Energy Security for
The 21st Century

Energy Security for The 21st Century

by Anco S. Blazev

LONDON AND NEW YORK

First published by Fairmont Press in 2015.

Published 2020 by River Publishers
River Publishers
Alsbjergvej 10, 9260 Gistrup, Denmark
www.riverpublishers.com

Distributed exclusively by Routledge
2 Park Square, Milton Park, Abingdon, Oxon OX14 4RN
605 Third Avenue, New York, NY 10017

First issued in paperback 2023

Routledge is an imprint of the Taylor & Francis Group, an informa business

Library of Congress Cataloging-in-Publication Data

Blazev, Anco S., 1946-
    Energy Security for the 21st Century / by Anco S. Blazev.
        pages cm
    Includes bibliographical references and index.
        ISBN 0-88173-738-0 (alk. paper) -- ISBN 978-8-7702-2322-5 (electronic) -- ISBN 978-1-4987-0966-8 (Taylor & Francis distribution : alk. paper) 1. Power resources. I. Title.

    TJ163.2.B594 2015
    333.79--dc23

                                                                2014038058

Energy Security for the 21st Century by Anco S. Blazev

While every effort is made to provide dependable information, the publisher, authors, and editors cannot be held responsible for any errors or omissions.

ISBN 13: 978-87-7022-930-2 (pbk)
ISBN 13: 978-0-88173-738-7 (The Fairmont Press, Inc.)
ISBN 13: 978-1-4987-0966-8 (hbk)
ISBN 13: 978-8-7702-2322-5 (online)
ISBN 13: 978-1-0031-5191-3 (ebk)

Contents

Foreword

There is a lot of talk in Washington, DC, and other high places these days about energy—present and future. Energy security, energy independence, national security, game changers, new energy reality, energy imports and exports, etc. buzz words are heard every day on the streets and all over the media.

Energy is an important factor in our lives, no doubt, with a lot of good and bad things happening simultaneously in the energy sector that affect our lives. So, this seems a perfect time to clarify some of them and get a closer look at all things energy and their effects on our energy security.

What is energy security, and why is it so important? Are energy security and energy independence related, and if so, how? Is our national security affected by any of these developments?

While some of the answers might seem simple and logical, the related subjects are incredibly complex and interwoven in a way that docs not lend to a superficial reasoning, or a skimpy knowledge of the technologies and other important details of the energy sector.

Energy security is the task of ensuring an uninterrupted and sufficient energy supply today and tomorrow, by employing efficient and safe internal and external risk prevention measures.

So, it is logical to reason that total energy security can be achieved only by a completely self-contained and self-supporting energy system, thus eliminating all internal and external interferences or risks.

But completely self-contained and self-supporting also means complete isolation from the outside world. Is this even possible in today's global economy? No, is the brief answer, so we are forced to deal with incomplete, or varying at best energy security, which involves very complex technological, logistical, social, political and other issues of national and international importance.

Energy is very important to our daily lives, and is the engine of our economic activities. As such it is a major wealth creator and a great political tool. Everybody is affected by the energy issues of the day, and yet there is ignorance and misunderstanding among politicians, regulators, and users alike.

For example, while focusing on our energy security, we might be (intentionally or unintentionally) overlooking the personal safety of the average citizen and/or ignoring more direct paths to energy independence.

Even more pertinent and bothersome is the possibility that while focusing on our present energy security (making sure that we have enough energy today), we might be ignoring our long-term energy survival, and that of the future generations.

It seems that this is exactly what is happening todays; we are doing everything possible to ensure plentiful (and in some cases excessive) energy supplies—fossils in particular—disregarding the fact that they are in finite supply. The use and overuse goes so far, in fact, that we are willing to dig out and pump out every last piece of coal and drop of oil for short-term benefits, regardless of the long-term consequences.

This reckless drive would leave the next generations without any fossils, which will compromise their energy security and negatively affect their way of life. Just imagine living without crude oil—life without vehicle fuels, heat, lubricants, plastics, fertilizers, medicines, cosmetics...

All this affects our energy security one way or another. Every single issue related to energy security and energy independence is important and very complex from technological and political points of view. The situation is made even more complex by the fact that the different issues are interwoven in an integral way, so to understand the entire energy scenario, we must address each subject in minute detail.

Ignoring, or misunderstanding, a single aspect of a particular issue might lead to wrong conclusions. This in turn might jeopardize our very lives and those of future generations.

In this text we evaluate, dissect, and analyze the different technical, logistical, regulatory, social, political, and financial aspects of the energy issues at hand—in light of our energy security.

The ultimate goal of this text is to bring out, understand, sort out, explain, and clear up the deep meaning and effects of energy on our lives in the 21st century and beyond.

Chapter 1

The New Energy Reality

The starting point for energy security today, as it has always been, is diversification of supplies and sources.
—Daniel Yergin

INTRODUCTION

The U.S. is the world's largest producer and consumer of energy. We produce huge quantities of coal, natural gas, crude oil, and nuclear materials, which are used to generate electricity and/or to fuel vehicles.

We are also a major producer and consumer of renewable energy, which is becoming more important as time goes on. The renewables, however, have different technological, political and regulatory problems that are limiting their progress. They also have to deal with steep competition from the conventional energy sources, so the short-term future of the renewables is unclear at best.

The renewables, however, are the best—if not the only—hope the future generations have for uninterrupted energy supplies. Alas, this is something that is missing in today's energy plans, and something we must be aware of as we move along the slippery slope of energy overuse.

Recently, we have discovered new energy supplies and have increased the production of natural gas and crude oil—so much so that today the quantity of *coal and natural gas* produced in the country meets the entire national demand, while the increasing excess is exported in huge quantities.

The additional quantities of *crude oil* produced today, however, are still not enough to meet our needs 100%. Because of that, we still import about half of the crude oil we use on daily bases. This is the case with nuclear materials too.

The U.S. now has plentiful coal and natural gas supplies. Crude oil remains in short supply and is the Achilles Heel of our energy security.

THE ENERGY SOURCES

The fossils are mostly burned as fuel for power generation or transport. A significant portion is also used for making chemicals, medications, fertilizers, cosmetics and other consumables, some of which are exported.

The major energy technologies and fuels powering the U.S. economy are shown in Table 1-1. The major sectors of the U.S. economy that use energy are shown in Table 1-2.

Table 1-1. Key energy sources in the U.S. (2013)

- Crude oil—37% (mostly used for transportation and production of consumables)

- Coal—21% (mostly used for electric power generation and in industrial processes)

- Natural gas—26% (mostly used for electric power generation and transportation)

- Nuclear—8% (mostly used for electric power generation)

- Hydro—5% (mostly used for electric power generation)

- Renewables—3% (mostly used for electric power generation)

Simply put, the energy sources (the fossils in particular), providing the major share of our electric power, vehicle fuels, and raw materials for the industry, are the foundation of our personal and business daily function.

Houses will not be comfortable places to live without enough power for lighting, heating, and cooling. Businesses will not be able to operate efficiently, or at all, without electric power, natural gas, and crude oil products.

Table 1-2. U.S. energy use by key fuels and sectors (2013)

- Transportation—a $535 billion business where:
  94% is powered by crude oil products, and
  6% is powered by natural gas and other fuels
Note: This is the Achilles Heel of our energy security since we import about 50% of the crude oil we need from unstable countries, transporting it through some of the most dangerous areas of the world.

- Electricity generation—a $357 billion business of which:
  48% is powered by coal,
  21% is powered by nuclear power,
  19% is powered by natural gas, and
  12% is powered by renewables (hydro and alternative technologies)
Note: This ratio is changing dramatically, as coal is being replaced by natural gas, and the renewables' contribution increases.

- Industrial uses—a $217 billion business in which:
  52% is powered by crude oil products,
  34% is powered by natural gas, and
  14% is powered by renewables
Note: Some crude oil and coal are used for production of consumables and in different industrial processes.

- Heating and cooling—a $125 billion business in which:
  58% is powered by natural gas,
  21% is powered by electricity,
  11% is powered by renewables, and
  10% is powered by crude oil products.

People and goods would not be able to move freely around without gasoline and diesel. Medications, plastic products, fertilizers, chemicals, paints, and many commodities would not be available, or would be extremely expensive, if we didn't have the fossils.

And yet, we don't have complete control over the supply chain of these materials—and especially that of crude oil.

Lack, or even temporary disruption, of crude oil supplies would have a chain-reaction effect on the economy with disastrous results. Airplanes, trucks, trains, and most other vehicles would have to be parked for the duration.

This would affect our personal lives, since we would not be able to move around—even going to work would be a challenge. The economy would slow down, causing increasingly serious damage to companies and entire sectors over time.

This doomsday scenario is not very likely today, but it shows a gap in our national energy security. It pro-vides a level of uncertainty and vulnerability that can be attributed to a number of internal and external risks. Ever changing domestic and international political, regulatory, and economic developments make it hard to predict and control events, thus adding to the energy uncertainty.

There are also threats of natural disasters, internal and external terrorist attacks, and other maladies that could disrupt the energy supply chain at any time.

*But most importantly, there is an eminent danger of **running out of the key energy sources** (coal, oil, and gas) in the near future.*

This is a key issue that has been a topic of discussion in energy circles since the 1970s. There have been many conferences, discussions, negotiations, and suggestions lately, but the results show that the world is still going in the wrong direction.

As a matter of fact, most people (and entire countries) tend to turn a blind eye and forget about the pending energy doom. Others think that we are already too close to it and that it is too late to reverse the events no matter what we do, so they can only predict different doomsday scenarios.

Until recently, the predominant thinking in the U.S. was that we will run out of fossil fuels in the very near future. But lo and behold, suddenly things have changed drastically for the good, making all previous doubts, prediction, and plans irrelevant.

Instead of importing massive amounts of fossils, we now see the U.S. getting ready to export huge quantities of all kinds of energy products: coal, oil, natural gas, and refined crude oil products. This is an unexpected and very welcome bonanza for the oil companies from which the consumers will benefit too. But the nagging question is still here, "How long will the new supplies last?"

Unfortunately, we also see glimpses of doomsday scenarios in the midst of all this newly acquired energy bliss. We see the energy prosperity lasting for awhile but not forever!

This is the key: fossils are a finite resource, so they are here today but gone tomorrow. We only don't know when that "tomorrow' will come.

We also see that the short-term benefits do not translate into long-lasting energy bliss. Everything has limits—and so do fossils. The gains of late do not assure us that we will have fossils forever.

Blinded by the new energy boon, we are failing to consider a reasonable balance between a) the amount of fossils we are using, b) the total amount available, and c) what to do when they are depleted. And, yes, they will be depleted someday soon. Not today, but soon enough to worry about it.

The energy frenzy in the U.S. is underway, and unless we manage to put it under control, it might lead to some unexpected negative consequences affecting us and tomorrow's generations.

These are the pertinent issues of today's energy and energy security, and those are what we are aiming to discuss from a technical point of view, and resolve (at least on paper) in this text.

THE ONGOING ENERGY DILEMMA

Our energy problems started in the 1970s, when the U.S. and the world economy as a whole was heavily affected by developments caused by foreign powers. Everyone in the U.S. at the time was affected by the gasoline shortage and the sharply rising prices at the pump.

The problems actually started around the end of the 1960s, when crude oil production and use in the U.S. and other countries peaked. This forced the industrial powers to accept and deal with real and perceived oil supply issues. A major factor was the reality of the West depending on oil supplies from unpredictable and potentially unfriendly Middle Eastern and other oil producing and exporting countries.

The fears finally materialized in 1973 unexpectedly and with unprecedented force, when OPEC decided to flex its muscle to see how far it could push the U.S. and the West. They decided to cut the supplies and raised oil prices as a protest in response to political developments in the area.

This created a serious global oil crisis that hit hard, mostly because the U.S. and the Western countries did not expect and were not prepared to handle such a drastic move. They always suspected that it could happen, but they did not even think that it could happen so soon and with such ferocity.

The impact was immediate and serious, with stagnant economic growth in many countries as oil supplies dwindled and oil prices climbed. There was also confusion and fear among the citizenry, when the long lines at the gas pumps became a daily event. This eventually grew into a general dissatisfaction with the government's handling of the situation.

Issues with the oil supply during that time where overblown by an overwhelming perception of crisis, which played a major part in the rise in prices.

Note: We saw a similar situation in 2008, when the oil prices rose to unprecedented levels, although there were no serious problems to account for the drastic rise.

The stagnant economic growth and excessive price inflation during the 1970s led to the invention of the term *stagflation*, which is today thrown around too, among other such threatening word combinations.

U.S. Energy History

The U.S. was getting back on its (oil) feet by the end of the 1970s, when a new disaster hit world energy markets. The 1979 Iranian Revolution started the second oil crisis in the United States. It started with massive protests in Iran, which forced the Shah of Iran, Mohammad Reza Pahlavi, to flee his country in early 1979. The Ayatollah Khomeini assumed power as the new authoritarian leader of Iran.

The internal turmoil disrupted the Iranian oil sector, production was greatly reduced and finally all exports were suspended. Oil exports resumed under the new regime soon thereafter, but they were inconsistent and at a lower volume. The new reality contributed to drastically increased world oil prices.

The Iran-Iraq War later on in the 1980s reduced world oil exports once again when Iran nearly stopped oil production, and Iraq's oil production was severely cut as a result of the war.

The U.S. government was anticipating another 1973 gas shortage, and printed coupons which were to be used in an attempt to reduce the gas line problems. The coupons were never used, because the crisis did not last long enough to affect the motoring public.

Figure 1-1. Gas coupons, following the 1979 energy crisis

OPEC nations compensated the reduction in exports from Iran by increasing production, so the overall drop in worldwide production was only about 4 percent. Nevertheless, the world was in panic resulting from the political chaos in Iran culminating with the taking of American hostages.

All this contributed to oil prices rising beyond what would be expected under normal circumstances. In response, President Carter began a phased deregulation of oil prices in the U.S., which allowed the U.S. oil output to rise sharply from a number of oil fields.

Those efforts did not prevent long lines from appearing once again, and several states even implemented odd-even gas rationing, as in 1973, but the issue never reached the levels it did in '73.

The average 1970s vehicle—V-8 large sedans in most cases—consumed more than a gallon of gas per hour during idle. With millions of gas-guzzlers waiting and idling in the long lines at the gas pumps, during several months, Americans wasted over 200,000 barrels of oil per day.

Americans at the time believed the rumors that oil companies were responsible for artificially created oil shortages, intended to drive up prices. Although the amount of oil sold in the U.S. in 1979 was only 3.5 percent less than the previous years, 54% of Americans believed the energy shortages were created by the oil companies. Only 37% of Americans thought the energy shortages were real, and 9% had no opinion.

President Carter outlined his plans to reduce oil imports and improve energy efficiency in his "Crisis of Confidence" speech (sometimes known as the "malaise" speech). Carter encouraged citizens to do what they could to reduce their use of energy. He had already installed solar hot water panels on the roof of the White House and a wood-burning stove in the building.

The solar panels were removed in 1986 by Ronald Reagan, reportedly for roof maintenance. The incident remains a symbol of the teeter-totter of government's energy policies (fossils vs. solar) that still persist.

In January 1980, Carter issued the Carter Doctrine, which declared that any interference with U.S. oil interests in the Persian Gulf would be considered an attack on the vital interests of the United States. Carter also proposed removing price controls (imposed during the administration of Richard Nixon before the 1973 crisis), as part of the deregulation effort. He suggested removing price controls in phases, an effort finally dismantled in 1981 by President Reagan. Carter also suggested imposing a windfall profit tax on oil companies.

The internally regulated price of domestic oil was kept to $6 a barrel, while the world market price at the time was $30 a barrel. The oil companies were not happy, and this unsustainable situation did not last very long.

President Carter called the 1979 oil crisis "the moral equivalent of war," as a justification of his proposals and mandates. The critics argue that his proposals did not serve any practical purpose on the international front and only made the bad situation worse.

In 1980, the U.S. government established the Synthetic Fuels Corporation to produce an alternative to imported fossil fuels. A number of other measures were introduced and some implemented. Millions of dollars were also poured into solar, wind and other alternative energy technologies. Many new companies started work in these areas, and some showed amazing progress within a short time. Examples of the solar technologies of those days are still operating in the U.S. deserts.

This euphoria, however, did not last long either. As soon as oil prices went down, all work on these technologies was abandoned. A trend we have seen repeated numerous times throughout the last several decades.

Detroit was affected by the sudden rise in fuel prices, with smaller cars imported from Japan, Italy, and Germany starting to dominate the domestic car market. The imported cars used novelty fuel-saving devices such as fuel injection and multi-valve engines replacing the conventional carburetor used in most American cars at the time. American automakers learned from the competition, and all this contributed to overall increase in fuel economy. This also was one of the factors leading to the subsequent 1980s oil glut.

Thirty-three years after President Carter warned against "outside" control of the oil-rich Persian Gulf we see that the U.S. effectively enforced a corollary to that doctrine, thus preventing control by a regional power during the Iraq wars of 1991 and 2003. Today, however, the Carter Doctrine must be revisited in light of Iran's nuclear ambitions and threats in the Gulf. This is perhaps the greatest immediate threat to U.S. energy and national security and a menace to U.S. economic interests, the least result of which could be long-term spike in oil prices.

The Carter Doctrine is as relevant today as it was thirty years ago, especially as the Persian Gulf's oil production capacity and transport activities are concerned. The global demand for crude oil is roughly the same today, about 30 percent, as it was in 1980. In case of a hostile takeover of, or serious interference with the

Gulf activities, the U.S. and China would be immediately and seriously affected, and crude oil prices would rise accordingly.

Figure 1-2. U.S. Navy warships patrol in the Gulf.

A major threat to the oil market is Iran's nuclear ambition. The U.S. Fifth Fleet and other military assets in the region are busy keeping the free flow of oil through the Persian Gulf for now. A nuclear-armed Iran, however, would gain *de facto* immunity from a conventional attack from another country, significantly limiting the effectiveness of U.S. forces influence in the region. What would happen then? Hard to tell, but nothing good could be expected for the Gulf crude oil supply.

The overwhelming conclusion is that price fluctuations and subsequent anomalies usually started with some type of crude oil manipulation—be it by foreign governments and companies, by public frenzy, or by stock trading abuse—and usually one anomaly feeds the other, so the problem grows bigger without a root cause.

We need to disentangle and control these major and interwoven external and internal political, social and economic drivers to avoid similar energy and economic disasters in the future.

The key to achieving this is optimizing our energy security and achieving total energy independence, which starts with understanding the issues at hand—and which is the object of this text.

The result of the 1970s crisis was that oil-rich countries in the Middle East benefitted tremendously from increased prices and slowed production in other areas of the world. Some other countries, such as Norway, Mexico, and Venezuela, benefitted as well by becoming world class exporters of this crucial commodity. In the U.S., Texas and Alaska, as well as some

other oil-producing areas, experienced major economic booms too, due to high demand and soaring oil prices.

The above mentioned foreign nations and U.S. states thrived while most of the rest of the nation and most of the world struggled with oil shortages and stagnant economies. The economic boon of the oil producing nations and U.S. states, however, was short lived and came to a abrupt halt as soon as prices stabilized and dropped in the 1980s.

The 1970s oil deficit and other oil-related problems were all but forgotten in the 1980s, when suddenly we saw ourselves floating in oil. Reduced oil demand and overproduction resulted in a glut on the world market. Oil consumption in major countries was down about 15% from 1978 to 1982, due in part the large increases in oil prices by OPEC and other oil exporters. This caused a six-year-long decline in oil prices, culminating with a 46 percent price drop in 1986.

Several years of peace were enough to rebuild the American public's confidence in continuous oil supplies and reasonable prices...until another war put our energy security and oil supplies in jeopardy. In 1990, Iraq invaded Kuwait, which started a 7-month armed conflict ending with occupation of Kuwait by U.S. military.

Iraq officially claimed that Kuwait was stealing oil, but its true motives seem to have been much more complicated. Iraq owed Kuwait $14 billion in debt, loaned during the Iraq-Iran war. Iraq also felt that Kuwait was producing too much oil, thus lowering market prices, which hurt Iraqi oil profits.

The war caused interruption of oil production and exports in the area, which coupled with threats to Saudi Arabian oil fields, led to a rise in prices. The price went from $21 per barrel to $28 per barrel within a month after the invasion. By the end of the invasion, oil prices rose to nearly double that, and went as high as $55 per barrel on the world energy markets. Soon after, however, prices dropped significantly, albeit temporarily.

The war and its energy consequences did not last long. After less than a year, the world oil supplies were back to normal, but this time oil prices remained high relative to pre-war levels. There were no long lines at the gas pumps in the U.S. this time either, but oil prices have been steadily rising ever since, as a result of persisting political turmoil and economic problems worldwide.

ENERGY ACCIDENTS

While energy is a precious commodity, which we cannot live without, it is also a vicious villain, that kills mercilessly when given a chance. Just in the last half century we have witnessed a number of serious energy related man-made disasters that have caused huge property damages, human suffering and death.

We've witnessed several coal mine collapses killing miners, several oil spills killing wildlife, and several nuclear accidents, the results of which are still outstanding.

Following are accounts of several of the worst disasters involving energy production.

Chernobyl

The city of Pripyat in Russia was built in 1970 to house scientists and workers of the nearby Chernobyl Nuclear Power Plant. Pripyat had over 50,000 residents and was brimming with life. It served well the locals during the 15 years of its existence, and during Chernobyl's construction and operation… until that infamous day in 1986, when the nuclear reactor melted down and took the plant with it.

Twenty-eight years after the accident, the buildings still stand tall as if waiting for their old inhabitants to return. The city is, however, nothing but a blood chilling image of a ghost town, devastated by a nuclear power gone seriously wrong. Bad went to worse during the critical time after the explosion, when the authorities failed to warn the residents of the nuclear accident in time, instead issuing an evacuation notice on the 3rd day after the explosion. Too little too late…

Figure 1-3. Pripyat… a ghost town.

This was 3 days more than most of the people exposed to the cloud of radiation could tolerate. Thousands got sick and many died shortly thereafter. Less than 2 miles away from the reactor, the ghost town is still littered with the remnants of the people who once lived there, the signs of their hasty abandonment frozen in time.

Officials downplayed the magnitude of the catastrophe, and the Soviet-era type politics of misinformation continue to this day. Official estimates of 4,000 people expected to eventually die from cancer-related illnesses as the result of the accident are an underestimate, to say the least. Major environmental organizations have accused the report of whitewashing Chernobyl's impact and state that more than 100,000 people have already died as a consequence of the disaster, and many more are expected to die.

The environmental damages of the area around the plant are permanent. Damages also extend all through Russia and parts of Europe and Asia, albeit most of them are temporary and remediation measures have erased the effects in most cases.

Note: The author was born in that part of the world and had a chance to visit the region in 1988. The devastation two years after the accident (even hundreds of miles from Chernobyl) was palpable, but what was most shocking was the fear in people's eyes every time Chernobyl was mentioned.

Even people who knew nothing about nuclear power, nor what exactly happened at Chernobyl in 1986, had acquired an animal-like fear from the vicious invisible killer that spread all over the place and that might be close enough to hurt them.

The fear was real—fear of the unknown fueled by the uncertainty imposed by the then-communist system, where people were not allowed to ask questions—that fear alone had very likely made many people sick and even shortened the lives of many others.

There was a large hole in the ground, where the nuclear reactor stood just hours before—a hole full of deadly radiation that would hurt many people and dislocate many more with time.

Today, the Chernobyl Nuclear Power Plant sits inside a fenced area, the famous Exclusion Zone. Radioactive remnants of the failed reactor still linger inside the 24-story concrete and steel encasement, which was hastily erected after the accident in an attempt to bury and forget the problems.

As a symbol of one of the final deeds and a lasting reminder of the totalitarian system, it still sits there full of cracks, leaky, and structurally unsound. Emitting enough radiation to cause a second disaster of similar magnitude, it is reminding us day after day that nuclear power is a force to be reckoned with, and that we must be very careful how we handle it.

New encasement is planned to slide over the existing cracked sarcophagus to seal in the remaining nuclear fuel. At an estimated cost of $2 billion, it is another

Figure 1-4. Chernobyl...or what is left of it.

reminder of the way things were, and that it pays to do things right the first time around.

Within the Exclusion Zone there are also dozens of abandoned villages, where collapsed houses and huts are disappearing under vegetation overgrowth. Hundreds of old people have returned to their village homes, and live in the radiation, ignoring the dangers. How many of them will survive is unclear, but nobody cares anyway, so those who die from it will just join the ranks of the undercounted and unaccounted victims of this huge unprovoked disaster.

Chernobyl is undeniably a lasting symbol of the mighty power of nuclear energy. It is awesome in its viciously dark glory, and a symbol of unstoppable devastation. Nuclear power can be a good friend who works for us day and night, but in a split second it can turn into an ugly monster that would destroy everything in its wake.

Horizon Deep Water Accident

9:45 P.M., April, 20, 2010. A geyser of seawater erupted from the marine riser onto the large oil rig, shooting 240 ft into the air, during the final phases of the drilling operation. An eruption of mud, gas, and water followed. The gas ignited, creating a series of explosions which turned into a firestorm on the platform, killing several people and hurting others.

The blowout preventer failed, and the stream of the wicked mixture coming from the depths of the ocean continued to flow uncontrollably for days. The Gulf was a mess, and local marine life paid the ultimate price.

At the initial stages of the disaster there were 126 crew members on board. Eleven of them were killed instantly by the initial explosion. Several were thrown, or jumped, into the water below but were recovered with some injuries. The rig was evacuated soon after, and the injured workers were airlifted to nearby medical facilities.

The oil rig continued to burn and emit copious amounts of smoke for the next day and a half, and finally sank two days later, resting on the seafloor close by. Then reality hit, and the nightmare began. The oil well continued to spew large amount of crude oil into the Gulf of Mexico for three months.

After several attempts, it was finally capped and then sealed permanently. The well was declared *dead* two months later.

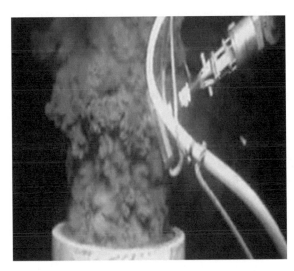

Figure 1-5. BP oil pipe spewing raw oil.

The world watched in disbelief the drama developing before their eye. The locals, and the local economy, went into a limbo at first, and then switched to damage control. What followed was simply dramatic. Birds covered with oil, workers cleaning the beaches at night, oil floating in the water and on the cost line. One bad report after another...day after day, week after week...

Then, the U.S. EPA, under BP advice, used millions of gallons of Corexit EC9500A and EC9527A as dispersants, to mitigate the oil damages. These chemicals are toxic, and not the most effective on EPA's list of approved dispersants. Although other products have better toxicity and effectiveness ratings, BP decided to use

Corexit because it was available immediately. Millions of gallons were injected in and sprayed over thousands of acres of ocean area.

Information on the composition and safety of the Corexit products was withheld by the EPA. Only after a lawsuit later on, was it disclosed that the dispersants contain propylene glycol, 2-butoxyethanol, and dioctyl sodium sulfosuccinate.

Test results showed that the dispersants contain cancer-causing agents, hazardous toxins and endocrine-disrupting chemicals. Several of these chemicals are severely toxic, as follows: 5 are associated with cancer, 33 with skin irritation ranging from rashes to serious burns, 33 are linked to eye irritation, 11 are suspected of being potential respiratory toxins or irritants, 10 are suspected kidney toxins, 8 are suspected or known to be toxic to aquatic organisms, and 5 are suspected to have a moderate acute toxicity to fish. That is a large menu, where anyone affected can pick and choose what to suffer or die from.

Thousands of gallons of this mixture were sprayed into the ocean. But wait, Corexit was banned from use on oil spills in the United Kingdom a decade ago for obvious reasons. More importantly, no toxicity studies have ever been conducted on this product, although 2-butoxyethanol in Corexit EC9527A was identified as a causal agent in the health problems experienced by cleanup workers after the 1989 *Exxon Valdez* oil spill.

According to the environmental scientists, the dispersants can cause genetic mutations and cancer and expose marine life (like sea turtles and bluefin tuna) to an even greater risk than the crude oil alone. The dispersants poured into the Gulf were picked up by the current and the winds, and were washed on shore throughout the entire Gulf coast. Only the locals know the extent of the damages to their environment and livelihood.

In May, 2010, a letter was sent to BP outlining the concerns related to potential dispersant impact on Louisiana's wildlife and fisheries, environment, aquatic life, and public health. Officials requested that BP release information on their dispersant effects. After three underwater tests, the EPA approved the injection of dispersants directly at the leak site to break up the oil before it reached the surface. Later that month, independent scientists suggested that underwater injection of Corexit into the leak may have been responsible for the oil plumes discovered below the surface.

EPA then gave BP 24 hours to choose less toxic alternatives to Corexit from the list of dispersants on the National Contingency Plan Product Schedule, or alternatively to provide a detailed reasoning why the approved products did not meet the required standards.

By end of May, BP was using about 26,000 gallons a day of Corexit. After the EPA directive, the daily average of dispersant use dropped to about 23,000 gallons a day. In the end, it was estimated that over 2 million gallons of this chemical mixture were dumped in the Gulf. USCG's attempts to verify if BP was exceeding approved volumes failed.

Only one thing is for sure: the total damage to the Gulf waters and the coast line will not be known for decades to come. BP was forced to pay billions of dollars for environmental mitigation and to the locals, but no matter how much money they pay, the Gulf won't be the same for a long, long time.

And now, television commercials describe the Gulf coast as heaven on Earth—even better than before. As if the millions of gallons of poisonous oil and dispersant chemicals dumped in the ocean somehow fixed something that was wrong in there, and improved the water quality and the entire coast line…

Kuwait

One of the most spectacular sights in oil well history was that of the flames and dense smoke bellowing from many burning oil wells in Kuwait during and after the 1990-1991 war. Iraqi armed forces set fire to 789 individual Kuwaiti oil wells, which were left burning for weeks, resulting in catastrophic consequences. The local ecology in the entire Persian Gulf was damaged, some permanently. The damage caused by the burning oil wells will take generations to repair. Some of the region will never be the same.

Amazingly, little attention was paid to the potential impact of a sustained damage to the regional

Figure 1-6. Iraq wars damaged the oil infrastructure

environment, but by the end of 1990 the experts (finally) realized that burning millions of barrels of oil per day is something to be concerned with. In early 1991, estimates were made that more than six million barrels of oil were burning daily in Kuwait, and the initial assessment of the environmental impact was staggering.

A variety of environmental disasters were feared, with the amount of soot generated daily being a major problem. We know that one gram of soot can block out 2/3 of the light falling over an area of 10 square meters, so the soot generated by burning 4-5 million barrels of oil per day would generate a plume of smoke able to cover the entire United States and part of Canada. Not a small thing, but what to do?

Uncontrolled weather patterns carried the smoke plumes great distances and hampered agricultural production in several distant areas of the world. In addition, about 250 million gallons of oil flowed in the Gulf, more than 20 times the amount spilled by the *Exxon Valdez*, Alaska. This caused irreparable harm to the biological diversity and physical integrity of the Gulf. There was oil covering 440 miles of Saudi Arabia's coastline, and it will take years before the oil can be cleaned by men, or swept away by natural forces.

By the end of 1991, the burning oil wells were capped, but the damage to the Kuwaiti economy and the Persian Gulf environment was done. The assessment revealed that hundreds of miles of the Kuwaiti desert were left uninhabitable, due to the newly created oil lakes and soot from the burning wells. The oil spillage changed the ecology, putting it at serious risk. Nearly two million migratory birds visit the Gulf shores each year on their way north, and during the war thousands of comorants died as a result of exposure to oil and smoke from the leaking and burning oil wells.

The fishing industry in the Gulf was brought to a halt. From 120,000 tons of fish a year prior to the war, now the numbers were several times lower. The Gulf coast people depend on fishing for subsistence, and the oil spillage and smoke disrupted their normal activities permanently because the spawning of shrimp and fish patterns changed.

Many species, like green and hawksbill turtles, are classified as endangered species, but the oil well leakage and smoke did not give them much of a chance. As a result, their populations, together with those of leatherback and loggerhead turtles, dugongs, whales, dolphins, cormorants, flamingoes, and sea snakes were decimated.

This was a man-made environmental disaster, which was unjustifiable, and made no sense whatsoev-er. Nevertheless, the local environment and the people living in it are innocent victims, who will be paying for it with their livelihood, health and lives for some time to come.

Ecuador

This is one of the best kept secrets of the oil industry. At a closer look, it is also the most revealing of the way oil business is done around the world.

Over 50% of Ecuador's national budget is funded by oil exports, so continued and expanding oil production is of utmost importance. It is the lifeblood of the country. Future government plans call for an increase in production and infusion of foreign investment. This is the good part.

The bad part is that this dependence on oil revenue is hindering Ecuador's full development as an industrialized nation. It has also created environmental enforcement blindness with damaging consequences to the Amazon region and the eastern part of the country. For the last 20 years the locals have been resisting the expanding oil exploration, in order to protect their rights to ancestral lands.

The oldest Indian tribes in the Amazon region, which not very long ago numbered in the thousands, have been reduced to hundreds. This human devastation is mostly attributed to the pollution brought upon the region by oil production. Careless and cheap production methods have caused serious water contamination, which has led to increased cases of cancer, abortion, dermatitis, and many other maladies. Drinking, bathing, and fishing water contains a level of various toxins that is much higher than any safe limits.

In addition, the oil companies are responsible for deforestation and seriously damaging thousands of acres of rain forest.

Their actions, including dynamiting the earth, spilling vast amounts of oil, destroying habitats, and fouling rivers are responsible for exterminating the fish population in the local streams and rivers. The wild game is also gone, which has forced many tribes to retreat deeper into the jungle.

It was documented that Texaco has already spilled nearly 20 million gallons of crude oil in the jungle and has abandoned hundreds of hastily installed toxic waste storage ponds. Oil used for road cover in order to cut dust is flowing into rivers in uncontrolled manner. The newly constructed roads for the oil production activities have devastated more than 2.5 million acres, and the deforestation process continues at a rate of over 350,000 hectares a year.

Oil production is responsible for a number of environmental problems in the Amazon region. The Sierra highlands have been almost completely deforested, and in the Orient numerous mammals are in danger of extinction.

Oil waste generated in the past was placed in holes in the ground, contaminating forests and rivers, while ruptures in pipelines have discharged over 18 million gallons of oil into the Amazon in the past two decades. This compared with the 10 million-gallon *Exxon Valdez* spill is a major disaster in progress, with no end in sight.

Any such activities in the U.S. would certainly put the guilty in jail, but out there in the jungle, anything goes...for as long as needed. Nobody is held responsible for some of the greatest environmental damage on Earth, and locals continue paying for it with their livelihood, health, and lives.

Fukushima

Fukushima Daiichi was one of the largest nuclear power complexes in Japan. Located at the peaceful Pacific cost line it adorned it with its presence and supplied millions of people with electric power.

All was well at the Fukushima nuclear plant until a powerful earthquake and a wicked tsunami surprised Japan in March 2011. The Fukushima Daiichi nuclear power plant on Japan's eastern coast was shaken up and flooded, causing its major operational and backup systems to fail.

Their instantaneous and simultaneous failure resulted in nuclear reactors running out of control, and used fuel storage tanks evaporating, triggering three meltdown incidents in a short time.

Assisted by hydrogen explosions at the site, the meltdowns caused massive escape of radioactive material in the environment.

Initially several workers were severely injured or killed by the disaster. Some drowned, some were hurt by falling equipment, mostly as a result of the earthquake. There were no immediate deaths due to direct radiation exposures, but at least six workers have exceeded lifetime legal limits for radiation and more than 300 have received significant radiation doses.

Mass evacuations, tests and checks followed, but the damage was already done, and now we could only watch, sympathize, and expect to see more disasters and negative consequences. Some 300,000 people evacuated their homes in the prefecture after the disaster caused multiple meltdowns at the Fukushima Daiichi nuclear

plant; over 1600 people died in the process.

Two years after, predicted future cancer deaths due to accumulated radiation exposures in the population living near Fukushima are inconclusive. But it is certain that it would take decades to decontaminate the surrounding areas, to decommission the plant, and get life in the area somewhat near normal.

Experts today predict the Fukushima disaster victims to include up to 600,000 deaths, over 100,000 still births, and over 100,000 children with genetic deformations now and in the future. There are also reports of wild animals—bears, seals, fish, and birds—hurt or killed by the radiation.

The damages and deaths were not limited to Japan, however. There were estimated 14,000 deaths in the United States by 2012 that are linked to the radioactive fallout from the disaster at the Fukushima nuclear reactors in Japan, according to the first peer-reviewed study published in a medical journal documenting the health hazards of Fukushima.

After three years of looking for solutions, the Fukushima nuclear power plant is work in progress. Radiation is still leaking out in the local soils and the ocean, and contaminated water is stored in huge tanks—which also leak.

Figure 1-7. Fukushima work in progress after three years.

Over 4,000 workers are busy 8-10 hours a day in the plant's decontamination and disassembly effort. This work is estimated to continue for the next 30 years, but the end results are still hard to predict.

The locals were affected the most, but the radiation spread quickly to distant areas too. Over 3,000 individuals from a town over 15 miles away from the disabled nuclear plant also bore detectable levels of radioactive cesium in their bodies.

Their total dose of less than one milliSievert is considered safe, and no radiation sickness was observed. Nevertheless, the exposed men, women and children need to be watched for the long-term effects of the radiation...for the rest of their lives.

The damages are noticeable in the local wildlife as well. One example of the continuing damage is the *pale grass blue* butterfly; a delicate local insect, famous for the way the colors and patterns of its wings change in response to environmental changes. Butterflies caught in the Fukushima region six months after the meltdowns (a generation in butterfly lifetime terms) had different colors and patterns than before. Some of the butterflies also had badly deformed legs, antennae, wings and even eyes. The deformities persisted and got even worse in the second generation of offspring. The same deformities were found in butterflies collected from the wild at a greater distance from the failed plant.

Such results are found in research labs, where normal butterflies are exposed to radiation from cesium particles like those escaping from Fukushima Daiichi. This leads to only one answer: the damage is due to radiation.

How many more changes to wildlife and the human population will be encountered in the future remains to be seen. The affected, however, will live in fear of developing cancer, or some other radiation-related illness, for the rest of their lives, asking, "Is it worth it?"

Chernobyl and Fukushima are the largest and most damaging nuclear power accidents since the inception of commercial nuclear technology. They, however, are very different in causation, response, and results.

The damages from both accidents are similar, but this is where the similarities end. The Chernobyl accident happened during the reign of communist awkwardness, and was caused by inferior technology and clumsy operation. There were no emergency response plans in place, so the actions of participants were erratic for the most part, further exacerbating the situation.

Chernobyl was a man-made accident, caused by ignorance and negligence, for the containment of which the main actors were not prepared.

The Fukushima nuclear accident was caused by natural events, triggering an earthquake, followed by a large tsunami. Faced with that awesome force, even the superior *Made in the U.S.A.* technology, and U.S. trained and well paid technicians and engineers could not prevent, nor even contain for months, the nuclear

meltdown. The nuclear plant was dead in the water... literally. What followed was a disaster that no one has ever even thought possible.

This was an accident caused by very powerful natural forces, which the existing technology and personnel simply could not handle. And yet, the designers should have provided additional layers of protection to prevent complete failure and devastation.

Lesson learned: now all nuclear power plants are going through inspections and redesign, geared to prevent such accidents. But the question persists, How many lessons do we still need to learn until nuclear energy is completely safe? Is that even possible?

Since we cannot go too far in the future, we work with what we have so far. So, here we have two different accidents: one 100% man-made and the other 100% nature-made. Both had some help from imperfect humans, and both had the same final results— devastation of environment and human life. We were not able to prevent either disaster, even though we were aware of the dangers and took some measures (albeit inefficient and ineffective) to prevent them.

These two accidents open a large hole in the safety record of nuclear power use. The world fears that a third accident could easily occur—even if we were advised ahead of time of the possibility.

The fear was so great for awhile shortly after the Fukushima blowup, that some governments decided to shut down all their nuclear facilities (albeit temporarily) as the only feasible way to prevent another Chernobyl or Fukushima. Now the memories are fading, and the fears are being replaced by the daily worries about energy supply. As a result, many nuclear plants were restarted after a quick inspection. Others are still shut down for more extensive repair work and upgrades.

There is a nuclear power plant close to many large population centers in the western world, so the burning question—fueled by the memories from the two greatest and totally different nuclear disasters we discussed above—is *who is next? And when?*

More importantly, it is time for the different governments—and the world community as a whole—to take a much closer look at the pros and cons of using nuclear power and come up with a long-term plan that minimizes the environmental and health risks while providing enough power for normal life. This is not an easy task, but it is one that is urgent in nature and that we cannot/should not ignore.

It is also part of the energy security quest too. We want our energy, including nuclear, to be safe, before anything else. It must be safe to produce, safe to use in all stages and aspects of its cradle-to-grave cycle. What is the point of having plentiful energy if it kills us at whim?

Climate Change

The greatest overall effect, directly or indirectly caused by energy and energy materials and products production and use, is measurable increase in global pollution and the resulting slowly changing climate conditions.

Scientists are still debating the details of energy production vs. global warming vs. climate change, but one thing is certain: energy from fossil fuels is at least partially responsible for these new phenomena. How *partial* their contribution is remains to be seen, but we should not ignore the fact that we are dealing with major pollution sources.

Millions of tons of CO_2 and other GHGs are pumped into the atmosphere every day by fossil-burning power plants and fossil-powered vehicles. Similarly so, millions of tons of solid and liquid waste from fossil materials and energy production facilities are stored or discarded somewhere, and the resulting air, water, and soil pollution is contaminating our fragile environment too.

The final results—in addition to air, water, and soil pollution affecting the environment, wildlife, and humans—is sharply changed global conditions. Increased polar warming, noticeable sea level rise, draughts, floods, as well as increased frequency and strength of hurricanes and tornadoes is becoming the norm.

An adequate energy supply is needed for our energy security, but it should not be achieved at the expense of human health and life.

There are many factors affecting our energy supplies, which in turn affect us and our environment, so we will review below the natural and man-made events that are related to energy production, and use. These are also components of our energy security, since their individual and combined effects play a significant role in the integrity of our energy supplies and human wellbeing in general.

To start with, there are many events around us that are natural; they are part of the Earth's natural life cycle. We understand most of them, and can partially account for their effects on the environment.

We have to remember that we live in a closed cycle world with many tightly interwoven natural events, which we humans affect one way or another. These events affect us in turn, or as somebody said, "a sneeze on one side of the world causes a tornado on the other side."

Those who cause changes in natural events may or may not care about the consequences, but we all must be aware of the entirety of the situation in which we live.

THE NATURAL CARBON CYCLE

Many natural events affect our overall energy security, the environmental picture, and our wellbeing and safety, so they must be taken into consideration.

Some of the most important natural events follow.

The Earth's Carbon Cycle

Carbon is a major constituent in the *carbon cycle* of our Earth, which is a very important biological, geological, chemical, and physical phenomena that regulates the generation and consumption of carbon in all natural and some man-made systems.

The carbon cycle basically determines and controls the exchange of carbon among the biosphere, pedosphere, geosphere, hydrosphere, and atmosphere of the Earth.

In combination with the nitrogen and water cycles, the carbon cycle creates and controls a number of daily events that help the Earth support its life systems—flora and fauna—including that of humans. The carbon cycle basically consists of the movement of carbon and carbon-related species in and out of the different environmental systems, including its reuse and recycling throughout the biosphere.

Figures 1-8a and 1-8b represent the movement of carbon (and carbon species) between land, ocean and the atmosphere in billions of tons of carbon equivalent each year. Here the global carbon cycle is divided into the major reservoirs of carbon interconnected by pathways of exchange in:

- the atmosphere,
- the terrestrial biosphere,
- the oceans, (including dissolved inorganic carbon and living and non-living marine biota),
- the sediments (including fossil fuels,
- the freshwater systems and non-living organic material, such as soil carbon),
- the Earth's interior (including carbon from the Earth's mantle and crust).

These carbon stores interact with the other components through geological processes, where the carbon

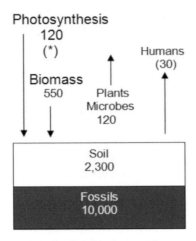

Photosynthesis
120
(*)

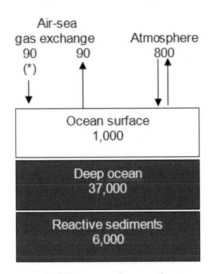

a. The land carbon cycle

b. The sea carbon cycle

Figure 1-8. Carbon cycle—land and sea

Note: Human contribution to the overall environmental gasses exchange.

exchanges between reservoirs occur as the result of various chemical, physical, geological, and biological processes.

The ocean contains the largest active pool of carbon near the surface of the Earth. The natural flow of carbon between the atmosphere, ocean, and sediments is fairly balanced, so that carbon levels would be roughly stable without human influence.

Note the key words, "without human influence." With the human influence—and it is extreme today—everything changes. Billions of tons of gasses, liquids, and solids are generated as products and byproducts around the world and are sent into the air, soil, and water. Every hour, every day, every month, every year…nonstop!

One doesn't have to be a scientist or a genius to de-

duct that the effect on the environment and us humans is significant and increasing by the day. This affects our energy security in a number of ways as well, as we will see later on in this text.

Carbon Use

Carbon is the building block of life, and major part of our daily lives. It is all around us; in the air we breath, in the water we drink, in the foods we eat, in our houses, our furniture, cars, shoes, clothing, and basically in almost everything we breath, touch, eat, and use.

Just take a look at some of the chemical compounds it can create: CO, CO_2, CO_3^{2-}, COS, CS_2, $CClN$, CCl_2O, CCl_2F_2, CCl_3F, CCl_4, CF_2O, CF_4, CN^-, CNS^-, $CHCl_3$, CHI_3, CHN, CHO_3^-, CH_5N, CH_6N^+, CH_2Cl_2, CH_2F_2, CH_2O, CHO_2, CH_2O_3, CH_3Cl, CH_3I, CH_4, CH_4N_2O, CH_4O, $C_2O_4^{2-}$, $C_2HBrClF_3$, $C_2HCl_3O_2$, $C_2HF_3O_2$, C_2H_2, $C_2H_2O_4$, C_2H_3Cl, $C_2H_3ClO_2$, C_2H_3N, $C_2H_3O_2^-$, C_2H_4, $C_2H_4Br_2$, $C_2H_4Cl_2$, C_2H_4O, $C_2H_4O_2$, $C_2H_4O_2$, C_2H_5Cl, C_2H_5NO, $C_2H_5NO_2$, C_2H_6, C_2H_6O, $C_2H_6O_2$, C_2H_6S, C_2H_7NO, $C_2H_8N_2$, $C_3H_3O_3^-$, C_3H_4, $C_3H_4O_3$, $C_3H_5N_3O_9^-$, $C_3H_5O_2^-$, C_3H_6, C_3H_6O, $C_3H_6O_2$, $C_3H_6O_3$, C_3H_7Cl, $C_3H_7NO_2$, $C_3H_7NO_2S$, $C_3H_7NO_3$, $C_3H_7O_2$, C_3H_8, C_3H_8O, $C_3H_8O_3$, C_3H_9N, $C_3H_{10}N^+$, $C_4H_2O_4^{2-}$, $C_4H_4N_2O_2$, $C_4H_4O_4$, $C_4H_4O_5^{2-}$, $C_4H_5N_3O$, C_4H_6, $C_4H_6NO_4^-$, $C_4H_6O_5$, C_4H_8, $C_4H_8N_2O_3$, $C_4H_8O_2$, $C_4H_9NO_3$, C_4H_{10}, $C_4H_{10}O$, $C_5H_5N_5$, $C_5H_5N_5O$, $C_5H_6N_2O_2$, $C_5H_8NO_4^-$, $C_5H_9NO_2$, $C_5H_{10}N_2O_3$, $C_5H_{11}NO_2$, $C_5H_{11}NO_2S$, C_5H_{12}, $C_6H_4Br_2$, $C_6H_4Br_2$, $C_6H_4Br_2$, $C_6H_5NO_2$, C_6H_6, C_6H_6O, C_6H_7N, $C6H_8N^+$, $C_6H_8O_6$, $C_6H_8O_7$, $C_6H_{10}N_3O_2^+$, $C_6H_{12}O_6$, $C_6H_{13}NO_2$, $C_6H_{13}NO_2$, C_6H_{14}, $C_6H_{15}N_2O_2^+$, $C_6H_{15}N_4O_2^+$, $C_7H_5NO_3S$, $C_7H_5N_3O_6$, $C_7H_5O_2^-$, $C_7H_6O_2$, $C_7H_6O_3$, C_7H_8, C_7H_{16}, $C_8H_6O_4$, C_8H_8, C_8H_{10}, $C_8H_{10}N_4O_2$, C_8H_{18}, $C_8H_{18}O$, $C_9H_8O_4$, $C_9H_{11}NO_2$, $C_9H_{11}NO_3$, C_9H_{20}, $C_{10}H_8$, $C_{10}H_{14}N_2$, $C_{10}H_{22}$, $C_{11}H_{11}N_2O_2$, $C_{12}H_{22}O_{11}$, $C_{16}H_{32}O_2$, $C_{18}H_{36}O_2$, $C_{27}H_{46}O$, C_{12}, C_{14}, C_{18}, C_{60}.

And this is only a very small number of products carbon is part of. These, and thousands of other carbon compounds are found in fuels, lubricants, plastics, textiles, paints, solvents, medical equipment, medicines, cosmetics, and much more.

It is getting hard to even imagine an area of our lives which is totally void of carbon-related materials.

Global Carbon Budget

The global carbon budget is the balance of the exchanges (incomes and losses) of carbon between the carbon reservoirs or between one specific loop (e.g., atmosphere ↔ biosphere) of the carbon cycle. An examination of the carbon budget of a pool or reservoir can provide information about whether the pool or reservoir is functioning as a source or sink for carbon dioxide.

Figure 1-8a shows the human contribution of 30 GT/year of carbon species in the atmosphere, which might seem small compared with the larger numbers of the natural carbon reservoirs. The man-made carbon emissions, however, represent the annual gas release, which when multiplied by 50 (for the last 50 years of rapid industrial development) we get an impressive 1,500 gigatons* of total carbon related accumulation in the atmosphere.

*Note: One gigaton is 1,000,000,000 tons, or 1,000,000,000,000 kilograms, or 2,200,000,000,000 lbs of volatile—some toxic—gasses. Fifty years of pollution produced 50 times that amount, or 110 trillion lbs of toxic gases are already floating in the Earth's atmosphere. During the next 50 years this amount will be doubled to 220 trillion lbs.

Such a huge amount of carbon gases is overwhelming the natural gas exchange cycles, and the overall environmental balance. So, we see the signs of drastic weather and climate changes coming upon us. It is logical to conclude that unless we reduce the amount of carbon released in the atmosphere, we are going to choke the planet's natural cycle, and ourselves, to death.

The most tragic fact is that even if we stop all emissions today, it might already be too late to do any good, because some of the accumulative effects are long lasting, and some of the serious damages are permanent and some are irreversible.

Since we are certain that the human carbon release related activities will not be reduced anytime soon, we are certain that the damage will continue. It is actually expected to get worse with time, as the developing countries are planning increase in fossils use for energy generation in the near future.

When the fossils are gone sometime this century, pollution emissions will be reduced and even stop to a large extent, but we can only guess what the environment will look like then. It is also hard to imagine how the people of tomorrow will provide energy to sustain their personal and economic lives.

CO$_2$ Lifetime

We agree that there is a lot of CO$_2$ in the atmosphere (like it or not), and it is now part of the carbon cycle. Some of it is emitted by natural events, but some comes from thousands of coal and gas power plants, and millions of trucks, cars, trains, boats, etc.

In all cases, most of the CO$_2$ flows up—and stays up—in the air. Some of it is absorbed or remains suspended in the local areas, while the majority is blown by the wind and taken to far away places and high in the atmosphere.

Then what? How long does it stay suspended in the air? What happens during and after its temporary residence in the locality, or high in the atmosphere? There is actually a great deal of confusion on this subject. This is because there are multiple physical and biological processes helping retain and remove CO$_2$. These processes are complex and behave differently and at different rates in different situations, depending on local conditions.

More confusion comes from the fact that the leading scientists, including the Intergovernmental Panel on Climate Change (IPCC), have not made up their minds yet regarding the lifetime of CO$_2$—or how long it stays in the atmosphere before disintegrating or changing into something else.

One report defined the lifetime of CO$_2$ as 50-200 years, while later it was reduced to 5 years. Which is it—5, 50, or 200 years? This is a huge range, and the difference it makes is so great that we cannot help but wonder if there is any science to back these findings, or if these numbers were just plucked out of thin air.

To make things worse, the final report changed the language entirely, saying that carbon dioxide cycles vary between atmosphere, oceans, and land biosphere, so its residence and removal from the atmosphere involves a range of processes with different time scales. And since the air at different heights moves around constantly, it does not lend easily to calculations or even estimates.

One thing scientists agree on, however, is that as a result from all these processes, 20% of CO$_2$ may remain in the atmosphere for hundreds or even thousands of years.

Now the lifetime range has been increased from 5 to thousands of years. This tells us that we know as much as the experts do…which is almost nothing.

Changes in the Summary for Policymakers have a lot to do with increasing the confusion on the issue, since many politicians and the media depend on the summary for reliable information. The exceptionally wide range of the estimates unfortunately contributes to increased confusion worldwide as well.

New studies attempt to clear the confusion over the lifetime of CO$_2$.

We know that the main CO_2 removal from the air processes are:

a) vegetation growth,
b) reactions with calcium carbonate in the oceans,
c) reactions with igneous rocks like granite and basalt, and
d) reactions with other gasses and substances in the atmosphere.

Vegetation absorbs CO_2 the fastest, but it's not well understood, since additional plant growth (and decomposition) also release CO_2, depending on local climate, so it is possible that the combination of all these natural processes produces as much CO_2 as is absorbed. In addition, the land area covered by vegetation also changes, as do the types of vegetation, so this adds more uncertainty.

The oceans are credited with only a limited amount of CO_2 absorption, since sooner or later, ocean water reaches an equilibrium state with the atmosphere and can't absorb any more. And since the oceans have done this for a long time, it is likely that the saturation limit has been reached long ago.

Considering the above, there are only two major and longest duration effects that can be attributed to CO_2 absorption—reactions with calcium carbonate in ocean sediments, and in igneous rock.

The calcium carbonate reactions take a very long time, so it is estimated that it will take tens of thousands of years to remove half of the dissolved CO_2 from the oceans. At the same time, igneous rocks reaction takes even longer—up to hundreds of thousands of years.

Using the "half-life" model is not very practical in the case of carbon dioxide because the physical system has too many variables and components, so a single atmospheric lifetime is not applicable. Analogous to radioactive waste, a variety of different radio-nucleotides with a wide range of half-lives would not allow the determination of one single lifetime estimate. Similarly, the half-life (lifetime) of CO_2 in the atmosphere is too complex a variable to be expressed in exact terms or numbers.

There are published estimates that CO_2 would stay in the air for up to 200-300 years, but these claims are not proven.

We know that the natural geological processes (without human intervention) that determine the return of CO_2 back to Earth could take tens to hundreds of thousands of years...again depending on many factors, including local conditions.

At the same time, the ocean and the biosphere are dependent on how much CO_2 is emitted into the atmosphere, which makes the CO_2 lifetime also a function of the amount of CO_2 gas emitted in the first place.

The conclusion we can make from all this is that a short-term (5-10 years) removal of CO_2 from the air by plants and the ocean is not a dominant factor in reducing CO_2 quantities. Instead, it seems that different chemical processes of long duration (up to hundreds of thousands of years) are most likely what controls the process. It also seems probable that a certain percentage of emitted CO_2 will remain in the atmosphere forever.

In all cases, we must draw the conclusion that the majority of every gram of CO_2 we put in the atmosphere will remain there at least for the duration of our lifetime, and that the zillions of grams of CO_2 will be there to poison us and to change our environment—its weather and climate—for many, many years to come.

How all this affects our energy security is another issue that requires thorough understanding of the overall energy-environment situation. So let's continue...

The Other Major Natural Events

When discussing our environment we must always remember that the events taking place in it daily and seasonally are many and very complicated. There are thousands of different parts and aspects of these. Because of that, it is not easy to understand what happens in any given place and time—even if we had all the facts—unless we know what's going on all over the world. In most cases, the interactions among the different components make the situation dynamic and even more unpredictable as more locations and events are involved.

Because of the importance of these events, however, we need to understand as much as we can and include them all in the overall picture. Some of the key events and factors that affect our energy supplies, the environment, and all life on Earth are given below.

Water Vapor

There is a lot of water vapor in the atmosphere. It is most likely the highest concentration of any gas species in the lower altitudes of some regions of the world—and surely over the oceans. Since the oceans are 2/3 of the Earth's surface, it is simply amazing that we don't pay more attention to the huge quantity of water vapor over and around them.

It has been confirmed that water vapor is a significant greenhouse gas (GHG). It effects the environment and plays a large role in our lives—from making it hard to drive in the LA's morning fog, to reducing the efficiency of solar panels, and to its great impact on global climate.

The distribution of atmospheric water vapor varies across the globe. Equatorial and tropical latitudes are especially saturated by water vapor, especially during the respective summer months. This is especially evident in South Asia, where monsoon thunderstorms sweep the water vapor gas some 2 miles above the Earth's surface for extended periods of time.

It is common knowledge that water vapor is abundant, since it is visible in as fog and clouds. It also lets us know its presence in the sky by dumping buckets of rain, or truck loads of snow, on our heads. So how can we miss including water vapor in our GHG and global climate change calculations?

The excuse scientists give for this omission is that the measurement methods used thus far have not been precise enough to provide reliable data. This is because the old space-based instruments could not measure water vapor at all altitudes in Earth's troposphere, and especially 1-10 miles above ground level.

Today, new instruments used in space finally give us solid proof that water vapor affects the environment and is a major cause of climate change.

NASA satellites have been giving some data about the heat-trapping effect of water vapor in the air, officially validating its role as a critical component of climate change. The scientists still don't know the exact extent of its contribution to global warming, so the debate will continue, but at least it will now be included in the discussions and estimates.

Water vapor has a heat-amplifying effect, which is powerful enough to double the global climate warming caused by increased levels of GHGs in the atmosphere. There are also scientific experiments that confirm the estimates of different climate models. Data from the Atmospheric Infrared Sounder (AIRS) on NASA's Aqua satellite continuously measure the humidity (water vapor concentration in the air) in the lowest 10 miles of the atmosphere.

AIRS is the first space instrument able to distinguish differences in the amount of water vapor at all altitudes within the troposphere. AIRS data are used to observe how atmospheric water vapor reacts to shifts in surface temperatures. From these data, changes in the average global strength of the water vapor feedback can be calculated and modeled. When combined with sur-

face-based measurements of humidity and temperature, the satellite data are used to build models of the interactions between water vapor, GHGs, and other gases in the atmosphere at different locations on Earth.

NASA's data are now filling the gaps in the global warming theories, by estimating the magnitude of water vapor feedback. It has been determined that increasing water vapor concentration leads to warmer temperatures, which in turn generate more water vapor into the air, thus creating a vicious cycle of water vapor and warming effects.

On the other hand, places with concentrated water vapor, such as ocean and lake shore communities, are unsuitable for solar energy, due to significant reduction of sun intensity. These locations, however, have significant wind currents, which can be used to generate wind power.

Water vapor interactions are complex, but now we know that water vapor plays a significant role in our energy supply and affects the climate change cycle. There is a big difference between the effects of dry air vs. those of air that is saturated with water vapor. Water vapor stops sunlight from reaching the Earth, while amplifying the warming effect of other GHGs.

The prevention of solar light from reaching the Earth's surface and allowing more water vapor to enter and remain longer in the atmosphere contributes to further increase in the global climate warming effect.

Researchers have determined that if Earth warms by an additional 1.8 degrees Fahrenheit, the associated increase in water vapor will trap an extra 2 Watts of energy per square meter. Keeping in mind that the Earth's surface is approximately 510,072,000 square kilometers, or 510,072,000,000,000 square meters, we see that the increase is equivalent to over 1.0 trillion kW of energy.

This is an enormous amount of energy (heat mostly) that water vapor traps, thus increasing the global warming with time. NASA scientists now think the tropospheric water vapor feedback is an extraordinarily strong event, capable of doubling global warming, compared with carbon dioxide acting alone.

This is another confirmation of previous model predictions that Earth's GHGs will contribute to a temperature rise of a few degrees by the end of the 21st century. The only new component in this estimate is the large presence, numerous effects, and great influence, of water vapor in environmental processes.

So, it is now confirmed that water vapor is one of the biggest contributors to atmospheric phenomena that influence global warming. Its role has been underestimated and even ignored until now, but considering the vastness of the world's oceans and other water bodies, which evaporate huge amounts of water non-stop, we must reconsider the role of water vapor in all this.

Including global water vapor in our climate change measurements and prognosis will give us a much better picture of what is going on in the atmosphere and what to expect in the future. One thing is for sure, the role of water vapor in global warming is much bigger than we previously thought.

The good news here, in addition to the fact that now we can measure the water content in the troposphere precisely, is that the volume of global water evaporation in the atmosphere is fairly constant, which would make predictions more accurate. Or would it?

Earth's Magnetosphere

One amazing thing that few people are aware of—although their lives literally depend on it—is the presence of a magnetosphere around our Earth. A magnetosphere is an area of space near the Earth that has magnetic properties, in which charged particles (such as high velocity photons and energetic particles (radiation from the sun) collide and are controlled by the force of the magnetic field.

The Earth's magnetosphere is the outer layer of the ionosphere, where in places the ionosphere and magnetosphere blend, and in places they are separate.

Near the surface of the Earth, the magnetic field lines are symmetrical and parallel and resemble those of an ideal magnetic dipole, while farther away, the field lines are significantly chaotic and distorted by external currents, such as collisions with charged particles hurling at Earth with the solar winds. Over the Earth's equator the magnetic field lines become almost horizontal, then return to connect back again at high latitudes, while at high altitudes, the magnetic field is significantly distorted.

Looking at a 24-hour span,

- On the dayside of the Earth, the magnetic field is significantly compressed by the solar wind to a distance of approximately 40,000 miles. The Earth's bow shock is about 11 miles thick and is located about 56,000 miles from the Earth.

The magnetopause exists at a distance of several hundred kilometers off the surface of the earth, and has been compared to a sieve, as it allows particles from the solar wind to enter our atmosphere. Kelvin–Helmholtz instabilities occur when large swirls of plasma travel along the edge of the magnetosphere at a different velocity from the magnetosphere, causing the plasma to slip past.

This results in magnetic reconnection, and as the magnetic field lines break and reconnect, solar wind particles are able to enter the magnetosphere.

- On the night side of the earth, the magnetic field is relaxed, since the sun influence is reduced, and extends in the magneto-tail at over 3,900,000 miles in length, and is the primary source of the polar aurora; the amazingly colorful and pretty polar lights we see in the northern hemisphere.

So in practical terms, the magnetosphere with all its components, complex composition and strange behavior is a powerful shield that filters some of the most harmful energetic particles coming from the sun, space, and other celestial bodies. Only harmless, lower power particles can go through, thus we get relatively low power sunlight and mild weather patterns.

Without this shield, the amount and intensity of the solar winds and the matter from the sun's great explosions—which emit particles that are thousands of degrees hot and travel at thousands of miles per hour—would damage the Earth and all life on it.

This is an amazing and quite spectacular phenomena, without which we would not even be here. Just think that a slight change in its state or size would make the Earth vulnerable to cosmic energies that could cause unprecedented damage to the environment and all life on Earth.

And yet, how many people are aware of the presence of this life-saving phenomena? We don't see many hands up in the air…

The Rain Forest

Oxygen is the engine of life. Without it there would be no life on Earth. Period. If there is not enough oxygen, or if it is contaminated with significant amounts of impurities, life would be full of hardship, illness, and death.

90 percent of our nutritional energy comes from oxygen, and only 10 percent is derived from the food we ingest.

Oxygen is the supreme element, the primary nutrient without which life couldn't exist. We can live without food for weeks and without water for three to seven days, but without oxygen we have between 3 and

5 minutes. So, how important is it?

Yes, we understand this, and also know that oxygen comes from plants, which absorb CO_2 and release oxygen. We know that we must value the forests for this gift. But...did you know that most, if not all, oxygen generated by the enormous Amazon rain forest, which is considered to be the greatest oxygen generator in the world, is re-absorbed by the rotting vegetation and the soil as soon as it is emitted?

One would think that's the end of hope for clean air, but the amazing natural process doesn't stop there. Instead, it takes the forest's vegetation and soil in the runoff waters and carries the oxygen locked in the oxygen rich substances to the oceans. Plankton in the nearby waters feeds on the substance and releases the oxygen in it, thus supplying a major part of the oxygen we breath.

Because rivers and streams are responsible for carrying the runoff, damming rivers stops the flow of oxygen into the oceans.

Reduced runoff reduces the total amount of oxygen emitted in the atmosphere by the plankton colonies. So, if building new hydro dams in and around the rain forest continues, we might experience accelerated oxygen depletion in the air we breath. This complicated twist of events of great proportion still remains somewhat poorly understood.

Regardless of how oxygen is generated and regenerated, the rain forest is a net oxygen producer and essential our daily resupply. And regardless of its benefits, we are destroying it at a fast pace, and with that we are reducing its capability to provide us with oxygen. Isn't that like cutting the branch we are sitting on?

Yes, it will take a long time to destroy the rain forest and all the world's forests, so this is a long-term problem, but it is something that should not be ignored, as we are cutting our own oxygen supply.

But this is a long-term problem, so we will let somebody else (the future generations) worry about it. They must simply come up with some ingenious way of making a lot of oxygen, or if that fails, replant the forests.

Volcanoes

Inside the earth, there are many volcanoes, which sometimes cause great damage upon eruption. This is the only thing we see and know about most of them, because the rest is hidden deep underground and is well beyond our control.

In addition to destroying property and hurting people, volcanoes also emit a lot of dust and soot, which hurts people and hinders air traffic.

Excess amounts of black dust and soot emitted by large volcanic eruptions are now being found to cover large areas of the ice cap too. The black color absorbs sunlight, which warms the ice cap and makes ice melt faster, thus further contributing to global warming and rising ocean levels.

But volcanoes also do a lot of good. Most of them emit gasses during eruption, or in smaller volume continuously, which are blown into the sea. Here, they feed sea life, including large sea plankton colonies, which also generate large amounts of oxygen.

Would you believe that half of the oxygen we breath every day is generated by different types of plankton in the world's oceans via the above processes?

Unfortunately, most people cannot see any of these processes, but Earth-orbiting satellites do see and document these events and their effects...every day, day after day.

Lightning Strikes

Now think of lightning. We have all seen its spectacular and intensive light displays in the sky, but few of us have a good idea of its effects.

Over 3 billion lightning strikes hit the Earth's surface every day.

This is a lot of energy, created by the friction of water and ice particles in the clouds. The friction creates electric charge, and when the charge grows large enough it is discharged, directing its enormous energy towards the nearest object in its path.

The temperature of a thunderbolt is 5 times hotter than the sun, and there is enough energy in an average thunderbolt to power a city like Denver, CO, for 10 hrs. This enormous amount of energy is discharged into a very small area, usually only several square centimeters, which is why people are killed and trees set on fire when hit by a lightning.

If we could capture this enormous energy, we could power entire cities, which would be a major boost to our energy supply system and our energy security.

During the electrical discharge, the enormous electrical voltage and current splits the N_2 molecules in the air. The free nitrogen molecules get quickly oxidized into nitrates. The nitrates are then dissolved by the rain drops and fall onto Earth to feed the plants as a heavenly fertilizer.

13,000 tons of nitrates are dissolved and fall with the rain onto the rain forest alone every year.

These are essential for the survival of life, for they feed the vegetation, which shelters and feeds the wild animals.

Wild Fires

Wild fires are feared since they damage property and kill wildlife and people. Now let's look at them from a different perspective. Wild fires basically destroy the old, but provide an opportunity for new life too.

Wild fires renew the forest by burning trees and animal remains, and deposit carbon rich ashes on the ground, where new vegetation can feed on them.

Nine tenths of the forest's energy is stored in the leaves and tissues of the trees themselves.

The average forest floor is a porous mass that prevents minerals and nutrients from being washed away and lost. As soon as a tree falls, or a creature dies, decomposer begin to turn it into a food source and mulch. The vegetation tries to renew the cycle quickly and absorbs the nutrients that are released.

Forest fires help the renewal process by accelerating the normal aging processes, whereby a tree can be burned in minutes, instead of slowly decomposing for years. This way its nutrients are delivered to the soil more quickly and efficiently.

This is an example of the tightest and most efficient ecosystem in nature. Damaging one part of the system, however, like destroying too much of the rain forest, can bring negative change and even destruction to the entire system.

There are an amazingly high number of wild fires around the world. We cannot see them all, but the satellites do. A satellite picture of the Earth during wild fire season makes it look like it is burning from end to end. About 19 million square miles of forests and brush are burned around the world annually according to their observations and calculations.

Wildfires occur on every continent except Antarctica. Wildfires are a common occurrence in Australia especially during the long hot summers usually experienced in the southern regions such as Victoria. Due to Australia's hot dry climate, wildfires (commonly referred to as bushfires in Australia) pose a great risk to life and infrastructure during all times of the year, though mostly throughout the hotter months of summer and spring. In the United States, there are typically between 60,000 and 80,000 wildfires each year, burning 3 to 10 million acres of land.

On average, over 200,000 acres are burned every day around the world, or over 150 acres every minute. Experts also estimate that 130 species of plants, animals, and insects are lost every day.

Today, more than 20% of the Amazon rainforest has been destroyed and is gone forever. The land is being cleared by fire for agricultural uses, and some forests are being burned to make charcoal to power industrial plants.

There were an estimated ten million Indians living in the Amazonian rainforest five centuries ago. Today there are less than 200,000. More than half of the world's rainforests have been destroyed by fire and logging in the last 50 years. At the current rate of destruction, the last remaining rainforests could be destroyed during the next 40-50 years.

Fossil records and human history contain accounts of wildfires, as wildfires can occur in periodic intervals. Wildfires can cause extensive damage, both to property and human life, but they also have various beneficial effects on wilderness areas. Some plant species depend on the effects of fire for growth and reproduction, although large wildfires may also have negative ecological effects.

Note: Ash and particles from large wild fires increasingly cover large portions of the surface of the ice in some areas of the world. This darker matter absorbs sunlight, which heats the ice surface and contributes to faster melting of the ice cap. This phenomena contributes further to quicker ice melting and the accompanying negative climate warming effects.

The Ocean's Air-conditioning System

Now here is something that very few of us know of. It involves special properties of water during freezing. The majority of this event happens at very cold weather at the ice-covered shores of Antarctica. Every winter, the ocean waters at the cost line start freezing. Salt ocean water resists freezing, but at temperatures below –110 degrees F (40 degrees lower than the North Pole) the water finally gives in.

During the slow freezing process, the water does something unexpected. The freezing ice crystals release brine, which is squeezed from them like from a sponge. Brine runs out of the ice at a rate of 1 trillion gallons per hour.

The concentrated brine is heavier than the diluted water, so it slowly sinks to the bottom of the ocean. Trillions of tons of brine are released this way every winter and as the underwater pool grows larger, the brine sinks deeper and deeper into the 2 mile deep abyss, where it will remain for centuries if undisturbed.

Ocean currents pick up some of the brine, and take it along for their never ending trips around the world. The cool brine lines up along the current paths around the globe and plays a major role in controlling water and air temperatures.

At the equator, the ocean waters warm and the cold brine rises, keeping the ocean temperature constant. This brine current regulates the ocean temp to ±1 degree, which affects the air above and with that the entire global climate. After each trip around the world, some of the brine eventually gets back home, where it is dissolved in the water, only to repeat the cycle next winter.

Just imagine how it would be without this little known effect of global proportions. The surface temperature of the equatorial waters would rise every year. Without the cooling effect of the cold brine, the rising ocean temperature would damage the marine and land environment, and increase the ice melting process.

So the Antarctica brine and its amazing effect is another thing we must consider in the overall discussions about our environment.

Note: The increased climate warming is contributing to increased melting of the ice cap, which is threatening to disrupt the natural cycle—including the ocean waters' cooling effects. As global temperatures rise, ice is melted. As the ice melts, the overall temperature at the poles increases, contributing to further and faster ice melting. This creates a catch 22 situation with the ice caps melting, the ocean air conditioning decreasing, and the ocean levels increasing worldwide.

Wind and Storms

The most noticeable events in our weather pattern are wind, rain, and snow. Water vapor, megatons of it, rises up a mile or more over the oceans and other water bodies every day under the heat of the sun.

High in the sky, some of the water vapor cools and condenses in the cooler air. Some of the mass continues to move around, agitated by the changes. These activities create disturbances, which result in low and high pressure areas, which develop into wind currents. If the wind speed and intensity increases, these disturbances build and swirl under the influence of the Earth's rotation, eventually evolving into storms.

The storms' power is expressed in very strong winds, hail, thunder and lightning, heavy precipitation, freezing rain, dust storms, blizzards, sandstorms, etc. It is estimated that a large storm system carries enough mechanical and electrical energy to provide 200 times the world's electrical power.

Hurricanes are a type of very strong storm that forms in some tropical areas. Their power is also ominous and devastating if a population center is hit. Hurricane Katrina hit New Orleans in 2010 and devastated parts of the city, killing 1,833 people, and dislodging many others. It broke the retaining walls around the city and flooded parts of it. It carried and dumped over 2 trillion tons of water.

We humans are defenseless against this type and scale of destructive power. All we can do is learn and understand what causes these events with the hope of predicting and avoiding them. Weather satellites are quite useful in this respect. They can see the developing disturbances in the atmosphere and warn us of their approach.

The Sun's Role

We should've started this section with a discussion about the role of the sun, because it, and its energy reaching the Earth, are the driver of all the activities around us. The sun's energy radiates equally around it, so the amount of energy reaching the Earth is only a very small portion of the total sun energy emitted in space. Yet, we still get over a billion terawatts of raw energy, which is many times the energy used daily around the world. The sunlight drives photosynthesis, creates winds, and helps humans with their daily tasks, providing light and warmth.

But sunlight can be harmful too. A set of satellites, orbiting up to 22,000 miles above the Earth, collect data on the sun's activities daily. They monitor and record the spectacular Coronal mass ejections on the sun's surface, where billions of tons of matter, heated to thousands degrees Celsius are ejected in all directions in massive explosions.

Some of these explosions have been recorded to release the power of over 14 million Hiroshima atom bomb blasts. Hurling towards Earth at a speed of over a million mph, this large amount of energetic matter (solar winds) could strip our atmosphere and dry up the oceans within minutes.

But another miracle is at play here, and watching over us all the time—the magnetic field surrounding the Earth—mysterious in itself. It is our magnetosphere, which shields and protects us from this ominous destructive power.

The strong solar winds hitting the magnetosphere distort it 120,000 miles across in all directions, but it manages to deflect or stop most of the sun's particles. During very strong solar winds, some of the most energetic particles penetrate the outer magnetic field, but the inner magnetic shield channels most of the particles to the pole regions.

Here the sun's originating particles interact with particles in the atmosphere, the result of which is an event we see reflected in a magnificent Aurora, or the polar lights. This amazingly beautiful light show is a reminder of the awesome power of the sun, and the incredibly important role it plays in our lives—albeit going mostly unnoticed.

The natural events have a major effect on our lives, in a complex way that is hard to comprehend, but we are aware of the combined results. As if this were not enough, there is another set of events that are not natural. We call these man-made, because humans—intentionally or unintentionally—play some role in these.

Following are some of the key man-made events.

THE MANMADE CARBON CYCLE

While by definition it is easy to separate natural from man-made events, these are sometimes interwoven and get confused in the rhythm of nature's and man's daily activities. It is hard to tell where some of the events originate, where they are headed, and their overall effect.

Since the environmental movement of the 1970s, the way we look at environmental issues has changed. While the initial emphasis was on conventional air and water pollutants, which were the most obvious and easily measurable problems, newer issues are long-term problems that are not easily discernible and can be surrounded by controversies.

For example, we know that the effects of our energy sources are significant, but we don't know exactly how and to what degree. We know that all energy sources and facilities emit some type and amount of pollutants—gasses, liquids, or solids (particles and solid residue)—during their cradle-to-grave lifecycle*.

*Note. The cradle-to-grave life cycle begins from the initial idea, and runs through the production of raw materials (coal, oil, gas, metals for equipment manufacturing, silicon for solar cells, etc), and through all stages of planning, design, permitting, installation, O&M, decommissioning and disposal of all materials

and buildings. This also includes the efforts to mitigate the environmental and health damages done for the duration, and the final effort of returning the area to its original condition.

The amount of GHG (carbon dioxide and other harmful gasses) emitted depends on the specific energy source, the method of power generation, and the equipment and facilities used in the process. In all cases we can measure, or at least estimate, the emissions from each location in grams of emission per kilowatt of generated power. These numbers can then be used to calculate the damages done by the particular energy source or facility, and figure out ways to reduce or eliminate the harmful effects.

The following events are a mixture of natural and man-made conditions which we will review keeping in mind that the boundaries might be too fuzzy to distinguish one from the other. These are also major contributors to present-day environmental issues at different stages of human understanding and mitigation attempts.

Let's take a closer look at what man is accused of doing wrong today:

CO_2 Emissions

Of all environmental culprits, CO_2 is the most notorious. It is the evil of all evils, and the reason for everything bad happening to the environment—or at least this is the present-day consensus. So let's take a look at CO_2 as the all-encompassing and most urgent concern as far as the global warming goes. We need to examine it also because PV installations are measured in terms of tons of CO_2 emissions saved, as compared with those generated by fossil fuel plants of the same size during a certain period of time.

A gallon of gasoline, which weighs about 6.3 pounds, could produce 20 pounds of carbon dioxide (CO_2) when burned. Impossible? Yes, possible! The $C+O_2$ combination, expressed in CO_2 units, is a wicked one, with some very special properties.

When gasoline or other carbon-containing hydrocarbon fuels burn, the carbon and hydrogen separate, the hydrogen combines with oxygen to form water (H_2O), and carbon combines with oxygen from the surrounding air to form carbon dioxide (CO_2).

Methane, which is one of the simplest hydrocarbon formations, CH_4, burns easily in an oxygen environment to produce CO_2 and water.

$$CH_4 + 2O_2 = CO_2 + 2H_2O$$

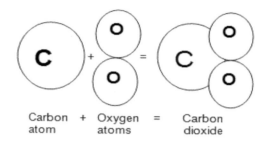

Figure 1-9. CO_2 formation during burning (i.e., coal)

Burning methane generates over two times less CO_2 than pure carbon, thus natural gas (which is mostly methane) is cleaner burning than coal. Some of the other harmful emissions from coal (particulates, SO_2, etc.) are also absent during natural gas burning. For this and other reasons, natural gas is the preferred fuel today for power generation and for powering vehicles.

At present, coal is responsible for 30-40% of world CO_2 emissions from fossil fuels. Carbon in coal and oil generates large amounts of CO_2 upon burning—four times more (in weight) than the original carbon content itself. This is because there is twice the amount of oxygen in the CO_2 molecule, and while O_2 is not harmful on its own, it is heavy and creates large amounts of CO_2.

Oxygen just happens to like the bad company of carbon and sticks to it tightly, thus making things even worse for coal. The resulting CO_2 gas is the main constituent of the harmful GHG gasses that cause global warming.

$$C + O_2 = CO_2$$
$$1.0 \text{ gr.} \quad 3.67 \text{ gr.}$$

As can be seen, one unit of carbon (in coal or oil) produces almost four times the amount of CO_2. Who knew? And we now know for sure that CO_2, chlorofluoro carbons (CFCs), methane (CH_4), and nitrous oxide (NO_x) emissions cause greenhouse effects. Since global warming has been increasingly associated with the contribution of CO_2, we need to pay attention and understand the phenomena at play in this process.

The contribution of the different gasses to global warming in the form of GHG generation is believed to be as follows:

CO_2	55%	GHG effect
CFCs	24%	"
CH_4	15%	"
N_2Ox	6%	"

If and when the *natural gas hydrates* (NGH), or *methane hydrates* (see below for more details) become a primary energy source, this ratio will be changed in favor of methane, which is several times more capable of producing the GHG effect, because it produces a large quantity of CO_2, as follows:

$$CH_4 + 2O_2 = CO_2 + 2H_2O$$
$$1 \text{ gr.} \quad 2.75 \text{ gr.}$$

Either by raw gas escapes, or during burning, methane will create large amounts of CO_2, if and when a large quantity of NGHs are extracted from the permafrosts and processed into useful fuels.

Other hydrocarbons, actually most of them, also produce a large quantity of CO_2 simply because they react with oxygen to produce heavier molecules, as follow:

$$2C_4H_{10} + 13O_2 = 8CO_2 + 10H_2O$$
$$1.0 \text{ gr.} \quad 3.03 \text{ gr.}$$

The reaction is $C + O_2 = CO_2$, where a carbon atom with an atomic weight of 12, and an oxygen atom with an atomic weight of 16 combine, producing a single molecule of CO_2 with an atomic weight of 44 (12 from one carbon atom and 32 from 2 oxygen atoms).

So, to calculate the amount of CO_2 produced from a gallon of gasoline, the weight of the carbon in the gasoline is multiplied by 3.7 (44/12). Since gasoline is about 87% carbon and 13% hydrogen by weight, the carbon in a gallon of gasoline weighs 5.5 pounds (6.3 lbs. x .87). Then we multiply the weight of the carbon (5.5 pounds) by 3.7, which equals approximately 20 pounds of CO_2 produced by a gallon of gasoline.

Amazing! The gallon of gasoline we burn every 20 miles of driving creates 20 lbs. of poisonous CO_2. So this means that an average car leaves approximately 1 lb. of poison with every traveled mile. This is doubled for SUVs, tripled and quadrupled for larger trucks and RVs, and multiplied many times over for jet planes, boats and other large vehicles. Who knew?

Consider the case of family van that burns one gallon of gasoline every 10-15 miles (or every 5 miles if we drive a Hummer, or RV). Every 10-15 miles, we leave 20 lbs. of CO_2 behind us, which goes up in the atmosphere to accelerate the global warming process.

Every 100 miles driven in the family van generates 400 lbs. of CO_2—not a commendable footprint. A summer vacation trip of 1,000 miles will leave 4,000 lbs. (over 2 tons) of CO_2 footprint behind our happy family's van. And 100,000 families with similar summer trip plans will load the atmosphere with an additional 400,000,000 lbs. (200,000 tons) of CO_2 within weeks… 1 million families… you get the picture.

How about a 2 GW coal-burning power plant bellowing dense clouds of smoke all day? Or a cruise ship navigating aimlessly around the Gulf of Mexico burning thousands of gallons of diesel? Multiply that by thousands of such activities and you'll see that millions of pounds of harmful CO_2 and other gasses are emitted into the atmosphere every day. All these poisons travel up into the atmosphere, where they accumulate and do their damaging job non-stop.

What we see in Table 1-3 confirms that coal is the worst GHG emitter. The rest of the fossils are not far behind, with nuclear, bio-, geo- and hydro-power showing the most respect for the environment. Of course, wind and solar are at the same level too, which means that they are our hope (and that of future generations) for energy independence and clean environment.

The Practical Aspects

Carbon dioxide emissions can be calculated as

$$Q_{CO2} = c_f/h_f \times C_{CO2}/C_m$$

where

Q_{CO2} = specific CO_2 emission (CO_2/kWh)
c_f = specific carbon content in the fuel (kg_C/kg_{fuel})
h_f = specific energy content (kWh/kg_{fuel})
C_{CO} = specific mass Carbon Dioxide (kg/mol CO_2)
C_m = specific mass Carbon (kg/mol Carbon)

Note: There is 55-75% heat lost during electric power generation. This energy loss is not included, but must be considered in the final calculations.

As total worldwide emissions do not seem to be going down, we must understand that as more CO_2 is emitted, more will remain behind in the atmosphere for a long, long time. Assuming that 20% of emitted CO_2 stays in the atmosphere forever, and that the ocean and biosphere are already saturated, we are faced with ever increasing levels of CO_2, quantities that cannot be reduced by the natural cycle. The only option is to make sure that we take serious steps to de-carbonize our world ASAP.

We all agree that there is a lot of CO_2 generated during the conversion of all fossils (and their derivatives) into electricity. A simple calculation shows that a 1.0 GW coal-fired power plant will emit 370 tons of CO_2 per hour (1,000,000 kWh x 0.37 kg). This amounts

Table 1-3. GHG emissions from power generators
*Including all cradle-to-grave emissions
**CCS—carbon capture and sequestration

Technology	Grams CO_2 per kWh*
Coal, conventional	1100
Coal w / CCS**	950
Diesel, oil	770
Fuel Cell, H2 gas	650
Natural gas	440
Nuclear	60
Geothermal	40
Biomass	30
Biogas	15
Hydro, river run	15
Hydro, dammed	10
Wind, onshore	25
Wind, offshore	15
Solar, PV modules	20
Solar thermal, CSP	40

Table 1-4. CO_2 emissions per kWh generated

Fuel	CO_2 emissions in kg per kWh
Coal	0.37
Gasoline	0.27
Light Oil	0.26
Diesel	0.24
LPG	0.24
Natural Gas	0.23
Crude Oil	0.26
Kerosene	0.26
Wood	0.39
Peat	0.38
Lignite	0.36
Bio energy	0

to 8,880 tons per day and to 3.2 million tons per annum. *This is from one average coal-fired power plant.*

Multiplied by the thousands of such power plants around the world this would bring the amount of CO_2 emitted in the atmosphere annually into many billions of tons—billions with a B—a very large number for our relatively small Earth and its fragile environment.

We must realize that it might be too late to return to pre-industrial CO_2 concentrations even if we stopped emitting CO_2 today. The present state of the art also would not allow us to neutralize or absorb the generated great quantities of CO_2 on a daily basis, so short of developing some exceptional (extremely efficient and cost-effective) absorption technology, we will be witnessing a rapid increase in CO_2 levels, and must be ready to accept the consequences.

Radically altering land use, designed to absorb as much CO_2 from the atmosphere as possible, is another, albeit unlikely possibility to combat extreme CO_2 increase.

And last, most of the CO_2 lifetime studies arbitrarily stop calculating the global warming potential (GWP) of CO_2 at around 100 years. But if the CO_2 lifetime were calculated over geologic time, it would increase by 50-100x, at the same time lowering the GWP of all other GHGs by an equal amount.

This only confirms the importance of considering and implementing immediate CO_2 emissions reduction, if we are to avoid great climate calamity in the near future. But that is quite unlikely to happen anytime soon, so we must keep in mind the aggregation of CO_2 and other GHGs in the atmosphere.

We can only hope that we will not see more air pollution-choked places like Beijing or Los Angeles. Or that the polluted air will not accelerate the global warming to the point where we are forced to fight for our basic daily survival.

Nitrogen Oxides

Nitrogen (N_2) is a predominant gas in the air we breath. Over 78 percent in dry air, to be exact. It is relatively inert element, which means that it doesn't participate readily in chemical reactions. At the high temperatures, however, nitrogen combines with oxygen (at about 1,100°C) and forms a variety of nitrogen oxides (NO_x).

The resulting gasses range from nitric oxide, or nitrogen monoxide (NO), nitrogen dioxide (NO_2); dinitrogen pentoxide (N_2O_5); and nitrous oxide, or "laughing gas" (N_2O).

The dominant component of NO_x gasses, NO, is not particularly harmful in itself, but NO_2 is toxic, and long-term exposure may result in lung damage and other illnesses. The reddish-brown color of NO_2 is why the polluted urban air has a distinct dirty brown appearance.

The worst part, however, is that NO_x gasses undergo further chemical reactions in the atmosphere with time and changing conditions. The most important reaction is that of NO which is readily converted into the more harmful NO_2.

Other important processes result in photochemical reactions, driven by the energy of sunlight. The result is the decomposition of NO_2, during which reaction NO and an isolated oxygen atom are produced. This oxygen atom is very reactive and is responsible for the formation of ozone, which is another story, and which we review in more detail.

Fossil-fueled power plants, industrial use of fossils, and transportation are the major sources of NO_x emissions worldwide. The NO_x emissions do not come from the fossil fuels, but from the air in the combustion chambers, which is needed for the burning to proceed. So there's no way to prevent NO_x emissions by altering the fuels or by removing nitrogen from the air supplied for combustion.

Nevertheless, there are a number of techniques available to help reduce nitrogen oxide production in power plants. Some of the practical methods consist of modifying the combustion chamber, controlling the flame geometry, using low-NO_x burners, air-fuel staging, catalytic combustion, adjusting the fuel-to-air ratios, and optimizing the fuel and air flow speed.

Recycling the exhaust gases through the combustion chamber is another efficient method of reducing NO_x emissions. The recycling reduces combustion temperature and with it the amount of NO_x produced. The problem is that by reducing the combustion temperature the thermodynamic efficiency is lowered, which also lowers the total useful energy generated by the power plant.

So, any of the NO_x reduction methods either increases the cost of operation, or reduces the efficiency and the total power output of the plant. The installation of additional scrubbers and their operation is an expensive undertaking, which costs millions of dollars in initial construction and annual O&M procedures. Some countries have regulations and can afford such modifications, but the majority of the world is not concerned with NO_x emissions.

Similarly, NO_x gas emissions from transportation vehicles (diesel trucks especially) are controlled to a degree in the developed countries, but not so much in the developing world. Because of that, global NO_x emissions are presently rising uncontrollably.

Sulfur Emissions

Sulfur is present in coal and, to a lesser extent, in oil. Unless it's removed before combustion (a technologically feasible but very expensive process) when burned sulfur produces sulfur dioxide (SO_2). However, the native sulfur content of fuels varies substantially, so sulfur reduction also can be accomplished by changing the fuel source.

For example, coal from the western United States typically contains less than 1 percent sulfur, whereas coals from the Midwest may contain 4 to 5 percent sulfur.

Sulfur dioxide itself is detrimental to human health. As usual, the health burden falls most heavily on the very young, the very old, and those already suffering from respiratory diseases. Acute episodes of high SO_2 concentrations have resulted in outright mortality. For example, in a 1952 air pollution incident in London, some 4,000 people perished when atmospheric SO_2 rose sevenfold.

Once in the atmosphere, SO_2 undergoes chemical reactions that produce additional harmful substances. Oxidation converts SO_2 to sulfur trioxide (SO_3), which then reacts with water vapor to make sulfuric acid (H_2SO_4):

$$SO_3 + H_2O = H_2SO_4$$

These reactions take place when the sulfur-containing gasses hit either water vapor or water droplets. In the liquid water, sulfuric acid dissolves into positive hydrogen ions and negative sulfate ions. Sulfate ions can join with other chemical species to make substances that, if the water evaporates, remain airborne as tiny particles known as sulfate aerosols.

These particles are typically well under 1/micron in size and are very efficient scatterers of light. As a result, they reflect sunlight and thus lower the overall energy reaching Earth. Consequently, sulfate aerosols have a cooling effect* on climate.

*Note: There are suggestions today to spread huge amounts of sulfur salts (in fine dust form) high up in the atmosphere by pumping them through large straws up in the sky. Once up in the air, they would travel around the world and serve as a reflector, which would reduce the sun's heating effects. This could reverse the climate warming trend with time, but the side effects are unpredictable.

Any large-scale project that has not been proven efficient and safe on the long run carries risks and contains unpredictable variables. In this case, increased acid rain, negative effects on solar power plants' operation, weather pattern changes, etc. are just some of the things that need to be considered before blowing tons of sulfur into the atmosphere.

Acid Rain

Acid rain and dry deposition (particles) of contaminated matter onto populated centers and other critical areas of human activities is primarily the result of SO_2 and NO_2 being emitted into the air. These gasses are generated in large quantity by fossil burning power plants. They travel freely, whichever direction the wind blows, and land in different places. Upon contact with different surfaces the chemicals in these gasses could change the acidity of the water or the soil they fall onto.

High temperatures created by the combustion of petroleum cause nitrogen gas in the surrounding air to oxidize, creating nitrous oxides. Nitrous oxides, along with sulfur dioxide from the sulfur in the oil, combine with water in the atmosphere to create acid rain. Acid rain is a condition where water droplets in the atmosphere meet SO_2 gasses and form sulfuric acid in the process. Thus formed dilute sulfuric acid falls to Earth during a rain storm, resulting in acid rain. The rain drops now are significantly acidic—much more than normal rain drops.

Acid rain is formed via different mechanisms, as follow:

1. Carbon dioxide reacts with water to form carbonic acid.

$$CO_2(gas) + H_2O(liq) = H_2CO_3(liq)$$

Carbonic acid then dissociates to give a hydrogen (H^+) and a hydrogen carbonate ions (HCO_3^-) H_2CO_3 produces H^+ which makes it an acid, thus lowering the pH of the solutions and causing corrosion.

$$H_2CO_3(liq) = H^+ + HCO_3^-(dissociation)$$

2. Nitric oxide (NO), which also contributes to the natural acidity of rainwater, is formed during lightning storms by the reaction of nitrogen and O_2, contained in the air.

$$N_2(gas) + O_2(gas) = 2NO(gas)$$

In air, NO is oxidized by the O_2 contained in it to nitrogen dioxide (NO_2)

$$2NO(gas) + O_2(gas) = 2NO_2(gas)$$

NO_2 in turn reacts with water to give nitric acid (HNO_3).

$$NO_2(gas) + H_2O(liq) = 2HNO_3(liq) + NO(gas)$$

This acid dissociates in water to yield hydrogen ions and nitrate ions (NO_3^-), similar to the dissociation of carbonic acid shown above, which similarly lowers the pH of the solution, thus making it acidic and corrosive.

$$HNO_3(liq) = H^+ + NO_3^- (dissociation)$$

3. SO_2 in the emitted gasses combines with H_2O and O_2 in the air to produce sulfuric acid, which is one of the principal bad guys in acid rain damage that can affect any structure and formation exposed to it.

$$SO_2 (gas) + H_2 (liq) + O_2(gas) = H_22SO_4(liq)$$

In all cases,

$$Emissions(gas) + Rain(liq) + Air(gas) = Acid Rain(liq)$$

Acid Rain Effects

Falling rain with acidity of pH 5 and below creates a number of unwanted effects, where it:

- Affects the aquatic life causing reproduction in fish to falter. Death or deformity are widespread among young fish living in acidic waters.

- Amphibians and invertebrates suffer similarly, and because of that, highly acidic water bodies and lakes do not support normal life. There are many such examples in the northeastern United States which has been particularly affected by acid rain created by the coal-burning power plants in the area. The same phenomena is observed in northern Europe and a number of other places around the world with a high concentration of coal-fired power plants.

- Many high-altitude lakes, even those in protected wilderness areas, have pH levels well below 5 and are considered "dead."

- Acid precipitation affects not only water quality; it also damages terrestrial vegetation, especially high-altitude trees.

- It also causes increased corrosion and weathering of buildings. Limestone and marble are particularly susceptible to damage from acid precipitation, as are some paints.

The effect of acid rain depends in part on geology; lakes or soils on limestone rock suffer less harm because the limestone can neutralize the excess acidity, but granite and quartz-based rocks lack this buffering effect.

Note: Technically speaking, pH is a measure of the concentration of the hydrogen ion p[H] in a water-based solution. Pure water has a pH of 7 +/- 0.1 at 25°C. Any solution with a pH less than 7.00 is acidic, while that with pH greater than 7.00 is basic or alkaline. Vinegar, for example, is strongly acidic (pH=2 or 3), while bleach or ammonia are basic (pH=10-12) depending on the concentration.

The pH level of aqueous solutions is usually measured with a special pH meter, or by using paper or liquid indicators which change colors with change of pH.

In the U.S., acid rain conditions in the Northeast from the burning of coal, and in the West from gasses generated by utilities and motor vehicles, have been causing problems since the beginning of the 20th century. The situation was partially exacerbated by the Clean Air Act, which forced coal power plants to use taller smoke stacks, resulting in transmission of acid rain gasses to much longer distances. This, of course, resulted in the contamination of larger land areas.

This is a good example of the complexity of the environmental issues, and how sometimes the best ideas, geared to solve problems, produce disastrous results.

During the Carter Administration, a risk-averse policy was undertaken through the EPA and Council on Environmental Quality (CEQ), aimed to research and control the pollutants suspected to cause acid rain even in the face of scientific uncertainty and growing opposition by interested parties. The Reagan administration, however, was not that concerned and was much more environmental risk tolerant. It believed that the scientific uncertainties surrounding environmentally caused exposure levels did not justify the necessary expenditures for their remediation. Therefore, any serious effort in that direction would unnecessarily curtail energy security and economic growth.

George H.W. Bush called for new Clean Air Act legislation to curtail SO_2 and NO_2 emissions. In 1990 he enacted amendments to the Clean Air Act where emissions were to be cut by over 12 million tons per year. A market-like system of emissions trading was implemented, and a cap on emissions was set during 2000, which was partially achieved by the installation of industrial scrubbers on the large emitters.

According to environmentalists, the initial costs in cutting emissions levels, to be paid by the utilities, was expected to be over $4.6 billion, resulting in a 40% rise in electricity costs. Instead, the total cost impact was only about $1 billion, which resulted in only 2-4% rise in electricity costs. The main reason for this "discrepancy" can be attributed to the fact that low-sulfur coal was used extensively in coal fired plants, so no major facilities upgrades were needed to reduce the emission levels.

Other groups, however, insist that trillions of dollars have been spent since the 1970s on emission reductions and other environmental measures, which have resulted in little improvement of air quality and instead have created other, even bigger, problems.

Note: In 1990 the U.S. Congress decided that a 50% reduction in SO_2 and NO_x emissions from power plants was a necessary first step to address the acid rain problem.

This resulted in the 1990 Clean Air Act Amendments which created the U.S. Acid Rain Program. This started one of the most drastic changes in the way pollutants are handled in this country, and especially SO_2 emissions which were now strictly regulated. Rather than relying on command-and-control MO, Congress set a strict emissions cap for power plants and, in return, provided them with unprecedented flexibility in determining how to meet that cap.

The Acid Rain Program set the cap at 8.95 million tons of SO_2 per year, to be implemented in two phases. In Phase I, which began in January 1995, the largest, highest emitting electric utility generating units were required to reduce emissions. Phase II, which began in 2000, required all electric utilities to reduce their emissions to roughly one-half of 1980 levels or the 8.95 million ton cap.

Under the program, sources are required to hold allowances equal to each year's emissions. Participants are free to choose how they want to comply, but at the end of each year they have to surrender enough allowances to EPA to cover their actual emissions. Failure to comply with this provision results in automatic penalties, which are enforced by EPA.

Although the U.S. Acid Rain Program is not perfect and is not expected to resolve the issues at hand, it is one of the most robust, functional, and practical environmental programs ever, which should be used as a model of how to handle the threat of excess pollution.

The biggest problem here, again, is that while such program, or programs, can be implemented in the U.S. and Europe, the rest of the world—and especially the developing countries—is not even considering it. And because of their rapid growth today, including increase of pollutants, we cannot expect any great reductions in the global GHG emissions—regardless of what the U.S. and EU countries do.

Gas Scrubbing

Sulfur emissions are controlled in the U.S., but since sulfur pollution is in a form of gas, it cannot be removed by the standard particulate pollution controls. So the most widely used sulfur-removal technique is *flue gas desulfurization*, or scrubbing.

The process is done in a wet scrubber, where the exhaust gases are forced through a water spray containing some base chemicals (usually $CaCO_3$ or $MgCO_3$. When the SO_2 gasses hit the water, they form sulfuric acid, which immediately reacts with the bases in the scrubber water, thus neutralizing the acids.

$$2SO_2 + 2CaCO_3 = 2CaSO_4 + 2CO_2$$

In a dry scrubbing process the exhaust gas is run through a pulverized limestone, which neutralizes the acids. In each case, the chemical reactions yield solid calcium sulfate ($CaSO_4$), which can be separated from the gas stream and is removed along with other particulate matter via standard filtering or centrifuge action.

Wet scrubbers can remove the majority (over 98 percent) of all sulfur from the exhaust gases, the price rising with the removal rate. Scrubbing costs about 15 percent of the total cost of a coal-burning power plant. In addition there is energy cost that could be close to 10 percent of the plant's total power output. On top of that, 3-4 percent of the power is used for particulate removal.

A great portion of this wasted energy is used to reheat the exhaust gases after the wet scrubbing so they can rise up the smokestack and be emitted into the atmosphere. Such a waste of energy and money, but there is no way around it—at least in the U.S., although the utilities have found a way to avoid using gas scrubbers. Instead they meet the sulfur emissions regulations with a number of strategies, such as switching to lower sulfur fuels, or trading so-called emissions credits.

By using these techniques, the coal-fired power plants are emitting less polluting gases when using more expensive low-sulfur coal and oil, and in some cases they can just pay emission credit penalties and keep on polluting as much as they want. How does this benefit the environment?

Gas scrubbers are widely used in Europe and Japan, but most other countries are much more flexible about using scrubbers, or any pollution reduction equipment, which is one reason for increased global air pollution.

Particulate Matter

The effects of small particles in the air have been of concern and are well known, but when it comes to predicting climate change (and their role in it) the picture is not that clear. The role of GHGs is quite clear, while the role of atmospheric aerosols and dust particles, as well as their interfacial chemistry's effects on the climate, and their interactions with the GHGs and water vapor are not that clear. The reactions up in the troposphere, where a lot of these interactions occur, are quite complex and not easy to observe, measure and document.

We have known for awhile now that the ways in which atmospheric particles, such as mineral dust, affect climate are important, but now we know that they are poorly understood. There are numerous processes where particles play some role in the troposphere, such as heterogeneous chemical interactions with the atmospheric gases, clouds formations, optical interferences, and combinations of these. All contribute to the radiative balance (or forcing), which is the net difference between incoming and outgoing radiation.

The role of particulate matter in the atmosphere and its chemistry is oversimplified by the existing climate models. At the same time, field researchers find the real world's complexity overwhelming, so better understanding of particles and their impact on climate, the environment, and human health is needed.

Particulate matter is emitted in great quantities on a daily basis by coal and gas power plants, airplanes, diesel engines, and small volcanoes. Periodically, equally large amounts of particles are emitted by the eruption of large volcanoes. The overall effect is complex, but we think of it as basically negative, because it contributes to the formation of acid rain, which damages buildings and metal structures.

But how about positive effects? On September 11, 2001, all U.S. based jet planes were grounded for three days. Ground climate tracking stations measured a significant increase of temperature at several points. We now attribute that anomaly to the fact that the jets' contrails were absent from the sky during that period of time. Without the particles in the contrails, the sun's rays were allowed to more easily penetrate the atmosphere and heat the ground than under normal flying operations.

On the other hand, the eruption of Krakatoa volcano in 1883 spewed ash and soot 50 miles into the air, and sent a cloud of particles around the world. It has been documented that the sun was hidden from the smoke cover, and that the global temperatures dropped by several degrees.

So, no doubt, the particle content of our atmosphere is a major factor in the daily environmental events. How this affects us is still debatable, but it must be included in future decisions affecting our environment and global warming in particular.

Figure 1-10. Many things to keep in mind...

Surely, the overall environmental picture has been complicated by man and nature recently, and we are now having a hard time figuring it all out. Our atmosphere is packed with a mixture of different chemicals (CO_2, SO_2, NO_x, VOC), and substances (water vapor, particulate matter, and who knows what else) during the last century of intense industrial progress and humanity increase.

Nevertheless, modern science is putting the pieces of the puzzle together...one by one...slowly but surely. There is no doubt that some day we will have the entire picture figured out, and decisive action will be taken to solve the major problems.

Photochemical Smog

Looking at the air above some densely populated centers (Los Angeles, Beijing, etc.) we see a brown-yellow cloud that consists of a complex soup of chemicals. Most of these gasses come directly or indirectly from fossil fuel burning in stationary power plants and industrial enterprises, and also from transport vehicles.

Urban smog formation is due mostly to high concentration of nitrogen oxides, as discussed above; hydrocarbons from incomplete combustion; production, handling, and storage of gasoline and diesel fuels. Significant amounts also come from evaporation of organic solvents in paints, inks, dry-cleaning fluids, and similar chemicals in daily use around the world.

These pollutants are trapped under some weather conditions, the most important of which is air inversion. This is a complex phenomena, based on the fact that air temperature usually decreases with altitude, because the atmosphere is transparent and doesn't absorb much energy from sunlight.

Earth's surface, however, is not transparent and heats up during the day. It then transfers energy to the air above it at a rate at which the temperature declines with altitude (lapse rate), which is normally measured at 6.5°C per kilometer of altitude.

Air in contact with the hot Earth's surface, or mixed with hot gases from fossil fuel combustion, is hot and less dense than its surroundings, so it tends to rise. On its way up, the hot air expands and cools, using its internal energy as it goes, pushing against the surrounding air.

Most of the time, this cooling is not as great as the lapse rate so the hot air remains warmer and less dense than its surroundings as it continues to rise, carrying away any embedded pollutants. In an inversion condition, however, the air temperature either increases with height, or decreases more slowly than a rising mass of air would cool. Thus, the hot air cools and becomes denser than its surroundings, which cause it to slowly sink back down.

The air in this case is stable, and the layers closest to the Earth's surface are *trapped* in place. Trapped with the air are any pollutants it contains and many more are attracted and locked in place during the day. Nearby topographical conditions facilitate and accelerate the trapping mechanism. Los Angeles basin is a good example of how nearby mountains trap polluted air, which assisted by the cool Pacific ocean create and maintain inversion conditions.

While idling in the trapped air of the inversion, the polluting gas molecules are activated by sunlight that impacts enough energy upon them to start chemical reactions among them. A key reaction in this situation is the decomposition of nitrogen dioxide into nitric oxide, which also produces an isolated oxygen atom (ion):

$$NO_2 + solar\ energy = NO + O\ (free\ ion)$$

The free oxygen ions are very reactive and react with molecular oxygen (O_2) to create another highly reactive chemical, ozone (O_3):

$$O + O_2 = O_3$$

On one hand, ozone in the higher stratosphere protects the Earth's surface from harmful UV radiation, but ozone in the lower atmosphere is harmful to human health. It is also a climate-changing greenhouse gas (GHG), which is blamed for climate warming and other environmental problems.

Ozone and O ions also are capable of initiating and supporting a number of additional reactions and effects. A number of other harmful substances can be created in the process, because O_3 molecules and O ions facilitate reactions among hydrocarbons (HC) and nitrogen oxides that result in a host of substances, such as volatile organic compounds (VOC) and so-called peroxyacyl nitrates (PAN), which are another set of highly reactive and harmful substances.

This is what photochemical smog is. It causes not only eye irritation, but much more serious respiratory problems. They also damage plants and many materials, such as paints, fabrics, and rubber.

In more detail its formation and behavior can be described as follows:

At night the levels of all air-suspended polluting substances are quite low, but the morning LA rush hour quickly increases the emissions of both nitrogen oxides and VOCs as people drive to work. Later in the morning, traffic dies down, and the NO_x and VOCs begin to form NO_2, quickly increasing its concentration.

As the sunlight becomes more intense later in the day, NO_2 is broken down and its by-products form increasing concentrations of O_3. At the same time, some of the NO_2 reacts with the VOCs to produce toxic chemicals such as PAN.

As the sun goes down, the production of O_3 slows down and the ozone that remains in the atmosphere is then consumed by several different reactions.

A number of meteorological factors can influence the formation of photochemical smog, including: Rain can reduce the photochemical smog as the pollutants are washed out of the atmosphere with the rainfall; winds can blow photochemical smog away replacing it with fresh air. The smog is then blown to distant areas where it can cause similar problems.

There is no quick or easy way to reduce or eliminate photochemical smog, short of reducing or eliminating the emissions of NO_x, HC, and other polluters. The power plants and vehicles NO_x emission standards in the U.S. and other developed countries are playing a major role in keeping them low, but that is not the case in most developing countries, so here again, we don't see a reduction in photochemical smog around the world in the near future.

Ozone Depletion

There has been talk about ozone depletion for decades now, and we know that it means reduced concentration of ozone in the Earth's stratosphere (the ever shrinking ozone layer). We also know that ozone is needed to block part of the sun's harmful UV radiation. But we still don't know the size of the problem and its actual effect on the Earth's environment, nor do we know exactly what causes the ozone layer problems.

Chlorofluorocarbons (CFCs), which have been used extensively in the 20th century, were blamed for much of the depletion of the ozone layer, so EPA and FDA banned CFCs in aerosol cans in the late 1970s and cars later on. Did that solve the problem? Not sure...

In the 1980s we learned that the problem was much worse than before, and a massive (albeit controversial) hole in the ozone layer over Antarctica was identified. International agreements were made to reduce the ozone-damaging substances, i.e., the Vienna Convention, the 1987 Montreal Protocol, and a third agreement in 1990 in London. The 1990 Clean Air Act Amendments phased out production of CFCs in the US and required recycling of CFC products.

So, the phase-out of CFCs and other similar policies are seen as successful, and a crisis seems to be averted, and yet, the ozone layer is still depleted. This, according to the proponents, is due to the longevity of CFC particles in the atmosphere, so the ozone layer will, hopefully, start showing signs of recovery by 2025.

A lot of time, effort and money was spent on fixing the hole in the ozone layer. We don't doubt that most of these efforts are of some benefit, but we doubt that we have done everything just right and on time. So, it is unlikely that the ozone layer will become intact, or even close to what it was in the 19th century.

Heavy Metals

Not surprisingly, fossil fuels contain a number of different heavy metals, which are released in one or another form in the exhaust gasses. There are many other ways for heavy metals to enter the environment. One of the most obvious is through wear in engines, turbines, and other machinery.

The heavier metals act as biological enzymes, which have a wide range of toxic effects on humans. Especially dangerous are the deleterious impacts on brain development in young children. For example, lead which until recently was a common component of gasoline and paints, is a particularly serious contaminant with levels that remain high in urban air, soil and other surfaces.

Mercury is another key toxin particularly harmful to children and animals. It is found in coal, and is emitted during coal burning at power plants or home heating. The dominant sources of mercury contamination are coal-burning power plants. Mercury pollution in surface waters collects in the food chain and ends up at very high levels in larger fish. Many lakes of the central and eastern U.S. are so highly contaminated that fish are unsuitable for human consumption.

A recent study shows that 1/5 of all Americans have mercury levels exceeding EPA recommendations, which are maximum 1 part per million (PPM). Mercury level presently is in serious violation of the federal standards and although there are efforts to control it, its concentration is still rising in some areas of the U.S. and even more so in developing countries with high levels of coal-fired power generation.

Radiation

Radiation is usually not associated with fossil fuel combustion (at least in our minds), but coal contains small quantities of uranium and thorium and their radioactive decay products, such as radium. The uranium content is minuscule, about 1 ppm, but it varies widely to over 1,800 ppm of uranium in some U.S. coal mines.

Radioactivity ends up in the air during the burning of coal, and in fly ash in the particulate-removal devices. Direct radiation from fossil fuels is not a major concern, but there is a large quantity of coal being burned every day worldwide, so the cumulative effects are unknown as yet.

Note: The radiation emitted during coal burning in some cases is higher than that emitted by a nuclear power during its normal operating cycle. Although it is not fair to compare coal with nuclear power, the fact is that 4-5 times more radiation is emitted from some coal-burning power plants than by nuclear plants.

Cement Production

Power plants are not the only polluters and energy guzzlers. There are other major culprits in the global energy and environmental pictures. The major ones are the cement and iron production industries.

The cement industry contributes about 5-8% to global anthropogenic CO_2 emissions, making it an important sector for CO_2-emission mitigation strategies. Large quantities of CO_2 are emitted during cement production, mainly from the calcination process of limestone, from combustion of fuels in the kiln, and from power generation.

Estimated total carbon emissions from global cement production are 350 million metric tons of "carbon" (MtC), 160 MtC from process carbon emissions, and 147 MtC from energy use. The top 10 cement-producing countries account for over 60% of global carbon emissions from cement production.

900 kg of CO_2 are emitted for every 1000 kg of cement produced, mostly CO_2. So there is nearly as much CO_2 produced as cement.

There were 3.4 billion tons of cement produced worldwide in 2011, accompanied by about 3.1 billion tons of CO_2 produced at the same time. In 10 years, this is 31 billion tons, in 20 years 62 billion tons, and over 90 billion tons in 30 years of cement production.

Globally, both cement production and steel production are indicators of national construction activity, with cement mainly used in building and road construction, and steel also in the construction of railways, other infrastructure, ships, and machinery. CO_2 emissions are generated by carbonate oxidation in the cement clinker production process, the main constituent of cement and the largest of the non-combustion sources of CO_2.

The world's cement production is heavily dominated by China, with an estimated share of 57% in global emissions from cement production, followed by India with a more than 5% share. The United States, Turkey, Japan, Russia, Brazil, Iran and Vietnam have shares of between 1.5% and 2%.

With a continuing trend in China, global cement production increased by 6% in 2011. China increased cement production by 11% and was responsible for 57% of the world's cement produced, while production in Germany, Brazil and Russia increased by 10%, 6% and 3% respectively during the same time period, according to USGS estimates.

However, emissions are not directly proportional to cement production levels, since the fraction of clinker—in this industry the main source of CO_2 emissions—in cement tends to decrease over time. Recent study by the World Business Council on Sustainable Development has shown that the share of blended cement that has been produced in recent years in most countries has increased considerably relative to that of traditional Portland cement.

Consequently, average clinker fractions in global cement production have decreased to between 70% and 80%, compared to nearly 95% for Portland cement with proportional decrease in CO_2 emissions per ton of cement produced. Both non-combustion and combustion emissions from cement production occur during the clinker production process, not during the mixing of the cement.

This has resulted in about a 20% decrease in CO_2 emissions per ton of cement produced, compared to the 1980s. At that time, it was not common practice to blend cement clinker with much other mixing material, such as fly ash from coal-fired power plants or blast furnace slag. According to EDGAR 4.2 data, this yielded an annual decrease of 250 million tons in CO_2 emissions, compared to the reference case of Portland cement production.

A similar amount has been reduced in fuel combustion for cement production and related CO_2 emissions. So this leaves the cement industry with the task of figuring out how to stop emitting the other 29.6 billion tons per year. Fat chance, you say. We agree; this is not going to happen anytime soon, and especially now with the construction boom in China, India, Russia and many other developing countries.

Emission mitigation options in the U.S. and other developed countries include energy efficiency improvement, new processes, a shift to low carbon fuels, application of waste fuels, increased use of additives in cement making, and, eventually, alternative cements and CO_2 removal from flue gases in clinker kilns. But even these improvements have a long way to go before making a major difference.

Iron and Steel Production

Looking at steel production, with related non-combustion CO_2 emissions from blast furnaces used to produce pig iron and from conversion losses in coke manufacturing, China accounted for 44% of crude steel production in 2011, followed by Japan (8%), the United States (6%), Russia and India (each 5%), South Korea (4%), Germany (3%) and the Ukraine, Brazil and Turkey (each 2%).

According to WSA (2012), global crude steel production rose 6.5% in 2011, compared to 10% in 2010. The 9% increase in China equated to almost one-third of the global increase in production in 2011. Production plummeted in Australia (-12%) and South Africa (-22%). In 2011, it strongly rose in Turkey (+18%), South Korea (+17%) and Taiwan (+15%). Other significant increases were seen in Italy (+8%), India (+8%), the United States (+7%), Brazil (+7%) and Russia (3%).

As with cement production, the steel industry produces a lot of CO_2.

2.0 kg. CO_2 are emitted in the production of 1.3 kg. of steel.

The total world production of steel is about 1.5 billion tons, so the steel industry produces over two billion tons of CO_2 annually. This is 20 billion tons in 10 years, 40 billion in 20 years…you get the picture.

In steel production, most CO_2 is generated in iron- and steel-making processes that use coke ovens, blast furnaces, and basic oxygen steel furnaces. However, the share of electric arc furnaces and direct reduction in secondary and primary steel making, which generate much less CO_2 per ton of crude steel produced, is increasing. Lime and ammonia production and other industrial sources of CO_2 emissions increased globally by an average of 5% lately.

China produces almost half of the world's steel and cement, and is approaching that for coal-fired power generation; therefore, China is also responsible for at least half of the world's emissions from these industries.

There are efforts to account for and reduce CO_2 emissions from these industries, but again, most of the money and effort in this area are spent in the developed countries. Cement factories in the Third World usually make a mess of the local environment, with some extreme cases of entire towns covered in cement dust.

There are significant reductions in CO_2 emissions as a result of technological improvements and structural changes in steel production in industrialized countries during the past 40 years. Substantial further reductions in those emissions will not be possible using conventional technologies. Instead, a radical cutback may be achieved if, instead of carbon, hydrogen is used for direct iron ore reduction.

This technology and the CO_2 generation emitted during the production of hydrogen as a reducing agent from various sources are the next goals ahead of the steel industry.

Waste Incineration

Municipal waste is processed in several ways, the most harmful for the environment of which is incineration. This is a waste treatment process that involves the combustion of residential and commercial products contained in waste matter.

This is a high-temperature treatment process designed to convert waste materials into flue gases, ash, and heat. The organic matter in the waste stream is reduced to gasses, which are exhausted in the atmosphere. Ash is formed by the inorganic content of the waste and is discharged from the incinerators in the form of solid lumps or small particulates which are carried into the atmosphere—together with other hazardous emissions—by the flue gas.

In some cases, the heat generated by incineration can be used to generate electric power. Incineration with energy recovery is one of several waste-to-energy (WtE) technologies. The others are gasification, pyrolysis and anaerobic digestion. The energy product from incineration is high-temperature heat, with combustible gas as the main energy product.

The main benefit of waste incineration is reduction in the solid mass of the original waste by 80-85%. So, although incineration does not completely replace land-filling, it significantly reduces the volume of waste for final disposal.

Incineration is especially suitable for the treatment of certain waste types, such as clinical wastes and certain hazardous wastes where pathogens and toxins can be destroyed by high temperatures. In industry, incineration is used in chemical plants that produce a large number of toxic wastes which cannot be processed in conventional wastewater treatment plants.

Waste combustion is particularly popular in countries where land is scarce. Japan, Denmark, and Sweden, for example, are the leaders in using incineration with heat recovery. Waste incineration produced about 5% of the electricity consumption and nearly 14% of the total domestic heat consumption in Denmark. Luxembourg, the Netherlands, Germany, and France also rely heavily on incineration for reducing municipal waste and using the generated heat.

The big problem with waste incineration on such a large scale is its toxic output. Huge quantities of ash and gas emissions are released in the flue gases. Particulate matter, heavy metals, dioxins, furans, sulfur dioxide, methane, and hydrochloric acid are just some of the toxic chemicals released in the atmosphere. Some plants are equipped with emission controls and release a small amount of pollutants, but in many cases the incinerator plants do not have emission controls. In such cases, and if the controls are inefficient, the toxic emissions add significant air pollution in the respective areas.

For an equal amount of produced energy, incineration plants emit fewer particles, hydrocarbons and less SO_2, HCl, CO and NO_x than coal-fired power plants, but emit more of these pollutants than natural gas-fired power plants.

It is generally agreed that waste incinerators reduce the amount of some atmospheric pollutants by substituting power produced by coal-fired plants with power from waste-fired plants. Yet, in the end they add significant amounts of toxic and GHG gasses to the atmosphere.

The worst part of waste incineration is that it adds dangerous toxins that the other plants do not emit. Significant amounts of dioxin and furan emissions are produced at these facilities and some are released in the atmosphere.

Dioxins and furans are serious health hazards and accumulate in the areas around incinerators. Added to the other sources—domestic waste burning, fireplace fires, etc.—these emissions add a significant threat to human health for the locals, and the global environment in general.

Livestock Farms

Giant livestock farms, which can house hundreds of thousands of pigs, chickens, or cows, produce vast amounts of manure, often generating the waste equivalent of a small city. A problem of this nature and scale is tough to imagine, and pollution from livestock farms seriously threatens humans, fish and ecosystems.

The agricultural sector emits about 15% of the world's green house gasses, and this amount is expected to double by 2030.

Cows in large-scale livestock farms release the greatest amount of GHGs in the form of methane gas. Cows are such methane factories because of the food they are fed. Farmers switched from natural feed to perennial ryegrass, because it grows abundantly and quickly, and is thus more profitable. Ryegrass, however, is the "fast food" of grasses, lacking in essential nutrients. It's also difficult for cows to digest and ultimately causes them to generate more gas—methane.

The cow's digestion process, which takes place in four separate stomachs, is another factor. Cows are "ruminants' (heavy-duty chewers) that chew and ingest their food, then regurgitate it as cud, to chew and ingest it again. This process is repeated several times because the microorganisms in the cow's stomach can't break down the food entirely in one pass. Instead, it is sent back to the stomachs to let the microbes finish their work.

These microorganisms give off gases which make the cows belch, expelling methane. The food then moves among the rest of the cow's four stomachs, until it is finally digested. This process is accompanied by a lot of flatulence too, which releases methane gas.

Agricultural activities which include cows are also a major source of nitrate pollution in more than 100,000 square miles of polluted groundwater in California alone. There is an established link between spontaneous abortions, increased risk of fatal "blue-baby syndrome," and high nitrate levels in drinking water wells close to feedlots.

Animal waste also contains disease-causing pathogens, such as *Salmonella, E. coli, Cryptosporidium,* and fecal coliform. More than 40 diseases can be transferred to humans through manure.

In the U.S., about 30 million pounds of antibiotics (roughly 80 percent of the nation's total antibiotics use) are added to animal feed every year to speed livestock growth. This contributes to the rise of resistant bacteria, making it harder to treat human illnesses. At the same time, the huge amount of antibiotics excreted by cattle contaminate the water supply and make water unsafe to drink.

Large hog farms emit hydrogen sulfide, a gas that most often causes flu-like symptoms in humans, but at high concentrations can lead to brain damage.

Huge open-air waste lagoons at large livestock farms are prone to leaks and spills. Spills kill fish and contaminate the local water supplies.

Nutrients in animal waste cause algal blooms, which use up oxygen in the water, contributing to a "dead zone" syndrome. These processes cause serious changes to the environment.

Energy Terrorism

Terrorism is an external factor which impacts our energy security. Terrorist attacks targeting oil facilities, pipelines, tankers, refineries, and oil fields are referred to as "energy industry risks," because they are part of daily life in the energy sector. The energy infrastructure is extremely vulnerable to external sabotage, with oil transportation and its exposure at the five ocean chokepoints, on the top of the risks list.

The Iranian controlled Strait of Hormuz is a prime example of a chokehold, where one attack on a Saudi oil field, or on tankers in the Strait of Hormuz, could disturb the oil supply. A prolonged conflict in the area, would surely throw the entire world energy market into chaos.

We take a close look at the choke points in the following chapters.

International terrorism affecting the world's energy reserves is of great concern lately as well, as evidenced by NATO leaders meeting in Bucharest in 2008, where international terrorism against energy resources was one of the key subjects. The group discussed the

Figure 1-11. Terrorist attack on French oil tanker *Limburg* **in 2002**

possibility of using military force to ensure the energy security of the region. One of the possibilities discussed included strategic placement of NATO troops in the Caucasus energy fields to police and protect oil and gas pipelines from terrorist attacks.

The U.S. energy supply and energy infrastructure are in the terrorists' sights too, and attacks are, no doubt, planned daily in terrorist dens around the world. But how would they do that? We are too far away, and too powerful, to invade by sea or air, so recently they have been choosing different tactics.

Nuclear power represents one of most dangerous risks to our energy security and safety. Computer warfare is the name of the game today, so a terrorist sitting in a cave in Afghanistan, or in a high-rise in Beijing, might gain access to a computer terminal at a power plant or refinery and simply put it out of commission, or worse.

Nuclear reactors are potential terrorist targets, since they are not designed to withstand attacks by large aircraft, rockets and other air-born weapons. A well-coordinated attack, using powerful weapons, could damage the reactors in a nuclear plant, which would in turn have severe consequences for human health and the environment.

A recent study concluded that such an attack on the Indian Point Reactor in Westchester County, New York, could result in high radiation within 50 miles of the reactor, and cause 44,000 deaths from acute radiation sickness. An additional 500,000 long-term deaths from cancer and other radiation-caused illnesses would be expected.

Terrorists could also target a spent fuel storage

facility by using high explosives delivered by ground or air vehicles. This would also results in radiation contamination of the immediate area. A terrorist group may infiltrate the personnel of a nuclear plant and sabotage it from inside. They could, for example, disable the cooling system of the reactor core, or drain water from the cooling storage pond. An internal attack is perhaps the most likely, and most dangerous, terrorist attack on a nuclear-power reactor.

There is also an inextricable link between nuclear energy and nuclear weapons, which pose the greatest danger related to nuclear power. The problem is that the same process used to manufacture low-enriched uranium for nuclear fuel, can also be used for the production of highly enriched uranium for nuclear weapons.

Expansion of nuclear power generation could, therefore, lead to an increase in the number of rogue states with nuclear weapons produced by their "civil" nuclear programs. The use of nuclear power also increases the risk that commercial nuclear technology will be used to construct clandestine weapons facilities, as done by Pakistan in the recent past.

THE NEW ENERGY REALITY

In the fall of 2008 we saw the U.S. and then the entire world economy collapse overnight. This was accompanied by another energy crisis, but this one was quite different. It was not triggered by a war and the related disasters. It was, instead, the result of a much more complicated combination of events, political developments, fears of oil reserves depletion, and financial inadequacies.

The price of crude oil was a major part of the panic that followed, and a reflection of the confused global socio-political situation. Until September 2003, the inflation-adjusted price of a barrel of crude oil on NYMEX was under $25 per barrel. In 2003, oil prices rose above $30, doubling by 2005, and peaking at over $150 per barrel (bbl) in the summer of 2008.

Then, within a few short months global oil prices dropped from over $150 to under $30/bbl again. Talk about an oil price teeter-totter; this is the biggest one ever. Why?

The gradual rise of oil prices has been attributed to many factors, such as the falling value of the U.S. dollar, a decline in petroleum reserves, worries over peak oil (end of oil global supplies), Middle East tensions, and a new phenomena—oil stock price speculations. The sharp increase of oil prices in 2008 was a reflection of all these factors, some of which were exaggerated to the maximum.

We can write another large book on the reasons for the 2008-2009 economic problems, which brought up the energy anomaly (or was it the other way around) but we will have to wait for the economic specialists to figure out and put together the entire puzzle.

Table 1-5 shows that the U.S. uses about 25% of all the energy used worldwide each year. The numbers vary from year to year, of course, but the ratio of the U.S. vs. global energy use remains fairly constant. Nevertheless, people and businesses in the U.S. depend on oil and are negatively affected by shortages and price variations.

Table 1-5. Energy consumption in the U.S. and the world (in TWh equivalent)

Fuel Type	U.S.	World
Oil	11,000	50,000
Gas	6,000	31,000
Coal	6,600	37,000
Hydroelectric	840	8,710
Nuclear	2410	8,140
Renewables	950	1,380
Total	**27,800**	**136,230**

Although things have stabilized recently, and we don't expect great problems anytime soon, our energy security is directly affected by crude oil supplies, its use and price. So, crude oil is a major part of our lives and a major problem for our energy security.

The primary factors affecting our energy (crude oil) supplies today are as follow:

- *Oil Demand* grew an average of 2% per year until 2006, peaking at 3.4% in 2004. The global oil demand is projected to increase almost 40% over 2006 levels by 2030, according to the U.S. Energy Information Administration (EIA). This is nearly 120 million barrels per day—up from 86 million barrels in 2006. The main driver of this rise is the increase in the transportation sector mostly in developing countries.

- *Oil Supply*, or rather the problems related to it, is another important contributor to price increases. There has been worldwide slowdown in oil supply growth lately, as is reflected in the statistics on oil production which is greater than new oil discoveries. There is a fear among the specialists that global oil production will decline soon, leading to an inadequate oil supply.

The continuing unrest in the Middle East, as well as wars around the world, also affect oil prices. The conflicts in oil producing countries are not going away anytime soon, and are therefore expected to contribute to price variations, not only of oil but many other commodities.

The situation has changed drastically in the U.S. lately too, where new oil and gas discoveries and the introduction of new exploration and production technologies have led to abundant oil and gas production. We are now at a point where switching from oil import to oil export mode is a feasibility. How that might affect the U.S. economy, the global environment, and overall oil prices, is yet to be determined.

- *Fuel subsidies* in many countries have been responsible for keeping oil prices level lately, but the bonanza is nearing an end, so global oil prices are entering an adjustment period. China has been reducing energy subsidies, and hiking retail gasoline and diesel prices steadily. Some Asian countries have followed the example and have seen prices increase by nearly 50 percent.

Half of the world's population enjoys fuel subsidies, and the estimates show that almost a quarter of the world's oil derivatives have been sold at less than the fair market price. So retail oil prices depend on government subsidies, but would cutting the subsidies reduce global prices? This is another question that will be answered in the near future.

- *Oil stock manipulation* (or trading) is a new, and quite powerful, vehicle for controlling oil prices. This is one of the major reasons for the huge increase in oil process in 2008.

Price manipulation occurs when investors purchase futures contracts to buy crude oil at a set price for future oil deliveries—which may be months away from the purchase date. Gambling comes to mind in this case, but gambling that involves our energy supplies and their prices.

Speculators are not buying any materials, just papers. They pay for (or contract for) a certain amount of oil, and wait until the contract matures, hoping it increases in value. Then they settle (making or losing money in the process, depending on current oil prices in relation to the purchase price), or sell the contract to other consumers. A pyramid scheme-like scenario comes to mind in this case.

Oil traders (financial speculators) have everything to lose when oil prices go down. Conversely, they gain when prices go up. So they want to see them go up, and do everything possible to facilitate the price increase, because this is the only way they can make money.

The oil stock traders have been blamed for the excess price increases during the 2008 energy crises, when oil went up to $150/bbl. The expert opinion is that most of that price increase (to high of $150/bbl) had nothing to do with supply and demand, or global pricing, but instead was due to speculator frenzy.

The oil gambling game continues, albeit in a less noticeable form. A testimony given to the U.S. Senate indicated that "demand shock" from "Institutional Investors" had increased by 848 million barrels over the last five years. Similar increases in demand came from China, to the tune of over 900 million barrels.

Institutional investors were very influential in the manipulations, and experts suggested that oil price increases were caused by commodities manipulations by these investors. It was estimated that for every $100 million in new inflows, the price of U.S. oil increased by 1.6%.

It was also observed in May 2008 that oil futures transactions on the NYMEX stock exchange mirrored closely the price of oil increases for the past several years. It is not clear if the increased investment was just following the rising prices, or if it was causing them.

According to the critics, huge investments in the oil market may not drive prices up simply because speculators are not buying any actual crude oil. Their transactions, therefore, do not disturb the supply and demand ratios in any way. It also must be noted that for the same period of time the prices of some commodities (which are not openly traded) have actually risen faster than oil prices.

Nevertheless, the world consumption of oil at about 100 million bbl/day is far exceeded by the crude oil traders "paper market," which equals about 1.5 billion bbls/day. This is over 15 times the actual market demand, so one must question the effects of such huge and disproportionate numbers.

The American dream is alive and well!

As a result of the energy and economic crisis of the past, the U.S. and most other nations are looking very seriously at their energy status and the terms "energy security" and "energy independence now have new meaning.

Although to the average Joe these words mean the same today as in 1973, the new world order requires clarification and even redefinition of them, which is what we are setting out to do in this text.

One of the most significant ways for the U.S. and other nations to protect themselves from oil emergencies in the past was the creation of national *strategic petroleum reserves* (SPRs). These are huge crude oil storage facilities, or stockpiles, held for the purpose of providing economic and national security during an energy crisis.

According to the U.S. Energy Information Administration, approximately 4.1 billion barrels of crude oil are held in strategic reserves, of which 1.4 billion is government-controlled, and the rest belongs to private corporations or entities.

The US Strategic Petroleum Reserves are one of the largest strategic reserves today, with much of the remainder held by the other 26 members of the International Energy Agency. Other non-IEA countries have begun creating their own strategic petroleum reserves as well. China is the largest holder of oil reserves.

What this means, however, is that in the best of cases, at the current consumption levels (over 100 million barrels a day), U.S. and global oil reserves would last only a few short months in case of a dramatic emergency situation. The estimates vary from 90 to 120 days, but that is the maximum. These reserves, therefore, are far from what is needed for long-term energy survival of the U.S. and the world.

We will take a closer look at the Strategic Petroleum Reserves and other efforts to ensure our energy future and obtain energy independence, later on in this text.

Presently, the U.S. is in the midst of an energy boon, which is shaping our new energy reality. This amazing phenomena is attributed to new discoveries, use of new technologies, and increased production of natural gas and crude oil via *hydrofracking*. Some call it *energy revival*, and some *energy revolution*. Whatever it is called, it is a new way of looking at everything related to energy, and it came upon us almost overnight.

How did it all happen? This new way of drilling and extracting oil and natural gas was unthinkable and impossible until recently. It is a new method of breaking through solid rocks to extract the crude and oil and natural gas deposits locked in them millions of years ago.

How can one process bring such drastic change to the way of life in the greatest country in the world, and to the entire world?

We will take a close look at this new phenomena—the processes and their results—in the following chapters.

The New Reality...

The U.S. is in the midst of new developments in the energy sector, with unexpected and unprecedented (albeit still somewhat not well defined) energy abundance and prosperity.

The new reality is much brighter than the situation we were forced to deal with during the energy and economic uncertainty of 2008-2009, when oil prices rose above $150 a barrel. During that same time, we also saw billions of dollars fly from the U.S. government coffers into different areas of the U.S. economy in a desperate attempt to stop it from a total collapse.

The energy sector also benefited from generous government subsidies and loan guarantees. The solar and wind industry were major recipients of billions of dollars, but the actual results do not reflect the massive spending. Billions of dollars were wasted on dubious companies and projects, and even more was spent on foreign companies that were importing their technologies in the most technologically advanced country in the world.

In the end, the US recovered slowly, and now we are suddenly on the other side of the equation. We see unexpected energy manna falling upon the nation in form of millions of barrels of oil and natural gas daily.

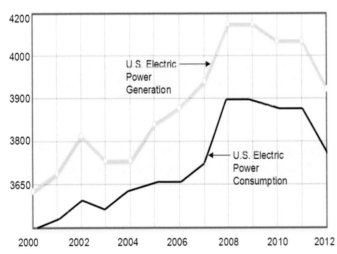

Figure 1-12. Electric power generation and use (in billion kWh)

At the same time, while we are getting more oil and natural gas than ever, the consumption of electric power in the U.S. dropped, leading to a decrease in the need for power generation. This was mostly due to the economic turmoil of 2008, followed by an increased awareness among U.S. consumers.

This slowdown in energy use also offered the opportunity to many coal-fired power plants to be con-verted to somewhat cheaper natural gas-fired plants to reduce air pollution and reduce cost.

The abundance of oil and natural gas changed everything, and the changes are still ongoing. We predict that the U.S. economy will gain unprecedented speed during this decade—fueled by huge oil, gas, and coal surplus—in addition to significant increase in use of renewables.

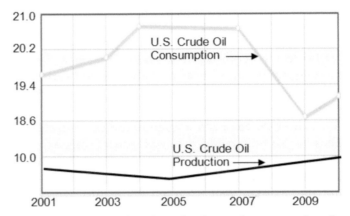

Figure 1-13. U.S. crude oil production and consumption (in million bbl/day

It took the U.S. a while to get to this stage, after going through several energy crises during the last 50 years. Now we are confident that there will be no more energy (electric power generation) crises, at least for the foreseeable future, unless something extraordinary happens to change the situation drastically.

Energy crisis is a key word that has been thrown around like a rock through a window in the past by anyone who wanted to make an important point with damaging results.

A convincing point was made every time the words *energy* and *crisis* were mentioned together, because who can forget the 1973 Arab oil embargo? Or the 1979 oil crisis after the Iranian revolution, or the Gulf war in the 1990s, or 9/11 and the following invasion of Iraq... and of course the 2007-2008 energy and economic debacle?

Emphasis and overemphasis on energy security and careful implementation of procedures designed to get us close to energy independence is the best approach to prevent serious energy crisis.

We are all aware of the importance of energy in our lives, so we just need to be better educated and more involved in the process. This might help prevent some energy crises in the future.

There is, however, one serious energy crisis that we will not be able to avoid. It is the impending, albeit distant, complete depletion of the fossils.

Even though this is a very serious, if not catastrophic event, all the indicators point to the fact that we will start worrying about crossing that bridge when we get to it.

The U.S.—A Net Energy Exporter

There is a lot of talk in U.S. political and energy circles today about the new energy bonanza. One of the biggest questions before politicians and regulators is how much liquid natural gas (LNG) and petroleum products we can and should export.

The U.S. has been a huge crude oil importer for decades, so how could it consider increasing exports? We are going almost overnight from sounding the energy-deficit alarms, to exporting energy products. How did this happen, and what are the results and consequences of this new energy revival, or as some call it, energy revolution?

With the abundance of energy derived from shale and other sources, the United States is becoming a net exporter of energy. Expectations are that huge amounts of oil, natural gas, and coal will be exported to China, Europe and all over the world in the near future.

We have a great excess of coal production, because the use of coal for electricity generation has been curtailed for a number of reasons. Coal is being replaced by natural gas, which is so plentiful now that we have excess of it too. So, both coal and natural gas are slotted for greatly increased exports.

U.S. law does not allow export of raw crude oil, but does not forbid the export of refined and otherwise processed petroleum products based on crude oil. So, we are ramping the exports of these products as well.

*The U.S. is now a net energy exporter. Coal in its natural form, liquefied natural gas (LNG), and processed petroleum products are headed to overseas markets as never before. The upward trend of energy exports will continue for the foreseeable future. **Most likely until we run out of them...***

The newly evolved situation radically changes the supply and demand forecasts. While producing, using, and exporting large quantities of oil and gas, coal use in the U.S. will be greatly curtailed as climate change initiatives dictate using other energy sources. At the same time, however, coal will be produced in even larger

quantities too...and most of it will be exported.

As the United States becomes a net exporter of energy, mostly fossils—coal, gas, oil derivatives—many new opportunities have been created. The abundance and relatively low cost of fuels and electric energy will encourage the industrial sector to rebound and bring back manufacturing jobs that were exported abroad not long ago.

But there are some negative sides to this new situation. By reducing domestic coal use, we reduce pollution, which is great. But by exporting such large quantities, we create new pollution centers somewhere else. The net result from these activities is an increased in total air pollution levels around the globe. How good or bad is this?

We will review this situation in more detail later on.

Crude Oil Status

We must agree that the key energy sources—coal, natural gas, crude, nuclear and hydro power—are the main drivers of our economy. Without them we will be thrown back into the dark ages...or at least back into the 19th century—using kerosene lamps and horses.

Of these drivers, crude oil is the most important to some sectors of our economy, the transportation sector and the petro-chemical industry in particular. It is the lifeblood of the nation, for over 90% of the vehicles run on crude oil derivatives.

Crude oil allows us to move from place to place and transport goods from shore to shore. It is also responsible for the production of millions of tons of plastics, fertilizers, cosmetics, and an endless number of chemical products.

Without crude oil, our lives will very quickly change in a drastic way, and it is hard to even imagine what could happen in the long run.

It is critically important to note that a significant part of the total crude available in the country (about 1/3 for now) is processed into different fuels and chemicals. A significant portion of these products is exported as finished goods, which translates into profits and thousands of jobs in many industries around the country. But it also raises the question of how long can we do this, knowing that this is a finite commodity?

Figure 1-14 tells us that we are consistently closing the import deficit gap, and that we might be able to reach a solid equilibrium point, where we import as much as we use and export.

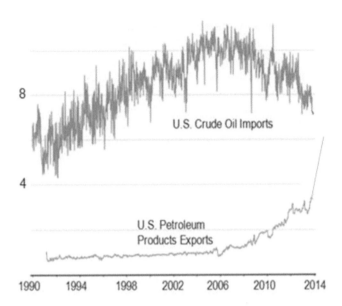

Figure 1-14. U.S. crude oil imports and exports (in million barrels/day)

The crude oil equilibrium, as needed for achieving our energy independence is reflected in the following simple equations:
On the input side:

Total Oil Use => Oil imports +
Domestic Oil Production

And on the output side:

Total Oil Use => Transportation +
Chemicals Production + Petrochemical Exports

In summary, we produce about 50% of the crude oil we use, and import the other 50%. We use 60-80% of the total amount of available crude oil for fueling our transportation sector. The rest is processed into different petrochemical products, a significant part of which are exported.

As a matter of fact, things are looking very good for the U.S. petrochemical sector and its exports. For example, in October 2013 the U.S. averaged exports of 7.7 million barrels of oil per day while taking in just 7.6 million barrels per day (bpd), according to the Energy Information Administration. The United States actually exported more oil than it imported for the first time since 1995, in a growing trend sped by hydraulic fracturing and other technological improvements.

So, on the output side we see a new situation:

Domestic oil production (50%) =>
Transportation + Petrochemicals

Crude Oil Imports (50%) => Processed for Exports

Could this be right? It means that all the crude oil we import is processed into different petrochemicals and then exported overseas. That situation is totally different from what we are used to, so it seems that we can live without crude oil imports. What's wrong with this picture?

And it gets even better. The experts forecast the U.S. oil production rising to 8.88 million bpd in 2015, at an average of 8.49 million bpd throughout the year. Imports, meanwhile, are expected to fall to as low as 6.0 million bpd by the same time and average about 6.5 million bpd for the year. Forecasts also predict that crude oil production will outstrip net imports every month for the near future.

Other extraordinary news is that between January and May, 2013, the U.S. satisfied over 86 percent of its own needs for crude oil. Whether this is a temporary or a permanent situation remains to be seen, but good news is good news, and we can only hope that it remains so.

This is due to the discovery of new oil and gas supplies, which with the help of new technologies makes it possible to extract fossils from depths and places that were unreachable, or financially unfeasible until recently. But how long can we continue pumping, using, and exporting so much oil non-stop?

One billion dollars every day, or over $350 billion just in 2103, is the rough estimate of the benefit the U.S. economy is receiving thanks to the development of new supplies of oil and natural gas, according to a new analysis.

A Merrill Lynch study attempted to quantify the new supplies of natural gas and oil from unconventional sources like shale and from the offshore production in the Gulf of Mexico. The study found that the gross gains from domestic energy supplies were $900 million per day in the spring of 2012. This is about 1,300 percent increase from $70 million per day for the prior years. As production increases, daily gains from domestic supplies are expected to rise well over $1.5 billion. It also pegs that new contribution at about 2.2 percent of U.S. gross domestic product, which is a truly astounding number for the huge $15 trillion U.S. economy.

The predominant thinking today is that we can and should export as much fossil fuels and products as possible while we can. But *can* and *should* mean different things to different people. It means monetary gains to the oil companies, and prestige for the government, while the rest of us are left wondering what will follow after the fossils are depleted.

We are also witnessing massive land, water, and air contamination as a result of the new widespread exploration expansion, so a fierce battle rages as a consequence of the new energy reality in the U.S. and abroad.

Oil Products Exports

1975 U.S. law prohibits U.S. companies from exporting crude oil anywhere and under any conditions. However, it *does not* prevent the export of refined oil products, such as gasoline, diesel, lubricants, and various chemicals.

So, U.S. refineries are shipping record amounts of gasoline around the world today and plan to ship even more in the future. In some cases we are exporting the fruits of the country's shale revolution to some of the same countries from which we get the crude oil. Tankers full of gasoline and diesel fuel—made from shale oil pulled out of places such as North Dakota and Texas—are being shipped to the Middle East, South America, Africa, and Asia.

Figure 1-15. American oil products transport

The U.S. shipped a record 3.2 million barrels a day of gasoline, diesel, and other refined petroleum products in 2012. That number is about 65 percent more than the U.S. shipped in 2010, before the peak of the shale revolution.

Just think: only three years before that, the U.S. was a net importer of crude oil and its products. Today, it's a net exporter. How about that?

But why and how did this happen?

U.S. refineries are enjoying the double-barreled advantage of 1) a quickly rising domestic oil supply and

domestic gasoline demand, and 2) the ban on crude exports of the 1970s is in reality a subsidy for U.S. refiners. So, the millions of barrels of oil being pulled out of new shale plays has nowhere to go except for export.

The U.S. also has another huge oil supply coming from the Canadian oil sands, which is actually in trouble now, due to the quickly changing dynamics of the U.S. oil market, leading to a "bitumen bubble."

A glut of oil in the U.S. Midwest caused the price of bitumen from Alberta's oil sands to trade as low as $50 a barrel—below the going rate in the rest of the world. The price of benchmark U.S. crude, similarly, is trading at a discount to world prices—anywhere from $10 to $25 a barrel.

With that kind of price advantage on feedstock and regulations, U.S. refiners are in hog heaven and are looking forward to a bright and profitable (albeit temporary) future.

European refineries, in contrast, are in trouble. They are paying high world prices for oil, but also their traditional business of exporting surplus gasoline to the U.S. is shrinking by the day. They're seeing their product (and profits) become consistently and increasingly displaced in the world markets by cheaper gasoline from the U.S.

Unfortunately, the happiness (and the discounts) from the bonanza have not reached American drivers yet, and despite record gains in domestic oil production, a gallon of gasoline is still above $3 (and diesel is even higher) and expected to remain there for a while, and even go higher in the future.

Why? Because even though the U.S. oil production is up by 2 million barrels a day from 2011 levels, most of the U.S.-made gasoline is now being loaded on tankers and shipped overseas.

What conclusions should we come to?

Could it be that the U.S. does not really need crude oil for domestic use? That we need it only to refine and produce numerous petrochemicals for export? The overall import-export numbers suggest that we are moving close to 50% imports and 50% exports with crude oil and its products.

Could it be that things have changed so quickly lately that crude oil—which until recently was the major problem in our energy security struggles, and a major obstacle to achieving total energy independence—is no longer an important strategic commodity? Could it be that we import such great quantities of crude only to create jobs and generate more profits by exporting its byproducts?

Could it be that we are building an economy based

on import and export of huge quantities of crude oil even though we know that it is a finite commodity that has a beginning and an end? And though we are acutely aware that the oil end is near, we continue pumping and using it as if there were no tomorrow.

These are serious questions that need to be answered, because if this is so, then we are foolishly wasting the most precious natural resource available today—crude oil. We cannot possibly be so blind, or greedy, as to shoot ourselves in the foot by wasting the few remaining drops of crude oil that took millions of years to create.

Natural Gas Exports

Natural gas has a life of its own too, albeit quite different from that of crude oil. It is good for power generation, because it is cheaper than coal and oil (per unit and MW generated), and emits less pollutants than they do. Producing and using natural gas is better for our energy security, the economy, and the environment than most of the alternatives—up to a point.

Exporting natural gas is a new phenomena that is also good for our economy, but presents a number of issues and creates some problems. It is a technological and political dilemma, and a touchy economic issue.

Exporting natural gas has good and bad sides that we all should become familiar with as Congress tackles the issues. The path we take is so important!

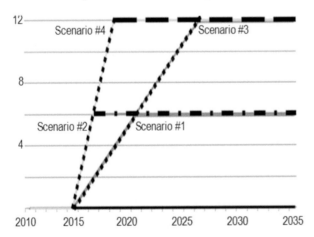

Figure 1-16. Four scenarios for future U.S. natural gas exports (in billion cubic feet/day)

Since there is no way to predict with certainty what will happen tomorrow, U.S. officials and experts on the subject are looking at four scenarios for future natural gas exports.

- Scenario #1 is the most conservative and, in our opinion, the most realistic—suggesting that the ex-

ports will increase slowly from 2015 through 2020, reaching about 6.0 billion cubic feet/day (bcfd). The slow increase would be due to political and regulatory restrictions, fueled by the continuing debate of exports vs. domestic only use of natural gas.

- Scenario #2 suggests a very quick increase of exports during 2015-2016 to the same 6.0 bcfd export level.

- Scenario #3 predicts slow increase to 12 bcfd, more than doubling the previous scenarios' predictions.

- Scenario #4 is the most aggressive, and most unlikely (in our humble opinion), where an almost overnight increase to 12 bcfd would be achieved. This is difficult to imagine, simply because there is no infrastructure to support such rapid increase.

Which path the U.S. takes—pending Congressional decisions—will determine to a large extent which scenario is most likely to develop. Here is where the U.S. taxpayers and voters have a great chance of a *say so*.

But the overwhelming question remains, "What do all these energy exports mean to our energy security, and especially its long-term faith?"

Note: Keep in mind that natural gas, like all natural resources, is a) in limited quantity, and b) the end will come quicker with the increase of exports.

U.S. Gas Bonanza

In addition to increased exports, which are bringing income and creating jobs in the U.S., the natural gas success story has been playing out in cities across America, where lower prices have reinvigorated existing businesses and enriched entire areas.

And there are other positive developments too. International firms are finding the "U.S. gas bonanza" an appealing proposition, and their investment is being felt in new jobs creation and many project starts around the country.

German gas-and-engineering company, Linde AG, said it would spend $200 million to build a new air-separation unit in La Porte, TX—joining dozens of other international firms taking advantage of low U.S. natural gas prices. Incitec Pivot, an Australian fertilizer company is joining the investment boon with an $850 million ammonia plant in the works in Louisiana for the same reason. And many more are being drawn in as officials in Washington are looking to drive additional foreign investment through a Commerce Department program called SelectUSA.

Ongoing investment promises to have a real impact on the American economy, with natural gas investments contributing to more than 300,000 new jobs and driving half a trillion dollars of production through 2025.

The domestic energy growth potential is significant and the U.S. Department of Energy (DOE) is paying special attention to it. The DOE has been also considering granting conditional approval to more natural gas export facilities which will facilitate the loading of gas for export.

The renewed interest in energy exports has two sides, and there are many people—including politicians—on each side of the key issues. With that, the debate has started, and we expect a real fight to unfold in the U.S. in the near future.

The issues on both sides of the energy export vs. energy security question are many and very serious. They include complex technical, logistic, regulatory, and political matters and considerations that need to be fully understood and analyzed before any major decisions are made. There is a delicate balance created by this unique moment for the American energy sector and the national economy, so we all should be involved in the discussion.

A compromise is needed between the possibility of making lots of money by exporting energy products (coal, natural gas, and crude oil derivatives) on one hand, vs. keeping the excess at home for future use, on the other.

Gas Export Politics

The export of U.S. natural gas is one of the most urgent issues on the energy export agenda. It is surrounded by a number of controversies, which U.S. politicians and regulators are trying to resolve. Recent congressional votes suggest that the tide may be turning on the issue, as the Energy Department authorizes more companies to sell the fossil fuel overseas. DOE recently also approved multiple applications to export natural gas to non-free trade agreement (FTA) nations, which might have unintended consequences in the near future.

Other major political developments of the gas exports thus far have been as follow:

— On Feb. 15, 2012, the House voted 173-254 to reject an amendment from then-Rep. Ed Markey, D-Mass. that would block oil and gas carried by the Keystone XL pipeline and resulting fuel products made from it from being exported. Twenty-four Democrats joined 230 Republicans in voting against the measure.

— On June 21, 2012, the House voted 161-256 to reject a Markey amendment that would bar the export of oil and gas produced under new leases covered by an underlying bill. Thirty-one Democrats joined 225 Republicans in voting against the proposal.

— On July 25, 2012, the House voted 158-262 to reject a Markey amendment that would bar companies from exporting any gas resources produced from leases sold that would be sold under the underlying legislation. Thirty Democrats joined 232 Republicans in voting against the amendment.

— In November 2013, the House of Representatives voted 142-276 to reject a plan from Rep. Peter DeFazio, D-Ore., to block exporting overseas large quantity natural gas produced on public lands. According to DeFazio, foreign sales threaten a resurgence in domestic manufacturing, as companies move operations back to the United States to take advantage of cheap power supplies and feedstocks from surging U.S. natural gas production.

This is because many experts believe that by expanding the marketplace for natural gas harvested inside the U.S., exports could cause the domestic price to rise—but this has not been proven and specialists' opinions vary widely on just how, when, and how much.

Presently many manufacturing companies are bringing production back to the U.S. because of the plentiful and cheap natural gas, but if we begin to export it in great volume, then we are part of the international market, which might mean a dramatic increase in domestic natural gas prices. With this we will instantly lose our competitive advantage for domestic manufacturing.

On the other hand, hoarding U.S. natural gas is not something that will be agreed on by most people. The experts believe that increasing the volume of natural gas exports would provide our allies with an alternative and reliable source of energy, helping to strengthen our economic and geo-political partnerships, while restraining U.S. natural gas exports would hurt our abilities to bolster strategic partnerships and create domestic jobs.

Ultimately, 56 Democrats joined 220 Republicans in voting against DeFazio's plan, while 140 Democrats and 2 Republicans voted for it. This is a slight uptick in the number of lawmakers who support energy exports, or at least don't want to ban them outright. This also shows growing momentum in Congress for increased natural gas production and exports.

But the Energy Department has approved four applications to sell natural gas to countries that do not have free-trade agreements with the United States, and increased the planned exports from the Freeport LNG facility in Quintana Island, Texas.

Does all this mean that we are ignoring our energy and national security in favor of making a quick buck?

Critics of expanded gas exports—like DeFazio—worry that the United States is moving too quickly to authorize foreign sales of liquefied natural gas, without a clear understanding of how that might affect domestic prices and the international political scenario in general.

The recent Energy Department decisions have been predicated on a third-party study that concluded in 2012 that broader natural gas exports would have net economic benefits for the United States, but some large natural gas users have been skeptical.

In November, 2013, America's Energy Advantage, a coalition including the Dow Chemical Co., other manufacturers and gas distribution firms, released a memorandum for policymakers that highlights the high stakes. "Unchecked LNG exports will cause domestic gas prices to spike, harming consumers, manufacturers and adversely impacting the economy," the analysis said, citing a raft of studies that predict higher electricity prices and energy costs if export levels tip above six billion cubic feet per day.

The group also faulted the Energy Department's reliance on the 2012 third-party exports study, saying its data are outdated now.

"The natural gas marketplace has evolved so rapidly that the conclusions made in the Department of Energy's commissioned report by NERA Economic Consulting have been rendered obsolete," America's Energy Advantage said. "Nearly every respected energy market analysis has forecast natural gas demand—domestically and internationally—to dramatically exceed DOE projections."

Oh, my...damn if we have gas, damn if we don't.
Double damn if we export it, and triple if we hoard it.

Subsequently, The U.S. Department of Energy conditionally approved more exports of liquefied natural gas from Freeport LNG in Texas, a move that could lead to increased shipments of the fuel in coming years. The approval is the fifth by the U.S. government since 2011 to countries with which it does not have a free trade agreement.

Subject to final regulatory approval, the facility is conditionally authorized to export an additional 0.4 billion cubic feet per day (Bcf/d) for a total rate of up to 1.8 Bcf/d, for a period of 20 years, the DOE said.

The agency initially granted Freeport approval to export 1.4 Bcf/d of natural gas a day of LNG from this facility on May 17, 2013. The oil industry's main trade group, the American Petroleum Institute, welcomed the announcement and urged Energy Secretary Ernest Moniz to approve more projects at a faster pace. "LNG exports will significantly reduce our trade deficit, grow the economy and support thousands of U.S. jobs," according to API.

While the U.S. natural gas boom has led to a long list of applications to export the fuel, the Obama Administration is weighing how fast to roll out approvals in order to keep domestic gas prices in check. The last approval was for Dominion Resources Inc. in Cove Point, Maryland, which was in line with the expectations.

President Barack Obama's nominee to be Assistant Secretary of the section of the DOE that oversees LNG, told a Senate hearing that the agency had no plans to pause its economic impact studies.

The chairman of the Senate energy and natural resources committee welcomed the news with caution, but praised the energy department for "proceeding in a deliberative manner" and considering applications on a "case-by-case basis."

Congress is weary of the potential of the gas exports to have a significant impact on domestic prices for families and manufacturers, and in turn harm America's energy security, growth and employment. Some manufacturers are worried that excessive exports of natural gas could lead to higher fuel bills and make it harder to open new domestic businesses.

The fear is that the Department of Energy continues to rely on obsolete data and ill-defined standards to justify continued LNG exports, according to the politicians, since uncontrolled LNG exports might threaten America's manufacturing renaissance, double or even triple prices for consumers, and negatively impact investment and job creation.

The National Association of Manufacturers on the other hand praised the decision. According to them, the principles of free trade and open markets should govern whether or not companies can move forward and construct LNG export terminals on U.S. soil.

We all agree that America's newfound abundance of natural gas is powering a remarkable economic revival. It is bringing an unexpected manufacturing renaissance, which to date has generated more than $110 billion of investment in over 120 different manufacturing projects. That prosperity has created nearly 100,000 manufacturing jobs in 2013 alone.

Manufacturing is the foundation of the U.S. economy, so this news is quite welcome after the doom and gloom of the 2008 financial crisis. The natural gas abundance and low prices are so significant that U.S. companies are beginning to bring foreign operations (which were exported in the 1980s and 1990s) back home. This trend leads to additional job creation, which is estimated to bring millions of new direct and indirect jobs by 2020.

On top of that, a large number of foreign companies are bringing their manufacturing operations to the U.S., and the jobs that go with them. Recent expert analyses claim that the growth of natural gas supplies and bringing businesses back home will have 15-20 percent cost advantage over the advanced economies of Europe.

The biggest driver of this cost advantage is the low cost of natural gas and electricity. The new developments combined are projected to create 3 to 5 million U.S. jobs by 2020, which could reduce the unemployment rate by several percentage points.

As always, however, each party defends its own position, to protect its own interests, and with so much information it is becoming increasingly hard to see the forest for the trees.

To export or not to export is the question before Washington lawmakers today, and the battle is just beginning.

The final decision of our lawmakers will determine to a large extent the role, magnitude, and effect of U.S. gas exports on world markets.

In the best of cases the impact might be huge, and will determine our energy future as related to energy security and the global environment. Let's hope that our politicians have access to the most relevant information on the subject, so that they can make the best decisions for the country, the world and our environment.

Still, if increased gas production, use, and exports are such a good thing, as it appears to be from the U.S. administration and international community points of view, why do so many people oppose the increased natural gas production and the approval for U.S. LNG export terminals?

Could it be that politics simply always has two sides? Nevertheless, there are a number of organizations in the U.S. that want to slow down or stop exporting gas to other countries, regardless of the benefits this activity brings home.

An important issue to consider is the fact that the more energy we export, the sooner we will reach its end. All fossils—including natural gas—are in limited quantity. Once they are depleted, there will be no more of them to use in the U.S., let alone export.

Also, exporting more natural gas will increase the global GHG emissions—not a small thing to consider. Do we really want to do this?

But the most pressing issue right now is the fact that hydrofracking is creating excess air, soil, and water pollution which are hurting some of the locals. The problem has been downplayed and even blatantly ignored until recently, but things are changing and some action is needed.

How we are going to resolve that issue is one of the big questions of the day, which will determine the future of natural gas production in the country. This in turn would affect exports too.

To Frack, or Not to Frack?

To frack or not to frack; that's the question. Fracking is good for the producers, but not that good for others. There is a strong movement in the U.S. opposing the new development in the energy sector—especially gas exploration via hydrofracking—where many people object to the pollution and property damage done by the increased and widespread hydrofracking activities.

The anti-fracking movement in the U.S. and around the world is growing by the day. It is a powerful movement that is capable of making some changes in the oil and gas hydrofracking industry as we know it. Its agenda varies according to local priorities and group composition. Public consultation is the preferred method in some countries like France. In Australia rural conservation issues dominate the battles.

Generally speaking, the movement is divided into four broad camps of people who:

1. Desire a better deal from the gas industry
2. Advocate further study into the environmental and economic impacts of unconventional gas development
3. Demand a complete ban on hydraulic fracturing
4. Demand tighter regulation of gas development

Some factions of the anti-fracking movement are not opposed to hydraulic fracturing *per se*, but simply want to get a better economic opportunity, taxation, compensation deals, and raise rebellion against the existing system. In the U.S., among other things, there is an opposition to energy companies from Texas and Oklahoma dominating the industry in Pennsylvania and New York.

The anti-fracking movement in Pennsylvania, for example, is questioning the state's low taxes on unconventional gas development, calling for both a severance tax (i.e. based on volumes extracted) and royalty rates in line with those in other parts of the country.

The internationalization of the anti-shale and anti-coal seam gas activist movements in defense of "the world's underground water supplies" was an important development in 2012. On a declared World Day Against Fracking/CSG in July, for example, a key activist group opposing Australian coal gas development spoke at a major national anti-fracking rally in Washington, DC, situating itself explicitly within the international anti-fracking movement.

On top of that, there are a number of lawsuits against U.S. oil and fracking companies, all of which will determine the way the industry develops in the future.

Gas Export Battles

Some groups object to exporting natural gas, which is in finite quantity, since exporting large quantities will exhaust the domestic supplies much sooner. One of the strongest organized anti-export movements objects the U.S. Department of Energy's (DOE) recent approval of multiple applications to export natural gas to non-free trade agreement (FTA) nations. This, according to them, is pushing the U.S. into a danger zone that will bring uncertainty to the U.S. and global energy markets.

It might actually increase prices for consumers and harm manufacturers who rely on low prices. All this might also eventually impact negatively capital investment, job creation and the overall health of the U.S. economy.

In support of the growing importance of natural gas export policy to America's manufacturing renaissance, a large manufacturer (a fertilizer manufacturer based in Australia, who is working on a new $850 million ammonia plant in Louisiana) has joined the protest. Their decision to invest in the U.S. points to the fact that the new energy boon is promoting investment, development and employment in the U.S.

The protest is led by a coalition of U.S. manufacturers and consumers insisting on a balanced approach to LNG exports to countries which have not negotiated FTAs with the US. The coalition is educating the public and policymakers about the importance of having an available and affordable domestic natural gas supply, since it is a major driver in the manufacturing renaissance presently underway in the U.S.

A reasonable LNG export policy and debate are seen as key to the economic viability and manufacturing renaissance in the U.S. They see this unique opportunity to revitalize the nation's economy and create millions of American jobs as much more important than exporting for profit and enriching a few corporations in the process.

The group vision is that of balanced approach to exports, taking into consideration the public interest like consumer prices and the potential for vast job creation, and welcoming all interested to join. Instead, DOE sees it as making decisions without legal standards to guide it as to when export applications and rescission of export authorizations are in the public interest.

DOE is currently reviewing other applications, which if authorized would raise the cumulative volume of authorized exports of LNG to 8.31 billion cubic feet per day (Bcf/d), which would go beyond the "low export scenario" level identified in a NERA report, which DOE used to grant three previous LNG export applications.

The group insists that, pending development of adequate standards, the Department should suspend its disposition of liquefied natural gas (LNG) export applications and assess the implications of further approvals on the public interest, before lasting harm is done to our economy.

One of the group's members has filed a formal motion to intervene in the DOE proceeding for the Freeport LNG Expansion, L.P. and FLNG Liquefaction, LLC (together "FLEX") export application (FE Docket No. 11-161-LNG).

The motion is a request for a more formal rulemaking process based on current data and assessments of today's supply and demand environment, since current applications are being granted based on guidelines developed for gas imports in the 1980s.

The motion indicates that the legal standards that DOE used to analyze the public interest in two previous grant applications were not "adequate, appropriate, or sustainable," and that American consumers of natural gas deserve as much say in the process as producers.

Coal Exports

Coal has been, and still is (to an extent), the engine of the global industrial revolution. It still provides nearly half of the U.S. electricity, and an even larger percentage in many other countries. We simply could not have achieved the present level of technological development, nor maintain our comfortable lifestyle, without coal.

But coal has problems. It is dirty and emits a lot of pollutants, which are causing major environmental and health problems around the world. Today, abundant, cheap, and cleaner natural gas is replacing coal as a major electric energy generator.

Coal-fired power plants cannot meet the stringent green house gasses (GHG) emission restrictions, and can no longer be built in some countries, including the U.S. At the same time the existing coal-fired power plants are being replaced by gas-fired plants.

According to an international banker, "banks won't finance coal, insurance companies won't insure it, the EPA is coming after it, and the economics to make coal clean simply do not make sense. Coal is a dead man walking..."

Wow...this banker is quite convinced that we are looking at the end of coal as we know it. He might be right, at least partially, since it is becoming obvious that North America's appetite for coal is waning by the day. But the coal companies are far from throwing in the towel. Not yet, for sure, and probably never!

Instead, they are saying, "America, you don't want our coal? Well, heck, we will sell it overseas instead, so try to stop us." So, they are looking to China and other Asian and European markets to buy US-mined coal. The mines and the entire coal infrastructure is in place, so they only have to change the final destiny of their product.

The only problem they face is that before large volumes of coal can be loaded on ships and sent overseas, new shipping terminals would need to be built at port sites in the Northwest. Building the terminals is easy, if it were not for some barriers in the permitting and resistance from the locals and some environmental groups.

In brief: five coal export terminals are proposed to be constructed on the Oregon and Washington state shores, which would be capable of dramatically increasing coal shipments to China and many Asian countries. The proposed terminals are the solution to the U.S. coal industry problems, so their importance is growing proportional to the decrease of coal use in the U.S.

The newly proposed export terminals are:

- Cherry Point, Washington, is to be built and operated by SSA Marine. The proposed Gateway Pacific Terminal, is a new shipping facility north of Bellingham, capable of handling 48 million tons of coal per year. Peabody Energy, the world's largest private sector coal company, has already agreed to supply 24 million tons of coal to be exported from this terminal.

- Longview, Washington, will be the host of the proposed Millennium Bulk Terminals, a subsidiary of the Australian coal mining company Ambre Ener-

gy, which purchased a port site on the Columbia River. Arch Coal, a major American coal mining company, has a 38 percent stake in the site and hopes to export 44 million tons of coal annually, starting with 25 million tons annual exports during the first phase.

- Port of St Helens, Oregon, is Kinder Morgan's Port Westward project and proposes to build and operate a coal export terminal near Clatskanie, Oregon, capable of handling 30 million tons of coal per year, stating with 15 million tons during phase one.

- Port of Morrow, Oregon, is the brain child of Ambre Energy, who plans to construct the new facility on the Columbia River in eastern Oregon. Coal from the mines will be transferred from rail cars to barges to be towed downriver to the Port of St. Helens and loaded onto ongoing vessels. The new terminal would be able to handle about 8 million tons annually.

- Coos Bay, Oregon, is the proposed site of "Project Mainstay," a new export terminal capable of handling up to 10 million tons of coal annually.

The Coal Exports Battle

All in all, about 140 million tons of coal could be shipped eventually from these new export terminals. This is more than the U.S. has ever exported in a single year, which means that the plan is to double U.S. coal exports.

But there are problems with these plans. The primary objections to the new coal export terminals, by a number of groups, are the spread of coal dust, possible environmental damage, the noise and vibrations of the trains, the traffic obstructions, loss of hunting grounds, and other problems the export terminals might cause to the locals.

Exporters store coal in large piles at the terminals, and these piles can feed prolific quantities of dust to the wind, when coal is loaded and unloaded. This makes coal terminals active sources of fugitive dust, which offends and sickens wildlife and people living and working near these facilities.

As a result, Seward, Alaska, residents have sued the local terminal operators because coal dust blowing off the terminal's stockpiles regularly coats the nearby harbor, fishing boats and neighborhood houses with dust and debris. The residents' suit also alleges that the conveyor system used to load ships drops coal dust into Seward's scenic harbor, violating the Clean Water Act.

In 2010, the State of Alaska fined the railroad

company that delivers coal to the terminal $220,000 for failing to control dust as needed for safe operation. Residents of Point Roberts, a beachfront community three miles away from British Columbia's Westshore coal terminal, which ships about 21 million metric tons per year, complain that coal dust blackens their homes, patio furniture, and boats moored in the local marina.

In 2001, a study of coal dust emissions in Canada found that the Westshore Terminal emits roughly 715 metric tons of coal dust a year. Another study conducted by researchers at the University of British Columbia found that the amount of coal dust in the vicinity of the terminal had doubled from 1977 to 1999.

The Lamberts Point Coal Terminal in Norfolk, Virginia, which ships 28 million tons of coal annually, is legally permitted to release up to 50 tons of coal dust into the air each year. Black grit from the coal piles commonly coats cars, rooftops, windows, and plants in neighboring communities. Neighbors worry that the dust is responsible for the vicinity's elevated asthma rates as well.

In Newport News, Virginia; Charleston, South Carolina; and on the Mississippi River, coal dust routinely blankets neighborhoods and local waterways. And coal dust is widespread near terminals in Australia, India, and South Africa.

The scale of likely dust emissions at the export facilities planned for the Northwest is unclear, but is suspected to be significant. The developers are promising to install mitigation devices to control dust, but promises are easy to make and hard to stand by. Coal is stored in huge piles outdoors, exposed to wind and weather, and the piles are shoveled into new positions by giant bulldozers and other machinery from time to time. All this makes it highly unlikely that the coal dust can be contained to an acceptable level.

Environmental Concerns

Environmental damage is one of the greatest objections on the list of the locals and environmentalists. Each of the planned coal export facilities would occupy hundreds of acres of mostly waterfront land with large capacity for storage of raw coal. Large quantities of dust and runoff would be generated day and night and might harm fish and other wildlife in the affected areas.

The Lummi Nation Indians have maintained the largest Native fishing fleet in North America around Cherry Point fisheries for thousands of years. They have officially objected to the Gateway Pacific export terminal construction, since it would certainly affect negatively their way of life.

The Lummi Nation sees the new terminal as a direct violation of their treaty rights to fish, gather, and hunt in the usual places, where they have been doing this for generations. They also fear that the pollution from the terminal would significantly degrade the already fragile and vulnerable crab, herring and salmon fishery, thus devastating the economy of the local community.

To them the choice is clear: preserving the way of life in a clean environment vs. exporting millions of tons of dirty coal, which will change and contaminate the local area, after which it will also spread more pollution on the other side of the world.

Traffic Issues

The other problem posed by the export terminals is the large number of coal trains that need to be added to supply the new export terminals in the Northwest. The estimate is for 56 trains per day to be added to the existing railroad traffic, or about 20,500 additional train compositions will rumble each year across the Northwest, each more than a mile and a half in length. Each railroad car carries about 100 tons of coal and there are over 100 cars in each composition. This is a lot of steel and coal rolling on the rails, making a lot of loud noise and vibrations—much more than any other trains, owing to their super size and extraordinary weight.

A statement issued by health professionals in Whatcom County, Washington, documents a number of health-related problems with coal exports. In addition to the risks of coal dust, doctors raise concerns about the impacts of the trains themselves, which generate noise, create collision hazards, and delay emergency medical response by impeding rail crossings. Trains are also responsible for hazardous air pollution from diesel engines, a documented threat to health in Washington.

Of course, the additional coal trains will worsen traffic congestion and impair truck freight in the area. The large number of trains required during full operation of the coal terminals means that coal trains traveling at 35 miles per hour would obstruct the railroad crossings more than 10 percent of every day.

In urban locations, where train speeds are much slower, the problem is likely to be much worse. A series of traffic analyses has found that coal train traffic will result in serious congestion and delay in many Northwest cities. In Marysville, Washington, for example, a single coal train passing through would delay traffic on the city's central arterials by the equivalent of three to four continuous red light cycles. Fifty-six trains crossing the city—in addition the present railroad traffic—points to a potential nightmare scenario where all of the city's access points are obstructed simultaneously and for long periods of time.

Health Concerns

There are also environmental effects and related health considerations. Powder River Basin coal is lower in ash and sulfur than some other kinds of coal, but it also produces less energy per pound than the coals more commonly burned in modern power plants. To produce the same amount of energy, 50% more coal is needed to be mined, shipped, and burned. Even if it were cleaner, coal is coal and an extremely polluting form of energy.

During burning, coal releases numerous hazardous substances, in addition to ash, CO, CO_2, and SO_2, including the radioactive materials uranium and thorium. Estimates are that U.S. coal plants have released hundreds of thousands of tons of uranium, and that radiation from coal plants is a greater threat to Americans than radiation from nuclear plants.

The true costs of coal in the U.S. alone are estimated to reach nearly $60 billion in health costs annually, and a grand total of $350 billion in harm to public health, mining damage, pollution, and government subsidies. The National Research Council concluded in 2010 that non-climate related damages from burning coal are 20 times higher than the damages from natural gas. This makes natural gas the second dirtiest and costliest fossil fuel in use, but that's another story…

Global Climate Change

Finally, coal is a serious contributor to global climate change, and there is very little variation in the carbon intensity of coal types on an energy-adjusted basis. This is because the amount of energy produced is simply a very close reflection of the carbon content of the coal.

All in all, coal has a much heavier carbon footprint than any other fossil, or non-fossil fuel used today. Sub bituminous coal, for example, produces 32 percent more GHG emissions than diesel and 82 percent more than (or almost twice as much as) natural gas. Other coal types are much "cleaner," if "clean" and "coal" can even be used in the same sentence.

Burning the 140 million tons shipped from the Northwest coal export terminals will release roughly 250 million tons of heat-trapping CO_2 into the atmosphere every year. This is the equivalent of the emissions from nearly 60 million cars.

An average large coal-fired power plant emits about 10 million tons of CO_2 per year, so the Northwest USA coal exports will power several dozen coal-fired power plants in China and throughout Asia and Europe.

Basically, we are exporting our dirty coal, thus freeing the North American continent from GHG pollution and toxic emissions, thus meeting and beating the national standards and international climate warming agreements. At the same time, however, we are allowing and even encouraging increase of the same polluting emissions in Asia and Europe.

Yet, we will be affected too; air pollution and toxic emissions won't stay in Asia or Europe, but will eventually spread all over the world. We all will be affected by the rise of pollution in any one part of the world.

Coming generations will be the judge. For now, the plan of the U.S. coal companies is to ramp up coal exports as much as possible. After all, they are just trying to preserve their way of life—at any cost.

Who is right, who is wrong and who is going to win the battle is still to be determined, but for now a long and difficult battle is underway on the U.S. Northwest shores.

Electric Power Imports

We all know by now about the imports and exports of fossils, because discussions about them become louder and overheated. There is one energy source, however, that is kept in the background only because it is an old, business-as-usual thing that not many of us are even aware of.

We are talking about the export and import of electric power.

After all, we burn millions of tons of coal, and billions of cubic feet of natural gas to generate electric power—which in turn emits thousands of tons of GHGs that poison our air in the process. So we'd better know and care how much of the stuff is imported and exported, how, why, and by whom.

Every day for dozens of years now our good neighbors, Mexico and Canada, have been trading electric power with us—importing some and exporting some. Figure 1-17 shows that the U.S. has been steadily increasing the balance of its electric power imports. Mexico supplies about 1/10 of the total, with Canada supplying the rest.

Starting with 6 TWh electric power imports in 2002, today we import almost 10 times that amount. The actual imports balance (imports minus exports) in 2012 was nearly 50 TWh. This is equivalent to the total power generated by several coal-fired power plants every year.

The rising U.S. electric power imports constitute a medium level risk to our energy security.

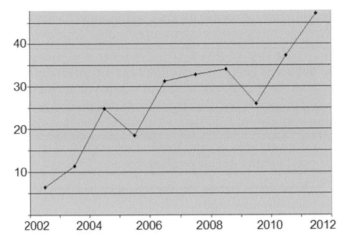

Figure 17. Balance of US electricity trade (imports in thousand TWh)

Mexico and Canada are friendly counties, so we don't expect them to intentionally sabotage the relations with the U.S., and yet there are a number of socio-political and economic factors that can affect the power generation in these countries. There are also natural events that might have a hand in electric power generation failure in these countries.

We have no control over these variables, so they could affect the U.S. electric power imports in a negative way, which in turn will affect our overall energy security.

Note: Amazingly, most European countries also import large quantities of electric power. Italy, for example imports almost as much as the U.S., which constitutes over 10% of the total electric power use in the country. Because of that, and other energy expenses, Italy has the highest electricity costs in Europe.

Another fact worth noting in Italy is that the electric power prices go up with increased consumption, which is one way to encourage energy conservation. This may be one approach that we might see used in the U.S. in the near future.

ENERGY SECURITY AND INDEPENDENCE.

The global energy market is undergoing major changes today, no doubt. Everything looks good today...but so it did on that beautiful day in the fall of 1973 when the Arabs decided to cut down our oil supplies. The result was a world in turmoil, which lasted several years, and which we still talk about and are affected by.

It was an equally beautiful day when the Chernobyl nuclear power plant blew up in the spring of 1986.

It changed the lives of the locals in an instant. The radiation cloud also affected millions in Europe, and some of those who did not die are still recovering from property and personal damages.

On a similarly beautiful day in September 2001, terrorists crushed planes loaded with unsuspecting passengers into the Twin World Center towers and the Pentagon. That also changed us and the world forever, so that we wonder constantly what to expect next...

Things like that happen often around the world, and could happen anytime anywhere, without warning. Internal and external threats to our security are nothing new, nor is the security of energy sources, and the related infrastructure—the preservation of which is critical to our survival as a nation.

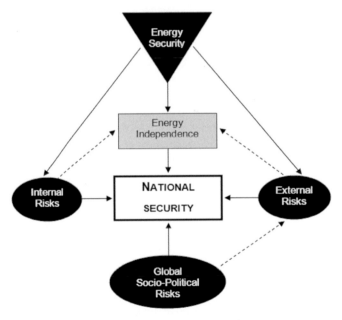

Figure 1-18. National security factors

From the necessity to preserve the integrity of our energy sector at all cost, we derive the term *energy security*, which simply means anticipating, planning, and preventing any interruptions in the energy sector—regardless of their source, magnitude, or intentions.

Internal and external risks alike could disrupt the production, transport, and/or use of energy, which may have disastrous results for our economy, and the personal safety of the population.

Energy security ensures a healthy and undamaged energy sector as a precursor of energy independence, which in turn contributes to a healthy national security.

In broader terms, *energy security activities* are intended to foresee, guard against, and prevent internal and external risks that otherwise might have negative effect on our drive to energy independence, thus preserving our national security in the process.

Today, as we are getting closer to energy independence, we are increasingly concerned with our energy security, mostly due to the increased terrorist activities worldwide. So our energy security and independence are at crossroads, where they are challenged by a number of internal and external risks.

In the following chapters we take a close look at the existing state of the art and future development of the energy technologies and the associated risks. We also review and analyze some of the possible solutions to ensuring energy security and independence.

But first, let's take a closer look at the meaning of the word *energy*.

Today's Energy Basics

What is energy? Generally speaking, energy is what makes everything move, change, grow and even exist. More specifically, translated from Greek energy means "activity, and/or operation." It is a quantity that exists, but is indirectly observed, for it is invisible and immeasurable in its purest form.

In practical terms, it is represented as the ability of a physical system to do work on other physical systems, or on itself.

Work is defined as force acting through a distance (a length of space), where energy is always equivalent to the ability to exert pulls or pushes against the basic forces of nature, along a path of a certain length.

The important facts here are that:

1. *Potential energy* remains an "indirectly observed" entity while any of its two main constituents "force" and "distance" are missing. This type of energy is called potential, because it has the potential to do work, if it is activated by its components force and distance, and

2. *Kinetic energy* is "visible," or measurable entity when both of the variables (force and distance) are present. In this case the potential energy has been "activated" and has become kinetic energy. Then it does work, and can be observed and measured.

Also, another two very important conditions to remember about energy:

1. Energy that is visible and measurable in an object, or a person, can be identified with and measured by its mass, and

2. Energy cannot be created or destroyed.

Finally, in this text we observe and measure the impact of energy—including power generation and use—on the global economy, and where we distinguish three different types of energy (or electric power) generating systems:

1. Conventional energy generators, such as coal, oil, and natural gas, also called fossils;

2. Renewable energy generators, such as solar, wind, and bio-fuels; and

3. A third category of energy generators, which don't fit in the above systems, or are hanging on the fringes, are nuclear, geo-, and hydropower generators.

Energy

Energy is measured in joules, but other units, such as kilowatt-hours, Btus, and kilocalories, are used too. All of these units translate into units of work, which are always defined in terms of force and distance that the forces act through. In the electric field we use the word power to replace energy, but for all practical purposes it is one and the same.

When ordinary material particles are changed into energy (such as energy of motion, or radiation), the mass of the system does not change through the transformation process. However, there may be mechanistic limits as to how much of the matter in an object may be changed into other types of energy and thus into work, on other systems. Energy, like mass, is a scalar physical quantity.

A system can also transfer energy to another system by simply transferring matter to it (since matter is equivalent to energy, in accordance with its mass). However, when energy is transferred by means other than matter-transfer, the transfer produces changes in the second system, as a result of work done on it.

This work manifests itself as the effect of force(s) applied through distances within the target system. For example, a system can emit energy to another by transferring (radiating) electromagnetic energy, but this creates forces upon the particles that absorb the radiation. Similarly, a system may transfer energy to another by physically impacting it, but in that case the energy of motion in an object, called kinetic energy, results in forces acting over distances (new energy) to appear in another object that is struck.

Transfer of thermal energy by heat occurs by both of these mechanisms: heat can be transferred by electromagnetic radiation, or by physical contact in which direct particle-particle impacts transfer kinetic energy.

Energy may be stored in systems without being present as matter, or as kinetic or electro-magnetic energy. Stored energy is created whenever a particle has been moved through a field it interacts with (requiring a force to do so), but the energy to accomplish this is stored as a new position of the particles in the field. This is a configuration that must be "held" or fixed by a different type of force (otherwise, the new configuration would resolve itself by the field pushing or pulling the particle back toward its previous position).

This type of energy "stored" by force-fields and particles that have been forced into a new physical configuration in the field, by doing work on them through another system, is referred to as *potential energy*. A simple example of potential energy is the work needed to lift an object in a gravity field, up to a support, while the object is resting. It is a measure of what *could happen*, rather than of *what happened*?

Each of the basic forces of nature is associated with a different type of potential energy, and all types of potential energy (like all other types of energy) appear as system mass, whenever present. For example, a compressed spring will be slightly more massive than before it was compressed. Likewise, when energy is transferred between systems by any mechanism, an associated mass is transferred with it.

Any form of energy may be transformed into another form. For example, all types of potential energy are converted into kinetic energy when objects are given the freedom to move to different positions (for example, when an object falls off a support). Kinetic energy then can be measured by the amount of work done during some activity—in this case, during the process of the object falling from the support.

When energy is in any form other than thermal energy, it may be transformed with good or even perfect efficiency to any other type of energy, including electricity or production of new particles of matter. With thermal energy, however, there are often limits to the efficiency of the conversion to other forms of energy, as described by the second law of thermodynamics. Thermal energy (heat) tends to be reduced or lost by a number of mechanisms, which complicates the conversion process.

Nevertheless, during all such energy transformation processes, the total energy remains the same, and a transfer of energy from one system to another results in a loss to compensate for any gain.

Although the total energy of a system does not change with time, its value may depend on the frame of reference. For example, a seated passenger in a moving airplane has zero kinetic energy relative to the airplane, but high kinetic energy (and even higher total energy) relative to the Earth.

The concept of energy and its transformations is useful in explaining and predicting most natural phenomena. The direction of transformations in energy (what kind of energy is transformed to what other kind) is often described by entropy (equal energy spread among all available degrees of freedom) considerations.

All energy transformations are permitted (and explainable) on a small scale, but certain larger transformations are not permitted because it is statistically unlikely that energy or matter will randomly move into more concentrated forms or smaller spaces. The fact that many forces and events control the outcome of most energy transformations also contributes to their complexity.

Energy Applications

Energy, in all its forms and variations is widely used in the sciences. For example:

1. *In physics*, energy is considered a quantity that exists, but is indirectly observed (or invisible) and immeasurable in its purest form. It comes to life, and is measurable when its other components (force and distance) are considered. In that case energy becomes work, and can be measured as a physical entity. The work then could be observed, measured and expressed as heat, electric power, mass, speed, etc. variables.

   For example, photons traveling from the sun through space have potential energy stored, which upon impact with a solar panel is released and converted into heat (heating the panels) and electricity, which can be extracted and used.

2. *In chemistry*, energy is an attribute of a substance as a consequence of its atomic, molecular or aggregate structure. Since a chemical transformation is accompanied by a change in one or more of these kinds of structures, it is invariably accompanied by an increase or decrease of energy in the substances involved. Some energy is transferred between the surroundings and the reactants of the reaction in the form of heat or light; thus the products of a reaction may have more or less energy than the reactants.

   Chemical reactions are invariably not possible unless the reactants overcome an energy barrier known as the activation energy. The speed of a chemical reaction (at given temperature T) is related to the activation energy E, by the Boltzmann's population factor $e^{-E/kT}$, which represents the probability of a molecule to have energy greater than or equal to E at the given temperature T. This exponential dependence of a reaction rate on temperature is known as the Arrhenius equation. The activation energy necessary for a chemical reaction can be in the form of thermal energy.

3. *In biology*, energy is an attribute of all biological systems from the biosphere to the smallest living organism. Within an organism it is responsible for growth and development of a biological cell or an organelle of a biological organism. Energy is thus often said to be stored by cells in the structures of molecules of substances such as carbohydrates (including sugars), lipids, and proteins, which release energy when reacted with oxygen in respiration. In human terms, the human equivalent (H-e) (Human energy conversion) indicates, for a given amount of energy expenditure, the relative quantity of energy needed for human metabolism, assuming an average human energy expenditure of 12,500kJ per day and a basal metabolic rate of 80 watts.

    For example, if our bodies run (on average) at 80 watts, then a light bulb running at 100 watts is running at 1.25 human equivalents (100 ÷ 80) i.e. 1.25 H-e. For a difficult task of only a few seconds' duration, a person can put out thousands of watts, many times the 746 watts in one official horsepower. For tasks lasting a few minutes, a fit human can generate perhaps 1,000 watts. For an activity that must be sustained for an hour, output drops to around 300; for an activity kept up all day, 150 watts is about the maximum. The human equivalent assists understanding of energy flows in physical and biological systems by expressing energy units in human terms: it provides a "feel" for the use of a given amount of energy.

4. *In geology*, continental drift, mountain ranges, volcanoes, and Earthquakes are phenomena that can be explained in terms of energy transformations in the Earth's interior, while meteorological phenomena like wind, rain, hail, snow, lightning, tornadoes and hurricanes, are all a result of energy transformations brought about by solar energy on the atmosphere of the planet Earth.

    For example, an erupting volcano releases its energy to create land slides, start fires etc., phe-

nomena related to the release of the thermal and mechanical energies stored in it through the millennia.

5. *In cosmology* and astronomy the phenomena of stars, nova, supernova, quasars and gamma ray bursts are the universe's highest-output energy transformations of matter. All stellar phenomena (including solar activity) are driven by various kinds of energy transformations.

    Energy in such transformations is either from gravitational collapse of matter (usually molecular hydrogen) into various classes of astronomical objects (stars, black holes, etc.), or from nuclear fusion (of lighter elements, primarily hydrogen).

Note: For the purposes of this text, the term "energy" is used with the understanding that it is contained in fuels, or devices, which generate or consume energy, or electric power. The energy contained in these fuels or devices is converted from one form to another—usually from potential to kinetic—by burning coal, oil, or natural gas, or from solar panels (converting solar energy into electricity).

The chemical and other energy types contained in these is converted into mechanical movement of and changes in their particles, which results in heat or electric flow. Thus produced heat boils water into steam, which drives steam turbines and generators that produce another form of energy, electric power, which can be measured and used to power our lives. In the case of solar panels, the solar energy falling on them is converted directly into electric power.

Similarly, energy (electric power) can also be generated by converting wind energy into mechanical motion and then into a flow of useful electrons coming from the generator, which can be used as electric energy (power).

In more practical terms, energy is all around us and can be observed and measured in its different forms—from light and sound, to the less obvious radio, TV, and satellite communication waves.

Energy Conversions

Energy in the universe is sometimes quickly and easily changed from one state (or type) into another. At times the conversion process is not that quick, or easy, depending on the type of materials and processes involved.

Here are some examples of the major energy conversion cycles:

1. Radiative to/from thermal energy conversion:
   — Incandescence is an example of radiative to thermal energy conversion
   — The solar collector is an example of thermal to radiative energy conversion

2. Radiative to/from chemical energy conversion:
   — Photosynthesis is an example of radiative to chemical energy conversion
   — Chemi-luminescence is an example of chemical to radiative energy conversion

3. Radiative to/from electric energy conversion:
   — The solar (PV) cell is an example of radiative to electric energy conversion
   — The fluorescent light is an example of electric to radiative energy conversion

4. Thermal to/from electrical energy conversion:
   — The thermo-cell is an example of thermal to electric energy conversion
   — The electrical resistance of wires is an example of electric to thermal energy conversion

5. Thermal to/from mechanical energy conversion:
   — Heated steam is used to turn the turbines in power stations
   — The mechanical energy from rubbing two sticks together is converted into heat

6. Thermal to/from chemical energy conversion:
   — Endothermic reactions are examples of thermal to chemical energy conversion
   — Exothermic reactions are examples of chemical to thermal energy conversion

7. Chemical to/from electric energy conversion:
   — The car battery is an example of chemical to electric energy conversion
   — Electrolysis is an example of electric to chemical energy conversion

8. Chemical to mechanical energy conversion process is observed in human muscles, where the chemical energy stored in the body is converted into mechanical, as needed to activate the different body muscles.

9. Mechanical to/from electric energy conversion:
   — The car alternator is an example of mechanical to electric energy conversion
   — The electric car engine is an example of electric to mechanical energy conversion

10. Mechanical to/from thermal energy conversion:
    — The car brakes are examples of mechanical to thermal energy conversion
    — The locomotive is an example of thermal to mechanical energy conversion

11. Mechanical to/from sound energy conversion:
    — A running car engine is an example of mechanical to sound energy conversion
    — Aretha Franklin breaking a glass with the high pitch in her voice is an example of sound to mechanical energy conversion

12. Nuclear energy, stored in nuclear materials, is converted to thermal energy in the nuclear power plants, and the resulting heat is used to generate electricity.

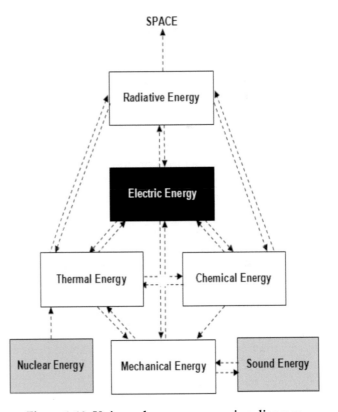

Figure 1-19. Universal energy conversion diagram

Energy transformations in the universe over time are characterized by various kinds of potential energy that have been available since the Big Bang, and are later being "released" (transformed to more active types of energy such as kinetic or radiant energy), when a triggering mechanism is available.

Familiar examples of such processes include nuclear decay, in which energy is released that was originally "stored" in heavy isotopes (such as uranium and thorium), by nucleosynthesis. This is a process ultimately using the gravitational potential energy released from the gravitational collapse of supernovae, to store energy in the creation of these heavy elements before they were incorporated into the solar system and the Earth. This energy is triggered and released in nuclear fission bombs. In a slower process, radioactive decay of these atoms in the core of the Earth releases heat.

Thermal energy also drives plate tectonics and may lift mountains, via orogenesis. This slow lifting represents a kind of gravitational potential energy storage of the thermal energy, which may be later released to active kinetic energy in landslides, after a triggering event. Earthquakes also release stored elastic potential energy in rocks, a store that has been produced ultimately from the same radioactive heat sources. Thus, according to present understanding, familiar events such as landslides and earthquakes release energy that has been stored as potential energy in the Earth's gravitational field or elastic strain (mechanical potential energy) in rocks. Prior to this, they represent release of energy that has been stored in heavy atoms since the collapse of long-destroyed supernova stars created these atoms.

In a similar chain of transformations beginning at the dawn of the universe, nuclear fusion of hydrogen in the sun also releases a store of potential energy which was created at the time of the Big Bang. At that time, according to theory, space expanded and the universe cooled too rapidly for hydrogen to completely fuse into heavier elements.

This meant that hydrogen represents a store of potential energy that can be released by fusion. Such a fusion process is triggered by heat and pressure generated from gravitational collapse of hydrogen clouds when they produce stars, and some of the fusion energy is then transformed into sunlight. Such sunlight from our sun may again be stored as gravitational potential energy after it strikes the Earth, as (for example) water evaporates from oceans and is deposited upon mountains (where, after being released at a hydroelectric dam, it can be used to drive turbines or generators to produce electricity).

Sunlight also drives many weather phenomena, save those generated by volcanic events. An example of a solar-mediated weather event is a hurricane, which occurs when large unstable areas of warm ocean, heated over months, give up some of their thermal energy suddenly to power a few days of violent air movement.

Sunlight is also captured by plants as chemical potential energy in photosynthesis, when carbon dioxide and water (two low-energy compounds) are converted into the high-energy compounds carbohydrates, lipids, and proteins. Plants also release oxygen during photosynthesis, which is utilized by living organisms as an electron acceptor, to release the energy of carbohydrates, lipids, and proteins.

Release of the energy stored during photosynthesis as heat or light may be triggered suddenly by a spark, in a forest fire, or it may be made available more slowly for animal or human metabolism, when these molecules are ingested, and catabolism is triggered by enzyme action.

Through all of these transformation chains, potential energy stored at the time of the Big Bang is later released by intermediate events, sometimes being stored in a number of ways over time between releases, as more active energy. In all these events, one kind of energy is converted to other types of energy, including heat.

Energy Measurement Units

The energy value of fuels such as coal and oil is usually given in Btus. As a result, most of the energy data in English-speaking countries are presented in Btus. The calorie, watt, horsepower, ton of oil equivalent, and joule units are used also in some scientific measurements and for simplicity in some conversions and calculations.

Table 1-6. Key energy conversion units
Note: toe means tons of oil equivalent

Unit	Measure
1 joule (J)	$1N.m$
	$1kg.m/s^2$
	0.2388 cal
	9.4782^{-4} Btu
	2.7778^{-4} Wh
	10^7 ergs
1 calorie	4.1868 J
1 kWh	3.6×10^{13} ergs
	3600 kJ
	860 kcal
	8.6^{-5} toe *
1 toe *	10^{10} cal
	41.8 GJ
	11.63 MWh
	1.28 tons coal equivalent
	39.68 million Btu
1 million Btu	1.0551 GJ
	2.52^{-2} toe
	0.2931 MWh
1 Watt	1 J/s
1 HP	746 W
1 GWh	86 toe

1 joule is equal to 1 Newton-meter and can be expressed as:

$$1J = 1kg\left(\frac{m}{s}\right)^2 = 1\frac{kg/m^2}{s^2}$$

where

kg is in kilograms
m is in meters, and
s is in seconds

The US units for both energy and work are the foot-pound force (or 1.3558 J), the British thermal unit (Btu, or about 1055 J), and the horsepower-hour (or 2.6845 MJ).

The unit used to represent everyday electricity, as in utility bills, is the kilowatt-hour (kWh), where one kWh is equivalent to 1,000 Watts used during one hour, or about 3600 kJ.

Another useful unit is kilowatt-hours per year (kWh/yr)—the total annual electric energy (power) consumption divided by 365—or the average rate at which energy is used during the period of one year.

Types of Energy

There are several major types of energy that we need to take a close look at, especially those that can be used to generate power on Earth, to understand them well, and to see if we can do something to delay, or even eliminate the doomsday scenario:

1. Energy can be divided into three major categories:
    a. Gravitational energy.
        This is the energy (or force) that attracts all objects, humans and animals to the Earth's center.
    b. Electromagnetic energy.
        This is the energy emitted and absorbed by charged particles. It exhibits wave-like behavior as it travels through space. It has both electric and magnetic field components in equal proportions and force. In a vacuum it moves at the speed of light. Sunlight is a type of electromagnetic energy.
    c. Nuclear energy.
        This is the energy that is hidden in the atoms and molecules of the elements and substances that make this world. This energy plays a major role in chemical and other reactions on the molecular level.

2. We can also divide energy, as briefly mentioned before, into two important types, according to their state and the work they do or intend to do:

    a. Potential energy is any energy that is stored and is idle; waiting to be released.
        A good example of potential energy is a large tank of water perched on a tall hill. The water in the tank is stored and idle, but has "potential" to do some work if and when released, at which point it will obtain kinetic energy. But until then its energy is "potential," or idle.
    b. Kinetic energy is energy in action. It is what moves things.
        Any object that is in motion has some amount of kinetic energy. The amount of kinetic energy at a certain time can be expressed as power.

3. Other major types of energy divisions are:
    a. Mechanical energy.
        This is the energy needed to move, bend, break and otherwise change material objects. Slamming a hammer onto a piece of metal, or breaking a twig require mechanical energy.
    b. Chemical energy.
        Reacting different substances (chemicals) to obtain a new or modified substance requires chemical energy.

In the physical and more practical aspects of our lives, there are several forms of energy, the main of which are:

*   Physical energy is the energy that moves things
*   Chemical energy is actually a number of energy types that drive chemical reactions
*   Electrical energy makes our lights, radios, and electrical equipment work
*   Thermal energy, or heat, is used to elevate the temperature of objects
*   Biomass energy is generated by plants
*   Heat energy is what we get when burning coal and other fuels
*   Geothermal energy comes from hot springs deep in the ground
*   Fossil fuels are the old energy sources—coal, oil and natural gas
*   Solar energy is the energy we obtain by capturing and converting sunlight
*   Hydropower is obtained from water turning a wheel, which generates energy
*   Ocean energy is the energy generated by ocean waves and tides
*   Nuclear energy generates heat by nuclear reactions
*   Solar energy comes from the sun and can be converted into heat or electricity

- Wind energy makes the wings of a wind mill rotate to generate electricity
- Transportation energy facilitates moving large loads
- Magnetic energy is created by permanent magnets or electromagnets
- Sound energy is created by increasing the noise level
- Cosmic energy is the different types of energy contained in the universe

Let's take a look at some of the major types that would help us clarify the concepts of energy and its use on Earth:

1. Physical Energy

Energy exists in its physical forms, though in the physical realm it is invisible and immeasurable in its purest form. When another component is added to it, however, such as force, mass, distance, speed, etc., it comes to life, and is easily measurable. In conjunction with its other components (force and distance), it is truly a physical entity that we use in our daily lives.

In those cases, energy becomes physical work, and can be measured as a physical unit. The work then could be observed, measured and expressed as heat, electric power, mass, speed, etc. variables.

For example, a large body of water sitting behind the secure walls of a dam contains energy which is hidden from view while the water is just sitting in the reservoir lake. The energy is released immediately when the intake gate is opened and the water is allowed to run down the penstocks. At the end of its travel it hits the turbine blades with a great force to generate electricity.

The water hitting the turbines converts the potential energy stored in the lake into kinetic energy (work) and eventually into electric power, which is measured in kilowatts (kW) and megawatts (MW) of electric power generated by its fall.

2. Chemical energy

Energy can be changed by rearranging the atoms in a molecule or by combining free atoms, at which point chemical energy becomes available. The reverse of the photosynthesis process that created the huge forests discussed above is a good example of that process. When the carbon, hydrogen, and oxygen molecules in carbohydrates and sugars (as in the fossils) are oxidized, either by burning or by the process of digestion, CO_2 and H_2O are formed and energy is released (such as when burning fossil fuels). Free carbon (in coal, or oil)

combines with oxygen during the burning process to form CO_2 and release heat energy too.

Chemical energy can be generated also when strips of lead (Pb) are dipped into sulfuric acid (H_2SO_4), where the Pb replaces the H_2 in the H_2SO_4 to form $PbSO_4$ and H_2O (like in a lead battery). This process goes in reverse during charge and releases energy during discharge, which can be used to generate electrons in a wire connected between the electrodes, which are then extracted in the form of electric current. This is the process of conversion of chemical potential energy into electrical energy, which takes place in a car battery. Many other combinations of metals, liquids, and gases can be arranged and forced to release energy in a similar way.

Chemical potential energy is one of our most important forms of energy known to man, since it is the energy of fossil fuels formation and use, as well as many other chemical reactions, which provide our food and power our society.

The fossil fuels are becoming too expensive to burn, however, so we must find other ways to generate electric energy, since it consumes the largest amount of fossils.

3. Solar energy

We now know how solar energy is created at the sun, where immense amounts of matter are converted to light and heat via fusion reactions of hydrogen converted to helium. The sunlight then traverses the 93 million miles to Earth in a matter of minutes, and arrives at our atmosphere as radiant energy contained in particles called photons. This is pure energy, riding on its own electromagnetic fields through space. The photons also contain potential energy which can be released upon impact.

Radiant energy (the photons) has properties that are wavelike, but at the same time it has particle-like properties. These wavelike, particle-like units are delivered in discrete chunks that have two distinct properties—wavelength and frequency.

The wavelength is measured by the distance between the peaks of the waves, while the frequency is the speed of its vibration, which, in turn determines its energy. The higher the frequency of the photons, the higher the amount of energy they carry. For example, X-ray photons are vibrating at very high frequencies, and carry enough energy to do physical damage to body tissues and organs. Radio waves vibrate several times slower, and do no damage.

Sunlight radiation waves have a wide range of

vibrations and carry enough energy to cause chemical changes in molecules that they encounter on the way. Most of the incoming radiation is visible light, and life forms have developed in response to those wavelengths. Reflected or converted solar radiation leaves Earth at the much longer infrared (IR, or heat) radiation.

4. Radiation energy

Radiant energy interacts with matter by 1) reflection from surfaces it encounters on its way, 2) absorption, where it sets molecules or atoms into vibration, which could cause re-radiation in the same or at longer wavelengths, 3) absorption and then dissipation within a substance as heat, and 4) absorption and production of a chemical change.

Heat absorbed by land provides the major heat input into the lower atmosphere. It is this converted solar energy that provides the kinetic energy for the great trade winds and all the other winds of the "atmospheric engine." Windmills are driven, indirectly, by solar energy which creates wind conditions. The combination of unequal heating, the Earth's gravity, and the spin about its axis are responsible for the atmospheric and oceanic motion.

Much of the solar technology depends on this radiant-to-heat energy conversion. Solar collectors for space and water heating, and other solar collectors rely on this interaction.

Radiant energy absorbed by the ocean and converted to heat ultimately provides humankind with another useful form of energy. If water molecules receive enough energy to become hot enough, they can escape from the liquid. This is the process we call evaporation. This water vapor is then lifted by rising heated air and is carried by the winds until it falls as rain. By this lifting of great masses of water away from the surface of the Earth, some of the solar energy is turned into gravitational potential energy.

Some amount of this gravitational potential energy is converted to mechanical energy when water is forced to run through a turbine on its way back to the sea. Thus, hydropower is a second indirect consequence of solar energy.

The selective absorption of radiant energy in the atmosphere (interaction 2 above) also has important consequences. Without the absorption at the top of the atmosphere, most of the ultraviolet radiation and other short-wavelength radiation life on Earth would be impossible. If this energetic radiation were not stopped by the thin layer of ozone, no living thing could exist in the sunlight.

5. Nuclear energy

The nuclear energy we discuss here is the energy that is stored in the nuclei of atoms and can be released by rearrangements of the protons and neutrons in it. Examples of this energy are the thermonuclear reactions in the sun (fusion), and the atom bomb (fission).

In fusion reactions, the lighter elements (with nucleus made up of a few protons and neutrons) can combine to make heavier ones, during which process a great amount of energy is released, as four protons fuse to make a helium nucleus, which contains two protons and two neutrons. Since the helium nucleus has less mass than the combined mass of the four protons, the mass difference is converted to energy.

Energy can be released in fission nuclear reactions too, by elements with heavy nucleus, such as uranium. It can be split into two medium-mass nuclei, during which a lot of energy is released. The two medium-mass nuclei have less mass than the original uranium and in this reaction the missing mass is also converted into energy.

For our everyday energy generating purposes we must be able to control the nuclear reactions. Thus far we have learned how to control the fission nuclear reactions, which are used in our many nuclear power plants. We still don't know how to efficiently initiate and safely control fission nuclear reactions. This challenge is going be left for the next generations to resolve, and no doubt they will, since it might be the only choice for energy production they'll have.

6. Gravitational energy

Energy is contained in and released during the moon's travel around the Earth. Potential energy, in the form of gravitational energy, is stored in the space between the Earth, moon and sun. At the same time all three bodies have a lot of kinetic energy, due to their rotation around each other and around their own axis.

Some of this energy is converted into mechanical energy by the mechanism of ocean waves and tides creation. Tides are formed as a results of the gravitational push-pull mechanism of the moon, and to a lesser extend by the sun. Since the oceans' surface is flexible, the waters beneath it can move up and down according to the moon's position. The converted energy causes the rise and fall of the tides, while some is dissipated as heat energy through friction.

The possibility of converting waves and tidal energy into useful electric energy has been on the agenda for many decades. Turbines can be turned by letting seawater run through them as it comes and goes in and out with the tides, or up and down with the waves. Such projects are feasible in a number of places, but technical difficulties and expense have not allowed their wide deployment.

Nevertheless, we do believe that this is another unexplored, and yet very plausible alternative for the next generations, which won't have the many less complicated and less expensive choices we have today.

7. Geothermal energy

Geothermal energy is one of the most interesting and elegant solutions to energy generation. The nucleus of the Earth consists of molten lava and some of the heat leaks through cracks in the Earth's core and crust and escapes to the surface by forming hot springs, geysers, or volcanoes. This is free heat in its basic form—lots of it and ready to use.

The Earth's heat can be captured as hot water, steam, or high-pressure steam. Different energy generators are used in each case, with the overall result of using this free (waste) energy to generate useful electric energy. In some places, the hot water or gasses are used to heat homes, or in commercial applications, but these are exceptions.

Geologists are not exactly sure how the Earth's molten core was formed, but we know that the inner heat is maintained by the slow decay of radioactive materials in it. The concentration of radioactive material in the Earth is small, yet enough to balance the heat loss through the surface.

As we remove significant amounts of the radioactive materials in the Earth's crust, and as we use more and more of its heat, it remains to be seen if geothermal energy would remain renewable energy for very long. Increasing the use of geothermal energy complicates the environmental picture, and seems to be a knife with two edges, so we must be careful using it.

The next generation will be forced to take a closer look at the inner and outer heat energy balance of the Earth and decide if this energy source can be expanded to serve their daily needs.

8. Wind energy

Wind power is another form of solar energy, where billions of photons arriving from the sun hit the Earth's surface and the air around it, and the energy in them is converted into mechanical and heat energy. The air molecules are heated to the point where they start rising. When enough air is warmed and moved up, an air current is created, which interacts with other environmental elements and surface structures and creates turbulence. Increased turbulence gives rise to wind currents, which can travel at a high speed across great distances.

Wind energy is a major electric power generator. Large-scale wind farms are built in many areas of the world, and provide a significant portion of global commercial electric power.

9. Cosmic energy

There is a lot of radiation in the universe that bombards everything in its way. Celestial bodies, such as stars, gas giants, and black holes are the best examples of massive radiation. Stars, planets, and black holes are continuously producing high energy radiation, which in most cases comes from nuclear fusion. The combining of separate atoms produces high levels of energy that are dispersed into space in the full electromagnetic spectrum. Gas giants can also produce large fields of radiation. This is because most gas giants are celestial bodies that failed to fully become stars, not having sufficient mass to complete fusion.

There is also radiation from the universe as a whole, which is called cosmic microwave background radiation. Luckily, each of the cosmic radiation sources is far beyond the safety of Earth's atmosphere. Unfortunately, to be of use to people, cosmic energy must be captured and controlled, something we are simply unable to do at this point.

Nevertheless, cosmic energy is available in immense quantities in high space, so the next generations will be charged with the task of figuring out how capture it, and to generate useful power for their use.

Other Energy Effects

The atmosphere does not absorb very much radiation at the visible wavelengths, since it is transparent to them. When this radiant energy is absorbed at the Earth's surface, however, it is reradiated at longer wavelengths, in the infrared region.

CO_2 and other gasses act as a barrier which traps some of the wavelengths, which helps keep the Earth warmer at night. Without this effect, the Earth would radiate back out into space most of the energy gathered during the day, and the nights would be very cold. Instead, the absorbed energy is reradiated so that about half of it comes back to Earth, maintaining a much milder night temperature.

This "green house effect" is quite sensitive to the content of the type and amount of gasses in the barrier layer, and we'll discuss it in more detail in the following chapters.

The process of photosynthesis is another interaction that provides not only food and fossil fuels, but also the resource of biomass in general, and the life-giving things—wood, water, grains, etc.—which we increasingly count on to provide food and energy.

Photovoltaic reactions are another example of this interaction. In the text below, we will take a close look at this process, as well as some of the new ways to get usable energy from sunlight in general.

A number of other effects are important, when painting a full picture of the energy cycles. Some of these are the shielding action of ozone (O_3), where it is formed when photons strike O_2 molecules on top of the atmospheric layer to make an additional layer of O_3 where it is beneficial to the Earth's environment. O_3 is also produced in the lower atmosphere by a similar reaction, but there it is harmful to humans.

In more practical terms, energy in its different forms—chemical, fuels, electricity, heat, mechanical, etc.—is used in all sectors of our lives; from residential to transportation, and in all industries, as can be seen in Table 1-7.

Practical Applications

Energy is contained in all substances around us, where in some cases we can easily observe or feel it, such as waterfalls, fire, explosions, etc. In most cases, however, it is locked in the molecular structure and can be released only under certain circumstances.

For example, wood has a lot of energy stored in, but it will stay there until the wood is ignited. This is true for the different solid and liquid fuels too—coal, natural gas, gasoline and diesel. Nuclear materials have the highest amount of energy locked in their molecules that is practically obtainable.

Different substances have different amounts and intensities of energy stored in them. This difference is the specific energy contained in each substance, where specific energy is actually the energy density (quantity) per unit mass or volume.

The specific energy of different substances is shown in Table 1-8.

Table 1-7. Ratios of daily energy uses

Sector	Energy uses
Industrial	22% chemical production
	16% petroleum refining
	14% metal smelting/refining
Transportation	61% gasoline fuel
	21% diesel fuel
	12% aviation
Residential	32% space heating
	13% water heating
	12% lighting
	11% air conditioning
	8% refrigeration
	5% electronics
	5% clothes washing
Commercial	25% lighting
	13% heating
	11% cooling
	6% refrigeration
	6% water heating
	6% ventilation
	6% electronics

Table 1-8. Specific energy of different substances in kWh per kg.

Water*	0.0004
Gunpowder	0.8
Battery	0.8
TNT	1.5
Sugar	5
Wood	5
Coal	7
Gasoline	13
Propane	13
H2 (liq)	35
U235	3,300,000

*Energy of water falling from 100m height

So, what we see here might be a bit confusing, and for sure surprising. Keep in mind that each substance in Table 1-8 weighs 1 kilogram (approximately 2 pounds).

The first thing that pops up from the table is the exceptionally high amount of energy achievable by an equivalent mass of nuclear materials—frightening amounts of energy that can be released under special conditions.

But there are other surprises too; for example, a kilogram of TNT has almost 5 times less energy than the same amount of coal, and almost 10 times less energy than gasoline. And it has even less energy (several times less) than the same amount of sugar.

How could that be? We all know how powerful and destructive TNT is, so comparing it with a handful of coal lumps makes no sense. Yet, this is the reality, and it is so because while coal has more energy, it cannot release it quickly. Upon ignition, the surface molecules start burning first, and when they are burned and gone up in smoke, the next row of molecules under them starts burning. Then the next, and the next, and so on, until the entire lump has gone up in smoke. This process takes a lot of time and releases a little energy at a time.

In contrast, TNT "burns" immediately and instantaneously upon ignition. Or better yet, it explodes violently with a great bang and lots of smoke in a split second. All molecules "burn" at the same time and the reaction is over faster than a blink of an eye.

That quick release of energy is what differentiates the energy in the loud and violent TNT from that in a quietly burning lump of coal, which could continue smoldering for hours.

Diesel fuel's energy density is 12% greater than gasoline, and so diesel engines get about 12% better mileage than gasoline engines of comparable efficiency.

Gasohol and ethanol have lower energy densities than gasoline, so they accomplish less work per unit of volume. In fact, a car burning one unit of ethanol will travel only 70% as far as the same vehicle burning the same volume of gasoline.

All in all, energy is a mysterious thing. Now you see it, now you don't! When you kick a ball, energy is transferred and you know it, because the ball flies away as a result of the energy transferred to it by the kick. But you see nothing else being transferred. The foot transfers some of its energy to the ball, but it does not transfer any objects, material, or anything physical.

Sound waves make air molecules move some, but they end up at exactly the same location where they started. They vibrated for awhile, transferring the sound wave energy from one to another, but they did not transfer or change their mass—they only changed their state from that of rest to that of a more excited state of a temporary vibration.

So where did the energy come from and where did it go? One way to explain it is by taking a close look at heat. At room temperature, the air molecules are vibrating with an average instantaneous velocity of about 770 miles per hour. Some are moving slower, some much faster—all moving in a random pattern. And although air is still and quiet, the molecules are moving and there is a lot of energy contained in them.

Energy Security Drivers

Energy security is defined as the uninterrupted supply of efficient, reliable, and affordable energy sources. It is also a critical link to achieving the ultimate goal of energy independence.

As such, it is an important part of the national security of any nation.

The drivers governing the secure supply of energy—which determine the level of the energy security of a given country—are many and of diverse nature. Below we take a cursory look at some of the key energy security drivers are:

- *Availability of energy resources* is of critical importance to any energy security system, since it determines how much of the energy requirements can be met by exploiting local resources, vs. looking for and procuring imported energy. Imports of energy resources limit the energy security to the level of reliability of the importers.

- *Energy suppliers*—their number and location—is of utmost importance to the energy security of energy importing countries, including the U.S. for we import crude oil and electric power from other nations. The reliance on imported fuels from a limited number of suppliers (especially the unstable and unfriendly countries we deal with) increases the risk of supply interruptions and energy crisis. The location is also of great importance, because due to transport charges, energy prices vary with the distance from the suppliers.

- *Political instability* in some regions of the world causes the energy supply system to be vulnerable to disruptions caused by internal or external political interests and terrorist attacks. Suppliers from politically unstable countries add an increased risk of supply disruption. Due to increased security precautions, the price of these fuels tend to rise too.

- *Transport issues* are a serious threat to the energy security of importing countries. Energy must be readily available at all times, and a number of transport issues hinder the ease and safety with which fuels are transported across the world's oceans and land expanses.

Domestic transport issues, such as lack of adequate highway systems, also hinder the reliable transport of energy sources. India, for example, has a great amount of coal deposits, but lacks the infrastructure to transport coal from the mines to the power plants.

- *Power distribution* (via power grids and distribution networks) is another issue that contributes to energy security nightmares. Aging power grids in the developed countries, and inefficient and inadequate power distribution infrastructure in the developing countries, limit the access of many customers to reliable electric power and fuels.

- *Diversification of power generation capacity* is required by any efficient energy system. It is comprised of various efficient power generation technologies which have enough suitable capacity to meet the needs. The overall system must be also able to maximize the advantages of each technology and coordinate their activities so that the energy delivery and prices are stable all the time.

- *Diversification in the uses of different fuels* is also important for a stable energy security, especially in emergency and crisis situations. Switching from coal to gas, converting gas to liquids, and coal liquefaction, for example, are necessary tools in meeting energy demand, especially when normal operating conditions are affected.

- *Interconnection* of the energy systems is another great factor to consider in terms of energy security. A limited interconnection will increase the risk of supply disruption and/or power generation by reducing the options available to meet the rising energy demand.

- *Technical expertise and experience* are absolutely needed to create and maintain a diverse and efficient energy mix. Once access to different energy sources has been secured, then expertise is required to set up and run the entire infrastructure: supply chain, power generation, fuel handling, delivery systems (pipelines and supply routes, and ports), electricity interconnections and transmission lines, etc.

- *The level of investment* in the energy sector is a significant factor in ensuring energy security, since significant investment is needed to support the growth in energy demand. The availability of that investment, as needed for new and existing infrastructure is key to keeping up with growing energy demand.

  This is especially important for developing countries, since their need for energy is greater, while at the same time the investment opportunities are limited. Investment or lack thereof will remain a significant factor in energy security.

- *Energy prices* are key to providing affordable energy. The final price the consumer pays depends on the cost of the energy supplies, power generation, transmission and distribution. The interruption of supply networks, for example, can impact the consumer prices and create economic difficulties for countries exposed by over-reliance on one energy source. Sustained price rises and short-term spikes in oil, gas or electricity can trigger inflation and recession.

Energy Security...The Beginning

So, we have many different types of energy that are used in many different sectors of our economy and our daily lives. Their effects and interactions are complex and although we may not account for all of them—and maybe don't even understand some—we depend on them for the comfort of our daily lives.

At least this is what we here in the U.S. were used to and did not question until that faithful day in October, 1973. It was a beautiful, sunny, mid-week day in Southern California, where millions were driving their large 8-cylinder cars to and from work on Los Angeles freeways, while many others were headed to the beach or the nearby mountains.

There was not a trace of worry in the air, and all were enjoying the weather and the carefree ride in their luxury cars. The American dream was at its peak. What can go wrong?

Around noon that day, from the other side of the country, Mr. James Schlesinger got a cable from the Saudi oil minister, "According to his Majesty King Faisal," it read, "I am instructed to cut off all oil supplies to the Sixth Fleet in the Mediterranean and to your armed forces in Europe."

This was the beginning. The U.S. energy problems started, and the meaning of the American dream changed from a care-free drive in a luxury car, powered by cheap gasoline, to waiting for hours in gas lines. When your turn came, you could get only a couple of gallons of gas. The American dream was quickly becoming a nightmare.

For the U.S. in general, the oil cutoff would soon morph into a full-fledged Arab oil embargo, which grew into a nationwide energy crisis. Gasoline prices spiraled skyward, and even then there was severe gas shortage spawning long lines at the pumps.

During the 1973 Arab-Israeli War, the Arab members of the Organization of Petroleum Exporting Countries (OPEC) imposed an embargo against the United States in retaliation for the U.S. decision to re-supply the Israeli military and to gain leverage in the post-war peace negotiations. Arab OPEC members also extended the embargo to other countries that supported Israel including the Netherlands, Portugal, and South Africa.

The embargo both banned petroleum exports to the targeted nations and introduced cuts in oil production. Several years of negotiations between oil-producing nations and oil companies had already destabilized a decades-old pricing system, which exacerbated the embargo's effects.

Figure 1-20. Gas lines all over the U.S. were the norm in 1973-1974.

The 1973 Oil Embargo acutely strained the U.S. economy that had grown increasingly dependent on foreign oil. The administration's efforts to end the embargo initiated a complex shift in the global financial balance of power to oil-producing states. It also triggered a flurry of U.S. attempts to address the foreign policy challenges emanating from long-term dependence on foreign oil.

Even before the Arab oil embargo, between 1971 and 1973, OPEC members repeatedly met with the major oil companies to demand higher export prices, higher tax rates, and greater equity in their local subsidiaries. The companies had little choice but to agree, because the economic terms of the world oil trade have shifted dramatically.

In late 1970, the White House began to notice that the increased dependence of the United States on OPEC oil, combined with the heightened power of the OPEC governments, made the United States vulnerable to an oil embargo.

In April 1973, President Richard Nixon announced a package of new energy policies designed to alleviate fuel shortages that had broken out around the country and reduce U.S. dependence on imported oil. They included an end to the system of import quotas, the deregulation of natural gas prices, and incentives to expand domestic energy production. The policies were almost exclusively focused on boosting production; they contained only token, voluntary measures to promote energy conservation.

A November 1970 National Intelligence Estimate, declassified in 2011, concluded that partial interruptions of oil flow from the Middle East to the United States and NATO countries "will probably occur during the next five years" and that "in the event of major Arab-Israeli fighting, some interruption of oil shipments seems almost certain."

Basically speaking, the U.S. was awaking to the reality, but was not fast enough to prevent the disaster. And as a result, we were caught with our pants down. Nobody— from the President down to the last citizen—suspected that our good "friends" at OPEC, who were getting billions of American dollars daily, would hurt us that badly. But they did it without blinking an eye, and the world has not been the same since…

The only good thing that came out from that episode of rude awakening from several decades of euphoria, was that the U.S. will never be caught again with its energy pants down. Never! The threats are still there, and some are even increasing, but we are now watching, anticipating, and planning ahead for all possible scenarios.

We have also developed and put into action enough preventive measures to ensure the uninterrupted supply of energy into the homes and businesses all around the country.

The long-lasting effect of the 1973 Arab oil embargo was a constant upward spiral in oil prices with global implications. The price of oil per barrel first doubled, then quadrupled, and kept on going up, imposing skyrocketing costs on consumers and challenges to the stability of the national economies.

Since the embargo coincided with a devaluation of the dollar, a global recession seemed imminent. U.S. allies in Europe and Japan had stockpiled oil supplies, and thereby secured for themselves a short-term cushion, but the long-term possibility of high oil prices and recession precipitated a rift within the Atlantic Alliance.

European nations and Japan found themselves in the uncomfortable position of needing U.S. assistance to secure energy sources, even as they sought to disassociate themselves from U.S. Middle East policy.

The United States, which faced a growing dependence on oil consumption and dwindling domestic reserves, found itself more reliant on imported oil than ever before, having to negotiate an end to the embargo under harsh domestic economic circumstances that served to diminish its international leverage. To complicate matters, the embargo's organizers linked its end to successful U.S. efforts to bring about peace between Israel and its Arab neighbors.

In response to these developments, in November, 1973 the Nixon administration announced Project Independence, which was designed to promote domestic energy independence. Intensive diplomatic efforts among our allies were underway, promoting a consumers' union that would provide strategic depth and a consumers' cartel to control oil pricing. These hasty efforts, born of desperation, were mostly unsuccessful.

The U.S. administration recognized the constraints of the (Arab-Israel war) peace talks coupled with negotiations with Arab OPEC members to end the embargo and increase production. They also recognized the link between the issues in the minds of Arab leaders, so the Nixon administration started parallel negotiations with key oil producers to end the embargo, and at the same time with Egypt, Syria, and Israel to arrange an Israeli pullout from the Sinai and the Golan Heights.

Initial bilateral negotiations started in November 1973 and culminated with the First Egyptian-Israeli Disengagement Agreement in January, 1974. A final peace deal did not materialize, but the prospect of a negotiated end to hostilities between Israel and Syria proved sufficient to convince the parties to lift the embargo in March 1974.

The oil embargo exposed one of the key challenges of the U.S.' Middle East policy. The strains on U.S. bilateral relations with Saudi Arabia revealed the difficulty of reconciling those demands. The U.S. response to the events of 1973-1974 also clarified the need to reconcile U.S. support for Israel to counterbalance Soviet influence in the Arab world with both foreign and domestic economic policies.

The 1973 oil crisis shocked the American people because it was a rebuke and a stark contrast to the growing prosperity of the postwar era, which was built on an ocean of cheap energy. Since the end of World War II, the real price of oil had steadily declined; a barrel of crude cost less in 1970 than at any time since the Great Depression.

Until 1971, the U.S. government was worried more about the dangers of too much foreign oil than too little. To protect domestic oil interests, oil imports to the United States were limited by a quota system established in 1959 under President Dwight Eisenhower. Foreign leaders from oil-exporting allies such as Canada, Iran, and Venezuela lobbied the White House for permission to sell more oil to U.S. consumers.

When the supply of oil from other countries was disrupted—due to the Western embargo against Iran in 1953, the Suez Crisis in 1956, and the Arab embargo that followed the Six-Day War in 1967—the United States boosted its own production, compensating for any shortfall and keeping global prices steady.

In October 1970, U.S. oil production peaked and there was no spare capacity left in East Texas. Suddenly, the only countries with unused capacity, and hence the power to manipulate global oil markets, were the members of OPEC. Just a few years earlier, those OPEC governments would have been unable to take advantage of their new market power.

Since the 1930s, a cadre of U.S. and European companies, the so-called Seven Sisters, controlled the vast majority of the extraction, export, and shipping of oil in the noncommunist world. The Seven Sisters brought stability to global energy markets, but only through a deeply unjust system.

Both Iraq and Libya, for example, had vast untapped oil fields that their governments were eager to develop. When the Seven Sisters refused to support production in those new fields, the Iraqi and Libyan governments had no recourse and were forced to leave them idle.

But the Seven Sisters' grip on oil markets loosened in the late 1960s and early 1970s, when the oil-exporting states of the Middle East, Latin America, Africa, and Asia won greater control over their own supplies. They did so thanks to three developments: the emergence of independent oil companies that were willing to offer better terms to the oil-exporting governments, the tightening of global petroleum supplies, and the rising power of OPEC.

Although officially the embargo lasted less than half a year, its impact was long lasting. A period of high inflation followed, and stagnation in oil importers resulted from a complex set of factors beyond the embargo actions.

The leverage of the U.S. and European oil corporations (the "Seven Sisters") that had stabilized the global oil market was declining.

The oil shock induced brought sweeping changes to global energy policies in the 1970s and 1980s, prompting world leaders to prepare for imminent depletion of global oil and gas reserves, which turned out to be illusory, but is still engraved in people's minds. In retrospect, it is easy to see the signs that global energy markets were on the cusp of a revolution.

For a century, the United States had simultaneously been the world's largest oil producer and its largest oil consumer. Until 1947, it produced more than it consumed and was a net oil exporter to the rest of the world. After 1947, the United States became a net importer as growing consumption outpaced slowing production. Despite its reliance on imports, the United States remained the pivotal actor in global petroleum markets, thanks to its policy of limiting production in the vast oil fields of East Texas.

That strategy allowed the United States to function as a "swing producer," able to boost or trim production to stabilize global supplies and prices, just like Saudi Arabia does today.

But now there was also a decline of capacity of East Texas oil fields, which combined with the decision to allow the U.S. dollar to float freely in the international exchange, played a major role in exacerbating and prolonging the crisis.

These factors together were the reason for new measures that focused on energy conservation and development of domestic energy sources. These measures included the creation of the Strategic Petroleum Reserve, a national 55-mile-per-hour speed limit on U.S. highways, and other such national efforts to conserve energy.

President Gerald R. Ford's administration later imposed new fuel economy standards, and prompted the creation of the International Energy Agency (IEA). As a result of these measures and to encourage domestic energy generation, the government issued billions of dollars of grants and loans to energy companies, including solar energy R&D and manufacturing operations. Hundreds of new companies and projects were sponsored by the U.S. government and the U.S. seemed to headed into a new direction—that of a bright energy future based on energy conservation and powered by renewable energy.

That period of energy renaissance did not last long. Soon after crude oil (and gas pump) prices went down, the energy conservation and renewables pipe dream was shut down. The renewable technologies were shoved into a closet, and many companies went bankrupt.

We have seen this trend a number of times during the 1980s and 1990s and the latest period which started in 2007 and ended in 2012. For fairness' sake, however, we must mention that this time, the renewables revival is somewhat different.

The world financial situation and technological advances are somewhat more favorable this time. The consumers are also much more aware of the need for re-

newables, so support is increasing. This alone will keep them going strong in the future.

Although hundreds of renewable energy companies went out of business lately, and many more were sold to larger competitors, the renewables are not going to be shoved back in the closet. As a matter of fact, we see them as a major factor in our energy future, but we will discuss this in more detail later on in this text.

Note: The International Energy Agency (IEA) was established in 1974 and is now one of the world's most authoritative sources for energy statistics. It provides annual analysis and data on oil, natural gas, coal, electricity generation and use, and renewables, which are indispensable tools for energy policy experts. Some of the data issued by IEA has also been used in this book, although we usually compare information from different sources. The information from the different sources varies widely at times, so we then perform our own analysis and comparative studies before coming to a final decision.

In any case, we do thank the EIA for their tireless work in the energy sector, and for the invaluable information contained in their publications.

All in all, some lessons were learned and some measures taken, but the resolve to reduce energy use, and/or find new ways to generate electricity did not last very long. The use of energy increased two-fold from 1973 to 2010 from 6,700 million tons oil equivalent (Mtoe) to 12,700 Mtoe.

Figure 1-21 shows that the worldwide use of crude oil had a dramatic drop from 1973 until 2010, while the use of most of the rest of the energy sources increased. This phenomena is due to a number of complex political and socio-economic factors, which we discuss in more detail later on in this text.

The increase of renewables is the most dramatic, where they rose from 0.1% of world energy sources in 1973 to 1.0% total participation in 2010, growing higher as we speak.

Nuclear power has also increased several fold, from 0.9% in 1973 to 5.9% in 2010, but since the Fukushima nuclear accident, its future is uncertain. As a matter of fact, a decrease in the global nuclear power generation was experienced recently as countries shut down their nuclear power plants.

The drop of biofuels use in 2010 can be explained by the difficulties the industry has been going through recently. Their share is projected to increase dramatically in the near future. We discuss this in more detail later.

Natural gas is also going through a quick increase today, and we expect its share to increase significantly in

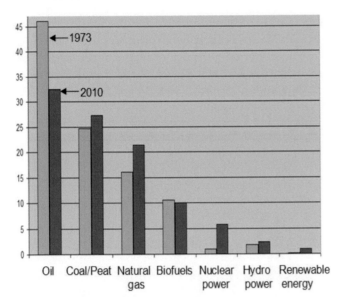

Figure 1-21. Worldwide energy use in 1973 vs. 2010 (in % of total world energy use)

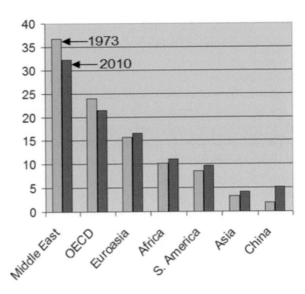

Figure 1-22. Major crude oil producing regions (in % of total world production)

the future, as we will see in the text below.

Crude oil's history is the most fascinating, since we depend on it so heavily for the daily transport of people and goods.

Figure 1-22 shows the rate at which the production in some major regions has decreased from 1973 through 2010, while at the same time, it has increased in other areas, including the U.S. It is hard to draw any conclusions from these numbers, so it suffices to say that the world is going through an adjustment period.

Note: There are 34 member countries of the Organization for Economic Co-operation and Development (OCED). These are: Australia, Austria, Belgium, Canada, Chile, Czech Republic, Denmark, Estonia, Finland, France, Germany, Greece, Hungary, Iceland, Ireland, Israel, Italy, Japan, Luxembourg, Mexico, Netherlands, New Zealand, Norway, Poland, Portugal, Slovakia, Slovenia, South Korea, Spain, Sweden, Switzerland, Turkey, United Kingdom, and the United States.

The new discoveries of huge oil and gas reserves in the U.S. and other parts of the world further complicate the picture, but the overall trend of increased crude oil production is quite clear.

Looking at Table 1-9, we see that the U.S. is the greatest user of this valuable resource. We import almost the equivalent of all the crude oil Saudi Arabia and Russia export put together. The situation is changing lately, where we produce more oil than we import.

The U.S. still imports almost half of the oil used daily, which represents the weakest link in our energy security.

Table 1-9.
Crude oil production, export, and import in Mt* (2010).

Country	Produce	Export	Import
Saudi Arabia	517	333	—
Russia	510	246	—
Iran	215	126	—
Canada	169	—	—
Norway	168	78	—
UAE 149	105	—	
Venezuela	148	87	—
Mexico	144	71	—
Nigeria	139	129	—
Iraq 110	94	—	
Angola	98	84	—
U.S.A.	346	60	513
China	255	—	235
Japan 3	—	181	
India	50	—	164
Korea	3	—	119
Germany	5	—	93
Italy	8	—	84
France	3	—	64
Netherlands	3	—	60

*__Note:__ Mt is million tons

At the same time, China is expected to surpass all other countries in crude oil use, and the projections are for a whopping 600 MT of crude oil by 2020. Is this even possible, let alone sustainable?

We will take a close look at this problem later on.

Natural gas has lately turned from the new kid on

the block into the ruler of the universe. The new discoveries, combined with new and much more efficient exploration methods, have brought it back on the world scene in a big way.

Natural gas is quickly changing the energy scenario in the U.S. and is expected to do the same in a number of other countries as well, in the very near future.

Take a close look at Figure 1-23. Windfall comes to mind, but miracle is even more appropriate when describing the unexpected and overwhelming bonanza of natural gas we have seen lately. It is an unprecedented transition from energy deficiency to energy oversupply.

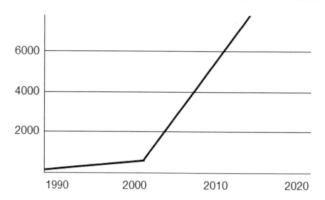

Figure 1-23. U.S. annual natural gas production (in trillion cubic feet)

The U.S. and the world are benefiting from this windfall in several ways. Excess natural gas is being exported to energy-poor nations, while at the same time making electric power generation in the U.S. more reliable and cheaper. All this brings more wealth into the U.S. economy.

As a bonus, natural gas is a cleaner alternative to power generation than coal, allowing the U.S. to meet its GHG obligations, and contributing to a cleaner global environment.

This is a win-win situation for the U.S. no doubt, but some nagging questions remain. We cannot help but ask, "How long will this bonanza last?" And what will happen when the gas is depleted?

The natural gas production increase of late is not limited to the U.S. Many other countries also report significant increases in domestic natural gas production.

One obvious thing that pops up in Figure 1-24 is the decrease of natural gas production by OECD countries, while it is on the increase in other parts of the world. The total world production of natural gas has increased from about 1,100 billion cubic meters (bcm) in 2000 to over 3,500 bcm in 2012.

Table 1-10 shows the unequal distribution of natural gas production and use around the world. The most worthwhile observation here is the fact that while most countries either only export, or only import natural gas, three countries do both. Russia, the U.S. and Canada are fortunate enough to have natural gas for domestic use, imports, and exports.

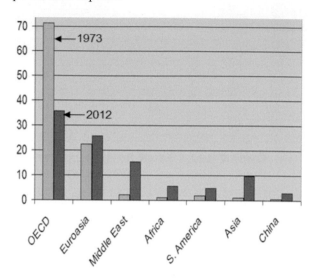

Figure 1-24. Natural gas producing regions in % of world production (2010)

Table 1-10.
Natural gas production, export, and import in bcm* (2011)

Country	Produce	Export	Import
Russia	680	195	40
U.S.A.	700	32	55
Canada	180	63	22
Qatar	150	120	—
Iran	150	126	—
Norway	110	100	—
China		100	—
30			
Saudi Arabia	95	—	—
Indonesia	90	50	—
Netherlands	80	35	—
Algeria	60	50	—
Malaysia	68	25	—
Japan	3.5	—	120
Germany	12	—	65
Korea		0.5	—
45			
Ukraine	20	—	45
Turkey	0.6	—	40
France	0.7	—	40
U.K.	0.1	—	35
Spain	0.6	—	35

*Note: bcm is billion cubic meters.

Energy Security Now

Energy security is vital for normal life in our society. Without energy, or in case of serious interruptions in the energy supply system, we would be vulnerable to internal and external threats. Energy security is one of the pillars of the national security of modern nations. Without energy, the industrial machine stands still and does not produce anything. Products that are produced cannot be moved to markets. Armies also stand still and cannot even get to the battle field.

Figure 1-25 shows how increasingly dependent we in the U.S. are on abundant and cheap electricity. The U.S., more than any other nation, would be badly hurt if the electricity supply were interrupted, since we depend heavily on many appliances, computers, and other gadgets in our daily lives.

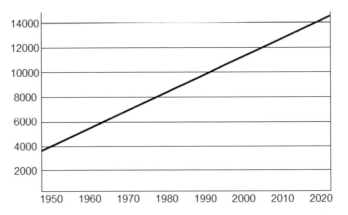

Figure 1-25. U.S. annual electricity use per capita (in kWh)

This is why the average U.S. citizen uses 10 times more electricity than the average person in most other countries. Because of that, it is imperative for our economy to have access to enough energy to function properly; interruption of the energy sector increases the risks of internal and external turmoil.

Energy security is not just making sure that our energy supplies and distribution networks are safe, but that the industries that depend on the energy sector function properly and efficiently without interruption and losses of profits or jobs.

Any interruption, no matter how minor or temporary, translates into some financial loss and damage to the affected companies and the overall national economy. In more detail, the basic energy security drivers can be summarized as follows:

- *Power generation and distribution diversification* is the foundation of any well-balanced energy system. It consists of power generating and distribution tech-nologies capable of adapting to changes in demand under normal and/or emergency situations. The diversification enables each technology as it also optimizes it, which allows energy price stabilization, and ensures a continuing energy supply to the nation.

Residential and business owners throughout the Northeast USA are now looking into energy diversification by evaluating different forms of distributed power generation, power distribution channels, and energy storage options for supporting building-critical loads during future grid outages. Such options must be included in the national energy security plans too, in order to prevent critical shortages in case of natural or man-made disasters capable of negatively affecting the energy supply and distribution.

- *Emergency planning* goes way beyond meeting the daily demand and lowering energy costs. For example, super-storm Sandy demonstrated once again the frailty of centralized power generation by disrupting the operation of dozens of oil rigs, refineries, and businesses in the area.

- *Energy prices* determine how affordable the energy sources are to the consumer. Prices dependent primarily on: a) the cost of the raw materials, as well as on b) cost of generation, c) transmission, and d) distribution. Any interruption of any of the supply or distribution networks impacts negatively the prices. In worst case scenarios, it can also cause major economic, logistic, and socio-political problems.

Countries that depend on one single source of energy are particularly vulnerable to interruption of the supply system. Prolonged price increases and spikes in the raw materials (oil, gas, or coal) and/or the electricity generation network can trigger inflation and recession.

- *The national energy supply chain* (including energy imports) must be diversified, because reliance on only a few energy suppliers increases the risk of negative market effects and could cause energy shortage and power outages. Many energy suppliers are from politically unstable countries, so there is always an increased risk of supply disruption from some parts of the world. This critical anomaly must be kept in consideration when addressing national energy security issues.

Fuel types diversification is an important element of national energy security. The ability to use different fuels, and even switch between fuels—such as coal to gas, gas to liquids, etc.—could ensure uninterrupted energy flow even in emergencies, or if the normal energy supply chain were temporarily disrupted, even when parts of it are totally eliminated.

- *Transportation* of raw materials to power stations, as well as the *distribution of electric power*, gasoline, and natural gas to the consumers determines the efficiency of any energy system. Lack of roads, railroads, or adequate fleets of vehicles could cripple any energy system. Adequate national power grid also must be readily available, thus assuring the delivery of electricity to all customers. Transportation and distribution are problematic in most developing countries, and must be resolved to provide a safe level of energy security for these nations.

- *Interconnection* of the national electrical network is key to supplying power (electricity, oil, and gas) to the entire population, and is an integral part of national energy security. Limited, or lack of, efficient interconnection jeopardizes the energy supply and distribution, and causes disruption of services to at least part of the population.

- *Technical expertise* is absolutely necessary for ensuring a diverse energy mix, as needed for efficient generation and use of energy. In addition to access to different energy sources, and a good transportation and distribution infrastructure, proficient technical expertise is needed to implement and run the different areas of the energy system.

The complexity of the generation technologies, fuel handling, pipelines, ports, interconnections, and transmission lines, to mention a few elements, is a complex undertaking that requires sophisticated equipment and technical expertise. Without it, the different elements of the energy system will not function properly or safely, and will eventually fall apart.

- *Investment* in the energy sector is an increasingly important driver, because significant investment is needed to keep the system operating properly and to add new capacity as needed by most countries. Such investment is generally available in devel-

oped countries, but is especially problematic in most developing nations. The lack of investment in the energy sector is becoming a significant factor in many areas of the world, threatening the energy security of the affected nations.

- *Political activities* are the cause of most major energy supply and distribution system disruptions. Political unrest, war, and terrorist attacks cannot be avoided, and must be considered in the design of a national energy security system, in order to make it less vulnerable to disruptions caused by these events.

The U.S. has its own energy security drivers, in addition to, or somewhat different from, those discussed above. We will be looking at these in the text below

Energy Security Risks

Today more than ever we rely on different energy sources to fuel everything from transportation to communication, to national security, health delivery systems, and all kinds of business enterprises. Perhaps most urgently experts claim that for every calorie of food produced in the industrial world, ten calories of oil and gas energy are invested in the forms of fertilizer, pesticide, packaging, transportation, and running farm equipment.

What Figure 1-26 tells us is that an average Chinese earns nearly 10 times less than the average American, and uses that many times less energy. This is not too difficult to understand, looking at the different life styles in these countries.

The average American wakes up in the morning in a 2300-square-foot house, turns on lights and a coffee

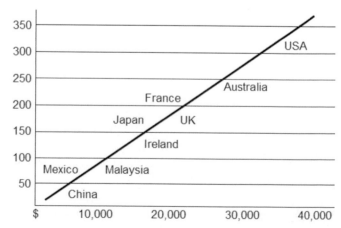

Figure 1-26. Annual income per capita (in $), vs. energy use (in GJ)

maker, takes a shower, then cooks breakfast on an electric stove. During all this time, a central air conditioning or heating system has been keeping the house at a comfort level temperature. Then we jump in a 300-horsepower SUV and drive 20 miles to work.

This is quite far from the life of the average Chinese, who would be lucky to have lights in the house to begin with, let alone air conditioning. They rode bicycles or mopeds to work—until recently. Now the Chinese are waking up to the idea that they deserve more, and want to have a better life, which includes air conditioning, SUVs, and other daily luxuries. Who can blame them?

All this, however, requires a lot of energy, which is why energy use around the world is increasing quickly. So, while the U.S. is at a plateau in both personal income and energy use, China and many other developing economies are pushing forward to higher income and more energy use.

Energy is an integral part of the national security of any country as it powers its entire economic engine. Some businesses rely on energy more heavily than others. The metallurgical industry, for example, would stop abruptly if huge amounts of fuel (electricity and natural gas) to power its many megawatt furnaces were no longer available.

The U.S. Department of Defense relies on fossil fuels for nearly 80% of its energy needs. If these are not available, then planes will be grounded, warships will be anchored, and troop movement around the world will stop. Lack of fuels would limit the response effectiveness of our armed forces, according to the severity of the energy shortage.

What would follow is anyone's guess, but one doesn't have to be a genius to deduce that many bad things would happen and a lot of pain and discomfort would be experiences by millions—not only in this country but around the world. Such a situation is a real threat to our national security, and must be avoided by all cost—not only now but in the future. This can be done only by providing reliable and uninterrupted long-term energy security.

The threats to our energy security are countless, but the major ones include the political instability of key energy producing countries, the manipulation of energy supplies by foreign powers, competition over energy sources with different countries, attacks on transport chain and supply infrastructure, as well as all kinds of accidents and incidents, natural disasters, and rising terrorism. All of this is exacerbated by the fact that the dominant countries of today rely heavily on a foreign oil supply, most of which comes from unstable areas of the world.

The major oil supplies are vulnerable to many types of disruptions to their production, which could come from internal conflicts, changes in exporters' interests, external effects on supply, and transportation of fossils. War, political unrest, and economic instability at times disrupt the proper functioning of the energy industry in some supplier countries.

Prior to 1970, only a handful of oil-producing countries had nationalized oil production: the Soviet Union, Bolivia, Mexico, Iran, Iraq, Burma, Egypt, Argentina, Indonesia, and Peru. Even though the oil production was nationalized, most of the key industries in these countries were still held by foreign firms, and of these countries only Mexico and Iran were significant oil exporters at the time.

Case Study: Venezuela

In Venezuela, the nationalization of oil production started in 1976. Major oil companies operating in Venezuela have had difficulty with the new situation and would walk away from their large investment in the Orinoco heavy-oil belt, rather than accept tough new contract terms that raise taxes and oblige all foreign companies to accept minority shares in joint ventures with the state oil company, Petróleos de Venezuela (PDVSA).

Contracts offered to foreign investors were often those that entailed high costs for extraction, leading to lower implicit tax rates. In the 1990s, private investment substantially increased, adding 1.2 million BD of production by 2005. While private investors were producing more oil and PDSVA decreased oil production, Venezuela still managed to increase its oil fiscal take for each barrel.

Continued shortcomings for PDSVA spurred an effort to eliminate the company, leading to strikes that severely reduced investment and production. This gave an opportunity for the government to seize control and, as a result, the contractual framework of the oil opening has been significantly changed, considerably increasing the government's take and control over private investments.

The nationalization of oil in Venezuela has triggered strikes and protests from which Venezuela's oil production rates have yet to recover. The country's political or economic incentives dictate its oil exports and often cause disruptions in the supply chain. And President Hugo Chávez threatened to cut off supplies to the United States on several occasions.

Other oil producing countries act the same way. The 1973 oil embargo against the U.S. is a historical example in which oil supplies were cut off to the U.S. due to U.S. support of Israel during the Yom Kippur War. This was done to apply pressure during economic negotiations—such as during the Russia-Belarus energy dispute.

Terrorist attacks targeting oil facilities, pipelines, tankers, refineries, and oil fields are so common they are now simply referred to and accepted as "industry risks." Infrastructure for producing fossil fuels and electric power is extremely vulnerable to sabotage as well. Oil transportation involves loading and transporting huge amounts of oil in gigantic vessels that are exposed to the risks of navigating the world's oceans. They also must cross the five most dangerous ocean chokepoints, like the Iranian controlled Strait of Hormuz.

Because of that, experts fear that it may take only one attack on a Ghawar Saudi oil field or tankers in the Strait of Hormuz to throw the world oil markets into turmoil and long-lasting chaos.

New threats to energy security are emerging. These come in the form of increased competition among the major countries for energy resources, which is due to the increased pace of industrialization. Countries such as India and China are racing towards new industrial and economic development, to support rising living standards of their large populations.

The competition over energy resources may also lead to the formation of security compacts in an effort to enable advantageous distribution of oil and gas between oil producers and/or major powers. This would happen at the expense of less developed economies.

Another concern is the possibility of price rises resulting from the pending oil peak conundrum. This is the time when the world oil production will start declining due to exhaustion of the oil fields.

This is a major future event that is not given the proper attention it needs. As a result, no proper measures to prevent or at least mitigate the consequences of an oil-less future have been taken.

Short-term Energy Security

In the short-term, the energy security of each country is determined by its readiness to deal with interruptions in energy supplies.

Petroleum

Petroleum, or crude oil, or oil is the energy resource that is most used by the developed countries worldwide, and their economies depend on it. But the oil fields where this oil comes from are located around the world, and most of it in politically and economically

unstable areas, so energy security has become a serious issue, and a lot of work and money has been spent on ensuring the safety of the petroleum that is being harvested there.

Oil fields and transport routes are targets for sabotage to manipulate the oil supplies, so many countries have opted to hold strategic petroleum reserves as a buffer against the economic and political impacts of an energy crisis. For example, all 28 members of the International Energy Agency hold a minimum of 90 days supply of oil surplus to respond to emergencies.

The European strategic petroleum reserves readiness was tested by the 2007 Russia-Belarus energy dispute, when Russia indirectly cut exports to several countries in the European Union. The result was a relative lack of disruption caused by the chaos surrounding the disruption, although prices increased some.

Natural Gas

Natural gas has been a viable source of energy, and today it is becoming a leading energy source in some countries—including the U.S. Natural gas is mostly methane, and is produced using two methods: biogenic and thermogenic.

Biogenic gas comes from methanogenic organisms located in marshes and landfills, whereas thermogenic gas comes from the anaerobic decay of organic matter deep under the Earth's surface. In all cases, natural gas is produced by drilling deep wells in the ground and letting the gas escape to the surface, where it is stored or transported via pipelines. It is sometimes compressed and stored in pressurized tanks as liquid natural gas (LNG).

One of the biggest problems facing natural gas providers—in the midst of the natural gas boom—is the inability to store and transport it. It is difficult to build enough pipelines in North America to transport sufficient natural gas to match demand, and is very expensive to compress and liquefy it, so the huge gas supply in the U.S. is limited by the transport channels.

Reliance on imported natural gas by many countries—including major world powers—creates significant short-term vulnerabilities. For example, many European countries saw an immediate drop in supply, coupled by sharp price hikes in some regions, when Russian gas supplies were halted during the Russia-Ukraine gas dispute in 2006.

Russia is now leading in the production of natural gas, and many neighboring countries—including most of Western Europe—depend on the Russian gas exports.

Nuclear Power

Nuclear power is a sustainable energy source that provides reliable power generation and reduces carbon emissions. Aside from the risks related to nuclear accidents—which also present a serious energy risk—nuclear power generation depends on a reliable fuel source.

Currently, nuclear power provides nearly 15% of the world's total electricity. Important applications of nuclear power are its use in powering aircraft carriers, cruisers, and submarines, which have been increasingly nuclear-powered for several decades now. These ships provide the core of the U.S. Navy's power, and as such are not dependent on fossil fuels supplies interruptions.

Uranium is mined and fuel is manufactured significantly in advance of need. Nuclear fuel is considered by some to be a relatively reliable power source, being more common in the Earth's crust than tin, mercury, or silver. Nevertheless, the debate over the timing of uranium's peak—or when the world's supplies would be exhausted—is just beginning.

This is a serious concern, because without uranium the world's electric power generation would be seriously reduced—not to mention that all nuclear vessels would be anchored indefinitely, compromising national security and the wellbeing of the affected countries.

Uranium for nuclear power thus far is mined and enriched in a number of stable countries. These include Canada (23% of the world's total in 2007), Australia (21%), Kazakhstan (16%) and more than 10 other countries spread worldwide.

The European Union depends on nearly 98% of its uranium supplies being imported from abroad, making it vulnerable to supply interruptions.

Table 1-11. EU uranium suppliers

Export Country	*%*
Russia	25.0
Canada	18.0
Niger	17.0
Australia	15.0
S. Africa	5.0
Uzbekistan	5.0
Other	15.0

Interruption in the nuclear fuel supply chain could be catastrophic for the EU community, and represents a serious energy security risk. This combined with an interruption of energy supplies from major but unreliable producers could bring Europe to its knees in a hurry.

Although a very viable resource, nuclear power

brings a number of controversies, due to the danger of nuclear accidents. Because of that, most people and companies simply do not want nuclear plants in their neighborhoods—another limiting factor in nuclear power proliferation.

Long-term Energy Security

While there is a concerted effort and emphasis on certain measures to ensure the short-term energy security of our country, the long-term measures are spread over a wider, albeit often less tended (and often ignored) spectrum of measures.

In the U.S., the long-term security measures are focused on:

- Reducing dependence on imported energy (crude oil primarily),
- Increasing the number of energy suppliers,
- Diversifying the import channels and methods,
- Exploiting local fossil fuel,
- Increased use of renewable energy resources, and recently
- Reducing overall demand through energy conservation measures.

On an even longer-term basis, the U.S. and its allies are attempting to enter into international agreements intended to reinforce international energy trading relationships. An example of that effort is the Energy Charter Treaty in Europe. The treaty covers some aspects of commercial energy activities including trade, transit, investments and energy efficiency. It is legally binding, including dispute resolution procedures in case of failures to comply.

The Energy Charter process was based on integrating the energy sectors of the Soviet Union and Eastern Europe at the end of the Cold War into the broader European and world markets. Later on, its role extended farther and now it strives to promote principles of openness of global energy markets and non-discrimination to stimulate foreign direct investments and global cross-border trade.

The effort's integrity is supported by legally binding instruments, such as settlements of the international arbitrations in case of breaking the law of the Energy Charter Treaty. The penalties could be quite steep—on the order of hundreds of millions of dollars. High-profile law firms often represent investors and the states in these cases.

Analyzing Figure 1-27 in light of the developments of late, we see that the emphasis of our long-term energy security measures is focused on the transportation sector, domestic infrastructure, and threats to reliable and cheap crude oil supplies.

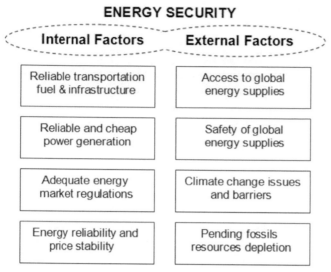

Figure 1-27. Key elements of our long-term energy security

We have enough coal and natural gas for power generation, so our efforts are focused on reducing future risks and cost of oil imports by not allowing any disruption or harm to the oil production or transport. This is easier said than done, however, so we need to look to the past for lessons in reading the warning signs.

The 1973 oil crisis came unannounced and totally unexpected. It caught us with our pants down. The emergence of the OPEC cartel was an ominous (after the fact) warning signal of increasing danger that prompted many countries to increase their energy security.

As a result, Japan, which was almost totally dependent on imported oil, steadily introduced the use of natural gas, nuclear power, and high-speed mass transit systems, and implemented energy conservation measures. At the same time, the UK started exploiting the North Sea oil and gas reserves, and became a net exporter of energy in the 2000s.

In many other countries, energy security has historically been (and in some remains) a lower priority. In most cases, the lack of stable economy and political balance force countries to ignore the potential threats to their energy security.

The example of the United States to date is different, and yet not perfect. Yes, we built significant national strategic reserves, which could help temporarily in extreme emergencies, but we continued to increase our dependency on imported oil. As a result, oil prices kept climbing.

The development of new oil and gas fields via hy-

drofracking and the introduction of solar, wind, biofuels and other renewables have helped to fuel our power generation and have also reduced the oil imports significantly.

Our dependence on energy imports, including imported oil, has decreased significantly although we still import over 50% of all crude oil used in the country. This critical import remains a gaping hole (perhaps the only serious one) in our energy security, and the only barrier to a total energy independence.

Crude oil is extremely important to the efficient daily function of the U.S. economy. Shortages would bring back the long gas lines of 1973. Things will get much worse if our power shortages are ever due to permanent depletion of global crude oil supplies.

There is not much talk today about reduction and "permanent depletion" of crude oil reserves, but it is a key point in our long-term future and that of the future generations.

We will take a closer look at the new situation in the following chapters.

The Major Risks

In more detail, today's energy security risks and their effects are many and fit a number of different classifications.

The major risk categories can be divided into:

a.   Physical and political security threats such as internal and external terrorism events, political unrest, wars, sabotage, theft or piracy, and

b.   Natural disasters (earthquakes, hurricanes, volcanic eruptions, the effects of climate change, etc).

Any of these can affect negatively any part of the energy supply chain including:

•   Power stations, sub-stations and transmission lines;
•   Oil and gas exploration, extraction and refining installations;
•   Oil and gas-fired plants, pipelines and storage facilities;
•   Rail or road networks; stations, terminals and ports;
•   Individual planes, oil tankers, trains or road vehicles.

The results of these acts could be expressed in:
•   Global energy market instabilities caused by un-

foreseen changes in geopolitical or other external factors, or compounded by fossil fuel resource concentration. Energy supply constraints may occur due to political unrest, conflict, trade embargoes or other countries successfully negotiating for unilateral supply deals.

Such supply constraints rarely result in physical supply interruptions thanks to the flexibility of the energy transport, storage, transformation and distribution systems as well as international mechanisms.

Nonetheless, they do have consequences for price developments in fossil energy markets—immediately in the case of oil, with a time lag also in natural gas and coal markets.

The impact on energy market volatility of such geopolitical threats is heightened by the uneven global distribution of fossil fuel resources. The world's proven conventional oil and gas reserves are concentrated in a small number of countries.

Taken together, OPEC countries account for 75% of global conventional oil and natural gas reserves. OECD countries only account for about 10% of the total production, but consume close to 60% of the world total.

The inequality of concentration of fossil fuel resources and their usage is the most enduring energy security risk. Experts project that oil demand will become increasingly insensitive to price, which reinforces the potential impact of a supply disruption on international oil prices. Transport demand is price-inelastic relative to other energy services. Since the heavy dependence on global oil consumption is projected to rise, oil demand will become less responsive to movements in international oil prices.

Thus, prices are expected to fluctuate more than previously in response to short-term demand and supply shifts. The relative weight of the impact of price fluctuations varies according to the robustness of economies and businesses, and the variation is amplified by a number of real or perceived risks.

Technical and logistical failures such as power outages (blackouts and brownouts) caused by grid or generation plant malfunctions are a major problem for the energy sector.

Faults in energy supply systems caused by accidents or human error may cause a temporary supply interruption. Due to network complexity and the immediate loss in network stability which has to be established system-wide, such failures have particularly sharp and wide-ranging effects if they occur in large interconnected systems as observed during recent power outages in India, South Korea, California, Italy, Germany and elsewhere.

The effects of any major incident result in technical failures, which can then spread to affect many areas of the national or global economies. Large-scale incidents may cause longer outages and have deeper impacts and longer-lasting effects on energy markets. The costs of security measures needed to prevent or mitigate technical failures affects energy prices, network stability and provision of energy services, and may have significant effects over the long term.

Oil platform and refinery closures off the south coast of the USA following Hurricane Katrina in 2005 and the Fukushima nuclear power plant incident following an earthquake and tsunami in 2011 exemplify the threat to energy supply infrastructure posed by extreme weather events.

According to the climate change models, such extreme weather phenomena are expected to increase. Terrorist attacks are similarly unpredictable, and the threat never goes away. As a matter of fact, they are also expected to intensify.

The probability and the negative impact of internal and external energy security related incidents and accidents can be reduced only by additional effort, increased investments, consistent watch, and efficient control measures.

Table 1-12 tells us that energy market issues would affect the transport sector most, simply because it uses the most crude oil and because crude oil (import) is seriously affected by interruptions in the production in different countries and restrictions in the delivery channels around the world.

Table 1-12.
Energy supply and use issues (in % of affected areas)

Major Issues	Electrical Supply	Lo-Temp Heating	Hi-Temp Heating	Transport
Energy Markets Issues	20	10	20	50
Technical Issues & Failures	50	20	20	10
Terrorism & Natural Disasters	50	10	10	30

The technical issues and failures affect mostly the power generation sector, since it is the most complex from engineering and technological points of view. It is also the most prone to breakage and failures. These failures vary from a break in a hydro dam (which could be remediated), to a nuclear power plant accident, which could bring a serious disaster, taking years to recover from.

Terrorism and natural disasters are events that would affect both the electrical generation and transport sectors, since they can be spread broadly across the world and the entire energy production, distribution, and fuel delivery networks.

Energy security risks can be further separated into *internal and external*, which we encounter on a daily basis. Some of these go unnoticed, while some are brought to international attention. Regardless of their source or nature, however, they all present some threat to our energy security.

We will review these in terms of *internal risks*, or those that occur in-country, and *external risks* or those that are caused by outside factors and powers.

Internal Energy Security Risks

There are a number of internal risks to our energy security. Unforeseen man-made accidents in mines, transport routes, refineries, and power plants threaten to disrupt the energy supply chain. Mine accidents are still happening in the U.S. and around the world. They take lives and shut down mining operations temporarily or permanently.

Railroad and transport ship accidents and spillage can close a transport route indefinitely and do significant damage to the environment. Hurricanes damage and shut down refineries almost every year, but the worst and most dangerous of them are nuclear power plant accidents.

As the related energy technologies mature and we learn how to handle natural and man-made accidents, the internal risks are getting more manageable. With that, the number of mine, railroad and transport ship accidents is significantly reduced. Refineries, on the other hand, are vulnerable to the force of hurricanes and other natural events, and not much can be done to improve that situation because of their complex outdoor infrastructure.

Nuclear accidents are also rare, but when they happen the local devastation is total and its effects are felt over broad areas. This is a critical component in considering our energy security and safety. One nuclear plant accident can not only disrupt the energy

supply, but can destroy the entire area, kill and sicken thousands. This is a serious issue that needs to be taken into consideration as we consider our energy options for the 21st century.

Today computers run everything—including our energy infrastructure. Every step of our energy supply chain, from mining operations to transporting, processing, refining, and using energy sources is monitored and controlled by computers. Internal terrorism is one way to gain control and damage a refinery or a power plant. There are a number of reported incidents, and an even greater number unreported.

In January 2013, critical control systems in two US power plants were found infected with computer malware, spread intentionally by computer virus brought in via USB drives. The infected computers controlled critical systems controlling power generation equipment. Intentionally planted malware poses a real threat by allowing the attackers to disable key equipment, thus disrupting its operation, or destroying it altogether.

In 2012, a "backdoor" in a piece of industrial software used to control power plants allowed hackers to illegally access a New Jersey company's internal heating and air-conditioning system. One of the infections was accidentally discovered after an employee called in an IT technician to troubleshoot the USB drive. A simple virus check discovered sophisticated malware, capable of doing a lot of damage to the plant equipment.

Since this technology is so new, defense mechanisms have not been fully developed and deployed, as in this case. The workstations lacked backup systems, so they were lucky to discover the malware before it was activated.

Another intentional malware attack, spread by a USB drive, affected 10 computers in a steam turbine control system of the power plant. The incident resulted in downtime for repair of the impacted systems, which delayed the plant restart by three weeks.

These examples are a lesson in safety precautions and a call to action by owners and operators of critical infrastructure, who must develop and implement serious security policies to maintain up-to-date antivirus mechanisms, and manage system patching and the use of removable media.

External Energy Security Risks

A number of risks can come from outside—outside the country and beyond. When we talk about energy production and use, natural events and man-made events must be kept in mind.

The Natural Events

The weather and other natural events have a profound influence on our way of life. In winter we need energy to keep warm, while in summer we need it to keep cool. At the same time, extreme natural events such as storms, hurricanes, earthquakes, etc. can disrupt the energy balance and our lives in general.

To get a batter idea of our energy arsenal we need to get a close look at the broad energy picture in its natural form.

Figure 1-28 suggests that the list of energy segments and interactions on Earth is long; they are overwhelmingly complex, where one malfunction in one area could lead to a chain reaction with unexpected negative results to the entire system. Because of the importance of these energy segments, scientists are constantly looking into them, finding new phenomena, and revising their previous findings and conclusions.

We will keep all these complexities in mind as we go along, while focusing on man-made power generation (be it electricity, heat, or other) and transportation.

In parallel we review the impact of the related technologies and processes on our energy security, the overall Earth energy balance, and the troubled global environment.

The solution to any problem starts with recognizing the problem, learning its nature and behavior, and then looking for ways to solve it. We agree that we have serious environmental problems, some of which are created by power generation (burning coal, oil, and natural gas).

We hear about the disaster caused by climate change every day, so now is the time to learn all we can about it all—including the origins and possible ways to reduce the negative impact on the environment and our life on Earth. This is an integral part of our energy and national security, since it affects our way of life in many ways.

In this text we are aiming to expose the issues at hand, presenting a detailed account of the effects of the energy sources and the related environmental problems with our national energy security.

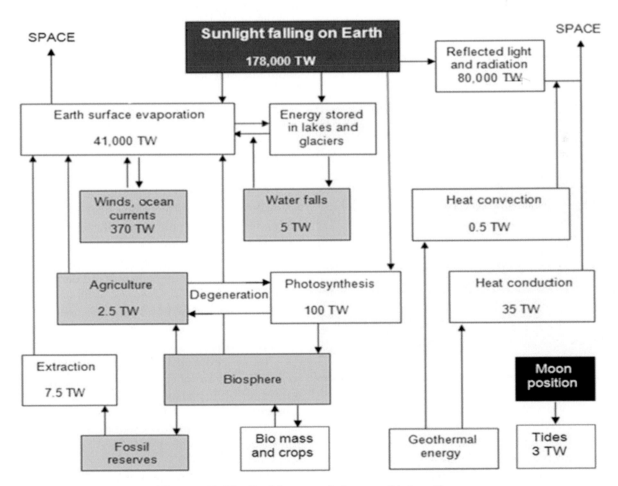

Figure 1-28. The Earth's energy balance and interactions

The Satellites

Most helpful in deciphering the mysteries around us are hundreds of satellites orbiting the Earth, looking from every possible angle at its oceans, ice caps, rivers, mountains, volcanoes, temperature changes, winds, etc. Satellites monitor weather, make meteorological observations, maps, etc. They operate on altitudes above 600 km, where the air-drag created by Earth's rotation is much less, and its effects are negligible.

Different satellites are designed for different types of detection—visual, thermal, UV, etc. They monitor different areas of the Earth and see things that we cannot even imagine. They can observe very large areas for extended periods of time, or focus on very small areas to see details that we can only guess.

Weather satellites are monitor the weather and climate. They see city lights, fires, pollution damage, storms, snow and rain, ice formations, and the ocean currents. This, and many other types of environmental information is continuously collected by the satellites and sent to weather and other stations on Earth for analysis.

Weather satellites have helped monitor volcanic ash clouds, smoke from fires around the world, and other important events that might influence the weather and life on Earth in general.

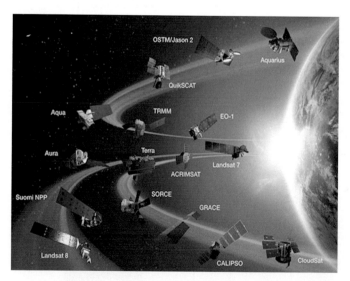

Figure 1-29. NASA's Earth observation and weather satellites

Some key events such as El Niño and La Niña are monitored daily from satellite images. Other satellites are looking at air quality, including the state of the ozone layer, and map the ozone hole above us. Hundreds of weather satellites owned by the U.S., Europe, India, China, Russia, and Japan provide continuous observations

via visible light, infrared rays, lasers and other tools.

Other satellites are designed to assist with environmental monitoring. These are set to detect changes in the Earth's vegetation, the atmospheric trace gas content, the overall state of the oceans, the ice cover and many other variables. By monitoring vegetation changes over time, for example, droughts can be monitored by comparing the current vegetation state to previous averages.

These types of satellites are almost always in sun synchronous orbits. Sun synchronous orbits keep the satellites close to the poles, to get the desired global coverage by maintaining relatively constant geometry to the sun. Other satellites are in "frozen" orbits, which are the closest to a circular orbit that is possible in the gravitational field of the Earth. This way they can keep a constant eye on environmental variables of a defined area of interest.

This colorful field of different satellites at different orbits and altitudes provides a constant 24/7 watch over us and our environment. Observing, measuring, and keeping track of the variables of life on Earth, they provide us with invaluable information.

There is still a lot we don't know, but our body of knowledge is increasing daily. We get a better and more complete picture of things that we could only guess about not long ago.

One overwhelming and undeniable conclusion we are reaching is that the Earth and its environment are going through major, unprecedented and unforeseen changes. Man-made activities are contributing to these changes through air and water pollution.

Electricity generated from fossils (coal and natural gas) as well as crude oil products used for transportation and other needs are being blamed for the greatest amounts of emitted pollution in the form of green house gasses (GHG). These gasses are affecting us by creating a number of abnormalities, the final result of which is climate warming and other strange and dangerous effects. In this text we take a close look at all events related to electricity generation and their consequences.

Natural Disasters

Today's educated person has enough information to be fully aware of the environmental problems and our responsibility to look at nature and the earth's environment as endangered. There is no doubt that humans have damaged the environment during the last century. It only remains to be determined to what degree the harm extends.

We also need to determine: a) how much of the damage is man-made, and b) what part of it is irreversible. Then, we could start remediation efforts.

Closing our eyes, or proclaiming helplessness and defeat, is not an option. We need to understand the problems; their sources, the related major issues, and then design solutions.

There is nothing that should stop us from acting as responsible 21st century people. We have the understanding and the tools, so all we need is to agree on the most appropriate course of action, and implement it. Are we responsible enough to do all this unselfishly, efficiently, and quickly?

Figure 1-30. BP Deepwater Horizon, 2011

Are man-made disasters like the BP's oil spill in the Gulf in 2010, or the Dust Bowl phenomena of 1934-1940 (yes, that was a large-scale, man-made) going to happen again? Yes, very likely, but regardless of their nature, we are more prepared to protect life and property.

We are more and more aware of the difference between man-made disasters and those of natural origin. This gives us more control over the environment, our actions, and our own health and destiny.

Figure 1-31. The Dust Bowl, 1936

Note: Environmental damage in a broad sense is a misnomer, because the environment only changes from one state to another, all we humans can do is watch and learn. It will surely survive in one form or another, as it has for millions of years.

There have been many ups and downs through the years, and each time the environment has recovered splendidly, in most cases evolving to a better state. However, those changes were all at the expense of the living things.

We may not like the upcoming changes or even survive them, if this pattern continues, but the environment will survive in one or another form—with or without humans.

Since the serious environmental changes we are presently experiencing might have devastating effects on the human race, we need to consider the preservation of the environment (in its present form) as a prerequisite for sustaining life on Earth, and as a vehicle to our own wellbeing.

Human activities have been blamed for recent negative environmental changes and, because there are no human activities that are absolutely pollution-free and totally environmentally friendly, there will always be some environmental effects and damages brought about by them. We must be fully aware of consequences and take measures to balance good and bad activities.

These are the major issues we will be tackling during the 21st century, and because most environmental changes and damages are being blamed on power generation, we will focus on its effects on the environment and human health.

The goal of this text is to make a detailed analysis of the cradle-to-grave processes of power generation—and how it affects the environment during its different stages. We will make some conclusions and provide recommendations, but it is you, the reader, who will make the final call, and all of us need to agree on a plan of action.

Man-made Events

We can separate the man-made events also in internal and external forms, where the internal can be controlled. External man-made events are harder to control, and require major effort and investment.

In the following chapters, we dedicate significant space to man-made events—international terrorism especially—that affect the most critical crude oil distribution channels, so we just mention it here as part of the external risks.

Other significant threats to our energy sector lately

have been cyber attacks originating in countries like China, Russia, and Iran. The Stuxnet worm and the Flame malware, for example, were developed by the US and Israel to spy on, and even control, critical systems in power plants. In the summer of 2012, these programs successfully disabled an entire enrichment facility in Iran. These programs relied on USB drives to store the commands, propagate attack codes, and carry intercepted communications over computer networks.

The same worms can be used against U.S. power plants and distribution networks. And although Microsoft has patched some of these vulnerabilities on Windows computers, there are additional steps to be taken by power plant and other energy sector operators.

The battle on the international front continues and seems to be getting more complex and dangerous.

The Future of Energy Security

We need to consider the way we have used energy and have impacted the environment—intentionally or unintentionally—in the past, and how we are continuing to impact it now.

The environment is an integral part of our energy and national security, for it would be ludicrous to have plenty of fuel while choking in its emissions—the beginning of which cycle is quite obvious in some large cities today.

With enough information, we can easily predict the future, provided we continue the present rate of energy production and use. It is easy to extrapolate from the intensive exploration of coal, natural gas, and crude oil that we are slowly but surely running out. And in parallel with the excessive use of the fossils, we are also increasingly polluting the environment.

It is easy to see in Figure 1-32 that as things are going, fossils will be depleted by the end of this century (some even sooner), while energy use increases with the increase in population.

As the demand for energy increases, more coal, oil, and gas are burned, which in turn creates new and more complex environmental issues.

Burning of coal, oil, and gas is giving us comfort, technical and economic progress, while at the same time causing global warming and other harmful effects, which are posing great threats to the environment and all life on Earth. The dangers seem to be growing greater as we intensify coal, oil and gas burning.

What are the plusses and minuses of these process and their consequences? This is what we aim to look at in detail in this text.

Renewable Energy

When talking about energy security we must always think of what is going on today from the perspective of our long-term energy survival, and that of the future generations. To them, a major subject in the energy sector that overshadows any other would be the status of the renewable technologies. This is because we will not leave any fossils for them to use, and the renewables will be all they will have for power generation and transportation.

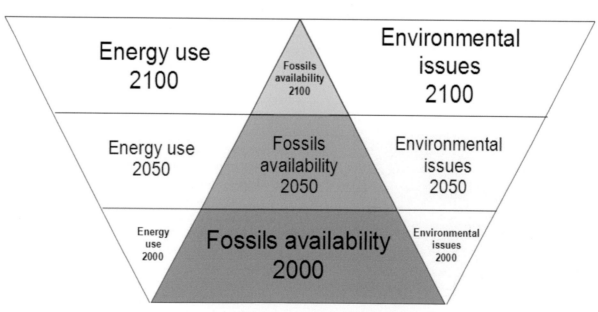

Figure 1-32. Energy use and supply vs. environmental issues of the 21st century

Energy security is the actual connection between, and dependence on, the availability of natural resources and our ability to access them for energy generation and transportation fuels.

Uninterrupted and constant access to cheap energy is essential to the functioning of our global economy. The major problems, however, still remain: uneven distribution of energy supplies among countries, interruption in deliveries, and ultimately running out of fossils, could lead to significant vulnerabilities, blackouts, discomfort, and economic and political disasters. This is especially concerning from the point of view of the energy fate of future generations.

Renewable energy resources might be the solution to us and them.

Renewables, in contrast to other energy sources, are different in that they are: a) inexhaustible, b) non-polluting, and c) not concentrated in small geographic areas. They are readily available and accessible to all, today and tomorrow.

Deployment of renewable energy technologies could be done on a mass scale worldwide, resulting in significant energy security and economic benefits for all. Renewables cannot be easily controlled by large groups or companies, so all people have a fair chance to participate in their use.

The deployment of renewable technologies usually increases the diversity of electricity sources and, through local generation, contributes to the flexibility of the energy system and its resistance to central shocks.

For those countries where growing dependence on imported fossils is a significant energy security risk, for example, renewable technologies can provide alternative sources of electric power and vehicle fuels. They can also displace electricity demand through direct heat production in some regions.

Biofuels, for example, are mostly used for transport, which is a major issue as oil prices rise. Biofuels have been researched using many different sources including ethanol and algae. These options are substantially cleaner than the consumption of petroleum. Perennial and ligno-cellulosic crops can be used to produce biofuels which can then supplement anthropogenic energy demands and mitigate GHG emissions to the atmosphere. Large quantities of biofuels can be produced in most countries, which offers diversification from petroleum products.

Geothermal energy sources can potentially lead to other sources of energy. The heat from the inner core of the earth is used to heat water, or the steam from the heated water is used to power machines. This option is one of the cleanest and most efficient available today.

Hydro-electric power which has been incorporated in many dams worldwide produces a lot of energy, and it is easy to produce the energy as the dams control the water that is allowed through seams which power turbines located inside of them.

The amount of sunlight that falls on Earth in one hour contains enough energy to power the world for one year. By adding solar panels around the world we can power our lives without the need to produce more oil and without polluting the atmosphere.

As the fossil resources start declining in numbers and rise in price, we begin to realize that the need for renewable fuel sources is vital. With the production of new types of renewable energy, including solar, geothermal, hydro-electric, biofuel, and wind power we could add an element of security to our energy systems.

We will look at all renewable energy options in this text, and consider their role in ensuring our energy security and bringing us closer to energy independence.

Energy Independence

Depending on energy (oil in particular) from other countries is the major U.S. energy security risk at the moment. Our economy is still exposed to manipulation of energy supplies, such as the OPEC-orchestrated oil crisis of 1973, and the extraordinary jump of oil prices in 2008.

Starting in 1973, the American government realized that we have a big problem and initiated a feverish race towards developing new energy sources to obtain energy independence. Soon after the crisis was averted, however, the energy independence urgency gave way to importing oil again.

Efforts to implement new renewable energy sources fizzled by the mid-1970s and were shelved until the next energy crises… including the last one during 2008-2012.

The Arab Oil Embargo temporarily reversed the steady increase of oil imports. It shocked the U.S. into the grim reality that we are not energy independent. We still depend heavily on the whims of other governments, who can turn the spigot of crude oil off any time they want, or raise the prices as much as they want.

Note in Figure 1-33 that the cross point—where equal oil production and oil imports volumes intersect—was reached in the mid-1990s. From that time on, the imports started increasing sharply, a trend that lasted until 2008.

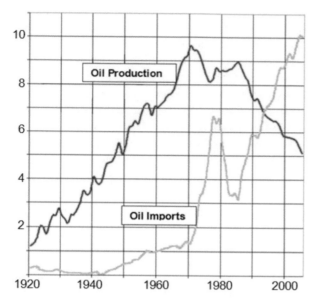

Figure 1-33. Historical U.S. oil imports and exports (in million barrels/day)

Today we are back at the cross point of the mid-1990s—producing almost exactly the same quantity of crude oil as we import. With that, our energy security is compromised by external factors, such as unstable governments and international events.

Depending on oil imports means that our energy security is compromised and also that we are far from achieving total energy independence.

Wars, political conflicts and many other factors, such as strikes in oil producing countries, can prevent the proper flow of energy supplies. Venezuela is a prime example of how things can go wrong with the oil supply chain. First it was the nationalization of the oil industry, followed by strikes and protests. It is several years after the fact, and Venezuela's oil production is yet to recover fully.

Exporters like Venezuela are driven by economic and political incentives, which at times force them to limit their exports. There is also increased competition for energy resources around the world, due to the increase of population and standard of living in India, China, and other developing countries. This creates energy price rises.

A major concern is the pending oil peak, which is lurking on the horizon. On the mid-term this means reduced amounts of oil at gradually steeper prices, which will create another wave of competition among the major importers. On the long term this means the end of crude oil. Period.

A big problem with the energy independence movement is that it has become a mantra of large companies, environmentalists, and others interested in pushing their agendas: renewable energy, climate change, etc. as needed to achieve their goals.

It is also used by interest groups as a weapon for defending technological and economic reasons for using unreliable energy sources and unrealistic projects that are unable to compete on the market, in order to justify subsidizing them.

A close look at the energy independence scenario reveals that all interest groups think that theirs is the only way of achieving it. The fossils are indispensible for providing reliable and cheap power, so they must be supported. Wind and solar are the solution for the future, so they must be supported too. Often supporting one technology is done at the expense of the others.

The problem is that the support structure is limited and crumbling, so the different energy sources must find a way to support themselves *and work together*, while taking the country closer to energy independence.

This—especially the working together part—requires a thorough understanding of the issues on both sides of the debate, and efficient leadership, both of which are simply missing, or inadequate at best.

INTERNATIONAL ENERGY SECURITY

Globally, the energy debate continues, but it is quite different from that in the U.S. Most European countries and Asian, for example, lack natural energy resources and are dependent on imports. This is forcing them to re-evaluate their energy policies and rely more and more on alternative energy sources.

The share of renewables—hydropower, biomass, wind and solar—in the global electricity supply is expected to reach 30-40 percent, surpassing natural gas and even coal as the leading fuel for power generation in 2050.

This scenario might not apply to the U.S., if the present abundance of hydrofracked oil and gas continues, but it is surely an achievable reality for most European and some Asian countries, including Japan and China.

The energy sources distribution and use around the world is unequal and quite interesting.

Table 1-13 shows the total energy picture of the different countries—all energy involved in power generation and transportation fuels. It clearly shows that a number of major countries, like S. Korea, Japan, and Germany import over half of their total primary energy resources. This includes all kinds of energy imports—coal, natural gas, and crude oil—which means that they depend totally on the international energy markets which are dominated by financial, logistical, and political issues.

Table 1-13. Global energy use, imports, and exports, 2010 (in TWh equivalent)

Country	Use	Import	Export	Import %
South Korea	10,341	9,169	0	89
Japan	20,725	17,124	0	83
Germany	13,900	8,792	0	63
France	11,053	5,568	0	50
UK	8,541	2,554	0	30
USA	93,575	22,190	7,500	24
Saudi Arabia	6,615	0	15,575	0
Russia	27,088	0	22,148	0
Canada	10,676	0	5,987	0
Australia	5,275	0	8,374	0

It is obvious from Table 1-13 that the European Union countries have long way to go toward energy security, and even farther to energy independence.

In stark contrast, countries, like S. Arabia, Russia, Canada, and Australia, do not depend on any imports, and in fact export more than they use. Saudi Arabia, for example, exports about 3 times more energy than it uses.

The major Asian economies, Japan and South Korea, are having problems with their energy supplies. Japan, for example, imports nearly 85% of its primary energy and has framed energy policies in light of this vulnerability. Its policy since the 1970s has been to have a balance among coal, gas and nuclear. Japan also set out to diversify its electricity generation, including a significant portion of nuclear power, alongside coal and gas, which has been increasing since 2000. However, since the Fukushima accident, this balance is being reviewed.

South Korea imports almost all of its energy and had a 12-fold increase in electricity demand from 1980 to 2009. It now gets 35% of its electricity from nuclear power while planning to increase this to 43% by 2020 and 59% by 2030.

The U.S. is the most interesting case, for it is one of the few countries which imports and exports large quantities of energy resources.

It is also worth noting that the U.S. imports only crude oil in significant quantities. U.S. imports represent about half of the total crude oil in the country. Then we refine the imported crude oil, use half of it to supplement the domestic supplies, and export the other half as petrochemicals.

Only about 25% of the U.S. total energy input depends on imports (crude oil), since the rest of the energy is obtained from domestic sources.

How long can this be sustained? The fossils are going fast.

European Union

The Europeans are having the hardest time with their energy supplies, and are trying hard to establish some semblance of energy security.

France imports half of its net primary energy, and this is a significant justification for its heavy reliance on nuclear power for electricity. The policy of having three quarters of its electricity from nuclear power was set in 1973.

Note: When the price of oil on world markets increased dramatically in 1973, several energy importing countries reviewed their energy policies and took steps to reduce their vulnerability to political and economic uncertainties. France was one of them. Now France is a world leader in nuclear energy reactors building and operation.

Germany imports more than half its net primary energy too, and in the past it has addressed this vulnerability with one third of electricity from nuclear and recently by providing major incentives for renewables.

The UK imports less than 20% of its net primary energy, but this is set to increase with depletion of North Sea gas. Continuing a high reliance on gas would make it vulnerable to supply interruptions from Russia and the Middle East.

Italy is the world's largest net importer of energy in the form of electric power—about 44 TWh net in 2010, which is about 15% of its consumption and which has been typical for a decade. Most of the imported power comes from French nuclear plants.

The European Union has had an energy policy for many years now, which actually evolved out of the European Coal and Steel Community arrangement. It is based on the concept of introducing mandatory and comprehensive rules of energy generation and use.

Today's European Energy Policy was approved at the meeting of the informal European Council in 2005 at Hampton Court. Later, the 2007 EU Treaty of Lisbon legally included solidarity in matters of energy supply and changes to the energy policy within the EU.

Prior to 2007, the EU energy legislation was loosely based on its authority in the area of the common market and environment on the continent. In practice, however, the actual energy policy competencies remained at national member state level. Changes or enforcement of the energy policy at the European Community level required voluntary cooperation by the different members states. As a matter of fact, little has changed since then, and the states still have the final word over all energy and environment related changes and enforcement efforts.

How this affects the energy security of the different countries is a complex matter. Because of the lack of uniformity, the different states have different approaches, with corresponding successes and failures of their energy security efforts.

Unfortunately, most European countries are fully dependent on energy imports. In 2007, for example, the EU imported 82% of its oil, 57% of its gas, and 75% of its uranium. This made EU the world's leading importer of fossil fuels, and Russia is a major exporter of energy materials and fuels to Europe.

In an attempt to mitigate some of the major risks, the EU legislation has been trying to design programs that could help it avoid getting into tight spots with its energy supplies. Nevertheless, energy exporters have succeeded in causing a number of serious disruptions.

Some of the measures taken to ensure energy supplies to the continent are summarized below.

The SET Plan

The European Strategic Energy Technology (SET) plan establishes (or rather suggests) a uniform energy technology policy for Europe. It's aim is to accelerate the development and deployment of cost-effective low carbon technologies.

The SET plan comprises measures relating to planning, implementation, resources development and international cooperation in the energy sector.

SETIS is the Information System for the SET plan and is led by the Joint Research Center. It supports the strategic planning and implementation of the SET plan initiatives. It provides information and guidance for technology options and sets priorities for monitoring and reviewing progress regarding implementation.

SETIS also assesses the impact of these developments on policy trends, and identifies corrective measures when needed.

Implementation of the SET-Plan started with the establishment of six European Industrial Initiatives (EIIs), which bring together industry, the research community, Member States and the European Commission in risk-sharing, public-private partnerships. These are aimed at the rapid development of key energy technologies at the European level, for which the barriers, scale of investment and risk can best be tackled collectively.

In parallel, the European Energy Research Alliance (EERA) has been working since 2008 to align the R&D activities of individual research organizations with SET-Plan priorities and to determine a joint programming framework at the EU level.

Implementation of the SET-Plan is also supported by a series of industry-led European Technology Platforms (ETPs). These help define research and technological development objectives and lay down concrete goals for achieving them.

Another pillar of support for implementing the SET-Plan is provided by a number of Joint Technology Initiatives. These are public-private partnerships, funded by the European Commission, along with Member States and industry, under the Seventh Framework Program (FP7) for large-scale research initiatives. SETIS is particularly involved with the Fuel Cells and Hydrogen (FCH) Joint Technology Initiative.

The SET plan includes a number of initiatives:

European Wind Initiative is part of the European Technology Platform for Wind Energy (TPWind). It is a joint initiative (public-private partnership) of the European Commission, representing the European Union, and the industry. Its goal is to improve the competitiveness of the European Union in the field of wind energy.

The effort is focused on the development and deployment of large wind turbines and the construction of large wind systems (wind power plants) through validation and demonstration of on- and off-shore wind technologies and power fields. TPWind is working to identify areas for increased innovation, new and existing research and development tasks. These will then be prioritized on the basis of "must haves" versus "nice to haves."

The primary objective of this effort is overall (social, environmental and technological) acceptance of the wind technologies that could lead to cost reductions. This will help to achieve EU energy objectives by increasing renewable electricity production.

TPWind is developing coherent recommendations, detailing specific tasks, approaches, participants and the necessary infrastructure, in the context of private R&D, as well as EU and Member State Programs, such as FP7. TPWind is also in charge of assessing the overall funding from public and private sources, as needed to carry out this work.

The strategic objectives of the EU wind power program are to improve the competitiveness of wind energy technologies, to enable the exploitation of the offshore resources and deep waters potential, and to facilitate grid integration of wind power.

The industrial sector objectives are to enable a 20% share of wind energy in the final EU electricity consumption by 2020.

The technological aspects of the wind technology have been evolving very fast recently. The trend in wind power is moving towards ever larger wind turbines. Since the first commercial turbine in the 1980s, their size has evolved from 0.022 MW to multi-MW machines of up to 5-6 MW today. For this to be achieved, rotor diameters grew from 10 meters to around 100 meters (over 300 feet). Larger machines are required today because of a booming offshore wind market.

Currently, the average turbine size in the EU is around 2 MW onshore and 3.5 MW offshore. By 2030, average turbine sizes of 3 MW and 10 MW are expected for on- and off-shore respectively, with offshore wind farms likely measured in the low GW range.

This recent scaling-up of turbine size is driven primarily by the move to take wind technology offshore. Due to land-use constraints, there are more suitable sites offshore than onshore, and wind speeds are higher away from land, leading to more wind energy generation.

Larger wind turbines lead to new challenges in the field of load control and turbine construction materials. While offshore sites require increased technological focus on foundations and materials adapted to the marine environment.

The further deployment of wind farms will also need to be accompanied by developments in storage technologies and increased grid flexibility to accommodate increasing levels of wind energy in the electricity network.

The speed of the tip of the rotor blade is limited by acoustic noise, operating at reduced speed in noise-sensitive areas, but above 80 m/s can be acceptable for offshore machines.

The specific wind technology objectives are:

- *General turbines and components development*
  — To develop large-scale turbines in the range of 10-20 MW especially for offshore applications.
  — To improve the reliability of the wind turbine components through the use of new materials, advanced rotor designs, control and monitoring systems.
  — To further automate and optimize manufacturing processes such as blade manufacturing through cross industrial cooperation with automotive, maritime and civil aerospace.
  — To develop innovative logistics including transport and erection techniques, in particular in remote, weather hostile sites.

- *Offshore technology development (focus on large-scale turbines and deep waters)*
  — To develop new stackable, replicable and standardized substructures for large-scale offshore turbines such as: tripods, quadropods, jackets and gravity-based structures.
  — To develop floating structures with platforms, floating tripods, or single anchored turbine.
  — To develop manufacturing processes and procedures for mass-production of substructures.

- *Grid integration techniques for large-scale penetration of variable electricity supply.*
  — To demonstrate the feasibility of balancing power systems with a high share of wind power using large-scale storage systems and high voltage alternative current (HVAC) or high—voltage direct current (HVDC) interconnections.
  — To investigate wind farms management as "virtual power plants."
  — Resource assessment and spatial planning to support wind energy deployment.
  — To assess and map wind resources across Europe and to reduce forecasting uncertainties of wind energy production.
  — To develop spatial planning methodologies and tools taking into account environmental and social aspects.
  — To address and analyze social acceptance of wind energy projects including promotion of best practices.

Solar Europe Initiative focuses on large-scale demonstration for photovoltaics and concentrated solar power (CSP) equipment and systems (power plants).

A major driver in the solar sector in Europe is the European Photovoltaic Industry Association (EPIA). It has several dozen members that are active along the whole solar PV value chain: from silicon, cells and module production to systems development and PV electricity generation as well as marketing and sales.

EPIA is also a member of the European Renewable Energy Council (EREC), the Alliance for Rural Electrification (ARE), PV CYCLE, the International Energy Agency Photovoltaic Power Systems Program (IEA-PVPS), and the European Forum for Renewable Energy Sources (EUFORES).

EPIA's mission is to give its members a distinct and effective voice in the European and world's energy markets, with emphasis on EU activities.

EPIA basically coordinates the European Photovoltaic Technology Platform and co-organizes the European industry event PV-SEC. It has been quite successful in achieving its goals and as a result of this (and other factors) Europe is one of the world leaders in solar installations.

There are two main areas of solar development: solar thermal and photovoltaics.

- *Solar thermal power*

After about a decade of low development, the concentrated solar thermal power sector (CSP) is now reviving, notably due to a favorable supporting framework in Spain and increasing investments in the USA. A CSP plant consists basically, of a solar concentrator system made of a receiver and collector to produce heat and a power block (in most cases a Rankine cycle).

Three main CSP technologies are under development: trough, tower/central and dish. Today CSP technologies are in the stage of a first commercial deployment for power production in Europe.

The most mature large-scale technology of these is the parabolic trough/heat transfer medium system. The first CSP parabolic trough power plant with a power capacity of 50 MWe and 7.5 hours of storage, Andasol 1, was installed in Granada, Spain, and two other plants of 50 MWe are also operational in Spain.

Central receiving systems (solar tower) are the second main family of CSP technology. An 11 MWe saturated steam central receiver project, named PS 10, is operating since March 2007 in Andalusia. This is the first commercial scale project operating in Europe. Solar Tres is another project under development in Spain based on a molten salt central receiver system.

Parabolic dish engines or turbines (e.g. using a Sterling or a small gas turbine) are promising modular systems of relatively small size (between 5 to 50 kWe). These are still in the development phase, and are primarily designed for decentralized power supply.

The solar-only average load factor without thermal storage of a CSP plant is about 1,800 to 2,500 full-load hours per year. The level of dispatching from CSP technologies can be augmented and secured with thermal storage or with hybrid or combined cycle schemes with natural gas, an important attribute for connection with the conventional grid. For instance, in the Solar Tres project, 15 hours molten salt storage is included leading to a capacity factor of 64% without fossil fuel power back-up.

Several Integrated Solar Combined Cycle projects using solar and natural gas are under development in Algeria, Egypt, India, Italy and Morocco.

- *Photovoltaics power*

Solar photovoltaic (PV) energy is a large, rapidly developing industry. Photovoltaic technology is poised to help Europe make good on its goal of sourcing 20% of its energy needs from renewable energy by 2020.

Recent progress has led to rising efficiencies, better reliability and falling prices. Improving solar cell efficiency while keeping costs low is the main challenge for further maturity and uptake of the technology.

In 2010, about 40 GW of grid-connected solar photovoltaic power were installed globally, which can produce over 45 TWh of electricity yearly. Europe has a cumulative installed capacity of more than 25 GW, making it the largest world market.

Crystalline silicon solar cells are by far the most common, maintaining a market share of almost 80%. These cells are made from thin slices cut from a single crystal of silicon (mono-crystalline) or from a block of silicon crystals (polycrystalline).

Commercial crystalline-silicon solar cells have efficiency ranges between 13% and 22%. The remainder of the market is principally provided by different thin film technologies.

Thin film modules are less efficient, but are more adaptable to mass production, which can potentially result in lower production costs compared with the more material- and labor-intensive crystalline technology.

Concentrating Photovoltaics (CPV) is an emerging market with two main tracks—either high concentration of 100-1000 suns (HCPV) or low to medium concentration with a concentration factor of 2 to 100.

HCPV systems are already exceeding 40% efficiency, but to maximize the benefits of CPV, the technology requires high direct normal irradiation (DNI), such as in the desert, but these areas have a limited geographical range in Europe.

The market share of CPV is still small, but an increasing number of companies are focusing on it. Market estimates of 100-200 MW of CPV installations are expected by 2020 in Europe.

Note: The widely advertised Desertec initiative, which planned the development of huge solar and wind power plants in North Africa to feed the European continent with terawatts of electricity, is now falling apart for a number of reasons.

The Europeans would do their future generations a great favor if they complete Desertec, or a similar project, because with the limited energy resources on the continent, the energy future of most EU countries looks quite grim.

• *Solar heating*

Solar thermal (passive) heating technologies are predominantly based on glazed flat plate and evacuated tube collectors. The vast majority of the European capacity (90%) of this technology comprises single-family house units used for the supply of domestic hot water. The remaining capacity consists of an equal share of domestic hot water—multi-family house units and single-family house combination systems that deliver both hot water and spatial heating.

There are a few large-scale systems installed in Denmark, Sweden, Germany and Austria which deliver heat to larger district heating networks. Some of these are coupled with seasonal heat storage.

There are also a small number of installations in industrial sites for the provision of low temperature process heat. The pumped heat systems installed in Central and Northern Europe can meet about 50-70% of the hot water needs for a house, generating 500-650 kWh of useful heat for each kW installed.

Thermo-siphon systems, typically in Southern Europe, provide up to 70-90% of the hot water requirements for buildings, and generate 700-1,000 kWh for each kW installed.

The passive solar-heating technology that can provide all heating and cooling needs of a building with good insulation has already been demonstrated. Further technological developments are anticipated in the near term, which will improve the competitiveness of the technology and facilitate the expansion of the solar-thermal market.

These technology improvements include the development of new systems that will incorporate superior collectors based on advanced polymeric materials, vacuum insulation and sophisticated heat storage media, combined with intelligent heat management controls.

These systems will be integrated in new and retrofitted buildings with new insulation, such as in facades, to provide hot water, and spatial heating and cooling.

In addition, the technology of concentrated collectors will be further developed for use in systems that will provide low and medium temperature process heat to the industrial sector.

If the solar-thermal capacity in Europe continues to expand, it is expected that system costs for small-scale forced circulation units installed in central Europe will be decreased to $300-400 per each kW installed by 2030.

Due to limited sunshine and land availability, solar alone cannot supply a large part of Europe's electric energy needs. Across the Mediterranean Sea, however, lie some of the world deserts, which have unlimited sunlight and land. This area will most likely supply a significant amount of energy to Europe in the future, and bring much needed prosperity to the African continent as well.

Note: The EU community declared war on Chinese imports in 2013 and in June imposed a tariff of 11.8% on imported solar cells and wafers. It threatened to raise the tariffs to 47% starting in August unless a settlement was reached.

European solar manufacturers and installers accused China of unlawfully dumping solar panels in Europe, i.e., selling them at up to 88% below the fair price in order undercut their European rivals.

The sanctions were modified in November 2013 when both governments endorsed an agreement that sets a minimum price and a volume limit on European imports of Chinese solar panels until the end of 2015.

So, participating Chinese manufacturers will be spared EU tariffs for now. The new solution—a two-year trade protection—is based on a provisional accord negotiated in July, 2013.

For the next two years, Chinese solar-product imports into Europe are fixed at a minimum price of 56 euro cents a watt.

Annual imports from China are now limited to maximum of 7 giga watts. Chinese companies willing to take part in the preliminary punitive EU import duties will be exempted from the new agreement.

The new agreement covers more than 90 Chinese exporters that presently cover about 60 percent of the EU solar-panel market. These include the solar giants Yingli, Suntech, Trina, Jiangsu Aide Solar Energy Technology Co., Delsolar (Wujiang) Ltd., ERA Solar Co., Jiangsu Green Power PV Co. and Konca Solar Cell Co.

The participating Chinese exporters will be exempted from anti-dumping and anti-subsidy duties being fixed definitively. Depending on the company, the two-year definitive anti-dumping levies range from 27.3

percent to 64.9 percent, while the anti-subsidy rates are from 3.5 percent to 11.5 percent.

In addition, in November 2013 EU imposed maximum 42.1% anti-dumping tariff on Chinese solar glass exporters. The provisional duties vary between 17.1% and 42.1% depending on each Chinese company's cooperation with the investigation. The tariffs are due to take effect immediately.

In response, the China's Commerce Ministry (Mofcom) initiated anti-dumping and anti-subsidy probes into European polysilicon imports.

Between solar cells and solar glass import tariffs, and the polysilicon exports, we are looking at a long-term struggle.

Bioenergy Europe Initiative focuses on development and deployment of biofuel production and use for transportation needs mostly. Its emphasis is on bringing the "next generation" biofuels to the world markets, within the context of an overall EU bio-energy use strategy.

Bio-ethanol and biodiesel are the most common biofuels used in transport, although other biofuels are also in use, such as pure vegetable oil and biogas. The main drivers for the production and use of biofuels are the security and diversification of energy supply, reduction of oil imports and dependence on oil, rural development and the reduction of greenhouse gas (GHG) emissions.

The European Biofuels Technology Platform (BiofuelsTP) was launched at a conference in June 2006 as part of the European Seventh Framework Program initiative, geared to improve the competitive situation of biofuels technologies in the EU community.

Its mission is to coordinate the development of cost-competitive world-class biofuels value chains, to create a healthy biofuels industry in Europe, and to accelerate the sustainable deployment of biofuels in the EU through a process of guidance, prioritization and promotion of research, technology development and demonstration.

It is a joint initiative (public-private partnership) of the European Commission, in representing the European communities, and the biofuels industry in particular on the world energy market. It is actively engaged with biofuels stakeholders (researchers, academia, civil societies, industry), EC-funded research projects and initiatives, related European TPs and global biofuels organizations in a wide range of activities in development and implementation of sustainable advanced biofuels technologies and projects in Europe.

The Biofuels' Future

The EU's Biofuels Technology Platform has set as a priority the development of ligno-cellulosic materials as a feedstock for biofuels production, which is also known as second generation biofuels.

Such biofuels currently under investigation include a direct production pathway for bio-ethanol production, which consists of mobilizing the cellulosic components of different plants through a saccharification stage prior to the fermentation process.

Biomass-based DME (dimethylether) is also currently under development. This can be produced from the gasification of biomass or black liquor and is currently being demonstrated as a transport fuel in heavy-duty vehicles.

Third-generation biofuels, including hydrogen produced from biomass, are expected to make a significant contribution to passenger car and urban transport markets as of 2030.

Biofuel production from algae is presently at the research and development stage, focusing on evaluating the optimum strains of algae, investigating process development and oil extraction.

One major aim for the technology platform is to bring to commercial maturity the currently most promising technologies and value chains to promote large-scale, sustainable production of advanced biofuels. For instance, feedstock-flexible thermochemical pathways, characterized by the use of high-temperature transformations, and biochemical pathways are to be developed.

For thermo-chemical pathways, research aims include the optimized use of advanced catalysts, the improvement of gas cleaning technologies and the quality and stability of bioliquids.

Within the biochemical pathways, three value chains will be optimized for the production of gas and liquids from biomass, including feedstock pre-treatment and downstream processing and the optimized use of advanced enzymes.

As part of the overall plan, Europe is in the process of assessing its biomass availability and plans to develop technologies and logistics for sustainable feedstock production, management and harvesting.

Europe's Nuclear Research Program focuses on activities in the areas of research, technological development, international cooperation, dissemination of technical information, and exploitation and training activities. Nuclear energy has an important role in a future low-carbon energy mix of Europe and is on the forefront of the energy and energy security plans of the EU community.

The two major specific efforts in this program are:

- Fusion energy research (in particular ITER), nuclear fission, and radiation protection;

- Activities of the Joint Research Centre (JRC) in the field of nuclear energy, including nuclear waste management, and environmental impact, nuclear safety, and nuclear security. JRC also carries out research in a number of other areas to provide scientific and technological support to EU policy making

Nuclear fission energy is a competitive and mature low-carbon technology, operating to high levels of safety within the EU. Most of the current designs are light water reactors (LWR), capable of providing baseload electricity often with availability factors of over 90%. The ageing of Europe's nuclear reactors and the requirements for secure, cost efficient and low-carbon energy systems will demand a substantial investment in construction and development of nuclear reactors.

One of the main features of nuclear energy is its very high energy density, whereby nuclear energy can provide a very large share of future energy needs. Nuclear energy today is used almost uniquely for base-load electricity production but heat application is another very large potential market that is completely dominated by fossil fuels.

Today's Generation II light-water reactors produce 16% of the world's electricity and 30% of the electricity in the EU. The related *sustainable nuclear fission initiative* focuses on the development of Generation IV reactors technologies.

In the Blue Map scenario of the IEA for a low-carbon energy system, the capacity should increase to 650 GWe in 2030 and 1200 GWe in 2050. Most of the expansion in the coming decades is expected to come from Generation III light-water reactors.

The levelized cost for electricity from newly built reactors is expected to be $45 per MWh. The capital cost represents typically 60-70%, operation and maintenance 20-25% and the fuel 10-15% (with only 5-7% for the natural uranium). The levelized cost for nuclear electricity is therefore very stable but the high investment cost is a problem in a completely free market.

Most of the natural uranium resources are in politically stable countries and due to its high energy density, reserves of nuclear fuel for several years of operation can be stockpiled, which guarantees security of supply. The further expected global expansion of the LWR fleet has repercussions on the availability of uranium resources.

Current estimated exploitable reserves at $130/kgU are 15 million tons (Mt). The present global rate of consumption of about 67,000 tonnes per year (t/y) will rise to an anticipated 100,000 t/y in 2025 for an installed world nuclear capacity of 550 GWe.

Assuming the installed nuclear capacity increases linearly up to 1200 GWe in 2050, the 15 million tonnes would be completely earmarked for the fuel needs of the light-water reactors by this date.

Nuclear energy also produces radioactive waste which is of great public concern. These observations underline the needs to develop sustainable nuclear technologies with respect to better use of the uranium resources and reduction of the ultimate radioactive waste.

- *Nuclear fusion* is the most attractive long-term energy solution, although, due to its complexity, it is unlikely that the technology will be ready for commercial power generation anytime soon—at least not during the next 40-50 years or beyond.

Nevertheless, fusion energy has made significant progress over the last few decades and is now considered as a credible option for clean, large-scale electricity generation.

Fusion is the process that produces the light and heat of the sun. Hydrogen nuclei collide in the sun's core and release huge amounts of energy as they fuse into helium atoms. On earth, fusion reactors heat gas to extreme temperatures to produce a plasma similar to the conditions found within a star.

Fusion's many benefits include an essentially unlimited supply of fuel, passive intrinsic safety and no production of CO_2 or atmospheric pollutants. It is one of the very few candidates for the large-scale, carbon-free production of base-load power. Compared to nuclear fission, it produces relatively short-lived radioactive products, with half-lives limited to less than 50 years.

The most efficient fusion reaction to use on earth is different from that in the sun: the reaction between two hydrogen (H) isotopes: deuterium (D) and tritium (T), produces the highest energy gain at the lowest temperatures.

Fusion power plant conceptual studies, including full lifetime and decommissioning costs, suggest that fusion could indeed be economically competitive with other low-carbon sources of electricity.

The Joint European Torus (JET) project has successfully demonstrated nuclear fusion technology, producing 16 MW of fusion power. This represents an energy output of 70% of the energy put in—the best results so far for a fusion reactor.

The ITER Agreement, signed in 2006 between the EU (via the Euratom Treaty) and six other countries including Japan, Russia and the USA, was heralded as a major step forward. This will see the construction of the ITER fusion reactor to demonstrate the technical and scientific feasibility of a "burning" plasma on the scale of a power plant.

ITER will produce the fusion reaction in a tokamak device, using magnetic fields to contain and control the hot plasma. The fusion between deuterium and tritium (D-T) will produce one helium nucleus, one neutron and excess energy. The helium nucleus carries an electric charge which responds to the magnetic fields of the tokamak, and remains confined within the plasma. The neutron has no electrical charge, however, so will carry some 80% of the energy away from the plasma.

Another important step is the Broader Approach agreement, signed between the EU and Japan in 2007, which also includes final design work and prototyping for the International Fusion Materials Irradiation Facility (IFMIF), a device that will subject small samples of materials to the neutron fluxes that will be experienced in fusion power plants.

These neutrons will be absorbed by the surrounding walls of the tokamak, transferring their energy as heat. In ITER, the neutrons are absorbed in the surrounding lithium blanket, producing heat which will be dispersed through cooling towers. The next fusion plant prototype DEMO and future industrial fusion installations will use this heat to produce steam and, by way of turbines and alternators in the conventional way, generate electricity.

Major challenges still remain, however, in making magnetic confinement fusion work reliably on the scale of a power plant. For example, how to sustain a large volume of extremely hot plasma (millions of degrees F) for long periods of time, at pressures that will allow for a large net energy gain from the fusion reaction.

Such a plant needs exceptional materials and components capable of resisting the extreme conditions required for continuous high power output. These are materials that simply do not exist presently, so fusion has long way to go before becoming a reality. Nevertheless, the work continues and a lot of money and effort is spent on this technology in the EU, the U.S., Japan and other countries.

European CO_2 capture, transport and storage initiative focuses on the whole system requirements, including efficiency, safety and public acceptance, to prove the viability of zero emission fossil fuel power plants at industrial scale.

Carbon capture and storage (CCS) technologies can be applied to energy production wherever carbon dioxide is produced in large quantities. This includes but is not limited to power generation and promises near zero emission electricity from fossil fuels.

Since we will continue to rely on fossil fuels for some significant time to come, CCS is the single action with the greatest potential to combat climate change. It may well be able to address almost half of the world's current carbon dioxide emissions, by preventing the gas emitted by large stationary sources from entering the atmosphere. For example, CCS can capture at least 90% of carbon dioxide emissions from power plants and heavy industry before transporting it by pipeline or ship and storing it at least 700m below the earth's surface.

CCS consists of three major steps:
- Capture and compression of the CO_2 at the emission site;
- Transport of the CO_2 to a storage location; and
- Storage of CO_2 in geological formations, usually deep underground.

Each step has several technology options, with different levels of performance and maturity, which can be combined in various ways. Capture can use post-combustion, in which CO_2 is removed from exhaust gases through absorption by selective solvents, as well as pre-combustion, whereby fuel is pre-treated and converted into a mix of CO_2 and hydrogen.

The CO_2 is then separated and the hydrogen is used as fuel. A third option is oxy-fuel combustion which burns the fuel with oxygen instead of air, producing a flue stream of CO_2 (which can be easily removed) and water vapor without nitrogen. Large-scale CO_2 transportation is mainly by pipeline, with some shipping and types of road tanker transport.

CO_2 storage options include deep saline aquifers (saltwater-bearing rocks) and depleted oil and gas fields (which includes the potential of enhanced oil recovery).

The portfolio of technologies currently being developed applies to both newly built power plants and also to retrofits of existing plants. Internationally, up to 20 pre-commercial implementation projects are aiming to demonstrate various combinations of CCS technologies. The European Commission has committed to support up to 12 projects to be operational by 2015. Funding will come from individual governments, the EU and industry.

CO_2 capture can be used with fossil-fuel power plants using pulverized coal, natural gas turbine combined cycle or integrated gasification combined cycle (IGCC). The first commercialized generation of such plants are expected to have efficiencies of 35%, 49% and 35%, respectively. The first CCS plants are expected to be coal-fired. Technological developments should reduce the efficiency gap between zero-emission plants (ZEP) and others by 2030.

- *Advanced fossil power* generation is one technique that promises better efficiency and reduced GHG emissions. Fossil fuel power plants produce the majority of electricity in the EU, mainly through pulverized coal (PC) combustion. But many of these plants are 15-20 years old and are relatively inefficient.

As fossil fuel power generation is the biggest contributor to carbon dioxide emissions, any gains in conversion efficiency would translate to substantial carbon dioxide savings. Using the best available technologies, such as "advanced supercritical plants" and "ultra-supercritical plants," can increase efficiency by allowing higher steam conditions (temperature and pressure).

Combined cycle plants using natural gas or biomass in pulverized coal power plants, and integrated gasification combined cycle (IGCC) plants, which turn coal into gas, can potentially reduce emissions even further, especially with carbon capture.

Most pulverized coal steam plants in Europe operate with sub-critical steam parameters and efficiencies between 32-40%. Supercritical plants with steam conditions of 540°C and 300 bar have been in commercial operation for a number of years, with efficiencies of 40-45%. When best technologies are used, as in "advanced supercritical" plants, with steam conditions up to 600°C and 300 bar, net efficiencies of 46-49% should be achieved. These technologies require stronger and more corrosion resistant steels, but potential efficiency savings offset their extra cost.

The next step for the use of coal, under development since the 1990s, is the "ultra-supercritical" (USC) power plants. Currently, steam conditions of 600°C and 300 bar can be achieved, with efficiencies of 45% and higher, for bituminous coal fired power plants. Future USC plants are planned which use 700°C and 350 bar or higher and yield net efficiencies of about 50-55%.

In 1998, a major group of power industry players started a 17-year demonstration project, supported by the EC THERMIE program, called the "Advanced (700C) PF Power Plant." The main aim is to make the jump from the use of steels to nickel based super-alloys for boilers to achieve the highest temperatures and increased efficiency (50-55%). However, it is not known when a 700°C steam coal power plant will become a commercial reality.

Many of the technology objectives and actions for advanced fossil fuel power generation are the same as for the development of carbon capture and storage (CCS). These include the establishment of an R&D program that will address fossil fuel conversion technologies aimed at improving power plant efficiency in all main fossil fuel power generation routes to better compensate for the efficiency penalty imposed by CO_2 capture.

- *Fossil power cogeneration* is another technique where the production of heat and electricity occurs in a single process or power plant. This technology is part of the European plan because of its increased efficiency and GHG reduction potential.

A modern fossil-fuel power plant transforms about half the primary energy content of its fuel into electricity and rejects the rest as "waste" heat. Cogeneration or combined heat and power (CHP), uses a part of that heat to satisfy a heat demand which would otherwise require energy from another source, usually a fuel.

The heat is often in the form of hot exhaust gases, steam or hot water. CHP thus improves the overall efficiency of fuel utilization and saves on primary energy in comparison to the conventional separate production of power and heat. This results in better operating efficiency and pricing, as well as significant GHG reduction per MW generated.

European electricity grid initiative focuses on the development of the smart electricity system, including storage, and on the creation of a European center to implement a research program for the European transmission network.

Electricity generated by renewables, such as wind and solar, is variable since it depends on the availability of sun and wind. Electricity generated by these energy sources, as well as by some conventional base load power plants, has to be stored for better utilization.

Large-scale energy storage can optimize the overall energy flows between supply and demand and therefore enable a higher contribution of renewable energy in our electricity mix. Renewables may also not be fully available at the moment when demand is higher or they may supply an excess when the demand is lower, thus creating an imbalance.

Electricity storage can overcome the mismatch between output and demand (the so-called time-shifting) and it can smooth out fluctuations in supply without calling on other back-up capacities. It can also save a supplier from penalties when forecast supply cannot be met (the so-called forecast hedging).

The principal electricity storage technologies include hydropower with storage, compressed air energy storage, flow batteries, hydrogen-based energy systems, secondary batteries, flywheels, super capacitors, and superconducting magnetic energy storage.

The energy storage technologies are still in their infancy, but due to their potential in bringing mass quantity of renewable energy into the mainstream energy supply, they are on the priority list of many companies and governments.

A wide range of energy storage technologies is available to store electricity, including those based on mechanical, chemical and physical principles. Ultimately, the main services that the storage has to provide will dictate the best-adapted technology.

For example, there are energy-related applications, where the electricity storage system is designed to discharge for several hours, with a nominal storage capacity of 10 to 500 MW and a time response of 1 to 5 minutes.

On the other hand, there are power-related applications, such as maintaining grid frequency, suppressing fluctuations and stabilizing voltages, that discharge for between a few seconds to less than an hour and require a response time of a few milliseconds.

The most mature storage technology is hydropower, where reservoir storage or hydro-pumped storage are used. The basic principle is to store energy as the potential energy between two reservoirs at different elevations.

The wide deployment of hydropower in Europe offers a significant technology base for regulating variable electricity production. The average plant size in the EU-27 is about 270 MW, but can reach 1800 MW, as in France (Isère) and Wales (Dinowig).

Most of any increase in pumped hydropower storage in Europe will come from retro-fitting of existing installations, or adding pumped storage to conventional reservoir-based facilities.

Compressed air energy storage systems (CAES) are another form of promising energy storage. The hybrid forms of this type of storage are already commercially used for large-scale energy storage.

In a CAES system, the compression cycle of a gas turbine is decoupled from its expansion cycle over time. Air is pre-compressed and stored separately in a geolog-

ical formation, prior to its use in the gas turbine.

Despite its reliance on mature technologies, CAES systems are not widespread around the globe. Their full implementation depends on a number of factors, the major of which are the trends that drive the rise in renewable technologies deployment.

The Framework Programs for Research and Technological Development, also called Framework Programs or abbreviated FP1 through FP8, are funding programs created by the European Union in order to support and encourage research in the European Research Area (ERA). The specific objectives and actions vary between funding periods.

The Framework Program presently in force, and until PF8 is designed, approved, and implemented (in 2014), is FP7. The FP7 research program has a moderate amount of funding for energy research, although energy has recently emerged as one of the key issues of the European Union. A large part of FP7 energy funding is also devoted to fusion research, a technology that will not be able to help meet European climate and energy objectives until beyond 2050. The European Commission is redressing this shortfall with the SET plan.

The specific programs constitute the five major building blocks of FP7:

- Cooperation
- Ideas
- People
- Capacities
- Nuclear Research

- *The Cooperation program* is the core of FP7, representing two thirds of the overall budget. It fosters collaborative research across Europe and other partner countries through projects by transnational consortia of industry and academia.

  Research is underway in ten key areas:
  — Health
  — Food, agriculture and fisheries, and biotechnology
  — Information and communication technologies
  — Nanosciences, nanotechnologies, materials and new production technologies
  — Energy
  — Environment (including climate change)
  — Transport (including aeronautics)
  — Socio-economic sciences and the humanities
  — Space, and
  — Security

- *The Ideas program* will support "frontier research" solely on the basis of scientific excellence. Research may be carried out in any area of science or technology, including engineering, socio-economic sciences and the humanities. In contrast with the cooperation program, there is no obligation for cross-border partnerships. Projects are implemented by "individual teams" around a "principal investigator." The program is implemented via the new European Research Council (ERC).

- *The People program* provides support for researcher mobility and career development, both for researchers inside the European Union and internationally. It is implemented via a set of Marie Curie actions, providing fellowships and other measures to help researchers build their skills and competences throughout their careers:
  — Initial training of researchers (Marie Curie Networks)
  — Industry-academia partnerships
  — Co-funding of regional, national and international mobility program
  — Intra-European fellowships
  — International dimension (outgoing and incoming fellowships)
  — International cooperation scheme, reintegration grants
  — Marie Curie Awards

- *The Capacities program* strengthens the research capacities that Europe needs if it is to become a thriving knowledge-based economy. It covers the following activities:
  — Research infrastructures
  — Research for the benefit of SMEs
  — Regions of Knowledge
  — Research Potential
  — Science in Society
  — Specific activities of international cooperation

- *FP-8 (2014-2020)*

The FP7 program (theoretically) ended in 2013, and FP8 is now shaping the future of European energy research and activities. It will basically continue the work started by PF-7. The final draft of the new program, however, was not available at the time of this writing.

China

The energy policy of China is dictated by the Central Government which manages the energy resources and the overall energy generation and use. China is currently one of the largest consumers of energy and the world's largest emitter of greenhouse gases.

The already large population of the country is increasing rapidly too, so China's per capita emissions are still far behind those in the U.S. and some of the other developed countries. This does not diminish the overall effects of unlimited fossil fuels use and increasing GHG emissions.

On the positive side, China is shaping up as the world's leading renewable energy producer. Figure 1-34 shows the incredible increase of electric power generation and use in the country. It amounts to nearly 500% increase in 10 years.

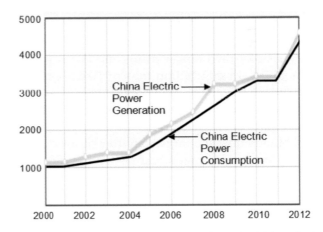

Figure 1-34. Electricity generation and use in China (in billion kWh)

China's total annual electricity output is about 4.0 trillion kWh, and the annual consumption about 3.75 trillion kWh, which makes it the second largest producer-consumer of electric in the world. The total installed electricity generating capacity in the country is nearly 1,000 GW, and the increase in demand is forcing China to undertake substantial long-distance transmission projects with record breaking capacities.

The goal now is to achieve an integrated nationwide grid in the period between by 2020, which means thousands of new transmission lines and hundreds of new coal-fired power plants. This also means millions of additional tons of pollution sent into the air of the large cities and the global environment.

It is obvious from Figure 1-35 that China is importing more than twice as much crude oil than it can produce.

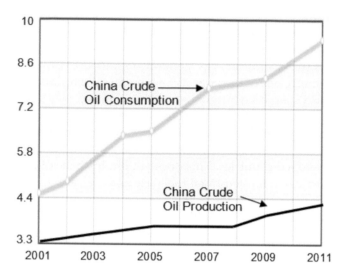

Figure 1-35. China oil production and consumption (in million bbl/day)

There is a two-fold increase in pollution emitted by China's coal-fired power plants and increased use of automobiles—from 4 billion tons of GHGs emitted in 2000 to over 8 billion tons emitted last year. The increase in both coal-fired power generation and automobile use is projected to continue rising in the foreseeable future.

China is now the largest emitter of toxic and GH gasses in the world, generating nearly four-fifths of its electricity from coal-fired power stations. It constructs about two new coal-fired power plants every week, or more than 100 every year.

There are over 2,300 coal-fired power stations worldwide, containing 7,000 power generating units. About 1/3, or 650 of these power stations are in China.

Nuclear Power

There were 15 nuclear power units operating in China in 2012 with a total electric capacity of 11 GW and total output of 54.8 billion kWh. These provided nearly 2.0% of the country's total electricity output. Two new reactors were added in 2013, and 32 reactors are under construction, which is the highest number of new nuclear reactors in the world.

The plans call for a significant increase in nuclear power capacity, bringing the total electricity output to 86 GW or 4% of total power generated in the country by 2020. Even more ambitious plans follow, calling for increase to 200 GWe by 2030, and 400 GWe by 2050.

After the Fukushima Daiichi nuclear disaster in Japan, China announced in 2011 that all nuclear plant approvals were being frozen. In addition, full safety

checks of existing reactors were mandated, which means additional safety-related costs and increased public awareness, all of which could cause a delay or even reduction in the planned nuclear expansion.

There is a trend recently in favor of expanded renewable energy programs, which could replace some of the planned nuclear plants construction. In fact, a revision of the existing policy in 2011 forced the cancellation of the construction of nuclear power plants in marine areas until further notice.

As a result, the official target of a capacity of 40 GW by 2020 is unchanged but earlier plans to increase this to 86 GW have been reduced to 70 GW. The reason for this change is shortages of equipment and qualified personnel, in addition to unresolved safety concerns.

The big question in front of China's policy makers and regulators is: coal, or nuclear power in the future? They will have to weigh the pros and cons of each technology in order to determine the risks and benefits of each.

Renewables

China leads the world in renewable energy products manufacturing (solar and wind) and recently user, with over 150 GW of renewable power plants around the country.

China's government has been investing heavily in the renewable energy field in recent years. This made China into the largest producer of wind turbines and solar panels in the world. In 2007, for example, the total investment in the renewable energy sector was about $15 billion, second only to Germany.

In 2012, China invested over $65 billion in clean energy, an increase of 20% over 2011. This is about 30% of the total investment by the G-20 countries put together. The investment portfolio at the time consisted of about $31 billion in solar energy, about $27 billion in wind energy, and over $6 billion in a number of other renewable energy technologies, such as small hydro, geothermal, marine, and biomass.

During the same time, over 23 GW of clean generation capacity (mostly wind and solar) was installed in the country. During 2013 wind power increased by nearly 10 gigawatts, solar rose by 5 gigawatts and nuclear by 2.2 gigawatts. Hydro electric power accounted for the remainder.

China is expected to build more renewable power plants through 2035 than the U.S., the European Union, and Japan combined. Wind and solar will account for half of the global power generation increase by 2035.

The major renewable energy source in China is still hydropower. Hydropower generates about 620 TWh annually, and provides over 17% of all electricity generated in the country. This constitutes the greatest hydro-electric capacity in the world.

The new Three Gorges Dam is the largest and most sophisticated hydro-electric power plant in the world, with a total capacity of 22.5 GW.

The great developments to date are a step in the right direction as far as reducing GHG emissions too, but are not going to solve the pollution problem created by hundreds of coal-fired power plants which number and emissions are increasing daily.

China plans to add 10 gigawatts of solar capacity each year in the 2013-15 period, with the goal of having 35 GW of installed solar power capacity by the end of 2015. This is 7 times the installed capacity in China of 5 GW it had in 2012, 5 times more than what the U.S. had, and more than the 32 GW that Germany—the world's largest player—had at the same time.

China is pushing hard its solar installations, in part using subsidies as an incentive, and plans to be third in the race for new solar installations by 2015. This scenario is very possible, because the present solar rise is mostly due to extremely generous government subsidies, which resulted in huge surplus, which can be used to increase the domestic solar installations.

The huge amount of power needed by the rising needs of the increasing population will be difficult to produce with renewables, so China may focus on developing nuclear power, hydropower, and shale gas as primary power generators.

Solar Problems

The development of solar power in the most productive areas of China might be limited by the poor infrastructure, though there is work underway. Still, it takes a long time and great expense to construct new power lines and substations, and those factorscould limit, or at least delay, the construction of significant new solar capacity.

There are also technical issues to be considered when undertaking such a large project, since most solar installations behave very differently from the established power plants.

We address these issues in more detail in the following chapters.

Russia

The energy policy of Russia is reflected in the Energy Strategy document, which sets out the direction of the energy sector up to 2020. The Russian government approved the main provisions of the energy strategy, and in 2003 the Energy Strategy was confirmed by the government.

The Energy Strategy document outlines the main priorities: an increase in energy efficiency, reducing impact on the environment, sustainable development, energy development and technological development, as well as improved effectiveness and competitiveness.

The Russian national energy policy was approved in 1992. At the same time the Energy Strategy of Russia was developed for which purpose the Interagency Commission was established.

In 1994, the Energy Strategy of Russia (Major Provisions) was approved, followed by the presidential decree confirming the first Russian energy strategy On the Main Directions of Energy Policy and Restructuring of the Fuel and Energy Industry of the Russian Federation for the Period up to the Year 2010, and the governmental decision from the 13 October 1995 approving the Main provisions for the Energy Strategy of the Russian Federation.

The strategy was changed somewhat by Putin, and in 2000 the Russian government approved main provisions of the Russian energy strategy to 2020. In 2002, the Russian Ministry of Energy elaborated on the main provisions and the new Russian energy strategy was approved in 2003.

The main objective of the Russian energy strategy is defined to be the determination of ways to reach a better quality of fuel and energy mix and enhance the competitive ability of Russian energy production and services in the world market. For this purpose the long-term energy policy focuses on energy safety, energy effectiveness, budget effectiveness and ecological energy security.

The major components of Russia' energy sector are given below.

Crude Oil

Russia has vast oil deposits, a major portion of which have not been exploited.

As Figure 1-36 suggests, Russia produces several times more crude oil (left hand side of the graph) than it can use (right hand side of the graph). For example, 11.25 million barrels per day were produced in 2011, and only 2.2 million were used. This is 4.5 times more production than consumption—it is the envy of the world. So no doubt, Russia is not worried about oil shortages, for now.

As the population becomes more prosperous, oil consumption will increase. For now, Russia has plenty

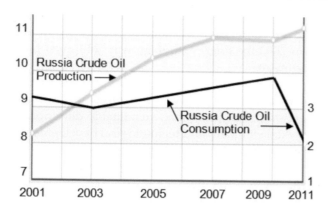

Figure 1-36. Russia's crude oil production and use (in billion bbl/day)

of oil, and as the Arctic ice cap shrinks due to global warming, the prospect of oil exploration and producing additional huge amounts of crude oil in the Arctic Ocean is becoming a reality.

In order to be able to develop any newfound deposits, in 2001 Russia submitted documents to the UN Commission on the Limits of the Continental Shelf, claiming expansion of the borders of the Russian continental shelf beyond the previous 200-mile zone in the Russian Arctic sector. In 2002 the UN Commission recommended that Russia should carry out additional research, which commenced in 2007.

There are estimates that the area may contain 10 billion tons of gas and oil deposits. This opens huge great opportunities for the Russian energy industry. The new deposits would put Russia firmly and permanently on the list of major world energy exporters.

Natural Gas

Russia's gas reserves, considered to be the largest in the world, are estimated at nearly 45 trillion cubic meters. The gas sector in Russia has not yet been developed to its highest potential as a result of technological deficiencies, declining fields, stringent governmental regulation, Gazprom's industry domination, and technical limitations on export capacity (measured by the throughput of Russian pipelines).

Gazprom has a monopoly for the natural gas pipelines and has exclusive rights to export natural gas, as outlined by the Federal Law "On Gas Exports" of 2006. Gazprom also has control over all gas pipelines leading out of Central Asian countries and controls their access to the European market.

Other main natural gas producers in Russia are gas companies Novatek, Itera, Northgas and Rospan,

and vertically integrated oil companies Surgutneftegaz, TNK-BP, Rosneft and LUKOIL.

Gazprom sets special subsidized prices for the power-generating industry under a contract with the now-disbanded Unified Energy System, which was the monopoly supplier of electricity in Russia. The increase in price as part of the state's initiative to deregulate the industry could prove to be unpopular.

Figure 1-37 shows that Russia produces more natural gas (left hand side of the graph) than is used for internal consumption (right hand side of the graph). About 20-30% of the natural gas production (100-200 million meters annually) is available for export to neighboring countries via pipelines.

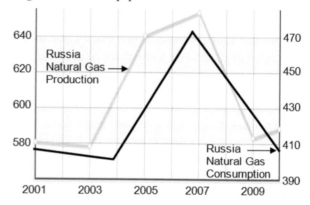

Figure 1-37. Russia natural gas production and use (in million cubic meters)

The main export markets of Russian natural gas are the European Union and the CIS. Russia supplies a quarter of the EU's gas consumption, mainly via transit through Ukraine (Soyuz, Brotherhood) and Belarus (Yamal-Europe pipeline). The main importers are Germany (where links were developed as a result of Germany's Ostpolitik during the 1970s and also Ukraine, Belarus, Italy, Turkey, France and Hungary.

Russia has used Central Asia's gas producers, primarily from Turkmenistan, on occasions where it has found itself unable to meet all its delivery obligations from its own production. Such circumstances led to Gazprom allowing Turkmenistan to use its pipelines to supply gas to the Russian domestic market.

This allows Gazprom to fulfill its obligations towards European customers at all times and maintain its position of major energy exporter to the continental Europe.

Note: The interference in Ukraine's internal affairs and the annexation of Crimea in 2014 put Russia an awkward situation. This will have a long-lasting effect on relations between Russia and the G7 countries.

Coal

Russia also has the second largest coal deposits in the world, estimated to contain 157 billion tons of reserves. Because of the plummeting demand in the 1990s, the Russian coal industry faces numerous problems.

After a brief downturn in 2002, coal production began to recover in 2003, and by 2008 Russian mines were producing 327 million tons of coal annually. The forecast for the years to come envisions that Russia's production in 2020 will be well over 400 million metric tons annually.

Russia's communist-era practices still effect the coal sector. The Kemerovo region, which accounts for more than half of Russia's coal production, for example, cannot sustain, let alone increase, production since it is still recovering from substantial environmental damages caused during that era.

The ecosystems of more than 200 rivers in the region have been heavily, and some irreparably, damaged by mining activities.

At the same time, Russia's adherence to international obligations pursuant to the Kyoto Accords is expected to lower the use of coal by public utilities, so the coal industry is trying to adjust to all these changes.

Electric Power

Generation of electricity in Russia is accomplished by a mix of fuels and technologies. The major fuels are gas, coal, hydropower, and nuclear energy. Today, gas accounts for approximately half of the power generated in Russia.

The power generating sector consists of over 450 thermal power plants and hydro-stations. Around 80 plants use coal as the source of fuel. Russia also has 31 nuclear reactors. A small number of generating facilities in the Eastern part of the country are not linked to Russia's electricity-delivering network. The power-generation facilities have the ability to produce 250 gigawatts. At its height, 2009-2010, the Russian system generated nearly a 1,000 terawatt hours of electric power, but there have been significant fluctuations in the sector during the last decade.

Russia has been using natural gas for electric power generation for awhile now. Natural gas comprises nearly 55% of the national power generation. Oil accounts for nearly 20% of the country's energy consumption, coal is third with less than 15%, while nuclear energy and hydropower share almost equally the remaining 10% of the national power generation mix.

Russia is also a major exporter of electricity, where the surplus of approximately 100 billion kWh is sent to neighboring countries. As of 2009, Russia's energy

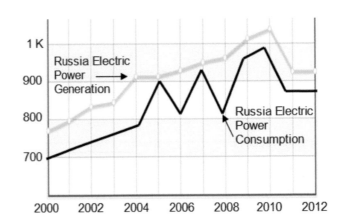

Figure 1-38. Russia's electric power generation and use (in billion kWh)

exports accounted for nearly 40 percent of electricity generated in the Eastern and Central European regions, which is expected to go down to 25-30 percent during this decade.

Nevertheless, Russia will remain a net exporter of electricity to adjacent states in continental Europe and Asia. The level and direction of its energy exports depend on political situations in the region.

Japan

Japan is a very special case from an energy point of view. With very limited natural resources, it has been forced to find ways to power its economy. Not an easy task, considering the scarcity of energy sources and their remote locations.

Japan was one of the largest importers of coal and natural gas in 2010—with 20% and 12% of the total world coal and natural gas imports, respectively.

Until 2011 Japan relied heavily on nuclear power to meet about 25% of its electricity needs. After the Fukushima Daiichi nuclear disaster, however, all nuclear reactors were shut down for safety concerns and the future of nuclear power generation in the country is uncertain. Nevertheless, as of January 2013 most power plants were considered safe and are being restarted. This will alleviate the energy problem, but while the problems at the Fukushima nuclear plants persist and continue to rise, the future of nuclear in the country remains uncertain.

Figure 1-39 shows a huge difference in production vs. consumption. Four to five million bbl of crude oil (left hand side of the graph) are needed every day to power the Japanese economy—transport and power plants—but only a total of 120,000 bbl/day (right hand side of the graph) are produced in the entire country.

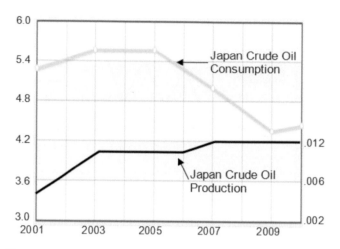

Figure 1-39. Japan crude oil production and consumption (in million bbl/day)

This means that slightly over 2 percent of the total amount of crude oil needed for daily operations is produced domestically. It also means that Japan is a poor country as far as natural energy reserves are concerned, so huge amounts of imports are needed daily.

Where would the energy come from? Imports, of course. Japan has some coal, and yet about 50% of its total energy needs (transport and power generation together) depend on imports. This is one of the world's largest economies. What if the supply is interrupted even temporarily?

Notice in Figure 1-40 the enormous disproportion of natural gas consumption (left hand side of the graph) vs. gas production (right hand side of the graph). Note the difference in the numbers on both sides of the graph. While the consumption is close to 100 bcm, the production is 0.003 bcm. This is over 30,000 times ratio in favor of the consumption.

Japan needs 100 **billion** cubic meters of natural gas

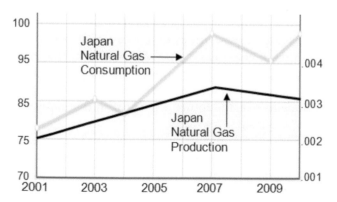

Figure 1-40. Japan's natural gas production and use in billion cubic meters (bcm)

annually for power generation, fueling transportation vehicles, but it produces only 3 **million** cubic meters, or less than 0.003 percent of its needs. Where is the rest coming from? What if the supplies stop?

Japan's energy security presents real challenges. It depend so heavily on imported fuels that if the world supply sneezes, Japan could catch a serious cold. So how do the Japanese manage that threat to their basic survival as a developed nation…and as people in general?

Even amidst the energy deficiency during 2008-2009, Japan generated 1.2 TWh electricity, ranking third in the world in electricity production after the United States and China. Per capita electricity consumption in the recent past was about 8,000 kWh, compared to 14,000 kWh for the average American, or 18th among the countries of the world per capita electricity consumption.

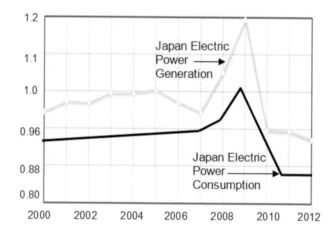

Figure 1-41. Japan Electric power generation (in TWh)

Japan has about 300 GW of total installed electricity generating capacity, also ranked the third largest in the world behind the United States and China. The Fukushima disaster changed the equation so that now Japan has just over 240 GW of power generation capacity.

Japan has 53 active nuclear power generating reactors, and in 2010 ranked third in the world in that respect, after the United States (104 reactors) and France (59). Prior to the Fukushima disaster, almost 25% of its electricity was generated by nuclear plants, as compared to over 75% for France and nearly 20% for the United States.

A large book could be written on ways to ensure the Japanese energy supply and its energy security, which then could be used as a reference and training manual for the rest of the world. As a matter of fact, the book is being written now, and we have a lot to learn from the Japanese—from frugal living to use of alternative fuels and ingenious approaches to power generation and energy use.

Still, the fact remains that Japan's energy security is in clear and present danger of interruptions imposed by foreign governments and factions, and other external factors.

At this point, Japan cannot even think of *energy independence* until some energy source of production is discovered and implemented.

International Developments

The recent crude oil and natural gas production boon in the U.S. has had a remarkable impact on our economy and our future energy outlook. The international impact is actually even greater. An important result from the new U.S. energy revolution is the elevation of America's status on the international energy playing field from a predominant importer to a energy exporter.

And what spectacular news this is! Global energy data issued in 2013 indicate that the U.S. is on track to surpass Russia as the world's largest producer of oil and gas combined. In 2012 U.S. companies produced more natural gas than Russia, as methods for extraction and an abundant supply made natural gas production more affordable than ever.

The numbers are amazing and quite surprising: the U.S. produced the equivalent of about 22 million barrels a day (mbd) of oil, natural gas and related fuels in 2013, according to the EIA and IEA. This, compared to 21.8 mbd for Russia.

Another unexpected effect of increased U.S. energy exports is the potential of helping our European partners become more energy independent. Now our allies can depend on a helping hand from the U.S. in the form of huge coal and natural gas imports.

The new supply (or rather over-supply) is quickly changing the equation for America's energy present and future. Abundant and (still) cheap natural gas is fueling a new generation of manufacturers producing American-made goods and creating new jobs in the process.

The question is no longer if we have enough supply, but rather about demand and the cost of production. Those are the only two drivers remaining, which bring an array of other questions, one of which is "Shall we export, and how much?"

As the Department of Energy weighs additional applications to export natural gas, it should consider the impact those cumulative exports could have on our natural gas advantage—and therefore our entire economy. If the U.S. exports too freely, the price could become volatile, which will put our manufacturing investment at significant risk.

Now that the full benefits of natural gas are in sight, we cannot lose track of the real benefits natural gas promises right here at home.

The Debate

Delegates from 10 countries—including Hungary, Haiti, India, Japan, and Thailand—argued in 2013 for increased U.S. exports of liquefied natural gas (LNG). They see this move as a step toward increasing global energy security and strengthening diplomatic ties.

According to Hungary and the Czech Republic, expanding LNG exports would bring U.S. gas into competition with Russia, the major supplier of oil and gas to most of European countries. Increased competition from U.S. gas will drive down gas prices in Europe to the benefit of all consumers.

Japan and some Asian countries also call for increased U.S. LNG exports, which would help diversify domestic energy markets and potentially lower energy bills. Many Asian countries import most of their energy from the few major world-class producers.

Singapore, for example—another country without any energy resources—gets its natural gas from Malaysia and Indonesia. Expanding the number of countries it buys gas from will enhance its energy security by avoiding over-dependence on a single energy source.

Singapore invested significant amounts of money in a new LNG terminal in 2013, and is making other such investments that are necessary to have the LNG supply to diversify their energy needs.

The other argument in favor of expanded U.S. LNG export activity is that it would improve U.S. diplomatic relations abroad, and open the possibility of strengthening relations between the U.S. and other countries. India, for example, expects that that new LNG imports from the U.S. would create a strong and mutually rewarding energy partnership and further the ties to the benefit of both countries.

There are a number of concerns that the international community is aware of and considering solutions for. One is that the importing nations may have proper and safe infrastructure to receive and distribute the large quantity of gas imports. Operating without proper and safe infrastructure might have adverse effects, such as spills and other accidents, creating unsafe conditions for the locals, and at the same time increasing the global environmental pollution.

This is unacceptable.

Nevertheless, the international community supports the U.S. administration's determination to increase LNG exports, which would be a welcome boon to the U.S. for both economic and diplomatic reasons. The exports will also bring other benefits to all participants. New jobs will be created, in addition to continued strengthening of the energy security of the participants.

The Obama Administration has already approved four LNG export terminals to ship natural gas abroad, with the last three LNG export terminals approved in 2013.

The long-term picture remains unclear, because we are treating the fossils—a finite commodity—as if they have no end. But they do, so we fear the day when the oil and gas wells run dry. It will be a sad, cold, and dark day in the land of plenty and around the world...

Conclusions

Security of the energy supply, as an integral part of the national energy security, is the key objective of most governments as they try to meet the goals of national economic growth and provide prosperity for the population.

For that purpose, following a period of stable and reliable energy supplies after World War II, many countries invested in roads and infrastructure. They counted on the flow of cheap and readily available energy supplies to continue indefinitely, so the roads and the related infrastructure were designed to provide many years of uninterrupted transportation.

The hopes and dreams were, however, shattered several times during the 20th century, starting with the Arab oil embargo in 1973. The 21st century started with the economic crisis of 2008, which led to oil price increases unheard until then. The world stood silent, expecting the worst... But soon enough—as every time before—things went back to almost normal...with a new energy reality shaping our destiny.

The new energy reality today is shaped by a continuing war on terror and political turmoil in many places around the world. With that, the number of threats to the continuation of conventional energy supplies is growing. This is forcing many governments to develop energy security policies to ensure the energy supply chain.

Efforts involved in mitigating the U.S. and the world energy risks presently could be categorized as:

- *Immediate* energy risks mitigation efforts, consisting of an effort to reduce the risks to energy (mostly crude oil) supply interruptions, be it internal, or external, natural, or man-made;

- *Mid-term* energy risks mitigation efforts, which consist of increased levels of fossil extraction, such

as hydrofracking, tar sands development; and

- *Long-term* energy risks mitigation efforts, consisting of development and deployment of renewable technologies, such as solar-, wind-, and geo-power, biofuels, ocean-power, etc. new and promising technologies.

In More Detail

The immediate concerns and efforts of most countries are to ensure the energy supply chain, both internal and external. The internal supply chain is not that easy to secure in some countries with unstable political and economic regimes. Crude oil imports interruptions, via inter- or intra-national conflicts, and/or global oil transport routes blockage are another immediate (and rising) threat to the energy security of most countries.

This threat is taken very seriously, because of the potentially disastrous consequences that could afflict some countries. Imagine Japan's crude oil deliveries discontinued for a period of 6 months or more. Since about 98% of the oil used by Japan is imported, disruption in the oil supply would be disastrous.

If national oil reserves were exhausted (within 60-90 days of an oil imports interruption), the entire transport sector would be paralyzed. This would trigger reduced electric power generation in some locations, since coal trains and delivery trucks would be parked. Gradually, the entire economy would slow down and possibly come to a complete stop if the crude oil supplies are not restarted soon enough.

This is a scary, but entirely possible, scenario, and not just for Japan; there are many other countries which would have to deal with a similar national disaster if left without oil imports for a long period of time.

The mid term effort to stabilize the energy supply is that of patching the energy sector while looking for a more permanent solution. We must agree that while the U.S. and some other countries are drowning in oil and natural gas for now, this is not a permanent solution. Even the largest bucket has a bottom, as the Saudis are learning.

After almost a century of uninterrupted pumping of millions of barrels of oil every day, they are having to drill deeper and deeper, while the quality of oil is deteriorating by the day. They realize that the day is coming when the wells will run dry. The same is true for the thousands of oil and gas wells in the U.S. and worldwide.

So, we in the U.S. are now pumping a lot of oil and gas. In fact, we have so much oil and gas today that we are starting to export a lot of it. What a cruel joke—exporting the lifeblood of our economy...our very well-being. We simply don't believe that the unthinkable is around the corner, that the oil and gas wells will dry up someday.

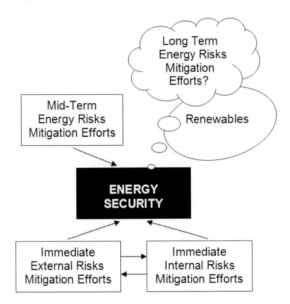

Figure 1-42. Energy security and energy risks mitigation efforts

The long-term risks to our energy security have been discussed and some have been even addressed, but a serious direction is still lacking. Judging by the ever increasing production and use of fossil fuels by the major countries, with the developing countries following the example, the world in general is not ready to face the long-term challenges.

For awhile during the last decade or two, we here in the U.S. were quite concerned about energy issues, and some serious effort was put into looking for alternative forms of energy. Billions of dollars were spent during 2008-2011 by the Obama Administration on R&D and implementation of renewable energy resources—solar, wind, biofuels and others—but that didn't last long and now we are back to drill, baby drill.

The renewable energy sources are the only sure path to a secure energy future. They alone have the potential to reduce the risk of energy supply disruptions and reduce the reliance by many countries on imported fuels. They are also widely distributed in many locations and could provide alternative, cheaper and more reliable (in the long run) choices for generating electricity, producing heat, and producing transport fuels in most countries.

Also, significant greenhouse gas reductions and various other benefits can be obtained by using renew-

able sources of energy, thus reducing the risks of energy supply interruptions, reinforcing the national energy security, and ultimately leading to energy independence—while at the same time cleaning the environment from pollution. Yet, the use of renewable energy is not a priority now. Not since we saw the abundance of oil and gas produced by the infamous hydrofracking process.

The renewables are not risk free either, nor are they the immediate solution to the energy issues now or in the foreseeable future, but we must keep in mind their importance for our future.

The major concern at the present stage of energy mix development is the fact that the key renewable technologies have variable output, which depends on their natural availability—the availability of sun and wind in the case of solar and wind power generation, for example. Because of that, their costs can be relatively high compared with traditional energy supplies.

In recent years however the costs for renewables have been falling, while the costs for fossil fuels have been increasing. This tilts the equation in favor of the renewables, since they have become more cost competitive. Because of that, and some other factors, the world growth in capacity for wind and solar until recently was around 20% per year for 10 years or longer.

During the last several years we witnessed the potential of solar and wind installations of providing significant amounts of energy. The quick expansion of solar and wind power plants, however, was unsustainable, since it was driven mostly by government subsidies. Countries like China, Germany, Spain and others spent huge amounts of money and effort on subsidizing renewable energy companies and projects. This ultimately resulted in creating a failure mechanism, which led to bankruptcies of hundreds of companies and projects.

The recent rise and fall of renewables—the last of the series of such occurrences since the 1970s—proved again that they can be a significant contributor to the energy supply system, but that government subsidies are not the right vehicle. Time after time, governments have failed to follow-through on the full development of the renewable technologies and their introduction into the national energy mix.

On many occasions, as soon as fossil prices exceeded the tolerance level of the U.S. government it jumped into developing new technologies and approaches—including the different renewables. But then, when energy (crude oil usually) prices went down—well below the tolerance level—then the renewables were shelved again. We are now going through a similar stage, where many governments are basically withdrawing their support for renewable technologies.

Also, and very importantly, more work is needed to optimize the key renewable technologies, to make them more efficient and reliable in the long run. Most of them are simply not ready for the global energy market. And those that are ready are still too expensive or difficult to operate.

For governments to obtain greater security of energy supply, and to help meet their climate change policy targets, greater uptake of more energy efficient technologies, demand reduction, and adding more renewable energy systems to the national portfolio are absolutely necessary.

Co-ordination and collaboration among nations and between the public and private sectors is essential if renewable energy technologies are to be successfully developed. This will help meet sooner the goals of sustainable development and climate change mitigation as well as to reduce the risk of continuing disruptive energy supplies.

Notes and References

1. U.S. Fuel Exports http://www.eia.gov/dnav/pet/hist/Leaf-Handler.ashx?n=pet&s=mgfexus1&f=m
2. U.S. Natural Gas Exports http://www.americanprogress.org/issues/green/report/2013/11/05/78610/u-s-liquefied-natural-gas-exports/
3. Contribution of Renewables to Energy Security, IEA
4. Oil Supply Security, IEA
5. Energy Security Initiative, Brookings
6. Energy Technology Perspectives 2012, IEA
7. Key World Energy Statistics 2012," IEA
8. Global Oil Security, Center for Strategic and International Studies.
9. *Photovoltaics for Commercial and Utilities Power Generation*, Anco S, Blazev. The Fairmont Press, 2011
10. *Solar Technologies for the 21st Century*, Anco S Blazev. The Fairmont Press, 2013
11. *Power Generation and the Environment*, Anco S. Blazev. The Fairmont Press, 2014

Chapter 2

The Game Changers

(Oil, Hydrofracking, Coal, and the Renewables in Brief)

*The energy "game changers" today are unprecedented in
type and size, resulting from the hard work of thousands
of dedicated and capable people around the world.*
Anco Blazev

The U.S. has gone through a number of game changing scenarios, with its energy supplies and their use. Some of these were perceived and some were real, depending on the opinions of those who were affected most.

Each time—starting with the Arab oil embargo, which was the biggest energy game changer ever—we have learned a number of lessons and have tried not to repeat the mistakes. Yet, although a lot has changed since those days, we continue with the status-quo—importing a lot of oil, using much more than we produce, and still failing to find the proper energy balance.

For fairness sake, the problem has been complicated by the fact that each time we encounter a new game scenario, new conditions and new players change the game rules, which makes our previous experience somewhat outdated and inappropriate. The developments of late are no exception.

THE U.S. ENERGY GAME

The latest development in the U.S. energy sector—the natural gas and crude oil bonanza—seems to be one of those cases without a precedent, where the old game rules do not apply and new rules must be established and implemented.

As a result of the newly created conditions, we have enough domestic coal and natural gas to ensure our energy tomorrow. This guarantees the electric power supply of the country for the time being and for the immediate future.

Crude oil, however, is still a problem; with over half of our needs being met by imports. As a result, new rules are now being made in those areas to fit the energy game of today.

*Again, as always, government policies and regulations
dictate the rules of the game.*

Crude oil has a monopoly over the transportation sector, thus we still depend on foreign powers and international events for the proper function of a critical sector of the U.S. economy. International oil companies and foreign oil producers have a monopoly over our crude oil supplies, which affects the global transportation sector as well. This way, they hold the U.S. and the other major global economies hostage.

*Unless and until the foreign crude oil monopoly is broken,
we will be far from obtaining complete energy security.*

Natural gas is a big factor in the new energy game. It is becoming a major, if not predominant, energy source in the U.S. and a number of other countries. Its importance is increasing with the discovery of new gas reservoirs and with the quick development of new and more efficient production techniques, such as hydrofracking, all of which contribute to abundant and cheap gas production. All other technologies and approaches (including renewables) are now more or less on the back burner.

Natural gas can also be used to supply fuel for the transportation sector, either as a direct fuel in the form of LNG (liquefied natural gas), or as an energy source in producing other fuels, such as methanol, or electricity for charging electric cars and trucks.

This fact increases further the importance of natural gas, but even then, reliance on one energy source, or approach, is not the best way to ensure long-term energy security.

Diversification as a principle of success in any undertaking is a common sense rule that applies as never before to our energy mix.

Adding other energy sources, like solar, wind, and other renewables would provide a more solid energy foundation, and a faster way to break crude oil's monopoly.

Of course, there are many other factors that play a major role in how the energy game is played. One is the fact that by making us energy secure today (by using excessive amounts of fossil fuels) we are depriving future generations of the benefits the fossils bring.

Fossils are useful and even indispensible for making a number of important commodities, from medications to fertilizers and many other things we need in our daily lives. But we burn over 90% of them for power generation or transportation. This drastically reduces the amount of fossils that future generations will have access to for those and all other purposes.

Here we need to make sure that we see clearly the line that separates the needs from the wants. We need to see, for example, the difference between the need to drive to work at the cost of 500 gallons of gasoline annually, from the want to go on a cruise ship and burn 500 gallons of diesel (per person) in a week, going purposely in circles around the world's oceans.

There are limits to all we do and many (written and unwritten) rules outline the limits and determine our best *modus operandi*. The limits and rules, however, are somewhat unclear and scrambled in the personal lives of the American people, and some are reflected in government policies.

Speaking of government policies, that is where the energy policies start and end in most cases. Regulators, oil companies, and utilities have a significant say in shaping the energy reality. This makes the entire energy situation very complex, and hard to grasp. So, this is a good time to take a closer look at the whole picture...or as much as we can.

Looking at the system diagram in Figure 2-1 we see the relationships between energy characteristics, energy solutions, and energy technologies. It is a product of a study by a team of West Point researchers (see reference 5). It shows that smart energy is one sure way of bringing energy solutions that produce good results.

Smart grids and other alternative energy solutions, represented by emerging technologies—in addition to energy efficiency and conservation—provide solutions to the growing concerns of security and resource depletion.

It also shows the complex effects of government policy and economics that provide solutions in the energy mix. This is a complex interpretation of all factors playing some role in the new energy game. It offers some solutions, which paired with government policies and consumer preferences impact the energy characteristics that create various energy profiles.

It concludes that creating a methodology for political and economic solutions that influences changes to the energy system through partnerships and investing in new energy sources, combined with consumer preference maximization should be the focus of our future drive to energy security and independence.

The developments in the energy sector of late (hydrofracking in particular) have a profound effect on our energy security in determining the path to energy independence. This new situation, however, comes with a number of problems, which makes it important to take a close look at the developments and understand the issues.

Let's start with the national and global discussion on the energy issues at hand.

The Energy Debate

The energy debate today has many facets, but the role of renewable energy sources is overwhelmingly becoming most important. The renewables have a number of advantages and disadvantages, so when it comes to power generation via renewable energy sources, such as solar and wind, or increasing the use of biofuels in the transportation sector, the debating parties are cleanly divided into three groups:

1. Those who are strongly against renewables,

2. Those who are strongly pro renewables, and

3. Those who don't care, or don't want to get involved.

The third group consists of a substantial number of people who don't care about or do not understand the issues, but their presence does not affect developments much, so we will ignore them for now. This third group, however (especially millions of energy-deprived people in developing countries who have no voice) is an important part of the growing energy and environmental crisis because those people are slowly waking up from their energy stupor and ignorance.

Slowly becoming aware of their rights and recognizing the injustice bestowed upon them by the developed countries throughout the 20th century, they will very likely play an important role in the energy and environmental debates and battles of the near future.

As can be seen in Figure 2-2, on the left side we have the defenders of renewable energy sources, which can be classified as follow:

- *The renewables extremists* are people, companies, and groups who believe, and vehemently insist, that renewables are the only reasonable, or at least the most important, solution to our energy and environmental problems. They claim that using any other energy sources is leading us and future generations to complete and irreversible doom. Their one-sided and at times misguided activities do more harm than good to the renewables cause.

- *The renewables scientists* are the brains of the industry and lead its technological development. They are responsible for developing new renewable technologies, and for improving the efficiency, reducing the cost, and optimizing the quality of the existing ones.

- *The renewables activists*, which sometimes include both groups mentioned above—the renewables extremists and scientists—are the muscle and face of the movement. The group is driven by the principle, "Enough of fossils; they are killing us, so it is

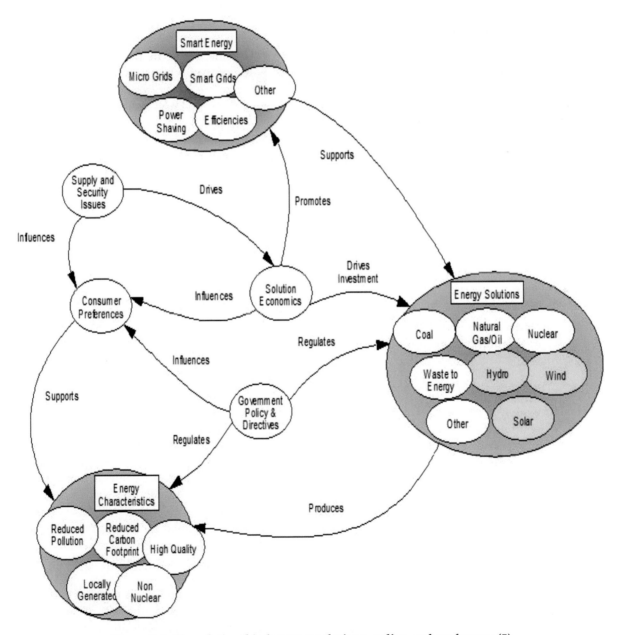

Figure 2-1. The relationship between solutions, policy, and preference (5)

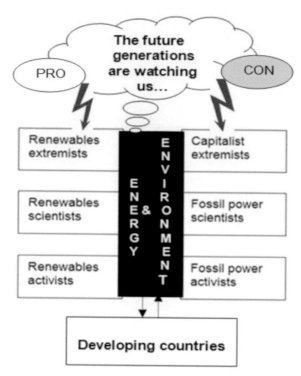

Figure 2-2. The energy and environmental battlefield

time to replace them once and for all with renewables." This group publishes magazines and articles, gets engaged in debates and public demonstrations, and organize protests against fossil and nuclear projects and companies they consider damaging to their cause and the environment.

On the other side of the energy and environment related issues we have interested groups led by the subgroup of extremists and opportunists.

- *The extremists* group usually includes the people who have the most to gain or lose from the matter at hand. Most often it is their representatives that we see on the front lines, since the beneficiaries are wealthy factions who don't like to get their hands dirty—such as oil and gas companies, utilities, and their investors. This sub-group is most damaging to the progress of renewables, and sometimes defends actions that cannot be defended—such as oil spills, excess smoke pollution, and other fossil-created disasters.

- *The fossil power scientists* are the brain of this group. Their findings are very often manipulated to suit the owner's whims and needs. The findings and results presented by scientists employed with oil and

gas companies are often biased, which is understandable. They are protecting their jobs and their bosses' interests. We saw similar scenarios during the asbestos and nicotine debacles of the previous decades, so we know what to look for.

- *The fossil power activists* are the lobbyists and large companies' representatives who are paid to protect the interests of their bosses. This group also sometimes includes people from the above two groups—the capitalist extremists and fossil power scientists. Some of these people have done more harm than good to the renewables industry.

- On the bottom of Figure 2-2 we see the developing countries, home of the world's poorest people. Some of these countries are quickly becoming the most active battleground of the pertinent energy and environmental issues.

The developing countries are an integral part of the discussion, and will soon become even more so. Their populations are growing, as are their energy needs. They are mostly interested in the survival and the wellbeing of their people, so they will do anything and everything to pull out of the misery.

They blame the developed countries for misusing fossils for their own benefit, and feel that they deserve the same, so they will do whatever is necessary to achieve these goals. And who can blame them...at least from that point of view.

The developing countries are simply unwilling (and some are unable) to sacrifice in the face of their blatant (past and present) overuse and misuse of fossils. A political compromise is crucial today, so it is up to the developed countries—and the U.S. especially—to lead the world in the right direction and eventually find one.

As for the battle between the fossil and renewable energy supporters, the renewable energy supporters blame the fossils for polluting the environment and for causing climate change that hurts people. On the other side fossil fuels—coal, oil, and natural gas account for nearly 90 percent of current US fuel use, so the fossil industry could be blamed for many of the damages caused to people, property, and the environment.

The major complaints in this area are:

- *Hidden costs.* Some of the costs of using these fuels are obvious, such as the cost of labor to mine coal or drill for oil, of labor and materials to build energy-generating plants, and of transportation of coal and oil to the plants. These costs are included in our electricity bills or in the purchase price of gasoline.

But some costs are not included in consumer utility or gas bills, nor are they paid for by the companies that produce or sell energy. These include human health problems caused by air pollution from the burning of coal and oil; damage to land from coal mining and to miners from black lung disease; environmental degradation caused by global warming, acid rain, and water pollution; and national security costs, such as protecting foreign sources of oil.

- *Environmental Impacts.* Many of the environmental problems our country faces today result from our fossil fuel dependence. These impacts include global warming, air quality deterioration, oil spills, and acid rain.

- *Global Warming.* Over the last century, burning fossil fuels has resulted in more than a 25 percent increase in the amount of carbon dioxide in our atmosphere. Fossil fuels are also implicated in increased levels of atmospheric methane and nitrous oxide, although they are not the major source of these gases. The global average surface temperature has risen 0.5-1.1 degrees Fahrenheit, and if it continues at that rate, the world will undergo damaging changes.

- *Air Pollution.* Many pollutants are produced by fossil fuel combustion: carbon monoxide, nitrogen oxides, sulfur oxides, and hydrocarbons. In addition, total suspended particulates contribute to air pollution, and nitrogen oxides and hydrocarbons can combine in the atmosphere to form tropospheric ozone (the major constituent of smog) and particulates, including dust, soot, smoke, and other suspended matter, which are respiratory irritants and contribute to acid rain formation.

- *Water and Land Pollution.* Production, transportation, and use of oil and natural gas can cause water and land pollution. Oil spills, or hydrofracking fluids, for example, leave waterways, their shores, and properties uninhabitable. Sulfur in coal is released as water washes through mines and forms a dilute acid, which is then washed into nearby rivers and streams, where it harms wildlife.

Materials other than coal are also brought to the surface in the coal mining process, and these are left as solid wastes. Coal burning creates huge amounts of ash residue which is also left as a waste product.

- *Thermal Pollution.* During the electricity-generation process, burning fossil fuels produces heat energy, some of which is used to generate electricity. Because the process is inefficient, much of the heat is released to the atmosphere or to water that is used as a coolant. Heated air is not a problem, but heated water, once returned to rivers or lakes, can upset the aquatic ecosystem.

- *National Security.* Our dependence on fossils means that we are forced to protect foreign sources of oil and the related transport routes, such as in the Persian Gulf. Wars are also started in the name of protecting oil sources and routes. US troops are to the Gulf often to guard against a possible cutoff of our oil supply. We pay for protecting oil supplies with our taxpayer dollars and our young men's lives.

If our energy policy is not changed, we may be relying on the Middle East for even more of our crude oil supply in the future.

The fossil power supporters often site a number of issues with renewable technologies. Some of the complaints against renewable energy development are:

- Uncertainties of renewable energy costs on the long run.
- Variability, due to dependence on weather patterns (sun and wind availability).
- Uncertainty about efficiency, performance and longevity (not enough track record).
- More work is needed to optimize materials, processes and final product.
- Much more land is needed to produce equivalent amount of energy
- Lack of solid government policies, subsidies, and other help.
- Lack of standardization and poor optimization of the best practices procedures
- Difficulties in scalability of the mass-production processes
- Serious permitting issues related to land, ROW, and connectivity.
- Too diffused—needs a lot of materials and land for unit of power.

- Solar does not work at night, and wind is weak during the day.
- Solar does not work in cloudy locations, and in many northern countries
- Wind doesn't work in many inland areas, and
- Solar and wind farms are usually far from population centers, thus in need of special and very expensive transmission infrastructure additions or modifications.

While at first glance there is some truth to all these claims and accusations, the simple explanation to all these concerns is that the renewables are too new—not fully developed, tested, explored, and not well understood—so they need some time to mature. In the meantime, it seems reasonable to lower the expectations at least temporarily in most cases.

We cannot expect a 2-year-old toddler to become an Olympic level athlete overnight just because we want him to. Given a chance, and proper training, he might become one.

Similarly, solar, wind, and the other renewables must be given a chance—today, tomorrow and for a number of years—to mature and develop into equal competitors to their older and more experienced cousins, the fossils.

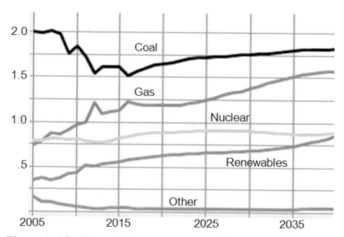

Figure 3. The U.S. future power generation mix (in TkWh)

Figure 2-3 says it all and confirms the upcoming changes we are discussing herein. According to the information in it, gathered from different sources, natural gas's position in the power generation mix is increasing, while coal is kept at a steady level. Nuclear power is also held at a steady level (provided that there are no new nuclear accidents), while the renewables are at a steady rise, projected to overcome nuclear power generation by 2040.

This is the short- to mid-term scenario, but the game will change again as the fossils are exploited very quickly and start coming to an end. The energy game is going to change then in a spectacular way.

For that reason we should always remember, that the fossils are in limited quantity and will eventually be depleted. The question is not if, but when. There is no way around it; every natural resource regardless of type or size will eventually be partially or totally exhausted and erased from the face of the Earth.

So, we must look at the fossils as a temporary solution to our energy needs, while the renewables are the only reasonable solution to our long-term survival—and especially that of future generations.

The battle between the renewables (solar, wind, and bio-fuels) and the conventional fuels (fossils, and nuclear) is more than 50 years old, with the fossils winning convincingly. Since 2008, however, the renewables have been putting up a real and very serious fight. The battle lines are clearly drawn, the stakes are great, but the outcome is unclear due to the complexity of the different technologies and the ever changing political and economic conditions in the U.S. and worldwide.

We all know that the fossils have passed their prime, and no matter how much we like them and want them to stick around, their end is coming. At the present rate of exploitation, there will be no more crude oil within the next 40-50 years, no more natural gas soon thereafter, and coal will be gone too by the end of the century, or soon thereafter. And then what?

At that time nuclear and hydropower maxed up with renewables and some other unconventional technologies (methane hydrates most likely) would be the only energy sources with some potential to grow and replace the fossils. Thus, they do deserve our respect and support.

Crude Oil—The Main Energy Security Risk

Energy security is defined as a secure way to produce, transport, and use energy, and (in the U.S. case) crude oil in particular, since it is the weakest link in our energy security portfolio.

For the U.S.:

True energy security can be achieved by decreasing the reliance on crude oil imports in the short term, and increasing the reliance on renewable energy in the long run.

This is the only way to ensure our comforts today

and our long-term survival as a civilized and progressive nation. Common sense also tells us that energy security is about safe and reliable access to not just one, but a diverse number of energy resources and procedures. It is very important that we have a choice how and when to produce and use these resources.

According to IEA, energy security is "the uninterrupted availability of energy sources at an affordable price." And that energy security is enhanced by having a diverse, efficient and flexible energy mix.

Simple definition that makes a lot of sense, right? It, however, contains a number of hidden details that are not that obvious and which change with time too. This makes it hard to figure out exactly what is going on, even at the highest government levels, which then makes it impossible to have complete control over the developments in the energy sector.

In idealistic terms, complete energy security could be viewed as a plentiful supply of all fuels, so that all we need to do is switch from one to the other as needed. We turn the solar power in the morning, add wind power later on when the sun goes down, and switch to natural gas or nuclear power when wind dies down. In other words, having a choice of energy sources means total independence in producing and using fuels for power generation.

In reality, however, this is not that easy, and only a few countries are close to it—and the U.S. is one of them. But then, we need to add the transportation sector, which uses crude oil, and which (crude oil) has no replacement. This makes the energy situation in most countries—including the U.S.—very difficult from an energy security point of view.

Energy security is essential for the U.S. since it is a critical component to crafting policies that ensure we can continue to enjoy basic services, such as driving cars, heat and cool air in our houses and businesses, access to abundant and clean water, efficient power grid function, as well as reliable medical supplies and services.

It also means that in times of need or disaster, police, hospitals, fire stations and other critical infrastructure will continue to operate and help us get back to normal by providing us with their life-saving services. All of these require fuels—crude oil mostly. Police cars, ambulances, fire trucks, and all other vehicles won't run without plentiful crude oil supplies. It is imaginable that they all could run on natural gas or biofuels some day, but the infrastructure is not here yet, so we depend heavily on crude oil.

As far as our energy security and independence are concerned, we must look at them from two separate points of view: a) electric power generation, and b) fuels for the transportation sector.

It appears that presently our electric power generation is fairly secure with the new natural gas bonanza. The supply of crude oil is the main threat to our energy security at present, since we import 50% of our daily needs.

Coal and natural gas are also not directly affected by international events or any external activities, while crude oil is affected by even the slightest political developments and depends heavily on overseas developments.

- The production of crude oil depends on the whim of foreign governments and the socio-political situation of the major producers.

- The transportation of crude oil depends on a number of factors, such as natural events, terrorist attacks, etc. which threaten the maritime delivery channels.

Because of the above facts, many experts equate the level of our energy security to our dependence on crude oil imports and the total quantity of crude oil we import.

The U.S. Energy Security Council declares that oil's status as a strategic commodity undermines U.S. national security and weakens the U.S. economy. Reducing oil's strategic importance requires breaking its virtual monopoly over transportation fuel.

We all agree; enough is enough. Oil must go, or be replaced. But how do we do that?

What Figure 2-4 tells us is that our energy security is affected by internal and external events and risks—both of which we must deal with daily. Both require different measures and approaches.

On the internal side, coal mine accidents, damage to properties and human life by mining and hydrofracking sites, workers' strikes, political currents and support, and other variables determine the status of our energy supplies.

The internal events and risks, however, are easier to handle and most are preventable to a certain extent. Thus far, they have not presented a great threat to our energy security.

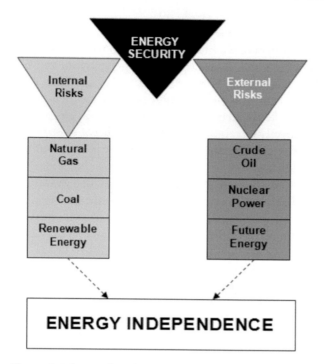

Figure 2-4. Internal and external energy security factors

Because of that, we now have enough coal and natural gas, and are even exporting significant quantities of these fuels.

On the external side, however, crude oil and nuclear materials are imported from many areas of the world—some of which are located in unstable countries. Restrictions on these imports, for whatever reason, would affect our energy security accordingly.

Nuclear power depends on imported raw materials, although presently that dependence is limited and cannot be construed as a significant risk unless something unexpected and very serious happens. Some external factors (terrorism, social and political unrest, etc.) in key producing countries might play a more significant role in the future, as the supplies dwindle, but for now the danger is small.

Crude oil is a critical commodity, which we import in huge quantities...every day. The problem is that most of the crude oil in the U.S. (both domestic production and imports) is ultimately used in transportation.

This means that threats to our energy security due to oil supply interruptions could be expressed mostly as unscheduled and unpredictable disruptions in the U.S. transportation sector.

Any serious crude oil shortage (like that of the Arab oil embargo of 1973) would paralyze our trans-

portation sector, which in turn would paralyze the entire economy. The duration of such unthinkable event would determine the short and long damage to our economy. A long duration might have extremely serious long-term economic effect.

A very long shortage—or permanent disruption—of the crude oil imports could bring the 'ol mighty U.S. to its knees....for the duration and beyond.

The future energy supplies, such as methane hydrates etc., are also dependent on external events and risks, since many of them are located overseas. The challenges of providing future energy, therefore, will be similar to what we are going through with crude oil supplies and transport today. For now these worries should be considered, but are still on the back burner.

So let's look at the present energy reality.

Crude Oil...Again!

The main and immediate problem—and the major threat to our energy security today—is that of stable crude oil supplies. Our goal is to reduce the importance of oil to our energy security. But how?

Let's look to history for some the answers.

During the early part of the 19th century, salt was a major strategic commodity worldwide. It was fundamental to national economies in its role of food preservative and for other uses. Salt deposits meant wealth and national power, so wars were fought over them. Countries that controlled salt aimed to keep uninterrupted production and prices as high as possible, to benefit as much as possible from the goods, as long as they lasted.

That frenzy, however, did not last long. Competing technologies of preserving food, such as canning and refrigeration ended salt's monopoly and put an end to its strategic value. The transformation was so quick and so dramatic that today a ton of salt is worth as much as a kilogram in those days.

Crude oil today is at the same strategic place salt occupied in the 19th century. Because the U.S. consumes over 25% of the world's oil, while producing only about 5% of it, we are most affected by its availability and price fluctuations.

We are forced to import over half of our oil from different countries, which is about half of our trade deficit, and which increases our energy security risks.

The problem, in addition to wasting hundreds of

billions of dollars every year, is that the majority of the world's oil is produced in counties that are undemocratic and/or hostile to the U.S. This presents serious challenges to U.S. energy and national security. There are also serious concerns about the negative impact of China and India's growing demand for energy on the global energy price structure.

These, and many other countries' policies are increasingly driven by the need to secure their energy supply, often at the expense of vital U.S. interests. This impacts our economy, and we see how oil crises are usually followed by economic downturns.

Oil imports amount to more than $1 billion daily, or well over $350 billion annually. This is money that could be much better spent domestically by creating new jobs and new investment opportunities.

Just as salt had a huge strategic importance in the 19th century, which was derived from its monopoly over food preservation, oil derives its importance from its virtual monopoly over transportation fuels. No crude oil, no transportation; leading to economic stagnation and worse.

Oil's Strategic Status

We have been trying unsuccessfully to get rid of oil dependency for a long time. This is mostly because we have been focusing on reducing the level of oil imports, while the real problems are:

a) oil's global status as a critical strategic commodity,
b) its overuse and abuse by consumers.

We have allowed the oil monopoly to grow to a world status and now crude oil has a virtual monopoly over transportation fuel, which we all depend on to move around and to grow our economy.

Oil is no longer used for generating electricity in nearly 99% of the U.S. energy sector. Crude oil use in transportation is now the main source of oil's global importance, a significant part of which is our (U.S. government, businesses, and citizenry) fault.

The special place oil takes in the economy means that no amount of coal or natural gas production, nor electricity generation from solar, wind, nuclear, and other power sources, will ever be able to displace oil in any perceptible manner in its reign over the transportation sector.

Presently, U.S. policies and efforts are geared towards increase in crude oil production, while increasing the efficiency of its use at the same time. These approaches are good, but they are tactical rather than strategic.

So, in this case, reducing oil demand through fuel economy—absent competitive markets in transportation fuels and transportation modes—serves to reduce the trade deficit (some) and the global emissions (even less).

All measures thus far have been totally incapable of changing the most important problem—the strategic status of oil as indispensible and irreplaceable transportation fuel. Changing that status is the first step in changing the oil's dominance of the global energy markets.

Today, when at times oil-consuming countries increase their domestic production or reduce net demand, OPEC (which controls 80% of global reserves) responds swiftly by throttling down crude oil supply to drive prices back up. And they do this at will—regardless of the damage their actions could cause.

Game Changer

Crude oil has been since the 1970s, and still is, the most serious threat to our energy security. The huge quantity of crude oil imports from unstable and unfriendly countries, and the well established status of crude oil as an indispensable and irreplaceable commodity, are major obstacles in breaking its dominance over the transportation sector.

The only way to reduce the strategic importance of oil is for the energy market to have viable choices that enable consumers to respond quickly to changes in oil availability and prices. These responses, however, should not be at the cost of constricting economic activities, but by substituting oil with other fuels, means of transport, and/or competing goods and services.

Introducing new transportation fuels such as LNG and biofuels would allow fuel competition and give drivers choices. This is already happening on a large scale in countries like Brazil, where drivers are given a choice to fill up and drive with gasoline, gasohol, or pure alcohol.

Actually, Brazil's biofuels story is not new. Gasohol was introduced there over 40 years ago. It has been a major vehicle fuel ever since, and its use is increasing in size and importance.

The U.S. response to this solution has been weak. It has been held at adding 10-15% or so of alcohol to gasoline during certain seasons of the year. Obviously, this is not a solution of scale, because we still import over 50% of the oil we use—even with the new crude oil bonanza and the biofuels addition.

A competitive transportation fuels market (with the introduction of alternative fuels) would place a ceiling on the price of oil once market penetration of vehicles that enable fuel competition is sufficiently high. If and when oil's prices surpass the threshold price at which competing fuels are economic (on a cost per mile comparison), then consumers whose vehicles are capable of switching fuels (as in Brazil) will choose to fuel with cheaper substitutes. That will result in more choices, more flexibility, and more control over the final motor vehicle fuel prices.

Automakers can produce flex fuel vehicles (FFV) at a very low cost—as demonstrated by the automotive industry in Brazil. The FFV vehicles are capable of running on any combination of gasoline and a variety of alcohols such as methanol and ethanol, made from a variety of raw materials.

Natural gas is another vehicles fuel option. Recent MIT studies show that as the economics of natural gas in the U.S. remain favorable due to progress in shale gas extraction, personal and commercial FFVs would be run most economically if natural gas is used to fuel them, or if it is used to produce methanol.

FFVs could provide a significant contribution to the effort to solve the oil dependency problem easily by giving drivers a choice of abundant and cheap fuels. This, in effect, would introduce competition at the pump by variety of fuels—with crude oil derivatives being part of the mix.

This will also let the market determine the winning fuels based on their availability and economics. Let the price in terms of mpg and dollars per mile be the driver of the competition among the fuels. This concept has been proven in Brazil, where in 2008, when oil prices were at record highs, more ethanol was used across the country than gasoline. This kept gas pump prices fairly leveled for the duration.

None of this, can happen, however, without government intervention and direction. Nor can it happen without the manufacturers' full acceptance of the rules of the game changers.

Economic theory clearly shows that market forces alone are incapable of breaking cartels and monopolies. It is the role of government and the business community to fight anti-competitive behavior, dismantle monopolies and cartels, and unleash (and support) free market forces to take their course.

It will require committed government leadership and active participation of the business community (including the large automakers) for the U.S. to break oil's virtual monopoly over transportation fuel. The oil companies have a huge role in this effort, but their participation is far from ensured, so it might take a long time and effort to bring them on board.

Placing oil into competition with other energy commodities will not only drive down its price, but it will alter the geopolitical balance of power in favor of net oil importers and countries with resources to become non-petroleum fuel producers. This scenario will also take time to develop and mature, but it is what the future will require as the fossil supplies dwindle.

For the U.S., crude oil affects mostly the transportation sector, since very little of it is used for electric power generation. Only some states (Hawaii, Washington DC, and Alaska) use crude oil and its derivatives for electric power generation. If push comes to shove, crude oil use in these states could be easily replaced by natural gas.

So, the battle of the electric power generation energy sources here is mainly between coal and natural gas, and we know who is going to win in the short run. The long term is less obvious, but in any case we see no threat to our energy security coming from the domestic power generation sector.

The battle for transportation fuels dominance, however, is just now starting, and its outcome will determine the level of our energy security, and if and when we can achieve energy independence. The new discoveries and technologies allow huge quantities of oil to be pumped out every day. As a matter of fact, the U.S. presently produces about 6.5 million barrels of crude oil daily from domestic sources—about 50% of the oil we use.

Note: Pumping out 6.5 million barrels, or nearly 275 million gallons, of crude oil every day is a lot of oil. It is about 2.4 billion barrels, or nearly 100 billion gallons, of crude oil every year. In 10 years this amount grows to 24 billion barrels, or 1 trillion gallons of crude oil. In 20 years...

*The estimates of how much oil remains, and how much more is available in the U.S., vary, but one thing is for sure: whatever that amount is, the oil reserves will **not** last forever.*

We have already noticed a decline in production at some of the major oil producing sites here and abroad. As demand increases, we will see more decline in domestic oil production. It is likely that more wells will be drilled then and more oil will be pumped up...and more and more...until the reservoirs dry out completely.

Crude oil, and all natural resources, are finite

commodities. Logic tells us that everything that has a beginning has an end. This is the absolute scientific truth about the fossils too. Their beginning was millions of years back, at which point a finite amount was created and stored in the Earth's bosom.

It took us only one century to dig and pump out most of the fossils, so now the end is only a few decades away.

What happens to our energy security then? Could we even survive a post-fossils reality without enough efficient, cheap, and safe alternative energy sources?

The Hydraulic Fracturing Process

Hydraulic fracturing (hydrofracking, or fracking) is a new technology using a special process that is mostly responsible for the recent unexpected and significant boom in natural gas production. Hydrofracking is used widely today since it makes it feasible to produce oil and natural gas from certain underground rock formations that are *not* productive under any other conditions.

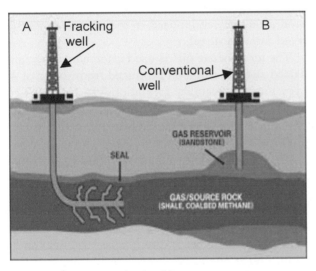

Figure 2-5. The fracking process setup

To produce oil or gas, usually a well is drilled to an underground rock formation that: a) contains oil or gas pools, or in some cases, b) the oil or gas can flow through the rock itself by moving from one pore or crack to the next, through interconnections between the pores. Rock through which such fluids flow easily has high permeability, and in both of these cases, some of the conventional methods of oil and gas extractions are used.

If there is no pool and if the rock is solid, with the interconnections between pores and cracks too narrow and too few, then the rock will have low permeability.

In that case, the oil or gas will not flow out easily, or at all. It is not economical, and in some cases impossible, to produce oil or gas from such formations with conventional methods.

Here is where the new hydrofracking methods comes into play. Production from these low permeability formations has become economical, thanks to fracking technology, which creates cracks or fractures in the rock so that oil or gas can flow through.

Fracturing wells started in the US in the 1860s, not very long after the first oil well was drilled in Pennsylvania in 1859. The earliest methods of fracturing were primitive, using explosive fracturing. Here an explosive charge—called a "torpedo"—was dropped into a well and detonated. The explosion would fracture the surrounding rock, and usually provided a higher rate of production.

In the late 1940s, hydraulic fracturing was developed, where water was pumped into a rock formation at extremely high pressure. At these high pressures, rocks fracture, and water moves into the fractures, forcing them to open even more. The water displaces the gas, which then escapes to the surface.

To prevent the fractures from closing when the high-pressure water is withdrawn, companies use "proppants" to prop them open. Proppants are small particles—typically sand—though small ceramic particles or sintered bauxite are sometimes used. The proppants are mixed into the fracturing water before it is pumped into the formation that will be fractured.

As the water moves into fractures, it carries proppants along with it. When the water is removed at the end of the fracturing process, the proppants stay behind, preventing the fractures from closing. For oil or gas to travel through the fractures, it must travel through the proppant particles, but this is not a problem because the proppants have high permeability.

By creating new pathways, hydraulic fracturing can exponentially increase oil and gas flow to the well. For example, a single fracture job can increase the pathways available for fluid migration in a formation by as much as 270 times in a vertical well, and much more in a horizontal well.

Usually about 99.5% of the fracturing fluid consists of water and proppants, but operators also add various other substances, including biocides (to inhibit microbial growth) and corrosion inhibitors to protect the well's piping. They also add other chemicals to assist the hydraulic fracturing process in other ways.

The use of fracking processes—the largest game changer—has been expanding lately for reasons most-

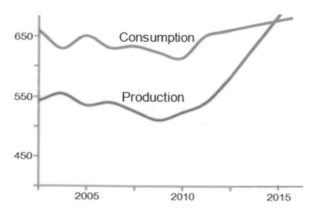

Figure 2-6. U.S. natural gas consumption and production (in billion m³)

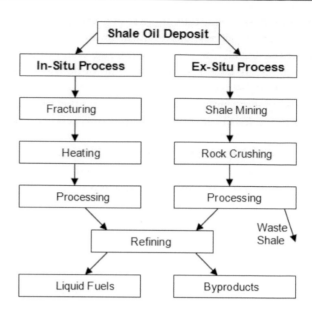

Figure 2-7. Shale oil production process

ly of economic and environmental natures. There are estimates of more than a million fracking wells being drilled and exploited in the U.S., since the process was commercially developed in the late 1940s.

In addition to being used in low permeability rock formations, hydraulic fracturing is also used in coal beds to facilitate the production of coaled methane, further expanding its availability at reduced production costs.

The new gas bonanza is reflected in the increased use of natural gas for power generation and as transportation vehicles fuel. It has also contributed to the increase of natural gas exports, which are projected to grow significantly in the near future.

As with any product or technology used on a mass scale, there are problems which deserve our attention. We will take a closer look at the fracking technology and its advantages and disadvantages later on.

Shale Oil

Shale oil, also known as kerogen oil, or oil-shale oil (not to be confused with tar sands oil) is a type of unconventional oil produced by extraction from the oil shale reservoirs and treatment by a number of chemical processes.

Oil shale, in which the shale oil is contained, is also known as kerogen shale. It is a fine-grained sedimentary rock, rich in organic matter (kerogen which is a solid mixture of organic chemical compounds). Upon extraction of this matter from the rock, liquid hydrocarbons or shale oil can be produced.

Pyrolysis, hydrogenation, or thermal dissolution processes are used to convert the organic matter within the rock into synthetic oil and gas. The resulting oil can be used immediately as a fuel or upgraded to meet re-

finery feedstock specifications by adding hydrogen and removing impurities such as sulfur and nitrogen. The refined products can be used in the same way as those derived from crude oil.

The term "shale oil" is used interchangeably, since it is used also for crude oil produced from shales of other very low permeability formations. To avoid the confusion of shale oil produced from oil shale with crude oil in oil-bearing shales, the IEA suggests using the term "light tight oil" for the latter.

The shale oil deposits are different from conventional oil-bearing shales, which contain crude oil that occurs naturally in the oil shale deposits.

Deposits of oil shale occur around the world, including major deposits in the United States. Examples of oil-bearing oil shales in the U.S. are the Bakken Formation, Pierre Shale, Niobrara Formation, and Eagle Ford Formation. The western states of Idaho, Wyoming, Utah, and Colorado seem to be the richest shale oil-rich basins in the U.S.

The potentially recoverable oil from western U.S. oil shale deposits is estimated at more than 800 billion barrels. This is nearly three times the total proven oil reserves of Saudi Arabia. Saudi Arabia times three…

Just think what this amount of oil could do to ensure our energy security. It would certainly be enough to put us on the path to complete energy independence.

Many countries have well-established oil shale industries. Estimates of the total global deposits are about

3 trillion barrels of recoverable shale oil. The amount of recoverable oil from these areas, however, is much less than that in the U.S.

Shale oil production is part of the overall oil-bonanza (or frenzy) going on in the U.S. during the last decade. Presently there are thousands of oil wells and shale oil operations around the country, delivering a total of over 6.5 million barrels of crude oil per day.

Currently, more than 70% of western U.S. oil shale deposits are on federal lands. Unfortunately, most federal oil shale locations are closed to development despite the growing need for new sources of oil supply.

Despite the inefficient permitting and regulatory processes, which are stifling the development of shale-oil industry in the U.S., shale gas production has been increased enormously lately. It is part of the new energy revolution in the country.

Tar Sands

Tar sands, or oil sands (not to be confused with shale oil) are petroleum deposits with a technical name of bituminous sands. These usually come in the form of loose sand, or sandstone that is saturated with a dense and extremely viscous form of petroleum derivatives (bitumen, or tar).

Bitumen is a thick, sticky form of hydrocarbon, so heavy and viscous (thick) that it will not flow unless heated or diluted with lighter hydrocarbons. At room temperature, it is much like cold molasses.

The World Energy Council (WEC) defines natural bitumen as "oil having a viscosity greater than 10,000 centipoises under reservoir conditions and an API gravity of less than 10° API.

The Orinoco Belt in Venezuela is sometimes described as oil sands, but these deposits are non-bituminous, falling instead into the category of heavy or extra-heavy oil due to their lower viscosity.

Natural bitumen and extra-heavy oil differ in the degree by which they have been degraded from the original conventional oils by bacteria. According to the WEC, extra-heavy oil has a gravity of less than 10° API and a reservoir viscosity of no more than 10,000 centipoises.

The bitumen in tar sands cannot be pumped from the ground in its natural state, since it is very dense and combined with the soil. Instead, the tar sand deposits are mined via the conventional mining methods, usually using strip mining or open pit techniques. In some cases, just like in the shale oil case, the oil is extracted by un-derground heating followed by additional processing. This, however, is a very expensive way to produce fuels, which can be used in certain locations and under certain conditions. Because of that, the cost of energy (electricity and heat) used during the extraction process often exceeds the value of the produced oil.

Ex- and in-situ processes used depend on the type and location of the deposits. Presently the oil sands in Alberta, Canada, are the largest such deposits with practical application to the U.S. oil industry.

There is a lot of oil potentially available in these sands, and the methods for its extraction are being perfected, but the transport from the origin in Canada to the refineries in the U.S. is problematic. It has stirred a huge controversy, known as the Keystone XL pipeline.

Case Study: Keystone XL Pipe System

The Keystone XL System is a large undertaking—a huge pipeline network intended to transport crude oil from the oil sands of Alberta, Canada, and from the northern United States, to the refineries in the Gulf Coast of Texas.

These pipelines will carry synthetic crude oil (syncrude) as well as diluted bitumen (dilbit) from the Western Canadian Sedimentary Basin in Alberta, Canada. They will also carry synthetic crude oil and light crude oil produced from the Williston Basin (Bakken) region in Montana and North Dakota.

The project consists of several phases, two of which were completed and are in operation, while the third phase—a long-distance, large diameter, pipeline from Oklahoma to the Texas gulf coast—is still under construction.

The fourth and last phase of the project is awaiting U.S. government approval.

When completed, the Keystone Pipeline System would consist of the completed 2,151-mile (3,462 km) Keystone Pipeline (Phases I and II) and the proposed 1,661-mile (2,673 km) Keystone Gulf Coast Expansion Project (Phases III and IV).

The fourth phase of the project is most controversial, and is creating a lot of discussion among politicians and professionals alike. It is proposed to start at the oil distribution hub in Hardisty, Alberta, and extend 1,179 miles (1,897 km), to Steele City, Nebraska. Environmentalists and some of the states which the pipeline is supposed to cross are putting up a fight.

The major controversy about the Keystone XL pipeline is focused on its impact on the local environment (soil and water contamination mostly), as well as on the way it affects the entire North American energy security efforts.

But how does the Keystone XL Pipeline fit into the North American energy security picture?

• *Safe and reliable access to energy* defines energy security, so the Keystone XL will provide safe and reliable access to North American crude oil. Pipelines are the safest and most efficient method of transporting crude oil over long distances, so the Keystone XL promises to be the safest and most advanced pipeline ever constructed in North America.

• The pipeline will be built with stronger, more flexible steel, featuring anti-corrosion and cathodic protection and other advanced safety features and operational controls. The pipes will be buried deeper beneath river beds and road crossings, thus ensuring safety of the environment, wildlife and people.

• Over 14,000 data and control points will feed real-time data to a control center, monitoring the condition of the pipeline and controlling the crude oil flow in it. It is the most advanced pipeline in the world. No other line has this degree of safety and sophistication.

• Keystone XL will bring us diverse and abundant energy resources from the largest strategic oil reserves in North America—the Canadian oil sands. Since Canada is our largest, most stable, friendly, and reliable trading partner, there will be no risks associated with the production or transport of the crude at any time for the duration.

The Keystone Pipeline legs presently in operation have already successfully shipped more than 600 million barrels of oil to the U.S. since 2010. Since the U.S. Gulf Coast refineries refine over 50 percent of all the oil imported to the U.S., it makes sense to provide a direct supply line from the Canadian and U.S. Bakken crudes.

These supply sources and routes are well insulated from any risks related to any disruptions and/or supply restrictions, so the Keystone project and a number of related pipelines have the support from major shippers. This has been instrumental in assuring that they will succeed, both in regulatory approval, in construction, and operation.

A big IF for these projects is the fact that there is active opposition from a number of local and environmental groups. So, while all sounds peachy from a distance, people who live nearby the pipelines are not happy with having such a serious potential problem in their

Figure 2-8. Oil transfer pipeline depot

backyards. Many of these people are actively involved in fighting against the pipeline's approval and stopping its construction.

There are a number of local issues to deal with, as well as national and international environmental groups, over the increasing array of issues.

Regardless of what the future brings, and how the Keystone XL pipeline and other such projects fair, crude oil will continue to flow across North America. If not to the U.S., oil will flow into China and other Asian countries. The only questions are how much, in which direction, and what changes, if any, will the existing oil flow to the U.S. incur.

Oil pipelines in the United States are already undergoing a historic realignment and upgrades in response to new production in the Eagle Ford development in south-central Texas, the Bakken flats in North Dakota, redevelopment of older production in the Permian Basin and new flows of oil from the Midwest and Canada that have oversupplied Midwest markets.

There are several dozen pipeline projects in the United States, in addition to the controversial Keystone XL pipeline and the related additions and extensions. Let's take a look at these, since it is important to understand the magnitude and the dynamics of the development in this area.

These projects are significant in size and expense, and play a significant role in ensuring our energy security and bringing us close to achieving energy independence.

Below are descriptions of some of the most important U.S. pipeline projects.

New Pipeline Projects

There are a number of completed, or near completion, pipeline projects in the U.S. as follow:

- *Allegheny Access Pipeline.* Connecting Midwest to eastern Ohio and western Pennsylvania with initial capacity of 85,000 bpd (barrels per day), expandable to 110,000 bpd.

- *Bakken Access Program.* Western North Dakota pipeline expansion by adding 26 miles of 16-inch pipeline between Enbridge stations in Beaver Lodge near Tioga, North Dakota. Also 29 miles of new 16-inch pipeline between Stanley and Berthold terminal; and expansion of Berthold with rail loading capability to handle three transport trains at a time. Added is 145,500 bpd pipeline capacity, plus additional 80,000 bpd of rail export capacity for a total of 120,000 bpd at a cost of $560 million for the pipeline and $145 million for the rail expansions.

- *BridgeTex Pipeline.* A 450-mile pipeline from Colorado City in the Permian Basin to Houston-area refineries, with access to Texas City and the Houston Ship Channel with 300,000 bpd capacity at a cost of $1.0 billion. Includes construction of 2.6 million barrels of crude storage.

- *Cline Shale Pipeline System.* From Irion, Sterling, Tom Green and Mitchell Counties in West Texas to Centurion's existing Colorado City, Texas, station. This is 100 miles of new pipeline and several origination stations, able to receive crude via truck or pipeline with total capacity of 75,000 bpd.

- *ConocoPhillips'* central delivery facility in Karnes County.

- *Eagle Ford Shale pipeline.* From Frio, LaSalle and McMullen counties NuStar's 600,000-barrel storage terminal at Oakville in Live Oak County, via 110-mile, 12-inch and 8-inch pipeline; Eagle Ford crude to NuStar's 1.6 million-barrel Corpus Christi North Beach terminal via an existing 16-inch pipeline. The cost of $325 million for acquisition of TexStar Midstream Services LP's 140 miles of Eagle Ford crude pipeline and gathering lines, and 643,400 barrels of storage assets. Also $65 million for integration and completion of the gathering and terminal assets.

- *Double Eagle Pipeline.* Connects to 50-mile, 14- and 16-inch existing pipeline to enable delivery of Eagle Ford condensate from Three Rivers, Texas to Magellan's marine and storage terminal in Corpus Christi. 140 miles of new 12-inch pipeline is connecting to the existing line at a capacity of 100,000 bpd initially, expandable to 150,000 bpd at a final

cost of $150 million.

- *Eaglebine Express.* Reversal of underused refined products pipeline from the Eaglebine and Woodbine shale plays in Hearne, Texas, to Sunoco Logistics' 22 million-barrel storage hub in Nederland, Texas.

- *Eagle Ford* crude-condensate pipeline and condensate processing facility. From Eagle Ford shale formation to Galena Park, Texas, on the Houston Ship Channel. Pipeline $225 million; processing facility $360 million; and $107 million pipeline expansion. 300,000 bpd of both Eagle Ford shale crude and condensate in 65 miles of newly built pipe and 113 miles of a converted natural gas pipeline. Will also process 100,000 bpd of Eagle Ford condensate and provide 1.9 million barrels of storage capacity. 31-mile pipeline expansion will connect Kinder Morgan's station in DeWitt County, Texas, to ConocoPhillips' central delivery facility in Karnes County, Texas.

- *Eagle Ford Pipeline.* 140-mile pipeline carrying crude oil and condensate from Eagle Ford production in Gardendale, Texas, to refineries in Three Rivers and Corpus Christi. A new 35-mile segment from Three Rivers to Enterprise Products Partners' Lyssy station in Wilson County and a 350,000 bpd take-away from western Eagle Ford to Three Rivers/Corpus Christi, plus a marine terminal facility at Corpus Christi and 1.8 million barrels of operational storage capacity across the system.

- *Enterprise Crude Houston Oil* (ECHO). Terminal and Houston area pipelines in the Houston Ship Channel area. Expansion of storage capacity to more than 6 million barrels, with access to Enterprise's marine terminal at Morgan's Point on the Houston Ship Channel. To be linked by pipeline to Eagle Ford shale to the west. 55 miles of 24- and 36-inch pipeline to connect the terminal with major refineries in the southeast Texas market with an aggregate capacity of about 3.6 million barrels per day, including plants in Baytown, Beaumont, Port Arthur and Texas City.

- *Galena Park to Houston Gulf Coast crude distribution.* Pipeline and terminal system at Galena Park, Texas, to deliver crude from Magellan's pipeline system Houston and Texas City refineries at a cost of $50 million.

- *Gardendale Gathering System expansion.* Consists of

several crude oil gathering pipelines, a total of 90 miles in length, from Dimmitt and La Salle counties to Plains' Gardendale Terminal in South Texas. Connecting at Gardendale to long-haul pipelines that deliver crude to refineries in Three Rivers, Corpus Christi and the Houston area. Also a new 40-mile Gulf Coast crude oil pipeline originating from Plains' Ten Mile terminal in Mobile, Alabama, with 115,000 bpd of incremental gathering capacity and a condensate facility of 80,000 bpd at a cost of $190 million for the latter three projects.

- *Gulf Coast Pipeline Project* (southern leg of Keystone XL). A 485-mile, 36-inch pipeline from Cushing, Oklahoma, to Nederland, Texas, and a 48-mile lateral pipeline to Houston with initial capacity of 700,000 bpd, expandable to 830,000 bpd at a cost of $2.3 billion.

- *Houma-to-Houston pipeline reversal.* From Houma, Louisiana, to Houston, Texas, allows deliveries of crude from connecting pipelines and terminals in Houston to Nederland and Port Arthur. Phase II extends the reversal to move crude from Texas to Louisiana. Phase III expansion brings additional capacity for a total of 250,000 bpd capacity at a cost of $100 million.

- *Longhorn Pipeline reversal.* To reverse flow of Crane-to-Houston segment of Longhorn Pipeline, which carries refined products from Houston to El Paso, Texas, and convert the line to 135,000 bpd, expanding to 225,000 bpd at a cost of $375 million.

- *Mississippian Lime pipeline.* 135-mile pipeline from Alva, Oklahoma, to Plains' storage facility in Cushing, Oklahoma, with capacity of 175,000 bpd. And a 55-mile extension moving crude oil to Alfalfa County from Comanche County, Kansas, with capacity of 75,000.

- *Pecos River Pipeline.* From Pecos, Texas, to Crane, Texas, where a 16-inch, 70-mile line will connect to Magellan Midstream Partners' reversed Longhorn pipeline. Intended to move Permian Basin crude oil to the Gulf Coast. A new 95-mile extension moves crude from southern New Mexico to Pecos with a total capacity of 150,000 bpd.

- *Permian Basin Expansion Projects.* Consist of several links extending and expanding crude oil lines in Permian Basin, West Texas, total of additional 145 miles and with a new capacity of 150,000 bpd in capacity in Texas, plus projects in southeast New

Mexico at a cost of $250 million.

- *Permian Express, Phase I.* From Wichita Falls, Texas, to Nederland, Texas, with initial capacity of 90,000 bpd. Phase II is 200,000 bpd pipeline from Colorado City, Texas, to Nederland and farther east to St. James, Louisiana, and the Louisiana refineries.

- *Pony Express Pipeline.* A 430-mile converted natural gas to crude oil pipeline and 260-mile new pipeline to carry North Dakota Bakken crude from Guernsey, Wyoming, to Cushing, Oklahoma, with capacity of 320,000 bpd.

- *Seaway pipeline reversal* from Cushing, Oklahoma, to Houston, Texas. $300 million for initial reversal, US$2 billion for final expansion with new parallel loop pipeline. 150,000 bpd initially, with first expansion to 400,000 bpd and further expansion to 850,000 bpd by 2015.

- *Sweeny Lateral.* Consists of 27-mile, 12-inch lateral pipeline moving oil from Kinder Morgan's crude and condensate to Phillips 66's 247,000 bpd refinery in Sweeny, Texas. The capacity is 100,000 bpd. Truck offloading capability in Dewitt County, Texas, and four storage tanks in Wharton County with 480,000 barrels of capacity are included.

- *Silver Eagle Pipeline.* 200 miles renovation of the existing 10- and 12-inch crude oil pipelines between Longview, Texas, and Houston, Texas.

- *South Texas Crude Oil pipeline expansion.* 100,000-barrel terminal near Pawnee in Karnes County, Texas, connected to the existing 12-inch pipeline system between Pettus and Three Rivers. Will also connect existing 12-inch pipeline to Oakville terminal for crude delivery to the NuStar North Beach terminal. Also included are new truck-receiving facilities at the Pawnee and Oakville terminals and a new ship dock in Corpus Christi. The total cost of this project is $120 million.

- *Toledo Pipeline* (Line 79) expansion. From Stockbridge, Michigan, to Toledo, Ohio, with capacity boosted to 180,000 bpd total at a cost of $198 million

- *Western Oklahoma Extension.* 95-mile extension of Plains' Oklahoma pipeline system from Orion, Oklahoma, to Reydon, Oklahoma, to provide access to the Granite Wash and Cleveland sands oil plays in western Oklahoma and the Texas Panhandle with capacity of 75,000 bpd.

- *West Texas crude system*; three different projects to bring Permian basin crude to Gulf Coast market. West Texas to Houston line of 40,000 bpd capacity is expandable to 44,000 bpd, intended to carry West Texas Sour and West Texas Intermediate at Midland. West Texas to Longview Access of 30,000 bpd, to carry Permian crudes to the Mid-Valley pipeline to the Midwest, and West Texas to Nederland Access of 40,000 bpd carrying capacity.

- *White Cliffs Pipeline*. A 527-mile, 12-inch crude oil pipeline from Platteville, Colorado, to Cushing, Oklahoma, with 150,000 bpd capacity.

Planned Pipeline Projects

A number of pipeline additions and renovations are planned for initiation during or after 2015. Some of these are:

- *Cactus pipeline*. A 20-inch crude pipeline from McCarney Texas to Gardendale, Texas, with a capacity of 200,000 bpd, expandable as demand warrants at a cost of $350 million.

  Westward Ho. From St. James, Louisiana, to Houston, Texas, with initial capacity of 300,000 bpd, expandable to 900,000 bpd depending on market conditions.

  Southern Access Extension. From Flanagan, Illinois, to Patoka, Illinois, with a 300,000 bpd capacity at a cost of $800-million.

- *Eastern Access/Line 6B Expansion*. From Griffith, Indiana, to Sarnia, Ontario, with 230,000 bpd to 490,000 bpd projected capacity at a cost of about $400-600 million.

  Southern Trails Pipeline conversion. A 485-mile line from San Juan Basin in New Mexico to Southern California near Essex, California. Additional 96 miles west leg section from Whitewater, California, to oil terminal in Long Beach. The projected capacity is 120,000 bpd.

- *Eastern Gulf Crude Access Pipeline* (formerly Trunkline Conversion). 700 miles pipeline intended to convert and reverse the 30-inch natural gas pipeline in order to carry Bakken and Canadian crude from Patoka, Illinois, to St. James, Louisiana, to the refineries along the Louisiana Gulf Coast. This includes new lateral near Boyce, Louisiana, to St. James and 574 miles of converted natural gas pipeline, about 40 miles of new 30-inch pipeline from the Patoka hub to the northern end of the converted trunkline, and 160 miles of new 30-inch pipeline to the St. James hub. Other lateral connections for deliveries to refineries might be included too. The total carrying capacity of this pipeline complex is 420,000 bpd at a cost of $1.5 billion.

- *Keystone XL, northern leg*. 1,200 miles, 36-inch pipeline from Hardisty, Alberta, to Steele City, Nebraska, with capacity 830,000 bpd at a cost of $5.3 billion.

- *Sandpiper Pipeline*. From the Bakken shale region to Superior, Wisconsin, with a capacity of 375,000 bpd and at a cost of $2.5 billion.

- *U.S. Mainline/Line 62 Expansion*. From Flanagan, Illinois, to Griffith, Indiana, with 235,000 bpd capacity at a cost of $500 million.

- *U.S. Mainline/Line 61 Expansion*. From Superior, Wisconsin, to Flanagan, Illinois, with a projected capacity of 550,000 bpd at a cost of $1.3 billion.

A number of pipeline projects to the tune of nearly $5 billion were cancelled in 2013, due to changing market conditions and lack of interest on the part of users and shippers. Some of these might be reconsidered if the situation changes and the market warrants the effort and expense.

Shale Gas

Shale gas is actually natural gas found within shale formations, and has been found in great quantities in the U.S. lately, and the interest in this commodity has grown quickly here and abroad. In 2000, shale gas was less than 1% of U.S. natural gas production, but by 2010 it was over 20% and has been steadily increasing since.

The predictions of the experts are that by 2035, 45-50% of the U.S. natural gas supply will come from shale gas.

Shale gas is also contained in the oil shales, but is much easier to produce than shale oil, so the extracted quantities are increasing very quickly worldwide.

Increased shale gas production in the US and Canada could help to eventually prevent Russia and OPEC countries from dictating increasingly higher prices for gas exports to Europe and Asia.

The Marcellus Shale gas formation is one of the largest shale regions in the U.S., and is estimated to

Project	Capacity (Barrels per day)	Destination	Year In-Service
Enbridge Line 9 Reversal	300,000	Eastern Canada	[2014]
Enbridge Clipper Expansion	800,000	Mid-West United States	[2015]
Enbridge Flanagan South	585,000	Gulf Coast United States	[2014]
Enbridge Southern Access Expansion	1,200,000	Mid-West United States	[2014]
Enbridge Southern Access Extension	300,000	Mid-West/Gulf Coast United States	[2015]
Enbridge Northern Gateway	525,000	West Coast Canada	[2018+]
Kinder Morgan TransMountain Expansion	890,000	West Coast Canada	[2017]
TransCanada Energy East	500,000 to 1,000,000	East Coast Canada	[2017/2018]
TransCanada Gulf Coast Project	700,000	Gulf Coast United States	[2013]
Enbridge Seaway Expansion	850,000	Gulf Coast United States	[2014+]

Table 2-1. Projects with binding contractual support from shippers.

be the second largest natural gas find in the world. It stretches across the states of New York, Pennsylvania, West Virginia, Ohio and Maryland. Its estimated total area is around 95,000 square miles, where the shale deposits range in depth from 4,000 to 8,000 feet.

The Marcellus shale formation is estimated to contain more than 410 trillion cubic feet of natural gas and could supply U.S. consumers' energy needs for hundreds of years. Technological advances such as horizontal drilling paired with hydraulic fracturing have allowed energy companies to unlock the energy resources from the Marcellus Shale, while trying to protect the environment.

China has one of the largest shale gas deposits in the world, which are now mostly idle, but are expected to be explored intensively. There are estimates that increased shale gas use (instead of coal) will help to reduce greenhouse gas emissions. It is believed that for that reason the U.S. GHG emissions dropped to a 20-year low in 2012, so it is likely China to follow our example.

The critics allege that the increased extraction and use of shale gas may result in the release of more greenhouse gases than conventional natural gas, simply because more of it is being used.

Another problem of the new energy revolution is that the supplies are unpredictable, and cannot be relied upon as a single solution to all our energy problems.

A recent study already points to high rates of decline of some major shale gas wells. This could be interpreted as an indication that the global shale gas content may actually be much lower than is currently projected.

Coal vs. Natural Gas

Natural gas is taking the energy sector by storm, and coal is slowly losing out. Natural gas is slowly but surely replacing coal in electric power generation, with a much smaller portion used for vehicle fuel, although this application is growing steadily.

We must note here that the battle between the different energy sources has never been so intense, and that they are going through some important changes right now, so it too early to identify the winners and losers.

The major battle fields are in the power generation and transportation sectors.

Power Generation

The biggest and most important battle for domination of the energy sector today is that between coal and natural gas. Coal has been the preferred fuel for power generation for almost a century, and its dominance has been unchallenged until recently.

With the discovery of new fossil reservoirs in North America and the wide implementation of hydrofracking methods, however, natural gas has been challenging its older cousin.

The most important development in this sector was the recent and much stricter EPA regulations.

On September 20, 2013, the U.S. Environmental Protection Agency announced its first steps under President Obama's Climate Action Plan to reduce carbon pollution from all power plants. EPA proposed carbon pollution standards for new power plants built in the future, and such for existing power plants.

Table 2-2. GHG emissions in the U.S.

Sector	%GHG
Power generation	33
Transportation	28
Industry	20
Commercial & residential	11
Agriculture	8

EPA-issued data show that power generation is the major emitter of greenhouse gases (GHG). The coal-fired power plants have been blamed for a major part of these GHG emissions, which prompted EPA's action.

According to the new, much lower, EPA emission standards, new coal-fired plants cannot be built in the U.S. unless equipped with some sort of carbon-capture or neutralization equipment. Appropriate technology has not been fully developed yet, and whatever is available is extremely expensive, so with the stroke of a pen, the EPA put an end to the coal dominance of the power generating sector.

After a century's undisputed reign over the power sector, King Coal was removed from the throne and King Gas was installed as the new ruler of the U.S. energy sector.

The EPA's new rules put coal in such a huge disadvantage that in addition to not being able to build new coal-fired plants, carbon credits and related penalties are making using coal so much more expensive, that many existing coal-fired power plants cannot maintain a profitable operation. Because of that some had to be shut down, and some were converted to natural gas. And the trend continues.

This has created a new and very different situation in the U.S. and world's energy sectors. Natural gas is now the king of the power generation. It is credited with being abundant, clean, safe, and cheap.

As we will see later on in this text, natural gas is also a polluter, but of different kind, so it remains to be seen how much cleaner and safer it is in the long run.

Still, many U.S. states and countries are plunging head first into massive natural gas production and use as if there is no tomorrow. The immediate, short term, benefits are clear and so natural gas is winning, but we suggest waiting for some more data before announcing the final results.

Figures 2-9a and 2-9b show that the increase in use of one commodity causes (or results in) decrease in use of the other, and vice versa. This is most likely due to the fact that states with coal-fired power plants prefer to keep using coal. Equally so, states with predominant natural gas production use more natural gas for electric power generation.

In both cases, the electricity prices per kWh use seem to suggest that using more coal is somewhat cheaper for the consumers in the predominantly coal-using states. There are, however, other factors to be considered here, such as the increasing use of other energy sources like wind, solar, hydro and nuclear power.

For example, only 2% of the electricity in the state of Connecticut is generated by coal and 45% by natural gas, but the remaining—almost half—of the electricity is generated by nuclear. Yet, the state has the highest electricity rates in the country, surpassed only by Hawaii, which generate electricity mostly by burning crude oil products.

The conclusion here is that power generation by crude oil as fuel has the highest cost, followed by nuclear power.

The battle in the power generation sector today is between coal and natural gas, with the other power sources filling the gaps.

Natural gas is shaping as the preferred fuel for power generation, with increased production levels and falling prices dictating the rules of the new game. Another major benefit of using natural gas is its (albeit questionable) advantage as a less polluting fuel.

Natural gas is a major contributor to the new energy reality—the key energy game changer—in the U.S., so below we take just a brief look at the newest energy miracle—the hydrofracking process—which is credited with producing large quantities of natural gas in the U.S. and other countries.

The abundance of natural gas, its low prices, and lower pollution are bringing a huge change in the way we generate electricity in the U.S. This is also changing our energy security status—at least the part of it that deals with electric power generation—providing us with an almost unlimited (for now) energy source,

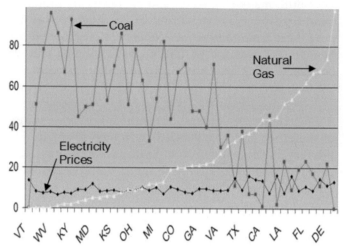

Figure 2-9a. Coal and natural gas use (vs. electricity prices) in the major producing states (sorted by percent increase of natural gas use in each state)

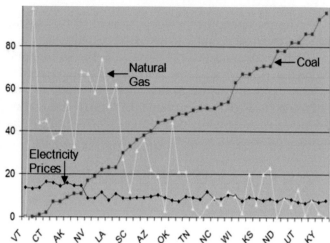

Figure 2-9b. Coal and natural gas use (vs. electricity prices) in the major producing states (sorted by percent increase of coal use in each state)

which is safer and cleaner than King Coal, who reigned for almost 100 years.

How long the natural gas supplies will last and the lasting consequences of their use are unknowns. We can only present the facts as they stand today.

Clean Coal

Now here is an oxymoron—clean and coal in the same sentence. How could coal ever be clean? This is like saying *clean dirt*, and much worse—dirt that emits poisonous smoke. Even babies know that **clean and coal** just don't go together, and that just touching coal gets you dirty and in trouble.

For the sake of the thousands of people working in the coal industry, we must admit that while we are so quick to turn a cold shoulder to "anything coal," we still depend on their efforts to bring coal to our homes in the form of electricity and heat.

We also must realize, and come to terms with the fact that no matter what we do, we will never get rid of coal.

Coal's importance and use in the U.S. is decreasing, but it is still the foundation of our electric power generation. It is still the major baseload provider, and although its use may be reduced even further, it will still remain as the most reliable and cheaper energy source in the U.S.—and even much more so in developing countries.

If the coal industry could only "clean" the coal it uses, it might take back its lost advantage. But try as they may, scientists have been only partially successful,

since it is technologically very hard to clean coal, in addition to being extremely expensive.

Amidst all the talk, the efforts, and the money spent, the jury is still out on the subject. Different experts would tell you different things—depending on which side of the debate they sit—but they all agree that the experiments must continue.

This reminds us of a joke coming from a virtual radio station, Radio Erevan. When asked if men can get pregnant, the radio host responded, "No, but the experiments continue." Similarly so, coal is not clean and it is very hard to make it somewhat cleaner, yet the experiments continue.

We must to wonder which would happen first: coal becoming clean, or men getting pregnant. In our humble opinion, the odds are in favor of male pregnancy methods being implemented on a mass scale first.

This is because coal is by nature one of the dirtiest things Nature has ever made. It is dirty to look at, dirty to touch, dirty to mine, dirty to transport, dirty to process, dirty to burn, and dirty to dispose of its waste products. During each step of the cradle-to-grave process, from coal mining to its final disposal as ash, it creates large quantities of dust, dirt, sludge, grime, and lots and lots of dirty, stinky, and toxic smoke. All this causes immense air, land, and water pollution in its proximity all through its lifespan.

But the largest negative contribution of coal to the environmental problems of late—including the deadly and finite climate warming—are the huge air emissions of GHG and toxic gasses belched non-stop 24/7 by thousands of coal-fired power plants around the world.

There is equipment at some coal-fired power plants designed to reduce air pollution, which does just that—reduces it somewhat. By no means, however, can any technology neutralize the smokestack gasses completely without breaking the bank first.

And there is also a lot of solid, liquid, and air pollution produced during the mining, transport, storage, refining, coal-burning and disposal, which won't go away even if the smoke stacks are clean as a whistle.

There are huge quantities of GHG pollutants already in the air to last several generations. Even if we stop pollution right now, we will suffer their consequences for decades. Since the GHG emissions will only increase, the future generations will live in a toxic gas chamber.

We are sure that Nature did not intend to make so much coal so we could pollute the atmosphere thoroughly and poison ourselves. As a matter of fact, we believe that coal was meant to be left untouched for many more hundreds of centuries to complete its final mission by being converted into diamonds. Or at least to be used for more noble purposes, like assistant in other industries and in producing different consumer products.

Coal is making ill and even killing thousands of people around the world directly or indirectly. We are afraid that before all is said and done, many more will suffer from coal's production and use and its intended and unintended consequences.

Still, we know quite well that coal is here to stay for a long while, and that its use around the globe will not decrease. It will actually keep increasing ad infinitum, because millions of people need it badly for electricity and heat production. This trend is expected to continue until there is no more coal to mine and burn.

So the only value-added question today, since we cannot reduce, let alone, stop using coal, is, "How do we reduce the damages created by coal?" Coal cannot be made cleaner during mining, processing, transport, and storage, so there is not much chance, or even efforts in those areas. Because of that, the negative consequences in these areas of the coal lifetime cycle will continue until the very end.

Although this problem is kept well hidden from the general public, its effects are horrendous. Just ask the people living near the coal mines. They will tell you horror stories of sludge disposal lagoons full of bubbling liquids spilling over pristine meadows and streams.

Removing entire mountain tops is another huge effect coal mining has on the environment. Although strict regulations require restoration of the environment upon completion of the mining process, the original environment cannot be restored completely just because you put the dirt back in its original place.

Instead, efforts today are focused solely on cleaning coal during and after burning so that its emissions do not exceed the established standards. The other pollutants left behind by coal—liquid, solid, and gaseous wastes—will remain unchanged and in ever increasing quantities for now.

Since we are using so much coal—half of our electric power is generated by coal—it is responsible for the majority of the total GHG emissions. So it behooves us to find a way to reduce its gaseous emissions, because they are most damaging...but how?

In technical terms, *clean coal* today means the use of some special technology that reduces emissions of carbon dioxide (CO_2) and other greenhouse gasses (GHG) that arise from the burning of coal for electrical power and other uses. It refers to technologies that can be used to reduce emissions of ash, sulfur, CO_2, and heavy metals from coal combustion at coal-fired power plants.

Lately, the term *clean coal* in practice refers mostly to carbon capture and storage (CCS), which consists of redirecting and storing CO_2 emissions underground instead of releasing them in the atmosphere. On the surface this seems a more doable and cheaper alternative to "scrubbing" the exhausts.

There is a huge amount of such emissions, however, so this is not an easy or cheap task. Especially so on the long run, because with such huge quantities to sequester and store, we will eventually run out of space to store them.

Still, the experiments continue...

Today there are coal treatment plants using an integrated gasification combined cycle (IGCC) during which coal is first gasified (turned into clean burning syngas) and then burned. During the IGCC process, CO_2 and most other polluting components are removed prior to combustion of the gas for electric power generation.

So coal could be made "clean," or rather made to burn much cleaner, but the cost is prohibitive, and as a result commercial scale IGCC plants are few and far between, and not many commercial such are planned. Especially not in the U.S., where coal is a truly dirty word today, and getting even dirtier (as compared to natural gas and the renewables) by the day.

So coal is being pushed out of the way as an electric power generator and replaced by natural gas. No new coal-fired power plants can be built in the U.S., due to very tight EPA regulations.

As a matter of fact, the "clean coal" technologies

in the U.S. are being developed in response to regulations in EPA's Clean Air Act, and in response to climate change legislations. The situation is similar in Europe and other countries. At the same time, most developing countries are just not interested in doing much, for financial and many other reasons.

Developing countries lack funds and the will to do anything, so clean coal is far down on their list of priorities. By the time they reach to it, we will witness billions of tons of GHG toxins belched into the atmosphere, without a chance to reduce, let alone stop them—regardless of what we do or intend to do here in the U.S. and the West in general. This is a done deal, which nothing and nobody can stop.

China is the exception in this case. Although there are more coal-fired power plants in China than in any other country, and more are being built every day, there are at the same time significant government sponsored efforts nationwide to reduce excess emissions coming from the coal-fired plants.

The air in some of China's major cities is becoming thick with pollutants, endangering the health of the population. Since the abnormality is blamed mostly on coal-fired power plant emissions, the government is obligated to do something about it.

For now China leads the world in the effort to clean coal emissions. The Tianjin IGCC power plant is an example of what can be done to clean coal. This, however, is not an easy, or cheap undertaking, nor is it a done deal since many technical issues remain to be resolved.

Case Study: Tianjin IGCC Power Plant

The Tianjin Integrated Gasification Combined Cycle Power Plant is a project in the city of Tianjin, China, approved in the beginning of 2010 by the China Central Government. It involves constructing and operating a 250 MW coal-fired power plant with added integrated gasification combined cycle (IGCC) module for removal of GHG during burning of coal.

This is the only IGCC clean coal project in the PRC, and the first IGCC power plant in developing Asia, using the most efficient and least-polluting technology currently available commercially.

The plant is expected to generate about 1,500 GWh of electricity annually, to be sold to the Northern China Grid Company Ltd. The power is to be transmitted via 220 kW transmission line to a substation in a nearby industrial park for distribution to customers in the area.

In more detail, IGCC technology uses a gasifier to turn coal and other carbon based fuels into gas—the so-called synthesis gas, or syngas. The syngas is then sent

Figure 2-10. Tianjin IGCC Power Plant

for purification to remove most of the impurities in it before it is burned for electric power generation.

Some of the pollutants, such as sulfur, can be turned into re-usable byproducts. The end result is power generation with much lower emissions of sulfur dioxide, particulates, and mercury. Additional process equipment is needed for the carbon in the syngas to be processed into hydrogen via the water-gas shift reaction. The resulting fuel is nearly carbon free, since the carbon dioxide from the shift reaction is separated, stored, or compressed for industrial use.

Excess heat from the primary combustion and syngas fired generation can be used for additional heat in a steam cycle. This is similar to what happens in the common combined cycle gas turbine and results in improved efficiency compared to conventional pulverized coal turbines.

Another such project in China is the Huaneng GreenGen IGCC Project, the first phase of which consists of a 250 MW integrated gasification combined cycle (IGCC) power plant and a pilot unit, designed to capture 100,000 tons per annum of carbon dioxide (CO_2) for use in the food and beverage industry.

The project is part of the Tianjin Lingang Industrial Zone Circular Economy Plan's GreenGen which plans to research, develop and demonstrate a coal-based energy system, with hydrogen production, through the use of coal gasification and electricity generation for widespread uptake of low emissions technology in China.

The full-scale CCS plant is planned to be built in 3 stages, including a 400 MW demonstration IGCC power plant that will capture up to 2 million tons of CO_2 per annum, which in turn can be used in enhanced oil recovery. The power output for the amount of CO_2 being captured is estimated to be in the range of 101-250 MWe. The captured materials will be transported to storage facilities 150-200km away. Secondary storage is planned at an Onshore Depleted Oil and Gas Reservoir.

Approvals were granted by the National Development and Reform Commission (NRDC) and the Pilot IGCC facility was completed and put in operation in 2012. The second phase is expected to be completed by 2016.

The overall goal of the project is to achieve generation efficiency of 55 to 60 percent by 2020 with more than 80 percent of the CO_2 being separated and reused. Expected full CCS operations date is 2020. The expectations are to capture and reuse about 50 million tons of CO_2 over the lifetime of the project.

Around the world: Poland's Kędzierzyn Zero-Emission Power & Chemical Plant is showing a promise to operate even more efficiently by combining coal gasification technology with Carbon Capture & Storage (CCS). The installation has been planned and work has started, but no information is available since 2009.

Some of the other IGCC plants operating or planned around the world are the Alexander plant in the Netherlands, Puertollano power plant in Spain, and JGC power plant in Japan.

IGCC in the U.S.

There are several IGCC plants in the U.S. too; the oldest ones are the Wabash River Power Station in West Terre Haute, Indiana, Polk Power Station in Tampa, Florida, and Piñon Pine in Reno, Nevada. These were constructed with the help of the U.S. DOE Clean Coal Demonstration Project.

The conclusions from the work at the Reno project showed that the then-current IGCC technology would not work well at locations more than 300 feet above sea level. The Wabash River and Polk Power stations are currently operating, following resolution of demonstration start-up problems.

Unfortunately, the Piñon Pine IGCC project encountered significant problems, which could not be fixed with the existing budget and the project was abandoned.

The first generation of IGCC plants emitted less pollution than regular coal-based technology, but created extensive water contamination. The Wabash River Plant, for example, could not comply with its water permit until 2001, because it emitted arsenic, selenium and cyanide.

Today the IGCC demo projects in the U.S. are touted as capture ready, with a proven potential to capture and store carbon dioxide.

General Electric's IGCC plant design is an example of greater reliability for the budding technology. It features advanced turbines optimized for production of coal syngas. Eastman's industrial gasification plant in Kingsport, TN, uses the GE Energy solid-fed gasifier. It was built in 1983 without any state or federal subsidies and has been profitable for awhile.

There are several disadvantages of the IGCC technology when compared to conventional post combustion carbon capture. The main problem for any IGCC plant is its high capital cost. It costs over $3,500 for each kilowatt installed of capacity. So a 500 MW power plant would cost over $1.75 billion, or almost 3 times as much as a normal cycle coal-fired power plant (estimated at $1,250 per kilowatt installed capacity). Optimistic estimates in the neighborhood of $1,500/kW installed capacity are being discussed, but these cost estimates have been shown to be incorrect thus far.

Estimates for the cost for each megawatt-hour (MWh) cost of IGCC plants vs. the cost of a pulverized coal plant is $56 vs. $52/MWh. IGCC might become even less expensive if the costs of carbon capture and sequestration are included. The IGCC cost is then estimated at $79/MWh vs. $95/MWh for normal pulverized coal-fired power plant.

However, recent results show that in the best of cases, the cost of IGCC is still about twice the predicted costs. The addition of carbon capture and sequestration at a 90% rate is expected to have additional $30/MWh cost.

Then there are the technical issues, some of which have been very serious. The Wabash River plant, for example, has been shut down repeatedly while fixing gasifier problems. The problems persist, but recently the plant has been running reliably, with availability comparable to that of other coal technologies.

The plant was turned over to commercial operation in 1999, and shows emissions that are far better than the targets. The Wabash River complex generates almost three times more power than the original unit, and yet the total emissions are a fraction of the pre-powering values, and particulate emissions are negligible, due to the IGCC system operation.

The Polk County IGCC had to overcome a number of design problems, from corrosion in the slurry pipeline, to thermocoupler problems that forced its replacement, to unplanned down time due to refractory liner problems. All these problems took a lot of time and money to correct.

The success of GE's IGCC design at Eastman's power plant in Kingsport, TN, led to a decision to expand the plant to over 1 billion pounds per year, thus meeting all of Eastman's needs for power generation while meeting the standards for reduced emissions.

If this is such a successful and profitable enterprise, why don't more companies adapt the model? We can no longer build normal cycle coal-fired power plants in the U.S. (due to DOE restrictions of emissions), so why don't we build them with IGCC technology? That would solve the excess emissions problem and we can continue burning coal, which we have in abundance.

This seems like a reasonable way to proceed in light of the huge coal deposits in the U.S., so what is stopping us? It is logical to assume that under today's reality (excess of natural gas) the high initial price of IGCC technology and maintenance expenses are issues that need to be resolved first.

More importantly, it seems that the technological complexities and the related financial risks are difficult to accept, when compared to the ease of building and operating a natural gas-fired power plant.

THE RENEWABLES

We cannot talk about energy security and energy independence without a close look at the new kid on the block—the renewable energy technologies. Although they—solar, wind, ocean, geo, and bio-energy technologies—are not that new, since they have been around for a long, long time, they have just recently been put on the list of promising energy sources that can help ensure our energy security.

Starting with the 1973 Arab oil embargo, the U.S. government made these a priority as part of our national energy security. Some exceptionally useful work was done in these areas, sponsored and financed by the government. Soon after the embargo was lifted, however, the priority to develop renewables to replace the fossils, was also lowered to non-emergency status. With that, the government funds dried up, the focus was shifted to coal and nuclear power and the work in the renewables sector just stopped.

This situation was repeated a number of times since the 1970s, with different nuances. During the last boom-and-bust cycle of 2007-2012, the renewables were again spread on the national energy security table, dissected, redesigned, refinanced, redeveloped, and redeployed as fast as possible. Some impressive results followed, accompanied by some not so impressive and even embarrassing consequences.

Still, the most important fact remains: solar and wind power generation in the U.S. have increased to unprecedented and unexpectedly high levels. Some amazing developments have also been documented in the bio-energy sector, where biofuels are taking a large place in the national fuel mix.

All renewables are expected to grow in the future, but the growth levels are unpredictable, since the urgency is once again gone. We now also have so much natural gas, and it is so cheap, that it makes no economic sense to spend money on the less efficient and much more expensive renewable energy technologies. Or does it?

Note: We will take a much closer look at the renewables in the following chapters, so here we will only outline the major points of their specifics and use in the national energy mix, as part of our energy security package.

Solar Energy

Solar energy is one of the oldest energy sources known to man. Since the beginning of time, humans have recognized, and even worshipped, the amazing source of light and heat coming from above. They have appreciated its magical powers, and we now know that without sunlight we wouldn't exist.

Background

Today, and for purposes of this text, we view solar power in terms of capturing and converting sunlight into heat or electricity that can be used by humans. Electricity from sunlight can be generated directly using photovoltaic (PV) cells and panels, or indirectly using concentrated solar power (CSP) systems.

PV devices—solar cells and modules—convert sunlight falling onto them directly into electric power using the photoelectric effect. Thus produced DC electricity is then converted into AC to be used immediately at location, or is sent into the grid. We will take a closer look at the PV technologies in the following chapters.

CSP systems use lenses or mirrors mounted on trackers to focus a large area of sunlight onto a target. The sunlight heats the liquid in the target, which is then converted into steam and used to turn the blades of a turbine. The turbine shaft is attached to a generator, which generates electric power to be sent into the grid.

PV systems are more flexible, since they generate DC power directly, and can be built in different sizes and for different applications, since they can be installed on roofs or in the fields. Solar cells and panels are used to power small and medium-sized applications, from calculators powered by a single very small solar cell, to off-grid homes powered by a few photovoltaic panels, and to huge megawatt solar power plants. PV technology is an important and relatively inexpensive source of electrical energy, frequently used where grid power is inconvenient, unreasonably expensive, or unavailable.

Solar power is increasingly being used in large-scale grid-connected power plants as a way to replace or supplement the conventional power generating technologies by feeding low-carbon energy into the grid.

A number of smaller PV installations were built in the U.S. in the past, but the real progress of this technology started in the mid-2000s. Today the 290 MWp Agua Caliente Solar Project in the United States, and the 221 MWp Charanka Solar Park in India, are some of the world's largest PV power plants. Even larger PV power plants are coming online soon, such as the 550 MW Topaz Solar Farm in California.

Commercial CSP power plants, like the 354 MWp SEGS CSP installation in California's Mojave Desert, were developed in the 1980s. It is still one of the largest solar power undertakings in the world. Other large CSP plants include the 280 MW Solana CSP power plant in Arizona, 150 MWp Solnova Solar power plant and the 150 MWp Andasol solar power plant, both in Spain.

U.S. Solar

The growing significance of the U.S. solar market, and especially the fast growth of large-scale photovoltaic (PV) installations, pushed the U.S. into third place in the world PV energy market. The PV projects pipeline in the US grew over 43.0 GW during 2013, consisting of nearly 2,500 projects of 50 kW and larger. This is double-digit annual demand growth, which is surpassed only by China and Japan.

NOTE: The U.S. PV projects pipeline is a list of PV projects (installations) that are underway, or planned for execution in the U.S. It does not include PV projects that have been already completed and are under full operation.

The breakdown of the 43 GW pipeline of U.S. PV installations in 2013 in shown in Table 2-2.

The main driver of new solar PV projects in 2013 was *the planned and under construction* projects, with double-digit annual growth forecasts for PV installations. 20 MW and larger installations dominate the pipeline in terms of installed capacity, driven by state renewable energy mandates.

The trend of declining equipment prices is making large-scale PV installations feasible and profitable, so PV projects above 100 MW are dominating the utility-scale market in the US.

As a matter of fact, the 10 largest PV projects underway in 2013 account for over 5.0 GW of new PV

Table 2-2.
U.S. PV pipeline of projects, 2013 (in percentage of total).

•	Planned:	63.0%
•	Pre-planning	27.5
•	Installed	5.5
•	Delayed	4.0

Small and medium-size PV installations		
•	Over 100 MW	4.7%
•	50-100 MW	3.8
•	20-50 MW	7.0
•	1.0-20 MW	40.8
•	50 kW-1 MW	43.7

capacity and are expected to come online by 2016.

Another noticeable trend is that of PV projects under 20 MW. Since 2012 the number of such installations increased by over 30%, encompassing over 2,100 projects. Many companies choose to focus on these smaller projects because they have shorter planning and implementation phases. They can be constructed within several weeks or months, instead of many years as in the large installations. This reduces the highs and lows of periodic revenue streams for the installers. In addition, the smaller projects are easier to finance in most cases.

The transition to smaller projects is also driven by the approaching deadline to qualify for the full U.S. Investment Tax Credit (ITC) of 30%, which will be reduced significantly by the end of 2016. So the rush is on.

Project developers in the U.S. are in a hurry to complete, or at least have a significant portion of their projects under construction, prior to the deadline. Still, and despite the urgency, less than 10% of the total 43.0 GW PV project pipeline projects were under construction in 2013, and over 4.0% were delayed (mostly for permitting and financing reasons). Over 63% of the PV projects pipeline are at the "planned" stage, with over 27.5% of the total still in "pre-planning" stages.

Note: The lower (darkest) portion of the bars in Figure 2-11 denote residential installations, the middle portion is the commercial, and the top (lightest portion of the bars) the large-scale, or utility type, PV power generating installations.

Obviously, the utility solar installations are growing the fastest, and are expected to grow even faster in the future as well. This is so because it is the most efficient and cheapest way of generating electricity from the sun. It is also a matter of numbers—one single utility scale installation is as large as several hundred, or sever-

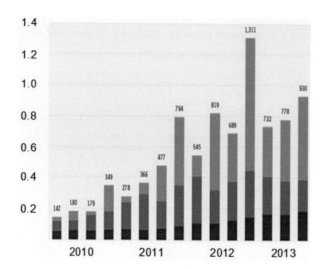

**Figure 2-11.
U.S. solar (PV) installations (in GW per quarter)**

al thousand, home installations.

Overall, the U.S. PV installations have been doubling every year since 2009, and the rush now is to complete the projects in the pipeline before the ITC changes in 2016. Unfortunately, some of these projects will be cancelled beforehand, if they seem likely to miss the deadline. In such cases, the projects will be delayed until (if and when) new financing structures are found in the post-ITC era.

It seems to us that things will change drastically—and for the worst—in 2017 with the ITC gone. We will just have to watch and hope for the best.

The overall effect remains to be seen, and we hope that the new (post-ITC) situation will pave a new way of growth for the solar industry by weeding out the inefficient technologies and companies and those that are surviving solely with the help of the ITC today. This, in turn, might open wide the doors to a new and healthier U.S. PV industry.

How do all these developments affect our energy security? In the short term, if and when implemented, the 43.0 GW of ongoing and planned PV installations in the U.S. is double the power generation lost by shutting down the 60+ coal-fired power plants that did not meet the new EPA standards.

43 GW is a lot of power, for the generation of which we'd need to permit, build, and operate 20-30 mid-size nuclear or gas-fired power plants. Because nuclear power safety is under greater scrutiny today, and because natural gas power plants also emit significant amounts of GHGs, 43 GW of new and clean solar power generation might make the big difference.

In the long term, solar energy—in conjunction with wind and other renewable energy sources and fuels—will replace most of the fossils and pave the road to a clean and sustainable future…down the road.

Wind Energy

Wind energy has been used for thousands of years as well. Humans have been harnessing the wind's energy for use in many areas of their lives, from sailing, to pumping water and grinding grains. Today, the modern wind turbine—an upgrade of the old wind mills—uses the wind's energy to generate huge amounts of electricity.

Modern wind turbines are quite sophisticated and are equipped with state-of-the-art electronics, mechanical, and energy engineering. These modern technological marvels usually consist of blades mounted on a tower and installed in windy areas, where they capture the wind's energy and send it to a generator to be converted into electric power.

At 100-200 feet above ground, these giants can take advantage of the faster and less turbulent wind prevailing at those heights. Wind turbines catch the wind's energy with their propeller-like blades, where two or three blades are mounted on a center shaft of the turbine body to form a rotor.

The turbine blade operates like an airplane wing when the wind blows, where wind hitting the blade (called the drag) creates a pocket of low-pressure on the back (downwind) side of the blade. That low-pressure air pocket creates a pulling force that forces the blade and rotor to turn. This action is known as the lift, which force, amazingly enough, is actually much stronger than the drag force. This force anomaly is what allows heavy airplanes to lift of and remain in flight against all odds. One can actually think of an airplane wing as a wind turbine blade in reverse; it creates its own wind by moving fast in the air, instead of being pushed by the wind as is the wind turbine blade.

This combination of lift and drag is also the force behind the wind turbine's operation, since it causes the rotor to spin like a propeller, which in turn spins a generator to make electricity.

Wind turbines can be used as stand-alone applications, or they can be connected to a utility power grid or even combined with PV and other power generating systems.

Stand-alone wind turbines are typically used for water pumping or for providing power for appliances, communications, etc. Homeowners and farmers in windy areas also use wind turbines as a way to cut their electric bills by either using the power directly, or sending it into the national grid for which they get paid by the utility companies.

Small wind turbines and small installations have potential to provide distributed power, which combined could improve the operation of the large local and national electricity delivery system.

For large-scale power generation, a large number of wind turbines are installed close together to form a wind power plant (or wind farm). Many utility companies today use wind power plants to supply large amounts of power to their customers.

Each large wind turbine generates about 2 MW (megawatts) of electric power, which is enough to power 200-250 average U.S. homes. A field of 250-300 wind turbines generates as much power as a mid-size coal-fired power plant.

Global wind power generation more than quadrupled between 2000 and 2006 and has been growing exponentially ever since. In 2013 the global wind power generating capacity was about 300 GW (gigawatt), with over 250,000 turbines operating in over 80 countries worldwide.

Note: These numbers mean that the average wind turbine size today is 1.2 MW. As the wind power generation technology improves, the size of the wind turbines is increased too, so we expect the average wind turbine size to increase to 1.5-1.8 MW in the near future.

And speaking of averages, there is currently about 25 MW of wind power capacity installed per 1,000 km of land area on the European continent. The highest wind power generating densities are in Denmark and Germany, with Germany having the most installed wind energy capacity, followed by Spain, the United States, India, and Denmark. Denmark is generating more than a quarter of its electricity from wind.

Note: Denmark has one of the highest densities of wind power generation per person in Europe, but it is a small country, so the total installed wind capacity is not that large.

Wind power is also growing quickly in over 80 countries which are using it for local use or to supply their national power grid. Over the last several years, the most intense wind growth moved from Europe and North America to Asia, which emerged as the global leader. China became the global leader in terms of total installed capacity in a very short time, overtaking the United States in 2010.

A number of recent important policy measures and programs have emerged in support of expanded wind markets, and many of the new policy developments concern offshore wind. Ten European countries have agreed to develop an offshore electricity grid in the North Sea to enable offshore wind developments.

Figure 2-12 shows unprecedented and truly amazing growth of wind power installations worldwide. Nearly 300 GW represents the power generated by over 500 mid- to large-size coal-fired power plants. These numbers also mean that millions of tons of GHGs were *not* emitted into the atmosphere by the replaced coal-fired power plants.

Industry experts predict that at this pace of growth at least one third of the world's electricity needs will be met by large-scale wind power generation by 2050, thus avoiding huge amounts of GHGs from being emitted into the atmosphere.

We will review and analyze in more detail the wind power pros and cons in the following chapters, but for the purposes of this section, we must agree that wind is undoubtedly one of the most serious and permanent game changers.

Wind power is clean and cheap, and is reaching LCOE (levelized cost of energy), the competitive price levels that the fossils have enjoyed for a long time. This is a major achievement, and one that signifies that wind is here to stay, so we better get used to it.

Wind enjoys huge and unprecedented success, where mega fields of off- and on-shore wind power plants worldwide are generating a lot of power, and where many are planned for the near future. Wind power is making the headlines on a daily basis, which is a major achievement that promises success for wind

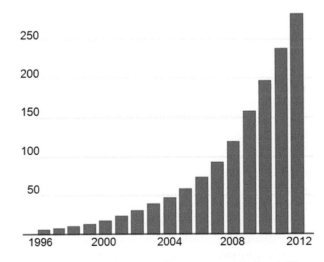

Figure 2-12. Global wind power growth (in GW)

power far into the future.

However, as its expansion increases, and most of us are happy to see low electricity prices, the outcry of the locals is also increasing. Many of them are negatively affected by the huge structures, and the visual and audio effects resulting from large-scale wind power use.

Wind is a growing element of our national energy security. At the same time, it is hurting some of us in the process. So we must take this seriously and determine how the energy security provided by wind power is improving the life of some people while hurting others.

We must figure out what can be done about it so that all involved benefit in the drive for more energy and energy independence.

Energy Storage

Solar and wind are *variable power* sources, since they depend on sunshine and wind, which are simply unavailable at times and intermittent at others. There is no sun at night and no wind part of the day, thus no electricity is produced during major parts of the time.

A major obstacle in the battle between solar and wind, on one hand, and the fossils, on the other, is **energy storage**.

The only way to correct this deficiency is to introduce *efficient, reliable, and cheap* means of storing excess energy. The stored energy can then be used during the non-productive periods, thus providing a continuous level of electric power on a 24/7 basis.

Well, there is a lot of effort in this area, and some progress can be reported, but *efficient, reliable, and cheap* energy storage is still far in the future. Because of that, at least for now, solar and wind will have to settle for a secondary role in the energy industry—that of gap-fillers, instead of base-load providers.

Presently, U.S. energy security does not depend on solar energy much, since it is still a small percentage of the country's overall energy generation. And as the domestic oil and gas production increases, the role of wind and solar will remain small for now.

Solar and wind, however, are shaping as drivers in the renewable energy arena, which is expanding in size and importance. And let's not forget that the value of solar, wind, and the other renewables should not be measured only in dollars; they are also our environmental responsibility.

And as an extension, they is our moral responsibility to future generations, which weighs heavily on our national conscious and which must be supported by effective actions.

In light of the achievements of late, and the increasing energy needs of the world, solar and wind power are shaping as major contributors to our future energy security. For that reason, we take a much closer look at these important energy sources in the following chapters.

Ocean Energy

There is an ocean full of energy out there—literally! It is ready to serve us day and night…if and when we find efficient, reliable, and cost-effective ways to harness the immense power contained in its waves, currents, and tides.

The world will never run out of energy as long as the oceans are still around, but the key to harnessing their power has not yet been discovered. Several different technologies look promising in extracting enough energy from the oceans, and those are under development.

- Ocean *tides* contain huge potential energy that can be harnessed by building a barrage or other forms of structures across an estuary. The powerful currents created in these could turn large turbine wheels, which in turn could generate large amounts of electric power.

- *Ocean currents* have similarly huge potential energy which can be harnessed using modular systems. Some of the very powerful ocean currents are found very close to shore, which makes the construction and operation of power plants easier and cheaper.

- *Wave power* offers readily available and never ending energy from the oceans. The kinetic and potential energy associated with ocean waves is immense, and can be harnessed by devices that move up and down with the waves while driving a pump or a generator that produces electric power.

- *Temperature gradients* in the world's oceans offer a huge potential energy source as well. The temperature differential between the sea surface and deep water (or warm currents in cold waters, for example), can be harnessed using ocean thermal energy conversion (OTEC) processes to generate electricity.

- *Salinity gradients* are found at the mouth of rivers, where freshwater mixes with saltwater. There is energy associated with the salinity gradient that can be harnessed using the pressure-retarded reverse osmosis process, and other salinity gradient related technologies. These then can be used to convert the salinity gradient difference into electric power.

Unfortunately, none of these technologies is fully developed, nor widely deployed in large-scale installations presently. Instead, they are mostly used in small-scale and demo projects. A lot of work remains to be done for their full commercial deployment, but the potential exists and the work continues.

The experts believe that the ocean-based energy technologies could start to play a sizable role in our electricity generation mix around 2025-2030. The worldwide resource of wave energy has been estimated to be greater than 2.0 TW, but limited to a number of locations. Some of these are the western coast of Europe, the northern coast of the UK, as well as the Pacific coastlines of North and South America, Southern Africa, Australia, and New Zealand.

Case Study: UK Tidal Power

Tidal power is seen as the most likely candidate for large-scale power generation from the oceans. One such project, planned in the U.K. could provide the long-awaited boost to this technology.

The river Severn flows from Wales to the Bristol Channel and has a tidal range of 40-50 feet, the second highest in the world. Scientists have been looking at this location as a potential site for tidal electric power generation.

Due to the need for more energy, pending climate change, and energy security, politicians are taking a second look at the potential under their noses. Three such barrages—giant dams—are under consideration. Two tidal lagoons will be built also to retain water when the tides come in and release it at exit.

Tidal energy is considered the most consistent of all renewable sources. Unlike wind or wave power, or hydropower, all of which depend on variable natural events, tidal energy never changes. The size of the potential power generation is impressive too.

For example, a ten-mile barrage running from Weston-super-Mare to Cardiff could generate nearly 10 GW of electricity. This is about 1/6th of Britain's peak consumption and more than every other renewable-electricity source combined.

The average output from such barrage would be below the peak, and yet it could still supply around 5-6% of Britain's total annual electricity demand—not to mention the potential of substantial GHG emissions reduction.

The environmentalists, however, will put up a fight. The Severn estuary is an important habitat for birds and large barrages would damage much of it. They will interfere with fish stocks in the river. To avoid these problems, offshore lagoons and installations have been suggested as a compromise.

The cost of such undertaking is about £25 billion, and even more if modifications are needed, so we need to wait and see how the Brits will handle this potential power source.

Geothermal Energy

Geothermal energy is another ocean of energy tucked deep into the Earth's core. In places, it is closer to the surface, where its heat is stored in rocks and in trapped vapor or liquids, such as hot water or brines. These are geothermal resources that contain large amounts of energy, which can be used for generating electricity and/or for providing heating and cooling.

The experts estimate that there is potential to increase the global production of heat and electricity from geothermal energy 20 times between now and 2050. The focus in this area will be on learning more about the different types of geothermal energy resources, which vary from the well-known high-temperature geothermal sources, to the lower temperature sources found in aquifers and other surface areas. Of special interest is the geothermal heat found in rocks worldwide, which lack sufficient water and permeability for natural exploration.

Through a combination of actions that encourage the development of untapped geothermal resources and new technologies, geothermal energy could account for 3-4% of the total global electricity production and over 5% of all energy used for heat by 2050.

Geothermal energy in hot springs and geysers has been used for a number of purposes—from bathing to cooking—since the beginning of time. The oldest known spa, a stone pool on China's Lisan mountain, was built in the 3rd century BC.

In the first century AD, the Romans used Bath, Somerset, in England, to provide hot water to public baths and also for underfloor heating of buildings. The world's oldest public geothermal heating system in Chaudes-Aigues, France, has been in use since the 14th century. Later on, in the 19th century, the first industrial exploitation of geyser steam was used to extract boric acid from volcanic mud in Larderello, Italy.

Around that time, America's first public heating systems in the U.S., those in Boise, Idaho, and Klamath Falls, Oregon, were powered directly by geothermal energy. A geothermal well in Boise was used to heat greenhouses, while natural geysers were used to heat greenhouses in Iceland and Tuscany at about the same time. The first downhole heat exchanger was developed in 1930 to heat a residence.

Steam and hot water from geysers have been used to heat homes in Iceland since 1943.

In the 20th century, geothermal energy was used for electricity generation, and the first geothermal power generator was put in operation in 1904 in Italy, and in 1911 the world's first commercial geothermal power plant was built there. Soon after that it produced about 3 MW of electricity for local use.

It was the world's first and only industrial producer of geothermal electricity until New Zealand built a plant in 1958, which produced nearly 500 MW in 2012.

In the 1940s the geothermal heat pump was designed and the first commercial geothermal heat pump was used to heat the Commonwealth Building in Portland, Oregon. Later in the decade, a residential open loop version was built and operated in Ohio.

Geothermal technology took off in a big way in Sweden after the 1973 Arab oil crisis. It is growing slowly worldwide with the development of polybutylene pipe, which greatly augments the heat pump's economic viability.

In 1960, PG&E installed the first geothermal electric power plant in the United States at The Geysers in California. The original turbine was rated at 11 MW and generated electric power for 30 years.

The latest improvement in geothermal technology—the binary cycle power plant—was first installed and operated in Russia in 1967 and introduced to the US in 1981. The advantage of this technology is that it allows the generation of electricity from low temperature resources. In 2006, the binary cycle technology was used at the geothermal plant in Chena Hot Springs, Alaska, where it produces electricity from a record low fluid temperature of 135°F.

In 2010, the United States led the world in geothermal electricity production with over 3.0 GW of installed geothermal capacity from 77 power plants, more than any other country in the world.

In 2012, U.S. geothermal energy accounted for roughly 0.3 percent of total installed operating capacity and about 1% of all new renewable energy projects brought online at the same time. Geothermal is a significant portion of renewable electricity generation in the states of CA and NV. The largest group of geothermal power plants in the world is located at The Geysers, a geothermal field in California. Its power is sold at $0.03

to $0.035 per kWh, and more during peak demand periods.

Geothermal power plants are also operating or under construction in Alaska, Hawaii, Idaho, Oregon, Utah, Washington and Wyoming. A significant amount of additional geothermal capacity (574-620 MW) could become operational by January 2016 if all goes according to plan.

In 2013 there was over 12 GW of geothermal power online in 70 countries, with nearly 700 geothermal power projects. This, compared with 24 countries in 2005, and 46 countries in 2007, shows the industry's great progress.

Globally, the Philippines is the second highest producer, with 1,904 MW of capacity online, and geothermal power makes up approximately 27% of Philippine electricity generation.

Another GW of geothermal power is expected to be added to the energy mix by the end of this decade. This is not much power, compared with the conventional energy sources, but these developments show the potential of this technology, and point the way to the future.

As most nations consider fossil fuel replacements, geothermal energy's unique benefits make it a worthy contender in the changing power sector. At the same time, projects abroad are providing the industry's top companies with opportunities to stay competitive by using cheap and reliable energy.

Regional financial institutions, country and provincial governments, as well as executives in both small and large businesses who are considering energy choices are taking a closer look at geothermal power as a potential supplement and even replacement of the fossils.

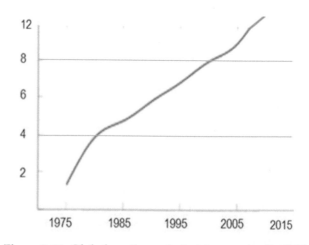

Figure 2-13. Global geothermal electric capacity (in GW)

Once a geothermal well is tapped and fully operational, it provides power 24/7/365, which consistency and reliability can be matched only by the conventional energy sources—without the environmental problems and depletion threats these present. Geothermal power provides energy that is cleaner, consistent, and much more reliable than most of the renewable energy sources.

The saying goes that if and when geothermal power is depleted, then life on Earth will be no more too. So we need not worry about a pending geothermal peak.

Although geothermal power generation is increasing very slowly, due to technological difficulties, it is truly renewable, clean, and reliable energy. As such, it will play an increasing role in our energy future, contributing to strengthening our energy security.

For this reason, in the following chapters we will take a much closer look at this promising technology as one of the vehicles to energy independence, during the transition to the post-fossils energy reality of our not-so-distant future.

Bio-Energy

Here we need to provide clarification of the nature of this energy source, and the terms used in this area of our energy mix.

- Technically speaking, bio-energy is a renewable energy source. However, it depends on climate change and weather variations. It also competes with the food supply of this hungry world, so we prefer to keep it in a separate category—that of semi-renewables.

- Biomass is any organic, i.e. decomposing, matter derived from plants or animals available on a renewable basis. This includes wood and agricultural crops, herbaceous and woody energy crops, municipal organic wastes as well as manure.

- Bioenergy is energy derived from the conversion of biomass where biomass may be used directly as fuel (bio-power), or processed into liquids and gases (biofuels).

- Biofuels, as we refer to them today, are any liquid or gaseous fuels that are produced from processing solid biomass, such corn and other so-called fuel crops.

- Cellulosic biofuels are produced from crop waste products and food manufacturing byproducts, such as corn husks, grasses, and other vegetation.

- Bio-products are different consumables that are made by processing and converting biomass into chemicals for making products that typically are made from petroleum or other raw materials.

Biomass accounts for roughly 12% of the world's total primary energy supply.

Most of the biomass used worldwide is used very inefficiently, mostly in developing countries for heating and cooking via inefficient open fires or simple cook stoves. These processes cause health problems by excessive smoke inhalation, and also harm the environment by the smoke emissions and by deforestation of the world's forests by the locals.

Modern bio-energy supply has been growing steadily recently, with a total of 300 TWh of bioenergy electricity. This amounts to about 1.5% of the total world electricity generation in 2012. Also, 10 EJ of bioenergy was produced as heat, which was used in the global industry sector.

Bioenergy used for heat and electric power can help us to achieve significant emission reductions in the energy sector. Sustainably produced bioenergy will play an increasing role in the future with demand expected to increase three-fold by 2050.

There are a number of bioenergy technologies already, but new technology options are still needed to optimize their efficiency and practical use. Co-firing biomass with coal, for example, could be an important option to achieve short-term emission reductions, and an example of efficient use of standing assets.

As Figure 2-14 shows, the U.S. has ambitious plans to double the biofuels production within the next decade. Using advanced processing methods for conversion of cellulosic matter into fuel seems to be the vehicle, but this technology is still under development. It is too complex and expensive for now, so some barriers must be overcome before it can become competitive enough.

So, the plan is not impossible to achieve, but looking to the past, we see the failures and the lessons in this area, so we must remain cautiously optimistic.

New crops production methods and schedules will have to be developed in order to not repeat the bitter lessons of the past ambitious attempts to increase biofuels production. New, dedicated bioenergy production plants—using new technologies—will also have to be built to meet the growing demand for bioenergy elec-

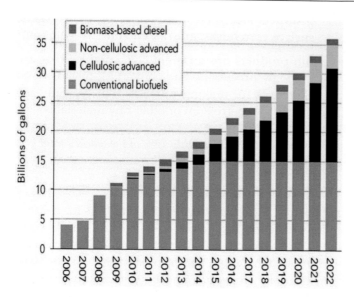

Figure 2-14. Biofuels in the U.S. energy mix

tricity and heat, and as a tool for reducing GHG emissions.

Increased use of biofuels is a no-brainer, and there are many methods for using them in the fight for reducing fossil fuel dependency and containing carbon emissions for a healthier and more eco-friendly future. Corn-produced ethanol has been one of the biofuels used for mixing with gasoline but there have been side effects like engine parts corrosion from the ethanol in the mix. More importantly, large tracts of precious farmlands need to be diverted for corn production, which has created some serious problems with basic foods availability and prices in the past.

New research, however, shows that common algae can be used for biofuel production instead of food crops. This is like hitting two birds with one stone; algae will make our waters clean and will also provide energy from matter that is presently not used for human consumption. This is especially important today as we are faced with a number of energy, environmental, and food shortage problems.

Several companies have been working on this for some time, rather quietly patenting efficient new techniques for getting fuels and even hydrogen from clean algae cultures. They also can offer specialized algae systems to companies looking to reduce carbon dioxide emissions.

Biofuels from grain and beets in temperate regions are relatively expensive and their energy efficiency and carbon dioxide savings vary at best. Biofuels from sugar cane and other highly productive tropical crops are much more competitive and beneficial, but they require

special climate conditions and a lot of water. Because of that only a few countries, like Brazil, have been able to produce large amounts of biofuels from sugar cane.

Most importantly, all first-generation biofuels compete with the production of food for land, as well as for water, fertilizers, and other resources. Commercialization of the second-generation biofuel technologies, such as biorefineries and cellulosic ethanol, is the key to enabling the flexible production of biofuels and related products from waste parts of the plants.

Cellulosic ethanol commercialization, for example, could allow ethanol fuels to play a much larger role in the future than previously thought. Cellulosic ethanol can be made from plant matter composed primarily of waste (inedible) cellulose fibers that form the stems and branches of most plants, and which is usually left unused in the fields or fed to animals.

Dedicated energy crops, such as switchgrass, are also promising cellulose sources that can be produced in many regions of the United States and other countries.

Biomass and biofuels energy sources are truly renewable, albeit their quantities are not unlimited and depend on land allocation, weather patterns, etc. Though they will never be able to provide a huge amount of our energy needs, they are an important part in our energy mix. This makes them especially important for providing stability to our drive for energy security and independence.

THE MOST PROMISING GAME CHANGERS

In addition to the technologies discussed in detail in this text, we are now witnessing the development and implementation of other technologies that produce or use energy in a different and/or more effective way.

There are actually too many of these to mention, so here are several that we feel have most potential to succeed and which could have greater impact on the energy sector.

The Renewables... Again

We need to clarify from the onset that the renewables are any and all technologies that are inexhaustible—such as solar, wind, ocean, and geo-power.

Note. Although hydro, bio, and nuclear power sources are technically classified as "renewables," we do not include them in the list of renewables, because they depend directly on natural resources that are limited and variable.

Hydropower and biofuels, for example, depend

on periodic water replenishment which is scarce due to droughts, and the uncertainty is increasing daily in the face of the present-day climate change. Thus, we cannot count on them in the long run.

Nuclear power depends heavily on importing raw materials which are fossils by definition and which will eventually be exhausted. Nuclear's future is also veiled in uncertainty, due to the extreme risks of huge devastation and human life loss in case of nuclear accident.

We must understand and remember that the renewables—solar, wind, ocean, and geo-power—are the only branch of the energy sector that has a long-term future.

We dare say this, fully aware of the controversy surrounding the subject. The facts, nevertheless, support this statement because the fossils will be depleted within the next several decades, while the other technologies—hydropower, biofuels, and nuclear—are not reliable, or safe enough. Because of that, we cannot count on them for uninterrupted power in the long run.

Access to cheap energy has become essential to the functioning of modern economies. However, the uneven distribution of fossil fuel supplies among countries, and the increasing need to access energy resources, has led to significant vulnerabilities and uncertainties.

These uncertainties bring threats to global energy security, aggravated by political instability of energy-producing countries, manipulation of energy supplies by different parties, competition over energy sources among the countries, terrorist attacks on supply and transport infrastructure, as well as accidents and natural disasters that negatively affect the supply and the transportation routes.

The Chernobyl and Fukushima nuclear accidents are great examples of how the entire national energy system can be affected by man-made mistakes and/or natural disasters. As terrorism and climate change are increasingly unpredictable, and their after-effects are more noticeable, the types and number of threats to our old energy systems are also more noticeable as they increase proportionately in number and size.

Increasing the renewable energy sources' share in our economies can help us to meet the dual goals of a) reducing greenhouse gas emissions, thus limiting future extreme weather and climate impacts, and b) ensuring reliable, timely, and cost-efficient delivery of energy, which is largely unaffected by terrorist threats or natural disasters.

The positive effects of these factors show that renewable energy can have significant dividends for our

energy security, especially in the area of providing fuels for the transportation sector, which is quite vulnerable to supply disruptions and price increases.

Rising petroleum demand by developing countries is accentuating our vulnerability in these areas, and biofuels used for transport fuels represent a key source of diversification from petroleum products.

In many countries, the growing dependence on imported fuels is a pressing energy security risk. As these fuels become less available and more expensive, only renewable energy technologies can provide alternative—locally produced—sources of electric power and heat.

The direct contribution of renewable energy in providing direct heat to domestic or commercial use—space heating, cooking, and industrial process—is a feasible alternative to the more expensive electric generation technologies. Heat from solar, geothermal sources, and ground heat pumps is increasingly economic but is often overlooked, so a second look is needed in the near future.

Electric generation from renewables is quickly becoming a major source. In 2010, four German states, totaling 10 million people, relied on wind power for over 50% of their annual electricity needs. Denmark isn't far behind, supplying 22% of its power from wind in 2010 (or 26% in an average wind year).

The Extremadura region of Spain is getting up to 25% of its electricity from solar, while the whole country meets 16% of its demand from wind. Just during 2005-2010, Portugal vaulted from 17% to 45% renewable electricity. Minnkota Power Cooperative, the leading U.S. wind utility in 2009, supplied 38% of its retail sales from the wind.

Other means of managing the use of energy also help. For example, in 2005, Cuba reorganized its electricity transmission system into networked microgrids which cut the occurrence of blackouts to zero within two years, limiting damage even after two hurricanes.

Networked islandable microgrids are where energy is generated locally from solar power, wind power and other resources and used by super-efficient buildings. When each building, or neighborhood, is generating its own power, with links to other "islands" of power, the security of the entire network is greatly enhanced.

The combined power plant, a project linking 36 wind, solar, biomass, and hydroelectric installations throughout Germany, has demonstrated that a combination of renewable sources and more-effective control can balance out short-term power fluctuations and provide reliable electricity with 100 percent renewable energy.

In other words, ensuring a reliable energy supply

cannot be done with just one technology or approach. A versatile mix is needed if we are to ensure the long-term security of our energy supply.

Clean Coal

We review the clean coal technology in the following chapters, so here we will only mention that coal can be made to burn "clean," or at least cleanly. This requires expensive equipment and procedures, but it can be done. As the technologies improve this day may not be too far.

Coal can also be converted into other, cleaner and equally efficient products. This is also expensive, so we don't expect it to happen overnight, but the possibility is there and we expect it to become a reality in the near future.

Water Desalination

Water is the new gold in many places worldwide. Clean water is becoming a luxury as the growing world population is demanding more water, while at the same time we are experiencing serious draughts and water contamination issues.

One way of solving the clean water problem is by distilling water from the ocean, rivers, lakes, and contaminated wells. This is a process that filters dirty or salty water through special membranes. Heat can be used to evaporate the dirty water and collect the condensate as clean water, thus separating the contaminants and salt from it.

The demand for fresh water is expected to outstrip supply by 40-45 percent by 2030, and as the natural water sources diminish, filtration and/or desalination is one way to meet the need if it is energy efficient and affordable.

All methods, however, require enormous amounts of energy to push the water through the membranes or to heat it to its boiling point. Since the major need for fresh water is in remote locations, or poor countries, there is not enough cheap energy for this process.

With the improvements in renewable technologies, solar and wind power could be used to provide the energy needed for the desalination process worldwide.

Other innovations include "capacitive deionization," which does away with separate membranes and uses porous electrodes that both attract and trap salt. Another is "multiple-effect distillation," which produces water vapor and leaves salt behind at 30 percent greater energy efficiency over the conventional methods.

Recent research MIT in Massachusetts have found that graphene, in the form of thin sheets of carbon-like material just one molecule thick, can separate clean water from the salts by applying very little pressure, which saves a lot of energy.

These and other new technologies might allow us to bring enough fresh water to the thirsty world regardless of the whims of Nature. But we are not there yet, and a lot of work still remains to be done before we can provide enough fresh water worldwide.

Energy Storage

The development of the electric vehicles (EV) industry might bring us faster to resolving the energy storage problem than initially thought. EVs are the new rage for motorists around the globe, and Tesla Motors is addressing it with its new model S electric car.

This is what *Consumer Reports* has to say about the Tesla model S electric car, "Sure, you can talk about this electric luxury car's blistering acceleration, razor-sharp handling, compliant ride, and versatile cabin, which can fit a small third-row seat. But that just scratches the surface of this technological tour de force. The Tesla is brimming with innovation. Its massive, easy-to-use 17-inch touch screen controls most functions. And with its totally keyless operation, full Internet access, and ultra-quiet, zero-emission driving experience, the Tesla is a glimpse into a future where cars and computers coexist in seamless harmony. Its 225-mile driving range and 5-hour charges, using Tesla's special connector, also make it the easiest, most practical, albeit pricey, electric car to live with."

Tesla cars come with a choice of 65 or 85 kW power platforms—a lot of power coming from a bank of batteries mounted in the bottom portion of the car's frame.

Tesla's specialized solar-charging network is available across some parts of the U.S. and Europe, which makes using this car very practical albeit expensive.

At a retail price for Tesla model S over $80,000 and model X even more, not many people would be able to experience the future anytime soon. Yet, Tesla expects to ship 10,000-15,000 cars every year from now on. 13,000 model X cars already have been reserved.

Amazingly, most of the growth of these sophisticated luxury vehicles won't be in North America, it says, but in Europe, China, and other countries. This is one indication that Americans have outgrown the "car enthusiast" phase, while other countries are just now getting into it. It is also a reflection of the difference in driving habits in different countries. Americans are used to driving hundreds of miles without stopping, so the 225-miles range and 5-hour charge time of Tesla cars is a nuisance—regardless of the other goodies.

Tesla's future plans are an extreme example of what our transportation future could look like. But since batteries are 50% of the cost (and the headaches) in EV manufacturing, Tesla must solve the battery (and energy storage) problem.

A new Tesla production plant is planned and expected to transform the car industry and bring the electric car technology to new highs. The goal is to lower the cost of batteries by 30% by 2018 and to produce a functional, exciting, and affordable electric car at half the present cost. This would be a real game-changer for the electric car and the energy storage industry.

The planned gigawatt production plant is a gigantic, 10-million-square-foot plant that employs some 6,000 people. The entire operation is expected to run predominantly on solar and wind power. $2 billion are needed to start the operation; an additional $4-$5 billion will be raised to operate through 2020.

The vertically integrated facility would bring battery manufacturing under one roof, starting with raw and precursor materials, and cell, module and battery pack production.

This could also be a great achievement for the solar industry too, because energy storage is its biggest problem today.

Producing cheap batteries for use with solar panels would solve the energy storage problem and unleash solar power's hidden potential. Deployed in solar power plants these batteries would ensure reliable and uninterrupted power delivery into homes and the grid.

But for all this to happen on a grand scale, the battery prices must be at least 70-80% lower than today, and even more importantly; their reliability must be several times higher. Is this possible? Maybe, and Tesla would be a the best indicator of the progress in the field.

Presently SolarCity and other solar companies sell Tesla's batteries for energy storage, so there is already some progress in this area. Tesla is no longer just another auto company. Instead, it is a real game-changer, jumping head first into the most disruptive intersection of technology, manufacturing, innovation, and capital ever experienced by the auto and solar industries.

We expect great things to happen with Tesla's EVs and especially with their batteries development efforts.

Food Energy

Food—its production, processing, and use—is also part of our energy security. It requires a lot of energy to produce, transport, refrigerate, and cook food. Some food crops are also used to make energy—biofuels—which are becoming increasingly important.

Food also provides energy to our bodies so that we can function properly. Worrying about energy security would be pointless If we have not enough food to keep us going strong and healthy.

Without proper and cheap food, our lives—especially those in the developed countries—would be quite different. In our small world, with its ever increasing population and its growing demands, food is increasing in importance.

Most countries, including the U.S., support biofuels production with tax breaks, mandated use, and subsidies. The unintended consequence of these efforts is diverting resources from food production, which in turn leads to rising food prices.

Fuel used by farmers is usually not taxed and they also receive many subsidies for growing fuel crops.

Farmers use more than a gallon of diesel to produce one gallon of biofuels, and still make a significant profit.

On a large scale this could lead to distortion of the energy markets. What is the point of importing a gallon of one fuel just to make a gallon of different fuel?

The increased use of agricultural crops for making fuels has also contributed to sharp food price increases. Food prices in some countries rose by 35-40 percent at the peak of the "ethanol revolution" between 2002–2008. Most of this anomaly was attributable to increased biofuels production.

The U.S.-led drive for biofuels has had the biggest impact on food supply and prices, due to increased production of biofuels in the US and Europe, supported by subsidies and tariffs on imports. Without these policies, food price increases at the time would have been much smaller.

There is also the misconception that biofuels contribute to reduction of GHG emissions. The experts estimate that the current biofuel production and use methodologies and policies have a marginal effect, and will reduce GHG emissions from transportation fuel by less than 1% by 2015.

In contrast, Brazilian ethanol fuel made from sugar cane reduces GHG emissions by 75-80% compared to fossil fuels. This discrepancy is due to the fact that great quantities of conventional fuel are used to make biofuels from crops such as corn and even more from cellulosic materials (which is the current rage). Because of that, a lot of fossil fuels are used and huge quantities of GHGs are emitted during the cradle-to-grave process of such biofuels.

Brazilian ethanol made from sugar cane, in contrast, is much less energy-consuming, thus is much cleaner and cheaper.

The demand for biofuels has also resulted in massive forest devastation in some countries and the destruction of natural habitats in huge areas around the globe.

The major battle today is for securing more land for growing crops that can be used for making vehicle fuels—biofuels. In many cases, this is done at the expense of reduced acreage for food crops, creating uncertainties in the global food market.

Case Study: Biofuels vs. Food Crops

In the early 2000s, the cultivation and use of jatropha and castor bean were the new hope for cheap and plentiful biofuels. These crops were thought to be sustainable sources of fuel, as they would reduce dependency on fossil fuels and reduce carbon emissions.

The experts argued at the time that the production of these seeds would have a positive impact on the economies of the producing countries too (Ghana and Ethiopia originally). As important export products they would boost the countries' GDP, create jobs, and lead to the development of new and more efficient infrastructure.

As an added and very important benefit, jatropha and castor bean were to be cultivated on marginal arid lands and wastelands. This way local farmers would be able to earn additional income by cultivating the waste lands. Too, food security would not be threatened, as these lands could not be used for the cultivation of food crops anyway.

As a result, several foreign biofuels companies started to operate in these countries and foreign investment in biofuels production increased. The European Union was very interested in this development and its foreign investment in the area increased.

Alas, the cultivation of jatropha and castor bean did not prove a clear-cut, win-win scenario. Scientific studies showed that the cultivation of biofuels crops actually increases carbon emissions and threatens biodiversity.

These crops also proved to be not as suitable as expected for cultivation on arid lands, and needed much more intensive irrigation than predicted. This increased the costs for local jatropha and castor bean farmers, which forced them to cultivate the crops on more fertile lands.

This move required use of land previously used for food crops, so the cultivation of the biofuel crops ultimately pushed up food prices and posed a threat to local and global food security.

Consequently, global interest in jatropha and castor bean has decreased significantly. The EU, once a significant investor in the cultivation of biofuels, decided to limit the use of biofuels.

The local enthusiasm for these new fuel crops soon evaporated too. Farmers cultivating jatropha or castor bean started complaining that they were better off cultivating other crops, given the irregular and decreasing earnings they received for cultivating biofuel crops.

In addition, farmers in both countries complained that foreign companies were making large-scale land grabs for the cultivation of biofuels stock, which ultimately results in tensions.

And in the end, the new hope became a nightmare and moved onto new and more profitable crops and land areas.

Methane Hydrates... Fuel of the Future

Natural gas hydrates, a relatively unknown to the public energy source, are also called methane hydrates, or clathrates. They are a variation of natural gas, where combining water and gas molecules under certain conditions forms non-stoichiometric, ice-like compounds. These molecules have crystalline (ice-like) structure, which is formed when temperature, pressure, gas saturation, water salinity, pH, etc. were favorable millions of years ago.

They are also referred to as flammable ice or combustible ice because the ice-crystal formations burn when defrosted and ignited. We take a much closer look at this very promising fuel of the future in the next chapters.

Very large quantities of this unusual material were formed millions of years ago and are usually found in the depths of the oceans, or in the permafrost of the North Pole. Their chemical composition is such that the water molecules in the hydrate combine into polyhedral cages through hydrogen bonds. The gas molecules are, therefore, trapped in and enclosed within the prison walls of the die cages forever...or until released by some form of energy.

There are three types of methane hydrate structure. They all include pentagonal dodecahedra of water molecules enclosing methane. This geometry arises from the happy accident that the bond angle in water is fairly close to the 108° angle of a pentagon. Generally, the dodecahedra are slightly distorted so that three dodecahedra can share an edge. This requires a dihedral (inter-face) angle of 120 degrees, whereas the dihedral angle of a true dodecahedron is 116.5 degrees. Between the dodecahedra are other cages of water molecules with different shapes. In practice, not all cages are occupied by hydrocarbons, but occupancy rates of over 90% occur.

Significant energy is needed to release the gas from its icy prison, and since the conditions deep down below on the ocean bottom, and the permafrost wilderness, don't change much with time, the gas will stay in there until the end of time, or until a natural disaster rescues it. But most likely it will be man who someday finds a way to break the prison walls and release large quantities of gas for use.

This type of fuel is made of colorless hydrocarbon and water molecules, so it is usually white or colorless. Some gas hydrates in the Gulf of Mexico, however, have yellow, orange, or even red tint. Other samples from the Atlantic deposits are gray or blue. The colors are most likely due to impurities mixed in the crystals and have no other meaning.

In nature, methane is the most common "guest" molecule to form natural gas hydrate. If 99% or more of the guest molecules are methane, then the natural gas hydrate is called methane hydrate. It, and most other hydrocarbon gases, always burns when ignited. Methane hydrate is regarded as an ideal clean fuel energy for the future.

Methane hydrates are common in sediments deposited in high latitude continental shelves and at the slope and rise of continental margins with high bio-presence. Biological products provided the organic matter that was buried in the sediment many millions of years ago, which under the pressure and temperature and after exhausting the oxygen supply, sulfate, and other electron acceptors, eventually generated methane through fermentative decomposition and/or microbial carbonate reduction. The resulting gas was then frozen in the depths of the Earth, crystallized, and perfectly preserved as methane hydrate.

The properties of sediment-hosted gas hydrates are strongly determined by texture, structure, and permeability of the sediment and the mode of supply of methane. The water molecules from a well-defined crystal lattice (the host lattice) contain cavities into which small gas molecules (guests) may be adsorbed; under appropriate conditions the adsorption energy may then reduce the free energy of the hydrate sufficiently to make the hydrate phase more stable than either pure water or ice. In the crystal lattice of hydrates, guest-guest interactions are negligible.

The host lattice is considered merely to give rise to an environment in which the guest molecules evolve, and the thermodynamic properties of the hydrate result from the classical behavior of individual guest molecules within the cavity potential.

The hydrate formation process can be described as gas absorption, primary and secondary nucleation, growth, agglomeration, and breakage. The interaction between hydrate and the host sediment at the grain level has been highlighted in recent years as laboratory and analytical investigations have shown that hydrate is not restricted to forming in a unique way in the pore space.

Parameters affecting gas hydrate formation and dissociation include temperature, pore pressure, gas chemistry and pore water salinity. Any change in the equilibrium of these parameters may result in dissociation and/or dissolution of the gas hydrate.

At 30 atmospheres pressure, methane hydrate begins to stabilize at temperatures above 0°C, and is stable at 100 atmospheres and 15°C. This has an important practical implication, because it is a nuisance to the gas company, since they have to dehydrate the natural gas pipes thoroughly to prevent methane hydrates from forming in their high-pressure gas lines. This is an expensive and time consuming process, so the gas company would have a hard time doing it.

Still, as time goes on and the technology gets more established, new techniques will resolve all issues, at which point methane hydrates will be produced and handled in a more efficient and cheaper manner.

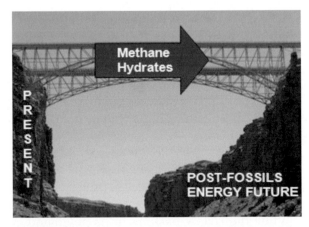

Figure 2-15. Methane hydrates in out path to post-fossils energy future

The methane hydrates will become increasingly important as the fossils dwindle, and new technologies appear to make it easier and safer to deal with them. Due to the expected importance of this type of fossil fuel in our path to the post-fossils transition, we will take a closer look at it in the next chapters.

Environmental Concerns

Methane hydrates are stable on the sea floor at depths below a few hundred meters and will be solid within sea floor sediments. Chunks occasionally break loose and float to the surface, where they are unstable and effervesce as they decompose.

The stability of methane hydrates on the sea floor has a whole raft of implications, the most important of which is that they may constitute a huge energy resource in the future. Also, natural, or man-made disturbances might suddenly destabilize sea floor methane hydrates, triggering submarine landslides and huge releases of methane. This would spell a catastrophic disaster, because methane is a very effective greenhouse gas, and large methane releases might spell large disaster for Earth and it inhabitants.

Such large releases may explain sudden episodes of climatic warming in the geologic past. The methane would oxidize fairly quickly in the atmosphere, but could cause enough warming to affect other mechanisms, such as release of carbon dioxide from carbonate rocks and decaying biomass, which could keep the temperatures elevated.

On the bright side, many scientists believe that methane hydrates are one of the major (if not the only) long-term energy sources of our future. The estimates are that at least 200 years of energy reserves are stored in that form, awaiting exploitation and use by man.

Their global abundance and distribution suggest that they may become energy resources of the future.

With increasing energy demand and depleting energy resources, gas hydrates may serve as a potentially important resource of future energy requirements for several centuries.

We take a much more detailed look at this new, potentially huge game changer (#2) in the last chapter of this text.

INTERNATIONAL DEVELOPMENTS

While the U.S. is basking in its newfound energy wealth, other countries are not that fortunate. The struggle in the developing countries continues, and the hope for better life shines brighter than ever for some. For most of them, however, energy security is only a far-away dream, and they cannot even dream of energy independence.

Developed countries have established energy security measures, yet for many of them the energy future looks quite shaky. Developed economies depend on plentiful and cheap energy supplies. But most European countries, Japan, and others import huge quantities of crude oil and natural gas from the Middle East and Russia. At times, many European citizens wonder if they would be able to survive the freezing winter months if Russia decides to reduce, or cut off, the life-giving natural gas supply at the critical time.

How long can Europe depend on Russia, and can a shale gas boom happen in Europe, are questions heavy on the mind of every European official. What will happen to the Japanese nuclear industry, and can Japan replace the lost power with something else?

The short answer to those questions is that neither Europe nor Japan has any plausible alternatives and will continue to rely on huge imports, controlled by unstable and unfriendly governments."

The U.S. shale gas and oil boom is a unique—once in a lifetime—development that cannot be duplicated at will here or anywhere. Favorable geological conditions, advanced technology, easy access to infrastructure, substantial public support resulting in attractive mineral rights legislation, available and efficient service industry, broad political support, a large market, and a favorable fiscal climate are special conditions that cannot be found anywhere else.

Europe

Europe has a large energy market, no doubt, but the shale oil industry has not developed even close to that in the U.S for a number of reasons. A major problem has been the fact that test drilling operations in a number of European countries have shown that the geological conditions in most parts of Europe are not as favorable as those in parts of the U.S. where the bulk of shale gas and tight oil extraction is currently taking place.

Also, government support has been limited and the economic factors largely ignored. Instead, public debate is almost exclusively focused on present-day environmental concerns of shale gas extraction. In Germany, the Netherlands and the Czech Republic, for example, the public has demanded further government studies on the potential damages, while in France, Bulgaria and Spain, there have been calls for a ban on the currently preferred technology altogether.

In the United Kingdom, Lithuania and Romania, the governments have cautiously moved ahead, but public opposition has remained strong. In Poland, public support has been large, but a wide range of other issues plague the commercial extraction of shale gas, making its future uncertain.

So European shale gas development has been fragmented, with different countries taking different positions on the issues. The European Commission is claiming to be "neutral" on the matter, and the environment (ENVI) committee in the European Parliament has more concerns than the industry (ITRE) committee about shale gas extraction in the European member states.

The big difference between the U.S. and Europe is that the latter has a large offshore industry, but no onshore industry. In 2012 the U.S. had 2,000 rigs working shale oil deposits, compared to about 70 in Europe, and the situation there is not changing much either way.

Europe is not going to see an energy revolution as in the U.S. It is expected, instead, to undergo a slow and bumpy energy evolution. The main reason is that the resource estimates in other parts of the world are much more promising than those in Europe. Estimates suggest that there are large basins of shale gas and/or tight oil all around Europe—with Algeria, Russia and Ukraine as the most prominent. When these are extracted they would not alter European import dependence as it has evolved over the last decades.

There are many uncertainties in Europe's shale oil and gas supplies future, as follow:

- *Technological Developments*

Energy resources in Europe are trapped in shale rock layers that are very deep into the subsurface, which raises the costs of extraction above the reasonable. Also, the existing infrastructure is unsuitable for large production, even if there was one. Historically, many EU countries have built their economies on domestic coal and imported oil, so substantial investments in infrastructure would be required in order to switch to domestic distribution and use of natural gas.

These are long-term projects, and although some governments have stepped up efforts in this area, much more and expensive work remains to be done. This includes the new construction of distribution grids or improving the inter-country interconnect facilities.

Basically speaking, shale oil and gas are too expensive to extract today, but future technological development is likely to make this process cheaper.

Excess methane emissions during shale oil and gas production is another great concern, which could also be resolved by a new green completions technology.

Excess use of fresh water for the drilling and production processes could be replaced by use of saline waters, as currently done in parts of Texas.

- *Unbundling the Pipeline*

There is a very important concept at play in the U.S. called "unbundling in the pipeline," where the owner of a pipeline in the U.S. does not own the oil or gas flowing through it. This way, the pipeline owner has no influence over the use of the pipeline to transport its natural gas. This in turn lowers the entrance barrier to different producers.

This rule alone has facilitated the emergence of smaller independent producers and the increase of venture capital in the area of investing in unconventional energy resources. This cannot happen in Europe anytime soon, which in addition to the apparent lack of successful initial drilling results and the departure of several private companies, further restricting the justification of these investments.

- *Regulation and Legislation*

A number of member states have been very hesitant when dealing with the hydrofracking process. France, Bulgaria and Spain have introduced outright hydrofracking bans, while most of the rest have started studies of the associated environmental impacts.

Policymakers in Brussels are revising the existing regulatory frameworks, and whether more stringent regulations are required. The dysfunctional European emissions trading scheme and post-2020 carbon and renewable targets for Europe are also on their agenda.

In contrast, the U.S. federal government has provided active support to the industry in the form of funding for research, starting in the early 1970s during the development of hydraulic fracturing technologies. The trend continued in the 1980s by favorable fiscal terms to stimulate the exploration of unconventional gas under the Natural Gas Policy Act, which deregulated wellhead sales prices of natural gas from Devonian gas shales.

- *Investments*

For these and other reasons, the European gas system is lacking investment and is not functioning properly. Investment in infrastructure, such as pipelines, interconnection facilities, reverse flow options or storage facilities, and a lack of implementation of existing legislation are hindering its proper development and efficient operation.

Poland is an example of how these factors can have a negative influence on the possibility of extracting unconventional energy resources.

U.S. exports of LNG to non-Free Trade Agreement countries is another uncertainty that stops investment, for it remains to be seen if LNG can compete effectively with pipeline gas from Russia. All this cripples domestic gas production in Europe.

- *The Environmental Debate*

The debate mainly focuses on environmental concerns related to shale gas extraction, and has risen on the political agenda. Energy security is also often cited as a major concern as are the lack of benefits for local communities and the challenge of population density. The European debate is completely polarized, with predominantly gross exaggeration of the pros and cons.

Environmental concerns are focused on water pollution, methane leakage and induced seismicity. Opponents point to some incidents in the U.S. as an example of what might happen to them. While this is the bare truth, there are thousands of wells that have been successfully drilled without problems, but are ignored in the European debate.

In contrast, even as concerns regarding shale gas and tight oil extraction in the U.S. have been rising, public support is staying high. So, the prevailing economic arguments in favor of this type of energy extraction drive the trend even in states where extraction takes place.

- *Industrial Slowdown*

The European Commission recently announced that industrial gas prices in the U.S. were about one quarter of those in Europe. If this price difference proves structural, heavy industry in Europe will suffer a serious blow, as several industry representatives are warning.

Some analysts warn that trade talks between the U.S. and EU could prove disastrous for European industries, if this price difference is not taken into account and included in the trade barriers to be taken away.

- *Energy Security*

Eastern European countries are especially worried about their energy security—perhaps a product of the Cold War. Most of these countries use very little natural gas in their energy mix, thus dependence on Russia is not that significant. Poland is especially adamant about moving away from Russia, and yet there has been no effort to obtain gas from Denmark or Germany. This is simply because there is no substantial gas market in Po-

land to justify such an effort and investment.

The lack of tangible benefits for local communities that might host hydrofracking activities is used as one of the core arguments in the French ban on the process. The ban was reconfirmed by the French supreme court and is stopping or delaying gas exploration.

In Summary

Many uncertainties make it difficult to predict the future of unconventional energy resources in Europe. What does seem certain is that an energy boom comparable to the one that has unfolded in the United States is not going to happen on the other side of the Atlantic Ocean.

Our major partner (and friendly competitor) Europe will be, at least for the foreseeable future, depending on imports from unpredictable neighbors and Arab countries. Maybe the U.S. could will help with reliable and cheap energy export in the future.

Japan

Japan is one of the world's largest economies with great energy challenges. Japan has no significant domestic energy supplies, and is far away from the exporting countries. This make its energy security seem like an oxymoron. How can you have energy security if you don't have much energy and if you depend totally on imports from unstable regions halfway around the world, let alone running a large economy efficiently?

Yet Japan has managed to do just that with the help of nuclear power and alternative energy sources. But nuclear power played a cruel joke on Japan's energy sector in 2011, with the Fukushima nuclear plant's damage, killing people and contaminating a large area in the locale.

As a consequence, all 50 nuclear plants were shut down. That's about 30 percent of the total electricity generation gone dark overnight. And all this happened amidst plans to increase nuclear power in the country by 50% by 2050.

This is a lot of lost electricity—now for the future. With most of its nuclear power plants shut off, Japan was forced to change plans and find other ways to supply power to its population and its growing economy.

The first thing was to increase the imports of all kinds of fossils, which increased the total GHG emission in the country by 10-15%. To help offset a looming environmental disaster (i.e., Beijing killer smog), Japan established a number of additional (albeit temporary) measures.

One of these measures was a lucrative feed-in tariff for solar power that received overwhelming public support. Within a short time it lead to rapid growth in solar installations around the country, and now Japan is one of the world's leaders in solar power. The number of solar installations is also projected to grow significantly in the near future.

But replacing nearly one third of their energy supply to provide enough electricity (especially during peak summer periods) cannot happen overnight—no matter what technology is chosen. Japan is also an island nation, so the area available for solar is limited. And on top of that, the coastal areas—which cover most of the country—are cloudy and not that suitable for solar power generation. Not to mention that solar is still an expensive alternative, with marginal returns.

Wind power was also considered and many new wind power plants were constructed in record time. But wind also has problems. It is a variable power source, thus cannot be relied on for baseload power generation. The land area for wind farms is limited, and wind is more expensive than the conventional fuels for now.

So, rolling blackouts were to be expected at least for awhile. The government then considered another option, that of planned energy efficiency and conservation by residential and commercial consumers. A new power saving program called "setsuden" was implemented and promoted as a solution to the energy deficiency.

The Japanese people are disciplined and rule-obeying for most part, so the program is working. By voluntary reduction of electricity use at night and rotating air-conditioning schedules, most of the blackouts were averted.

These measures, however, are perceived as a short-term solution to a permanent problem, and in the long term, demand for electricity would return to where it was, thus taxing the system beyond its limits.

So far, Japan has surprised the experts and the naysayers by replacing half of its lost nuclear power capacity through energy efficiency and conservation measures. *It took only three years for Japan to show the world what serious energy saving efforts can do.*

This was achieved by increasing the people's awareness of energy use and energy efficiency. The effort was initiated by the government and is now conducted by large companies who run high-profile efficiency programs.

Even with these significant achievements in energy efficiency, Japan is still far from 100% efficiency and there is a lot of room for improvement. And as always, the majority of Japanese people support these measures, so there will be more progress in the future.

This way, the need for building new coal-fired power plants, which was considered at first, is totally eliminated. The need for the nuclear power plants is also reduced, so the pressure is off at least partially.

Could this happen in other countries? How about the U.S.? What would we do if all nuclear power plants—providing 1/5th of the total electric power are shut down today? Analysis shows that following the Japanese example, the U.S. could cut its total energy use by 20-25% within a decade. In 2011 the U.S. government issued a call for achieving 20% improvement in energy efficiency in buildings by 2020 and double U.S. energy efficiency by 2030.

Energy efficiency has been seen as the best solution to our energy and environmental problems, so it makes sense to focus on energy efficiency. Could energy efficiency measures be implemented in the U.S. today? It's a question that we are hesitant to answer; we are just not that desperate…yet.

Researchers also have estimated that the world could save almost 3/4 of its energy through efficiency measures. Better building construction, insulated walls and windows, reducing heating and air-conditioning use, and turning off appliances when not in use could save a lot of energy.

Basically speaking, we have part of the energy solution in our hands, but we are not as willing to embrace it as our Japanese counterparts did. Maybe if forced, we would follow their example. Maybe…

China

China is one of the most significant and impressive 21st century phenomena—a giant rising from the ashes of the communist era. This is an unprecedented development, since most of the post-communist societies, including that of the former Soviet Union, sank into political inequity and economic misery.

China, by contrast, is progressing economically, and will surpass the U.S. in the near future. Its political system is still based on old semi-communist principals, mixed with capitalist intentions and methods, but that has not prevented China from achieving great industrial and economic progress.

What we need to be aware of, however, is the huge amounts of fossils China is consuming and the millions of tons of pollutants its power plants, factories, and cars are emitting into the atmosphere, soil, and water.

China's population is increasing—doubling from 750 million in 1970 to nearly 1.5 billion today. With the industrial growth of late, the new generations of Chinese have achieved an unprecedented economic wealth.

They insist on achieving the Chinese Dream, which similar to the American dream requires material goods and benefits that make life easy and pleasant—large houses, cars, vacations—all of which require a lot of energy. To keep the population happy—which is the main goal of the ruling class—China is increasing electric power generation by huge amounts. There are, for example, several coal-fired power plants constructed and put in operation around the country every month.

Please note that the power generation capacity in Figure 2-16 was added in just one year, and that most of it represents power generated by newly constructed coal-fired power plants. The previous years are similar, and the following years will see even greater power generation additions in the country—again mostly coal-fired power plants.

Most of China's total energy supply is used for electric power generation via coal 70%, with crude oil second at 18%, hydro power 6%, natural gas 5% and the renewables at about 1%. That makes China the greatest consumer of coal in the world.

And as of 2014, China is also importing as much crude oil as the U.S.

The U.S. and China together import over 13 million barrels of crude oil daily. This is almost as much as the oil imports of all other countries combined. And while the U.S. is reducing imports by adding domestic production and efficiency measures, the projections for China's energy future point to more coal-fired power plants and more cars on the roads.

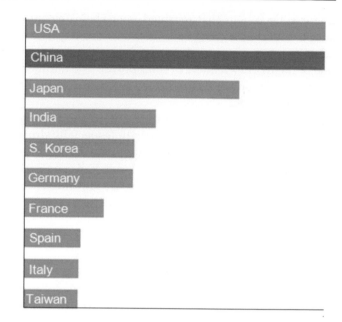

Figure 2-17. Global crude oil imports and use, 2014

China is now the greatest fossils user and largest GHG emitter in the world.

It is clear from Figure 2-18 that China's energy imports and their use are projected to double and triple within the next few decades, and there is no way to stop China from achieving these goals.

The situation is quite similar in most other developing countries. Their population is increasing and demanding better life, which requires more energy. With

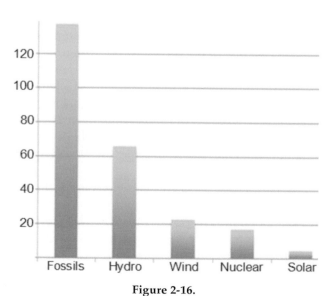

Figure 2-16.
Power generation added in 2013 (in thousand TWh)

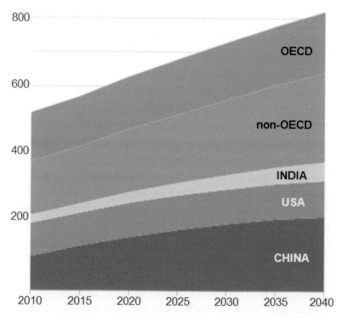

Figure 2-18. Global energy consumption (in quadrillion Btu)

that, the demand for energy is growing faster than ever before.

The 1.2 billion people in India are closely following the Chinese path to economic progress, albeit not as successfully. India, like China, is fed by many coal-fired power plants, which add to the already enormous amount of GHGs pumped into the atmosphere.

The economic prosperity of the developing countries leads to increased demand for energy and fossils more than ever before. The result is more pollutants than ever in the short run, and an imminent depletion of the fossils in the near future.

Above we have two extreme examples:

a) Japan's responsible energy efficiency increase in response to the energy crises, and

b) China's and other developing countries unstoppable increase of fossils use in response to their growing populations and their needs.

The increase of fossils use by China and the developing world is much larger than any counter-measures taken by Japan and the all developed countries put together, including the U.S.

Our Energy and Environmental Paths

So, the world slowly but surely is sinking into a deeper imbalance in energy supplies, energy security, and environmental protection. Energy related measures led by the U.S. and the developing countries cannot change the situation. Not in our lifetime, and very likely never!

Figure 2-19 shows the path we have been walking thus far and where we are headed in the future. They are both very clear—both leading to steadily increasing environmental pollution and quickly approaching fossils depletion.

As things are going now, there is no chance for reversal, and the feasible alternatives are too few and far between.

Some experts go as far as to insist that even if we start what we should've started 50 years ago, it is too late to do any good. The Earth and its atmosphere are polluted and poisoned to the point of no return, so we can only sit like the proverbial duck on ice, waiting for things to change.

And we have seen some of the signs: melting ice cap, increased draughts and floods, more violent storms and hurricanes, and temperature extremes.

And, as the old wise men say, "We must lie in the

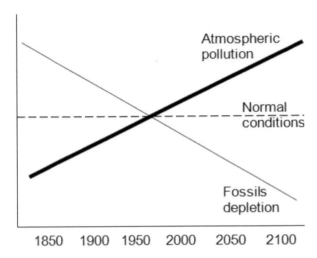

Figure 2-19. The global energy/environment imbalance

bed we have prepared for ourselves."

Looking objectively at all of this, one cannot help but notice that we humans are slowly but surely cutting the branch we are sitting on. Why worry about energy security if soon enough there will be no more fossils? Why worry about more fossils if soon we will not even be able to breath the air around us?

Far fetched? Maybe, but spending a week breathing Beijing or Los Angeles summer smog is enough to remind us that we are getting close to that day.

And looking closely at the technical specs, one can easily see that 50 years ago oil in the Middle East was easily accessed, but today it is miles beneath the surface—as much as 10 miles down. Too, deep-well oil is also heavier and dirtier, so it is much more expensive to pump out and refine.

One must be blind to not see that the branch we are sitting on is getting more crowded and heavier by the day, so the deeper we cut, the closer we come to falling.

We can blame the developing countries for pushing us down the energy cliff, but we must remember that we—the developed countries—created that cliff and didn't do much to avoid it.

ENERGY SECURITY CHALLENGES

The risks to our national and global energy securities today are dominated by a number of challenges, as follow:

• Increasing growth in energy consumption and demand worldwide is the major issue affecting the global energy reserves. In the last half century, energy use tripled. Over the next several decades, if current trends continue, energy use will exceed that of all energy used to date combined.

If current consumption trends and government policies continue, we will see over 50% increase in global energy consumption by 2035, with 70% coming of demand from developing countries.

- Energy supplies are limited, and yet we continue to use them as if they were unlimited. The competition for energy resources is increasing—again mostly by developing countries. Each nation seeks to secure its own energy supply at all cost and by trying various strategies—all resulting in energy use increase.

Some countries focus on diversifying their energy suppliers in order to create a buffer against fluctuations in fuel market prices, and as protection against political sanctions from oil and natural gas producing nations.

This is not a new challenge, but the competition seems to be increasing as the developing nations seek to achieve modernization. China, India, Russia and others are competing for the same limited energy resources. The stakes are high, because those countries which fail to secure adequate energy supplies will fall short of achieving "the dream."

- Environmental impact is playing an increasingly important role in the energy security scenario as well. Carbon emissions from fossil fuels bring pollution and climate change, which could lead to higher global temperatures, rising sea levels that would threaten to submerge coastal regions, prolonged droughts, more frequent violent storms, and threats to many wildlife land- and water-based species.

Because of that, any use of fossil fuels is under scrutiny, and while it won't be stopped, people who use fossils will need to pay for the cleanup. This complicates the global logistics and financial pictures and will eventually determine the path taken by many developed countries.

The developing economies are not paying much attention to the facts. Most of them are determined to use as much fossils, and emit as much pollution as they see fit to keep the population satisfied.

This will further complicate the world energy distribution, and is becoming a great threat to our energy security, and even more challenging to the health of our environment.

The energy game is fast changing. Energy sources, power generating technologies, transport methods, environmental changes, economic development—all play great roles in this change.

The notion of a "balanced energy" use in the world is just a dream that will take a long time to materialize…if ever.

The Game Changing Issues

There is no question that we are living a world in which energy plays a major role. In the U.S., we have huge amounts of fossils—coal, natural gas, and crude oil. We also have equally large amounts of renewables—hydropower, biofuels, geo, ocean, solar, and wind energy.

While the benefits are quite clear—energy abundance and cheaper prices—the hidden costs are just now being calculated. All these technologies come with some problems and risks.

Our short-term energy security is affected in a positive way by the availability and lower cost of the fossil technologies and fuels. In the long run, however, human health, lives, and the global environment in general are affected in a negative way.

Most of these factors and the related cases have not yet been fully recognized, accepted, calculated and applied to the energy security formula. But, we must consider drawing a line in the sand, where the drive for energy security intersects the borders of human health, safety, and wellbeing. Having energy today, only to suffer tomorrow, is not what energy security is about.

This critical line is something we need to understand and discuss more openly, finding solutions that ensure our energy security with minimal negative effects on humans and the environment.

It is easy to see from Table 2-3 that all energy generating technologies and fuels have problems. What is more difficult to see is the degree to which each of these problems is related just to energy output, and which affect the humans and the environment. It is even harder to determine what must done about it.

In more detail:

- Coal is dirty from the beginning to the end of its life cycle, but we need its energy, so its use will only increase. That means that we simply must be aware of the problems and take the necessary precautions. This is happening increasingly in the developed nations, including the U.S.

- Natural gas is cleaner, but still a serious polluter. Its production and use will also only increase, so we need to implement safety measures that control and limit the damages from hydrofracking and power generation—some of which have been ignored in the midst of the present energy frenzy.

- Crude oil is a heavy polluter, which sickens and kills many people worldwide. We need it to drive our cars and trucks, so we need to be aware of the harmful effects and implement procedures to reduce them.

- The issues of nuclear power can be summarized by one word; Fukushima. That word contains the risk of devastation of property and life that most people are not willing to take. Because of that, the future of nuclear power hangs in the balance—one more serious nuclear accident might put an end to global nuclear power as we know it.

- Hydropower is a clean technology if you are not personally being displaced in order for the hydro dam reservoir to be built. Or, God forbid, you are not living downstream when the dam wall breaks.

- Biofuels are good and much larger quantities are needed, but we must be aware of the damages done to humans and lands by their production, use, and especially their expansion into the food production sector.

- Geo- and ocean-power are still in their infancy and have a long way to go before becoming major energy sources. They hold great promise, however, for future generations.

- Solar and wind suffer from the same serious problem—variability. If there is no sun or wind, there is no power. Because we cannot control the sun and wind, we cannot count on those for reliable energy. For now, they will remain gap-fillers, instead of becoming major baseload power providers.

In summary, the U.S. and the world are witnessing an energy renaissance driven by new 21st century game changers, summarized as follow:

Table 2-3. Today's energy sources and their major issues

Technologies and Fuels	Pertinent Issues
Coal	Mine accidents and local air, soil, and water contamination. Power plants air, soil, and water contamination.
Natural gas	Well accidents and local air, soil, and water contamination. Power plants air, soil, and water contamination.
Crude oil	Well accidents and air, soil, and the related water contamination. Large scale GHG emissions by transportation vehicles.
Nuclear	The threat of nucelar accidents hinders the future development of the global nuclear power.
Hydropower	Destruction of large land areas, population centers, and cultures by hydro dams and their use.
Biofuels	Competition with food crops the related human hardshop. Large land areas deforestation.
Geo power	Available mostly in remote areas and in small quantity.
Ocean power	The technology is not developed for large scale power generation.
Solar	Power variability—depending on sun availability—and lack of energy storage. Some solar technologies contain toxic and carcinogenic materials.
Wind	Power variability—depending on wind availability—and lack of energy storage. Visual and audio effects and damages to human health.

- Solar and wind power ramp-up around the world
  — Replacing fossil-fired power plants with wind and solar

- Coal use decrease (in the U.S.)
  — Replacing coal-fired with gas-fired power plants
  — Increased coal exports to Asia and Europe (resulting in global coal use increase)

- Nuclear power on-and-off shifts around the world
  — Increased coal- and gas-power use to compensate for nuclear power decrease.

- Hydrofracking ramp-up (in the U.S.)
  — Reduced volume of crude oil imports (50% of total oil use)
  — Natural gas and oil products exports (resulting in increased global gas and oil products use)

The other major energy sources—nuclear power, hydropower, geo- and ocean-power, and the biofuels— also play a significant role in the new energy game, but they cannot be considered *"game changers"* for reasons we will discuss in detail later on in this text.

Notes and References

1. The four energy changers in the energy sector. http://www.eia.gov/countries/cab.cfm?fips=ch
2. China Energy. EIA. http://www.eia.gov/countries/cab.cfm?fips=ch
3. Game changers: Five opportunities for US growth and renewal. http://www.mckinsey.com/insights/americas/us_game_changers
4. The Four "gamechangers" of the energy sector, http://www.energypost.eu/many-gamechangers-energy-sector/
5. *Methodology for Prioritization of Investments to Support the Army Energy Strategy for Installations,* by George Alsfelder, Timothy Hartong, Michael Rodriguez, and John V. Farr Center for Nation Reconstruction and Capacity Development United States Military Academy, West Point, NY 10996 http://www.usma.edu/cnrcd/CNRCD_Library/Energy%20Security%20Paper.pdf
6. Energy game changers-IQ2 talks. http://www.energypost.eu/many-gamechangers-energy-sector/
7. *Photovoltaics for Commercial and Utilities Power Generation,* Anco S, Blazev. The Fairmont Press, 2011
8. *Solar Technologies for the 21st Century,* Anco S Blazev. The Fairmont Press, 2013
9. *Power Generation and the Environment,* Anco S. Blazev. The Fairmont Press, 2014

Chapter 3

Crude Oil

(The Weak Link)

*The U.S. energy security problems today are not due to energy shortage per se—we have plenty of domestic energy—but **oil imports** to excess.*

—*Anco Blazev*

The latest energy revival, attributed to discovery of large oil and natural gas reserves, and the use of new extraction techniques as discussed in this text, is great for our country and (albeit partially) for the entire world.

Because of the importance of these developments, we will take a very close look at what it takes to produce, distribute, and use the major energy sources, starting with the fossils—crude oil, natural gas, coal, and nuclear power. We then review the renewable energy technologies, in their drive to compete with the fossils, and as the greatest promise in the upcoming post-fossils energy future.

Note: Yes, we put nuclear in the fossil energy category, because uranium is dug out of underground mines and transported, just like coal and oil. And just like the rest of the fossils, uranium is also being depleted quickly. At this rate of exploitation, it may not see the end of the 21st century.

The U.S. nuclear industry and military import a great quantity of uranium—about 18,000 tons annually—or about 30% of world production, thus we heavily depend on the availability and distribution of uranium around the globe.

So, obtaining our key imports, crude oil and uranium, depends on the risks in the variability of the producer countries' production capabilities and their willingness to export, as well as on the security of global transport routes.

The immediate, and biggest, problem for our energy security, however, is the availability, price, and transport of crude oil.

Global competition for crude oil is increasing, prices are going up, and the threat of terrorist attacks in transport routes is growing.

All this makes crude oil a major stumbling block in achieving energy security, which in turn renders the dream for energy independence a practical impossibility.

We still import half of the oil we use every day, thus exposing our energy and national security to the vulnerability of oil imports from unstable and unfriendly nations. Then, we process and export half of the imported oil in order to keep jobs and industries running, which adds to the crude oil imbalance.

Without a doubt, the fossils have been and still are the foundation of our energy sector and a main factor in our economic prosperity. Just imagine for a moment life without coal, gas, or oil. What would this world be without them? Where would the American Dream be without them? It is actually hard to imagine, but later on in this text we will dare take a peak at a fossil-less existence. Not a pretty sight…

The fossils are our friends and enemies. They consistently provide abundant and cheap energy, but at the same time they pollute the air, soil, and water with dangerous chemicals. In addition to contaminating the environment, the different polluters also damage wildlife and human health.

Worse, they can kill people. A lot of things can happen during the long lifetime of the fossils, and when things go wrong the consequences are dire. A number of intentional or unintentional incidents and accidents await us every step of the way in mines, on drill platforms, on transport routes and in processing facilities.

The fossils go through a long cradle-to-grave process. They are dug, or pumped, out of the ground, transported many miles, burned, and finally used as electricity or gasoline fuel to provide power to homes and businesses, or run our vehicles.

Each step of this process has its special advantages and disadvantages as far as safety, efficiency and environmental impact issues are concerned. Each step also has its vulnerabilities, which if not properly managed could create a disruption in the supply chain, and in some cases even cause environmental disasters and deadly accidents.

Let's dig deeper into the fossils by looking at their technical, logistic, financial, political, and regulatory details—starting with *crude oil*, since it is the most important as well as the weakest link in our energy security portfolio.

How and where is crude oil produced? How is it transported, processed, distributed and used? What are its market characteristics? What is its impact on our economy and the world's environment? What is its effect on our energy security and its role in our energy independence?

We will review all these subjects in the light of the present and future energy security concerns and the related issues.

CRUDE OIL

Crude oil is second to none in importance in a number of key areas. It is widely used in the U.S. and around the world:
a) primarily, as a transportation fuel,
b) partially, as a convenient power generation fuel in some areas where other fuels are not available, and very importantly,
c) as a raw material in a number of major industries.

Our dependence on its magical qualities during the last century has been total and unprecedented, albeit unsustainable in the long run. Within a single century, we managed to burn, and otherwise use, a major part of the oil reserves which it took Nature millions of years to create.

There is no possible way to create more crude oil, so we must be very careful how we handle it now and in the future. The situation is very serious and deserves our full attention and utmost care. Using less oil now, will leave more for the basic needs of future generations.

Unfortunately, reducing the use of crude oil would mean significantly slowing down the economic development in many countries, and most of them are not willing to accept that. Overuse, at the present rates, will exhaust the already semi-depleted oil supply within several decades and force us into an era of oil-less exis-

tence. This is truly a catch 22 situation…

Crude oil is the lifeblood of our nation's economy and is what drives our technological and social progress. Just like blood flows in our veins, so does the vehicular traffic flow across the country, non-stop all day and all night, all through the year.

Vehicles of all types and sizes travel the roads and highways, the waterways and the skies, carrying people to work or play. A fleet of large trucks, boats, and trains is carrying millions of tons of goods across the country. From the agricultural fields of Imperial Valley, California, to the streets of New York, they carry fruits and vegetables. From Main to Arizona they haul seafood.

Cars, trucks, boats, trains, and planes fill their tanks with oil products—gasoline and diesel—every day. This continuous flow—millions of tons of vehicles and cargo—must keep going in order to keep the economy going and the people happy. This is something only crude oil can do.

Got that? No oil: no vehicles on the roads. No vehicles on the roads: no people at work and no vegetables and fish on the dinner tables.

Electric and hybrid vehicles, fuel cells powered vehicles, and all such hi-tech wanna-be-substitutes cannot compete with the power, speed, reliability, and flexibility of gas and diesel cars and trucks. And even less can they compete with the fossils for driving our 18-wheelers, trains, and planes. Not for now, and very likely never.

With the finality of fossil fuels we see a simple and scary reality—no crude oil, no traffic on the roads! Simple and final…the lifeblood flow stops. Period. End of story, and end of the American Dream. The consequences to the economy would be disastrous, even if the vehicular flow is slightly reduced for lack of crude oil.

Here's hoping our political and economic powers consider the importance of oil today and what could happen, if and when we run out of it. The consequences of a fossil-less future are very serious and they must think clearly before deciding to drill another oil well or exporting crude oil products abroad.

In the Beginning…

Immediately after the Big Bang there were no coal, oil, or natural gas deposits on Earth. It took millions of years to complete the conversion of organic matter into coal, crude oil, and natural gas. As a matter of fact, the process is not fully completed yet. This is evidenced by the different coal layers which show different stages of coal development—from peat (the most recent) to anthracite (the oldest) coal products.

The oil and natural gas formation process, in most cases, were somewhat different from that of coal. The origins of oil and natural gas can be traced to the much shallower and warmer oceans, where dead animals, vegetation, and other organic matter was continuously falling and piling on the bottom.

The bulk of the organic matter were tiny, surface-dwelling organisms called plankton, which includes several animal species (zooplankton) and plant species (phytoplankton). The phytoplankton uses photosynthesis to capture the solar energy that is the basis of marine food chains.

The organic materials would sink to the floor eventually, to join with the inorganic matter, consisting of sand and dirt, carried in by the rivers and land runoff. The different materials slowly mixed up on the ocean floor, resulting in a thick layer of ocean-floor sediments.

As the ocean levels rose, the pressure of the overlying water compacted the sediments. A number of natural processes, such as crystallization and cementation of inorganic minerals eventually converted the soft sediments into solid material.

As time went by, the ocean-floor sediments got buried deeper and deeper, and the water got deeper, so the sediment layers were subjected to ever-increasing, extremely high pressures and temperatures. This resulted in the formation of the layered solids under the ocean floor, called sedimentary rock.

During the long exposure to high pressure and elevated temperature, organic material in the sedimentary layers got *cooked*, very similar to preparing food in pressure cookers, but at much higher pressure and temperature. Of course, the process took a very long time, which is the key to producing the large quantities of fossil materials with proper quality that we use today.

Initially, the sediments *cooked* to a waxy substance called kerogen. Kerogen, as in tar and shale oil deposits, is similar to the peat segment of the coal formation process. It represents a not-yet completed process of conversion to crude oil—sort of half-cooked crude oil.

Like peat, it has different (inferior) heating properties than the respective final products, crude oil and coal. As the cooking processes continued, the sediment was eventually converted into a combination of organic liquids and gases which we call *petroleum*.

Note: Although the term *petroleum* is usually used for liquids (i.e. crude oil), officially petroleum refers to both liquids and gases. So when we talk about *petroleum*, we do refer to a group of fossils which contains both

crude oil and natural gas and their derivatives and by-products.

Petroleum contains some of the energy (sunlight) that was captured many, many years ago by the plankton on the ocean surface. This energy is what we need badly today, and which our comfort and progress depend on. Because of that we now extract millions of gallons from the large petroleum deposit areas, which were formed by migration and pooling of the liquid oils and gaseous natural gas into large underground reservoirs.

There are also small petroleum deposits in many sedimentary rock formations around the world, but reaching them, due to the insufficient quantities and the great effort and expense involved, makes most of them technologically difficult and economically infeasible to extract with the present state of the art. Many of them, however, could and will be exploited at the right time—most likely when petroleum prices increase enough to make the process economically feasible.

Note: Remember, when we talk about *petroleum*, we do refer to a group of fossils, which contain both *crude oil* and *natural gas*. Below we will take a close look at the present day crude oil production and use, followed by natural gas.

In the Recent Past...

We know that oil in the large deposits comes mainly from strata of coarse sandstone, which was formed by sand deposited by water ages and ages ago. We've also noticed that there is no oil in flat strata, so we must deduce that the oil flowed from the sediment layers into the sandstone strata, where it was held as in a sponge, and where it remained throughout the millennia.

After the initial Big Bang and the following overheating at the very beginning, the Earth mass grew cooler. At that time and later on, it went though a number of processes, such as *folding*, during which horizontal movements press inward and move the rock layers upward into a fold or anticline.

Note: *Faulting* resulted where the layers of rock crack and one side shifts upward or downward. *Pinching out* happened where a layer of impermeable rock was squeezed upward into the reservoir rock.

While undergoing these processes, the Earth's surface layer—the crust—twisted and wrinkled, so that oil and gas are now found in the large sandstone wrinkles—with the gas on top and oil on the bottom.

If cracks develop in the strata, the oil would usually run away by gravity to find another place to settle. This is how large reservoirs are formed.

There is also trapped gas in the mix, which is usually under pressure, and is constantly looking for a way out. If we know where it is, we could easily drill to it and pipe it for use in a house or business.

Not very long ago, there was a gas deposit close to Niagara Falls that constantly released gas. One day somebody drove a pipe down in the ground and lit the gas. The flame was quite high and was visible from far away. This, however, is a rare occurrence, since the oil and gas deposits are usually much deeper in the ground.

The presence of gas is often an indication that there is oil underneath too. This is how the first colonists discovered petroleum in New York. They, however, didn't know what to do with it, while the native Americans nearby used oil found in the local springs as medicine. Later on they figured out that oil can be sold, so they would soak their blankets with oil from the springs and wring them out to sell the oil. This was the first commercial oil extraction process.

Later on, the first "gusher" was discovered by accident. Workmen had drilled 500 feet down when a high pressure stream of crude oil burst forth, hurling tools and people high up into the air. They just let the oil gush out until the pressure was reduced enough. They did not know how to stop it, nor what to do with it, so a lot was wasted that way.

A large gusher in Lakeview, California, threw up 50,000 barrels of oil a day, to a height of 350 feet for a long time. The steady oil column sprayed the country for a mile around. The oil flowed away as a river, and no one could stop it or store it. Finally, a large storage tank was built around the well with stones and sand bags.

Yet, oil could not be efficiently collected, processed, or used for a long time.

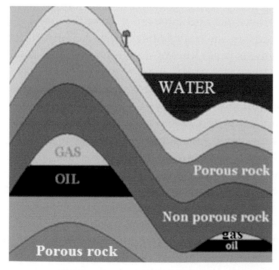

Figure 3-1. Crude oil deposits

The high pressure that oil comes up with sometimes is due mostly to the high pressure of natural gas in the deposit mix. This is confirmed by the fact that the pressure is reduced with time, after the oil is pumped out for awhile, and/or when the gas pressure is reduced.

So in the past, the lucky land owner who thought that there was oil underneath would build a derrick over it with the intention of collecting some of the oil and selling it. The derrick principle is used today too, and it usually consists of four wooden or metal beams, 30-100 feet high, firmly held together by crossbeams. It is positioned over where the oil well is to be dug. An engine house nearby provides power for the drilling machine.

An iron pipe eight or ten inches in diameter is driven slowly down in the ground until it comes to the oil-containing rock. At this point regular drilling operations begin. A pulley mounted on top of the derrick holds a heavy-duty rope to one end of which are attached the drilling tools. The drilling tools go down in the pipe and drill a hole in the rock day and night. While the drill is busy making the hole, a sand pump sucks out water and loose bits of stone from the drilling hole.

As soon as the drill reaches the bottom of the strata, the sides of the bore hole are cased to keep the water out, and the drilling continues. This time the drill penetrates the oil-bearing sandstone and if there is oil in it, it will gush out and the lucky land owner will join the ranks of oil millionaires.

Often, however, drill as much as you may, oil is nowhere to be found. That's because although it is underground, it is near the surface in some places, but more often than not it is deep down to 1,000-5,000 feet depth. To reach these depths, special and very expensive equipment and advanced procedures are usually needed.

In some cases, the oil is not under much pressure, so it is hard to determine where the oil deposit is even if it is hit directly. And this is where the modern oil drilling industry comes in handy.

Modern drilling rigs with their sophisticated drilling tools, computerized controls, and GPS navigation can find and reach oil and gas many thousands of feet under the surface—be it land or sea—with amazing precision and speed.

A new, more sophisticated and efficient, method of horizontal drilling is promising extraction from a depth of several miles, and from locations that were impossible to even imagine accessing a few years back.

Hydrofracking is a new process which has revolutionized the oil and gas industry. Later, we will take a closer look at this process, its advantages and issues.

Crude Oil Properties

Crude oil is actually a mixture of many (dozens and even hundreds) of different species of organic chemical compounds, most of which are classified as hydrocarbons. Hydrocarbons are chemicals that contain both hydrogen (H-hydro) and carbon (C-carbon).

Table 3-1. Hydrocarbon species in crude oil and natural gas

Formula	Name	# C atoms	MP	BP	State
CH_4	Methane	1	-183	-162	Gas
C_2H_6	Ethane	2	-172	-89	Gas
C_3H_8	Propane	3	-187	-42	Gas
C_4H_{10}	Butane	4	-138	0	Liquid
C_5H_{12}	Pentane	5	-130	36	Liquid
C_6H_{14}	Hexane	6	-95	69	Liquid
C_7H_{16}	Heprane	7	-91	98	Liquid
C_8H_{18}	Octane	8	-57	126	Liquid
C_9H_{20}	Nonane	9	-54	151	Liquid
$C_{10}H_{22}$	Decane	10	-30	174	Liquid

Methane and ethane (one and two carbons respectively) are the main components in natural gas.

Propane (three carbons) is most commonly used as gas, but can be liquefied under modest pressure. Its different forms are used in countless applications, where petroleum products are used for energy; cooking, heating, and transportation.

Butane (four carbons) is added in the winter to motor fuels, because its high vapor pressure helps with cold starts of car and truck engines. It is also used to fill cigarette lighters at slightly above atmospheric pressure. It is also a main fuel source in many developing countries.

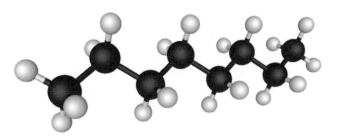

Figure 3-2.
Octane is one of the hydrocarbons found in crude oil.

Octane, a main component of crude oil, has 8 carbon molecules, and is one of the most useful components of the motor fuels. It is part of the alkanes group, which start with pentane (five carbons). The alkanes are refined into gasoline, while hydrocarbons with a higher carbon number from nonane (nine carbons) to hexadecane (sixteen carbons) are refined into diesel, kerosene, and jet fuel.

Alkanes with more than 16 carbon atoms are not used for fuel, but are refined into fuel oil and lubricating oils. At the high end of the alkanes range is paraffin wax which has 25 carbon atoms, and asphalt with 35 and more carbons. Most of these fractions are usually cracked (refined) by modern refineries into lighter and more valuable products.

Some other types found in crude oil are paraffins, naphthenes, aromatics, or combinations of these, such as alkyl naphthenes and aromatics, and the polycyclic compounds. Mercaptan, thiophene, and others are also present in high concentration as well. There are also olefins, which usually are not present in crude oil, but are formed during decomposition of the crude oil components during high temperature distillation processes.

The shortest molecule alkanes, those with four or fewer carbon atoms, are the petroleum gasses, which remain in a gaseous state at room temperature. Depending on demand and the cost of recovery, these gases are either flared off, sold as liquefied petroleum gas under pressure, or used to power the refinery's own processes.

The cycloalkanes, also known as naphthenes, are saturated hydrocarbons which have one or more carbon rings to which hydrogen atoms are attached according to the formula CnHn. Cycloalkanes have similar properties to alkanes but have higher boiling points.

The aromatic hydrocarbons are unsaturated hydrocarbons which have one or more planar six-carbon rings called benzene rings, to which hydrogen atoms are attached with the formula CnHn. They tend to burn with a sooty flame, and many have a sweet aroma. Some are carcinogenic.

To complete the soup mix, a number of other chemicals, some of which are inorganic (non carbon and hydrogen containing) are found in raw crude oils as well.

Sulfur produces the most troublesome byproducts of the oil refining process, in the form of free (elemental) sulfur (S) and hydrogen sulfide (H_2S). They cause corrosion and produce sulfur dioxide (SO_2) which is toxic and creates acid rain when burned. The other major sulfur compound, H_2S, is a vicious poison, which paralyzes the olfactory nerves so that its victim is unaware of its presence and can choke to death.

On the other hand, the petroleum industry produces large quantities of sulfur by converting the H_2S by-product to elemental sulfur, which is quite useful in other industries. A lot of nitrogen is produced also, which also can be captured and used in other applications.

Salts, such as chlorides, sulfates, and carbonates (compounds of sodium, calcium, and magnesium) are found in crude oil in liquid or small particles of these salts, which are expressed in pounds of salt (NaCl equivalent) per thousand barrels of crude.

These salts cause corrosion and deposits in the heating and heat transfer equipment, by adhering to the surfaces of boilers and pipes. Worse, when in the presence of H_2S and H_2O, they are extremely corrosive and can damage all metal equipment, even stainless steel.

Of course, and very importantly, crude oil and all its organic components are highly flammable. They have different boiling point ranges.

It can be easily seen in Table 3-2 that the boiling point increases as the compounds get heavier—their MW increases.

Table 3-2. Crude oil fractions

Compound	Boil°F	MW
LPG	-44–31	44–58
Gasoline	160	100–110
Jet fuel	380–520	160–190
Diesel fuel	520–650	245
Gas oil	650–800	320
Residuals	800–1000	—
Vacuum gas	800–1000	430
Crude coke	2,000	2,500

where:
Boil°F   is the boiling point in degrees F, and
  MW   is the molecular weight of the different compounds.

Most crude oil compounds can be ignited with open flame and at low temperatures, which can be good, but could also be quite dangerous. For this purpose, we use a standard called auto-ignition temperature (AIT), which is used for proper use and safety purposes.

AIT is basically a measure of the temperature at which a vapor from a particular compound will ignite spontaneously (in the absence of a flame).

Note: AIT should not be confused with the boiling point of the respective substances. Gasoline and naphtha, for example, boil at 160°F or less, but could ignite spontaneously on coming into contact with a hot surface at 600°F or above, even if in the absence of an open flame.

Classification

Another important crude oil measure is the American Petroleum Institute gravity, or ***API gravity.*** API is a measure of how heavy or light a petroleum liquid is compared to water. American Petroleum Institute has an inverted scale for denoting the lightness or heaviness of crude oils and other liquid hydrocarbons. Calibrated in API degrees (or degrees API), it is used universally to expresses a crude's relative density in an inverse measure. The lighter the crude, the higher the API gravity, and vice versa, because lighter crude has a higher market value. Wow. Who knew...

If its API gravity is greater than 10, it is lighter and floats on water; if less than 10, it is heavier and sinks. API gravity is thus an inverse measure of the relative density of a petroleum liquid and the density of water, but it is used to compare the relative densities of petroleum liquids. For example, if one petroleum liquid floats on another and is therefore less dense, it has a greater API gravity.

Although mathematically, API gravity has no units (see the formula below), it is nevertheless referred to as being in "degrees." API gravity is gradated in degrees on a hydrometer instrument. The API scale was designed so that most values would fall between 10 and 70 API gravity degrees.

$$API = \left(\frac{141.5}{S.G.}\right) - 131.5$$

where:
S.G. is the specific gravity, which is the ratio of the density of the material at 60°F to the density of water (accepted as 1) at that temperature.

For example, water at 60°F has

$$API = \left(\frac{141.5}{1}\right) - 131.5 = 10° \text{ API.}$$

Crudes can be classified as "light" or "heavy," a characteristic which refers to the oil's relative density based on the American Petroleum Institute (API) Gravity. This measurement reflects how light or heavy a crude oil is compared to water. If an oil's API Gravity is greater than 10, it is lighter than water and will float on it. If an oil's API Gravity is less than 10, it is heavier than water and will sink.

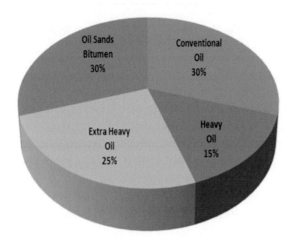

Figure 3-3. Global crude oil reserves by type

- Light crude oil is defined as having an API gravity higher than 31.1°API. Brent Crude has an of 35.5°, and gasoline's API is 50°.

- Medium oil is defined as having an API gravity between 22.3°API and 31.1°API.

- Heavy oil is defined as having an API gravity below 22.3°API.

- Asphalt is the heaviest of the petrochemicals with average API gravity of 8° or less.

Lighter crudes are easier and less expensive to produce and bring the highest prices. They generally have a higher percentage of light hydrocarbons that can be recovered with simple distillation at a refinery.

Heavy crudes *cannot* be produced, transported, and refined by conventional methods because they have high concentrations of sulfur and several metals, particularly nickel and vanadium. Heavy crudes have densities well above that of water. Some heavy crude oils are also known as "tar sands" because of their high bitumen (tar) content.

Crude oil quality is also measured and qualified by their sulfur content. Crude oils with low sulfur content are classified as "sweet." Those with a higher sulfur content are classified as "sour."

Sulfur content is generally considered an undesirable characteristic with respect to both processing and end-product quality. Sweet crudes are usually more desirable and valuable than sour crudes.

Unfortunately, the light and sweet crude oil supplies are being exhausted at a high pace, and will be eliminated the fastest. What we have left now are the heavy, sour, and much dirtier oils, which also will be gone soon enough.

Toxicity

Toxicity of crude oil and its derivatives refers to how harmful these materials might be to humans and other living organisms. Generally, the lighter the oil, the more toxic it is. EPA has classified crude oils in four categories that reflect how the oils would behave in a spill and its aftermath:

- *Class A* crude oils are light and highly liquid, these clear and volatile oils can spread quickly on impervious surfaces and on water. Their odor is strong and they evaporate quickly, emitting volatiles. Usually flammable, these oils also penetrate porous surfaces such as dirt and sand and may remain in areas into which they have seeped. Humans, fish, and other biota face the danger of toxicity to Class A oils. These high quality light crudes and the products produced from them are in this class.

- *Class B* oils are considered less toxic than Class A, these oils are generally non-sticky but feel waxy or oily. The warmer it gets, the more likely Class B oils can be to soak into surfaces and they can be hard to remove. When volatile components of Class B oils evaporate, the result can be a Class C or D residue. Class B includes medium to heavy oils.

- *Class C* oils are heavy, tarry oils (which include residual fuel oils and medium to heavy crudes). They are slow to penetrate into porous solids and are not highly toxic. However, Class C oils are difficult to flush away with water and can sink in water, so they can smother or drown wildlife.

- *Class D* are non-fluid, thick oils that are comparatively non-toxic and don't seep into porous surfaces. Mostly black or dark brown, Class D oils tend to dissolve and cover surfaces tightly when they get hot, which makes cleanup much harder. Heavy crude oils, such as the bitumen found in tar sands, fall into this class.

The new planned Keystone Pipeline is scheduled to transport tar sands oil from Canada through several U.S. states to Gulf Coast refineries. The battle between pipeline operators and the locals (supported by environmental groups) is presently centered on the danger of soil and water table contamination resulting from spills of these Class C and D oils.

In any case, all types of crude oils and their derivatives are considered hazardous compounds that can harm living things and humans under some circumstances. Because of that, special care must be exercised when handling, or otherwise dealing with, all types of oils and petrochemicals.

Combustion Risks

An important measure of the oil quality and safety is its vapor pressure, also called raid vapor pressure (RVP). Different oils have different RVP, which determines the volatility, or how fast the substance evaporates, and the related explosion risks. Crude oils with higher RVP are usually more prone to explosions during storage and transport than those with lower RVP.

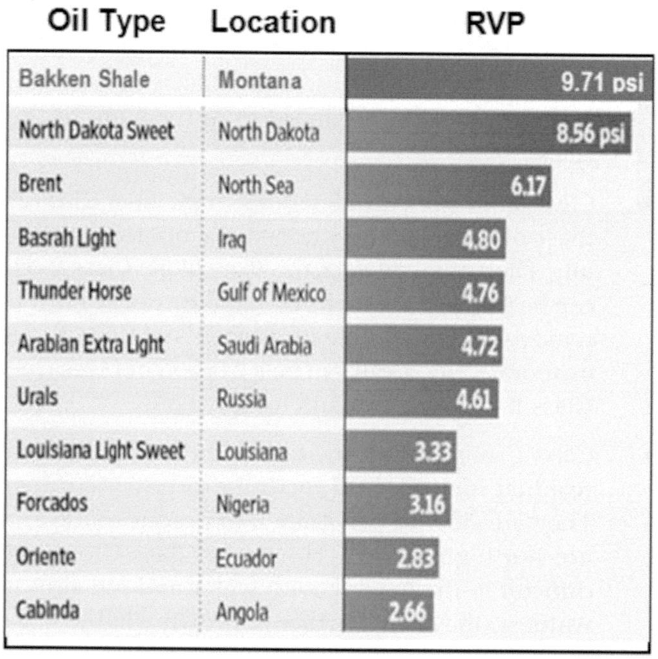

Oil Type	Location	RVP
Bakken Shale	Montana	9.71 psi
North Dakota Sweet	North Dakota	8.56 psi
Brent	North Sea	6.17
Basrah Light	Iraq	4.80
Thunder Horse	Gulf of Mexico	4.76
Arabian Extra Light	Saudi Arabia	4.72
Urals	Russia	4.61
Louisiana Light Sweet	Louisiana	3.33
Forcados	Nigeria	3.16
Oriente	Ecuador	2.83
Cabinda	Angola	2.66

Figure 3-4. RVP of different oils per location

Crude oil from the Bakken Shale formation, for example, contains several times the combustible gases as average crude oil from other places. This raises questions about the safety of producing and shipping it by rail or pipes across the U.S.

The volume of oil moving by rail from the Bakken Shale had soared to nearly a million barrels daily at the end of 2013; way up from about 300,000 barrels a day in 2010.

The rapid growth of oil production has increased the transport of oil by rail from other locations too, and with that the number of accidents have increased.

In the summer of 2013 an oil-tanker train loaded with 72 cars of crude oil exploded near Lac-Mégantic, Quebec, leveling the downtown area and killing 47 people. Later that year, trains derailed and exploded in Al-

abama and North Dakota, creating huge fires, complete with giant fireballs in the sky.

Tanker car derailments are usually caused by track problems or some other equipment failure, so the crude oil cannot be blamed for the accidents. But crude oil is considered hazardous, and although it usually does not explode, it did on several occasions recently.

Bakken crude is a mixture of oil, ethane, propane and other gaseous liquids, which are in a higher concentration than is usually found in conventional crude oils. Unlike conventional oil, which sometimes looks like black syrup, Bakken crude tends to be very light in color. Some people claim that it smells like gasoline, so you can just fill your tank with it and run the engine with no problems.

How true this is depends on whom you are talking to, so we just need to remember that crude oil presents a number of dangers, which we must be aware of and account for.

We also should know that it is hard to separate the liquids from the gasses in the raw oil, and since there are no particulate federal or state regulations on the matter, the oil companies just fill the railroad tanker cars with whatever comes out of the oil well spigot.

The standards are upcoming, but for now it is a gamble, so you don't always get what you see or want. Because of that, oil tanker accidents accompanied by oil spills, explosions, and fires will continue. Properties and human life along the oil tanker railroads are also in danger.

Crude Oil Production

Crude oil is usually pumped out in a liquid from underground deposits, where it could vary in density and other physical and chemical properties. There are also other forms and shapes that crude oil can be found in. These, of course, require different production and processing methods, some of which we will take a closer look later on.

According to OPEC, at the end of 2011, world proven crude oil reserves stood at 1,481,526 million barrels, of which 1,199,707 million barrels, or 81 percent, were in OPEC Member Countries. The picture today is changing, however, and a number of non-OPEC member countries—including the U.S.—are increasing their production levels significantly due to new discoveries and more efficient technologies and processes.

One major change in the oil equation today is the fact that the U.S. now produces domestically over 50% of the oil used on a daily basis. Other countries are following its example.

Table 3-3. Global oil production and use, 2011.

Country	% of world production	% of world use
Saudi Arabia	13.1	2.4
Russia	12.3	3.3
USA	8.0	4.1
China	4.7	9.0
Canada	3.9	2.5
Iran	5.4	1.1
Mexico	4.7	1.4
Venezuela	3.7	1.0
Kuwait	3.4	1.0
Norway	3.3	1.0

Crude Oil Production

What do we need to find and extract crude oil? In general terms, the cradle-to-grave oil production and use process consists of:

- Oil search and location
- Proposed area survey
- Environmental studies
- Oil rights and agreements
- Permits and legal issues
- Drill area preparation
- Drilling rig construction
- Daily O&M operations
- Oil transport
- Oil use (burning)
- Decommissioning and area reconstruction

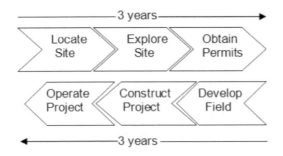

Figure 3-5. Project timeline

We'll take a close look at these steps, to determine the total energy cost and environmental impact of the cradle-to-grave oil production and use process.

Oil Search and Locating

The first step of the oil extraction process is locating significant oil deposits that are worth the time and expense to drill and extract the oil from them. Geologists employed by oil companies or under contract from a private firm, are the ones that look for and find oil.

Their efforts are focused on recognizing the signs and indications of oil deposits. Some of the signs could be a source rock, reservoir rock, or entrapment.

In the past, geologists had to read the surface features, surface rock and soil types, and then drilled small core samples to confirm their findings. A variety of other methods were applied in different cases too. Sensitive gravity meters were used to measure small deviations in the Earth's gravitational field, which could be translated into gaps of oil deposits. Magnetometers were also used to measure changes in the Earth's magnetic field that might be caused by flowing oil.

Detecting the smell of escaping gas by sensitive electronic sniffers is another good way, but most commonly seismology is used. In this method shock waves were created close to the surface and traveled through the rock layers, where some are reflected by the different materials.

Such shock waves are created by a compressed-air gun which shoots pulses of air into the water (for exploration over water). Thumper trucks are used to slam heavy plates into the ground (for exploration over land). Explosives can be detonated in holes drilled into the ground (for exploration over land) or thrown overboard (for exploration over water) to create shock waves.

The waves reflected back to the surface travel at different speeds depending on the type and/or density of rock layers through which they must pass. Sensitive microphones or vibration detectors detect the reflections (hydrophones over water, and seismometers over land are used). Seismologists interpret the readings, looking for signs of oil and gas reservoirs. Once a prospective oil strike is found, they mark the location using GPS coordinates on land or by marker buoys on water.

Today, satellite images and other sophisticated, computerized technologies do most of the job of detecting, interpreting and locating oil deposits.

Oil drilling can be separated as land and off-shore drilling. Land drilling is done on solid ground, while off-shore drilling is done miles from the shore, and sometimes at great depths in the middle of the ocean.

Conventional Oil Drilling

Once the oil has been located and the drilling site has been selected, the area must be surveyed, to determine its exact boundaries. Environmental impact studies are conducted as well in most cases.

If everything checks out and the decision is made to proceed with drilling at this site, then the legal work starts. Lease agreements, or land titles, rights-of-way accesses, legal jurisdiction determination and other conditions must be determined and met.

After the legal issues are settled, the crew goes about preparing the land, in land drilling cases. The land is cleared and leveled for derrick erection, pumping equipment, and support structures construction. Access roads also need to be constructed for transporting equipment and products to and from the oil well.

A lot of water is used during the drilling process, so a source of water must be located nearby. A water well has to be drilled in many cases prior to initiating oil drilling. In isolated cases, water truck-tanks are used instead.

A reserve pit must be dug nearby as well, to deposit rock cuttings and drilling mud during the drilling process. The pit is a temporary structure, which is usually lined with heavy plastic to prevent ground contamination and the related cleanup expense. In ecologically sensitive areas, such as a marsh or wilderness, a pit might not be allowed, so the waste materials must be disposed of by trucking them offsite.

The next step is to dig several holes in the place of the oil rig and the main hole, and a rectangular pit, or cellar, is dug around the drilling hole area. This is needed to provide a work space around the hole for the workers and drilling accessories.

The main hole drilling can be then started, first with a small drill truck, since the top part of the hole is larger and shallower than the main portion. The drill hole is then lined with a large-diameter metal pipe, and additional holes are dug off to the side for temporary equipment storage.

The main rig components can be set up now, starting with the power system, which usually is a diesel engine coupled with a generator, to provide electric power for drilling process. The components of the mechanical system, a hoisting system, and a turntable are setup next. The hoisting setup is used for lifting heavy loads and consists of a mechanical winch with a large steel cable spool, a block-and-tackle pulley and a receiving storage reel for the cable. The turntable, which is part of the drilling mechanism is installed next.

The main part of the drilling mechanism is the rotating equipment, which consists of a swivel, which is a large handle that holds the weight of the drill string and allows it to rotate. It also makes a pressure-tight seal on the hole. A kelly is four- or six-sided pipe that transfers rotary motion to the turntable and drill string, while the turntable drives the rotating motion using power from the electric motors. The drill string consists of a drill pipe (connected sections of about 30 feet) and drill collars con-

sisting of larger-diameter, heavier pipe that fits around the drill pipe and places weight on the drill bit.

A special drill bit is mounted at the end of the drill that actually cuts up the rock. These come in many shapes and materials, such as tungsten carbide steel, diamond, etc. Drill bits are usually specifically designed for the different drilling tasks and rock formations. Finally, the casing is large-diameter concrete pipe that lines the drill hole and prevents it from collapsing and leaking. It also allows the drilling mud that is pumped into the smaller drill pipe to circulate to the surface.

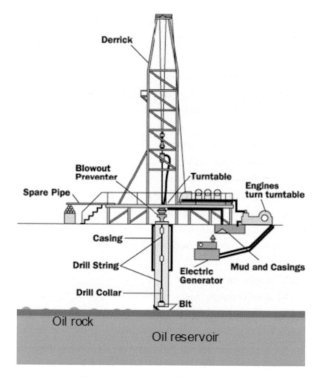

Figure 3-6. Land oil drilling rig

A thus constructed oil rig, or derrick, is tall enough to allow new sections of drill pipe to be added to the drilling apparatus as drilling progresses and the drill bit goes farther down into the ground. A blowout preventer is added to prevent explosions. It consists of several high-pressure valves that seal the high-pressure drill lines and relieve pressure when necessary to prevent a blowout. This prevents uncontrolled gush of gas or oil to the surface and eliminates fires.

The circulation system, when activated, pumps the drilling mud through the kelly, the rotary table, the drill pipes and collars. A mixture of water, clay, and other materials and chemicals are pumped down in the drill hole and used to carry rock cuttings from the drill bit to the surface.

The pump sucks mud from the mud pits and pumps it to the drilling apparatus and down the hole. Pipes and hoses connect the pump to the drilling apparatus, while the mud-return line returns mud from the hole to the surface and into a shale shaker which separates rock cuttings from the mud.

The shale slide pumps send the waste rock cuttings into the reserve pit, where they are separated from the mud. The cleaned mud is sent into a mud pit to be mixed and recycled in a mud-mixing hopper for another use.

The drilling starts by lowering the drill bit into the ground and rotating it. The drill bit goes down and is stopped when is very close to hitting the oil deposit. Technical and safety checks are completed at this point, and a *surface hole* is then drilled from the starter hole to a pre-set depth, just above the oil deposit. The surface hole is lined with a casing pipe, which is cemented in place in the center of the hole, as needed to prevent it from collapsing in on itself. The casing pipe is centered by spacers around the outside to keep it centered in the hole. The cement is allowed to harden and is then tested for hardness, alignment and a proper seal.

Drilling in the main hole continues for awhile, and then stops again for the new surface hole area to be lined and cemented. Then the drilling continues farther down, repeating the lining and cementing of the surface hole until the rock cuttings in the mud show oil sand, which is a signal that the well's final depth is very near.

Now the drilling mechanism is removed from the hole and a number of tests are performed to confirm the oil presence. These tests are done by lowering different sensors in the main drill hole. Some of tests are well logging, measuring the rock formations, drill-stem testing for measuring the pressures in the hole, and core samples for taking samples of rock to look for characteristics of oil deposit bedrock.

If the tests look good, the hole is drilled farther until oil flows into the casing in a controlled manner. A perforating gun is lowered into the well to the depth of the oil deposits, using explosive charges to create holes in the casing through which oil can flow. A smaller-diameter pipe is then run in the hole as a conduit for the oil and gas to flow up through the well and to the surface.

A packer is then run down the outside of the tubing and is allowed to expand to form a seal around the outside of the tubing. A multi-valve structure called a Christmas tree is mounted on the top of the tubing and is cemented to the top of the casing. It allows control of the flow of oil from the bottom of the well to the surface.

The flow of oil into the well is started by acids pumped down into the well and out the perforations.

Thus dissolved channels in the limestone allow oil into the well. If the rock is sandstone, then a specially blended fluid containing a mixture of sand, walnut shells, and aluminum pellets is pumped down the well. The pressure makes small fractures in the sandstone that allow oil to flow into the well, while the mixture holds these fractures open.

At this point the oil rig is removed from the site and a set of production equipment is installed for continuous extraction of the oil from the well to the surface. Several methods are used for this. The pump system uses an electric motor to drive a gear box that moves a lever. The lever pushes and pulls the pump rod, attached to a pump. The up and down action creates a suction that draws oil up through the well.

If the oil is too heavy to flow freely, an *enhanced oil recovery* method is used, where a second hole is drilled into the oil reservoir, and steam is injected into it under pressure. The heat thins the oil and the pressure helps push it up the well.

Offshore Oil Drilling

While oil extracted by the methods described above is usually done on solid ground, there is a lot of oil under the deep waters of rivers, lakes and the oceans too. Extracting oil deposits covered by deep water is a much more complex and dangerous effort. Done correctly, it can be efficient, safe and profitable, but if things go wrong, the results can be deadly for the oil workers and devastating for the surrounding environment. We only need to mention Deep Horizon to get a feel of the consequences from such a scenario.

The search, location, and exploration of potential oil deposits is similar to those described for land use. When all preliminaries are completed, a mobile offshore drill-

Figure 3-7. Offshore oil rig

ing unit (MODU) is brought in and installed over the potential oil hole. It is then used to dig the initial well and sometimes is converted into production rigs. More often, however, the MODU rigs are replaced by permanent oil production rigs for long-term oil extraction.

There are several different types of MODUs:

Submersible MODU is used in shallow and calm waters. It is a barge supported on the sea floor, and on the deck are steel posts that extend above the water line. A drilling platform rests on top of the steel posts and is used to drill the oil hole, similar to the methods described above.

Jackup MODU is a rig that sits on the deck of a floating barge, which is towed to the drilling site. The jackup can be used in depths of up to 525 feet, by extending its legs down to the sea floor and resting them on it without penetrating the floor. The jackup is then ratcheted up so that the platform is kept above the water level to keep it safe from high waves. Drilling can commence and proceed in a similar fashion as described above

Drill ships are special ships, designed for deep water oil drilling. They are equipped with a drilling rig mounted on the top deck, with a drill setup operating through a hole in the hull. Once the drilling starts, the ship uses a combination of anchors and propellers to maneuver as needed to correct for waves and currents.

Semi-submersible drilling rigs float on the surface of the ocean on top of huge, submerged pontoons, using propulsion systems to navigate to drilling sites and to maneuver over the hole. Computers control the anchor chain tension and engine power to correct for waves and currents.

During the drilling process, a blowout preventer (BOP) is installed at the ocean floor. It is equipped with a pair of hydraulically powered clamps that close off the pipe to the rig in the case of a blowout.

When the hole is drilled and ready for production, the well is sealed by a pair of plugs. The bottom plug sits near the oil deposit, and drilling mud or seawater keeps it in place, while the top plug is placed to cap the oil well. Then the well is hooked to a production rig, which operates in a similar way to the land-based oil rigs.

Below are the most promising oil exploration technologies. We take a closer look at hydrofracking in the next chapters.

The Hydrofracking Process

Hydraulic fracturing, hydrofracking, or fracking, is a major process that has played an increasingly important role in the development of America's oil and natural gas resources for over half a century. Until recently, few wells

were developed and exploited by hydrofracking, simply because the technology was not fully developed.

Advances in the hydrofracking state of the art have allowed the technology to be used safely, cheaply and efficiently where it was unthinkable just a few years ago. Presently, there are over 35,000 wells processed and operated with the hydraulic fracturing method. Estimates are that over one million such wells have been developed during the last half century.

There are differences in each well development and operation, and with each well the industry takes a step ahead in its progress. This allows the development of new best practices, to increase safety and minimize the environmental and societal impacts associated with new well development.

At the present levels of oil and gas extraction, up to 80 percent of oil and natural gas wells drilled in the near future will require hydraulic fracturing. This process is essential for oil and gas production from hard-to-reach formations and for production optimization of existing wells.

Horizontal drilling is a key component in the hydraulic fracturing process.

Horizontal Drilling vs. Vertical Drilling

In traditional drilling, a well is drilled more or less vertically downward. When the target formation is reached, drilling continues for some distance into the target formation. The operator then uses a special tool to create perforations in a portion of the wellbore that is within the target formation. Oil or gas can then flow into the well through the perforations.

The longer the length of perforated pipe, the faster oil or gas can flow into the well, but with vertical drilling the length of pipe that can be perforated is limited by the vertical height of the target rock formation.

On the other hand, a formation that may be only a couple of hundred feet or less in height might extend horizontally for miles. Horizontal drilling takes advantage of this. In horizontal drilling, the operator drills vertically downward toward the target formation, then turns the drill bit at an angle to drill in a horizontal direction. The drilling might then proceed horizontally for a mile or more within the target formation, providing a long horizontal "leg" that can be perforated and exploited.

Hydraulic fracturing makes it feasible to produce oil and gas from shale and other low-permeability formations from which production would otherwise not be feasible. Such production and the activity associated with it are beneficial for several reasons.

First, the activity has substantial economic benefits.

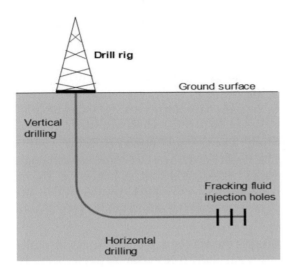

Figure 3-8. Horizontal (directional) drilling and fracking

Economists have estimated that shale gas development has created more than 200,000 jobs—direct and indirect—in the United States. Some of the new jobs are in the oil and gas industry itself, while other jobs are with companies that supply products, materials or services to the oil and gas industry. These include companies that mine the sand or manufacture the ceramic particles used as proppants; transport water, sand, and equipment to drilling sites; manufacture the high-pressure pumps used in fracturing; operate pipelines; perform construction; and operate the hotels, restaurants and caterers that house and feed workers.

State and local governments benefit from increased tax revenue, and sometimes mineral royalty revenue. In northwest Louisiana, where Haynesville Shale is located, some local governments have seen their sales tax revenue double over the course of a few years, enabling those governments to pay cash for the construction of numerous capital improvements, even while state and local governments elsewhere are struggling. State and local governments in other areas, including Texas, Pennsylvania, and North Dakota, also have benefitted.

Second, the increased production of oil and gas bolster our national security by reducing our country's dependence on imported oil and gas, some of which comes from areas that are politically unstable.

Third, hydraulic fracturing can have environmental benefits because it is often used to facilitate production of natural gas, which is the cleanest-burning of all fossil fuels. For a given amount of energy production, the combustion of natural gas produces only half as much carbon dioxide as coal, and about one-third less than oil. The combustion of natural gas also produces smaller, albeit significant, amounts of other pollutants.

Note: The glass in this case is always half empty or half full, depending on your point of view and your personal interests. Fracking has produced the most controversial issues and debates on very important socio-economic and political subjects to date.

On one hand we are happy that we now have plenty of clean, affordable natural gas, which can replace coal burning, thus killing two birds with one stone—providing cheap energy and cleaning the environment at the same time.

On the other hand, however, there are documented reports of environmental disasters caused by large-area fracking wells—water table contamination, ground cave-ins, earthquakes, toxic gas emission, etc. People and animals have been hurt by the liquid and gas emissions, and property values have been decreased because of them as well.

The Issues

In addition to causing local traffic congestion, dust and other problems for the locals, fracking can contaminate large land areas with waste water from wells and chemical storage ponds. At times the process goes wrong—both below and above ground. In some cases, the drilling wells are not built sturdily enough, or a piece of equipment malfunctions, causing leaks that contaminate the adjacent soil and the groundwater in the area.

"Flowback" water (process water used during the production cycle) has been documented at a number of sites to travel to the surface or into water sources, thus contaminating water tables, lakes, and streams.

Once the contamination enters the fresh water supplies, it shows up at people's dinner tables. In some cases, it can be seen bubbling out in water wells, and even coming out of kitchen faucets. These, albeit rare, occurrences are reason enough for us to consider the consequences of large-scale water contamination.

Luckily, at least for now, such incidents have been recorded in remote locations, affecting people living in small towns. Such contamination would be disastrous in a densely populated centers.

The fracking frenzy is increasing and proceeding quickly—well ahead of appropriate environmental safeguards. For example, toxic fracking fluids, including known cancer-causing chemicals like benzene and toluene, are not controlled and are exempt from federal regulation, according to the old Safe Drinking Water Act.

More importantly, there are no requirements for the drilling companies to disclose the contents of their fracking fluid, which are proprietary in most cases.

This regulatory gap lets millions of gallons to seep into the ground around each drilling site and eventually contaminate the water table. Because of that, property values in many areas near fracking sites have fallen, and human health has been adversely affected.

This is a precursor of a large-scale disaster, and since no one knows what is in the contaminated areas, except the drilling owners, nobody knows what can happen next. The drill operators are not going to reveal the toxic components for fear of competition and lawsuits...unless forced to.

At the same time, for state and local governments, oil and gas are cash cows so they are not going to make too many changes, let alone stop, or even limit, the hydrofracking activities anytime soon.

Another danger of the fracking overdevelopment is the fact that, just as happened in the Rust Belt in the recent past, large industries coming into unsuspecting areas, use the local resources and leave gross toxic waste, and eventually high unemployment behind. This has happened thousands of times; everyone is aware of it, and yet such disasters happen time after time.

Short-term solutions to the key points of the hydrofracking controversy are to:

a)   open the communication channels between producers and locals,

b)   accelerate the R&D and testing of new proppants, chemicals and procedures for use in fracking, and

c)   introduce new legislation that is acceptable to all parties involved.

The latter option might be easier said than done, but it is the only fair and most efficient solution. If properly designed and executed, said legislation could be the best long-term solution, where the interests of all parties could be protected.

For now, however, crude oil and natural gas—and by association the hydrofracking process—rule over the energy sector in the U.S. and many other countries. The positive economic effect of this new development is so great that the entire U.S. economy is starting to depend on it and its products—abundant crude oil and cheap natural gas.

While nobody can deny the short-term benefits of the hydrofracking process, it is extremely likely that as the fracking sites multiply in numbers, so will the number and intensity of long-term negative effects. It is also very likely that eventually they will get so big (and so many people will be negatively affected) that the state

and federal governments will need to step in and take appropriate—even drastic—measures to protect people and their properties.

Until then, it will be "Drill, Baby, drill." Oil and gas drilling are the most important (albeit short-term) solutions to the national energy security issues, after all...

Case Study: The Energy Country of Texas

The Eagle Ford Shale deposits are bringing a new oil bonanza to Texas and the U.S. in general. The massive production coming out of Eagle Ford and other shale deposits in Texas is significantly enhancing our energy security.

Texas is the energy capital of the world, with Houston as its home base. It is the nerve center of the U.S. energy sector too, with more energy-related jobs than most other places on Earth.

200,000 people are directly involved in the energy business, with thousands additional oil- and gas-field service industry jobs. Most oil and gas companies are headquartered in Houston, with thousands of employees in the city.

Factoring all indirect and induced jobs that service the industry, like hotels, restaurants, gas stations, and supermarkets, the numbers are in the millions. Putting all this together, we see that Texas' economy is larger than that of many countries.

If Texas were a country, it would rank as the 14th largest oil and 3rd largest natural gas producing nation on Earth.

Texas produces about 30% of US natural gas and about 30% of our oil production. As far as natural gas is concerned, only Russia and the other 49 states put together would rank ahead of Texas. The Eagle Ford Shale alone tripled natural gas production in Texas.

The Eagle Ford Shale in East Texas is estimated to be the largest oil field ever discovered in the lower 48 states. If this were not enough, there are now estimates that the Cline Shale formation in West Texas is even larger than the Eagle Ford. Who knew?

The implications of the Texas shale revolution are equally important for the entire country and the world. There are estimates now that due to these and other developments the U.S. will surpass Saudi Arabia in total oil production by 2020. And that was before the potential of the Cline Shale was added to the total. It is now also projected that, if things go as planned, the U.S. could become completely independent of imports by 2020-2025.

One amazing fact is that there are about 400 major rigs in the US drilling for natural gas, while four years ago, there were over 1,600 such rigs. While the rig count has dropped by 75%, the total natural gas production has continued to rise steadily.

This is because technologies and rig operators are more efficient and because the major oil shales—the Eagle Ford, the Bakken, and the Permian Basin deposits—contain a lot of associated natural gas. So, natural gas is becoming more available whether the companies are drilling for it or not.

There is also a huge amount of excess drilling capacity in the system, waiting to come on line as soon as natural gas and crude oil prices stabilize (and preferably rise some more).

This true-to-life energy revolution started only 6-7 years ago, when most of the major new shale plays today were discovered.

The best part is that new shale plays are still being discovered on a daily basis, and we are far from having discovered all of them. In addition, new technologies promise to bring even more deposits, and larger quantities will be extracted from each.

This is one of the best news for the U.S. oil companies and partially for the consumers. It is, however, bad news for future generations. If we proceed as planned, we will discover and pump out all the oil and gas in the U.S. within a short time. That would leave those who come after us with a serious energy deficit, and we clearly see them scratching their heads, saying, "What were those people thinking?"

Tar Sands Oil

Tar sands deposits occur naturally in many places around the globe, but are found in very large quantities in Canada, Kazakhstan, and Russia. Global tar sands deposits are estimated to contain the equivalent of over 2 trillion barrels of crude oil, although some have not even been discovered yet.

Today, most of the proven tar sands reserves, estimated at about 250 billion barrels, are located in Canada. This is about 80 percent of all global reserves.

Oil sands reserves have been largely ignored until recently, mostly due to technical difficulties in extracting the oil from them. There are also logistic problems, due to the remoteness of the deposits from point of use. Recently, however, they have been included in the world's oil reserves, as higher oil prices and new technology enable profitable extraction, transport, and processing.

Note: Oil produced from bitumen sands is often referred to as unconventional oil or crude bitumen, to distinguish it from liquid hydrocarbons produced from traditional oil wells.

Due to the nature of the raw materials and their locations, producing useful crude oil from tar sands is a grandiose undertaking. Large land areas must be bulldozed to remove the sand and transport it for processing. Equally large areas are needed for its processing too.

As Figure 3-9 shows, an entire city needs to be built in order to obtain enough oil from the tar sands. At this particular location, bitumen recovered from oil sands can be upgraded through various processes to a lighter oil (also called syncrude), and to other products such as naphtha, diesel, and gas oil.

Figure 3-9. Tar sands oil producing facility

Alternatively, the bitumen can be mixed with dilutents like naphtha to form the so-called dilbit, which can then be transported by pipeline or rail for further processing.

Dilution is necessary, because raw bitumen has a very high consistency ranging from molasses to tar. Because of that, it has to be upgraded and diluted for long-distance transport via pipes.

It can be also heated for pipe transport, but this option is very expensive and impractical for long-distance oil transport. However, it is used for short-distance transfers in refineries and processing facilities.

Here is a close look at the different tar sands production processes:

Ex-situ Process

The first step in the tar sands production process starts in a surface mine, similar to those in which coal is mined. The sand is scooped out by bulldozers and other heavy equipment, loaded on huge trucks and transported to nearby extraction plants for processing.

The object of the processing is to separate the oil (bitumen) from sand, water, and minerals. The separation takes place in large separation cells, into which the sand is dumped. Hot water is added to it, and the resulting slurry is piped to the extraction plant where it is agitated for thorough mixing.

The combined action of hot water and agitation releases bitumen from the oil sand, and causes tiny air bubbles to attach to the bitumen droplets. The oil droplets float to the surface of the separation cell with the help of the air bubbles.

Eventually, most of the bitumen gathers on the surface and can be skimmed off. Further processing removes residual water and solids. The bitumen is then transported and eventually upgraded into synthetic crude oil.

After oil extraction, the spent sand and other materials are returned to the mine where they are eventually reclaimed. The mine surface can be restored to pre-exploration condition too.

About two tons of tar sands are required to produce one barrel of oil, and only a total of 75% of the oil can be recovered from its sandy grave.

Although there is a lot of oil-containing sand, its mining, transport and processing requires a lot of energy—diesel fuel, electricity, and heat. The tar sand production process uses much more energy, and generates 12-15% more GHG per barrel of final product, than the extraction of conventional oil.

In-situ Process

In-situ production methods are used on some bitumen deposits that are buried too deeply underground to be economically recovered by mining. These techniques include **steam injection**, solvent injection, and firefloods, in which oxygen is injected and part of the resource burned to provide heat. So far, steam injection has been the favored method. All of these extraction methods require large amounts of both water and energy for heating and pumping the oil sludge.

The two primary methods of *in situ* bitumen production are Cyclic Steam Stimulation (CSS) and steam assisted gravity drainage (SAGD).

• *Cyclic Steam Stimulation (CSS)*, or the "huff-and-puff" method, was first used commercially in Alberta by Imperial Oil at Cold Lake in 1985. This technique involves the injection of steam into the formation for a period of time, followed by an extraction period in which the oil is pumped out. When the oil flow slows to a certain point, steam is once more injected. This cycle continues until the well is no longer economical.

• *Steam-assisted gravity drainage (SAGD)* is another *in situ* method, enabled by the same horizontal drilling improvements that enabled the hydraulic fracturing revolution. SAGD was first commercialized in 2001 by Cenovus at Foster Creek, and its commercial application was the single biggest reason that Canada's oil reserves more than quadrupled in the past 20 years.

SAGD involves drilling a pair of horizontal wells, one about 5 meters above the other. Steam is injected into the upper well for months to heat the bitumen. I learned from Cenovus that its initial projects required the company to inject steam for 18 months before producing oil, but as the engineering progressed, the time has been reduced to 3 months of steam injection. Once the wells start to produce, they have tended to produce almost without depletion for 10 years (a situation very unlike fracking, where wells initially deplete rapidly). The water that condensed when the steam was injected is also returned, separated from the oil, and reused in the process.

The horizontal wells can be drilled for miles in many directions from a single well pad, and as a result, a large land area can be accessed without a huge environmental impact on the surface. A normal well pad can produce nearly 20,000 bpd of bitumen for 10 years before depletion begins to curtail production.

It is obvious from Figure 3-10 that the extraction of the oil from the tar sands deep underground is not a simple process, nor cheap. First, two wells must be drilled and properly reinforced. Then the aboveground facilities must be constructed and equipped with steam generators, pumps, storage tanks, and tar sands processing capabilities.

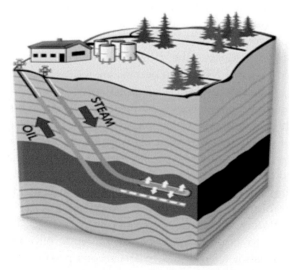

Figure 3-10. Tar sand *in-situ* process

Once a technique makes it both technically viable and economical to produce a resource, it can be placed in the reserves category. This is a similar situation to fracking, where resources in places like the Bakken and Eagle Ford became reserves when fracking made them economical to produce.

In all cases, during the production cycle, large quantities of super-heated steam are generated and pumped underground into the tar layer. The steam dislodges and dissolves the oil from its sandy embrace and the pressure sends the oil up the other well into the processing facility. Here, it is separated from the sand slurry and further processed. Thus obtained oil is stored in aboveground tanks where it awaits transport to refineries.

The vast majority of future oil sands growth is expected to come from *in situ* production. As an example, in 2013 there were 127 operating oil sands projects in Alberta, all of which were *in situ*. Production from both methods is expected to continue to grow, but the vast majority of the oil sands resource is too deep to be mined. Thus, most future production growth will be *in situ* production. There are only 5 shale mining projects in the area and no others have been planned thus far.

Currently, oil is not produced from tar sands on a significant commercial level in the United States; in fact, only Canada has a large-scale commercial tar sands industry, though a small amount of oil from tar sands is produced commercially in Venezuela and some other countries.

The Canadian tar sands industry is centered in Alberta, and more than one million barrels of synthetic oil are produced from these resources per day. Currently, tar sands represent about 40% of Canada's oil production, and output is expanding rapidly.

Approximately 20% of U.S. crude oil and products come from Canada, and a substantial portion of this amount comes from tar sands. The Canadian tar sands are different than those in the U.S.; they are water wetted, while U.S tar sands are hydrocarbon wetted. As a result of this difference, extraction techniques for the tar sands in Utah will be different than for those in Alberta.

As crude oil prices rise, tar sands-based oil production in the United States will become commercially attractive in the near future. The government and the oil industry are interested in pursuing the development of tar sands oil resources as an alternative to conventional oil.

For now, the vast Canadian tar sands are the focus of the U.S. energy industry. The bitumen in these sands is actually rich in petroleum in semi-solid or solid phase. Plans have been made to extract and transport the tar oil to the U.S. for processing and use.

Growth in the Canadian oil sands industry, however, will depend on the construction of a major new pipeline, the disputed Keystone XL, stretching hundreds of miles across the United States. The new pipeline's approval would mean rapid increase in tar sands production, but that also means surge in global greenhouse gas emissions, which are already dangerously high.

So the major part of the debate over the proposed Keystone XL now is whether the project would worsen global warming. There are also serious concerns being raised by nearby localities and entire states about potential contamination of land and water supplies by oil leaking from the pipe.

The IEA has concluded that there is significant danger of contamination and air pollution, while the U.S. State Department's high-profile draft environmental impact study disregarded any such concerns.

Tar oil is the dirtiest of all carbon products in terms of producing, handling, and burning it. It also sinks in water, which makes it impossible to clean or remove from contaminated river or ocean floors, which might create new and very serious problems for our aquatic resources and their wildlife.

The 1,700-mile TransCanada pipeline would carry as much as 830,000 barrels of oil a day, mostly from Alberta's tar sands, to Texas refineries. Tar sands production would more than double to 4.3 million barrels a day in 2035, up from 1.8 million in 2012.

The State Department also insists that tar sands production would increase with or without the Keystone XL, and therefore the pipeline wouldn't affect the GHG pollution one way or another. So now the decision hinges on the net effects of the pipeline's impact on the local environment and climate.

Light tight oil and heavy crude from the oil sands, according to a recent IEA report, are more than making up for a slow decline in Alberta's conventional crude oil. Total Canadian oil output is projected to rise by 62 percent to 6.1 million barrels a day from 3.8 million in 2035. The share coming from oil sands will rise to 70 percent of the total, from just under half today. Cumulative tar sands production could add up to a total of nearly 30 billion barrels of oil by 2035.

The Keystone XL and two export pipelines to the British Columbia coast could easily spur an additional one million barrels a day of tar sands production—most of it going to the U.S. But rising production of tar sands oil is the biggest driver of the increase in Canada's greenhouse gas emissions. It is also the main reason why

Canada says it will probably miss its 2020 goal of cutting emissions 17% below 2005 levels.

So, the environmentalists see the Keystone XL as a test of the U.S. government's commitment to tackling global warming. The battle is on.

Environmental Impact

Canada's tar sands undertaking is a very large project, and its environmental footprint is growing by the day. Turning tar sands into oil destroys the locale and harms the environment.

To start with, the Canadian Boreal forest's rich ecosystem will altered for a very long time, in order to expose the tar sands sludge underneath. The enormous diggers and trucks, some five stories high, will dig up and move millions of tons of sand.

The process requires 2-4 tons of earth to be dug out and processed for every barrel of tar sands sludge they extract. Imagine the mess, dust, mud, and smoke coming from such a colossal effort.

Next, the resource-intensive process uses heat and very hot water to dissolve and separate the sludge from the sand. More water is also used for the sludge to be further cleaned of unwanted toxins. The water-intensive processes create huge quantities of water pollution, not to mention that heating massive quantities of water requires massive quantities of energy too.

Fossil fuel consumption from the tar sands is Canada's fastest growing source of pollution.

Extracting tar sands oil poses a number of problems, but that is just the beginning. Transporting the toxic oil by pipelines, tankers, or rail, poses real and unacceptable risks to the involved communities. The highly corrosive oil is especially difficult to clean up in case of a spill, since it is heavier and contains more chemicals than regular oil.

Not surprisingly, however, the big oil companies are proposing a 700% increase in the number of tankers carrying toxic tar sands oil on the pristine West Coast. In response, a number of environmental groups are launching multinational campaigns in opposition to the tar sands development.

Recently, the federal government of Canada, which supports the tar sands development, has made it unreasonably difficult for Canadians to have a say at hearings for these proposed projects. Because of that, a number of lawsuits have been filed against the government.

The August, 2013 lawsuit filed in Toronto calls for the Federal Court of Canada to strike down provisions of the National Energy Board Act, which unreasonably restrict public comment on project proposals. The new rules created by the National Energy Board (NEB) to prevent any discussion of the wisdom of tar sands development at the upcoming hearings are also challenged.

The National Energy Board regulates the oil, gas, and electricity industries and approves pipeline construction, coal and uranium mining, liquefied natural gas projects, and tar sands development in Canada. Its decisions have massive environmental and health implications. Under the new rules, many Canadians are blocked from participation in the discussions on the issues.

Recent legislative changes mean that many Canadians with something to say about these projects will not have a chance to be heard. The NEB now requires that anyone wanting to submit a letter of comment to the NEB must complete a nine-page application for a chance to speak at NEB hearings. The NEB then decides who can and cannot provide testimony. The NEB reserves for itself the right to exclude anyone except for those that it considers to be "directly affected" by the proposed project.

"The amendments not only restrict who can speak to issues before the National Energy Board, but they also limit what those individuals are allowed to say," according to advocacy members, "Canadians deserve a fair public debate about the future of our economy and energy systems. Right now, they aren't getting it."

Under the new rules enacted by the government, people are not allowed to submit letters or petitions at the hearings.

"Tightening the rules around public participation to the extent that any citizen of this country—regardless of expertise or geographical location—cannot express their concerns is an extraordinary and profoundly dangerous affront to our democracy. I love my country and my beliefs call on me to respect our environment. That is why I chose to join this lawsuit," says one of the lawsuit participants.

This is the first time the amendments have been challenged in court. This suspension of citizen speech was evident in 2012, when over 1,500 people spoke at NEB hearings for the Enbridge Northern Gateway pipeline project. But now only 175 are permitted to speak at hearings this year for the Enbridge Line 9B reversal project. What happened to the freedom of expression in Canada?

With a set of lawsuits against the Canadian tar sands development in Canadian courts, and another set filed against the XL Keystone pipeline in the U.S., we must wonder what's next for these projects? Whatever it is, the future won't be free of troubles on either side of the Canadian-U.S. border.

What this means to U.S. energy security and how the outcome of the legal battles will affect our energy supplies are unknowns that depend on political and legal developments.

Shale Oil

Shale oil is another type of very special oil that is constantly in the news. It is contained (locked) in solid rock formations, so getting it out is quite different, and much more difficult, than pumping liquid oil from underground reservoirs. Shale oil extraction is a technologically difficult, energy and labor intensive, and very expensive process, even more so than tar sands oil extraction in most cases.

Nevertheless, as global crude oil prices keep rising and the technology improves, shale oil extraction is becoming more attractive and is the present-day reality in the U.S. It is also the biggest game changer in the energy sector.

Note: Shale oil is defined as synthetic crude oil derived from oil shale. It is also known as tight oil, or light tight oil (LTO), which is petroleum that consists of light crude oil contained in petroleum-bearing formations of low permeability, such as shale or tight sandstone.

Economic production from tight oil formations requires hydraulic fracturing, similar to the horizontal well drilling and exploitation technology used in the production of shale gas. Shale oil is different from oil shale, since the oil shell is a just a shale (rock or sand layer) rich in kerogen, or shale oil.

In layman's terms, shale oil is the actual crude oil-like substance produced from oil shale formations.

The International Energy Agency recommends using the term "light tight oil" for oil produced from shales or other very low permeability formations, while the World Energy Council uses the term "tight oil" for the same purpose. We prefer using the term shale oil in this text, unless a differentiation is needed.

The North American oil production surge led by unconventional oils—mainly light tight oil produced in the U.S. and oil from the Canadian tar sands—has created a global supply abnormality that is still to be sorted and finalized, but which inevitably will shape and reshape the way crude oil is produced, transported, stored, refined, marketed, and used.

Shale Oil Production

Shale oil is usually trapped (incorporated) in rocks or sandy oil shell formations deep underground. Oil

Table 3-5. Global shale oil reserves (2013 estimates)

Country	Reserves
Australia	95 billion barrels
Russia	75 billion barrels
United States	55 billion barrels
China	32 billion barrels
Argentina	27 billion barrels
Libya	26 billion barrels
Venezuela	13 billion barrels
Mexico	13 billion barrels
Pakistan	9 billion barrels
Canada	9 billion barrels
Indonesia	8 billion barrels
World	345 billion barrels

shale formations are heterogeneous and vary widely from location to location, and even over short distances within a location.

It is not uncommon to have totally different amounts of recoverable oil in different areas of a single horizontal drill hole. This makes evaluation of the quantity of recoverable oil and the overall profitability of each well an almost impossible task.

For example, profitable production of oil from some oil shale formations requires at least 15-20% natural gas content in the reservoir pore space in order to drive the oil toward the borehole.

Note: Shale oil formations which contain oil without sufficient gas content cannot be economically exploited at the present oil market prices. For example, one of the largest and most promising shale oil formations in the U.S., the Bakken oil shale formation,

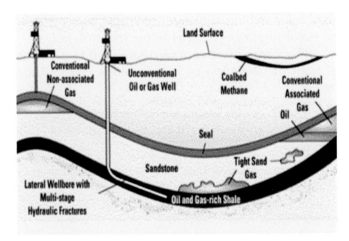

Figure 3-11.
Different oil formations and exploration methods

contains an estimated 20 billion barrels of shale oil in its 200,000-square-mile area. Of these, however, only a small portion—about 3 to 4 billion barrels—are economically recoverable at this time.

Nevertheless, the Bakken formation is now producing about 10% of the U.S. domestic crude oil via hydrofracking methods. We reviewed the hydrofracking technology earlier in this text.

The Future

Only a portion of the shale oil in each well can be extracted from the oil shale via hydrofracking, due to rock porosity and oil density variations. As we saw above, the Bakken formation is presently exploited via hydrofracking, but that process can extract only 5-10% of all recoverable oil in the oil shale.

Oil that is tightly held by the rock or sand formation cannot be hydrofracked out. Instead, the more difficult to recover oil shells require chemical and thermal processes, such as pyrolysis, hydrogenation, and thermal dissolution to release the oil.

Heating the oil shale to a sufficiently high temperature causes the chemical process of pyrolysis to yield an oil melt or vapor. The melt can be pumped out, while the vapor is cooled to produce liquid shale oil that is mixed with the melt.

Thus produced oil is then separated from the oilshale gas and is sent to additional refining.

Note: *Shale gas* also occurs naturally in oil shales and can be extracted by the conventional methods.

The organic matter within the melt (kerogen) is refined to be converted into synthetic oil and gas. The resulting products can be used immediately as lubricants or fuel. They can be also upgraded to meet refinery feedstock specifications (suitable for vehicle fuels and such) by enriching with hydrogen and then removing the major impurities such as sulfur and nitrogen.

There are a number of processes used to extract the oil from the shale layers, but the major classifications are:

- *in-situ*, or processing the oil in its shale reservoir and then extracting it for further processing and use,

- *ex-situ*, or taking the shale matter out and then extracting the oil from it.

In-Situ Process

The actual *in-situ* shale oil extraction process is fairly simple: find the shale oil vain, drill holes into it, insert pipes along its width, heat the pipes—which in

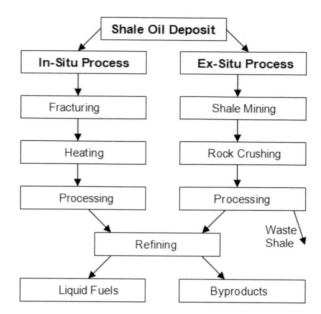

Figure 3-12. Shale oil production processes

turn heat the frozen shale oil around them—and pump the liquid oil out.

At around 650-700 degrees F, the kerogen, or crude oil trapped within the rock formation starts to liquefy and if kept hot enough for awhile it could be separated from its centuries long rock embrace and floated out. The result is an oil-like substance that can be further refined into different petroleum products, such as synthetic crude oil.

Thus obtained oil is then stored on site or transported for further refining and use.

There is only one problem: heating large areas of frozen soil deep underground is not simple, not cheap. Bringing a large area of frozen ground to a temperature of 700 degrees Fahrenheit and maintaining it requires a lot time and energy. A lot! This, however, is the only way to separate the oil and gas from their icy grave, where they have been resting for millions of years.

The heating pipes deep underground must be fed electricity or steam constantly for days, months, and sometimes years, until the oil and gas reach high enough temperature to be extracted. Supplying so much energy uses more electricity than the entire power supply for a small town. If the electricity or steam are expensive, then the produced shale oil would be expensive too.

In most cases other fossil fuels such as natural gas, oil, or coal are burned to generate the needed heat, but this is an expensive way to produce oil—by burning expensive oil. So, some new experimental methods are in development mode, which use electricity, radio waves, microwaves, or reactive fluids for this purpose.

Other strategies used to reduce, and even eliminate, external heat energy requirements use the oil shale gas and other by-products generated in the extraction process. These are burned as a source of energy, which is much cheaper than the previously mentioned method. The heat contained in hot spent oil shale and oil shale ash may also be used to pre-heat the raw oil shale, thus saving additional energy. Solar and wind power are also evaluated as a potential solution to this problem, but this is just an idea for now, and far from practical implementation for a number of reasons.

Always, *in situ* processes take several weeks or months and a lot of energy to heat the shale layers underground. The *in-situ* oil extraction may be conducted at temperatures as low as 480°F, and must be kept below 1,110°F, to prevent decomposition of surrounding lime stone and dolomite in the rock. This also limits carbon dioxide emissions and energy consumption.

The other methods of hydrogenation and thermal dissolution (reactive fluid processes) are used to extract the oil using hydrogen donors, solvents, or a combination of these. Thermal dissolution involves the application of solvents at elevated temperatures and pressures, increasing oil output by cracking the dissolved organic matter. Different locations and methods produce shale oil with different properties

Obviously, cheaper sources of energy to keep the elevated underground temperature up is the key to achieving a low-price product. This is a problem, which considering the enormity of the U.S. oil-shell reservoirs and our increasing thirst for energy, will be fully resolved in the near future.

There are major incentives for finding solutions to the efficient and cheap shale oil extraction, so the U.S. scientific community and the oil companies are feverishly working on these problems. There is no doubt that large quantities of shale oil will flow in the U.S. energy system in the very near future; just like there is already enough gas flowing into it.

Ex-Situ Process

For *ex situ* processing, the oil shale is dug out and broken into small pieces, which are then crushed into even smaller pieces in order to increase the overall surface area as needed for thorough extraction. The rocks are placed into a processing vessel to be heated. The temperature at which decomposition of oil shale occurs depends on the time-scale of the process. In all cases, a minimum of about 570°F process temperature is needed but the process proceeds more rapidly and completely at somewhat higher temperatures.

The pyrolysis of the rock is performed in a retort, or distillation column, filled with the crushed shale rock formation. The temperature of kerogen decomposition into usable hydrocarbons varies with the time-scale of the process. The *ex-situ* retorting process decomposition starts at 570°F. The process proceeds more rapidly and completely as the temperature is elevated, and the decomposition is most efficient between 900 and 970°F. At these temperatures the process is most efficient, and the amount of oil produced is the highest.

Hydrogenation is a another reactive process used to extract the oil using hydrogen donors, solvents, or a combination of these. Thermal dissolution involves the application of solvents at elevated temperatures and pressures, increasing oil output by cracking the dissolved organic matter. Different methods produce shale oil with different properties.

EROEI...again

The most important measure of the usefulness of the shale and tar oil production is the ratio of the energy produced by the final product, as compared to the energy used in its mining and processing of a similar quantity of the same product.

This ratio is known as *Energy Returned on Energy Invested* (EROEI). This is an important variable that determines the usefulness of an energy product or technology. It is often ignored, or intentionally obscured to hide facts that companies prefer to keep hidden.

The estimated EROEI of the various known oil-shale deposits vary between 0.7 and 13. This means that, for example:

- EROEI of 0.7 means that 7 gallons of oil are produced by using energy equivalent to burning 10 gallons of crude oil, in addition to all other consumables and labor. This also means that the site is operating at a nearly 30% financial loss.

The only justification for continuing the production in this case would be if low quality or contaminated fuel that is useless for any other purpose is used to produce a higher quality product.

- EROEI of 13, on the other hand, means that 13 gallons of shale oil are produced by burning just one gallon of crude oil. This ratio indicates good return on energy invested, which could result in profitable operation while the supplies last.

Recent estimates of some global oil shales put their EROEI at 1.1, 1.2 when the cost or internal energy is excluded and only purchased energy is counted as input. Actual oil company EROEI results of 1.3 to 1.4 have been reported from *in situ* development.

The amount of oil that can be recovered during the recovery process is another variable that varies with the type and location of the oil shale formation and the specific technology used. For example, 1/6th of the oil shales in the Green River Formation have a relatively high yield of 25 to 100 US gallons of shale oil per ton of oil shale. This is 2.5 to 10% yield.

One third of all oil shales in that region produce from 10 to 25 US gallons of oil per ton shale rock, or 1-2.5% yield. This is approximately 1 to 3 tons of oil produced from each 100 tons of shale. The rest, or over half of these oil shales deposits, however, yield less than 10 US gallons of oil per ton of shale rock, or less than 1% yield. This is only about a ton of oil from 100 tons of shale rock.

It takes a lot of equipment, labor, and fuel to dig out, crush, and process 100 tons of rock. The shale oil production from such low oil content, therefore, is economically unfeasible under normal conditions...unless crude oil prices are high enough.

Reports from major shale oil producing companies, however, are encouraging. As an example, Fushun Mining Group reports producing 300,000 tons per year of shale oil from 6.6 million tons of shale, a yield of 4.5% by weight.

Petrobras produces in their Petrosix plant 550 tons of oil per day from 6,200 tons of shale, a yield of 9%. And VKG Oil produces 250,000 tons of oil per year from 2 million tons of shale, a yield of 13%.

The Challenges

The immediate problem is that the extraction and processing of kerogen add two extra steps to the conventional extraction process in which liquid oil is simply pumped from the ground. In the shale oil process, in addition to the mining steps, there's also retorting and refining of the kerogen into synthetic crude. Both are energy consuming processes, which make the price of the final product quite high.

Presently, at global crude oil prices below $50 per barrel shale oil production becomes economically unjustifiable.

There is no danger, for now, of global oil prices falling that low, but crude oil is a big business and we should not be surprised at anything coming from any big business in the capitalist reality. This is even more true for the big guns of the global oil production business.

Oil shale production presents environmental challenges as well. It takes two barrels of water to produce one barrel of oil shale liquid. And without cutting-edge water treatment technology, the water discharge from oil shale refining increases the salinity in surrounding water, and poisons the water table in the local area.

There's also the matter of the waste materials—chemicals, rocks, dirt, etc. Every barrel of oil produced from shale leaves behind about 1.2 to 1.5 tons of rock. Just imagine the mountain left behind by producing the estimated 1 million barrels of shale oil. 1.2-1.5 million tons of rock materials piled up high, or spread around... every single day in several U.S. states.

This is a huge amount to deal with on daily basis. What should be done with this remaining rock? There are certainly projects that require loose rock—like covering ground beneath highway overpasses to discourage homeless settlements. But the supply may exceed the demand several fold if oil shale production continues on a massive scale.

Then, there is the never ending controversy of chemical pollution of the water table, which we take a closer look in this text.

Case Study: In-Situ Conversion Process

Royal Dutch Shell Oil Company is trying a new *In Situ* Conversion Process (ICP). Here, the rock remains where it is; it's never excavated from the site. Instead, holes are drilled into an oil shale reserve, and heaters are lowered into them, deep underground.

Over the course of two or more years, the shale is slowly heated and the kerogen seeps out. It's collected on-site and pumped to the surface. This cuts out the mining aspect, and further reduces costs since there's no need to transport or dispose of spent rock.

Shell's design includes a freeze wall—essentially, a barrier around the oil shale site where cooled liquids are pumped into the ground. This freezes any groundwater that may enter the site and keeps harmful byproducts like hydrocarbons from seeping out.

Because of current obstacles, oil shale hasn't been commercially produced oil on a large scale. Simply put, it's currently more expensive and environmentally harmful than conventional drilling. But as the supply of crude oil diminishes and the price of petroleum rises, oil shale—especially under Shell's plan—is becoming increasingly attractive. Read about some of the positive and negative global consequences of emerging oil shale production on the following pages.

The advantages of the *in-situ* heating process are:

- No open-pit or subsurface mining

- No thousands of tons of shale waste, as the traditional mining method does

- No groundwater contaminants via a "freeze wall" between the oil shale and water sources

- Minimizes water use and unwanted byproducts

This is, nevertheless, a very expensive process presently, but as usually happens with new manufacturing processes, operating costs can be expected to decrease over time via design enhancements and improved efficiency.

The huge amount of power that is required for this process and its cost won't go down, so we need to wait and see how Shell will resolve this problem.

The international interest in the new technologies is growing. In Jordan, for example, Shell pledged to spend $500 million in exploration of the country's oil shale resources in return for the right to develop these resources if and when the exploration begins. What does Jordan have to lose? This is a big gamble for Shell, which can turn into a big gain. Shell will acquire a huge resource to be exploited for many years at a fairly small price (if the effort is successful.) We will keep our fingers crossed for them…

After refining, the petroleum products can be used for the same purposes as those derived from crude oil. Oil shale can also be burned directly in furnaces as a low-grade fuel for power generation and district heating or used as a raw material in chemical and construction-materials processing.

Oil shale is now recognized as a potential abundant source of oil, but its production is financially feasible only when the price of crude oil rises above a certain level. Although shale oil is a substitute for conventional crude oil, extracting it from the oil shale is much more costly than the production of conventional crude oil, both financially and in terms of its environmental impact.

The major problem that needs to be addressed and resolved globally is the fact that oil-shale mining and processing raise a number of environmental concerns, such as land use, waste disposal, water use, waste-water management, greenhouse-gas emissions, and air and other pollution.

This is an important factor when assessing our energy security, because uncontrolled and unregulated large-scale contamination of our land, water, and air is not an option. Because of that, future environmental regulations might restrict, and somewhat change, the shale oil production in the U.S. and the world.

U.S. Shale Oil Production and Our Energy Security

The development of new exploration methods in the U.S. was put on the back burner by the Obama Administration recently, when it withdrew the research and development of oil shale leases that the Bush administration had offered in the Energy Policy Act of 2005.

Private sector research and development is now needed to bring these resources to market. Without these leases and the potential to commercialize the energy resource, companies will not invest the hundreds of millions of dollars required to develop the necessary technology.

Nevertheless, the six major shale deposits in the U.S.—Bakken, Eagle Ford, Haynesville, Marcellus, Niobrara and Permian—are well and very productive. They account for nearly 90 percent of the total growth in our domestic oil production, according to U.S. Department of Energy.

We are now producing about 6 million barrels of crude oil every day, and the aim is to exceed the 8 million barrel crude oil production per day mark by 2015. This means that we are relying less and less on foreign suppliers to meet our energy needs.

As a matter of fact, the United States is expected to soon pass Saudi Arabia in terms of projected crude oil production. Wow! What a twist of events.

The new energy bonanza in the U.S. might be the most important development in the world's energy sector ever! Its consequences will realign the world energy markets.

The surge in output from U.S. shale basins (and its tight oil formations) is responsible for much of that momentum. The Bakken, Eagle Ford and Permian basins are the bright spots now and account for more than half of the expected production. Production from reserve areas in North Dakota and Texas are expected to account for the bulk of forecast production growth over the next two years.

But wait. DOE also warned that the energy gains of late are misguided as far as the drive for energy security is concerned. Why is that? DOE warns that the increase in U.S. oil production and the frenzy around excess shale oil production doesn't mean that we are shielded completely from the global marketplace and its variations.

As a matter of fact, increased oil production, according to the DOE secretary, does very little to strengthen our energy security, as long as the U.S. and global economies are significantly dependent on crude oil imports.

According the DOE Secretary Moniz:

"It would be a misconception to think because of our increased domestic (oil) production that somehow we have become free of the global oil market, the global oil price, and global oil price volatility."

Who knew? So seriously, could we ever get free from 1973-like worries about our energy present and future? The U.S. domestic oil production has cut imports almost in half, which is good, but there are other alternatives to consider as well.

Looking beyond the shale oil and gas boom, we see $300 billion in investments in the energy efficiency markets globally in 2011 alone and some more thereafter, which is also a good thing, the results of which are significant.

Energy efficiency is a new concept, and according to IEA it is a huge "hidden fuel" that is driving the global energy markets. It is even greater than all the oil findings in the U.S. and Canada.

A low-carbon future would provide both enough cheap oil and a cleaner environment. So, while increased oil production in the U.S. is a good thing as a relief from constant energy worries or geopolitical turmoil, the best, if not the only, way to reduce oil dependence is to not use oil all...or at least use much less of it.

Not using oil, or even using less, is somehow not American, so we don't see it happening anytime soon. Oil used for transportation is one thing that cannot be replaced by renewables or any other technology or product. Yes, there will be solar hydrogen and biofuels and such, but at least 80-90 percent of all transportation fuel for the foreseeable future will still depend on crude oil for a long time.

In conclusion, we must emphasize, again, that our energy security consists of two totally different segments:

1. Power generation, and
2. Transportation.

This is a good place to make a distinction between fuels used for power generation vs. transportation.

We can easily replace most of the coal with less polluting, cheaper natural gas, as we are doing now, and eventually replace the natural gas with renewables.

Crude oil needed by the transportation sector, however, cannot be replaced by any of the existing technologies or products...for now and for the foreseeable future.

While we could internalize and fully control the power generation segment by use of domestically produced coal, natural gas, and renewables, we cannot by any means internalize and control completely the production, import issues, and transport of crude oil, which is absolutely needed for transportation.

So, unless we witness some serious technological breakthroughs in producing and/or using crude oil, or replacing it with renewable fuels, we will always need to import it from wherever we can and anyway we can.

*Since **only** crude oil can drive our cars, trucks, busses, trains, and planes, and since we need to import over 50 percent of our daily crude oil needs, we, and our energy security, are and will remain dependent on unstable, unfriendly foreign powers for oil production and transport for a long time to come.*

Oil Rigs Decommissioning

Oil production has problems that start with the initial stages of discovery, planning, and drilling of oil reservoirs, and continue all through their productive years. After 20, 30, or 50 years of non-stop operation, the oil deposits are depleted, and/or the oil rig is getting too old for reliable and safe operation. At that time the oil rigs go into the last phase of their lifespan—the decommissioning process.

Decommissioning of oil rigs is a complex and expensive effort. It can cost $5-$10 million per rig in the shallow water the Gulf of Mexico (GOM), and several times that much for decommissioning of deep-water rigs.

The U.S. Department of the Interior, Bureau of Ocean Energy Management, Regulation, and Enforcement (BOEMRE) Gulf of Mexico's OCS Region, issued a new decommissioning regulation in September 2010, which tightens further the requirements and increases the costs.

The NTL 2010-G05 required oil and gas wells that have not been used for the last five years to be classified as permanently abandoned, temporarily abandoned, or zonally isolated by Oct. 15, 2013.

If wells are zonally isolated, operators have 2 additional years to permanently or temporarily abandon the wellhead. Platforms and supporting infrastructure that have been idle for five or more years must be removed within 5 years as of the same date.

This means that the new NTL on top of the typical volume of decommissioning work in the GOM, will increase demand for contractors and, in turn, the expenses.

The Process

There are 10 steps to the decommissioning process:

- Project management
- Engineering and planning
- Permitting and regulatory compliance
- Platform preparation
- Well plugging and abandonment
- Conductor removal
- Mobilization and demobilization of derrick barges
- Platform removal
- Pipeline and power cable decommissioning
- Materials disposal
- Site clearance and final test

Project management, engineering and planning for decommissioning an offshore rig usually starts three years before the well runs dry. The process involves review of contractual obligations engineering analysis, operational planning, and derrick barges contracting.

Due to the limited number of derrick barges, many operators contract these vessels two to three years in advance. In addition, much of the decommissioning process requires contractors who specialize in a specific part of the process. Most operators contract out the project management, hardware cutting, civil engineering, and diving services.

Permitting and regulatory compliance requires obtaining permits to decommission an offshore rig, which can take up to three years to complete. Often, operators will contract a local consulting firm, which is familiar with the regulatory framework of their region, in order to ensure that all permits are in order prior to decommissioning.

An execution plan is one of the first steps in the process. Included in this plan is environmental information and field surveys of the project site. The plan describes a schedule of decommissioning activities and the equipment and labor required to carry out the operation.

An execution plan is required to secure permits from federal, state, and local regulatory agencies. The BOEMRE (a federal agency) will also analyze the environmental impact of the project and recommend ways to eliminate or minimize those impacts.

Federal agencies often involved in decommissioning projects include BOEMRE, National Marine Fisheries Service, US Army Corps of Engineers, US Fish and Wildlife Service, National Oceanic and Atmospheric Administration, US Environmental Protection Agency, US Coast Guard, the US Department of Transportation, and the Office of Pipeline Safety.

Platform Preparation

To prepare a platform for decommissioning, all holding tanks, and processing equipment and piping must be flushed and cleaned, and the waste waters and solid materials must be disposed of.

Then the platform equipment has to be removed which includes cutting pipe and cables between deck modules, separating the modules, installing padeyes to lift the modules; and reinforcing the structures in order to withstand the disassembly and transport efforts. Underwater workers usually prepare the jacket facilities for removal, which includes removing marine growth.

Well Plugging and Abandonment

Plugging and abandonment is one of the major costs of a decommissioning project and can be broken into several stages.

The planning phase of well plugging includes:

- Data collection
- Preliminary inspection
- Selection of abandonment methods
- Submittal of an application for BOEMRE approval

In the GOM, the rig-less method, which was developed in the 1980s, is primarily used for plugging and abandonment jobs. The rig-less method uses a load spreader on top of a conductor, which provides a base to launch tools, equipment and plugs downhole.

Actual well abandonment involves:

- Well entry preparations
- Use of a slick line unit
- Filling the well with fluid
- Removal of downhole equipment
- Cleaning out the wellbore
- Plugging open-hole and perforated intervals(s) at the bottom of the well
- Plugging casing stubs
- Plugging of annular space
- Placement of a surface plug
- Placement of fluid between plugs

Plugs must be tagged to ensure proper placement or pressure-tested to verify integrity.

According to BOEMRE, all platform components including conductor casings must be removed to at least

15 ft below the ocean floor or to a depth approved by the regional supervisor based upon the type of structure or ocean-bottom conditions.

To remove conductor casing, operators can chose one of three procedures:

- Severing, which requires the use of explosive, mechanical or abrasive cutting

- Pulling/sectioning, which uses the casing jacks to raise the conductors that are unscrewed or cut into 40-ft-long segments.

- Offloading, which utilizes a rental crane to lay down each conductor casing segment in a platform staging area, offloading sections to a boat, and offloading at a port. The conductors are then transported to an onshore disposal site.

Mobilization/Demobilization and Platform Removal

Mobilization and demobilization of derrick barges are key components in platform removal. According to BOEMRE, platforms, templates and pilings must be removed to at least 15 ft below the mudline.

First, the topsides are taken apart and lifted onto the derrick barge. Topsides can be removed all in one piece, in groups of modules, reverse order of installation, or in small pieces.

If removing topsides in one piece, the derrick barge must have sufficient lifting capacity. This option is best used for small platforms. Also keep in mind the size and the crane capacity at the offloading site. If the offloading site can't accommodate the platform in one piece, then a different removal option is required.

Removing combined modules requires fewer lifts, thus is a time-saving option. However, the modules must be in the right position and have a combined weight under the crane and derrick barge capacity. Dismantling the topsides in reverse order in which they were installed, whether installed as modules or as individual structural components, is another removal option and the most common.

Topside can also be cut into small pieces and removed with platform cranes, temporary deck-mounted cranes, or other small (less expensive) cranes. However, this method takes the most time to complete, so any cost savings incurred using a smaller derrick barge will likely be offset by the day rate.

Removing the jacket is the second step in the demolition process and the most costly. First, divers using explosives, mechanical means, torches or abrasive technology make the bottom cuts on the piles 15 ft below the mudline. Then the jacket is removed either in small pieces or as a single lift. A single lift is possible only for small structures in less than 200 ft of water. Heavy lifting equipment is required for the jacket removal as well, but a derrick barge is not necessary. Less expensive support equipment can do the job.

Pipeline and power cable decommissioning can be done in place if they do not interfere with navigation or commercial fishing operations or pose an environmental hazard. However, if the BOEMRE rules that it is a hazard during the technical and environmental review during the permitting process, it must be removed.

The first step to pipeline decommissioning in place requires flushing it with water followed by disconnecting it from the platform and filling it with seawater. The open end is plugged and buried 3 ft below the sea floor and covered with concrete.

Site Clearance and Materials Disposal

Clearance and disposal of platform materials is used to ensure that all materials are refurbished and reused, scrapped, recycled or disposed of in specified landfills.

To ensure proper site clearance, operators need to follow a four-step site clearance procedure.

- Pre-decommissioning survey maps the location and quantity of debris, pipelines, power cables, and natural marine environments.

- Post decommissioning survey identifies debris left behind during the removal process and notes any environmental damage

- ROVs and divers targets are deployed to further identify and remove any debris that could interfere with other uses of the area.

- Test trawling verifies that the area is free of any potential obstructions.

How much each step of this effort costs depends on the location and size of the rig. In all cases, many millions dollars is a good estimate...for each step of the process. The total cost can be a mind boggling number, which the oil companies are usually reluctant to release.

In case of an accident, such as a rig collapse, or human injury and fatality, the final total costs can be even more unbelievable. This is another reason why crude oil will never be cheap, and why its cost will only go up—regardless of how much more oil we discover.

OIL TRANSPORT

Crude oil is the life blood of the global transportation and petro-chemical sectors. It is also vital to the international energy market. Since most of it is found in areas that are far away from the point of use, it has to be transported—sometimes at great distances. The transport is done via pipelines, or transport vehicles—tankers, trains, and trucks.

After crude oil is pumped out of the ground, it has to be transported to a refinery for conversion into useful final products. The different oil transport methods include pipelines, marine vessels, tank trucks, and rail tank cars to transport crude oils, compressed and liquefied hydrocarbon gases, liquid petroleum products and other chemicals from their point of origin to pipeline terminals, refineries, distributors and consumers.

Oil can be transported to the refinery by a pipeline, which can be hundreds, even thousands, of miles long. The pipelines are either connected directly to the oil wells, or are supplied by storage tanks at tanker ports and terminals.

The U.S. pipeline system totals about 223,000 miles. Following is a closer look at the main components of this important chapter of crude oil's cradle-to-grave life cycle.

Pipeline Transport

Transporting oil via pipes is the safest, albeit not 100% safe, method of moving oil from one location to another—even if these are thousands of miles a part. The U.S. pipeline system totals about 223,000 miles. From the refineries, gasoline, fuel oil, and such, go by truck or train tank car to wholesalers or large consumers. The U.S. tank truck fleet numbers about 160,000, and the number of rail tank cars is slightly larger, over 165,000.
The risks related to pipe transport are few, but serious in nature if and when the pipe system fails.

Oil spills are the most common result from pipeline failure, and result in contamination of large land areas. These spills are difficult to clean and are the cause of a number of law suites and protests from locals and environmental organizations—some of which have contributed to shutting down pipeline sections and prevented others from being built.

Background

Pipelines are widely used for transport of crude oil from the oil wells to the refineries for processing. Thus processed products are also often delivered via pipe-

lines to the point of use.

There are aboveground, underwater and underground pipelines, varying in size from several inches to several feet in diameter. These pipelines move vast amounts of crude oil, natural gas, LHGs and liquid petroleum products across the US and most other countries.

The first successful crude-oil pipeline, a 2-inch-diameter wrought iron pipe, 6 miles long with a capacity of about 800 barrels a day, was built in Pennsylvania in 1865.

During WWII large pipe networks were built in the US, to move oil from coast to coast. The Keystone XL pipeline, planned to deliver Canadian syncrude oil to Gulf Coast refineries, is stalled in the permitting and political debate stages. If and when completed, it will transport millions of gallons of crude oil and petrochemicals across the U.S.

Many locals and a number of environmental groups are fighting the project, because of oil spill dangers, but it is unlikely that they can stop it. Oil is too important to our energy and national securities; so, if not this one, the next administration for sure, will make the Keystone XL pipeline happen.

Figure 3-13. Section of the Alaska pipeline

The lower 48 states receive 1.5 to 1.6 million barrels of oil per day, courtesy of the Alaskan pipeline. This pipeline, 789 miles of 48-inch steel pipe, curving up and down and around hills and valleys, carries great quantities of oil from the rich North Slope field to the port of Valdez in southern Alaska. Here the oil is loaded on tankers for shipment south. The Alaskan pipeline was built in controversy which is still unresolved and is a subject of high-level debates and legal battles.

Today, liquid petroleum products are moved long distances through pipelines at speeds of up to 6 miles per hour, assisted by large pumps and compressors along the way, at intervals ranging from 60 miles to 200 miles.

Pipeline pumping pressures and flow rates are controlled throughout the system to maintain a constant movement of product within the pipeline. Sensors, control mechanisms, and safety devices are installed along the length of the major pipelines to prevent spills, fires, and other possible disasters.

Pipelines run from the frozen tundra of Alaska and Siberia to the hot deserts of the Middle East, across rivers, lakes, seas, swamps and forests, over and through mountains and under cities and towns.

There is a network of over 95,000 miles of petroleum product pipelines in the United States. This network delivers finished petroleum products to the end customers. It is separate from the network of crude oil pipelines, and balances the demand and supply conditions in each region.

The initial construction of pipelines is difficult and expensive, but once they are built, properly maintained and operated, they provide one of the safest and most economical means of transporting these products.

Types of Pipelines

There are four basic types of pipelines in the oil and gas industry—flow lines, gathering lines, crude trunk pipelines and petroleum product trunk pipelines.

- *Flow pipelines* move crude oil or natural gas from producing wells to producing field storage tanks and reservoirs. Flow lines may vary in size from 5 cm in diameter in older, lower-pressure fields with only a few wells, to much larger lines in multi-well, high-pressure fields. Offshore platforms use flow lines to move crude and gas from wells to the platform storage and loading facility. Lease lines carry all of the oil produced on a single lease to a storage tank.

- *Gathering and feeder pipelines* collect oil and gas from several locations for delivery to central accumulating points, such as from field crude oil tanks and gas plants to marine docks. Feeder lines collect oil and gas from several locations for delivery directly into trunk lines, such as moving crude oil from offshore platforms to onshore crude trunk pipelines. Gathering lines and feeder lines are typically larger in diameter than flow lines.

- *Crude trunk pipelines* move natural gas and crude oil long distances, from producing areas or marine docks to refineries and from refineries to storage and distribution facilities by 1- to 3-m-diameter or larger trunk pipelines.

- *Petroleum product trunk pipelines* move liquid petroleum products such as gasoline and fuel oil from refineries to terminals and from marine and pipeline terminals to distribution terminals. Product pipelines may also distribute products from terminals to bulk plants and consumer storage facilities, and occasionally from refineries direct to consumers. Product pipelines are used to move LPG from refineries to distributor storage facilities or large industrial users.

Batch Intermix and Interface

Although pipelines originally were used to move only crude oil, they evolved into carrying all types and grades of liquid petroleum products. Because petroleum products are transported by pipelines in successive batches, there is co-mingling or mixing of the products at the interfaces.

The product intermix is controlled by one of three methods: downgrading (derating), using liquid and solid spacers for separation, or reprocessing the intermix.

Radioactive tracers, color dyes and spacers may be placed into the pipeline to identify where the interfaces occur. Radioactive sensors, visual observation, or gravity tests are conducted at the receiving facility to identify different pipeline batches.

Petroleum products are normally transported through pipelines in batch sequences with compatible crude oils or products adjoining. One way to maintaining product quality and integrity, downgrading or derating, is by lowering the interface between the two batches to the level of the least affected product. For example, a batch of high-octane premium gasoline is typically shipped immediately before or after a batch of lower-octane regular gasoline. The small quantity of the two products which has intermixed will be downgraded to the lower octane rating regular gasoline.

When shipping gasoline before or after diesel fuel, a small amount of diesel interface is allowed to blend into the gasoline, rather than blending gasoline into the diesel fuel, which could lower its flashpoint. Batch interfaces are typically detected by visual observation, gravitometers or sampling.

Liquid and solid spacers or cleaning pigs may be used to physically separate and identify different batches of products. The solid spacers are detected by a radioactive signal and diverted from the pipeline into a special receiver at the terminal when the batch changes from one product to another. Liquid separators may be water or another product that does not co-mingle with either of the batches it is separating and is later removed and reprocessed. Kerosene, which is downgraded (de-rated) to another product in storage or is recycled, can also be used to separate batches.

A third method of controlling the interface, often used at the refinery ends of pipelines, is to return the interface to be reprocessed. Products and interfaces which have been contaminated with water may also be returned for reprocessing.

Here's a closer look at the most controversial pipe-line in the U.S., the Keystone XL pipeline.

Case Study: Keystone XL Pipeline Debate

There is a big question before the American con-sumer—especially the people in the states that are af-fected by the Keystone XL pipeline. Carrying oil from the Canadian oil sands to Texas—crossing the U.S. from border to border—has a lot of pros and cons.

Those who care about energy security support the pipeline, and those who care about the environment oppose it. In that number are the locals, closest to the pipeline route.

And if you care about both, then you pick a side, or design your own battlefield. Maybe there is a middle ground: that of supporting the Keystone XL pipeline as long as its construction and use are regulated by strict environmental protections which are enforced.

Here are the problems from different points of view:

The Pros

As far as our energy supply and national security are concerned, increasing domestic oil production in-stead of importing petroleum from the Middle East is a win-win situation. It is in our best interests, and for our best friend and neighbor, Canada, too, while imports of petroleum from regions like the Middle East not so much. By buying their oil, we support governments and philosophies that are contrary to our values and our de-mocracy.

Pumping dollars into the Middle East and other volatile regions of the world also supports unfriendly currents and terrorist organizations in those areas. Some of these groups are fighting against us and would not hesitate to attack us and our allies. The Keystone XL pipeline helps mitigate this problem.

The Keystone XL pipeline also has a number of economic advantages to the United States and the state of Texas in particular. Canada is our largest trading part-ner, so the richer Canada gets, the more we benefit too, since Canadians buy a lot of U.S.-made goods.

The pipeline would also reinforce the technical cooperation with Canada, since we have the advanced technologies and skills that are needed to get oil from the oil sands. Increasing the oil imports from stable Can-ada, would decrease imports from unstable countries like Nigeria and Venezuela. This will have a leveling effect on oil prices too.

The Cons

Water supply protection is a major issue that won't go away no matter what happens now or later. Like all oil and gas pipelines, the Keystone XL pipeline poses some real and present risks and dangers. The construc-tion and operation of a large-diameter pipeline, with thousands of gallons of oil flowing every day cannot possibly be free of accidents.

The following land contamination, property de-valuation, and in particular poisoning of local water supplies are unavoidable. The only question is when, where, and who will be affected and how? For those who live near the line, this becomes a life and death is-sue…thus the persisting battles.

These major risks are at, a) the sand oil production sites in Canada, and b) along the pipeline's path from Canada to Texas. The risks in Canada are under the con-trol of Canada's environmental agencies. The risks on this side of the border, however, are our own problem, which is under the control of EPA and other government bodies.

Oil leakage along the pipeline in water bodies such as the Ogallala aquifer are a reality that is under con-sideration and a subject of lively discussion. We already have tens of thousands of miles of pipelines carrying oil and gas across the U.S. over sensitive aquifers and other environmentally delicate areas.

The new pipeline increases the risk of leaks pro-portionally, which is actually a small number compar-atively speaking. Looking from the eyes of the locals potentially affected by accidents, that number is huge. Opposing the pipeline won't do much if and when a decision is taken to proceed. A better approach would be to raise the standards of pipeline integrity and safety inspections. Imposing a fee on pipeline owners and op-erators would ensure regular inspections, repairs and mitigation of risks.

The other big problem under evaluation is the additional carbon emissions that the sand oil project brings. Due to process complexity, oil produced from oil sands is much more energy-intensive. This, in turn, makes it much more carbon-intensive than conventional oil production.

Therefore, while we increase oil production from the Canadian oil sands, we also proportionally increase the carbon emissions at the production sites. We cannot change the minds of Canadian companies and government, who are determined to develop the process to its maximum. Because of that, the oil sands will be exploited and oil will be shipped somewhere no matter what. If it is not shipped to Texas, it will be surely shipped to China, which will use much more energy (and emit much more GHGs) than shipping it to Texas.

Also, the oil shipped to China will be refined in outdated refineries, which emit many times the GHGs of comparable facilities in the U.S.

All in all, sending the oil to the U.S. will produce fewer carbon emissions, in addition to the other benefits like jobs, stable energy prices, etc. So, it is obvious that Canada oil sands and the Keystone XL pipeline offer a serious economic and national security advantage, provided that we design and implement proper protection procedures to ensure the safety of the environment and the people living nearby.

Oil-tank Trucks

From the wells, or refineries, oil, gasoline, fuel oil, and other petrochemical products go by truck or train tank car to wholesalers, or to large consumers. The U.S. tank truck fleet numbers about 160,000, and the number of rail tank cars is over 165,000.

Tank trucks are normal heavy-duty trucks, modified to carry large metal oil tanks. The oil tanks are typically constructed of carbon steel, aluminum, or plasticized fiberglass material, and vary in size from 1,900 liter tank wagons to jumbo 53,200 liter capacity. The optimum capacity of tank trucks is governed by regulatory agencies, and usually is dependent upon highway and bridge capacity limitations and the allowable weight per axle or total amount of product allowed along the scheduled routes.

There are pressurized and non-pressurized tank trucks, which may be non-insulated or insulated depending on their service and the products transported. Pressurized tank trucks are usually single-compartment, and non-pressurized tank trucks may have single or multiple compartments.

Regardless of the number of compartments on a tank truck, each compartment must be treated individually, with its own loading, unloading and safety-relief devices. Compartments may be separated by single or double walls. Regulations may require that incompatible products and flammable and combustible liquids carried in different compartments on the same vehicle be separated by double walls. When pressure testing compartments, the space between the walls is also tested for liquid or vapor.

Tank trucks have either hatches which open for top loading, valves for closed top- or bottom-loading and unloading, or both. All compartments have hatch entries for cleaning and are equipped with safety relief devices to mitigate internal pressure when exposed to abnormal conditions. These devices include safety relief valves held in place by a spring which can open to relieve pressure and then close, and hatches on non-pressure tanks which pop open if the relief valves fail and rupture discs on pressurized tank trucks.

A vacuum relief valve is provided for each non-pressurized tank truck compartment to prevent vacuum when unloading from the bottom. Non-pressurized tank trucks have railings on top to protect the hatches, relief valves, and vapor recovery system in case of a rollover. Tank trucks are usually equipped with breakaway, self-closing devices installed on compartment bottom loading and unloading pipes and fittings to prevent spills in case of damage in a rollover or collision.

Railroad Tank Cars

Railroad tank cars are constructed of carbon steel or aluminium and may be pressurized or unpressurized. Modern tank cars can hold up to 171,000 liters of compressed gas at pressures up to 600 psi. Non-pressure tank cars have evolved from small wooden tank cars of the late 1800s to jumbo tank cars which transport as much as 1.31 million liters of product at pressures up to 100 psi.

Non-pressure tank cars may be individual units with one or multiple compartments, or a string of interconnected tank cars, called a tank train. Tank cars are loaded individually, and entire tank trains can be loaded and unloaded from a single point. Both pressure and non-pressure tank cars may be heated, cooled, insulated and thermally protected against fire, depending on their service and the products transported.

All railroad tank cars have top- or bottom-liquid or vapor valves for loading and unloading and hatch entries for cleaning. They are also equipped with devices intended to prevent the increase of internal pressure when exposed to abnormal conditions. These devices include safety relief valves held in place by a spring which can open to relieve pressure and then close; safety vents with rupture discs that burst open to relieve pressure but cannot reclose; or a combination of the two devices.

A vacuum relief valve is provided for non-pressure tank cars to prevent vacuum formation when unloading from the bottom. Both pressure and non-pressure tank cars have protective housings on top surrounding the loading connections, sample lines, thermometer wells and gauging devices. Platforms for loaders may or may not be provided on top of cars.

Older non-pressure tank cars may have one or more expansion domes. Fittings are provided on the bottom of tank cars for unloading or cleaning. Head shields are provided on the ends of tank cars to prevent puncture of the shell by the coupler of another car during derailments.

Oil Train Accidents

A number of accidents involving oil trains have dominated the news recently. Although isolated, their effect is significant. Public opinion, shaped by these accidents, is ringing the alarm bells and the oil industry is considering the alternatives (pipelines).

The more significant accidents of the last several years were:

January 7, 2012. A 122-car Canadian National Railway train derailed in New Brunswick, Canada. Three cars containing propane and one car transporting crude oil from Western Canada exploded after the derailment. This created intense fires that burned for days. About 150 residents of nearby Plaster Rock were evacuated.

January 20, 2012. Six CSX train cars containing oil from the Bakken region of North Dakota, derailed on a bridge over the Schuylkill River in Philadelphia, near

Figure 3-14. Burning oil-tanker train

the University of Pennsylvania, a highway and three hospitals. No oil was spilled and no one was injured. The train originating from Chicago was more than 100 cars long.

July 5, 2013. A parked train from the Montreal, Maine & Atlantic Railway had been left unattended and sped down the hill for unknown reasons. It finally derailed, spilling oil and catching fire inside the town of Lac-Megantic in Quebec. Forty-seven people were killed and over 30 downtown buildings were damaged. About 1.6 million gallons of crude oil from the Bakken oil wells, transported to a Canadian refinery was spilled during this accident. The cleaning efforts continue.

November 8, 2013. An oil train from North Dakota derailed and exploded near Aliceville, AL. There were no deaths but an estimated 749,000 gallons of oil spilled in the surrounding area from 26 leaking tanker cars.

December 30, 2013. A fire erupted on train cars loaded with oil on a Burlington Northern-Santa Fe train after a collision with another train about a mile from Casselton, ND. No injuries were reported, but more than 2,000 residents were evacuated as emergency responders struggled to put out the intense fire.

April 30, 2014. A train carrying crude oil derailed in Lynchburg, VA. Seventeen cars were involved in the derailment en route from the Bakken oil reserves to Virginia. Three cars tumbled into the James River and spilled unknown quantities of crude oil into it. One of the cars breached and caught fire, sending flames and black plumes of smoke into populated areas along the James River.

Accident investigation boards in the U.S. and Canada are calling for tougher regulation of trains carrying crude oil. They are warning that experience shows that accidents in populated areas could cause major property and environmental damages and even loss of life, similar to that in the town of Lac-Megantic. Shipping crude oil by train is now seen as a dangerous undertaking.

Oil Tanker Transport

A great portion of the world's crude oil and about half of the oil headed to the U.S. is transported via ocean tankers. Oil transport is very big business. The world tank ship fleet numbers more than 5000. On any given day as much as 750 million barrels of crude oil and products may be in tankers on the world's oceans.

These undergo major risks every step of the way as they traverse the world's oceans, and especially in the most vulnerable places—choke points.

The issues related to energy security of oil transport could be summarized as follow:

- Global problems, such as political changes and turmoil, can affect the oil routes and the safety of the tankers and the people working on them.

- Weather related problems, such as storms and hurricanes on the tanker route could cause serious damage to the vessels and their personnel.

- Terrorist attacks are the most threatening part of global oil transport routes. This is the worst form of damage to the oil transport tankers and their personnel. There are several "choke points" around the world, where the ocean transports are most vulnerable.

Oil tankers and barges are vessels designed with the engines and quarters at the rear of the vessel and the remainder of the vessel divided into special compartments (tanks) to carry crude oil and liquid petroleum products in bulk. Cargo pumps are located in pump rooms, and forced ventilation and inerting systems are provided to reduce the risk of fires and explosions in pump rooms and cargo compartments.

Modern oil tankers and barges are built with double hulls and other protective and safety features required by the United States Oil Pollution Act of 1990 and the International Maritime Organization (IMO) tanker safety standards. Some new ship designs extend double hulls up the sides of the tankers to provide additional protection. Generally, large tankers carry crude oil and small tankers and barges carry petroleum products.

- *Oil tankers* are ocean traveling vessels, which in addition to ocean travel can navigate restricted passages such as the Suez and Panama Canals, shallow coastal waters and estuaries. Large oil tankers, which range from 25,000 to 160,000 SDWTs, usually carry crude oil or heavy residual products. Smaller oil tankers, under 25,000 SDWT, usually carry gasoline, fuel oils and lubricants.

- *Barges* carrying oil products operate mainly in coastal and inland waterways and rivers, alone or in groups of two or more, and are either self-propelled or moved by tugboat. They may carry crude oil to refineries, but more often are used as an inexpensive means of transporting petroleum products from refineries to distribution terminals. Barges are also used to off-load cargo from tankers offshore whose draft or size does not allow them to come to the dock.

- *Supertankers* are the least expensive way for long-distance shipment of oil and oil products. These modern ocean-going vessels are huge floating oil tanks that make their slow cumbersome way from the giant oil fields of the Middle East to ports in the industrial countries. But there are no supertanker ports in this country, so they are usually unloaded off-shore into smaller tankers. Supertankers are huge. The largest ones presently in service have a DWT (deadweight tonnage, i.e., cargo and fuel capacity) of more than 546,000 tons. They are up to 400 meters long (the length of about five football fields laid end to end) and can carry about 40 million barrels of oil.

- *Ultra-large and very large crude carriers* (ULCCs and VLCCs) are restricted to specific routes of travel by their size and "draft" (depth of water to which a ship sinks according to its load). ULCCs are vessels whose capacity is over 300,000 SDWTs, and VLCCs have capacities ranging from 160,000 to 300,000 SDWTs. Most large crude carriers are not owned by oil companies, but are chartered from transportation companies which specialize in operating these super-sized vessels.

Their draft is very important for their navigation. A fully loaded draft of over 90 feet prevents such a large ship from going into any of the U.S. ports. Our deepest port, Los Angeles, cannot handle ships of more than 100,000 deadweight tons. As a result supertankers now unload offshore in the Caribbean and transfer their cargo to smaller tankers for delivery to the United States.

The ocean transfer method costs a lot of additional energy and money, and for this purpose an alternative LOOP facility was built. *LOOP* is the Louisiana Offshore Oil Port (LOOP), which is a deepwater port in the Gulf of Mexico off the coast of Louisiana, near the town of Port Fourchon. It provides offloading and temporary storage services for crude oil transported on some of the largest tankers in the world, since most of them cannot enter US ports.

LOOP presently handles 13% of the nation's oil imports, or about 1.2 million barrels a day. It is connected by pipeline to the refineries, thus feeding 50% of the U.S. refining capability.

Tankers offload at LOOP by pumping crude oil through hoses connected to a single point mooring (SPM) base. Three SPMs are located 8,000 feet from the marine terminal. The SPMs are designed to handle ships of up to 700,000 deadweight tons. The crude oil then moves to the marine terminal via a 56-inch diameter submarine pipeline.

The marine terminal consists of a control platform and a pumping platform. The control platform is equipped with a helicopter pad, living quarters, control room, vessel traffic control station, offices and life-support equipment. The pumping platform contains four 7,000-hp pumps, power generators, metering and laboratory facilities. Crude oil is handled only on the pumping platform where it is measured, sampled, and boosted to shore via a 48-inch-diameter pipeline.

The Issues

The majority of the world's crude oil is transported by tankers from producing areas such as the Middle East and Africa to refineries in consumer areas such as Europe, Japan and the United States. Oil products were originally transported in large barrels on cargo ships. The first tanker ship, which was built in 1886, carried about 2,300 SDWT (2,240 pounds per ton) of oil. Today's supertankers can be over 300 meters long and carry almost 200 times as much oil.

Gathering and feeder pipelines often end at marine terminals or offshore platform loading facilities, where the crude oil is loaded into tankers or barges for transport to crude trunk pipelines or refineries. Petroleum products are also transported from refineries to distribution terminals by tanker and barge. After delivering their cargoes, the vessels return in ballast to loading facilities to repeat the sequence.

Oil tankers also pose a significant danger of oil spills due to accidents or terrorist attacks.

Millions of tons of crude oils are transported across world's maritime channels every day by tankers which are exposed to considerable risks.

Oil spills from transport tankers account for less than 8 percent of oil spills in the ocean. Although not that frequent, these are usually large spills caused by accidents, and the public is well acquainted with most of them. For example, everybody knows about the *Exxon Valdez* spill in Alaska, the effects of which are still visible.

Recently published list of oil spill effects by the International Tanker Owners Pollution Federation shows that the most harmful result of oil-related incidents is the negative effect these have on marine animals. Toxic chemical components in the oil smother and kill marine life. Less-than-lethal levels of toxicity have long-term effects on marine animals' ability to feed and reproduce. In the long run, oil spills contaminate the entire marine food chain, causing a crippling and even deadly domino effect on all species.

Aquatic birds and mammals are particularly affected. Even a slight, or momentary exposure to oil products is fatal for many of these animals, and ingested oil poisons them. Birds with oil in their feathers lose their ability to fly, and the waterproof coating that protects their vital organs is damaged, which causes health problems and even death. Older and sick mammals often suffer hypothermia due to oil-related complications.

The *Exxon Valdez* spill in 1989, for example, killed more than 40,000 birds and thousands of other sea mammals after the spill and during the clean-up efforts.

Following the *Exxon Valdez* spill, increased government regulations were implemented. One of these is the International Safety Management Code of 1998, which requires tankers to conform to new, much tougher, standards of quality and their owners are held fully accountable for all damages.

Individual states have crafted their own laws and methods of preventing oil-related accidents.

In California, for example, the regulatory agencies require oil tankers and transport ships to have active contingency plans for handling oil spills, as well as an additional $300 million in accident insurance.

Terrorism

The total world oil production amounts to approximately 80-90 million barrels per day. About one-half of this quantity is moved by oil tankers on fixed maritime routes. The global economic downturn of 2008 reduced the world oil demand, and the volumes of oil shipped to markets via pipelines and along maritime routes.

There are several chokepoints around the world, through which large quantities of oil float daily and

Figure 3-15. Oil transport tanker spill and fire

which are under constant threat of terrorism. By volume of oil transit, the Strait of Hormuz leading out of the Persian Gulf, the Strait of Malacca linking the Indian and Pacific Oceans, Bab-el-Mandeb at the entrance to the Red Sea, and the Strait of Gibraltar at the exit from the Mediterranean Sea, are the world's most strategic and dangerous chokepoints.

Oil tankers crossing those chokepoints are exposed to political unrest in the form of wars or hostilities and are vulnerable to theft from pirates and terrorist attacks. All this can lead to shipping accidents, resulting in disastrous oil spills and or blockage of the waterways. The blockage of a chokepoint, even temporarily, can lead to substantial increases in total energy costs.

Over half of America's oil is imported, so terror organizations like al-Qaeda and its affiliates see the disruption of oil transportation routes as one of the ways to hurt the U.S. and its allies. And sure enough, disruption of scheduled oil flow through any of those routes could impact negatively the global oil prices.

Several attacks against oil tankers in the Arabian Gulf and Horn of Africa have been planned in the past, starting with June 2002, when a group of al-Qaeda operatives suspected of plotting raids on British and American tankers passing through the Strait of Gibraltar were arrested by the Moroccan government. In October of the same year, a boat packed with explosives rammed and badly damaged a French supertanker off Yemen. al-Qaeda claimed responsibility for that attack, and for plans to seize a ship and crash it into another vessel or into a refinery or port.

The terrorists' goal is to cut the economic lifelines of the world's industrialized societies, but such attacks would also weaken and perhaps topple some of the Gulf oil monarchies. This would have even greater and long lasting effect on the world's oil supplies and transport routes.

There are approximately 4,000 tankers crossing the world's oceans every day. Each of those slow and vulnerable giants can be attacked at the narrow passages, where a single disabled or burning oil tanker and its spreading oil slick could block the route for other tankers.

This constant threat and the increased risks contribute to further increase in oil prices. Insurance carriers are raising the premiums to cover tankers in risky waters, as evidenced by the fact that the insurance for oil tankers entering Yemeni waters tripled since the attack in Yemen a decade ago.

A typical supertanker with two million barrels of oil on board would pay about $450,000 instead of the usual $150,000 for insurance of the ship alone. The cargo requires a separate and equally expensive insurance policy each trip. This adds about 15-25 cents a barrel to the cost of the oil.

Even the best insurance cannot ensure complete safety and reliability of our energy supply under those conditions. The only way to achieve energy independence, therefore, is to eliminate the need to import and transport oil across the globe. This, however, is not going to happen anytime soon, so the new situation has created a new niche market for individuals and companies that provide insurance and ensure the physical safety of the oil tankers.

Some oil tank owners are resorting to placing security personnel on board, especially when crossing pirate-infested international waters. This measure, in addition to equipping the tankers with special equipment, is one of the services that is expected to grow. The number of complexities of those services, as well as their costs, will increase as the pirate attacks are on the rise, at least until we figure a way around it.

The Choke Points

Choke points are narrow channels along widely used global sea routes, some so narrow that restrictions are placed on the size of the vessel that can navigate through them. They are a critical part of global energy security due to the high volume of oil traded through their narrow straits.

Global crude oil production is at approximately 85 million barrels per day, half of which must be transported by oil tankers all over the world. Oil tankers traveling along these fixed routes cross some of the most dangerous chokepoints, which present especially high risk for piracy that might result in hazardous oil spills.

Even a temporary blockage of a strategic chokepoint leads to substantial economic damage to the affected countries, and overall increase in energy costs.

In 2011, total world oil production amounted to approximately 87 million barrels per day (bbl/d), and over half was moved by tankers on fixed maritime routes. By volume of oil transit, the Strait of Hormuz, leading out of the Persian Gulf, and the Strait of Malacca, linking the Indian and Pacific Oceans, are two of the world's most strategic chokepoints.

The international energy market is dependent upon reliable transport. Blockage of a chokepoint, even temporary, can lead to substantial increases in total energy costs. In addition, chokepoints leave oil tankers vulnerable to theft from pirates, terrorist attacks, and political unrest in the form of wars or hostilities as well as shipping accidents that can lead to disastrous oil spills. The seven straits highlighted in this brief summary serve as major trade routes for global oil transportation, and disruptions to shipments would affect oil prices and add thousands of miles of transit in an alternative direction, if even available.

Chokepoints for crude oil transport are a critical part of the global energy security dilemma, since most of the world's crude oil and petroleum production move on maritime routes.

Table 3-6.
The major choke points (in million barrels per day)

Choke Point	Volume
Panama Canal	1.6
Danish Straits	3.0
Suez Canal	3.0
Bab el Mandab	3.5
Turkish Straits	3.8
Strait of Malacca	15.0
Strait of Hormuz	18.0
Total	**47.9**

Due to the importance of oil supplies to the U.S. economy, and since about half of the crude oil used daily is imported, some of the world chokepoints are of especial interest.

The Strait of Hormuz

About 18 million barrels of crude oil are transported along the Strait of Hormuz in the Persian Gulf every

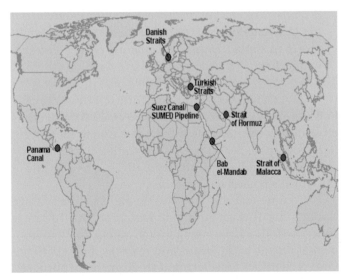

Figure 3-16. The major oil transport choke points.

day. Located between Oman and Iran, the Strait of Hormuz connects the Persian Gulf with the Gulf of Oman and the Fabian Sea. Flow through the strait accounts for 30-40 percent of all seaborne oil. This is almost 20 percent of all oil traded worldwide.

More than 85 percent of these crude oil exports go to Asian countries, with Japan, India, South Korea, and China as the largest customers.

Qatar also exports about 2 trillion cubic feet of liquefied natural gas (LNG) through the Strait of Hormuz annually, which is about 20 percent of all global LNG trade. Kuwait, on the other hand, imports huge LNG volumes that also travel through the Strait of Hormuz. This is a total of about 100 billion cubic feet per year passing through one single spot.

At its narrowest point, the strait is 21 miles wide, but the width of the shipping lane in either direction is only two miles, separated by a two-mile buffer zone. The strait is deep and wide enough to handle the world's largest crude oil tankers, with about two-thirds of oil shipments carried by tankers in excess of 150,000 deadweight tons.

The Strait of Hormuz is the world's most important oil chokepoint due to its daily oil flow of about 18 million barrels—up from 16 million barrels per day during 2010.

Even temporary closure of this chokepoint would immediately and seriously affect the U.S.' energy security.

There are no options for oil tankers to bypass Hormuz presently. Iraq, Saudi Arabia, and UAE, however, have pipelines for shipping crude oil outside of the Gulf, bypassing the strait at least partially. Only Saudi Arabia, and UAE have enough pipeline capacity to bypass Hormuz completely. The total available, but not utilized, pipeline capacity from these two countries combined is approximately 1 million bbl/d, which is to be increased to 4.3 million bbl/d, as both countries try to increase their pipeline capacity to bypass the strait if and when needed.

Iraq's major crude oil pipeline, the Kirkuk-Ceyhan (Iraq-Turkey) pipeline transports oil from the North Iraq to the Turkish Mediterranean port of Ceyhan. This pipeline has a nameplate capacity of 1.6 million bbl/d, but has been the target of sabotage attacks and is operating at much lower levels. It also cannot send additional volumes to bypass the Strait of Hormuz unless it receives oil from southern Iraq via the strategic pipeline, which links northern and southern Iraq. Portions of that pipeline have been closed on and off during war activities, and renovations are underway.

Saudi Arabia's 745-mile-long petroline (the East-West Pipeline), runs across Saudi Arabia from its Abqaiq complex to the Red Sea. The petroline system consists oft two pipelines with a total nameplate capacity of about 5.0 million bbl/d. This means that Saudi Arabia's spare oil pipeline capacity—able to bypass the Strait of Hormuz—has a capacity of 3.0 million bbl/d at full capacity.

The UAE constructed a 1.5 million bbl/d Abu Dhabi Crude Oil Pipeline in order to bypass Hormuz. It runs from the Habshan collection point (serving Abu Dhabi's onshore oil fields), to the shipping port of Fujairah on the Gulf of Oman. The pipeline is capable of carrying nearly 2.0 million bbl/d, which is over half of UAE's oil exports at full capacity

The Strait of Malacca

The Strait of Malacca, located between Indonesia, Malaysia, and Singapore, links the Indian Ocean to the South China Sea and Pacific Ocean. This is the shortest route between the Persian Gulf and the Asian markets. Most oil tankers headed to China, Japan, South Korea, and the Pacific Rim pass through Malacca.

Crude oil makes up about 90 percent of the daily cargo transiting Malacca, while petroleum products make the remainder of the traffic.

The Strait of Malacca is the key chokepoint in Asia, with over 15 million bbl/d headed to China and Indonesia—the major Asian oil consumers. This flow is up from 13 million bbl/d in the mid-2000s.

A temporary closure of this chokepoint would not affect the U.S.' energy security. A permanent closure, however, might result in global chaos and overall energy markets instability.

At the Phillips Channel of the Singapore Strait (its narrowest point), Malacca is only 1.7 miles wide, which creates a natural bottleneck. It also presents a great potential for collisions, grounding, and oil spills. Over 60,000 vessels navigate through the Strait of Malacca every year, where piracy, including attempted theft and hijackings, are a constant threat. The number of attacks has dropped since 2005, due to the increased patrols and interventions.

In case of a blockade, half of the world's fleet will need to be rerouted around the Indonesian archipelago and through Lombok Strait, between Bali and Lombok.

Another long route is through the Sunda Strait, between Java and Sumatra.

Options to bypass Malacca have not materialized, so China is working on the 400,000 bbl/d Myanmar-China Oil and Gas Pipeline bypass. These are two parallel (crude oil and natural gas carrying) pipelines from the ports in the Bay of Bengal, Myanmar to the Yunnan province of China. This addition will transport crude oil imports from the Middle East in order to bypass the Strait of Malacca, thus ensuring steady oil supply to China and the neighboring countries.

Turkish Straits

The Turkish Straits are located at the Bosporus and the Dardanelle, which divide Asia from Europe. The Bosporus is a 17-mile passage, between the Black Sea and the Sea of Marmara. The Dardanelles is a 40-mile waterway, linking the Sea of Marmara with the Aegean and Mediterranean Seas. Both straits are located on Turkish territory and serve tankers supplying Europe with oil from the producers of the Caspian Sea Region.

Almost 3 million barrels of crude oil flow through the Turkish Straits daily. The ports of the Black Sea are one of the primary oil export routes for Russia and other former Soviet Union republics. Oil shipments through the Turkish Straits vary as Russia shifts crude oil exports toward the Baltic ports and as crude production and exports from Azerbaijan and Kazakhstan rise.

Oil exports from the Caspian Sea region to Western and Southern Europe make the Turkish Straits one of the busiest and most dangerous chokepoints in the world.

A temporary closure of this chokepoint would affect the U.S.' energy security. A permanent closure might result in global chaos and overall energy markets instability.

At its narrowest point, the Turkish Straits are only half a mile wide. This makes them one of the world's most difficult waterways to navigate due to their peculiar geography. 50,000 vessels, including 5,500 oil tankers, pass through the straits annually, so it is also one of the world's busiest chokepoints.

Commercial shipping has the right of free passage through the Turkish Straits in peacetime, but Turkey is concerned about the navigational safety and environmental threats to the straits and reserves the right to impose regulations for safety and environmental purposes.

Bottlenecks and heavy traffic present dangers and create problems for oil tankers in the straits, but there are no alternate routes from the Black and Caspian Sea region westward. As a solution to the problem, there are several pipeline projects near completion and others in various phases of development underway. These will help reduce the congestion and dangers of oil tankers passing through the straits.

Bab el-Mandab

Bab el-Mandab is located between Yemen, Djibouti, and Eritrea, and connects the Red Sea with the Gulf of Aden and the Arabian Sea. It is a major chokepoint of oil shipments between the Horn of Africa and the Middle East.

It is also a strategic link between the Mediterranean Sea and Indian Ocean. Tankers from the Persian Gulf crossing the Suez Canal and SUMED Pipeline also pass through the Bab el-Mandab.

Four million barrels of oil flow through Bab-el-Mandab annually, headed to ports in Europe, the United States, and Asia. The flow varies from 4.5 million bbl/d in 2008, down to 2.9 million bbl/d in 2009 as a result of the global economic downturn and the decline in northbound oil shipments to Europe.

Decline in northbound traffic through the Suez Canal and SUMED Pipeline in 2009 also reflect the adverse affects of the latest global economic crisis. Over half of the northbound oil shipments through Bab el-Mandab, about 2.0 million bbl/d, move en route to the Suez Canal and SUMED Pipeline.

Closure of the Bab el-Mandab could keep tankers from the Persian Gulf from reaching the Suez Canal and Sumed Pipeline, with the only alternate route being around the southern tip of Africa.

Even a temporary closure of this chokepoint would immediately and seriously affect the U.S.' energy security.

Bab el-Mandab is 18 miles wide at its narrowest point, which is limited to two 2-mile-wide channels in the inbound and outbound directions—not much for the large amount of the huge oil tankers crossing the chokepoint.

Restricted traffic flow in the Bab el-Mandab chokepoint limits oil entering the Red Sea from Sudan and other countries, as ships have no access to the most direct route to the Asian markets. Instead, the tankers must go into the Mediterranean Sea and through the other chokepoints of the Suez Canal and SUMED Pipeline.

A large French oil tanker was attacked off the coast of Yemen by terrorists in 2002, causing damage to the vessel and chaos around the world. Since that accident, security is a major concern of tanker owners, operating in the region. During the last several years Somali pirates (off the northern Somali coast in the Gulf of Aden and southern Red Sea, including the Bab el-Mandab) have attacked a number of tankers, capturing their crews and demanding large sums of ransom money.

The Suez Canal

The Suez Canal is located in Egypt and connects the Red Sea and Gulf of Suez with the Mediterranean Sea. Crude oil and LNG shipments account for 30 percent of the total annual Suez Canal business. The canal is too small for ultra large crude carriers (ULCC) and also for fully laden very large crude carriers (VLCC) class crude oil tankers. In 2010 the Suez Canal Authority extended the depth to 66 feet to allow more tankers to go through, so now over 60 percent of all world tankers can use the canal.

About 3 million bbl/d of oil tankers transit the canal in both directions. This is about 6-7 percent of the total worldwide seaborne oil. The majority of the oil goes north toward the European and North American markets. The remainder flows south toward Asian markets.

The oil volume varies according to economic and other conditions. For example, it increased in 2012 due to the restart of oil production in Libya, following the civil war. At the same time, southbound oil flow from Libya quadrupled.

SUMED Pipeline

A 200-mile long SUMED Pipeline, or Suez-Mediterranean Pipeline, was constructed to serve ULCC and VLCC type vessels that are too large to transit through the canal. These vessels are unloaded at Ain Sukhna terminal along the Red Sea coast, and the oil is transported via pipelines to the Sidi Kerir terminal on the Mediterranean. Here it is loaded on other tankers for shipment to its destination points.

Some fully laden VLCCs trying to cross the Suez Canal also use the SUMED Pipeline for reducing their weight and draft by offloading some of their cargo. A portion of the crude is offloaded at the SUMED Pipeline at the Ain Sukhna terminal and pumped to the Sidi Kerir terminal. The lighter VLCC now can go through the Suez Canal, then pick up the offloaded portion of their loads at the other end of the pipeline at Sidi Kerir terminal.

The crude oil is pumped through two parallel, 42-inch-diameter, pipelines with a total pipeline capacity of around 2.5 million bbl/d. The SUMED pipeline is owned by the Arab Petroleum Pipeline Co., a joint venture between the Egyptian General Petroleum Corporation (EGPC), Saudi Aramco, Abu Dhabi's National Oil Company (ADNOC), and Kuwaiti companies.

In 2012, around 1.54 million bbl/d of crude oil was transported through the SUMED pipeline. SUMED crude flows decreased somewhat in 2012, the total crude oil transited northbound from Suez and SUMED combined increased to 2.44 million bbl/d in 2012 from 2.20 million bbl/d in 2011.

The SUMED Pipeline is the only alternative route to transport crude oil from the Red Sea to the Mediterranean, for tankers too large to navigate through the Suez Canal.

Closure of the Suez Canal and the SUMED Pipeline would force oil tankers to travel around the Cape of Good Hope, or 2,700 additional miles from Saudi Arabia to the United States, adding time and cost to the transport.

Even a temporary closure of this passage would immediately and seriously affect the U.S.' energy security.

Navigating around the southern horn of Africa adds 15 days of travel time to Europe-bound tankers, and adds 8-10 days to the tanker going to the United States. The oil cost goes up accordingly too.

LNG also flows through the Suez Canal in both directions, amounting to about 1.5 trillion cubic feet in 2012, or around 13 percent of total LNG traded worldwide.

Southbound LNG transit originates in Algeria and Egypt and is headed mostly for the Asian markets. The northbound transit originates mostly from Qatar, destined for European markets.

U.S. LNG imports from Qatar fell by around 63 percent in 2012 compared with 2011, mostly due to sharp increase in domestic supply in the U.S., and a decrease in LNG demand in some European countries. This is also an indication of increased complexity and competition in the global LNG marketplace.

Danish Straits

The Danish Straits are a 300-mile-long waterway passage that connects the Greenland Sea (an extension of the Arctic Ocean) and the Irminger Sea (a part of the Atlantic Ocean). This waterway is 180 miles wide at its narrowest point, and 625 feet deep at the shallow end, so it does not present nearly as much danger as narrower chokepoints, except that the cold East Greenland Current passes through the strait and carries icebergs south into the North Atlantic. The icebergs are the only danger in the path of oil tankers in the straits, but we are far from the *Titanic* days, so the danger is minimal.

The Danish Straits are a major route for Russian oil exports to Europe, where about 3 million bbl/d flow mostly westward. Russia is increasing its oil production and is using its Baltic ports for crude oil exports to Europe.

The new port of Primorsk accounts for nearly half of the exports through the Danish Straits.

Additional 500,000 bbl/d of crude oil from Norway flow eastward to the other Scandinavian markets. About one-third of the westward shipments through the straits are of refined petroleum products coming from Baltic Sea ports of Tallinn, Venstpils, and St Petersburg.

Panama Canal

The United States is the top user of the Panama Canal (as country of origin and destination) for all commodities going to/from the U.S. It is, however, not a significant route for U.S. crude oil or petroleum transports.

The Panama Canal is an important route connecting the Pacific Ocean with the Caribbean Sea and the Atlantic Ocean. It is 50 miles long, and only 110 feet wide at its narrowest point, Culebra Cut on the Continental Divide.

Over 14,000 vessels transit the canal annually, of which more than 60 percent (by tonnage) represent United States coast-to-coast trade, along with United States trade to and from the world that passed through the Panama Canal.

Closure of the Panama Canal would greatly increase transit times and costs adding over 8,000 miles of travel. Vessels would be forced to reroute around the Straits of Magellan, Cape Horn and Drake Passage under the tip of South America.

The Panama Canal is not a significant route for U.S. petroleum imports, so a closure would not affect the U.S. energy security much.

Roughly one-fifth of the traffic through the canal is oil tankers. 755,000 bbl/d of crude and petroleum products were transported through the canal in 2011, of which 637,000 bbl/d were refined products, and the rest crude oil. Nearly 80 percent of total petroleum, or

608,000 bbl/d, passed from north (Atlantic) to south (Pacific).

The relevance of the Panama Canal to the global oil trade has diminished, as many modern tankers are too large to travel through the canal. Some oil tankers can be nearly five times larger than the maximum capacity of the canal.

The largest vessels that can transit the Panama Canal are the PANAMAX-size vessels—ships ranging from 60,000 to 100,000 dead weight tons in size, and no wider than 108 ft.

In order to make the canal more accessible, the Panama Canal Authority began an expansion program to be completed by the end of 2014. Many larger tankers will be able to transit the canal after 2014, and yet some ULCCs will still be unable to make the transit. These vessels will need to find an alternative route.

The Trans-Panama Pipeline

The Trans-Panama Pipeline (TPP) is located outside the former Canal Zone, near the Costa Rican border. It runs from the port of Charco Azul on the Pacific Coast, to the port of Chiriquie Grande in Bocas del Toro on the Caribbean.

The pipeline's original purpose was to facilitate crude oil shipments from Alaska's North Slope to refineries in the Caribbean and the U.S. Gulf Coast. In 1996, the TPP was shut down as oil companies began shipping Alaskan crude along alternate routes.

In 2009, TPP completed a project to reverse its flows in order to enable it to carry oil from the Caribbean to the Pacific. The pipeline's current capacity is about 600,000 bbl/d.

Today BP is leasing storage located on the Caribbean and Pacific coasts of Panama and uses the pipeline to transport crude oil to the U.S.' West Coast refiners. BP has leased 5.4 million barrels of PTP's storage and committed to east-to-west oil shipments through the pipeline averaging 100,000 b/d.

This route significantly reduces transport time and costs of tankers navigating around Cape Horn at the tip of South America on their way to the U.S. West Coast.

The Oil Exporters

While worrying about safe ways to transport oil across the world, we also need to take a look at the global oil producers. One thing most of them have in common is that they have nothing in common with us.

Their political, religious, and social systems are often far from what we consider and value as democratic and fair. In many cases, these are countries ruled by

monarchs who are shamelessly imposing their will on the people, which is as undemocratic as undemocratic can be.

And some of these monarchs are as anti-American as they could be. Many of these support terrorist organizations, and/or are openly declaring their animosity towards America and everything it stands for.

And yet, we call many of these people "friends," and engage in "friendly" negotiations, while behind our backs they are planning how to hurt us. All this in the name of crude oil.

It is a game of broken promises, desperate moves, conflicts, and wars which has been going for a century and will continue like this for awhile longer. However, the world's energy, financial, and political systems change—sometimes very fast and in unpredictable manners—so we must be ready to react and counter-react to protect our energy system.

This is vital for our survival as the most powerful nation in the world. Without enough energy—and crude oil in particular—our military strength would be diminished, our economic growth crippled, and our safety compromised.

For that purpose, the national strategic petroleum reserves were established in response to the 1973 Arab oil embargo.

Strategic Petroleum Reserves

The fourth principle of President Carter's 1977 energy policy was that we must reduce our vulnerability to potentially devastating embargoes.

The policy was to be implemented by 1985…albeit partially.

President Carter's policy was designed to protect us from uncertain supplies by reducing our demand for oil, by a) making the most of our abundant resources, such as coal, and b) developing a strategic petroleum reserve.

Note: After the Arab oil embargo, reducing oil use and looking for other energy sources was on the national agenda for awhile, and for a number of times after that. Today, however, it is on the back burner…again, mostly due to the new crude oil and natural gas bonanza we are experiencing.

Reducing oil use does not seem to be a priority now, and the alternatives are too expensive and bothersome. The good old happy days of 8-cylinder cars and trucks are back again, and only somewhat higher prices at the gas pump keep us from getting full blast into the 1950-1960s energy oblivion.

The strategic petroleum reserves (SPR) plans were

implemented as scheduled, to the tune of many billions of dollars. Now the U.S. strategic petroleum reserve is a federally owned stockpile of oil with a capacity of nearly 750 million barrels stored at different strategic locations.

The national strategic oil reserves are actually crude oil and petroleum products stored in different underground storage reservoirs and/or aboveground tanks at different areas of the country.

Thus great quantities of crude oil and petroleum products sit in storage, but are available for use only in case of emergency, or sudden disruption in oil supply.

The storage facilities vary in type and size as follow:

Oil Storage Tanks

There are different types of vertical and horizontal aboveground atmospheric and pressure storage tanks in tank farms, which contain crude oil, petroleum feedstocks, intermediate stocks or finished petroleum products. Their size, shape, design, configuration, and operation depend on the amount and type of products stored and company or regulatory requirements. Aboveground vertical tanks may be provided with double bottoms to prevent leakage into the ground and cathodic protection to minimize corrosion.

Horizontal tanks may be constructed with double walls or placed in vaults to contain any leakage.

- *Atmospheric cone roof tanks* are aboveground, horizontal or vertical, covered, cylindrical atmospheric vessels. Cone roof tanks have external stairways or ladders and platforms, and weak roof-to-shell seams, vents, scuppers or overflow outlets; they may have appurtenances such as gauging tubes, foam piping and chambers, overflow sensing and signaling systems, automatic gauging systems and so on.

  When volatile crude oil and flammable liquid petroleum products are stored in cone roof tanks there is an opportunity for the vapor space to be within the flammable range. Although the space between the top of the product and the tank roof is normally vapor rich, an atmosphere in the flammable range can occur when product is first put into an empty tank or as air enters the tank through vents or pressure/vacuum valves when product is withdrawn and as the tank breathes during temperature changes. Cone roof tanks may be connected to vapor recovery systems.

- *Conservation tanks* are a type of cone roof tank with an upper and lower section separated by a flexible membrane designed to contain any vapor produced when the product warms and expands due to exposure to sunlight and to return the vapor to the tank when it cools and condenses at night. Conservation tanks are typically used to store aviation gasoline and similar products.

- *Atmospheric floating roof tanks* are aboveground, vertical, open top or covered cylindrical atmospheric vessels that are equipped with floating roofs. The primary purpose of the floating roof is to minimize the vapor space between the top of the product and the bottom of the floating roof so that it is always vapor rich, thus precluding the chance of a vapor-air mixture in the flammable range.

  All floating roof tanks have external stairways or ladders and platforms, adjustable stairways or ladders for access to the floating roof from the platform, and may have appurtenances such as shunts which electrically bond the roof to the shell, gauging tubes, foam piping and chambers, overflow sensing and signaling systems, automatic gauging systems and so on. Seals or boots are provided around the perimeter of floating roofs to prevent product or vapor from escaping and collecting on the roof or in the space above the roof.

  Floating roofs have legs which may be set in high or low positions depending on the type of operation. Legs are normally maintained in the low position so that the greatest possible amount of product can be withdrawn from the tank without creating a vapor space between the top of the product and the bottom of the floating roof. As tanks are brought out of service prior to entry for inspection, maintenance, repair or cleaning, there is a need to adjust the roof legs into the high position to allow room to work under the roof once the tank is empty. When the tank is returned to service, the legs are readjusted into the low position after it is filled with product.

- *Aboveground floating roof storage tanks* are further classified as external floating roof tanks, internal floating roof tanks or covered external floating roof tanks.

  External (open top) floating roof tanks are those with floating covers installed on open-top storage tanks. External floating roofs are usually constructed of steel and provided with pontoons or other means of flotation. They are equipped with roof drains to remove water, boots or seals to prevent vapor releases, and adjustable stairways to reach the roof from the top of the tank regardless of its position.

These may also have secondary seals to minimize release of vapor to the atmosphere, weather shields to protect the seals, and foam dams to contain foam in the seal area in case of a fire or seal leak. Entry onto external floating roofs for gauging, maintenance or other activities may be considered confined-space entry, depending on the level of the roof below the top of the tank, the products contained in the tank, and government regulations and company policy.

- *Internal floating roof tanks* usually are cone roof tanks which have been converted by installing buoyant decks, rafts or internal floating covers inside the tank. Internal floating roofs are typically constructed of various types of sheet metal, aluminum, plastic or metal-covered plastic expanded foam, and their construction may be of the pontoon or pan type, solid buoyant material, or a combination of these.

   Internal floating roofs are provided with perimeter seals to prevent vapor from escaping into the portion of the tank between the top of the floating roof and the exterior roof. Pressure/vacuum valves or vents are usually provided at the top of the tank to control any hydrocarbon vapors which may accumulate in the space above the internal floater. Internal floating roof tanks have ladders installed for access from the cone roof to the floating roof. Entry onto internal floating roofs for any purpose should be considered confined-space entry.

- *Covered (external) floating roof tanks* are basically external floating roof tanks that have been retrofitted with a geodesic dome, snow cap, or similar semi-fixed cover or roof so that the floating roof is no longer open to the atmosphere. Newly constructed covered external floating roof tanks may incorporate typical floating roofs designed for internal floating roof tanks. Entry into covered external floating roofs for gauging, maintenance or other activities may be considered confined-space entry, depending on the construction of the dome or cover, the level of the roof below the top of the tank, the products contained in the tank, and government regulations and company policy.

Tank Farms

Tank farms are groupings of storage tanks at producing fields; refineries; marine, pipeline and distribution terminals; and bulk plants which store crude oil and petroleum products. Within tank farms, individual tanks or groups of two or more tanks are usually surrounded by enclosures called berms, dykes or fire walls. These tank farm enclosures may vary in construction and height, from 45-cm earth berms around piping and pumps inside dykes to concrete walls that are taller than the tanks they surround.

Dykes may be built of earth, clay or other materials; they are covered with gravel, limestone or sea shells to control erosion; they vary in height and are wide enough for vehicles to drive along the top. The primary functions of these enclosures are to contain, direct and divert rainwater, physically separate tanks to prevent the spread of fire from one area to another, and to contain a spill, release, leak or overflow from a tank, pump or pipe within the area.

Dyke enclosures may be required by regulation or company policy to be sized and maintained to hold a specific amount of product. For example, a dyke enclosure may need to contain at least 110% of the capacity of the largest tank therein, allowing for the volume displaced by the other tanks and the amount of product remaining in the largest tank after hydrostatic equilibrium is reached. Dyke enclosures may also be required to be constructed with impervious clay or plastic liners to prevent spilled or released product from contaminating soil or groundwater.

The History of Strategic Oil Storage

Following the 1973 Arab oil embargo, the U.S. started looking into ways to protect itself from similar disasters. In 1975, the U.S. founded the strategic petroleum reserves (SPR) in order to mitigate future temporary supply disruptions.

In July, 1977, approximately 412,000 barrels of Saudi Arabian light crude were delivered for a first time to the SPR caverns in the Gulf. This was the beginning, and the flow continued until a total of about 600 million barrels were completed by the end of 1994.

Then direct purchase of crude oil was suspended during 1995-1999 due to budget resources being redirected to refurbishing the SPR equipment and other work intended to extend the life of the complex through the first quarter of the 21st century.

The filling activities were resumed in 1999 using a joint initiative between the Departments of Energy and the Interior to supply royalty oil from federal offshore tracts to the Strategic Petroleum Reserve, known as the Royalty-in-Kind (RIK) program. It continued in phases from 1999 through 2009, when the Department of the Interior discontinued the RIK program for good.

The first direct purchase of crude oil from the SPR since 1994 was in January 2009 using revenues available from the 2005 Hurricane Katrina emergency sale. DOE purchased 10.7 million barrels at a cost of $553 million.

In 2012, the SPR conducted an emergency exchange with Marathon Oil following the Hurricane Isaac disruptions to the commercial Gulf Coast oil production, refining, and distribution systems. Marathon Oil repaid the one million barrels in 3 months.

Today, the U.S.' SPR is the largest emergency supply in the world, with the capacity of over 700 million barrels, worth some $20-30 billion. Depending on how much oil is used, the U.S. SPR represents a 60- to 90-day supply of oil.

But the maximum total withdrawal capability from the SPR is only slightly over 4.0 million barrels per day, so it would take nearly 200 days to use the entire inventory. During this time, provided there were no other fuel supplies, the U.S. transportation sector would function on 50% capacity.

The U.S. oil reserves are stored at four sites on the Gulf of Mexico, located near a major center of petrochemical refining and processing. The storage areas are actually a number of artificial caverns created in salt domes located below the surface. These structures were built at a cost of $4 billion by drilling into the salt deposits and then dissolving the salt with water.

Each cavern can be up to 3,500 feet below the surface, and capable of holding 10-35 million barrels of oil. The caverns-concept was used to reduce costs, since it is 10 times cheaper to store oil below the surface than in large aboveground tanks. The added advantages are those of no leaks, and there is also a constant natural churn of the oil due to a temperature gradient in the caverns, which keeps it fresh.

Decisions to withdraw crude oil from the SPR are made by the U.S. president under the authority of the Energy Policy and Conservation Act. In the event of an energy emergency, SPR oil would be distributed by competitive sale.

The SPR has been used under these circumstances only three times, most recently in June 2011 when the president directed a sale of 30 million barrels of crude oil to offset disruptions in supply due to Middle East unrest.

The United States acted in coordination with its partners in the International Energy Agency (IEA). IEA

Figure 3-17. U.S. SPR aboveground delivery system

countries released altogether a total of 60 million barrels of petroleum. The SPR's formidable size (capacity of well over 700 million barrels) is a significant deterrent to oil import cutoffs and a key tool of foreign policy.

The U.S., however, has no significant gasoline or diesel reserves. So, while we are somewhat protected from disruptions in oil supplies, we depend on other sources in emergencies in case of major disruption to refinery operations. This is a problem, since no new refineries have been constructed in the US for thirty years, thus there is little excess refining capacity. Hurricane Katrina made that point, when many Gulf coast oil refineries were disrupted for the duration.

Update 2014

There have been some more than dramatic developments in the U.S. energy sector recently, marked by an unprecedented increase in domestic crude oil production. This has brought up significant changes in the overall national energy system, which include pipeline expansions, and construction of new energy infrastructure.

New methods of handling our energy sources (crude oil in particular) have been developed, including reversed flow of existing pipelines, and increased and diversified use of the domestic crude oil terminals.

The government is also assessing the strategic petroleum reserves system's capabilities and evaluating the appropriate responses in the event of emergencies and oil supply disruptions.

For that purpose, in March 2014, the U.S. Department of Energy initiated a test of the system by drawing and selling about 5 million barrels of sour crude oil from its strategic petroleum reserves. This exercise was done solely for operational purposes, and was designed to have a relatively minimal market impact. The release was a "test sale," according to DOE, designed to evaluate how the reserve system works when releasing and selling the oil.

Refineries that were interested in buying crude oil from the reserve did so in preparation for the summer driving season. This is when the U.S. refineries switch over to summer grade gasoline, which is more expensive than the winter types.

As a result, the oil markets responded by extending the losses, due to an unexpected flood of crude-oil supplies, and the accompanying announcement of selling a significant amount of oil from the U.S. strategic petroleum reserve. Light, sweet crude for April delivery fell $2.05, or 1.5%, on the New York Mercantile Exchange. Brent crude on ICE Futures Europe slipped 33 cents, or 0.3%.

The last time the U.S. DOE conducted a similar test sale was in August 1990 when 4 million barrels of crude oil were put on sale. Such test sales are required by law and are done periodically, but there is no specific order or schedule of when and how these tests are conducted.

Crude oil was also released from the U.S. strategic oil reserves in 2011, as discussed above.

The biggest shortcoming of the strategic oil reserves system is that it can keep the U.S. economy going at full speed for only 60-90 days—maybe longer with some modifications and restrictions. It cannot keep the economy going much longer, let alone forever, and things can get even tighter in case of a more serious, unexpected problem.

We can easily see how such problems could occur, but we are well prepared to handle most of them. And yet, there is another, even bigger, problem that even the mighty U.S. of A. cannot handle. It is the pending depletion of the global fossil reserves. The so-called "Peak Oil" is approaching and there is not much we can do about it. As a matter of fact, most of us—including oil companies, regulators, and politicians—are unaware of, or turn a blind eye to, its existence

It is still largely ignored by the world's energy markets too, but will become a major factor in the near future, most likely when oil prices increase, and/or oil supplies decrease to certain levels.

CRUDE OIL PROCESSING

Crude oil is the stuff that comes from the ground when a crude oil deposit is found. It is the *unprocessed* oil, which is also known as *petroleum*, or liquid *fossil fuel*

(coal is the solid, and natural gas is the gaseous form of the fossil fuels).

Crude oil was made naturally a long time ago from decaying plants and animals (or fossils) living in ancient seas millions of years ago. In most cases, crude oil is found under the oceans, or in places which were once sea beds.

Crude oils vary in color, from clear to tar-black, and their viscosity varies from water-like, to molasses-like, to almost solid. Crude oil is an extremely valuable material. It is for making many different substances, which also contain hydrocarbons.

Hydrocarbons (HC) are organic compounds that contain hydrogen and carbon atoms in their molecules. They come in various sizes and structures;, from single atoms to straight molecular chains, to branching chains, and rings.

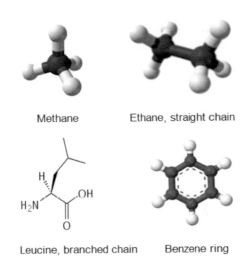

Methane   Ethane, straight chain

Leucine, branched chain   Benzene ring

Figure 3-18. Different HC configurations

The smallest hydrocarbon molecule is methane (CH_4). With its one carbon atom, it, like all HCs with up to 4 C atoms, is a gas. Longer HC chains with 5 or more C atoms are liquids, like gasoline and diesel. Very long HC chains, with 6 or more carbon atoms are solids, or semi-solids, like wax and tar.

The two most important characteristics of crude oil and its derivatives are:

• Their high energy content. Many of the things derived from crude oil, like gasoline, diesel fuel, paraffin wax and many others, have very large energy content.

• Hydrocarbon chains are very versatile and can take on many different forms and properties, which makes them very useful.

- Via special chemical reactions we can react and cross-link hydrocarbon chains to obtain consumer goods—anything from synthetic rubber, to nylon, to different types of plastics.

The major classes of hydrocarbon types contained in crude oil include:

- *Paraffins* with general formula C_nH_{2n+2}, where n is a whole number, usually from 1 to 20. These are straight- or branched-chain molecules, which can be gasses or liquids at room temperature depending upon the molecules. Some of these are methane, ethane, propane, butane, isobutane, pentane, hexane—all of which are gasses at room temperature.

- *Aromatics* with general formula C_6H_5 are ringed structures with one or more rings. Each ring contains six carbon atoms, with alternating double and single bonds between the carbons. These hydrocarbons are typically liquids, like benzene and napthalene

- *Napthenes or Cycloalkanes* with general formula: C_nH_{2n}, where n is a whole number usually from 1 to 20 are also ringed structures with one or more rings. Their rings contain only single bonds between the carbon atoms. These compounds are typically liquids at room temperature, like cyclohexane and methyl cyclopentane.

- *Alkenes* with general formula C_nH_{2n}, where n is a whole number, usually from 1 to 20, are linear or branched chain molecules containing one carbon-carbon double-bond. These hydrocarbon materials can be liquid or gas, such as ethylene, butene, isobutene

- *Dienes and Alkynes* with general formula C_nH_{2n-2}, where n is a whole number, usually from 1 to 20, are linear or branched chain molecules, containing two carbon-carbon double-bonds. These hydrocarbons can be liquid or gas, such as acetylene and butadienes.

Now that we know what's in crude oil, let's see what we can make from it.

Types of Oil Products

Crude oil contains hundreds of different types of hydrocarbons and impurities, all mixed together in different proportions in the bulk. To be useful, the different fractions must be separated by type of hydrocarbons and impurities. This is done by refining the crude oil in special refineries.

Different hydrocarbon chain lengths have progressively higher boiling points, so they can all be separated by distillation. This is what happens in an oil refinery—in one part of the process, crude oil is heated and the different chains are pulled out by their vaporization temperatures and separated for later use. Each different chain length has a different property that makes it useful for different purposes.

There is great diversity contained in crude oil, which is why refining it is so important to our society.

The following is a list of some key products that can be refined from crude oil:

- Petroleum gas is used for heating, cooking, and making a number of plastics. It contains small alkanes molecules (1 to 4 carbon atoms), commonly known by the names methane, ethane, propane, butane with a boiling range of less than 104 degrees Fahrenheit. These can be liquified by compressing them to a high pressure to create LPG (liquefied petroleum gas). This process uses a lot of energy and requires special container and transport, so it is used only in special circumstances usually for export purposes.

- *Naphtha or Ligroin* is an intermediate product that will be further processed to make gasoline. It is a mix of 5 to 9 carbon atom alkanes with a boiling range of 140 to 212°F.

- *Gasoline* is a liquid motor fuel, which is a mix of alkanes and cycloalkanes, containing 5 to 12 carbon atoms. Its boiling range is 104 to 401°F.

- *Kerosene* is fuel for jet engines and tractors and is also used as a starting material for making other products. It is liquid mix of alkanes, containing 10 to 18 carbons, and aromatics. Its boiling range is 350 to 617°F.

- *Gas oil or Diesel distillate* is used for diesel fuel and heating oil, as well as a starting material for making other products. It is liquid mix of alkanes, containing 12 or more carbon atoms. The boiling range is 482 to 662°F.

- *Lubricating oil* is used for motor oil, grease, other lubricants. It is a liquid long chain, containing 20 to 50 carbon atoms, alkanes, cycloalkanes, and aromatics. Its boiling range is 572 to 700°F.

- *Heavy gas or Fuel oil* is used for industrial fuel, as well as a starting material for making other products. It is a liquid long chain, containing 20 to 70 carbon atoms, alkanes, cycloalkanes, and aromatics. Its boiling range is 700 to 1112°F.

- *Residuals* are materials like coke, asphalt, tar, waxes. They are also used as a starting material for making other products. These materials are usually solid multiple-ringed compounds with 70 or more carbon atoms. Their boiling range is greater than 1112°F.

All these products have different sizes and boiling ranges, which is the basis for their refining. They can be easily separated by heating, boiling and condensing each fraction, as we will see below.

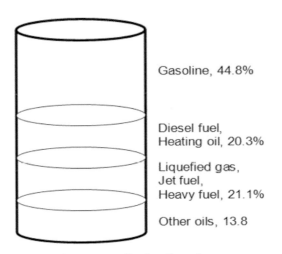

Gasoline, 44.8%

Diesel fuel, Heating oil, 20.3%

Liquefied gas, Jet fuel, Heavy fuel, 21.1%

Other oils, 13.8

Figure 3-19. Crude oil products

As seen in Figure 3-19, most of the oil fractions produced from each barrel of crude oil are fuels (gasoline, diesel and such) intended to be burned in a number of applications, while about 7-8% of the crude oil is turned into road-making products, lubricants, and other petrochemical feedstocks.

A smaller amount of the crude is converted into oil derivatives, many of which go into making a vast array of the commonplace products we use today. Plastics are of the largest quantity of these products, followed by medicines, contact lenses, insecticides, toothpastes, cosmetics, fertilizers, paint, food preservatives, etc.

What a waste of a precious commodity—to be burned with no trace left except for deadly, climate warming gasses. What will future generations think of us, burning their inheritance without remorse or thought about a tomorrow without fossil reserves?!

The Oil Refinery

We refine crude oil to produce a variety of fuel products, ranging from the heaviest oils for use in industrial boilers, to fuel oil used in home heating, to diesel oil that powers most trucks and some cars, to jet aircraft fuel, to gasoline for our cars. Each of these products contains a mix of different molecules, so no single chemical formula describes a given fuel. Molecules typically found in gasoline, for example, include heptane, octane, nonane, and decane. These molecules consist of long chains of carbon with attached hydrogen molecules.

Higher grades of gasoline have a greater ratio of octane to heptane which makes them burn better in high-performance engines. This is the origin of the term high octane at the gas pump. The energy content of refined fuels varies somewhat, but a rough figure is about 45 MJ/kg for any product derived from crude oil.

Figure 3-20. Oil refinery

Oil refineries are huge developments—entire cities of buildings, towers, piping, electrical substations, wires, storage tanks, railroad stations and ports. The life of a refinery is that of never-ending loading and unloading, processing and re-processing of oil products in gaseous, liquid and solid form.

Oil refining is an energy-intensive process; some 8% of the total U.S. energy consumption is used to run the fleet of oil refineries around the country.

The actual quantitative breakdown of the different fractions produced in a refinery varies with demand. For example, the percentage of crude oil refined into heavier fuel oil for heating increases significantly in the winter months.

The Oil Refining Process

The crude oil is pumped from the storage tanks by charge pumps located at one of the oil storage tanks. Each of them can pump 350 gallons of oil per minute at 240 PSI outlet pressure.

The crude is then mixed with "make-up water," which is fresh water pumped from a well into an automated water desalination system, to purify it prior to mixing. After mixing with "make-up water," the crude is pumped through two heat exchangers, connected in series and mounted about twenty feet above the ground, in close proximity to the crude distillation tower.

The crude goes through the tubes of the heat exchangers and gets heated to over 400°F by indirect contact with the hot liquids coming from different stages of the process. It is then pumped into a condenser unit to separate one of the low-boiling fractions contained in the oil product.

Then the crude is mixed with more "make-up water" to form a water-in-oil emulsion. Salts, of the type of magnesium chloride and calcium chloride are in this emulsion and must be removed. The goal is to obtain salt concentration less than 1 pound of salt per 1000 barrels of crude. This is achieved by an electrostatic device and by adding demulsifying chemicals to the mix.

Thus purified crude is then pumped into the atmospheric distillation unit where the crude oil is distilled (separated) into low boiling point fractions. It is then pumped into a vacuum distillation unit, which distills the higher boiling point fractions remaining after the atmospheric distillation.

The most important processes take place in the distillation column, or vacuum distillation units. This is a method of separating the different fractions by reducing the pressure in the vessel, which together with the high temperature of the oil causes evaporation of the volatile liquids. Vacuum distillation can be used also without heating the mixture, but in such case the high boiling point fractions will not be separated.

Some of the fractions go through a naphtha hydro-treater unit where hydrogen gas is used to de-sulfurize the naphtha fraction left from the previous atmospheric distillation. The naphta is then hydro-treated to separate other fractions and impurities, before sending it to a catalytic reformer unit, where the naphtha fractions are converted into higher octane products. Hydrogen is released during the catalyst reaction, and is used either in the hydrotreating or the hydrocracking process steps.

The different fractions go through a number of additional steps and pieces of equipment, such as fluid catalytic cracker, hydrocrackers, visbreaking, merox,

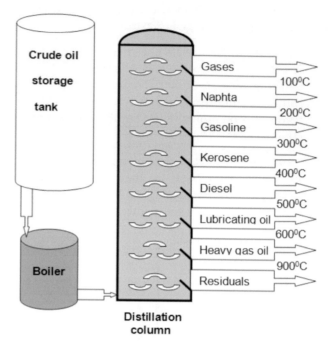

Figure 3-21. Distillation of the different oil fractions

coking, alkylation units, dimerization, isomeration, steam reforming, amine gas treater, claus unit, and other units and steps during the long and complex refining process.

The final products, obtained and separated during the different process steps are petroleum based products, that can be grouped into light distillates (LPG, gasoline, naphta); middle distillates (kerosene and diesel); heavy distillates (misc. products); and residuum (heavy fuel oil, lubricating oils, wax, asphalt).

In more practical terms, the refinery produces are liquefied petroleum gas, gasoline, naphtha, kerosene, jet aircraft fuel mixes, diesel fuel, fuel oils, lubricating oils, paraffin wax, asphalt and tar, petroleum coke, and sulfur. A number of gasses, like propane and others, are also produced.

All products are stored and eventually shipped for use to different customers in different industries.

Synfuels

An important part of our drive towards energy security is finding new products to replace crude oil and its derivatives. Synfuels, or synthetic fuels, are a good example of such products.

Synfuel is basically a liquid fuel produced by processing some types of organic matter, such as coal, natural gas, oil shale, oil sands, or biomass into useful fuels.

Synfuels can be also derived from other solid materials, such as some types of plastics, rubber waste, municipal waste, and others. The production of these fuels also includes gaseous fuels, which are usually produced as by-products of the synfuel production process.

Synthetic fuels are usually produced by the chemical process of conversion, which could be direct conversion into liquid transportation fuels, or indirect conversion. In the latter case, the source substance is converted into syngas, which then goes through additional conversion process to become liquid fuels. Basic conversion methods include carbonization and pyrolysis, hydrogenation, and thermal dissolution.

The most common production processes for synfuels are:

* The Fischer Tropsch conversion,
* Methanol to gasoline conversion, and
* Direct coal liquefaction.

Presently, there are about 250,000-300,000 barrels of synfuels produced daily around the globe. The effort in this area continues and the total production volume is increasing.

Types of Synfuels

The race to find the next big energy solution is definitely on. There may not be a magic bullet to solve the energy crisis, or a perfect fuel that is infinitely available and doesn't pollute the environment, but one option, synthetic fuels—or synfuels—offers some advantages and some drawbacks when compared to conventional oil-based fossil fuels.

Synthetic fuels are produced from coal, natural gas or biomass feedstocks through chemical conversion into useful high energy fuels.

These types of fuels are often called Fischer-Tropsch liquids, after the more common process used to create them. The synfuels category also includes fuels derived from synthetic crude, a substance similar to crude oil that is synthesized from natural resources like bitumen or oil shale.

Chemically, synfuels are similar to the gasoline and diesel fuels we use today and can be used in existing engines, but producing them requires complex chemical conversions.

Governments and energy companies have been paying more attention to synthetic fuels in recent years, as rising oil prices and political instability in oil-producing countries have created incentives to seek out alternatives. Synfuels are also quickly becoming an important part of our energy security portfolios.

The main benefit of synfuels is that they can be produced from coal, natural gas and even biomass, and other widely available raw materials. Many synfuels also burn cleaner than conventional fuel, thus they help in the battle for environmental improvement.

Like anything else, there are also disadvantages. Although they burn cleaner, during production they emit just as much, and even more, GHG pollution than producing or burning regular gasoline.

Synfuels are also much more expensive to produce than conventional fuels, mostly because of the extensive use of fossil fuels during their production. So, much more research, development and investment are necessary to make synfuels production economically viable.

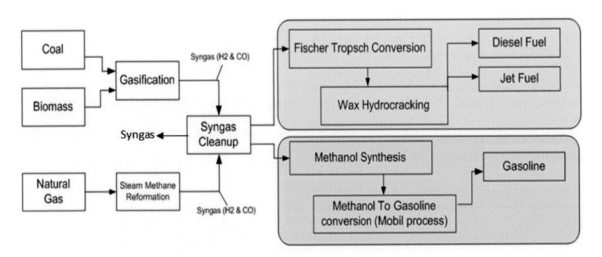

Figure 3-22. Synfuel production process flow (indirect conversion method)

Some of the important types of synfuels are:

Syncrude is a type of synthetic fuel that is quite similar in properties to crude oil. It is produced from extra-heavy oil, which occurs naturally, and is formed when conventional crude is exposed to bacteria that breaks down the hydrocarbons and changes the oil's original physical and chemical properties.

Due to its high consistency, that of molasses, this type of oil cannot be pumped out directly. Instead, it can be recovered through open pit mining or by piping hot steam or gas into a well to liquefy the heavy oil and collect it through a second well. All existing methods have their limitations and are very expensive.

Open pit mining (*ex-situ*) can only be used to collect extra-heavy oil near the surface. This process damages the environment by destroying forests and animal habitats. Huge amounts of water, required during the process, must be disposed of as waste after use.

In situ collection involves piping hot steam or gas into a well to break up the heavy oil and collecting the fluid through a second well.
Both methods have their limitations. They are not well developed for now, and need a lot of external energy to produce large amounts of heavy oil, which makes them very expensive.

In situ methods are less damaging, but much more expensive. They also need further research to produce large amounts of heavy oil efficiently and economically.

Syncrude is a synthesized crude oil that is unsuitable for use in its as-is state. Instead, it must be further refined before it can be used commercially. In its natural state, extra-heavy oil is basically a more viscous form of crude, resembling the consistency of molasses.

To get the extra-heavy oil into a useful form, it is processed in refineries by a process very similar to that used to refine crude oil, but it is in fact much more technologically complex and expensive.

The oil is typically exposed to heat and gases that break down the hydrocarbons into those that can be burned as fuel and those that can't. The different fractions are then separated and stored for use or further processing. This is similar to the process of refining crude oil into fuels, but is more complex and expensive.

Gas-to-Liquids (GTL) synfuels are produced by converting natural gas to liquid, petroleum-based fuels. GTL products usually require less complex and expensive processing than the extra-heavy oil. Some types of GTL synfuels can be used as fuel (in gaseous form) as-is and without further processing.

The most widely used method for converting gas to liquid fuels is the Fischer-Tropsch process (F-T synthesis). Here, natural gas is mixed with air before burning in a chamber with a catalyst (cobalt or iron.) The catalyst, under high heat and pressure in the chamber, triggers a chemical reaction that forms long chains of hydrocarbons.

The gas is then condensed into a liquid form, and depending on type catalysts used, different hydrocarbon structures can be obtained.

F-T synthesis can produce diesel fuels and naphtha, which can then be converted into gasoline, and a number of industrial lubricants.

The GTL process is mostly used to produce diesel fuels, which like other Fischer-Tropsch fuels, produces fewer emissions when burned. This chemical separation process creates purer fuels, because impurities can be filtered out easily, resulting in a higher quality and cleaner burning fuels.

Another benefit of GTL synfuels is that the chemical reactions involved in converting the gas to liquids create electricity and steam as byproducts. Those resources can be reused to save costs and reduce the environmental impact. Or they could be sold on the commercial market, as needed to make the process more cost effective.

Coal-to-liquids (CTL) synfuels are produced by converting coal into liquid or gaseous synthetic fuel that can be used for electric power generation or in internal combustion engines.

There are several production methods for that conversion; indirect and direct coal liquefaction.

- Indirect coal liquefaction (ICL) uses the same Fischer-Tropsch process as gas-to-liquids fuels. It has one additional step of converting the solid coal into gas that can feed the F-T reaction. First, solid coal is crushed and then burned in a high-temperature, high-pressure chamber, along with steam and oxygen. The reaction of the species in the reaction chamber produces gas, which is a mixture of carbon monoxide, hydrogen, and other gases. Thus obtained gas mixture is then fed into a Fischer-Tropsch reaction to create liquid synfuels.

- Direct coal liquefaction (DCL) requires coal to be pulverized, and then burned in hydrogen atmosphere at high temperature and pressure levels to produce liquid syncrude oil. The syncrude is then collected and processed as described above. This is a complex and expensive method, which is not as widely used as its ICL cousin.

CTL synfuels' advantage is that they generally burn cleaner than conventional gasoline or diesel. CTL manufacturing also produces byproducts, such as electricity and metals. Some of these can be reused or sold to offset the processing costs, thus making the process more sustainable.

In all cases, the CTL production has drawbacks, the most serious of which is cost. There are estimates today that the cost of a small-size, 100,000 barrels per day, modern CTL synfuels production plant will be about $5-6 billion.

On top of that, a number of environmental issues are to be resolved. Consumption of huge amounts of water and excess CO_2 emissions are the main concerns. The process also produces huge amounts of solid waste, or "slag," left from the burning of the coal.

The high cost—way above the cost of any of the conventional technologies—and the environmental concerns promise to keep the CTL technology in the closet for a long time to come.

- Biomass-to-Liquids (BTL) advantage is that here biofuels are produced by processing natural organic materials, most of which are renewable, instead of using fossils (organic material that has been decomposed and compressed over millions of years).

BTL fuels can be made from grass, wood, crops leftovers, straw and grains. Here the great advantage is that only the parts of plants that are not useful for food or other purpose are used.

First, the biomass is burned in a low oxygen environment, producing syngas. This step requires less energy than other synfuels, so potentially it could produce cheaper fuel. The following liquid fuel production from syngas is similar to the other synfuels, using the Fischer-Tropsch reaction.

Unfortunately, it takes very large quantities of raw biomass feedstock to make fuel.

Approximately 5 tons of biomass, grown in 3-5 acres of crops, and a lot of energy, are needed to produce about 1 ton of BTL synfuel.

Adding transportation and other costs, BTL costs exceed those of CTL or GTL production. Biomass also takes up much more space than other synfuel feedstocks, which adds to the final cost.

So, for now BTL is not nearly as widespread as other forms of synfuels, so we need to invest a lot of money to get BTL programs up and running.

Nevertheless, BTL uses renewable resources and is much better for the environment in the long run. Plants grown to produce the fuel could cancel out some of its CO_2 emissions, which alone might be able to bring BTL on the world energy markets sooner than the rest.

- Solid waste in municipal waste sites can also feed the synfuels production process. Solid waste that can be used for this purpose includes old tires, sewage and waste from landfills. Any material in the waste that contains organic matter, represented by high levels of carbon, can be used to create some form of fuel.

Municipal waste sorted for this purpose could be used for feedstock in the same process as other synfuel feedstocks discussed above. The waste materials are burned under special conditions to produce syngas, which then goes through the Fischer-Tropsch process to be synthesized into liquid fuel.

As an alternative, the gas that landfills naturally emit as waste decomposes (methane gas mostly) can be captured and added to the process, which could result in significant cost reduction.

The Synfuels Future

In 1984 President Ronald Reagan cut off government funding for the new U.S. synthetic fuels industry because he believed U.S. synthetic fuels would be too expensive, relative to falling oil prices. Reagan's National Energy Policy was crafted by his economic advisors who believed world oil prices would remain low. This established a precedent, that kept repeating, although oil prices kept going steadily higher.

Oil prices remained low for several years, mostly due to increased oil imports. This, however, established another precedent—that of complacency that led to increased dependence on foreign oil—from 25-30% in 1984 to over 50% today.

Today—even in the midst of an unprecedented energy boon—the U.S. is still unable to fuel 50% of its economy.

Twenty years later, President George W. Bush attempted to reverse Reagan's policy, when in 2005 he asked Congress to fund a synthetic fuels plant for the U.S. Air Force. It was supposed to be an example of using a secure domestic supply of jet fuel for military aircraft. Congress failed to support his request.

Today, the U.S. Air Force, several private companies, and universities are making an effort to develop synthetic fuels. Due to environmental and economic reasons, however, these efforts have not been supported by Congress and other key groups.

So, the short-term future of a serious domestic synfuels production is on hold, while its long run success depends on political winds and market conditions. Yet, synfuels might be the ultimate solution to our long-term energy security problems.

Refinery Logistics

During World War II, the US Department of Defense established the "Petroleum Administration for Defense Districts" (PADD) to control and facilitate oil allocation. At first, the refineries processed crude oil and distributed petroleum products for use in the local areas only. Soon, however, it became necessary to construct the Virginia and Colonial product pipelines, linking the Gulf Coast with the Northeast United States. This led to a network of crude oil and petroleum product pipelines that interlinks the PADDs, making them interdependent.

The PADDs Development

PADD 1, or the East Coast refineries, process crude oil shipped from all over the world, while PADD 2, Midwest, and PADD 4, Rocky Mountains, depend on crude oil produced and moved by pipeline from Canada and the Gulf Coast.

PADD 3, the Gulf Coast, is the largest refining region in the U.S., and obtains crude oil from the Gulf Coast outer continental shelf, Mexico, Venezuela, and the rest of the world.

Note: A large pipeline project, planned to deliver Canadian syncrude oil to the Gulf Coast Refineries—the XL Keystone pipeline, which we review in more detail in this text—is stalled in the permitting and political debate stages.

PADD 5, West Coast, gets crude oil delivered by tankers from Alaska and California oil wells and imports.

US refineries increased in number and capacity through the last century, reaching a peak during the mid-2000s, when the US and world's demand for gasoline and other fuels were increasing by the day. Then, due to the economic crisis, the U.S. gasoline consumption declined by almost 9 million barrels in 2008, and another 10 million barrels in 2009.

Amazingly, the US imported 81 million barrels in 2009, while the production of ethanol peaked at over 256 million barrels, plus nearly 4.6 million barrels of ethanol imports. It's complicated, the decision makers and oil companies would say. No doubt, it is!

The market volatility and overall confusion increased at the time, which led to record high crude oil prices, peaking at $150/bbl in 2008. The oil companies and other related parties' profit margins rose accordingly during the period. The U.S. consumers stood by watching the developments with their mouths wide open. What else could they do?

As a result of it all, the U.S. refineries were faced with the possibility of a long-term decrease in fuel demand, which forced them to cut costs, reduce capacity. A number of refineries were shut down for good.

At the golden age of the 2000s there were 158 fully operational refineries in the US and the territories. Now we have about 124 refineries that process crude oil into fuels, and 13 that produce lubricating oils and asphalt.

The refineries' production volumes go up and down according to market supply and demand conditions, and the political and economic factors. Today the U.S. refineries are operating at full capacity with short shutdowns for maintenance and upgrade work.

Overall, while the number of refineries is going down, their efficiency is increasing and so is their total operable refining capacity. From 16.5 million barrels/day total capacity a decade ago, it is now over 18 million barrels/day. This is an amazing achievement and a good example of producing more with less.

Case Study: The Oil Country of Texas

Texas is an amazing place. One can drive many, many hours without seeing any people or houses. The most frequent sighting on these long trips would be an oil rig or natural gas well, evidenced by pumps, well heads, or the related activities. Lately, lots of wind mills have been added to the Texas countryside as well, thus completing the energy-filled travel experience.

Texas was the leading crude oil-producing state in the nation in 2013, exceeding the production levels from any other offshore areas. The 27 Texas refineries had a capacity of over 5.1 million barrels of crude oil per day, which is almost 30% of the total U.S. refining capacity.

West Texas Intermediate (WTI) is a grade of crude oil produced in Texas and southern Oklahoma, which is traded domestically, but is used as a global benchmark for oil quality and pricing.

Texas is also the U.S. leader in wind-powered generation capacity, with over 12 GW of installed wind power capacity, which generates nearly 40 million MWh of electricity annually. Yet, the average annual electricity cost per Texas household is over $1,800, which is among the highest in the nation.

Texas's gasoline prices have been among the lowest in the nation for many years. Recently, however, they rose to just about average levels. With the long distance most Texans must drive, this is another penalty for the citizens of the country's leader of energy production.

Energy Security and Oil Refining

The issues related to energy security (of oil production, transport, and use) start at the oil wells. A number of risks exist at each well and extend throughout the oil distribution paths across the country and the world. Similar problems exist at the oil refineries too.

The risks related to oil refineries operation can be summarized as follow:

- Pipe transport from the source to the refinery, resulting in oil leaks, spills, explosions, fires, property and human health damage;

- Local problems, such as geophysical and political changes that can affect the performance and safety of the refinery and its workers. These include strikes, regulations, civil war—events that can interrupt, or permanently shut down a refinery;

- International conflicts, which could limit the flow of oil and/or physically harm a refinery, causing it to shut down;

- Weather-related problems, such as storms and hurricanes hitting the refinery infrastructure and causing damage to property and hurting workers. Golf Coast hurricanes often shut down many of the refineries in the area.

The U.S. refineries are well designed and protected against most events, but even then, we have witnessed production disruption after earthquakes, storms, hurricanes, etc. The major effect on U.S. oil refining is the fact that it is one of the most heavily regulated industries in the U.S.

EPA has recently settled enforcement actions with oil companies, comprising over 30% of U.S. domestic refining capacity, and has been also engaged in settlement negotiations with companies comprising an additional 20%.

Nearly 50% of the U.S. refineries are under some sort of notice or legal action, as a result of, or related to, safety or environmental impacts of daily operations.

In addition to the EPA restrictions and sanctions, the U.S. refiners face complex regulatory issues involving their products. Gasoline formulations, for example, are restricted differently depending on the season of the year, and on the geographical location of the market in which they are sold at the time.

The dominant environmental impacts from refinery operations currently involve air quality. About three-fourths of the EPA's Toxic Release Inventory (TRI) of GHG emissions reported by the oil refineries are releases into the air, so the industry is under strict scrutiny. This is changing the status quo and will determine the future development of the industry. These developments contribute to availability issues and price fluctuations.

The major classes of processes typically carried out by refineries are desalting, distillation, reforming and extraction, and waste recovery and treatment. Each step of the process is accompanied by consumption of a lot of energy (electric power and steam mostly) and huge GHG emissions.

The major impacts associated with the oil refinery operations include:

- Air emissions: volatile hydrocarbons from crude oil, SO_x from crude oil, process heat related and particulates, NO_x, and H_2S from sulfur recover operations

  Note: About 75% of total TRI emissions by weight is released into the atmosphere. Refinery emissions contain several major ozone precursors, and the associated impacts would be most significant near and downwind of each refinery.

- The average breakdown of GHG emissions are volatile organic compounds (VOC) about 200,000 tons per year, nitrogen oxides (NO_x) about 300,000 tons per year, and other hazardous air pollutants (HAPs) about 200,000 tons per year.

- Water discharges: process wastewater from desalting, distillation, cracking, and reforming operations, as well as large quantities of cooling water used at different steps of the process.

  Note: About 24% of total TRI emissions is released to wastewater, which has the potential for contaminating the local environment from leaks and spills at the facilities.

- Solid wastes: desalter sludges, spent catalysts, other process sludges, storage tank bottoms.

   Note: Environmental contamination of the local environment is from spills and leaks from storage areas, and during loading, unloading and transport.

Because of the air emissions and other releases, oil refining operations are blamed for global warming. Oil refining is an energy-intensive operation—the most energy intensive in the U.S.

At an average efficiency of 85%, U.S. refineries use (the equivalent of) 15 gallons of fuel to produce about 100 gallons of gasoline.

In More Detail

The net energy content of crude oil of average quality is approximately 132,000 Btu per gallon. At approximately 45 gallons of refined product per barrel, it can be estimated that about 21,000 Btu (15%) of energy is lost per gallon of gasoline.

$$(1 - 85\%) * \frac{45}{42} * 132,000 \approx 21,000 \text{ Btu/gallon}$$

This is roughly equivalent to 6 kWh of electric energy, which is what refineries use for power.

But this is only part of the story. In addition to the energy used directly for the refining (conversion of oil into gasoline) there are other, secondary, operations that also use a lot of energy.

- For example, even before the crude arrives at the refinery, a lot of energy is used during its discovery, well drilling, pumping, transport, and storage. These activities would add another 2-3% to the total energy lost during the entire crude oil's cradle-to-grave lifespan.

- Then, there is the energy used to generate the electric power used in the refining process. Here we need to look deep into the coal mines, and figure out the energy used to dig, transport, process, store, and burn the coal as needed to generate electricity.

- On top of that, there are huge losses in efficiency during some of these steps, with the largest one in the coal-fired power plants, most of which operate at 30-40% efficiency. So basically more than half of the initial energy content is lost in the electricity production, and more during its transmission to the refinery—even before the refining process has started.

Putting all these losses together, we come up with a total loss of over 40,000 Btu per gallon, or about 1/3 of the energy content of the original crude oil (or the equivalent of 12 kWh per gallon of gasoline lost in its production process).

Since the U.S. uses about 12 million barrels of crude oil annually, and gasoline is about 45% of the crude oil yield, we get 5.4 million barrels of annual U.S. gasoline production. About 227 million gallons of gasoline are used annually for powering our vehicles in the U.S.

This requires about 2.7 TWh of electric power, which is equivalent to the electric power generated by several coal-fired power plants. That much, or more electric power would be needed to refine the rest of the crude oil to produce other products.

The energy used in the actual refining process, then, has to be multiplied by 2 or 3 in order to get the total amount of energy used during the cradle-to-grave process of crude oil production and refining.

The actual dollar and environmental cost of each step of this process is significant, and has to be entered in the overall formula, which at the end would look something like this:

$$Pt = Op + Ot + Or + Cp + Ct + Pg + Ec$$

Where:
   Pt is the total cost of petrochemical production
   Pp is the cost of oil production
   Ot is the cost of oil transport
   Or is the cost of oil refining
   Cp is the cost of coal* production
   Ct is the cost of coal transport and processing,
   Pg is the cost of power generation, and
   EC is the environmental cost of the above steps.

*Note: This refers to coal used to generate power used in oil refining

It is obvious here that there are many steps in the process of producing and refining crude oil. Each of these steps is expensive and polluting. The cost of each step is usually calculated in the final price of the finished petroleum products, but there are a number of hidden costs and subsidies that obscure the picture.

Most of the energy used at the oil refinery is consumed as process heat, some of which is generated from by-products of the process. This energy source cannot be replaced by cleaner renewables, or other non-CO_2-intensive sources. This means that regardless of what happens, oil refineries will operate as usual—stinky polluters—in the long run.

Note: We take a detailed look at all costs—including the hidden costs—of crude oil production and use in our book on the subject, *Power Generation and the Environment*, published by the Fairmont Press in 2014.

Because of the prevailing socio-political conditions, the oil refining industry is likely to face ever increasing pressure to expand capacity as rapidly as possible. This is likely to bring resistance from local and environmental groups, which will force tightening of the emissions, which would lead to price hikes and shortages.

The potential future expansion of the industry is also very likely to exacerbate the ongoing regulatory struggle regarding the distinction between existing energy sources and the new such. The regulations are getting more stringent by the day under the Clean Air Act, which makes adding new refining capability in the U.S. almost impossible (or at least very expensive). Because of this new situation, the long-term expansion of the industry is uncertain.

There are additional potential regulatory consequences of the global warming and climate change problems as well, which may strongly affect the operations of this highly energy-intensive industry. This, in addition to increased volatility in the global oil markets and oil transport, will affect the availability (and demand) for oil refineries' products, and their prices.

2014 Update

The U.S. has been the world's largest importer of oil for decades, so American refineries are predominantly fitted to refine the heavy (lower-quality) crude imported from Mexico, Venezuela and the Middle East rather than the sweet-light domestic crude.

As a result, analysts are forecasting that in 2014-2015, the U.S. refineries will reach their limits to process excess sweet crude oil. That leaves them with two choices: a) export excess production, or b) shut down domestic production. Neither of these choices is acceptable, thus far, so new alternatives are sought presently.

With nearly 50% of our oil supplies coming from overseas, and with rising internal pressures on energy production and consumption, we must be prepared to change our habits and/or switch to using less oil in the near future.

For these and other reasons, the U.S. energy security—as far as crude oil is concerned—is uncertain in the long run, even though we see significantly increased domestic crude oil production.

While living with less oil is possible and is suggested as one of the solutions, it is an unwelcomed event, which might hurt our economy badly and affect our way of life negatively. On the other hand, if we don't change our ways—regardless of what the short term brings—we will continue to depend on unreliable foreign imports and live with never-ending energy price increases and increasing shortages.

OIL-FIRED POWER PLANTS

Crude oil-fired power plants are the exception to the rule today. They are few and far between and are used in some special circumstances; remote locations and such lack other energy resources.

Washington, DC, Alaska, and Hawaii, for example, are powered mostly by crude oil, which is also the case, albeit to a lesser extent, in some other states.

The furnaces of these types of power plants are powered directly by unrefined crude oil, or more often by some of its derivatives. The oil-fuel is transported to the power plants by ship, pipelines, truck, or train, where several methods can be used to generate electricity from the oil.

One way is to burn the oil in boilers to produce heat and steam, which is used by a steam turbine to generate electricity. Another common method is to inject the oil in combustion turbines, which are similar in operation to jet engines.

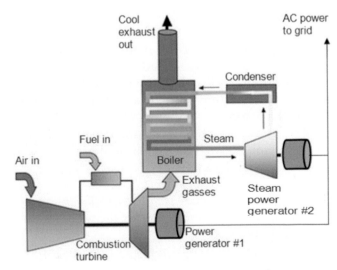

Figure 3-23. Combined cycle power plant

A more efficient method, or "combined cycle" technology, where oil is burn in a combustion turbine, but the hot exhaust gasses is then used to make steam and drive a steam turbine and generator to generate electricity. This technology is much more efficient because it uses the same fuel source twice. What makes this setup efficient is that the waste steam generated in the combustion turbines is send to a boiler to generate steam, which drives the turbine and generator sets.

Another use of crude oil derivatives (gasoline and diesel) is to fuel internal combustion engines, which then drive the turbine-gen sets to generate electricity.

Saudi Arabia's Shoaiba is one of the largest oil-fired power plants in the world. Following completion of stage 1 in 2003, stage 2 consisting of six additional 400 MW units was completed, bringing the total output of the plant's stage 1 and 2 to 4400 MW.

Stage 3 was completed recently, adding 1.2 GW of electric power generation, bringing the total power generation to 5.6 GW. The plant uses oil furnaces to burn crude oil and generate steam, which drives the steam turbines and generators to produce electric power.

This monster crude-oil-wasting plant alone provides over 40% of the country's western region's current power requirements. It is also a record breaker in terms of size for any oil-fired plant in the world. During the construction of stage 2, several records were broken in terms of construction and commissioning time for a steam power plant.

A multi-stage flash distillation water desalination plant was constructed as part of the plant complex. It has desalination capacity of 50 million cubic meters of water per year. A second desalination plant with another capacity of 50 million cubic meters of water per year was constructed later on as well.

The clever feature of this setup is that the waste steam generated in the power plant turbines is send to the desalination plant to be used to heat sea water. This provides cheap(er) clean water, while at the same time cools the steam before reuse, thus reducing the waste heat cooling losses.

The crude oil-fired power plant life-span cycle consists of several phases, very similar to those of a gas- or coal-fired power plant, as follow:

Plant site and structure planning and design
- Selecting, testing and exploring the proposed power plant location
- Engineering design of plant facilities, equipment and process
- Estimating the construction and operating cost and

the related tasks
- Applying for and obtaining the necessary federal, state and local permits

Plant construction and setup
- Plant building construction
- Plant equipment procurement and setup
- Labor training

Power plant daily operation
- Petroleum fuel oil delivery
- Oil burning and energy generation
- Electric power distribution

Power plant decommissioning and land reclamation
- Plant shut down and disassembly
- Plant waste disposal
- Surface land reclamation

Since the different steps of the power production process are similar to those used for coal-fired plants construction and use, we'd refer the reader to the section on COAL for more details.

Energy Security and Oil-power Plants

Crude oil and oil derivatives-fired power plants are few and far between in the U.S., but they are very important where they are. Currently, there are many oil-fired (oil or diesel mostly) power plants in many countries—Saudi Arabia, Japan, Israel, etc.—and in many remote locations, which lack any other fuels. Because of that, these power plants are very important for the locals and the local economies.

There are over 740 oil-fired power plants in the U.S. alone, which is several times the number of such plants around the world. Washington DC, Hawaii, Alaska, and other states depend on crude oil and its distillates to generate a significant portion of the electric power.

One could imagine what will happen in Washington, DC, for example, if crude oil is suddenly unavailable. Chaos will overcome the city at first, followed by disaster after disaster all around the affected area. You think DC is a disaster now, just imagine then…

Oil-fired power plants depend exclusively on the oil supplied to their burners, so the risks here are limited primarily to timely and plentiful oil supplies. Another set of factors that can affect the sector are unscheduled power generation events—due to accidents, weather phenomena, strikes, abnormalities.

Oil-fired power plants also cause environmental damage from GHG exhausts, which could be significant.

There is also the possibility of oil spills and leaks that could contaminate the local water table and soil.

All these factors—in addition to the fact that huge quantities of precious crude oil (which could be used for more noble purposes) is being burned in these facilities—must be taken into consideration when evaluating the existing and planning new oil-fired power plants.

Lack of crude oil supplies in these areas would be a direct hit on our energy and national security. The impact of a prolonged oil supply shortage would be truly disastrous for these areas and have a pronounced effect on the entire country's economic and political life.

Environmental Effects of Oil-fired Power Plants

Constructing and installing an oil rig and pumping millions of barrels of oil requires many pieces of large equipment. Building a oil-fired power station also requires large pieces of equipment, in addition to huge amounts of concrete, steel, and many other materials. All these materials are made with the help of fossil fuels—lots of them. During the production and transport of these materials, a lot of gaseous, liquid and particulate pollution was generated as well, including huge amounts of CO_2 pollution.

It has been estimated that the steel and concrete production and transport for the construction of a 1 GW power station have a carbon footprint of roughly 300,000 tons CO_2. Spreading the CO_2 over a 25-year reactor life we come up with almost 1.5 grams of CO_2 emitted per each kWh of electricity generated.

According to industry estimates, the total carbon footprint of an oil-fired power plant, including construction, fuel processing, decommissioning, etc.) is about 40 grams of CO_2 per kWh.

Amazingly, there are still many oil-fired power plants around the world, including the U.S. A number of states generate their electricity the old fashioned way—by burning crude oil, or diesel. Washington, DC, gets almost all of its electric energy from oil-fired power plants. The states of Alaska and Hawaii follow closely the DC example.

Nearly half of all toxic air pollution in the U.S. comes from coal- and oil-fired power plants.

The largest emitters are the power plants located in the states of Alaska, Arizona, Colorado, Delaware, Florida, Georgia, Hawaii, Illinois, Indiana, Kentucky, Maryland, Massachusetts, Michigan, Missouri, Nebraska, Nevada, New Hampshire, New Jersey, New Mexico, New York, North Carolina, Ohio, Pennsylvania, South

Carolina, Tennessee, Virginia, West Virginia, and Wyoming.

Note that Alaska, Washington DC, and Hawaii—all using oil-fired power generation—are among the largest polluters.

Figure 3-24. The future of oil-fired power plants.

There are hundreds of oil-fired power plants in the U.S., which are as bad, in terms of GHG pollution, as their coal-fired cousins. New EPA rules make it almost impossible to construct new oil-fired power plants, so there are no plans for such undertakings in the U.S.

At the same time, the existing oil-fired power plants will be forced to meet the tightened emission standards, and will be burdened by additional carbon and GHG exhaust charges. In many cases their conversion to natural gas or other alternative fuels is inevitable.

Because of that, and to save precious crude oil, there are calls from high places for the demolition of oil-fired power plants and/or conversion to natural gas. Conversion is easy to do in most cases and a number of such efforts are under way.

We are not sure what will happen to the oil-fired power plants in Washington, DC, Alaska, and Hawaii, where oil-to-gas conversions would not be as easy due to lack of local supplies. The responsible parties better get busy and figure what to do soon, because things won't get any better in the future.

Practical Uses of Crude Oil

Crude oil in the U.S. is used predominantly in the transportation sector. Smaller quantities of crude oil are used for electric power and heat generation, and for the production of different petrochemical products.

The average use of crude oil in the U.S. is about 0.07 bbl/day per person, which is nearly 1/10th of a barrel. Some countries use over 0.80 bbl, or almost a barrel full of oil per person every day, while other countries use about 0.0001 bbl/day, and some use none.

The global crude oil use average today is over 90 million barrels per day. This is a huge lake of oil we burn and waste every single day…non-stop!

This amount comes to over 30 billion barrels of oil use annually around the world. The actual oil use has increased significantly since the 1950s, so assuming 20 billion barrels average annual use, we come up with a staggering number of 1.3 trillion barrels used during the last half-century.

Figure 3-25. Lake Erie

This is over 54 trillion gallons. Or in just over half a century we have used about half of Lake Erie. As things are going we will dry Lake Erie by the middle of this century, and half another Lake Erie by the end of the century.

Note: Lake Erie is one of the Great Lakes, containing about 120 cubic miles of water. This is a cube with 120 miles on all sides, containing about 120 trillion gallons of water. The surface area is close to 10,000 square miles with nearly 900 miles of shoreline. The lake's drainage basin covers parts of the states of Indiana, Michigan, Ohio, Pennsylvania, New York and Ontario. A small ocean…

You get the picture…huge amounts of water…or oil, which brings a simple question to mind, How many Lake Eries are there floating in oil under the Earth's surface? And how long would it take us to drain them all? It's not like we aren't trying…

Let's take a closer look at the crude oil use in different sectors:

Transportation
The world transportation sector consumes more oil than all other sources combined. In the U.S. alone more

than 13 million barrels of oil-equivalent are used every day to fuel cars, trucks, trains, boats, and planes. This is roughly two-thirds of total U.S.' daily oil use.

Gasoline, diesel, and jet fuels account for more than 95% of all fossil fuels used for transportation in the U.S.

Oil products power virtually every mile we drive, and we drive a lot of miles in our gas guzzling cars and trucks. And then, millions of gallons of oil are used every year for motor oil and lubricating oils and greases.

There are more than 250 million cars and trucks on the U.S. roads today, traveling a whopping 3 trillion miles annually. This is enough miles for one car to make more than 14,000 round-trip voyages to the sun—enough to fill a lake with oil used for lubrication.

The U.S.' transportation system requires about 5 billion barrels of crude oil to meet the annual needs of all cars, trucks, trains, and boats on the road, and planes in the sky. At the same time, the number of vehicles and miles driven is expected to rise.

Oil companies and a handful of foreign countries benefit from this trend, while consumers are usually the losers.

However, there's a better, cheaper, and cleaner way to power America's transportation system. By increasing the use of clean biofuels and creating the next generation of advanced vehicles that no longer rely exclusively on oil, we can decrease our reliance on petroleum for fuel.

By improving the fuel efficiency of our cars and trucks, we can dramatically reduce the amount of oil we need in the first place.

Clean vehicle and fuel technologies are discussed and some designs are implemented, but much more work is needed to reduce the amount of oil we use for transportation.

If we act now, we can cut America's projected crude oil used for transportation in half during the next 20 years.

This would save us billions of dollars at the gas pump, slash oil consumption, move us toward a cleaner environment, and ensure our energy security. Yes, this is just a pipe dream for now, but it will certainly become a reality when crude oil hits $250/bbl. The time is coming…

Transportation Efficiency
Crude oil is the life blood of the transportation sector. High oil prices and reduced oil volumes put

pressure on the transport sector, and affect negatively the consumers and the entire economy. One way to put some order in the unstable oil supply situation is by optimizing the transport sector efficiency.

For that purpose, the 2007 Energy Independence and Security Act (EISA) amended the "corporate average fuel efficiency" (CAFE) standards.

CAFE mandates that by 2020, a manufacturer's combined fleet of passenger and non-passenger vehicles must achieve an average efficiency of 35 mpg.

According to the American Council on an Energy Efficient Economy, when implemented, this standard will save 2.4 million bbl/d of oil by 2030. This is a significant number, equivalent to nearly 15% of the current U.S. refinery output.

The EPA and the National Highway Transportation Safety Administration (NHTSA) recently published final rules to implement the first phase of these new standards. They also announced their goal, in addition to improving fuel efficiency, to reduce greenhouse gas (GHG) emissions for commercial trucks. Later on, they intend to adopt the second phase of GHG and fuel economy standards for light-duty vehicles.

Usually, improving energy efficiency releases an economic reaction that partially offsets the original energy savings, by causing a rebound effect. This means that improving a vehicle's fuel economy reduces its fuel cost per mile driven.

This is the good thing. The bad thing is that reduced per-mile cost of driving will encourage some drivers to increase the amount of driving they do, just because it is cheaper to do so. The magnitude of such response is uncertain, but most experts agree that imposing stricter fuel economy standards will eventually increase the total miles driven. In such case, the increased fuel efficiency is not only not doing any good, but is actually harming the environment by increase of total GHG emissions.

Research on the magnitude of the rebound effect in light-duty vehicles done in the early 1980s concluded that a statistically significant rebound effect—due to increased driving habits—usually occurs with every increase of vehicle fuel efficiency.

Recent evidence shows that the rebound effect declines over time, and may decline even further if income rises faster than gasoline prices. In light of the various study results, NHTSA elected to use a 10% rebound effect in its analysis of fuel savings and the overall benefits from higher CAFE standards for MY2012-MY2016 vehicles.

The EPA prefers to use a more conservative 5% effect.

The legal mandate for increased ethanol content in motor fuels further complicates the effort to improve vehicle fuel efficiency. This is because, based on the energy content, it would take roughly 1.4 gallons of E85 (85% ethanol, 15% gasoline mix) to move a vehicle the same distance as one gallon of pure gasoline.

EPA's partial Clean Air Act waiver allows the sale of E15 (15% ethanol, 85% gasoline mix) which might cause further decreases in vehicles' advertised mile-per-gallon ratings.

Obviously, great brains are at work in the U.S. and some other countries, analyzing and designing the ways and means to reduce oil use. How fast and how far these efforts will take us into the fossil-less future remains to be seen.

Unfortunately, this is not the case in most developing countries, where with the number of cars and trucks on the roads is increasing daily. In parallel with that, the use of crude oil in these areas is increasing, as is the pollution from millions of vehicles and thousands of power plants.

The Refineries' Effect

The fuel blends changes, discussed above, might be an obstacle for the automobile manufacturers to meet the new CAFE standards. This also might lead to reduced demand for refined petroleum products, the unintended effect of which could be a damaged petroleum refining industry.

The petroleum refining industry operates in cycles which are driven by political and economic developments. For example, the recent economic downturn was followed by the "golden age" of refining. Cycles in the industry are also related to changes in the price of crude oil, which is the primary cost element in refinery operations, and will likely remain as such.

Recent regulations are causing reduction in the demand for the industry's output. Higher mileage standards, higher ethanol content in fuel blends, use of bio-fuels and the influx of electric vehicles mean that regardless of the economic conditions and the increased demand for transportation fuels, the need for refined petroleum products will decrease proportionately.

The government policies were intended to accommodate the growing demand for refined petroleum products, but the declining motor-fuel demand means that the use of the alternatives could hurt the refining industry.

The reduced demand will force the refining operations to reduce production or idle. Some may choose to consolidate, or even permanently close refineries. This explains why most refiners do not plan to expand, or even maintain, production capacity in the United States. This is forcing some refineries to move out of the country as well.

The new market forces, technological changes, and regulatory pressures on the refining industry will continue the trend and we might see more refineries shutting down. At the same time, some of the more technologically advanced and more efficient refineries will most likely stay and even expand.

A trend toward larger refineries is also likely to emerge, which in turn could lead to concentration of the industry on a national level. New competitive conditions in oil refining will be in force, which will eventually have an impact on fuel prices.

Heating Oil

Home and business oil-fired furnaces are a major user of heating oil, which is derived from refining crude oil. Home oil heating is widely used in developed countries, especially during the winter months.

Heating oil is usually delivered by tank truck to residential, commercial and municipal buildings and stored in above-ground storage tanks. These are located either outside, near the building, or in empty areas like basements and garages. When used in larger quantities the oil is stored in underground storage tanks. Heating oil is sometimes used as a fuel in industrial applications, and for small or remote power generation.

Heating oil's properties are very similar to that of diesel fuel, and consist of a mixture of petroleum-derived hydrocarbons in the 14- to 20-carbon atom range. During oil distillation, heating oil condenses at between 482 and 662°F. It condenses at a lower temperature than the heavy (C20+) hydrocarbons such as petroleum jelly, bitumen, candle wax, and lubricating oil, but it condenses at a higher temperature than kerosene, or over 500°F.

Heating oil has a high heating value, producing 138,500 Btus per gallon—close to the same heat per unit mass of diesel fuel—and is known in the U.S. as No. 2 heating oil. It must conform to ASTM standard D396, which is somewhat different from that of diesel and kerosene.

Heating oil is widely used during the winter months in parts of the United States and Canada where natural gas is not available and propane is priced higher. The northeastern United States and Canada are the most likely users for heating coming mostly from Irving Oil's refinery located in Saint John, New Brunswick, which is the largest oil refinery in Canada.

In addition to the toxic gases emitted during the burning process, leaks from tanks and piping are a well known environmental concern. Various federal and state regulations are in place regarding the proper transportation, storage and burning of heating oil, which are classified as hazardous material by U.S. federal regulators.

With about 4% of the U.S. total consumption of crude oil going into heating oil, this is a significant issue, which is on the radar of the environmentalists, awaiting prompt resolution.

Consumable Products

Transportation fuels and heating oil are only some of the many products derived from petroleum. Americans consume petroleum products at a rate of 3½ gallons of oil and more than 250 cubic feet of natural gas per day—per person!

The problem in Figure 3-26 is obvious; we use three times more oil than we produce. The effect is also obvious; we are spending a lot of money to support sheikhs and their artificial kingdoms, instead of using the money to build our own economy.

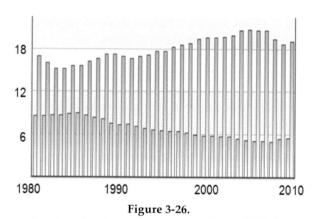

Figure 3-26.
U.S. oil production and use (in millions of bbls/day)

This unreasonable and unjustifiable imbalance has been going on for almost a century, during which time we made several nations very rich, and have pumped and used most of the oil, which took billions of years to form. How long would this uncontrolled pumping and burning of a limited resource continue?

Petroleum is not only used as a fuel; there are many other uses of this precious material. One 42-gallon barrel of oil creates 19.4 gallons of gasoline and 9-10 gallons of diesel and heating oil.

The rest (over half of the total content) is used to make things like:

• ammonia, anesthetics, antifreeze, antihistamines, antiseptics, artificial limbs, artificial turf, aspirin, awnings, balloons, ballpoint pens, bandages, basketballs, bearing grease, bicycle tires, boats,

- cameras, candles, car battery cases, car enamel, car tires, cassettes, caulking, CD player, CDs & DVDs, clothes, clothesline, cold cream, combs, cortisone, crayons, curtains,

- dashboards, denture adhesive, dentures, deodorant, detergents, dice, diesel fuel, dishes, dishwasher parts, dresses, drinking cups, dyes,

- electric blankets, electrician's tape, enamel, epoxy, eyeglasses,

- fan belts, faucet washers, fertilizers, fishing boots, fishing lures, fishing rods, floor wax, folding doors, food preservatives, football cleats, football helmets, footballs, footballs,

- gasoline, glycerin, golf bags, golf balls, guitar strings,

- hair coloring, hair curlers, hand lotion, heart valves, house paint,

- ice chests, ice cube trays, ink, insect repellent, insecticides,

- life jackets, linings, linoleum, lipstick, luggage,

- model cars, mops, motor oil, motorcycle helmet, movie film,

- nail polish, nylon rope,

- oil filters, oils and lubricants

- paint, paint, brushes, paint rollers, panty hose, parachutes, percolators, perfumes, petroleum jelly, pillows, plastic wood, purses, putty,

- refrigerants, refrigerators, roller skates, roofing, rubber cement, rubbing alcohol,

- safety glasses, shag rugs, shampoo, shaving cream, shoe polish, shoes, shower curtains, skis, slacks, soap, soft contact lenses, solvents, speakers, sports car bodies, sun glasses, surf boards, sweaters, synthetic rubber,

- telephones, tennis rackets, tents, toilet seats, tool boxes, tool racks, toothbrushes, toothpaste, transparent tape, trash bags, TV cabinets,

- umbrellas, upholstery,

- vaporizers, vitamin capsules,

- water pipes, wheels,

- yarn, and many, many more.

This is only a partial list of products made from petroleum—only 150 of the over 6000 different items made every day from this miraculous liquid. Our lives are so wrapped in it that it is absolutely impossible to even imagine living without it. Just think…no gas for the car, no heat for the house, no plastic bags, no cosmetics…

Yet this is exactly what is going to happen by the mid 21st century when most petroleum deposits in the ground will be pumped up, refined, used and otherwise wasted. Just think…

Cost of Petroleum Production

As other industries, oil rigs and refineries use a lot of expensive equipment and employ many specialized engineers and technicians. The cost of equipment and labor adds to already high prices of the raw materials, which in turn keeps the price of the final products high.

Too, there is a significant cost for transporting crude oil from the oil rigs, some of which are thousands of miles away (in the Middle East, South America and Asia). Shipping, which is done mostly by huge ocean-cruising oil tankers, is a complex and expensive proposition. The cost of new oil tankers is major, and has been increasing lately, due to increased requirements and rising cost of materials, energy and labor involved in tanker building. Tankers in the 32,000-45,000 DWT, 80,000-105,000 DWT, and 250,000-280,000 DWT cost approximately $18, $22, and $47 million respectively.

Oil tankers are often sold and bought second-hand too. In 2005, 27.3 million DWT worth of used oil tankers were sold. As an example, in 2006, First Olsen paid $76.5 million for *Knock Sheen*, a 159,899 DWT tanker.

After refining, the oil products are again loaded on trucks and trains to be transported to the end users at power plants, gas stations, manufacturing facilities, etc.

All this loading, unloading and wheeling hundreds and thousands of miles is expensive, since it depends on fossils as well. So when the price of oil goes up, the cost of transportation goes up. Catch 22…

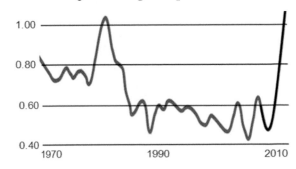

Figure 3-27.
Historical oil refining and distribution prices (in $/gallon)

The approximate breakdown of the costs of materials and services required to produce and sell oil products is as follows:

74% - Cost of the crude oil
11% - Taxes
10% - Refining costs
5% - Distribution and marketing

As an example, paying $100 for a barrel (42 gallons) of crude oil, the cost for a gallon of gasoline would be about $2.38. At a gas-pump the price will go up to $4.00 per gallon, where 44 cents goes to pay for taxes and 20 cents for distribution and marketing expenses.

This leaves $3.36 for the oil company, who pays for the cost of the gallon of crude oil itself which was $2.38 and also the 40 cents to refine it into gasoline. This leaves $0.58 profit per gallon of gasoline for the oil company. In reality, the reported profit-margin is in the 30¢- to 60¢-per-gallon range.

As crude oil prices go up, fuel consumption goes down, and oil companies find it harder to pass the increases to customers, so they do often lose money.

Global Oil Markets

The global oil market is the most important of the world energy markets because of oil's dominant role as an energy source. Oil is a commodity, as any other, but much more important than most. Contracts for its supply are usually traded through commodity exchanges, and are designed to allocate efficiently and fairly resources between those who supply and those who demand a particular oil product.

There are two economic concepts that are important to understanding how supply and demand function in global energy markets: the marginal unit and elasticity. We will take a close look at those in the next chapters.

One of the standards in the marketplace has been West Texas Intermediate (WTI) crude oil, which is high quality crude oil with a 39.6° API gravity. It is considered a "light" crude oil, and with a 0.24% sulfur content it is also a "sweet" crude oil.

North Sea Brent crude oil, at 38-39° API gravity is also light sweet crude oil, but with higher sulfur content than WTI, which makes it a global standard for other types of crude oil grades and widely used to determine global crude oil prices. Brent is typically refined in Northwest Europe, and also is exported to the U.S. Gulf and East Coasts.

Gasoline, diesel and other petroleum products' prices vary according to supply and demand, political situations, etc. One major factor in the pricing of fuels is also the pricing of crude oil futures contracts traded on the New York Mercantile Exchange (NYMEX) and the Tokyo Commodity Exchange (TOCOM).

Crude oil futures are standardized, exchange-traded contracts in which the contract buyer agrees to take delivery from the seller of a specific quantity of crude oil (in 1000-barrel lots at NYMEX and 50,000-liter lots at TOCOM) at a predetermined price on a future delivery date. So traders and producers' traders buy and sell virtual oil, as part of their daily business.

What drives the oil markets crazy is speculators and risk-taking moves by large traders. When the speculators think that oil prices are going up for some reason, as happened in 2008, they continue to outbid each other, driving prices of the actual crude oil higher and higher. We saw an increase to over $150/bbl in 2008, which was explainable only by the speculators' betting on the prices increasing indefinitely.

Stupid and unfair, you say? Maybe, but this is capitalism, after all, so everyone is afforded the right to make a quick buck. And the speculators will continue driving the energy markets in the future.

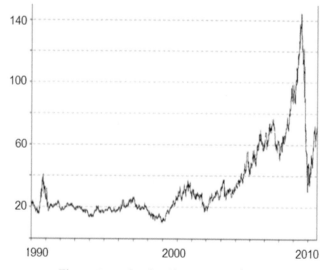

Figure 3-28. Crude oil prices (in $/bbl)

No wonder we are panicking. Look at the jump in oil prices to near $150/bbl in 2008 and the resulting repercussions during 2007-2009. For a short period in 2008, diesel at a truck stop near Bakersfield, CA, rose to $6.95 a gallon—a U.S. record of sorts, in this author's experience. Filling a 50-gallon diesel tank came to $350. This same tank just a year or two earlier was filled with $75 at the same gas station. Panic we did…oh, yes!

But of course, this is nothing new to most European and Asian drivers, since gas and diesel prices there have been at least twice as higher than those in the U.S.

for the longest time. There are various reasons for the high fuel prices, but one thing is for sure: things won't get any better...ever. On the contrary; they will only get worse, as oil prices will continue rising...forever.

The simple dynamics of the capitalist enterprise driver—the supply and demand ratio—is showing increasing demand (in countries like China and India) and reduced supplies (due to depletion of major deposit sites), so prices are inevitably headed up. Fasten your seat belts for some big surprises in the near future as well, when most producing nations (none of them friends of ours) would not hesitate to initiate another oil embargo or two just because they can.

Figure 3-29 is a rough representation of the relation between crude oil production and demand vs. price. Excluding the glitch in 2008, when prices went sky high, the general trend thus far has been increased demand, production and prices. The big difference in the future is the projected production levels and gradually increasing prices.

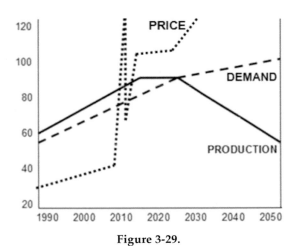

Figure 3-29.
Crude oil demand, and production vs. price (in $/bbl)

The demand and price will continue to increase, while global crude oil production will decrease sharply, due to well depletion and increased technical difficulties. The availability of other energy sources, such as natural gas and renewables like solar and wind will contribute to the high oil prices by the mid-21st century.

PRESENT OIL ISSUES

In addition to its other problems—limited quantity and fast exploitation, dealing with shady governments, risky transportation routes, and excess emissions—crude oil has another big problem. It is toxic, flammable,

and dangerous in many aspects of its lifetime journey.

In its natural form, crude oil is a substance composed of a number of organic and inorganic chemicals and chemical compounds. It is this complex composition—mixture of different chemicals—that makes it dangerous to all living things, including humans. A number of the chemicals in crude oil and its derivatives have been identified and listed as dangerous to human health due to their toxins and carcinogens.

Since different life forms react differently to different chemicals, it is impossible to predict what the consequences of a contact with the different petrochemical products would be. And since there is a mixture of these in crude oil, it is likely to affect in a dangerous way anything and anybody.

Benzene and the polycyclic aromatic hydrocarbons, in particular, are the most dangerous of the bunch in crude oil, because they are well-known toxins that can harm life forms and humans. Benzene is a known carcinogen too. Given a chance, it is able to readily affect living organisms. The worst part is that the harmful effects are sometimes not instantly recognizable and sometime even remain undetected for years after exposure.

Many chemicals in oil have a low BP (boiling point) and can easily vaporize and move from the oil into the air very quickly. Exposure to crude oil can occur either by air inhalation, accidental ingestion, and/or skin contact.

Crude oil toxicity in all its forms causes physical, chemical, and mental changes and damages, such as fatigue, dizziness, headaches, and some more serious (or fatal) illnesses. In the summer months especially, when petrochemicals are most volatile, these symptoms can be mistaken for heat exhaustion and are often not properly treated.

Gasoline Evolution

Since most of us drive cars and breath the gasoline and diesel vapors coming from them and from filling stations, we are affected by the exhausts of benzene and other chemicals contained in the common vehicle fuels around us.

The conventional gasoline refined today has changed considerably since the Clean Air Act of 1970 prohibited lead additives. Oxygenated gasoline and reformulated gasoline (RFG) are in use today, where each of the three formulations of gasoline (conventional, oxygenated and reformulated) are available in at least three grades (87, 89-mid grade, and 91+ super) gasoline.

The volatility of these fuels is adjusted for winter/summer and northern/southern driving conditions by the addition of additives and other chemicals such as alcohols.

The Clean Air Act of 1990 directed the EPA to designate areas not complying with national ambient air quality standards (NAAQS) as ozone "nonattainment areas." Cities with the worst smog pollution are required to reduce harmful emissions that cause ground-level ozone by using reformulated gasoline (known as RFG), which is blended to burn cleaner, thus reducing smog-forming and toxic pollutants during the summer ozone season.

Reformulated gasoline undergoes additional processing to remove volatile components that contribute most to air pollution, and to make it less prone to evaporation. It also contains chemical oxygen, known as oxygenate, to further improve combustion and reduce emissions.

Since then, a growing number of distinct types of gasoline ("boutique fuels'") have entered the supply chain. Currently, 15 distinctly formulated boutique fuels are required in portions of 12 states. In addition to the federal RFG standards, State Implementation Plans to improve air quality require low-Reid Vapor Pressure (low RVP) type gasoline, which requires special processing and handling.

California mandates a cleaner fuel than federal RFG (referred to as California RFG, or CaRFG), while the Midwestern states require a unique ethanol-blended RFG.

Between 1992 and 2005, EPA also mandated oxygenated fuel blends to reduce ground-level ozone and smog. Much of the gasoline sold in the U.S. during that period was blended with up to 10% methyl tertiary-butyl ether (MTBE), which was produced and added at the refineries, as the oxygenate in almost all RFG outside of the Midwest, while ethanol was used in the Midwest.

Both MTBE and ethanol served several functions: as an oxygenate in RFG, as an octane booster, and as a volume extender in conventional gasoline. Groundwater contamination concerns and the State of California's ban on MTBE as a gasoline additive left ethanol as the most popular fuel oxygenate.

However, ethanol's corrosive nature makes long-distance shipment of ethanol mixed into gasoline impractical. In consequence, ethanol (produced mostly from corn fermentation) is blended with gasoline at the storage terminal where the fuel is dispensed to the fuel tank truck. The shift from MTBE to ethanol thus contributed to a significant reduction in refinery production.

Environmental Effects of Oil Production and Use

Similar to the mining process, a number of large and small pieces of equipment are used during the oil rig and oil-fired power plant construction and operation. Starting with the well and plant's design and construction, we see the pollution taking place in the shape of ground leveling, digging and bulldozing, with heavy plumes of smoke coming from the many trucks at the site.

Then the oil rig and power plant equipment pieces arrive on trains and trucks. All this machinery was made somewhere else at some time in the past, where and when their manufacturing processes and transport created a significant pollution footprint.

As with coal mining, we would assign 5-10% of the manufacturing and transport of the oil-well and power plant's equipment to the initial construction step. The main pieces of equipment, rig structure, boilers, turbines, generators, etc. arrive in wooden boxes or on the backs of enormous trailers or railroad cars.

These are then assembled with the help of other equipment and put to work after creating another set of pollution problems. In addition, there are many other items and small pieces of equipment, tools and consumables for personal or specialized operations. These are shovels, helmets, computers, boots, shoes, overalls, first aid kits, goggles, etc.

All these were also made somewhere else at sometime in the past and transported to the mine. During their manufacturing processes and transport they also left a significant pollution footprint.

To complete the pollution footprint picture, we must add that of the people involved in the entire process—engineers, technicians, and operators involved in equipment manufacturing, transport, installation, setup, and operation. These people drive cars and travel by air, all of which leaves another significant set of pollution footprints.

So, the total pollution footprint of the power plant's equipment cradle-to-grave process is expressed by:

$$P_f = A + B + C + D + E + F$$

where

P_f is the total pollution footprint
A is the equipment manufacturing and transport to the well and power plant
B is equipment assembly, test and installation at the well and power plant
C is the manufacturing and transport to the well and power plant of tools and consumables,

D is the pollution footprint of the equipment personnel during the life of the project,

E is the actual pollution from daily operations, and

F is the pollution during well and power plant decommissioning

Combining the pollution footprint of A (the plant equipment manufacturing and transport) with B (plant equipment assembly, test and installation), adding C (the manufacturing and transport of the tools and consumables), plus D (the pollution footprint of the equipment personnel) gives us the environmental footprint of the plant equipment *before* even starting plant operations.

Here again, the pollution footprint is significant, to be sure! It goes deep and wide, stretching from one part of the country to the other... from end of the world to the other.

We must add to that the even greater pollution coming from the daily operations during the lifetime of the well or power plant (E). And finally, there is the pollution created during the decommissioning stages (F).

To put a value on all of this we must take inventory of all items delivered to the plant and needed for its design, construction and operation. We must take a close look at the manufacturing of the oil rig and power plant components, rig structure, boiler, steam turbine, generator, and other pieces of equipment used at the wells and power plants.

These are also very large pieces of equipment, and it takes hundreds and thousands of tons of metals to make them. Each of these started as an iron ore, dug out from a mine somewhere in the world, which was transported to a smelter. The molten metal was shaped in different forms and shipped to the different parts manufacturers, who drilled, welded and otherwise constructed parts for these vehicles.

Remember, we are talking about huge pieces of metal—some as big as an entire building. These parts are then packed and loaded on railroad cars or trucks to be shipped to the assembly facility. The parts of a large dump truck or a dragline might require a dozen of railroad cars for transport. The parts are then sent to the assembly plant, where they are assembled into major components, or entire units. After one more loading and unloading operation, they finally arrive at the plant for final assembly, installation, and final testing, before being put to work.

This is a long and winding process, with many stop-and-go, loading and unloading steps.

Putting a dollar value to all this would be too complex for our purposes, so it suffices to say that the pollution effects—in dollars spent and hidden costs—during and after all these steps are significant.

Upon starting the well and/or power plant operation, the production equipment continues to generate secondary pollution. Spare parts for these monsters are periodically arriving from the parts manufacturers, after leaving their pollution footprint. We will attempt to assign a number—quantity of air and land pollution—and related dollar amounts to these activities and their environmental effects in the following chapters.

After 20-30 years of non-stop operation, the oil well and the oil-fired power plant are finally declared obsolete, at which point they must be shut down and decommissioned. Decommissioning and land reclamation are major, expensive and polluting undertakings as well. The equipment and materials used during the decommissioning and waste disposal process create additional air and ground pollution which is not to be ignored.

Additional, specialized equipment and personnel (environmental specialists and inspectors), are usually brought in to assist and/or coordinate the effort. With that, more pollution footprints are left at the already exhausted mine.

When the final tally is made, a significant part of the emissions and overall pollution are to be attributed to the well and plant's equipment and support personnel. This number is significant too.

The Regulations

In addition to the obvious health hazards, gasoline and diesel used in transportation are blamed for causing excess GHG emissions in the U.S. and the globe. In 2007, the United States Supreme Court ruled that EPA has the authority under the Clean Air Act to regulate carbon dioxide (CO_2) emissions from automobiles, and directed EPA to conduct a thorough scientific review.

EPA issued a finding in April 2009, stating officially that greenhouse gases contribute to air pollution that may endanger public health or welfare. Though the finding pertained to automobile emissions, it has wide ranging implications.

In response to the FY2008 Consolidated Appropriations Act (H.R. 2764; P.L. 110-161), EPA issued the Mandatory Reporting of Greenhouse Gases Rule. It requires suppliers of fossil fuels or industrial greenhouse gases, manufacturers of vehicles and engines, and facilities that emit 25,000 metric tons or more per year of GHG emissions to submit annual reports to EPA.

The rule includes final reporting requirements for

31 of the 42 emission sources listed in the proposal. The rule establishes the basis for future legislation and regulations that could cap GHG emissions from refineries as well as other industrial sources.

The American Clean Energy and Security Act of 2009 H.R. 2454 amends the Clean Air Act by establishing a "cap-and-trade" system designed to reduce greenhouse gas emissions (GHG) and would cap emissions from refineries and allow trading of emissions permits ("allowances").

Over time, H.R. 2454's provisions would reduce the cap to 83% of current emissions, forcing industries to reduce emissions by that amount (cap) or purchase allowances from others who would have reduced emissions more than required or offsets from eligible entities not covered by the cap (trade). The bill would allocate the refining industry only 2% of the total emission allowances for the entire U.S. economy.

U.S. petroleum refineries emit approximately 205 million metric tons of CO_2 annually, which (according to the new EPA rule) represents approximately 3% of the U.S. GHG emissions. The cost of complying with the new EPA rule could be minimal, but the cost of complying with "cap and trade" provisions of H.R. 2454 or similar legislation could be disruptive to the refining industry according to recent expert analyses.

As proposed, H.R. 2454 would require U.S. refiners to purchase emission credits for both their stationary emissions and the subsequent combustion of their fuels (predominantly consumed in the transportation sector). U.S. refiners could face competitive disadvantages with refined petroleum products imported from countries where refinery greenhouse-gas emissions are treated differently.

U.S. refiners would need to purchase roughly 2,000 million credits in 2015, whereas European Union refiners who export their products (predominantly gasoline) to the United States would only need to purchase 3 million allowances.

LNG Vehicles

The Clean Air and Oil Accountability Act of 2010, S. 3663, was introduced in August 2010, to establish a Natural Gas Vehicle and Infrastructure Development Program to promote natural gas as an alternative transportation fuel, to reduce domestic oil use. The program would also offer incentives to convert or repower conventionally fueled vehicles to operate on compressed natural gas (CNG) or liquefied natural gas (LNG).

Natural gas is abundant in the United States and has already been used as a transportation fuel for in-

tra-city buses, principally as means of reducing air emissions. U.S. automobile manufacturers marketed passenger vehicles modified to run on compressed natural gas in the 1990s.

U.S. refineries currently produce over 1,400 million barrels of diesel fuel annually, so about 5.3 trillion cubic feet (tcf) of natural gas would be needed to replace this fuel, as S. 3663 proposes. To displace its diesel fuel with natural gas, the U.S. would need to increase natural gas production by more than 25% above today's levels, on top of the levels mandated by policies aimed at replacing coal-based electricity generation with natural gas. This also does not consider the lost efficiency in converting the more energy efficient diesel fuel in diesel engines to natural gas.

Note: A refined barrel of 350 API crude can yield about five gallons of diesel fuel, or 12% of the barrel. Displacing all diesel fuel consumption with natural gas represents about two million bbl/day in refining capacity.

One of the problems is that U.S. refineries cannot cut back present levels of diesel production without cutting back on production of gasoline and other refined products. Assuming no decreased gasoline demand, refiners would likely export the excess diesel, or market it as heating oil.

Energy Security and Crude Oil

Above we took a fairly close look at crude oil production, transport, and use. There is much more to it, of course, but we feel that this is enough to give us a good understanding of the technological and logistic complexities involved in the crude oil lifespan.

The fact that crude oil is a key component of our energy security, and because of the complexity of its cradle-to-grave cycle, demands that we be very careful with how we use it.

The issues related to energy security, oil production, transport, and use start at the oil wells. A number of risks have been identified and are present at each well from its initiation until decommissioning.

Then we have unresolved issues related to the transport, refining, and use of crude oil. These overall oil life cycle risks are as follow:

- Transport equipment, refineries, power plants, components of the crude oil infrastructure (their design, construction, installation, and operation which are technologically and logistically complex), with associated risks at each step of the process;

- Local problems, such as geophysical changes that can affect the performance and safety of the installations and their personnel;

- Weather related problems, such as storms and hurricanes can cause property damage and/or hurt workers;

- Outside accidents, incidents, and terrorist attacks, especially affecting off-shore rigs and transport tankers, could damage and even disable the structures and hurt personnel.

The reliance on excessive crude oil imports from different countries around the globe is especially damaging to our energy security.

According to the U.S. Department of Energy, the increase in U.S. oil production and the excess enthusiasm over large internal oil production doesn't mean that the U.S. economy is shielded completely from the global marketplace.

True, because the U.S. still imports over 50% of its oil, which means that at least 50% of our energy and national security depends on unreliable oil imports. Granted, we don't need coal and natural gas…for now…but crude oil is still in short supply, even with the amazing bonanza in the U.S. oil fields of late.

Forty years after the Arab oil embargo, DOE warns that even sharply increased oil production does very little for our energy security as long as the U.S. and the key global economies still run on oil imports.

The International Energy Agency estimates that the increased domestic oil production in the U. S. could cut imports by more than half by the end of the decade. An important indicator of the changing times is the $300 billion in investments attracted to energy efficiency markets globally in 2011 alone. This is a new wave in the energy markets, where energy efficiency is a "hidden fuel" driving markets even faster to energy independence and a low-carbon future.

"It would be a misconception to think because of our increased domestic production that somehow we have become free of the global oil market, the global oil price and global oil price volatility," said the Energy Secretary Moniz in 2013.

According to him, huge oil production is a welcome, albeit temporary and partial relief for our oil anxieties and the policymakers who are wary of geopolitical turmoil, but it is not the solution. The only way to avoid revisiting the 1973 Arab Oil Embargo calamity is to reduce our oil dependency. Period!

Obviously, the best, if not the only, sure way to ensure our long-term energy security is to not use oil at all. Not using oil at all is not possible for now, especially in the U.S., but reducing its use is. The more we reduce our use, the more secure we'll be.

We must become fully and painfully aware that while our oil and gas bucket is still half-full today, we are using a lot of it, so it won't last very long. Yet, most of us prefer to rejoice and enjoy today with all its benefits, totally forgetting about tomorrow, when the bucket will be mostly empty.

There is no way around it—the bucket will be empty sooner or later. The only rational solution is to find ways to make what we have last longer. This would give us more time to find energy sources that are capable of replacing crude oil and natural gas—before we tumble head-first off the steep fossil-less energy cliff of the future.

EROEI

In physics, EROEI is defined as "energy returned on energy invested" for a particular process. In oil production this is basically the ratio of the amount of usable energy obtained from a type of oil, as compared to the amount of energy used to drill and pump out an equal amount of oil.

$$EROEI = \frac{A}{B}$$

where
  A   is the usable energy obtained from an amount of produced oil, and
  B   is the energy used to obtain that amount of oil

At EROEI less than or equal to one, crude oil production is unprofitable, because it uses more energy to pump out and transport it than can be produced from it once delivered to the point of use. Such oil is more expensive than the global market and *cannot* be used as a primary source of energy.

As global crude oil supplies dwindle, it is getting harder to reach the oil deep underground, so it becomes more expensive to produce it. With that, crude oil's EROEI increases steadily and proportionately with time.

Figure 3-30 shows that EROEI in the first half of the 1900s was over 100. It dropped to 25 by the end of the century and is now even lower. This drastic decrease of EROEI happened in less than a decade, and the digression continues at the same rate.

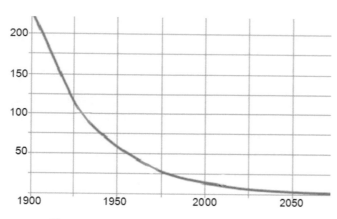

Figure 3-30. U.S. crude oil EROEI since 1900.

There is no change in sight, since the world is determined to use as much oil as possible and for as long as possible.

Because of that, we must assume (or calculate and predict) that the crude oil EROEI will continue dropping and get to near zero by the end of this century.

It is quite obvious from Table 3-4 that biodiesel and corn ethanol production uses as much energy as it produces. The profit margin from these alternative fuels, if any, depends on the global fossil fuel prices, since fossil fuels are used in their production.

Table 3-4. EROEI of different fuels and technologies

EROEI	FUEL / TECHNOLOGY
1.3	Biodiesel
1.3	Corn ethanol
2	CSP trough
2	Solar water heater
3	Tar sands
5	Sugarcane ethanol
5	Shale oil
10	PV power
10	Nuclear, diffusion enriched
12	Oil imports, 2013
15	Natural gas, 2010
15	Oil production, 2010
18	Wind power
30	Oil production, 1970
35	Oil imports, 1990
35	World oil production
75	Nuclear, centrifuge enriched
80	Coal
100	Hydropower

At the same time, coal and some types of nuclear power seem to be the most profitable of all fuels. The EROEI of oil production has dropped by half during the last 30 years and is dropping further as we speak. EROEI of oil imports dropped 3 times during the last 20 years.

Presently, the major crude oil production technologies in the U.S. are based on hydrofracking, tar sands, and shale oil extraction processes. These are energy-hungry processes, which affects the EROEI and increases production costs.

Nevertheless, they are profitable enough, and as global fuel prices go up, the profit difference is increasing in their favor. As global crude oil competition increases, the new oil exploration technologies are a welcome solution for many countries, including the U.S.

But there is another problem: it is the biggest and most worrisome of all. It is the fact that the amount of fossils in the earth's bosom is decreasing quickly, and soon most of these precious commodities will be gone.

The Hubbert's Peak

As far back as 1956, Mr. Hubbert, an oil engineer, proposed that the global fossil fuel production would look roughly like a bell-shaped curve. The highest point in the curve is the highest level of fossil production, and it is where crude oil production is at its peak. After that point is reached, production levels would slope down to zero within 30-40 years.

The curve in Figure 3-31 and the logic behind it later on became known as the Hubbert's peak theory. The highest point on the Hubbert's curve denotes "peak oil," or the time when oil production is at its maximum volume. It is basically a scientific projection of the time when future petroleum production (whether for indi-

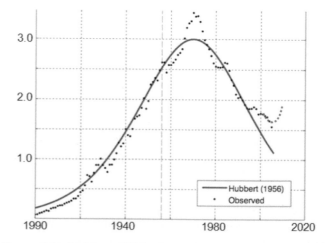

Figure 3-31. Hubbert Oil Peak, predicted and actual (in million barrels/year)

vidual oil wells, entire oil fields, whole countries, or worldwide production) will eventually peak and then start a slow decline at a similar rate to the rate of increase before the peak, as those reserves are exhausted.

Looking at Figure 3-31 we see clearly that we have passed the point of no return (the Hubbert peak). And although we can now rejoice (albeit temporarily) in a new significant "peak" in domestic oil and natural gas production, it only shifts the inevitable by several years into the future.

Considering all past oil discovery and production data, a Hubbert's curve that attempts to approximate past discovery data can be used to provide estimates for future production, as follows:

$$Qt = \frac{Qmax}{1+AB^{-bt}}$$

The year of maximum annual production (peak) Tmax then is:

$$Tmax = \frac{1}{b} \ln (a)$$

This is were the cumulative production Qt reaches half of the total available resource:

$$Qt = Qmax/2$$

Where:

Qmax is the total recoverable crude oil
Qt is the cumulative production volume
a and b are constants

Hubbert's peak oil theory accurately predicted the peak of U.S. oil production at a date between 1966 and 1970, based on data available at the time of his publication in 1956. Using the same data, Hubbert predicted world peak oil by 2006. That date came and went, but the global peak oil date is still debated and seems to be shifting further into the future.

The same is true for the U.S. oil volume. As Figure 3-31 shows, we are experiencing a significant deviation from the original curve today, so it may need to be redrawn. This anomaly is because Mr. Hubbert had a fairly limited amount of data, which considered only the easy ways of obtaining oil, without advanced technologies and new discoveries, as we know them today. As time went on, new oil reserves were found and the new technologies, such as hydrofracking, shale, and tar sands, are now pumping oil from places that were

unknown to Mr. Hubbert and even unimaginable at the time.

Nevertheless, even with the new discoveries, oil reserves in general are more difficult to extract, and the alternatives are too expensive for some companies and especially the developing countries.

According to the International Energy Agency (IEA), the global production of conventional crude oil peaked in 2006, as predicted by the Hubbert's model. Since then, however, things have been changing so rapidly that don't know exactly what the curve looks like.

Even if it is extended by decades, the end of oil as we know it is imminent.

There is a finite quantity of crude oil in the Earth, and we have already extracted most of it.

So whether the crude oil will last 10, 20, 30, or 50 years, its end is approaching fast. This path is clearly seen, but efforts to find solutions are not that clear and vary from time to time and from country to country.

Oil Stock Trading

Earlier, we took a fairly close look at the new phenomena—crude oil stock and commodity trading. Many large crude oil stock traders were to be blamed for the excess price increases during the 2008 energy crises, when oil went up to $150/bbl. It is well known now that most of the oil price increases at the time had nothing to do with oil supply and demand, or variations in global pricing. Instead, the variations were due in large part to speculator frenzy.

A testimony given to the U.S. Senate indicated that "demand shock" from "Institutional Investors" had increased by 848 million barrels over the last five years. Similar increases in demand came from China, to the tune of over 900 million barrels.

Institutional investors were very influential in the manipulations, and experts suggest that oil price increases were caused by commodities manipulations by these investors. It was estimated that for every $100 million in new inflows, the price of U.S. oil increased by 1.6%.

The crude oil gambling game on the world's stock and commodity markets continues, albeit in a less noticeable form.

Crude Oil and Energy Independence

Energy independence sounds good, right? And this is what we all want to see. However, it is much easier said than done, and although we have tried hard to achieve it, the U.S. has been consistently one of the largest importers of crude oil since the late 1940s, and far from energy independence.

The problem is that regardless of the numerous statements on the subject by the oil industry, its paid consultants, and the think-tank academics the facts are simply not supported by the actual numbers. Crude oil in all its forms is a finite commodity and once a well is producing, it eventually slows down and stops. One by one, the wells dry up. It is unavoidable.

Since we rely on the newly found shale rock oil in these estimates, we must consider its present volumes and estimated decline rate. At a conservative 10% per year decline rate for existing U.S. crude oil production, in order to simply maintain current U.S. crude oil production, we would need the productive equivalent of 10 new Bakken plays over the next 10 years, to maintain current crude oil production.

According to Citi Research, a Citigroup research unit, the decline rate for existing U.S. natural gas production is about 24% per annum. This would require the industry to replace about 100% of current U.S. natural gas production in four years, just to maintain current production. In other words, it would need the productive equivalent of 30 new Barnett Shale plays over 10 years, to maintain current natural gas production.

But there is only one (huge as it is) Barnett Shale and no others found. Nor is there a new Bakken-sized play discovered each year. So, the U.S. natural gas production has been flat since 2012.

At the same time, U.S. crude oil production continues to grow steadily, slowly outpacing most projections. But, at the present levels of consumption, the U.S. would need to more than double its future output to supply all of the country's crude oil needs.

It is critical to remember that the U.S. energy independence is now all about oil, since we have all other energy sources we need.

From 1990 through 2012, the United States imported about 54 percent of its crude oil and petroleum products such as gasoline and diesel fuel. In 2012 the imports were cut down to 48.3 percent. That's good, but we are still importing almost 50% of the oil we use.

This reminds us of the 1980s when the huge oil find around Prudhoe Bay in Alaska briefly boosted U.S. oil production. Prudhoe Bay peaked in 1988 and has been on the decline ever since. The total U.S. production followed that pattern until recently, when new discoveries and techniques brought a new oil and gas bonanza.

Yet, the experts believe that we are only temporarily out of trouble. High global oil prices make it possible to recover tight oil from deep shale deposits (at a high cost), so the U.S. production could grow for a time.

At some point, however, the production decline rates for tight oil wells (around 30-40 percent per year) will be too much of a barrier, and total U.S. crude oil production will begin to decline once again. This could happen even sooner if global oil prices fall. In that case, we will go back to importing most of our oil.

Common sense tells us that regardless of what the experts know or think, crude oil is in limited quantities. It has a beginning and an end. We found the beginning long ago and have been living as if there were no end.

We might have shifted the peak oil point several years into the future for now, but we cannot ensure continuous oil supplies forever. And we have not even started to consider the dark consequences of getting at the oil's end without being prepared for that.

The amount of our petroleum product *exports* has been rising slowly, compared with the total crude oil *imports*. In 2012 the U.S. petroleum products exports were—for a first time ever—more than the total oil imports. So we import oil to export oil (products). This is done for economic gain, without concern for the long-term energy future or future generations.

So the bottom line is that for now, the total oil imports represent half of all the crude oil we use. This means that half of our energy security depends on other countries—some of which are not that friendly. At this rate, energy independence is not going to happen.

The Energy (post-fossils) Transition

All things considered, we know that since virtually all economic sectors rely heavily on petroleum, and since its supply is limited, a peak will occur and eventually lead to partial or complete failure of the energy sectors and others.

In 2007, the U.S. GAO issued a report that provided an urgent call for action due to eminent reduction and approaching collapse of global oil supplies. The report reaffirmed that peak oil is real and inevitable by 2040, which might be followed by a quick decline in world oil production without warning.

New and more efficient exploration methods and production techniques allow new discoveries and improved production, which might delay the peak oil estimates, but oil is a finite resource, so whether it is 2040, 2050, or 2060, its end is eminent. Rising demand for oil from developing countries also shows sharply increased risks of significant disruption to oil supplies.

For now, the U.S. fossil fuel industry (oil, gas, and coal) is encouraged to grow. It is good for the country and it is good for all of us, the TV commercials proclaim, so in the short term we are looking at more consumption of crude oil and other fossils, accompanied by more damaging carbon emissions.

Due to its priority status, the fossil industry has been receiving subsidies since the very beginning to the tune of over $72 billion in subsidies between 2002 and 2008, according to the Environmental Law Institute.

The nuclear industry gets significant subsidies also, but the exact number is only a guess, for it is part of the national security package, and as such it enjoys special privileges.

The renewable energy industry has been getting subsidies on and off through the years, depending on which direction the political and energy winds blow. Lately, however, it has been receiving massive amounts of subsidies too. Proponents claim that the renewables are more deserving to receiving subsidies than the fossil fuels, because they are much more versatile and environmentally friendly, thus they are good for the environment, our energy security, and for other reasons, including fuel diversification.

In 2009, the G20 agreed to phase out energy price subsidies in principle, which alone could result in a 10% drop in global GHG emissions by 2050. These, however, are only words, while in reality many countries are increasing the subsidies and the use of fossil fuels now and in the future.

While officially agreeing to reduce subsidies and cut emissions, most governments are in fact encouraging fossil fuel use and expansion by increasing the subsidies, which will lead to quick depletion of the natural resources and an increase of GHG emissions.

This is a new phenomenon, where not only China and India, but also developed countries like Germany, are building new coal-fired power plants and are switching from gas to coal at the same time.

These new developments go against all international agreements and all efforts to clean the environment. As a result, the global GHG emissions are projected to increase in the future, if this trend continues.

This will eventually lead to:

- A worldwide economic stagnation and collapse, which if no decisive measures are taken might be as bad as the U.S. recession in the 1930s, or

- If enough influential people and countries decide on a radical solution, it might lead to a speedy and orderly transition to 100% renewable energy within a very short time. In the best of cases, experts predict that at least a decade is needed for a drastic transition.

The Energy Gap

The transition from A (fossil economy) to B (fossil-less economy) is not easy to design, let alone agree to among the nations. The ultimate choice of what is the best way to get from A to B will most likely be different for different areas of the world, and can be influenced by many variables.

As far as crude oil is concerned, however, the transition would be most difficult and painful. It is one commodity that cannot be easily replaced, unless a lot of effort and money is poured into finding renewable and sustainable solutions…none of which would be quick, easy, or cheap.

Because of the uniqueness of crude oil and its derivatives, we predict a large gap forming in the global energy system. The gap would be formed and would grow as crude oil supplies dwindle and the world hesitates, and/or is incapable of finding the proper solutions.

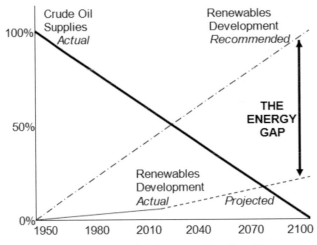

Figure 3-32. The global energy (crude oil) gap

It is impossible to foresee what will happen in the long run, because of the complexity of the energy sector and the large number of variables. In the short term, however, we see a number of developments taking place to fill the crude oil gap—evidenced by increasing prices.

Some of the most plausible developments in filling the oil gap would be as follow:

- In the people transport sector, which uses the most crude oil, free shuttle buses running frequently along the most travelled routes may persuade many to leave the car home and take the bus.

This won't happen in America anytime soon, because of the American mentality and the way life is structured here. In many other countries, however, most people would take the free bus service even if they had access to cars and other means of personal transportation.

- Safe designated bicycle lanes and parking racks at both ends of traffic areas may persuade many to ride bikes to work. Again, not many Americans would take this option, but the numbers will increase significantly elsewhere.

  Note: Some European cities even have free public bicycles programs, where people ride a bikes from one point to another and leave them at designated racks for others to use. There are also inexpensive bike rental systems where users can pay by the hour using a credit or debit card.

- Parking space availability and price often determines the choice of transport too, where the car-alternative options become more attractive with price increases.

  Note: City planners in some European cities are rationing the use of private cars in city centers by making parking scarce and expensive. With a few exceptions, Americans do not follow this rationale because often they have no choice but to drive a car from their suburban residences to their work. On the other hand, most of the residents of big, congested cities such as New York, London, or Tokyo do not drive cars, for lack of parking and congestion of inner-city traffic.

- Road congestion is another big factor, for which there are dual occupancy lanes and other measures that help traffic flow.

  Note: In many European cities there are special dedicated bus lanes, where buses run at normal speed while other traffic is stuck in a jam. This option is available in some places in the U.S. as well, but it will be a long time before it becomes a standard, due to the Americans' love affair with the automobile and the lack of well-developed mass transit systems.

Anyway, a big change in the way we look at and use crude oil and its products is coming our way. Our choice is to ignore it (and the upcoming Hubbert's peak) until it hits us in the head, or to start planning ahead. This is the right time to plan ahead for the transition to a fossil-less existence, because when the inevitable comes,

it might be too late for any plans or actions. Sadly, for now, efficient planning for the fossil-less future is on the back burner.

Update 2014: It is now estimated that the U.S. could produce over 10 million barrels of oil per day by 2020. Basically, in a decade, the shale revolution in the U.S. has reversed a four-decade-old trend of increasing oil imports, to where we are now a major oil exporter of petrochemicals.

If this happens, then we would be well on our way to energy independence. While we don't doubt that this is possible, we must consider the technical and logistical complexities the industry is faced with, which might slow down and even derail some of these estimates.

In best of cases, we must never forget the fact that no matter how much oil we have today, or how much we can produce in the future, oil is a finite commodity and its end is approaching.

AGAIN: oil's end is predictably near, and at the rate we are pumping it out of the ground, it will come even faster than presently estimated. Meanwhile, as the oil wells are emptied, the oil quality is decreasing, extracting it gets harder, and its prices keep rising

Regardless of whether it will be next year, next decade, or next century, oil in the U.S. and around the globe will eventually become very expensive to the point where we would not be able to afford it. Of course, soon after that it will cease to exist altogether.

Think about this horrendous possibility every time the subject of oil production and use pops up. It is very important that we understand the long-term issues and their impact on our lives and those of future generations.

Imagine a cold winter in the year 2050, when heating oil supplies hit record prices and are not readily available. Imagine rigid temperatures hitting the Northeast, where minus 20°F temperatures hold the area in a deep freeze and high snowfall stops all traffic for weeks at a time.

With about half of the population using heating oil to heat their homes, this freezing cold weather means that the furnaces will be running longer trying to keep up…if there is enough oil available and if we can afford the rising prices.

Much more natural gas and heating oil are needed during the cold winter months—five-six times more than usual in the northern states. As demands from each home rise, increased usage would result in declining fuel reserves across the nation, to where the U.S. inventories fall below the emergency threshold and the spigots are turned off. Prices jump sky high, so only a few can afford enough fuel to heat their homes.

But these are still the good times. Now imagine people in the year 2150, when all oil and gas supplies are absolutely gone. With weather forecasts predicting even lower temperatures in the future, and all fossil fuels gone, we can only imagine the hardship of the people in these freezing states, unless other energy forms are found.

How likely is it that new energy sources will be available 50-100 years from today is uncertain, but with the focus on "drill, Baby, drill," there is not much effort in this area.

The longer we rely on fossils, the wider the energy gap gets, which increases the energy uncertainty of the post-fossils society.

Ensuring the nation's energy security, its economic development, and the comfort of the people in the short run is the priority today. The future will take care of itself... We can only wonder what future generations would think of our lack of foresight and selfishness.

INTERNATIONAL DEVELOPMENTS

It's a funny world we live in! It is full of opposites and inequalities. Energy is one of the greatest inequalities, surpassed only by global food distribution.

The "haves" use almost a barrel of crude oil (equivalent) daily, living in luxury, while the millions of have-nots have access to very little energy, if any.

Some of us live in huge air conditioned and brightly lit palaces, while others live in huts with only wood and brush scraps used as light and energy for daily survival.

Crude oil is becoming increasingly important in the global development, since it is needed to fuel most of the vehicles on the roads, in the water, and in the air. Here again we see even bigger inequalities. Some of us drive 8-cylinder gas guzzlers, while others can afford only a bicycle, if that. The discrepancy is amazingly deep and can be easily seen by just observing the city and rural traffic patterns in the different countries.

Another significant difference that can be easily observed around the world is that in cities with very

heavy traffic, the air is also very heavy. It gets so heavy at times, that it is hard to see around the corner, and sometimes even hard to breath. At the same time, in cities without motor vehicles, the air is clean and light and one does not need a face mask to breath.

Nevertheless, crude oil drives the world's economies and everyone wants it. As the world population increases, the demand for oil increases proportionally too. And that reveals another amazing discrepancy—some countries have more oil than they know what to do with, while others have little to none. Because of that, we live in a world full of ever changing energy discrepancies and increasing energy problems.

Here are some of the countries that reflect the greatest global energy discrepancies.

Saudi Arabia

Saudi Arabia has been for a long time one of the world's largest producers of crude oil, most of which is exported. A significant part of the domestic oil is used locally for power generation.

Note: Most other countries abandoned oil-fired power plants long ago in favor of gas, nuclear and renewable energy, so in this respect the Kingdom is still in the dark ages. Things may be changing, as the Saudis are looking for ways to overcome this long addiction to generating electricity from their own oil and export it instead. If and when that happens, it might help ease the increasing supply concerns on the global oil market, which is heavily reliant on free-flowing and cheap Saudi Arabian crude oil exports.

Over the summer months of 2013, the country reversed a trend of burning increasing amounts of its oil for domestic electric power production. The excessive oil-burning was needed to provide power to cool homes and businesses during the hot summer months, but thanks to more readily available and cheaper natural gas supplies, greater energy efficiency, and some cooler weather, the Kingdom was able to ramp down oil use for the duration.

Saudi Arabia has been relying on crude oil to generate peak power in the summer as needed by surges in demand for air conditioning, which has triggered alarms and warnings about the Kingdom's ability to keep up its oil exports. So, the good news now is that the volume of oil burned for power production in the world's leading crude exporter fell to 689,750 barrels per day (bpd) during the summer of 2013. This, compared with a record 763,250 bpd in the same period of 2012 represents almost 10% of oil use decrease.

This cooler weather, however, may not be the case

in the summers to come, so crude oil use for power generation may be increased again.

The other good news is that the plans call for decrease in oil burning over time...which remains to be seen. As a matter of fact, similar predictions were made in 2012, when oil-fired power generation was supposed to be reduced by using output from the new Karan gas field. However, an abnormally hot summer forced more oil use and the plans for decrease of oil use were pushed back.

Nevertheless, the domestic consumption is expected to go down or at least stabilize in the near future, as the economic growth is expected to remain strong.

About half of Saudi Arabia's natural gas supplies come from its vast oil fields, so increase of crude oil production also boosts natural gas production. The Kingdom increased crude production by around 2 million bp/d since the summer of 2010 as needed to make up for lost supplies from Libya, Syria and Yemen, and Iran, due to the political and social abnormalities in these countries.

Nearly 10 million bpd of crude oil were produced in a four-month period in 2013, which is the highest since 2002. With that, the gas supplies increased accordingly. The Saudi crude oil exports were increased for the duration to nearly 8 million bpd for the same four months period, which is the highest four-month average since the winter of 2005/06.

This means that the Kingdom used about 2 million barrels of crude oil per day for its own power generation and transportation needs during that time period. With a population of 27 million, this amounts to about 27 bpd of oil use per person...every day for the duration. Wow! This must be a new world record.

This leads us to the conclusion that Saudi Arabia is responsible for the largest waste of crude oil in the world.

A long-term problem for the Kingdom is the fact that the global demand for its oil might decrease. Reasons for this change in oil exports could be:

a) Return of large volumes of Iranian, Iraqi, and other countries oil exports, if domestic and international situations stabilize, and

b) Decrease in the demand for imported crude oil in North America, due to rising tar sands and shale oil production in the area.

c) Decrease in oil exports will lead to reduction in crude oil production, which will effectively reduce

the natural gas production, so that the Saudis cannot rely on it for power generation and must substitute it with more crude oil.

To balance the energy use and electric power baseload, the Kingdom is making ambitious plans to build nuclear, wind, and solar power plants. Unfortunately, these programs are still on the drawing tables, making very slow progress.

But electricity use has been rising in the country by 6-7% annually over the last decade, with summer peak demand more than doubling from 24 GW in 2002 to over 50 GW in 2012.

Until those new (planned) power plants start production, the Saudis need to find a way to keep pumping crude (and natural gas) at record high levels to avoid wasting oil by burning it in their power plants in the hot summer months.

The short-term future of Saudi Arabia is clear—pumping as much oil as possible, to keep the world markets happy, while burning tons of crude to keep its people cool in summer. As the oil production gets more expensive, the global market's reaction is unclear and cannot be predicted from day to day. The high global oil prices mean that the Saudis are losing money by providing cheap electricity to their subjects by burning expensive crude oil. Still, this is a manageable situation...for now.

In the long run, however, the energy security of Saudi Arabia is uncertain, because as soon as its oil wells dry up—and they will dry up some day—the Kingdom will have no energy resources. No energy resources, no income and no fuel for the power plants will lead to internal unrest.

Let's hope that by that time, they would have had implemented new ways to generate electric power and find income to maintain their lifestyle.

More importantly, we must hope that by the time the Saudi oil wells dry up, the entire world will have switched to other fuels. If not, this world will be a very sad and even dangerous place.

Saudi Arabia also has a number of serious environmental problems. Desertification, depletion of underground water resources, and the lack of perennial rivers and other permanent water bodies have prompted the development of extensive seawater-desalination facilities.

Saudi Arabia has the third highest per capita freshwater consumption in the world, despite being one of the world's driest countries. Water supply is becoming an increasingly major problem, the solution of which requires a lot of energy for water desalination, distillation, and recovery.

Since a major portion of the electric power in the country is generated by burning oil, air pollution and waste management are also becoming increasingly important issues. Carbon dioxide emissions from burning crude oil, water desalination, and cement production are on the increase.

Today CO_2 emissions in the country are estimated at nearly 20 tons per capita, or almost as much as the U.S. With a growing population of nearly 30 million, the GHG pollution over Saudi Arabia's sunlit skies amounts to about 600 million tons annually.

Numerous oil spills have caused additional and very serious coastal pollution, while natural threats from frequent sand and dust storms are ravaging the interior of the country.

So, not everything is peachy in Saudi Arabia. The large oil reserves are decreasing, while the air and soil pollution is increasing. Things are getting to a point where something drastic must be done. But what? We will be watching.

Japan

Japan is one of the advanced Western democracies that depends heavily on energy imports. Japan had the world's third largest nuclear fleet, and nuclear power was the hope for getting less dependent on energy imports.

In March, 2011, the Fukushima nuclear disaster destroyed within minutes what took the Japanese many years and billions of dollars to build. Japan's entire nuclear industry and the global energy markets went into a phase of uncertainty and confusion.

This changed the socio-political climate in the country drastically, and now nuclear—and with it the entire energy security of one of the world's largest economies—is hanging by a thread that could break at any point.

Use of coal, fuel oil, and direct-burn crude for power generation in Japan increased dramatically after the 2011 earthquake that triggered the Fukushima nuclear power demise. Many nuclear reactors were shut down, driving the Japanese power utilities to use oil fuel, reaching an 18-year high within months. About 336,000 barrels per day (bp/d) were burned in March 2013 alone across Japan.

Japan imports nearly 5 million bpd of crude oil and nearly that much LNG; most of which are used for electric power generation.

Around 2013 the power companies started cutting oil use, ramping up coal and liquefied natural gas use instead. Tepco, Japan, for example, cut its consumption of the high-cost oil products for power generation by over 41 percent to 71,566 bpd over the nine months of 2013, as compared with the same period the previous year.

The use of coal and LNG reached record rates soon after, and that trend looks to be extending with new gas- and coal-fired units coming online. Japan now imports nearly 40% of the global LNG market. Nevertheless, Japan saw lower rate of use of fuel oil for power by mid-2013, but the ultimate fuel oil use decline will depend on the nuclear power reactors restarts...if and when they occur.

In all cases, coal and LNG-fired power plants will take an increasingly larger share in Japan's power generation sector for the foreseeable future.

Fuel oil will still be the cushion for peak-demand fluctuations, at least until nuclear power comes back fully. When this is going to happen and how long it will last depends on the reliability of the nuclear reactors and the public's sentiment on the matter.

The energy security of Japan is one big question mark that will take a long time to sort out. Until then, Japan's economy will depend on ships full of crude oil and LNG arriving at its harbors every day.

With that, the Kyoto Protocol agreement, to which Japan was a host country and a principal instigator and supporter, is on hold for now and until needed. The GHG emissions in Japan are presently 15-20% above the pre-Fukushima levels and rising by the day.

This is simply another confirmation that energy is a priority for any economy. That nice words about the environment are just that—words. Energy supply is the number-one priority. The energy sources flow must be maintained...

Jordan's Shale Oil

Jordan has been surviving on handouts from the United States and the Persian Gulf monarchies for a long time...as it sits on top of an estimated 100 billion barrels of shale oil. The only problem is taking them out of the ground. If and when these deposits are developed and fully explored, Jordan could become the fourth-largest shale oil producer in the world, after the United States, China and Russia.

Now, however, Jordan depends on imported energy, which is a growing problem, mostly due to rising oil prices, and the recent turmoil in the Arab world. These developments hit Jordan hard, and were emphasized by the loss of low-cost Egyptian natural gas via a pipeline across the Sinai Peninsula, which totaled about 80% of Jordan's electricity generating fuel needs.

Jordan had to replace the cheap, local, natural gas with expensive crude oil imports. This cost Jordan the equivalent of over 25 percent of the kingdom's gross domestic product. Energy issues have gained even greater political importance recently as political and social unrest in the country continues to grow.

The influx of over 400,000 refugees from the civil war in neighboring Syria has complicated the situation and overwhelmed its already stretched resources. This is now fueling domestic discontent and demands for democratic reform that are increasingly posing a challenge to the monarchy.

Because of that, and to improve its energy problems, Jordan is planning to build the Middle East's first shale oil-fired power plant. This would be a major step toward achieving energy security by the energy-poor kingdom. There were actually attempts to develop the shale deposits as early as in 2006, but the prohibitively high price of the project at that time, as well as cheap oil and gas global prices, put an end to that project.

Now they are looking again at shale oil as a plausible solution to at least some of the most pressing energy problems. So, Royal Dutch Shell and British Petroleum will be developing the shale-oil fired power plant, to meet 14 percent of the kingdom's energy needs from shale deposits by 2020.

The power plant is scheduled to start operations in 2017, with a planned capacity of 500 megawatts, which is estimated to cut the cost of Jordan's electricity generation by over $500 million annually.

Upon completion, the shale power plant will be operated by Estonia's Enefit and Jordan's Ministry of Energy and Mineral Resources. There is a good indicator that, if successful, similar power plants will be built in the future, which will help further Jordan's energy crisis.

The first step in the process now is to make sure that enough shale oil can be extracted from the shale play, for which purpose several exploration contracts were signed in 2012 with different companies.

The Energy Ministry also signed an agreement with Canadian Global Oil Shale Holdings in 2012 to assess oil shale resources across 86 square miles of the Attarat Um Ghudran and Isphere al-Mahatain regions of southern Jordan. Work on the project is ongoing, but the success of even this first step is not guaranteed.

The feasibility of extracting enough shale oil in Jordan depends on a number of factors. The global crude oil prices at the time are the most important factor, since shale oil makes sense only when the global oil and gas prices are high enough.

The vulnerability of Jordan's economy to external events makes the shale oil exploration a necessity, and yet, Jordan cannot afford shale oil, if crude oil is twice as cheap. It just makes no sense.

As if to prove the point, the Jordan shale oil project is threatened by the increasing imports of cheap natural gas from Israel's offshore gas fields in the eastern Mediterranean. The first of these wells began production in 2013 and Israel will be exporting 40% of all gas produced.

Supplying Jordan with cheap natural via a relatively short pipeline seems to be a reality and a challenge for the domestic shale plays development and power generation. With the abundance of cheap natural gas, we will need to say goodbye to Jordan's shale oil…at least until the cheap gas flows from Israel's wells.

SUMMARY

Importing huge quantities of crude oil was the biggest problem in the U.S. transport sector during the 20th century. We have learned the lessons of the past and know that the game has to change. We know that we should find a way to stop relying on unstable regimes and fighting terrorists across the world's oceans to meet the energy needs of our transport sector. We've learned all these lessons, but what have we done to change the situation?

Luckily, we now produce a lot of oil, but we still buy over 50% of what we need from foreign nations. Then, the oil must travel cross dangerous oceans and travel long distances before we can use it.

Even if we could ensure its safe transport, foreign oil production levels cannot be relied on for very long. Global crude oil supplies simply won't last forever.

Oil is a commodity of limited quantity on this Earth, and we have already used most of it.

So, what shall we do? Reducing the amount of oil used in the U.S. sounds good, but history tells us that it is not easy to do, and most likely it is not going to happen. Replacing it with other energy sources sounds even better, but it is also impossible, because it is too expansive, especially while oil is so cheap. So we are looking at a long time of drilling as usual in the U.S. and a lot of imports.

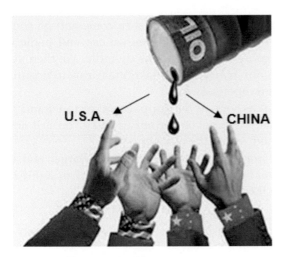

Figure 3-33. Me too, me too...

Generally speaking, our energy security is determined primarily by the integrity of the power generation and the transportation sectors.

While our power generation is 100% secure presently, mostly due to use of abundant coal and natural gas resources, our transportation sector is less than 50% secure. This is due to uncertainty in the supply chain and constantly increasing prices of crude oil imports.

Crude oil is the foundation of the U.S. transportation sector, and since most of it is imported, the short-term energy security of our transportation sector is questionable.

The long-term future is even more uncertain, as the global oil resources are getting depleted by the day and will eventually become exhausted. At that time prices will increase dramatically and we will be forced into a chaotic and even dangerous existence, unless measures have been taken well in advance.

Presently, the U.S. is not low on fossil fuels, only on transportation fuel—crude oil in particular. Since efficient and cheap transportation is absolutely necessary for a healthy economy, we dare not think of the possibility of another Arab oil embargo. It would have devastating consequences on our way of life.

Oil shortages, and eventually running out of oil would leave a mark on our lives, but its effects on the U.S. military would be most serious. This will also present serious challenges to our national security.

The fear of running out of fossils—and crude oil in particular—forces us to consider alternative fuels, like biofuels, synfuels, natural gas, shale oil, etc.

Biofuels are a promising fuel for transportation energy security, and their use will increase as crude oil

supplies dwindle. New technologies—in addition to corn ethanol—must be developed for biofuels to compete. Cellulosic matter is the most promising alternative biofuel source.

Synfuels are another promising practical and important fuel. When fully developed, they should be cheap enough if, for example, they can be made from cheap natural gas. Synfuels, in theory, are one the few energy sources that have a chance to eventually compete with compressed natural gas.

Shale oil is also shaping up as a potentially promising new energy source for the transportation sector. The U.S.' shale reserves are enormous, and successful extraction technologies have been developed. Shale oil production is cheaper than synfuels, and will eventually become a major competitor.

Other important factors in the long-term energy security of the transportation sector are improved automobile mileage and efficient use of cars and trucks.

Car and truck manufacturers are working on more efficient engines, but this is not a priority for them, so they need a push from the politicians from time to time to come up with plausible solutions.

Smarter driving habits is the best way to save fuel and protect the environment.

Walking, biking, and mass transport are widespread in many countries, but getting the Americans (this one included) out of their luxury cars won't be easy. No amount of regulation would make much differ-

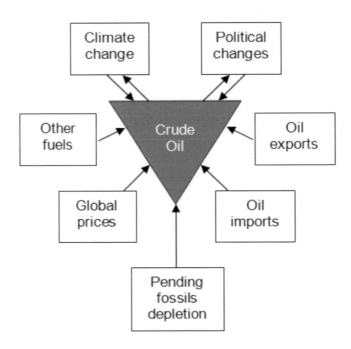

Figure 3-34. Crude oil risks

ence unless gasoline at the pump is $20-30 a gallon...or more!

Hybrid cars have some promise, but plug-in hybrids and all-electric automobiles are losing their attraction very quickly as fuel prices go down. Complexity, high cost of repairs and battery replacement, short range, and sluggish performance (per American standards) have been some of the reasons of their slow U.S. market penetration.

As a confirmation, while Volts and Prias abounded at the International Auto show in Detroit, there were only a few of these types of vehicles at the 2014 show, and even fewer were the interested visitors. Muscle and sports cars, large trucks and SUVs took the spotlight and are now headed to the American roads in huge numbers. Goodbye Volts and Prias...until the next time oil rises over $150/bbl.

Not that Volts and Prias will make a big dent in the energy situation, but they are one drop less in the bucket of energy solutions we need to combat the huge energy problems ahead of us.

So we are back to where we started. Oil is plentiful again, so we will use as much as we want for as long as we want, and to heck with tomorrow. This is a reoccurring theme, which we have seen come and go so many times. We use as much oil as we can and panic when prices go up, or the supplies dwindle. We then vow to change our ways and to take measures to ensure this doesn't happen again.

But as soon as the supplies are restored and prices go down, we forget our promises and go on as if nothing happened...until the next time.

A major energy disaster—much larger and worst than ever before—is the only thing that can significantly change this pattern.

Notes and References

1. IEA, Crude oil statistics, http://www.iea.org/topics/oil/
2. Oil news, http://oil.com/
3. Oil prices, http://www.oil-price.net
4. Bloomberg, Energy & Oil Prices. http://www.bloomberg.com/energy/
5. CME, Crude Oil Futures, http://www.cmegroup.com/trading/energy/crude-oil/light-sweet-crude.html
6. IEA, http://www.eia.gov/cfapps/ipdbproject/IEDIndex3.cfm?tid=5&pid=53&aid=1
7. *Photovoltaics for Commercial and Utilities Power Generation*, Anco S. Blazev. The Fairmont Press, 2011
8. *Solar Technologies for the 21st Century*, Anco S. Blazev. The Fairmont Press, 2013
9. *Power Generation and the Environment*, Anco S. Blazev. The Fairmont Press, 2014

Chapter 4

Natural Gas

(The New Hope)

The new natural gas bonanza is quickly reshaping the global energy markets and reinforcing U.S. energy security.

Anco Blazev

NATURAL GAS

The natural gas industry, and with it our entire energy sector, is undergoing unprecedented energy revival. It is much more spectacular than anyone could have ever imagined just several years back. We have so much natural gas today, thanks to new large deposits discoveries and use of new production techniques, that we simply cannot use it all.

Who would've thought just several short years back that the U.S. would become a major natural gas exporter? Yet, this is exactly what is happening today. We are now planning to export large quantities of surplus natural gas and crude oil products to other countries.

In addition to being the world's largest importer (of crude oil), we would now become a major exporter (of LNG) in the global energy market. Maybe even grabbing the title of the largest natural gas exporter someday soon...

We'll let the reader be the judge if this is possible, how right or wrong it is, and what it would do to our energy security after we present (as complete as possible) a picture of the energy products, processes, markets, and related events.

What is natural gas?

In the beginning...

Natural gas, like coal and crude oil, was created millions of years ago when large quantities of plants and animals were buried amidst sand and rock deposits deep underground. The events that led to the death of such large quantities of life vary, but the important thing is that layers of mud, sand, rock, plants, and huge amounts of animal matter built up, while the heat and pressure in their deep graves underground increased, slowly turning the once-living matter into fossils—coal, oil and natural gas.

Another theory states that the Earth was made up of primordial materials that combined in space billions of years ago when the basic structure of our Earth evolved. These materials are still buried far below the earth's crust where they have been trapped for 4.5 billion years, some of them slowly turning into coal, oil, and natural gas.

No one really knows exactly when and how this all happened, since no one was there to record the events. And because this happened so long ago, we simply lack the tools and ability to understand and analyze the situation in detail. Nevertheless, we must at least consider the theories provided by the specialists, and it is very likely that there is more than just one mechanism responsible for the formation of fossil fuels.

What we know for certain is that the formation of natural gas was driven primarily by three main physical-chemical processes, those of thermogenic, biogenic, and abiogenic natural gas formation.

- *Thermogenic methane* is formed from organic particles that are covered in mud and other sediment. Over time, more sediment, mud and other debris are piled on top of the organic matter. This sediment and debris put a great deal of pressure on the organic matter, compressing it. This compression, combined with high temperatures found deep underground, breaks down the carbon bonds in the organic matter.

  As we scan deeper under the earth's crust, the temperature gets higher. At low temperatures (shallower deposits) more oil is produced than natural gas. At higher temperatures, however, more natural gas is created than oil.

  That is why natural gas is usually associated with oil in deposits that are 1 to 2 miles below the earth's crust. Deeper deposits, very far underground, usually contain primarily natural gas, and in many cases, pure methane.

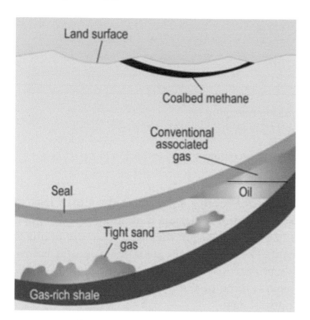

Figure 4-1. Natural gas deposits

- *Biogenic natural gas* formations are created through the transformation of organic matter by tiny microorganisms. This type of methane is referred to as biogenic methane. Methanogens, tiny methane-producing microorganisms, chemically break down organic matter to produce methane. These microorganisms are commonly found in areas near the surface of the earth that are void of oxygen. These microorganisms also live in the intestines of most animals, including humans.

  Formation of methane in this manner usually takes place close to the surface of the earth, and the methane produced is usually lost into the atmosphere. In certain circumstances, however, methane can be trapped underground, recoverable as natural gas. An example of biogenic methane is landfill gas. Waste-containing landfills produce a relatively large amount of natural gas from the decomposition of the waste materials that they contain. New technologies are allowing this gas to be harvested, adding to the supply of natural gas.

- *Abiogenic natural gas* formation processes start extremely deep beneath the earth's crust, in the presence of hydrogen-rich gases and carbon molecules. As these gases gradually rise towards the surface of the earth, they may interact with minerals that also exist underground, in the absence of oxygen.

This interaction may result in a reaction, forming elements and compounds that are found in the atmosphere (including nitrogen, oxygen, carbon dioxide, argon, and water). If these gases are under very high pressure as they move toward the surface of the earth, they are likely to form methane deposits, similar to thermogenic methane.

NOTE: Natural gas is informally referred to simply as "gas," especially when used in conjunction with the other fossil energy sources, coal and oil. This is how we are referring to it most of the time in this text as well.

Types of Natural Gas

There are several different types of commercially available natural gas, some of which are as follow:

- *Pipe natural gas* or natural gas that is transported via pipeline, is known in commercial terms as sale gas. Sale gas is mainly composed of methane. It is transmitted to customers to be used as fuel at power generation and industrial plants. It is not used for residential purposes, but instead 24% of the energy generated in the US comes from natural gas, and its share is increasing as new deposits are discovered and exploited.

- *Coalbed methane* is basically a type of natural gas that is extracted from coal beds. Coal seams that contain some coalbed methane and are isolated from other fluid units are also called coalbed methane reservoirs. By analysis and comparison of the typical coalbed methane reservoirs, these can be divided into hydrodynamic sealing coalbed methane reservoirs, and self-sealing coalbed methane reservoirs.

  Currently, hydrodynamic sealing reservoirs are the main target for coalbed methane exploration and development, since self-sealing reservoirs are unsuitable for profitable coalbed methane extraction.

- *Natural Gas for Vehicles (NGV)* is the form of natural gas used as fuel for vehicles. NGV is purified natural gas, primarily composed of pure methane, which is transported through a pipeline, or via truck (under pressure) to gas stations. At the gas stations, low pressure gas will be compressed and stored at high pressure (3000-3600 PSI), and can be then injected into the vehicles' gas tanks.

- *Liquefied Natural Gas (LNG)* is natural gas that is liquefied by lowering its temperature to $-160°C$. At this temperature it becomes liquid and is 600 times smaller in volume. It is then stored at atmospheric pressure in specially designed vessels (tanker trucks, railroad cars, or cargo ships) and transported to end users. The cost of waterway transport today is most convenient for distribution to many areas of the world, and could be less than transportation through pipeline.

- Cooking gas (LPG) has a commercial name of liquefied petroleum gas or LPG which is a product from the oil refineries, or the gas separation plants. LPG is a mixture of several hydrocarbon gases, with propane and butane as the major constituents. LPG can be in any ratio or purely propane or butane. LPG is sold in pressurized bottles and can be used as fuel in homes, industry and transportation. Natural gas is not used directly for domestic cooking and heating purposes.

The Imitators

There are several products that are similar to natural gas, but have different origins and should not be confused with it. Some of these are town gas, biogas, landfill gas and methane hydrate.

- *Town gas* is a synthetically produced mixture of methane and other gases, which contains the highly toxic carbon monoxide. Town gas is produced by treating coal chemically, and can be used the same way as natural gas. This technology is not economically competitive with other sources of fuel gas today, but there are still some specific cases where it is the best option, and it may be even more so in the future.

  There were town "gashouses" in the eastern US in the late 19th and early 20th centuries, which were simple by-product coke furnaces, where bituminous coal was burned in air-tight chambers. The gas driven off from the coal—the town gas—was collected and distributed through networks of pipes to residences and other buildings where it was used for cooking and lighting.

  The coal tar (or asphalt) that collected in the bottoms of the gashouse ovens was often used for roofing and other water-proofing purposes. When mixed with sand and gravel, it was used for paving streets. These methods were replaced by natural gas heating by the end of the 20th century.

- *Syngas* is similar to town gas in its production and final properties. It is a fuel gas mixture consisting primarily of hydrogen, carbon monoxide, and very often some carbon dioxide. It is produced by steam reforming of natural gas or liquid hydrocarbons with the final goal of enriching the gas with hydrogen. Gasification of coal or biomass is another method for syngas production used today.

- *Biogases* are methane-rich gases produced by the anaerobic decay of non-fossil organic matter (biomass), and are also called natural biogas. Biogas was formed mostly in swamps, marshes, and landfills. It is also found in sewage sludge and cow manure where the conversion is done by way of anaerobic digesters or enteric fermentation, as is the case with cattle-based biogas.

- *Landfill gas* is a type of methane gas which, as the name suggests, is formed in landfills by the decomposition of organic matter. Landfill gas is already used in some areas, but its use could be greatly expanded. Landfill gas is a type of biogas, but biogas usually refers to gas produced from organic material that has not been mixed with other waste, so landfill gas has its own place.

- *Methane hydrate* is a type of gas that is formed on the bottom of oceans and other water bodies. It is simply molecules of methane trapped into ice (water) pockets, which must be heated to release the methane. We will take a very close look at this type of gas in this text, because there are huge amounts of it locked in the ocean floor and in permafrosts, which hold the promise of becoming primary energy sources of the distant future.

NOTE: When methane from any source is released directly into the atmosphere, it is considered a pollutant, because it is oxidized by the oxygen in the air and produces carbon dioxide. This is a dangerous process that, considering the great amount of gas used today, has severe environmental consequences.

The conversion methane to carbon dioxide in the atmosphere is a very slow process. Methane gas has a half life of seven years, which means that of 1,000 kg of methane emitted today, 500 kg will have broken down to carbon dioxide and water after seven years. This also means that methane released in the atmosphere today will be slowly but consistently damaging the environment by releasing carbon dioxide at a slow rate for decades to come.

We take a closer look at this and other types of natural gas and its imitators in the following sections.

Natural Gas Properties

Since we will be talking about natural gas in more detail, and because it has its own measurements, terms and nomenclature, we'd like to familiarize the reader with them for future use:

Nomenclature

Natural gas measurement units:

Volume units:

    CCF—one hundred cubic feet
    Mcf—one thousand cubic feet of natural gas
    Mmcf—one million cubic feet of natural gas
    Bcf—one billion cubic feet of natural gas
    Tcf—one trillion cubic feet of natural gas
    Mmcf/d—millions of cubic feet of gas per day

Energy equivalents:

    Boe (barrel of oil equivalent)—equal to 6,000 ft^3 natural gas
    Mboe—one thousand barrels of oil equivalent
    Mmboe—one million barrels of oil equivalent
    Mmcfe—one million cubic feet of natural gas equivalent
    Bcfe—one billion cubic feet of natural gas equivalent
    Tcfe—one trillion cubic feet of natural gas equivalent

NOTE: The energy contained in one barrel of crude oil equals that contained in 6,000 cubic feet of natural gas.

General Heat Measurement Units

Btu: British thermal unit, or the amount of energy required to raise the temperature of one pound of water by one degree Fahrenheit.

— One Btu is equivalent to 252 calories, 0.293 watt-hours or 1,055 joules.

Calorie: The energy needed to increase the temperature of 1 gram of water by 1°C at standard atmospheric pressure.

— One calorie is equal to 4.18400 joules.

Joule: A derived unit of energy, work, or amount of heat in the International System of Units.

— One Joule is equal to the energy expended (or work done) in applying a force of one Newton through a distance of one meter (1 Newton meter or N·m), or in passing an electric current of one ampere through a resistance of one ohm for one second.

The Properties

Raw natural gas is a mixture of methane, higher hydrocarbons (primarily ethane), and some noncombustible gases. Some other constituents, principally water vapor, hydrogen sulfide, helium, and other petroleum gases are present in different ratios. These are usually removed from the mix prior to distribution and use by the public.

Table 4-1. Natural gas components

Gas species	Content
Methane, CH_4	70 to 96%
Ethane, C2H$_6$	1 to 14%
Propane, C_3H_8	up to 4%
Butane, C_4H_{10}	up to 2%
Pentane, C_5H_{12}	up to 0.5%
Hexane, C_6H_{14}	up to 2%
Carbon dioxide, CO_2	up to 2%
Oxygen, O_2	up to 1.2%
Nitrogen, N_2	4 to 17%

Some of the practical natural gas measurements are as follow:

    1 cubic foot (CF) natural gas averages 1,000 Btus
    A typical Gas Quality Spec allows 950 to 1,100 Btus per CF
    1 hundred cubic feet (CCF) average 100,000 Btus
    1 CCF is about 1 therm
    1 Therm averages 100,000 Btus
    1 dekatherm (10 therms) averages 1,000,000 Btus
    1 MCF (10 CCF) = 1,000 CF = 1,000,000 Btus

The composition of natural gas depends on its location and source. Commercial gas is usually a mixture of gas drawn from various sources, so its composition can vary slightly. Nevertheless, a fairly constant heating value is usually maintained for control and safety.

Heating values of natural gases vary from 900 to 1,200 Btu/ft^3; but the usual commercial range is 1,000 to 1,050 Btu/ft^3 at sea level.

Natural gas is a nearly odorless and colorless gas that accumulates in the upper parts of oil and gas wells. For safety purposes, odorants (such as mercaptans) are added to natural gas and LPG before distribution to consumers to give them advance warning.

LPG

Liquid petroleum gas (LPG) is another natural gas imitator, also called propane (C_3H_8) or butane (C_4H_{10}) on the commercial markets. When used as a vehicle fuel, it is referred to as autogas.

LPG is derived from fossil fuel sources; i.e., during the refining of crude oil, or extracted from petroleum or natural gas streams as they emerge from the ground. It is used primarily for heating, cooking, and as an aerosol propellant and a refrigerant, replacing chlorofluorocarbons which damage to the ozone layer.

LPG's heating value is approximately 10% lower than natural gas (methane vs. propane). There is about 4% propane in the commercial natural gas used today.

The major differences between LPG and natural gas are:

Natural gas is:
- Very safe as it is lighter than air, and when leaked it will float up and dissipate quickly.
- Ready to use as it is in gas form.
- Its ignition point is 593 degree C.
- The concentration at which natural gas would ignite or explode is 5-15% gas in air.
- Colorless and odorless, and burns completely, without emissions of soot or sulfur. Storage tanks are usually not needed for storing natural gas.

LPG is:
- Less safe, as it is heavier than air, and when leaked it will pool on the ground.
- A liquid, so it needs to be converted to gas.
- Colorless and odorless, but odor is normally added for safety reasons.
- It burns completely, without emissions of soot or sulfur.
- Storage tanks and advance ordering are needed in most cases.
- Its ignition point is lower than natural gas; 410-580 degree C.
- The concentration at which LPG gas would ignite or explode is 2.0-9.5% gas in air.
  NOTE: The great variation of ignition points in LPG is due to the fact that there are a number of different petroleum compounds in it, each of which has a different ignition point. Basically, LPG is more dangerous to use than natural gas. It is easier to ignite or explode LPG gas than natural gas, since its ignition point is significantly lower.

Also, and very importantly from safety point of view, is the fact that LPG is heavier than air, so it tends to settle down and concentrate in corners, where it could explode in the presence of a spark or open flame.

Now we know what natural gas is and what it can do, so let's take a close look at its production process.

Natural Gas Production

Lately, natural gas production has reached sky-high levels of volume and efficiency, accompanied by low prices. The trend is very strong and there is no stopping it—come hail or high water—for now. The use of natural gas as fuel for electricity generation is climbing faster than all other energy sources combined, so it is now king of the energy sector.

Other uses of natural gas such as in the production of fuel for vehicles and making different chemicals are also on the rise. This is leading to further expansion of the U.S. natural gas drilling and production now, and even more is planned for the future.

Table 4-2. Global natural gas production and use, 2010.

Country	% of world production	% of world use
Russia	21.3	15.1
USA	18.5	22.0
Canada	6.5	3.4
Iran	3.7	3.7
Norway	3.0	3.0
Algeria	2.9	2.9
U.K.	2.8	3.2
Saudi Arabia	2.6	2.6
Netherlands	2.2	2.2
Ukraine	2.3	-
Japan	3.0	-

Table 4-2 tells us that the U.S. was number 2 in natural gas production just a few years ago, but recently we surpassed Russia in total natural gas production. United States natural gas production in 2012 was 619 million tons of oil equivalent, or 20.4 percent of the world. At the same time, the Russian Federation produced 533 million tons of oil equivalent, or 17.6 percent of the world total.

Russia's gas production today is nearly 15% less than that in the U.S.

On the usage side of the equation, the U.S. has always been, and still is, the undisputed number 1 user of natural gas. This is a big business in the U.S., and getting bigger by the day. Natural gas is used for cooking and heating homes and businesses, powering transportation vehicles, making different products; both as process fuel and raw material.

So, how is natural gas produced? In this section we take a look at the usual, on-shore (on land), natural gas production process and its different stages.

Natural gas cradle-to-grave lifespan is quite similar to that of crude oil, and consists of.

- Deposits search and location
- Proposed area survey
- Environmental studies
- Gas rights and agreements
- Permits and legal issues
- Drill area preparation
- Drilling rig construction
- Daily O&M operations
- Gas transport
- Gas use (burning)
- Decommissioning and area reconstruction

Well Site Search and Location

The practice of locating natural gas and petroleum deposits has been transformed dramatically in the last 20 years with the advent of extremely advanced, ingenious technology. In the early days of the industry, the only way to locate underground petroleum and natural gas deposits was to search for surface evidence of these underground formations.

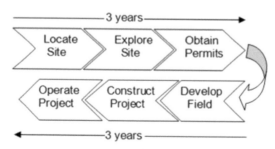

Figure 4-2. Gas well development

In the past, those searching for natural gas deposits were forced to scour the earth, looking for seepages of oil or gas emitted from underground before they had any clue that there were deposits underneath. However, because such a low proportion of petroleum and natural gas deposits actually seep to the surface, this was a very inefficient and difficult exploration process. As the demand for fossil fuel energy has increased dramatically, so has the necessity for more accurate methods of locating these deposits.

Modern technology has allowed for a remarkable increase in the speed and success rate of locating natural gas reservoirs. Today, geologists and geophysicists use advanced technologies and their knowledge of the properties of underground natural gas deposits to gather data that can be used to find natural gas deposits and estimate their size.

However, the process of exploring for natural gas and petroleum deposits is still characteristically an uncertain one, due to the complexity and difficulties related to finding deposits that are hundreds or thousands of feet below ground.

Exploration for natural gas typically begins with geologists examining the surface structure of the earth with the help of various instruments to determine where it is geologically likely that petroleum or gas deposits exist.

By surveying and mapping the surface and sub-surface characteristics of an area, the geologist can extrapolate which areas are most likely to contain a petroleum or natural gas reservoir. The geologist uses techniques that range from outcroppings of rocks (on flatland, or in valleys, and gorges) to geologic information obtained from rock cuttings and samples from the digging of irrigation ditches, water wells, and other oil and gas wells.

This information is then processed to make decisions on the fluid content, porosity, permeability, age, and formation sequence of the rocks beneath the surface of a particular area.

Once the geologists have identified an area where it is geologically possible for a natural gas or petroleum formation to exist, further tests can be performed to gain more detailed data about the potential reservoir. These tests allow for the more accurate mapping of underground formations, most notably those formations that are commonly associated with natural gas and petroleum reservoirs.

Seismology is one of the greatest tools used to locate oil deposits. It refers to the study of how energy, in the form of seismic waves, moves through the Earth's crust and interacts differently with various types of underground formations.

- Onshore seismology involves artificially creating seismic waves, in the form of small explosions, the sound reflections of which are then picked up by sensitive pieces of equipment called "geophones" that are embedded in the ground. The data picked up by these geophones are then transmitted to a seismic computer which records the data for further interpretation by geophysicists and petroleum reservoir engineers.

Lately a new, non-explosive technology is used, which consists of a large heavy-wheeled or tracked-vehicle carrying special equipment designed to create a large impact or series of vibrations. These impacts or vibrations create seismic waves similar to those created by dynamite explosions. The signal is then used to locate deposits.

- Offshore seismology uses a slightly different method of seismic exploration, where instead of trucks and geophones, equipment on a ship is used to pick up seismic data, and hydrophones are used to pick up seismic waves under water. These hydrophones are towed behind the ship in various configurations depending on the needs of the geophysicist. Instead of using dynamite or impacts on the seabed floor, the seismic ship uses a large air gun, which releases bursts of compressed air under water, creating seismic waves that can travel through the Earth's crust and generate the seismic reflections that are necessary.

There are also other techniques for locating oil deposits today, using:

a) Magnetometers, which detect the magnetic properties of underground formations, and which are then measured to generate geological and geophysical data;

b) Gravimeters, which measure and record the difference in the Earth's gravitational field to figure out what is underground; and

c) Exploratory wells, which are used to dig deep into the Earth's crust to allow geologists to study the composition of the underground rock layers in more detail.

NOTE: There are important differences between the drilling and production methods for shale gas vs. coal bed methane (CBM), known as coal seam gas, as follow:

1. Shale strata (at 1,000 to 3,000 meters) are typically much deeper than coal deposits (100 to 1,000 meters). This reduces the likelihood that gas, fracking fluids and produced water could migrate from the shale formation to contaminate the water table.

2. Coal deposits are much more porous than shale rock. Nearly all shales require hydraulic fracturing to extract gas, whereas only a portion of coal deposits require fracturing. In Queensland, Australia, for example, only 10%-40% of CBM wells are estimated to ultimately require fracturing.

3. Shale is much denser than coal, requiring greater energy to fracture (25,000-35,000 horsepower, compared with 4,000-5,000 horsepower to fracture coal seams). Therefore, more and more powerful equipment is required to fracture shale layers than to fracture coal deposits.

4. Water must generally be pumped out of coal deposits to extract natural gas, while it must be pumped into shale to fracture the deposit and extract natural gas. Therefore, coal gas projects potentially put less immediate strain on surface water resources. However, they also have greater subsurface water risks insofar as water from adjacent aquifers could migrate into coal formations as reservoir pressures decline.

For both coal and shale gas wells, water produced from the well (also known as "produced water" or "gray water") is saline and toxic in high concentrations, requiring similar types of handling, special treatment and disposal.

Well Site Development

Once a potential natural gas deposit has been located by a survey team, the team of drilling experts takes over. Their job is to dig down to the natural gas deposit site and figure out a way to extract it. Natural gas deposits, like those of oil, can be both onshore and offshore.

Although the process of digging deep into the Earth's crust to find deposits of natural gas that may or may not actually exist seems daunting, the industry has developed a number of innovations and techniques that both decrease the cost and increase the efficiency of drilling for natural gas.

Figure 4-3. Natural gas drilling site

Determining to drill a well depends on many factors, including the economic potential of the hoped-for natural gas reservoir. It costs a great deal of money for exploration and production companies to search and drill for natural gas, and there is always the inherent risk that no gas will be found.

The exact placement of the drill site depends on many factors, including the nature of the potential formation to be drilled, the characteristics of the subsurface geology, and the depth and size of the target deposit.

After the geophysical team identifies the optimal location for a well, it is necessary for the drilling company to ensure that it completes all the necessary steps to legally drill in that area. This usually involves securing permits for drilling operations, establishment of a legal arrangement to allow the natural gas company to extract and sell the resources under a given area of land, and a design for gathering lines that will connect the well to the pipeline.

There are often a variety of potential owners and stakeholders of the land and mineral rights of a given area. The interests of these groups are at times in conflict with the well construction and proper operation, which could hinder or stop the exploration or well drilling activities.

If after solving all logistics problems a new well does in fact come in contact with natural gas deposits, it is developed to allow for the extraction of this gas, and is termed a "development" or "productive" well. At this point, with the well drilled and hydrocarbons present, the well may be completed to facilitate its production of natural gas.

However, if the exploration team was incorrect in its estimation of the existence of a marketable quantity of natural gas at a well site, the well is termed a "dry well," and production does not proceed. Such cases usually result in great financial loss for the company, the land owners, and the investors.

On-shore Drilling

There are several types of on-shore drilling

- *Percussion*, or cable tool drilling consists of repeatedly dropping a heavy metal bit into the ground, eventually breaking through rock and punching a hole through to the desired depth. The bit, usually a blunt, chisel shaped instrument, can vary with the type of rock that is being drilled. Water is used in the well hole to combine with all of the drill cuttings, and is periodically bailed out of the well when this "mud" interferes with the effectiveness of the drill bit.

- *The "springpole" technique,* used in the early 1800s, consisted of a flexible pole (usually a tree trunk) anchored at one end, and laying across a fulcrum, much like a diving board. The flexible pole, or springpole, would have a heavy bit attached at the loose end. In order to get the bit to strike the ground, workers would use their own body weight to bend the pole toward the ground, allowing the bit to strike rock. The tension in the pole would spring the bit free if it became stuck in the ground.

Many improvements have been made since these early percussion rigs. It was from cable tool drilling that one of the most important drilling advancements was made. In 1806, David and Joseph Ruffner were using the springpole technique to drill a well in West Virginia. To prevent their well from collapsing, they used hollow tree trunks to reinforce the sides and to keep water and mud from entering the well as they dug. They are credited as the first drillers to use a casing in their well—an advancement that made drilling much more efficient.

- Innovations, such as the use of steam power in cable tool drilling, greatly increased the efficiency and range of percussion drilling. Conventional man-powered cable tool rigs were generally used to drill wells 200 feet or less, while steam powered cable tool rigs, consisting of the familiar derrick design, had an average drilling depth of 400 to 500 feet. The deepest known well dug with cable tool drilling was completed in 1953, when the New York Natural Gas Corporation drilled a well to a depth of 11,145 feet.

- *Directional drilling* (or slant drilling) is used to drill non-vertical wells, which are preferred method today (vs. vertical wells). This method consists of three different groups:
  a) Oilfield directional drilling, which is used in areas with large number of oil wells. The full oilfield exploitation entails multiple production and exploratory wells (usually vertical wells) scattered across the area.
  b) Utility installation directional drilling or directional boring, is a steerable, trenchless method of installing underground pipes, conduits and cables in a shallow arc along a prescribed bore path. It uses a surface-launched drilling rig, with minimal impact on the surrounding area.
  c) Surface in seam (SIS) drilling, which horizontally intersects a vertical well target to extract coal bed methane.

• Horizontal drilling is the new, most successful, and most controversial drilling method. It is flexible in that it allows for the extraction of natural gas that had previously not been accessible. Although on the surface it resembles a vertical well, beneath the surface, the well inclines and switches to a 90 degree (or close to it) angle, so that it runs parallel to the natural gas formation.

These parallel to the surface well legs can go in different directions at different depths and can be more than one mile long and thousands of feet below the surface.

Horizontal drilling allows one surface well to branch out underground and tap many different natural gas resources. It basically allows the well to make contact with unprecedented larger areas within productive formations, increasing well production to quantities unthinkable by other methods.

NOTE: The terms directional and horizontal drilling are often used interchangeably, although directional drilling usually refers to drilling at a slight angle to increase contact with the resource, while horizontal drilling is a type of directional drilling which often uses a technique known as hydraulic fracturing to extract natural gas from geologic formations by running through and fracturing them.

Well Casing

Installing well casing is an important part of the drilling and well completion process. Well casing consists of a series of metal tubes installed in the freshly drilled hole. Casing strengthens the sides of the well hole, ensures that no oil or natural gas seeps out of the well hole as it is brought to the surface, and keeps other fluids or gases from seeping into the formation through the well.

A good deal of planning is necessary to ensure that the proper casing for each well is installed. The type of casing used depends on the subsurface characteristics of the well, including the diameter of the well and the pressures and temperatures experienced throughout the well. The diameter of the well hole depends on the size of the drill bit used. In most wells, the diameter of the well hole decreases, the deeper it is drilled, leading to a type of conical shape that must be taken into account when installing casing.

There are five different types of well casing: conductor casing, surface casing, intermediate casing, liner string, and production casing. They are used during the different cycles of the well drilling process, and are designed to provide a complete, efficient and safe well structure.

Well Completion

Once the casing is complete, the well is ready for production of natural gas. At this point, a decision is made on the characteristics of the intake portion of the well in the targeted formation. This process is called well completion, and the different types of well completions include: open hole completion, conventional perforated completion, sand exclusion completion, permanent completion, multiple zone completion, and drain hole completion. The use of any type of completion depends on the characteristics and location of the hydrocarbon formation to be mined. For example, an open hole completion is used when the surface is not in danger of collapse (cave in). In this case, the pipe end that is above ground is left as is, open without any protective hardware.

The type of completion used depends on the particular characteristics and location of the formation to be exploited.

The Wellhead

The wellhead consists of pieces of equipment mounted at the opening of the well, which are used to control the extraction of oil or gas from the underground formation. It prevents leaking of oil or natural gas out of the well, and it prevents blowouts caused by high pressure. Formations that are under high pressure typically require wellheads that can withstand a great upward pressure from the escaping gases and liquids. These wellheads must be able to withstand pressures of up to 20,000 psi.

Figure 4-4. Gas wellhead

The wellhead consists of three components:

a)   the casing head, which consists of heavy fittings that provide a seal between the casing and the surface;

b)   the tubing head, which provides a seal between the tubing, which is run inside the casing and the surface; and

c)   the Christmas tree, which is the piece of equipment that fits on top of the casing and tubing heads. The latter contains tubes and valves that control the flow of hydrocarbons and other fluids out of the well. It is the only visible—above ground—part of an operating gas or oil well.

As Figure 4-4 shows, the gas wellhead is a clean, uncomplicated, and unimposing piece of hardware. Underneath, however, the picture is much different. A close look underground reveals a complex network of pipes, casings, cement filling, etc. components that are put together professionally and with good intentions in mind.

The problem is that while we can control the gas wellhead hardware, we cannot control what's underneath, and here is where the problems start.

Well Treatment (hydrofracking)

In most cases, in the beginning of the exploitation of a new well, the pressure of the gas deposit is high enough to expel the gas to the surface without any external help. At that time all we need is a good connection to the gas wellhead and a long string of pipes to deliver thus produced gas to its final destination.

With time, however, the pressure in the deposit will diminish and the gas or oil must be pumped out by means of electric pumps. Eventually, even with the external pumping in place, the gas or oil flow would be reduced and even stopped. The well is pronounced depleted, and has to be abandoned or "treated."

In the past, most wells were abandoned at this stage, but recently new technological developments have opened new possibilities. So today abandoned wells, or low volume such, can be reopened and "treated."

Well treatment is a method of ensuring the efficient flow of gas from a formation after the natural pressure has been reduced and the free flow has diminished. The well treatment method, also called hydrofracking, or fracking for short has been perfected lately to produce amazingly huge quantities of gas from wells that were thought dead, or impossible to operate in the past.

Hydrofracking consists of injecting huge amounts of water, acid, different chemicals, and/or gas mixtures

into the well to break the formation (fracking) and allow the petroleum products and gas to flow through it. The use of this method is increasing with the help of different fracking mixtures and techniques.

Acidizing a well is a chemical method that consists of injecting acid (usually hydrochloric acid) into the well. This approach is especially effective in limestone or carbonate formations, where the acid dissolves portions of the rock in the formation, opening up existing spaces to allow for the flow of petroleum and natural gas trapped in the formations.

The fracturing fluids are injected into the well at very high pressure, which breaks or "cracks" new, or opens up existing, fractures in the formation. In addition to the water being injected, "propping agents" are also used.

The propping agents consist of a mixture of different materials and chemicals, such as acids, petrochemicals, sand, glass beads, epoxy, silica sand, organic chemicals, etc., which individual or combined properties assist in the opening of the newly widened fissures in the formation.

NOTE: Hydraulic fracturing involves the injection of water into the formation, while CO_2 fracturing uses gaseous carbon dioxide. Gaseous carbon dioxide can also combine with undersurface water to make H_2CO_3 acid, which is quite effective in breaking lime formations. Fracturing and acidizing can be used on the same well to increase permeability and widen the pores of the formation.

Hydrofracking is one of the most controversial subjects in the energy sector today. Claims for damage to properties, water table contamination, and human illness have been made recently, and the protests in some locations are spilling from the streets and the fields into the courts.

Offshore Gas Drilling and Production

Offshore drilling for natural gas is usually done on anchored or fixed platforms in lakes or the ocean, some of which are dozens to hundreds of miles away from the coastline. This creates challenges that just don't exist when drilling on-shore, although the actual drilling processes are very similar.

The key difference is that with offshore drilling, the sea floor can be hundreds or thousands of feet below sea level. An artificial, manmade drilling platform must be constructed to support the drilling equipment on the surface, and a very long drill must be lowered to the bottom before starting the drilling.

This is not an easy undertaking. There are many

different types of such platforms and drill rigs, depending on the type and location of well to be drilled. The most important factor in off-shore drilling is the depth of the underwater drilling target and the water conditions—high seas are much harder to work in than the waters of an inland lake.

A subsea drilling template is a piece of equipment that connects the underwater well site (the actual hole in the ocean floor that leads to the gas deposits) to the drilling platform on the surface of the ocean. It consists of an open steel box with multiple holes in it, dependent on the number of wells to be drilled. The subsea drilling template resembles a cookie cutter that is placed over the well site on the ocean floor, exactly where the hole(s) will be drilled.

A shallow hole is then dug and the drilling template is cemented into place, secured to the ocean floor and connected with cables to the drilling platform above. This setup provides stable and accurate drilling, while at the same time allowing the platform to move on the ocean surface some, as it will inevitably be affected by waves and currents.

A blowout preventer is installed on the sea floor next, to prevent any oil or gas from seeping out into the water. Remember the malfunction of the blowout preventer of BP's Deep Water Horizon in 2010, and you'll get a good idea of how important this piece of equipment is.

On top of the blowout preventer, a special "marine riser," a long pipe, is raised to the drilling platform above. It houses the drill bit and drill-string. It is flexible enough, via slip and ball joints, to accommodate the shifting of the drilling platform.

Types of Offshore Gas Drilling Rigs

The actual design of the platform and the rest of the equipment vary, so that ultimately there are several types of offshore drilling rigs, as follow:

- *Drilling barges* are usually used for drilling in inland water bodies—lakes, swamps, rivers and canals. Drilling barges are large, floating platforms which must be towed by tugboat from location to location. Suitable for still, shallow waters, drilling barges are not able to withstand the water movement of large open-water situations.

- *Drill-ships* are large ships designed to carry out drilling operations. These boats are specially designed to carry drilling platforms out to deep-sea locations. A typical drillship will have, in addition to all of the equipment normally found on a large ocean ship, a drilling platform and derrick located in the center of its deck. In addition, drillships contain a hole (or "moonpool"), extending right through the ship, down through the hull, which allows for the drill string to extend down into the water.

  Drillships are often used to drill in very deep water, which can be turbulent. They use what is known as "dynamic positioning" systems. They are equipped with electric motors on the underside of the ship's hull, capable of propelling the ship in any direction. These motors are integrated into the ship's computer system, which uses satellite positioning technology, in conjunction with sensors located on the drilling template, to ensure that the ship is directly above the drill site at all times.

- *Jack-up rigs* are similar to drilling barges, with one difference. Once a jack-up rig is towed to the drilling site, three or four "legs" are lowered until they rest on the sea bottom. This allows the working platform to rest above the surface of the water, as opposed to a floating barge. However, jack-up rigs are suitable for shallower waters, as extending these legs down too deeply would be impractical. These rigs are typically safer to operate than drilling barges, as their working platform is elevated above the water level.

- *Moveable drilling rigs* are two basic types of offshore drilling rigs: those that can be moved from place to place, allowing for drilling in multiple locations, and those rigs that are permanently placed. Moveable rigs are often used for exploratory purposes because they are much cheaper to use than permanent platforms. Once large deposits of hydrocarbons have been found, a permanent platform is built to allow their extraction. The sections below describe a number of different types of moveable offshore platforms.

- *Submersible rigs* consist of platforms with two hulls positioned on top of one another. The upper hull contains the living quarters for the crew, as well as the actual drilling platform. The lower hull works much like the outer hull in a submarine, so that when the platform is being moved from one place to another, the lower hull is filled with air, thus making the entire rig buoyant. When the rig is positioned over the drill site, the air is let out of the lower hull, and the rig submerses to the sea floor. This type of rig has the advantage of mobility in the water, but its use is limited to shallow-water areas.

- *Semi-submersible rigs* are the most common type of offshore drilling rigs, combining the advantages of submersible rigs with the ability to drill in deep water. A semisubmersible rig works on the same principle as a submersible rig: through the "inflating" and "deflating" of its lower hull. The main difference with a semisubmersible rig, however, is that when the air is let out of the lower hull, the rig does not submerge to the sea floor. Instead, the rig is partially submerged, but still floats above the drill site. When drilling, the lower hull, filled with water, provides stability to the rig.

  Semisubmersible rigs are held in place by huge anchors, each weighing upwards of 10 tons. These anchors, combined with the submerged portion of the rig, ensure that the platform is stable and safe enough to be used in turbulent offshore waters. Semisubmersible rigs can be used to drill in much deeper water than their submersible cousins.

Gas Production Platforms

Now that we have the drill setup figured out and set, we need to start the production process. There are several types of production platforms on which the entire extraction and production cycles are conducted, using a number of methods.

Generally, offshore production platforms are very expensive, semi-permanent structures used when exploiting large, commercially viable natural gas or petroleum deposits, and on which the entire gas extraction and production processes are conducted.

Some of the largest offshore platforms are located in the North Sea, where, because of almost constant inclement weather, strong structures able to withstand high winds and large waves are absolutely necessary.

A typical permanent platform in the North Sea must be able to withstand wind speeds of over 90 knots (103 mph), and waves over 60 feet high. Correspondingly, these platforms are among the largest structures built by man.

A real technological marvel, these platforms, resembling a small city built on top of the waves in the middle of the world's roughest and coldest waters.

Types of Offshore Production Rigs

There are a number of different types of temporary or permanent offshore platforms, each useful for a particular depth range. The major types of gas and oil production rigs are as follow:

- *Compliant Towers* consist of a narrow tower attached to a foundation on the seafloor and extend-

ing up to the platform. The support tower is flexible, however, vs. the rigid legs of a fixed platform. The flexibility of the support system allows the rig to operate in deep water, since it is capable of "absorbing" surface waves and currents, and yet it is strong enough to withstand even hurricanes.

- *Fixed Platforms* are usually found in shallower water, where they are attached to the sea floor via "legs" constructed with concrete and steel. The legs extend down from the platform and are fixed to the seafloor with piles. Large platforms mounted on heavy-duty concrete legs structures are so heavy that they are not attached to the seafloor, and the entire rig relies on its weight for stability.

  There are many possible designs for these fixed, permanent platforms. The main advantage of these types of platforms is their stability. As they are attached to the sea floor, there is limited exposure to movement due to wind and water forces. The limitation here, of course, is depth, for it is simply not economical to build very long legs.

- *Floating Production Systems* are essentially semisubmersible drilling rigs, or ships, like those used for drilling, except that these contain both drilling and production equipment. Surface platforms are kept steady via large, heavy anchors attached to the ocean floor. They can also be fixed in place via the dynamic positioning system used by drillships.

  The wellhead of this setup is attached to the seafloor once the drilling is completed, rather than being attached up to the platform. The extracted petroleum is transported via risers from this wellhead to the production facilities on the semisubmersible platform. These production systems can operate in water depths of up to 6,000 feet.

- *Seastar Platforms* are like miniature tension leg platforms. The platform consists of a floating rig, much like the semisubmersible type discussed above. A lower hull is filled with water when drilling, which increases the stability of the platform against wind and water movement. In addition to this semisubmersible rig, however, Seastar platforms also incorporate the tension leg system employed in larger platforms.

- *Tension legs* are long, hollow tendons that extend from the seafloor to the floating platform. These legs are kept under constant tension, and do not allow for any up or down movement of the platform. They do allow for side-to-side motion, which lets the platform withstand the force of the ocean and wind without breaking the legs off.

Note: *Seastar platforms* are typically used for smaller deep-water reservoirs, when it is not economical to build a larger platform. They can operate in water depths of up to 3,500 feet.

Tension leg platforms are larger versions of the Seastar platform. The long, flexible legs are attached to the sea floor, and run up to the platform itself. As with the Seastar platform, these legs allow for significant side-to-side movement (up to 20 feet), with little vertical movement. Tension leg platforms can operate at around 7,000 feet.

- *Spar Platforms* are among the largest offshore platforms in use. These huge platforms consist of a large cylinder supporting a typical fixed rig platform. The cylinder does not extend all the way to the seafloor, but instead is tethered to the bottom by a series of cables and lines. The large cylinder serves to stabilize the platform in the water, and allows for movement to absorb the force of potential hurricanes. The first Spar platform in the Gulf of Mexico was installed in September of 1996. Its cylinder measured 770 feet long and was 70 feet in diameter, and the platform operated in 1,930 feet of water.

- *Subsea Production Systems* are wells located on the sea floor, as opposed to the surface. The gas or oil is extracted at the seafloor, and then "tied-back" to an already existing production platform. The well is drilled by a moveable rig, but instead of building a production platform for that well, the extracted natural gas and oil are transported by riser or undersea pipeline to a nearby production platform. This allows one strategically placed production platform to service many wells over a reasonably large area. Subsea systems are typically in use at depths of 7,000 feet or more.

-

Now that we have produced a lot of natural gas, we need to see what we can do with it.

Natural Gas Refining and Processing

Natural gas in its "as is" state—and regardless of what formation it comes in—is not very useful for daily applications. It usually contains mixtures of other gasses, moisture, and other contaminants that are impractical and even dangerous for normal use.

So in all cases, natural gas from the wells must be processed to remove impurities and water. Thus purified gas meets the customers' specifications for use as marketable natural gas that is safe for use in normal domestic and commercial applications.

The by-products of natural processing include chemicals such as ethane, propane, butanes, pentanes, and higher molecular weight hydrocarbons. Hydrogen sulfide is another major byproduct which can be converted into and sold as pure sulfur. There are also carbon dioxide, water vapor, and sometimes helium and nitrogen gasses in the "as is from the wellhead" natural gas mixture.

The Refining Process

From the production site, the raw natural gas is transported, usually via pipes, to a processing facility, where it is treated as needed for its final use. It could be used as fuel in power generating plants, or for fueling cars and trucks.

A number of different processes and techniques are used to create pipeline-quality or other types of natural gas. These depend on the source and makeup of the wellhead production stream and its intended final use. Several of the steps of the usual process may be integrated into one operation, or be performed in a different order, while some are not required at all.

The usual stages of processing/treatment of natural gas are:

- Gas-oil separation
- Condensate separation
- Dehydration
- Contaminants removal
- Nitrogen extraction
- Methane separation
- Fractionation

In more detail:

- *Gas-oil Separation* is needed in some cases, where a multi-stage gas-oil separation process is used to separate the gas stream from the crude oil. These gas-oil separators are usually closed cylindrical shells, horizontally mounted with inlets at one end, an outlet at the top for removal of gas, and an outlet at the bottom for removal of oil. The actual separation of the liquid and gas phases is accomplished by alternately heating and cooling (by compression) the flow stream through multiple steps. Some water and condensate, if present, will also be extracted as the process proceeds.

- *Condensate Separation* is used to remove condensates from the gas stream at the wellhead with mechanical separators. Most often the gas flow into the separator comes directly from the wellhead, since the gas-oil separation process is not needed.

  The gas stream enters the processing plant at high pressure (600 pounds per square inch gauge [psig] or greater) through an inlet slug catcher where free water is removed from the gas and then directed to a condensate separator. Extracted condensate is routed to on-site storage tanks.

- *Dehydration Process* is needed to eliminate water which may cause the formation of hydrates. Hydrates form when a gas or liquid containing free water experiences specific temperature/pressure conditions.

  Dehydration is the removal of this water from the produced natural gas and is accomplished by several methods. Ethylene glycol (glycol injection) systems are used as an absorption mechanism to remove water and other solids from the gas stream.

  Alternatively, adsorption dehydration may be used, utilizing dry-bed dehydrator towers, which contain desiccants such as silica gel and activated alumina, to perform the extraction.

- *Contaminants Removal* is needed to remove contaminates from the gas. This includes the elimination of hydrogen sulfide, carbon dioxide, water vapor, helium, and oxygen. The most commonly used technique is to first direct the flow though a tower containing an amine solution. Amines absorb sulfur compounds from natural gas and can be reused repeatedly.

  After desulphurization, the gas flow is directed to the next section which contains a series of filter tubes. As the velocity of the stream reduces in the unit, primary separation of remaining contaminants occurs due to gravity.

  As gas flows through the tubes, separation of smaller particles occurs and they combine into larger particles which flow to the lower section of the unit. Further, as the gas stream continues through the series of tubes, a centrifugal force is generated which further removes any remaining water and small solid particulate matter.

- *Nitrogen Extraction* is performed after the hydrogen sulfide and carbon dioxide are processed to acceptable levels, and the stream is routed to a nitrogen rejection unit (NRU), where it is further dehydrated using molecular sieve beds. In the NRU, the gas stream makes a series of passes through a column and a brazed aluminum plate fin heat exchanger. Using thermodynamics, the nitrogen is cryogenically separated and vented.

  Another type of NRU unit separates methane and heavier hydrocarbons from nitrogen using an absorbent solvent. The absorbed methane and heavier hydrocarbons are flashed off from the solvent by reducing the pressure on the processing stream in multiple gas decompression steps.

  The liquid from the flash regeneration step is returned to the top of the methane absorber as lean solvent. Helium, if any, can be extracted from the gas stream in a pressure swing adsorption (PSA) unit.

- *Methane Separation* is the process of demethanizing the gas stream and can occur as a separate operation in the gas plant or as part of the NRU operation. Cryogenic processing and absorption methods are some of the ways to separate methane from NGLs (natural gas liquids).

  a. The cryogenic method is better at extraction of the lighter liquids, such as ethane, than is the alternative absorption method. Essentially, cryogenic processing consists of lowering the temperature of the gas stream to around –120 degrees Fahrenheit.

     While there are several ways to perform this function, the turbo expander process is most effective, using external refrigerants to chill the gas stream. The quick drop in temperature that the expander is capable of producing condenses the hydrocarbons in the gas stream, but maintains methane in its gaseous form.

  b. The absorption method, on the other hand, uses a "lean" absorbing oil to separate the methane from the NGLs. While the gas stream is passed through an absorption tower, the absorption oil soaks up a large amount of the NGLs.

     Thus "enriched" absorption oil, now containing NGLs, exits the tower at the bottom. This oil is fed into distillers where the blend is heated to above the boiling point of the NGLs while the oil remains fluid. The oil is recycled while the NGLs are cooled and directed to a fractionator tower.

     Another absorption method that is often used is the refrigerated oil where the lean oil is chilled rather than heated, a feature that enhances recovery rates somewhat.

NOTE: *Adsorption* is the binding of molecules or particles to the surface of a material, while *absorption* is the filling of the pores in a solid. The binding to the surface is usually weak with adsorption and, therefore, easily reversible.

- *Fractionation* is the process of separating the various NGLs present in the remaining gas stream, using the varying boiling points of the individual hydrocarbons in the stream (by now virtually all NGLs). The process occurs in stages as the gas stream rises through several towers where heating units raise the temperature of the stream, causing the various liquids to separate and exit into specific holding tanks.

The Refining Process

There are a great many ways to configure the various steps in the processing and refining of raw natural gas. The usual process flow can be described as follows.

To start with, the raw natural gas is pumped up from a well or a group of adjacent wells. It is then processed on-site usually, for removal of free liquid water and natural gas condensate. The condensate is trucked to a petroleum refinery and the water is disposed of as wastewater.

Processing raw natural gas yields several byproducts, such as natural gas condensate, sulfur, ethane, and natural gas liquids (NGL), such as propane, butanes and C5+ (commonly used term for pentanes plus higher molecular weight hydrocarbons). All these fractions must be separated and otherwise treated in separate streams, until the final product is pure enough.

The somewhat cleaned raw gas is sent from the wellhead via pipeline to a gas processing plant where it is treated first to remove the acid creating gases; hydrogen sulfide and carbon dioxide, using Amine gas treating. Polymeric membranes can be used also to dehydrate and separate the carbon dioxide and hydrogen sulfide from the natural gas stream.

Thus removed acid gases are sent to a sulfur recovery unit, where they are converted into elemental sulfur, using the Claus process in most cases. The residual gas from the Claus process is processed in a tail gas treating unit (TGTU) to recover and recycle residual sulfur-containing compounds.

The final residual gas from the TGTU is incinerated, so that the carbon dioxide in the raw natural gas ends up in the incinerator flue gas stack and is disposed

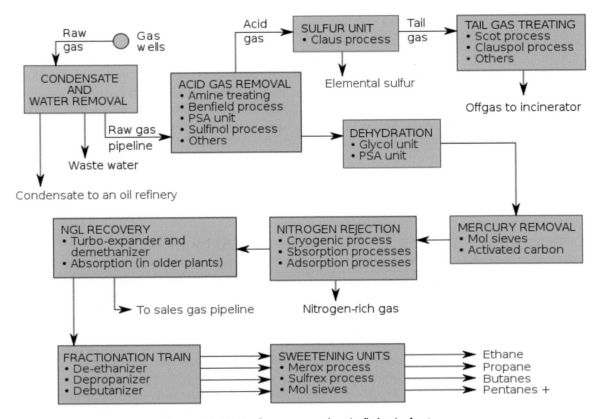

Figure 4-5. Natural gas processing (refining) plant

by venting in the atmosphere.

Removal of water vapor from the gas is an important step, which is done by using either regenerable absorption in liquid triethylene glycol (TEG), also called glycol dehydration, or by using a pressure swing adsorption (PSA) unit using a solid adsorbent.

Mercury is then removed by using adsorption processes using activated carbon or regenerable molecular sieves.

Nitrogen is removed next and rejected using one of these processes:

— Cryogenic process, using low temperature distillation.
— *Ab*sorption process, using lean oil or a special solvent as the absorbent, or
— *Ad*sorption process, using activated carbon or molecular sieves as the adsorbent.

The natural gas liquids (NGL) are recovered next, via cryogenic low-temperature distillation process involving expansion of the gas through a turbo-expander, followed by distillation in a de-methanizing fractionating column. Lean oil absorption process can be used here also, rather than the cryogenic turbo-expander process. The residue gas from the NGL recovery section is the final, purified "sales" gas which is pipelined to end-user markets.

The recovered NGL stream is processed through a fractionation train consisting of three distillation towers in series: a de-ethanizer, a de-propanizer and a debutanizer. The overhead product from the de-ethanizer is ethane and the bottoms are fed to the de-propanizer.

The overhead product from the depropanizer is propane and the bottoms are fed to the debutanizer. The overhead product from the debutanizer is a mixture of normal and iso-butane, and the bottoms product is a C5+ mixture.

The recovered streams of propane, butanes and C5+ are each "sweetened" in a Merox process unit to convert undesirable mercaptans into disulfides and, along with the recovered ethane, are the final NGL by-products from the gas processing plant.

NGLs

Natural gas liquids (NGLs) are liquid hydrocarbons, other than methane, that are separated from raw natural gas during the refining process at the refineries. These liquids include ethane, propane, butane and pentane in variable amounts.

Natural gas for an average well contains 11 percent ethane, 5 percent propane, 2 percent butane and about 2 percent of natural gasoline or drip gas (a low-octane fuel used mostly as a solvent.)

Other natural gas sources contain much higher percentage of methane and correspondingly smaller percentages of NGL—7 percent ethane, 4 percent propane, 1 percent butane, and other components including carbon dioxide and pentanes.

In most cases, ethane makes up about half of the NGL total, propane makes up about a quarter, and butane makes up 5-10 percent. Unfortunately, ethane cannot be used as is for vehicle fuel; only propane and butane can be used as such.

The problem here is that they are a small component (25-35%) of the total NGL volume, and since most of the as-is volume of NGL cannot easily be used as vehicle fuel, propane and butane will not become major vehicles fuels, due to their small fraction of the total natural gas volume. This makes their total availability limited by the total amount of natural gas extracted.

Within the last decade, the industry began to count the NGLs as part of our energy supply, and there is increasing talk about using some of these as vehicle fuel.

Some NGLs can also be used as feedstocks for chemical production, just as petroleum is. Nevertheless, although NGLs would not be able to displace crude oil in this market, as it is currently configured, they play an important role in it.

So, in the future NGLs will most likely be used as a fuel in gas-fired power plants and other heat-producing applications. Their use in the transportation sector as replacement for crude oil, however is most important to our economy, since it is one key element of our energy security and future energy independence that no other fuel can provide.

The more natural gas and NGLs we use for transportation fuels, the closer we get to our goal of reaching energy independence, so the efforts to include NGLs in the fuel mix will continue.

Natural Gas Distribution

Distribution of natural gas starts at the wellhead, where it has to be transported from the origination point to storage, refinery, or a customer. This involves a number of pipelines, combined with several physical transfers of custody, and multiple processing steps along the way.

Once the gas leaves the well, a pipeline gathering system directs the flow either to a natural gas processing plant or directly to the mainline transmission grid, depending upon the initial quality of the wellhead product.

The processing plant produces pipeline-quality natural gas, which meets the federal and states standards as needed for use in a variety of residential and commercial applications. This gas is then transported by pipeline to consumers or is put into underground storage for future use. Storage helps to maintain pipeline system operational integrity and to meet customer requirements during peak-usage periods.

Transporting natural gas from wellhead to market involves a number of pieces of equipment, a series of processes, and an array of physical facilities.

The list of these is long, but the key pieces are:

- *Gathering Lines* are small-diameter pipelines which move natural gas from the wellhead to the natural gas processing plant or to an interconnection with a larger mainline pipeline.

- *Processing Plant* is the operation that extracts natural gas liquids and impurities from the natural gas stream. Thus produced plant gas is then sent to the customers via mainline pipelines.

- *Mainline Transmission System* is a large-diameter, long-distance pipelines that transport natural gas from the producing area, or the refinery to the market areas.

- *Market Hubs/Centers* are locations where pipelines intersect and gas flows are combined or transferred.

- *Underground Storage Facilities* are special places around the U.S. where some large quantity natural gas is stored for later use. These could be depleted oil and gas reservoirs, aquifers, and salt caverns, where the excess gas is pumped in for future use.

- *Peak Shaving* is a system design methodology permitting a natural gas pipeline to meet short-term surges in customer demands with minimal infrastructure. Peaks can be handled by using gas from storage or by short-term gas line-packing.

NATURAL GAS USE

Natural gas is the fuel of the future and a leading fuel in ensuring our energy security. It powers large numbers of chemical and refining processes in the U.S., and is also the fuel of choice for our electric power generating network. It is a major feedstock for hydrogen production, hydrofracking, hydro-desulfurization, ammonia and other chemicals production.

Natural gas is also used for making syngas, methanol, and its derivatives—MTBE, formaldehyde, and acetic acid. The condensate derivatives, ethane and propane, are used as an advantageous raw material in a number of processes and products. Via ethylene and propylene, natural gas is also used in much of the organic chemicals production industries today.

Since at times there is much more natural gas produced than can be used, it is then stored underground inside depleted gas reservoirs from previous gas wells, salt domes, or in liquefied natural gas tanks.

In most cases, the gas is injected in a time of low demand and extracted when demand picks up. Storage near end users helps to meet volatile demands, but due to limited capacity and the volatility of the gas, storage may not always be practical.

Only 15 countries account for 85 percent of the worldwide extraction, so access to natural gas has become an important issue in international politics, and countries vie for control of pipelines. As an example, in the 2000s, the Russian Gazprom started disputes with Ukraine and Belarus over the price of natural gas and on several occasions shut off the supplies in the midst of freezing winter weather.

This situation has increased the concerns of the largest European economies which depend on Russian gas. Most European countries are looking for alternatives, which are not many.

Using larger quantities of U.S. coal and natural gas is one possibility, and is under discussion on different levels of the European governments.

Table 4-3 shows that residential home use of different types of natural gas, delivered by pipeline, (LNG and CNG), is almost half the gas used in the U.S. for power generation, and almost 20% of all gas delivered to customers in the country. This gas was used mostly for heating in winter and cooking by U.S. households.

Table 4-3. Gas use in the U.S. (2012)

GAS USE	$M ft^3$
Electric power	9,100,000
Industrial	7.225,000
Residential	4,150,000
Commercial	2,900,000
Vehicles	30,100
TOTAL	23,405,100

When the amount of natural gas used by commercial and industrial enterprises is added, we come up with 60% of the total natural gas delivered to customers being used by these businesses. The amount of natural gas used by U.S. homes, commercial, and industrial enterprises has kept steady during the last decade.

At the same time, natural gas used for vehicle fuel has increased by about 20 percent and is expected to continue increasing as these vehicles are becoming widely spread around the U.S. and the globe.

The largest increase of natural gas use is for electricity generation—a 30% jump, from 6,800,000 to 9,100,000 M ft^3 annually in 2012.

The total use of natural gas increased by another 10% from 2012 to 2013 to 26,034,000 M ft^3. This was due mostly to the increase of natural gas use for electric power generation. That trend is expected to continue for the foreseeable future, as natural gas is quickly replacing coal for electric power generation in the country.

Natural gas is mainly used for power generation, but it has many other uses as well. Some of the major uses of natural gas follow.

Electric Power Generation

Natural gas is a major fuel for electric power generation that is becoming more important by the day. The U.S. produces a significant amount of natural gas, which is increasing exponentially as we speak. Some of that amount is exported, while most of the rest is used for power generation.

There is a trend in the U.S. presently, where old coal-fired power plants are converted to gas-fired such. The trend is increasing as the benefits of cheap gas are added to the list.

EPA has also tightened the GHG emissions specs so that the operation of existing coal-powered plants is getting more expensive. At the same time, no new coal-power plant can be build, because they cannot meet the new GHG specs.

Because of that, and as a sign of the times, natural gas provides a major part of the electric power used in the country, and especially in California. While there are a lot of discussions, disagreements, and even fights over the renewable energies, natural gas is quietly taking the driver's seat of the California energy industry and other sectors.

Figure 4-6 points to the fact that California is leading the country in using natural gas for power generation. Over three times more natural gas-fired capacity than nuclear is used in the state—four times more than hydro, and four times more than all other energy sources put together.

The trend continues, and now California is on the path to double the use of natural gas by increasing its use in this and other sectors of its economy—transportation primarily.

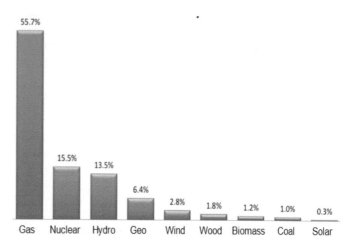

Figure 4-6 California electricity generation, 2010

Gas usage for electric power generation is also growing in many other states due to its abundance, low price, and (relative) environmental friendliness. Natural gas is literally taking over the electric power generation in the U.S. and a number of countries are following closely.

We reviewed drilling and extraction procedures and some of the uses of natural gas in the home and the industry, so now we will take a close look at its major uses.

LNG

Liquefied Natural Gas (LNG) is a form of natural gas in liquid state. It is obtained by compressing the natural gas to a very high pressure into specially designed tanks. It is an odorless, colorless, non-toxic and non-corrosive liquid. It is, however, dangerous as it can cause freeze burns and asphyxia, and is extremely flammable after vaporizing into its natural gaseous state.

LNG is made at a gas processing plant, where the gaseous natural gas is first purified by removing any condensates such as water, oil, mud, and other gasses such as CO_2 and H_2S. Special attention is paid to removing trace amounts of mercury from the gas stream to prevent mercury amalgamizing with aluminum in the cryogenic heat exchangers.

Thus purified gas is condensed into a liquid at close to atmospheric pressure at a transport pressure of around 4 psi by cooling it to approximately −260°F. The gas is then cooled in stages until it is completely lique-fied and stored in tanks, ready to be loaded and shipped.

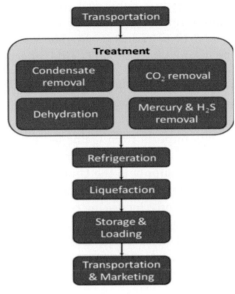

Figure 4-7. LNG production process

LNG has several times lower volume than compa-rable amounts of compressed natural gas (CNG). The volumetric energy density of LNG is 2.4 times greater than that of CNG, and is about half that of diesel fuel.

LNG is the only cost efficient method of trans-porting natural gas over long distances where pipelines do not exist, or over ocean transport routes. Specially designed cryogenic sea vessels, or LNG carriers, and cryogenic road tanker trucks and trains are used for its transport.

LNG is usually transported to energy markets, where it is re-gasified and distributed as pipeline nat-ural gas. It can also be used in natural gas vehicles, although CNG type vehicles are more common.

240 million tons of LNG were produced in 2013, which represents only 1/500th of the total volume of natural gas produced in its gaseous state.

The greatest increase of LNG production was ob-served between 2009 and 2011—about 60 million tons increase of LNG production—as compared with the previous periods. Production has been stable at 230-240 million tons annually since 2011. It is expected to in-crease rapidly during the next several years.

Production and transport costs of LNG are rela-tively high, which has hindered its widespread commer-cial use...until now. As global demand increases, LNG is becoming more and more the energy alternative of choice for many countries.

LNG Transport

Liquid natural gas (LNG) is the liquid version of natural gas that is becoming more popular worldwide by the day. LNG is stored and shipped as a cryogenic material in insulated tank trucks and rail pressure tank cars. Pressure tank trucks and rail tank cars for LNG transport have a stainless steel inner reservoir suspend-ed in an outer reservoir of carbon steel. The annular space is a vacuum filled with insulation to maintain low temperatures during shipment.

To prevent gas from igniting back to the tanks, they are equipped with two independent, remotely con-trolled fail-safe emergency shut-off valves on the filling and discharge lines and have gauges on both the inside and outside reservoirs.

LPG is transported on land in specially designed rail tank cars (up to 130 m^3 capacity) or tank trucks (up to 40 m^3 capacity). Tank trucks and rail tank cars for LPG transport are typically un-insulated steel cylinders with spherical bottoms, equipped with gauges, ther-mometers, two safety relief valves, a gas level meter and maximum fill indicator and baffles.

Rail tank cars transporting LNG or LPG should not be overloaded, since they may sit on a siding for some period of time and be exposed to high ambient tempera-tures which could cause overpressure and venting.

Bond wires and grounding cables are provided at rail and tank truck loading racks to help neutralize and dissipate static electricity. Truck and rail loading facilities are typically protected by fire water-spray or mist systems and fire extinguishers to be used in case of accidental fires.

LNG Ocean Carriers

LNG is the only accepted overseas transport meth-od of natural gas. A typical ocean LNG carrier has four to six tanks located along the center-line of the vessel. These are heavy-duty tanks, built to withstand any type of failure and accidents.

The LNG tanks are also equipped with contain-ment systems. Today there are four containment sys-tems in use for new vessels. Two of the designs are of the self-supporting type, while the other two are of the membrane type and today the patents are owned by Gaz Transport & Technigaz (GTT).

A typical containment system consists of a thick layer of foam insulation on the outside of the tank. It is fitted in panels or is wound around the tank height. Over the insulation is a layer of tinfoil which keeps the insulation by a slight pressure of nitrogen. The outside of the tank is checked every 3 months for any cold spots that would indicate breakdown in the insulation and potential leakage.

The LNG tank is supported around its circumference by an equatorial ring which is supported by a large circular skirt which distributes the weight of the tank onto the ship's structure. This skirt allows the tank to expand and contract during cool-down and warm-up operations, corresponding to charging and discharging the tanks.

During these cool-down and warm-up periods the LNG tanks expand or contract about 2 feet. Because of this, all piping is connected to the top of the tank, and is connected to the ship's lines via flexible bellows. Inside each tank there is also a set of spray heads, mounted around the equatorial ring. They are used to spray LNG onto the tank walls as needed to reduce their temperature.

A series of ballast tanks, cofferdams, and voids surrounds the LNG tanks, to provide double-hull type design.

Three submersible pumps are located inside each LNG tank, contained in the pump tower, which is attached to the top of the tank and runs its entire depth. The tank gauging system and the tank filling line are also mounted in the pump tower, near the bottom of the tank.

Two main cargo pumps are used in charge and discharge operations, while a smaller pump, referred to as the spray pump, is used for either pumping out liquid LNG to be used as fuel (via a vaporizer), or for cooling down cargo tanks.

In membrane-type vessels there is also an emergency pump tower, which contains an empty pipe with a spring-loaded foot valve that can be opened by weight or pressure. In case of pump failure, the pump can be removed and an emergency cargo pump can be lowered down to the bottom of the tank and pump the cargo out.

The cargo pumps from all LNG tanks discharge into a main pipe, which runs along the deck of the vessel. The main pipe branches off to either side of the vessel cargo manifolds, which are used for charging and discharging of the frozen liquid.

All tanks have vapor spaces, which are linked via a vapor header pipe running parallel to the cargo header. This pipe also has connections to the sides of the ship, in close proximity to the charge and discharge manifolds.

Some of the major international exports of liquefied natural gas (LNG) go in both directions through the Suez Canal. Southbound LNG transit mostly originates in Algeria and Egypt and is largely destined for the Asian markets and China. The northbound transit orig-

inates mostly from Qatar, and is largely destined for the European markets.

About 1.5 trillion cubic feet (tcf) of natural gas crossed Suez Canal locks in 2012. This was over 13 percent of total LNG traded worldwide, but was down from its peak of 2.06 tcf in 2011. This reflects the fall in northbound LNG flows and is consistent with LNG import data for the United States and Europe, which show that total LNG imports into both areas decreased, particularly from Qatar.

U.S. LNG imports from Qatar fell by around 63 percent in 2012, as compared with 2011. This reflects the growing domestic supply in the United States, in addition to some decrease in LNG demand in Europe, and even more importantly, increasing competition for LNG on the global energy market.

Northbound LNG traffic was also curtailed by decreased LNG exports from Yemen due to sabotage attacks on a major gas pipeline in the area.

Syngas

Synthesis gas (Syngas) is another of the natural gas imitators. It is a fuel gas mixture consisting primarily of hydrogen, carbon monoxide, and very often some carbon dioxide. The name comes from its use as an intermediate in creating synthetic natural gas (SNG) and for producing ammonia or methanol.

Syngas is not natural gas, and as the name suggests, it is a synthetic fuel. It can be produced via production methods such as steam reforming of natural gas or liquid hydrocarbons to produce hydrogen, and gasification of coal or biomass. It is also used in some waste-to-energy gasification facilities.

Each individual component in Syngas can be isolated and used for other purposes, as follow:

- hydrogen for use in electricity generation and transportation fuels;

- nitrogen for fertilizers, and pressurizing agents;

- ammonia for fertilizers;

- carbon monoxide for feedstock and fuels in the chemical industry;

- carbon dioxide to be stored by injection into sequestration wells, and

- steam used to drive turbines for electricity generation.

Today, syngas from large-scale coal gasification plants is primarily used for electricity generation, such as in integrated gasification combined cycle power plants, for production of chemical feedstocks, or for production of synthetic natural gas.

The hydrogen in syngas can also be used for various other purposes such as powering a hydrogen economy, or upgrading fossil fuels. Alternatively, some syngases can be converted into transportation fuels such as gasoline and diesel through additional treatment via the Fischer-Tropsch process or into methanol which itself can be used as transportation fuel or fuel additive, or which can be converted into gasoline by the methanol to gasoline process.

Coal gasification is one of the processes of producing syngas. Here, coal, water, air and/or oxygen are combined to produce syngas. Historically, coal was gasified to produce coal gas (also known as "town gas"), under oxygen depleted, high pressure, high-heat and/or steam conditions. The resulting syngas is a combustible gas which was traditionally used for municipal lighting and heating before the advent of industrial-scale production of natural gas.

The energy density of Syngas is only about 50 percent that of natural gas and is therefore mostly suited for use in producing transportation fuels, and manufacturing chemical products. It is also used as an intermediary building block for the final production (synthesis) of various fuels such as synthetic natural gas, methanol, and synthetic petroleum fuel (dimethyl ether, or synthesized gasoline and diesel fuel).

Heating Gas

The two most common forms of home and business heating systems in the U.S. are electric and natural gas furnaces. Depending on the geographic area and local climate each type of furnace offers advantages and disadvantages.

In the U.S. and much of the developed world, natural gas is supplied to homes and businesses via special gas pipes. It is used for many purposes including gas-powered ranges and ovens, gas-heated clothes dryers, and central heating. Home and commercial buildings heating may also include boilers, furnaces, and water heaters.

Compressed natural gas (CNG) is used in rural homes without connections to piped-in public utility services. It is also popular for use in portable grills. However, CNG is less economical, so instead, LPG (or propane) is the dominant source of rural gas today.

One of the advantages of natural gas furnaces is that they provide heat almost instantaneously, while the resistance coils in electric furnaces take time to build up heat. In most cases, heating with electricity is more expensive than heating with natural gas, but prices can depend on the market at the time and other external factors.

During the winter months gas heating is a necessary evil for many people in the Northeast who have experienced problems with high prices and irregular deliveries.

A big problem with gas furnaces is that each furnace must have its gas supply resupplied (at great expense) and turned on every winter, since it is idling the rest of the year. This is another hassle and waste of time and money, but a necessary evil without which many homes, and the people in them, would become frozen statues.

Natural gas burning on stovetops can generate up to 2000°F heat, which alone makes it a powerful domestic cooking and heating fuel.

So, natural gas is the most popular form of home heating in most of the U.S., with about 70 percent of the new single-family home market in the recent past. This is up from the 1980-1990s, when natural gas and electricity virtually split the heating share in the new home market.

Today's heating systems offer incredible choices to contractors—from top-of-the-line furnaces that achieve efficiency levels of more than 90 percent, to moderately priced units that meet or slightly exceed the minimum efficiency standard of 78 percent, so that customers don't have to pay for more efficiency than they need.

Many people also prefer natural gas, because it feels warmer than heat produced by an electric heat pump. Natural gas heat is delivered from forced-air systems at temperatures ranging from 120-140 degrees Fahrenheit. In contrast, the air from an electric heat pump is typically delivered at 85-95 degrees Fahrenheit—warm enough to heat a room, but cooler than the average human skin temperature.

Case Study: Winter of 2014

Exceptionally cold winter weather caused natural gas stockpiles in the U.S. to decline significantly in January 2014, as furnaces in the Northeast fought the blast of arctic weather.

The inventories dropped by 231 billion cubic feet to 191 trillion by the end of January. At the same time, gas futures topped $5 for the first time since 2010; a 22% increase—the largest ever in the Standard & Poor's GSCI commodity index. Spot gas prices in the Northeast surged to all-time highs as well. Some people complained about paying 4-5 times more for gas in 2014 than the previous years.

Due to the inclement weather patterns throughout the region, deliveries were delayed by days or weeks. So, high price or no high price, people were in desperate need to get the life-giving fuel.

Nevertheless, the shortage and price increase are somewhat unbelievable, considering the fact that the U.S. now has unlimited natural gas supplies at the best prices ever.

This also means that we still have serious natural gas delivery and price problems at times...today—amidst the natural gas explosion across the country.

If this is so today, imagine what the scene will be by the end of the century when natural gas supplies will be dwindling and prices rising to unaffordable levels.

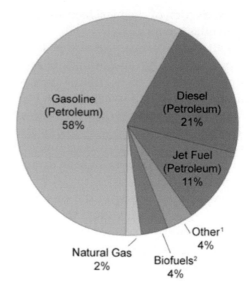

Figure 4-8. Natural gas use in the U.S., 2010

Gas for Transportation

Natural gas has been around for a long time, and has been considered as an alternative fuel for the transportation sector since the 1930s.

There are presently over 150,000 natural gas vehicles (NGVs) on U.S. roads, and more than 5 million NGVs worldwide. But this is a small number, compared to the large picture of over 250 million cars and light-duty vehicles, 11 million heavy-duty trucks, and almost 1.0 million busses—in the U.S. alone. These numbers are several-fold when the global number of vehicles is included.

Figure 4-8 shows that natural gas powers only 2% of the transportation in the U.S. The use of natural gas in the U.S. transportation sector has increased, and now over 3 percent of all natural gas used in the country (comprising 2% of all transportation fuels) powers the transportation sector. The rest 97% of the natural gas produced in the U.S. is used for electric power generation and making industrial products.

In recent years, the natural gas abundance, its low price, technology improvements, and environmental concerns have played an important role in allowing the proliferation of natural gas vehicles. For now, natural gas is mostly used for powering fuel-intensive vehicle fleets, such as buses and taxis.

There are also many types of natural gas powered vehicles in production today available to the public, and many new types of passenger cars, trucks, buses, vans, boats, off-road, and heavy-duty utility vehicles are in the development stages.

NGVs use compressed natural gas (CNG) to power their engines. The compressed gas, just like a gasoline tank, is attached to the vehicle's frame in a tube-shaped

storage tank, which can be filled in the same way as a gasoline tank.

At the same time, there are a number of limitations of this alternative fuel that prevent the growth of the NGVs market. Some of these are limited range, lack of adequate trunk space, and much higher initial cost.

Natural gas also contains much less energy per unit than most of the other fuels.

Figure 4-9 points to the fact that CNG has very low density, which means that the fuel tanks have to be much bigger and that the range the vehicles can travel would be limited. LNG has more potential to compete with the conventional fossil fuels, but it is more expensive and more delicate to store and use.

The natural gas fueled engines also require serious modification to allow them to run efficiently. Some natural gas vehicles are fueled by liquefied natural gas

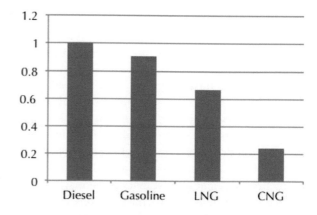

Figure 4-9. Energy density of different fuels

(LNG), while some are multi-fuel vehicles, which allows switching from gasoline or natural gas and vice versa. This allows for greater flexibility in fuel choice, but the conversions to either of these types is costly, and usually results in less efficient use of natural gas.

There is also the perceived fear of explosion of the high pressure CNG and LNG tanks during an accident or malfunction. There are no data to support such fears, yet, driving with a high-pressure tank under one's seat has a negative effect on some people, which influences their driving preferences.

More importantly, there is a very limited refueling infrastructure, which limits the use of personal vehicles to certain areas of the country and the world. These are serious considerations that need immediate solutions if natural gas is to penetrate the personal and light-duty vehicle market.

The future will bring some changes in this sector of the transportation industry. New, more efficient, and cheaper engines will bring costs down and improve performance of NGV vehicles. At the same time, expanding fueling infrastructure will make driving these vehicles less burdensome.

Emissions

Very importantly, natural gas is a much cleaner burning fuel, compared to the competitors—coal, gasoline, and diesel. We already see drastic transition from coal-fired to natural gas-fired electric power generation.

This pattern is extending to crude oil and its derivatives. As the environmental standards and government incentives for NGVs increase, supplying natural gas as a vehicular fuel will become increasingly common and more attractive as well.

Table 4-4 shows the amount of carbon in grams of CO_2 per unit energy (1 MJ) produced by each fuel during their entire cradle-to-grave life cycle. Obviously, CNG has the lowest carbon content of any of the other fuels, except the H_2 fuel cells.

Table 4-4. Carbon content in different fuels

FUEL	Carbon
Gasoline	95.85
Diesel	94.71
LNG	83.13
CNG	68.00
H_2 Fuel cells	42.00

LNG is also relatively low in carbon content, so these two fuels are preferred from an environmental point of view, which is one of the reasons their use will be encouraged in the future—regardless of the prices and other changes in the energy sector.

Table 4-5 shows that cars, vans, pick-ups and trucks emit the largest portion of the vehicular pollution. LNG and CNG could be used for powering all types of vehicles in Table 4-5 except aircraft, so with a little imagination (and some luck) we can see our transportation sector moving towards NGVs in the near future.

**Table 4-5.
GHG pollution ratios of different vehicles**

VEHICLE TYPES	GHG %
Light duty	61.0
Medium and heavy-duty	22.0
Commercial aircraft	7.0
All other	10.0

This would not only help to clean the environment, but would also close the energy security gap, which presently is forcing us to import over 50% of the crude oil we use—most of which is refined to fuel cars, trucks, boats and planes.

NOTE: Please keep in mind looking at all these figures that natural gas is composed mostly of methane, which after release into the atmosphere is removed over about 10 years by gradual oxidation to carbon dioxide and water by the reaction $CH_4 + 2O_2 \rightarrow CO_2 + 2H_2O$.

The lifetime of atmospheric methane in the atmosphere is relatively short, as compared to carbon dioxide, but it is much more efficient at trapping heat in the atmosphere while it's up there.

One unit of methane has 62 times the global-warming potential of the same unit of CO_2 when calculated over a 20-year period, 20 times over a 100-year period, and over 8 times during a 500-year period.

This makes natural gas a much more potent GHG than carbon dioxide due to its greater global-warming potential.

EPA estimates that global emissions of methane are 3 trillion cubic feet annually, or just above 3.0 percent of the entire global emissions. Direct emissions of methane represented 14.3 percent of all global anthropogenic greenhouse gas emissions by the end of the 20th century.

The extraction, storage, transportation and distribution processes have been known to release methane into the atmosphere, with the extraction process emitting the most methane. The high rates of methane leakage compromise the global warming advantage of natural gas over coal use. The only way to solve this is to reduce the methane leakage rates during all stages of the natural gas cradle-to-grave lifetime.

Heavy Duty Vehicles

Natural gas presently has wider use in large, high-horsepower (HHP) NGV vehicles, such as busses, boats, large trucks, and such. The numbers of these vehicles, however, are still only a small fraction of their full potential. In 2013, for example, natural gas accounted for less than 4% of the total fuel consumption in the HHP NGV transportation sector.

There are now about 85 LNG-powered HHP marine vessels operating around the world, compared to a total of 40,000 such vessels operating across the US alone. There are plans to invest in NGVs by the marine, mining and rail industries, but there are still a number of unknowns and barriers to the full implementation of these plans.

The immediate problem with this sector is the high initial capital expenditure, which is justifiable only if natural gas prices are low enough. There is also a reasonable concern with the development of new engine technologies, which have not reached their peak. Competing technologies and alternative fuels are adding to the puzzle, which is completed by the limited LNG supply and infrastructure in some regions.

Because of these concerns, natural gas use as a fuel for HHP NGVs is slow and sporadic. A good example here is the fact that the natural gas industry, which has the best familiarity with the natural gas fuels, is slow in adapting natural gas as a fuel to their vehicles and production equipment. As a matter of fact, in 2013 only 1% of U.S. drilling rigs and no well drilling and operating units were powered by natural gas.

Recently, a series of new programs in the HHP NGV sector have been announced, which indicates that the interest in substituting diesel with natural gas is rising. The HHP sector is split between pioneers already adopting LNG such as BC Ferries, Shell and Apache and those still in the testing phase such as Union Pacific, CSX and BNSF.

LNG adoption efforts are expected to continue in 2014-15, but HHP sector experts expect it to accelerate from 2016 on.

Natural gas fuel in the transportation sector in general is expected to grow about 3% annually during the next 20 years. Even then, transportation will remain the smallest sector by consumption at least until 2035. Energy consumption in the transportation sector is expected to remain flat over the next 30 years at 27.1 quadrillion Btu, ending 36 years of 1.1% year-on-year growth.

Marine, air, and rail transport sectors are expected to increase use of total energy consumption through 2040, with heavy-duty vehicles exceeding the total consumed by these sectors. Railroads are expected to consume 27% more energy by 2040, while marine traffic will increase its energy use 10 times during the same time.

LNG, however, is expected to provide about 35% of the power for the freight rail industry, but only 2% of the marine industry. Yet, the marine market potential is huge, with 5,000 towing vessels and 27,000 barges navigating the American waterways. Fueling even only 2% of such a huge fleet is still a lot of fuel saved and GHGs not emitted.

Keep in mind that just one bulk cargo barge burns up to two million gallons of diesel fuel annually—and there are hundreds of them, using 2% of the fuel supplied to the entire marine fleet. This is equivalent to hundreds of millions of gallons of diesel that can be replaced by natural gas.

One area that is left untouched, however, is fueling aircraft—commercial and military—that use millions of gallons of crude oil derivatives around the world every day. There are some successful attempts to replace aviation fuel with CNG, and some are plausible. While this may be the case with small airplanes, it is theoretically impossible to power a jumbo jet with natural gas, or any other gas, at least for now.

Although the alternative is quite attractive—20% lower price and 30% less emissions—the energy density of natural gas in the form of LNG, or CNG is 2 to 5 times less than that of aviation fuels. So lifting several hundred tons of metal off the ground with that power would take time to perfect, if ever.

For now, fixed-wing aircraft, helicopters, and some very heavy and specialized machinery—tanks, submarines, etc.—will depend on crude oil derivatives for fuel.

Mining and Well Use

On- and off-road transportation and HHP vehicles presents the largest hope for expansion of natural gas fueling the transportation sector in the U.S. during 2020-2040. This is followed closely by the potential of its use in mining, drilling and fracking industries, in which 99% of the natural gas fueling potential remains unrealized.

Large haul trucks in strip mining and well drilling operations burn a lot of diesel, which could be easily replaced by LNG. The 28,000 haul trucks, each 100 tons of live cargo, operated by the 10 largest companies in the world, for example, burn over 2 billion gallons of diesel every year. Replacing even 1-2% of that fleet's fuel with natural gas represents a lot of diesel.

The oil and gas drilling sector used about 700 million gallons of diesel in 2012 to pump liquids into the ground during hydrofracking operations at a cost of nearly $2.5 billion.

Using field gas (gas produced at the well) is a no brainer that could replace diesel fuel and reduce fuel costs by up to 70%. This billion-dollar opportunity, in addition to the environmental benefits of using cleaner natural gas instead of dirty diesel, is another no brainer.

Why that is not done on a large scale is a question that only the oil companies can answer. Our guess is that it is more expensive to retrofit the existing internal combustion engines to run on natural gas.

Also, the gas as it comes from the ground—and used without additional processing and refining—contains contaminants that can damage the engines. For now at least, field gas is not a widespread option and diesel fuel use is the only answer.

LNG Exports

During 2012, U.S. domestic Liquefied Natural Gas (LNG) production increased to nearly 2.5 million barrels per day (mmbd). LNG is now expected to become a critical component of the U.S. industrial complex, to take advantage of the newly found bonanza According to the insiders, this might play a significant role in our drive toward energy security and independence.

Large-quantity LNG production is the key for sustained production and healthy revenues for the gas producers. This is absolutely needed not only for ensuring their financial feasibility, but also for maintaining production levels during crisis and low natural gas prices.

While production is increasing steadily, it is often hindered by lack of distribution pipelines and processing facilities, which are delayed by regulatory and permitting problems.

Estimates are that LNG production will account for roughly 25 percent of all U.S. energy liquids supply. This, in turn, is expected to increase the LNG exports significantly, which is important for reducing overall price volatility and incentivizing further production.

There is confusion, coming from the fact that many experts consider LNG production as part of the overall energy (oil) production—energy is energy, right? Watch out; 2.5 mmbd equivalent of LNG today is counted as "oil" pro-

duction, but in most cases this fuel CANNOT be used as substitute for crude oil—especially in the most affected by oil imports transportation sector.

The increased potential for U.S. exports of LNG has its advantages, since we benefit from the exports in terms of financial gains and recognition as a serious energy player. The big problem is that:

- In the long run, we export our liquid gold while it is available, without considering the long-term energy security of the country, and

- In the short term, we export LNG to some countries that do not have a free-trade agreement (FTA) with the U.S.

The LNG export procedures call for the DOE to review each application, but it can deny an application only if the exports are not in the "public interest." This is a vague definition and requires added clarity. But the U.S. has a FTA only with South Korea, which means that the DOE must have some way to determine which country is suitable for LNG exports.

DOE relies on the Natural Gas Act to make these decisions, but the Act presumes that exports are in the national interest unless it is proven otherwise. This raises a question about our national security, which especially today with political turmoil around the world must include a better defined criteria as to whom and why we export our energy. These decisions also affect the cost of constructing export facilities, and changes in the LNG prices according to changes in the gas market.

In order to ramp-up LNG exports, DOE gave conditional approval in 2013 to the Freeport LNG export terminal to ship LNG to countries that do not have a free-trade agreement with the U.S. In addition to the Sabine Pass terminal, this is the second export point for such dubious exports.

Natural Gas Today

Natural gas is a versatile fuel—or better said, it can be used in high volume in a number of different and very important applications. It is becoming increasingly important, because it is well suited for many residential and commercial uses.

Its use for electric power generation presently is of utmost importance to our energy security. Power cogeneration, and combined use in association with renewable energy sources such as wind or solar have a bright future.

Lately gas is also finding increasing use for alimenting peak-load power stations functioning in tandem with hydroelectric plants. This wide use of a plentiful fuel leads the way to our energy security and is a major contributor to the dream of energy independence, no doubt!

Many grid-peaking power plants and some off-grid generators use natural gas for multiple reasons. Very high efficiencies can be achieved through combining gas turbines with a steam turbine in combined cycle mode. Natural gas burns more cleanly than other hydrocarbon fuels, such as oil and coal, and produces less carbon dioxide per unit of energy released.

For an equivalent amount of heat, burning natural gas produces about 30% less carbon dioxide than burning petroleum and about 45 percent less than burning coal.

Coal-fired electric power plants emit around 2,000 pounds of CO_2 for every MW/h generated, which is almost double the CO_2 released by a natural gas-fired electric plant for the same MW power generation.

Because of the higher carbon efficiency of natural gas generation, as the fuel mix in the US reduces coal and increases natural gas generation, CO_2 emissions have fallen well below the expected levels. Those measured in the first quarter of 2012 were the lowest on record for the first quarter of any year since 1992. Of course, there are other factors that affect emissions, so we will be watching the future developments carefully.

Combined cycle power generation using natural gas is currently the cleanest available source of power using hydrocarbon fuels. This technology finds wide acceptance as natural gas can be obtained easily and at increasingly reduced costs.

Locally produced electricity and heat using a natural gas powered combined heat and power plant (CHP or cogeneration plant) are considered energy efficient and as one efficient and cheap way to cut carbon emissions.

In May 2013 a number of energy-related organizations reported that gas prices will continue to ease, and some countries are expecting their gas import bills to fall by half, mostly thanks to lower shale gas pricing and extraordinary availability.

During the same time, however, the International Energy Agency issued a warning that any attempt to push gas prices up beyond $5/MMBtu could prompt the U.S. and other countries to step up use of coal, after years of cutting back on its consumption in favor of cleaner-burning gas.

So we now have a theoretical ceiling on gas prices.

It is $5/MMBtu. How long it will last, and what is next for natural gas remain to be seen.

The Future

Regardless of the unresolved issues, low prices and increasing availability lead us to the conclusion that natural gas will play an even more important role in many aspects of our immediate energy future. It is perceived as an environmentally friendly fuel, which, at the same time, could bring us closer to energy independence.

The use of natural gas as a motor fuel is the most interesting and important facet of our future energy security. It could also play a most critical role in achieving energy independence.

A. Eliminating crude oil imports is the only path to energy independence. Only natural gas could replace crude oil in the transportation sector, thus bringing us closer to the goal. Nothing else could do it…for now!

This attractive possibility is a major factor in many decisions about the future production and use of natural gas. Gas to oil transition, however, is not going to come quickly or easily, and a lot of work must be done to accomplish it, if we are ever to have energy independence.

Still, we must always remember that—regardless of its advantages and disadvantages—the more natural gas we use, the less there is of it left in the ground. As things are going, we are planning to increase the production and train the entire global economy to depend heavily on natural gas.

B. Increasing dependency on, and huge rates of use of, natural gas mean that we will run completely out of it by the end of this century.

Looking at the above statements, A and B, we see that we are getting into a Catch 22 situation. On one hand we need a lot of natural gas to replace crude oil, but on the other we don't want to use so much that we run out of it soon.

Today the trend around the globe—including the U.S. and the developed countries—is to use as much natural gas as possible, to achieve the goals of the present: energy security and independence. These goals must be achieved at all cost, the common wisdom goes.

Using huge quantities of natural gas for electric power and motor fuel could get us close to the goals, no doubt. But we would not be able to enjoy the achievement for very long, simply because there is not enough natural gas in the ground. So, eventually we would run out of it...

Then the urgent question that begs an answer is, "What would the new generations do without natural gas? How would they heat their homes, generate electricity, and/or drive their cars, if they have no natural gas in an economy run mostly by natural gas?" Good question, you say, but the answer is still in the wind...

THE GAS-FIRED POWER PLANT

Presently, natural gas is bringing a revival to the electric power generation sector. Many coal-fired power plants are replaced with, or retrofitted for, natural gas-fired operation. This trend is expected to increase exponentially in the future in the U.S. and the globe as well.

Natural gas-fired power plants are similar to their coal-fired cousins. The only difference is the fuel, which is in gas form and is fed via gas pipe. Here, natural gas is fed into the combustion turbine where it is burned (as in a jet engine) to turn the turbine blades. The turbine in turn rotates the rotor of a generator, and the produced electricity is sent into the grid.

In modern, more efficient designs, called co-generation power generators, the exhaust heat from the combustion turbine is sent into a boiler, where water is heated and turned into steam. The steam is sent into a second turbine, which turns and rotates the rotor of a second generator. The additional electricity is sent into the grid as well. See Figure 4-10.

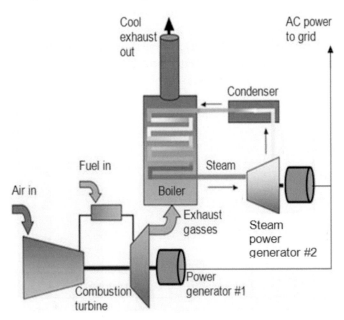

Figure 4-10. Cogeneration power plant

We've seen lately a large-scale conversion of coal-fired plants into gas-fired. The main reasons for this are that *natural gas emits less GHG, and it benefits from less strict regulations.*

A compromise, oil- and gas-fired mixed fuel power generation can be used in some areas where both fuels are readily available and their prices are compatible.

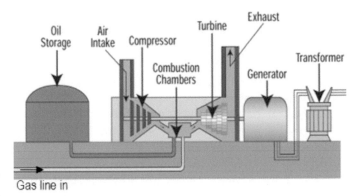

Figure 4-11. Mixed fuel power plant

In this setup, dual fuel (oil and gas) is fed into a turbine, where both fuels burn simultaneously to rotate the turbine and the generator's rotor, which generates electricity.

This setup is convenient since allows versatility, and also benefits from lower emissions, in addition to more lenient regulations.

Gas Power Plant Life Cycle

The gas-fired power plant life-span cycle consists of several phases, very similar to those of coal- and oil-fired power plants, as follow:

Plant site and structure planning and design
- Locating, testing and exploring the proposed power plant location
- Engineering design of plant facilities, equipment and process
- Estimating construction and operating cost and related tasks
- Applying for and obtaining the necessary federal, state and local permits

Plant construction and setup
- Plant building construction
- Plant equipment procurement and setup
- Labor training

Power plant daily operation
- Gas delivery

- Gas burning and energy generation
- Electric power distribution

Power plant decommissioning and land reclamation
- Plant shutdown and disassembly
- Plant waste disposal
- Surface land reclamation

Since the steps are very similar to those used for coal-fired plants construction and use, we refer the reader to the section on COAL for more details.

Natural Gas Cost

Natural gas prices have varied through the years, driven by the overall energy markets and local political and socio-economic conditions.

Large gas reserve finds and better extraction technologies have already begun pushing gas production costs much lower than they were in previous years, as shown in Figure 4-12.

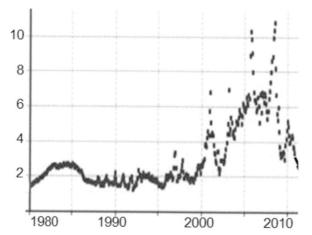

Figure 4-12. Natural gas cost of production (in \$/thousand ft³)

Nevertheless, there are many other factors influencing the production cost of natural gas. A major one is the fact that unlike crude oil, where the production spigot can be turned off when needed, natural gas wells have no spigot. Once turned on, the well cannot be turned off easily. That creates the necessity of identifying long-term markets and contracting purchasers long before a well is up to full capacity. Else, a lot of gas could be wasted, which won't sit well with the investors. This advanced thinking and planning makes natural gas prices somewhat stable, and at least somewhat more stable than oil.

Enron lobbied for the linking of gas prices to oil prices, even though gas competes with and is mostly compatible with coal. But this suited the Texan lobby, as the state produced large quantities of both oil and gas. Oil prices soared after the Middle East formed a cartel in the 1970s (the OPEC embargo). That too suited Texas' oil oligarchy.

Even today, there are charges of price rigging, which both the US and the EU are investigating. One glaring example is Qatar, which had been selling LNG to the East Coast of the US at \$15 per million metric British thermal unit (MMBtu), when in actuality the average global price at the time was \$3. It seems that someone was trying to push prices higher, as often happens with energy sources. But who is to say and judge; it is capitalism after all, which allows making a top buck anytime you can within the law.

Everything about natural gas changed with the advent of hydrofracked shale gas, and by 2010 the world had over 200 trillion cubic meters of natural gas available on top of the existing reserves.

But with more shale gas being found everyday, especially in China, which has more than the combined reserves of the US and Canada put together, shale gas could exceed other energy reserves. This then might bring prices even lower than expected. The biggest shale gas production centers today, in addition to those in the U.S., are in Australia, China and Russia.

Figure 4-13 shows that:

- The cost of power generation using natural gas equals that of coal, as far as the entire cradle-to-grave production and power generation cycles are concerned.

- We also see that the material cost of natural gas is higher than coal, but its cost of operation and maintenance (O&M) is much lower. The total of the sums of base materials and O&M of both fuels—coal and natural gas—is almost the same today.

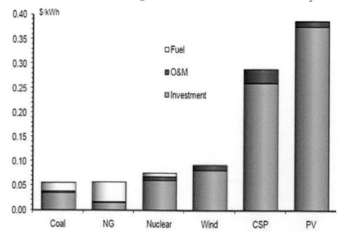

Figure 4-13. Energy generation cost per fuel (in \$/kW/h)

But while coal's cost is level and cannot be reduced any further, natural gas production costs are going down and will become even lower in the future, which will make it cheaper than coal.

There are other advantages to natural gas, such as its flexibility in use as a transportation fuel, which is very important, since it is the only replacement of crude oil. Gas is also used in large quantities as raw material and fuel for chemicals, metals, and other products manufacturing.

As the price of fuel gas decreases in the very near future, it might become the cheapest and one of the cleanest fuels in the world—until it is depleted or until a large-scale environmental disaster (due to hydrofracking anomalies, for example) changes the face of the entire energy industry.

End of Life Decommissioning

After 20, 30, 50 or more years of non-stop operation, a gas well is tired and has to be decommissioned. Decommissioning gas wells and rigs is a complex and expensive effort.

There are several steps in the oil and gas rigs decommissioning process:

- Project management, engineering and planning
- Permitting and regulatory compliance
- Platform preparation
- Well plugging and abandonment
- Conductor removal
- Mobilization and demobilization of derrick barges
- Platform removal
- Pipeline and power cable decommissioning
- Materials disposal
- Site clearance and final test

- *Project Management*, engineering and planning for decommissioning an offshore rig usually starts three years before the well runs dry. The process involves review of contractual obligations, engineering analysis, operational planning, and contracting. Due to the limited number of derrick barges, many operators contract these vessels two to three years in advance. In addition, much of the decommissioning process requires contractors who specialize in a specific part of the process. Most operators contract out the project management, cutting civil engineering and diving services.

- *Permitting and Regulatory Compliance* consists of obtaining permits to decommission an offshore rig

and can take up to three years to complete. Often, operators will contract a local consulting firm to ensure that all permits are in order prior to decommissioning. Local consulting firms are familiar with the regulatory framework of their region.

An execution plan is one of the first steps in the process. Included in this plan is environmental information and field surveys of the project site. The plan describes a schedule of decommissioning activities and the equipment and labor required to carry out the operation. An execution plan is required to secure permits from federal, state, and local regulatory agencies. BOEMRE will also analyze the environmental impacts of the project and recommend ways to eliminate or minimize those impacts.

Federal agencies often involved in decommissioning projects include BOEMRE, National Marine Fisheries Service, US Army Corps of Engineers, US Fish and Wildlife Service, National Oceanic and Atmospheric Administration, US Environmental Protection Agency, US Coast Guard, and the US Department of Transportation, Office of Pipeline Safety.

- *Platform Preparation* starts with the preparation of tanks, processing equipment, and piping that needs to be flushed and cleaned. Residual hydrocarbons must be disposed of; platform equipment must be removed (which includes cutting pipe and cables between deck modules, separating the modules, installing padeyes to lift the modules), and reinforcing the structure. Underwater, workers prepare the jacket facilities for removal, which includes removing marine growth.

- *Well Plugging and Abandonment* is one of the major costs of a decommissioning project and can be broken into two phases. The planning phase of well plugging includes:
  — Data collection
  — Preliminary inspection
  — Selection of abandonment methods
  — Submittal of an application for BOEMRE approval

In the GOM, the rig-less method, which was developed in the 1980s, is used primarily for plugging and abandonment jobs. The rigless method uses a load spreader on top of a conductor, providing a base to launch tools, equipment and plugs

downhole. Well abandonment involves:
— Well entry preparations
— Use of a slick line unit
— Filling the well with fluid
— Removal of downhole equipment
— Cleaning out the wellbore
— Plugging open-hole and perforated intervals(s) at the bottom of the well
— Plugging casing stubs
— Plugging of annular space
— Placement of a surface plug
— Placement of fluid between plugs

To remove conductor casing, operators can chose one of three procedures:
— Severing requires the use of explosive, mechanical or abrasive cutting
— Pulling, or sectioning uses the casing jacks to raise the conductors that are unscrewed or cut into 40-ft-long segments.
— Offloading utilizes a rental crane to lay down each conductor casing segment in a platform staging area, to offload sections to a boat and for offloading at a port. The conductors are then transported to an onshore disposal site.

• *Mobilization/Demobilization and Platform Removal* of derrick barges is a key component in platform removal. According to BOEMRE, platforms, templates, and pilings must be removed to at least 15 ft below the mud line.

First, the topsides are taken apart and lifted onto the derrick barge. Topsides can be removed all in one piece, in groups of modules, reverse order of installation, or in small pieces.

If removing topsides in one piece, the derrick barge must have sufficient lifting capacity. This option is best used for small platforms. Also the crane size and capacity at the offloading site are very important. If the offloading site can't accommodate the platform in one piece, then a different removal option is required.

Removing combined modules requires fewer lifts, thus is a time-saving option. However, the modules must be in the right position and have a combined weight under the crane and derrick barge capacity. Dismantling the topsides in reverse order in which they were installed, whether installed as modules or as individual structural components, is another removal option and the most common.

Topsides can also be cut into small pieces and removed with platform cranes, temporary deck-mounted cranes, or other small (less expensive) cranes. However, this method takes the most time, so any cost savings incurred using a smaller derrick barge will likely be offset by the dayrate.

Removing the jacket is the second step in the demolition process and the most costly. First, divers using explosives, mechanical means, torches or abrasive technology make the bottom cuts on the piles 15 ft below the mudline. Then the jacket is removed either in small pieces or as a single lift. A single lift is possible only for small structures in less than 200 ft of water. Heavy lifting equipment is required for the jacket removal as well, but a derrick barge is not necessary. Less expensive support equipment can do the job.

• *Pipeline and Power Cable Decommissioning* can be done in place if they do not interfere with navigation or commercial fishing operations or pose an environmental hazard. However, if BOEMRE rules that it is a hazard during the technical and environmental review, during the permitting process, it must be removed.

The first step to pipeline decommissioning in place requires flushing it with water, followed by disconnecting it from the platform and filling it with seawater. The open end is plugged and buried 3 ft below the seafloor and covered with concrete.

• *Materials Disposal and Site Clearance* ensures that all platform materials are removed to be refurbished, reused, scrapped and recycled or disposed of in specified landfills. To ensure proper site clearance, operators need to follow a site clearance procedure consisting of:
— Pre-decommissioning survey maps the location and quantity of debris, pipelines, power cables, and natural marine environments.
— Post-decommissioning survey identifies debris left behind during the removal process and notes any environmental damage.

If the rig is off-shore then:
— ROVs and divers target are deployed to further identify and remove any debris that could interfere with other uses of the area.
— Test trawling verifies that the area is free of any potential obstructions.

This is obviously a long, complex, and very expensive process, full of safety hazards. It is imperative that each step be executed properly, efficiently, and safely. One misstep, one hesitation, and the entire project is in jeopardy.

Here we need to mention also the complex and expensive process of decommissioning gas-fired power plants, which is quite similar to that used for decommissioning coal- and oil-fired power plants. The costs and environmental impact of these operations are serious, so they need to be kept in mind when analyzing the entire cradle-to-grave cycle of gas production and use.

THE ISSUES

Living close to natural gas wells has become a way of life for many U.S. ranchers and people in rural areas. Raising cattle and taking part in the natural gas production is how many of them make a living.

Natural gas operators pay the land owners generous monthly payments for using the land. This way, the farmers can increase the size of their farms, modernize their equipment and afford more luxuries.

In many economically depressed areas, natural gas development is breathing new life into family farms and the local economies, thus ensuring that the locals will continue to thrive for a long time.

The official version goes something like this, "Natural gas wells are drilled thousands of feet down into shale formations—well below local drinking water supplies. The wells are reinforced with multiple layers of steel and cement casing that seals off natural gas from aquifers. All of this work is done by skilled operators, and supervised by watchful state regulators, who are trained to protect local groundwater supplies and air quality. These efforts result in wells that safely produce natural gas without harming the local environment. Many farmers think that gas production and farming can coexist very well."

Sounds great, and is actually true in most cases, where the locals do not have to choose between economic prosperity and protection of the land and environment. They can have both: income from natural gas production and a healthy environment...in most cases.

Then there are a number of cases, where this ideal situation is far from the true reality. These are the exceptions to the rule, but stark reality for those affected. As their numbers and the related problems grow, we must pay close attention to what is going on.

We hear in the news and the media that there are serious problems related to natural gas drilling, extraction and use, so we need to know all about it in order to take the necessary precautionary and regulatory measures.

Ignorance or denial will not work! Educated reasoning and decisive action are needed in order to handle the situation properly now and prevent a major disaster(s) from happening in the future.

We often hear talking heads, company executives, and government officials speak enthusiastically about natural gas—its benefits and its bright future. We hear how great things are and how much better they will get as we increase the production and use of natural and shale gas.

Very seldom, however, do we hear people discussing the problems experienced by the locals. Even less do we hear about any efforts to improve the situation for them. It is the best kept secret of the industry.

More importantly, even less often do we hear discussions on what will happen when the gas and oil wells run dry? Where would we get the energy to power our lives? What could we replace gas and oil with?

Is that silence due to the fact that we are so blind that we do not see the obvious? Or is it that we are so self-centered that everything and everybody else (the future generations in particular) just don't matter?

This calls for an open and honest national discussion on the issues, but we just don't see it happening anytime soon.

Let's take a close look now at the evils of the natural and shale gas production—and the new hydrofracking process in particular.

Above we discussed the actual production process and mentioned that the different companies use different proppants. These are chemicals in liquid form that are injected into the wells under high pressure to break the shale and to push the gas to the surface.

The problem with this process is that regardless how careful the work is done, there is always an occasional mishap. And then, these toxic chemicals could travel freely underground, contaminating the water table, and surfacing in people's farms and backyards.

And mind you, these proppants are a vicious mixture of many toxic and carcinogenic chemicals. So, it is quite easy to call the hydrofracking process safe from the third floor of your office, 1,000 miles away from the leaking proppants. It is different, however, if you have to live with them. Imagine green and brown liquids bubbling in your pasture, poisoning the cattle, or spurting up in your backyard, making you and the entire family sick day after day. And even worse, proppants have been found in the communities' drinking water supplies.

Not a big deal, you say from your third floor office, because you don't have to watch your property values drop, nor drink the poisoned water, nor watch your children suffer from ingesting and breathing the poisons.

So, let's take a close look at these proppants and what they are doing to those less fortunate people living close to less-than-perfect natural gas production operations. Since perfection cannot be expected in every case, we must agree that the problems exist and we need only to understand and analyze them and their effects.

Fracking Chemistry

Proppants are chemicals, water, oil, and a large number of other organic and inorganic compounds that are used in large quantities during hydrofracking operations. These chemicals are pumped at a high volume and under high pressure into the oil and natural gas wells to increase their production.

The proppants work miracles in releasing huge quantities of gas and oil trapped in rocks and other underground formations. This was simply impossible just several years back, and is the main reason for the present-day bonanza in the U.S. and many other countries.

Intentionally or unintentionally, however, prop-

pants have leaked up into pastures, playgrounds, and people's back yards. Water tables have been contaminated by escaping from their containment proppants, and have hurt vegetation, animals, and humans as well.

Even the best of intentions and the best methods of operation cannot prevent occasional spills and leaks. It is the nature of the beast. So, properties, the environment, wildlife, and even humans have been hurt by this process—all in the name of natural gas—pumping it as much and as fast as possible out of the ground.

Human life and wellbeing should not be sacrificed in the name of energy security. We should not pay for our energy security and independence with human suffering. And yet, this is exactly what is increasingly happening today...

Over 700 chemicals have been identified and listed as proppant additives for hydraulic fracturing in a report to the US Congress in 2011. But the manufacturers are not required to officially disclose the contents of their proppant liquids. So, they basically use whatever amount and combination of chemicals they want, regardless of what the final effect of these might be on the local environment and human life.

Table 4-6 is a partial list of some of the chemical constituents used in fracturing operations, which have been identified by the New York State Department of Environmental Conservation.

Table 4-6. List of chemicals and substances in proppants

1,2-Benzisothiazolin-2-one / 1,2-benzisothiazolin-3-one
1,2,4-trimethylbenzene
1,4-Dioxane
1-eicosene
1-hexadecene
1-octadecene
1-tetradecene
2,2 Dibromo-3-nitrilopropionamide, a biocide
2,2-azobis-{2-(imidazlin-2-yl)propane}-dihydrochloride
2,2-Dibromomalonamide
2-Acrylamido-2-methylpropane sulphonic acid sodium salt polymer
2-acryloyloxyethyl(benzyl)dimethylammonium chloride
2-Bromo-2-nitro-1,3-propanediol
2-Butoxy ethanol
2-Dibromo-3-Nitriloprionamide (2-Monobromo-3-nitriilopropionamide)
2-Ethyl Hexanol
2-Propanol / Isopropyl Alcohol / Isopropanol / Propan-2-ol
2-Propen-1-aminium, N,N-dimethyl-N-2-propenyl-chloride, homopolymer
2-propenoic acid, homopolymer, ammonium salt

2-Propenoic acid, polymer with 2 p-propenamide, sodium salt
2-Propenoic acid, polymer with sodium phosphinate (1:1)
2-propenoic acid, telomer with sodium hydrogen sulfite
2-Propyn-1-ol / Propargyl alcohol
3,5,7-Triaza-1-azoniatricyclo[3.3.1.13,7]decane, 1-(3-chloro-2-propenyl)-chloride,
3-methyl-1-butyn-3-ol
4-Nonylphenol Polyethylene Glycol Ether Branched / Nonylphenol ethoxylated / Oxyalkylated Phenol
Acetic Anhydride
Acetone
Acrylamide—sodium 2-acrylamido-2-methylpropane sulfonate copolymer
Acrylamide—Sodium Acrylate Copolymer or Anionic Polyacrylamide
Acrylamide polymer with N,N,N-trimethyl-2[1-oxo-2-propenyl]oxy Ethanaminium chloride
Acrylamide-sodium acrylate copolymer
Aliphatic Hydrocarbon / Hydrotreated light distillate / Petroleum Distillates / Isoparaffinic solvent / Paraffin Solvent / Napthenic Solvent
Aliphatic acids

Table 4-6 (*Continued*). List of chemicals and substances in proppants

Aliphatic alcohol glycol ether
Alkyl Aryl Polyethoxy Ethanol
Alkylaryl Sulfonate
Alkenes
Alkyl (C14-C16) olefin sulfonate, sodium salt
Alkylphenol ethoxylate surfactants
Amines, C12-14-tert-alkyl, ethoxylated
Amines, Ditallow alkyl, ethoxylated
Amines, tallow alkyl, ethoxylated, acetates
Ammonia
Ammonium acetate
Ammonium Alcohol Ether Sulfate
Ammonium bisulfate
Ammonium bisulfite
Ammonium chloride
Ammonium Cumene Sulfonate
Ammonium hydrogen-difluoride
Ammonium nitrate
Ammonium Persulfate/Diammonium peroxidisulphate
Ammonium Thiocyanate
Aromatic hydrocarbons
Aromatic ketones
Aqueous ammonia
Bentonite, benzyl(hydrogenated tallow alkyl) dimethylammonium stearate complex/organophilic clay
Benzene
Benzene, 1,1-oxybis, tetratpropylene derivatives, sulfonated, sodium salts
Benzenemethanaminium, N,N-dimethyl-N-[2-[(1-oxo-2-propenyl)oxy]ethyl]-, chloride, polymer with 2-propenamide
Boric acid
Boric oxide/Boric Anhydride
Butan-1-ol
Ethoxylated Alcohol
Alcohol, Ethoxylated
Carboxymethylhydroxypropyl guar
Cellulase/Hemicellulase Enzyme
Chlorine dioxide
Citrus Terpenes
Cocamidopropyl betaine
Cocamidopropylamine Oxide
Coco-betaine
Copper(II) sulfate
Crissanol A-55
Crystalline Silica (Quartz)
Cupric chloride dihydrate
Decyldimethyl Amine
Decyl-dimethyl Amine Oxide
Dibromoacetonitrile
Diethylbenzene
Diethylene glycol
Diethylenetriamine penta (methylenephonic acid) sodium salt
Diisopropyl naphthalenesulfonic acid

Dimethylcocoamine, bis(chloroethyl) ether, diquaternary ammonium salt
Dimethyldiallylammonium chloride
Dipropylene glycol
Disodium Ethylene Diamine Tetra Acetate
D-Limonene
Dodecylbenzene
Dodecylbenzene sulfonic acid
Dodecylbenzenesulfonate isopropanolamine
D-Sorbitol/Sorbitol
Endo-1,4-beta-mannanase, or Hemicellulase
Erucic Amidopropyl Dimethyl Betaine
Erythorbic acid, anhydrous
Ethanaminium, N,N,N-trimethyl-2-[(1-oxo-2-propenyl)oxy]-, chloride, homopolymer
Ethane-1,2-diol/Ethylene Glycol
Ethoxylated 4-tert-octylphenol
Ethoxylated alcohol
Ethoxylated branch alcohol
Ethoxylated C11 alcohol
Ethoxylated Castor Oil
Ethoxylated fatty acid, coco
Ethoxylated hexanol
Ethoxylated octylphenol
Ethoxylated Sorbitan Monostearate
Ethoxylated Sorbitan Trioleate
Ethyl alcohol/ethanol
Ethyl Benzene
Ethyl lactate
Ethylene Glycol-Propylene Glycol Copolymer (Oxirane, methyl-, polymer with oxirane)
Ethylene oxide
Ethyloctynol
Fatty alcohol polyglycol ether surfactant
Ferric chloride
Ferrous sulfate, heptahydrate
Formaldehyde
Formaldehyde polymer with 4,1,1-dimethylethyl phenolmethyl oxirane
Formamide
Formic acid
Fumaric acid
Glassy calcium magnesium phosphate
Glutaraldehyde
Glycerol/glycerine
Guar Gum
Heavy aromatic petroleum naphtha
Hemicellulase
Hydrochloric Acid/muriatic acid
Hydrogen peroxide
Hydroxy acetic acid
Hydroxyacetic acid ammonium salt
Hydroxyethyl cellulose

Table 4-6 (*Continued*). List of chemicals and substances in proppants

Hydroxylamine hydrochloride
Hydroxypropyl guar
Isomeric Aromatic Ammonium Salt
Isoparaffinic Petroleum Hydrocarbons, Synthetic
Isopropanol
Isopropylbenzene (cumene)
Isoquinoline, reaction products with benzyl chloride and quin-
    oline
Kerosene
Lactose
Light aromatic solvent naphtha
Light Paraffin Oil
Magnesium Silicate Hydrate (Talc)
methanamine, N,N-dimethyl-, N-oxide
Methanol
Methyloxirane polymer with oxirane, mono (nonylphenol)
    ether, branched
Mineral spirits/Stoddard Solvent
Monoethanolamine
N,N,N-trimethyl-2[1-oxo-2-propenyl]oxy Ethanaminium chlo-
    ride
Naphtha (petroleum), hydrotreated heavy
Naphthalene
Naphthalene bis(1-methylethyl)
Naphthalene, 2-ethoxy-
N-benzyl-alkyl-pyridinium chloride
N-Cocoamidopropyl-N,N-dimethyl-N-2-hydroxypropylsulfo-
    betaine
Nitrogen, Liquid form
Nonylphenol Polyethoxylate
Organophilic Clays
Oxyalkylated alkylphenol
Petroleum distillate blend
Petroleum Base Oil
Petroleum naphtha
Phosphonic acid, [[(phosphonomethyl)imino]bis[2,1-ethanedi-
    ylnitrilobis(methylene)]]tetrakis-, ammonium salt
Pine Oil
Polyethoxylated alkanol
Polymeric Hydrocarbons
Poly(oxy-1,2-ethanediyl), a-[3,5-dimethyl-1-(2-methylpropyl)
    hexyl]-w-hydroxy-
Poly(oxy-1,2-ethanediyl), a-hydro-w-hydroxy/Polyethylene
    Glycol
Poly(oxy-1,2-ethanediyl), α-tridecyl-ω-hydroxy-
Polyepichlorohydrin, trimethylamine quaternized
Polyethlene glycol oleate ester
Polymer with 2-propenoic acid and sodium 2-propenoate
Polyoxyethylene Sorbitan Monooleate
Polyoxylated fatty amine salt
Potassium acetate
Potassium borate

Potassium carbonate
Potassium chloride
Potassium formate
Potassium Hydroxide
Potassium metaborate
Potassium sorbate
Precipitated silica/silica gel
Propane-1,2-diol, or Propylene glycol
Propylene glycol monomethyl ether
Quaternary Ammonium Compounds
Quinoline,2-methyl-, hydrochloride
Quinolinium, 1-(phenylmethl),chloride
Salt of amine-carbonyl condensate
Salt of fatty acid/polyamine reaction product
Silica, Dissolved
Sodium 1-octanesulfonate
Sodium acetate
Sodium Alpha-olefin Sulfonate
Sodium benzoate
Sodium bicarbonate
Sodium bisulfate
Sodium bromide
Sodium carbonate
Sodium Chloride
Sodium chlorite
Sodium chloroacetate
Sodium citrate
Sodium erythorbate/isoascorbic acid, sodium salt
Sodium Glycolate
Sodium Hydroxide
Sodium hypochlorite
Sodium Metaborate .8H$_2$O
Sodium perborate tetrahydrate
Sodium persulfate
Sodium polyacrylate
Sodium sulfate
Sodium tetraborate decahydrate
Sodium thiosulfate
Sorbitan Monooleate
Sucrose
Sulfamic acid
Surfactants
Tall Oil Fatty Acid Diethanolamine
Tallow fatty acids sodium salt
Tar bases, quinoline derivs., benzyl chloride-quaternized
Terpene and terpenoids
Terpene hydrocarbon byproducts
Tetrahydro-3,5-dimethyl-2H-1,3,5-thiadiazine-2-thione (a.k.a.
    Dazomet)
Tetrakis(hydroxymethyl)phosphonium sulfate (THPS)
Tetramethyl ammonium chloride
Tetrasodium Ethylenediaminetetraacetate

Table 4-6 (*Concluded*).
List of chemicals and substances in proppants

Thioglycolic acid
Thiourea
Thiourea, polymer with formaldehyde and 1-phenylethanone
Toluene
Tributyl tetradecyl phosphonium chloride
Triethanolamine hydroxyacetate
Triethylene glycol
Trimethylolpropane, Ethoxylated, Propoxylated
Trisodium Ethylenediaminetetraacetate
Trisodium Nitrilotriacetate
Trisodium orthophosphate
Urea
Vinylidene Chloride/Methylacrylate Copolymer
Xylene
Misc. fillers and additives

Wow! The list in Table 4-6 must be the most extensive and elaborate mixture of vicious chemicals for use in one single operation ever concocted. Taking a close look at the list, we see so many problems with using these chemicals that it is hard to even figure where to start.

Here are the problems with any mixture containing any of these chemicals, in no particular order of importance or gravity:

1. Most of the proppant chemicals, in varying concentrations and exposure time, can, and usually do, damage the environment, and hurt wildlife and humans, when given a chance.

2. Many of these chemicals are toxic, and some are carcinogenic.

3. Often, there is not enough information as to the degree of harm these could cause and under what circumstances.

4. In most cases, the information available in the MSDS sheets has not undergone human toxicity and carcinogenic tests, so the exposure levels are uncertain.

5. Of the information that can be extracted from the MSDS sheets, most of it cannot be applied directly in any particular case, because the manufacturers do not specify the concentrations of the chemicals in the mixtures.

6. Even if we knew the toxicity levels and the concentration of each chemical:
   a. We do not know what the interaction among the chemicals in the vicious soup would be, since there is no previous experience, and/or any research done on the matter, and
   b. In the best of cases we cannot possibly know the reactions occurring underground between the different chemicals and the varying rock formations.

Because of these, and many other reasons we simply are not able to predict the effect the different proppant formulations might have on the environment or humans.

NOTE: We would like to officially challenge anyone who can explain and demonstrate from a technical—chemical, geological, and biological—point of view, what exactly interactions and results can be expected from a mixture of any of these chemicals used as proppants.

What exactly do they do to the underground layers? How can the chemical reactions and mechanical (pressure) action be controlled in all directions and at all times? How can we prevent chemical over-reaction, over-pressurizing in certain areas, and the related unwanted cracks creation, and excess penetration of the soup into the surface layers as a result of the chemicals' reactivity under high pressure and increased temperature?

What reactions are expected in/with the upper soil levels? What would these chemicals do to the local vegetation and agricultural crops? What effects are expected from the chemical mixture, or parts of it, entering the local water supply? What would that do to wildlife and humans who depend on it?

And finally, the most important question, "How do we prevent all possible negative effects?" Anyone?

The Failures

So, what we have here (the proppants) is magic, the effects of which nobody can even predict. It is magic, because it performs miracles underground. It is also a miracle that we don't have more incidents and accidents of proppant damage to property and human health.

The chemicals used in hydrofracking create serious problems by leaking into places where they do not belong. Since some of the chemicals' properties are not well known, and the different manufacturers use different chemicals in different concentrations, we cannot imagine how we will ever be able to put this chemical debacle under complete control.

Since the number of negatively affected people is not that great (yet), but steadily increasing, this problem is becoming a scenario of many possibilities and of very long duration.

While this might sound abstract, it is a scary proposition if it affects you personally. Imagine waking up one morning to find a green-brown pool of unknown chemicals bubbling in your back yard. What would you do?

Worse, imagine turning on the faucet to get a glass of drinking water, but seeing instead a yellow blub floating in it. And then you smell natural gas all over the house. What is the next step you'd take? Call the owner of the nearby drill? Or the state authorities? EPA?

Many people have gone through these experiences and none of them is pleasant. Many people have lost their property values and even their health in similar situations. We can only hope that the number doesn't grow much faster...although we are sure that it will grow some.

things. Contrary to what we are told, no production material, or process setup is perfect.

So, there is always a chance of:

1. Mechanical failure and breakage of the piping system, the well casings, pipe connections, etc. critical components, and

2. Cracks or pore imperfections in the ground could expand, extend, and propagate uncontrollably in whichever direction under the high pressure and chemical action of the proppants.

Any of these failures and malfunctions could cause unpredictable, and sometimes hazardous, below- or above-ground incidents and accidents.

Figure 4-15 shows what we could see in the end after many days of pushing proppants into the rock formation. Proppant fluids and/or gas under high pressure could be forced to flow all the way into the water table and to the surface through the existing or newly formed cracks.

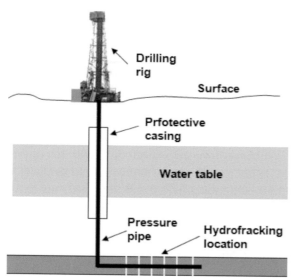

Figure 4-14. Normal hydrofracking setup

In most cases of normal daily operation, there are no mechanical problems, nor is there any presence of ground or surface abnormalities. All of the drilling rig components do their job as planned and no problems are encountered. The proppants are pumped into the pressure pipe and sent deep down under to initiate the hydrofracking process as they are designed to do deep and safe below the surface.

After many days, weeks and even months of pumping huge amounts of pressurized proppants into the formation and shuttering it piece by piece, enough gas is released, at which point the proppant flow is stopped and the gas production starts.

"So what can go wrong?" you ask. A number of

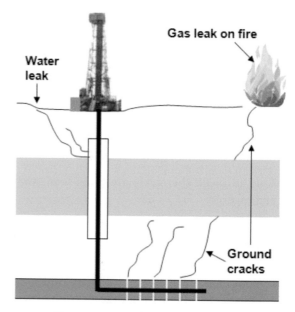

Figure 4-15. Hydrofracking failures

Thousands of gallons of proppants are pushed down the pipe and into the rock formation day after day. The pressure and the chemical action of the proppants is tremendous and there is no way to control what it does deep underground.

With time, the proppants could penetrate any existing pore and crack in their way, or create new ones and travel into the ground water and onto the surface areas around the drilling rig.

If the well casing fails too, then the water table will be surely contaminated and there is no man-made process to stop a disaster from happening in such case.

These are not planned events, mind you, and don't happen all the time, but when they happen, the consequences vary from a mild nuisance of dealing with small spills and gas leaks, to major damage of property and health problems related to the chemicals in the proppants.

These events might be quite small in the overall scheme of things, but they are huge for the locals who are affected by them. There are now a number of cases where environment quality has been destroyed, properties have been made uninhabitable, and wildlife and people have been made sick and killed.

What we see often, as result of the damages caused by hydrofracking, is one or a combination of the following:

- Ground water and aquifer contamination
- Air quality degradation
- Migration of gases and hydraulic fracturing chemicals to the surface
- Risk of gases in potable water wells, creating "flammable water"
- Tremors and earthquakes within surrounding geographic areas
- Potential waste water spillage risks
- Loss of land and agricultural production
- Property damage
- Loss of land value
- Human and animal health risks
- Birth defects
- Miscarriages
- Cancers
- Attention deficit disorder
- Hyperactivity
- Learning deficiencies
- Neurological damages

The chemical mixture in the proppants is blamed for a number of these anomalies, so controlling the mixture's contents and use would require a number of things to happen. Such an effort, if ever undertaken, will take many, many years, and billions of dollars, so we don't see it happening anytime soon.

One of the principals of our energy security efforts is to ensure the safe production and use of energy for all people involved. Hydrofracking violates this principle for some of us because the proppants used in the process have not been proven safe for use in the way they are presently used.

For now we have to assume that we are on our own and depending on the good will of the manufacturers and well operators. We also trust that the U.S. regulators will become more active in this area and come up with an acceptable standardized way to deal with the issues at hand.

Our hope and prayers are that this will happen before the vicious underground chemical attack has a chance spread beyond the point of return for some people.

Land Damage

Hydraulic fracturing is a controversial process to say the least. On the positive side, oil and gas companies are now able to reach previously inaccessible reserves, which bring financial prosperity to many states and increase domestic energy production.

The downside is the creation of huge waste chemicals ponds which damage local lands and water tables. Millions of gallons of salty, chemical-infused wastewater, or brine, are used during drilling and fracking at each well site. Drillers inject the brine deep underground, but some of it doesn't make it that far.

In N. Dakota alone there are more than 1,000 accidental releases of oil, drilling wastewater, and other fluids annually. Many more illicit releases go unreported, according to the state regulators. These are instances when companies dump truckloads of toxic fluid along the road or drain waste pits illegally for different reasons.

Brine laced with carcinogenic chemicals and heavy metals, dumped in rivers and streams has wiped out aquatic life and damaged wetlands and sterilized farmland. These effects are not temporary; they can last decades.

The enforcement of drilling and fracking operations in the U.S. is inadequate. In some cases there are no regulations to enforce.

Most companies clean up spills voluntarily, but not all and not in all cases. As the activities increase, so does the number of these "exceptions." Over 1,000 wells are drilled every month across the country, where millions of gallons of brine are pumped into them, and that much is stored in waste brine pits. Who can figure out and keep track of the number and size of these "exceptions."

Oil companies report accidental spills to the Department of Mineral Resources, which then works with the Health Department to investigate, assess, and eventually mitigate the incidents. The Department of Mineral Resources requires companies to report brine use and how much they dispose of as waste, but usually do not audit the numbers.

So, short of catching bad operators in the act, the regulators have no way to detect, let alone stop illegal dumping. As a result, the states have no real estimate for how much fluid spills out accidentally from waste pits, storage tanks, pipes, trucks and other fracking equipment.

About 30% of all spill reports involve hydrofracking brine, where in most cases the operators were not able to contain a leak and some waste materials leaked into the ground or waterways.

There are also many unofficial and uninvestigated reports of truckers dumping their wastewater load at the roadside instead of waiting in line at the injection wells.

Extensive damage to pastures, cattle watering holes, rivers, streams, and roads have been reported.

Amazingly, even when the authorities catch a well operator in illegal action—such as dumping large quantities of brine—they either can't or don't bother to impose sanctions. So, the intense contamination of the U.S. land and water continues unhindered.

Pollution Footprint

A number of large and small pieces of equipment are used during gas drilling, production, and use. Starting with the gas rig and power plant design and construction, we see pollution taking place in the shape of heavy plumes of smoke from large trucks and bulldozers all around the gas site. There are also trains and trucks used for transporting the production equipment to the site, the exhausts of which are significant.

Don't forget that all this machinery was made somewhere, sometime in the past, where and when their manufacturing processes and transport created a significant pollution footprint as well. Since most of these pieces of equipment are used only temporarily, we cannot assign the entire footprint to them, so we would consider assigning 5-10% of it to the manufacture and transport of gas rigging and power plant equipment (A).

The equipment arrives in wooden crates on enormous trailers or railroad cars. The different pieces are assembled with the help of other equipment and put to work after creating another set of pollution problems (B).

In addition, there are many other items and small pieces of equipment, tools and consumables—some for personal use, and some for specialized operations. These are shovels, helmets, computers, power drills, boots, shoes, overalls, goggles, first aid kits, chemicals…

the list is long. All of these were also made sometime in the past, and transported to the gas site. During their manufacture and transport, they also left a significant pollution footprint. (C).

To complete the pollution footprint picture, we must add the footprint of the people involved in the entire process—engineers and technicians involved in equipment manufacturing, transport, installation, setup, and operation. These people drive cars, or travel by air, trains and busses, all of which leave another significant set of pollution footprints (D).

So, the total cradle-to-grave pollution footprint of the well drilling and exploitation equipment is expressed by:

$$Pf = A + B + C + D$$

where

Pf is the total pollution footprint

A is the gas rigging equipment manufacturing and transport to the drill site

B is gas production equipment assembly, testing and installation at the drill site

C is the manufacturing and transport of tools and consumables to the new well

D is the pollution footprint of the equipment personnel during the design, setup, drilling, and production phases of the operation.

Combining the pollution footprints of A (equipment manufacturing and transport), B (production equipment assembly, test and installation), C (manufacturing and transport of tools and consumables), and D (pollution footprint of equipment personnel) gives us the environmental footprint of the gas rigging and production equipment before even starting drilling operations.

This pollution footprint is significant, to be sure! It goes deep and wide, stretching from one part of the world to the other.

To put a value on it, we must take inventory of all items delivered to the site, as needed for its design, construction and operation. We must take a close look at the manufacturing of a gas rig, a bulldozer, a dump truck and a number of other pieces of equipment used at the gas well site.

Keep in mind that some of these pieces are quite large, and that it takes many tons of metals to make them.

Each of these pieces started as an iron or aluminum ore dug from a mine, and which was transported to a smelter. The molten metal was then shaped into different forms and shipped to various parts manufacturers who drilled, welded and otherwise constructed parts for the equipment.

Again, remember that we are talking about huge pieces of metal—some of the gas well pipes are 20-30 feet long, and longer. One single bolt from the gas rig might be over a foot long, and the drill bore might need 5,000 feet of large- and small-diameter tubing for casing.

After the equipment is manufactured, which is usually an energy-intensive and polluting process, these parts are packed and loaded on railroad cars or trucks to be shipped to the assembly plant where they are assembled into major components, or entire vehicles. After one more loading and unloading operation, they finally arrive at the gas well site, or the power plant, for final assembly and testing, before being put to work.

This is a long and winding process with many loading and unloading steps, all of which require a lot of energy and generate a lot of toxic gasses.

Putting a dollar value to all this would be too complex for our purposes, so it suffices to say that the energy use and pollution effects during and after these steps are significant. We will take a closer look at these steps and their environmental effects in the following chapters.

Production Equipment Environmental Impact

Now, we have started the drilling operation and see another set of significant pollution footprints that are somewhat hidden. As an example, at the remote gas rig, a number of large diesel generators provide power for drilling operations and personnel accommodations. During an average work day and under a full load, these will burn 1,000 gallons of diesel...every day. Multiply this number by several thousand such rigs around the U.S. and many more around the world and you get the picture. Or at least part of it...

Several bulldozers, water carriers, and dump trucks are serving the rig for months during the drilling of the gas well. They also burn a lot of fuel in their never-ending runs, so the entire area is fogged with dust and soaked in diesel fumes and lubricating oils, leaving their pollution imprint day after day, hour after hour—for the duration!

Spare parts for equipment periodically arrive from the parts manufacturers, after leaving their footprint around the country and the globe too. This is to be added to the pollution footprint left during the manufacturing of all well drilling and operation equipment.

A number of work related activities require the use of different pieces of equipment during the 20-30 years

of non-stop operation of the gas well. These include people driving to and from work in cars and trucks, pumps running day and night, water and chemicals hauled to and from the site in huge trucks, etc.

After most of the gas has been produced, the well is finally declared exhausted. At this point it must be shut down and decommissioned. Decommissioning and land reclamation are another set of major, expensive, and polluting undertakings. The equipment and materials used during the decommissioning and waste disposal process create additional air, soil, and water table pollution which is not to be ignored.

Specialized equipment and personnel (environmental specialists and inspectors) are usually brought in to coordinate and supervise the effort. Some of the equipment and chemicals residue must be disposed of as hazardous waste, so it is loaded on trucks and transported to a waste disposal facility. With that, more pollution footprints are left at the area around the well site.

When the final tally is made, a significant part of the emissions and overall pollution are to be attributed to the gas drilling and production equipment, chemicals, and support personnel. This effect is significant and needs to be well understood and included in all calculations of gas well design, drilling, and exploitation.

Externalized Cost of Natural Gas

Natural gas is the new priority of many governments because of its low cost and abundance. Low prices and abundance, however, cannot change the fact that gas drilling is a hazardous operation, from both environmental and human health points of view.

Although safety is paramount in gas drilling and production operations (especially in the US), there are still problems which often result in accidents and injuries—even in the US. In addition to the large accidents that make news, there are smaller ones daily, in which people get hurt or sick as a result of the effects of gas drilling and related operations. Among the damages:

- Burn injuries occur at the site of a gas drilling operation, where the victims are likely to suffer burn injuries from gas or chemicals, accompanied pain, suffering and loss of limbs or life.

- Chemical spills and leaks of toxic chemicals can cause injury to people close to a spill by direct exposure, or contact or inhalation of fumes, which could lead to chemical burns and respiratory illness. If the water supply becomes contaminated, it could lead to long-term health problems such as cancer.

- Drill site fires at oil and gas rigs are a constant threat. The source could be negligence, lack of experience, or defective equipment, but the consequences can be grave.

- Explosions could occur too, although the oil and natural gas industry is tightly regulated to ensure the safety of workers. An explosion in a gas drilling rig is a major event that could hurt and kill people, so precautions are the norm, but even then, accidents do happen.

- Gas field accidents—small and large—are not unheard of. They are rare, but when they do happen, those affected can be badly injured and even killed.

- Gas truck and train accidents are exceedingly dangerous, given the combustible nature of the cargo. Injuries can vary from broken bones to severe burns and death when a tank is ignited.

- Property damage happens during gas drilling accidents. Homeowners and individuals who live close to a drilling rig could be indirectly hurt by damage to their property, caused by fire, chemical spill or air contamination.

- Groundwater contamination is one of the most pervasive forms of property damage from gas drilling today.

- Toxic vapors and chemical exposure might cause damages that appear immediately or are not evident for months or years. Common examples include chemical burns, respiratory complications, and various forms of cancer related to contaminated air or water at the drill site. In these cases, gas rigs and processing workers, as well as local people, are sickened by pollution from the gas emissions and liquids escaping from the work sites.

Hydrofracking is a main cause of contamination of soil and the water table, causing permanent property damages, and where fish, wildlife, and people are poisoned by harsh chemicals dumped intentionally or unintentionally into their habitat.

The externalized costs of gas production and use are estimated to be in the billions. Since the gas drilling business is growing daily, estimates vary, but soon costs could be even larger than those of coal and oil.

As a result (of the "externalities"), future generations will be heavily impacted by global warming from the exhaust gases and liquid wastes that gas drilling, production and use spew into the air and spill into the soil.

We would like to emphasize here that the complex, expensive, and dangerous gas drilling, exploitation and natural gas use (fuel for power generators and cars) have some components that are very complex and difficult to understand, sort, and estimate accurately.

Nevertheless, while we must be aware of the problems and work towards their solution, we must also keep in mind that natural gas is a necessary evil. It is something we need badly. So, we need to treat natural gas and the people who work in the industry with respect and admiration, because their efforts and sacrifices make our lives comfortable and prosperous.

The Solutions

Natural gas production is quickly expanding across the nation because the bonanza was allowed by new technologies that make it easier to extract gas from previously unknown and/or inaccessible sites.

Recently, the industry has drilled thousands of new wells in the central and south USA, and is expanding operations in the eastern side of the country, the most promising being a 600-mile-long rock formation called the Marcellus Shale, which stretches from West Virginia to western New York.

Ninety percent of all natural gas production today is done via hydrofracking—a technique that was perfected recently. Here, many (some dangerous) chemicals are mixed with large quantities of water and sand and are then injected into the drilling wells at extremely high pressure.

Hydrofracking is suspected to have polluted drinking water in Arkansas, Colorado, Pennsylvania, Texas, Virginia, West Virginia and Wyoming. In many of these places, residents have reported unusual events and water quality deterioration. In most cases, these findings coincide with the commencement and expansion of hydrofracking operations.

Natural gas producers have been ignoring the complaints of the communities across the country with their extraction and production activities for too long. Meanwhile, the number of cases of contaminated water supplies, dangerous air pollution, destroyed streams, and devastated landscapes is increasing daily.

The industry deals with a number of problems related to weak safeguards and inadequate oversight, which fail to protect the locals from harm. We shouldn't have to accept unsafe drinking water just because we have a lot of natural gas, and because it burns more cleanly than coal.

The rules of the game are unclear, and many companies ignore even the few rules set thus far. The indus-

Figure 4-16. Hydrofracking site—waste water and chemicals storage areas.

try in general prefers to use its political power to escape accountability for its actions, which leaves the locals unprotected.

New safeguards are needed, as follow:

- Sensitive lands, including critical watersheds, must be off limits to fracking.

- A lot of independent, in-depth research still remains to be done to understand and determine exactly the effects of the different chemical mixtures on vegetation, wildlife, and humans.

- New clean air standards are needed to ensure that natural gas leaks from wells are under control, and to prevent local air pollution and reduce global warming.

- New and safer well drilling and construction standards must be implemented to ensure the strongest well siting, casing and cementing, as well as other best practices during drilling and operation of the gas wells.

- Closing Clean Air, Clean Water and Safe Drinking Water loopholes is needed to reduce toxic waste, and hold toxic oil and gas waste to the same standards as other types of hazardous waste.

- Robust inspections and enforcement of the safety programs, including full disclosure of fracking chemicals, are needed.

- Affected communities close to fracking sites must be allowed to protect their environment and themselves by having a voice in comprehensive zoning, planning, and management of fracking sites.

There is no doubt that having enough natural gas is a good thing, which is critical to our energy future. We cannot and should not stop the developments, but we must make sure that they are not hurting and killing us in the process. Efficient regulations are needed to make the process fully transparent and protect the environment, and property and lives of the locals.

INTERNATIONAL DEVELOPMENTS

The natural gas bonanza has transformed North American gas markets and shows big potential internationally in Russia, China, Poland, Argentina, and elsewhere. At the same time, LNG exports from Qatar and Australia are expected to ramp up over the coming years. In parallel, intense exploration activity in West Africa and the Caspian region are promising even greater gas supply.

But what does all this mean as far as the global energy situation is concerned?

- In the U.S., the geographic location of other energy sources (coal, nuclear, hydro-, wind and solar power) determines the winner in many cases, which shows how close the competition is.

- In Asia, coal prices are rising which is creating a new pricing dynamic between the fuels that is too early to define, so the winner is still to be announced.

- China is a huge country with very large population, whose demands for energy are growing by the day. China is where Europe was in the 60s and 70s, so now they have to first clean the mess they've made—cleaning the GHG emissions. Natural gas is one of the alternatives, so huge amounts of it will be produced and imported in the future.

- Currently gas and oil prices are linked in Asia, and are foreseen to stay linked at least for the next 10 years, so no major game change is expected. Only if the shale gas production increases enough would there be a chance for gas prices to de-link from oil. If China, for example, ramps up its shale gas production over 100 Bcm/year (billion cubic meters) within the next 10 years, that huge amount would reduce China's need to import LNG and pipeline gas.

- In Europe, new coal plants cannot be built without carbon capture and storage, but this is not going to happen soon, so gas is the king for now and for the foreseeable future.

- Japan has reduced sharply its nuclear power generation, to the tune of about 12-15 Mt/annum of LNG equivalent nuclear capacity, or the equivalent of over 150 million barrels of crude oil. This is a huge loss for any country, so the energy security of Japan is in dire straights. With minimal natural resources, Japan has to import most of its energy from abroad. LNG is shaping as the temporary stopgap, so there will be increased LNG demand from Japan for the foreseeable future.

- The other industrial powers in the region, South Korea, Taiwan, and China are taking a close look at the safety of their nuclear power infrastructure. They are analyzing the risks and looking into the alternatives, and LNG is one of them.

- Australia is shaping as a serious potential LNG supplier to the Asian countries and Japan, but their shale gas prices are still too high and unpredictable, depending on raw materials availability and cost inflation. Nevertheless, Australia has clearly shown that LNG from shale gas is technically feasible, and will become a major LNG exporter as soon as the shale gas prices make it feasible.

Europe

Europe is a very special case as far as energy security is concerned. At this time most Europeans are in a difficult energy situation, an energy pickle of sorts. Many of these countries depend on Russia and some other not-so-stable neighbors for their energy supplies.

This vulnerability becomes most palpable during the freezing winter months, when Europeans depend on Russian natural gas imports to heat their houses and run their businesses.

Mr. Putin stated in 2013 that he was worried about two issues related to the long-term natural gas trade:

1. The "take-or-pay" principle that has recently faced certain legal challenges; and

2. The problems related to long-term contracts being indexed to the U.S. dollar.

Therefore, Moscow sees the rulings against Gazprom—its second loss—in the arbitration cases opened in Europe as a disturbing nuisance. Gazprom lost $1.3 billion, and was forced to revise its agreements by determining gas prices according to market prices.

In the summer of 2014, Russia signed a long-term gas export contract with China worth $400 billion. The 30-year gas-export deal is a unilateral move by President Putin to shift Russia's commercial interests towards the East. This, in response to the mounting sanctions from the U.S. and Europe, and the accompanying embarrassment.

EU member states have reached a consensus about the oil-indexed prices, the sustainability of which will be tested during the coming months and years. Generally speaking, oil market prices are the major indicator in determining the natural gas prices in Europe. This is one way of ensuring some predictability for both the producer and the consumer. The oil prices' tendency to rise has created considerable pressure in consumer markets, which in turn shapes consumer habits that have been reflected in the natural gas trade as well.

NOTE: In Europe, there is a significant disconnect between oil and gas prices, especially in northwest European countries, thus gas prices tend to stay low. Southeast European countries are also seeing lower gas prices than traditional prices in oil-linked contracts, but not as much as the northwest European states.

Enter the U.S. shale gas revolution—the new kid on the gas exports block. The U.S. has been successful in taking a significant share of the Middle Eastern LNG markets, which have been historically Russian-dominated. This changes the long-term picture in Europe and Asia, with the U.S. becoming an important exporter with the potential of changing the global price-supply balance.

The U.S. influence on the natural gas markets might affect Europe more than any other place, thus reducing the pressure from dealing with unstable suppliers and establishing a more competitive gas market.

In summary:

- Generally speaking, availability is factor number one, because countries with large gas and oil deposits will benefit in the long run, while those with little or no gas and oil reserves will continue to be dependent on the global markets' shifts and glitches.

- The future of natural gas in the *developed* countries—including the U.S. and Europe—depends on many factors, the major of which are:
  a) The environmental impact of gas production and use is a major issue, the solution to which depends on the implementation of efficient

and cheap carbon capture and storage solutions. This would put natural gas well ahead of the competition since it would be much easier and cheaper to decarbonize natural gas than coal. Nevertheless, the technological and high production cost issues have to be resolved first with this technology, which might take several years, or even decades, and

b) The price of the competing energy sources is also very important, because if at a certain point the cost to produce and use natural gas increases substantially (for whatever reason), then its reign might come to an abrupt end too.

- Natural gas, however, is shaping as a destination fuel in some developing countries, including China. As the volumes are increasing and the benefits of using natural gas become more obvious and well known, the demand in regions like Africa, Asia, and Latin America is increasing as well. In all of these countries, the gas cost is the only factor under consideration…at least for now.

China's Gas Future

China is another very special case. In 2013, the production of natural gas in China increased to 121 billion m^3. This represents a trend of approximately 10% annual increase. Of this total, 118 billion m^3 came from conventional natural gas, 3 billion m^3 from coalbed methane, and only 200 million m^3 from shale gas.

This, however, is short of the demand for natural gas, which has been increasing significantly as the country is retrofitting old coal-fired power plants to burning natural gas.

China is now the third largest consumption country of natural gas in the world. The domestic energy market demand for natural gas will continue the upward trend as the economy grows, the population increases, and the energy policies are adjusted accordingly.

The demand for natural gas is increasing, with its proportion to the total energy consumption rising to over 7% in the next several years. The domestic reserves are not enough, and the production of natural gas is simply unable to meet the rising demand.

At the same time, the government introduces measures and regulations that encourage increase in the proportion of clean energy consumption. Since there are no alternatives right now, the demand for natural gas increases proportionately.

So China imports a lot of natural gas to meet domestic demand. China imported 54 billion m^3 in 2013, over 25% more than the previous year. The growth rate of pipeline gas and LNG were 24% and 27% respectively.

The Chinese government sets the factory price of natural gas, while the final retail price is set by the respective local governments. In many cases the retail price is well below the factory price.

The demand for natural gas and LNG imports will increase rapidly as the Chinese economy develops and urbanization accelerates. The market potential of natural gas and LNG is huge, as over half of China's households still use firewood or coal as a fuel. Using natural gas would be a great improvement to their lifestyle, which is the goal of the government.

So what is a government to do? Enter the Russian Bear.

The Deal of the Century

After a decade of talks and plans, in the summer of 2014, Russia made a deal to sell natural gas to China. With a long-term contract in hand, China is now set to become the world's leading consumer of natural gas.

The new 30-year import contract calls for 38 billion m^3 a year (or 3.8 billion cubic feet a day) to be piped from Siberia starting in 2018. The Russians still have to build thousands of miles of gas pipeline from the frozen tundra of Siberia to the Chinese distribution centers. This is a huge, multi-billion dollar project, estimated for completion by 2018.

Anything can happen between now and then, but for now we need to assume that this is a done deal and plan accordingly.

China already consumes about half of the world's coal, copper and iron ore and 4 percent of its gas, is to become the biggest natural gas user by 2035.

This truly is the deal of century for China, Russia and the world, for it establishes one of the most important gas benchmarks in decades. This development is very important because it also establishes a new market dynamic—not only for China and Russia, but the entire world.

If, for example, liquefied natural gas (LNG) prices fall below Russian import parity levels in the future, then the Chinese market could absorb them.

In the future, global gas markets will probably converge to the Russian export price, with spot Asian LNG cargoes likely from the Chinese floor of $11 per million Btu to the Japanese ceiling of about $16 per million Btu.

LNG supplies may rise by 18 billion cubic feet a day by 2020 to 43 billion cubic feet a day, an amount that can be "relatively easy" to absorb by emerging markets such as China, with the help of other developing countries.

The gas price of the new contract was not disclosed, but experts believe it is in the range of $10.50 to $11 per million Btus. The new contract also sets (unofficially) a long-term price floor of $4 per million Btu for U.S. gas. This is because re-gasification, liquefaction, and transport costs of as much as $7 per million Btu from the U.S. to Asia will become a "key component" of the global Henry Hub pricing.

This also means that increased demand from Asia is will likely going to keep Europe's gas prices steady, albeit on the high side, since most European countries are forced to bid on the global energy markets.

This is good news for America's natural gas industry and its plans for increased LNG exports. The U.S. natural gas prices are presently heavily discounted for both the European and Asian markets, so the new Chinese demand would provide some stability, new export opportunities, and a floor for our LNG export prices.

Natural Gas Cartel?

The second summit of the Gas Exporting Countries Forum (GECF) was held in the summer of 2013 in Moscow. It was attended by all 13 member, and 4 non-member states. The members are the United Arab Emirates, Bolivia, Algeria, Equatorial Guinea, Iran, Qatar, Libya, Egypt, Nigeria, Oman, Russia, Trinidad and Tobago, and Venezuela, while. Iraq, Kazakhstan, the Netherlands, and Norway all have observer status.

Wait…what is missing in this picture? The U.S., and its huge natural gas industry are not represented at GECF…yet. But this might change soon.

The official goal of the GECF is to "support the sovereign rights of member countries over their natural gas resources and their abilities to independently plan and manage the sustainable, efficient, and environmentally conscious development, use and conservation of natural gas resources for the benefit of their peoples.

Although there have been many unexpected and unprecedented advances in the natural gas sector during the last several years, GECF is not OPEC. It is nothing more than a platform for the interested parties to exchange views.

As yet, ECF has no official agenda or plan of action, but this might be changing soon too, since it became obvious at the summit that gas exporting countries—realizing their growing advantages—are beginning to act as a new cartel—the new OPEC of natural gas.

Russia insists officially that the member states do not intend to be a cartel, but Moscow is known to use energy (and natural gas exports in particular) as a political instrument.

In the recent past, Russia cut supplies to several neighboring countries in the middle of severe winter freeze. Yet Russia realizes that energy is one sure way to balance its budget, so it favors long-term natural gas agreements and is looking for ways to solidify trade relations with importing countries.

The different GECF member countries follow different strategies depending on their national characteristics, and there is no coordinated effort to introduce stability in the global natural gas market.

The remarkable fact is that the 13 member states own 70 percent of the world's natural gas reserves, which includes 85 percent of LNG production, and 38 percent of pipeline-based trade. Iran, Qatar, and Russia are thus far the largest constituents with almost 50 percent of the world's natural gas reserves.

The difference in modus operandi among the member states is also quite large. For example, in 2012 Russia exported 93 percent of its natural gas via pipelines and only 7 percent in form of LNG, while Qatar exported 85 percent of natural gas as LNG, and only 15% via local pipelines. There is similarly disparate data from other states, which stem from geographical and historical factors.

Due to its geography, finance, and population, Qatar, for example, is much better equipped to deal with fluctuations in the export market. Russia, in contrast, must protect its long-term interests and the competition of the existing structure, since gas exports problems could negatively affect its fragile internal economic and political balances.

So, among the socio-political changes worldwide—including those in the energy sector—we are now witnessing the birth of a new global power. The power of the states with excess natural gas. Would they become another OPEC? Would the U.S. join in the new game?

Update 2014

With all the talk about energy revolutions, the new shale deposits and their developments take the cake… with a cherry on top. We now live in an amazing period of unprecedented energy (mostly shale gas) prosperity that boggles the imagination in character and scope.

This bonanza will most likely never happen again…or could it? The good news continue coming and now the expert analysis of the global prospects of shale gas and oil in countries like China show that there might be significantly more shale gas and oil reserves than in the U.S. There are other countries too, such as Argentina and Canada that might be hiding megatons of productive shale deposits too.

Most countries will not be able to get into the shale game anytime soon—at least not during the next 20-30 years. Once they do, however, the global impact on oil and gas production and the reserves distribution would be unprecedented. Another wave of discoveries and gas and oil production increases...maybe?

US accomplishments in the shale oil and gas industry to date are quite impressive. Even disregarding to an extent the possible future U.S. achievements in the area (which are fully expected to come), we see a huge increase in the U.S. energy supply.

With a roughly 10% share of global shale gas resources and a 5% share of shale oil resources, the U.S. has expanded its gas output by almost 200 mtoe (million tons of oil equivalent) and oil output by almost another 200 million tons annually.

So, comparing these amounts with those available in the rest of the world—if and when accessible—would yield a shale gas output of 2,200 mtoe of natural gas (11 times the U.S. output) and a shale oil output of 4,650 mt (23 times the U.S. output) by 2030-2040.

These facts are important in many ways, but their overall practical impact would be stunning. The 20-year output growth projections for global shale gas are twice as large as the global production rise in the preceding 20 years. Oil production will increase more than five times.

Future conventional output increases, such as those following from recent discoveries of gas in the eastern Mediterranean, must be added to those from shale gas and shale oil too, in order to obtain the global 2035 aggregates.

This means that the total change over the coming 20 years would be revolutionary—an energy game changer for the gas industry that would revolutionize the entire global energy market.

Energy Security and Natural Gas

Natural gas can significantly enhance the U.S. energy security by providing the bulk of fuel for electric power generation. Its potential as transportation fuel is extremely significant as well, and all this adds to a significant increase in our energy security.

We need to take a close look at the energy security concept under the loupe of the new situation in the U.S. and the world:

Energy security can be defined as an umbrella term that links energy with economic growth and political power.

The energy security concept varies depending on the perspective of the beholder:

- Consumers focus on cheap energy on demand and without disruptions.

- Oil and gas companies focus on access to new reserves, developing new infrastructure, and access to investment.

- Power companies focus on the integrity and operating cost of the power network.

- Oil producing countries focus on security of revenue and increased demand.

- Developed countries focus on securing uninterrupted energy supplies at the best prices.

- Developing countries focus on their ability to afford energy for economic growth without price shocks.

- Policymakers focus on ensuring energy supply, the infrastructure integrity, and the strategic reserves.

Generally speaking, for most countries today energy security is:

- The availability of energy needed for stable economic and social development;

- Freedom from interruption of the energy supply; and

- The affordability of energy prices.

In the past, one of the possible (and at times preferred) instruments for achieving energy security was direct or indirect military intervention. In today's civilized and peaceful "global village," however, energy security is primarily determined by social and geopolitical factors. It is a combination of national elements and policies affecting the control of energy development and transportation worldwide. A clear distinction must be made between these two approaches, because there is still the possibility of a mixture of both approaches at times.

So, when viewing a nation's energy sources in light of national energy security, the policymakers and regulators must consider all the national elements of power—military, diplomatic, informational, and economic—in addressing the interplay with numerous international actors under different socio-political conditions.

The U.S.' energy security is directly affected by our relations with oil exporting countries. This, in turn, depends to a large extent on the global developments. Entering into a conflict with one nation, for example, could trigger a response from another—oil exporter—as it happened during the Arab oil embargo in 1973.

Recently, new developments in the U.S. are dictating a close look at our relations with coal and natural gas producing and importing countries, due to the large quantity of gas produced in the U.S. territory. This is a great bonus, giving us a chance to become a major LNG exporter. We have also reduced coal use, thus ensuring a cleaner environment, in addition to freeing huge quantities of coal for export.

The new developments complicate the overall energy security picture, because they add new elements to consider. The possibility of increased exports of LNG to Europe, for example, might affect negatively our relations with Russia, while exporting millions of tons of coal to China would certainly increase the global GHG emissions.

As a result of the newly increased natural gas production, the U.S. is becoming a major exporter of coal and LNG.

And yet, there are still uncertainties and risks in natural gas production and use.

Figure 4-17 depicts some of the risks playing a major role in the natural gas production and use lifecycle. In the short term, we see risks related to internal issues—excess water use, environmental and landowners' law suites, and political changes (regulatory restrictions)—as the major factors that might affect and even alter the path of natural gas in the U.S.

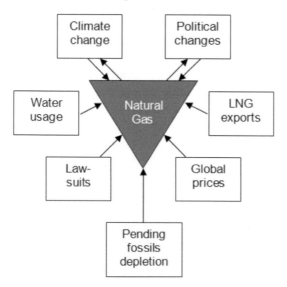

Figure 4-17. The risks of natural gas production and use

In the long run, global political changes, LNG export competition, global LNG prices, and climate change will also play a major role in the future of natural gas production.

The greatest (and most ignored) risk to the future of natural gas, and our future energy security, is the imminent depletion of the fossil energy sources. This is a problem without a solution. Or better said it is a problem that—for many reasons—is not looked at as such, and, at least for now, there is no problem.

As a result, the excess gas drilling and pumping trend in the U.S. and worldwide continues. Many developing, and most of the developed, countries are making plans for increased reliance on natural gas for electric power generation and vehicle fuel. The major producers are planning to supply as much as they can for as long as they can.

Obviously, nobody sees that the fuel gauge has long passed the half-way mark to empty. So for now the party is on and the frenzy is increasing. The future can wait...so back to the present:

Electric Power

Natural gas is steadily gaining solid ground as a reliable power generator. New studies assessing the reliability of the natural gas supply system during electric power grid outages concluded that it is highly resilient to the loss of electricity provided by the conventional power grid.

This is due to the fact that natural gas supplies would continue to flow with minimal risk of interrupted deliveries even during electric power grid outages of three months or more. The natural gas supply network is distributed over large areas across the country, thus it has few single points of failure that can lead to a system-wide propagating failure.

A large number of wells, pipe and storage systems are widespread across the country, so the transmission system can continue to operate at high pressure and normally even with the failure of the majority of the support network. Basically, it can continue running unattended and without external power.

In contrast, the present-day power grid cannot operate unattended, or without power, since it has fewer generating points, and requires oversight to balance load and demand on a tight timescale. Its transmission and distribution networks can be affected by shutting down a single point, which can result in cascading and catastrophic failures.

So, having more natural gas ensures more reliable grid operation by providing an uninterrupted energy supply to the power grid. This in turn ensures the integrity of our energy sector and our energy security.

Natural gas use for vehicle fuel is growing too,

and can also be counted on for uninterrupted supply of vehicle fuel in case of serious crude oil emergencies.

NGVs

Using more natural gas for transportation can make the biggest difference in our energy security as well.

Natural gas could replace crude oil, which is the weakest link in our energy security portfolio, for use as motor fuel. This could also bring us closer to energy independence.

Replacing 3.5 million heavy-duty vehicles running on fossil fuels with natural gas vehicles would save more than 1.5 million barrels of oil per day, or more than we import from either Venezuela or Saudi Arabia.

The NGC Efforts

Over 50 North American producers and distributors of natural gas have joined together to advance a common vision of enhancing our national energy security by promoting the development of natural gas vehicles (NGV) and infrastructure across the continent.

The goal is to formalize the shared commitment of natural gas producers and utility companies to advancing the development and utilization of natural gas vehicles and fueling infrastructure in the North American marketplace.

These joint activities will focus on infrastructure development, vehicle production, marketing and education for clean transportation solutions, and targeted advocacy. The final objective is to work in a cooperative and complementary fashion with other stakeholders who share our commitment to promoting natural gas vehicles and clean American transportation solutions.

Natural gas vehicles outperform conventional fuels with a significantly higher octane rating, providing better fuel efficiency and lower operating costs, while at the same time offering dramatic reductions in emissions.

As a result, adoption and augmentation of this clean transportation alternative, in addition to significantly improving the air quality across the country, adds flexibility to our vehicle fleets and reduces the risks of interrupted energy supplies.

NGV Facts

- Transportation accounts for 30% of U.S. CO_2 emissions, and natural gas vehicles run 25% cleaner than vehicles powered by traditional gasoline or diesel.

- Natural gas vehicles reduce smog-producing pollutants by up to 80-90%.

- There is tremendous potential in converting heavy-duty commercial vehicles from diesel to natural gas. A number of technical and economic problems still need to be resolved for this to become a universally accepted alternative.

- Natural gas vehicles are 25% cleaner than those that run on diesel or traditional gasoline.
  NOTE: Diesel exhaust includes more than 40 substances that are listed by the EPA as hazardous air pollutants. In contrast, natural gas vehicles emit only 2-3 of these substances.

- Businesses and cities have found that natural gas vehicles pay for themselves in fuel savings. Fleets report 15-30% savings compared to gasoline and diesel fleets.

- The lifecycle costs of natural gas trucks are competitive with traditional fuels, and many companies have found that these vehicles are cheaper to maintain.

- Replacing 3.5 million medium- and heavy-duty trucks and buses with CNG-powered counterparts by 2035 would save at least 1.2 million barrels of oil per day.

- Natural gas has the potential to reduce transit and school bus oil use by 80,000 barrels a day. If 420,000 diesel buses in the country are replaced with NGVs by 2035, then natural gas transit and school buses could eventually displace half of the diesel used by buses by 2035.

- America can lead the world in the effort to help clean the global air, while at the same time advancing its energy security.
  NOTE: Switching to NGVs cannot be done easily in many countries, due to high initial cost of engine conversion, adding or retrofitting fuel stations, and technical and economic issues.

The biggest problem in the use of natural gas for transportation fuel presently, in addition to the high cost of engine conversion, is the shortage of fueling stations. The U.S. natural gas vehicle fueling infrastructure is on the rise, however, thanks to a combination of a) company efforts to run cleaner vehicles, b) more affordable fleets, c) local governments concerned about air quality, and d) a growing commitment to vehicles that reduce our reliance on foreign oil.

There are presently over 150,000 natural gas vehicles on the U.S. roads, and over 1,000 natural gas fueling stations.

Looking forward, we see legislation in Congress which would provide tax credits for the purchase of natural gas for vehicles. In addition, companies that invest in natural gas refueling infrastructure would get up to $100,000 in subsidies per station.

Another encouraging sign for NGV infrastructure is expanding natural gas corridors. South Coast Air Quality Management District and UPS are completing a 700-mile natural gas corridor from Las Vegas to Ontario, California. In addition to fueling more than 200 UPS heavy-duty vehicles, the corridor would be accessible to all private NGVs too.

Similarly, Utah Governor Jon Huntsman has designated I-15 from Idaho to Arizona as a natural gas vehicle corridor. Seattle is also among the growing number of cities making fueling stations for municipal fleets available to the public.

Under, the American Recovery and Reinvestment Act of 2009, the Department of Energy funded 25 different projects for alternative fuel, infrastructure and advanced technology vehicles, and 19 of these 25 projects included natural gas. These commitments include support for 140 new fueling stations.

Figure 4-18. LNG pumping station

In summary, natural gas' potential to close the energy security circle is tremendously important.

Natural gas promises 100% energy security to our domestic power generation. Increased use of NGVs could reduce crude oil imports significantly, thus increasing our energy security proportionately as well.

If successful, NGVs could eliminate a significant portion of crude oil imports, which are presently used mostly for fueling transportation vehicles. Crude oil imports represent the weakest link in our energy security, and the greatest threat to our energy independence.

The fastest way to do this, and as needed to have an immediate impact on our energy use, is to focus on the conversion of larger vehicles—commercial fleets, trash trucks, transit and school buses, delivery vehicles, taxis and shuttles. These vehicles are the heaviest and busiest on the road, and need a well established regional fueling infrastructure, which is lacking presently.

Nevertheless, the U.S. energy future looks bright for now, mostly thanks to a large supply of clean and cheap natural gas. It is closing some of the gaps in our energy security, leading the way to long-term energy independence. Can it be done? Will we ever be able to produce all the energy we need domestically? The answer is a definite maybe.

Conclusions

Natural gas is a miracle fuel that provides more than half of the power we use. It is increasing in volume and importance in the U.S. energy sector and our entire economy. Yet, it is not the solution to all our energy or economic problems—although some insist that this is the case.

Instead, natural gas is only a partial—and temporary—solution to some of the problems. It has a place in the present stage of our development, with advantages and disadvantages which must be well understood to properly use the gas bonanza in a responsible and efficient manner.

The advantages and disadvantages vary from location to location and from time to time, so there is no "cookie cutter" approach that can be used universally.

When all advantages and disadvantages are combined in a particular case or project, the picture gets extremely complicated. Deep understanding of the technical, political and economic aspects, as well as logistics, is absolutely necessary to grasp the situation and make the right decisions in every case.

Natural gas has a number of advantages over the conventional and alternative fuels.

Some of the key advantages are:

- Natural gas production in the U.S. and many other countries is rising to unprecedented levels, which, together with its other qualities, puts it in a very special class—that of a priority fuel. The new gas "bonanza" brings many benefits to the U.S. economy, and if done properly, it might be the greatest thing that ever happened in the U.S. energy sector and our energy security.

- Natural gas has a competitive advantage over renewable energy and nuclear power in almost every country because it is similarly environmentally clean and cheap, and yet does not require subsidies.

- A great advantage of natural gas is that gas-fired power plants can easily be built in a modular fashion by adding indefinite amounts of modules and capacity as needed.

- Another serious benefit of natural gas is its potential to provide fuel for the transportation sector, thus replacing crude oil. Although 100% replacement is not possible, any smaller contribution would have great positive effect on our energy security by reducing oil imports.

- Coal presents a real and complex challenge to gas' dominance of the electric power generation sector. It is equally abundant and cheap, so here the environmental emissions (SO_x, NO_x, and CO_2) are the only difference, which plays a significant role when comparing both fossil fuels. Natural gas is winning for now, and so many coal power plants have been shut down or retrofitted to gas-burning, mostly because of environmental issues and requirements.

Unfortunately, natural gas also has some disadvantages. The key disadvantages are:

- The growing environmental problems associated with shale gas production are becoming more obvious and serious. If this problem grows enough to cause reduction in shale gas production, there won't be enough gas to produce electricity in the U.S. and some other countries. This could grow into a national and even global energy and economic disaster.

The sense of responsibility is slowly catching up with the gas volumes, but the related actions are still lagging and as a result, the incidents of environmental damage and people suffering are increasing.

- Almost every day we hear new scare stories about damages inflicted by hydrofracture operations, which are the main source of shale gas. This means that until the environmental obstacles are resolved, the future of gas in the U.S. and most of the globe is uncertain.

- Natural gas also emits GHGs at several stages of its life cycle, albeit in smaller quantities than coal. Some of the gas emitted GHGs (i.e. methane) are many times more dangerous and long lasting than those emitted by coal.

- The natural gas GHG emissions will increase significantly in the future, due to the present and planned increase of shale gas production and gas-fired power generation. This will require special measures to be taken to reduce the effects of the increased global emissions.

- A lot of fresh water is used during hydrofracking operations. Since water is scarce and is getting even more so in many areas of the U.S. and the world, the production model has to change to reduce water usage.

For example, instead of operating one huge well by conventional methods, many smaller wells can be drilled from one production pad, each using a closed-loop water system. This method uses significantly less water, which combined with safe disposal of the smaller waste-water streams, would resolve the majority of the environmental concerns.

All modification and process improvements, however, would increase shale gas prices accordingly. The closed-loop model has not been tested on a large scale, so it is not certain if it is technically and economically feasible. If it is not, then use of large quantities of water would continue to be a serious issue that might eventually limit gas production and/or make it prohibitively expensive.

- A big danger in the natural gas' future is overhyping its benefits—emissions reduction and lower price in particular.

  — We know that natural gas emits less GHGs, but they are different types from those emitted by coal or oil. Because of that, we simply don't know what exactly the overall long-term global effect would be. Some experts claim that the global environment would suffer more as a result of the transition to natural gas power generation. Time will tell, but we need to be vigilant.

  — We also hear that natural gas is a "cheap" fuel. While it is a fact that it is more abundant and somewhat cheaper than some fuels, it still costs a lot of money to produce and transport.

It also has a lot of growing challenges to overcome, so it will never be "cheap" *per se.*

A global price range of $7-8/Mcf (million cubic feet) could be acceptable in the future, since we already see $8/Mcf shale gas prices in Europe, and since global energy prices tend to rise. Higher prices, above $4-5/Mcf, would challenge natural gas' dominance of the power generation sector.

• Finally, and most importantly, we must watch for the excesses in the present and planned over-exploitation of natural gas reserves. With the recent huge increases in gas production in the U.S., China, and Russia, and similarly large production plans in these and other countries, natural gas reserves won't last very long.

We must always remember that natural gas, just like the other fossils, is a commodity of limited quantity. Exploit-ing it as if the supplies are unlimited is simply irrespon-sible. Continuing this trend will damage our Earth and hinder the economic development of the future generations.

Notes and References

1. BP http://www.bp.com/sectiongenericarticle.do?category-Id=3050046&contentId=3050873
2. NaturalGas.org http://naturalgas.org/overview/uses_transportation.asp
3. Exxon Mobile http://www.exxonmobilperspectives.com/2012/03/22/natural-gas-cars-a-look-under-the-hood/?gclid=CPC32OHOo7wCFclkKgodDX-kAAA&gclsrc=ds
4. C2ES http://www.c2es.org/publications/natural-gas-use-transportation-sector
5. Cool, Clean Fuel. http://www.lngfacts.org/
6. Natural Gas. http://www.eia.gov/naturalgas/
7. Natural Gas Facts. http://www.aga.org/Newsroom/factsheets/Pages/NaturalGasFacts.aspx
8. *Photovoltaics for Commercial and Utilities Power Generation,* Anco S, Blazev. The Fairmont Press, 2011
9. *Solar Technologies for the 21st Century,* Anco S Blazev. The Fairmont Press, 2013
10. *Power Generation and the Environment,* Anco S. Blazev. The Fairmont Press, 2014

Chapter 5

Coal
(Friend or Foe)

This island (Earth) is made mainly of coal and surrounded by fish.
Only an organizing genius could produce a shortage of coal and
fish at the same time.

Aneurin Bevan

Wow! This man, Mr. Bevan, living over half a century ago was able to foresee what we are doing today—something that many of us are unable or unwilling to see.

Mr. Bevan is well qualified as an expert on our island Earth and its problems. His father was a coal miner, so he grew in the Earth-energy environment. Later on, as Minister of Health, he was able to put his experience in action by focusing on the negative consequences of coal mining. This prompted him to spearhead the establishment of the National Health Service, which was instrumental in elevating the health care on his small island to world standards.

Amidst the chaos of WWII—10 days before D-Day—he was able to capture in the above quote the essence of what we are experiencing today, 70 years later. He was able to see the role of the "organizing genius" on a large scale and project it in the future.

Looking around us now, we see how prophetic his statement was and still is. It is so much so, that if Mr. Bevan could see what we have done, he himself would be surprised at the accuracy and timeliness of his statement. This is how clear the *organizing genius'* actions, and their disastrous effects on our lives, are today.

A close look at today's delicate balance of energy, environment, politics, and socio-economic development reveals a mixture of greed, ignorance, negligence, and irresponsible behavior that promises to bring Mr. Bevan's predictions to fruition... soon!

We see increased demand for and accelerated exploitation of fossil fuels—the fastest route to their premature and complete depletion. In parallel, we also see the signs of increased GHG emissions, global warming, natural disasters and other events that threaten not only the fish, but all wildlife and humans as well.

These developments are usually driven by corporate greed, political mediocrity and economic over-ambition; all parts of the organizing genius. As a matter of fact, the "organizing genius" has already succeeded in getting us very close to coal (and all fossils) shortage, and has reduced the fish and other wild animals well beyond Mr. Bevan's imagination.

One thing Mr. Bevan did not foresee is the fact that, in addition to the fish, these changes would be causing great property damages and hurting people too. This is a new development that was simply unthinkable half a century ago.

At the rate we are going now, we will reach a point soon when most fossil fuels—including coal—together with fish and many other wild animal species—will be no more.

Future generations will only be able to see pictures of the fossils and some animal species in the history books... if they even survive the transition to non-fossil and non-fish reality.

We know that this worst-case scenario can be avoided. Humans' survival instincts are very strong and people are very smart, so they usually are able to figure out how to handle any situation. Even that which the "organizing genius" is cooking now can be resolved successfully someday. We hope that this happens soon, but we just don't know when and how.

So let's take a close look at the present-day reality first, starting with a brief look at the fossils in general, and then a closer look at coal.

THE FOSSILS

The fossil fuels—coal, crude oil, and natural gas—are the major primary energy sources, responsible for over 90% of the global energy generation, transportation, chemicals production, etc. There are other fossil products, such as peat and bitumen (and potentially methane hydrates in the future) that can also be used for power generation, consumer goods, and other applications.

Right now, however, due to availability and price differences, coal, crude oil, and natural gas are by far the dominant fossil fuels used for power generation and transportation in our modern society.

- Coal consists largely of carbon, although lower coal grades contain significant quantities of other organic and inorganic contaminants such as different types of hydrocarbons, sulfur, mercury, etc.

- Crude oil and its petroleum relatives are even more complex as a chemical mix of different organic and inorganic substances. The liquid component of the petroleum reserves is crude oil, which contains typically about 85% carbon by weight and a mix of different hydrocarbons with various amounts of sulfur, oxygen, and nitrogen mixed in. The exact composition of the crude oil reserves varies with location, as well as the age and the conditions of its formation.

- Natural gas consists mostly of methane gas—the simplest of the hydrocarbon family—with the rest being a mixture of small amounts of different hydrocarbons and impurities in different proportions.

Natural gas coexists with coal and crude oil, so the gas-to-solid or liquid in coal and crude oil reserves respectively is quite important, for it determines the quality of the final products and their extraction, transport, and processing methods.

Note: An important fact to remember is that we don't burn all products that are made from coal, crude oil, or natural gas. Nearly 10% of crude oil, for example, is turned into consumer goods, road making products, lubricants, and feedstock used in the petrochemical industry. Similar quantities of coal and natural gas are also used in different industrial processes, both as fuels and feedstock.

This feedstock is an important derivative that goes into making a vast array of products we use today. Without these, there would be no plastics, some medicines, inks, insecticides, pesticides, toothpaste, perfumes, fertilizers, lipstick, paints, steel, food preservatives…just to mention a few.

The remarkable thing is that we have built our

lives around the fossils and their byproducts, so that a shortage of any of them would bring a number of serious problems. Imagine life without gasoline or diesel—with no cars and trucks on the highways, no trains and boats, and no planes. Imagine fields without fertilizers, insecticides, and other sprays to prevent crop attacks by bugs and parasites. Imagine life without plastics…here goes ¾ of the Chinese gadgets-producing industries.

More realistically, imagine all these products costing 10, 50, 100 times the present value. Impossible, you say? Yes, possible, we say, for this is what will happen 50-100 years from today when the fossils are mostly, and some completely, gone.

We foresee a future, as fossil supplies dwindle, where plastics and some of the above-mentioned commodities would become as scarce, precious and expensive as gold. This might be the incentive for our society to look elsewhere for energy sources. Or would it?

We clearly see future generations looking back at us confused; scratching their heads and asking incredulously, "Did they really burn all that precious stuff?"

But we will let the future generations worry about it, at least for now, and get back to the fossils.

The Fossil Energy Science in Brief

When we burn fossil fuels, the carbon in them reacts with atmospheric oxygen to create carbon compounds, during which process a lot of heat and smoke are generated. In coal-burning, the heat released in the burning process sustains the combustion. We can use the generated heat for heating our homes and generating electricity.

The bad part is that because fossil fuels consist largely of carbon and hydrocarbons, the combustion process produces a lot of carbon dioxide (CO_2) gas and some water vapor (H_2O). Both are nontoxic, common substances, which are present in the atmosphere, but when their quantities increase above certain limits, they become harmful climate-changing greenhouse gases.

CO_2 is one of the most, if not the most dangerous greenhouse gases, blamed for serious environmental damage and climate warming.

And here lies the dilemma: we need heat and electricity from the fossils, but we don't need their greenhouse gasses with toxic and harmful effects on the environment. Since we cannot have one without the other, we are stuck with a big problem that needs a quick resolution. Hence, the need to gain a complete understanding about what we are dealing with.

Different fossil fuels have different chemical compositions, thus different combustion processes, which result in differing amounts of CO_2 emission. Since coal consists of simple molecules, which are mostly carbon, CO_2 is the main product of coal burning. Sure enough, coal burning power plants are the major emitter of CO_2 gasses, where millions upon millions of it are pumped into the atmosphere every day.

Note: Carbon emissions—and CO_2 gas especially—from coal-burning is a major global environmental problem. CO_2, however, is not incidental, or a by chance by-product of coal fuel combustion. No, it is actually exactly what we want and need to produce when burning coal. This is because CO_2 represents the lowest possible energy state of carbon in the Earth's atmosphere.

So by burning high energy value coal, extracting its energy and letting the less energy value CO_2 escape, we've achieved our purpose of efficient extraction of the maximum possible amount of energy from the coal we burn for energy. This is also the major point in the present-day dilemma of coal power generation. We can't live with it, and we can't live without it...

Crude oil has a different, much more complex content. It actually consists of different long hydrocarbon molecule chains, which contain at least twice as many hydrogen as carbon atoms. Because of that, when oil burns, most of the released energy comes from the formation of H_2O so, for a given amount of energy, less CO_2 is emitted—much less than equal amount equivalent of coal.

Crude oil is what makes transportation as we know it possible. Millions of cars, trucks, trains, boats, planes and other types of vehicles are fueled every day, using millions of gallons of crude oil. They are then crisscrossing the country and the world leaving their deadly gasses behind and filling the atmosphere with them.

Vehicular pollution from gasses emitted by crude oil derivatives (gasoline, diesel, heating oil, jet fuel and other petrochemicals) is the number two source of pollution in the world. It is also blamed for a number of ills, together with its cousins coal and natural gas emissions.

Finally, natural gas is largely methane gas (CH_4), so it contains much less carbon than hydrogen and as a result produces much less CO_2. Almost two times less CO_2 is emitted for an equal quantity of coal.

Considering the average efficiency and other coal-to-electricity conversion factors and variables, an average coal-burning power plant emits CO_2 at a rate of approximately 500 lbs. per second. Got that? Every second there are 500 pounds of CO_2 billowing from the plant's stacks.

Each coal-fired power plant produces about 40 million lbs. of CO_2 every day, or close to 16 billion lbs. CO_2 annually.

All this smog from just one coal-fired power plant. Multiply that number by the thousands of such plants worldwide and you'll see the magnitude of the problem. All this, because we need a lot of energy and enhanced energy security. It's a global frenzy of sorts.

A similar size natural gas-burning power plant has better conversion efficiency and less carbon content, so it will emit half the amount of CO_2 of the coal-burning unit. But even then it blasts 8 billion lbs. CO_2 annually (half of the coal-fired plant), which is still 8 billion lbs. more than we need. A significant number that cannot and should not be ignored.

When multiplied by the thousands of similar power plants around the world, we could clearly see a dense cloud of smoke full of GHGs that our atmosphere is getting saturated with, forcing changes in our weather patterns and causing global warming.

Also, studies have shown that the fracking process, as used today emits large amounts of gasses, raw methane being one of them, which are even worse than CO_2 as far as their GHG content and overall effect on the global climate are concerned. So we must wonder if in this case we are not jumping from the frying pan into the fire.

Nevertheless, there is a drive presently to shift from coal and oil to natural gas. It is presumed that this switch to a cleaner fuel would slow the process of anthropogenic climate change in the long run. While it might slow down the emissions in the short term, it cannot halt—let alone reverse—the harmful effects in the long run. This is because natural gas too produces greenhouse gasses, both during its production and during the burning processes, as discussed in this text.

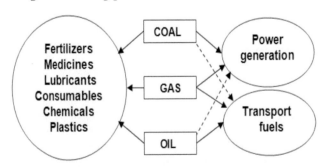

Figure 5-1. Fossils use in the 21st century

In summary, fossil fuels are the primary energy source in the US and most other countries. They also provide some priceless commodities and products that we simply cannot live without.

We use the fossil fuels and their products every day and in every area of our lives. At the same time, however, we complain of the increasing cost of the energy they provide, and the pollution they create. Well, as the old folks say, "You cannot have your cake and eat it too."

So instead of complaining, we must get serious about understanding the different fossil and renewable technologies and figure a way to use them all—a way that preserves the energy balance and the environment now, while searching for the alternatives to be used in the future.

This is exactly our goal in this text—to present the technical, logistical, and economic details of all energy generating technologies, in order to show their good and bad sides. Having a view of the complete picture is a good start. It will allow us to recognize the problems and help us look for solutions. Whining and complaining won't do!

Closing our eyes to the problems, or remaining ignorant on the subject, is a crime. Smart thinking and hard work are needed, if we are to stop the destruction of the world as we know it, in our drive to ensure temporary energy security.

Fossils Now and Then

The fossil energy sources—coal, crude oil, and natural gas—were created and deposited deep underground millions of years ago. They have also been used occasionally and in very small quantities throughout the centuries, but their large-scale exploitation and use started just 100 years ago.

- *Coal* has been used as a fuel, mostly for personal use—cooking, heating, etc.—for over 2,000 years. Other uses were also known as early as 1,000 BC, where the Chinese were using coal to smelt copper, and coal cinders from 400 BC Roman-occupied Britain show traces of other use, but the written record of coal use dates back to the 13th century.

  From the 17th century on, coal was a major heating fuel in England, and there are a number of famous buildings that are blackened from coal smoke. The industrial revolution in Europe made coal a driving force, and the steam engine invention put coal on a pedestal. By the mid-1800s, coal mining was well underway in the Eastern U.S.

too, supplying coal for industrial uses and steam locomotives that were crossing the country on the newly developed railroads.

- *Crude oil* seeps naturally to Earth's surface in some locations, where it has been used for variety of purposes, including medicinal use, as early as 5000 BC. Later, crude oil became a weapon, when Persians and Arab tribes used it quite successfully in oil-soaked arrows and other incendiary weapons. Oil use increased in the 19th century, when it was used for lighting and heating.

  In the mid-1800s the first commercial oil well was drilled in Pennsylvania, where oil was struck at 70 feet below ground, and the rest is history. Today, oil is by far the largest and most important natural resource that provides many of the essential necessities of modern society.

- Natural gas is the new kid on the block. It has very humble beginnings as a waste gas, until the early 20th century, when it was widely used for heating and cooking in industrialized countries. Today it has taken a new form and risen to unimaginable levels of importance.

  There was very little use of it in the distant past and as a matter of fact until recently it was simply vented, or burned, at oil well sites in a process known as "flaring."

  Flaring is now mostly illegal for environmental reasons, in addition to being a great waste of money. Instead, natural gas is treated with due respect and is quickly turned into a leading fuel.

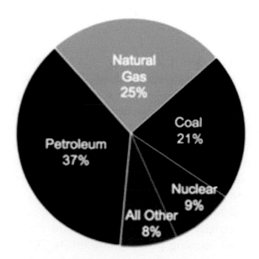

Figure 5-2. Use of fossil fuels in the U.S.

Natural gas is now an increasingly important energy resource. With a number of new and very large gas deposits discoveries, and the improvement of horizontal drilling and fracking processes, it will become even more so. Gas rules the U.S. energy markets, and no doubt it will continue so for a long time.

In the process of taking over the energy sector, natural gas is making coal obsolete in some areas of our economy, such as electric power generation. No new coal-fired power plants can be built in the U.S., due to stricter EPA emission standards. At the same time, the existing coal-fired plants are becoming more expensive to operate, and as a result, there is an increasing incentive to convert coal- to gas-fired power generation.

Yet, coal is the solid foundation of our power generation, and is even more valuable as such in the developing countries. So, natural gas may be the new king of power generation, and crude oil still remains the king of transportation, but coal is here to stay in a close head-to-head competition with all other energy resources. This a new and complex situation that requires a deep knowledge of the properties and logistics of the energy sources. We just need to get used to the new situation, understand it well, and then make appropriate decisions.

The fossils contain a lot of carbon. Here coal has no competition, because it is mostly carbon. Thus, carbon is what determines the properties, advantages, and disadvantages of the fossil fuels. Important, no? So let's see what is carbon.

CARBON

Carbon is a miracle element. It is in the structure of all living things, animals and people. Without carbon there would be no life on Earth. It is also the main constituent of coal, and although it has a fairly simple physical structure, it is a chemically complex element in the way it behaves and in the never-ending versatility of its use. It is found in many different compounds; it is in the food we eat, the clothes we wear, our cosmetics and the gasoline that fuels our cars.

Carbon is the sixth most abundant element in the universe, and is number one in usefulness. It is a very special element because it plays an essential role not only in thousands of products, but it is a dominant component in the chemistry of life. Life as we know it cannot exist without carbon. All living matter is made out of carbon with other elements added to it. Amazing…

Carbon was known in ancient times, when it was made by burning organic material in cooking fires, and later on intentionally produced during charcoal making processes. Today it is the foundation of power generation and transportation, and many industrial processes and products. Let's see why.

Background

There are four known allotopes of carbon: amorphous, graphite, diamond and fullerene. A new (fifth) allotrope of carbon was recently found. It is a spongy solid that is extremely lightweight and, very unusually, attracted to magnets. This new magnetic addition to the carbon family, nanofoam, may someday find important medical applications that might change our lives in a drastic way.

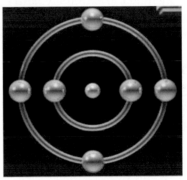

Figure 5-3. The carbon atom

Carbon has four electrons in its valence shell (the outer shell), or the energy shell, which plays major role in chemical reactions, and which in this case can hold a maximum of eight electrons. Each carbon atom, therefore, can share electrons with up to four different atoms, be it carbon or other elements. Since carbon can combine with other elements as well as with itself. This allows carbon to form an amazing number of different compounds of varying size and shape.

Carbon alone forms the familiar substances graphite and diamond. Both are made only of carbon atoms. Graphite is very soft and slippery, while diamond is the hardest substance known to man, and although both are made only of carbon atoms what gives them different properties is the different ways the carbon atoms form bonds with each other in the different materials.

In the diamond structure the carbon atoms are arranged very symmetrically and are compacted very closely together, so that they cannot move at all. This gives the diamonds their very special physical and chemical properties—exceptional shine, hardness and inertness—not to mention great price, for they are very rare in nature, which makes them hard to find and mine.

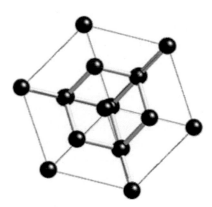

Figure 5-4. Diamond structure

Cutting and polishing the extremely hard raw diamonds increases their price even more.

Carbon's key physical and chemical properties are:

Atomic Number	6
Atomic Mass	12.011
Melting Point	6422°F
Boiling Point	6917°F)
Density	2.267 g/cu.cm.
Hardness Scale:	0.5 Mohs
Stable Isomers:	2

The most important property of carbon, at least for our purposes, is its ability to form long chains of interconnecting C-C, C-H, and other bonds, called catenation. Carbon-carbon (C-C) bonds are strong, and stable and allow carbon to form an almost infinite number of compounds. There are more known carbon-containing compounds than all the compounds of the other chemical elements combined.

Hydrogen is the exception, but only because in combination with carbon it is in almost all organic compounds, too.

When combined with hydrogen, carbon forms various hydrocarbons which are important to industry as refrigerants, lubricants, and solvents. They are used as chemical feedstock for the manufacture of plastics and petrochemicals and are the core of the molecule of all fossil fuels as well.

The hydrocarbons are a large family of organic molecules that are composed of hydrogen atoms bonded to a chain of carbon atoms. It is the simplest organic molecule, which is part of all organic matter. The difference between the different hydrocarbons is the chain length, the number and size of the side chains, as well as the number and type of the functional groups, all of which determine the properties of the different organic molecules.

ALL organic chemicals and substances are based on carbon and hydrocarbons.

Carbon is part of ALL known organic life, so no living organism can exist without carbon. Carbon is also the basis of organic chemistry, including fossils chemistry. Its reaction with oxygen produces CO_2, which is causing so many problems lately.

When combined with oxygen and hydrogen, carbon can form many groups of important biological compounds including sugars, lignans, chitins, alcohols, fats, and aromatic esters, carotenoids and terpenes.

When reacted with nitrogen it forms alkaloids, and with the addition of sulfur forms antibiotics, amino acids, and rubber products. Adding phosphorus to these other elements, forms DNA and RNA, the chemical-code carriers of life, and adenosine triphosphate (ATP), the most important energy-transfer molecule in all living cells.

Carbon Uses

Carbon on Earth is most abundant in coal, which is usually burned to generate electric power. There are thousands of other uses and products that of which carbon is a part.

Some of our daily encounters with, and major uses of, carbon are:

- Graphite combined with clays form the "lead" used in pencils.

- Diamonds are used for decorative purposes, and also as drill bits.

- Carbon is added to iron to make steel products.

- Carbon is used for control rods in nuclear reactors.

- Graphite carbon in a powdered, caked form is used as charcoal for cooking, artwork and other uses.

- Charcoal pills are used in medicine in pill or powder form to adsorb toxins or poisons from the digestive system.

- Carbon compounds are used in a great number of medical preparations.

- Basically, all compounds in the so-called organic chemistry branch are carbon based. This includes, coal, crude oil, natural gas, and all their derivatives.

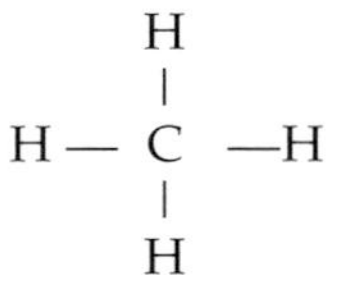

$$H$$
$$|$$
$$H — C — H$$
$$|$$
$$H$$

Figure 5-5. Methane molecule

The simplest carbon based molecule is that of methane, which is the major component in natural gas. Crude oil and coal are much more complex. By adding more hydrogen atoms we can change the methane molecule to others—ethane, hexane, octane, propane, etc.

Adding oxygen and other atoms would take us in the depths of the organic chemistry compounds, which are too deep and complex for our purposes, but which are the origins of most life matter and many things (plastics, cosmetics, and such) around us.

Now we know all about carbon and its role in the function of fossils, so it is time to look at our subject in this chapter—coal.

COAL

Coal is one of the major fossils. It is also the greatest friend humanity have ever had, since it provided 100% of the electric power we used most of the last century. It alone started and was primarily responsible for the technological and economic revolutions in the recent past that made our lives so comfortable and efficient.

Today, its contribution is down significantly, and yet it is still impossible to think of reliable power generation without coal. At least 50% of the electric and heat energy we use every day of our lives comes from coal. As a matter of fact, electric power base load would not be even possible to maintain, or at least not as reliably so, without coal. Period. And this won't change anytime soon. It is just not physically possible to replace coal 100%—not with what we have available today.

But coal is also our big foe, because its pollution is killing us. There is no way to put it in mild terms—it is a killer of vegetation, animals, and people in the local areas around mines and coal-fired power plants. It is also filling the atmosphere with green house gasses that promise to gradually make our lives more difficult, and even kill many of us—directly or indirectly. These Earth-choking gasses could eventually kill us all, if we don't take the necessary measures.

In any case, coal is here to stay, we need it regardless of its bad sides, and as such it deserves our respect and understanding.

A Close Look

Coal deserves a very, very close look, since it is mostly responsible for the economic prosperity we enjoy today. Coal needs to be fully understood, because we cannot live without it, and cannot live with it.

Coal provides cheap and reliable power to our homes and allows uninterrupted business activities, while at the same time it is killing us with its smog and pollution.

This makes it a major segment in the energy security puzzle, which we are trying sort out and clarify in this text.

We need coal, but it can hurt us. Since it will be around us for a very long time, we must respect it and try to understand all about its production and use.

We must understand the issues related to coal and the other fuels in order to form an educated opinion and make the right decisions.

Where do we start? Let's start from the very beginning:

In the Beginning...

How were beds of coal created in a way so convenient for mining? What forces were responsible for their creation, and support their existence? One theory claims that many centuries ago strange forces made a number of weird things happen very quickly. The skies became cloudy and the air so moist that even the sun's rays had a hard time going through the mush, according to this theory. The weather was very warm, and the crust of the earth was still new and not as thick as we know it now. Enormous amounts of heat were coming from the Earth's core, where nuclear reactions were going wild, contributing energy to the flesh and vegetation cooker on the surface of the Earth.

The vegetation and trees grew wildly in the moist air, and very large plants, trees and animals developed at the time. Forty- to fifty-foot-high vegetation and trees that grew extremely fast were the norm, especially in swamps.

When time came for them to die, others took their place within weeks. Dead leaves and trees fell and heaped up all over the place. Stumps stood up for awhile and decayed slowly and every year the soft, black, decaying mass of vegetation and animal carcasses grew deeper and deeper.

The crust of the fairly new Earth was still very thin and pliable, so it bent and wrinkled easily. Some swampy places rose up, and some sank, so that water rushed over them, pressing down on the deposits with its great pressure, while sweeping in sand and clay.

Under normal condition, the carbon in the wood burns completely, with only a pile of ashes left as a reminder. But if you cover the wood pile before lighting it, and make sure that only a little oxygen reaches it after it is lit, it will smolder slowly with much of the carbon left in it in a form of charcoal.

When wood decays, the process is similar to the smoldering, where carbon reacts with small amounts of oxygen in the air to create charcoal—provided not much oxygen reaches the decaying mass.

This is the big difference between the forests of today where leaves and trees fall and decay in the open air with nothing remaining. In the times when coal was forming, the water and soil covering the vegetation and dead animals restricted air access to the decaying mass, so that much of the carbon was left to form our useful coal deposits. High heat and pressure from above and below finished the job.

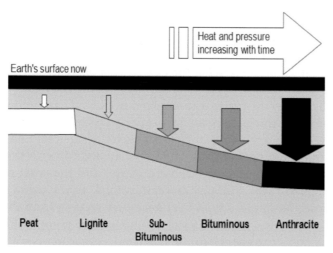

Figure 5-6. Coal formation with time.

Anthracite is the oldest coal. Formed so long ago and given a chance to mature, it is the highest quality coal and has the highest heat content. In contrast, peat is the youngest member of the coal family. It is still a baby that needs centuries to become mature coal.

Some of the layers, or strata, of coal are fifty or sixty feet thick, while some are very thin. Most, however, have a layer of sandstone or dark-gray shale on top, which was created by the sand and mud which covered the decaying mass. This layer, as the lid of a pressure cooker, sealed the mass under and kept the heat and pressure in until the process was complete.

The Earth's crust was thin in the beginning, so it

wrinkled under the weight of the mass; the pressure cooker lid was cracked open slightly. This brought some of the strata of coal to the surface, which we now find as surface coal deposits. These deposits are mined today in the surface (or pit) mines.

Some of the mass, however, travelled (together with its pressure cooker lid) deep underground, as if trying to hide. Today we can easily find it, but have to dig long and deep shafts to get to it. This is where miners go every day to do their dirty and dangerous job of digging the coal and taking it to the surface to be shipped to coal-burning power plants to make electricity for our comfort.

Sixty percent of U.S. coal is scraped and ripped out from the earth in surface mines. The rest is dug out from underground mines. Coal companies often remove entire mountaintops to expose the coal below, which is then easily excavated and loaded on trucks and railroad cars for transport to the power plants. Underground mines consist of large and very long holes dug deep into the ground, where miners descend into deep shafts to scrape and load the coal on carts and transport it to the surface.

It took billions of years to create this precious, albeit limited, coal, but it took us only a century to dig out and burn most of it.

Coal Properties

Coal has a large and complex molecule. It consists mostly of carbon combined with hydrogen, oxygen, nitrogen and sulfur atoms. The actual numbers of different atoms in each molecule vary, but basically speaking its chemical formula can be expressed as

$$C_{100}H_{73}O_8NS.$$

What this means is that for every 100 atoms of carbon in the coal molecule, there are 73 atoms of hydrogen, 8 oxygen, one nitrogen and one sulfur atom. The distribution of the atoms and links to each other are too complex for this text. Only organic chemists, specialized in this subject, understand exactly and can deal with the coal structure and the relations within, so we will leave this task to them.

For our purposes, it suffices to say that coal's chemical structures consist of a large number of carbon, hydrogen and oxygen atoms, which are different for each type of coal. The molecules are interlocked into long chains, which, in addition to the quantity of hydrogen and oxygen atoms, determine the overall properties and qualities of the different types of coals.

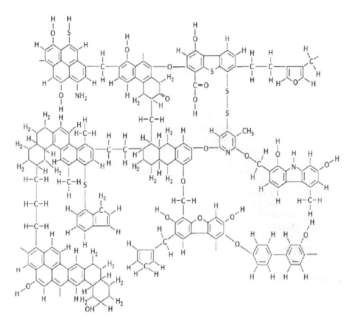

Figure 5-7. Segment of a coal molecule.

Coal Grade	(Btu/lb)
Anthracite	12910
Semi-Anthracite	13770
Low-volatile bituminous	14340
Medium-volatile bituminous	13840
High-volatile bituminous A	13090
High-volatile bituminous B	12130
High-volatile bituminous C	10750
Sub-bituminous B	9150
Sub-bituminous C	8940
Lignite	6900

Figure 5-8. Heating values of coal

value) is the factor that determines what coal is going to be used for, how much, and at what price. Anthracite and its sub-categories are the most valuable from that perspective and bring the highest price per unit as compared with most other coal types.

Another way of evaluating the heat content in coal is by looking at its calorific value Q, which is the actual heat liberated by its complete combustion at different oxygen levels. The Q value is determined experimentally by using special test equipment, such as calorimeters.

The approximate formula for determining Q at an oxygen content of less than 10% is:

$$Q = 337C + 1442(H - O/8) + 93S$$

Where:
 C is the mass percent of carbon in the coal sample,
 H is the mass percent of hydrogen,
 O is the mass percent of oxygen, and
 S is the mass percent of sulfur.
 Q in this case is given in kilojoules per kilogram.

Physical Properties
 Moisture is one of the most important properties of coal, which determines in part its commercial value. All coals contain some amount of moisture. External moisture is known as adventitious moisture and is readily evaporated during mining and transport. Moisture that is held within the coal itself is known as inherent moisture and is analyzed.

Moisture may occur in four possible forms within coal: a) surface moisture is water held on the surface of the coal particles, b) hydroscopic moisture is water held by capillary action within the microfractures of the coal pieces, c) decomposition moisture is water held within the coal's decomposed organic compounds, and d) mineral moisture is water which comprises part of the

Although the different types of coals have somewhat similar formulas, they have different qualities, which are determined by the arrangement of the atoms in the molecules. This arrangement varies with coal type and location and is usually an indication of the time and way the coals were formed.

One important thing to remember is that some coals have more hydrogen and oxygen atoms in their molecules than others, which determine their qualities. Basically, coals that were formed in the more recent past tend to have more hydrogen and oxygen atoms in their molecules. They are also of lower commercial quality and monetary value, mostly due to their lesser heating value.

There are four types or ranks: lignite or brown coal, bituminous coal or black coal, anthracite and graphite. Each of these has a certain set of physical characteristics which are expressed by moisture, volatile content (in terms of aliphatic or aromatic hydrocarbons) and carbon content in the coal.

A closer look of coal's properties:

Energy Content
 The most important (from commercial point of view) property of coal is its energy content. It is expressed in Btu/lb, or the amount of heat in Btus that a pound of coal can generate.

The energy or fuel content is the amount of potential energy contained in the coal that can be easily converted into actual heating ability. This heating ability (or heating crystal structure of hydrous silicates such as clays.

Volatile matter are all other components of coal, except for moisture, which can be freed (evaporated) at high temperature in absence of air. These are usually a mixture of short- and long-chain hydrocarbons, aromatic hydrocarbons and usually some sulfur. The volatile matter of coal is determined under rigidly controlled standards. In Europe this involves heating the coal sample to $1650 \pm 10°F$ in a muffle furnace. The US procedures involve heating the coal samples to $1740 \pm 45°F$ in a vertical platinum crucible and weighing the before and after samples. The difference is the volatile matter.

Fixed carbon is the carbon found in the material which is left after all volatile materials have been evaporated. This is different from the *ultimate* carbon content of the coal, because some carbon is lost with the volatiles. Fixed carbon is an estimate of the amount of coke that will be yielded from this type of coal. Fixed carbon is determined by removing the mass of volatiles determined by the volatility test from the original mass of the coal sample.

Ash is the non-combustible residue which is left after coal is burnt. It is the bulk mineral matter left behind after all carbon, oxygen, sulfur and water have been evaporated during combustion. Ash content is determined by burning the coal thoroughly, and the remaining ash material is weighed and expressed as a percentage of the original weight.

Coal is sorted and classified by rank, according to its physical properties, which is basically a measure of the process of its formation and present state. It represents the amount of alteration (the amount of heat and pressure) that the coal has undergone during its formation. The increase in rank describes an increase in temperature and pressure which results in the coals having a lower volatile content, therefore increased carbon content. That also determines its heat content, which in turn determines its commercial value, with anthracite having the highest heating and commercial value.

Coal is also classified according to the content of its contaminants, sulfur, phosphorous, volatile and ash contents, which generally vary according to, and determine, its rank. Consecutive stages in evolution of rank, from an initial peat stage, are brown coal (or lignite), sub-bituminous coal, bituminous coal, and anthracite.

Coking coal is used in the steel making industry where the coal requires specific qualities such as low sulfur and phosphorous contents. Approximately 630 kg of coal are used for every ton of steel produced. Electricity generation normally uses lignite (thermal) coal, which is ground to a fine powder prior to combustion.

Coal Types

Coal is classified into four general categories or "ranks." They range from lignite through subbituminous and bituminous to anthracite. These ranks reflect the formation process in terms of time and the progressive response of individual deposits of coal to increasing heat and pressure.

The carbon content of coal determines its heating value (or most of it), but other factors also influence the amount of energy it contains per unit of weight.

Note: The amount of energy in coal is expressed in British thermal units per pound. A Btu is the amount of heat required to raise the temperature of one pound of water one degree Fahrenheit.

About 90 percent of the coal in the U.S. falls in the bituminous and subbituminous categories, which rank below anthracite and, for the most part, contain less energy per unit of weight. Bituminous coal predominates in the eastern and mid-continent coal fields, while subbituminous coal is generally found in the western states and Alaska.

Lignite ranks the lowest and is the youngest of the coals. Most lignite is mined in Texas, but large deposits also are found in Montana, North Dakota, and some Gulf Coast states.

The coal types are determined and priced by their carbon and energy content, as determined during testing procedures. In practice, the different coal types are:

Commercial Classification

The commercial classification of coal is generally based on the content of volatiles. However, the exact classification varies between countries.

So, depending on the different variables, specific uses and application we now have several types of coal, as follow:

Anthracite

Anthracite is coal with the highest carbon content, between 86 and 98 percent, and a heat value of nearly 15,000 Btus per pound. Most frequently associated with home heating, anthracite is a very small segment of the U.S. coal market. There are 7.3 billion tons of anthracite reserves in the United States, found mostly in 11 northeastern counties in Pennsylvania.

Steam Coal

Steam coal is a grade between bituminous coal and anthracite, which was widely used as a fuel for steam locomotives in the past. In this specialized use, it is sometimes known as "sea-coal" in the US. Small steam

Class	Volatile matter weight in %
Anthracites	< 6.1
Dry steam coals	9.1 - 13.5
Cooking steams coal	15.1 - 17.0
Low volatile steam coal	19.1 - 19.5
Prime cooking coal	19.6 - 32.0
Heat altered coal	19.6 - 32.0
Strong coking coal	32.1 - 36.0
Medium coking coal	32.1 - 36.0
Weak coking coal	32.1
Very weak coking coal	32.1 - 36.0
Non-coking coal	32.1 - 36.0

Figure 5-9. Coal types per volatile matter content

coal, or dry small steam nuts, was used as a fuel for domestic water heating in the past as well. Today steam coal is used in a number of applications that can afford its somewhat higher cost.

Bituminous Coal

Bituminous coal has a carbon content ranging from 45 to 86 percent carbon and a heat value of 10,500 to 15,500 Btus per pound.

The most plentiful form of coal in the United States, bituminous coal, is used primarily to generate electricity and make coke for the steel industry. The fastest growing market for coal, though still a small one, is supplying heat for industrial processes.

Sub-bituminous Coal

Subbituminous coal contains 42 to 52% carbon (on a dry, ash-free basis), and its heat (calorific) values ranges between 19 and 26 MJ per kilogram, or about 8,200 to 11,200 Btus per pound. It is also characterized by greater compaction than lignite and has greater brightness and luster. The lower quality, woody-like structure lignite is not found in subbituminous coal, which exhibits alternating dull and bright maceral bands composed of vitrinite in patterns similar to those found in bituminous coals. Some subbituminous coal is macroscopically indistinguishable from bituminous coal.

Subbituminous coal is the type of coal used most extensively in coal-fired power plants, so it deserves a closer look. It is also called black lignite, and has dark brown to black color. Its qualities (and heating value) are between lignite and bituminous coal according to the coal classification used in the United States and Canada. In many other countries subbituminous coal is considered to be a brown coal and is used for domestic and commercial heating and cooking purposes.

There are reliable estimates that nearly half of the world's proven coal reserves are made up of subbituminous coal and lignite. There are large deposits of these coal types in Australia, Brazil, Canada, China, Russia, Ukraine, Germany, other European countries, and the United States.

Most subbituminous coal is considered to be relatively young from a geological point of view, dating from the Mesozoic and Cenozoic eras, or about 250 million years ago. Although age is important, the quality of coal is determined primarily by the pressure and temperature reached during the "cooking" cycle.

Subbituminous coal usually also contains less water, around 10 to 25% and is harder than lignite. This makes it more valuable and easier to transport, store, and use.

One characteristic of great importance to the energy generation sector is the fact that although subbituminous coal has lower heat value than bituminous coal, it is lower in sulfur content, usually less than 1 percent, so it is preferred in some cases. Since it has a lower heat value, however, more subbituminous coal must be burned to obtain an equal amount of energy.

Recently a number of coal-fired power generating plants have switched from burning bituminous coal to subbituminous coal and lignite. The main reason is the relatively low sulfur content.

Lignite

Lignite is a geologically young coal which has the lowest carbon content, 25-35 percent, and a heat value ranging between 4,000 and 8,300 Btus per pound. Sometimes called brown coal, it too is mainly used for electric power generation, but its low energy content makes it less desirable than any of the other types.

Peat

Peat is a fairly young coal, actually considered to be a precursor of coal. It has some industrial uses as a fuel for some special applications. One of the wide applications is its dehydrated form. Since dehydrated peat is a highly effective absorbent, it is used in great quantities for fuel and oil spills on land and water. It is also used as a conditioner for soil to make it more able to retain and slowly release water.

Graphite

Graphite is technically the highest rank of coal with the highest carbon content, but its physical structure and chemical properties make it difficult to ignite and is not commonly used as fuel. It is instead mostly used for making pencils, electrodes and, when powdered, it is widely used in lubricants.

Uses of Coal

In addition to its wide use in electric power generation and domestic heating and cooking, coal has a number of other, quite versatile applications. Some of the most important uses of coal follow.

Sea Coal

Finely ground bituminous coal, known in this application as sea coal, is a constituent of foundry sand. While the molten metal is in the mold, the coal burns slowly, releasing reducing gases at pressure, thus preventing the metal from penetrating the pores of the sand. It is also contained in "mould wash," a paste or liquid with the same function applied to the mould before casting.

Sea coal can be mixed with the clay lining (the "bod") used for the bottom of a cupola furnace. When heated, the coal decomposes and the bod becomes slightly friable, easing the process of breaking open holes for tapping the molten metal.

Chemicals Production

Coal is used extensively as feedstock to produce chemicals using processes which require substantial quantities of water, and which release a number of toxic gasses and liquids. Because of that, presently most of the coal-to-chemical production is concentrated in China, where environmental regulation and water management policies are weak to non-existent.

In coal-to-chemicals, synthesis gas (syngas), which is a gaseous mixture of primarily carbon monoxide and hydrogen gas is produced by gasification of coal. The syngas can then be used as a chemical building blocks in a number of chemical processes, such as in making methanol or acetyls.

Ammonia and urea are products of coal-to-chemicals for use in fertilizers. The syngas composition, or the ratio of hydrogen to carbon monoxide, is important for some downstream processes, so a water-gas shift reactor is sometimes used to change this balance.

Coking Coal

Coking coal (coke) is a solid carbonaceous residue derived from low-ash, low-sulfur bituminous coal from which the volatile constituents are driven off by baking in an oven in the absence of oxygen at 1,832°F in order to fuse the fixed carbon and residual ash together.

Coke from coal is grey, hard, and porous and has a heating value of 24.8 million Btu/ton (29.6 MJ/kg). Some coke-making processes produce valuable byproducts, including coal tar, ammonia, light oils, and coal gas.

Metallurgical coke is used as a fuel and as a reducing agent in smelting iron ore in a blast furnace. The result is pig iron, and is too rich in dissolved carbon, so it must be treated further to make steel. The coking coal is low in sulfur and phosphorus, so that they do not migrate into the metal to deteriorate its properties.

The coke is also strong enough to resist the weight of overburden in the blast furnace, which is why coking coal is so important in making steel using the conventional route. An alternative route is direct reduced iron, where any carbonaceous fuel can be used to make sponge or pelletized iron.

Petroleum coke is the solid residue obtained in oil refining, which resembles coke, but contains too many impurities to be useful in metallurgical applications.

Gasification

Coal gasification can be used to produce syngas, which is basically a mixture of carbon monoxide (CO) and hydrogen (H_2) gas. This syngas can then be converted into transportation fuels, such as gasoline and diesel, through the Fischer-Tropsch process. This technology is currently used by the Sasol chemical company of South Africa to make motor vehicle fuels from coal and natural gas. Alternatively, the hydrogen obtained from gasification can be used for various purposes, such as powering a hydrogen economy, making ammonia, or upgrading fossil fuels.

During gasification, the coal is mixed with oxygen and steam while also being heated and pressurized. During the reaction, oxygen and water molecules oxidize the coal into carbon monoxide (CO), while also releasing hydrogen gas (H_2).

This process has been conducted in both underground coal mines and in the production of town gas.

$$C \text{ (as Coal)} + O_2 + H_2O \rightarrow H_2 + CO$$

If the refiner wants to produce gasoline, the syngas is collected at this state and routed into a Fischer-Tropsch reaction. If hydrogen is the desired end-product, however, the syngas is fed into the water gas shift reaction, where more hydrogen is liberated.

$$CO + H_2O \rightarrow CO_2 + H_2$$

In the past, coal was converted to make coal gas (town gas), which was piped to customers to burn for illumination, heating, and cooking.

Liquefaction

Coal can also be converted into synthetic fuels equivalent to gasoline or diesel by several different processes. In the direct liquefaction processes, the coal is either hydrogenated or carbonized. Hydrogenation processes are the Bergius process, the SRC-I and SRC-II (Solvent Refined Coal) processes and the NUS Corporation hydrogenation process. In the process of low-temperature carbonization, coal is coked at temperatures between 680 and 1,380°F. These temperatures optimize the production of coal tars richer in lighter hydrocarbons than normal coal tar. The coal tar is then further processed into fuels.

Alternatively, coal can be converted into a gas first, and then into a liquid, by using the Fischer-Tropsch process. Coal liquefaction methods involve carbon dioxide (CO_2) emissions in the conversion process. If coal liquefaction is done without employing either carbon capture and storage (CCS) technologies or biomass blending, the result is lifecycle greenhouse gas footprints that are generally greater than those released in the extraction and refinement of liquid fuel production from crude oil.

If CCS technologies are employed, reductions of 5-12% can be achieved in Coal to Liquid (CTL) plants and up to a 75% reduction is achievable when co-gasifying coal with commercially demonstrated levels of biomass (30% biomass by weight) in coal/biomass-to-liquids plants. For future synthetic fuel projects, carbon dioxide sequestration is proposed to avoid releasing CO_2 into the atmosphere.

Sequestration, however, adds to the cost of production. Currently, all US and at least one Chinese synthetic fuel projects, include sequestration in their process designs.

Refined Coal

Refined coal is the product of a coal-upgrading technology that removes moisture and certain pollutants from lower-rank coals such as sub-bituminous and lignite (brown) coals. It is one form of several pre-combustion treatments and processes for coal that alter coal's characteristics before it is burned.

The goals of pre-combustion coal technologies are to increase efficiency and reduce emissions when the coal is burned. Depending on the situation, pre-combustion technology can be used in place of or as a supplement to post-combustion technologies to control emissions from coal-fueled boilers.

COAL MINING

Coal is found in many places around the U.S. and the globe. Not all coal deposits are physically or economically feasible to exploit, so many are sitting deep underground awaiting their day under the sun.

The most economical method of coal extraction from coal seams varies, depending on the depth and quality of the seams, the local geology, and a number of environmental factors. Coal mining processes are mainly differentiated by whether they operate *on the surface* or *underground*.

As the name suggests, *surface mining* entails gathering coal from, or close to, the surface. Underground mining, on the other hand, is what we are more familiar with since it is the oldest method of coal mining.

For obvious reasons, surface mining is cleaner and safer than underground mining, but it causes serious visual and environmental damages. Underground mining is equally harmful to the environment, but the scars and damage are much less visible in most cases. It is also one of the most dangerous human undertakings on Earth.

Mine Planning

To determine the technical and economic feasibility of the mine, a number of pieces of information are needed, such as:

- *Market analysis*. Potential customers, contract agreements, size and location of the markets, etc. data needed to figure out the size and other peculiarities of the potential markets.

- *Transportation*. Size and type of loads, property access, elevation of the mine and the customer sites, road system, etc.

- *Utilities*. Availability and distance to electric power and substations, location, rights-of-way and the related costs for new transmission lines.

- *Water*. Type, quantity, and quality of potable and process water, and source location, type, and the cost of aqueducts, etc.

- *Labor*. Local availability, type and quality, rates and trends, and the local labor history.

The technical and economic feasibilities of coal mining are evaluated according to the following criteria:

- Regional geologic conditions

- Overburden characteristics

- Coal seam continuity, thickness, structure, quality, and depth

- Condition of the ground above and below the seam for use as roof and floor

- Local topography, especially altitude and slope

- Local climate and weather conditions

- Land ownership as it affects the availability of land for mining and access

- Surface drainage patterns

- Ground water conditions

- Availability of labor and materials

- Coal purchaser's requirements in terms of tonnage, quality, and destination

- Capital investment requirements

Based on these analyses, experts can determine if mining coal in a particular location is feasible and profitable. If yes, then they have to decide what type of mining process to use.

Surface, or strip, mining is usually preferred for extracting coal that is less than 200 feet under the surface. So, today we have two distinct methods of mining coal: a) surface, or strip, mining and, b) underground mining.

For example, coal that is found at depths of 180 to 300 ft is usually mined in underground mines, but in some cases surface mining techniques can be used. For example, some western U.S. coals that occur at depths in excess of 200 ft are mined by the open pit methods, due to thickness of the seam of 60–90 feet.

Coals deposits below 300 ft are usually mined in underground mines. Although there are open pit mining operations working on coal seams up to 1000-1500 feet below ground level, for instance in some regions in Germany, they are exceptions.

Below we will take a close look at the cradle-to-grave coal mining and burning operations. "Cradle-to-grave" in this case means every single step of the process, and every single material and effort used from the very beginning of the coal mine plan to the very end of its existence and burial. All these materials and efforts have to be accounted for, because they have some impact on the overall process.

The cradle-to-grave coal mining, cleaning, transport and burning process basically consists of:

Planning
- Locating, testing and exploring the coal reserves in the proposed mine location
- Estimating the construction and operating cost and the related issues
- Applying and obtaining the necessary federal, state and local permits

Design
- Mine area design
- Equipment design
- Process design

Construction
- Mine construction
- Surface facilities
- Infrastructure

Daily operations
- Coal digging and surface transport
- Coal preparation and waste treatment
- Coal and waste transport

Decommissioning and land reclamation
- Mine shutdown and evacuation
- Mine reconstruction
- Surface land reclamation

In addition, a major portion of thus mined coal is transported to a coal-fired power plant to be burned into furnaces. The heat of these furnaces is used to boil water and generate steam, which is run through turbines attached to electricity generators. Thus generated electric power is sent into the national grid for use at residences and businesses.

Let's start with a close look at coal mines' development and operation:

Surface Mining in Brief

When coal seams are near the surface, it may be economical to extract the coal using surface (also referred to as open cut, open cast, open pit, or strip) mining methods. Surface coal mining recovers a greater proportion of the coal deposit than underground methods, as more of the coal seams in the strata may be exploited.

Large surface mines can cover an area of many square kilometers and use very large pieces of equipment. This equipment can include draglines which operate by removing the overburden, power shovels, large trucks which transport overburden and coal, bucket wheel excavators, and conveyors.

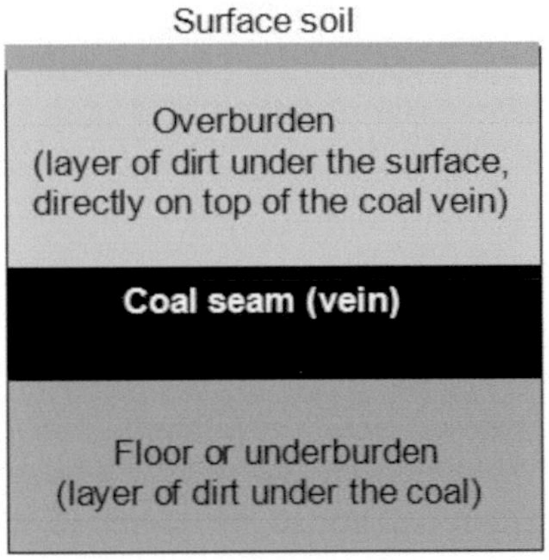

Surface soil

Overburden
(layer of dirt under the surface,
directly on top of the coal vein)

Coal seam (vein)

Floor or underburden
(layer of dirt under the coal)

Figure 5-10. Coal deposit, cross section

In surface mining, explosives are first used to break through the surface soil and the overburden, of the mining area. The overburden is then removed by draglines or by power shovel and truck. Once the coal seam is exposed, it is drilled, fractured and thoroughly mined in strips. The coal is then loaded onto large trucks or conveyors for transport to either the coal preparation plant or directly to where it will be used.

Most open cast mines in the U.S extract bituminous coal. In Australia and South Africa open cast mining is used for both thermal and metallurgical coals. In New South Wales open casting for steam coal and anthracite is practiced. Surface mining accounts for around 80 percent of production in Australia, while in the U.S. it is used for about 67 percent of production. Globally, about 40 percent of coal production involves surface mining.

Surface mining is the most widely spread mining method, with most obvious—in your face—environmental damages. Underground mining has its problems too, but they are hidden and harder to observe.

Underground Mining in Brief

Most coal seams are deep underground, and since opencast mining is impractical, underground mining is used instead. It accounts for about 60% of world coal production. Underground mining offers a number of totally different challenges extracting coal from the Earth. Since coal is very deep in some locations, and strip mining would not work there, deep holes (shafts) are dug into the ground down to the level of the coal veins. Miners are lowered down into the hole, where they dig

the coal and bring it to the surface.

Figure 5-11. Underground mining operations

Let's follow the miners in their daily work. They arrive early morning by the shaft's opening and get into an elevator, which lowers them down hundreds and sometimes thousands of feet underground. The cage stops at the bottom of the mine shaft, and the miners spill out to their assigned places for the daily toil.

Some of them dig with picks, some drill holes, some drive coal trains, but all are covered with, and breathe, great amounts of black coal dust. Most miners wear tiny lamps on their caps, since it is dark in there and not many lamps are allowed. Today large machines do most of the digging and loading, but manual work in different areas of the mine operations is still a major part of the miners daily effort.

Looking around you'll see a low roof, usually held up by pillars of wood and coal. It is noisy and dusty in there. From time to time even greater noise approaches and a light moving towards the workers signals the approaching coal train, dragging many cars loaded with coal. It soon rushes back with empty cars, taking them to be refilled. And explosions thunder at certain periods, when new veins are exploited.

There are small rooms and larger chambers everywhere you go. They were created by digging out the coal and taking it up to the surface. You can only imagine the amount of earth and rock above the ceilings of these rooms; hundreds of feet in thickness. And you also start imagining what would happen, if the roof in one of these rooms caved in. The rooms are usually small enough to not allow caving of the roofs, but you never know. Stuff happens…

Note: In many cases the "room-and-pillar" mining methods are used, where "pillars" of coal are created by digging the coal around the columns, pillars. that serve as supports for the rooms' ceilings. This method accounts for a significant portion of the total mineral production in the United States. Well in excess of $6 billion worth of mineral commodities are produced each year by this method. A substantial portion ($3.55 billion) of coal production still comes from room-and-pillar mining.

Coal mines experience large-scale, catastrophic pillar collapses, if the strength of a pillar in a room-and-pillar mine is exceeded. When one pillar collapses, the load that it supported is transferred to neighboring pillars. The additional load on these pillars may lead to increased stress of their structures and the ceiling above them.

This mechanism of pillar overloading, load transfer, and continuing pillar failures can lead to the rapid collapse of very large areas of a mine. In some cases, only a few pillars might fail; however, in extreme cases, hundreds, even thousands, of pillars can fail.

This kind of failure has many names, such as progressive pillar failure, massive pillar collapse, domino-type failure, or pillar run. A special term, "cascading pillar failure" or CPF has been coined to describe and study these rapid pillar collapses. There are over 21 instances of large-scale pillar collapses in room-and-pillar mines, mainly in the United States, in the recent past.

There are many other dangers of underground mining, which we will review in the next chapters.

Coal Mines Logistics

Large-scale industrial process operations are expensive, and most of them cause significant environmental damage—usually negative—which must be taken into consideration and calculated in terms of economic and health terms. Coal mining is no exception.

When all costs are considered, the process of locating the coal, applying for and getting the necessary permits, designing the mine structure (including facilities, equipment, process, and labor), hiring and training engineers, technicians and engineers, and the mining operations process (its daily operation and maintenance) is overwhelming and expensive. There are a number of steps in the overall mining process that cause significant air, soil or water table damage.

Let's take a close look at the entire cradle-to-grave mining process and estimate the cost and environmental and other damages at each step of the process.

It all starts with an idea, "Hey, there is lots of coal in this area. Let's go and get it." When the principals agree and decide that a mine is justifiable in the reference area, they start the cradle-to-grave coal mining process. This consists of cleaning, transport and burning—a number of consecutive steps, which vary in type and magnitude from state to state and from country to country. Below, we explore these processes in more detail.

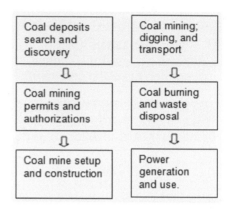

Figure 5-12. Cradle-to-grave coal power generation

Preliminary Mine Design Considerations

Coal mine sites vary in type and size, but typically involve large land areas, especially when surface mining methods are used. Coal seams vary in size, depth and quality, and the mining methods are chosen on the basis of physical feasibility, economical viability and safety. The size of the mining operation will depend on the site characteristics, the coal reserve, and the mining method.

For coal mined by surface methods (the mine plant) should be located off the outcrop, as in a visible exposure of coal deposits on a hill, if possible.

The mine plant itself consists of coal handling and storage facilities, offices, shops and laboratories, equipment storage buildings, and waste disposal areas.

Access to coal deposits at a surface operation involves the use of large equipment such as bucket-wheel excavators, draglines, and shovels to remove overburden from the coal, so extraction can begin. As mining progresses, development consists mainly of extending paved roads and power lines, and constructing new roads for access to the coal deposit.

The mined coal typically goes on a conveyor belt and on small cars to a preparation plant that is located close to the mining site. At the plant the coal gets cleaned and otherwise processed to remove dirt, rocks, ashes, sulfur, and other unwanted materials. After that, the coal is sorted by quality and size, according to the customer's needs. This is needed to increase the heating value of the coal.

Once the coal is processed, it is shipped typically by rail, but also by truck or barge or even a coal-slurry pipeline, to the coal burning power plant. Transportation methods depend on the distance to be traveled, as well as the access to existing transportation systems.

Coal is delivered to coal-fired power plants and burnt to boil water. Thus produced steam is injected in steam turbines that turn generators to produces electricity. The electricity is sent into the system that consists of electric transmission lines , towers, substations and other components. Coal accounts for over 50% of the electricity produced in the US.

The process, however, starts with finding the coal seam, or vein.

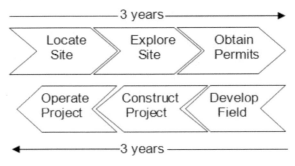

Figure 5-13. Project timeline

Coal Reserves Locating, Tests and Estimates

Before a coal mine can be planned and built, a number of tests and surveys must be conducted to estimate the type and quantity of coal in the particular location. Then engineering and financial estimates and calculations must be made to ensure the project's technical and economical feasibility. The environmental aspects of the entire cradle-to-grave process are plugged in as well, to complete the picture.

These activities include historical research, mapping, drilling (to obtain geological samples), and geophysical exploration. The latter includes a number of field and lab tests, such as aerial photography, airborne geophysical surveys (magnetic, radiometric, and electromagnetic); on-the-ground geophysical surveys (drillhole logging ; electrical, magnetic, electromagnetic, radiometric, gravimetric, and refraction-seismic surveys; and induced polarization surveys using exposed electrodes). Field surveys for identifying cultural resources , paleontological resources , and ecological resources (habitats, species, etc.) in the project area should also be conducted during this phase.

Reclamation of exploratory areas that will not be part of the project area would occur during this phase as well.

A number of engineering disciplines and scientific methods are applied when evaluating all activities related to mine design, and estimating its technological and economic potential. The cradle-to-grave cost and environmental impact are major parts considered during every step of the process.

At the end, the obtained data and information about the different aspects of the potential mine are provided to the mine engineers, managers and investors to make decisions on plant design and financing. Some of the information needed for such decisions follows.

Geology

Complete geological information on the soil, overburden and coal deposits, obtained from historical data and from the preliminary exploration steps includes the overall geological structure and the relevant physical properties of the proposed mine, the thickness and variability of the different layers, the quality and quantity of the coal seams—all vital information.

Geometry and Geography

A complete picture of the location, its climate and earthquake zoning are needed, as well as the size shape, continuity, attitude, and drainage patterns of the different layers.

Hydrology

A complete picture is assembled of the permeability and porosity of the overburden and coal layers, as well as location of the aquifers.

Toxicology

A complete overview of the hazardous, toxic, reactive, and radioactive species found on the surface and in the ground is assembled. Best and worst case scenarios are developed for dealing with industrial accidents and hazmat situations during operation, as well as during the mine's decommissioning and land reclamation stages.

Surface Mines

The planning and design stages of a large size surface or pit mine in the U.S. is a complex and specialized undertaking, with many different and equally complex variables. The information collected in the evaluation stage, as well as all available project-related information, are reviewed as a first step, to develop the most appropriate plan for the particular case coal extraction plan.

Various mining methods and equipment set combinations and permutations are considered by the team of mining engineers, geologists, environmentalists, and economists. After the technology, process and equip-

Figure 5-14. Surface mine

ment set are determined, an economic and market analysis are performed of the best case scenario, as needed to determine the project viability from a financial point of view.

The preliminary mine plan usually goes through a number of iterations, which finally result in a technically feasible and economical operating plan, subject to the specific contractual, legal, environmental, and other constraints of the specific property and the related equipment and operating procedures.

The process of strip mining is quite simple and self-explanatory. It basically consists of a) carefully clearing the topsoil (mountaintop removal in some cases) and stockpiling it for reclamation, b) removing the overburden and stockpiling it for the same reasons, and then c) mining the coal seam.

The coal is removed from the ground layers by a number of methods, depending on type of formation etc. It is the loaded on dump trucks or conveyor belt for transport to the preparation plant. Here it is processed as required by the customer and loaded on railroad cars for transport to the coal burning power plant.

As the mining of the surface seam moves forward, the mined area is simultaneously reclaimed by replacing and re-contouring the overburden and replacing the topsoil.

In open-pit mines, the mining begins by drilling and blasting waste rock and clearing the overburden and debris to expose the coal seam. The removing of the coal is done one layer (bench) at a time, which forms terraces. The mined bench continues to get wider as the mining continues, and goes deeper with each bench.

This is a long and complex process, which, together with obtaining the necessary permits, etc., could take 10-15 years, or more, in addition to millions of dollars of related expenses—long before a single lump of coal is removed from the mine.

Here are some of the lengthy, complex, and expensive steps in this process:

Environmental impact package
- Initial site evaluation
- Scope of work program
- Environmental impact report
- Environmental monitoring program
- Reclamation bonds

Coal mining package
- Lease or buyout rights acquisition
- Mapping the area
- Site drilling and sample analysis
- Coal type and quantity evaluation
- Drilling coal samples and analysis
- Development drilling and sample analysis

Land rights
- Land requirements, surface land and minerals ownership, oil and gas wells location and ownership, etc.

Taxes and royalties
- Federal, state and local taxation, royalties payments, zoning, and operating and reclamation requirements.

Process design and equipment set planning
- Concept mine design
- Mining process design
- Equipment selection and ordering
- Economic evaluation
- Overall mining plan development

NEPA procedures
- Lead EIS agency communications
- EIS draft and reviews
- EIS hearing and federal review
- CEQ filing and approvals

Permits
- Surface drilling rights acquisition
- Federal permits
- USFS land use
- State water use and mining
- State industrial siting
- Local permits

Mining preparation

- Stripping equipment setup
- Loaders and conveyors setup
- Support equipment setup
- Labor hiring and training
- Production ramp-up procedures
- Full production specs and documentation

A quick look at these steps reveals that:

- The environmental studies of the proposed site, combined with post-test monitoring, are a major part of the process, and could take a long time—sometimes 10 years or more.

- The initial process design and equipment ordering are strictly engineering disciplines. Their completion is also of long duration and might take 2-3 years at a minimum.

- The design and construction of the site facilities and support structures are specialized tasks that would take another 2-3 years or more.

- Completing the National Environmental Policy Act (NEPA) requirements and the related procedures could take 5-6 years.

- The necessary federal, state, local, tribal and other permits and negotiations take the project well over the 10-year mark, and might even extend it *ad infinitum*.

- If all goes well with the above tasks and sub-projects, and if their design and implementation are well coordinated and properly and efficiently executed, the initial setup and limited coal production could be started around year 10 from the beginning of the project.

It is important to point out that most of the tasks and sub-projects mentioned above can be conducted in parallel, thus saving time.

There are a number of additional complex details that go into the mine development and exploitation process. The design and equipment selection, for example, are very special and vary with individual sites. In some cases, these are considered "exceptional."

Some of these exceptions are:

- *Equipment limitations*

  In many cases the size and type of equipment limits the full exploration of the coal seams. Land instability might also limit the activities in some areas. Some of these limitations cannot be foreseen, and so they must be dealt with by planning on eventual equipment or strategy changes, to obtain the maximum amount of coal from the area.

- *Mining losses*

  As in any process, mining also suffers losses during full operation. Some mines are different than others. The most common losses are those in the top, bottom, and rib coal layer. These are losses, where the loaders, for a number of reasons, cannot grab all the coal from the top, bottom and the sides of the coal seam.

  There are also fly rock losses from blasting, and transportation losses from dump trucks and conveyor belts. These are so-called barrier losses, and can amount to over 5% of the mine gross, which is a significant number and which must be taken into consideration in the preliminary designs.

When finally a green light is given, a year or more will be needed for final setup, operators training, and ramp up of the production cycle. Another year will be needed to establish the final, full production cycle and all of its steps, specs, procedures, training, and the related documentation. Then the mine is in full production until there is enough coal to dig out at full speed.

A day comes many years later, when the coal deposits in the mine seams diminish and the mine production volume is reduced. This requires special, albeit temporary, measures, which eventually lead to reduced profits. Eventually the mine has to be shut down, and special procedures have to be followed to accomplish that process successfully and safely.

After shutdown of mining operations, the mine area has to be reconstructed as close as possible to its original state. This is seldom possible, so there are a number of mines that have been simply abandoned, while others are partially reconstructed.

Note: As part of the permitting process, new mine constructions in the US are required to purchase Reclamation Bonds, as required by the Bureau of Land Management (BLM) and various state environmental agencies. These are long-term surety obligations, which are like insurance that the site will be returned to its original condition upon termination of the mining operations.

Once issued, the bonds cannot be canceled. Adequate performance can be highly subjective, and bond penalties and losses in some cases can be large.

Surface Mine Operation

Surface mining, "strip mining," is used where large coal deposits are near the surface. This way the coal can be reached and extracted easily by digging deep trenches and loading it directly on large trucks. This method has its advantages; it is much safer and more efficient than underground mining. It also has disadvantages; it damages the Earth's surface and changes the environment and life in it for miles around.

Surface mining is simply working on the surface of a mountain or a hill, where huge machines, called power shovels, scoop large amounts of soil just above the coal seam. The seam, also called overburden is then dug out by the same machines in similarly large quantities.

This technique has vast economic advantages since manual labor is replaced by machines. Some of these power shovels have buckets that hold over 200 cubic yards, or over 300 tons with each scoop, which is enough to fill 15-20 regular dump trucks.

This level of mechanization and automation of strip-mined coal is responsible for the low price of coal, which is estimated at almost two times lower cost than underground mined coal. And the gap is increasing due to increasing labor costs and new mine safety regulations for underground mines.

There are several types of surface mining:

- *Open pit mining* is used on very thick coal seams,

- *Contour mining* is used in mostly hilly terrain, and

- *Auger mining* usually accompanies contour mining.

- *Mountain top removal mining* involves removal of large land areas.

- *Area Mining* is a surface mining method where the overburden is removed in mile-long, 100-ft-wide strips. Holes are drilled in the exposed coal seam, filled with explosives and blasted. The coal is scooped and loaded onto dump trucks or conveyors for transport to the coal preparation (or wash) plant.

- *Area mining* is the preferred technique in most areas,

Once this strip is empty of coal, the process is repeated with a new strip being created next to it. At the same time, but with some delay, a parallel trench is dug close to the first trench and the overburden from it (the new trench) is dumped in the old trench as the machines move forward simultaneously.

The coal from the new trench is also removed, and a third, parallel trench is dug. This process is repeated until all the coal on the surface is removed.

If the vein is deep enough, the process might be taken to a deeper level. The depth at which coal can be profitably reached has been increasing as the equipment has become larger and more efficient. The maximum overburden that could be handled at first was about 70 feet, increasing to 125 feet with time, and now it is close to 200 feet.

The equipment used in strip mining depends on the local geological conditions. For example, to remove overburden that is loose or unconsolidated, a bucket wheel excavator might be used as the most efficient and productive piece of equipment.

Some area mines may be productive for more than 50 years.

The problem of the area mining method is that flat farmland is replaced by a series of ridges and gullies, which brings the usefulness of the land and its value for crops, recreation, or anything else, to practically zero. Elaborate reclamation processes can be used to restore the land to its original shape and quality, but this is a very expensive, and rarely completely successful undertaking.

Area mining is widespread in the Midwest, and the huge coal reserves in Montana, Wyoming, the Dakotas, and the Southwest. About 1/3 of the estimated 440 billion tons of U.S. coal reserves lies in beds that are less than 100 feet below the surface, thus suitable for strip mining.

- *Open Pit Mining* is similar to area mining except that larger and often deeper trenches of up to 1000 feet wide, are usually dug out. This method is used on thick beds of coal, and similarly when the coal is removed from the first trench, it is filled with overburden from the second trench. The land is left in a similarly useless shape and quality for any practical purpose.

- *Contour mining* is used in mountainous coal deposits, and consists of removing overburden from the seam in a pattern following the contours along a ridge or around a hillside. It is most commonly used in areas with rolling to steep terrain. It is quite different from area strip mining, and is much more destructive.

The process begins with the power shovel cutting at the area where the coal seam reaches the surface. The resulting overburden is pushed down the mountainside and the coal is removed in long strips, in a way very similar to paring an apple.

The haul-back or lateral movement method is widely used and consists of an initial cut with the overburden (or spoil) deposited down slope or at some other site. Spoil from the second cut usually refills the first. A ridge of undisturbed natural material 15 to 20 ft (5-6 m) wide is often intentionally left at the outer edge of the mined area. This barrier adds stability to the reclaimed slope by preventing spoil from slumping or sliding downhill

The shovel moves toward the center of the mountain, with each cut removing more overburden. A wall over 100 feet high is created and is too thick to remove, and an auger is used at that time to drill out more coal.

The overburden is stacked on the edge or thrown down the slope, and this is what causes the major damage to the surrounding area. The loose soil in an area with no trees or grass to anchor it is easily eroded and washed down the hill and into the streams below. Heavily travelled access roads, with heavy dump trucks and other equipment roaring up and down day and night, add even more to the erosion and overall area damage.

But erosion is not the only problem. Loose boulders and landslides are responsible for even larger and more permanent damage. Entire towns in Wales and Virginia have been buried by such landslides.

The remaining high walls circle the mountains, making them useless and inaccessible. There are over a million such disturbed acres in the U.S., with an additional 30,000 acres added every year. Most of this land remains unreclaimed.

Adding insult to the injury, rainwater floods the abandoned coal seams and reacts with the sulfur-containing pyrites and other minerals in them. Leachate, a corrosive and destructive pollutant such as sulfuric acid is formed in many areas and runs downhill killing all vegetation. It then runs into the streams where it kills aquatic life by increasing the acidity.

Mine drainage also contributes to increased amounts of sediments, sulfates, iron, and hardness in the streams and lakes, which also changes their environment and affects the life forms there.

Surveys in some post-contour stripping areas in the U.S. show that about 5% percent of the hillsides had a pH less than 3 (highly acidic), and 80% with pH of 3-5—which is also unacceptable. Over 6000 miles in of the Appalachian streams are affected by acid mine drainage, and over 11,000 miles of streams are affected by other mine pollutants as well.

The limitations on contour strip mining are economic and technical. When the operation reaches a predetermined stripping ratio (tons of overburden/tons of coal), it becomes unprofitable. Also, depending on the equipment used, it may not be technically feasible to reach a certain height of high wall. Producing more coal with the auger method is possible today.

- *Auger mining* is a method for coal extraction by boring into a coal seam at the base of strata exposed by excavation. Augering is usually associated with contour strip-mining, recovering coal for a limited depth (up to 1,000 feet) beyond which stripping becomes uneconomical because the seam of coal lies so far beneath the surface. It is also limited to horizontal or slightly pitched seams that have been exposed by geologic erosion.

In this process, auger drills mounted with cutter heads cut and fracture through both overburden and coal, operating very similarly to a drill machine. The augering differs from other types of coal cutting machines such as continuous miners in that it tends to exploit the lower tensile strength of coal rather than trying to over compensate for its high compressive strength.

The power of the auger as well as diameter of the cutterhead are the two features that govern an auger drill's performance. The greater the power of the machine, the greater the depth of the coal seam into which it is able to bore, producing a higher rate of coal.

Auger drills used in auger mining can range from 60 to 200 feet (18 to 61 m) in length, and two to seven feet (0.6 to 2.1 m) in diameter. The cutter head on the auger bores a number of openings into the seam, similar to how a wood drill produces wood shavings. The coal is then extracted and transported up to the surface.

As the depth of the bored hole is extended, coal production is most likely to decrease. The auger drill will continue to penetrate into a high wall until the maximum torque of an auger is reached, usually at a depth of 492 feet (150 m). Once the coal arrives at the surface, it is lifted up to a dump truck for hauling by a conveyor or front-end loader.

Recent auger drill technology has led to the introduction of a new type of auger drilling machine called the thin-seam miner (TSM). It is actually a type of continuous miner that can cut an entry up to eight feet (2.4 m) wide and up to five feet (1.5 m) high into a coal seam situated under a high wall in surface mines.

One of the drawbacks of this method is that once the cutter head enters the coal seam, the operator is unable to view the cutting action directly and must rely more on a sense of feel for the machine, to control it and its performance, as well as to detect potential problems.

- *Mountaintop removal mining* is a surface mining method that relies on the removal of entire mountaintops, to expose large coal seams. Mountaintop removal is a combination of area and contour strip mining methods. In areas with rolling or steep terrain with a coal seam occurring near the top of a ridge or hill, the entire top is removed in a series of parallel cuts, with the overburden deposited in nearby valleys and hollows.

This method usually leaves ridge and hilltops as flattened plateaus. The process is highly controversial since it creates drastic changes in local topography. It is accompanied by the creation of head-of-hollow-fills, or filling in valleys with mining debris, and for covering streams and disrupting ecosystems.

In preparation for filling the overburden disposal area, vegetation and soil are removed and a rock drain is constructed down the middle of the area to be filled, thus replacing natural drainage. Upon completion of the fill, the underdrain forms a continuous water runoff system from the upper end of the valley to the lower end of the fill. Typical head-of-hollow fills are graded and terraced to create permanently stable slopes.

Always, the change of many acres of the local area is dramatic and permanent.

Underground Mines

Until recently, mining was a fully manual job, where miners cut into the walls and ceilings of the rooms with special pickaxes. They reached as far as the pickaxe would reach, and then a hole was bored into the top of the coal, and a cartridge of dynamite was exploded to release a ton or two of coal. The coal was then shoveled into a car and pushed out of the room to join the long string of cars going to the surface. The digging and exploding work continued all day.

Today, miners cut and dig the coal with machines that are as sophisticated and safe as the state of the art and economics allow. The machines grind the coal, and leave a deep cut all along the sides of the room. Then another group of miners bore holes in the walls for blasting. The holes are made with powerful, compressed air-driven drills, and dynamite charges are placed in the holes.

After the explosions are set off, and the dust has settled, the coal is loaded manually or with machines into the cars. Then it is taken to the surface and made ready for market.

Modern, fully automated operations are the most efficient and safest in deep mining. This type of mining is done along the seam with machines, and pillars and timbers are left standing to support the mine roof.

When the seam mining is complete, for reasons of geology formations or economics, a supplementary version of the room and pillar mining (second mining) takes over. It consists of removal of the coal in the support pillars in the dugout, thus recovering the maximum amount of coal possible from the seam.

Modern methods for coal pillar sections removal use remote-controlled equipment, which includes large hydraulic machines that are used to support the roof during the pillar removal process. The mobile roof supports look like a large dining-room table with hydraulic jacks for legs. After the coal pillars are removed, the legs of the table are shortened and it is withdrawn to a safe area.

This efficient and safe method is used to prevent cave-ins until the miners and their equipment have left a work area, because the unsupported roof of the room usually collapses when the roof supports are removed.

If the coal is to be mined by underground methods, the mine plant is constructed near the main portal or entrance.

Access to coal deposits at an underground operation is provided by drifts, slopes, or shafts. The coal bed is developed for further operations by driving entries. Although terminology varies, the following system of entries is universal in the industry.

Main entries are extensions of access openings and often run several miles in one direction. Three or more parallel entries, 12 to 22 feet wide and 40 to 100 feet between centers, are driven in a given direction and connected at intervals by crosscuts to provide proper air circulation. These are the major routes of underground transport and access, and serve for the life of the mine.

Panel entries are driven from the main entries, resulting in a subdivision of the coal bed into blocks or panels having dimensions that may be as large as 1 by 1/2 mile. Panel entries serve as routes from the main entries to the working places, and for air circulation. Although coal is removed during the driving of the main and panel entries, the production cycle begins upon completion of the panel entries.

Underground Mine Planning and Design

The main activities during planning and design of mine construction and during operation phases of either surface or underground mining are focused on the efficient and safe function of the proposed mine as well as any auxiliary facility (e.g., shaft construction) and coal transport system (e.g., access roads, rail lines, pipelines, conveyor systems).

As with the surface mine, we need to go through a number of steps, some of which are:

1. Coal mine
    *   Planning and design
    *   Permitting
    *   Land preparation
    *   Facilities construction
    *   Equipment purchase, delivery and installation

2. Daily operations
    *   Coal digging
    *   Onsite transport
    *   Coal preparation (on-site)

3. Coal transport
    *   Loading
    *   Unit train transport
    *   Unloading

4. Coal preparation (off-site)
    *   Sorting
    *   Washing
    *   Treating

5. End of life decommissioning and waste disposal
    *   Mine or plant shutdown and disabling
    *   Equipment disassembly and demolition
    *   Facilities demolition
    *   Waste disposal
    *   Land decontamination and reconstruction

Here, as in the surface mines' initial steps, we need to go through the different steps of site location, testing, and exploration. Then many months and thousands of dollars will be spent obtaining the necessary federal, state and local permits and approvals. Design of the actual mine, the mining equipment, and the support infrastructure (buildings, roads, railroad, etc.) follow. When all these tasks are completed, the miners go to work.

Underground Mine Development

The process of underground mining includes cutting into the coal deposit and removing it from the coal face via room-and-pillar methods using a continuous mining machine, or through longwall methods using a longwall cutting machine. In either method, once the coal is removed, the supports or pillars can be removed and the roof of the mine is allowed to collapse. The mined area is then abandoned, and later the land around them is reconstructed, if and as much as possible.

Underground mine planning and design is a unique engineering discipline. It involves the development of infrastructure and working conditions that are very sophisticated, highly specialized, and quite different from other industrial processes.

This type of mine design consists of the three basic engineering phases, conceptual, preliminary, and final design. While the second and third steps are common for many industrial processes, the development of the conceptual design is different for mining and is the key to the success of the entire operation. An error in interpreting the results from preliminary tests, for example, could lead the entire process in the wrong direction, cause technical and financial difficulties, and even disaster.

The goal of underground mine planning and design is integrated with mine systems design, with the final result being efficient and safe extraction of coal. The coal is then prepared to desired market requirements, at a minimum cost, while meeting social, legal and regulatory constraints.

A number of engineering disciplines are needed for successful mine planning and design process. Mining is a complex undertaking, so proper planning leads to the correct selection and implementation of all subsystems. Proper design, by the same token, ensures the implementation of traditional engineering subsystems.

An underground mining operation is a system which, due to the diversity of the technological processes, facilities, personal skills, and large capital investment, must consider and coordinate the behavior of, and interactions between, the different subsystems.

Advances in several fields used in mining operations have the potential for making a significant impact on mining. These, therefore, must be taken into account during the planning and design process as well.

First Things First

The initial process of evaluation and exploration of the particular area considered for underground mining is similar to that used for surface mining, although the objectives and the results are different.

After this step, the planning and design process is also similar to that of surface mining, although additional steps and safety precautions are added for obvious reasons.

Following is a list of steps taken in the initial evaluation of underground mine sites.

Baseline Assessment

This is an essential process, and encompasses the evaluation of all available data, prior to starting the actual planning efforts. This is a comprehensive review of all available information gathered through historical materials and by actual site tests.

This process is somewhat more complicated than the surface mine evaluation, where one can dig a shallow hole to find some of the results. A number of specialized tests and measurements have to be taken and properly evaluated when looking for coal deep underground.

Underground mine data evaluation also includes the review of all geographic, geologic, environmental, technical, economic, and other data available. Hopefully, the available data contain enough accurate historical and present-day information for proper planning and mine design.

Preliminary Planning

Most plans start with a feasibility study—an overview of the project—making reasonable assumptions and estimates of the physical and other key operating factors of the mine. The intent is to figure out as quickly as possible if the project justifies further effort.

A life-of-mine plan must be developed to determine the reserve's type and size, and other mining parameters, including the costs of site reclamation. Reclamation costs are often great, so those could put the project's feasibly in question.

Regulatory and Legal Factors

The planning process must also review the state of current regulatory affairs. These play a significant role in the overall mining operation, and these must be faced in the very beginning and attacked in a proactive manner, rather than addressing them after the fact.

Each sub-system and step of the process is subject to compliance, and often needs to be submitted to the various agencies for inspection and approval. At a minimum, these may include mine layout, strata and roof control plan, shaft ventilation plan, fan stoppage plan, medical and emergency evacuation plan, fire control evacuation plan, and escape route plan. It is of utmost importance to ensure that the latest regulations, policies, and proposed rulemaking have been incorporated.

Geologic/Geotechnical Factors

The most important part of the data collection is the information on the coal deposits. Coal variability can be mathematically defined, if enough accurate information is available.

Exploration permits are needed to start the data collection, and must be first on the plan schedule.

Understanding the regional geology and features of the deposit are of utmost importance. Potentially adverse geologic conditions, such as faults, wants, rolls, low cover, or water inflow must be located and well defined. Seam or horizon conditions are important also.

A thorough review of the land lease is needed to determine the requirements and compliance provisions, which may be excessive for a profitable and safe mine operation.

Reserves Data

A complete and accurate coal reserves inventory is needed. Since composing the geological model is a free interpretation of the available data, it depends heavily on the experience of the geologists working on it. Exploration efforts do not rely on actual core recovery, but on indirect conclusions.

Changes in the data interpretation are usually made as new data come in, and might change the entire geologic model. Geophysical logging, core photography, and petrographic identification are used with the most success. There are different approaches for geologic modeling: accepting the geology and developing the plan around it, or considering the geologic model incomplete and incorporating flexibility in the planning for potential changes.

The evaluation and calculation of the coal reserve is one of the most crucial factors to the long-term success of a mine. The reserve type, magnitude, grade, depth, inclination, geometry, etc. are key to proper mine design.

There are a number of methods applied in proper estimation of the variables and the overall mine design characteristics. Mathematical methods are of utmost importance. This involves taking data, such as drilled samples, and extrapolating the data into blocks or grids to make the appropriate calculations. Mapping, determining reserve classification, leasehold boundaries, etc. are required for the final calculations. The data are then processed via different mathematical techniques, such as polygonal, inverse distance weighing, and others.

Geographic and Economic Factors

Geographic factors include the location, transportation infrastructure, type, size, and skill level of the local work force, private and public facilities available locally, local climate, local power availability, etc.

Economic factors include the local political and tax environment, government stability (if a foreign country), socio-economic conditions, and availability of support networks. Economics usually favor starting with the "lowest hanging fruit" approach, which in this case means extraction of the best-grade material, or starting with the lowest mining cost areas. While this approach might maximize the return on investment in the short term and shorten the payback period, it might also create a compromise in the mine's design and operation.

Environmental Factors

Environmental data gathering is very important from feasibility and economic points of view. Because of that, sometimes as much as 5 years' data are necessary, especially if an environmental impact statement is required. The minimum necessary baseline environmental data required for planning include a) topsoil, subsoil, and overburden analysis, b) hydrologic studies, c) vegetation and land use surveys, d) air quality analyses, e) wildlife surveys, and f) archeological survey.

Tasks, such as core holes sealing and site reclamation, and their impacts on the local environment, must be considered and included in the planning stages. Some of the impacts are aesthetics, noise, air quality, vibration, water discharge and runoff, subsidence, and process wastes. Surface and groundwater quality during operation and through the remedial and treatment stages must be developed to meet supply and discharge standards.

Planning is basically responsible for environmental protection, from the initial exploration to final reclamation. It is to alleviate or mitigate potential impacts of mining to a) minimize the cost of environmental protection by proper steps in the overall design, thus eliminating remedial measures, and b) minimize negative publicity or poor public relations which may have severe economic consequences.

Technical Factors

From an engineering point of view, the technical aspects of mining operation planning and design are the most extensive and detailed. Data from regulatory, geologic, and environmental analysis must be evaluated and translated into technical specifications. This information is used to determine which process to use, and then to outline and develop each step.

The layout of the mine is determined by the size and shape of the coal deposits, and these features are used to calculate the mine reserves and determine the best way to extract them. Access to the reserves can be by vertical shafts, inclined slopes and drifts, or horizontal entries, and the production levels will determine the number and size of the access openings made.

The technical parameters will form the conceptual basis for the plan, from which the detailed plan will be drawn. The larger and more extensive the area, the more complicated the plan.

Each item and step of mine construction and the production process are defined by the available key assumptions, the physical factors, the equipment, mine facilities and infrastructure, and transportation. These are detailed as much as possible in the beginning, and then modified as each item is executed, according to the new data and the "best-fit" technical models.

Equipment

The type and size of coal deposits, including their hardness, will determine the types of equipment to use. The seam and working height, mining dilution limits, production rates, and property extent, ventilation, size constraints, regulations, and floor pressures may impact the choice of equipment. For example, a large flat-lying coal seam may allow the use of longwall mining equipment. Floor condition plays a big part in the equipment type. Equipment productivity is also a factor that might prove essential in the final decision.

Maintenance, equipment overhaul and replacement schedules must be developed accordingly, to ensure continuous production. Transportation of the product may be by rail, truck or a combination.

Transportation

Movement of materials, personnel, and equipment into and out of a mine is a critical part of mine operation. Workers must reach their designated work area in an expeditious manner. Supplies must get to their points of use before need becomes critical. The equipment itself must be transported through the mine to the working area. Then the coal must be transported from the working face to the processing facility.

A smooth flow of people and materials is critical to the efficient operation of a mine, so the various transport vehicles and pathways must be properly selected for efficient operation.

Underground Mine Operations

Upon obtaining the necessary permits, and after a series of inspections and modifications, the mine is ready for exploitation. There are basically several methods of underground mining:

Room and Pillar Mining

Here coal deposits are mined by cutting a network of rooms into the coal seam. Pillars of coal are left behind to hold up the roof. The pillars can make up to 30-40% of the total coal in the seam, as needed to provide space for head and floor coal.

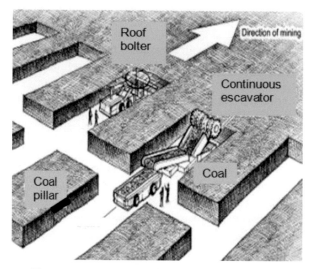

Figure 5-15. Room and pillar mining operation

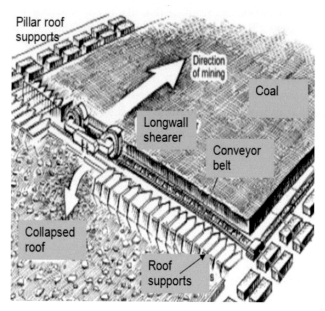

Figure 5-16. Longwall mining operations

There is evidence from recent open cast excavations that 18th century operators used a variety of room and pillar techniques to remove 92% of the *in situ* coal. The coal in the pillars can be extracted at a later stage by the retreat mining method, where a large machine with a self-supporting roof digs out the coal in the pillars and lets the roof collapse, as it pulls out of the area.

Longwall Mining

This process accounts for about 50% of underground production. The longwall method utilizes a large machine, a shearer, with a cutting face of 1,000 feet (300 m) or more. It is a sophisticated machine with a rotating drum that moves mechanically back and forth across a wide coal seam. The loosened coal falls onto a pan line that takes the coal to the conveyor belt for removal from the work area.

Longwall systems have their own hydraulic roof supports which provide safety, and can advance with the machine as mining progresses.

As the longwall mining equipment moves forward, overlying rock that is no longer supported by coal is allowed to fall behind the operation in a controlled manner. Sensors detect how much coal remains in the seam, while robotic controls enhance the efficiency of the process. Longwall systems allow a 60-100% coal recovery rate when surrounding geology allows the use of this method. Once the coal is removed from approximately 75% of a section, the roof is allowed to collapse in a safe manner.

Continuous Mining

This process utilizes a continuous-miner machine with a large rotating steel drum equipped with tungsten carbide teeth that scrape coal from the seam. Operating in a "room and pillar" (also known as "board and pillar") system—where the mine is divided into a series of 20- to 30-foot (5-10 m) "rooms" or work areas cut into the coal bed—a mine produces as much as five tons of coal a minute. This is more than a non-mechanized mine of the 1920s would produce in an entire day.

Conveyors transport the removed coal from the seam. Remote-controlled continuous miners are used to work in a variety of difficult seams and conditions, and robotic versions controlled by computers are becoming increasingly common. Continuous mining is truly a misnomer, as room and pillar coal mining is very cyclical. In the US, one can generally cut 20 ft (or a bit more with MSHA permission, or 40 ft in South Africa) before the continuous miner goes out and the roof is supported by the roof bolter. After this, the face must be serviced, before it can be advanced again.

During servicing, the continuous miner moves to another face. Some continuous miners can bolt and dust the face (two major components of servicing) while cutting coal. A trained crew must be able to advance ventilation to truly earn the "continuous" label. However, very few mines are able to achieve it.

Most continuous mining machines in use in the U.S. lack the ability to bolt and dust. This may be partly because incorporation of bolting makes the machines wider, and therefore, less maneuverable.

Blast, or Conventional, Mining

This is an older practice that uses explosives such as dynamite to break up the coal seam, after which the coal is gathered and loaded onto shuttle cars or conveyors for removal to a central loading area. This process consists of a series of operations that begins with "cutting" the coal bed so it will break easily when blasted with explosives. This type of mining accounts for less than 5% of total underground production in the US today.

Shortwall Mining

This method currently accounts for less than 1% of deep coal production. It involves the use of a continuous mining machine with movable roof supports, similar to those used in the longwall method. The continuous miner shears coal panels 150 to 200 feet (40 to 60 m) wide and more than a half-mile (1 km) long, keeping in mind the local geological strata and other factors.

Retreat Mining

This is a method in which the pillars, or coal ribs, used to hold up the mine roof are extracted, allowing the mine roof to collapse as the mining works back towards the entrance. This is one of the most dangerous forms of mining, owing to the unpredictability of ceiling behavior, and possibility of collapse, which could crush or trap miners.

In all cases, and no matter what type of mining is used, underground mining is complex, expensive and dangerous work, where experience counts. That often determines the difference between loss and profit—even life and death.

Major Coal Mines

Coal use in the U.S. is decreasing, due to tightening EPA regulations, but the worldwide demand is increasing by the day. There are presently nearly 1,200 proposed coal plant projects in 59 countries around the globe, with most targeted for the Pacific market. China and India top the list with over 800 coal-fired plants planned for implementation in the next several years.

So while new coal-fired plant construction has slowed somewhat and is under increasing fire in the U.S., the rest of the world is poised to go all-in with a dirty coal future. The expert conclusion is that coal demand could rise more than 20% by 2035.

This will increase the coal mining activities proportionately and we expect that the US will be a major player in this new game by increasing coal exports.

Coal production is growing fastest in Asia, while in Europe it has declined lately. The top coal mining nations in Table 5-1 (in millions of tons) are:

Table 5-1. Coal mining, 2010

Country	Mining
China	3,050
United States	973
India	557
Australia	409
South Africa	250
Russia	298
Indonesia	252
Poland	135
Kazakhstan	101
Colombia	72
TOTAL	6,097

Usually, a major part of the coal production is used in the country of origin, with only around 15% of hard coal production being exported. The U.S. is the only country that is planning to drastically increase its coal exports in the future. This controversial move promises to bring interesting developments, so we will watch carefully.

Global coal production is expected to more than double to over 13,000 Mt/yr by 2030. At that time, steam coal production is projected to reach around 5,200 Mt/yr; coking coal 620 Mt/yr; and brown coal 1,200 Mt/yr.

There were 1,325 mines in the U.S. in 2011, spread all over the U.S.

Table 5-2.
Major U.S. mines production (in million tons/annum)

Mine	Type	State	Annual
North Antelope	Surface	WY	98,279,377
Black Thunder	Surface	WY	81,079,043
Cordero Mine	Surface	WY	39,380,964
Antelope Coal Mine	Surface	WY	33,975,524
Jacobs Ranch Mine	Surface	WY	29,021,485
Belle Ayr Mine	Surface	WY	28,395,952
Enlow Fork Mine	Underground	PA	11,092,684
Bailey Mine	Underground	PA	10,232,360
Mcelroy Mine	Underground	VA	9,863,588
Foidel Creek Mine	Underground	CO	7,827,079

Coal mining produced over 1.1 billion tons in 2011, with northeastern Wyoming contributing the largest amount produced in any state in the US. It also presently produces more coal than any other region in the world.

Note the larger amount of coal produced at surface mines. This shows the relative ease of extraction by this method, which ultimately determines the total amount and price of thus produced coal.

The only problem is the massive destruction of local environment and huge scars left in the surface. This is justifiable to an extent by the need of coal for base-load power generation. It is totally unjustifiable for digging massive amounts of coal for export.

What does this do to our energy security? It simply punishes the future generations of Americans by depriving them of this valuable commodity, in addition to contributing to environmental pollution around the globe.

All this for the sake of making a quick buck today. Hardly fair!

Cost of Mining Operations

Underground mining currently accounts for about 60% of world coal production, although surface is prevalent in some regions of the US and some countries. For example, surface mining accounts for around 80% of production in Australia, while in the USA it is used for about 67% of production.

The cradle-to-grave mining operation is an expensive undertaking. The cost of a large mine—from concept to exploitation—is in the billions of dollars.

• The site location, exploration and data collection alone is a major task, consisting of a number of sub-tasks, involving a number of professionals and specialized firms.

• The construction of the surface facilities, the local infrastructure, and mine structure is another grand undertaking worth millions of dollars and involving huge equipment and hundreds of people.

• The equipment procurement and installation is a considerable effort and expense as well.

• Mine exploitation is, of course, the main goal of this undertaking and it's day-to-day operations are a never-ending stream of expensive equipment, projects and sub-projects.

• The mine decommissioning and land reclamation are also major, and equally lengthy, labor intensive and expensive undertakings.

The initial process of mine location, development and construction is estimated on the average at about

$150 million per each million tons of annual production, or, a 5 million tons per annum coal mine would have an initial cost estimate of $750 million. This is almost a billion dollars needed, if everything goes well, to plan, design, set up, and start operations.

Then the mining operations start—an equally expensive and complex process, the cost of which is determined mostly by the initial cost and coal cleaning. In the eastern USA, for example this cost ranges from $15 to $45 per ton of clean coal (in 2010 dollars).

Assuming that the overall weighted average cost per ton of clean coal is $25, our 5 million tons of annual production would spend about $125 million annually for daily operations—not including emergencies, accidents, etc.

The retail value of coal depends on the type, quality, quantity, the season, and the location of the order, so it varies up and down accordingly. Due to increasing transport charges, taxes, broker fees, etc., the retail prices could be twice the cost of mining, or more. They also vary with the type of coal, the period of the year, etc., and fluctuate in the $25-$125/ton range.

Mining Labor

As in any industry, labor pay in the mines is different from person to person and from mine to mine. Labor remuneration in the mines is categorized according to the level of expertise and experience. It is also dependent on the mine location. Different states have different pay scales. And of course, underground workers are paid higher wages, due to the increased level of difficulty and danger, which requires additional training and expertise.

Table 5-3. Labor rates in different states

Worker	Pay $/hour
Laborer - surface mine	7.00-22.00
Laborer - underground mine	14.00-27.00
Mill equipment operator	12.00-30.00
Stationary equipment operator	19.00-26.00
Mechanic - surface mine	11.00-30.00
Electrician - underground mine	14.00-32.00
Equipment operator	11.00-32.00
Production truck driver	9.00-28.00
Heavy equip. operator	9.00-30.00

Miners are not the best paid workers in the world. Considering the dangers, the dirt and misery in underground mines, and the other hazards these people are exposed to on a daily basis, they might be the most underpaid workers in the world.

Equipment Cost

Mine operations require a number of specialized pieces of equipment for digging, loading and transport of coal from the mining site to the surface and beyond.

Mining equipment is usually very large, really large, and not cheap. When all additional expenses related to equipment transport, assembly, maintenance, disassembly, EOL disposal, etc. are added, we get many more billions of dollars spent on mining equipment (after the initial purchase) through its long life in the mines.

acid. The acid in the leachate is sometimes neutralized with limestone, as part of a long-term land management process.

The acid can be also neutralized with ashes, which are a serious solid waste from coal burning. Using these ashes to fill the trenches and neutralize the leachate acidity would, however, be practical only at mine mouth plants (power plants operating close to the mine). Transport of large quantities of ashes to mines that are far away from the power plant would be prohibitively expensive and has never been tried.

In other cases, sewage sludge and even liquid sewage from the water reclamation plants can be spread

Table 5-4. Mining equipment specs and prices

Equipment	Specifications	Weight in lbs.	Cost in $
Dragline	55 cu yd bucket, 250 ft. dump height	16 million	$100 million
Shovel, hydraulic	5.2 cu yd bucket 23.6 ft (7.2 m) dump height	131,000	$925,000
Loader, wheel	9.0 cu yd bucket, 12'1" dump height	114,000	$720,000
Truck, rear-dump	60 ton, 46 cu yd, mechanical drive	22,000	$120,000
Drill, rotary (crawler)	5.13" to 7.88" hole, 25 ft drill length	30,000	$600,000
Tractor, crawler (dozer)	13.7' maximum blade width	39,100	$300,000
Grader, road	14 ft blade width	52,200	$450,000
Truck, water	5,000 gallon water tank	20,000	$255,000
Truck, service	Off-road tire service truck	15,000	$55,000
Truck, shot loader	1,000 per minute capacity	15,000	$75,000

Land Reclamation

After many years of exploitation, the coal in any mine would be depleted. The law requires that the land a mine occupies is brought back, as closely as possible, to its original condition. Restoring the land 100% is very seldom possible, so enforcing the law is subjective at best.

Land reclamation is the preferred post-strip mining method for bringing it back to original conditions. It's expensive, however, and at a cost of about $10,000 per acre, it would cost $2-3 billion to reclaim all damaged land in the U.S. This is possible but improbable to happen anytime soon.

So, what happened in the past more often than not is that mine companies would just abandon the mine, file for bankruptcy, or find another way to evade the expensive land restoration process.

There are, however, examples of great success in this area. In the Rhineland, Germany's coal fields, they store both the topsoil and the subsoil during excavation, to be replaced later. The land is fully refilled, graded and then fertilized and seeded. Drains carry the water away before it can be contaminated by forming sulfuric

over strip-mined areas to help restore the fertility and soil texture. The land can then be revegetated, thus slowly returned to its original state.

The most difficult reclamation is in the North Central U.S. regions, where grassy ranch land is being destroyed in large areas. Restoring this arid land to usable status would be almost impossible from an economics point of view. As a result, no coal company has ever gotten its reclamation performance bond (which is deposited at the beginning of the mine process) returned to a Montana coal company.

Note: Meeting the requirements of the federal surface mining controls adds to the cost of the coal. The cost of reclaiming western surface mines is $1,000 to $5,000 per acre. A more useful comparison is provided by the additional cost per million Btu (MBtu) of energy obtained from coal.

In the mid-1980s coal energy was available to electric utilities at an average cost of about $1.50 per MBtu. It is estimated now that federally mandated reclamation adds about $0.02/MBtu to western coal (where the seams are deep, but the energy content is relatively low), about $0.05/MBtu to Midwestern coal, and $0.11/MBtu to Appalachian coal.

Existing requirements already cost an additional $0.10 MBtu for Appalachian coal. Added to these costs is a tax, equivalent to $0.02/MBtu, assessed on the coal companies for the reclamation of abandoned strip-mined land.

Thus, the highest total of additions, Appalachia's $0.23/MBtu raises the cost of coal to the utility, and ultimately of electricity to the consumer, by over 15%.

About 1.1 million acres of coal mined land currently needs reclamation and new land is being disturbed at a rate of about 65,000 acres/year. At this pace, in 20 years we will have an additional 1.2 million acres of pit mining land in need of surface restoration and reclamation.

Reclamation laws need careful enforcement, but are not up to par, and cannot keep up with the fast pace of proliferation of coal draglines around the country. The situation is even worse in many other countries.

The Impacts

Typical activities during the decommissioning and site reclamation phase include removing infrastructure, such as structures, conveyors, rail lines; filling in the mined area or shafts; recontouring the surface; and revegetation . Potential impacts from these activities are presented below, by the type of affected resource. Depending on the mining method, some reclamation activities occur while the coal mining continues, such as in strip mining.

The following potential impacts may result from decommissioning and site reclamation:

- *Acoustics (Noise)* sources during decommissioning would be similar to those during construction and mining, and would include equipment (rollers, bulldozers, and diesel engines) and vehicular traffic. Whether the noise levels exceed guidelines established by the U.S. Environmental Protection Agency (EPA) or local ordinances would depend on the distance to the nearest residence. If near a residential area, noise levels could exceed the EPA guideline, but would be intermittent and occur for a limited time.

- *Air Quality* affecting global climate change and carbon footprint during decommissioning activities is determined by vehicle tailpipe emissions; diesel emissions from large construction equipment and generators ; and fugitive dust from many sourc-

es such as backfilling, dumping, restoration of disturbed areas (grading, seeding, planting), and truck and equipment traffic. Permitting would be required (as during construction and mining), and therefore these emissions would not likely exceed air quality standards or impact climate change .

- *Cultural Resources* would be unlikely to be affected during decommissioning because these resources would have been removed professionally prior to mining, or would have been already disturbed or destroyed by prior activities. Collection of artifacts could be a problem if access roads were left in place and the area was not monitored.

- *Visual impact* of the coal mine would be mitigated if the site were restored to its preconstruction state. However, despite the physical removal of any surface facilities, the impact of a scarred landscape on an area would likely remain.

- *Ecological Resources* impacted by the decommissioning activities would be similar in nature to impacts from construction and mining, with a reduction or elimination of blasting activities. Negligible to no reduction in wildlife habitat would be expected, and injury and mortality rates of vegetation and wildlife could be lower than they would be during mining. Impacts resulting from acid mine drainage could continue if not properly managed. Restoration of the mine site would reduce habitat fragmentation. Following site reclamation, the ecological resources at the project site could return to preproject conditions.

- *Environmental Justice* could result from significant impacts in any resource areas, and when these impacts disproportionately affect the populations. The environmental justice impact issues that could be of concern during decommissioning are noise, air quality, water quality, loss of employment and income, and visual impacts from the project site.

- *Hazardous Materials* and Waste Management of industrial wastes, such as lubricating oils, hydraulic fluids, coolants, solvents, and cleaning agents would be treated similarly to wastes generated during mining activities (that is, put in containers, characterized and labeled, possibly stored briefly, and transported by a licensed hauler to an appropriate permitted off-site disposal facility). Impacts could result if these wastes were not properly handled and were released to the environment. Additional solid and industrial waste would be generated during the dismantling of any ancillary facilities . Much of the solid material from dismantling facilities could be recycled and sold as scrap or used in road building or bank re-stabilization projects; the remaining nonhazardous waste would be sent to permitted disposal facilities.

- *Human Health and Safety* are potential impacts to worker and public health and safety during the decommissioning and reclamation of a coal mine, and would be similar to those from any construction-type project with earthmoving, crushing, large equipment, and transportation of overweight and oversized materials. Added risk may be involved with the reclamation of underground mines due to the potential for mine subsidence . In addition, health and safety issues include working in potential weather extremes and possible contact with natural hazards, such as uneven terrain and dangerous plants, animals, or insects.

- *Land Use*, upon decommissioning of the mine site and rectifying the impacts of coal mining, would be largely reversed. Future subsidence of underground mines could be a long-term issue. Open pit mines could have lasting land-use impacts; the land may be irreversibly altered if reclamation to pre-development condition is not possible. Alternate land uses may be established.

- *Paleontological Resources* during decommissioning activities would not be impacted, because these resources would have been removed professionally prior to mining, or would have been already disturbed or destroyed by prior activities. Fossil collection could be a problem if access roads were left in place and the area was no longer periodically monitored.

- *Socioeconomics impacts* of decommissioning of the mine and reclamation would include the impacts resulting from the cessation of mining activities, including job loss and revenue loss, and also the creation of new jobs for workers during reclamation activities and the associated income and taxes paid. Indirect impacts would occur from both the loss of economic development created by the loss of mining jobs and new economic development that would include things such as new jobs at

businesses that support the reclamation workforce or that provide project materials and associated income and taxes.

No adverse effect to property values is anticipated as a result of decommissioning. Site reclamation could result in economic values of residential properties adjacent to the coal mine becoming equivalent to similarly developed residential areas that were not affected by the coal mine. The loss of royalty and tax revenue could adversely impact the local and regional economies.

- *Soils and Geologic Resources* (including Seismicity/ Geo Hazards) activities during the decommissioning/reclamation phase, include removal of access and on-site roads and heavy vehicle traffic. Surface disturbance, heavy equipment traffic, and changes to surface runoff patterns can cause soil erosion. Impacts of soil erosion include soil nutrient loss and reduced water quality in nearby surface water bodies. Disturbed areas would be contoured and revegetated to minimize the potential for soil erosion.

- *Transportation impact* is reflected in short-term increase in the use of local roadways, occurring during the reclamation period. Heavy equipment would remain at the site until reclamation is completed. Overweight and oversized loads, when removing the heavy equipment, could cause temporary disruptions to local traffic.

- *Visual Resources* during decommissioning would be similar to those from construction and mining. Restoring a decommissioned site to preproject conditions would entail recontouring, grading, scarifying , seeding and planting, and perhaps stabilizing disturbed surfaces. Newly disturbed soils would create visual contrasts that would persist at least several seasons before revegetation would begin to disguise past activity. Restoration to pre-project conditions may take much longer. Invasive species may colonize newly and recently reclaimed areas. Nonnative plants, not locally adapted, could produce contrasts of color, form, texture, and line.

- *Water Resources* (surface water and groundwater) might be trucked in from off-site or obtained from local groundwater wells or nearby surface water bodies, depending on availability. It would be used for dust control for road traffic and mine filling and for consumptive use by the decommissioning/site reclamation crew.

- *Water Quality* could be affected by continued acid mine drainage if not effectively managed, activities that cause soil erosion, weathering of newly exposed soils leading to leaching and oxidation that could release chemicals into the water, discharges of waste or sanitary water, and pesticide applications. Upon completion of decommissioning, disturbed areas would be contoured and revegetated to minimize the potential for soil erosion and water quality related impacts.

- *Water Flow* would be affected by withdrawals made for water use, wastewater and storm water discharges, and the diversion of surface water flow for access road reclamation or stormwater control systems. The interaction between surface water and groundwater could also be affected if the two resources are hydrologically connected, potentially resulting in unwanted dewatering or recharging of any of these water resources.

Thus produced coal cannot be used as is, so it is treated for its final journey, loaded on tracks or trucks, and transported to a treatment facility.

PRE-BURNING TREATMENT

Coal straight from the ground, known as run of mine (ROM) coal cannot be used in its as-is state. It usually contains unwanted impurities such as rocks and dirt. It also comes in a mixture of different-sized fragments, which are not easy to use as-is in most commercial operations.

Coal users, with very few exceptions, need coal of consistent quality and size, so coal preparation—also known as coal beneficiation or coal washing—and/or other treatment of ROM coal is performed to ensure consistent quality and to enhance its suitability for particular end uses.

Coal can be transported to a special pretreatment processing facility, but that requires an additional step of loading and unloading, so instead, so it is cheaper to process it at, or close to, the mine.

The Coal Treatment Process

The mined coal deep down in the mine shafts is loaded on special cars and transported to an elevator. The cars are then hoisted to the surface, where the coal is dumped. Surface mines, of course, omit this step since the coal is near the surface. Once on the surface, the coal is piled in great piles, awaiting its turn in the preparation process.

The preparation work is done in large buildings, called *breakers,* which reach heights of 150 feet or more. Here the coal is taken to the top of the breaker by a conveyor belt system and undergoes several transformations on its way down.

The coal is crushed between rollers, and is then sifted over sorting screens into different shapes and sizes, as required by the customers. On the way down, the smaller pieces fall through the sifters and are sold as a lower quality coal. The coal that remains is what gets sold to the regular customers.

One efficient way of sorting coal is by putting it and the accompanying slate into moving water. The slate is heavier than the coal, and sinks in the water, while the coal floats on the surface and can easily be separated and carried away.

The different sizes and types of coal have different names. For example, an "egg" must 2 to 2-5/8 inches in diameter, a "nut" is usually between 3/4 and 1-1/8 inches; and a "pea" is between one 1/2 and 3/4 of an inch.

Crushing and cleaning of mine-run coal is also referred to as beneficiation or preparation. Often, crushing and sizing is all that is required, but many coal seams, especially those in eastern and mid-western states, contain enough impurities to necessitate further cleaning. Whether the cleaning process is wet or dry, it is commonly referred to as "coal washing."

The dry washing method uses high-pressure, pulsating airflow to blow dust from the coal. Wet washing starts with breaking and screening the coal to remove the large, hard pieces of impurities.

Larger material is usually treated using "dense medium separation," where the coal is separated from other impurities by floating in a tank containing a liquid of specific gravity, usually a suspension of finely ground magnetite.

As the coal is lighter, it floats and can be separated off, while heavier rock and other impurities sink and are removed as waste. The smaller size fractions are treated in a number of ways, usually based on differences in mass, such as in centrifuges. A centrifuge is a machine which turns a container around very quickly, causing solids and liquids inside it to separate.

Equipment can include any of the following: jigs, screens, landers, heavy-medium cyclones, tricone separators, concentrating tables, froth flotation, cells, filters, and driers.

Additional cleaning depends on the amount, size, and nature of impurity, how it is dispersed in the coal, and how the coal is to be used. The treatment depends on the properties of the coal and its intended use.
It may require only simple crushing or it may need to go through a complex treatment process to reduce impurities.

Alternative preparation/washing methods use the different surface properties of coal and waste. In "froth flotation," coal particles are removed in a froth produced by blowing air into a water bath containing chemical reagents. The bubbles attract the coal but not the waste and are skimmed off to recover the coal fines.

Recent technological developments have helped increase the recovery of ultra fine coal material too, which reduces waste and profit loses.

Coal Transport

After the coal is dug out from the ground and taken up to the surface, it is processed per consumer specifications of size, cleanliness, etc. before being shipped to the customer for burning. Transportation is a major step in the overall coal production-use process.

The coal transport system will depend on site-specific and project-specific factors and could be a conveyor system within the mine site to the coal preparation plant or a rail system. A system of haul roads is also likely to be present. Transporting coal off-site may be accomplished by rail, truck, barge, or some combination thereof. A coal-slurry pipeline also may be used to send coal off-site.

Most of the 4,325 coal mines in the U.S., however, produce less than 50,000 tons per year. It is impractical, and/or economically unfeasible to process the coal at the smaller mines, so it is usually trucked to a central processing facility. On arrival it is unloaded, cleaned, sorted by size and otherwise processed to customer specs before shipping to the point of use (POU).

If the mined coal is to be processed at a dedicated facility, away from the mine, then coal transport includes, a) loading on railroad cars at the mine and moving them to a central point for unloading, treatment, and upgrading (cleaning, sorting, refining, etc.), and b) loading and shipping thus processed coal to a power plant where it is unloaded again and burnt for electricity generation.

A close look at the coal transportation systems shows that coal mines are growing larger and their production increases exponentially. There were 316 mines in the mid-1970s with an annual output of above 500,000 tons. These large mining complexes execute all coal preparation operations onsite. From there the coal moves, usually by train, to the end use site—utility power plant or a large industrial consumer.

Most of the coal in the U.S. is shipped by "unit train." These are special train compositions, containing 100 or more coal hopper cars. The unit trains shuttle back and forth between large coal mines and coal-fired power plants non-stop, and sometimes on dedicated railroads lines.

The typical coal train is 100-120 hopper cars long, with each of the cars holding 100-115 tons. This is almost a mile of coal, which can feed a large coal burning power plant operating about a day, or two maximum.
The larger surface mines load two or three unit trains of coal a day.

Note: Two coal trains a day is 100 cars x 100 tons = 20,000 to 30,000 tons of coal daily. Imagine the noise, dust, stink, and expense that accompanies the loading, transport, and unloading operations.

There are approximately 80 trains leaving Wyoming mines every day, or about 26,000 trains annually. This is 26,000 miles of coal, or more than the circumference of the earth. If the unit trains from the other coal producing states are added, they can be wrapped around the Earth several times.

Figure 5-17. Unit Train

There is a trend of building "mine mouth" plants, in order to avoid the transportation costs. These power plants are built and operated right at, or very close to, the mines, because the coal is cheap at the mine (about $5 per ton).

Transportation makes 50-80% of the total cost of coal, depending on location, type, quantity and other variables.

Environmental and capital expense considerations, however, make the mine mouth plants option impractical and unprofitable in some locations, such as the West Coast. But there are several large complexes in the Four Corners area (where Colorado, New Mexico, Arizona, and Utah join) and in the Dakotas and Montana, which are good examples of mine mouth plants.

Coal can be also shipped by water—by barge on the nation's inland waterways, or by ocean freighters, to coal's export customers.

Overseas transport (coal for export) is done by large ocean cargo ships. A huge transoceanic coal export facility is located at Norfolk, Virginia. Coal is currently shipped also from seaports of Baltimore, Philadelphia, New York City, New Orleans, and Los Angeles.

Coal can be shipped by another water method, *slurry pipelines*. In this method, powdered coal and water are mixed at the mine and pumped through a long, large-diameter, pipeline to the coal burning power plant.

An example of that was the 273-mile-long pipeline from the Peabody mine on the "Black Mesa" in Arizona to the 1500-MW Mohave power plant near Page, Arizona. The Black Mesa slurry pipeline delivered about 8 tons of coal per minute along with 2700 gallons or 11 tons of water. At the plant the coal was dried and burned.

Note: This slurry pipeline is an example of lack of consideration for the natural resources and the local environment. Built on Native American land in the harshest and driest desert on Earth, it used 2700 gallons of water a minute, or close to 4 million gallons of per day. This was 1.5 billion gallons—an entire lake—taken from one area of the desert and pumped to another, where it was discharged as waste water after the coal had been filtered out.

What brilliant mind conceived such a monstrosity? And where did they get these lakes of water to waste in the middle of one of the most arid deserts in the world? Did they think of the consequences?

Similar slurry pipelines are proposed for the north central coal fields of the Dakotas, Montana, and Wyoming to feed the power plants of the Midwest (Chicago, St. Louis, etc.). Although studies show that slurry transport is less expensive than rail—after the pipeline is installed—environmental and economic controversies have so far frustrated pipeline projects.

Transportation Cost

The cost of transport via unit train today is approximately $0.020 per ton/mile, or a 1,000-mile trip would

cost approximately $20.00 per ton of coal transported from the mine to the power plant. This multiplied by 100 tons per car and 100 cars per unit train gives us the grand total of $200,000, which is the amount a large-scale power plant pays *every day* for coal transport alone. Remember that coal can start with a humble price of $5 per ton at some mines.

The transportation cost is even higher when considering other methods, such as barges, trucks, etc. This, compared with the cost of mining the coal, which is approximately $5-15 per ton in the U.S. (but much higher in other countries), is true highway robbery. There is no way around it, except in the case of "mine mouth" versions, where the power plant is located at the mine site, so the coal goes directly into the furnace to be converted into electricity.

There is a loss of power during the transmission of the generated electric power from the remote power station to the populated centers, but this is a small price to pay, compared with shipping millions of tons of coal across country.

COAL-FIRED POWER PLANTS

Coal produced in the U.S. and most other countries is used primarily for burning in power plants for electric power generation. Since this is a huge industry (and expanding in some countries like China and India), coal burning in coal-fired power plants is considered to be, a) vitally important to providing a steady baseload, and b) the most serious contributor to air pollution.

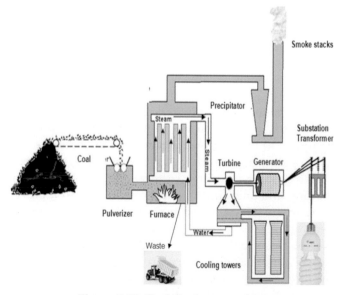

Figure 5-18. Coal-fired power plant

So let's go through the entire cradle-to-grave lifespan of coal-burning power plant and see where the problems start and end.

Life Cycle

The coal-fired power plant life cycle consists of several phases, very similar to those of gas- and oil-fired power plants, as follow:

Plant site and structure planning and design
- Locating, testing and exploring the proposed power plant location
- Engineering design of plant facilities, equipment and process
- Estimating the construction and operating cost and the related tasks
- Applying for and obtaining the necessary federal, state and local permits

Plant construction and setup
- Plant building construction
- Plant equipment procurement and setup
- Labor training

Power plant daily operation
- Coal unloading and transport
- Coal burning and energy generation
- Coal waste disposal
- Electric energy distribution

Power plant decommissioning and land reclamation
- Plant shutdown and disassembly
- Plant waste disposal
- Surface land reclamation

Planning and Design

In a coal-burning power plant the chemical energy stored in coal is converted into thermal energy (steam), which is converted into mechanical energy by turning a turbine. The energy of the turbine turns a generator, which produces electrical energy. Multiple generating units may be built at each power plant, for more efficient use of land, natural resources and labor.

The development of a coal-burning power plant is a complex and expensive process. It begins with the planning and design of the plant, starting with the main power plant building, including the main power generating unit, and related equipment and processes. Facilities and equipment must be designed to guaranty long-term efficient, profitable and safe operation of the plant.

The coal-burning power plant planning process starts with the need for electric power. A utility company decides that it needs more power, and a new power plant needs to be built. The executives decide what type of fuel will be used. If there is a coal mine nearby, and/or if a railroad node is available to transport large quantities of coal, then a decision is made to use coal for fuel, necessitating the design and construction of a coal-burning power plant.

Then a search is undertaken for an appropriate location for the plant buildings and infrastructure. The plant must be located close to railroad tracks, yet far enough from population centers. A water source must be located nearby (a well, or enough underground water) for the cooling cycle. Access to roads and electrical distribution lines must be considered.

When the location is chosen, environmental tests and analyses are conducted for state and federal permits. The permitting stage might take years, and construction can be started only upon obtaining the necessary permits.

The technical aspects of the coal burning power plant equipment and process design includes: The coal handling system—designed to provide the equipment required for the entire cycle of unloading, conveying, preparing, and storing the coal delivered to the plant. The scope of the coal handling system includes everything from the transport vehicles to storage areas. This includes the operation of the trestle bottom dumper and coal receiving hoppers, including the slide gate valves on the outlet of the coal storage silos.

The steam generator design—the most critical part of the planning and design effort. The coal-burning power plant is usually designed to operate non-stop as a base-load unit for the majority of its life, with some weekly cycling the last few years.

The heat and mass balance of the main plant steam power cycle is of utmost importance to planning and design. As an example, a plant using a 2500 psi and 1000°F single reheat steam power cycle has a high-pressure turbine that uses 2,734,000 lb/h steam at 2415 psi and 1000°F. The cold reheat flow is 2,425,653 lb/h of steam at 600 psi and 630°F, which is reheated to 1000°F before entering the intermediate-pressure turbine.

These are extremely high pressures and temperature regimes, so the proper heat and mass balance (steam generated vs. coal burned) is critical. So, the proper design requires the effort of a lot of experienced engineers with the appropriate tools.

The limestone handling and reagent preparation system is designed to receive, store, convey, and grind the limestone delivered to the plant. The scope of the system is from the storage pile up to the limestone feed system. The system is designed to support long-term operation, and roadways, turnarounds, and unloading hoppers are usually included in the overall plant design as well.

Dry scrubber, using electrostatic charge or other dry technology, can be used instead of limestone in the combustor for sulfur capture. NO_x control in both cases can be accomplished by a selective non-catalytic reduction (SNCR) system.

The electric power generator is an integral part of the system, so its design and operation are done according to established industry procedures. The generators are large units, usually made in the US or Germany. Their maintenance is critical as well, so detailed O&M procedures are usually part of the plant design and operation.

The pollution emission systems and controls must comply with 1990 CAAA imposed two-phase capping of SO_2 emissions in the U.S. For a new greenfield plant, the reduction of SO_2 emissions that would be required depends on possessions or availability of SO_2 allowances by the utility, and on local site conditions. In many cases, Prevention of Significant Deterioration (PSD) regulations will apply, requiring that Best Available Control Technology (BACT) be used. BACT is applied separately for each site, and results in different values for varying sites.

The flue gas desulfurization (FGD) system is part of the pollution emission control system. It is designed to scrub the boiler exhaust gases to remove at least 90% of the SO_2 gas in these, prior to release to the environment. The FGD system design includes the outlet of the induced draft fans to the stack inlet. The system is designed to support long-term operation with minimum maintenance.

The ash handling system is designed to provide the equipment required for conveying, preparing, storing, and disposing of the fly ash and bottom ash produced on a daily basis by the boiler. This includes the precipitator hoppers, air heater hopper collectors, and bottom ash hoppers to the ash pond (for bottom ash) and truck filling stations (for fly ash). The system is designed to support long-term operation with minimum maintenance

The support facilities and infrastructure are major elements of the design process. Proper design of buildings, roads and a rail spur within the plant fence line is of great importance, since thousands of tons of coal are needed to run the plant every day. The rail spur design

Figure 5-19. Control room

includes coal receiving and handling, crushing, storing, drying and shipping. There is also a limestone leg of the spur that includes facilities for receiving, crushing, storing, and feeding the fresh limestone needed for the operation.

The waste disposal system is dedicated to solid waste disposal, flue gas desulfurization, wastewater treatment and related equipment, as needed for an efficient, safe facility capable of a full 30-year life cycle.

The U.S. power regulations are based on equipment manufactured in the United States, Germany, or England, all of which comes with the standard manufacturer's warranties. Power plant designs are usually based on a referenced design approach to engineering and construction, where all facilities, process equipment and procedures are designed and procured in accordance with the applicable codes and standards, such as ASME, ANSI, IEEE, NFPA, CAA, and many other local, state, and federal regulations. OSHA codes are adhered to at all times and phases of the plant planning and design stages.

The coal power plant planning and design process is a lengthy one and starts immediately after the design and permitting processes are completed. In some cases, permitting and construction processes can be undertaken in parallel, but in most cases permitting and related steps are full of uncertainty, so proceeding with construction might result in great financial loss.

There are several methods of power plant design and construction, the most popular being stick and modular. Stick-built design power plants incorporate conventional design and construction from the ground up. In this case, each area and piece of equipment is designed and installed as a separate entity in a predetermined sequence of events.

Modular design and construction uses modules of shop assemblies, sub-assemblies, and full-scale modular packages. By maximizing the use of modular design and construction, significant cost and schedule savings can be realized. This method, however, is not widely accepted, so most power plants are constructed with the old, stick-built method.

Plant construction process consists of:
- Land clearing and preparation
- Support structures and infrastructure construction
- Plant building construction
- Plant equipment procurement and setup
- Labor training

The overall construction process is also a very a lengthy one, as outlined in Table 2-9. Here again, the different steps can be taken in parallel, the proper execution of which requires exceptional planning and management.

Daily Operations

Just like any business, coal-fired power plants follow a daily schedule, complete with routine, periodic, and emergency procedures.

In brief, the coal-fired *power plant daily operations consist of*
- Coal unloading and transport
- Coal preparation and burning
- Energy generation
- Coal waste disposal
- Energy distribution

A coal-fired power plant operates primarily by burning coal to generate electricity. There are a number of different steps and procedures which allow this process to proceed. Coal cannot be efficiently burned in its natural form, because it comes in large chunks, which make the combustion process inefficient.

Instead, the coal has to be pulverized into extremely small particles, as fine as baby powder in most cases. The powder is then mixed with hot air, and the mixture is blown into a furnace called a "firebox." Here the coal powder is burned in suspension, and before being able to settle on the bottom or the walls of the furnace. This space burning results in the most complete combustion of the coal, which produces the hottest flame and heat which can be obtained from coal.

There are also rows of pipes in the furnace walls, through which water flows. The heat from the burning coal heats the water quickly and turns it into a high tem-

Figure 5-20. Coal-fired power plant control room

perature (1,000°F) and high pressure (3,000 PSI) steam. The superheated steam is injected into the blades of a turbine, where the extreme pressure is enough to turn the turbine blades fast. When the turbine blades turn fast enough, they engage a generator, which produces electric power by rotating magnets on its axles into wire coils. Thus generated electric power is sent to a transformer in the substation, where it is conditioned and sent into the transmission lines for use at a distant location.

After the steam exits the turbine compartment it is sent into a condenser in the basement of a power plant, where the steam is cooled by running it between rows of pipes in which flows cool water which can be returned to the boiler, where the entire process is repeated. The cooling water (which upon exiting the condenser is very hot) could be used for heating in industrial processes, or is discharged into a local river, lake, or the ocean.

The daily operation of the different stages of coal burning and power generation are complex, but are yet standardized operations, where the personnel is highly skilled (properly trained), and where the execution of the different steps is controlled and documented at all times, with safety as a priority.

Standard operation and maintenance and safety procedures, the replacement of critical parts and major equipment overhaul, and personnel training are important parts of daily operations and must be taken into account during plant planning and design.

Operational Efficiency

A 500 MW power plant theoretically produces a maximum of 500 million Watts of power. This could be expressed as 500,000 kW of power. The power demand of a typical household is about 2 kW; so, this plant could easily supply 250,000 households.

Suppose this plant operates at 100% capacity for 12 hours per day, 7 days a week. In a year, it would produce 500,000 kW x 12 hours/day x 365 days = 2.2 billion kWh. Theoretically, the plant would supply 2.2 billion kWh/250,000 households = 8,800 kWh per year to its average residential customers.

In reality, our plant would produce about the same amount of power if it operated 24 hours a day, simply because its efficiency is about 50%, so half of the power is lost in heat and other wastes.

So, a typical 500 MW plant burns enough fuel to release 1,500 MW of thermal power, where:

- 1500 MW is released by burning coal (input)
- 500 MW appears as electric power output
- 1000 MW is dumped as waste heat
- Efficiency = output power/input power = 1/3 = 0.33 = 33 %

A 50% efficient plant would require much less coal (2/3) to produce the same electric power:

- 1500 MW is released by burning coal (input)
- 750 MW appears as electric power output
- 750 MW is dumped as waste heat
- Efficiency = output power/input power = 1/2 = 0.50 = 50 %

More efficient power plants emit less pollution and carbon dioxide.

- A 33% efficient coal plant releases 2.1 lb of CO_2 for every kWh generated.
- A 50% efficient coal plant releases 1.4 lb of CO_2 for every kWh generated.
- A 50% efficient natural gas plant releases 0.76 lb of CO_2 for every kWh generated.

Natural gas plants have almost 2 times less emissions than coal power plants of the same size.

Initial Costs

The overall cost for the planning, design and construction of a power plant in the US is approximately $1,500 per kW installed, so a 500 MWp coal-burning power plant would cost about $750 million in 2010 dollars. This includes all steps of the concept-to-implementation process. Of course, running the plant is a different matter that requires additional expense for materials, power, labor, etc.

We need to emphasize here that it is very expen-

sive to start and operate a coal-fired power plant, so only very large companies or governments can undertake and complete such tasks.

Now even they cannot get the necessary permits, mostly due to the new EPA emission standards which do not allow any coal-fired power plant to be built in the U.S.

The cost of construction has been sky rocketing lately as well. For example, the cost to build a new 300 MW coal-fired power plant in Wisconsin, which would generate enough power to supply 150,000 homes, is now projected at $1.1 billion (if it were built in southwestern Wisconsin) and $1.2 billion if it were built in Ohio. Only several months prior, the projected cost of this plant was $850 million to $950 million, but the cost of building power plants has risen very quickly lately, as the prices of steel, concrete and other materials have escalated.

Note: The price of constructing a gas-fired power plant, and the related conditions, are similar.

Decommissioning

After several decades of operation, the power plant has to be shut down and decommissioned. The land usually has to be brought back to its original condtion too.

The major steps in the coal burning power plant decommissioning and reclamation are:

- Plant shutdown and disassembly
- Turbine and generator dismantling and removal
- Smoke stack demolition and dismantling
- Rigging and removal of support equipment
- Construction and equipment waste disposal
- Surface land decontamination and reclamation

Coal burning power plant decommissioning is similar to that of any power plant, and basically consists of safe removal of the different buildings from service by demolition, and dismantling and disposal of their components.

The decommissioning effort usually starts with disassembly and removal of key electro-mechanical components, including breaker boxes, transformers, conduit, wring, and electrical components. The rest of the equipment, furnace components, piping, turbine rotor and blading, generators, static exciters, frequency changers, heat exchangers, and other related mechanical equipment, is disassembled and removed from the facilities as well.

This effort might also involve selective dismantling and rigging of some power equipment pieces, which are in good condition and could be reused or sold. When all equipment has been removed from the facilities, the buildings are demolished and the waste, together with the demolished equipment, is loaded on dump trucks for removal and disposal.

Last, but not least, the land that was occupied by the plant buildings must be decontaminated, leveled, and/or otherwise brought to its original state. This is usually also a very complex, lengthy, and expensive undertaking.

All decommissioning, rigging and dismantling work is done in accordance to all safety rules and regulations and is quite labor intensive and expensive as well.

Air, Liquid and Solid Pollution

The coal mining, transport, and burning process consists of several steps, each of which generates air, liquid, and solid pollution:

• *Air pollution* is generated at all stages of the coal's cradle-to-grave process—from the mine site development, to subsequent mining operations, and during the processing and transport of coal. Most of the air pollution throughout the entire coal lifetime cycle, however, is emitted during the burning of coal in coal-fired power plants.

Coal-fired power plants have been blamed for emitting most of the GHG pollutants that are causing environmental problems—including global warming— today. The numbers are so large that it is hard to grasp the magnitude of the related problems.

Get this: *The average GHG emission rate in the U.S. for each MWh of coal-fired electric power generation is 2,249 lbs CO_2, 13 lbs SO_2, and 6 lbs of NO_x.*

Please note that this represents several thousand lbs. of GHG emissions with every MWh of power generated in the U.S., while the average emission rates are much higher in other countries.

But staying with the U.S., we see that in 2012, the United States generated about 4,000 billion kWh of electricity, 68% of which was generated from fossil fuels (coal, natural gas, and petroleum), with 37% of the total generated by coal-fired power plants across the country.

So, 4,000 billion kWh electricity is 4 billion MWh, which multiplied by the coal GHG emissions (as 37% of

the total) gives: 3.3 trillion lbs. of carbon dioxide, 19 billion lbs. of sulfur dioxide, and 9 billion lbs. of nitrogen oxides emitted annually in the U.S. alone.

Multiply this number by six to obtain the total global emissions and you will see well over 20 trillion lbs. of GHGs emitted annually around the world. This number is so huge that most people would give up comprehending the magnitude of the problem.

Yes, the U.S. emits 1/6th of the total global GHGs, while China emits twice as much today, with India in third place emitting half as much GHGs as the U.S. All in all, we are choking the atmosphere with so much toxic gas that our fragile closed system is going to burst at the seams one of these days.

Because of that, the air in some large cities around the world is becoming increasingly contaminated, and the locals are paying with their health and lives.

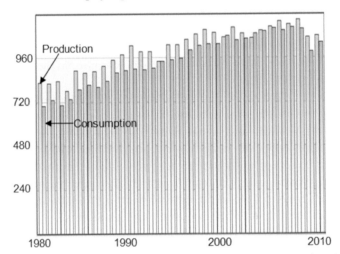

Figure 5-21. U.S. coal production and use (millions of tons/year)

• *Water pollution* is also a big problem all through the coal mining, processing, and burning cycle. To process coal, mines and processing facilities use huge amounts of water which is dumped after use. Stored in large pools, the waste water is contaminated and often finds its way into the local water table or other bodies of water.

Coal-fired power plants alone generate about 140 million tons of fly ash, scrubber sludge, and other wastes every year. Coal processing (crushing and washing) facilities generate additional quantities of similarly harmful materials.

These wastes contain some deadly pollutants, including toxic chemicals, heavy metals, particulate matter, and other substances that kill wildlife and cause cancer and neurological damage in humans.

Coal combustion waste sites are known to have contaminated groundwater, wetlands, creeks, or rivers. This could easily have been prevented with sensible safeguards such as phasing out leak-prone ash ponds and requiring the use of synthetic liners and leachate collection systems.

Amazingly enough, ash and other coal combustion wastes are not subject to strict federal regulations, and a certain amount is allowed to be dumped in water bodies.

Recently the EPA slapped a $210 million fine on a large coal company for dumping large amounts of unauthorized wastes and contaminating the water systems of five U.S. states. Six thousand violations have been documented in this single case, where sloppy methods of waste water disposal have led to mass contamination of the water system of the affected states.

There are many more such violations across the country, and the problem is not going away, simply because the coal companies don't have many options. They seem willing to pay the penalties instead of putting more money in developing new technologies and recycling facilities.

Coal-fired plants also use a lot of water for cooling the steam prior to being reused. This process creates additional air and water pollution. Large amounts of cooling water are evaporated from the heat exchangers, and together with the exhaust gasses change the local environment significantly. Some of this water is dumped from time to time in the environment too, which creates additional contamination and other problems in the local areas.

- *Solid pollution* is also a problem during coal production and use, because the process demands movement of large amounts of coal, which generate a lot of dust and solid waste. Removal of large amounts of soil from surface mines, the overburden, is part of the mining process that is accompanied with the generation of a lot of dust and soil waste. Depending on the nature, attitude, and grade of the deposits, often much more waste soil is removed from the surface mine than the total quantity of ore mined during the entire life of the mine.

The solid waste is classified as either *sterile*, or *mineralized* and the movement and stacking (or dumping) of this material forms a major part of the mine planning process. In cases where the mineralized package is determined by an economic cut-off, the soil waste is dumped separately (if it contains some minerals) with view to treatment if and when it becomes economic viable to treat this material.

Civil engineering design parameters are used in the design of the waste dumps, and special conditions apply to high-rainfall areas, e.g. Brazil or Venezuela, or where the dumps are created in seismically active areas like Chile, Peru, and parts of Canada.

During the coal-burning process, coal-fired power plants also generate immense amounts of ash and solid waste which is stored in huge mountains on site, or transported for disposal off-site.

All in all, coal production, processing, transport, and burning are dirty processes, where large quantities of toxic gasses, liquids, and solids are emitted, spilled, piled up and otherwise created. All of these issues must be taken into account when discussing the coal process.

What does all this environmental stuff have to do with energy security, you ask? The simple answer is that energy security means *nothing* to those dying of lung cancer due to polluted air from energy production, overproduction, overuse and abuse.

As we continue to ignore the connection between energy and the environment, more of us will die from the environmental impacts of pollution from energy production.

Pollution Footprint of Plant Equipment

Similar to the mining process, a number of large and small pieces of equipment are used during power plant construction and operation. Starting with the plant's design and construction, we see pollution taking place in the shape of ground leveling and digging by large bulldozers with heavy plumes of smoke and by large trucks invading the area.

Then the power generating equipment arrives on trains and trucks. All this machinery was made somewhere on the planet some time in the past, where and when their manufacturing processes and transport created a significant pollution footprint.

Just like in the mining case, we would assign 5-10% of the manufacturing and transport of power plant equipment to the initial construction step. The main pieces of equipment, boilers, turbines, generators, etc. arrive in wooden boxes or on back of enormous trailers or railroad cars. These are assembled with the help of other equipment and put to work after creating another set of pollution problems. In addition, there are many other items and small pieces of equipment, tools and consumables for personal or specialized operations. These are shovels, helmets, computers, boots, shoes, overalls, first aid kits, goggles, etc. All these were also made somewhere around the world, sometime in the past, and transported to the mine. During their manufacturing processes and transport, they also left a significant pollution footprint.

To complete the pollution footprint picture, we must add to that the footprint of the people involved in the entire process—engineers and technicians involved in the equipment manufacturing, transport, installation, setup, and operation. These people drive cars, ride on planes, all of which leaves another significant set of pollution footprints.

So, the total pollution footprint of the power plant's equipment cradle-to-grave process is expressed by:

$$P_f = A+B+C+D$$

where

P_f is the total pollution footprint

A is the equipment manufacturing and transport to the mine

B is equipment assembly, test and installation at the mine

C is the manufacturing and transport to the mine of tools and consumables, and

D is the pollution footprint of the equipment personnel during the design and setup phases.

Combining the pollution footprint of A (the plant equipment manufacturing and transport) with B (plant equipment assembly, test and installation), adding C (the manufacturing and transport of the tools and consumables), plus D (the pollution footprint of the equipment personnel) gives us the environmental footprint of the plant equipment *before* even starting plant operations.

The equipment pollution footprint is significant, to be sure! It goes deep and wide, stretching from one part of the country to the other...and from end of the world to the other.

In order to put a value on it we must take inventory of all items delivered to the plant and needed for its design, construction and operation. We must take a close look at the manufacturing of a boiler, a turbine, or a generator, as well as a number of other pieces of equipment used at the plants.

These are also very large pieces of equipment, and it takes hundreds and thousands of tons of metals to make them. Each of these started as an iron ore, dug out from a mine somewhere in the world, and transported to a smelter. The molten metal was shaped in different forms and shipped to parts manufacturers who drilled, welded and otherwise constructed parts for these vehicles.

Again, remember that we are talking about huge pieces of metal—some as big as an entire building. These parts are then packed and loaded on railroad cars or trucks to be shipped to the assembly facility. The parts of a large dump truck or a dragline, for example, might require a dozen railroad cars for transport. The parts are then sent to the assembly plant, where they are assembled into major components or entire units. After one more loading and unloading operation, they arrive at the plant for final assembly, installation, and testing, before being put to work.

This is a long and winding process, with many stops, starts, loading and unloading steps.

Again, putting a dollar value to all this would be too long and complex a procedure for our purposes, so it suffices to say that the pollution effects during and after all these steps are significant.

Upon starting the power plant operation, the equipment continues to generate secondary pollution. Spare parts for these monsters are periodically arriving from the parts manufacturers, after leaving their pollution footprint all over the place. We will attempt to assign a number—quantity of air and land pollution—and a related dollar value to these activities and their environmental effects in the following chapters.

After 20-30 years of non-stop operation, the power plant is finally declared obsolete, at which point it has to be shut down and decommissioned. The decommissioning and land reclamation are major, expensive, and polluting undertakings as well. The equipment and materials used during the decommissioning and waste disposal process create additional air and ground pollution which is not to be ignored.

Additional, specialized equipment and personnel (environmental specialists and inspectors), are usually brought in to assist and/or coordinate the effort. With that more pollution footprints are left at the already exhausted mine.

When the final tally is made, a significant part of the emissions and overall pollution are to be attributed to the plant's equipment and support personnel. This number is significant and we will go through the calculations in the next chapters as well.

Summary

Coal is a fossil fuel formed from the decomposition of organic materials that have been subjected to geologic heat and pressure over millions of years. Because of that, coal is considered a nonrenewable resource because it cannot be replenished on a human time frame.

The activities involved in generating electricity from coal include mining, transport to power plants, and burning of the coal in power plants. Initially, coal is extracted from surface or underground mines. The coal is often cleaned or washed at the coal mine to remove impurities before it is transported to the power plant—usually by trains, barges, or trucks.

At the power plant, coal is commonly burned in a boiler to produce steam. The steam is then sent into a turbine, which turns to generate electricity.

The biggest problem is that coal emits huge amounts of pollutants—GHGs—which are blamed for causing environmental problems around the world—including climate changes, global warming, and damages to human health and well-being. This has led to increased emissions standards in the U.S. which are having a huge impact on the energy industry, and coal in particular.

Environmental Impacts

Although power plants are regulated by federal and state laws to protect human health and the environment, there is a wide variation of environmental impacts associated with power generation technologies.

The list below is intended to give the reader a better idea of the specific air, water, solid waste, and radioactive releases associated with coal-fired generation.

Air Emissions

When coal is burned, carbon dioxide, sulfur dioxide, nitrogen oxides, and mercury compounds are released. For that reason, coal-fired boilers are required to have control devices to reduce the amount of emissions that are released.

Mining, cleaning, and transporting coal to the power plant generate additional emissions. For example, methane, a potent greenhouse gas that is trapped in the coal, is often vented during these processes to increase safety.

Water Resource Use

Large quantities of water are frequently needed to remove impurities from coal at the mine. In addition, coal-fired power plants use large quantities of water for producing steam and for cooling. When coal-fired power plants remove water from a lake or river, fish and other aquatic life can be affected, as well as animals and people who depend on these aquatic resources.

Water Discharges

Pollutants build up in the water used in the power plant boiler and cooling system. If the water used in the power plant is discharged to a lake or river, the pollutants in the water can harm fish and plants. Further, if rain falls on coal stored in piles outside the power plant, the water that runs off these piles can flush heavy metals from the coal, such as arsenic and lead, into nearby bodies of water. Coal mining can also contaminate bodies of water with heavy metals when the water used to clean the coal is discharged back into the environment. This discharge usually requires a permit and is monitored.

Solid Waste Generation

The burning of coal creates solid waste, called ash, which is composed primarily of metal oxides and alkali. On average, the ash content of coal is 10 percent.[3] Solid waste is also created at coal mines when coal is cleaned and at power plants when air pollutants are removed from the stack gas.

Much of this waste is deposited in landfills and abandoned mines, although some amounts are now being recycled into useful products, such as cement and building materials.

Land Resource Use

Soil at coal-fired power plant sites can become contaminated with various pollutants from the coal and take a long time to recover, even after the power plant closes down. Coal mining and processing also have environmental impacts on land. Surface mining disturbs larger areas than underground mining.

To summarize all this, we must agree that coal is critical for use as a power generator, but beyond that, it is not doing us any favors. While our energy security depends on coal, it is slowly killing us…

COAL'S FUTURE

Coal is largely responsible for bringing our civilization to the level of technical and economic development we enjoy presently. It is also the fuel that might help us with the transition to a fossil-less energy future. But now it is blamed for excess pollution…a catch 22 that we will have to deal with for a long time—hopefully in a responsible manner.

While we are thankful for all the good things coal has done for us and for our economic progress, we are at the same time blaming it for great environmental evil, property damages, as well as for making people sick and even killing them.

Before we delve into the future, let's see what we can do with coal today:

Present-day Coal Use

Coal is still the primary energy source for a number of countries worldwide, and provides about 1/3 of the world's primary energy and over half of the power generation. Coal is the main fuel for the generation of electricity since its price is low, compared to other fuels. Unfortunately, it is also the highest polluting source of electricity.

Other major uses of coal are in the production of steel and synthetic fuels. Bituminous coal is also used to produce coke for making steel and other industrial process heating. Coal gasification and coal liquefaction (coal-to-liquids) are used also to produce synthetic fuels.

The primary use of coal, however, is to generate electricity or heat, with electric generation surpassing any other use. Coal generates over $1.0 trillion annual revenue, and its use is rapidly growing around the world.

In contrast, coal use in the U.S. is decreasing, driven mainly by the conversion of coal-fired power plants to natural gas fuel.

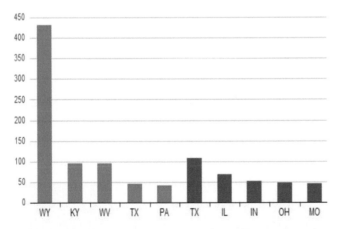

Figure 5-22. Coal producing states (in millions of tons)

Obviously, we produce more coal than we need for internal use. Only 4-5% of the coal mined in the United States was exported in the past, mostly for making steel. Export plans, however, call for a major jump in coal exports in the near future. Coal will be exported to Asia and Europe mostly for electric power generation.

Coal is mined commercially in over 50 countries. Over 7,036 Mt/yr of hard coal is currently produced, a

substantial increase over the past 25 years. In 2006, the world production of brown coal and lignite was slightly over 1,000 Mt, with Germany the world's largest brown coal producer at 194.4 Mt, and China second at 100.6 Mt.

Coal production has grown fastest in Asia, while Europe's has declined. The top coal mining nations in 2010 were, unsurprisingly, China and the U.S.

Table 5-5. Global coal production and use (millions of tons).

Country	Annual production	% of world production	% of world use
China		3,050	39.4
38.6			
US	960	19.3	18.4
India	557	6.8	7.7
Australia	415	6.6	1.7
S. Africa	250	4.7	3.0
Russia,	310	4.9	3.6
Indonesia	250	4.7	3.2
Poland	150	2.2	1.9
Kazakhstan	125	1.8	1.1
Japan	-	-	3.9
S. Korea	-	-	1.8

Most coal production is used in the country of origin, but some of it is being exported. U.S. coal production in 2011 increased slightly from 2010, driven by export demand, to roughly 1.1 billion short tons. Production in the western region, which includes Wyoming, totaled 587.6 million short tons, a 0.7 percent decline from 2010.

Productive capacity of coal mines increased by 2.5 million short tons to 1.3 billion short tons. At the same time, the average number of employees in U.S. coal mines increased 6.3 percent to 91,611.

Domestic coal consumption of metallurgical coal by the coking industry rose 1.6 percent to 21.4 million short tons. The average sales price of coal increased 15.2 percent to $41.01 per short ton.

The average coal use in the U.S. is estimated at 18 lbs. per person per day (p/d). The global average is 6.4 lbs p/d, while in Australia it is over 33 lbs. p/d.

We can't help but wonder what the Aussies are doing with so much coal.

Case Study: Loy Yang Coal Mine

Loy Yang coal mine in the Latrobe Valley in Victoria, Australia, is the biggest coal mine in the southern hemisphere, excavating over 30 million tons of brown coal annually. At the same time it burns 65,000 tons of brown coal every day for power generation, releasing the same amount of carbon dioxide as 60 million cars. This is about 50-75% of the mine's daily coal production, so the excess coal is transported to other power plants across the country.

Estimates are that there is enough brown coal easily accessible in the Loy Yang mine to continue the present-day level of operation for the next 1200 years.

Note: Global coal production is over 7,000 Mt/yr and by 2030 is expected to reach 12-13,000 Mt/yr, with China accounting for most of the increase. Steam coal production is projected to reach over 7,500 Mt/yr; coking coal 1,500 Mt/yr; and brown coal over 3,500 Mt/yr.

Coal reserves are available in almost every country worldwide, with recoverable reserves in around 70 countries. At current production levels, proven coal reserves are estimated to last 147 years. However, production levels are by no means level, and are in fact increasing and some estimates are that peak coal could arrive in many countries such as China and America by around 2030.

The estimation of hypothetical coal resources in areas where geologic, thickness, rank, and areal size data are sparse or absent is necessary to promote exploration for poorly known and undiscovered coal areas.

Currently about 2.24 trillion short tons of the U.S.' 3.68 trillion tons of remaining coal resource inventory are classified as undiscovered (hypothetical). Much additional unknown coal may be concealed in the central parts of basins and is not as yet included in the Nation's coal inventory.

This additional unknown coal must be identified, as must the 2.24 trillion tons currently remaining in the inventory, because knowledge of the quantity, quality, and rank of the unknown and hypothetical coal could influence the nation's energy usage plans.

Coal reserves are usually stated as either:

1. "Resources," which consist of the total of all "measured" + "indicated" + "inferred" resources,

2. "Run of Mine" (ROM) reserves (or the actually recoverable coal), which is usually much less than the general "reserves" estimates, and

3. "Marketable reserves," which after sorting of the produced coal may be only 60% of ROM reserves.

The standards for reserves are set by stock exchanges, in consultation with industry associations. For example in ASEAN countries reserves standards follow the Australasian Joint Ore Reserves Committee Code used by the Australian Securities Exchange.

Because of increased pollution concerns, coal has fallen on hard times lately and the U.S. coal industry is forced to adapt. The axe is mercilessly falling on many coal-related businesses around the country. EPA regulations issued in the last several years limit the emissions from power plants to a point where coal can no longer compete.

The last round of EPA regulations put the coal power generation in an absurdly awkward position. It simply made coal-fired power plants things of the past. No new plant can be built and expected to operate at a profit. Period. The same is true for the existing coal-fired power plants.

Coal mining will be hurt, but temporarily. Coal exports are expected to reach unprecedented levels, and U.S. coal mines will most likely increase their output for the foreseeable future. Most of the coal will be exported to Asia and Europe.

Coal Production

Coal is a very big business worldwide, and getting bigger by the day. With nearly 25% of the world's reserves located in the U.S. it will remain big business in this country for a long, long time as well.

Coal in the United States is found in great quantities in several regions, such as the:

- Appalachian region in the states of PA, OH, WV, KY, TN, and AL;

- Midwest region in the states of IL, IN, and KY;

- Gulf Coast region in the states of TX, LA, AR, and MS; and in

- West in the states of UT, CO, AZ, NM, WY, SD, ND, and MT.

Figure 5-23. Monster truck

Different types of coal can be found in different areas, as follow:

- *Lignite* is found in the states of MT, TX, and ND;

- *Sub-bituminous coal* is found in the states of MT and WY; *bituminous coal* is found in the states of IL, KY, and WV; and *anthracite* is found in the state of PA.

To begin with, we need to emphasize that there are different types and qualities of coal, which determine the design and operation of the coal mine, as well as the use of the coal produced at each mine.

Coal varies in composition from deposit to deposit, and sometimes even within the same deposit.

There are four major and very different types of coal. Each of these is characterized by a differences in appearance, but more importantly in energy output. These are determined by the different pressure, heat, and time the different coal reserves have endured.

The different types of coal are:

- *Lignite* is a brownish-black coal with high moisture and ash content, which has the lowest heating value of the four types of coal. It is considered an "immature" coal that is still soft. It is used for generating electricity.

- *Sub-bituminous* coal is a dull black coal with a higher heating value than lignite, and is used principally for electricity and space heating.

- *Bituminous coal* is the most common type in the United States, accounting for over 50% of the demonstrated reserve base. It is the most commonly used type of coal for electric power generation in the United States. It is a dark, hard coal that has a higher heating value than lignite and sub-bituminous coal, but a lower heating value than anthracite.

- *Anthracite* is also known as "hard coal" that was formed from bituminous coal under increased pressures in rock strata during the creation of mountain ranges. In the United States, it is located primarily in the Appalachian region of Pennsylvania. It is very hard and shiny. This type of coal is the most compact and therefore, has the highest energy content of the four levels of coal. It is used for space heating and generating electricity. It makes up only 1.5% of the demonstrated reserve base for coal in the United States.

There were 1,325 mines in the U.S. in 2011, with total coal production of roughly 1.1 billion short tons. Production in the Western Region, which includes Wyoming, totaled 587.6 million short tons. The productive capacity of the U.S. coal mines was approximately 1.3 billion short tons.

The average number of employees in U.S. coal mines at the same time was 91,611. Domestic coal consumption of metallurgical coal by the coking industry was 21.4 million short tons. The average sales price of coal was $41.01 per short ton.

Approximately 25% of the world's recoverable coal deposits are in the US; approximately 270 billion tons, with Russia following at 176 billion tons, except that most of the deposits are in areas that are very difficult to mine.

There are estimates of over 1.0 trillion tons of coal available globally, equivalent to over 120 years of current production and 100% of the world total, with the largest reserves located in the US, Russia, China, Australia and India.

The major coal producers today are China, the USA, India, Australia, Russia, Indonesia and South Africa. The global coal production lately has been in the 7-7.5 billion tons annually. China accounts for nearly 50% of the world's coal consumption, followed by the US, India, Japan and Russia as major coal consuming nations.

Let's take a brief look into the regulations that affect coal production and use.

The Regulations

Coal is not clean by any means, and attempts to clean it are few and far between, not to mention that it is very expensive to do that. So, recently the U.S. government agencies in charge of the energy sector gave up on any attempts to clean coal.

Instead, they decided to tighten the GHG emission regulations for power plants, which meant, a) no new coal-fired plants can be build in the U.S. ever, and b) many old plants were, or will be, forced to shut down.

CSAPR

In July, 2011, the EPA issued a rule intended to protect the health of Americans by helping states reduce air pollution and attain clean air standards. This rule, known as the Cross-State Air Pollution Rule (CSAPR), requires states to improve air quality by reducing power plant emissions that contribute to ozone and/or fine particle pollution in other states.

In a related, regulatory action, EPA finalized a supplemental rulemaking on December 15, 2011 to require five states (Iowa, Michigan, Missouri, Oklahoma, and Wisconsin) to make summertime NO_x reductions under the CSAPR ozone season control program.

CSAPR requires a total of 28 states to reduce annual SO_2 emissions, annual NO_x emissions and/or ozone season NO_x emissions. The goal is to attain the 1997 ozone and fine particle and 2006 fine particle National Ambient Air Quality Standards (NAAQS).

In February and June, 2012, EPA also issued two adjustments to the CSAPR, which replace EPA's 2005 Clean Air Interstate Rule (CAIR).

Note: A court decision in 2008 kept the requirements of CAIR in place temporarily but directed EPA to issue a new rule to implement Clean Air Act requirements concerning the transport of air pollution across state boundaries. This action is in response to the 2008 court's decision.

MATS

In December 2011, EPA issued the first national standards for mercury pollution from power plants (MATS). This is the first national standard intended to limit power plant emissions of mercury, arsenic, acid gas, nickel, selenium, and cyanide. The new standards also slash emissions of these dangerous pollutants by relying on widely available, proven pollution controls that are already in use at more than half of the nation's coal-fired power plants.

EPA estimates that the new safeguards will prevent as many as 11,000 premature deaths and 4,700 heart attacks a year. The standards will also help America's children grow up healthier—preventing 130,000 cases of childhood asthma symptoms and about 6,300 fewer cases of acute bronchitis among children each year.

Table 5-6. U.S. coal-fired power plants shut down due to tightened EPA regulations

State	Plant	Owner	MW
CO	Clark	Black Hills	42
GA	Harlee Branch	Georgia Pwr	581
IL	Hutsonville	Ameren	150
	Meredosia	Ameren	203
KY	Green River	LG&E	189
	Cane Run	LG&E	618
	Big Sandy	AEP	278
	Tyrone	LG&E	135
ME	R. Paul Smith	FirstEnergy	110
MA	Salem Harbor	Dominion	806
	B.C. Cobb	CMS	312
	D.E. Karn	CMS	515
MI	J.R. Whiting	CMS	345
MN	Black Dog	Xcel	294
MO	Meramec	Ameren	924
NM	Four Corners	APS	633
NC	LV Sutton	Progress	604
	Cape Fear	Progress	323
	Weatherspoon	Progress	171
	HF Lee	Progress	397
OH	Miami Fort	Duke	163
	WC Beckjord	Duke	1,222
	Muskingum	AEP	840
	Conesville	AEP	165
	Bay Shore	FirstEnergy	499
	Avon Lake	GenOn	732
	Lake Shore	FirstEnergy	300
	Niles	GenOn	217
	Ashtabula	FirstEnergy	256
	Eastlake	FirstEnergy	1,289
	Pickway	AEP	100

State	Plant	Owner	MW
OR	Boardman	Portland	601
PA	Armstrong	FirstEnergy	326
	New Castle	GenOn	330
	Shawville	GenOn	597
	Titus	GenOn	243
	Portland	GenOn	401
	Elrama	GenOn	460
SC	Jefferies	Santee	306
	McMeekin	SCE&G	294
	Urquhart	SCE&G	100
	Canady's Sta.	SCE&G	490
	Grainger	Santee	170
SD	Ben French	Black Hills	25
TX	Monticello	Luminant	1,186
	Welsh	AEP	528
	J.T. Deely	CPS	897
VA	Yorktown	Dominion	376
	Chesapeake	Dominion	813
	Clinch River	AEP	235
	Glen Lyn	AEP	335
WV	Albright	FirstEnergy	278
	Kammer	AEP	630
	Rivesville	FirstEnergy	110
	Willow Island	FirstEnergy	213
	Kanawha	AEP	400
	Phillip Sporn	AEP	600
WI	Alma	Dairyland	45
WY	Neil Simpson	Black Hills	22
	Osage	Black Hills	35
	Total		**24,459**

All these actions are done in good faith, no doubt. And while these may be justifiable measures, they also mean that nearly 30,000 workers are losing their jobs, millions of consumers will be paying more for their electricity and the reliability of our electricity supply is being compromised.

The coal related problems were (partially) eliminated overnight by not burning more coal. As a result, many coal-fired power plants were shut down or converted to natural gas fuel. What is not clear yet, however, is how much GHG savings these moves bring, since natural gas is also a polluter.

A total of 24.5 GW of coal power generation has been eliminated thus far, and much more will be eliminated in the future as a result of the new EPA emission rules. Over 60 power plants were shut down in recent years—most of them to never be restarted. Thousands of jobs were lost in the process, with the accompanying financial devastation for the local economies.

The amount of lost power generation thus far is more than the total power generation of most countries in this world. The future loss of power generation will be even greater.

The Environmental Protection Agency (EPA), however, seems to have grossly underestimated the impact of its mandates on coal-based electricity generation. Instead of 4.8 to 9.5 GW of electric plant retirements, as predicted by EPA recently, 63 power plants with over 24 GW of generating capacity are already shut down, or on the chopping block, due to the new regulations.

The EPA rules also resulted in higher monthly bills for most electric power customers. Arizona, for example, the solar capital of the world, gets 40 percent of its electricity from coal, 22% from natural gas, 29% from nuclear, and 9% from renewables. It is now the 18th most expensive state for energy prices in the country.

The average retail price in the U.S. is 10 cents per kWh. During summer months, the price during peak hours (1-8 PM) rises to 25-30 cents per kWh. Go figure! Additional EPA regulations might send the prices even higher.

Much stricter EPA regulations are coming soon.

Update 2014

For the U.S. coal producing and power industries, the worst is yet to come. In the summer of 2014, EPA came out with new rules, directing the states to cut greenhouse gas (GHG) emissions from power plants.

The new rules call for 25 percent reduction of GHGs by 2020 and 30 percent by 2030 from 2005 emission levels. Considering the fact that the 2005 emission standards were already too low to be met by coal, the new standards basically mark the end of coal-fired power plants in the U.S.

According to the EPA, this move will provide the country with cleaner air and $90 billion in climate and health benefits, in addition to avoiding hospitalizations due to health issues such as asthma.

Environmentalists are welcoming the new rules as potential gain for the American customers. The hope is that the new rules will force people and companies to look into alternative power generation and introduce efficiency measures. The major gains could come from improved energy efficiency, as well as the health and environmental benefits of curbing climate change.

The already impossible to meet GHG standards, have made it very hard for coal-fired plants to operate profitably. As a result, many coal power plants have been closed or converted to natural gas.

The new rules make it virtually impossible to operate coal-fired power plants in the U.S., so we expect many more to shut down or be converted to natural gas. This also means that no new coal plants will be built in this country ever.

In May 2014, the U.S. Chamber of Commerce released a paper estimating that the expected rules would cost the U.S. economy $50 billion a year. This move would also eliminate a quarter—224,000 jobs—of the total 800,000 coal industry jobs. This is in stark contrast with the EPA estimate of over $90 billion saved in climate and health benefits. Who is right; who is wrong?

The new EPA standards give the states full flexibility in deciding how to meet the new limits—as long as they cut the GHG emissions 30% by 2030. Period.

Power plants can cut emissions directly, by switching to a fuel source with lower carbon emissions, such as natural gas. They can also make upgrades to equipment performance and improve the process efficiency.

The states also have a choice to meet the new standards by increasing the amount of energy generated by renewable sources such as solar, wind, biofuels, or hydropower.

There are successful programs based on similar principles already in place in some locales. The Northeast's Regional Greenhouse Gas Initiative is a good example of this effort. It is a nine-state agreement started in 2005 that uses a cap-and-trade system to reduce emissions. These states have been able to make greater cuts in GHG emissions at much lower costs than was initially expected.

This achievement has been made possible mainly by the increased use of natural gas, with some help from expansion of, and reduced costs for, renewables and other innovations in the energy market.

The new EPA mandate puts coal at a great disadvantage and makes natural gas officially the leader of the U.S. power sector, simply because no other fuel or technology can replace coal.

Here are the major problems with the new regulations, as we see them:

1. The U.S. coal industry won't give up. On the contrary; plans are made right now to produce more coal than before. If it is no longer needed in the U.S., then it will be exported to China, India, Europe and other energy-hungry places around the globe.

2. This will contribute to increased emissions in other countries, because nothing can stop China, India, and other countries from using more coal as planned presently. In the end, we are only moving the pollution overseas—and the global emissions will continue rising…with our help.

3. The U.S. is giving a good example of what can and should be done to reduce GHG emissions before it is too late. The U.S. is a rich country and can afford all these new rules and regulations. Most of the world is poor and cannot afford such drastic changes, and therefore will not follow our example. Period!

4. The new regulations put us at a disadvantage compared to most other countries, where coal production and use are unrestricted. The disadvantage can be expressed in terms of lost jobs, and a number of future problems—read on.

5. The very worst part of the new rules is that the U.S. will now rely on natural gas for most of its power generation needs. Our economy will become 70-80% natural gas based.

6. Here is one big disadvantage that can profoundly affect our national security: any serious interruptions in natural gas supplies would cripple an economy that relies exclusively on it. EPA regulators apparently have never heard that putting all your eggs in one basket can lead to disaster.

7. Another disadvantage is for the general public, which must pay more every month while all these changes take place.

8. The natural gas bonanza is good for the short term, but how long can it last? A decade, two, three? The production from existing and any new gas deposits will eventually decrease to a point where gas will become more expensive than coal. Would the EPA issue new, less strict, rules so we could start using coal again?

9. In any event, the gas deposits will be completely drained eventually. What standards would the EPA issue then?

Here is a new idea: How about cleaning coal for use with less pollution?

Carbon Cleaning…

We fully understand that coal is not going away. We also know that it is not clean, and yet it is a most important part of our energy supply. Because of that, we have accepted these facts of life, which we must deal with. But is it possible to clean coal, at least partially, so it stops killing us with noxious fumes emitted at high volume?

The clean coal technology is here and it is just a matter of implementing it and putting lots of money into it. Yes, it might be somewhat more expensive than the conventional processes, but it is important that we stop, or at least reduce, the CO_2 and other GHG gasses produced during coal burning.

IGCC Process

Integrated gasification combined cycle (IGCC) is a technology that uses special equipment, a gasifier, to process coal and turn it into gas. Thus produced synthesis gas, or syngas is then further purified before it is sent to the power plant for burning.

Some pollutants found in coal and released during the IGCC process, such as sulfur, can be turned into re-usable byproducts. This brings additional income to the processing facility, and reduces the emissions of sulfur dioxide, small particulates, and mercury.

Further processing of syngas via the water-gas shift process, shifts the carbon in it to hydrogen, which results in even cleaner, nearly carbon-free fuel. The reaction produces carbon dioxide, which can be compressed and stored.

Excess heat from the primary combustion and syngas fired generation can be re-used in a combined cycle gas turbine, which results in improved efficiency compared to conventional pulverized coal burning.

The IGCC process consists of several major steps:

- Coal and oxygen are combined in the gasifier and converted into syngas and steam.

- Hot syngas is sent into a refining unit to remove sulfur and other contaminants, such as mercury and particulate matter.

- Thus processed clean syngas is piped into a combustion turbine where it is burned to produce steam.

- The steam in turn rotates the blade of a turbine connected to an electric power generator.

- The generator spins, during which action its rotor and stator generate electric power.

- Thus generated electric power is used locally, or is sent into the national grid.

- Large amounts of steam generated in the combustion turbine are recovered and used to generate more steam.

- The steam recovered in the different steps of the process is sent into a heat recovery steam generator where the steam is conditioned and sent into another steam turbine attached to a generator to generate additional electric power.

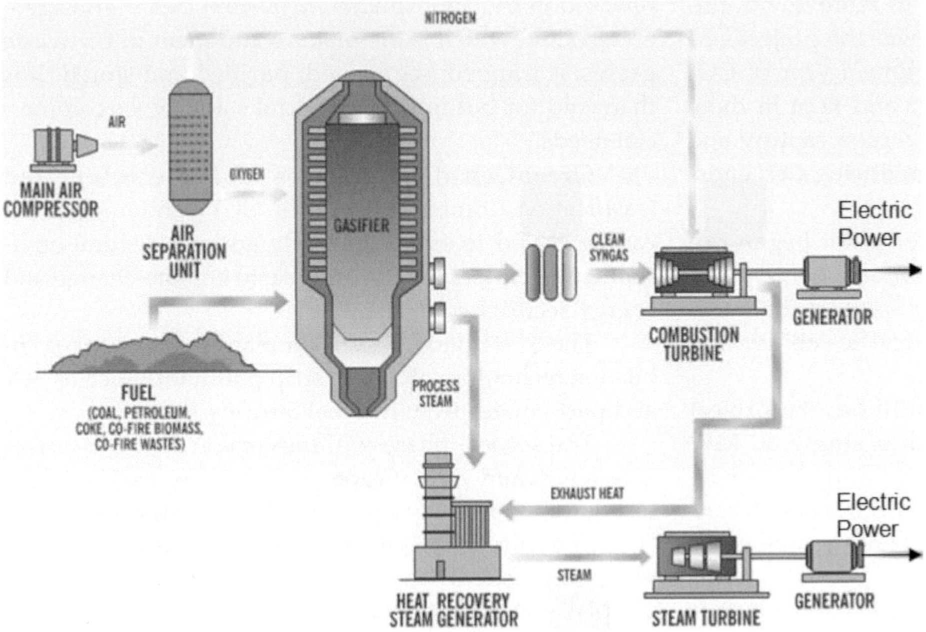

CCS Process

The carbon capture and storage (CCS), or sequestration, process is used to capture waste carbon dioxide (CO_2) from fossil fuel power plants, and store it in special underground reservoirs. This way, large quantities of CO_2 are prevented from entering the atmosphere, in an attempt to mitigate the contribution of fossil fuel emissions to environmental damage. CCS can also be used in a similar manner to remove large amounts of CO_2 from ambient air, but power plants' CO_2 CSS is the focus today.

Storage of the CO_2 can be done either as gas or liquid in deep geological formations, or in the form of mineral carbonates, but natural geological formations are the most promising sequestration sites for now. North America has enough storage capacity for more than 900 years worth of carbon dioxide at current production rates.

A major problem is the fact that any long-term predictions about submarine or underground storage security are uncertain, and there is always the risk of CO_2 leakage into the atmosphere.

Note: Deep ocean storage has been considered, but has been largely abandoned, because it brings further uncertainty to the problem of ocean acidification.

The industry has experience with injecting CO_2 into geological formations for different purposes, such as enhanced oil recovery. The huge amount of CO_2 released by today's power plants, and the need for it to be stored indefinitely are new concepts that need to be proven.

CCS applied to a modern conventional power plant could reduce CO_2 emissions to the atmosphere by approximately 80-90% compared to a plant without CCS. The economic potential of CCS could be between 30% and 50% of the total carbon mitigation effort until the year 2100.

Capturing, compressing and liquefying CO_2 is one method of CCS, but it increases the fuel needs of a coal-fired CCS plant by 30-40%, which ultimately increases the cost of the generated power.

Figure 5-24. IGCC process

Applying the CCS technology to existing plants would be even more expensive since they are usually far from a sequestration site.

There is no way to change the existing power generating infrastructure to alternative renewable sources of energy, so the best approach is to work with the existing technologies, while working on the alternatives until they become economically feasible. That process, however, could take many decades, so storing CO_2 and reducing greenhouse gas emissions might be one way to do this painlessly.

Coal-fired power generation plants produce a rather diluted flue gas, so coupling coal-fired power generation with CO_2 geological storage through carbon capture and storage would produce greener electricity, since there are much less GHG emissions emitted during power generation.

With successful research, development and deployment, and considering the related laws and regulation, CCS coal-based electricity generation may cost less than unsequestered coal-based electricity generation by 2030.

Case Study: German CCS Projects

One of first commercial CCS projects was started in 2000, where a 200-mile pipeline transfers CO_2 gas from the Weyburn Sask oil fields to a coal gassification plant in Beulah, North Dakota. Tons of carbon dioxide (CO_2) are transported through it daily to be injected deep underground for long-term storage.

This effort alone is equivalent to removing 8 million cars from the road every year, over the project's 35 years of operation. Over a million tons a year of CO_2 are being injected into the reservoir and kept in there as long as possible. This method of carbon capture and storage is seen as a practical way of reducing CO_2 emissions in the short term.

Another pilot-scale CCS power plant began operating in 2008 in the eastern German power plant Schwarze Pumpe, hoping to answer some of the pertinent questions about the technological feasibility and economic efficiency of CCS methods.

At a total investment of $96 million, the project consists of a steam generator with a single 30 MW top-mounted pulverized coal burner and the subsequent flue gas cleaning equipment (electrostatic filter/precipitator), wet flue gas de-sulfurization, and a flue gas condenser.

The CO_2 purification and compression plant is downstream of the flue gas condenser to produce liquid CO_2 and gaseous oxygen with 99.5% purity. This high level of purity required for combustion is supplied by a cryogenic air separation unit.

In 2009 the project achieved nearly 100% CO_2 capture, and in the middle of 2010, Schwarze Pumpe reported 6500 hours operation during 18 consecutive months without any major problems.

The pilot plant is designed to have flexibility in terms of construction and the ability to exchange key components like burners. The plant has undergone two remodeling/rebuilding periods to enlarge the project. During those times changes were made to three different burners for testing and optimization purposes.

Schwarze Pumpe, also known as the "CO_2-Free Power Plant Project," is scheduled to be operated for a ten-year period (until 2018), delivering base knowledge and validation data. The project boasts a complete process from fuel input to delivery of liquid clean CO_2 ready for storage. Upon completion, life cycle assessment of the entire CCS process, including the upgrades, will provide information for future practical applications of the CCS technology.

Case Study: IGCC-CCS Combo

How about using IGC and CCS at the same time? A good example for this technology is the new GreenGen power station near Beijing, China. It is a 400 MW coal burning plant using a coal-to-gas conversion and gas burning process, which is much cleaner than solid-coal burning. The plant is also equipped with efficient gas sequestration equipment, which ensures that the gasses released in the atmosphere are 90-95% clean. Thus generated CO_2, which is the major component in the waste gasses, is trapped, segregated, purified and stored. It is then sold for bottling of soft drinks and other commercial needs.

GreenGen demonstrates multiple Integrated Gasification Combined Cycle (IGCC) technologies that can be scaled to simultaneously address several environmental challenges in response to climate change and energy security concerns.

The $1.0 billion GreenGen plant features pre-combustion technology that will strip pollutants such as SO_2 and particulates from the coal burning process.

The second phase will implement fuel cell power generation and carbon capture and sequestration (CCS) technology for nearly zero-emission power generation.

The third phase of the project is planned for completion by 2016, when the plant would produce a total of 650 MW and 3,500 tons of syngas per day.

When completed and optimized, this model will be duplicated in many such IGCC-CCS plants. With

China continuing to build about 30 power plants a year, these technologies could have a significant impact on the air quality and greenhouse gas emissions not only in China but around the world.

So far so good, so what is stopping us? The answer is money. Lots of money is needed, as well as more work on the technologies as a whole.

In any case, even if all this works well, it is not the final solution to all our energy problems. Nevertheless, we have to agree that during this energy (fossil-less) transitional period, which is upon us already, we must look into all possible solutions, no matter how expensive or temporary they might be. Things may change tomorrow so that what is impossible or impractical today becomes a valuable solution tomorrow.

Little by little, we might just be able to add another drop to the bucket of emergency energy and environmental solutions. Any measure that reduces environmental pollution must be given a priority and implemented immediately—it is a matter of life and death at this point.

Coal Liquefaction

Converting coal into liquid fuels might be one of the most important uses of coal, as far as our energy security and independence are concerned. As discussed above, we have enough domestic fuels for power generation and heating, but we are importing half of our transportation fuel   crude oil—from abroad.

Making enough liquid fuels via coal liquefaction would eliminate the import needs, so this is one process/product which we need to watch in the future.

Coal liquefaction is a number of different technologies that are capable of producing liquid oils and fuels using hydrogenation and/or carbon-rejection processes. This is necessary because liquid hydrocarbons have a much higher hydrogen-carbon molar ratio than coal, so the coal molecules must be changed to become hydrocarbon-like.

The coal liquefaction technologies can be divided into:

- *Direct coal liquefaction (DCL) and*

- *Indirect coal liquefaction (ICL).*

Coal liquefaction is a high-temperature and high-pressure process, which requires a lot of energy input, and which increases its cost often well above the crude oil price. Also, the large-scale industrial coal liquefaction operations require multi-billion dollar capital investments. Because of these factors, coal liquefaction is only economically viable when crude oil prices are very high.

This presents a high investment risk, so we won't see many coal liquefaction operations until either, a) crude oil prices jump well above \$150/bbl, and/or, b) international or national disaster forces us to use coal liquefaction for transportation and other applications dominated by crude oil.

In more detail, the major coal liquefaction technologies in use today are:

Direct Coal Liquefaction (DCL) Process

Direct coal liquefaction (DCL) process converts coal into liquid oils and fuels directly without the intermediate step of gasification as used in the ICL process. This is done by breaking down coal's organic molecule via solvents and catalysts at very high pressure and temperature.

Some of the key DCL processes are:

- *The Bergius Hydrogenation process* is a DCL process for coal liquefaction by hydrogenation, developed by Friedrich Bergius in the beginning of the 20th

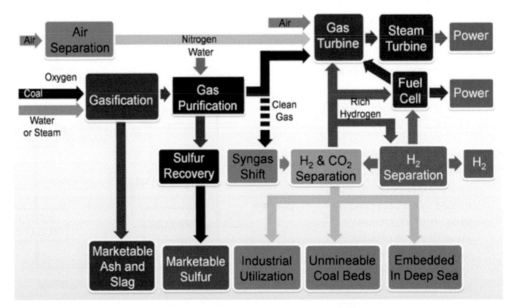

Figure 5-25. Green Gen function

century. Here, dry coal is mixed with heavy oils that have been used previously and recycled. The mixture (slurry) is pumped into a reactor, where catalyst is added to it. The catalyst enriched mixture is exposed to high temperature in the 800-950°F range and up to 10,000 PSI pressure in hydrogen atmosphere.

The resulting reaction is of the type:

$$xC + (x+1)H_2 \rightarrow C_xH_{2x+2}$$

or

$$6C + 7H_2 \rightarrow C_6H_{14} \text{ (hexane)}$$

The mixture is sent into a flash drum, where the liquid and solids are separated, and the liquid is vaporized. It is then sent into a distillation column, where the different fractions of oils and fuels are separated and collected individually. These liquids are then stored, and shipped to customers for use as lubricants and liquid fuels.

Figure 5-26 depicts a generic *coal* hydrogenation process, involving reactors and distillation columns, which is all that is needed to convert the single minded carbon molecule in coal into much more versatile hydrocarbon molecules in liquid fuels.

This relatively simple (but very expensive) petrochemical process amplifies the uses of coal from those of power and heat generation to thousands of applications that crude oil is dominating now. Most importantly, thus liquefied coal can replace the huge, expensive, and insecure crude oil imports.

There are several variations of this process:

•   *The H-Coal process*, developed in the 1960s uses pulverized coal mixed with recycled liquids, hydrogen and catalyst in an ebullated bed reactor. The dissolution and oil upgrading takes place in the single reactor at a fast pace and the products have high H/C

ratio. The main disadvantage of this process is high gas yield, since it is basically a thermal cracking process, high hydrogen consumption, and limitation of the produced oil usage as a boiler oil because of impurities.

•   *The NEDOL process* was developed in the 1970s, where coal is mixed with a solvent and a synthetic catalyst and the mixture is heated in a tubular reactor to 800-900°F and 3,000 PSI in H_2 atmosphere. The produced oil has low quality and requires expensive refining before regular use.

•   *The Solvent Refined Coal (SRC-I and II)* processes were developed in the 1960 and 70s. They use dried, pulverized coal mixed with molybdenum catalyst. The hydrogenation is conducted under high temperature and pressure via syngas, produced in a separate gasifier. This process yields synthetic liquid product Naphtha, some C3/C4 gas, light-medium weight liquids (C5-C10) suitable for use as fuels, small amounts of NH_3 and significant amounts of CO_2.

There are also a number of two-stage direct liquefaction processes; however, after the 1980s only the catalytic two-stage liquefaction process, modified from the H-coal process, the liquid solvent extraction process by British coal, and the brown coal liquefaction process of Japan have been developed.

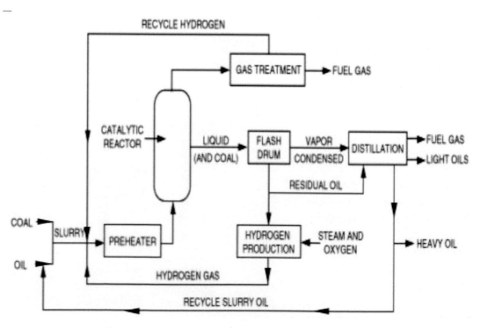

Figure 5-26. Coal liquefaction via hydrogenation.

Historically, there have been many small and large size coal liquefaction attempts, with the most famous dating back to WWII, when several plants were built in Germany to supply the German army with fuel and lubricants from coal.

The Kohleoel Process was used in the demonstration plant with the capacity of 200 tons of lignite per day, built in Bottrop, Germany, during 1981-1987. This process was also explored by SASOL in South Africa around the same time.

The NEDOL process was used by Japanese companies Nippon Kokan, Sumitomo Metal Industries and Mitsubishi Heavy Industries in the 1980s.

Case Study: Shenhua DCL Plant:

The Chinese coal mining company Shenhua built a DCL plant in Inner Mongolia in 2002. The process flow is as follows: a slurry of pulverized coal in recycled (heavy coal-derived oil) is premixed and pumped through a preheater along with hydrogen and catalyst into the first-stage reactor.

The effluent from the first stage undergoes separation to remove gases and light ends with the heavier liquid stream flowing to the higher temperature second stage. Effluent from the second stage, joined with overhead from the inter-stage separator, flows to the fixed bed in-line hydro-treater for enhanced upgrading to very clean fuels.

The effluent from the hydro-treater is the major liquefaction product, mostly diesel, naphtha, and a jet fuel fraction. Bottoms product from the second-stage separator is flashed, and the overheads are pumped to the in-line hydro-treater for upgrading.

The atmospheric bottoms stream containing solids is used as recycle with a portion going to a vacuum still and to solvent solids separation, with the resulting bottoms going to partial oxidation and the overheads to recycle.

The plant coal liquefaction line was designed for 12,000 tons of coal use daily, and produced around 3,000 MTD of gasoline, 18 tons ammonia, and about 53 MTD sulfur every day.

The economics of such a commercial coal liquefaction plant show 18.5% discounted-cash-flow rate of return, with 33.3% equity financing and a 10-year debt carrying an interest rate of 10.5%.

The process economics can be improved significantly by decreasing the investment cost, lowering the coal price, using cheaper natural gas for hydrogen production, and extending the plant's operating lifetime. But most importantly, the economic feasibility of a direct coal liquefaction plant depends heavily on the prevailing price of gasoline and diesel fuel products.

Indirect Coal Liquefaction (ICL)

The indirect coal liquefaction (ICL) process, on the other hand, involves gasification of the coal solids, where coal is reacted to produce a mixture of carbon monoxide and hydrogen (syngas) first. These are then treated via special processes, such as Fischer-Tropsch, which convert the syngas mixture into liquid oils and fuels.

The major ICL processes operate in two stages. First, coal is converted into syngas (a purified mixture of CO and H2 gas). In the second stage, the syngas is converted into light hydrocarbons using one of three main processes: Fischer-Tropsch synthesis, methanol synthesis with subsequent conversion to gasoline or petrochemicals, and methanation.

The first stage of the process—conversion of coal into syngas, or gasification—uses a gasifier, a cylindrical pressure vessel about 40 feet high by 13 feet in diameter. Feedstocks (coal in this case), water and oxygen are fed into the top of the reactor, and when the temperature and pressure produced are high enough, syngas and steam are produced. Recycled black water and slag are also produced and discharged from the bottom of the reactor.

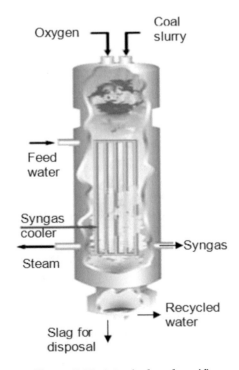

Figure 5-27. A typical coal gasifier

Any kind of carbon-containing material can be a feedstock, but coal gasification, of course, requires coal. A typical gasification plant could use 16,000 tons (14,515 metric tons) of lignite, a brownish type of coal, daily.

Coal gasifiers operate at high temperature and pressures—about 2,500°F and 1,000 PSI respectively. This causes the coal to undergo a number of thermo-chemical reactions, starting with partial oxidation, where the carbon in coal releases heat which further feeds the gasification reaction.

Pyrolysis reactions occur under these conditions, during which coal's volatile matter decomposes into several constituent gases. This process leaves char, a charcoal-like substance, as a main byproduct.

Reduction reactions following the pyrolysis step convert the remaining carbon in the char to a gaseous syngas mixture, with carbon monoxide and hydrogen as its two primary components.

The raw syngas is then run through a gas-cleaning process in a cooled chamber, where the various components accompanying it are separated. During this step, harmful impurities, including sulfur, mercury and un-converted carbon, are removed, separated and used in different steps of the process, or for other purposes.

Carbon dioxide can also be separated at this step, to be either stored underground or used in ammonia, methanol, or other chemicals production.

At this point, only hydrogen and carbon monoxide remain in the syngas, and it can now be burned cleanly and efficiently in gas turbines to produce electricity. Syngas can also be converted into a natural gas-like substance by passing the cleaned gas over a nickel catalyst. This causes carbon monoxide and carbon dioxide to react with free hydrogen and form methane gas.

Thus produced gas has the same properties and behaves just like regular natural gas, so it can be used in gas turbines to generate electricity, heat homes and businesses, or for fueling vehicles.

The second step of the ICL process could be the Fischer-Tropsch reaction, which converts the syngas into liquid hydrocarbons. The simplified form of this reaction is as:

$$CO + 2H_2 \rightarrow CH_2 + H_2O$$

This process was discovered in the 1920s and has been used for the production of liquid fuels and chemicals from syngas ever since. Fischer-Tropsch process was first used on a large technical scale in Germany during WWII, and is currently being used by Sasol in South Africa, and other companies around the world.

The Fischer-Tropsch process works by synthesizing hydrocarbons via chain reactions, where the length of the chain depends on the catalyst properties and the reaction conditions.

There are two types of Fischer-Tropsch conversion steps in use today. The first one uses a Slurry Phase Reactor to produce substances like waxes and distillate fuels, and the other uses the advanced synthol reactor to produce light olefins, gasoline, and other liquid fractions.

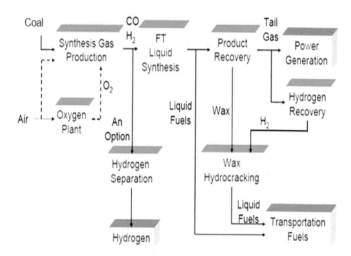

Figure 5-28. Fischer-Tropsch ICL process

In the Fischer-Tropsch process, syngas is preheated and fed into the reactor. Here it is spread into a slurry made out of liquid wax and catalyst particles. The gas is allowed to travel through the slurry, where it diffuses into it and is subsequently converted into even more wax by the Fischer-Tropsch reaction.

The heat released by the reaction is removed via cooling coils mounted inside the reactor, which results in steam generation. The wax is separated from the slurry containing the catalyst particles via special process. At the same time, the more volatile gas fractions and water leave the reactor as a gas stream and can be captured for later use by cooling the steam.

Thus separated hydrocarbon streams are collected and sent for upgrading to their final state as liquid oils and fuels. The water stream is recycled by treatment in the water recovery unit.

- *Methanol synthesis processes* use the reaction of converting syngas into liquid methanol, which is then polymerized into alkanes in special reactors in the presence of zeolite catalysts. This process, known as MTG (methanol to gasoline) method, was developed in the early 1970s, and was tested at a demonstration plant by Jincheng Anthracite Mining Group (JAMG) in Shanxi, China.

- *Methanation reaction* converts thus produced syngas into synthetic natural gas (SNG). This process is used at the Great Plains Gasification Plant in Beulah, North Dakota, which is a coal-to-SNG facility, known to produce 160 million cubic feet SNG per day. The plant has been in operation since the early 1980s.

Coal liquefaction processes, however, generate significant CO_2 emissions, which result from the gasification process and from the generation of heat and electricity needed for the proper operation of the reactors.

There is also high water consumption and discharge rates in some water-gas shift or methane steam reforming reactions, which cause adverse environmental effect.

Nevertheless, properly designed and implemented synthetic fuels produced by coal liquefaction processes are proven to be less polluting than naturally occurring crudes, as far as sulfur pollution is concerned. This is because heteroatom (sulfur) compounds are not synthesized during these processes, and/or are always separated from the final product before its final use.

In conclusion, we must admit that a full integration of coal liquefaction in the U.S. energy supply system for use in the overall economy won't happen overnight. The DCL and ICL technologies require a lot of energy to operate, and the infrastructure and operation are expensive undertakings, so it will take a while before any significant large-scale DCL or ICL plants are in production.

We need to remember, however, that this technology can bring us the much needed relief from expensive and unreliable foreign crude oil imports. We also must remember that as soon as crude oil prices hit the $150 mark, DCL and ICL will become instantly economically feasible, so we must be ready to scale up their production to avoid another economic disaster.

Coal Wars

President Obama initiated a major overhaul of the coal mining regulations recently, but the House of Representatives did not go along. Instead, bill HR 3409, Stop the War on Coal, was introduced which in effect prohibits the Secretary of the Interior from issuing regulations under the Surface Mining Control and Reclamation Act.

This measure was taken in order to avoid adverse affects on the production and delivery of coal and to preserve jobs in the coal industry. Also, the legislation compiled several bills already passed by the House and stalled in the Senate, intended to halt the onslaught of regulations intended to severely curtail the use of coal for electricity generation in the United States.

At the same time, layoffs and mine closures were announced, mostly due to tight regulations. This basically means that there would be less coal needed from the Wyoming mines, that could lead to layoffs in the Powder River Basin.

Wyoming produces about 40% of all the coal used in the U.S., the new regulations affect the state's coal mining operations, the workers and their families.

U.S. Congressman Cynthia Lummis (R-Wyo), issued the following statement after the bill's passage:

There is a clear anti-coal agenda in the President's so called "all-of-the-above" energy plan. In casting coal aside, this Administration is casting aside domestic energy security, high-paying jobs, and affordable abundant energy. President Obama has turned a blind eye to the real-life consequences of these actions in Wyoming and across the nation. We cannot sacrifice the jobs or energy supply these regulations take away, period. It's time the Administration takes the choke hold off America's most abundant energy supply and lives up to their promise of domestic energy security and job creation.

So, looking at this newly created coal-less situation we see a number of complex scenarios at play:

- On one hand, reducing the coal used for power generation would benefit the environment, but we just don't know how much. The problem is that coal is being replaced by natural gas, which also emits significant—if not equal—amounts of GHG during its entire cradle-to-grave life cycle.

- On the other hand, we see abandoned mines, lost jobs, and most importantly abrupt change that might affect our energy supplies and security.

- Natural gas production and delivery is not as stable as coal, so the reduction of the amount of coal used for power generation will result in increased energy costs, at least in the short term.

- Converting the entire U.S. power generating fleet to natural gas might have a significant impact on our energy security. A temporary—and God forbid long lasting—interruption in the natural gas supply would reduce the power generation and even shut down parts of the sector.

- In the long run, we see exports of large quantities of coal to China and other countries, which in effect nullifies the effect of reduced coal use in the U.S. Instead of eliminating the coal-related GHG emissions, we are simply transferring them from the U.S. to China. Is this a smart thing to do? How does that benefit the global environment?

This is a serious subject that is taking new turns every day with some surprising results at each turn. Here is one of these frightening turns of events.

Coal Workers

There is one group of Americans for whom "energy security" means losing jobs and benefits. It is a very large group too—workers in the coal industry. Thousands of these people working in the surface and underground mines and the coal-fired power plants all across the U.S. are the latest victims of the "green" revolution.

During the last 4-5 years the products and services they have been providing for over a century have been labeled "evil" and have been targeted for extermination. As a result of changing regulations, coal mines and coal-fired power plants are shutting down, and through no fault of their own, these people are left without jobs.

While the issues brought out as a reason for the shutdowns are legit, the way this was done is questionable. President Obama's re-election campaign slogan was "all of the above" energy strategy. "We need an energy strategy... for the 21st century that develops every source of American-made energy," Obama said in March, 2012. Voters in the coal states of Virginia, Ohio, and Pennsylvania, believed that this means fairness.

A year later—after his re-election—the President pushed for tighter environmental restrictions on GHG emitting power plants and businesses. As a result, EPA issued CO_2 regulations, which, in fact, made impossible building new coal-fired power plants. The new regulations also make the operation of most existing such plants more expensive.

During his 2014 State of the Union address, Obama confirmed that he meant what he said in 2013, and that he intends to finish what he has started—reduce GHGs to the maximum possible levels. Since coal is in the crosshairs of the anti-CO_2 movement, it is logical to conclude that coal will have to go away, one way or another.

In other words, coal mines and coal-fired power plants, and their workers, have no place in America's energy future, and many of them simply have no future at all. How does that serve our energy security and independence, if some of us are badly hurt in the process?

Things can get worse with the new EPA mandate on GHG emissions, which will surely affect the coal industry as dozens of coal-fired power plants shut down.

It is obvious that President Obama intends to make his final years in office an example of how to deal with climate emissions. He has decided to proceed—on his own if he has to—and do whatever is needed to make a drastic change in the way we do business. It will be his legacy: "When our children's children look us in the eye and ask if we did all we could to leave them a safer, more stable world, with new sources of energy, I want us to be able to say yes, we did."

This is a noble act, no doubt, but results would not match the initial intent. While we are shutting down the coal mines and power plants in the U.S., we are making plans to export more coal overseas, which will be burned in China and India, producing the same, and even worse, GHG pollution, but elsewhere around the globe. This means that the overall effect of sharply cutting coal production and use in the U.S. on the global environment would be a big round zero.

Instead of such a drastic change—reducing emissions overnight—the EPA could've and should've implemented a gradual reduction. This would give the coal producing and using industries enough chance to take measured actions, instead of throwing it in panic mode.

How fair or unfair all this is remains to be debated by the experts, but what we can clearly say is that abrupt changes like this are not a good way to do business. They hurt the industry, which has powered, and is still powering, most of our economy.

Those who suffer most and immediately as a consequence of these changes are the workers in the coal industry. At least 30,000 of them are to lose their jobs.

Our energy security is also compromised by such drastic and sudden changes. The widely spreading use of natural gas is not a proven alternative, and any problems in the plans might cost us dearly.

Energy security should not mean abundant, safe, and cheap energy for some, at the expense of others. But it is too late to cry over spilled milk, so we just need to find solutions to the issues at hand before we end up with bigger national and international problems.

Coal Cost

Coal is a predominantly domestic fuel—it is produced and used locally in 85% of the cases. Domestic energy markets are more stable, since they are not usually exposed to international prices. Yet the prices can vary significantly because of quality, geographic, contractual and regulatory aspects. Different types of coal and purchase conditions, including time and point of delivery, create a great variety of pricing schemes too.

We must consider the not-so-obvious fact that coal is not a single product. Instead, it is a family of many types of different materials. There are many coal classifications, but the main price determining types are: a) non-coking (steam or thermal coal and also lignite), and b) coking coal.

In more detail:

- Coking coal, which is used in iron making, is of a higher quality, since it has more desirable higher caking properties and strength, than non-coking coal. Because of that, its price is usually higher too. But this is oversimplification. Coking coal is also not a "homogenous" product, since there are a variety of qualities and other factors that determine its final price.

- Hard coking coal represents the highest quality, while the other types, such as semi-soft or high-volatile coking coal are usually sold at much lower prices.

- Some high-quality non-coking coals are also used in metallurgy for pulverized coal injection, which reduces total coke consumption in blast furnaces. The price of these types is related to coking coal, although still at some discount rate.

- Then, there are the different market niches. For example, high-grade anthracite can be used for a number of applications and would follow the usual prices in each of these. Ultra-high-grade anthracite can replace coke in blast furnaces, where it approaches coke price at a slight discount.

- The non-coking coals, such as steam and thermal coals are used for heat and power generation, where calorific value is the main factor in defining performance and price. Lower calorific value generally means a lower price, where the price falls faster than the energy value.

- The calorific values can also be referred to as "gross" or "net."

- The price-reporting agency Argus lists five different price indexes for one single source—Indonesian coal. Each index reflects the actual kilocalorie counts of 6500, 5800, 5000, 4200 and 3400 kilocalories per kilogram of different coals.

- Supply and demand is another determining factor of price formation.

- The different physical conditions—air dry, wet, etc.—of the as-received coal also determine the price per ton.

- Different geographic markets, although basically well integrated, depend on many factors, such as transportation.

- The different importing and exporting regions also have different prices, determined mostly by availability and production and transportation costs.

- Seaborne coal's price is determined mainly by the freight and insurance costs. Here, terms such as free-on-board (FOB), cost insurance freight (CIF) or cost freight (CFR) determine the final price.

- The price of coal also depends on the type of purchase instrument; contracted coal or spot purchase. Generally, coal is traded internationally on a spot basis, so most price markers and indexes refer to spot purchases. Some Japanese utilities are the exception. They buy most of their thermal coal through one-year term contracts, while coking coal is bought on quarter or monthly contracts.

- Coal futures, forwards, as well as swaps for different dates, different types of coal and different locations are also common on the coal marketplace.

So, our old, simple friend coal is a complex commodity—so much so that we really have to know very well what we need and expect in order to get the best coal quality and value.

In the U.S., coal prices vary from state to state, from mine to mine, and also depend heavily on the seasons and the overall market conditions.

Table 5-7.
Average sales price of coal (in \$/ton)

STATE	Under ground	Surface mine
Alabama	100.17	108.71
Illinois	51.43	46.60
Indiana	51.77	44.91
Illinois	51.43	46.60
Indiana	51.77	44.91
Kentucky Total	63.38	64.01
East Kentucky	78.63	70.86
West Kentucky	47.87	38.93
Ohio	47.86	43.41
Pennsylvania	78.67	82.89
Tennessee	66.27	77.27
West Virginia	89.40	77.39
North W. Virginia	60.91	65.74
South W. Virginia	114.25	78.15

These are only examples of the purchase costs of coal from mines in different U.S. states. The price would vary according to quality of coal, location, transport charges, and other fees. Compare these prices with the average cost of $15 per ton in 1950.

The official estimates show coal prices at $40-60 per short ton (2,000 lbs.). At the same time, Powder River Basin coal costs only $8.75 per ton. The variation is due to the quality of coal and the amount of heat as measured in Btu generated.

Figure 5-29 shows the relative prices of the fossil fuels during the energy crisis of 2008-2012. The fluctuating prices were caused and driven by faulting economy and rising crude oil prices and production uncertainties.

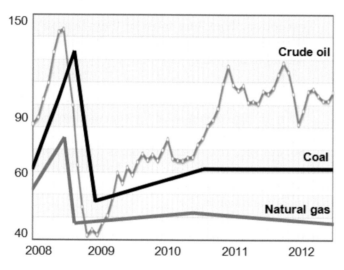

Figure 5-29. U.S. fossil fuels, comparative prices in US$

Note that while we can clearly see the price fluctuations in Figure 5-29, the actual dollar per unit comparison among the different energy sources is very difficult to do for a number of reasons.

- The physical properties of the different energy sources are so very different—oil is liquid, coal is solid and natural gas is gaseous;

- The heating values of the fuels are quite different; and

- The official market prices are expressed in different units; oil is given in $ per barrel (42 gallons), the price of coal in $ per ton (1,000 kilograms), and that of natural gas in $ per thousand cubic meters.

If we equate the energy of the different sources to that of crude oil (42 gallons), which contains approximately 1.7 MWh of power, we see that 1,000 m^3 natural

gas contains 10.5 MWh of energy, while 1 ton of coal contains 6.5 MWh energy equivalent.

This information reveals that a $90 barrel of oil contains only 1/4 of the energy contained in a $60 ton of coal, and 1/6 of the energy contained in 1,000 cubic feet of natural gas worth $40.

But this is not all. The above numbers are the theoretical content of energy, while the actual amount of electricity, or heat, produced when using the different fuels in different applications depends on additional factors such as the type and efficiency of the equipment, and the actual quality of the input materials—for example, all coal is not the same.

The prices also reflect not only the actual energy value of the energy sources, but other factors, such as the convenience of using them in different applications. For example, no matter how cheap coal is, we cannot pour it in gas tanks and drive away.

Coal Externalities

This is the skeleton in the closet of the coal industry. In economic terms, an external cost, or externality, is a negative effect of an economic activity on a third party or society in general. When coal is mined and used to generate power, for example, the external costs include the impacts of water pollution, toxic coal waste, air pollution, and the long-term damage to ecosystems and human health.

The list of such damages is long, and we cannot possibly look at all items and cases on that list, so we will limit this writing to the very basics, to underline the importance of this subject:

- Greenpeace released an analysis in 2008 showing that the global total cost of coal, including the externalities, was at least $450 billion annually and growing. This number was later updated to $500 billion in 2011 by a Harvard study group, and is now most likely much higher.

Researchers arrive at these figures by looking at CO_2 and other pollution damages, and the damages, resulting in health costs from mining and power plant accidents, air, soil, and water pollution.

The first report was released at the time when industry ministers from 20 big GHG emitting countries met in Warsaw with the world's climate-polluting industries, where it was agreed that the relentless expansion of the coal industry is the single greatest threat to increasing the damages of global climate change.

Coal is presently the most environment-polluting

fossil fuel, responsible for one third to one half of all CO_2 emissions, which are projected to increase to 60% of all emissions by 2030. Thus, reducing coal burning will benefit not only the global climate, but it also will reduce the other external impacts and the related costs which everybody else has to pay for.

In calculating the $450 billion annual global cost of coal figures, the focus was on the external costs of coal. This includes: the cost for rectifying damages attributable to climate change; human health impacts from air pollution; fatalities due to major mining accidents; and other factors for which reasonably reliable global data are currently available. The above dollar number was derived by taking into account 90% of the global emissions and looking at the damages, which are projected to rise significantly, due to the impacts of climate change. With these changes, the total global cost of coal is likely to increase sharply—unless its use is reduced and/or the climate change is stopped soon enough.

The projected impacts of climate change include billions of people facing water shortages, while others will be flooded by rising ocean levels. This also includes the hundreds of millions who will be threatened by food insecurity and hurt by exceedingly extreme weather events, such as storms, earthquakes, and tsunamis.

The Other View...

There are a number of specialists of the other kind, who insist that it is too late now. There are climate models showing that no matter what we do now—even if we stop all CO_2 and other GHG emissions immediately—we have already triggered a massive global warming and other environmental changes that can be reversed only by reducing the GHG gasses already in the atmosphere. Doing this is simply impossible to achieve, so no matter what we do, we cannot stop, let alone reverse the environmental damages.

But the negative impacts of coal are not only related to global climate change. Coal also pollutes local water resources, and dirties the local air, which causes black lung disease while large geographic areas are changed forever by blowing apart entire mountaintops. A dramatic increase of air pollution in China is a good example of such large-scale damages to environment and humans alike. Entire river systems and lakes are lost as a result in China and other countries as well.

- The new "clean coal" technologies have the potential to sharply reduce CO_2 emissions from coal-fired power plants, but industry experts claim that the newly developed carbon capture and storage

(CCS) techniques have their own problems that could lead to another set of dangers. The CCS technology is immature, unproven, and contains inherent risks. In addition, it comes with an enormous price tag and a long process and equipment development time period.

One thing everyone who is well informed on the CCS technology agrees on is that the global GHG emissions must start declining soon, and in the best of cases the new CCS approach is in no position to play a significant role in making this happen for the above-mentioned reasons.

- Greenpeace insists for the world governments to see and agree on a "climate vision" that is to address immediately the threat of serious global emissions peaking by 2015. Developed countries must agree on immediate reduction targets in the 30-40% range, if any significant results are to be achieved before it is too late. How this is to be done and how much will be done, if anything at all, is still uncertain, but the talks continue—yes, we are still in the talking stages.

The Environmental Damage

There are a number of environmental changes of late, some of which are attributed to coal mining and burning. We cannot list them all, let alone discuss each in detail in this text, but here is a sample of what is happening and what could be expected to happen in the future.

Habitat Destruction

The majority of the coal mining in the U.S. is surface mining, which includes strip mining and mountaintop removal. In these cases, the original ecosystem which once was on the surface is destroyed and removed in the process of removing the surface dirt and then mining the coal underneath.

Mining destroys fish and wildlife habitat, which has rippling effects not only on their populations, but also on the local population that relies on them. In addition to the mine site itself, coal mining affects the surrounding areas too. Air pollution, widespread coal dust, and mine runoff discharge into nearby water bodies are the usual results from coal mining operations.

Coal dust particles, chemicals, and sediment from coal mining discharged in the local rivers and lakes can reduce life expectancy of fish, damage their immune systems, and suffocate fish eggs.

After the coal in the mine is depleted, coal companies usually restore the surface to a certain level and plant vegetation in an attempt to restore the original ecosystem. This, however, is difficult and sometimes impossible, especially in wetlands areas.

It is very unlikely that any amount of effort or money could restore the environment and the wildlife in it exactly the way it was during pre-mining times. Because of that we must accept the fact that mining operations, if and when approved, will leave a permanent scar on the Earth's surface, which might take centuries to heal completely.

Health Effects

Mining and burning coal in coal-fired power plants releases a number of toxic pollutants, some of which rise in the air, and remain behind as liquid or solid waste materials and chemicals. These pollutants are responsible for a large number of illnesses and premature deaths to people directly involved in the industry, people who live nearby, and people worldwide.

Coal dust in mines and near storage and transport facilities contributes to serious respiratory illnesses such as asthma and Pneumoconiosis (black lung). Solid combustion wastes such as fly ash pollute groundwater near storage facilities, contaminating individual and community water supplies.

Airborne pollutants have a larger footprint. Despite air pollution regulations, toxic emissions (soot, sulfur dioxide, nitrous oxides) from coal-fired power plants are estimated to be responsible for thousands of deaths due to lung disease each year in the U.S. and Canada.

A government study in Ontario found that the coal-fired plants in that province alone were responsible for an average annual total of about 660 premature deaths, 920 hospital admissions, 1090 emergency room visits, and 331,000 minor illness.

Coal combustion emissions released into the atmosphere contain nitrous oxides which are responsible for industrial and urban smog, sulfur dioxide which is the primary reactive agent behind acid rain, mercury which accumulates in the food chain, and large amounts of carbon dioxide which is the most important greenhouse gas contributing to climate change. Coal mining itself also releases significant amounts of methane, another extremely potent greenhouse gas. Coal mining is responsible for over 25% of the energy-related methane emissions in the U.S.

Coal combustion wastes (CCW) include ash, sludge, and boiler slag left over from burning coal to make electricity. These wastes (120 million tons/year in the U.S.) concentrate toxins such as arsenic, mercury, chromium, cadmium, uranium and thorium. In addition, these wastes create an expensive storage problem "in perpetuity."

Coal-fired power plants are also a major source of atmospheric mercury, which accumulates in the food chain and can damage the developing nervous systems of human fetuses, as well as lead to reduced immune function, weight loss, reduced reproduction rate, mental defects and other neurological problems.

There are a number of health effects and illnesses attributed to the pollution released during coal mining, preparation, transport, combustion, and waste removal and storage.

Some of the negative coal-related health effects and illnesses in the U.S. are:

- Respiratory illnesses like asthma OCPD and such, caused by particulates, ozone, and sulfur dioxide emissions

- Black lung from coal dust

- Congestive heart failure, caused by particulates and carbon monoxide emissions

- Non-fatal cancer, osteoporosis, ataxia, renal dysfunction, caused by benzene, radionuclides, heavy metals, emissions and waste materials

- Chronic bronchitis and asthma attacks caused by particulates and ozone emissions

- Loss of IQ, caused by excess air and water pollution

- Nervous system damage, caused by mercury releases in the drinking water

- Terminal cancer, caused by air and water pollution

- Reduction in life expectancy, caused by particulates, sulfur dioxide, ozone, heavy metals, benzene, and radionuclides emissions and releases

- Degradation and soiling of buildings caused by excess sulfur dioxide, acid deposition, and particulates emissions that can effect human health

- Ecosystem loss and degradation with negative effects on human health and quality of life, caused by excess emissions and releases of gasses, liquids, and solids

- Global warming, caused by large quantities of carbon dioxide, methane, nitrous oxide, and sulfur dioxide emitted by coal-fired power plants. This is the greatest threat to the future of the human and wildlife population, for it will affect us in many negative ways.

Coal Related Illnesses

Coal-fired power plants emit pollutants that may act as triggers to a number of medical conditions in humans and animals. These pollutants include sulfur dioxide, nitrogen oxides, and particulate matter. In addition, the carbon dioxide emissions from coal accelerate global warming, which is likely to increase the concentration in air of pollen from some plants, such as ragweed, and thereby contribute to the development of additional asthma and other conditions.

Chronic inhalation of coal dust causes several lung disorders, including simple coal-workers' pneumoconiosis, progressive massive fibrosis, chronic bronchitis, lung function loss, and emphysema.

We will review some of the most important illnesses related to coal mining and use, such as:

Asthma

Coal dust is one of the primary causes of asthmatic conditions for coal miners, people living and working next to coal mines, or those living in areas with high pollution levels. There are a number of cellular actions, interactions, and conditions related to coal dust toxicity on human tissues and the lungs in particular. The dust particles are known to cause lung disorders on a cellular level like macrophages and neutrophils, epithelial cells, and fibroblasts.

Inhaling coal dust particles induces cellular and non-cellular activities in the lungs, and may be involved in the damage of lung cells as well as some macromolecules including α-1-antitrypsin and DNA.

Recent studies with coal dusts show its effects on important leukocyte recruiting factors, such as: leukotriene-B4, platelet derived growth factor. Coal dust particles also stimulate the macrophage production of various factors with potential capacity to modulate lung cells and/or its extracellular matrix.

In more practical terms, asthma is a chronic disease of the lungs characterized by inflammation and narrowing of the airways. Airway inflammation in asthmatics causes swelling in the throat that narrows a bronchial tree that has been previously sensitized to inhaled irritants, including many air pollutants.

Exposure to an inhaled irritant causes further narrowing of the airways and the production of mucus that makes airways even narrower, a vicious and very dangerous process for all affected.

Patients with asthma experience recurrent episodes of dyspnea (shortness of breath), a sensation of tightness in the chest, wheezing, and coughing that typically occurs at night or early in the morning. This can lead to hypoxia (low blood oxygen level), hypercarbia (high blood carbon dioxide level), and respiratory acidosis (acidification of the blood caused by carbon dioxide retention) that may, in turn, cause cardiac arrhythmias and other medical emergencies.

During severe attacks, the lungs also fail to perform their task of exchanging carbon dioxide, produced by metabolic processes in the body, for life-giving oxygen. This, combined with other conditions, and the resulting panic, could prove fatal in some cases.

There are about 25 million asthmatics in the U.S., including 6-7 million children. The Centers for Disease Control and Prevention report that the number of persons with asthma increased by 84% from 1980-2004. More than half of the U.S. states report that 8.6% or more of their inhabitants have asthma. These high-asthma states are clustered in the Northeast and Midwest.

Again you ask, what does this have to do with energy security? Again we say, who cares about energy security while lying in a hospital bed dying from lung cancer. The issues are intimately connected and separating them intentionally should be a criminal offense.

The Economics of Coal Externalities

All of the impacts of coal have a direct and indirect economic cost, ranging from the jobs lost by fishermen downstream from a coal mine, to the health care costs of the people sickened by coal-fired power plant pollution, to the cost of cleaning up spills of toxic coal waste. And there are a number of other consequences—the list of which is too long to list here.

Some of the indirect economic costs of coal are in the form of subsidies and tax breaks which are not reflected in the market price of coal. For example about $4.6 billion in coal-related subsidies was introduced in the 2009 stimulus package.

On top of that, in the state of Kentucky, for example, the government spends $115 million more on subsidies for the coal industry than it receives in taxes or other benefits.

Coal mining and power plant operation are expensive projects, and often require major investments. The risks and costs of those investments are often passed on to taxpayers via different financial vehicles such as infrastructure subsidies and loan guarantees. For example, the Healy clean coal plant (HCCP) cost the state of Alaska and the federal government nearly $300 million since its construction started in the mid-1990s and as yet has not produce any power in return.

As a result of this and other abnormalities in Alaska, the coal industry pays only 5% of its market value to the state, even though the nominal rates are much higher.

Taxpayers also pay the costs of cleaning up environmental disasters caused by the coal industry. These cases and the related expenses are not well publicized but many millions of dollars are spent on these activities every year.

Cleanup of the recent coal ash spill in Tennessee is estimated to cost up to $1 billion, not including pending litigation. Now that the cleanup at this site has been taken over by the EPA under the superfund law, most of this cost will be borne by the U.S. taxpayer. Congratulations, taxpayers, here is another gift from the coal industry.

The most serious impact of the coal industry is air pollution, which has enormous economic costs through health care costs and lost productivity. A recent study by the government of Ontario estimated these costs in the billions of dollars within the state alone.

Similarly, the cost associated with premature death due to coal mining in the state of Virginia is five times greater than all measurable economic benefits from the mining industry. Measuring human death in terms of economic benefits is not easy, but our esteemed economists find ways, and we agree that it is the best way to reflect the reality in our society.

A recent study found that coal mining in the state of Illinois resulted in a net cost to the state of almost $20 million in 2011, even before including any externalities. Including the externalities would put this number in the hundreds of millions of dollars.

Another direct effect of coal mining is that it also hurts the ecosystems that other industries and the population use and depend on. Recreational fishing, camping, commercial fishing, and tourism are important for the economy of most states. For example, there are over 55,000 full-time jobs in Alaska that are closely related to, and depending on, the health of the state's fragile ecosystems. These jobs are over a quarter of Alaskan employment and produce over $2.5 billion of income in the state.

At the same time, the proposed Chuitna coal strip mine in the Cook Inlet, Alaska, would create 350 jobs, which will impact the 55,000 ecosystem-dependent jobs in a negative way. These effects will affect negatively the economy by leading to worsening the health of the population, which in turn impacts the health care costs, thus compounding the economic impact.

At the same time, it is undeniable that coal pro-

vides cheap electricity, which helps economic development and contributes to improving the health of the population, thus lowering the total health care costs. While this is true in theory, it has to be applied in the same equation with the negative economic impacts.

When the total of all costs of air, soil, and water pollution, global warming, habitat destruction, and human health are added and compared with the benefits of cheap electricity, the overall economic and social impact of coal is negative by almost a two-to-one ratio:

$$\text{Economic benefit} = \frac{\text{Negative effects}}{\text{Positive effects}} = \frac{2}{1}$$

Could it be that the negative effects of coal mining and burning are twice as many and as serious as the positive? The coal industry proponents will surely argue with the above numbers, and the overall conclusions herein, so we just need to mention that using coal is not the only option.

So, even if the negative effects are not that great, the fact that coal is continuously emitting more and more air pollution is undeniable. The pollution in turn is causing illness and climate warming, which is also undeniable.

The other fact that needs to be considered in this context is that China, India and other large developing countries are planning to increase their use of coal mining and burning in the coming decades. That will cause an increase in coal pollution to levels we have never seen before.

What that increase in GHG emissions will do to the health of the people and to the global environment is anyone's guess, but we venture a guess that it won't be good… and might even have detrimental consequences to humans and Earth's tortured environment.

So, to get the real value in dollars of each energy source, we must know the actual application, its location, the type of process and equipment to be used, the type and quality of the fuels, and a number of other factors that might play a role in determining the final production cost.

Coal prices have historically been lower and more stable than oil and gas prices (before including the external costs). Coal is also very likely to remain the most affordable fuel for power generation in many developing and industrialized countries for decades to come.

In countries with energy intensive industries, the impact of fuel and electricity price increases is compounded. High prices can lead to a loss of competitive advantage and in prolonged cases, loss of the industry altogether.

Countries with access to indigenous energy supplies, or to affordable fuels from a well-supplied world market, can avoid many of these negative impacts, enabling further economic development and growth.

Those that have limited supplies have to import the needed quantities of fuels. They would be most affected, and most vulnerable on the growing global coal export markets.

And speaking of coal exports...

U.S. Coal Exports

Thus far, most coal produced in the U.S. has been consumed domestically at a fairly steady pace. At the same time, the U.S. coal exports have been going up and down, but now with decreased domestic production, there are serious plans to export huge amounts of coal to China and other countries.

The new regulations are forcing the coal companies to look at exporting their production to keep the mines open. As a result we foresee export volumes of coal to increase.

The proportion of coal production going toward exports has also increased, doubling from 5 percent in 2009 to 10 percent in 2011, sharply increasing for awhile in 2012, and projected to take off in the near future.

Several new shipping terminals have been planned in the U.S. and the coal companies are gearing up for exporting as much coal as they can produce—to whomever wants to buy the stuff.

But what does that mean to U.S. energy security—especially that of future generations? What would exporting so much coal do to their energy supplies? Isn't that similar to the proverbial "shooting oneself in the foot?"

The majority of the present-day coal production consists of thermal (steam coal) used mostly for electricity generation. Metallurgical (coking coal), although in smaller quantities, is also important for iron and steel production.

Seventy percent of global steel production depends on this type of coal. In 2011, metallurgical coal exports accounted for 77 percent of U.S. coal exports in terms of trade dollars and 65 percent in terms of volume.

In 2010, exports to Asian markets increased 176 percent from 2009 levels, primarily because of a surge in exports of metallurgical coal to China, Japan, and South Korea.[8] Metallurgical coal exports accounted for 83 percent of the growth in 2010 export volumes.

U.S. coal producers increased both export volumes and prices in 2010 because there were huge demands from China and India, due to reduced world coal supply caused by heavy rains and flooding in major coal exporting countries.

In 2012, export coal prices turned down sharply, due to decreased domestic demand, a slowdown in economic growth in both China and India, and the continuing weakness in some European countries, which promise to be a bottomless coal market.

Oil and natural gas prices in Europe are, and will remain, quite high for a long time, so relatively inexpensive U.S. coal will be in even greater demand by European utility companies. This, combined with the decline of carbon emissions permits in the European Union, makes coal even more attractive to the end users.

Today U.S. coal export volume and prices are influenced by the difference between domestic and international demand and gas prices, as well as by the coal use in developing countries and the Asian coal production levels.

While all indicators point to favorable trade winds for U.S. coal exports, the nagging question persist, "Is this the best we can do with such a key commodity, that is in limited supply?" Is this the best for our long-term energy security? The answer is also blowing in the wind...

Coal and Our Energy Security

Coal plays an important role in our energy security, since it meets the demand for a secure and cheap energy supply. It is abundant and widespread, with commercial mining taking place in over 50 countries.

Over 45% of global electricity is currently based on coal, and this number is growing as some developing countries ramp up coal-fired power generation. Coal provides the base load in most of the world's countries today, via well established coal-power generation technologies and know-how.

At the same time, ongoing research enables the coal-related technologies to meet the efficiency and safety demands. Lately, however, excess regulations are pushing the limits to the breaking point, which is causing changes in the coal market.

Coal resources are essential in enabling economic development by guarding against import dependence and large price variations. Coal prices have always been lower and more stable than any of the other fossil energy sources, so coal is likely to remain the cheapest and most available fuel for power generation around the world during this and part of the next century.

Coal is the most abundant and economical of the fossil fuels, and is estimated to be available at least for the next 120-150 years at current production and use levels. This, compared to just over half that time for oil and natural gas is a significant advantage in our future energy security as well.

Coal is also readily available from a wide variety of sources in a well-supplied worldwide market, and can be easily transported to points of use quickly and safely by ship and rail. A number of supplier countries ensure competitive and efficient functioning of the international coal markets.

Coal can also be stored in large quantities at power stations, for use during emergency situations, thus adding to the security of our energy supply.

Recent developments in coal to liquids conversion processes can be used to hedge against oil-related energy security risks. Using this approach could allow us to minimize our exposure to serious oil price volatility, by providing liquid fuels to the transportation sector.

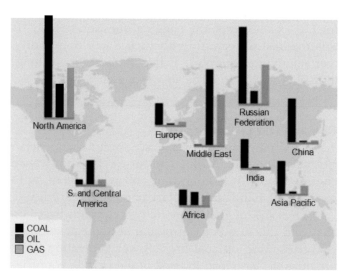

Figure 5-30. World's fossil fuel reserves

Global demand for energy continues to rise in most developing countries, so the energy security issues are becoming ever more important on a global level. Continuous energy security provides economic growth through readily available and cheap energy.

Our energy security continuity can be ensured only by the ability to provide reliable, affordable, and risk-free sources of power without risks, vulnerability to long- or short-term disruptions, or serious price variations.

Vulnerable energy supplies cause interruption of economic growth and financial losses. In more serious cases, such interruptions create havoc in national eco-nomic development and could even affect negatively the health and wellbeing of the population.

While some of us are somewhat aware of the short-term risks to our energy security and the environment, and are taking all possible precautions to eliminate them, long-term energy security and environmental integrity are mostly on the back burner. They are something that we could do tomorrow…

This anomaly has created a gap in the continuous global energy security and environmental integrity. It reflects the fact that we are using unreasonably large amounts of fossils today, but are not worried about tomorrow when these natural resources will be fully exhausted and/or the atmosphere would be totally clogged by air pollution.

It is logical that the fossils—including coal—will be gone in less than a century. We know also very well that economies cannot run efficiently, or at all, without the fossils. We know that the people of the next centuries will have to live in very cold (or hot) houses without fossils to heat or cool them, and will not be able to move from place to place as freely as today.

Yet, looking at the forecasts, we see the energy demand increasing, which means that we will need much more fossils—and coal in particular—in the future.

We also should always remember the relation between energy and the environment. It is impossible to think of energy and energy security without considering the environment. There is a very strong relation between these variables, and we must be fully aware of the present reality and the future possibilities.

We need air to breath, and water to drink in order to survive. We also need energy for our personal and business activities. Energy emits pollution that contaminates our air and water. So we need it and don't want it at the same time.

It's a catch 22 situation, the ultimate effects of which we are still trying to understand and avoid.

Update 2014: King Coal Still Reigns

As a confirmation of our observations that coal's consumption in the overseas energy markets is increasing, 2013 annual reports released by energy companies show that it jumped by 3%. Overall, coal accounted for 30.1% of total primary energy consumption around the globe, which is the highest proportion since 1970.

*Increased use of coal overseas suggests that the developing countries are looking primarily for cheap—**not clean**—fuels.*

The increased U.S. coal exports are partially responsible for the increased coal use overseas. The planned ramp-up of coal exports from the U.S. would reduce the global coal prices further and contribute even more to the increase of coal use around the world.

The increase in consumption made coal the world's fastest-growing fossil fuel, driven largely by increased use in developing regions.

Japan is now ramping up power generation from its coal-fired power plants in response to the shutdown of the nuclear power fleet after the Fukushima accident. Some European countries are also planning increased use of coal, since their energy options are limited and since Russia is threatening to shut down the natural gas flow to the continent.

In parallel with the increased coal consumption, we see an increase in the global GHG emissions, coming from increased numbers of coal-fired power plants worldwide. This abundant and cheap coal also contributes to a delay in the implementation of renewable and other alternative fuels, simply because they cannot compete.

Renewable energy sources, for example, accounted for a record share of global energy consumption last year too; about 2.7% of global consumption in 2013 was generated by renewables. This is up from 0.8% a decade ago, but is still 10% less than the coal's increase during 2013 alone.

Solar was the fastest growing technology in the renewables class, growing by 33% compared to wind's 20.7% during the same time. Overall, renewables accounted for over 5.3% of global power generation. This is about 10 times less than coal-fired power generation and the margin is increasing in favor of coal. This gradual increase is likely to continue *ad infinitum*. The biggest challenge for renewables is lack of subsidies and investment in areas where penetration is highest but economic growth sluggish; particularly in some European and most developing countries in Asia and Africa.

For now at least, the renewables cannot compete without subsidies. Short of massive increases in government subsidies and other benefits, as needed to level the playing field and allow the renewables to compete with the fossils, coal will continue to grow and dominate the global energy markets... until it is gone for good.

Summary
- Coal fuels almost half of the electricity generated in the U.S.

- Coal also emits half of the CO_2 gasses that harm our environment.

- Electricity generated from coal *costs* about 5-6 cents per kWh, which is among the lowest in the world.

- The U.S. national average *price* of electricity from all fuel sources—with 50% from coal—is about 10 cents per kWh, which is also one of the lowest in the world.

- Each person in the U.S. uses on the average over 7,500 pounds of coal annually.

- Coal accounts for 90 percent of America's fossil fuel reserves, and is enough to power the U.S. for the next 100-150 years, but this needs to be adjusted as coal exports are increasing.

   Note: The question that begs an answer here is, why are we in such a hurry to get rid of this precious commodity by exporting it? Is "selling the branch you are sitting on" (which is exactly what we are doing by exporting coal and other fossils), the smart and responsible thing to do? This is a very serious question that nobody is even trying to answer.

- Coal is a key element of our national security, since it is the lifeblood of our domestic energy supply, which depends on reliable electricity. It allows us to avoid further dependence on imports of energy from abroad.

- Coal provides thousands of long-term, well paid jobs. Each mining job also creates 3-4 additional jobs somewhere in the U.S. economy, or about 500,000 people (including over 125,000 coal miners) are involved in the U.S. coal industry. Many times more people are involved in the global coal industry.

- Coal mining jobs average $65,000 per year, or about twice as much as the average wage for other industrial jobs at the same technical level.

- Coal mining, however, is a dirty, hard, and dangerous job, so we must thank our miners and their families, who stand by them in good and bad times, every time we turn the lights on.

- 65-70 percent of U.S. coal production is via surface mining, which alone is responsible for over $5.0 billion of business activities every year and provides fuel to power more than 25 million American homes.

OK, we are totally convinced. We see clearly the pros and cons. Coal is indispensible, granted. But...coal is dirty, damages the environment and harms wildlife and people.

It is dirty in the ground and makes people (miners and innocent bystanders living near mines and coal power plants) sick. It is dirty above ground and makes people sick (during transport and processing).

It is even dirtier during burning for power generation by emitting tons of toxic and GHG gases that make people sick, pollute the environment, and cause global warming. It is dirty after use too, since its ashes are deposited in slurry ponds, which leak and create huge environmental disasters and illnesses.

Surface mining operations are also responsible for damaging land and local environment, as entire mountains and their habitat are erased from the face of the Earth and replaced by huge gaping holes in the ground.

So, coal is good and coal is bad. We cannot live with it, but we cannot live without it either. Where do we go from here?

INTERNATIONAL COAL

Coal has always been an international commodity, and is getting even more so as we speak. The U.S., for example, has been exporting coal for a long time, but is now planning to increase the coal exports to unprecedented levels.

China, India, and other developing countries, on the other hand, have been importing large quantities of coal for awhile, but are now planning to increase the imports to unprecedented levels.

Supply meets demand, you say. Right; the business model is clear; export as much as you can to make money. At the same time, however, the future of the fossils and our environment is also clear, albeit not in such positive terms. Increased levels of coal use mean that it will be gone sooner, and also that much more GHGs will be emitted in our environment.

If this situation continues for awhile, it will lead into an uncertain future, where global energy security will be compromised and human survival made more difficult.

Nevertheless, we cannot change the situation, so at least we need to understand it, in order to be able to do something about it if and when given a chance.

The global coal markets are going through major changes. Let's look at some of these:

China

The Intergovernmental Panel on Climate Change said recently that demand for coal power led to a spike in global carbon dioxide emissions between 2000 to 2010. Much of the new demand for coal seen over the past 10 years has come from China, which is building new coal plants at a rapid pace to keep up with electricity consumption.

At the same time, China has been praised for installing huge quantities of solar installations. As a matter of fact, it is about to become the number one in the world in "green" power installs. Unfortunately solar and wind are just a very small portion of the energy portfolio. As the need for energy in the country grows, coal is the fuel of choice. It is the answer to the energy problems in China—clean environment or no.

It is easy to see from Figure 5-32 that solar in China is almost irrelevant as far as the total of its new power generation is concerned. While wind and nuclear take a more respectful place in the energy mix, it is obvious that fossils play a huge role in the national power generation cycle. Is it possible, then, for solar, wind or any of the renewables to become major power sources in this huge and growing economy? The obvious answer is, "Not now! Maybe later."

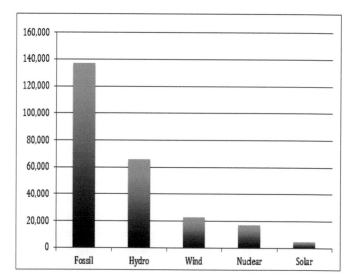

Figure 5-31. China's power generation added in 2013 alone (in TWh)

On the positive side, some of the new fossil power plants replaced older facilities with outdated and much more polluting equipment.

In 2014, China gave green energy advocates some hope when its energy administration announced that 1,725 small-scale coal mines would be closed by the end of the year. The closures are part of China's effort to reduce CO_2 levels in the atmosphere, and cut massive pollution problems in cities like Beijing.

But that was more of a PR move than a step towards a greener future. China hopes to cap its total coal production at 4.1 billion metric tons by 2015, which is actually up 10% from the 3.7 billion metric tons produced in 2013.

The picture is clear enough: While solar and wind are making dramatic strides in China, fossil fuels dominate by making even more dramatic strides. Fossils—coal in particular—will continue to dominate the Chinese energy markets for a long time.

What does this do to China's image as the world's largest emitter of GHG pollutants? While it is a major discussion point in all climate change debates, and especially as far as coal-burning issues are concerned, the issues are wide reaching:

- China is planning to continue burning dirty coal for a long time, since there is no other choice to generate enough electricity to power its growing population.

- Awareness of the fact that air pollution is choking and killing people in the major cities is increasing.

- Discussions on the subject are intensifying, but the actions are few and far between.

- There are some efforts by the government to reduce pollution, and though solar, wind, nuclear and hydroelectric dams are making some impressive gains, all this is small potatoes in the huge borsch pot of Chinese energy.

- At the same time, the country's economy is slowing down and getting more efficient, with a tendency to shift away from heavy, less energy demanding, industries.

- All this considered, experts estimate that China's coal use will peak by 2015, and might stabilize and even decrease after that.

- But there is another possibility: that of China's increasing use of coal to continue indefinitely. In such case, even if coal-plant construction is reduced, the thousands of existing power plants are going to increase emitting their poisons for many decades—at least until the middle of the 21st century.

- Even the most optimistic low-coal power generation forecasts show that coal will provide about 60 percent of China's electric power by 2020. This, in the best of cases, will still be many times more than all clean technologies put together.

So, while India and some other countries are foreseen to hit "peak coal use" in the near future, this is not the case in China. Instead, China is financially capable of, and is surely planning to, ride the coal band wagon for a long time. Because of that, China's overall coal consumption is predicted to increase during the next two-three decades, representing up to half of the world's coal use.

This, in turn, makes tackling climate change impossible, since as a result of China's increased coal use, the global GHG emissions will also continue to increase for the next two-three decades. This comes at a time when emissions need to start declining, if we want to avoid 2°C or more of global warming.

Decades of excess carbon emissions from China's increasing coal fleet will nullify all international efforts. So, short of implementing some ingenious carbon capture and storage technology for coal plants emissions, our environment is doomed and uncontrolled global warming will become a reality in the very near future.

The coal-rush in China is on, and only a miracle can stop it. We see no such miracle on the horizon, now or in the near future. So in China the GHG gasses will continue flowing, negatively affecting the global environment.

How much would the U.S. help China and the local environment by exporting millions of tons of coal there as planned? How much would that help our energy security? Or is it just the old, established way of making money?

India

Large amounts of cheap fossils are available today around the globe, while imported coal, oil, or gas are 60 to 80 percent more expensive than locally produced such. That is Finance 101, and while many countries are getting serious about getting a handle on this problem, India is not of them.

Even though India is home to one of the world's largest coal deposits, the Indian coal sector is in shambles. The country can't mine coal fast enough, and then it lacks the infrastructure to transport it to the points of use. While in the U.S., for example, unit trains (100 cars loaded with 100 tons of coal each) shuttle back and forth between mines and the power plants, in India coal transport is still done by horse carts and trucks.

Coal imports have been blamed as a major part

Figure 5-32. Coal transport in India.

of the energy problems in India. Because of their high cost, the Indian government relies heavily on them to plug the faltering domestic supply. And the discrepancy is growing, with official government estimates placing coal imports around 50-60 percent of total demand by 2030.

Putting these numbers in perspective, China's record-breaking coal imports account for only slightly over 7 percent of the country's total coal consumption. Imagine an increase to 50-60 percent imports, as in India, which has a similarly huge demand for energy. No wonder India's bank is broken.

One reason China doesn't increase imports to such high levels is that power plants and the economy they power are subjected to wild price swings driven by the exporting countries. Since some of these countries are not that friendly, or well-meaning, China's government is very careful with the imports.

India, on the other hand, is past the point of no return, so for now (and the foreseeable future) it must import huge amounts of fossils from abroad. Coal imports in particular pose a real threat to India's energy security and the current deficit levels, which could run up to 15 percent of total GDP by 2030. Fifteen percent is a lot of money for India or any other country.

Even worst-case scenarios estimate the deficit increase as high as nearly 40 percent. The rupee has fallen precipitously already, so such a huge drain on foreign exchange is putting India in a precarious and totally unsustainable position from energy and budget sustainability standpoints.

International advisers have suggested that India start planning for the oncoming "peak coal." Remember "peak oil?" It is when oil extraction around the world reaches a maximum and then starts declining. It is the

same with coal. It will eventually reach a peak point and then dwindle into oblivion…unless something drastic is done soon.

We have not considered "peak coal" as a serious problem until now, because it is still far in the distant future, at least for the U.S. and most other countries. Not so in India. And not because coal is running out *per se* (although in reality it is, but very slowly.)

The "peak coal" for India is coming from the fact that the current path of increasing coal imports is simply unsustainable economically. This also means that—since increasing coal production is not considered feasible—the government should dramatically ramp down the planned coal plant construction.

The current five-year plan proposes 51 GW of new coal-fired power plant construction, which would bring India into the "peak coal" reality. The experts consider that the only sustainable path is for new coal-fired power plant construction to be reduced to about 10 GW by 2030.

Replacing the potentially lost coal capacity would require 350-400 GW of new energy additions. This would require a dramatic ramp up of solar energy power generation: a jump from the estimated 2 GW solar power added annually today to 20-25 GW annually by 2020. This is quite unlikely due to "confusing" policies that change with time and from state to state.

The wind manufacturing also must be at least doubled, during the same time from the current 10 GW production level. But the existing policies are as confusing as in the case of solar power.

A drastic energy transition makes economic sense too, since new coal power generation is estimated at 25-29 cents/kWh, while new wind projects, for example, are estimated 8-10 cents/kWh, and solar is roughly 10-12 cents/kWh.

This transition, however, requires providing both energy sectors with stable political support and enough low-cost finance, which is particularly important. Amidst the political confusion and scarcity of capital for clean energy in India, however, this is not very likely to happen soon.

So, in India the problem is clearly visible and solvable; imported coal is expensive and unavailable, clean energy is available, but equally expensive. And equally importantly, the entire energy situation in the country is in a state of confusion.

If and when India's government and regulators get a good grasp on the problems at hand, they have clear choices to make: continue with unsustainable fossils imports and throw the country down the "peak coal" cliff (the first country to get there), or take the "green" path. Not an easy, quick, or cheap choice to be sure, but one that cannot be postponed...for the sake of the Indian people!

Continuing to count on huge coal imports increases will solve the immediate energy supply problem, but it does not guarantee the energy security of the country. It might even bankrupt the already struggling economy.

Taking the "green" path, on the other hand, would be harder in the short term, but promises to provide long-term energy security, at least as far as India's power generation is concerned.

In the best of cases, there is still the crude oil imports issue, which is as serious, and getting even more serious with time, but which is, unfortunately, not easily solvable. So what will the Indian government do?

Japan

Most surprisingly, a number of developed countries are following China's example by increasing the use of coal for power generation. Japan is the most controversial of them all. After the Fukushima disaster, Japan is headed to a near-term energy future powered mostly by coal.

The Japanese government released a new energy plan in 2014, which intends to increase coal's use in the energy matrix at home. The new energy plan puts coal in the role of an "important long-term electricity source" while the costlier "green energy" is conspicuously absent in the new plan.

The plan also gives nuclear power the same prominence as coal, despite ongoing concerns regarding nuclear fall-out from the 2011 meltdown at the Fukushima plant. Japan's nuclear reactors have been idled since 2011 for safety checks and upgrades. As a result, 10 power companies consumed a record breaking 5.6 million metric tons of coal in January 2014. This is 12% more than the same time a year earlier.

Japan is stuck between a hammer and a hard place now. Its huge nuclear power fleet, which was responsible for providing 30% of the nation's electricity (with plans for growth to over 40%), was completely shut down after the Fukushima accident in 2011.

Overnight, literally, Japan went from a leader in clean environment to a major polluter, along with China and the developing countries. After trying the renewables route for awhile, Japan realized that it will require a lot of time and money to achieve its energy goals with solar and wind, so they're back to using a lot of coal as a long-term (maybe permanent) solution to the lost 30% power generation capacity.

Yet, being the good steward of Mother Earth, Japan is now looking at the possibility of implementing "clean coal" technologies by improving on the latest gasification and carbon storage technologies that make carbon emit less carbon into the atmosphere. One step forward, two steps back...

Solar and wind are still considered, but are far down on the list of priorities per the new 2014 energy plan. This new situation draws a totally new path for Japan, the Kyoto Protocol instigator and host.

After making real progress in GHG emissions reduction until 2011, Japan is again a major polluter for the foreseeable future. The Japanese government seems to have decided to put the Protocol in the closet, where it will stay wait for better days.

Until then, King Coal rules the energy sector in the land of the Rising Sun.

Summary

Global demand for energy is on the rise, with the developing countries leading the race. With that, energy security concerns are becoming ever more critical. Economic development is fueled by energy (no pun intended), so energy sources must be readily available, reliable, and affordable. In that order!

A readily available and reliable source of power is key to keeping up the rapid pace of economic development. Power must be available without vulnerability to long- or short-term disruptions. Interruption of energy supplies and price fluctuations cause major disruptions in the economies and bring financial losses. The abnormalities could even create havoc in economic centers, which could also damage the health and well-being of the population.

Coal meets all these needs and requirements.

Coal is an abundant, reliable, and secure supply of energy. It provides short- and long-term energy security to the countries with sufficient coal reserves.

The problem with coal is its pollution, so the growth in energy demand means that we either use more coal, or find other, less polluting, sources of energy. A diverse mix of energy sources, each with different advantages, is the best way to provide security to an energy system. Energy diversity allows flexibility in meeting the individual needs of the different social and industrial sectors.

Coal is an integral part of that diversity by being part of the mix of different energy sources. Its role might be diminishing in some countries, but—pollution or no pollution—it will remain a major part of the mix for a long time to come.

Coal is Here to Stay

Coal is an important commodity. It is the most important in our energy security arsenal, for it has an important role to play in meeting the demand for a secure energy supply. It is abundant worldwide—present in almost every country—with large commercial mining taking place in over 50 countries on all continents.

Coal is the most abundant and most economical of all fossils, and all other fuels at present.

Another important factor of coal's significance is that at the current production levels it will be available at least for the next 100-150 years. This, compared to 50-60 years availability for oil and gas puts coal in the rank of the fuels we can count on and cannot do without in the long run.

Coal is also readily available in a well-supplied worldwide market. It can be transported to demand points quickly, safely and easily by ship and rail. A large number of suppliers are active in the international coal market, ensuring proper competitive behavior and efficiency.

Unlike gaseous, liquid or intermittent renewable sources, coal can be stockpiled at the power plants and stocks drawn on to meet demand.

Coal has been and still is one of the most affordable source of energy. Coal prices have historically been lower and more stable than oil and gas prices. It is likely to remain the most affordable fuel for power generation in many developing countries for several decades to come.

Coal-fired power plants are the most established, and famous for their high reliability. Over 50% of global electricity is currently based on coal. The electric power generation technologies are well-established and the related technical capacity and human expertise is widespread and readily available.

Ongoing research activities ensure that the coal power generation capacity is continually being improved and expanded, facilitating innovation in energy efficiency and environment performance.

One important fact is that coal could also be used as an alternative to oil in the near future.

The *future* is the key word here, because the development of an efficient coal-to-liquids industry is still in question. If and when developed, however, crude oil from coal could reduce oil imports and serve to hedge

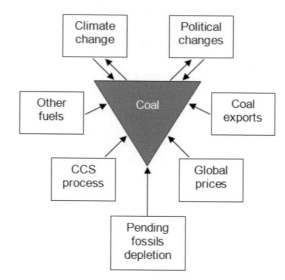

Figure 5-33. Coal's risks

against oil and other fuels energy security risks.

Yet, coal also comes with its own risks. Presently it is the excess GHG emissions from coal-fired power plants. In the long run, it is the pending depletion of the other fossils—crude oil and natural gas—which will force even larger increases in coal production and use.

This, in turn will bring coal closer to extinction, and will contribute to further increase of GHG pollution and the related climate changes.

On the positive side of things, using domestic coal reserves, and/or supplementing them with the relatively stable international coal market, will allow many countries to minimize their exposure to oil price volatility while providing the liquid fuels needed by their economies.

And also quite importantly, domestic coal resources enable economic development, which also can be used and/or transformed to guard against import dependence and future energy price shocks.

So what are we saying here? On one hand, we see that coal is bad because it emits large quantity of GHGs, but on the other hand we praise it for being so efficient, practical, and useful.

While both of these situations are true, we can always remember to weigh the benefits against the damages. One doesn't have to be a genius or great scientist to see the pros and cons in coal use.

Like anything in life, overuse is bad while controlled use is the preferred MO. The problem is that we now have an out-of-control situation where coal's use is increasing well above the limits of its benefits. And yet, there is no solution in sight, so we continue counting the bad effects, while hoping for a miracle.

Update 2014

We talk about "renewables" revolution, and saw predictions that solar will grow exponentially to unprecedented heights and will replace coal. While this is a (future) possibility, it is far from a done deal. Many thousands of acres of land have to be covered by solar and wind farms, and hundreds of billions must be spent before they can make a significant dent in the power generation sector.

So, coal is not going away. On the contrary, as the oldest energy source, it has a lot of experience and many supporters that know how to deal with newcomers.

Coal, nevertheless, is changing the global energy game. As an example of this new reality, we just need to take a close look at China. It is currently the darling of renewable energy, investing heavily in solar and wind power, and yet in 2013 new fossil fuel energy output in China (most of it coal) exceeded the renewables growth by a huge margin.

New coal power generation exceeded new wind energy output by more than six times. Solar was outpaced not by 50%, or even 100%, but by almost 3,000%.

This means that 30 coal-fired power plants were built and put in operation in China for each solar plant on-line in one single year.

And this ratio is increasing in favor of coal, as China is in a hurry to provide more energy to its burgeoning population. Building and running coal-fired power plants is much faster and cheaper than any of the renewables. Coal power plants are also many times more reliable than their renewable counterparts.

While the Chinese government is still supporting the renewable technologies—solar and wind in particular—in reality they are viewed as toys in the large arsenal of energy sources China controls. Coal is one of the main fuels in that arsenal and China intends to use it to the fullest extent. Nothing and nobody can stop China from accomplishing its energy plans, so King Coal is back on its throne, complete with a veil of copious GHG emissions.

One thing to keep in mind here is that Chinese coal plants are fairly new. Most of them were built during the last 20 years, so they have a long life ahead of them. That means also that huge amounts of pollution will be emitted for another 50 years or more.

As confirmation of all we have been saying in this text, China's government recently approved more than 100 million tons of new coal production capacity. This is 6-7 times more than all coal produced in previous years. The increase alone equals about 10% of U.S. total annual coal use.

The increase includes most major mines, and reflects Beijing's aim to put 860 million tons of new coal production capacity into operation by 2015, more than the entire annual output of India.

So no matter what we say or do, and even considering the most aggressive renewable development scenarios, more than 2/3 of China's power generation by 2020-2030 will come mostly from coal.

Short of coming up with efficient and cheap carbon capture and storage technologies sometime soon, the upward coal use trend in China is the largest and unavoidable contributor to unmitigated climate disaster.

According to the Chinese government, this is the best, if not only, way for China to ensure its energy security and provide energy for its growing population and economy. How can we argue that point? Energy security is good, right?

But it is not just China that is on the path of ensuring its energy security by increasing coal use. There are other large developing countries, like India and Indonesia, that are also ensuring their energy security by building coal-fired power plants on a massive scale. Amazingly, some of these plants are even less efficient and more polluting than the old Chinese coal power plants.

But even more surprisingly, Japan, Germany, and other developed countries are also looking into increasing coal use for their energy security. Japan is using a lot of coal while recovering from the Fukushima nuclear accident, which shut down the entire nuclear plant fleet in the country. Germany and other European countries are also increasing their use of coal in response to increased risk to Russian natural gas imports.

The role of the U.S. here cannot be understated. We plan to export a lot of coal in the future, which would help the economic growth of many developing (and even some developed) countries. At the same time, however, the increased GHG pollution would defeat our efforts to reduce it by reducing the domestic coal-fired capacity. Another catch 22...

The increase in alternatives—solar and wind in particular—helps, but it is just not enough to successfully drive the growth of these large economies. So

King Coal rules...still, but the environment will pay the ultimate price.

Energy supplies and temporary energy security is the goal now. What the future brings is anyone's guess...although as things look today, it (the future) is obscured by huge, dark clouds of GHG pollution coming from thousands of coal-fired power plants around the world.

Global energy security is ensured for the next several decades by increased use of coal. Our health and the environment, however, are in great danger. Something has to be done...soon!

Notes and References

1. DOE, Coal. http://energy.gov/coal
2. Coal Energy http://www.ifpaenergyconference.com/Coal-Energy.html
3. USGS, Coal Resource Classification System, http://pubs.usgs.gov/circ/c891/guidelines.htm
4. IEA, Coal. www.iea.org/topics/coal/
5. EPA. Coal. http://www.epa.gov/cleanenergy/energy-and-you/affect/coal.html
6. *Photovoltaics for Commercial and Utilities Power Generation*, Anco S, Blazev. The Fairmont Press, 2011
7. *Solar Technologies for the 21st Century*, Anco S Blazev. The Fairmont Press, 2013
8. *Power Generation and the Environment*, Anco S. Blazev. The Fairmont Press, 2014

Chapter 6

Nuclear and Hydro Power

Nuclear and hydropower fit in the very special category of energy sources in peril and on trial. As major power generators for decades, these respected power generators are well established, and most countries depend on their reliable, clean, efficient, and cheap power.

We cannot even imagine what this world would be without these major players. Yet, they have major problems, and their future depends on a number of factors mostly out of our control:

- Nuclear power has a poor safety record due to several major nuclear disasters. The circumstances around these incidents and their consequences have added a high level of uncertainty and fear, which threatens to curtail global nuclear power development. And things could get much worse, if another major nuclear accident happens anytime soon. Such an event could mark the end of nuclear power as we know it.

- Hydropower, on the other hand, is very safe, but recent extended and severe draughts are putting a big question mark on hydropower's reliability and its expansion plans across the globe. As a matter of fact, global warming is changing the world climate drastically to where a greater number and even more severe draughts are expected in many areas in the future. This could put hydropower in a very bad place. No water from the sky, no hydropower…it is that simple!

In the case of both nuclear and hydropower, there are a number of complex, serious, and out-of-our control forces that drive, and which will ultimately determine, their future development in the U.S. and the world in general.

Let's take a closer look at our old friends in peril.

NUCLEAR POWER (FRIEND OR FOE)

Like the genie in the bottle, nuclear power brings miraculously huge quantities of useful energy to its masters. When released by mistake or malicious intent, however, it becomes a monster bringing instant and merciless misery, *pain, and devastation to thousands of innocent and unsuspecting victims.*

Anco Blazev

BACKGROUND

Since the tragic Chernobyl nuclear disaster, nuclear power has been the Cinderella of the energy industry. Like Cinderella, it does good most of the time, but from time to time, things turn out badly. Usually it is the fault of the masters—their negligence or ignorance, that create the disasters. Nuclear power, however, is always blamed for the masters' mistakes.

Like Cinderella, it is never thanked for the good things it does, but is always blamed for anything bad that happens around it. As a result, some love, some envy, and some even hate her. In the end, most of us don't know for sure what to think and or do about our nuclear Cinderella.

The recent Fukushima nuclear accident elevated suspicion and fear surrounding nuclear power. Its veil of mystery, complexity, and danger became even more intimidating for the average person. Because of that, most of us don't know what to say about it for fear of being technically or politically incorrect. So, the controversy on high levels continues.

Nuclear power is a truly mysterious and misunderstood force. One frightening thing we have learned the hard way is that its ominous power is hard to control even with our advanced technologies and methods. This is a big problem, which, regardless of the assurances of energy companies and governments, puts nuclear power on the guilty list, where it will remain until proven innocent.

Nuclear power is unequalled in its fierce responses. No other disaster—natural or manmade—can get even close in magnitude and cruelty. Just whispering the word nuclear sends shivers down our spines and conjures images of explosions, fires, smoke, twisted metal, buildings with blown off roofs, and death.

This evokes respect and terror, inevitably reminding us of Hiroshima, Nagasaki, Three Mile Island, Chernobyl, and Fukushima. These are symbols of the fierce power, frightful widespread devastation, and merciless revenge of a wicked, out-of-this-world force. It is much more than we can even imagine, until we experience the effects ourselves.

There are awful signs of nuclear power damage around the world, and there will be more before we find similarly useful, but safer, form of large-quantity electric power generation. For now we have no choice, if we want to continue living in comfort provided by unlimited quantities of electric power...compliments of nuclear generation.

Many people see nuclear power as a serial killer, a mass murderer that can strike any moment and kill thousands in seconds. Yet, most of us cannot live without it. We are, therefore, willing to gamble.

Nuclear power generates a significant portion of our electricity, so we need the beast, which provides us with lots of reliable and cheap electric power on demand. We just have to tame and befriend this necessary evil to the best of our abilities, to control and use it efficiently and safely.

This is not an easy chore, and we are not there yet. So, from time to time, we have problems handling the overwhelming and finicky power contained in nuclear reactors...and sometimes we pay for it with our health and lives.

Nuclear power pushes our technical abilities to the limits, keeping us on our toes—afraid that we might make a mistake to which the beast would respond with unspeakable fury, widespread destruction, and deadly devastation.

Yet, we are fully engaged in the nuclear power generation game, and although many countries are making serious efforts to reduce their dependence on nuclear power, it will not happen easily or anytime soon.

As a matter of fact, we need nuclear power now more than ever before, because the world population is exploding and the fossil fuels are becoming too expensive. We are also acutely aware that the fossils are coming close to their end, and that nuclear is the perfect fuel to help us transition to a post-fossils future. A real catch 22, no?

Giving up on nuclear power now, and before we have lined up alternative energy sources, will cause unimaginable hardship during the transition to fossil-less energy reality in the future.

Nuclear Nightmare...

Before we get far into the future, a quick glimpse at the present energy situation in Japan provides a good picture of what nuclear power means to us, and what can be expected with and without it. It is the latest proof that we cannot live with it or without it.

During several tragic days in March 2011, we witnessed the dangers of nuclear power after the Fukushima accident, and now we are witnessing what it is to live without it. The entire Japanese nuclear power fleet was shut down after the 2011 accident, and the country entered, and is still meandering in, a period of energy and environmental crisis.

To replace the lost power, Japan started importing and burning more coal, natural gas, and crude oil than ever before. This makes it vulnerable to the risks and fluctuations of the global energy markets, so Japan's energy security is on a limb for the duration.

As a result of the increased fossil consumption, Japan now also emits more GHGs than ever before. The Kyoto Protocol agreements can wait...now is more important than tomorrow.

Now, imagine all global nuclear power generation shut down for a long while. Not possible? Yes, possible!

One more Fukushima-like accident in the near future anywhere in the world would seriously cripple the nuclear industry for a long time. Perhaps forever...

The global nuclear industry would not be able to survive the resulting panic and massive nuclear plants shutdowns around the globe that would most likely follow another major nuclear disaster. This is because people have not forgotten the Chernobyl and Fukushima tragedies, and another one on top of those would be too much to bear. Unprecedented chaotic responses and decisions might just make nuclear power unfeasible for a long time.

There are countries, like France and others, however, where nuclear power provides more than half of the daily power generation. What would France and the other nuclear power-dependant countries do if their power fleets were shut down? Hard to tell, but their energy situation would be even worse than Japan's is now. That, in turn, would increase the risks and compromise their energy security, in addition to increasing significantly the global GHG emissions.

Nevertheless, at least for the foreseeable future, we have to live with nuclear power, keeping our fingers crossed and dealing with its problems, trying to avoid another big surprises at all cost.

Germany and Japan tried to reduce and even eliminate nuclear power for awhile, but are now reconsidering their previous decisions. Germany's nuclear power is back up 100%, while Japan is considering doing the same soon, but gradually.

Although we all are afraid of nuclear power, we need it, since we simply cannot replace it easily with the other sources.

U.S. politicians and regulators were not impressed, let alone scared, by the Fukushima events, most likely thinking that such a thing cannot happen here. This *is* the U.S., after all, with the best nuclear technology and nuclear safety in place. So what can go wrong... go wrong... go wrong...?

We hope they're right, but we must remember that the Fukushima power plant was equipped with U.S. hardware and expertise, and look what happened to it. This cannot happen again, they say. But this is what they said before Fukushima too, so whom are we to believe?

Several months after the Fukushima disaster, when the damages and the victims were being still counted, the nuclear industry watchdog, the U.S. Nuclear Regulatory Commission, approved the applications for several new nuclear power plants for the first time since 1978. In defiance of the Fukushima disaster and its aftermath, the U.S. will add to its fleet of nuclear plants.

We will not let one accident stop us in our drive for more and cheaper energy. No sir! But it wouldn't hurt to keep our fingers crossed, because one Fukushima-like disaster in the U.S. would certainly shove a big rod in the nuclear power industry's spokes. The final outcome would be anyone's guess.

Now at least two new nuclear reactors are back on the schedule to be built in Georgia. Other states may follow soon.

Is this the first of the last batch of nuclear power plants to be built in the U.S., or is this a new wave of nuclear power increase? Is this the beginning of the nuclear renaissance, we have been talking about for decades? We don't know what will happen, because it could go either way fast.

Nuclear Facts

Obviously, nuclear power is very important, so let's see what it is and what it offers in terms of convenient, reliable, and plentiful energy generation, as compared to its cost, safety, and environmental effects.

Amazing fact: the specific energy of one kilogram coal (2.2 lbs.) is 24 MJ, while that of uranium-235 is 80,620,000 MJ...or 3.4 million times greater.

While one kilogram of coal can theoretically produce 8 kWh of electric power, 1 kg. of uranium-235 produces over 25 million kWh of electric energy. This is about 3,400 tons of coal, or 34 train cars full of coal, 100 tons each, generating as much power as one single lump of 1 kg. uranium.

It's hard to imagine, yet this is only the theoretical side of the equation. When all other factors and variables—such as labor, transport, power generating efficiency, and all other losses—are included, nuclear looks even better. Nuclear uses less labor, and less transport, it is more efficient in operation, cheaper, and much less polluting. These are huge advantages that should not be overlooked when comparing the different power sources.

No matter how one looks at it, however, nuclear contains a tremendous amount of power, which if properly used and controlled, can generate a lot of electricity—much more than any other sources. If not properly controlled, however, it could do a lot of damage...much more than any other energy source! One single picture from Chernobyl or Fukushima tells the whole story.

Most of us feel lucky that we were not nearby when those nuclear accidents happened, and hope that we will never witness such monstrous events in our lifetime.

Table 6-1 shows the energy content of different materials (in mega joules per kilogram) that can be used as fuel for different applications, including heat and electric power generation. The incredible energy content of uranium, results in a huge difference, as compared to the energy content of the fossils. This is an enormous advantage for the nuclear power generation, in addition to other benefits and conveniences.

- Nuclear power offers the convenience of transporting and handling much smaller amounts of raw

Table 6-1. Specific energy of fuels (MJ per kilogram)

Energy source	MJ / kg
Brown coal (lignite)	24
Firewood (dry)	16-18
TNT	4-5
Compressed hydrogen	142
Natural Gas	40-45
Crude Oil	45-50
Uranium (nuclear grade)	80,620,000

material; a truck load of uranium, for example, would keep producing power for many months and even years. This, compared to endless lines of coal cars loaded and unloaded 24/7 at the mines and coal-fired power plants respectively for the duration.

- And then, once the nuclear fuel is spent, only one truck load of nuclear waste is driven away from the nuclear plant. This, vs. several hundred train cars loaded with ash, soot, and slurry at the coal-fired power plant dumped periodically nearby, or transported to some distant, hazardous dump site.

- The truck full of nuclear waste, however, presents a number of specific and significant dangers too, but in a smaller package. This smaller package is more difficult to handle, because while coal byproducts can be just dumped in specially designated areas and left there unattended, nuclear wastes have to be packaged in special containers, deposited in special storage sites, maintained, and guarded ad infinitum.

This is one of the big problems the nuclear power industry has not been able to resolve thus far.

In addition to the safety of nuclear operations, the disposal of nuclear waste remains unresolved and very controversial issue.

One very important advantage of nuclear power from environmental point of view is the fact that nuclear plants produce less harmful gasses or liquids. At the same time, thousands of tons of toxic smoke, ash, and soot are produced at the coal-fired plants. Not to mention the acres of liquid and solid waste contamination around coal-fired plants and at remote dump sites.

Yet, nuclear is not 100% clean as we will see below. Since there is no perfect solution, we need to pick and chose carefully how we generate and use energy, keeping in mind that in all cases there would be some problems. Because of that, we must be well informed and educated on the issues, in order to make the right conclusions and take appropriate decisions.

Let's dive deeper into the world of nuclear power to see what it is, its role in our lives and its effects on our energy security.

In the beginning...

There are a number of theories about the creation and availability of uranium and other nuclear (radioactive) materials on Earth. Some of these theories claim that the radioactive materials were produced in one or more supernovae; exploding stars in which the radiated energy increases several billion times within, as a result of the catastrophic collapse of the star's core. Such events are hard to describe exactly, because of the enormity and complexity of the different energies generated and released for the duration.

Supernovae events are extremely luminous and noisy, and cause a burst of radiation that often briefly outshines an entire galaxy, before fading from view over several weeks or months. During this short interval a supernova can radiate as much energy as the Sun is expected to emit over its entire life span.

The initial explosion expels much or all of a star's material at high velocity of up to 20,000 miles per second, thus driving a shock wave into the surrounding interstellar medium. This shock wave sweeps up an expanding shell of gas and dust called a supernova remnant, accompanied by a flood of free neutrons. The main process that leads to the creation of uranium and other radioactive isotopes can be explained by the rapid capture of these free neutrons on seed nuclei at rates greater than disintegration through radioactivity during the explosive supernovae event.

For the sake of simplification, we will assume that the Earth's uranium and some of the other radioactive materials we find here were produced through similar processes in one or more supernovae, which must've occurred billions of years ago.

This is actually a crude oversimplification, since there were many other extraordinary events with even more spectacular effects around the time of the Big Bang and shortly thereafter. I.e., there is evidence that more than ten separate and distinctly different stellar sources were involved in the genesis of the solar system material. Thus the relative abundance of U-235 and U-238 at the time of formation of the solar system is most likely as a result of the explosive debris of many supernovae and the related interactions.

In any and all cases, the final result of all this is that we now have large deposits of uranium, thorium, and other radioactive materials here on Earth that allow us to extract and use them for generating electric power via nuclear power reactors.

The average abundance of uranium in meteorites is about 0.008 parts per million (ppm, or gram/ton), while the abundance of uranium in the Earth's "primitive mantle" (prior to the extraction of the continental crust) was 0.021 ppm. Allowing for the extraction of a core-forming iron-nickel alloy with no uranium (because of the characteristic of uranium which makes it combine more readily with minerals in crustal rocks rather than iron-rich ones), this still represents a roughly two-fold enrichment in the materials forming the proto-Earth compared with average meteoritic materials.

The present-day abundance of uranium in the "depleted" mantle exposed on the ocean floor is about 0.004 ppm. The continental crust, on the other hand, is relatively enriched in uranium at some 1.4 ppm. This represents a 70-fold enrichment compared with the primitive mantle. In fact, the uranium lost from the "depleted" oceanic mantle is mostly sequestered in the continental crust.

The processes which transferred uranium from the mantle to the continental crust are complex and consist of many consequent steps over long time period.

It took over 2 billion years to go through the:

- Formation of oceanic crust and lithosphere through melting of the mantle at mid-ocean ridges,

- Migration of the oceanic lithosphere laterally to a site of plate consumption (this is marked at the surface by a deep-sea trench),

- Production of fluids and magmas from the down-going (subducted) lithospheric plate and overriding mantle "wedge" in these subduction zones,

- Transfer of these fluids/melts to the surface in zones of "island arcs" (such as the Pacific's Ring of Fire),

- Production of continental crust from the island arc protoliths, through re-melting, granite formation and intra-crustal recycling.

- In nature, uranium ore is found as uranium-238 (99.27%), uranium-235 (0.72%), and a very small amount of uranium-234 (0.006%).

All through the crust-forming cycle, the lithophile character of uranium is manifested in the constancy of the potassium to uranium ratio in the rock range from peridotite to granite.

Keeping track of how uranium is distributed in the Earth, we see the abundance and isotopic characteristics of lead (a relative of U-235 and U-238) as a useful parameter. There is relatively low abundance of lead in the Earth's mantle and high uranium to lead ratio, compared with meteorites, which can be explained by lead's volatile nature and its tendency to combine with iron. Thus lead is being lost during terrestrial accretion and core separation.

One of the consequences of these high ratios is the comparatively high radiogenic/non-radiogenic content of Pb-207/Pb-204, and conversely Pb-206/Pb-204 in the Earth's crust and mantle compared with meteorites or the Earth's core.

Note: Pb-207 is the final stable decay product of U-235, and Pb-206 is that of U-238, while Pb-204 is non-radiogenic. Uranium decays slowly by emitting alpha particles. The half-life of uranium-238 is about 4.47 billion years and that of uranium-235 is 704 million years.

Therefore, the decay from strongly radioactive U-235 to its final state of inert and harmless Pb-207 takes several hundred million years. This is fairly uniform process, which makes it useful in dating the age of the Earth and the materials contained in it.

Nuclear Materials and Fuels

Materials that can sustain a nuclear chain reaction, or produce a nuclear explosion, are known as *fissile* materials, such as uranium (U). Some examples of the nuclear fuels are; ^{233}U, ^{235}U, ^{239}Pu, ^{237}Np, ^{243}Am, etc.

Energy is released when heavy elements like thorium, uranium, or plutonium undergo fission, during which process the heavy nuclide with atomic mass of 235 for ^{235}U, is split into two lighter nuclides like ^{90}Sr, or ^{137}Cs, including several neutrons, and occasionally a hydrogen atom.

Note: ^{233}U can also be presented as U-233, or uranium-233. This nomenclature is also valid for the other nuclear elements

Uranium is the preferred fuels for most nuclear reactors, and presently generates over 16% of all electricity worldwide. This makes it an important fuel, and a key chemical element, which also played an important role in the evolution of the Earth.

Uranium

Uranium is found in a number of places around the world, since it is as common as some more common metals, such as tin and germanium. As a matter of fact, uranium is a constituent of most rocks, dirt, and of the oceans—albeit in very small quantities. And this is a problem, because mining is economically feasible only in deposits of somewhat larger concentration of uranium.

The estimates show that, at the present rate of use there are enough uranium deposits around the globe to last about 100 years.

Today, uranium is economically recovered at a price of $100 to 150/kg. This higher level of assured resources is normal for most minerals, and a significant price increase (as expected), as well as improved methodologies, could create a correspondingly high increase in extractable resources over time as well.

Uranium's price constitute only 3-5% of the final cost of nuclear power, so we can count on cheap fuel for at least a century to come, or more.

Uranium supplies in the Earth depths are nevertheless limited, so nuclear power is not an eternal power source, nor is it a renewable. It is mined quite extensively today, so someday soon, probably some time in the 22nd century, it would become scarce and harder to mine. At that time, prices would increase to the point where nuclear power would be no longer competitive with the other—renewables most likely—energy sources.

Sporadic uranium shortages are foreseen in the near future, because over 95% of the global uranium ore deposits are located in a hand full of countries, most of which don't have our best interests in mind.

Because of that, and many other factors and variables, the uranium market, like all commodity markets, is volatile. It depends not only on the standard international forces of supply and demand, but also on regional geopolitical, socio-economic, and other global conditions and trends.

U.S. Uranium Reserves

At the end of 2008, U.S. uranium reserves totaled 1,227 million pounds of U_3O_8 at a maximum forward cost (MFC) of up to $100 per pound U_3O_8. At up to $50 per pound U_3O_8, estimated reserves were 539 million pounds of U_3O_8.

Based on average 1999-2008 consumption levels (uranium in fuel assemblies loaded into nuclear reactors), uranium reserves available at up to $100 per pound of U_3O_8 represented approximately 23 years worth of demand, while uranium reserves at up to $50 per pound of U_3O_8 represented about 10 years worth of demand.

Domestic U.S. uranium production, however, supplies only about 10 percent, on average, of U.S. requirements for nuclear fuel, so the effective years' supply of domestic uranium reserves is actually much higher, under current market conditions.

In 2008, Wyoming led the Nation in total uranium reserves, in both the $50 and $100 per pound U_3O_8 categories, with New Mexico second. Taken together, these two States constituted about two-thirds of the estimated reserves in the country available at up to $100 per pound U_3O_8, and three-quarters of the reserves available at less than $50 per pound U_3O_8.

By mining method, uranium reserves in underground mines constituted just under half of the available product at up to $100 per pound U_3O_8. At up to $50 per pound U_3O_8, however, uranium available through in-situ leaching (ISL) was about 40 percent of total reserves, somewhat higher than uranium in underground mines in that cost category. ISL is the dominant mining method for U.S. production today.

In summary, we import 90% of our uranium ore supplies, which poses a great threat to our energy security—even greater than crude oil, of which we import only 50% of our total needs. Depending on foreign countries and negotiating traitorous transportation routes daily cannot be considered a safe way of doing business. Especially not when it concerns critical energy supplies.

And the worst is yet to come—a day in the near future when the global fossil reserves get more and more scarce and prices rise astronomically. And then finally, one beautiful day we will wake up with the realization that even the friendliest countries have no more crude oil nor uranium ore to send us…

Uranium Peak

Just like the other fossils, uranium ore has beginning and end. It is a finite commodity that has been explored at a fairly large pace for over half a century. Because of these two factors—finite reserves and fast exploration— we must be well aware of the uranium, and other radioactive ores, status now and in the future.

At the present, there are 440 nuclear reactors in operation worldwide, while 50-60 new reactors are under construction, and over 150 nuclear power reactors around the world are in planning or design stages. Another 350 nuclear reactors have been proposed for construction in the next 2-3 decades. All new reactors, plus the existing ones, depend on uranium ore supplies.

It takes about 8 kg of uranium ore to produce 1 kg of enriched uranium for use in nuclear reactors.

Large nuclear reactors are loaded with 50-100 tons of enriched uranium, which requires 400-800 tons of uranium ore every 12-18 months. This number multiplied by the number of reactors operating worldwide produces a rather large amount of uranium ore needed every 12-18 months.

A rough estimate shows that approximately 500 reactors in the near future, using an average of 600 tons of uranium ore would require about 300,000 tons of uranium ore every 12 to 18 months. Since there are about 7 million tons of total proven uranium resources in the world, our rough (unproven) estimate shows that we have only 20-30 years of uranium ore supplies at the future rate of use.

Even if our estimates are way off, at the accelerated future rate of use, uranium ore deposits will not last long. Uranium peak—the point of maximum possible production—has been estimated by reliable sources to occur around 2035. A total depletion of uranium, therefore could be expected some by the end the 21st century at the latest.

In worst-case scenarios, rapid expansion of the world nuclear industry might increase the demand and the prices of uranium, which might make it cost prohibitive. Worse, increased use of uranium might lead to a more rapid depletion of this valuable resource than presently estimated.

On the bright side of things:

New technologies could add to the proven resources and facilitate their exploitation, thus reducing the price and increasing the amount of available uranium ore. Another great hope is closing the existing nuclear fission cycle—by recycling and reusing uranium—which might extend the uranium supplies by centuries.

Note: A new nuclear fusion technology promises to make the use of the present-day fission technology obsolete... sometime in the distant future, hopefully before we run out of uranium ore, and space to store the nuclear wastes. All efforts in this area have failed so far, which only shows that our fusion technological capability is in its infancy. Yet, it remains a distinct possibility for future generations. More on this technology and its issues can be found later on in this text.

Uranium Properties

Uranium (U) is a very heavy and hard, silvery-white metallic chemical element, of the rare earth/actinides series in the periodic table. Its chemical symbol is U, and corresponds to atomic number 92. Each uranium atom has 146 neutrons, 92 protons and 92 electrons in its atom. Six of the electrons are in the outer shell and are its valence electrons. It has a high melting point, of 2070°F.

Uranium metal has three allotropic forms:

α (orthorhombic) stable to up to 660°C

β (tetragonal) stable from 660°C to 760°C
γ (body-centered cubic) from 760°C to melting point. This is the most malleable and ductile state.

U is malleable, ductile, slightly paramagnetic, strongly electropositive and is a poor electrical conductor. It has very high density, being approximately 70% denser than lead, and only slightly less dense than gold. Unlike gold, it reacts readily with almost all nonmetallic elements and their compounds, with reactivity increasing with temperature.

Hydrochloric and nitric acids dissolve uranium, but non-oxidizing acids (other than hydrochloric acid) attack the element very slowly. When finely ground, it can react with cold water, and in air it gets coated with a dark layer of uranium oxide. Because of that, uranium in ores is extracted chemically and converted into uranium dioxide or other chemical forms usable in industry.

The Uranium Isotopes

Naturally occurring uranium ores contain one or more of the three U isotopes: uranium 234, 235 and 238. All three isotopes are radioactive, but only one of them, uranium 235 with 143 neutrons, is fissionable and can be used in the generation of nuclear power.

Pure uranium is slightly radioactive, and has the second highest atomic weight of the primordially occurring elements, lighter only than plutonium. Its density is about 70% higher than that of lead, but not as dense as gold or tungsten. It occurs naturally in low concentra-

Table 6-2. Uranium isotopes

Isotope	Half Life
U-230	20.8 days
U-231	4.2 days
U-232	70.0 years
U-233	159000.0 years
U-234	247000.0 years
U-235	7.0004xE8 years
U-236	2.34xE7 years
U-237	6.75 days
U-238	4.47xE9 years
U-239	23.5 minutes
U-240	14.1 hours

tions of a few parts per million in soil, rock and water, and is commercially extracted from uranium-bearing minerals such as *uraninite*.

The uranium atom decays slowly by emitting an alpha particle. The half-life of uranium-238 is about 4.47 billion years and that of uranium-235 is 704 million years, which makes them useful in dating the age of the Earth.

The uranium nuclear fuel is usually based on the metal oxides of uranium metal. The oxides are used rather than the pure metal itself simply because the oxide melting point is much higher than that of the metal, thus they are safer in case of reactor overheating and meltdown. Another benefit of the oxides is that they cannot burn, being already in the final oxidized state of matter of the uranium metal. This is another important operational and safety consideration.

Plutonium

The second-most used fissile isotope is *plutonium-239*. It can also fission on absorbing a thermal neutron, with the end product being *plutonium-240* (Pu-240). Pu-240 makes up a large proportion of *reactor-grade plutonium* used today, which is plutonium recycled from spent fuel that was originally made with enriched natural uranium and then used once in a light water reactor (*LWR*).

Note: Current light water reactors (LWR) make relatively inefficient use of nuclear fuel by fissioning only the very rare and expensive uranium-235 isotope. Nuclear reprocessing can make this waste reusable, and more efficient reactor designs allow better use of the available resources.

Pu-240 decays with a half-life of 6,561 years into U-236. In a closed *nuclear fuel cycle*, most Pu-240 will be fissioned (after more than one neutron capture) before it decays. However, Pu-240 discarded as *nuclear waste* will decay over thousands of years.

Recently, several countries have experimented with using thorium as a substitute nuclear fuel in nuclear reactors. The growing interest in a thorium fuel cycle is due to its abundance in some areas (3-4 times more abundant than uranium), its safety benefits, and absence of non-fertile isotopes.

Thorium

Thorium is a naturally occurring radioactive chemical element with the symbol Th and atomic number 90. In nature, virtually all thorium is found as Th-232, which has a half-life of about 14.05 billion years. Other isotopes of thorium are short-lived intermediates in the decay chains of higher elements, and only found in trace amounts. Thorium is mostly refined from monazite sands as a by-product of extracting rare earth metals.

Thorium undergoes a complete combustion in specialized nuclear reactors, vs. only 1% for standard uranium reactors using natural uranium. Thorium reactors are popular in India and will become more popular around the world when global supplies of uranium ore are near depletion.

Thorium reactors generate 3.6 billion kWh of heat per ton of thorium at 40% efficiency, which means that a 1 GW reactor uses about 6 tons of thorium per year. Worldwide thorium resources are estimated at 2 million tons, so the thorium supply (theoretically) could power the world for several centuries

India's three-stage nuclear power program is possibly the most famous, well funded, and most advanced thorium nuclear process development effort.

Metal Nuclear Materials

Metal nuclear materials have much higher heat conductivity than oxide types, but cannot withstand high temperatures. Metal fuels have the highest fissile atom density, and are normally alloyed, made with pure uranium metal.

Uranium alloys include uranium aluminum, uranium zirconium, uranium silicon, uranium molybdenum, and uranium zirconium hydride. Any of these can be made with plutonium and other actinides as part of a closed nuclear fuel cycle. Metal fuels have been used in water reactors and liquid metal fast breeder reactors, such as EBR-II.

Some of the metal nuclear materials are:

- *Uranium dioxide* (UO_2) *nuclear fuel* is a black solid material, which is prepared by reacting uranyl nitrate with ammonia to form a solid (ammonium uranate). It is then heated (calcined) to form U_3O_8 that can then be converted by heating in an argon/hydrogen mixture at 700°C to form UO_2. Thus obtained UO_2 is mixed with an organic binder and pressed into pellets, which are then fired at a high temperature again in argon/hydrogen gas mixture to sinter the pellets into a solid material with few pores.

- *Mixed oxide, or MOX*, is a blend of plutonium and natural or depleted uranium which behaves similarly to the enriched uranium feed, for which most nuclear reactors were designed. MOX fuel is an alternative to low enriched uranium (LEU) fuel used in the light water reactors (LWR) which are used in the global nuclear power generation.

- *TRIGA nuclear fuel* is used in TRIGA (training, research, isotopes, general atomics) reactors, which use uranium-zirconium-hydride (UZrH) fuel, which has a built-in safety, where as the temperature of the core increases, the fuel reactivity decreases. This pretty much eliminates the possibility of a meltdown. Most cores that use this fuel are "high leakage" cores where the excess leaked neutrons can be utilized for research.

  TRIGA fuel was originally designed to use highly enriched uranium, however in 1978 the U.S. Department of Energy launched its Reduced Enrichment for Research Test Reactors program, which promoted reactor conversion to low-enriched uranium fuel. A total of 35 TRIGA reactors have been installed at locations across the USA. A further 35 reactors have been installed in other countries.

- *Actinide nuclear fuel* is a by-product of fast neutron reactors, where minor actinides produced by neutron capture of uranium and plutonium can be used as fuel. Metal actinide fuel is typically an alloy of zirconium, uranium, plutonium and the minor actinides. It can be made inherently safe as thermal expansion of the metal alloy will increase neutron leakage.

Note: The minor actinides include neptunium, americium, curium, berkelium, californium, einsteinium, and fermium. The most important isotopes in spent nuclear fuel are neptunium-237, americium-241, americium-243, curium-242 through -248, and californium-249—252.

Ceramic Nuclear Materials

Ceramic nuclear materials, in addition to the oxides, also have high heat conductivities and melting points, but they are more prone to swelling than oxide fuels and are not understood as well.

- *Uranium nitride (UN)* is used in NASA reactor designs, because it has a better thermal conductivity than UO_2, since it has a very high melting point. UN fuel has the disadvantage that a large amount of 14C would be generated from the nitrogen by the (n,p) reaction. As the nitrogen required for such a fuel would be so expensive it is likely that the fuel would have to be reprocessed by a pyro method to enable the 15N to be recovered. It is likely that if the fuel was processed and dissolved in nitric acid that the nitrogen enriched with 15N would be diluted with the common 14N.

- *Uranium carbide* was used in the form of pin-type fuel elements for liquid-metal fast breeder reactors during their intense study during the 60s and 70s. Recently there has been a revived interest in uranium carbide in the form of plate fuel and most notably, micro fuel particles (such as TRISO particles).

The high thermal conductivity and high melting point makes uranium carbide an attractive fuel. In addition, because of the absence of oxygen in this fuel, as well as the ability to complement a ceramic coating, uranium carbide could be the ideal fuel candidate for certain Generation IV reactors such as the gas-cooled fast reactor.

Liquid Nuclear Materials

Liquid nuclear fuels are basically liquids that contain some percentage of dissolved nuclear fuel. Liquid-fueled reactors generally have large negative feedback mechanisms and therefore are particularly stable designs. Liquid fuels have the disadvantage of being easily dispersible in the event of an accident, such as a leak in the primary system.

Molten salts nuclear fuels have nuclear fuel dissolved directly in the molten salt coolant. Molten salt-fueled reactors, such as the liquid fluoride thorium reactor (LFTR), are different than molten salt-cooled reactors that do not dissolve nuclear fuel in the coolant.

Molten salt fuels were used in the LFTR known as the Molten Salt Reactor Experiment, as well as other liquid core reactor experiments. The liquid fuel for the molten salt reactor was a mixture of lithium, beryllium, thorium and uranium fluorides: $LiF-BeF_2-ThF_4-UF_4$ (72-16-12-0.4 mol%). It had a peak operating temperature of 705°C in the experiment, but could have operated at much higher temperatures, since the boiling point of the molten salt was in excess of 1400°C.

Aqueous solutions of uranyl salts are used in the aqueous homogeneous reactors (AHRs) in a solution of uranyl sulfate, or other uranium salt, in water. Historically, AHRs have all been small research reactors, not large power reactors.

An AHR, known as the Medical Isotope Production System is being considered for production of medical isotopes.

Recovery from Seawater

Uranium ore reserves are limited and estimated to last about 100 years, thus nuclear power is not a renewable energy source. Because of that, there are efforts to extract uranium, and other materials suitable for nuclear fuel, from other sources. Rocks like granite, contain minute amounts of uranium have been considered, but the process is cumbersome and very expensive.

Extracting uranium from sea water, as unlikely as it might seem, is actually more doable, so there are a number of efforts in that area. The uranium concentration in sea water varies but averages about 3.3 mg uranium per cubic meter of seawater. This is a very small amount by all means, but the quantity of this resource is huge, so even a small portion of the uranium in seawater could provide fuel for nuclear power generation for a long time.

A number of methods have been tried to extract uranium from seawater, including using a uranium-specific nonwoven fabric as an absorbent. In one case, the total amount of uranium recovered from three collection boxes containing 350 kg of fabric was >1 kg of yellowcake after 240 days of submersion in the ocean, which brings the cost of uranium to over $150/lb. Not bad for an infant process with a promising future.

A new absorbent material called HiCap was developed in 2012 and outperformed previous best adsorbents five to seven times. HiCap also effectively removes toxic metals from water, so it might have a dual use. It is possible that this and other efforts will bring the cost of uranium from sea water down to acceptable levels, thus significantly extending the time we can use nuclear power for generating electricity.

The major cost of nuclear power is in the construction and operation of the power station, so the fuel's contribution to the overall cost of the electricity produced is relatively small. Because of that, even a large fuel price escalation will have relatively small effect on generated electric power price.

Doubling the uranium price would increase the fuel cost for an LWR reactor by 26% and the electricity cost about 7%. In contrast, doubling the price of natural gas would add 70% to the price of electricity from that source.

If and when the price of uranium and other fuels rises high enough, extraction from sources such as seawater and other materials containing traces of uranium, might become economically feasible.

In 100 years or so, uranium as we know it will be gone, so we need to explore other avenues for producing radioactive raw materials.

All power plants require the use of large quantities of different fuels. Nuclear plants are no exception, and

need a constant large supply of high quality fuel. The most common fissile nuclear fuels, used in today's fission nuclear power plants, are uranium-235 (^{235}U) and plutonium-239 (^{239}Pu). These materials have the highest energy density (contain most Btus per given mass) of all practical fuel sources in use today.

Nuclear Fuel Life Cycle

All of the actions of mining, transporting, refining, purifying, using, recycling, reprocessing, reusing, and ultimately disposing of the nuclear fuel materials put together make up the nuclear fuel cycle.

It starts deep underground, where uranium and other radioactive ores are dug out for transport to a processing facility. Here the ore is converted into nuclear fuel, loaded in special canisters and transported to the nuclear plant to fuel the reactors to generate steam and electricity for several months.

When the fuel is exhausted, the reactor is shut down, the fuel bundles are removed, packed in special containers and taken to temporary storage.

This process sequence forms the life-cycle of the nuclear fuel. New bundles are loaded in the reactor and the power generating cycles starts anew.

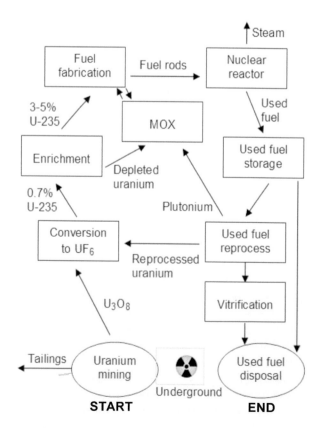

Figure 6-1. Nuclear fuel cycle

Uranium Ore

Uranium is found in seams under the Earth's surface, just like coal, and is dug out of these in the same way as coal. Uranium mining, however, is much more dangerous. It is actually one of the dirtiest and most dangerous jobs in the mining industry...if not in the world.

Here, in addition to the other dangers related to surface and/or underground mining operations, the miners digging the radioactive ores are exposed to high levels of radiation. As a matter of fact in the olden days, prisoners were forced to work in the mines while serving life or death sentences. Only a few survived the long work hours with minimum protection against the radiation. This inhumane punishment is still practiced in some countries today.

Although modern uranium mining is much more sophisticated and many times safer, it is still a dangerous occupation. Miners still get sick and hurt in the uranium mines, so we must appreciate their sacrifice. Every time we flip the light switch on we must remember that there are miners deep underground responsible for our comfort, for which we should thank them.

Uranium deposits are usually in sedimentary rocks that include those in sandstone (Precambrian unconformities, located in Canada), phosphate, Precambrian quartz-pebble conglomerate, collapse breccia pipes (Arizona Breccia Pipe Uranium Mineralization), and calcrete.

Sandstone uranium deposits are generally of two types; roll-front type deposits occur at the boundary between the up dip and oxidized part of a sandstone body and the deeper down dip reduced part of a sandstone body.

Peneconcordant sandstone uranium deposits, also called Colorado Plateau-type deposits, most often occur within generally oxidized sandstone bodies, often in localized reduced zones, such as in association with carbonized wood in the sandstone.

Precambrian quartz-pebble conglomerate-type uranium deposits occur only in rocks older than two billion years. The conglomerates also contain pyrite. These deposits have been mined in the Blind River-Elliot Lake district of Ontario, Canada, and from the gold-bearing Witwatersrand conglomerates of South Africa.

Hydrothermal uranium deposits encompass the vein-type uranium ores.

Igneous deposits include *nepheline syenite* intrusives found at *Ilimaussaq, Greenland*; the disseminated uranium deposit at Rossing, Namibia; and uranium-bearing *pegmatites*. Disseminated deposits are also found in the states of Washington and Alaska in the US.

The worldwide production of uranium in recent years has been in the 50,000-60,000 tons annually. Kazakhstan, Canada, and Australia are the top three producers and together account for 63% of world uranium production. Other important uranium producing countries in excess of 1000 tons per year are Namibia, Russia, Niger, Uzbekistan, and the United States.

India and several other countries have large deposits of thorium. India, for example, is sitting on over 30% of the world's thorium reserves, so they are trying desperately to develop thorium fueled reactors.

Known uranium ore resources that can be mined at low cost are estimated to be sufficient to produce fuel for about 100-150 years, based on current consumption rates.

Note: The estimates put uranium deposits at 100-150 years, which shows that we can count on steady nuclear power generation now and at least for a century more. After that...? No more nuclear power...at the same time when the last of the fossils—yes, coal included—will be also totally and permanently depleted.

Think about that for a moment. Flipping the light switch won't light the room, dialing the thermostat up won't bring heat in the house. What a sad and painful day this would be!

Uranium Ore Mining

The goal of uranium mining is to locate and extract uranium ore, suitable for nuclear fuel, from the ground. Just like coal mining, uranium is mined via open pit (surface mining), or underground.

In surface mining, the overburden is removed from the surface by drilling into and blasting the top soil to expose the ore layers. The ore is then mined by conventional blasting and excavation, with the help of large loaders and dump trucks.

The big difference here is that the entire area is contaminated by different levels of nuclear radiation. Because of that, special precautions are taken to protect all life forms in the local. The miners wear special clothing and safety equipment and spend most of the work day in enclosed, air conditioned cabins in order to limit the exposure to radiation.

The entire area is carefully monitored and special precautions are taken to limit the radiation levels and reduce its spreading in the environment. For example, large quantities of water are used all through the operation to suppress airborne dust, preventing it from spreading through or leaving the immediate mine area.

The underground mining method is another story

altogether. Here the workers are working in close quarters deep underground, often exposed to high levels of radiation from the rocks and from radon gas in tight quarters. Uranium is often mined in association with copper, gold, silver and other ores, although uranium-only mines are also in operation in some countries. In all cases, the dangers to the miners are present and significant.

Once the ore body has been identified, a shaft is sunk close to the ore veins, crosscuts are driven horizontally at various levels, usually every 100 to 150 meters, and tunnels are driven along the ore veins from the crosscuts. Drive tunnels, or raises, are driven through the deposit from level to level to the stopes, where the ore is mined from the veins.

"Cut and fill" or open stoping method used today consists of removal of the ore and refilling the space left behind with waste rock, sand, and cement.

The "shrinkage" method is used when sufficient broken ore is removed via the chutes below to allow miners working from the top of the pile to drill and blast the next layer to be broken off. This eventually leaves a large hole in the mine body.

Another method, known as *"room and pillar,"* similar to that used in coal mining is used for thinner, flatter ore bodies. In this method, the ore body is first divided into blocks by intersecting drives, removing the ore in some sections, and then systematically removing the blocks one by one, but always leaving enough ore for roof support.

Heap Leaching

Heap leaching is another prominent type of mining, which is suitable for oxide type ore deposits. Here, dilute sulfuric acid is used to extract the uranium from the ore. In this method, bulldozers level large areas of land with a small gradient, after which workers layer the area with thick plastic liner. Sometimes clay, silt, or sand are placed beneath the plastic liner for additional protection against leakage.

The ore is crushed, piled in heaps in the plastic container, sprayed with the leaching agent and left to react 1-3 months. The leaching agent breaks the uranium bonds with the rocks around it and dissolve them. The solution filters down along the gradient into collecting pools. As different heaps will yield different concentrations, the solution is pumped to a mixing plant that is carefully monitored. The properly balanced solution is then sent to a processing plant where the uranium is separated from the sulfuric acid. This method allows only about 70% of the uranium content

to be extracted.

Heap leaching has a number of advantages. It is more convenient and significantly cheaper than traditional milling processes, so lower grade ore can be economically mined.

The damages from this method, however, could be significant too, especially in countries with poor environmental regulations and enforcement practices. Leachate from the collecting pools could leak out in the ground and into the local water system. Radiation from the pool can also harm wildlife and humans.

In-Situ Leaching

This method is very similar to heap leaching technique, but here the ore doesn't even need removal from the mine shafts. *In-situ* leaching (ISL), also known as *in-situ* recovery (ISR) in North America, involves leaving the ore untouched where it is in the ground. The recovery of the minerals from it is done by flooding the shafts with acid, dissolving the uranium and pumping the solution to the surface where the uranium can be recovered.

In most cases, native groundwater in the ore body is used. It is fortified with complexing agents and in some cases by the addition of an oxidant. This chemical mix is pumped through the underground orebody, to dissolve and recover the minerals in it by leaching. Once the pregnant solution is returned to the surface, it is sent to a processing plant to separate and recovered the uranium in it.

The advantages of using this method are many; there is little surface disturbance and no tailings or waste rock generated. This method, however, poses the greatest environmental damage, so special precautions are taken prior to issuing the permits.

The oxidant used in most cases is hydrogen peroxide and the complexing agent is sulfuric acid. Some mines do not employ an oxidant but use much higher acid concentrations in the circulating solutions instead.

ISL mines in the U.S., for example, use an alkali leach, instead of acids, due to the presence of significant quantities of acid-consuming minerals such as gypsum and limestone in the host aquifers. The high carbonate minerals presence dictates the alkali leach use vs. the more efficient acid leach.

Here again, heavy contamination of soil and water table is unavoidable. While engineers and scientists can calculate and estimate the path of the leachate, once released they cannot control it. This could cause serious damage to the local environment.

Uranium Ore Processing

In nature, uranium is found as uranium-238 (99.2742% U-238) and uranium-235 (0.7204% U-235).

All U-based isotopes are radioactive, and pose serious health danger. They can be deadly upon exposure, when improperly handled.

At the mine, the uranium ore is dug out via mechanized tools, similar to those in coal mines. It is then loaded on special trucks or trains and transported to a processing facility. Here the ore is first crushed to a fine powder by crushers and grinders. The "pulped" ore is further processed by a treatment with concentrated acid, alkaline, and/or peroxide solutions, which dissolve and extract the uranium from the mix.

The resulting solution is further refined, filtered and dried to yield the final product, called yellowcake, or urania, which is actually brown or black in color.

Note: The yellowcake name is a remainder of the color and texture of the final product in the past, when it was considered to be ammonium or sodium di-uranate, and was quite un-uniform and unstable, depending on the refining process conditions. The natural uranium, yellowcake, is sold on the uranium market as U_3O_8. It is then processed into UO_2, as required for making fuel rods.

Yellowcake is a coarse powder which has a pungent odor, is insoluble in water and contains about 80% uranium oxide (U_3O_8), which melts at approximately 5212.4°F. It contains, among other things; uranyl hydroxide, uranyl sulfate, sodium para-uranate, and uranyl peroxide, and some uranium oxides.

Today yellowcake's quality is tightly controlled to contain about 90% triuranium octoxide (U_3O_8) by weight. It is produced by all countries in which uranium ore is mined. Yellowcake is used predominantly in the preparation of uranium fuel for nuclear reactors, for which it is smelted into purified UO_2 for use in fuel rods for pressurized heavy-water reactors and other systems that use natural un-enriched uranium.

Purified, pure uranium, metal (not UO_2) can be enriched into the isotope U-235, by combining the pure uranium metal with fluorine to form uranium hexafluoride gas (UF_6). The gas is then processed via gaseous diffusion, or through a gas centrifuge, where it undergoes isotope separation. This process produces low-enriched uranium containing up to 20% U-235, or the type used in most large civilian electric-power reactors.

Further processing produces highly enriched uranium, containing over 20% U-235, which is used in smaller reactors to power naval warships and submarines. Even further processing can yield weapons-grade uranium, which contains over 90% U-235, which is used for making nuclear weapons.

Since the end of the cold war in 1990, there is a worldwide surplus of highly-enriched uranium, and it is often diluted for use in some nuclear reactors.

Isotope Enrichment

Isotope separation is designed to concentrate (enrich) the fissionable uranium-235 for nuclear weapons and most nuclear power plants, except for gas cooled reactors and pressurized heavy water reactors. Most neutrons released by a fissioning atom of uranium-235 must impact other uranium-235 atoms to sustain the nuclear chain reaction. The concentration and amount of uranium-235 needed to achieve this is called a "critical mass."

To be considered "enriched," the uranium-235 fraction should be between 3% and 5%. This process produces huge quantities of uranium that is depleted of uranium-235 and with a correspondingly increased fraction of uranium-238, called depleted uranium (DU).

To be considered "depleted," the uranium-235 isotope concentration should be no more than 0.3%. The price of uranium has risen since 2001, so enrichment tailings containing more than 0.35% uranium-235 are being considered for re-enrichment, driving the price of depleted uranium hexafluoride.

The gas centrifuge process, where gaseous uranium hexafluoride (UF_6) is separated by the difference in molecular weight between $^{235}UF_6$ and $^{238}UF_6$ using high-speed centrifuges, is the cheapest and leading enrichment process. The gaseous diffusion process had been the leading method for enrichment for a long time. Here, uranium hexafluoride is repeatedly diffused through a silver-zinc membrane, and the different isotopes of uranium are separated by diffusion rate (since uranium 238 is heavier it diffuses slightly slower than uranium-235). Today, however, gas diffusion is becoming an obsolete technology that is steadily being replaced by the later generations of technologies, as the diffusion plants reach their ends-of-life.

The molecular laser isotope separation method employs a laser beam of precise energy to sever the bond between uranium-235 and fluorine. This leaves uranium-238 bonded to fluorine and allows uranium-235 metal to precipitate from the solution.

An alternative laser method of enrichment is known as *atomic vapor laser isotope separation* (AVLIS) and employs visible tunable lasers such as dye lasers.

The uranium enrichment facilities are usually designed, built, and operated under strict security, to which very few outsiders have access. Because of that, we will limit our discussion on uranium processing and handling to the very minimum.

Nuclear Fuel Production Process

During production, the final uranium dioxide (UO_2) product, in the form of a fine powder, is compacted into cylindrical pellets and sintered at high temperatures. The objective is to produce ceramic nuclear fuel pellets with a high density and well defined physical properties and chemical composition.

The pellets are machined to give them uniform and precise cylindrical shape. Thus obtained fuel pellets are then stacked into metallic tubes. The metal tubes type, size, and shape depends on the design of the reactor. Stainless steel used in the past is now replaced by zirconium alloy, which is highly corrosion-resistant and has lower neutron absorption.

The metal tubes with the fuel pellets inside (fuel rods) are sealed, and grouped into fuel assemblies which comprise the core of a power reactor. The outer layer of the fuel rods (cladding) is made of a corrosion-resistant material with low absorption cross section for thermal neutrons. It also protects the tubes from reacting with the surrounding media. Cladding also prevents radioactive fission fragments from escaping the fuel into the coolant and contaminating it.

The fuel rod assemblies are then shipped to the nuclear power plant for installation in the reactors. Since there are a number of different types of reactors, the fuel rod assemblies are also different in type and size.

Several nuclear fuel bundles are submerged in water inside a pressure vessel, where the water acts as a coolant. If the bundles are left unattended in the reactor, the uranium fission would accelerate beyond control and would eventually overheat the bundles, evaporate the water and melt the containment vessel. To prevent such uncontrolled conditions and overheating, a number of "control rods" made of a material that absorbs neutrons are inserted between the active rods in the uranium bundle.

While the uranium rods are stationary, and usually attached to the bottom of the reactor, the control rods are attached to a special mechanism in the ceiling of the reactor, that can be raised or lowered at will with great accuracy. By raising and lowering the control rods, operators can control the rate of the nuclear reaction and the overall temperature in the reactor.

In order for the uranium core to produce more

heat, the control rods are slowly lifted out of the uranium bundle. This reduces the surface area of the uranium rods in contact with the control rods, so that they absorb fewer neutrons, which results in greater heat generation.

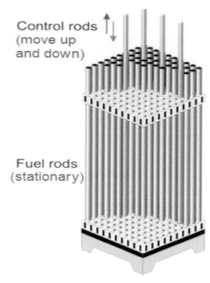

Figure 6-2. Uranium and control rods arrangement

To reduce the speed of the reaction and lower the generated heat, the control rods are lowered into the uranium bundle. This increases their surface area in contact with the uranium rods, which leads to the absorption of more neutrons, which slows the nuclear reaction and reduces the heat in the reactor.

The lower the rods, the more area to absorb neutrons, less power produced. If the rods are lowered completely into the uranium bundle, then all emitted neutrons are absorbed and the entire nuclear reaction stops. This is how a nuclear reactor is normally shut down for maintenance, or in the event of a malfunction in an attempt to prevent an accident.

If the rods are not lowered in time, however, the reaction can get out of control, causing overheating and even explosion in the reactor. This is one of the ways nuclear accidents start, and what follows could be anyone's guess. In worst-case scenarios, we end up with a Chernobyl-type nuclear accident.

Nuclear Fuel Assemblies

The type, quality and use of radioactive materials, as well as those of the nuclear fuel assemblies, is regulated by the federal Nuclear Regulatory Commission (NRC), as well as different state regulating agencies

The major fuel rod assembly types are:

- *Pressurized water reactors'* (PWR) nuclear fuel assemblies consist of cylindrical rods put into bundles. Here, uranium oxide ceramic is formed into pellets and inserted into the Zircaloy tubes that are bundled together.

The uranium oxide is dried before inserting into the tubes to try to eliminate moisture in the ceramic fuel that can lead to corrosion and hydrogen embitterment. The Zircaloy tubes are pressurized with helium to try to minimize pellet-cladding interaction which can lead to fuel rod failure over long periods.

The fuel bundles are usually enriched to several percent of U-235, with tubes of about 1 cm in diameter. The fuel cladding gap is filled with helium gas to improve the conduction of heat from the fuel to the cladding.

There are between 179 and 264 fuel rods per fuel bundle. Between 121 to 193 fuel bundles are loaded into each reactor core. Generally, the fuel bundles consist of fuel rods bundled in groups of 14×14 to 17×17 and about 4 meters long.

In PWR fuel bundles, control rods are inserted through the top directly into the fuel bundle.

- *Boiling water reactors'* nuclear fuel, as the name suggests is used in boiling water (BWR) reactors. It is similar to PWR fuel, except that the bundles are "canned" by means of a thin metal tube of appropriate diameter enveloping each bundle. This is primarily done to prevent local density variations (difference between different tubes) from affecting neutronics and thermal hydraulics of the reactor core.

Modern BWR fuel bundles consist of either 91, 92, or 96 fuel rods per assembly, depending on the manufacturer and reactor type. The reactor core contains a total of 368 assemblies for the smallest, and 800 assemblies for the largest U.S. based BWR reactors.

Each fuel rod is back filled with helium to a pressure of about three atmospheres, and is used similarly to the PWR rods.

- CANada Deuterium Uranium (CANDU) nuclear fuel bundles are about a half meter long and 10 cm in diameter. They consist of sintered (UO_2) pellets in zirconium alloy tubes, welded to zirconium alloy end plates. Each bundle weighs about 20 kg, and a typical core loading is in the order of 4500-6500 bundles, depending on the design. Modern types typically have 37 identical fuel pins radially arranged about the long axis of the bundle. Several different configurations and numbers of pins have been used in the past to figure out the ultimate, most efficient and safe design.

- *The CANDU FLEXible* fueling (CANFLEX) nuclear fuel bundle has 43 fuel elements, with two element sizes. It is also about 4 inches in diameter, 20 inches long and weighs about 44 lbs. It replaces the 37-pin standard bundle in use in some reactors.

This configuration was designed specifically to increase fuel performance by utilizing two different pin diameters. Current CANDU designs do not need enriched uranium to achieve criticality (due to their more efficient heavy water moderator), however, some newer concepts call for low enrichment to help reduce the size of the reactors.

A number of less common nuclear fuels are:

- Magnox *nuclear fuel* is a type of fuel used in reactors, which are pressurized, carbon dioxide cooled, graphite moderated reactors using natural, unreached uranium as fuel, and magnox alloy as fuel cladding.

- *Tristructural-isotropic (TRISO) nuclear fuel* is a type of micro fuel particle. It consists of a fuel kernel composed of uranium compounds (UOX, UC, or UCO) in the center, coated with four layers of three isotropic materials. The four layers are a porous buffer layer made of carbon, followed by a dense inner layer of pyrolytic carbon (PyC), followed by a ceramic layer of silicon carbide SiC to retain fission products at elevated temperatures and to give the TRISO particle more structural integrity. There is a dense outer layer made out of PyC.

- *Quad-structural-isotropic (QUADRISO) nuclear fuel particles* consist of europium oxide, erbium oxide, or carbide layer surrounding the fuel kernel of ordinary TRISO particles, which helps to better manage the excess reactivity.

- *RBMK nuclear fuel* was used in Soviet designed and built RBMK type reactors. This is a low enriched uranium oxide fuel. The fuel elements in an RBMK are 3 m long each, and two of these sit back-to-back on each fuel channel, pressure tube.

Reprocessed uranium from Russian VVER reactor spent fuel is used to fabricate RBMK fuel. As a result of the Chernobyl accident, the enrichment of nuclear fuel was changed from 2.0% to 2.4%, to compensate for control rod modifications and the introduction of additional absorbers.

- *CerMet nuclear fuel* consists of ceramic fuel particles (usually uranium oxide) embedded in a metal matrix. It is thought that this type of fuel is what is used in United States Navy reactors. This fuel has high heat transport characteristics and can withstand a large amount of expansion.

- *Plate type nuclear fuel* is commonly composed of enriched uranium sandwiched between metal cladding. It is used in several research reactors where a high neutron flux is desired, for uses such as material irradiation studies or isotope production, without the high temperatures seen in ceramic, cylindrical fuel.

- *Sodium bonded nuclear fuel* consists of fuel that has liquid sodium in the gap between the fuel pellet and the cladding. The sodium bonding is used to reduce the temperature of the fuel. It is often used for sodium cooled liquid metal fast reactors, and has been used in EBR-I, EBR-II, and the FFTF type reactors. The fuel pellets may be metallic or ceramic.

Uranium Use

Upon bombardment with slow neutrons, uranium-235 isotope divides into two smaller nuclei, releasing nuclear energy (which was used for the binding) and more neutrons.

Uranium has the ability to absorb thermal neutrons, during which time the reaction may go one of two ways:

1. Over 80% of the time it will fission;

2. 18% of the time it will not fission, and will instead emit gamma radiation, thus yielding U-236.

If too many neutrons are absorbed by other uranium-235 nuclei, a nuclear chain reaction occurs. If not controlled, the chain reaction could result in a burst of heat, and under some special circumstances it gets so violent it that could end with an explosion. The size of the explosion would depend on the type and quantity of nuclear material involved.

In a nuclear reactor, such a chain reaction is initiated under normal production regime, but it is strictly controlled. It can be slowed by a *neutron poison*, which absorbs some of the free neutrons that cause acceleration of the reaction. The neutron poison (or better said neutron absorber) could be made of a number of materials that make part of a reactor's control rods.

The main use of uranium in the civilian sector is to fuel nuclear power plants. One kilogram of uranium-235 can theoretically produce about 45,000 kWh of electric energy, assuming complete fission. This is as much energy as is contained in 10,000 kg oil, or 14,000 kg coal. This is a tremendous amount of power, which, if properly controlled, could be harnessed to generate a lot of electricity. If not properly controlled, however, it could do a lot of damage.

Commercial nuclear power plants use fuel that is typically enriched to around 3% uranium-235. The CANDU and Magnox reactor designs are the only commercial reactors capable of using unenriched uranium fuel. Fuel used for United States Navy reactors is typically highly enriched in uranium-235 (the exact values of this fuel are classified).

In the military, uranium is used in nuclear reactors to power ships and submarines. It is also used for making atomic bombs. Fifteen pounds of uranium-235 is all that is needed to make a small atomic bomb with huge destructive powers. The first nuclear bomb used in war, the Little Boy, was based on uranium fission, while the very first nuclear explosive (The Gadget) and the bomb that destroyed Nagasaki (Fat Man) were plutonium-based bombs.

NUCLEAR FISSION

Nuclear fission is what 100% of the global nuclear power industry uses to generate electricity today. There is no better way to use nuclear power at present, so a number of variations of nuclear fuels, reactors, and methods—all used in fission nuclear reactions—have evolved.

Below we take a brief look at the major products and processes used in the fission nuclear power generation.

Background

In a nuclear reactor, a nuclear chain reaction is initiated under normal production regime, but it is strictly controlled at all times. Redundant controls and safety mechanisms ensure that the nuclear reaction is under complete control.

It can be slowed down by a neutron poison, which absorbs some of the free neutrons that cause acceleration of the reaction. The neutron poison (or better said neutron absorbers) are made of a number of materials that make the reactor's control rods.

The process starts by pilling the control rods in the nuclear reactors up to initiate a nuclear reaction. At this point, bombarding uranium-235 isotopes in the uranium rods with slow neutrons causes the uranium atoms to divide into two smaller nuclei, releasing nuclear energy (which was stored in the atoms and used for the binding).

Large quantity neutrons are released at this point as well, and since uranium has the ability to absorb thermal neutrons, the reaction may go one of two ways:

- Over 80% of the time it will fission, or

- 18% of the time it will not fission, and will instead emit gamma radiation, thus yielding U-236.

When many neutrons are absorbed by uranium-235 nuclei, a nuclear chain reaction occurs. If controlled, the reaction heats the liquid in the reactor, sending it as steam into the steam turbine where it generates electricity. The steam is then condensed in the cooling towers and sent back into the reactor to repeat the heating-steam-power generation cycle.

To find the actual amount of energy released by uranium (U-235) we could use the famous formula:

$$E = mc^2$$

Where
　　E = energy released
　　c = speed of light (3.0×10^8 m/s)
　　m = nuclear mass

The fission of 1 kg of uranium fuel produces 9×10^{16} Joules of energy compared to 3×10^7 Joules of energy produced when 1 kg of coal is burned. So, theoretically speaking, the actual difference in energy contained in uranium vs. coal is about 3 billion times.

If a fission reaction releases 100 MeV of energy, then the different components of the released fission energy (assuming complete capture and accounting of the species) are as follow:

　85.00　MeV of kinetic energy of fission fragments.
　2.50　MeV of kinetic energy of neutrons.
　7.50　MeV of energy beta particles and gamma rays.
　5.00　MeV as energy of antineutrinos.

In practice, however, we must consider interferences and losses in many places during the entire power generating processes. These are different for the different fuels and processes, so we end up with different numbers.

One kilogram of uranium-235 can produce about 45 MWh of electric energy, assuming complete fission and no losses.

So, in practice, considering the losses, we would get less than 45 MWh, maybe 30, or 35 MWh, which is still a lot of energy. It is as much as contained in approximately 10 tons of crude oil, or 14 tons of coal. This is a tremendous amount of power, which, if properly controlled, could be harnessed to generate a lot of electricity. If not properly controlled, however, it could do a lot of damage.

A nuclear reaction out of control creates a chain reaction that could result in a burst of heat, and under some special circumstances it gets so violent that it cannot be stopped and could overheat the vessel and even explode. This is what happened at Fukushima's nuclear plant in March of 2011.

The magnitude of overheating or explosion conditions depends on the reactor type, and the type and quantity of nuclear materials involved.

Note: Commercial nuclear power plants use fuel that is typically enriched to around 3% uranium-235. The <u>CANDU</u> and Magnox reactor designs are the only commercial reactors capable of using unenriched uranium fuel. Fuel used for United States Navy reactors is typically highly enriched in uranium-235 (the exact values of this fuel are classified).

In the military, uranium is used in nuclear reactors to power ships and submarines. It is also used for making atomic bombs. Fifteen pounds of uranium-235 is all that is needed to make a small atomic bomb with huge destructive powers.

The first nuclear bomb used in war, the Little Boy, was based on uranium fission, while the very first nuclear explosive (The Gadget) and the bomb that destroyed Nagasaki (Fat Man) were plutonium based bombs.

Nuclear Fission Reactions

Now we know all there is to know about the nuclear fuels, so let's see what happens to them in the nuclear reactor. There are two types of nuclear reactions; fission and fusion. Presently fusion nuclear reactions are in a development stages, and only fission is used in nuclear power plants, so it is fission that we will consider in more detail in this text.

During uranium fission reaction, see Figure 6-3, neutrons are captured by the nucleus of the U-235 atoms and are absorbed within it. This briefly turns the nucleus into a highly excited U-236 atom, and very quickly it splits into two lighter atoms, Ba-141 and Kr-92, plus two or three neutrons. The number of ejected neutrons depends on the conditions under which the U-235 atom splits. This process (neutron capturing and splitting) is very fast and energetic.

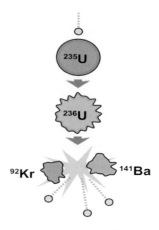

Figure 6-3. Nuclear fission

The decay of a single U-235 atom releases approximately 200 million electron volts (MeV). Since there are billions of atoms undergoing this reaction in a uranium fuel source, the energy release is enormous. For example, the energy released by just a pound of highly enriched uranium undergoing a fission reaction is equal to the energy released by over a million gallons of gasoline burning.

In addition to energy (heat) released during the splitting of the atoms, a large amount of gamma radiation (radiation made of high-energy photons) is released. The two atoms that result from the fission of each atom go in their own way releasing beta radiation (super-fast electrons) and gamma radiation of their own. These particles create the radioactivity of nuclear fuels, and they are what makes nuclear accidents so very dangerous.

For all this to work most efficiently, uranium used for nuclear fuel must be enriched so to contain 2 to 3 percent additional U-235. Three-percent enrichment is sufficient for nuclear power plants, but weapons-grade uranium requires at least 90% U-235.

Note: It is important to point out here that one neutron impinging onto the U-235 nucleus creates three neutrons, which can in turn impinge on three other U-235 nucleus, thus propagating and escalating the reaction *ad infinitum*. This spontaneous, exponential generation of neutrons is why nuclear reactions must be controlled to avoid overheating the vessel.

The fission process is accompanied by large amounts of kinetic (heat) energy that is generated during the U-236 fission. When a large amount of U-235 material is bombarded by neutrons in a containment vessel (as in a nuclear power plant) an enormous amount of heat is generated and consequently used to make electricity.

Nuclear Fusion

Nuclear fusion is a futuristic concept that promises to have great impact on the energy sector…in the distant future, if ever. Fusion power is the energy generated by nuclear fusion processes—not to be confused with nuclear fission processes, which we reviewed above.

In nuclear fusion, two or more atomic nuclei collide at a very high speed and join to form a new type of atomic nucleus. During this process, no matter is conserved because some of fusing nuclei is converted to photon energy. This is the opposite of the fission power, where neutrons are captured by the nucleus and are absorbed within it, which excites it and causes it to split into two lighter atoms and emit neutrons.

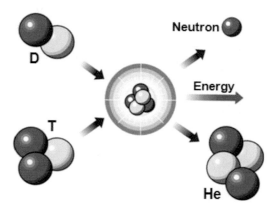

Figure 6-4. Nuclear fusion reaction

The fusion reaction needs a lot of energy to reach the operating state, at which point it releases a huge amount of energy—much more than the input energy—from overcoming the binding energy of the powerful nuclear forces. This results in a rapid and very large increase in temperature at the reaction site.

Fusion power is primarily an area of research in plasma physics, with only a few universities and government labs working it full-time. The term is commonly used to refer to potential (future) commercial production of net usable power from a fusion source.

The major designs for controlled fusion reaction use magnetic (tokamak) design, or inertial (laser) confinement of a plasma ignition. The heat released by the fusion reaction can be used to run a steam turbine and electrical generator setup, very similar to those in conventional power stations

Both approaches are many years away from success in the labs, and even many more away from commercial application. The temperature developed by full fusion reaction is so high that it makes the modern fission reaction seem like a toy.

In our estimate, if the Fukushima power plant's nuclear reactors were fusion based, the damage would've been many times faster, longer lasting, and more serious.

Fusion Reactors

Fusion reactors are still a dream that might power the future someday. Fusion nuclear reactions (vs. fission, which was described above) are the ultimate nuclear reactions in which two small particles, such as two hydrogen atoms, are combined, or "fused," to form one larger particle, releasing enormous energy in the process.

The fusion process is actually a mirror image of fission. Here two nuclei must be brought close enough and fused together to create different and heavier nucleus as a result. This can be done by overcoming the electrostatic repulsion by the attractive nuclear force, which is stronger at close distances. A lot of energy is required to overcome the electrostatic force barrier between the nucleus and initiate the fusion process.

If the nucleus is sufficiently heated, to the point of being plasma, the fusion reaction may occur due to collisions with extreme thermal kinetic energies of the particles. This process is called thermonuclear fusion, and is the only one that can be used for obtaining fusion energy for practical purposes. Fusion reactors presently are experimental types of nuclear reactors, built on the principle of controlled nuclear fusion reactions as the source of energy.

The amount of energy produced in fusion reactions is far greater than that obtained from fission reactions, since fusion reactions are the mechanisms by which stars (including the Sun) produce energy. The Sun's fusion reactions, which give us all the energy we need on Earth, are maintained by 620 million metric tons of hydrogen being fused every single second in the Sun's core.

Equally enormous energy produced during any fusion process is a fundamental problem, as far as the materials used for construction of fusion reactors and their containment is concerned.

There are no known materials thus far, that can contain the excessive heat generated during a fusion reaction. Nevertheless, the experiments continue.

Case Study: National Ignition Facility (NIF) at the Lawrence Livermore National Labs in Livermore, CA.

Note: The author was part of a team serving NIF during the last stages of the construction of the large, laser-based, inertial confinement fusion (ICF) research system.

Using 192 powerful lasers, NIF is the largest, and most energy consuming and releasing laser setup ever. The lasers are focused on a very small target (2mm in diameter), to create a nuclear fusion reaction within the target, which would then generate enormous amounts of heat.

This is how the theory behind this monstrous, multi-billion dollar project goes. It seems convincing enough for the U.S. Congress to authorize that much money for a single project with unproven theory and unknown outcome.

NIF's ultimate mission was to achieve fusion ignition with high energy gain, which would help to study the behavior of matter in nuclear weapons and reactions, and hopefully build a fusion reactor someday.

NIF construction began in 1997, but management problems and technical delays slowed progress, and the facility was completed five years behind schedule in 2009. It was also four or five times more expensive than the originally budgeted $1.1 billion.

After a number of tests, on July 5, 2012, the NIF laser system pointed its 192 beams at the small target, and delivered more than 500 trillion watts (terawatts, TW) of peak power and 1.85 megajoules (MJ) of ultraviolet laser light to the target. All this power was discharged into the miniscule, 2 mm diameter, target—(1.85 MJ is about 100 times the energy any other laser setup ever made can produce). *Nothing happened...*

Note: Five hundred TW of electric power is approximately 1,000 times more power than the U.S. uses at any instant in time.

This was a great achievement, nevertheless. And a lot of research will be done at NIF to further advance the US nuclear power generation and weapons technol-

Figure 6-5. NIF's laser bay

ogies. The initial idea to create a fusion reaction, however, was and still is a mirage. And many more billions of dollars will be needed to implode the pea-size target and release its enormous energy.

But that—even if achieved some day—it would be only the beginning. Billions more dollars will be needed to design a commercial reactor that could withstand the enormous temperature, and the accompanying problems, created during the fusion of a sizeable chunk of fusionable material.

As demonstrated by NIF's 500 TW laser blast, we need as much energy as the entire country uses at the time to initiate the fusion reaction in a small ~2 mm diameter target. With that in mind and after everything else we know about fusion, there a number of unanswered questions:

1. How much energy could be produced by a 2 mm target during its fusion process, and how long would it last?

2. How much of the energy released by the target can be captured and used?

3. Is it possible to use larger targets for practical power generation with the NIF setup?

4. How much energy is needed for the fusion of a 4 mm target? 4 cm? 4 foot?

5. How big of a target, and how much power, do we need to produce 1 GW of electric power via fusion reaction?

6. What would be the maximum practical amount of energy generated by a larger target?

7. Could the fusion reaction of a larger target be controlled?

8. What materials would be used to build the vessels needed to contain the enormous heat generated during thermonuclear fusion of larger targets?

9. Would a practical fusion power plant be as safe, or safer, than today's fission power plant of the same size?

Judging from how things stand today, there is a lot of work to be done before fusion becomes a practical nuclear power generator, and many more billions of dollars will be spent by NIF, and others, on tests, equipment modifications and upgrades.

One doesn't have to be a laser or nuclear scientist to figure out that NIF is the model T of the fusion industry, and that a lot of time is needed to build a practical,

functioning and safe fusion reactor on Earth.

Note: The latest achievement of NIF was that parts of the facility were used as the set for the starship Enterprise's warp core in the 2013 movie "Star Trek Into Darkness." Is this an indication of the path to darkness the NIF facility is headed?

Update 2014:

The NIF's plan was that upon successful completion, the fusion reaction triggered in the fuel cell would release net energy many times greater than the energy the 192 lasers use. The fuel cell made up of the two hydrogen isotopes tritium and deuterium was bombarded by the lasers on several occasions and produced enormous pressures and temperature for less than a billionth of a second.

Alas, this was still not enough to trigger the expected sustained fusion reaction. This is step one in the process of several steps, which ultimately requires that the target (fuel capsule) produces more energy than the input for the duration.

In addition, we must remember that even if and when fusion is achieved and sustained, the reaction temperature is so high that it would be impossible to contain it. The Sun's core temperature is over 28 million degrees F, which is what would be expected in a large-scale fusion reactor.

Such high temperature could melt any presently known material and drill a hole in the ground all the way to China. We simply have no materials on Earth today that can withstand even a fraction of this temperature. So, provided that somebody develops a fusion reaction, they should also know what materials would be used to build the reactors with.

After all the hoopla about the potential of the NIF's amazing technology, in 2014 the management quietly announced—as if this was not such a big deal—that the initial goal of igniting a fusion reaction was officially abandoned.

Effective immediately, NIF's $10 billion facility will be used for mundane basic research and possibly more futuristic movies. This is analogous to converting a rockets-building facility into a donut shop.

Another analogy comes to mind here, that of Solyndra, where poorly designed solar equipment was pitched to the U.S. government as the technology of the future. After half a billion taxpayer dollars were wasted, Solyndra went belly up overnight. Similarly so, after $10 billion wasted, NIF's high purpose was changed to a daily maintenance routine.

The biggest difference here is that Solyndra went

away quickly, while NIF is planning to continue operations for many years, and spend many additional billions. The most likely result of these efforts would be hundreds of theoretical papers justifying NIF's existence.

As with many grandiose ideas in the past, NIF did not even come close to achieving the goal of demonstrating a complete fusion reaction. Instead, it wasted about $10 billion on worthless hardware and finally gave up on the idea.

"The experiments at NIF are laying the groundwork to provide the nation with abundant clean energy by using lasers to ignite fusion fuel," was how the management sold the NIF project. Sounds good, and those who do not know what fusion is would fall for it. The U.S. Congress did!

Today, NIF touts the amazing power of its 192 lasers, and there are plans to shoot at and break a lot of things with them over time. At this price we could've fed half of the world instead of dumping so much money in a hole in the ground...literally.

Note: The author was part of a contractor team serving NIF at the latest stages of its development, and had first-hand experience with the project's equipment, procedures, managers and engineers.

Without going into the details, we can positively say that if fusion experiments are to be financed again with public funds, the money should go to private companies, not to a government entity. The inefficiency and waste of NIF's experience should serve as a lesson of how not to do things, and should never be repeated.

Note: A number of institutions and private firms are claiming that they are working on "the last stages of a fusion reaction." What they are saying, with no exception, is that they are close to creating a split-second fusion reaction in a target as big as a flea.

The questions they need to answer are:

- What would it take to expand the small-scale fusion reaction to large commercial scale?
- What materials would be used to contain the out-of-this-world heat in the reactors?
- What safety measures would be implemented to prevent nuclear fusion accidents?

Until all these questions are answered plausibly, nuclear fusion will remain just a pipedream and a way for institutions and companies to make money from government subsidies and private investment.

Nuclear Radiation

Nuclear radiation is defined as the energy and matter released during radioactive decay of nuclear elements and fuels. Nuclear radiation can take the following two principle forms:

Particulate radiation consists of actual subatomic particles being emitted from the nucleus of the atom.

Electromagnetic radiation is the energy emitted in wave form that possesses both electrical and magnetic characteristics.

The two types of particulate radiation are a result of alpha and beta decay. Electromagnetic radiation is a result of gamma decay.

Alpha (α) particles are typically ejected from heavy uranium or other atoms, which have excess of neutrons.

For example:

$$^{-238}U \rightarrow {}^{-234}Th + \alpha \text{ particle}$$

Alpha particles are massive, relatively slow moving, with a short travel length (mean free path) before losing their energy. At the end of the travel path they pick up two electrons and become stable helium atoms. Alpha particles are stopped within a few millimeters of travel in water or human tissue.

Beta (β) radiation originates at the shell of the atoms as a free electron (a negatively charged particle). Beta particles have energies ranging from a few keV to several MeV. The emission of a beta particle (with its negative charge) results in converting a neutron into a proton, which effectively raises the atomic number of the affected element by 1, but at the same time keeping the atomic mass unchanged.

For example:

$$^{-239}U + \beta \text{ particle} = {}^{-239}Pu$$

Here we see uranium emitting a beta particle, and is then converted into plutonium.

Gamma (γ) radiation is a photon (electromagnetic wave, not a particle), which originates in the nucleus. The gamma emission transfers energy without changing mass or charge. The fission of U-235 produces 2 or 3 neutrons each with several MeV. Since they have no charge, they can pass through matter, unless they collide with other nuclei. In case of collision with a proton (hydrogen ion), the energy is shared with the proton.

An example of gamma rays production is:

$$^{60}Co \rightarrow {}^{60}Ni^* + e^- + \gamma \text{ radiation}$$

$$^{60}Ni^* \rightarrow {}^{60}Ni + 2\gamma \text{ radiation}$$

In this case, ^{60}Co decays into excited ^{60}Ni during beta decay and emits an electron and gamma radiation. The excited ^{60}Ni then drops down to its ground energy state, emitting two gamma rays in the process.

High energy gamma waves have mean free paths in water of over three feet. Anti-gamma radiation shielding of about 20 feet of water, or 4-6 feet of cement slab, are required for human safety.

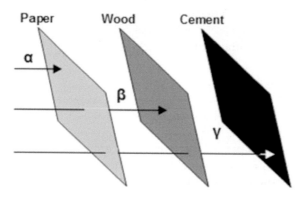

Figure 6-6. α, β, γ particles penetration potential

Neutron radiation arises from nuclear fission, such as the fission of U-235, which produces 2 or 3 neutrons each with several MeV of power. Since the protons have no charge, they pass through matter, unless they collide with other nucleus.

If they strike a proton, such as a hydrogen ion, the energy is equally shared. If a neutron strikes heavy nuclei it may be deflected with little energy loss, or it may be absorbed and incorporated into the heavier nuclei, as in the collision with a U-238 nuclei:

$$^{-238}U + neutron = {}^{-239}U$$

This results in a chain reaction of dangerous nuclear (neutron) radiation. Neutrons are electrically neutral, more penetrating than any other radiation, and could be more harmful over great areas.

NUCLEAR POWER PLANTS

A nuclear power plant is similar in design to a conventional power plant in which electricity is generated by the combustion of coal, oil, or natural gas. The most important, if not only, difference between the different power plants is the source of energy from which the electricity is generated; i.e. fossil fuels in a conventional power plant vs. nuclear material in a nuclear power plant.

The reactor of a nuclear power plant is where the nuclear reaction happens and where heat is generated. Its functional part is called the "reactor core." It contains a fissionable fuel, uranium or plutonium, from which energy is released.

The reactor core—a steel container with bundles of uranium and control rods in it—is surrounded by a large dome-shaped structure, made of concrete and reinforced steel. It is called a "containment" building, which purpose is to shield and prevent radioactive materials from escaping from the reactor core during operation and in case of an accident.

Along with the fuel rods, long and thin cylindrical control rods (made out of boron or cadmium) are placed in the core. They have the ability to absorb neutrons very efficiently, so that the position of the control rods with regard to the fuel rods determines the rate at which fission occurs within the reactor core.

When the control rods are completely inserted into the core, they block the surface of the fuel rods, so that large numbers of neutrons are absorbed, and the fission reaction can be stopped. If the control rods are removed completely, however, very large numbers of neutrons would become available, and fission would occur at a very rapid rate. By moving the control rods up and down, the reactor operators control the speed of the nuclear reaction.

Modern reactor cores are designed to prevent explosion within, under normal operation. The mass of fissionable material in the core is not enough, and the controls of the reactor are good enough to stop uncontrolled chain reaction.

A core meltdown is, however, an extremely serious event, and is quite possible as shown by the core meltdown at Chernobyl in 1986. Overheating and explosion under extraordinary circumstances, as in the Fukushima accident, are also possible consequences of abnormal operation.

Under normal operating conditions, the successful, safe, and efficient operation of a nuclear power plant depends on the exact control of the control rods, as needed to release enough energy to heat the water, but no more than that. And of course, there are many other operational and safety controls that ensure the proper and safe operation of a nuclear power plant, as we will see below.

The nuclear reaction starts by activating the uranium or plutonium atoms in the fuel. The fission reaction starts and releases energy. Once the nuclear reaction is under control, heat is generated in the reactor core, and water or some other (heat transfer) liquid is run via pipes in the core. Thus heated liquid is used to drive a turbine. The turbine in turn runs an electric generator, which generates electrical energy.

There are two major methods of transferring heat from the reactor core to the steam generator. These are: a) the boiling water system, and b) the pressurized water system.

In the boiling water system, water circulating around the reactor core is allowed to come to the boiling point. Thus generated steam is sent via pipes to the turbine where an electric generator attached to it generates electric power.

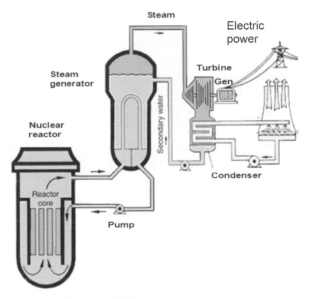

Figure 6-7. Nuclear power reactor

In the pressurized water system, the water in the reactor core is kept in pipes and under pressure, to prevent it from boiling. The super-heated water is then transferred through pipes to the external building where it is used to heat water in a second system, causing that water to boil.

Each type of nuclear power plant has its technical advantages and disadvantages. Today, about two-thirds of all nuclear power plants in the United States use pressurized water systems, and the other third use boiling water systems.

In addition to the type of water heating, the nuclear reactors used in nuclear power plants can be classified as: a) breeder reactors, and b) fusion reactors.

Breeder Reactors

A seemingly illogical concept, that any reactor can make more fuel than it consumes, is actually what the reaction that occurs between uranium-238 and neutrons

in a nuclear reactor core is based on. Strange, but fact.

Since uranium-238 is not fissionable, only uranium-235 atoms in a fuel rod undergo fission. Uranium-238, however, does react with neutrons to produce an isotope of the next heavier element in the periodic table, plutonium-239 (Pu-239), which is one of the isotopes that can undergo fission. It can also be removed from the "waste" products of the nuclear reactor, reprocessed, and used as fuel in another nuclear reactor.

A breeder reactor then is a nuclear reactor capable of generating more fissile material than it consumes because of its neutron interactions and fuel economy, which is high enough to breed fissile fuel from fertile material like uranium-238 or thorium-232.

Breeder reactors were considered for use in the past because of their superior fuel economy compared to light water reactors. That changed after the 1960s, as more uranium reserves were found, and new methods of uranium enrichment were used to reduce fuel costs.

Today, breeder reactors are again of research interest, since they offer means of controlling nuclear waste, thus closing the nuclear fuel cycle. Another major advantage of FBRs is the potential of using alternative fuels, like thorium, which are available in large quantities in some countries.

That, however, won't happen until the price of uranium generated power exceeds that of the breeder reactor operation. Experts estimate that this might happen sometime in the distant future, but it is hard to tell when.

Nevertheless, using FBRs is not to be discounted, because it is one of the potential nuclear power generating methods of the future.

Breeder Reactor Types

There are a number of breeder reactors, since a "breeder" is simply a reactor designed for very high neutron economy with an associated conversion rate higher than 1.0. So, in principal, almost any reactor design could possibly be modified in some way to become a breeder.

The evolution of the Light Water Reactor (LWR) is a good example. LWR is a very heavily moderated thermal design, into the Super Fast Reactor concept, using light water in an extremely low-density super-critical form to increase the neutron economy high enough as to allow breeding.

There are many types of breeder reactors possible, such as water cooled, molten-salt cooled, and liquid metal cooled designs that can be fueled by uranium, plutonium, minor actinides, or thorium, and may also be designed for alternative operations, such as creating more fissile fuel, long-term steady-state operation, or active burning of nuclear wastes.

We can divide the reactor designs into two broad categories based on their neutron spectrum: a) those designed to utilize primarily uranium and transuranics, and b) those designed to use thorium.

Fast breeder reactor or FBR uses fast (unmoderated) neutrons to breed fissile plutonium and possibly higher trans-uranics from fertile uranium-238. The fast spectrum is flexible enough that it can also breed fissile uranium-233 from thorium.

Thermal breeder reactors use thermal spectrum (moderated) neutrons to breed fissile uranium-233 from thorium (thorium fuel cycle). Due to the behavior of the various nuclear fuels, a thermal breeder is thought commercially feasible only with thorium fuel, which avoids the buildup of the heavier trans-uranics.

Here are the major types of nuclear reactors:

Fast Breeder Reactors

Recently all large-scale fast breeder reactors (FBR) based power plants have been of the liquid metal fast breeder reactors (LMFBR) type, which is distinguished from the rest by the fact that it uses liquid (molten) sodium for heat transfer and cooling.

There are two basic designs of LMFBR reactors:

1. *Loop type*, in which the primary coolant is circulated through primary heat exchangers outside the reactor tank (but inside the biological shield due to radioactive sodium-24 in the primary coolant), and

2. *Pool type*, in which the primary heat exchangers and pumps are immersed in the reactor tank.

All current fast neutron reactor designs use liquid metal as the primary coolant, to transfer heat from the core to steam used to power the electricity generating turbines. FBRs have been built cooled by liquid metals other than sodium—some early FBRs used mercury, other experimental reactors have used a sodium-potassium alloy (NaK).

Both have the advantage that they are liquids at room temperature, which is convenient for experimental rigs but less important for pilot or full-scale power stations. Lead and lead-bismuth alloy have also been used.

The new Generation IV FBR reactor types, still in development, are: Gas-Cooled Fast Reactor (GFR) which is cooled by helium; Sodium-Cooled Fast Reactor (SFR) which is based on the existing Liquid Metal FBR (LM-FBR); Integral Fast Reactor designs; and Lead-Cooled Fast Reactor (LFR) which is based on Soviet naval propulsion units.

FBRs usually use a mixed oxide fuel core of up to 20% plutonium dioxide (PuO_2) and at least 80% uranium dioxide (UO_2). Another fuel option is metal alloys, typically a blend of uranium, plutonium, and zirconium (used because it is "transparent" to neutrons). Enriched uranium can also be used on its own.

In many designs, the core is surrounded in a blanket of tubes containing non-fissile uranium-238 which, by capturing fast neutrons from the reaction in the core, is converted to fissile plutonium-239 (as is some of the uranium in the core), which is then reprocessed and used as nuclear fuel.

Other FBR designs rely on the geometry of the fuel itself (which also contains uranium-238), arranged to attain sufficient fast neutron capture.

The plutonium-239 (or the fissile uranium-235) fission cross-section is much smaller in a fast spectrum than in a thermal spectrum, as is the ratio between the $^{239}Pu/^{235}U$ fission cross-section and the ^{238}U absorption cross-section. This increases the concentration of $^{239}Pu/^{235}U$ needed to sustain a *chain reaction*, as well as the ratio of breeding to fission.

On the other hand, a fast reactor needs no moderator to slow down the neutrons at all, taking advantage of the fast neutrons producing a greater number of neutrons per fission than slow neutrons. For this reason ordinary liquid water, being a moderator as well as a neutron absorber, is an undesirable primary coolant for fast reactors.

Because large amounts of water in the core are required to cool the reactor, the yield of neutrons and therefore breeding of ^{239}Pu are strongly affected. Theoretical work has been done on reduced moderation water reactors, which may have a sufficiently fast spectrum to provide a breeding ratio slightly over 1. This would likely result in an unacceptable power derating and high costs in an liquid-water-cooled reactor, but the supercritical water coolant of the SCWR has sufficient heat capacity to allow adequate cooling with less water, making a fast-spectrum water-cooled reactor a practical possibility.

Thermal Breeder Reactors

Thermal breeder reactors operate on the principle of neutron absorption by fertile isotopes in a thermal spectrum. These reactions also produce more fissile fuel than they consume. Their absorption cross-section is an important factor in choosing fertile material for the core, so the fact that Th-232 breeds U-233 through neutron absorption and successive beta decays with higher neutron absorption cross-section than U-238 is an overriding factor for using thorium in thermal breeders.

Adding low enriched uranium provides another safety advantage to thorium fuel because the lower absorption cross-section for epithermal neutrons in pure Th-232, reduces the negative power coefficient in case of a power transient. Too much uranium in the fuel, however, might result in a higher concentration of plutonium produced by the fertile isotope U-238.

In a thermal breeder reactor, the low enriched fuel reaches high burn-up and achieves higher Pu-240/Pu-239 ratio for the fuel. The blanket with half the size of the driver also breeds only reactor-grade plutonium in lower quantity.

Advanced heavy water reactor (AHWR) is one of the few large-scale reactors that use thorium, and is presently developed for use in India and other countries with large thorium reserves, and which lack significant uranium deposits

The Shippingport Atomic Power Station, a 60 MWe reactor was a light water thorium breeder, which began operating in August 1977 and after testing was brought to full power by the end of that year. It used pellets made of thorium dioxide and uranium-233 oxide; initially the U233 content of the pellets was 5-6% in the seed region, 1.5-3% in the blanket region and none in the reflector region. It operated at 236 MWt, generating 60 MWe and ultimately produced over 2.1 billion kilowatt hours of electricity. After five years the core was removed and found to contain nearly 1.4% more fissile material than when it was installed, demonstrating that breeding from thorium had occurred.

The liquid fluoride thorium reactor (LFTR) can also be developed as a thorium thermal breeder. LFTRs have many advantages, such as inherent safety (due to their strong negative temperature coefficient of reactivity and their ability to drain their liquid fuel into a passively cooled and non-critical configuration), no need to manufacture precise fuel rods and the possibility of relatively simple continual reprocessing of the liquid fuel.

This concept was first investigated at the Oak Ridge National Laboratory Molten-Salt Reactor Experiment in the 1960s. It has recently been the subject of a renewed interest worldwide. Japan, China, the UK, as well as private US, Czech and Australian companies have expressed intent to develop and commercialize the technology.

Nevertheless, most efforts have been unsuccessful, delayed, and/or cancelled. In fact, after decades of tri-als, France decided to close its Superphenix fast-breeder reactor in 1998 following radioactive leaks. Japan's Monju reactor is idled due to technological challenges, including a sodium leak. In Germany, the United Kingdom, and the United States, breeder reactor development programs have been abandoned.

According to the International Panel on Fissile Materials, "After six decades and the expenditure of the equivalent of tens of billions of dollars, the promise of breeder reactors remains largely unfulfilled and efforts to commercialize them have been steadily cut back in most countries."

India is the exception here, where active research continues with the hope of using abundant domestic nuclear fuels suitable for FBR operation—thorium in this case. For that purpose, India has been developing its own FBR program since the 1980s, but the progress has been slow and full of problems and delays. A new 500 MWe FBR project—as part of India's three-stage nuclear program—is in the news now, but its destiny is unknown.

India will not stop the FBR research no matter what, because it has the world's largest reserves of thorium, which could provide power for 10,000 years, according to some estimates, and as long as 60,000 years according to others. The carrot at the end of the stick is too attractive for India to stop chasing it.

Another front runner in this area is Mitsubishi Heavy Industries, as the leader in FBR development in Japan since 2007. MHI continues the related basic FBR research as Mitsubishi FBR Systems even now. The ultimate goal of this lengthy and expensive exercise is to develop, optimize, and eventually use and even sell its new FBR technology around the globe.

Not an easy task, and one with unclear future, due to ever increasing pressure by the events of late, and the future nuclear plans of Japan and the world.

Nuclear Power Plant Life Cycle

On top of the regular permits and authorizations, as needed for the construction and operation of coal, oil and gas power plants, a nuclear power plant requires a number of additional analyses, permits, certifications, and licenses. These are needed to comply with safety rules and regulations, specific to the nuclear industry, and/or as required by local, state and federal authorities. This is truly a complex and expensive process.

A nuclear power plant's cradle-to-grave process includes:
- Nuclear plant site search
- Proposed area survey

- Environmental studies
- Local rights and agreements
- Federal and state permits
- Legal concerns and work
- Plant site preparation
- Power plant construction
- Initial tests, certification and licensing
- Reactor(s) ramp-up
- Power generation process optimization
- Daily power plant O&M operations
- Fresh and spent nuclear fuel logistics
- EOL plant decommissioning, and
- Local area reconstruction

The difficulties in every step of the way are enormous, which is one of the reasons why there have not been any new nuclear plants built in the US since the mid 1970s.

The problems start on day one; locating an area to construct a nuclear plant is a difficult task. The locals usually evoke their NIMBY (not in my back yard) rights. NIMBY demonstrations of local activists and neighbors accompany all plans for nuclear plant construction. We have seen these all around the country and the world since the beginning of nuclear power generation, and the movement is not going away.

Obtaining the needed local, state and federal permits, certifications and authorizations is another enormous task. It is actually the major reason of the stoppage and/or delay in new nuclear plants construction in the U.S. and most other countries.

Due to changes in regulations, and safety and environmental requirements, it takes forever (and a bank full of money) to complete the permitting process. Most planned nuclear plant plans are on hold presently for this reason alone.

Once a site for the new nuclear plant is chosen, the opposition is kept at a respectable and safe distance, and there is enough assurance of successful permitting that the site area survey and analysis can start. These consist of the usual geological analyses and historical data collection, as needed to determine if the site can support large structures, deep excavations, roads, and infrastructure as needed for the construction activities.

Several stages of environmental tests and studies are conducted, to determine the present environmental conditions, and as needed for the required reports. A long chain of events in the never ending permitting and licensing processes follows.

Once the necessary permits and licenses have been issued, after many years into it, plant construction starts in a fashion similar to that of coal, oil, and gas-fired power plants...although somewhat differently, as we will see below.

Since no new nuclear power plants have been built in the U.S. since 1974, we are unclear how the construction of the new plants will be handled. Most of them are now stuck in permitting and licensing stages, so we must wait and see how the situation develops in the near future.

Note: The above is actually a very condensed summary of what is needed to design, build, and commission a modern nuclear power plant. A complete description of the entire process and the related procedures would require an entirely new, and quite thick, book.

We believe, however, that this is enough to give the reader a good idea of the complexity and grandeur of a nuclear plant site.

Nuclear Power Plant Construction

Nuclear power plant (using fission nuclear processes for now) design and construction is a complex and expensive undertaking. It is different from a coal- or gas-burning power plant by the different (nuclear) fuel it uses, and the special requirements for efficient and safe use of that fuel.

Safe here must be underlined, emphasized, and over-emphasized, for the future of the entire nuclear energy industry depends on safe operation—now more than ever!

That makes a big difference in designing and building the front end of the plant (the nuclear reactors), while the rest (steam generation, turbines and electric generators) and other support infrastructure and equipment are quite similar in design to that of coal- and gas-fired power plants.

Figure 6-8. Nuclear power plant under construction

The fission nuclear plant consists of a number of key elements, each with a high level of complexity, expense, and size.

In more detail:

Reactor Pressure Vessels (RPV) are large pieces of equipment, over 20 feet inside diameter by 90 feet high, and can weigh up to 1,200 tons. Each GEN III+ unit, for example, has one RPV and one RPV head, so with everything included, the total weight of the entire unit would be well over 2,000 tons. This is about 4 million pounds of metal and other materials for one reactor.

Steam Generators are nearly 80 feet tall with an 18-foot-diameter upper section and a 14-foot-diameter lower section, and weigh about 730 tons.

Moisture separator reheaters are up to 100 feet long and 13 feet in diameter. Each moisture separator reheater weighs around 440 tons. Each GEN III+ unit uses either two steam generators, or two to four moisture separator reheaters.

Control Rod Drives and Fuel Assemblies are key elements of a nuclear reactor. One single 1100 MWe PWR reactor core contains 193 fuel assemblies composed of over 50,000 fuel rods and some 18 million fuel pellets. Once loaded, the nuclear fuel generates electricity for several years depending on the operating cycle. During refueling, every 12 to 18 months, some of the fuel—usually one third or one quarter of the core—is removed to storage, while the remainder is rearranged to a location in the core better suited to its remaining level of enrichment.

The removed fuel is considered "nuclear waste" and is disposed of via special equipment and procedures.

Steam Turbine Generators (STG) and Condensers are low pressure (LP) turbines with last-stage blades, about 52 inches long. The high pressure steam turbine weighs over 500 tons. Several low pressure rotors are used, each weighing about 250 tons. The generator stator weighs 500 tons and the generator rotor another 250 tons. Imagine the force needed to move and spin that mass...

The STG condenser lower sections each weigh over 650 tons with dimensions of 57 feet by 31 feet by 34 feet. Each STG would have up to three condensers.

Electric pumps are used to move liquids and gasses around the plant. Ten reactor coolant pumps, two turbine-driven feedwater pumps, and two motor-driven feedwater pumps are used for each reactor.

Each unit also has several very large (>400 HP) safety-related pumps, and over 100 smaller pumps designed for different functions. Some reactors have "passive safety" features and do not require that many safety-related pumps.

Valves are used to control the flow of liquids and gasses around the nuclear power plant. Each reactor unit has over 2,000 valves. Over 1,000 motor operated valves (MOVs) and air operated valves (AOVs) are used in each unit, with 700 valves of 3" size and larger. For example, different GEN III+ units have a total of 9,000 to 18,000 valves, with over 2,000 values used in the reactors alone.

Class 1E Switchgear and Equipment are used to assist the power plant operations and usually consist of two or three medium voltage switchgear panels, three 5 MW emergency diesel generators, nine 480 V motor control centers, four 125 VDC uninterruptible power supply systems, and three 120 VAC uninterruptible power supply systems. Some reactors have "passive safety" features and do not require many, if any, emergency diesel generators.

Control Equipment is used to control the power plant operations, and consists of 2,000 to 3,500 instruments, digital plant control systems, main control panels, reactor protection panels, local panels, and a plant simulator.

This short review reveals that thousands of tons of metals and other materials are used to make the different components of a nuclear power plant. These materials are then processed one way or another, and then machined before assembly and delivery to the plant.

Some of the equipment pieces used in the construction of nuclear reactors, as well as other plant components, are prefabricated. There are 500-600 prefabricated modules used in the assembly of each reactor, half of which are fabricated by third-party contractors, usually at their own offsite facilities.

These units include electro-mechanical equipment modules, piping, pipe supports, valves, controls and measurement instrumentation, tubing, conduit, cable tray, junction boxes, structural bases, and structural supports. These are then loaded on trains or trucks and transported to the construction site.

The maximum size of a module or sub-assembly fabricated off-site is 12'x12'x80' feet, to allow shipment by rail or truck. Larger structural and equipment modules are assembled on-site from scratch, or from multiple, smaller sub-assemblies delivered by the vendors.

Over 250 reinforcing steel modules and piping assemblies, and over 140 mechanical equipment modules are required for each plant of the most modularized design. In addition, over 60 structural modules and 20 electrical equipment modules are required for each unit.

The largest structural modules consist of numerous factory preassembled sub-modules that weigh up to 800 tons. Some of the structural modules include leave-in-place formwork for concrete placement.

Upon delivery of the different parts and components, they undergo final assembly and inspections onsite and are then mounted in place, to complete the plant construction. Huge cranes and other construction equipment move the parts in place, and workers finish the assembly by bolting or welding the different components.

This is a truly gigantic operation with many facets, complexities, and dangers, requiring an equally huge labor force.

Construction Labor

The construction work, from the first shovel to the initial plant test requires hundreds, and at times thousands, of engineers, technicians, and general labor personnel. The construction labor force is usually temporary, and is different for different power plant designs. The construction labor, and on-site labor support personnel includes shift supervision, warehouse personnel, clerical staff, security personnel, quality control inspectors, EPC contractors, engineers, schedulers, start-up personnel, and the ever-present Nuclear Regulatory Commission inspection staff.

All these people are highly trained and have specific daily tasks and overall goals. The labor force is divided into groups according to their specialty and work location. All jobs and tasks are managed, monitored and inspected.

Between 130 and 150 administrators, engineers, and loss control personnel are required at each unit site during the peak construction period. This does not include the work supervision, quality control, and system start-up personnel associated with vendors and subcontractors since these are different specialized groups that are counted separately.

From 40 to 50 quality control inspectors are employed during the construction phases, and are assigned to different areas of the plant. At the same time, NRC is also represented by 10-20 NRC inspectors onsite and at different off-site locations during construction, start-up, and testing periods. In addition, over 60 highly trained specialists are onsite during the critical start-up phases.

EPC contractors are involved and usually in charge of the plant design and construction, for which they need a staff of over 100 specialists at each site during different phases of construction. This includes project managers, engineers, schedulers, technicians, and other specialized technical and clerical personnel.

The plant owner usually is represented by operating and maintenance (O&M) staff of 200 specialists who support the commissioning, start-up, and maintenance of unit systems during the construction and start-up phases. The O&M staff would be over 650 for a single unit and 400 for the second unit of a twin unit plant, during normal plant operation.

Most of the permanent plant operating and maintenance (O&M) personnel and plant management, engineering, and security staff are also present onsite during the construction period, but are not included in most labor efforts. Instead, they observe the construction and undergo hands-on training for the duration.

Craft labor and onsite labor supporting construction and start-up during the peak construction period totals 2,400 people of different occupations and specialties. Basically, 800 onsite managers and specialists support 1600 personnel during daily labor activities for the duration of the plant construction and start-up phases.

Some of these people will be gradually transferred to, and/or interchangeably used in, the construction of the second and other consecutive units.

A five-year nuclear power plant construction schedule includes:
12 to 18 months for site preparation,
36 to 42 months for construction from first concrete to fuel load, and
6 to 12 months for commissioning, final equipment testing, and certification.

There are 10-12 million man hours of craft-labor estimated for the construction of one unit of an average nuclear plant in the United States. The peak construction craft labor of the same unit is estimated at 1,500-1,700 personnel working on each unit.

Most construction schedules have a 12- to 18-month period between the commercial operation dates of units in multiple unit plants. The cost of the consecutive units within the 12- to 18-month construction staggering period is usually significantly lower than the construction of the first unit for a number of obvious reasons.

There are preliminary estimates that additional GEN III+ units will be built after 2014 in some countries at the rate of eight units under construction at any time. By staggering the construction effort, two units would be entering commercial operation every year using this method. Such massive construction efforts would require a substantially large sustained labor force during the construction and deployment period.

Nuclear plant design, engineering, equipment set, construction procedures, testing and initial start-up are some of the world's most complex, sophisticated, efficient and safe operations. The same is true for the plants' O&M procedures, which are a model of precision and accuracy that is simply unthinkable in most other industrial operations.

Of course, this is absolutely necessary, because nuclear power is a monster that cannot be allowed to get out of control even for a split second...

And now, after we build the nuclear power plant, we have to run it. This is another complex operation, that involves sophisticated equipment and many well trained people.

Nuclear Power Plant Operation

A nuclear power plant is a 24/7, 365 days a year operation. It never sleeps, even when its power is not needed for awhile, at which time the power output is reduced as needed. The control of the power plant operation is done from a control room, where trained technicians and engineers control every aspect of plant function 24/7.

Nuclear power plants are shut down completely from time to time for refueling and scheduled, or preventative maintenance work. The power is also shut down partially or completely in emergencies to prevent accidents. Emergency shutdown is a major, complex, and dangerous undertaking that requires skilled operators, well executed emergency procedures, and some luck.

The most critical and dangerous operation in a nuclear power plant is the refueling of the reactor. This operation is done once or twice annually, when the spent fuel rods are removed and replaced with a fresh set. This work is done by licensed and well trained operators under strict supervision, using approved fuel handling operating and safety procedures.

A crane mounted above the reactor is used to execute this operation. The refueling is done by lowering a special lifter mechanism into the reactor core and attaching the fuel rod assembly to it. The assembly is then raised to the floor level and lowered into the nearby spent fuel storage pool. The spent fuel bundles remain in the fuel storage pool for a long time and until moved to a more permanent location.

Personnel

Nuclear power plants operate non-stop 24/7/365; fully staffed by operations and maintenance (O&M), technical, and administrative personnel regardless of the condition or stage of operation.

Each unit usually has a shift manager and several supervisors, several control room operators, and system operators whose job is to monitor and control the different pieces of equipment around the plant. There are also numerous maintenance technicians, general workers, and support personnel with varying qualifications and responsibilities.

Nuclear power plants have general managers who are ultimately responsible for operations of the entire site. Engineering staff are responsible for technical aspects of daily operation and maintenance procedures.

During normal operation, the operators perform many tasks such as support operation and maintenance activities, test safety and emergency equipment, perform minor maintenance, and ensure the proper handling of radioactive liquids and gases in the process.

During a refueling, which is conducted at least every 18-24 months, the O&M personnel handles the old fuel removal and refueling the reactor with fresh fuel by following carefully designed and strictly supervised fuel replacement and safety procedures.

The shift manager and supervisors are highly trained and licensed by the national regulatory agency. Mid- and high-level managers and supervisors are licensed as senior reactor operators, for which they have undergone special training and tests, to show superb accident assessment, supervisory, and team management abilities. They spend a lot of time working in plant simulators, where real life situations can be reproduced, followed by written and oral tests. In addition, supervisory personnel usually have many years of hands-on experience as plant operators or engineers, prior to assuming the higher responsibility.

The control room operators are licensed as reactor operators by NRC, and are in charge of and control different areas of the plant from the control room. For example, while one operator watches and controls the operating parameters of the turbine, generator, circulating water and related systems, another is in charge of the reactor, reactor cooling, and the emergency systems.

Figure 6-9. Nuclear power plant control room

The different plant systems are usually controlled from different areas of the control board in the room, but the operators communicate with each other and make coordinated decisions in different situations, with the plant efficiency and safety as the first priority.

Operators also get trained on simulators, where they learn their basic duties, in addition to practicing activities that they would be rarely required to perform. Plant startups, shutdowns, and emergencies are all included in the simulator training, and are needed to provide the operators with the maximum understanding and comfort level of managing the plant under all possible situations.

In addition, plant personnel undergo constant on-the-job training and re-qualification, via plant simulator training and testing, to maintain their skills and demonstrate competence.

Note: The plant simulator is a functional (but not operational) copy of the control room, with exact copies of all gauges, alarms, controls, and gadgets that function just like the real plant. It is driven by computers and can be programmed to simulate any possible situation—from refueling to a critical accident. The plant simulator is an integral part of plant operation, which, in addition to operator training, allows process improvements and provides solutions for unusual situations, and ultimately contributes to optimizing the plant's efficiency and safety.

It goes without saying that nuclear plants have the best trained and qualified personnel of the power generation industry. The complexity of nuclear reactor operation and maintenance is also unrivaled, as are the associated dangers. While incidents and accidents happen at all power plants, only those at nuclear power plants are the ones we hear about.

It is truly amazing that with hundreds of nuclear reactors operating non-stop around the world, there have been only a few minor and 2 major accidents during the last 60+ years. This is a truly remarkable achievement, for which we must thank the smart and brave people operating the nuclear power plants.

Spent Nuclear Fuel

Basically, 3% of the spent fuel's mass consists of fission products of ^{235}U and ^{239}Pu, which comprise the radioactive waste. These can be processed and separated further for various industrial and medical uses. The fission products include the second transition metals row; Zr, Mo, Tc, Ru, Rh, Pd, Ag, and the next in the periodic table; I, Xe, Cs, Ba, La, Ce, and Nd.

Many of the fission waste products are either non-radioactive or only short-lived radioisotopes, while a considerable number of these are medium to long-lived radioisotopes such as 90Sr, 137Cs, 99Tc and 129I.

The fission waste products can modify the thermal properties of the uranium dioxide, where the lanthanide oxides tend to lower the thermal conductivity of the fuel, while the metallic nanoparticles slightly increase its thermal conductivity. Different types of nuclear fuels produce different kinds of spent nuclear fuel (nuclear waste).

Spent low enriched uranium nuclear fuel, for example, is a type of nanomaterial. In the uranium oxide spent fuel, intense temperature gradients cause fission products to migrate. The zirconium tends to move to the centre of the fuel pellet where the temperature is highest, while the lower-boiling fission products move to the edge of the pellet. The pellet is likely to contain lots of small bubble-like pores which form during the fission cycle. Fission xenon migrates to these voids, and some of it decays to form cesium, C-137 (137Cs), so many of the bubbles contain a large concentration of 137Cs.

In MOX, the xenon tends to diffuse out of the plutonium-rich areas of the fuel, and it is then trapped in the surrounding uranium dioxide, while the neodymium is not very mobile.

Metallic particles of an alloy of Mo-Tc-Ru-Pd also tend to form in the fuel, while other solids form at the boundary between the uranium dioxide grains. The majority of the fission products remain in the uranium dioxide as solid solutions.

Successful efforts have been led to segregate the rare isotopes in fission waste including the "fission platinoids," Ru, Rh, Pd, and Ag as a way of offsetting the cost of reprocessing. Although this is possible, it is not economical enough, so it is not done commercially.

In all cases, spent nuclear fuel has only two possible paths: stored as waste, or shipped for reprocessing. Most of the spent nuclear fuels are stored in large onsite water pools at the nuclear plant, where they are left to cool down for a long time. Eventually the cooled fuel rods and bundles are placed into special canisters and shipped to an offsite storage facility.

Nuclear fuel stored as waste remains radioactive for thousands of years, and the fact that there are no permanent storage areas is becoming a huge issue. There are only a few options around the world, so this unresolved issue will continue throw a shadow of danger and uncertainty over the future of nuclear power.

Reprocessing is an expensive and uneconomical process in most cases, so it is done infrequently. Some fuels are showing promise, so the efforts in recycling, reprocessing, and reusing spent nuclear fuels will continue and even intensify in the future.

Nuclear Materials Transport

Uranium production and processing are dangerous undertakings, and so is the transport of the radioactive materials from place to place. The transport of nuclear materials is a complex process that involves a number of sophisticated equipment pieces, many different regulations, and equally numerous and complex procedures.

Unlike the other fuels, which can be dumped on any old truck, uranium ore from the mines, or spent nuclear fuel from the plant, is transported using special equipment and procedures, and following strict regulations. No exceptions!

These special precautions are needed because uranium and its products are radioactive, and people must be protected. Additionally, uranium presents security issues, so transport must be done in such a manner as to prevent its falling into the hands of terrorists and rogue nations.

Uranium ore is transported from the mines to the milling plant in produce the yellow cake. The yellow cake is then transported to the enrichment facility to make UF_6 which is transported to the fuel fabrication plant, where the final nuclear fuel is produced. Thus enriched UF_6 (nuclear fuel) is finally transported to the power plant to be used for generating electricity. Once used at the plant, the nuclear fuel is transported to a used-fuel storage facility, from where it is transported to a nuclear waste disposal site or to a reprocessing facility. From the reprocessing plant the transportation cycle repeats. These steps are full of dangers, precautionary and safety measures, expert technical personnel, and specialized security personnel.

There are an estimated 20 million shipments of radioactive material from one place to another every year around the globe. These vary from a single small package, or a number of packages sent from one location to another via common carriers, to large containers shipped on special trucks or trains.

Radioactive materials are used in many other areas (not nuclear fuel related) too, such as medicine, agriculture, university research, special manufacturing, non-destructive testing, and minerals exploration. As a matter of fact, only 5% of these annual shipments are nuclear fuel cycle related, so shipping of this commodity is a big business.

The shipment of any radioactive materials is governed by international regulations for the transport of radioactive materials, established as far back as 1961.

These regulations control shipment of radioactive material in a manner that is independent of the material's intended application.

Nuclear fuel processing facilities are located in various parts of the world and materials of many kinds are transported between them. Many of these are similar to materials used in other industrial activities. However, the nuclear industry's fuel and waste materials are radioactive, and it is these "nuclear materials" about which there is most public concern.

In the U.S. only 1 percent of the 300 million packages of hazardous materials shipped each year contain radioactive materials. This is still 3 million radioactive packages on the U.S. roadways every year.

Of these 3 million packages, about 250,000 contain radioactive wastes from US nuclear power plants, and only 50 to 100 shipments contain actual used fuel.

Decommissioning Nuclear Plants

Decommissioning costs of nuclear plants are very high, and significantly different than that of other power plants. This is due to the complexity of the technology, but also because of the added serious danger of nuclear radiation at every step of the way. This means that the demolition crew must be well trained and thoroughly familiar with, and trained in, dealing with radioactive components and materials. Every action could result in the unexpected release of radiation, which could be damaging and even fatal.

Decommissioning involves a number of administrative and technical actions, including inspection, clean-up of radioactivity, demolition of the plant, and removal and disposal of components. There are contaminated materials, parts and components that have to be handled with utmost care.

Engineering services and labor used in decommissioning of nuclear plants can be described by specialized services, starting with concrete and metal cutting and removal, including:

- Man access cuts
- Equipment hatch enlargement
- Containment wall penetrations
- Dog house cutting for SGRPs
- Elevated platform removals
- Fuel pool and canal segmentation
- Fuel transfer canal segmentation

And in more detail, there are a number of tasks performed by experienced professionals during decommissioning of a nuclear power reactor, such as:

- CRDM cutting
- Reactor nozzle cutting
- Heat exchanger cutting
- Ion guide tube removals
- Monitoring line cutting
- Steel plate stabilizer cutting
- Carbon containment liner cutting

Hands-on demolition services experience is needed here, but it has to be supplemented by thorough understanding of the nuclear plant components and their function. Only well trained and committed workforce, well attuned to personal radiation and toxic materials exposure concerns, and waste minimization awareness, could be allowed and successful in such undertaking. Anything less might result in environmental and personal damage and even death.

Note: The above is a very sketchy and superficial description of a very complex, dangerous, and lengthy operation. It is intended to give the reader just a brief glimpse of the entire nuclear fuel life cycle. This is because going into intimate technical detail would require another thick book.

Decommissioning costs represent about 5-10% of the initial capital cost of a nuclear power plant, but when discounted, they contribute only a few percent to the investment cost and even less to the generation cost. For example, in the US the decommissioning costs average 0.1-0.2 cent/kWh, which is about 5% of the cost of the electricity produced.

Decommissioning and land remediation is a complex and expensive undertaking. The dismantling could cost from $5 million to $10 million. Handling and removal of the fuel could cost another $5-10 million alone. The entire plant decommissioning could be in the range of $25 million to $1 billion. In some cases there is an additional charge for upkeep of the area, which could be up to $ 10 million annually.

One can only hope that all these costs were foreseen and estimated in the overall operating budget of the plant, or the owner might be faced with a seriously uncertain financial future, or worse.

Once a nuclear facility is decommissioned properly, there is no longer danger of radioactivity. At that time, and after official inspection and certification, the area and the owner of the plant are released from regulatory control, and the owner is no longer responsible for the nuclear safety of the area.

NUCLEAR POWER USE

The benefits of using nuclear power are obvious, so a lot of it is produced non-stop and routed into the national power grid for distribution to residences and commercial applications.

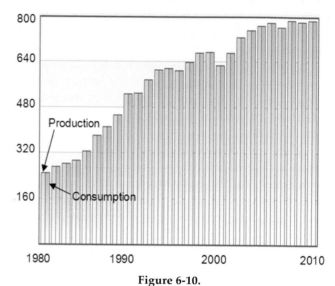

Figure 6-10.
U.S. nuclear power production and use (in TWh)

Nuclear power generation started in the 1960s, increased sharply by the 1980s, and quadrupled since then. Power generation and use are now stabilized at the 800 TWh level and very little change is expected in the near future.

Nuclear Power Generation

Nuclear power in the United States is generated by 104 commercial reactors, of which 69 are pressurized water reactors (PWR) and 35 are boiling water reactors (BWR). These are installed in and operate at 65 nuclear power plants spread strategically around the country, presently producing a total of over 800 TWh of electricity annually. This represents about 20% of the total national electrical generation, making the U.S. the world's largest supplier of commercial nuclear power.

Nuclear power provides about 6% of the total world's energy needs and over 13% of the world's electricity generated by about 440 operational nuclear power reactors in 31 countries. There are also approximately 140 naval vessels using nuclear propulsion in operation, powered by 180 reactors.

In 2012 there were 68 nuclear power reactors under construction in 15 countries around the world, 28 of which are in China. The newest nuclear power reactor was connected to the electrical grid in February 2013 in

Hongyanhe Nuclear Power Plant, China.

In the USA, two new Generation III reactors are under construction at Vogtle—the first after 34 year period of stagnation in the US nuclear power industry. Yes, true—no nuclear power plants have been put in operation in the U.S. since the 1980s.

So now, a big problem we face is that most nuclear plants are old. The first reactors were built in the 1960s and 1970s, but after the Three Mile Island accident in 1979, and following years of changing economics, many nuclear projects were discontinued and plans for others were canceled.

There have been no new nuclear power plants built in the United States since 1974, although a number of reactors started before 1974 have been completed since then.

Because of that, the licenses of most of the 104 power reactors presently in operation in the US have expired. Some have been given extensions to 60 years, instead of the original 30 years useful operation design. These extensions are close to expiration again, however, so many U.S. nuclear plant owners and the regulators are evaluating the options.

Plans for another series of extensions and building new nuclear power plants are under consideration and in negotiations stages, but their future is unclear.

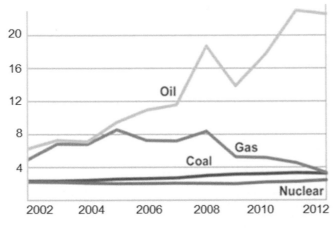

Figure 6-11. U.S. electric power generation cost (in $/kWh)

A talk about "nuclear renaissance" has been on the increase during the last several years, and the construction of several new reactors began in the early 2010s. However, as a result of the 2011 Japanese nuclear accident, and the persisting global economic crisis, most of the newly planned nuclear plant projects were reconsidered or canceled. Now, only five new reactors are expected to enter service by 2020.

This is a very shaky situation, to say the least. It almost seems like the nuclear industry is waiting for something to happen (or to not happen) before it decides to grow—or not—a teeter-totter it has been riding for the last 40 years.

Things won't get any better until the Fukushima nuclear disaster fades away in the memory of the people, and—hopefully—no other Fukushima-like disasters happen anytime soon.

This is a good time to stop, ponder the facts, and give due credit to the global nuclear industry.

—Nuclear power, under normal circumstances, is the most stable, reliable, cleanest, and cheapest electric power generator ever, and

—After over half a century of non-stop operation of hundreds of nuclear power plants around the world, generating zillions of watts of clean, reliable electric power, there have been only two major accidents.

Of course, we must admit that even one accident is one too many for the locals at Chernobyl and Fukushima, who paid with their properties, health, and lives for serious accidents that were no fault of their own.

Amazingly, most of us are still willing to take the gamble and prefer to ignore the dangers of nuclear power plants close by, in return for clean, reliable, and cheap electric power.

This won't change until and unless, another Fukushima-like accident or two, cause so much damage as to turn the general public's opinion totally and permanently against nuclear power.

Such a scenario, however, is highly unlikely, so we must assume that nuclear power, as we know it, is here to stay...at least until all uranium in the world is dug out and used.

Military Nuclear Power

The discussion herein is focused on civilian use of nuclear power, but it won't be complete if we don't mention the U.S. Armed Forces' use of nuclear power, since they are one of the largest, most versatile, and prolific users of nuclear energy in the world. As a matter of fact, the nuclear power generated and used by the U.S. Armed Forces exceeds that generated and/or used by most countries in this world.

The U.S. military is also planning to spend over $600 billion on nuclear weapons and the related programs during the next decade. This is more than the GDP of many countries, and more than most nuclear nations spend on their civilian nuclear power development.

$300 billion goes to nuclear weapons related spending, as needed to develop and maintain the nuclear weapons arsenal. Cleaning the environmental damages caused by nuclear process development and related blunders of the past is also part of this spending, but it is uncertain how much and how it will be used.

$100 billion goes for reduction of nuclear threat programs, focused on removing and securing nuclear materials and equipment from vulnerable locations. Another $100 billion is to be used for missile defense, designed to protect us from an incoming nuclear attack.

Nuclear power has one very important application that the other energy sources and fuels don't have; it comes in small packages and is portable, thus convenient for field use, in addition to being very powerful in these relatively small packages.

As a result, the U.S. Department of Defense is reducing the use of fossil fuels in its operations while proportionately increasing the nuclear use in certain applications.

Nuclear power represents a growing part of the military energy use, with important national security implications. Nuclear powered aircraft carriers, submarines, and other large vehicles are playing increasingly important roles in defending our national interests.

In all cases, we must clearly distinguish the military from civil uses. The most important difference in this respect is the fact that while we (the general public) have some say about the proliferation (or not) of nuclear energy for civil purposes, we have no say as to its use for military purposes.

The U.S. military has powerful friends in high places, and use of nuclear power as an integral (and increasing) part of our national security is usually not open for discussion.

For the DoD, energy security means having assured access to reliable supplies of energy and the ability to protect and deliver sufficient energy to meet "operational" requirements as needed.

It is implicit in this definition that military energy security enhances and does not sacrifice other operational capabilities. Operational energy is defined as the "energy required for training, moving, and sustaining military forces and weapons platforms for military operations. "This includes energy used by tactical power systems, generators and weapons platforms."

Approximately 75 percent of the energy consumed by DoD operations in 2009 was considered "operational" under this definition, while fixed installations accounted for the other 25 percent, used mostly for facilities and non-tactical vehicles.

In practice, the Department considers operational energy to be the energy used in:

- Military deployments across the full spectrum of missions;
- Direct support of military deployments around the world; and
- Training in support of unit readiness for military deployments.

Although most of the energy used by DoD operations today comes from fossils, a large (and very important) portion comes from nuclear power. The current trend in DoD energy use is to decrease fossils and increase nuclear power use, regardless of the risks associated with use of nuclear power in stationary or field power generation units.

Here we also need to give credit to the military machine for using nuclear power over half a century with so few nuclear related accidents. It is truly a great achievement, and one that makes nuclear power seem safe for military purposes.

This success might be due to increased discipline among the military personnel, better safety procedures, and/or to the fact that the nuclear plants used by the military are much smaller than the civilian giga-watt power plants, thus easier to control. Either way, good job!

Battle Ready Nuclear Power

DoD's operational experience with nuclear power started in the 1950s, when the U.S. Navy launched the *USS Nautilus*, the world's first nuclear powered submarine in 1954. The Navy currently operates over 100 nuclear power plants aboard submarines and aircraft carriers.

The Army also has significant operational experience with small land-based reactors too. The U.S. Army Corps of Engineers ran a nuclear energy program from 1954 to 1979. The small nuclear plants provided power to remote installations where connection to the power grid would have been difficult or impossible.

During this time, the Army constructed and operated nuclear reactors at Fort Belvoir, Virginia, and at Fort Greeley, Alaska. The Army also operated a nuclear reactor onboard the Sturgis, a barge used to supply electricity to the Panama Canal.

Small nuclear reactors were also located at Sundance, Wyoming; Camp Century, Greenland; and McMurdo Sound, Antarctica. Most of these reactors were decommissioned over time and the Army's participation in research and development in nuclear power stopped by 1980, around the same time that national interest in nuclear power began to wane.

Today the interest in stationary nuclear power is gaining acceptance among politicians, regulators, and the military.

Stationary Nuclear Power

Recognizing that nuclear power is a potential benefit to DoD facilities, Congress directed the DoD, in section 2845 of the National Defense Authorization Act (NDAA) of 2010, to "conduct a study to assess the feasibility of developing nuclear power plants on military installations."

Specifically, the study was to consider the following topics:

- Options for construction and operation
- Cost estimates and the potential for life-cycle cost savings
- Potential energy security advantages
- Additional infrastructure costs
- Effect on the quality of life of military personnel
- Regulatory, state, and local concerns
- Effect on operations on military installations
- Potential environmental liabilities
- Factors that may impact safe colocation of nuclear power plants on military installations
- Other factors that bear on the feasibility of developing nuclear power plants on military installations.

As a result, the military is developing detailed strategic energy plans to meet the goals established by presidential and DoD orders, as follow.

- The Navy has set a goal of meeting 40 percent of its energy needs for operations and shore installations, with alternative sources by 2020.

- The Army is incorporating sustainability into planning, training, equipping, and operations, and it has established a goal to reduce its greenhouse gas emissions by 30 percent by 2025.

- The Air Force is the largest consumer of energy in DoD. Like other services, it has made investments in sustainable energy. At the end of 2007, the Air Force was the number one purchaser of renewable energy in the federal government and number three in the United States. The Air Force continues to invest in renewable energy sources, including geothermal, wind, biomass, and solar power.

Nuclear power is part of this effort, where small size nuclear reactors are planned to be installed on military bases as needed to provide more reliable, cleaner, and cheaper power for daily operations. This in turn, would provide additional benefits to the effort of strengthening the national energy and national security.

Nuclear Renaissance

A pending nuclear renaissance (or not) has been in the news lately, describing ways for a possible return to the nuclear power industry's glory days. Global economics and rising fossil fuel prices would be the drivers, while rising concerns about greenhouse gas emission limits and global warming would provide the support for unprecedented nuclear progress. At least this is what the nuclear power proponents hope for.

Unfortunately, there are a number of unresolved issues on the way to a full nuclear renaissance. Some of the issues slowing down the progress of nuclear power today are:

1. Serious delays in permitting and siting of new power plants,

2. Unfavorable economics and safety record as compared to other sources of energy,

3. Industrial bottlenecks and personnel shortages in the nuclear sector, and

4. Uncertainty about what to do with nuclear waste and spent nuclear fuel,

5. The fear of additional nuclear accidents, such as Chernobyl,

6. Lingering national security issues related to nuclear plants, fuels, and weapons,

7. The increasing threat of nuclear terrorism, and

8. The increasing threat of nuclear weapons proliferation.

The nuclear accidents at Fukushima I Nuclear Power Plant, Chernobyl, Three Mile Island, and other nuclear facilities linger in people's minds and restrict the introduction of new nuclear programs around the world. These recent developments raise serious questions about the future of nuclear power as such, and the nuclear renaissance is on hold for now, with a nuclear reversal underway.

Germany led the nuclear reversal trend when following the 2012 Fukushima nuclear disaster they decided to shut down all nuclear power plants by 2020, while reviewing the safety of the nuclear energy in general. Several countries followed the example to a certain degree. Several new nuclear reactors planned for construction in several European countries (which were headed toward nuclear renaissance) have been delayed or cancelled.

There are estimates that 30 nuclear plants may be closed worldwide in the near future. Those located in seismic zones or close to national boundaries are the most likely to shut down first. At this time Switzerland, Israel, Malaysia, Thailand, United Kingdom, Italy and the Philippines are all reviewing the safety of nuclear power programs as well.

At the same time Australia, Austria, Denmark, Greece, Ireland, Latvia, Lichtenstein, Luxembourg, Portugal, Israel, Malaysia, New Zealand, and Norway remain unshakably opposed to nuclear power.

By 2025 over 100 older reactors will have to be decommissioned around the globe, with many nuclear power programs running over-budget and out of time.

As a result of these developments, the International Energy Agency reduced its estimate of additional nuclear generating capacity built by 2035 by more than 50%. Indonesia and Vietnam are some of the countries on the other extreme, with grandiose plans to build nuclear power plants in the near future. China is officially undecided and yet 27 new nuclear reactors are under construction in the country. A number of new nuclear power plants are being built in South Korea, India, and Russia as well.

A pro-nuclear France was planning to close at least two reactors to demonstrate political action and restore the public acceptability of nuclear power. That, however, was a publicity stunt, more than anything else.

Nuclear Failures

The turmoil in the global nuclear industry continues.

- As a confirmation of the upcoming debacle, and to reduce its bottom line risks and liabilities, Siemens (one of the giants in the industry) decided to withdraw entirely from the nuclear business.

- In the U.S., Exelon Corporation, the nation's largest nuclear operator, threw in the towel too on a planned twin-reactor project in Victoria County

in Texas. This in the face of opposition from locals who claim that there is not enough water in the area, and who also proved that the local ground is subject to subsidence which could wreck a cooling pond with time.

The Nuclear Regulatory Commission might have approved the site over these objections anyway, as it is known to do in the past, but Exelon admitted that on top of the local hassles, the economics were not that favorable so it would be better to give up on this project. So it did. 1:0 for the locals.

- At the same time, a panel of administrative law judges ruled that Électricité de France (EDF) could not proceed with a plant in Maryland's Calvert Cliffs. That plant was originally a joint venture between Constellation Energy, which owned the adjacent Calvert Cliffs 1 & 2, and the French EDF.

- In 2012 the Unistar consortium fell apart when it could not obtain a loan guarantee from the Department of Energy on terms acceptable to Constellation, which was later on bought by Exelon.

- Two new nuclear plants, each with two reactors, are in the design stages now, one in Georgia and one in South Carolina, but permitting and financial issues make us believe that no groundbreakings seem very likely anytime soon.

Even with significant political and regulatory support, the nuclear renaissance, in the U.S. at least, seems to be on hold for now. We will watch carefully the developments in the other countries as the nuclear power race goes through the different stages.

Another Show Stopper

The nuclear industry is also suffering from a number of technical and technological problems. These get more obvious as the plants' equipment and infrastructure deteriorate with time. This is evidenced by growing dilemma of a number of persisting problems with units 2 and 3 of San Onofre nuclear power plant in Southern California.

The entire plant's power generation was shut down in January 2012 after it was discovered that a generator might have been tampered with. Engine coolant was also found seeping in the oil system of the backup diesel generator.

Technicians later on found excessive wear on hundreds of tubes in units 2 and 3, which had been taken offline earlier for maintenance. The problems center on damage to alloy tubing in four steam generators that

Figure 6-12. San Onofre nuclear power plant.

were installed during a $670 million overhaul in 2009 and 2010. These and number of other abnormalities forced a temporary plant shutdown.

Although these issues pose no direct safety risk, since the plant is in shutdown mode, the bad news of abnormalities at the nuclear plant, and record-keeping errors stretching back to 1985, are another blow to San Onofre's reputation.

The work to rectify the problems, if allowed to proceed as planned, will cost over $300 million in additional repairs and other costs.

Two years late, the plant was still shut down due to safety issues. California Edison proposed in the summer of 2013 to run the plant at reduced power to avoid the safety issues and reduce financial losses. This haphazard solution for operating a crippled nuclear power plant is a worrisome development and an indication that the U.S. utilities are getting careless about the safety of nuclear power.

All California needs right now is a nuclear disaster à la Chernobyl at its San Diego coastline. That would be a lesson that the survivors would never forget.

Gregory Jaczko, ex-chair of the U.S. Nuclear Regulatory Commission expressed doubts about Edison's proposal to restart the San Onofre nuclear plant at 70 percent power. "When you're operating at a reduced power level, it indicates a lack of confidence," said Jaczko. "It raises a lot of questions."

Yes, there are lots of questions, because if they go unanswered and the genie locked in San Onfre's reactors is left out, a large part of Southern California would become a Chernobyl-like desert land. This is something that the locals and the U.S. nuclear industry do not need and must avoid at all cost.

Update: Finally, Southern California Edison (SCE) announced in June, 2013 that it will permanently retire Units 2 and 3 of the embattled San Onofre nuclear plant. SCE's decision is based on continuing uncertainty about when or if San Onofre Unit 2 might return to service. This uncertainty is not good for customers, the investors, and for the region's long-term electricity needs.

Actually, the blundered attempt to repair the two steam generators, to the tune of $670 million and a lot of embarrassment, had a lot to do with it. This, combined with increasing pressure from Senator Barbra Boxer and Friends of the Earth, a political activist group, who filed a legal petition with the Nuclear Regulatory Commission to stop San Onofre from ever starting up, were the major reasons for SCE's decision. Whatever the reasons, San Onofre as we knew it is no more.

Now SCE will use safety, stewardship, and full community engagement approaches to transform San Onofre's demise into a success story by making its decommissioning a model for the nuclear power industry.

But this is easier said than done. It will cost SCE over $4 billion and several decades to bring the project to 100% completion, as required by law. This is no peanut operation, but a very complex and dangerous operation, where a lot can go wrong... go wrong... go wrong during the long decommissioning period.

We can only wish all the best to SCE and its San Onofre project.

Note: A big problem revealed by the San Onofre case is that the facts are hidden in legal and technical secrecy. We (the public and even Congress) will never be allowed to see all the documents, thus we will never know exactly what went wrong and who's to blame.

The main issue for the locals remains: radioactive materials will lay on the beach for decades to come.

Lesson learned: as nuclear power plants in the U.S. age, they should use San Onofre as a case study of what to do, and not to do. The general public must be educated about, and pay attention to, these issues and take action whenever needed.

The abnormalities at a number of nuclear plants lately cast a shadow of doubt on the future of all global nuclear energy too, since it is apparent that even the most technologically advanced countries (Japan and the U.S.) cannot ensure the safety of their nuclear power generators.

There are second thoughts about using nuclear power on a large scale now as the issues are brought out in the open. They will surely have long-lasting consequences, and might even signal a major shift away from nuclear power for the most part. This was confirmed by the decision of the German government to shut down all nuclear plants by 2020. If that really happens, then the other energy sources and technologies like renewables, gas and coal will benefit most. Such a drastic development, however, will eventually put additional upward pressure on the availability and price of energy.

As always, nations with limited local energy resources will suffer most. Deprived of nuclear power the developing economies will have a hard time maintaining the increasing demand for electricity to keep up with their economic growth.

Nuclear Power and Energy Security

According to the Nuclear Energy Agency (NEA), "The continuous availability and affordability of energy and, in particular, electricity supply is an indispensable condition for the working of a well-functioning modern society. This is especially true for advanced industrial or post-industrial societies, where electricity provides the services essential for production, communication and exchange."

This is especially important for OECD countries, so NEA continues, "Unsurprisingly, governments of OECD countries are thus concerned with understanding the factors influencing the security of energy and electricity supplies and seek to develop policy frameworks and strategies to enhance them."

True, but developing policy frameworks lately (as a result of the Fukushima nuclear accident and the following disastrous events) has been a problem. A big problem that has been emphasized and overemphasized since that faith defining day of 2011 when news from Fukushima shook the world.

The world watched carefully, and what we saw was horrible. Three years after that day, the Fukushima region is still a deadly zone, and the prospects for it, and the affected nuclear reactors, are bleak at best. The battle at the destroyed nuclear power plants is just now starting, with over 4,000 workers fighting radiation and other damages daily.

There is still radiation contaminated soil and water in the area that promises to stay around and create even more problems with time. This is making the Japanese government very nervous and a lot of effort has been put into remediating the situation…unfortunately without much success. This situation is not embarrassing, but it is also exposing the hidden dangers of nuclear power; its wicked and merciless way of acting in a way we not only cannot control, but don't even understand well.

Yet, nuclear power is a critical component in most OECD countries' energy security plans. Just ask the French what would they do without nuclear power. They would have to import 80-90% of their energy supplies, a significant increase of the 50% present-day energy imports. With all that said, we must agree that nuclear energy has some major advantages that deserve serious consideration:

- Nuclear power plants produce electricity while using over 90% domestic capital, materials, equipment, and labor.

- Although the majority of OECD countries still import some of the of uranium used for power generation, a major part of these imports come from other OECD countries, thus the supply is safe from risks.

- But even if uranium ore supplies come from non-OECD countries, they are distributed all over the globe and have never been, nor are expected to be, a great risk to the supply chain in the past. This might change as global socio-economic and energy security situations change.

- Nuclear energy provides large amounts of clean and safe base-load electric power in many countries at stable costs.

- Nuclear power is unaffected by sudden changes in demand and/or tightening of regulations on the emission of greenhouse gases. As a matter of fact, many long-term sustainable GHG emissions scenarios include expansion of nuclear power.

There is no question that we need nuclear power as a major part of the energy supply mix. No one can deny the facts, but there are still different opinions on the matter. The developments in Europe are most poignant.

The highest proportions of proponents of nuclear energy are found in Bulgaria (90%) and Poland (88%), and there are plans for new nuclear power in both countries. This trend might be driven by the desperation of the locals, since they see no other (less painful) options. Depending on Russian natural gas deliveries, as was the case for decades, is becoming a bad option.

Several EU countries, such as Italy and the UK, shut down or put on hold their nuclear power programs in the near past, but recently changed their policy under the influence of about 75% public support for nuclear energy.

The situation in the U.S. is a bit more complicated, mostly due to the recent oil and natural gas discoveries and record-setting expansion of their production. Here safety is not a major concern; the mighty dollar is. As the extracted volumes of natural gas increase and their production costs decrease, nuclear power (with its huge capital expense) is left on the other side of the $$$ equation...for now.

The U.S. public seems split on the matter of nuclear power generation, where only the locals and environmentalists oppose nuclear power openly. Nevertheless, because of a number of complex facts and issues, nuclear power in the U.S. is not expected to increase significantly...at least not in the near future.

This is just a matter of dollars and cents (and a little sense) for now. All the facts to date lead to the conclusion that nuclear power will remain at its present level for now, with two possible scenarios for the future:

- In case of a great increase of the price for coal and natural gas, or in case of diminishing availability of the fossils, nuclear power would increase quickly in a short time period (thus entering a period of nuclear renaissance), and/or

- In case of another nuclear accident, nuclear power generation might be drastically reduced, or even put on standby for a long time around the world.

What will happen next depends on these and many other factors and variables. We just have to be ready for any eventuality. Our energy security depends on it.

THE ISSUES

No new nuclear plants have been built in the U.S. since the 1980s. This says it all. It is an unprecedented development, where a number of safety, political, and regulatory issues are throwing the shadow of a doubt on the entire future of nuclear power in this country.

Nevertheless, more than 20 companies have announced plans to build new nuclear plants in the near future, both regulated and unregulated. Planning is one thing, but starting construction is another, so the ultimate question is how many nuclear power plants will actually be built, where, and when?

In General

There are a number of issues that the nuclear pro-activists must find answers to before things change for the better. Here are the major ones:

Safety

The major problem with nuclear is its safety record—a perception that is getting more persistent after each nuclear accident. First, there was Three Mile Island in 1979, which was not as bad, but was a great surprise, because it happened in the U.S. This was a U.S. designed and built nuclear power plant, using the most advanced technology and materials in the world at the time. And yet it failed, which in turn engraved in our minds the perception that if this can happen in the U.S. it could, and most likely will, happen anywhere in the world.

Then came Chernobyl in 1986, which scared people all around the world, and the panic lasted for several years. To top it off, in 2011 the Fukushima nuclear power plant was shaken up and flooded by a powerful earthquake followed by a huge tsunami. Many aspects and issues of this disaster are still unresolved.

Regardless of the circumstances, the perception was confirmed—Fukushima used American technology and it failed. But nuclear technology—especially American is not supposed to fail, because when it does, very bad things happen.

So here we have several accidents and many humans hurt and dead—all because of nuclear power. How can you make this scenario look good? And how do you convince people that a nuclear power plant close to your home won't kill you someday, just like it did at Chernobyl and Fukushima?

Accident Insurance

Nuclear power has good friends in high places. They ensure unconditional government (and military) backup that makes nuclear power look like a kept concubine—preferred before all others. Nuclear power has benefitted by billions of dollars in government grants and subsidies through the years and continues to enjoy preferential treatment in many cases.

For example, most nuclear power plants are supposed to have some insurance—hundreds of millions of dollars in some cases—to cover any and all types of accidents. This is adequate coverage in most cases, such as with Three Mile Island.

But in a total melt-down, as with Chernobyil and Fukushima, this insurance is peanuts. Billions upon billions of dollars are needed to cover the damages for many years. Who is responsible for such huge amounts? You guessed it: you and I end up with the short end of the stick. The government steps in and takes care of every single problem, and pays every single extra dollar for as long as it takes.

Imports Dependence

The U.S. imports 90% of its uranium ore supplies. This is an important fact that is left (intentionally or unintentionally) on the back burner. We tend to forget that the raw materials used as fuel in nuclear power reactors are uranium ores found in and dug out of mines all over the world. What this does to our energy security is quite clear. The situation is very similar to that of crude oil imports.

We depend fully on unfriendly foreign governments and factions, who would not hesitate to harm us given a chance. We depend on the internal and international political scenarios to obtain the precious ore. Then we endure the risks on the world's oceans while transporting it through terrorist ridden waters. And in the process we spend billions of dollars.

Uranium Peak

Uranium, like all fossils, has a beginning and an end. It is in limited quantities and no matter what we do, its end is coming. We just do not know when. In the best of cases, Uranium will last for 100-150 years and then it will be no more. After that that the nuclear reactors will sit idle, or will be refueled with coal (if it is still around).

This is a major issue for future generations, but it is not something we are worried about today. Achieving our energy security at all cost and for as long as we can is our primary goal.

Health Problems

Work in uranium mines, nuclear research facilities and plants is associated with illnesses, and death. This is another best kept secret of the industry, which is often overlooked, and even given a positive spin by the proponents. There are a number of cases, however—in addition to the major nuclear accidents—where human health and life are affected. These are usually kept out of the media, so we don't hear about them. That way most of us don't know and don't care unless we are personally affected—which in most cases is too late.

Terrorism

One doesn't have to be a genius or an expert to see that nuclear power plants are sitting ducks, as far as cyber and air attacks are concerned. Cyber attacks have already proven successful in affecting operations of several power plants around the world, while there are no incidents of air attacks as yet.

There is also the distinct threat that nuclear facilities are vulnerable to theft of bomb-grade material, which could be used in the form of "dirty bombs" on the population.

Too, the risk of internal sabotage with the potential of causing a massive meltdown and release of radiation should not be ignored. It is quite easy to imagine that the prospect of a successful internal sabotage, in addition to cyber and air attack, on a nuclear power plant is quite attractive to many terrorist organizations. And we know that many of them are looking for such opportunities.

Defending a nuclear power plant from cyber attacks, and or internal sabotage, seems more doable than protecting it from air attacks. According to a Defense Department-commissioned report released in the summer of 2013, all 107 nuclear reactors in the U.S. are inadequately protected from terrorist attacks. It warns that the current security required of civilian-operated reactors fails to safeguard against airplane attacks, rocket-propelled grenades and such invasions.

911 proved what terrorists are capable of and willing to attack, so we must never think that they won't try attacking nuclear power plants.

So, all 104 commercial nuclear power reactors and the three research reactors operating presently in the U.S. are not protected against a 911-type terrorist attack. Of these, eleven nuclear reactors are most vulnerable, including those in Southport, NC, Port St. Lucie, FL, Columbia, MO, and Gaithersburg, MD. The latter is less than 25 miles from the White House, which makes it a very attractive target for internal sabotage or external terrorist attack.

It is very unlikely for the existing nuclear plant buildings to be strengthened to the point of withstanding a collision with a fully loaded 747. And how could we prevent a 911-type attack by airplanes or RPG rockets launched from a great distance?

It is not logistically, let alone financially, feasible to equip each facility with its own battery of anti-aircraft artillery, and even its own small air force, which is authorized to shoot down civilian aircraft that get too close and fail to respond.

Safeguarding against RPG attacks is equally difficult, if not impossible in some situations. Most RPGs today have a shooting range of over a thousand yards, which means that the ground around the nuclear plant must be strictly controlled a thousand yards in any direction. It also means that the local traffic must be closed, and nearby housing, industry, etc. infrastructure must be away from the restricted zone.

While this might be possible when designing a new nuclear power plant, this luxury is not possible for many of the existing units. San Onofre nuclear power plant, for example sits right between the busiest highway in California, I-5, and the ocean shore. To secure

1000 yards safety zone around the plant, a sea wall of some sort must be built in the ocean, and I-5 has to be rerouted through the mountains. While both approaches are possible, they would be prohibitively expensive and incredibly impractical. So what can San Onfre do? Luckily it was shut down for other reasons…

There are many ideas and suggestions, and many people are working day and night, to increase the safety of nuclear installations, but solutions are few and far between. It seems that nuclear-power sites themselves cannot provide adequate external protection, so thy must rely on the US military to provide airport security to prevent another 911-style attack—this one targeting nuclear reactors.

Controlling the national borders and with active support from the local communities we could prevent RPG attacks as well. But this is the best, if not the only thing that can be done to protect our nuclear power plants. Anything else is just not going to happen.

How good this is for the continuing safety of the nuclear power generation in the country remains to be seen, but we just don't see any other way of handling these risks.

So, in summary, when we talk about the great contribution of nuclear power to our energy security, we should always remember that the risks are equally great and never too far away. Nuclear accidents, and internal and external attacks can cripple any power plant, damage the local area, and the entire nuclear power industry with it.

We saw what happened in Japan after the Fukushima disaster, and the loud reverberation around the world, which resulted in shutting down many power plants, including total shutdown of nuclear power generation in Japan and Germany, and partial shutdown in other countries.

In other words, there is a lot of work left to do before we can call our nuclear power generation totally safe. On the contrary, it is even potentially very dangerous, especially for the people living nearby, who are always worried about accidents, attacks, and other evils attracted by, and lurking in and around the nuclear power plants.

Dirty Bombs

This is a long-term dream of the terrorists, which became reality after the fall of the Iron Curtain and the following disintegration of the Russian empire. Russian and Ukrainian nuclear power plants and many military nuclear installations were compromised, and nobody knows what happened there for several long years.

Some radioactive material was reported stolen during that time.

A dirty bomb, or radiological dispersion device, is a bomb that combines conventional explosives, such as dynamite, with radioactive materials in the form of powder or pellets.

The main purpose is to blast radioactive material into the area around the explosion. This could possibly cause buildings and people to be exposed to radioactive material. The expected result from a successful dirty bomb explosion is to frighten people and make buildings or land unusable for a long period of time.

Dirty bombs are easy to make by people who have experience with these things. They can be packaged as a regular explosive, containing some radioactive material, which is to disperse in the target area. The relative ease of constructing such weapons makes them a particularly worrisome threat. Even so, expertise matters.

Not all dirty bombs are equally dangerous; the cruder the weapon, the less damage it could cause. Terrorists could probably handle and detonate high-grade radioactive materials, and attempt to make dirty bombs, but there is a good risk of injuring themselves first.

Nevertheless, the hardest part in making a dirty bomb, and the reason we don't see more activities in this area, is acquiring the radioactive material. There are reports, however, that al Qaeda terrorists have stolen radioactive materials like strontium 90 and cesium 137, which could be used to make dirty bombs.

Can terrorists get enough quality radioactive material to make dirty bombs? This is not that easy, because the most harmful radioactive materials are found in nuclear power plants and nuclear weapons sites. These, however, have increased security, which makes obtaining materials from them more difficult.

Because of the difficulty of obtaining high-level radioactive materials from such facilities, there is a greater chance of getting them from low-level radioactive sources. These less secure sources of radioactive materials are found in hospitals, construction sites, and food irradiation plants. These materials are used to diagnose and treat illnesses, sterilize equipment, inspect welding seams, and irradiate food to kill harmful microbes.

Breaking into one of these facilities is much easier, but the materials that can be obtained there are of limited amount and of much lower quality. Still, they can be used to make a dirty bomb, which at the very least would contaminate the area and scare people enough to abandon it for a time.

In 2003, documents found in Afghanistan led to the conclusion that al Qaeda had successfully built a small dirty bomb. Also in 2003, U.S. homeland security officials and the Department of Energy sent dozens of undercover nuclear scientists with radiation detection equipment to key locations in five major U.S. cities. They were supposed to watch for, detect, and neutralize any attempt to use dirty bombs around the New Year's celebrations.

How safe we are from such attacks remains to be seen, but the efforts to make the nuclear materials safer continue, so we hope that we don't have to deal with this threat anytime soon.

We all should be aware of the dangers of nuclear radiation and be educated enough to know that nuclear radiation cannot be seen, smelled, felt, or tasted by humans. Therefore, if people are present at the scene of an explosion from a suspected dirty bomb, they will NOT know whether radioactive materials were involved at the time of the explosion. Expert radioactive measurements are required to determine the level of the radiation.

First Aid

In case of a dirty bomb attack, people in the immediate blast area should:

- Leave the immediate area on foot, avoiding panic.

- Not take public or private transportation such as buses, subways, or cars because if radioactive materials were involved, they may contaminate cars or the public transportation system.

- Go inside the nearest building away from the blast area. Staying inside will reduce further exposure to radioactive material that may be around the scene.

- Remove their clothes as soon as possible, place them in a plastic bag, and seal it. This is very important because a) removing clothing will remove most of the contamination caused by external exposure to radioactive materials, and b) saving the contaminated clothing would allow testing for exposure without invasive sampling.

- Take a shower or wash themselves as best they can. Washing will reduce the amount of radioactive contamination on the body and will effectively reduce total exposure.

- Be on the lookout for new information on the blast. Emergency personnel assessing the scene will release news as to what type and strength of radiation was involved, if any.

Note: Even if people do not know whether radioactive materials were present, following these simple steps after a terrorist bomb attack can help reduce injury from other chemicals that might have been present in the blast.

Potassium iodide, also called KI, is misunderstood and misused in many instances, so here are some guidelines:

- KI only protects a person's thyroid gland from exposure to and damage from radioactive iodine.

- KI will not protect a person from other radioactive materials, nor will it protect other parts of the body from exposure to radiation.

- KI must be taken immediately after exposure to be effective.

- KI taken before a nuclear explosion is most effective in protecting humans, but the uncertainty and health risks are too great for this approach to become a routine.

- KI can harm people and its use is not recommended, unless there is an imminent and unavoidable risk of exposure to radioactive iodine. Doctor's recommendation is advised in all cases.

The ultimate, and most serious, threat to survivors of a dirty bomb explosion is the possibility of developing cancers as a result of exposure to radiation released during the incident. Not all people close to the explosion would be exposed to enough radiation. In all cases, radiation scan of people's skin with sensitive radiation detection devices is a must, and is the only way to determine whether they were exposed or not.

Just being near a radioactive source for a short time does not mean that cancer is imminent. Only doctors can assess properly the cancer risks after the exposure level has been determined.

Technological Problems

As if to add to the risks, the complexity, and huge amounts of money spent, the newly proposed nuclear reactor designs don't have a proven track record. They look good on paper, even function well in a demo mode, but no one knows if they are efficient and safe in full-scale operation, since there is limited experience with them.

For example, the Advanced Boiling Water Reactor (ABWR) design had some problems in the past so a number of new designs have been considered as potential improvements. The power outputs of the new designs vary from 600 to 1800 MWe, with the most developed design being the ABWR-II, which started as an enlarged 1718 MWe ABWR.

The new ABWR-II was intended to make nuclear power generation more competitive in the late 2010s, but unfortunately, it has not been properly long-term field tested. So, the question of their usability, efficiency and safety still remains to be answered—hopefully before they are put in large-scale operation.

Another example, the EPR (European Pressurized Reactor, or Evolutionary Power Reactor, now simply named EPR) design, was supposed to be completed and ready for construction in the early 2000s, but remains unfinished.

The design has numerous flaws:

- The EPR is the first reactor design proposed that is to be controlled by fully computerized systems both during normal operation and during accidents. The original design for the computer systems has been found to violate just about every basic principle of nuclear safety, and many regulators are requiring an analogue back-up system. Using several complex software systems to control a nuclear power plant introduces an enormous amount of potential errors and unpredictable interactions.

- No approved design of the control systems exists as yet, even though lots of work has been done on this system for years. In addition, some EPR components are proposed to use off-the-shelf computer systems that do not comply with nuclear safety standards. The EPR design is not equipped to deal with a sustained blackout of the power supply to the reactor's emergency systems, a crucial design defect that caused the Fukushima nuclear disasters in March 2011.

- The EPR reactor's emergency diesel generators seem dangerously insufficient to power many crucial subsystems needed to cool down the reactor. If the diesel generators malfunction, the reactor is designed to prevent a meltdown of the reactor and the nuclear waste ponds for only 24 hours before risking meltdown. In Fukushima, the blackout lasted 11 days—11 times above the safety limits. Once cooling is lost, an accident can proceed fast: in the Fukushima reactors, the fuel was completely molten and the reactor was totally out of control only 11 hours after the meltdown started.

So, lots of work still needs to be done in this and all other areas of safe nuclear reactor design.

Nuclear Plants Standardization

Presently, nuclear plant's construction and operation is a black box, which no one is allowed to look into. This is done for a number of reasons, but their ultimate effect is hindering the progress of the nuclear industry.

Because of that black box mentality, we now have a multiplicity of specialized and customized reactor designs, governed by different regulatory approaches and licensing requirements. As in many other cases, this un-uniformity of equipment and procedures has the effect of increasing cost and uncertainty of the operations, and is far from being optimally conducive to nuclear plant efficiency and safety.

All this also affects investment and political decisions, which depend on the level of risk manageability. Future standardization, resulting in transparent and predictable licensing processes and oversight would contribute significantly to a stable investment framework and contribute to a rapid, efficient and orderly expansion of nuclear power worldwide. The concept of standardization does not have to cover every single detail in a nuclear plant's operation.

To start with, all that is needed is sufficient standardization to:

- Enable the owner to prepare standardized specifications for the procurement of new plant equipment, and to

- Establish standardized regulations in determining the adequacy of a nuclear facility's safety.

This may not be enough, but is much more than what is practiced today, and would limit the degree of individual nuclear power plant adaptation, while allowing enough flexibility to meet site-specific conditions and other local factors, as needed.

In other words, nuclear plant standardization would mean taking the nuclear power plant's equipment and procedures out of the black box, and making them uniform and acceptable to all.

This effort, although unlikely to materialize anytime soon, will eventually lead to:

- Developing much higher levels of detail for standard reactor designs, where special reactor designs will be the exception to the rule, and will need special, and different requirements;

- Harmonizing the nuclear industry standards and requirements, focusing on convergence of codes and standards applicable to key components affecting efficiency and safety;

- Clarifying and expanding the existing feedback sharing among utilities and participating parties, during power plant construction and operation;

- Enhancing cooperation between vendors and utilities by establishing efficient mechanisms for long-term design knowledge management, such as training materials, operator certification, and plant operation procedures;

- Information and expertise sharing among governments, vendors and regulators, which will eventually lead to wide adaptation of the standards.

Regulation

The nuclear power industry in the U.S. is regulated by the U.S. Nuclear Regulatory Commission (NRC). This is a huge body of commissioners, inspectors, and other technical and administrative personnel located in several U.S. states. NRC "…formulates policies and regulations governing nuclear reactor and materials safety, issues orders to licensees, and adjudicates legal matters brought before it."

With $1 billion annual budget, and over 4,000 employees in five different states, the commission oversees and ensures the proper and safe use of nuclear power and the related materials, procedures, etc. in the U.S. So far so good. Job well done; no Fukushima-type disasters on U.S. territory.

Internationally, there are presently efforts to identify the differences in different areas and develop international operational and safety codes and procedures, such as mechanical codes and instrumentation and control (I&C) procedures.

Many world-class companies and organizations, such as ASME, AFCEN, IEEE, and IEC are fully involved in this effort, and some general utility requirements for new reactor designs have already been developed by EPRI-URD in the U.S., and EUR in Europe.

In addition, a number of multinational regulatory initiatives have been created, such as the Multinational Design Evaluation Program (MDEP), which main objective is to establish convergent reference reactor design and regulatory practices.

Regional initiatives have been taken also by regulators and utilities, such as the Western European Regulators' Association (WENRA) and the European Nuclear Installations Safety Standards (ENISS) initiative in Europe).

WENRA, for example, has established common reference levels for reactor safety to be implemented in member countries and which, if accepted and implemented, will lead to further harmonization of the nuclear power cycle.

The International Atomic Energy Agency's IAEA's Integrated Regulatory Review Service (IRRS) provides reviews of national regulatory systems to identify and spread best practices in licensing and oversight. It also provides a reference point for states seeking to establish a nuclear infrastructure.

IAEA's Safety Standards specify safety requirements and guides, representing best/good practices, are increasingly used as reference for review of national safety standards and as a benchmark for harmonization in all countries utilizing nuclear energy for peaceful purposes.

Lots of work has been done, but it seems somewhat fragmented, so much more effort in coordinating the activities is needed to bring the global nuclear industry to the highest level of safe and efficient operation.

The Logistics

In addition to the key technological and regulatory problems of today, we see another set of issues that the U.S. and global nuclear industries must address and resolve:

Global Competition

The increased costs for today's new generation of nuclear plants are due, in large part, to a fierce worldwide competition for the resources and manufacturing capacity needed in the design and construction of new power plants. This competition has led to double-digit annual increases in the costs of key power plant commodities such as steel, copper, concrete, etc.

Worldwide demand is straining the limited capacity of EPC (engineering, procurement, and construction) firms. Since the industry was in decline, many of these firms went broke, or switched their areas of involvement to other industries. As a result, now we have only a few reliable EPC contractors able and willing to undertake nuclear power plant design and construction projects.

For the same reasons, there is also a small number of manufacturers and suppliers of specialized nuclear plant equipment. This could also cause significant bottlenecks in construction if, as expected, there are multiple orders for new power plants in the U.S. and abroad.

For example, there are only two companies that have the heavy forging capacity to create the largest equipment/components in new nuclear plants. Japan Steel Works and Creusot Forge in France are the remnants of this once thriving industry. The demand for heavy forgings for nuclear plant facilities is expected to increase significantly so the nuclear industry will be waiting in line alongside the petrochemical industry and its new refineries, for these products and services.

Twenty years ago, there were about 400 suppliers of nuclear plant components and 900 so-called nuclear stamp, or N-stamp, certifications from the American Society of Mechanical Engineers. Today there are fewer than 80 suppliers in the U.S. and fewer than 200 N-stamp certifications.

This means that there will be a great reliance on overseas companies to manufacture nuclear power plant systems and components—another unforeseen blow to our energy security.

So, the NRC would need to inspect the quality of the manufacturing programs of foreign firms if it is to ensure that substandard materials or equipment don't end up installed in U.S. plants. This may sound easy, but is in fact extremely difficult to do. So, it would take NRC much more time to inspect foreign-made components than it would to check quality control of U.S.-manufactured such.

Heavy reliance on overseas suppliers also will lead to cost increases due to the continuing weakness of the U.S. dollar relative to other currencies.

Standardization would be very helpful in this area too, for it will force all global manufacturers and providers of commodities and services to operate in a uniform and reliable manner. This, however, is something that can be expected in the future, so for now we have to do things the old and proven ways.

Cost and Finance

The long period of inactivity (lack of new construction) in the U.S. nuclear industry is taking its toll in unexpected ways. One of these is financing, or lack thereof.

Millions of engineering man-hours have been spent and many more will be spent before cost estimates become reliable enough to allow original equipment manufacturer (OEM) and engineering, procurement and construction (EPC) contract-cost negotiations to fully define cost and performance risk parameters of the new nuclear plant designs.

While simplified nuclear steam supply system (NSSS) designs and innovations in modular construction suggest the potential to build nuclear plants more economically, a confluence of largely uncontrollable forces are pushing preliminary factor cost estimates up-

ward. These include commodity price escalation, dollar fluctuations, engineering and craft labor shortages, as well as manufacturing and shipping constraints.

All these variables, combined with uncertainties about executing engineering and construction, overnight cost estimates, including costs of escalation and financing of proposed plants, are increasing so the range of cost/expense estimates is wide.

It is now estimated that it would cost between $4,000 and $5,500/kW to build a new nuclear power plant in the U.S., depending on a number of factors, including the date of the estimate. This initial investment is well above any of the existing power generating technologies—including the renewables.

Adding to the list of uncertainties, cost variability, and consequently financing uncertainty, threatens the overall economic attractiveness of nuclear development. This in turn means that it is unlikely, or at least very difficult, to ensure financing of new nuclear power plants, even if they were approved and ready for construction.

The government and most private investors would find it quite difficult to justify spending billions of dollars on a new nuclear power plant under present-day conditions, especially now, when natural gas is taking over the power generation field with its availability, low cost, and added domestic economic benefits.

Update

In the summer of 2014, a new 2 GW nuclear power plant was approved for construction in Hungary. The cost of the Russian designed and build plant was set at $13.6 billion. Assuming no overrun expenses (which always must be considered in such large projects) this is still a huge capital expense of about $6,800/kW ($6.8 per installed watt.)

This is several times more than any fossil or renewable power plant of this size—here, there or anywhere. Although the cost might be somewhat lower in the U.S. and other EU countries, it is becoming obvious now that the initial cost alone will be a major factor in the future of nuclear power.

Although this project using Russian technology is not an example of superb planning and most likely would be even less so for execution, it still sets a precedent. Adding to this abnormality, the other factors and issues make the future of nuclear power seem not that bright.

Personnel Problems

To the list of regulatory, logistical, financing, and technological problems we must add one more equally important one. It is the poor record of personnel running nuclear facilities in the U.S., no less! In the winter of 2013, about 30 out of 150 watch-standers at the Navy's Charleston Nuclear Power Training Unit in South Carolina were investigated for alleged cheating on written qualification exams.

This incident involves the written qualification exam in one of the 11 watch stations at nuclear facilities. It consisted of cheating on some portions of the periodic propulsion exam. The test sheets were simply copied and shared among senior enlisted personnel of military nuclear facilities as a routine procedure and without blinking an eye.

Wow! These people are supposed to be the best and brightest (and the most trustworthy?) among us. How is this even possible?

Navy officials were quick to take action by starting an investigation and applying corrective measures. The Navy claims that this is an isolated incident that does not affect nuclear weapons operations. Maybe so, and yet it is disruptive to the units operations and affects the security of our nuclear installations. Not to mention that it does no good to the overall morale of the rest of the military personnel and all of us!

The problem is that these are not new recruits, but the most senior enlisted personnel—the most experienced and trusted veterans—involved in the cheating ordeal, that makes this incident point to a set of internal issues and serious risks.

The guilty will be held accountable, no doubt, but what does this say about the overall reliability of the U.S. nuclear system and its safety? The burning question here is, "If highly trained and qualified U.S. military men—the elite of the U.S. nuclear arsenal—can be so sloppy and devious, then what can we expect from civilian nuclear power operators? What if they are equally sloppy and devious in their training, certification tests and their daily tasks?"

Remember that it took one sloppy, ill-trained technician to turn the wrong valve the wrong way and blow up the Chernobyl nuclear power plant, thus erasing the viability of the entire region forever, while at the same time hurting and killing many people.

Could this happen again…in the U.S. or anywhere else? Let's hope not, but let's be watchful.

NIMBY

Not in my back yard (NIMBY) movements are alive and well in the U.S. and around the world. We still remember clearly the powerful and painful images of the past. People demonstrating in front of a nuclear power plant site, others chained to fences, and some handcuffed and thrown on the backs of police cars.

The U.S. has seen many anti-nuclear protests, starting in the 1950s when PG&E planned to build the first commercial nuclear power plant in the US. The chosen site was Bodega Bay, a fishing village on the Pacific Ocean, fifty miles north of San Francisco. Bad choice, PG&E!

As soon as it was announced, the proposal was confronted by angry locals and anti-nuclear activists. From 1958 to 1963 large, non-stop demonstrations were a daily scene at the site, watched by the American public on nightly TV broadcasts.

PG&E didn't have a chance in front of the huge NIMBY lobby and national attention. It finally gave up in 1964, forced to abandon its plans for a nuclear power plant in the area.

Similar plans to build another nuclear power plant on the coast near Malibu, California, later on were also boycotted and cancelled in a fashion very similar to those at Bodega Bay.

The anti-nuclear and NIMBY movements grew stronger and captured national attention especially during the 1970s and 1980s. Clamshell Alliance protests at Seabrook Station Nuclear Power Plant, and the Abalone Alliance protests at Diablo Canyon Nuclear Power Plant are some of the most memorable. Thousands of seemingly peaceful locals and anti-nuclear activists were arrested at different locations and times, while the nation and the entire world watched in disbelief.

The Three Mile Island accident prompted another wave of similar protests starting in 1979, with the largest anti-nuclear demonstration at the time held in May 1979 in Washington, DC. Sixty-five thousand people, including some prominent politicians, attended a march and rally against nuclear power. Then in September, about 200,000 anti-nuclear demonstrators attended a similar protest against nuclear power as well.

These protests and their aftermath are engraved in the American psyche, and were very successful, as they are primarily responsible for the shutdown of the Shoreham, Yankee Rowe, Millstone I, Rancho Seco, Maine Yankee, and a dozen other nuclear power plants around the country at the time.

The wave of anti-nuclear protests in the U.S. has subsided since those days, but we expect it to surge if the construction of new nuclear power plants is announced. Would it be as successful as before remains to be seen, but the possibility should not be ignored.

Internationally, the anti-nuclear movement is on the rise. In India, Bulgaria, Poland, and some EU states people are sharply divided along the lines of pro- and con-nuclear power proliferation.

The latest such conflict was observed in a village near one of the newest nuclear plants, the Kudankulam nuclear power plant in India. Some of the villagers have been protesting for years, until a fringe group supporting the plant attacked them in the summer of 2014.

The conflict did not attract international attention, only because the small village has no access to the media. This, however, is not an isolated incident. NIMBY protests will certainly happen again and again—more violently in some areas than others—when a nuclear power plant starts construction.

The NIMBY movement is a powerful force, and should not be underestimated. It shapes people's opinions, which influences politicians, so it is capable of stopping construction of any power plant anywhere.

As such, NIMBY movements must be included in our plans and estimates anytime we consider building a power plant of any sort—and even more so when planning a nuclear one.

The Achilles Heel...again

Nuclear power has one major advantage, and one major disadvantage. In both cases, it is the immense power of the nuclear reaction. It is truly a blessing and a curse.

It is a blessing, no doubt, because of the huge amount of energy contained in a small quantity of uranium lumps. It is millions of times greater than the amount of energy contained in the same amount of coal, or any other conventional fuel.

At the same time, it is a curse, because when this huge amount of energy is released by accident, it is a vicious, silent killer that has no mercy and does not hesitate to bring total devastation to anything in its path. Fukushima's power plant accident is an example of this misguided power.

On top of that, the damage in the affected area continues for decades—even centuries. Unlike conventional explosives, which release their deadly power in a second, nuclear radiation remains in full power in the affected areas and renders them dangerous and unsuitable for humans forever.

Unlike Chernobyl, Fukushima reactors did not explode, although hydrogen gas buildup caused an explosion that blew up the upper building. The reactor, however, remained intact even after the tsunami hit. It even turned itself off as per specs, and was sitting idle and safe for several hours.

The nuclear chain reaction in it had stopped, but the radioactive materials in the core continued to generate dangerous levels of heat. The cooling system operated properly for awhile and kept the core from overheating.

Unlike most modern reactors, which use the natural convection of the cooling water to keep it circulating, the Fukushima reactor was of the older kind. It required outside power to keep the cooling water pumps running. In fact, they kept working well, even after the earthquake and the tsunami devastation of the surrounding infrastructure, and successfully kept the reactor core cool and out of danger of melting...for awhile.

The emergency cooling was designed to run about 8 hours, or until outside power was restored—whichever came first. In this case, outside power was not to be restored ever. Eventually, the emergency power ran out and the cooling pumps stopped circulating water through the core's cooling jacket. Soon the core overheated, and started melting under the tremendous temperatures developed within. So, the Fukushima accident was caused by lack of outside power, not by the earthquake or tsunami themselves.

The core meltdown exposed the nuclear fuel to the outside, which allowed a huge cloud of radiation to escape. The emitted radioactivity was much greater than any previous nuclear reactor accidents, except that at Chernobyl in 1986.

The intense heat in the Fukushima nuclear reactors melted the metal capsules that contain the nuclear fuel and nuclear waste, releasing copious volatile gases mixed with water vapor. Among them, radioactive iodine and cesium found their way out and up into the surrounding air.

Radioactive iodine's atoms have a very short half-life and decay rapidly, while releasing radiation as they travel through the air. Radioactive iodine decays rapidly; 50% is gone in 8 days, and most of it in 2 months.

They make the largest source of initial stages of radioactivity. Inhaling or consuming radioactive iodine collects and concentrates in the thyroid, where it usually induces cancer. However, within a week from the explosion, the amount of radioactive iodine in the area was close to zero.

But radioactive cesium was just then starting to play its vicious role of mass killer. Its radiation levels are low at first because it decays very slowly.

So while it was almost undetectable at first, the radiation levels, increased and remained for a long time at Chernobyl. During the first 30 years after the accident

50% of the initial radioactive cesium is still present, which is enough to kill anyone nearby.

Radioactive strontium, which is also found in significant amounts around leaking radiation sites, is also slow to unleash its radiation. When it does, however, it sticks around for many, many years. During this time, they circulate in the air and/or penetrate in the soil and water table. The radioactivity also settles on plants and vegetation, which when eaten by animals and humans becomes concentrated in organs and bones.

This is exactly what happened after the Chernobyl disaster. The authorities were too ignorant or careless, so the locals were not protected in time from the iodine, cesium, and strontium radiation, and they continued to drink milk and eat foods from contaminated sources.

This did not happen at Fukushima, because immediate evacuation of the locals was ordered and executed. At the same time, the consumption of any food grown in the region, and fish from the local waters was forbidden.

The only protection against radioactive iodine is potassium iodide, taken internally in the form of pills or solution, immediately after exposure.

Note: Iodide taken internally saturates the thyroid gland, stopping the absorption of its radioactive cousins. Unless, of course, the pills are not taken in time, or if the exposure to radioactive iodide is too great. In such cases, the overall health of the individual and luck play major roles in the outcome. There is no effective protection against radioactive cesium or strontium. None!

At the end, thousands of people new Fukushima were hurt and made ill as a result of the nuclear accident and its aftermath. Many of them are still living away from home, and some will never return to their homes, due to high levels of radiation, which are expected to remain for several generations.

There is an estimate of about 1,500 people killed by the accident, which is only 10% of the total number of people killed during the entire episode of the earthquake/tsunami/nuclear plant explosion. Yet, these are 1,500 people who are dead as a consequence of our drive to make cheap energy.

We must never forget that energy generation and the related energy security should not come at the expense of human suffering or loss of life. It is one thing to live comfortably, but totally another to lose your life for it.

Yet, what can we do? We cannot make everything—including nuclear power plants—absolutely safe. Can we design nuclear reactors to withstand everything imaginable? Anything that comes their way—be it RPG attacks, airplane strikes? Or how about large asteroids hurling at them?

There is no possible way to provide such absolute protection, so we must evaluate the risks and include the most practical designs in our nuclear plants, in order to ensure against the most dangerous negative effects.

Fukushima Update, 2014

The Fukushima disaster is a never-ending nightmare, unlike Chernobyl, which was packed in cement and left to brew silently for ages to come. Out of sight and out of mind; almost forgotten in its silent grave, Chernobyl is now a fading memory.

Fukushima, however, is very much alive and continues to create serious problems on a daily basis.

Water draining from the local mountains runs through or close to the plant, becomes contaminated with radioactive materials and runs into the ocean. Local fisheries are contaminated with ever increasing doses of radioactivity, spreading, threatening the globe.

Fukushima cannot be entombed like Chernobyl, so other alternatives to prevent further damage to the environment are sought. The latest one, approved by the Japanese government, calls for building a wall of iced soil around the plant, and containing the radiation within a "frozen wall."

Figure 6-13 shows the location of the proposed "frozen wall" to be built around reactors 1, 2, 3, and 4. The idea is simple; a frozen soil barrier deep underground would contain the contaminated water coming from the plant's leaking infrastructure, while at the same time not allowing the groundwater to enter the plant perimeter and get contaminated.

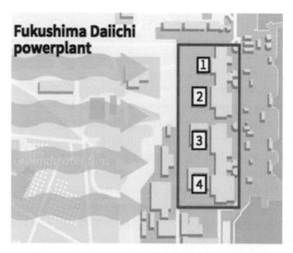

Figure 6-13. The Fukushima "frozen wall"

First a row of deep shafts will be drilled into the ground around the wrecked reactors. A special refrigeration plant will be then built nearby. A network of "freeze pipes" will be laid from the refrigeration plant, or plants, all around the "frozen wall" perimeter and into the underground shafts.

Special coolant, kept at subzero temperature will circulate in the freeze pipes, freezing the soil in the deep shafts around them. The cryogenic action will eventually create a deep and thick layer of frozen soil in the shafts along the "frozen wall."

If successful, the new underground wall would prevent some 300-400 tons of radioactive water from running into the ocean every day. Although the frozen soil method has been used in other, albeit much smaller projects in the past, it has never been attempted on such a large scale and under such complex conditions.

This is a huge construction project, expected to last 2-3 years (if no problems are encountered), at the cost of half a billion dollars. The project is faced with a number of technological, safety, and economic problems, and nobody knows to what extent it would be able to contain the radiation contamination of the ocean.

The biggest unknown is how long the "freeze wall" has to be operated and maintained. Radiation contamination can last hundreds of years. Can the "freeze wall" last that long too? And at what expense?

Nuclear Waste Transport

The waste nuclear fuel and its transport, storage, and disposal in particular have been of most concern to all involved—from the power plant owners to the investors and the customers. Because of that, special attention is paid to the technology and procedures used in this process.

There are two distinct types of nuclear wastes: a) low and intermediate, and b) high level radiation fuel wastes.

Low Level Radiation Waste

Low-level (LLW) radioactive wastes are a variety of materials that emit relatively low (yet unsafe) levels of radiation, slightly above normal background levels. They often consist of solid materials, such as clothing, tools, or contaminated soil. Low-level waste is transported from its origin to waste treatment sites, or to an intermediate or final storage facility.

LLW are generated throughout the nuclear fuel cycle and/or from the production of radioisotopes used in medicine, industry, research labs, and other such facilities.

The transport of these wastes is commonplace and they are safely transported to waste treatment facilities and storage sites.

A variety of radio-nuclides give low-level waste its radioactive character. However, the radiation levels from these materials are very low and the packaging used for the transport of low-level waste does not require special shielding.

Low-level wastes are transported in drums, often after being compacted in order to reduce the total volume of waste. The drums commonly used contain up to 200 liters of material. Typically, 36 standard, 200-liter drums go into a 6-meter large transport container.

Low-level wastes are moved by road, rail, and internationally, by sea. Most low-level waste is only transported within the country where it is produced.

The composition of intermediate-level wastes is broad, but they usually require some shielding. Much ILW comes from nuclear power plants and reprocessing facilities. Intermediate-level wastes are taken from their source to an interim storage site, a final storage site (as in Sweden), or a waste treatment facility. They are transported by road, rail and sea.

Intermediate Level Radiation Wastes

The radioactivity level of intermediate-level waste (ILW) is higher than that of low-level wastes. The classification of radioactive wastes is decided for disposal purposes, not on transport grounds. Yet, the transport of intermediate-level wastes take into account any specific properties of the material, so they usually requires some minimal shielding.

In the U.S. there had been over 9000 road shipments of defense-related trans-uranic ILWs for permanent disposal in the deep geological repository near Carlsbad, New Mexico, by 2010, without any major accident or any release of radioactivity. Almost half of these shipments were from the Idaho National Laboratory.

The repository, known as the Waste Isolation Pilot Plant (WIPP), is about 700 meters deep in a Permian salt formation. It is used for disposal and permanent storage of LLW and ILW nuclear waste streams.

High Level Radiation Waste

Used fuel unloaded from a nuclear power reactor is considered to be high-level radiation waste. It usually contains: 96% uranium, 1% plutonium and 3% of fission products (from the nuclear reaction) and trans-uranics. This waste emits high levels of both radiation and heat so is stored in water pools adjacent to the reactor to allow the initial heat and radiation levels to decrease.

Typically, used fuel is stored on-site for at least five months before it can be transported, although it may be stored there long-term as well. This method is preferred if there is enough space in the water storage pool, or if there is no available permanent storage.

From the reactor site, used fuel is transported by road, rail, or sea to either an interim (semi-permanent) storage site or a reprocessing plant where it will be recycled and reused.

The waste nuclear fuel assemblies are shipped in Type B casks which are shielded with steel, or a combination of steel and lead, and can weigh up to 110 tons each when empty.

A typical transport cask holds up to 6 tons of used fuel, which is what an average modern nuclear power plant uses in a year.

Case Study: Yankee Rowe Nuclear Power Station

Yankee Rowe Nuclear Power Station was the third commercial nuclear plant built in the U.S. It was a large (for its time) power plant located in Rowe, Massachusetts, rated at 185 MW power output, which operated non-stop from 1960 to 1992.

The capital cost of the entire plant was $45 million vs. an estimated cost of $57 million—almost 25 percent below budget, something that is unheard of today. It was quite an achievement then and an example for us now of what can be done with proper project planning, design, and management.

Amazing fact: the total construction cost of the Yankee Rowe nuclear plant was about $240/kW, or $0.24 per installed watt. Compare this to the latest $6.8 per installed watt estimates.

During its 32-year non-stop operation, the Yankee plant generated over 44 billion kilowatt-hours of electricity, and had a lifetime capacity factor of 74%. No other power generating technology (at the time) could show such remarkable numbers.

The plant was decommissioned primarily due to concerns with embritlement of the reactor core materials.

Another amazing fact: the spent nuclear fuel was stored on site and after 32 years of non-stop operation, it all fit in 16 casks. No other significant pollution was emitted for the duration.

By comparison, a 32-year-old coal-fired power plant of this size would have produced mountains of coal, ash, and slurry piled nearby, slowly destroying the natural landscape and contaminating the local water supply.

The plant would have also emitted millions of tons of GHGs during its life time. None of this damage can be attributed to nuclear power generation. Not a small difference that must be kept in mind when discussing the energy sources and making decisions of their use.

Nuclear Fuel Shipping

Since 1971 there have been some 7,000 shipments of used fuel (over 80,000 tons) over many millions of miles around the world with no property damage or personal injury, no breach of containment, and very low dose rate to the personnel involved.

This includes 40,000 tons of used fuel shipped to Areva's La Hague reprocessing plant, at least 30,000 tons of mostly UK used fuel shipped to UK's Sellafield reprocessing plant, 7140 tons used fuel in 160 shipments from Japan to Europe by sea and 4500 tons of used fuel shipped around the Swedish coast.

A typical transport consists of one truck carrying one protected shipping container. The container holds a number of packages with a total weight varying from 80 to 200 kg of plutonium oxide. A train shipment may have several cars with several containers each. A sea shipment may consist of several containers, each of them holding between 80 to 200 kg of plutonium in sealed packages.

Some 300 sea voyages have been made carrying used nuclear fuel or separated high-level waste over a distance of more than 6 million miles. The major company involved has transported over 4000 casks, each of about 100 tons, carrying 8000 tons of used fuel or separated high-level wastes. A quarter of these have been through the Panama Canal.

In Sweden, more than 80 large transport casks are shipped annually to a central interim waste storage facility called CLAB. Each 80-ton cask has steel walls 30 cm thick and holds 17 BWR or 7 PWR fuel assemblies. The used fuel is shipped to CLAB after it has been stored for about a year at the reactor, during which time heat and radioactivity diminish considerably. Some 4500 tons of used fuel had been shipped around the coast to CLAB by the end of 2007.

Shipments of used fuel from Japan to Europe for reprocessing used 94-ton Type B casks, each holding a number of fuel assemblies (e.g. 12 PWR assemblies, total 6 tons, with each cask 6.1 meters long, 2.5 meters diameter, and with 25 cm thick forged steel walls). More than 160 of these shipments took place from 1969 to the

1990s, involving more than 4000 casks, and moving several thousand tons of highly radioactive used fuel—4200 tons to UK and 2940 tons to France. Within Europe, used fuel in casks has often been carried on normal ferries, e.g. across the English Channel.

Plutonium is transported, following reprocessing, as an oxide powder since this is its most stable form. Plutonium oxide is transported in several different types of sealed packages, and each can contains several kilograms of material.

Risk of exposure is reduced by the special design of the package, limiting the amount within, and the number of packages carried on a transport vessel. Special physical protection measures apply to plutonium and all other nuclear consignments as well.

Nuclear Waste Storage

Ideally, a nuclear power plant runs at the maximum allowed power level from one refueling to the next. It has to be shut down for several hours, or days, during each refueling, and then restarted and brought up to the maximum power output.

Used nuclear fuel, which is removed from the reactor after a year or two of service, is a solid material that has to be stored safely, usually at nuclear plant sites. This storage is only temporary, which is one component of an integrated used fuel management system in use at all nuclear plants, that addresses all facets of storing, recycling and disposal.

The integrated used fuel management approach mandates that used nuclear fuel will remain safely stored at nuclear power plants for the near term, with safety as the key word. Eventually, the hope goes, the government will find a way to recycle it, and place the unusable end product in a deep long-term repository.

- Low-level wastes are byproduct remaining from uses of a wide range of radioactive materials produced during electricity generation, medical diagnosis and treatment, and various other medical processes. These could be liquids or solids, and are treated in a number of ways prior to disposal as hazardous waste, burning, or storage in temporary containers.

- Recycling of waste nuclear fuels is a program of the federal government, which includes plans to develop advanced recycling technologies in order to take full advantage of the unused energy in the used fuel, and to also reduce the amount and toxicity level of byproducts requiring disposal.

- Transportation of the waste nuclear fuel is the responsibility of the U.S. Department of Energy, which will transport used nuclear fuel to the repository by rail and road, inside massive, sealed containers that have undergone safety and durability testing.

- Repository for long-term storage is under review, and under any used fuel management scenario, disposal of high-level radioactive byproducts in a permanent geologic repository is necessary.

Note: The desperation of the nuclear waste storage is reflected in the roughly 53 million gallons of nuclear waste stored in 177 large underground tanks at DOE's Hanford Nuclear Reservation in Washington State. Of these, 149 are more than 40 years beyond their expected 25-year design life. And of these more than one-third are known or suspected to be leaking, releasing roughly 1 million gallons of waste to Hanford's surrounding soils.

Hanford lacks the storage capacity to retrieve the waste from these tanks until the waste treatment and disposal process is underway. Washington's $12.3 billion Waste Treatment Plant (WTP) continues to be designed and constructed to meet standards specific to the Yucca Mountain facility. Design and engineering for the WTP is 78% complete and construction is 48% complete.

In 2002, Congress designated Yucca Mountain as the nation's sole current repository site for deep geologic disposal of high-level radioactive waste and spent nuclear fuel. At that time, the Secretary of Energy concluded that, "The amount and quality of research the DOE has invested... done by top-flight people... is nothing short of staggering... I am convinced that the product of over 20 years, millions of hours, and four billion dollars of this research provides a sound scientific basis for concluding the site can perform safely."

Congress then directed DOE to file a license application for the Yucca Mountain site with the Nuclear

Figure 6-14. Airtight containers for nuclear waste storage

Regulatory Commission (NRC) and thereby commence a formal evaluation and licensing process overseen by the NRC.

But...in January 2010, President Obama, Secretary Chu, and DOE determined that they would withdraw with prejudice the application submitted by DOE to the NRC for a license to construct a permanent repository at Yucca Mountain, Nevada, for high-level nuclear waste and spent nuclear fuel.

Also that month, President Obama, Secretary Chu, and DOE chose to unilaterally and irrevocably terminate the Yucca Mountain repository process mandated by the Nuclear Waste Policy Act, 42 U.S.C. §§ 10101-10270. Several lawsuits were filed as a result, but these will take years to resolve, during which time, the US has no place to store its large pile of nuclear waste produced during power generation and nuclear weapons production.

Onkalo Nuclear Waste Storage Site

Now there are real plans to build an enormous bunker for permanent storage of the dangerous radioactive waste—not in the U.S., but in Finland. It consists of burying the nuclear waste deep underground, sealing the depository and throwing away the key. Literally. The intent is to keep everyone—including future civilizations— away by hiding the nuclear waste somewhere so unremarkable and unpleasant that nobody would ever think to go there, let alone dig into it.

Located on Olkiluoto Island, just off Finland's southwest coast, the underground facility known as Onkalo will hold all of the country's 5,500 tons of nuclear waste—all that is expected to be produced by the end of the century. It is designed to keep that waste secure for at least 100,000 years, taking measures and counting on making humans forget that Onkalo was ever there.

Onkalo is intended to safely and permanently store high-level waste (HLW), which consists of spent nuclear fuel and some equally dangerous decay products. This residue emits dangerous types and levels of radiation for tens of thousands of years. Over 300,000 tons of this stuff now exists around the world and about 12,000 more tons are produced annually. It is also expected that numbers will increase significantly in the years to come.

This maze of deep underground bunkers is being carved from impermeable rock, in geologically stable zones, where the waste can be permanently buried and then redundantly sealed.

At a $3 billion price tag thus far, Onkalo will start accepting nuclear waste in 2020, while construction of new tunnels will continue as needed until the facility is shut down and sealed forever.

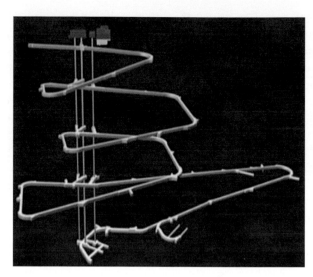

Figure 6-15. The Onkalo's maze of underground tunnels

Will this grandiose plan work? Who knows? It is too big, expensive, complex, and ambitious to be properly assessed with the available information, so we just have to wait and see.

Nuclear Waste Recycling

One of the best and most plausible solutions to reducing nuclear waste is its recycling and reuse. To this purpose, in 2000 the United States and Russia signed a bilateral agreement, committing to eliminate 34 metric tons of surplus military plutonium produced during the Cold War by recycling it as fuel for civil nuclear applications.

In 2008, the Department of Energy made an agreement with a joint venture created by the AREVA and SHAW groups for the construction of a mixed oxide (MOX) fuel production plant.

The effort consists of two parts:

1. Construction and operation of a pit disassembly and conversion facility (PDCF), where nuclear warheads are dismantled and where the recovered metal is converted into plutonium oxide, and

2. Construction and operation of a fuel fabrication plant, Mixed-oxide Fuel Fabrication Facility (MFFF), where plutonium oxide is mixed with uranium oxide to make MOX assemblies.

The 600,000 ft^2 MFFF plant currently under construction at the Savannah River Site near Aiken, SC, is on track to be completed by its target date of 2016. Its purpose, upon completion, is to reduce the surplus weapons grade plutonium and provide fuel for commercial plants. If successful, the MOX process might be the ultimate answer both politically and environmentally for a safer nuclear power industry.

It is estimated that when the MFFF operation is complete, enough electricity could be generated to power all households in South Carolina for up to 20 years. The project is overseen by the National Nuclear Security Administration (NNSA), via third-party contractors.

In addition, there will be a waste solidification building, and a pit disassembly and conversion unit that are pivotal in the attempt to shrink the waste plutonium mountain.

At the MFFF facility, surplus plutonium will be processed and blended with depleted uranium oxide to make mixed oxide fuel that will be used as new fuel for nuclear plants.

The waste solidification building and the pit disassembly and conversion unit are vital cogs in the wheel of using the plutonium as a resource, the waste unit is forecasted to treat 150,000 gallons of transuranic waste, and approximately 600,000 gallons of low level radioactive waste from the MFFF and pit disassembly buildings.

Note: As an additional benefit to national and international security, once the MOX fuel has been irradiated by the commercial reactors, the plutonium can no longer be used for nuclear weapons activity.

Most importantly, MOX facilities would provide environmental safety for future generations, by converting potentially environmentally dangerous radioactive materials into safe commercial nuclear fuel.

Abandoned Uranium Mines

Finally, we need to mention this important issue, which has been causing problems for the locals for decades. The uranium mining industry began in the 1940s primarily to produce uranium for weapons and later for nuclear fuel. Although there are about 4,000 mines with documented production, a database compiled by EPA, with information provided by other federal, state, and tribal agencies, includes 15,000 mine locations with uranium related activities in 14 western states. Most of those locations are found in Colorado, Utah, New Mexico, Arizona, and Wyoming, with about 75% of those on federal and tribal lands.

The majority of these sites were conventional (open pit and underground) mines. The mining of uranium ores by both underground and surface methods produces large amounts of bulk waste material, including bore hole drill cuttings, excavated top soil, barren overburden rock, weakly uranium-enriched waste rock, and subgrade ores (or protore).

At some abandoned mine sites, ore enriched with uranium was left on site when prices fell, while transfer stations at some distance from remote mines may contain residual radioactive soil and rock without any visible facilities to mark their location.

While most pose minimal radiation risk to the public, since exposure is most likely to be short and intermittent (e.g., visitation, recreation), they may pose other physical safety risks.

No single national AML program exists; rather, several authorities and multiple departments and agencies address AML sites as part of broader programs. Similarly, funding for remediation projects is spread among separate appropriations for participating departments and agencies. Federal and state agencies are doing more than ever before to display their spatial data and exchange information. Data from the former U.S. Bureau of Mines and the U.S. Geological Survey provide the foundation of most mine inventories.

Funding for remediation projects is spread among separate appropriations for participating departments and agencies. Federal and state agencies are doing more than ever before to display their spatial data and exchange information, but this is where the effort dwindles in the bureaucratic mix of other pressing issues.

There is a "polluter pays" principle that requires the federal government, where possible, to compel responsible parties to clean up their sites or help cover the costs. Priorities focus on water quality and sites involving release or potential release of hazardous substances. The enforcement of this policy is partial and little information exists on its compliance by the coal industry, as evidenced by the large number of abandoned mines that are just that—abandoned. No one is held responsible for the environmental damage, human health issues, and other consequences, resulting from abandoned mines

Basically, the uranium mining industry leaves its dangerous mark everywhere, even after operations have ceased. Abandoned underground and surface mines are the longest lasting, and least controlled, damaging the environment and the people in it.

Although there are signs of revival of the movement to secure abandoned mines, we do not see it being resolved anytime soon. So it will be up to the future generations to clean up our mess—abandoned uranium mines included.

Environmental Impact

The greatest environmental impact of nuclear plants is during and after a major accident such as at Chernobyl and Fukushima. Although such accidents do not happen every day, they cannot be ignored. The negative effects of their radiation releases are so widespread and so devastating that environmental damages from other power plants are negligible in comparison.

Figure 6-16 shows some of the negative effects caused by radiation releases during and after nuclear accidents. Air, soil, and water in the immediate location are no longer useful and are very dangerous for use for many, many years. Animals and humans directly or indirectly in contact with any of the contaminated areas become ill, some suffer extended illnesses and some never recover. Many of the affected animals and people die as result of the radiation.

But there are other environmental effects, during the entire cradle-to-grave life cycle of nuclear power plants.

- Building a nuclear power plant requires huge amounts of concrete, steel, and other materials. All these materials are made with the help of fossil fuels—lots of them. During the production and transport of these materials a lot of gas, liquid and particulate pollution was generated as well, including huge amounts of CO_2 pollution.

- Many pieces of large and expensive equipment are delivered and installed at a new nuclear plant. This equipment is also made out of metals and plastics, which use a lot of energy, and leave a lot of pollution during manufacturing and transport to the power plant.

- It has been estimated that the steel and concrete production and transport for the construction of a 1 GW nuclear power station has a carbon footprint of roughly 300,000 tons CO_2. Spreading the CO_2 over a 25-year reactor life we come up with almost 1.5 grams of CO_2 emitted per each kWh of electricity generated.

- According to industry estimates, the total carbon footprint of a nuclear power plant—including construction, fuel mining, transport, processing, operational emissions, fuel disposal, plant decommissioning, and operations—is about 40 grams of CO_2 for every kWh generated. This is exactly 10 times less than the fossil-fuel average of 400 grams of CO_2 emissions per kWh generation.

It may sound like not much, but considering that a typical 1 GW nuclear power plant generates 1,000,000 kWh every hour, this is the equivalent of 40 tons (or 88,000 lbs.) of CO_2 and other GHGs emitted every hour. This is over 2 million lbs every day, over 63 million every month, and over 760 million lbs every year.

Almost a billion lbs GHGs are emitted by one average-size nuclear plant annually. Who knew...?

This amount is 10 times much less than the emissions from a similar size coal-fired power plant, which emits 4-5 million tons GHGs annually. And yet, when multiplied by the number of nuclear power plants around the world, this becomes a significant number that cannot and should not be ignored.

External Costs

The report of a major European study of the external costs of various fuel cycles, focusing on coal and nuclear, ExternE, shows that in clear cash terms nuclear energy incurs about one tenth of the costs of coal, under normal use and in the absence of accidents.

The external costs are defined as those actually incurred in relation to health issues and environmental damage. These costs are quantifiable, but are presently not built into the cost of the electricity. If these costs were in fact included, the EU price of electricity from coal would double and that from natural gas would increase 30%. Nuclear power

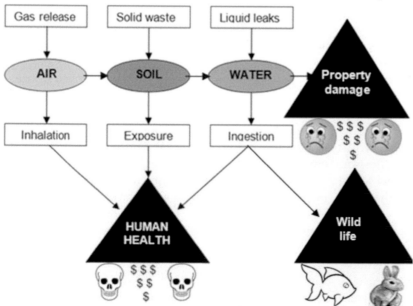

Figure 6-16. The effects of radioactive release

will also be affected in a great way by external cost increases, but even the most aggressive estimates fall short of including all the external costs of accidents.

Figure 6-17 shows the total estimated cost of recovery and remediation from the Fukushima nuclear plant disaster to rise to over $500 billion and those from the Chernobyl disaster are estimated to over $350 billion. The partial damages resulting from the Three Mile Island accident are estimated at $1.0 billion, while those from oil, natural gas, and coal-fired power plants are well below the billion range (thus cannot be seen in Figure 6-17 due to comparably small size).

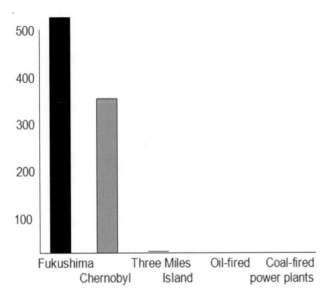

Figure 6-17. Total cost of damages and remediation of nuclear accidents (in $/billion)

Nuclear plant accidents are not well understood, nor is there a standardized way to handle them. This is because things are complicated by a number of factors, some of which are not included in the external costs calculations as used presently. The external cost of environmental damage, death, pain and suffering caused by nuclear plants accidents are hard to even estimate, let alone assign a hard number.

How do you estimate the cost of damages caused by the Chernobyl accident in Russia and surrounding countries. Or Fukushima? What is human suffering and loss of life worth? How do you calculate the value of being forcefully evicted from your home and moved to a strange place forever?

We still have no proven way of putting a value on all the total effects from nuclear accidents, since we are unable (and perhaps unwilling) to put a standard value on direct and indirect damages to property and human

suffering, life quality reduction, evictions, local and global environmental damages, global warming, and many other related factors.

A nuclear accident causes the complete devastation of a large land mass; damaging its animal life, crops, water sources etc. Most accidents caused by the conventional power generators are usually limited to the immediate land and easily localized and recovered. A nuclear accident not only damages the local land making it into waste land forever, but it also emits clouds of radiation that spread and damage vast areas around the plant—damages which last a long time. In most cases, the radiated areas are rendered useless for many, many years—maybe forever.

But the most important and un-reconcilable damage is the fact that thousands of animals and people have been made ill and killed during and after nuclear plant accidents. How do you estimate the cost of one human life...or thousands of them after Fukushima? Or the pain and suffering of over a million after Chernobyl?

If these enormous costs are somehow compiled in their totality and spread over the entire global nuclear industry, it may in fact prove to be the most expensive energy source in use today. This, however, is the dark secret of the industry, which—regardless of its noble intentions—must be taken out of the closet, discussed and resolved once and for all.

Considering the complexity of the matter, as well as the place and importance of the nuclear industry in our socio-political landscape, we don't see this issue completely resolved anytime soon.

Experts' conclusions on the pros and cons of nuclear power vary, but here are several things we can take from the above discussion:

- Nuclear power is the mightiest force on Earth generating unequaled amounts of electricity, but it can also bring unequaled destruction and death.

- Because of its huge and uncontrollable destructive potential, nuclear power is not cost-competitive when all costs of completely insuring against nuclear disasters are considered in the final power price.

- Nuclear disaster damages are socialized in most cases, except in the U.S. Usually, however, major nuclear disaster costs cannot be covered completely in many countries.

- Developing countries lack the finances to completely cover nuclear accidents, which could also spill into neighboring countries, thus increasing their liability. The U.S. is an exception in this case too, since there is only a very slight possibility of this happening here. Lucky devil…

- Nuclear power is a significant part of our energy and environmental present and future, but we cannot rely on it alone. Nuclear plant safety, as well as long-term availability and the rising price of uranium are major issues that must be considered in the nuclear discussion.

INTERNATIONAL DEVELOPMENTS

Figure 6-18 shows that the U.S. was the world's largest nuclear producer 30 years ago, generating almost half of the world's nuclear power. The U.S. share steadily declined and, if the present trend continues, it is projected to hit rock bottom by 2030. Things might get even more difficult after that.

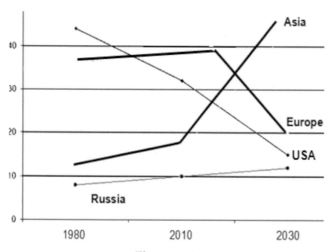

Figure 6-18.
Global nuclear power 1980-2030 (in percent of total)

The European nuclear power follows roughly the same path as the U.S. Russia has been slowly increasing its nuclear capacity, which is expected to continue until 2030 at a faster pace.

Asia's nuclear power, on the other hand, had a slow start, and is still on a slow rise, but is expected to increase exponentially by 2030, reaching much higher levels than the U.S., Europe and Russia put together.

By 2030 China is expected to have the major share of the nuclear power installations in Asia and only second in the world after Russia, with about 130 GWe, following Europe's 140 GWe.

Things may get quite bad for the U.S.'s nuclear in-

dustry in the near future, when due to financial, market, and public and regulatory pressures the industry may shrink to 50 GW. Some analyses go as far as predicting a complete collapse of the sector with less than 10 GWe nuclear reactors left standing and operating in the U.S. by 2050, in stark contrast to the growing nuclear ambitions of the other global nuclear power leaders.

Presently, the picture is a bit confusing, since the nuclear energy bandwagon rolled downhill in 2011 in the aftermath of the Fukushima disaster. Most nuclear power plants in Japan and Germany were shut down for a period of time, and now the nuclear plans of most countries are in revision mode.

Figure 6-19 shows the approximate nuclear power generation capacity (in GW) of the world's major nuclear powers in 2012-2013, and the percent of nuclear power generation in the domestic energy mix. Russia and China are still low on the charts, but the balance is changing fast, and as a result the new global nuclear lineup (the future estimates) are changing too.

Country	GW	Domestic
United States	102.1	19.0%
France	63.1	74.85%
Japan	44.2	18.1%
Russia	23.6	17.8%
South Korea	20.7	30.4%
Canada	14.1	15.3%
Ukraine	13.1	46.2%
China	12.1	2.0%
Germany	12.1	16.1%
United Kingdom	9.9	18.1%
Sweden	9.4	38.1%
Spain	7.6	20.5%
Belgium	5.9	51.0%
Taiwan	5.0	18.4%
India	4.8	3.6%

Figure 6-19 Nuclear power around the world

Presently Russia is building, or planning to build, about 37% of the world's nuclear reactors, followed by China with 28%, Korea with 10%, France and the U.S. with 8% and 7% respectively, India 4%, and the rest of the world with the remaining 6%.

These are interesting and plausible plans, but are they achievable? In more detail:

Russia

In 2003, Russia set a policy priority for reduction in natural gas based power generation by doubling its nuclear power generation capacity by 2020.

In 2006 the Federal Atomic Energy Agency of Russia (Rosatom) announced targets for future nuclear power generation; providing 23% of electricity needs by 2020 and 25% by 2030. This is to be done by increasing the number of reactors in operation from 31 to 59.

In 2007 the Russian Parliament adopted the law "On the peculiarities of the management and disposition of the property and shares of organizations using nuclear energy and on relevant changes to some legislative acts of the Russian Federation," which created Atomenergoprom—a holding company for all Russian civil nuclear industry, including Energoatom, nuclear fuel producer and supplier TVEL, uranium trader Tekhsnabexport (Tenex) and nuclear facilities constructor Atomstroyexport.

In 2010 the government approved the federal target program (FTP) "New-generation nuclear energy technologies for the period 2010-2015 and up to 2020" designed to bring a new technology platform for the nuclear power industry based on fast neutron reactors. It anticipated about $5 billion to 2020 out of the federal budget, including $2.5 billion for fast reactors. Subsequent announcements started to allocate funds among three types: BREST, SVBR and continuing R&D on sodium cooled types.

In 2012 total electricity generated by nuclear power plants in Russia was 180 TWh, or about 18% of all power generation. The installed gross capacity of Russian nuclear reactors is now nearly 25 GW.

Starting 2020-25 Russia envisions fast neutron power reactors playing an increasing role in the country, though these will probably be new designs such as BREST with a single core and no blanket assembly for plutonium production.

The latest Federal Target Program (FTP) envisions a 25-30% nuclear share in electricity supply by 2030, 45-50% in 2050 and 70-80% by the end of the century. Ambitious plans, no doubt, but there were and still are many obstacles. A number of projects were started, but then modified or even cancelled.

For example, in 2009 Siemens announced that it would enter into a joint venture with Rosatom. A memorandum of understanding was signed with the intent to set up a joint venture with Rosatom as majority shareholder; developing Russian VVER designs, upgrading existing nuclear plants at home, and building many new nuclear power plants around the world.

Mr. Putin hailed the JV as an important achievement in furthering Russia's nuclear ambitions around the globe. But the agreement was delayed pending Siemens disengaging from Areva, and finally in September 2011 Siemens announced that it would not proceed with the JV.

Rosatom's long-term strategy now includes an overly optimistic scenario with expansion to over 90 GWe nuclear capacity by 2050, focused on building inherently safe nuclear plants using fast reactors with a closed fuel cycle and MOX or nitride fuel.

Using the experience of Chernobyl as a selling point for its current safety measures and crisis management expertise, Russia is also pursuing an ambitious plan to increase sales of Russian-built reactors overseas.

So, in addition to the domestic expansion, the Russian government is now banking on developing its nuclear exports; nuclear materials, equipment and services. They envision undertaking many projects overseas built and operated with Russian components by Russian companies.

Through its membership in the ITER project, Russia is also participating in the R&D of nuclear fusion reactors.

China

Compared with most nuclear countries, China is well ahead of the pack. It is in the midst of an energy revolution, led by a strong nuclear renaissance. China is quickly becoming self-sufficient in reactor design and construction, as well as in other aspects of the nuclear fuel cycle management.

China is quickly shaping up as the emerging giant of the nuclear industry. The newly planned nuclear plants in China include some of the world's most advanced reactors, which will give a five- or six-fold increase in nuclear capacity to 58 GWe by 2020, then 130 (and possibly 200 GWe) by 2030, and a whopping 400 GWe by 2050.

The Chinese-made AP1000 reactor units have substantially lower manufacturing and installation costs, which is a reflection of the significantly lower labor rates in the country. For example, the cost for the first two AP1000 units under construction in China was $5.3 billion, or $2,750 per installed kW. Additional four AP1000 reactors constructed in China were estimated to cost the total of $8 billion, or about $2,000 per kW installed capacity. This represents over 30% in cost reduction, which reflects both the low labor rates and the value of mass production.

Construction costs in China are expected to fall

even farther once full-scale mass production is fully underway. Another domestic CAP1400 reactor design, based on the AP1000 model started construction in 2013 with scheduled completion by 2017. Once the CAP1400 design has been installed and proven, work is scheduled for a CAP1700 design with a target construction cost of $1000 per kW installed. This is a 50% decrease in the cost of today's nuclear plant construction.

Mainland China presently has 16 nuclear power reactors in operation, 27 under construction, and even more in the planning stages—some just about to start construction. This includes China's most ambitious nuclear project, started in the fall of 2012, when China resumed construction on a "fourth generation" nuclear power plant. The construction was started in 2011, suspended for awhile in the wake of the Fukushima disaster, and is now undergoing "upgrade" work.

The Shidao Bay nuclear plant in Rongcheng, a city in eastern China's Shandong province, resumed in November, 2012, and is planned to be China's biggest nuclear project. The 6.6 GW reactors will be cooled by high-temperature gas. It might also become the world's first successfully commercialized 4th generation nuclear technology, designed as the safest and most cost effective nuclear plant ever.

The official plans, however, do not specify the level of safety of this overgrown giant. Nuclear safety information, in China, is kept secret for as long as possible. Our major concern is that the rush for more power may compromise project quality.

Europe

Europe's nuclear future is quite clear—full steam ahead, carefully watching for and taking care of the threat of nuclear accidents. Europe is densely populated, and any type of nuclear disaster is unthinkable, for the devastation would be very serious. Nuclear power is tricky, so Europe seems to be getting ready to handle anything it throws at her.

Europe is also learning from the Fukushima disaster, which is far from over. Like Chernobyl, it had a beginning, but no end that we can see. As a matter of fact, we are just beginning to understand that the cost of using the immense power of nuclear energy for our comfort is sometimes too great to even consider in terms of damages to human life, properties, and financial loss.

There are estimates that the cost of Fukushima's damages to people's health and properties, which are still rising, might get close to mind boggling $500 billion.

This high cost, added to the similarly amazingly high cost from the Chernobyl disaster which is still rising too, could bring us to an astronomical number of damages from just two nuclear accidents to over $1 trillion in actual monetary expenses. The "cost" estimates do not take directly into consideration the human suffering and loss of life during and after these accidents.

And now, looking at the damages and the related expenses, the European Commission is considering amending the laws regarding insurance of nuclear power plants on European territory. In other words, the huge dollar amount of potential damages in the densely populated European continent, in the event of the unthinkable taking place in a European reactor, will have to be paid somehow, by somebody. But how and by whom?

The European conclusion is simple: "We will continue to operate nuclear power plants for the foreseeable future, and since there is no exemption from nuclear disasters, we have to decide on the best ways to recover from, and compensate victims."

This conclusion is not a precursor for secure, healthy and happy life, is it? Our common sense response, together with that of the grand majority would be, "Who cares about the expenses when my family is in the hospital dying from nuclear radiation overdose?" Nope, this is not the ultimate solution, EU community. There must be a better way, but for now, this is all they could come up.

The proponents of nuclear energy and some experts argue that nuclear energy is still among the cheapest means to produce electricity...in most cases. This is disputed by many, claiming that the initial cost of nuclear power plants is very high. Adding to that the cost of externalities and disaster-related costs, which could be much higher, we come up with numbers that make nuclear power the most expensive technology ever.

But Europe needs electric power...a lot of it, and nuclear is one of the available solutions. This, combined with the fact that CO_2 emissions from nuclear power generation are the lowest of all energy sources makes nuclear energy a sure way of providing reliable energy, while at the same time contributing to a low-carbon energy future. Safety aside, we must agree that it is a much better option than ramping up the coal use or Russian natural gas imports.

So, everything considered in March, 2013 (two years to the day after the Fukushima nuclear accident), the European Parliament issued a resolution on the European Energy Roadmap, 2050. In it, the EU dignitaries agree that "...nuclear energy will continue to play a large and significant role in energy production on the European continent for the foreseeable future..."

The potential for nuclear disasters cannot be ignored, and according to a senior engineer at the Fukushima Daiichi, "...we have to keep thinking: what if ... " He must know something...so we must listen carefully.

At present, the potential nuclear accident costs in Europe are partly covered by a compulsory disaster insurance that the nuclear operator carries. This insurance package, however, is usually very small, and fails to include the expected costs of a mass-scale disaster. The projected $500 billion in damages resulting from the Fukushima nuclear disaster is a good guideline for the European lawmakers.

The Japanese government was caught unprepared for all eventualities of such a large-scale disaster. As a result, it was forced to socialize the damages that were not covered by the original Fukushima power plant insurance. For that purpose, it partially nationalized (or rather took over) Tepco, using its proceeds to reimburse the victims and pay for the damages. But that is not enough to cover all the damages, so the Japanese government keeps on poring taxpayers' money into the large hole in the ground where Fukushima Daiichi proudly sat several years back—a very sad case with no end or solution in sight.

This is what the European Commission is trying to avoid by modifying the insurance of nuclear disasters. Since such disasters cannot be limited by national borders, reforming nuclear plant disaster insurance makes sense on the European level.

Presently, the European nuclear insurance is governed by the Paris and Brussels Conventions, which most European countries have signed and ratified. These conventions place full liability on the operator, who must also provide insurance, but they also limit the operator's maximum liability significantly.

Based on the 2004 amendments, both the operator and the country of origin cover €1.2 billion of all disaster costs. Of this, €700 million is to be paid by the operator or its insurance, €300 million is paid by the other European convention members, and the rest by the country of origin.

If the EU nuclear plants are insured for the whole potential damage, the nuclear energy price will increase up to €2.30 per kWh.

This is huge amount, which reflects the enormity of the problem. In France, which derives most of its electricity from nuclear power, the estimates show a price increase of €0.05 per kWh, which is a more reasonable but unverified number. It also seems too small to compensate for $500 billion in damages, or similar amount.

France is nevertheless the leading nuclear power generator in Europe, and is one of the most populated countries as well. In France, a serious nuclear plant disaster like Chernobyl could wipe out half of the country. It would also most certainly bring drastic changes in the rest of the country, the surrounding countries and the world.

Recently, more realistic estimates show that the insurance cost would raise current electricity prices for the French nuclear industry by 60%, and for the German industry by at least 100%. This, translated into electricity prices for households means an increase of 25-35% in France and 50-60% in Germany. Not small change, that won't go well with the average Joe electricity consumer.

A serious nuclear disaster with the present insurance coverage will not only shut down most nuclear power plants, but would also make the nuclear electricity uncompetitive in the long run. So, to protect the nuclear industry, the Europeans have limited the operator's maximum liability, which leaves open the question, "Who is going to pay for the rest of the damages?"

Only some countries, like Germany and Finland, have set unlimited operator's liability. If the operator's insurance is insufficient to cover the losses from an accident, then the operator would be fully liable with the total of their own corporate equity.

But which company has $500 billion lying around? Even the largest European nuclear plant operators have stock market value of less than €10 billion—after selling the shirts off their backs. This is very far from the said $500 billion, so then what?

The solution the European commission came up with, following Japan's example, is to partially socialize the costs of nuclear disasters. Since the operators are unlikely to be able to cover the total costs of a larger nuclear accident, the public of the resident country, partially assisted by member countries, will pay the rest.

Several European states signed the Convention on Supplementary Compensation for Nuclear Damage, which is similar to the Brussels Convention, and as a result provides up to an additional €360 million on top of everything paid by the operator in the event of a nuclear disaster.

Noble intent, yes, but still short of covering the total costs of larger disasters, as discussed thus far. But, this is what it is; the Europeans are moving towards socialization and internationalization of nuclear disaster losses and we just have to wait and watch how that will work in case of a large nuclear accident.

Note: The U.S. uses a slightly different approach, where the risk-sharing is kept within the national borders and within the industry. It is basically implemented through the Price-Anderson Act, where the nuclear power plant operators are required to carry €375 million damages insurance for each nuclear plant.

This is still not enough, so on top of that all US nuclear plant operators need to have an available pool up to $100 million per plant. With the current 104 US reactors in operation, we have about $12 billion available for use in case of a nuclear disaster.

With that money in the bank, in addition to the fact that the U.S. is not as densely populated as some parts of Europe or Japan, does the U.S. seem better prepared to handle nuclear disasters?

Europe is still searching for the right path to take with nuclear energy. The quick trigger reaction of the German government to shut down ALL nuclear power stations in the country after the Fukushima disaster, and the following equally quick-triggered decision to reopen ALL nuclear power plants shortly thereafter, is a clear indication that the nuclear power issue is a sensitive one on the continent—a teeter-totter of sorts…

It also shows that no proven nuclear accidents procedures are in place, and that the proverbial "seat-of-the-pants" management approach is used by the EU regulators and politicians.

This reminds us of a sail boat driven by uncertain and uncontrollable winds promising to blow the nuclear power sails into new and untested territory, hopefully not into the shore cliffs.

Case Study: UK's New Nuclear Plant

The UK has been planning to build a great new nuclear plant for awhile now, and things looked good for awhile. The project finally hit a brick wall of resistance on the part of the European Commission, questioning the project's legality.

The UK signed a deal with EDF—a French energy company—to build UK's first in 20 years nuclear power plant at Hinkley Point in Somerset. The new large nuclear plant could generate about 3.0 GW of electric power, enough to power over five million homes.

The new plant is seen as a key to the government's plans to reduce UK's energy sector emissions, while at the same time providing reliable and cheap electricity to its people.

The deal obligates the government to guarantee £90/MWh for power delivered by the plant, and to underwrite £10 billion in loans to build the plant. The loans would be basically subsidized by the public through a levy on their energy bills. That seems to pave a new way of doing nuclear business in Europe, since most renewable energy projects also get a similar deal across the continent.

But the EU has strict criteria for what and when a government is allowed to subsidize an industry—any industry, including nuclear power. This provision is known as the "state aid," the rules of which are intended to prevent governments from giving unfair advantage to some industries.

So, the European Commission is stalling over the deal as it is unconvinced that the plan fits the bill, no pun intended. It is not sure that the plan is fair, and/or that the new nuclear plant will help the EU meet its long-term economic and environmental goals—which is one way the Commission allows governments to subsidize particular industries and projects.

Basically, the Commission exempts projects from the state aid rules if they help the EU hit its emission reduction target set for 2020, or if they increase the security of the EU's energy supply, and/or only if they wouldn't get built without help from the government.

The Commission disagreed with the UK's arguments in the above areas, stating the UK could reduce emissions to the same extent, and at the same rate, in other ways. It is concerned that subsidizing Hinkley Point would interfere with the development of alternative energy technologies such as wind and solar, which is seen as unfair. The commission also objects to the claim that the Hinkley Point plant is needed to secure the EU's energy supply, among other things.

It will be interesting to follow the developments of this case, but whatever happens with the Hinkley Point nuclear plant, the arguments and counter-arguments in Table 6-3 reflect a changing attitude in Europe.

There is an increased awareness in Europe that nuclear power is not the best solution to the long-term energy and environmental problems of the continent. With its tight spaces and highly populated areas, Europe is the worst place for a nuclear disaster. Any serious nuclear accident close to a populated center would cause untold damage to everything and everybody around it. The EU nuclear industry would take a long, long time, if ever, recovering from such a doomsday scenario.

Japan

Japan's nuclear power history is one of the greatest stories of the 21st century. It is full of great successes and even greater failures. Japan was, and still is, one of the

Table 6-3. Hinkley Point nuclear plant arguments

UK arguments	EU Commission response
The Hinkley Point nuclear plant will help the UK to reduce its greenhouse gas emissions on the long run.	It's not clear the UK needs more nuclear power to do this, since renewable energy could do the same job and better
The Hinkley Point nuclear plant will help secure energy supply for the UK and EU.	The power plant won't come online until 2023, so the UK must be making other plans to help secure supply by then, and that it has also underestimated the extent to which European interconnection will help secure energy supply on the continent.
The new nuclear plant will help the UK and EU achieve the broader goals, so it should be allowed to receive state aid.	Nuclear power is not an 'immature' technology, so it doesn't qualify for state aid. Also, other government policies make nuclear power an attractive investment, so new nuclear plants can be built without state aid, like these in France and Finland.

world's largest nuclear power generators. March, 2011 changed all that. Just like the *Titanic*, proudly cruising the ocean, Japan's nuclear industry hit a huge iceberg that changed its faith forever.

Almost instantly, Japan sank into a state of nuclear nightmare. All nuclear power plants in the country were shut down from fear of another accident, and soon after that official government plans were revealed to mothball the entire nuclear fleet. But that didn't last long...

On April 11, 2014—three years and a month to the day from the Fukushima accident—Japan's government reversed its previous plans to systematically mothball all nuclear power plants in the country. The move was sudden and unexpected, so it is quite unpopular with the majority of Japan's wary public in the wake of the Fukushima disaster.

Even if the plan is accepted and implemented, it might be too little too late for Japan's nuclear industry. There are about $50 billion in losses, which are rising by the day as the recovery work at Fukushima and upgrades at most other nuclear power plants continue.

As a consequence, two utilities were forced to apply for government assistance a week or two before the government decision to restart the nuclear plants was made. Nuclear plant owners have already spent almost $90 billion on replacement fossil fuels to keep the grid running. In addition, an estimated $16 billion have been spent on nuclear plant upgrades, to meet the new safety guidelines.

Note: This is still peanuts, compared with Fukushima's ongoing debacle, where the radiation escape and the daily work of about 4,000 people to contain it contin-

ue non-stop day and night.

The precise value of the total damage to the local towns, agricultural lands, businesses, homes and other properties in the roughly 300 square miles of exclusion zones around the plant has not been established.

There are estimates that the total economic loss ranges from $300 to $500 billion. This is an unprecedented amount of money that is hard to imagine, let alone justify.

And on top of that, the human costs have not been tallied up yet either. In 2012, Fukushima officials estimated that about 160,000 people had been evicted from the exclusion zones. All these people lost their homes and virtually all their possessions forever.

Thus far, most of the these thousands of victims have received only a small compensation, in most cases enough to cover their costs of living as evacuees. And many are forced to make mortgage payments on the homes inside the exclusion zone, which they are not even allowed to visit. Amazingly, many of these people have not yet been told that they can never return to their homes.

Instead, radioactive cesium is inhabiting their homes in the exclusion zone, and is planning to stay there for hundreds of years, thus replacing the human inhabitants virtually forever.

Note: Cesium-137 has a half-life of 30 years, and since it takes about 10 half-lives for any radionuclide to disappear completely, it will keep the exclusion zone contaminated and void of wildlife and people for the next several centuries.

As a result of the increasing financial burden, about 33 (or 2/3) of the 48 idled nuclear reactors will not be restarted because of the high maintenance and upgrade costs.

Another big problem for many nuclear plants in Japan is the local opposition, which is afraid of high seismic risks. Recent polls put opposition to nuclear restarts at about two-to-one over support. An Asahi newspaper poll in June 2014 found that nearly 80 percent of those surveyed supported a gradual exit from nuclear power.

The Japanese utilities will have a hard time fighting the rising financial and opposition tides, so they will have to write off some (and even most) of their nuclear assets and find a different way to move on.

The new government energy plan defines nuclear as a "very important baseload power source," which is needed to feed power into the grid constantly to meet the national power demand requirement. But the plan does not specify the share of nuclear power in the nation's energy mix.

So, because of the slim prospects for nuclear power to provide a major share of the energy mix, flexibility will rule, and with it the nuclear baseload concept flies out the window. The government also named coal and hydro power as baseload power sources, so in the absence of nuclear power, the fossils are back on Japan's energy throne.

For now the government energy plan makes it clear that the reliance on nuclear power in the country will be reduced in the short term through a variety of measures, most likely by increased use of fossil fuels.

The government also plans to continue working on reprocessing nuclear fuel at the Rokkasho facility in northern Japan and will maintain storage facilities for used nuclear fuel. The Japanese government will evaluate the situation periodically and will come up with the "ideal energy mix" within the next two or three years.

The energy industry wants the plan to be implemented steadily, as a core national policy, with nuclear power use based on the premise of ensuring safety first. Meanwhile, to achieve the goals in the plan, Japan will increase fossils imports and accelerate building domestic renewable energy capacity.

Previous government plans had set a target for renewable energy sources to contribute 13.5 percent of total power generation in 2020 and around 20 percent in 2030. Renewable energy sources, including hydro power, solar, and wind contributed around 10 percent of the country's energy by 2012. The set targets remain Japan's policy...for the time being, but things might change one way or another in Japan's fluid energy future.

After all is said and done, nuclear power remains an integral part of the present and future of the world's energy security. Because of that, China, India, and many other countries are seriously contemplating significant increases in nuclear power generation in the short term.

The long-term future of nuclear power, barring any additional Fukushima-like disasters, looks even better. The embattled nuclear industry is here to stay and will fight to retain its position in the global energy mix.

Update 2014

Japan's nuclear problems started in March 2011 when the Fukushima power plant exploded after an earthquake triggered a huge tsunami that overwhelmed the plant's safety systems.

After that accident, the Japanese government shut down all nuclear plants in the country and started thinking of the alternatives. Life goes on in Japan even after the Fukushima devastation.

Renewables + coal (mostly coal) were thought to be a reasonable substitute for nuclear power. So, coal imports increased, along with the pollution, proving to be an expensive, environmentally damaging, and unsustainable solution. It simply didn't work quite as planned and brought more questions than answers.

Looking to the future, Japan is doubling efforts to upgrade, improve and restart the nuclear plants fleet. This is an expensive and long process with uncertain outcome. Would the public agree that the new and improved nuclear plants are safe enough? Would it believe the nuclear industry that no other Fukushima is coming to Japan's shores?

There was similar hesitation around the world at the time, but as soon as Fukushima images faded away, nuclear power plants resumed their proliferation globally. Presently, 60 new nuclear power plants are being constructed worldwide, most of them in Asia, and 160 new power reactors with a total net capacity of nearly 200 GW are planned for construction in the near future.

India is a good example of the new situation in the nuclear field. Here the vision of becoming a world leader in nuclear technology due to its expertise in fast reactors and the thorium fuel cycle is flourishing. India counts on its immense thorium ore reserves to become energy sufficient in the future.

Unfortunately, thorium-based nuclear power reactors are not fully developed as yet, so India's thorium nuclear future—as bright as it might look on paper—must wait several decades.

Instead, India started a new Russian-built nuclear

power plant at Kudankulam, which is claimed to be one of the safest nuclear power facilities in the world. It is so safe, according to the designers and owners, that it could resist even the strongest of tornadoes or even direct impact by an aircraft.

And this is just the start of a trend in the country of 1.2 billion people, most of whom desperately need electric power. So, India is planning to build almost 500 GW of new nuclear power by 2050. But wait...this is 200-300 new nuclear plants, or more than the entire nuclear power capacity in the world today, which is leveled at about 400 GW.

Crazy, right? This also means that India has to build 1-2 new nuclear power plant every month for the duration. And they haven't started yet...

But looking back we see that in one decade of the 1980s, about 220 new nuclear power reactors were started up around the world. This is an average of one power plant every 15 20 days. During that time, one new nuclear plant came online every 75-80 days in the U.S. alone.

This period of nuclear craziness did not last very long, but proved that crazy or not, nuclear power can be ramped-up very quickly. This also means that it is not unthinkable, or impossible, for India (and/or China) to build one nuclear power plant a month for a few decades.

Considering the fact that over 600 million people in India do not have access to electricity, which is almost twice the entire U.S. population, we can clearly see the need and the urgency.

For all these people to get electric power by 2050, India will have to massively expand its power production capacity. Nuclear is a major part of this equation for India and all other developing countries, since it is the most reliable, cleanest, and cost-competitive power source. This will also allow them to avoid becoming overly dependent on outside energy sources.

Nuclear power is very important in India and all over the world. Would it be the solution to our energy problems, or would it put an end to the world as we know it? This the (big) question.

HYDRO POWER
(Here Today, Gone Tomorrow)

Hydropower is the cleanest, safest, and cheapest form of electricity today. Its growth, however, is limited by location, water availability, and the need to construct huge dams and water storage reservoirs.

Anco Blazev

Introduction

Hydropower, in contrast with nuclear power, is the safest and friendliest of all power generators available today. Although it also can kill people when dams collapse and/or when other disasters happen around hydropower facilities, these accidents are much more predictable and avoidable in most cases. The electricity from hydropower plants is also the cleanest and cheapest available today.

Because of these very important reasons—large quantities of reliable, safe, clean, and cost effective electric power generation—hydropower is on the priority list of most politicians around the world.

The IEA's Energy Technology Perspectives 2010 BLUE map scenario (which sets the goal of halving global energy-related CO_2 emissions by 2050 as compared to 2005 levels) estimates that hydro "could" generate over 6,000 terawatt-hours of electricity in 2050, or roughly twice as much as it generates today. This would add to our energy security, while avoiding the emission of millions of tons of GHGs from coal- and gas-burning power plants.

This is the good news, but to complete the picture we must add, "God willing...," because the "could generate" in the above IEA estimate is not the same as "will generate."

Yes, technologically speaking, we are capable of doubling and tripling the hydropower generation around the world; no problem. But to do that we need large quantities of water—much more than is available. We cannot control the exact location and/or quantity of available water suitable for hydropower plants. Not today, not tomorrow!

Water is an essential element which we have not been able, and will never be able, to control. Those who have it, have it...while those who don't, just don't.

Recently, the number of fortunate people around the world who have enough water has been decreasing, while the number of those who don't have enough water is proportionately increasing. Worse, it is also becoming very likely that those who don't have water now will never have it.

Water is the liquid gold of the world, they say, and this is becoming more so by the day.

There is not enough water for drinking in some areas, let alone for hydropower use. There is too much water in other areas, but it is usually not suitable for hydropower. Go figure! And to top it off, future weather and climate forecasts predict increased frequency and duration of draughts, which threaten to curtail the existing, and severely limit new, hydropower plants construction.

The U.S. Southwest is a good example of the changing situation with deepening and prolonged draughts and water shortages. Other areas around the world are in even worse shape. In many places water is actually more expensive than gold. So building a hydro dam or any water-using installation is not only overly expensive, but is also practically impossible even for the richest of the rich; for they cannot make rain no matter how much money they have.

On the other hand, during the wet periods of the year, some areas get enormous amounts of water in the form of prolonged rainfall. While this could be good or bad for the locals, the form and periodicity of available water are usually simply unsuitable for hydropower generation.

Another big problem with hydropower is the fact that there are limited numbers of locations globally that lend themselves to building large hydro dams, or river damming, as needed for safe and efficient hydropower generation. Large hydro dams require huge areas of land to be submerged, which causes enormous environmental damage. It also forces thousands of locals to leave their homes. These people are transferred permanently to new areas, where they must start a new life, which leads to protests and painful existence and is confronted by increasing opposition.

Running river water hydropower generation is also restricted to locations that are suitable for hydropower generation. The river's surface level, contamination content, seasonal variations, currents flows and speed, and international regulations and requirements limit the number of river dams.

River dams also restrict commercial traffic along river routes and create navigational hazards, which further limits their number.

Background

We all agree that water is the liquid gold of the Earth. Fortunately, it is available in varying quantities all over the world. Unfortunately, it is getting severely limited in some areas, where even drinking water is becoming scarce, and where hydropower would be impossible.

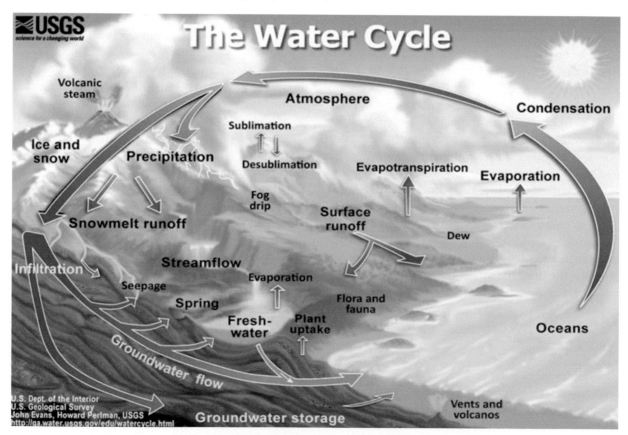

Figure 6-20. The natural water cycle

Although ¾ of the globe is covered by water, most of it is in the oceans.

Salty, ocean water is useless for human consumption, and/or most any human use. The type of water that is useful for human consumption and other practical uses—including hydropower—falls mostly from the sky in the form of rain and snow.

In areas low in rain and snowfall, water is getting steadily scarcer. Some areas, such as the deserts, are plagued by draughts which are getting increasingly longer in duration and severity.

The world's total water supply of about 332.5 million cubic miles of surface water is over 96 percent saline—ocean water. Of the total freshwater, over 68 percent is locked up in ice and glaciers, so we have no access to it. Another 30 percent of freshwater is deep in the ground.

Fresh surface-water sources, such as rivers and lakes, only constitute about 22,300 cubic miles, or about 1/150th of one percent (0.15%) of the total available water on Earth. Yet, rivers and lakes are the major source of available water people use everyday.

Water is not created nor lost in our close global system. It only changes from one state of matter to another (from liquid to gas and ice). It also travels from place to place, and we have no control over any of its properties or movements.

There is also a trend of deterioration of quality along with quantity of water available for human use, which is getting to a critical point in many places. The lack of water and its contamination cause illnesses and death to the locals in many places.

On top of the usual natural events that control the quantity and quality of water, recent climate and environmental changes are making our liquid gold even more scarce and unequally distributed, causing some areas to get flooded, while others do not see a drop of rain in years.

This is raising a number of questions about the future use of water for power generation in general, because all power plants (not just hydropower plants) need a lot of water—for cooling during the electricity generating process, in the case of coal- and gas-fired power plants.

Hydropower power plants need much more water than any other power generator, since water is their fuel; their energy source. Although water is not totally wasted at hydropower plants, a lot of it is diverted, evaporated, and otherwise made unavailable for human use in the huge storage reservoirs after a hydro dam has been constructed and put in operation.

Nevertheless, we still depend on hydropower to supply a major portion of our electric power. Since for the most part, the U.S. has already built its hydropower infrastructure, which is an integral part of our energy security, we just need to make sure that it is used safely and efficiently in the long run.

We will not see many, if any, new large hydropower projects in the U.S. anytime soon, so we need to do the best we can with what we have. We also need to be ready to face the reality of future water shortages affecting our rivers and hydro dam water storage reservoirs.

First, we must have a clear understanding of the advantages and disadvantages of hydropower, which is needed for making rational decisions.

Renewable, or Non-Renewable?

Water that fuels hydro-power generation is not considered (officially) a renewable energy source.

This is a major problem for the industry, so the issue was brought up before the U.S. Congress for a vote. The decision might affect the current hydropower industry, but will have a much more profound effect on the future hydropower generation in the U.S.

That water is a renewable energy source, however, is debatable regardless of the Congressional decision. There are many ways to look at the global water flow, storage and use, including the damage and waste that come as a result of these activities.

The most important issue today is the expected increase of frequency, duration, and severity of draughts around the world. Without enough water to fill the upstream (above the dams) water storage reservoirs, hydroelectric power plants are no more than a monument of human endeavor. A tourist attraction of sorts that is not capable of generating much revenue, or electricity.

This long-term uncertainty combined with the life-giving importance of water, makes all issues surrounding it controversial. One's opinion depends on which side of the equation s/he sits.

Power companies and some politicians are demanding hydropower to be recognized as a renewable energy source, while the locals, environmentalists and others argue that there is not enough water at the right places for it to be considered and used as renewable.

Hydropower Basics

Hydropower is a form of electrical energy generated by turbines driven by flowing water in rivers or from reservoirs behind man-made dams. Hydropower represents the largest share of renewable electricity production. It was second only to wind power for new-built capacities in 2010.

Note: Hydropower, usually refers to the potential energy contained in water (dams, rivers, and streams). The energy contained in ocean waters is subject to a different type of technology which we review later on in this chapter.

Hydropower can also be viewed as the energy created by the Earth's hydrologic cycle, which is ultimately driven by the sun, making it an indirect form of solar energy. Energy contained in sunlight evaporates water from the oceans and deposits it on land in the form of rain and snow, thus forming and maintaining thousands of streams, rivers, and lakes on Earth.

Some of the rain water is absorbed by the ground, while the part that is not absorbed runs off the land into the ocean by means of streams and rivers. Some of the water in the long journey to the ocean is lost to evaporation, irrigation, refilling the water table etc., while the water reaching the ocean repeats the evaporation cycle.

Most often, hydroelectric plants are built along the rivers, where they generate power by the flowing water, or by releasing water stored behind concrete dams built across the river to turn power generating turbines. The power plants capture the energy released by water falling through a turbine which converts the water's energy into mechanical power. The mechanical energy of the rotating turbines then drives generators to produce electricity.

Hydropower from dams accounts for approximately 75% of the U.S.'s total renewable electricity generation, making it the leading renewable energy source of electric power.

The annual hydropower output is equivalent to the energy produced from burning 200 million barrels of heating oil. There are estimates that more than 200 million tons of CO_2 emissions are avoided in the U.S. annually because of hydropower generation (as replacement for coal, oil, and gas power generation).

Let's take a close look at the details of hydropower generation and use, and let the reader be the judge.

Hydropower Generation

Hydroelectric power plants produce electricity by using a power source (water in this case) to turn a turbine, which then spins an electric generator that produces electricity. While a coal- or gas-fired power plant uses steam to turn the turbine blades, the hydroelectric plant uses falling water to turn the turbine and generate electricity.

The results are the same; electric power flows from the facility into the grid. A distinguishable exception in case of a coal-fired plant is that there is a significant emission of toxic gasses, liquid and solid waste, in addition to the electrical flow.

Hydroelectric power is generated by the gravitational force of falling water, so the capacity to produce energy is dependent on both the amount of flow and the height from which it falls. The water behind the dam accumulates huge amounts of potential energy, which is transformed into mechanical energy when the water

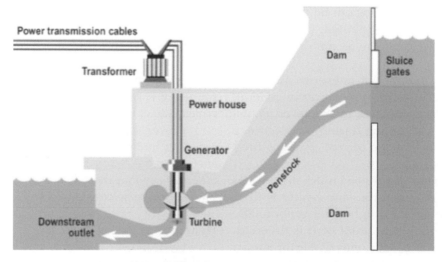

Figure 6-21. Hydroelectric power plant at a dam

rushes down the sluice and strikes the rotary blades of the turbine.

The power available from falling water can be calculated from the flow rate and density of water, the height of fall, and the local acceleration due to gravity, as follows:

$$P = \eta \rho Q g h$$

Where

P is power generated by falling water in watts

η is the dimensionless efficiency of the turbine

ρ is the density of water in kilograms per cubic meter

Q is the flow in cubic meters per second

g is the acceleration due to gravity

h is the height difference between inlet and outlet

The turbine's rotation spins electromagnets which generate current in stationary coils of wire. Finally, the current is put through a transformer where the voltage is increased for long-distance transmission over power lines.

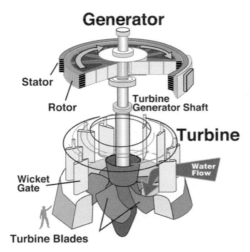

Figure 6-22. Turbine-generator set

Hydropower turbines are capable of converting more than 90% of available energy into electricity, making it the most efficient form of electricity generation ever. By comparison, fossil fuel plants are only 40-50% efficient.

In addition to providing low-cost electricity, multi-purpose dams provide water for irrigation, wildlife, recreation, barge transportation, and flood control benefits.

Note: While significant CO_2 emission reduction and 90% efficiency at the last leg of the hydropower generation cycle are undeniable facts, there are serious

inefficiencies, pollution, and environmental damage throughout its life cycle that must be taken into an account when comparing the different power generating technologies. We will take a close look at these here.

Types of Hydropower Turbines

Hydro-turbines vary depending on location and application, but the major types are: impulse turbine, reaction turbine, and Francis turbine.

Impulse Turbine

The impulse turbine generally uses the velocity of the water to move the runner and discharges to atmospheric pressure. The jet pushes on the turbine's curved blades which changes the direction of the flow. The resulting change in momentum (impulse) causes a force on the turbine blades. Since the turbine is spinning, the force acts through a distance (work) and the diverted water flow is left with diminished energy.

In an impulse turbine, the pressure of the fluid flowing over the rotor blades is constant and all the work output is due to the change in kinetic energy of the fluid. The water stream hits each blade on the runner and there is no suction on the downside of the turbine. The water flows out the bottom of the turbine housing after hitting the runner. An impulse turbine is generally suitable for high head, low flow applications.

Pelton Wheel has one or more free jets discharging water into an aerated space and impinging on the buckets of a runner. Draft tubes are not required for impulse turbine since the runner must be located above the maximum tailwater to permit operation at atmospheric pressure.

Turgo Wheel is a variation on the Pelton and is made exclusively by Gilkes in England. The Turgo runner is a cast wheel whose shape generally resembles a fan blade that is closed on the outer edges. The water stream is applied on one side, goes across the blades and exits on the other side.

Cross-Flow turbine is drum-shaped and uses an elongated, rectangular-section nozzle directed against curved vanes on a cylindrically shaped runner. It resembles a "squirrel cage" blower. The cross-flow turbine allows the water to flow through the blades twice. The first pass is when the water flows from the outside of the blades to the inside; the second pass is from the inside back out. A guide vane at the entrance to the turbine directs the flow to a limited portion of the runner. The cross-flow was developed to accommodate larger water flows and lower heads than the Pelton.

Reaction Turbine

A reaction turbine develops power from the combined action of pressure and moving water. The runner is placed directly in the water stream flowing over the blades rather than striking each individually. Reaction turbines are generally used for sites with lower head and higher flows than compared with impulse turbines.

Propeller turbine generally has a runner with three to six blades in which the water contacts all of the blades constantly. Picture a boat propeller running in a pipe. Through the pipe, the pressure is constant; if it isn't, the runner would be out of balance. The pitch of the blades may be fixed or adjustable. The major components besides the runner are a scroll case, wicket gates, and a draft tube.

There are different types of propeller turbines:

Bulb turbine is where turbine and generator are a sealed unit placed directly in the water stream.

Straflo turbine uses generator that is attached directly to the perimeter of the turbine.

Tube turbine uses penstock that bends just before or after the runner, allowing a straight line connection to the generator.

Kaplan turbine uses blades and the wicket gates that are adjustable, allowing for a wider range of operation.

Francis Turbine

Francis turbines have a runner with fixed buckets (vanes), usually nine or more. Water is introduced just above the runner and all around it and then falls through, causing it to spin. Besides the runner, the other major components are the scroll case, wicket gates, and draft tube.

Kinetic energy turbines, also called free-flow turbines, generate electricity from the kinetic energy present in flowing water rather than the potential energy from the head. The systems may operate in rivers, man-made channels, tidal waters, or ocean currents.

Kinetic systems utilize the water stream's natural pathway. They do not require the diversion of water through man-made channels, riverbeds, or pipes, although they might have applications in such conduits. Kinetic systems do not require large civil works; however, they can use existing structures such as bridges, tailraces and channels.

Types of Hydropower Plants

"Hydropower" refers to electric power generated by river flow, or from permanent water storage reservoirs. These power plants can be small and large in size and could be driven by running water through their turbines.

Hydropower plants range in size from large power plants that supply many consumers with electricity, to small and micro plants that individuals operate for their own energy needs or to sell power to utilities.

According to the U.S. DOE:

- Large hydropower plants are facilities that have a capacity of more than 30 megawatts.

- Small hydropower plants are facilities that have a capacity of 100 kilowatts to 30 megawatts, and

- Micro hydropower plants are facilities with capacity of up to 100 kilowatts.

Note: A small or micro-hydroelectric power system is usually designed to produce enough electricity for a home, farm, ranch, or a small village. In some cases, the excess power is sold to the local utility.

Hydropower can also be generated by ocean waves and currents, and we categorize these as special "ocean hydropower" generators.

The types of hydropower we will be discussing herein, are:
- Dam hydropower
- River hydropower
- Pumped storage hydropower
- Ocean hydropower

Hydropower Plants Cost

Compared to hydropower, *thermal plants* take less time to design, get approved, build and recover investment. However, they have higher operating costs, and typically shorter operating lives (about 25 years). They are also huge sources of air, water and soil pollution, and provide fewer opportunities for economic spin-offs.

Other renewable sources of power (solar, wind, etc.) are valuable options in addition to hydropower in specific contexts, but, even if major efforts were made to develop them, they will not be able to produce large amounts of energy in the coming decades. Nor will they be able to offer the same level of service, as they are intermittent sources requiring back-up power supply.

In assessing life cycle costs hydropower consistently compares favorably with virtually all other forms of energy generation. Hydropower is also successfully used in providing stability to large solar and wind power generation installations, which are notorious for their inconsistent, variable level power generation (during cloudy and windless days and nights).

The actual *initial* cost of hydropower plant design and construction are astronomical. While the Hoover

Dam's construction in 1936 was estimated at less than $50 million, the total cost of the Three Gorges dam in 2012 is estimated at $28 billion. This great difference makes comparing difficult, but even when currency variations are considered, it does emphasize the great changes in all aspects of dam building.

Large dams are things of the past, however, so we now can talk only about smaller dams and power plants. These are much cheaper, but still cost billions of dollars. The cost of the proposed Marvin Nichols reservoir in northeast Texas, for example, has been estimated at $2.2 billion, with no power plant included. It would flood 30,000 acres of rare bottomland hardwood forest, and another 42,000 acres of mixed forest, family farms, and ranches along the Sulphur River in northeast Texas.

The other thing to consider , always, is the external cost of the generated power. The suffering of many relocated families is heard in the residents' outcry, which is usually quite loud, as they are concerned with the damage to vital habitat for wildlife, and the local water quality. Long-lasting bad memories and negative publicity surround most large, and even smaller, hydro projects.

Note: Figure 6-23 shows the cost of the hydropower plant alone. This does not include the cost of building a dam and storage reservoir. These costs could be many, many times higher.

The actual cost of power plant construction (as related to the amount of produced power) varies with location, size, and type of power plant. The main variable in the power generation cycle is always the size of the plant, because it could take as many people to operate and maintain a small one-unit generator as it would to operate and maintain two larger generators.

This means that the cost of operation and mainte-

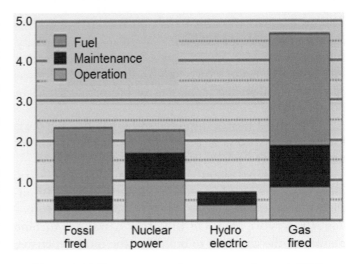

Figure 6-24. Power generation expenses in cents/kWh

nance per kilowatt produced would be higher (maybe twice) for the smaller plant.

A number of other variables also affect the final cost of the generated energy. The type, availability, quantity needed, and price of the different fuels, used in different power plants, is a critical component. Hydropower plants are the only power plants that have no allowance for fuel, because water is free (if the external effects are not considered).

Lately, however, water is becoming a precious commodity, so this issue is becoming more important, and we will eventually end up assigning a price to the water used by hydropower plants.

The maintenance and operation variables are also significant and are to be considered. Here again, hydropower requires very little daily maintenance, and only scheduled periodic maintenance and replacement of turbines and other components are adding cost.

Operational costs of hydropower plants are also quite low, so the entire fuel-maintenance-operation regime is quite stable, efficient and cheap. All this, however, provided there is enough water. And this is the main question before the global hydropower generating industry.

Hydropower plants' initial cost and production expenses (average estimates) are as follow:

Capital cost	$1700-2300/kW
Operation cost	0.4¢/kW
Maintenance cost	0.3¢/kW
Operating life:	50+ years
Capacity factor:	50-60%
Average size:	30-100 MW
Large size:	over 1.0 GW

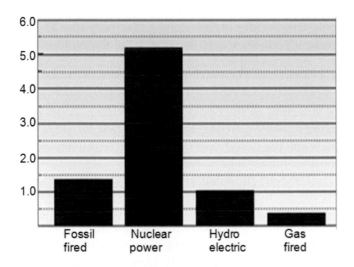

Figure 6-23. Power plants construction costs (in $/watt)

In general, the larger the hydroelectric plant, the cheaper the cost per kilowatt to produce the electricity. When compared to other means of producing electricity, hydroelectric production costs run about one third those of either fossil-fueled (coal, gas, or oil) or even nuclear power plants.

The main contributing factor here is the fuel costs for the other means of producing electricity (coal and natural gas-fired power plants).

The original plant cost for a small hydroelectric plant is somewhat cheaper than either fossil fuel or nuclear plants (not including the support infrastructure—water reservoir, etc.). On the other hand, gas turbine plants are the cheapest to build but the most expensive to operate.

As long as there is sufficient water to run the turbines, electricity can be produced very cheaply. Compared even to mature nuclear plants, large-scale hydropower costs less than half as much to produce, at under a penny per kWh.

Thus generated power is then transmitted to the national power grid and sold in the wholesale and retail markets at the same prices as electricity generated by other means, complete with premiums for peak demand production. This means a significant profit for the middle man and the utilities.

In summary, hydropower is a great energy source which will continue to provide a major part of the world's energy...if water availability does not alter its destiny. This is a major issue that has to be properly addressed and resolved, if we are to rely on hydropower in the 21st century.

Dam Hydropower

Dams on rivers and streams have been around for a long time, and are still in use in almost all regions of the globe. They have played a key role in the development of human activities, and are now used for irrigation, water supply, flood control, electric power generation, and improvement of navigation. They also provide recreation, such as fishing and swimming, become refuges for fish and birds, and slow streams and rivers so that the water does not carry away soil, thereby preventing erosion.

During the last two centuries dams have also played a key role in producing large-scale power and electricity. Hydropower is the only renewable resource currently used on a large scale to generate electricity.

Note: Although hydropower relies on the renewable water cycle, it is not officially recognized as a renewable energy resource in the U.S. There are efforts

presently in the U.S. Congress to include hydropower in the official list of renewable energy sources like solar, wind, bio- and geo-energy.

Hydroelectric plants range in size from several kW to many GW. For example, the Three Gorges Dam on the Yangtze River, the largest hydroelectric dam in the world, has installed capacity of over 20 GW, which is 25% more than the next largest hydropower plant, the Brazil-Paraguay Itaipu Dam, at 14 GW.

There are about 35,000 large dams in existence around the globe today. Their importance for boosting economic development has increased in recent decades, most significantly in developing countries.

Since there is not much more room (or river water) for large dams, the hydropower development in most countries is focused on building smaller dams. Refurbishing and upgrading existing hydroelectric plants and retrofitting dams constructed for other purposes is also underway in a number of places.

Small hydropower plants are quite different from their large-scale cousins. They usually depend on a back-up source of electricity, since most of them often do not have an adequate storage reservoir. Instead, they depend on the variable river flow.

Large- and intermediate-scale dams, however, will continue to be very important in developing countries, in Russia and China, and in some industrialized nations, such as Canada.

Case Study: Hoover Dam

Building a hydropower plant on a large dam is quite different from building any other power plant. Building a large dam with a storage reservoir is a huge undertaking of national proportions. It cannot be done by one company or entity alone. Instead, a number of companies, under the direction and supervision of government bodies are usually involved in the project.

As an example, here we take a look at the construction of Hoover Dam on the Arizona-Nevada border. It was approved by the U.S. Congress and executed under the direction of the Bureau of Reclamation. A decision on the massive concrete arch-gravity dam structure was made and the design was overseen by the Bureau's chief design engineer John L. Savage.

The Bureau issued bid documents available to interested parties, where the government was to provide the materials, but the contractor was to prepare the site and build the dam. The dam was described in minute detail, covering 100 pages of text and 76 drawings. A $2 million bid bond by the contractors accompanied each bid, and the winner posted a $5 million performance bond, allowing him 7 years to build the dam, or else.

There were three valid bids, and a bid of $48,890,955 was the lowest. It was amazingly only within $24,000 of the confidential government estimate of what the dam would cost to build. The best bid was also $5 million dollars lower than the next lowest bid.

An entire city was built in the desert near the dam site, which is still there as we know it today as Boulder City, Nevada. There was also a special railroad line constructed to join Boulder City to Las Vegas for transport of people and materials.

The dam building began by diverting the Colorado River away from the construction site. Four diversion tunnels (56 ft. in diameter and nearly 3 miles total length) were drilled through the canyon walls, on both sides of the proposed dam structure. This work had to be completed quickly in late fall and winter, when the water level in the river was low enough.

When the tunnels were completed, they had to be lined with concrete, using Gantry cranes running on rails through the entire length of each tunnel. Then the sidewalls were poured, using movable sections of steel forms, to create a concrete lining 3 feet thick.

The river was then diverted into the two Arizona tunnels, by exploding the temporary cofferdam protecting the Arizona tunnels, while at the same time dumping rubble into the river until its natural course was blocked and it started flowing through the two tunnels. The Nevada-side tunnels were kept in case of high water floods.

Upon completing the dam, the entrances to the diversion tunnels were sealed at the opening and halfway through the tunnels with large concrete plugs, while the downstream halves of the tunnels following the inner plugs are now the main bodies of the spillway tunnels.

Two cofferdams were constructed to facilitate the river's diversion, each 96 feet high, and 750 feet thick at the base, which is actually thicker than the dam base. Each contained 650,000 cubic yards of rock material and cement.

The site was then drained of water and the accumulated erosion soils and other loose materials in the riverbed were dredged until sound bedrock was reached for the dam foundation. This required the excavation and removal off-site of over 1,500,000 cubic yards of river bed material. Since the dam was an arch-gravity type, the side-walls of the canyon would bear the force of the impounded lake. The side-walls of the surrounding rock channel were excavated too, to reach virgin rock as needed for the load-bearing side walls and to eliminate water seepage.

The dam foundation was reinforced with grout and holes were driven into the walls and base of the canyon, as deep as 150 feet into the rock, and all cavities were filled with grout, to stabilize the rock. This would also prevent water from seeping past the dam through the canyon rock and limit the upward pressure of water seeping under the dam.

After the dam base and sides were secured the pouring of concrete into the dam structure was initiated. This was a complex undertaking in such an enormous structure, because concrete heats and contracts for a long time while it cures. The potential for uneven cooling and contraction of the concrete is a serious problem, so to avoid a very long curing process, the dam was built in sections.

The ground where the dam was to rise was marked with rectangles, and concrete blocks in columns were poured. These were 50 ft. square and 5 ft. high, strengthened by a series of 1-inch steel pipes through which first cool river water, then ice-cold water from a refrigeration plant was run.

Once each individual block had cured and had stopped contracting, the pipes were filled with grout. Grout was also used to fill the hairline spaces between columns, which were grooved to increase the strength of the joins.

Huge steel buckets (7 ft. high by 7 ft. in diameter, and weighing 18 tons when full), suspended from aerial cableways above the construction site, were used to pour the concrete for each block. The concrete was prepared at two large concrete plants on the Nevada side, and were delivered to the site in special railcars. A team of men worked the newly poured cement in each block throughout the form until achieving the desired uniformity.

A total of 3,250,000 cubic yards of concrete was used in the dam. An additional 1,110,000 cubic yards of concrete were used for the construction of the power plant and other works. More than 582 miles of cooling pipes were placed within the concrete. It was estimated that there is enough concrete in the Hoover Dam and the surrounding structures and infrastructure to pave a two-lane highway from San Francisco to New York.

Although the dam was completed in 1935, concrete cores removed from the dam for testing in 1995 showed that the concrete has continued to slowly gain strength. It was also confirmed that the dam is composed of a durable concrete having a compressive strength exceeding the range typically found in normal mass concrete.

Hoover Dam concrete is not subject to Alkali-Silica Reaction (ASR) as the Hoover Dam builders happened to use nonreactive aggregate, unlike that at downstream Parker Dam, where ASR has caused measurable deterioration.

Figure 6-25.
Upstream side of Hoover dam before filling the reservoir

The huge monolithic dam is 660 ft. thick at the very bottom and getting thinner as it goes up, ending with a 45-ft.-wide road, connecting Nevada and Arizona. It has a convex face towards the water level above the dam. It was estimated that the curving arch would transmit the water's force into the abutments of the rock walls of the canyon.

Following an upgrade in 1993, the total gross power rating of the Hoover hydropower plant, including two 2.4 MW Pelton turbine-generators that power Hoover Dam's own operations, is a maximum capacity of 2.1 GW. The annual power generation varies, according to water conditions and other factors. The maximum annual generation of 10.3 TWh/y was recorded in 1984, and the minimum was 2.6 TWh in 1956. The average has been about 4.2 TWh/year.

The dam reservoir has been very low since around 2005, and although still far from minimum level, it is watched closely and preventative measures are planned.

The upstream picture is much more impressive and even dramatic. 250 square miles (160,000 acres, or 28,537,000 acre/feet) in the Arizona/Nevada desert were permanently flooded, causing a radical change in the local environment. Good or bad, this change has to be taken seriously and evaluated for what it is, which is what we will do in the next chapters.

Labor

The dam and the power plant construction was very dangerous and exhausting work. 5,000 people on

Figure 6-26. Downstream of Hoover dam.

the average were involved in the daily dam construction. The most dangerous job was that of the "high scalers." These were people suspended from the top of the canyon with ropes, who climbed down the canyon walls every day. Suspended on the vertical rock wall, they removed the loose rock with jackhammers and dynamite.

There were falling rocks and other debris that hurt and killed workers. One high scaler was able to save a government inspector, who lost his grip on a safety line and began tumbling down a slope towards a certain death, when the high scaler intercepted him, risking his own life, and pulled him into the air and into safety.

The workers were under severe time constraints too, because the concrete pour was piling up and drying fast, and at times negligently ignored seepage and cavities in the wall. As a result, many holes were incompletely filled. This eventually caused unacceptable leaks in the completed dam, and the Bureau decided to fix the problem by drilling new holes from inspection galleries inside the dam into the surrounding bedrock. This work was done in secrecy, and at additional cost, taking over nine years to complete after the dam was already in full operation.

Although there are myths that men were caught in the pour and are entombed in the dam to this day, each bucket only deepened the concrete in a form by an inch, and Six Companies engineers would not have permitted a flaw caused by the presence of a human body.

Nevertheless, there were over 110 deaths during the construction of the Hoover Dam. One of the first victims was a surveyor who drowned in 1922, while looking for an ideal spot for the dam. Incidently, his son was the last man to die working on the dam's construction, 13 years to the day later. Ninety-six of the deaths occurred during construction at the site, and 91 of those were contracted employees. The rest were helpers and visitors.

In addition to the official fatalities due to accidents and incidents, there were a number of deaths due to illness, such as pneumonia. There were allegations, however, that this diagnosis was a cover for death from carbon monoxide poisoning of people overexposed to the gasoline-fueled vehicles in the diversion tunnels, which diagnosis was used by the contractor to avoid paying compensation claims.

River Hydropower

Hydropower plants based on the run of the river generate electricity on a smaller scale by simply using the water flow of a local stream or river. One way to generate electricity is to channel a certain amount of water through a pipe from the river to a turbine at the outlet. The water flow makes the turbine blades spin, which in turn spins a generator to produce electricity. This method is preferred to damming the river because it is much cheaper, and causes much less environmental damage by flooding large areas (as in damming the rivers).

Electricity can also be generated from turbines installed directly in the river. While this seems as a most practical and economic way to use the water flow, this method has not yet been commercially used, although there are a number of tests underway in the U.S. and Europe.

The problem here is that the water flow is usually low, so the power generator relies solely on the density and mass of the running water. The slow speed limits significantly the amount of generated electricity, while on the positive side it also reduces the risk to acquatic life.

River dams of the run of the river type are the new trend in hydropower generation. One of the most remarkable examples is the joint Romanian-Yugoslavian mega hydro project on the river Danube that started operation in 1972. The first version, Iron Gate I dam, had two dams with power plants each, containing 6 turbines at approximately 1.0 GW power generation each. Iron Gate II dam extension followed in the late 1990s by an upgrade and 10% increase of the total power, and two additional smaller power plants are still to be built as well.

The construction of these dams created a large reservoir in the Danube shoreline, and raised the water level of the river near the dam by 35 meters. Six villages, totaling a population of 17,000, were evacuated, and the villages, together with the local Orşova island were flooded. The locals were relocated and the settlements have been lost forever to the Danube. The dam construction had a major impact on the local environment, especially on the fauna. The spawning routes of several species of sturgeon were permanently interrupted.

The good news is that during post-dam construction, the geo-morphological, archaeological and cultural historical artifacts of the Iron Gates have been under protection by both nations. In Serbia the Dzerdap National Park was created in 1974 on 245.59 sq mi., and in Romania, the Porţile de Fier National Park was created in 2001, with 446.55 sq mi., both officially protected territories.

There are a number of similar examples of river hydropower, but nevertheless, a lot of work remains to be done on this technique, before its efficiency and cost-effectiveness can be assessed, and all problems resolved—before it finds wider commercial applications.

Small Hydropower Plants

Building a small hydropower plant on an existing lake, spillway or a river, is a straightforward process, which, like any other power plant, would go through the normal channels of development.

The only difference here is the fact that there is no fuel to be procured and transported, nor is there any waste or byproducts to be stored and disposed of. Of course, with that the emissions coming from the hydro power power during full operation are close to nil too.

A typical cradle-to-grave development process for a small hydropower plant includes:

- Hydropower plant site search
- Proposed area survey
- Environmental studies
- Local rights and agreements
- Permits and legal issues
- Plant site preparation
- Plant construction
- Daily plant O&M operations
- Decommissioning and area reconstruction

All steps of this process must be considered, to obtain a complete picture of the project, as needed to assess properly its technical and financial feasibility. As with all power plants, the survey, environmental studies and permitting processes are complex undertakings, in need of expert personnel. They also might take months, but most likely years, of intense work and negotiations with local and federal authorities and regulators.

After all studies have been completed, and the necessary permits secured, the construction of the plant can begin. Here we see again large pieces of equipment moving in, leveling the site and digging holes in the ground. Cement trucks dump their loads in the foundation and construction workers erect the buildings and complete the other elements of the plant infrastructure.

In most cases, small- to medium-size hydropower plants are built in remote locations—far away from populated centers and/or the power grid. In such cases, a substation is built nearby to transform the generated electricity into voltage suitable for transmission to a connection point in the national power grid, or to remote customers. In all cases, right-of-way for the transmission lines is needed, which is another great obstacle in building remote power plants.

After all the work has been completed, the power plant goes through a series of start-up testing and certification procedures. If everything checks OK, then the plan is given a clean bill of health and is switched into full operation, at which point it starts sending electricity to the grid, and normal O&M procedures take place on daily, usually 24/7, basis.

Case Study: Small Hydro Plant Construction

In the late 1980s this author was part of a team contracted for the design and implementation of small hydropower plants in Eastern Europe. The permitting and financing of the projects was the greatest hurdle, which had to be considered first. Due to the complexity of these procedures, several sites were rejected. Another set of potential sites had to be rejected because of their remoteness, which made connecting to any customers or the grid impossible (or at least prohibitively expensive). The remote sites also bore the risk of vandalism, which was not justifiable from the investors' point of view.

After the preliminary evaluation of dozens of proposed sites, only two were approved for development, and only one of these was eligible for government support. So, the team focused on the proposed site located on a large stream in a mountain region, 1.5 miles away from a large resort, which was the customer.

The stream is full, and water flow is fast 8 months out of the year, while for the other 4 months the level and flow vary from year to year. The stream also freezes occasionally for 2-3 months in winter, during which time the turbine is frozen and shut down.

A natural curve on a hill, where the stream zigzags for about a mile, as in Figure 6-27, was the proposed hydropower plant site. An intake at the top of the curved section was to be built and the water diverted from the natural stream. A catch box at the mouth of the intake was to filter out floating debris and fish, using an array of bars to keep large debris out, and a large mesh screen to remove smaller objects. An inspection gate was installed to divert the water for inspection and maintenance of the box.

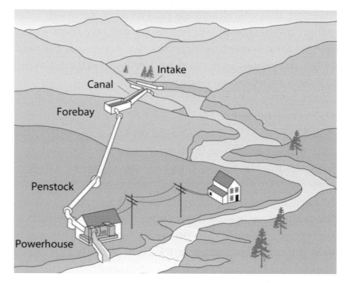

Figure 6-27. Small hydropower plant

A large bore cement pipeline was run ¼ mile from the intake to the turbine building. This was a major challenge, due to the mountainous terrain. An electro-mechanical control was installed in the turbine building to regulate the speed of the turbine and the electric generator attached to it. Thus generated power was sent to a transformer and then to the main breakers of the building for use on demand.

The power plant was rated at 150 kW and was equipped with a German turbine/gen set. The plant operation was quite simple, and consisted of periodic cleaning and inspection of the catch box, and normal periodic maintenance of the turbine/gen stack.

The only anticipated long-term problem was the availability of water in the stream during the low flow months in the summer and freeze in the winter. Also the late night use of power at the resort was drastically reduced, and since the power grid was far away, the turbine was idling several hours every night. The team could not come up with a better (economically feasible) solution for using the excess power generated during the night.

The total cost from concept to final inspection was about $1.2 million, or $8,000/kW. This somewhat higher cost was due to extensive work needed for the ¼-mile pipe on the mountain hills.

Pumped Storage Hydropower

There are several types of storage hydropower.

- *Pumped water* storage is basically storing water and reusing it for peak electricity demand. Demand for electricity is not constant. It goes up and down during the day, and overnight there is less need for electricity in homes, businesses, and other facilities. For example, in Phoenix, Arizona, on a hot August day, the demand for electricity to run millions of air conditioners is huge, but by midnight, it is drastically reduced.

  Some hydroelectric plants use "pumped storage" to generate power on demand anytime of day by reusing the same water more than once.

- *Pumped energy* storage is a method of keeping water in reserve for peak period power demands by pumping water that has already flowed through the turbines back up a storage pool above the power plant *at a time when customer demand for energy is low*, such as during the middle of the night. The water is then allowed to flow back through the turbine-generators at *times when customer demand is high*, like at noon on hot August days in Arizona. This reduces the heavy load placed on the electric system at peak hours, when the air conditioners are cranked up.

  The reservoir acts much like a battery, storing power in the form of water when demands are low and producing maximum power during daily and seasonal peak periods. An advantage of pumped storage is that hydroelectric generating units are able to start up quickly and make rapid adjustments in output. They operate efficiently when used for one hour or several hours. Because pumped storage reservoirs are relatively small, construction costs are generally low compared with conventional hydropower facilities.

- Compressed air hydro is used when a plentiful head of water can be made to generate compressed air directly without moving parts. In these designs, a falling column of water is purposely mixed with air bubbles generated through turbulence at the high-level intake. This is allowed to fall down a shaft into a subterranean, high-roofed chamber where the now-compressed air separates from the water and becomes trapped. The height of the falling water column maintains compression of the air in the top of the chamber, while an outlet, submerged below the water level in the chamber allows water to flow back to the surface at a slightly lower level than the intake. A separate outlet in the roof of the chamber supplies the compressed air to the surface. A facility on this principle was built on the Montreal River at Ragged Shutes near Cobalt, Ontario, in 1910 and supplied about 4.5 MW electric power to nearby mines.

Another method of compressed air storage is the use of excess electric energy during low power demand hours to run compressors, which compress air in special storage tanks. During high power demand hours the compressed air can be released to run turbines connected to generators. Thus produced power is sent into the grid to reduce the power generated by peaking power plants.

Compressed air storage suffers from low efficiency (less than 50%), so there are designs to improve the efficiency by using a mist of water sprayed into the air storage tanks, which acts to absorb and store the heat generated from the compression (and release it on expansion). This allows the improved system to achieve an 80-90% thermodynamic efficiency, and a 70% overall efficiency.

Large hydropower plants are the largest and most efficient of all power generation methods. Above, we took a close look at one of largest hydropower plants in the U.S., Hoover Dam power generation plant.

Ocean Hydropower

The potential energy stored in the world's oceans is huge. Calculating the energy stored in ocean tides, currents and waves, we come up with a very large number, and conclude that even if a small part of this energy can be captured and used, then we won't have to worry about energy crises or global warming.

Alas, the key components here, capture and use, are still in the early stages of planning and development. Capturing ocean energy won't be easy, for its powers are many and awesome. Powerful storms devastate everything in their way, wild waves can crush any mechanical device, and corrosion could eat through any material man has available today.

Nevertheless, the energy is there, we need it, and since it is human nature to try the impossible, we will be trying for as long as needed to capture this untamed giant and use its endless energy sources.

Ocean energy occurs in the form of tides, waves, currents, and heat. Tidal energy resources are modest on a global basis, and tapping them involves building major dams on inlets and estuaries that are prized for other purposes, so few tidal energy facilities have been developed. Harnessing waves and currents on a significant scale will involve designing turbine structures that are large, inexpensive, and can operate for long periods under the physical stresses and corrosive forces of ocean environments. For the most part, such systems are at the research stage today.

The largest but yet experimental form of ocean energy is ocean thermal energy conversion, which taps heat stored in the ocean to generate electricity. This process runs warm surface seawater through several different types of systems that use the water's stored heat to turn a turbine, then cools the resulting steam or vapor with cold deep-sea water.

Making this conversion work affordably on a large scale is technologically very difficult because it requires large structures and physical challenges associated with working in the ocean environment. It works most effectively in regions where there are large temperature differences between surface and deeper waters, mainly in the tropics. If ocean thermal energy conversion can be commercialized at some point, however, it could become an enormous new energy source.

Hydropower Use

Hydropower generation is probably the oldest method of producing mechanical power. We can even imagine the amazement of the first caveman who noticed that moving stream water can do work. He later on figured out how to make the water spin a wheel that crushed grain, thus feeding the entire village. This also must've been one of the first mass-service capitalist enterprises.

People have used moving water in increasing sophistication to facilitate their work throughout history. Presently, people make use of moving water mainly to produce electricity.

Most energy in the United States is produced by fossil-fuel and nuclear power plants, and only 7% of total power is produced by hydroelectric plants. Most of the hydropower comes from huge power generators placed inside dams.

A great limitation of this type of generation is the fact that there is no more room to build new hydropower plants. On top of that, available water is in short supply. Frequent draughts and other natural events make hydropower's reliability questionable for the long run.

Figure 6-29 shows that most U.S. states do not have major hydropower capabilities. Many use some type of small-scale hydroelectric power, but the major hydro-

power capacity is limited to the few states represented in Figure 6-29.

States with low topographical relief, such as Florida and Kansas, produce very little hydroelectric power. Some states, such as Arizona, Idaho, Washington, and Oregon use hydroelectricity as the main power source. And, most if not all of Idaho's electric power comes from hydroelectric plants.

In 2006, Texas had 23 dams with hydroelectric power plants, but hundreds of medium to large dams built for purposes other than power generation. The 23 dams have a total generating capacity of 673 MW, but the annual electricity they actually produce is well below their maximum potential. This is due to water level

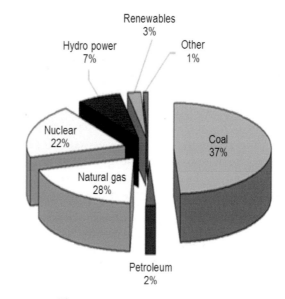

Figure 6-28. Power use in the U.S.

Figure 6-29. U.S. hydropower generation states

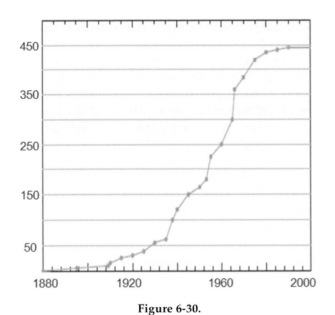

Figure 6-30.
U.S. hydropower reservoirs (in million acre-feet)

and demand variations, as well as other factors that reduce the plants' up-time and efficiency.

For example, in 2004 Texas hydropower plants operated at an average 22% capacity factor, and in 2006 the capacity factor averaged only 11 percent. Hydropower production is limited most drastically by droughts or other factors that affect surface water flows.

Hydropower Generation

No doubt, we use all the hydropower we generate since it is clean and cheap, and want and need much more too. Hydroelectric power is the best and cheapest power available. It is clean and reliable (to a point), and is totally clean at the customer end.

So the question is, "Why don't we use more of it

to produce a lot, or even all, of the electric power we need?" The answer is quite simple—there simply is not enough water, and no more can be produced due to lack of excess hydro resources. River damming is also an enormously long and painful process. It requires many years of permitting and negotiations, where the locals always lose.

Just ask the thousands of evicted from their homes—people in China and other countries—who were resettled into unfamiliar settings and deprived of making a living, in order to flood their homes as needed to build new dams on the local rivers.

Hydropower is good, but the beginnings of each large hydropower plant on a dammed river are bad and even ugly. This is due mainly to the fact that lots of water and a lot of land are needed to build a dam and create a large lake (reservoir) at the expense of the local environment and the people living in it.

The local environment covered by the lake is lost forever, and once flooded it cannot be restored to its original condition. Large dam hydropower projects also require a lot of money, time, construction materials and effort.

But all this is mostly in the past. Today, in the U.S. and most other countries, the good spots to locate large dams and hydro plants have been taken, so only smaller plants can be built now.

While in the early part of the 20th century hydroelectric plants supplied about half of the nation's power, that number is down to about 7% today and falling, as the other energy sources (gas and other) increase in use. So, the most feasible trend for the future of hydropower in most cases is to build small-scale hydro plants on riv-

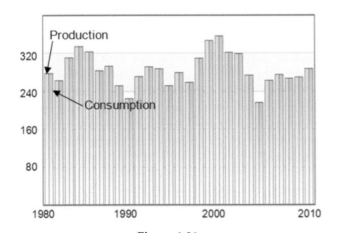

Figure 6-31.
U.S. hydropower generation and use (in billion kWh)

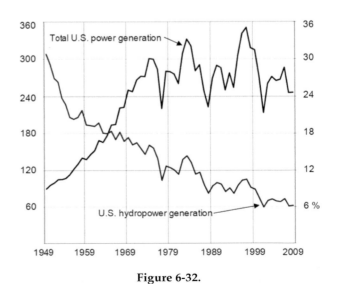

Figure 6-32.
Percent U.S. total power vs. hydropower generation (in GW)

ers and streams that can generate electricity for a single project or community.

Obviously, hydropower is taking a back seat in the overall power generation and use picture in the U.S. as compared to the overall amount of electricity generated in the country. This denotes its basic and terminal limitation—lack of additional natural resources for hydropower generation. It has basically reached its maximum volume and will stay there—regardless of the nation's needs and the related energy and environmental issues.

Water…Water…Water

The following examples point to some major problems facing hydropower in the U.S.; mostly lack of water. This seemingly simple and yet profoundly important problem is growing in size and importance and will determine hydropower's future.

Case Study: California's Water Woes

The present-day record drought in California has reduced availability of water, including that for hydropower generation. Nearly 60% of the state of California is in a state of extreme drought. Following the driest December on record in 2013, the drought is California's worst ever, including evidence that it may be the region's worst drought during the last 500 years.

The governor of California declared a state of emergency in 2014, and many communities are in danger of running out of water—even drinking water—in 2014 or shortly thereafter. Lack of water for drinking and agricultural needs is a real possibility, and with it the danger of reduction of water available for electricity generation is increasing too.

California gets most of its water from rain and snow precipitations in the winter months. Later on, the melting snowpack releases water slowly during the spring and summer months, a large part of which is used to generate electricity at hydropower stations. The snowpack in the Sierra Nevada Mountains is down to 10-12% of the normal levels. This is an unprecedented event, the results of which are hard to predict.

The absence of rain makes the problem even worse, especially as electricity demand rises during the warmer months of summer. But rain is usually much less likely in summer, and with the water reservoirs empty, the drought could create a real crisis in the near future.

The share of electricity from hydropower sources varies from year to year, from 10% to 30% of the state's total. Reduced water flow means reduced electricity generation, so California will be forced to import electricity from neighboring states—some form of hydro-

power generation from other states, mostly from the Pacific Northwest and Arizona.

Most Southwestern states are also in one or another state of prolonged and unprecedented draughts, due to lower levels of precipitation. This shapes the entire future of the area, since the wellbeing of the locals depends on what comes down from the sky. Little snow and rain means hardship is underway.

Hydropower is seriously affected in these areas. With the water levels in the major reservoirs down to historically low levels, hydropower is at low speed, or on standby in many locations. What will happen in the future is anyone's guess, but the scientists are predicting worsening of water availability in these regions, which means that hydropower cannot be relied on. It also means that building new dams in the near future is a risky undertaking.

At the present draughts trend, only a miracle would be able to retain, let alone increase, the amount of hydropower in the Southwest U.S.

Nevertheless, hydropower is not to be ignored. Where available, it offers large water and energy storage capacity and fast response characteristics that are especially valuable to meet sudden fluctuations in electricity demand. Hydropower is also successfully used to match and supplement power supply from less reliable energy sources and variable renewable sources, such as solar and wind power.

On the other hand, the environmental and social effects of hydropower projects need to be carefully considered. We have the example of large populated and cultural centers being erased from the face of the Earth, followed by huge environmental changes and damages to the affected areas.

Most countries are experimenting with integrated approach in managing their water resources, and hydropower development is part of their considerations, including the full life-cycle of the related projects.

A responsible approach to assessment of the benefits and impacts of hydropower projects and their use is a hot issue, with a lot of efforts going toward solving it, and yet we see a number of problems in these areas worldwide.

Case Study: U.S.-Mexico Water Wars

The situation is quite similar on the Mexican side of the Colorado river, which runs hundreds of miles on the U.S. side. Several dams and thousands of acres of agricultural irrigations on the U.S. side reduce the river's flow to a trickle by the time it gets to Mexico.

On top of that, agricultural leakage contaminates the Colorado River water, making it toxic and basically useless for human use as it goes down towards Mexico.

Raw sewage dumped into the river on the Mexican side further contaminates the slow running and already contaminated river, so by the time it gets to the ocean, the water quality is that of a sewage dump. It is not only useless, but is even harmful to wildlife and humans.

Finally, the filthy water flows into the ocean and contaminates the beaches of Tijuana, causing additional damage to wildlife, properties and people. The contaminated river water, full of hazardous waste is then taken by the ocean currents towards the U.S. border and dumped at or near San Diego's beaches.

The principle of cause and effect, derived from the biblical admonition that "what you sow is what you reap" is clearly reflected in this situation. Or in simpler words, "what goes around comes around…" We created the problem and we will suffer the consequences. It is as simple as that!

Solutions? None…yet. Although the trials have been going for dozens of years, and still continue…

The U.S. has already spent over $250 million on building a water desalination plant in Yuma, AZ, in an attempt to clean the Colorado water before it enters Mexico.

The Yuma desalting project is a delayed response to a 1944 bi-national treaty that promises Mexico 1.5 million acre-feet of water each year from the Colorado River, while U.S. farmers get more than 3 million acre-feet. Fair? You be the judge.

In the absence of other solutions, the desalination idea is plausible, and the technology available. The plan and the plant, however, have not worked at all thus far. The plant has been literally sitting idle most of the time since its inception in 1992, at a cost of $6 million annually—spent on keeping it idle. There are also additional and rising expenses for lawyers in several U.S. states and Mexico, fighting a war over the Colorado River water.

Recently, new plans were made to restart the plant and attempt new ways of using it, but we must remain skeptic as to their success. Let's hope that we are wrong. The locals on the Mexican side would appreciate it, and an international wrong would be made right. That, however, is not easily done, so we must wait and see.

Business Models

Administratively, there are six types of hydropower plants in the U.S.:

- Municipal and other non-federal public
- Private utility
- Private, non-utility
- Industrial
- Federal
- Cooperative

There are 2,388 licensed U.S. hydroelectric plants, assigned to one or more of the six owner classes. These plants represent the bulk of the U.S. hydroelectric capacity of approximately 75 GW in 1996. Of these, 69% of the plants are owned by private owners, as follow: private utility 31%, and private non-utility 27%, while industrial hydropower plants represent 9%, and cooperative 2%.

Table 6-4. Large U.S. dams

Dam name	Year built	MW
Grand Coulee	1942	6,809
Chief Joseph Dam	1958	2,620
Robert Moses Niagara	1961	2,515
John Day Dam	1949	2,160
Bath County PSP	1985	2,100
Hoover Dam	1936	2,080
The Dalles Dam	1981	2,038

Nearly 75% of the total hydropower capacity is owned by federal and public owners, as follow: federal 51% and public 22%. Different federal agencies are considered owners of different dams and hydro plants. Since an owner may own plants in more than one owner class (e.g., private non-utility and industrial) the total number is referred to as "presence" rather than owners.

The total number of owners in the U.S., therefore, is 1,134, while the total number of presences is 1,152. As with the distribution of the plant population by owner class, the distribution of the plant ownership shows that approximately 70% of the plant ownership is in the private sector.

There are several federal, state and local agencies that regulate different aspects of the hydropower projects. Many hydroelectric projects were built and managed by the federal government through the Army Corps of Engineers and the Bureau of Reclamation. Other projects are licensed by the Federal Energy Regulatory Commission (FERC).

Other federal and state agencies are usually involved in the process. Examples include: the National Marine Fisheries Service, U.S. Fish and Wildlife Service, National Parks Service, the Environmental Protection Agency, state fish and wildlife agencies, state water resource agencies and the state agency with Clean Water Act authority.

On top of this maze of government authority sit tribal governments, environmental and other non-profit groups who have significant but varying interests and concerns. Some of these groups are: American Rivers, the Sierra Club, Trout Unlimited, and a number of fishing and hunting associations, and boating groups.

Hydropower Pros & Cons

Hydropower has its own good and not-so-good sides. The good, however, prevails by a huge margin:

- The water resources are widely spread around the world. Potential exists in about 150 countries, and about 70 percent of the economically feasible potential remains to be developed. This is mostly in developing countries.

- It is a proven and well advanced technology (more than a century of experience), with modern power plants providing the most efficient energy conversion process (> 90 percent), which is also an important environmental benefit.

- The production of peak load energy from hydropower allows for the best use of baseload power from other less flexible electricity sources, notably wind and solar power. Its fast response time enables it to meet sudden fluctuations in demand.

- It has the lowest operating costs and longest plant life, compared with other large-scale generating options. Once the initial investment has been made in the necessary civil works, the plant life can be extended economically by relatively cheap maintenance and the periodic replacement of electromechanical equipment (replacement of turbine runners, rewinding of generators, etc.—in some cases the addition of new generating units). Typically a hydro plant in service for 40-50 years can have its operating life doubled.

- The "fuel" (water) is renewable, and is not subject to fluctuations in market. Countries with ample reserves of fossil fuels, such as Iran and Venezuela, have opted for a large-scale program of hydro development, recognizing environmental benefits. Hydro also represents energy independence for many countries—although the number of countries with adequate water supplies is limited.

Note: There is a question if water used for energy generation can be considered a renewable resource, which is becoming a focal point in U.S. politics. A resolution in Congress calls for officially recognizing it as renewable, but its destiny is uncertain.

Benefits of Hydropower

Hydropower provides unique benefits, rarely found in other sources of energy. These benefits can be attributed to the electricity itself, or to side benefits, often associated with reservoir development. Despite the recent debates, few would disclaim that the net environmental benefits of hydropower are far superior to fossil-based generation. In 1997, for example, was calculated that hydropower saved GHG emissions equivalent to all the cars on the planet (in terms of avoided fossil fuel generation).

While development of all the remaining hydroelectric potential could not hope to cover total future world demand for electricity, implementation of even half of this potential could have enormous environmental benefits in terms of avoided generation by fossil fuels.

Carefully planned hydropower development can also make a vast contribution to improving living standards in the developing world (Asia, Africa, Latin America), where the greatest potential still exists. Approximately 2 billion people in rural areas of developing countries are still without an electricity supply.

As the most important of the clean, renewable energy options, hydropower is often one of a number of benefits of a multipurpose water resources development project. As hydro schemes are generally integrated within multipurpose development schemes, they can often help to subsidize other vital functions of a project.

Typically, construction of a dam and its associated reservoir results in a number of benefits associated with human well-being, such as secure water supply, irrigation for food production and flood control, and societal benefits such as increased recreational opportunities, improved navigation, the development of fisheries, cottage industries, etc. This is not the case for any other source of energy.

Electrical System Benefits

Hydropower, as an energy supply, also provides unique benefits to an electrical system. First, when stored in large quantities in the reservoir behind a dam, it is immediately available for use when required. Second, the energy source can be rapidly adjusted to meet demand instantaneously.

These benefits are part of a large family of benefits, known as ancillary services. They include:

- Spinning reserve—the ability to run at a zero load while synchronized to the electric system. When loads increase, additional power can be loaded rapidly into the system to meet demand. Hydropower can provide this service while not consuming additional fuel, thereby assuring minimal emissions.

- Non-spinning reserve—the ability to enter load into an electrical system from a source not on line. While other energy sources can also provide non-spinning reserve, hydropower's quick start capability is unparalleled, taking just a few minutes, compared with as much as 30 minutes for other turbines and hours for steam generation.

- Regulation and frequency response—the ability to meet moment-to-moment fluctuations in system power requirements. When a system is unable to respond properly to load changes its frequency changes, resulting not just in a loss of power, but potential damage to electrical equipment connected to the system, especially computer systems. Hydropower's fast response characteristic makes it especially valuable in providing regulation and frequency response.

- Voltage support—the ability to control reactive power, thereby assuring that power will flow from generation to load.

- Black start capability—the ability to start generation without an outside source of power. This service allows system operators to provide auxiliary power to more complex generation sources that could take hours or even days to restart. Systems having available hydroelectric generation are able to restore service more rapidly than those dependent solely on thermal generation.

Avoided Emissions

Today 85 percent of the primary energy consumption is fossil (coal, oil and gas) or traditional (wood), with associated large-scale emissions to the atmosphere of greenhouse gases: carbon dioxide from combustion, and methane from processing coal and natural gas. It is well recognized at the international level that this is leading to major climatic changes, and will therefore also have consequences on the hydrological system (and thus on water supply and agriculture, as well as the sea level).

Recent research in North America confirms that the GHG emission factor for hydro plants in boreal ecosystems is typically 30-60 times less than factors for fossil fuel generation. Studies have also shown that development of even half of the world's economically feasible hydropower potential could reduce GHG emissions by about 13 percent, and the impact on avoided sulfur dioxide (SO_2) emissions (the main cause of acid rain) and nitrous oxide emissions is even greater.

Taking into account the fuel required to build hydropower stations, a coal-fired plant can emit 1000 times more SO_2 than hydropower systems. The magnitude of the impact of particulate emissions from fossil fuel is now also becoming recognized, particularly in connection with respiratory disease, and a recent estimate of the environmental cost of this form of pollution is put at $100-500 per ton annually.

Case Study: Tucurui Dam, Brazil

Research is continuing on dams, reservoirs and hydropower plants emissions, and it is recognized that more research is needed particularly regarding tropical reservoirs. A theoretical calculation has been done for the case of Tucurui in Brazil, including "worst case" assumptions, which include the decomposition of flooded biomass—assuming that 100% of the affected biomass would decompose over 100 years, and that 20% of its carbon content would be emitted as methane gas.

In this case, the emission factor for Tucurui would be 213 grams CO_2 equivalent per kWh, which is still five times lower than that for coal, and yet it is a huge number. Tucurui generates 8.4 GW of power, which means that if the 213 grams/kWh estimate is correct, then the project emits almost 16 million tons of CO_2 equivalent annually. Who knew…?

This is not a small amount, for sure, and it is not to be ignored, but the number of emitted CO_2 would be several times greater, if similar amounts of electricity were generated by coal, or natural gas-fired, power plants.

Generally, hydropower reduces the CO_2, SO_2, and particulate matter emissions, as compared to fossil-fired power plants, but its share is too small to make the big difference. Still, every little bit helps in the fight for cheaper and cleaner energy.

As with all things, there are some not-so-good aspects to hydropower. Let's see:

Environmental Impact

Any infrastructure development inevitably involves a certain degree of change. That change can be quite significant, as is the case with new dam construction. A new dam requires the impounding and inundation of a large area for a water reservoir. This always creates certain social and physical changes upstream and down.

There are difficult ethical issues, and ensuring that the rights of people and communities affected by a new project are respected, are usually on top of the list.

Social Aspects

As with other forms of economic activity, hydropower projects can have both positive and negative social aspects and effects. Social costs are reflected in transformation of large land areas, displacement of people living in the reservoir area, and effects on the lifestyle of people living downstream.

Relocating people from the reservoir area is, undoubtedly, the most challenging social aspect of hydropower. This usually leads to significant concerns regarding local culture, religious beliefs, and the effects of submerging homes and burial sites.

There is no good way to handle this, so compromises are commonly used. The countries in Asia and Latin America, where resettlement is a major issue, have developed comprehensive strategies for compensation and support for people who are impacted.

We should keep in mind that other power generating technologies also cause significant resettlement. For example, coal mining and processing, and coal ash disposal sites, also displace communities. And the long-term GHG-induced climate change may eventually cause the greatest population migrations, as sea levels rise.

Social effects of hydro schemes are variable and project specific. However, if anticipated and tackled early in the planning stage of a project with the required resources, the negative impacts can be addressed in a positive manner for local people, or in some cases avoided altogether. Whenever these impacts cannot be avoided or mitigated, compensation measures can be implemented.

During the construction phase of a hydro scheme (often several years) there may be a large workforce, and access roads can lead to a sudden influx of outside labor and the development of new economic activities, with resulting tensions if populations in the area in question are unprepared. Issues of resettlement, sustainable livelihoods, cultural impacts and flood control must be addressed.

Effective mitigation measures can be implemented if local authorities and project promoters acknowledge and address these issues. On the positive side, the additional economic activities create new employment opportunities.

During the operational stage, the hydro project may represent a significant source of revenues for local communities. The access roads, local availability of electricity and other activities associated with the reservoir are all possible sources of sustainable economic and social development. It is clear there must be good co-operation between proponents, authorities, political leaders and communities, and long-term benefits must be directed to affected communities.

Socially acceptable hydropower means that any proposal for a project must be discussed with stakeholders and adapted to their needs, and that successful negotiations must be concluded with affected local communities for a project to move ahead.

From a social point of view, the relative success or failure of a hydro project is determined by integrating social considerations early into the project design.

And there are of course a number of negative effects on wildlife in the area. Both up- and downstream fish, birds and other wildlife are affected by landscape changes. Since these changes are permanent, the change in wildlife habits and patterns is irreversible too.

Case Study: Farakka Barrage Dam

In 1970, the Indian government built the huge Farakka Barrage dam on the Ganges River. Located on the border with Bangladesh, the gated dam has had devastating effects on the environment and the population of both the Indian state of West Bengal and neighboring Bangladesh.

The dam is controlled by the Indian side, and has been creating its own environment, with regular floods in the summer and prolonged droughts in the winter months. The negative effects are felt mostly downstream in Bangladesh, where the locals are forced to live in flooded houses part of the time and without water the rest of the time.

Through the years, the dam forced the Ganges river to shift almost six miles eastward, eroding several villages in the process. Factories, sugar mills, hospitals, and government buildings were flooded by the river at the settlements. The Ganges's fast shift, however, made land that was once underwater reappear on the other side of the river bank. The land shift has created a territory dispute, so neither West Bengal nor Bangladesh recognizes thousands of displaced villagers.

The Farakka barrage was built at the narrowest point on the Ganges to divert water to Calcutta to the south and flush out the silt that was clogging up its port. But scientists say the project was ill-conceived from the start: Water upstream from the dam carried massive amounts of silt, dropping it directly behind the dam. The buildup—almost 700 million tons annually—has clogged the dam's gates and raised the river bed more than 20 feet, which is the reason the river was forced to change its course.

In that process the river is swallowing villages and buildings in its path one day and bringing absolutely devastating droughts the next. This abnormal man-made disaster has affected the locals living downstream from the dam, and has created a large-scale humanitarian crisis.

Negative Effects of Small River Dams

It is not just large dams that have such profound negative effects. A recent five-year study concluded that for certain environmental impacts the cumulative damage caused by small river dams is even worse than that caused by the large dams.

According to the researchers, hydropower may be renewable, but there are serious questions about whether or not the overall process is sustainable. There is obvious and not-so-obvious damage to streams, fisheries, wildlife, threatened species and communities.

The small hydropower projects are often located in poor areas of the world, which get no benefits from them since the small hydropower stations are almost always connected to the national grid, so the electricity is being sent outside of the local area. So the locals are left to endure the results from the small hydropower project construction and operation, which can be profound.

There are already 750,000 small (under 50 MW) dams in China alone, and about one new dam is being built every day, according to the study. The ecological concerns these projects raise, especially in "hotspot" areas of biological diversity have been ignored until now, but are surfacing slowly as the number of hydropower projects increases.

Per megawatt of energy produced, small tributary dams in some cases can have negative environmental impacts that are many times greater than large, main stem dams. Small dams can have significant impacts on habitat loss when some or the entire river flow is diverted into channels or pipes, leaving large sections of a river with no water at all.

Fish, wildlife, water quality and riparian zones are all affected by water diversion, and changes in nearby land use and habitat fragmentation can lead to further species loss. The cumulative effect on habitat diversity can be 100 times larger for small dams as compared to large dams.

China and other countries encourage more construction of small dams, but because they are usually built in remote areas, inhabited by poor people there is much less oversight and governance during the construction, operation and monitoring of small hydropower projects.

As a result, mitigation actions and governance structures that would limit social and environmental impacts of small hydropower stations are not adequately implemented. These conclusions are relevant to national energy policies in many nations or regions, and the locals are left to bear the brunt of the negative effects.

Large hydropower projects are usually well designed, organized, and managed, which is not the case with small hydropower projects, many of which are constructed and operate under the radar of the media and the regulators. There is much less control during the lifetime of the project and lots of bad things can happen unnoticed.

The overwhelming result of all this, according to the study, is that the negative biophysical and social impacts of small hydropower exceed those of large hydropower, especially where the local habitat and hydrologic changes are concerned.

Case Study: Columbia River Salmon Massacre

Salmon have thrived in the Columbia Basin since the beginning of time and have had tremendous impact on its flora and fauna. Salmon have also been central to the life of the local Indians in the Basin. Salmon sustained their lives and was even used as trade commodity—until the hydro dams changed everything forever.

In 1855 the U.S. government signed a treaty with the local Indians, allowing them full and unrestricted access to fishing in the local rivers in exchange for their lands. Fair deal...maybe...but good enough to sustain the people and keep the local environment healthy. Soon after that, however, dam construction on the Columbia river started, and everything changed.

Dozens of them were built on the Columbia river. As a result, in less than a century, the once-plentiful salmon is on the brink of extinction. The dams' structures stopped the salmons' annual migration to their traditional spawning grounds. Salmon cannot change their millenniums old routine, and now the number of salmon in the river has drastically decreased, and a number of a species are gone forever; never to be seen in this part of the country.

As a result, the Columbia River salmon population is now less than 3% of what it was only a century ago.

The government has been trying to manage the salmon population by raising millions of baby salmon in captivity, which are then released in hopes of keeping the species alive and well. The effort is not working as planned and is actually creating some even bigger problems in the area.

The salmon, however, are only one small part of the big picture, although their absence started the chain of events. The salmon's fate is the most obvious reflection of a dying ecosystem. The damage is now spread over the entire mighty Columbia Basin and the Pacific Ocean, also endangered with many other species undergoing changes, and some facing extinction.

Not only salmon, but also sturgeon, eels, and many other local fish species are threatened with extinction unless immediate and drastic changes are made in the management of the Columbia Basin and the Ocean.

One big lesson we must learn from all this is that every time we modify Nature's design, she responds in untenable ways.

The dams were built to generate power and to serve other purposes, and they do what is expected. However, they are also changing a huge part of our ecosystem—not only in the Columbia River Basin—in an unprecedented manner, the end result of which is uncertain. In all cases, very little good is going to come out of all this—at least for the salmon and the local environment—if we don't change the way we manage our rivers and oceans.

The problem here is that we put energy generation and all our other needs before the environment. The way things are done in the Columbia River Basin, we are basically trading the long-term health of our environment and all life in it for the short-term benefit of generating enough energy to make our lives more productive and comfortable.

This then is a war between our prosperity and comfort now vs. a dying environment in the future. We are ensuring our energy security today at the expense of a seriously damaged environment tomorrow.

Hydropower and Energy Security

Hydropower has been and still is a major energy source in the U.S. and around the globe. About 20% of the world's electricity is generated one way or another by using water. In the U.S., hydropower accounts for nearly 10 percent of the nation's electric power supply. It also provides over 90% of the total electricity generated by all renewable resources in the country—solar, wind, geo-thermal, bio-fuels, etc.

Hydropower plants produce nearly 100 GW of electricity annually, which is enough to provide power to about 35-40 million customers. Although we have over 80,000 dams around the country, only about 3% of them actually generate electric power.

- The most important benefit of hydropower is that it provides reliable, clean, and cheap energy.

- Its energy is produced domestically, so it insulates the United States from fluctuating and dangerous world energy markets and political conditions in supplier countries.

- Hydroelectricity is a "base load" power generator, which provides electricity to the grid as needed and when needed, which is an important means of regulating the flow of electricity throughout the national power grid. Unlike plants that use coal, hydropower generation can be quickly increased and decreased to maintain the exact balance of the grid. This power regulation ability is especially valuable during "peak" hours.

- Hydroelectric projects provide other benefits as well, in the form of flood control, irrigation, navigation, recreation, and steady water supply.

- The environmental and other concerns that are brought up by hydropower include barriers to upstream and downstream fish passage in the dammed rivers. This also brings changes to the river water quality, which affects wildlife habitat conditions.

- The varying flow rate and quality of river water moving downstream affects the locals in a serious way as well.

- Many hydropower projects in the U.S. are old, and their licenses are due for renewal during the next several years. This requires complete environmental impact studies that must include the public in the process. Choosing and following mitigation and enhancement strategies regarding these impacts is also part of the overall strategy of bringing hydropower into the energy mainstream.

- Continuous research and mitigation efforts are underway in an attempt to address the major concerns. However, after decades of work, the results are insufficient and much more work is needed, although some enhancement and mitigation efforts have been successful.

• These efforts are funded both by the federal government and project owners. Many of the large dam hydropower plants are huge—national size—projects, so the government will have to cover a substantial portion of the associated expenses. This is just another burden on our already crumbling infrastructure.

It is obvious from Figure 6-33 that there are many factors that affect hydropower generation both negatively or positively. Some of these are under our control, and others not so much. While we can work with and manage some of these, there is a third group of variables that are totally out of our control.

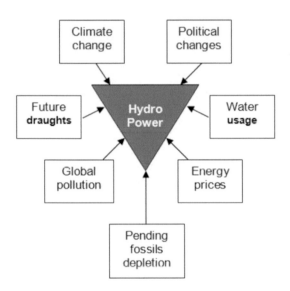

Figure 6-33. Hydropower's future

Water availability, or lack thereof, is the biggest problem. It will determine the future of hydropower. The pending climate changes and the accompanying long draughts in some areas will not help much.

Hydropower, like all other energy sources, can simply become unavailable. We know quite well when the fossils will be gone, and we know that solar and wind will be around as long as the Earth is. Unfortunately, we don't know what exactly will happen with hydropower in the future. It might continue as is for a long time, but if the global temperature increases by another 2-3 degrees (as predicted by the experts) then we must expect the unexpected and the unthinkable.

Unprecedented changes could significantly reduce, and even put an end to hydropower in many areas. We can hope for the best...but we should be ready for the worst.

INTERNATIONAL DEVELOPMENTS

Worldwide hydropower represents about 20% of total global electricity production. Hydro power supplies more than 50 percent of national electricity in about 65 countries, more than 80 percent in 32 countries and almost all of the electricity in 13 countries.

China has been building large hydroelectric facilities in the last decade and now leads the world in hydroelectricity usage. Canada and Brazil follow in their use of hydropower. Many other countries use hydroelectricity, and in many it is the most important widely used renewable energy source. In addition to large hydro dams, which are no longer an option in most countries, hydropower is produced by using the power of large rivers and natural drops in elevation.

While large-scale hydropower is almost fully developed in developed countries, the underdeveloped countries have untapped hydro resources, which are still abundant in Latin America, Central Africa, India and China.

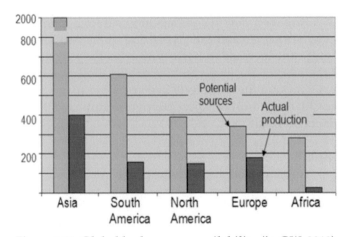

Figure 6-34. Global hydropower availability (in GW, 2010)

Obviously, the developing countries have much greater potential to add hydro power in the future. The developed countries, the USA included, have reached the limits of their hydro power resources.

The world's total technically feasible hydropower (the amount of hydropower that can be achieved if no other conditions were considered) is estimated at approximately 15,000 TWh/year, of which about 8,000 TWh/year is currently considered economically feasible for development. This is about 700 GW (or about 2600 TWh/year) already in operation, with more than 100 GW under construction. Most of the remaining potential is in Africa, Asia and Latin America.

Table 6-5. World-wide hydropower potential

Continent	Technically feasible potential	Economically feasible potential
Africa	1750 TWh/year	1000 TWh/year
Asia	6800 TWh/year	3600 TWh/year
N. America	1660 TWh/year	1000 TWh/year
S. America	2665 TWh/year	1600 TWh/year

It seems that although there are significant (potential) hydropower resources, using these is not that easy. A number of countries, such as China India, Iran and Turkey, are undertaking large-scale hydro development programs, and there are projects under construction in about 80 countries.

According to recent world surveys, a number of countries see hydropower as the key to their future economic development. Some of these are: Sudan, Rwanda, Mali, Benin, Ghana, Liberia, Guinea, Myanmar, Bhutan, Cambodia, Armenia, Kyrgyzstan, Cuba, Costa Rica, and Guyana.

Provided that the global climate does not hold too many surprises, and the rivers flow full, we might see significant increase in hydropower in the near future. If the predictions are even half correct, however, useful water around the world will become more scarce, which would make building hydropower dams and power plants more difficult.

China

China is planning a hydropower revolution. Right now! Its energy needs are increasing, and in order to meet them, as well as the renewable energy goals by 2020, the Chinese government plans to increase hydropower capacity to 300 GW by 2015 and to 420 GW by 2020. This is twice as much as the power generated in 2013. It is also much more generated power than that of all European countries combined.

Note the white marks on the map in Figure 6-35—dozens of hydropower plants that the Chinese government is planning to build during the next several years in this part of the country alone. Some of these use river flow to generate power, while some will be equipped with water storage reservoirs. Some of the water storage reservoirs will be contained by dams, while some will be using pumped storage.

This is truly an amazing undertaking. The world has never seen so much hydropower addition. While the official reason is improvement of the global environment, in reality it is driven by economic and political

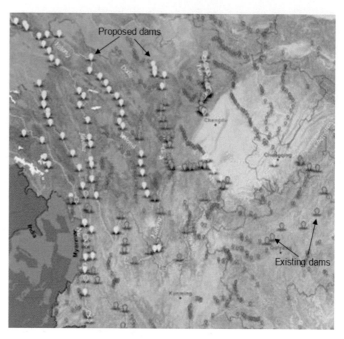

Figure 6-35. China's hydropower revolution (planned)

interests. The local environment cannot possibly be improved by adding hundreds of river dams.

And whatever good this development means for the global environment is nullified by the addition of hundreds of coal-fired power plants in other parts of the country emitting as much GHGs as the entire world put together. One good thing cannot rectify a thousand bad ones.

Chinese officials are biased toward the hydro projects, which trend has created a new hydro-industrial complex syndrome. This way the officials are able to cite their contribution to renewable energy and flood control, while using the dams to drive economic development, industrial expansion, construction, and increase the overall revenue streams.

All in all, China has the incentive, political support, and the money to add all this new capacity, and yet the projects are hitting a barrier. Tougher approval rules and permitting regulations are hindering the successful plan execution.

Only 4.82 GW of new hydropower capacity was approved for construction in 2013, and only 26.8 GW of new capacity has gone into construction since 2011.

Yet, a quarter of the planned capacity, or nearly 100 GW has been approved and is in different stages of design and construction on major rivers in the southwest. However, controversial projects on the Nu River in Yunnan and the Brahmaputra River in Tibet are not expected to get approval anytime soon, if ever. This might slow down, and even reduce, the ambitious government plans, but it is not clear by how much.

The resistance of the locals is increasing too, since they have learned the lessons from the previous hydro projects, which forced thousands of people to leave their homes. Over its 50-year history of developing hydropower, China has never addressed and resolved all issues related to involuntary resettlement caused by dams and other hydro projects construction. This may become an even larger issue as the new hydropower plants' construction expands.

There are also scientific arguments that dam reservoirs change the local seismic activity, which can be blamed for triggering earthquakes much earlier than they would have occurred naturally. This is a real danger in many regions that are already quake-prone. Cascading dams such as those planed on the Nu river at Burma's border, are capable of causing chain reactions that can amplify the impact of an earthquake.

Too, several of the rivers to be dammed are part of Burma, India, Bangladesh and other parts of Southeast Asia's river systems. This situation is triggering true international "water wars."

There is no doubt that China's ambitious plans to become the world's largest hydro-industrial complex are real and near realization. The only thing that is not certain is when and how much hydro power will be added as a result of the domestic socio-political and international issues the new developments have created.

India-Pakistan

They say that water is the liquid gold of the Earth, and while this is true, today it is becoming even more scarce and precious than gold. Gold is just a pretty yellow metal, that most of us can live without. Water is life itself, without which people, animals, and vegetation cannot survive.

The addition of hydropower plants on some rivers is adding serious uncertainties and risks in some cases that threaten to destroy the local's way of life.

Above the dams, thousands are forcefully relocated and their homes flooded to create a new water storage reservoir. Below the dams, the river flow is severely limited and even stopped for long periods of time. This converts previously productive agricultural lands into dry deserts.

Rivers cross many borders too, so water use in large hydropower projects has been on the international arena for many years. As global water conditions change, the related issues are becoming more serious as well.

One example of the seriousness of the situation is the Kishanganga power plant in India. It is a fairly small power plant (330 MW) with huge repercussions for neighboring Pakistan. The construction started in 2007, and is expected to be complete by 2016. The construction on the dam, however, was halted for awhile by the Hague's Permanent Court of Arbitration in 2011 in response to Pakistan's protest.

The claim is that the diverted river flow will have huge effects on the downstream flow of the Kishanganga River (Neelum River in Pakistan) that would affect negatively the Pakistan side of the project.

The Hague court ruled in 2011 that India could divert some (minimum) amount of water for power generation. This is not satisfactory for either party, so the water wars continue.

On one hand, Pakistan is mainly worried about the effects of the diverted water on the locals. On the other, however, it is concerned with the low river flow on its new Neelum–Jhelum Hydropower Plant downstream of the Kishanganga.

The problem is that the upstream Kishanganga power plant project will divert a portion of the Neelum river from Pakistan which will reduce power generation at the Pakistan's downstream Neelum–Jhelum hydropower plant.

India claims that the project will divert only 10 percent of the river's flow, but Pakistan estimates the diversion to be 33 percent or more. Worse, since both projects divert and restrict water flow downstream, the river flow through Pakistan's Neelum Valley is expected to be reduced to the very minimal. This has the potential to have adverse impacts in the entire Neelum Valley in Pakistan.

Pakistan is already in a state of extreme water shortage. Lack of rain has drained its water storage and groundwater supplies, and as these reserves run out, a water crisis looms. Recent rainfall has been far short of filling the reservoirs, so less water downstream means more problems for Pakistan.

So, the battle for life-giving water is shaping as a major international conflict. This is the case not only with India-Pakistan, but many other countries, including the U.S. See U.S.-Mexico water wars case study in this text.

Notes and References

1. World Nuclear Associations. http://www.world-nuclear.org/info/Country-Profiles/Countries-T-Z/USA—Nuclear-Power/
2. Natural Resource Defense Council. http://www.nrdc.org/nuclear/fallout/?gclid=CKKThPr8t7wCFVJcfgod6WsAZA

3. U.S. EIA, Nuclear & Uranium. http://www.eia.gov/nuclear/data.cfm
4. I.A.E.A., Nuclear Data. http://www-naweb.iaea.org/napc/nd/index.html
5. AE Hydropower. http://www.alternative-energy-news.info/technology/hydro/
6. National Hydropower Association. http://www.hydro.org/
7. Renewable Energy. http://www.renewableenergyworld.com/rea/tech/hydropower
8. *Photovoltaics for Commercial and Utilities Power Generation*, Anco S. Blazev. The Fairmont Press, Inc., 2011
9. *Solar Technologies for the 21st Century*, Anco S. Blazev. The Fairmont Press, Inc., 2013
10. *Power Generation and the Environment*, Anco S. Blazev. The Fairmont Press, Inc., 2014

Chapter 7

Renewables—The Future's (only) Hope!

*The argument must be made that the value of alternative energy is
directly related to energy security and environmental stewardship.*
—*West Point Team*

THE RENEWABLES

This quote is from a report (5) of West Point team of researchers, which concludes that alternative energy technologies are a small part of our energy present, and that their importance will (and must) grow proportionally with time. This includes their use in the U.S. military operations, as well as the entire U.S. economy.

And while it is not easy to imagine a solar-powered F-18 fighter jet, M1A3 Abrams-class battle tank, or Nimitz-class aircraft super-carrier, we must consider that this might be the only option in the 21st century and beyond, when all fossils will be depleted. Is this even possible? Not today, for sure. The question remains for future generations to answer, since today's world is not planning to make it easy on them.

At the rate we are digging, pumping out, and burning the conventional fuels—the fossils—there will be none left within a century or so. At that point, our children's children will have nothing to use for power generation and transportation, and will have to fully rely on alternative and renewable sources of energy. No choice.

The environmental consequences of excess fossils use are also important and must be kept in mind when discussing energy and energy security now and in the future. The relentless drive to achieving energy security is blinding us to the fact that we are running out of clean air to breath in the process.

So what do the renewables have to do with our present and future energy security and energy independence? The brief answer is, "Not much today, but everything tomorrow."

Sorting it Out...

This is a good place, before we delve deeper into the subject, to bring some clarity to the terms we hear every day: alternative energy, renewables, and cleantech. These are often used (intentionally or unin-tentionally) interchangeably, indiscriminately, and even incorrectly.

Alternative Energy

Alternative energy today denotes any energy source that can be used instead of conventional fuels—(coal, natural gas, crude oil, and uranium). The main purpose of using alternative energy is to prevent the depletion of the limited fossil resources, while at the same time saving society from the pollution those create.

Historically, today's conventional energy sources—the fossils—were at different times used as alternative fuels to replace other (than conventional) fuels.

Here are several examples of the evolution of alternative fuels in the past:

- Wood was the fuel that started it all. It was used since the beginning of time as source of heat and for cooking. It is hard to imagine the life of our ancestors without wood fire.

- In the late Middle Ages, *coal* was used as an alternative to wood. It saved society from overusing the dominant fuel, as the vast forests of Europe were being depleted from uncontrolled overuse.

People became so skilled at deforestation that by 1500 AD they were running short of wood. Europe was on the edge of a severe fuel disaster, from which it was saved in the sixteenth century only by the burning of coal. Coal still reigns in many areas of the world, it is indispensible in some industries, and dominates others.

- *Whale oil* revolutionized the lighting business, replacing candles, and was extensively used in oil lamps for a long time. It was also used for making a number of other products, like soaps and lubricants.

Whale oil was made by boiling strips of whale blubber in a process called "trying out." The boiling was carried out on land, in the case of whales

433

caught close to shore or beached. On deep-sea whaling expeditions, the trying-out was carried out on the ship itself so the waste carcass could be thrown away to make room for the next catch. This became a big business and millions of whales—the most magnificent of ocean creatures—were converted into lamp oil. It is truly the crime of the 19th century.

- *Crude oil* became an alternative to whale oil in lubrication and lighting, and later replaced coal in powering vehicles. Whale oil was the dominant form of lubrication and fuel for lamps in the early 19th century, but the depletion of the whale stocks by mid-century caused whale oil prices to skyrocket. This allowed crude oil to replace whale oil for these purposes. Crude oil was first commercialized in Pennsylvania in 1859, and it still dominates the transportation sector and the production of lubricants.

- *Grain alcohol* was used as an alternative to fossil fuels, when in the early 1900s, U.S. companies started making ethanol from corn, wheat and other foods, and used it as an alternative to coal and oil. Brazil took this idea further, and since 1970 it has had an ethanol fuel program, thus becoming the world's second largest producer of ethanol after the United States and the world's largest exporter. Brazil's ethanol fuel program uses modern equipment and cheap sugar cane as feedstock, and the residual cane-waste (bagasse) is used to process heat and power. All light vehicles in Brazil run on either pure alcohol or some gasoline and alcohol mixture.

- *Coal gasification* was developed and used for fueling vehicles in Germany during WWII. The Nazis demonstrated the feasibility of the process, and produced a lot of liquid petroleum products from coal during that time. These products kept the German war machine going for years, and yet, coal gasification has not been used commercially since.

Coal gasification was promoted as an alternative to petroleum in the 1970s by President Carter's administration, in an attempt to replace the scarce and expensive imported oil. The program created the Synthetic Fuels Corporation, which, however, was scrapped when petroleum prices plummeted in the 1980s. The carbon footprint and environmental impact of coal gasification are both very high, as are the costs. Because of that, it is not on the world's energy stage, and will be in development for awhile longer, according to the specialists.

- There are other, less known and less explored, alternatives to the fossils. Some of these are the so-called methane hydrates; frozen natural gas deep under the permafrost. There are huge quantities of this potential energy source, which could provide a major part of our energy balance if and when appropriate and safe extraction technology is fully developed.

- Solar and wind are touted as potential alternatives to using the fossils today, and as the most promising for future generations. We will review these energy sources in more detail later in this chapter.

Any energy sources that can be used instead of fossils could be called "alternative" energy sources.

Some of the alternative energy sources—but not all—are also considered to be renewable.

Renewable Energy

Renewable energy is energy provided by any source that is naturally renewable, and which is constantly or periodically reproduced. The renewables are supposed to retain their quantity and quality indefinitely.

For example, solar and wind are constantly present and reproduced non-stop day after day. Their quantity and quality is, and is expected to remain, constant on a global level for the millennia to come.

The most common and widely accepted renewable energy sources today are solar and wind energy.

- *Solar energy*—the use of sunlight, which can be converted into thermal (heat) energy and/or electric power.

- *Wind energy*—the power of the wind currents used to generate electricity by converting it into mechanical and/or electric power.

Note: The problems with wind and solar is their availability and variability, which properties are at times interchangeable. Solar energy, for example, renews itself constantly, if and when available. It is theoretically available all through the day at any place on the globe, but is often hidden by clouds, fog, and smog, so that its practical application could be severely limited. Clouds are the main reason for its variability as well.

Similarly so, wind currents theoretically exist all the time at anyplace, but their intensity and duration are not always suitable for conversion into electric power.

Because of their time and location inconsistencies, solar and wind are "variable" energy sources.

This makes them intermittent and not fully reliable for large-scale power generation. So, until new ways to store their energy for use 24/7 are implemented, solar and wind power use will be marginal.

- *Geothermal energy* uses the earth's internal heat to heat buildings or generate electricity. It is available at all times at anyplace where we access it; limited only by our technological capabilities.

- *Ocean waves and tides* are a truly renewable energy source that can be used for power generation. Expansion is limited by location and our technical capabilities...for now.

The Not-So-Renewables

There are several important energy sources in a class of their own, since they are not totally renewable, but do not fit in other categories either. These are hydropower, nuclear power and bio-energy.

- Hydropower depends on plentiful water supply all through the year. In theory, it is renewable, because rain and snow replenish the global freshwater supply. But in reality, the global climate warming is disrupting the natural water cycle, which results in long and severe draughts all over the world.

  Hydropower is therefore a variable (semi-renewable, if you will) power source, due to variations in the yearly snow cover and rainfall. During long draught periods, as in the U.S. Southwest, some parts of Africa, etc., the water reservoirs' level falls below optimum levels. At such times the hydropower plants operate at a reduced capacity. In isolated cases they cease power generation altogether.

  In the long run, global warming promises to bring more serious and longer lasting draughts, so we must question the future of hydropower altogether. Because of that, we hesitate to put it in the renewables category.

- *Nuclear power* is the cleanest, and most reliable power generator today. Its safety record, however, puts it in the "guilty until proven innocent" category. It will take a long time for nuclear energy to clear its name after the massive disasters at Chernobyl and Fukushima. Another Fukushima-like accident might cripple the nuclear power industry beyond repair.

  Also, uranium ore is a fossil material *per se*, in limited quantity, which will eventually be depleted. For these reasons nuclear power cannot be considered as a renewable energy source.

- *Bio-energy* is derived from agricultural and other plant materials which depend on a number of factors, such as rain, local soil conditions, human intervention, etc. While these energy sources are officially considered renewable, their dependence on water—under the threat of serious climate warming and increasing draught conditions—makes their long-term "renewables" status less than certain. At the very least, they're not completely unreliable but are variable from year to year.

- *Hydrogen fuel* is in a category of its own, since it is a byproduct from other energy sources, such as solar and wind. It can be used directly for power generation and as fuel to power spaceships and land vehicles via fuel cells.

  Its future is closely related to that of the renewables, because they can produce cheap hydrogen fuel. Thus produced fuel can be then stored, transported, and used for power generation and vehicle fuel—a win-win situation for the hydrogen and the renewables...in the not-so-distant future.

Finally, the broadest and most difficult of the terms—cleantech:

Cleantech

Cleantech, or "green tech," is defined as a diverse range of products, services, and processes that harness or produce clean and renewable materials, energy sources, and goods in order to:

a) reduce the use of natural resources,
b) cut or eliminate pollution and waste, and
c) improve people's lives.

So basically:

Any technology or product, part of the Cleantech sector, must be efficient, free of pollution, safe to use, and not causing any mess or damages all through its life span.

The Cleantech sector today includes wind power, solar power, nuclear power, hydropower, biomass, biofuels, "green" transportation, electric motors, green chemistry, efficient lighting and heating, grey water use, and many other energy efficient technologies and products.

Note please that some non-renewable technologies fit in the "cleantech" category because they are basically clean during power generation. None of them, however, is absolutely clean during the entire cradle-to-grave life cycle of the fuels—from raw materials production to use in power plants.

For example, nuclear power is often included in the "cleantech" category, simply because during use it emits much less GHGs than the other fossil fuels. Nevertheless, when the entire life-cycle is considered, we see a lot of pollution and many dangers coming from the mining, transport and use of nuclear materials. Not to mention the threat of nuclear accidents hanging over every nuclear power plant in the world.

So "cleantech" is basically a means to create electricity and fuels with a *smaller (not necessarily absolute zero)* environmental footprint in order to minimize or eliminate pollution of land, water, and air. The only question here is, how small is "*smaller*," so the debate on this subject continues.

Cleantech is the most misunderstood, misinterpreted, misused, and abused part of today's energy revolution.

The reasons for this discrepancy are many, but most often the confusion is due to mixing and lumping together, and using interchangeably the different elements in a, b, and c above.

Also, some technologies might be somewhat cleaner than others, and yet not part of the cleantech group. We must remember that "*cleaner*" is not always "*clean*." Not all alternative and renewable technologies are clean enough to be part of Cleantech, at least during parts of their life cycle. Some of these provide limited advantages, but are not truly Cleantech material.

While many technologies could qualify for the Cleantech label, they must be evaluated, judged, and classified per the level of their "cleanliness." For example, solar and wind energy sources are 95-96% clean, as far as the pollution footprint during their cradle-to-grave lifespan is concerned. This is so because, although they are nearly 100% clean during operation, there is significant pollution generated during the equipment manufacturing, transport, installation, and decommissioning stages.

There are also exceptions within the "cleantech" technologies. For example, Cadmium Telluride (CdTe) and SIGS solar cells and modules, which are part of

the solar cleantech sector, contain significant amounts of toxic, poisonous, and carcinogenic materials. While these are part of the "renewable" technology class, they contain harmful substances, and although they may not be dangerous in small installations, when millions of toxic solar modules are installed in a large-scale power plant under blistering desert sun, the danger is multiplied several million-fold. Since we have no experience with these products, and there are no precedents, we might be exposing ourselves to a solar Chernobyl someday in the future...under the name of "clean tech."

Biofuels are also clean in the gas tank, but a lot of pollution is created during the crop growing, harvesting, transport, and processing. In addition, the produced bio-fuels also emit significant pollution during their burning in power plants or vehicles. Although their pollution is much less than the fossils, millions of gallons are burned every day, so we see a lot of pollution coming from them too, which cannot and should not be ignored.

So, although we put the biofuels in the Cleantech category for being cleaner than the fossils, they are quite dirty—since there are huge amounts of energy used and pollution emitted during their lifetime.

Similarly, all Cleantech technologies must be ranked by the level of their efficiency in generating benefits vs. the pollution and other damages they create now or could create in the future.

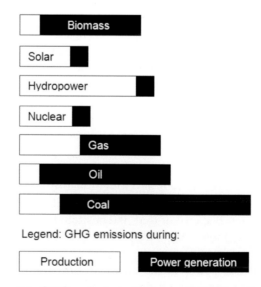

Figure 7-1. GHG emissions during production and power generation

Figure 7-1 shows the amount of GHG pollution emitted by the different technologies during different stages of their life span. Biomass, oil and coal, for example, emit a relatively low amount of pollution during their production, but a lot during use (power generation).

Solar, hydro- and nuclear-power, on the other hand, pollute mostly during their production (equipment manufacturing, transport and installation, hydro-dam construction, etc.), but do not emit much during use (power generation).

Natural gas is somewhat different, as it emits almost as much pollution during its production as during power generation. Yes, this includes damages done to property and the environment during hydrofracking operations, which in the long run might exceed all other damages combined.

The main question is what is "cleantech," and how much can we stretch the term to accommodate materials and products that are not completely "safe and clean." This clarification is absolutely needed, because while we drive towards energy security we also want to make sure that we would survive all impacts its products and services might have on us and our children.

There is no point in living in an energy secure society while gasping for breath as a consequence of energy materials poisoning or excess radiation. This important concept is sometimes lost on politicians and regulators, and is simply ignored by many corporate officials.

This, however, is not China, and we the people have a voice. We just have to understand the issues at hand and exercise our will. We should not be led blindfolded to another disaster by political and corporate self-interests.

Now, let's take a look at electric power generation using renewable energy sources.

Power Generation and Use

The energy generation and use in the U.S. during 2013 is reflected in Table 7-1. The power generation trend is depicted in Figure 7-2, which clearly shows the decline of coal-fired power generation, which is being replaced by natural gas, and the rise of thenewables.

Figure 7-2 shows clearly that, a.) crude oil dominates the transportation sector, b.) natural gas is quickly replacing coal for electric power generation, and c.) the renewable energy sources are increasing in size, but very slowly.

The "country of California" generates and uses more energy than most U.S. states, and even more than many countries. See Figure 7-3. Within a decade or two, the power generation in California switched to the cleaner technologies—natural gas, nuclear and hydropower—mostly due to the increased level of awareness

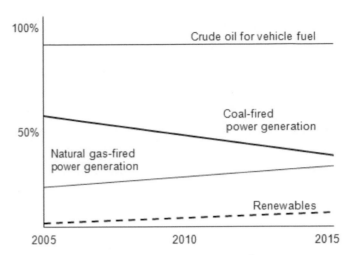

Figure 7-2. Power generation and use

in the state, with the emphasis on "clean" and even "cleaner" technologies and methods of power generation and use.

Note: California is in the midst of unprecedented long and severe draught, which is impacting all areas of its economy. Hydropower is also affected, and if the draughts trend continues, the state has to adapt accordingly, at which point hydropower might be the first vic-

Table 7-1. U.S. energy generation and consumption, 2013

Electric power generation (USA, 2013)
- Coal 36%
- Natural Gas 29%
- Petroleum 1%
- Nuclear 19%
- Hydropower 7%
- Renewables 8%
  — Biomass 1.48%
  — Geothermal 0.41%
  — Solar 0.23%
  — Wind 4.13%
  — Other < 1%

Total Energy consumption (USA, 2013):
- 40% as electric power generated by: 36% coal, 29% natural gas, 1% petroleum, 19% nuclear, 7% Hydropower, and 8% Renewables
- 29% in transportation used as: 95% crude oil, 3% renewables, and 2% natural gas
- 21% for industrial operations used as: 42% crude oil, 40% natural gas, 10% renewables, and 8% coal
  — 10% for residential and commercial use in form of: 78% natural gas, 18% crude oil, 2% coal, and 2% renewables.

tim (partially due to global warming).

At the same time, the renewables are growing fast, but much less than the major power generators. This balance will remain for the next several decades, with the renewables catching up after the mid-century mark.

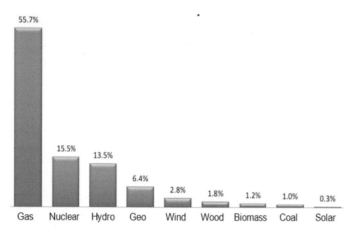

Figure 7-3. California energy generation

Energy generation and consumption in the country increased much faster than energy production during the 20th century, and now the difference is met through imports, especially crude oil used for transportation purposes.

Table 7-2. World energy consumption (in PWh)

Fuel	U.S.	World
Oil	11.71	50.33
Gas	6.50	31.65
Coal	6.60	37.38
Hydro	0.84	8.71
Nuclear	2.41	8.14
Renewables	0.95	1.38
Total	29.26	138.41

The U.S. consumes almost ¼ of the total global energy and much more than any other country in the world. That won't change, but there is a push in the U.S. energy sector towards replacing coal with natural gas, reducing crude oil imports, increasing renewables power generation, and introducing and enforcing carbon reduction and energy efficiency measures.

This is all good for our energy security, and as the awareness increases so do the benefits of using less fossil energy. The renewables have a huge role to play in all this too, but it will take some time to figure out all technical, administrative, and financial issues.

Let's start with a closer look at solar energy.

Solar Energy

Energy from the sun (sunlight) is one of the most abundant, clean, and cheap energy sources available to man. Sunlight has been used throughout the centuries for cooking, heating water, and other applications.

Only about 50 years ago did man start to use sunlight to generate electricity, albeit at a slow pace. Finally, after many false starts and hesitation by governments and companies, the global solar industry took off during the last 5-6 years.

While it is still in its infancy, solar energy used for power generation has the potential of becoming a major factor in our energy security. It also promises to be the power of the future generations...since we won't leave any fossils for them.

In order to produce electricity, sunlight must be captured and converted into thermal or electric energy suitable for human consumption. As we discussed previously, sun energy races toward the Earth at very high speed, and if its path were not obstructed by space junk or clouds and dust in the atmosphere, it would arrive to us at full power.

Nevertheless, enough sunlight reaches the Earth's surface which contains sufficient energy to allow generation of large quantities of electric power.

The sunlight falling on Earth every day is equal to the total power used by the entire world during an entire year.

We call the sunlight falling on the Earth's surface "beam" or "direct" radiation. Arriving "directly" from the sun it is most powerful and measures around 1367 W/m² (watts per square meter) just above our atmosphere.

Its power drops after crossing the atmosphere to approximately 900-1,000 W/m² as measured at noon in the deserts during the summer months, and much less than that in other parts of the globe and during different seasons of the year.

When the sunlight hits clouds, dust, or man-made gasses in the air, it gets scattered. We call what reaches the Earth's surface in this case "diffused" radiation. Diffused radiation has properties very different from those of the direct radiation.

It contains much less energy, thus PV modules will produce less power under diffused radiation, in some cases much less. Particularly, concentrating thermal or PV equipment is affected by this diffusion, and this will cause it to lose its focus and operate well below its maximum efficiency rating, if at all, under extreme diffused sunlight conditions.

Sunlight hitting the Earth is reflected from its surface, and we call this effect "albedo." Different materials have different reflecting properties, but most do reflect and some reflect a lot. Take fresh snow, for example. It will reflect almost 80% of the light falling on its surface. Water, on the other hand, absorbs most of the sunlight and gets heated in the process. Thus, reflected sunlight can be captured by our PV modules installed nearby as well.

The albedo always has an effect on PV module performance, so it should be taken into consideration, especially in areas with snow cover or other highly reflective ground surface cover.

Another factor that affects the amount of energy available for conversion into electricity is the time of day and location on the globe, both of which determine the distance that the sunlight travels once it enters the Earth's atmosphere.

At certain times of the day and year, the sun seems to be overhead at a 90-degree angle to the Earth's surface, and this is when sunlight travels the shortest path and is the strongest. We call this Air Mass = 1 (or AM 1).

AM 0 is measured above the atmosphere and is much stronger than AM 1.

In the early morning and later afternoon, the sun is at a sharper angle and sunlight travels a longer path through the atmosphere, so the angle decreases (approaching 45 degrees), the sunlight has a longer path to travel and the AM number increases: AM 1.15, 1.5 etc. depending on the angle.

The sharper the angle of the sun rays, the larger the AM number and the objects' shadows. AM 1.5, for example, is measured at an angle close to 45 degrees.

The air mass number can be determined by the formula:

$$AM = 1 + sh^2$$

where:

h is the object's height, and
s is its shadow length

The revolution of the Earth around the sun and its rotation on its axis produces seasonal and daily effects, which vary by location on the globe. The location is measured on the world map in terms of longitude (east-west direction), and latitude (south-north direction). The intersection of these provides us with a precise point on the map. Locations close to the equator are usually the closest to the sun, while the poles are the farthest. This is why the equator locations are always warm, and the poles are always cold.

All these components taken together represent what we call "global radiation," which is a very important factor in the proper design, installation and operation of solar energy generating systems. Solar professionals need to be very familiar with it, if a properly designed PV system is their goal.

The Solar Technologies

There are today a number of technologies capable of producing large amounts of electric power from sunlight, which can be used for residential or commercial applications.

The technologies that are suitable for large-scale photovoltaic (PV) power generation, and which are most important from an energy security point of view, have special characteristics and requirements which need to be thoroughly understood during the planning, design, installation, operation and maintenance stages of each solar power plant.

These are herein referred to as commercial and utilities power generation technologies in order to distinguish them from residential and small commercial types of solar power generating systems. They are also known as "large-scale," or "utility" type, usually grid connected installations.

Figure 7-4 shows the solar energy technologies in use today, divided into three major categories: thermal solar, solar photovoltaic, and the combination of solar and other technologies; the hybrids.

In more detail:

Solar thermal technologies:
• Flat plate water heater
• Stirling engine dish
• Parabolic troughs
• Power towers
• New and exotic solar thermal technologies

Solar photovoltaic technologies
• Crystalline silicon PV modules
• Thin film PV modules
• HCPV trackers
• New and exotic PV technologies

Solar hybrid technologies
• Thermal solar-fossil
• Thermal solar-wind

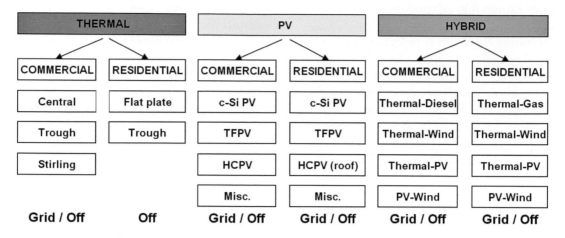

Figure 7-4. Major solar technologies and their uses

- Thermal solar-PV
- PV-wind

Let's take a brief look at some of these technologies.

SOLAR THERMAL TECHNOLOGIES

We will review here the solar thermal and thermo-electric technologies, focusing on the concentrated solar power (CSP) technologies, since they are most suitable for commercial and utilities power generation.

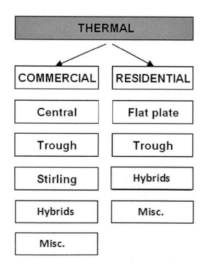

Figure 7-5. Solar thermal technologies in the 21st century

Flat Plate Solar Water Heater

Flat plate water heaters (FPWH) have been, and still are, used in residential, commercial and industrial applications, primarily for heating water in homes, laundromats, restaurants, public parks, car washes, and canning and bottling facilities.

These heating systems could be used practically anywhere where low temperature hot water is needed during the day. Adding a storage tank could provide water for use during the night and/or cloudy days. In all cases, they are truly "thermal" systems designed to provide hot water.

Flat plate water heaters are the simplest and cheapest energy conversion devices today, consisting of a frame onto which a heat exchanger plate is mounted. Water runs through the heat exchanger plate and absorbs the sunlight energy, thus heating the plate and the water (or other heat absorbing liquid) running through it.

The materials, as well as the manufacturing, installation and operation procedures are straightforward and relatively inexpensive. The return on investment (ROI) is one of the highest in the industry, if the systems are properly designed, installed and operated. There are also a number of incentives today which make it even more feasible and desirable to own and operate such a renewable energy system.

The simplicity and practicality of this type of solar power system are obvious, so there is an increased effort lately worldwide to encourage their wider use. Significant numbers of these systems could replace millions of barrels of oil, and add to a significant increase in our energy security.

There is renewed interest and increased financial support for solar water heating in the U.S., and we expect to see many new installations in the near future. Due to its simplicity and low price, this technology will also see increased demand and implementation in the developing countries. The combined effect of the expan-

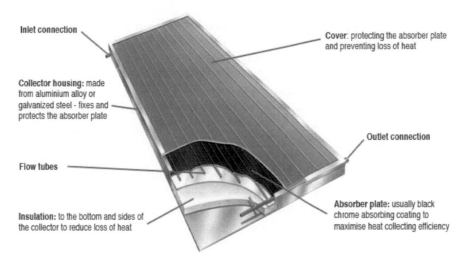

Figure 7-6. Flat plate solar water heater

sion of solar water heating would benefit the world by saving crude oil and limiting the GHG pollution.

CSP Technologies

Concentrated solar power (CSP) technologies can be divided into three major types.

- The Stirling engine-dish tracker
- The parabolic trough tracker
- The power tower (central receiver)

These three technologies have one thing in common: they all use trackers and optics of some type to optimize their efficiency. They also require direct solar insolation (deserts are the best) and relatively flat land with slopes not exceeding 3-5 percent to accommodate the solar collectors.

The area of land required depends on the type of plant, but it is about three-four acres per installed megawatt (MW). A commercial-scale CSP facility in the range of 100 MW or larger requires in excess of 500 acres for the collectors and whatever else is needed for the proper and safe infrastructure.

Unlike solar photovoltaic technologies, which use semiconductors to convert sunlight directly into electricity, CSP plants generate electricity by converting sunlight into heat first. Much like a reflective mirror, their reflectors focus sunlight onto a receiver. The heat absorbed by the receiver is used to move an engine piston (Stirling engine), or generate steam that drives a turbine to produce electricity (parabolic troughs and power tower).

Power generation after sunset is possible also by storing excess heat in large, insulated tanks filled with liquids or molten salt during the day and using the stored heat to generate power at night.

Since CSP plants require high levels of direct solar radiation to operate efficiently, deserts make the most ideal locations. As a matter of fact, these types of systems cannot operate efficiently in any other environment.

A 2010 study, indicates that about 90% of fossil fuel-generated electricity in the U.S. and the majority of U.S. oil usage for transportation could be eliminated by using solar thermal power plants. The total cost would be less than it would cost to continue importing oil, in addition to reducing the GHG emissions.

The land requirement for the CSP plants in this case would be roughly 15,000 square miles in the southwestern U.S. deserts, or the equivalent of 15% of the land area of the entire state of Nevada. While this may sound like a large tract of land, in the long run CSP plants use less land per equivalent electrical output than large hydroelectric dams, when flooded and wasted land is included. This is also less than the land used for the equivalent amount of coal-fired power generation, when factoring in the land used for mining, waste disposal, and power generation.

Another study estimates the possibility of using a combination of CSP and PV plants to produce 65% of U.S. electricity and 35% of total U.S. energy including transportation by 2050…overly optimistic, yes, but doable nevertheless. Something to think about—hopefully before we run out of the fossils!

The Stirling Engine

One of the most elegant and flexible solar thermal power conversion technologies today is the Stirling engine dish system. It consists of mirrors mounted on a frame which is continuously tracking the sun all through the day. The mirrors focus the reflected sunlight onto the receiver of the Stirling engine mechanism which is activated by the heat and turns on a shaft, connected to the rotor of an electric generator similar to that of the alternator of your car. The generator rotor turns with the engine shaft and generates electric power while the sun is shining and the receiver is hot enough to activate the engine and rotate the shaft.

A Stirling engine system is actually a solar electricity generator because the heat produced by the mirrors attached to it is converted into electric energy on the spot, so small installations of a few units are possible; something that is just not practical with the other CSP technologies.

The Stirling engine needs cooling just like a car engine for more efficient operation and to cool the engine walls, bearings and other moving parts.

The mirror, or mirrors, are mounted on a metal frame which is driven by two motor-gear assemblies (x-y drives), programmed to move the frame in such a fashion that it follows the sun's movement precisely all day long, thus providing accurate focusing of the sunlight onto the heating plate of the Stirling engine.

Figure 7-7. Stirling engine dish

When the plate gets hot enough, the air in one of the cylinders in it is compressed and forces the piston in it to move up. This action forces the piston in the other cylinder (which is simultaneously cooled) to move down.

Eventually, the compression in the second cylinder increases to the point that its piston is forced to go back, thus forcing the piston in the first cylinder to assume its initial position. The cycle repeats over and over while there is enough heat to maintain the process.

The Stirling engine function, under ideal operating conditions, can be represented by four cycles, or thermodynamic process segments, of interaction between the working gases, the heat exchanger, pistons and the cylinder walls.

The Stirling engine is a very ingenious and efficient piston engine, without the noise and exhaust of internal combustion engines. As a matter of fact, it can be classified as an "external combustion" engine. The gasses inside the cylinders are not exhausted, so there is no pollution and there are few moving parts with very little noise, so it can be used virtually anywhere.

A number of attempts were made during the last decade to build large-scale projects using Stirling engine design, but unfortunately they all were unsuccessful. Thus, the future of the Stirling engine as a large-scale power generator remains uncertain.

The Parabolic Trough

Parabolic troughs can focus the sunlight many times its normal intensity on a receiver pipe, where heat transfer fluid flowing through the pipe is heated and then used to generate steam which powers a turbine that drives an electric generator.

Figure 7-8. Parabolic trough

Parabolic trough solar systems consist of a frame in a parabolic trough shape in which glass, metal or plastic reflectors are mounted to focus the sun's energy onto a receiver pipe running above and in parallel with the trough's length.

The collectors are aligned on an east-west axis and the trough is rotated up and down (north-south), following the sun as needed to maximize the sun's energy input to the receiver pipe.

The receiver pipe, or heat collection element (HCE), is centered at the focal point of the reflectors and is heated by the reflected sunlight to very high temperatures. Liquid of some sort—usually mineral or synthetic oil—is pumped through the receiver pipe and is heated in the process.

The HCE of the parabolic trough units is usually composed of a metal pipe with a glass tube surrounding it, and with the space between these evacuated to provide low thermal losses from the pipe. The pipe is coated with a material that improves the absorption of solar energy.

Several improvements have been made or are underway to improve performance, the most significant of which is the seal between the glass and the pipe, which seal has not been as reliable as desired and development of better seal materials/seal configuration is still underway.

Parabolic troughs can focus the sunlight many times its normal intensity on the receiver pipe, where heat transfer fluid (HTF—usually mineral or synthetic oil) flowing through the pipe is heated. This heated fluid is then used to generate steam which powers a turbine that drives an electric generator. The collectors are aligned on an east-west axis and the trough is rotated north-south, following the sun as needed to maximize the sun's energy input to the receiver tube.

Parabolic trough power plants, also called solar electric generating systems (SEGS), represent the most mature CSP technology, with the most installed capacity of all CSP technologies.

The first SEGS solar trough plant started operating in 1984, with the last one coming on line in 1991. Altogether, nine such plants were built; SEGS I-VII at Kramer Junction and VIII and IX at Harper Lake and Barstow respectively. All SEGS plants are still operating.

Natural gas supply added to the plant converts it into a "hybrid," which allows it to generate electricity 24/7 and contributes up to 25% of the output.

There are a number of parabolic trough power plants operating around the world and many others are planned.

Power Tower

The power tower (or central receiver) power generation uses methods of collection and concentration of solar power based on a large number of sun-tracking mirrors (heliostats) reflecting the incident sunshine onto a centrally positioned receiver (boiler) mounted on the top of a high tower, usually in the middle of the collection field.

Eighty to 95 percent of the reflected energy is absorbed into the working fluid which is pumped up the tower and into the receiver. The heated fluid (or steam) returns down the tower and is fed into a thermal electrical power plant, steam turbine, or an industrial process that uses the heat.

Figure 7-9. Power tower

The difference between the central receiver concept of collecting solar energy and the trough or dish collectors discussed previously, is that in this case all of the solar energy to be collected in the entire field is transmitted optically to a relatively small central collection region rather than being piped around a field as hot fluid.

Because of this, central receiver systems are characterized by large power levels (100 to 500 MW) and higher temperatures (540-840°C) of the working fluids. This in turn allows the creation of high quality superheated steam which is more efficient for electricity generation.

The reflecting element of a heliostat is typically a thin, back (second) surface, low-iron glass mirror. This heliostat is composed of several mirror module panels rather than a single large mirror. The thin glass mirrors are supported by a substrate backing to form a slightly concave mirror surface. Individual panels on the heliostat are also canted toward a point on the receiver. The heliostat focal length is approximately equal to the distance from the receiver to the farthest heliostat. Subsequent "tuning" and optimization of the closer mirrors is done upon installation.

Another heliostat design concept, not so widely developed, uses a thin reflective plastic membrane stretched over a hoop. This design must be protected from the weather but requires considerably less expenditure in supports and the mechanical drive mechanism because of its light weight. Membrane renewal and cleaning appear to be important considerations with this design.

In all cases, the reflective surface is mounted on a pedestal that permits movement about the azimuth and elevation axis. Movement about each axis is provided by a fractional-horsepower motor through a gearbox drive. These motors receive signals from a central control computer that accurately points the reflective surface normally halfway between the sun and the receiver.

System design and evaluation for a central receiver application is performed in a manner similar to that when other types of collectors are used. Basically, the thermal output of the solar field is found by calculating the collection efficiency of the boiler and multiplying this by the solar irradiance falling on the collector (heliostat) field, minus some optical, transmission and other losses.

Power tower technology for generating electricity has been demonstrated in the Solar One pilot power plant at Barstow, California, which went into operation in 1982. This system consists of 1818 heliostats, each with a reflective area of 430 ft^2 covering 72 acres of land. The receiver is located at the top of a 298-ft-high tower and produces steam at 960°F at a maximum rate of 42 MW (142 MBtu/h).

The future of the power tower concept, similar to all CSP technologies, is determined by a number of factors, such as sunlight availability and intensity, efficiency, economics and the availability of cooling water.

SILICON PV TECHNOLOGIES

Photovoltaic (PV) systems are the most versatile and practical for all types of use solar power generating technologies. Because of that, their use has been increasing rapidly during the last several years. At the same time, their quality has improved, while the prices have decreased sharply.

As seen in Figure 7-10, there are a number of different PV technologies. Silicon-based PV technology is the most mature and most widely used today. It dominates the solar markets by far and will continue doing so in the foreseeable future

The most common, commercially available, and most preferred in PV products today are crystalline silicon (c-Si) solar cells and modules. Thin film (TF) PV cells and modules are also used in large quantities, but are limited by lower efficiency and unproven safety and reliability issues.

c-Si PV

Crystal silicon (c-Si) technology uses silicon material, made of sand which is melted then cooled and crushed into small pieces. Thus produced metallurgical-grade silicon is melted again, and shaped into ingots while cooling. Depending on the process, thus produced ingots can contain single or multi (poly) crystal silicon material.

The silicon ingots are then sliced into thin wafers, and processed into solar cells. The solar cells are tested, sorted, interconnected and encapsulated into PV modules.

This is a simple and yet long and technologically complex process using sophisticated materials, equipment.

Silicon Solar Cells Manufacturing

The poly- and sc-Si solar cell processing sequence, as used in the mid-1990s by Alpha Solarco, Inc. in Phoenix, AZ, is outlined below. This process, or a variation of it, is used by some world class c-Si PV cells and modules manufacturers today.

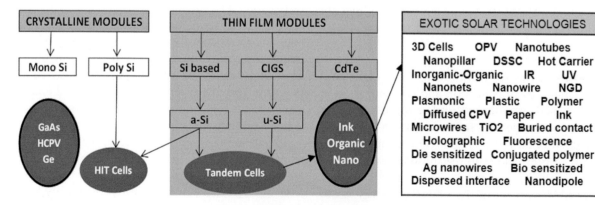

Figure 7-10. Major PV technologies

A number of variations of this process are used by other manufacturers as well.

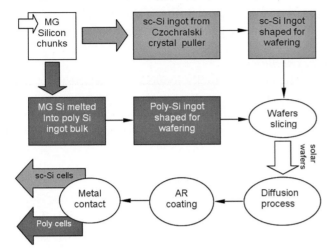

Figure 7-11. sc-Si and poly Si solar wafers and PV cells manufacturing process

The major process steps of the solar cells manufacturing sequence are as follow:

Wafers Inspection and Sorting

Incoming silicon wafers are placed on inspection tables and are inspected visually and with optical equipment. Any wafers with visible mechanical defects are rejected. The wafers are then tested with a 4-point probe, and are sorted according to their resistivity. Wafers, or wafer samples are sent to an outside lab for metal and organic contamination analysis. Results from these tests determine the level of quality of the finished cells.

Wafers Cleaning and Etch

The wafers that pass all initial inspections and tests go to the wet cleaning line and are chemically processed in special chemicals where they are cleaned and etched to remove damage and oxide formed on the surface. The wafers are then rinsed with de-ionized (DI) water and dried via spin dryer.

Surface Etch (Chemical Etch)

This process is used only on single-crystal silicon with 1-0-0 orientation. (Polycrystalline wafers cannot be textured, because the different strings of silicon have different orientation and the resulting surface is only partially and unevenly textured, if at all.) A controlled chemical solution (composition, concentration, temperature and time) etches the pyramid-like structures in the wafer surface and the surface takes on a dark-gray appearance. The pyramids blend into each other and block excess light reflection.

Each pyramid is approximately 4-10 microns high. This step has critical process parameters. The wafers are then rinsed, spin-dried, and stored in special containers for processing.

Note: In a variation of this process, wafers are loaded in a fixture two-by-two with the backs of each pair touching, so that the pyramid structure is formed only on their front surfaces.

Diffusion for P-N Junction Formation

The clean wafers are oven dried, placed in the diffusion furnace at 900-950°C in reactive carrier gasses to impregnate the wafers. POCl3 gas is used for n-type diffusion, which diffuses phosphorous atoms in the wafer's surfaces. This creates a p-n junction in the lightly boron-doped wafers.

Note: This method is easier to control and has more uniform distribution of dopant than using the spray-on diffusion liquid and belt furnace diffusion process used by many companies today. This step has critical process parameters, so any compromise will be reflected in the cells' overall performance and longevity. In a variation of this process, wafers are loaded in a fixture two-by-two with the backs of each pair touching, so that the diffusion layer is formed only, or mostly, on their front surfaces. This facilitates the processing of the back surface later on.

Mass production solar cell operations use a different method in which the wafers are sprayed with a dopant chemical and run through a conveyor belt type furnace, where the dopant is diffused in the wafers' surface. Both methods have advantages and disadvantages.

Plasma Etch for Removal of Edge Layer (diffusion etc.)

The diffusion process implants P dopant in the wafers' side edges, causing an electrical short circuit between the top and bottom (negative and positive) surfaces of the cell, so it is necessary to remove the dopant with a wet chemistry or plasma etch. Wafers are coin-stacked and etched for a brief period in an RF plasma etch reactor; only the edges of the wafer are exposed to the plasma which removes the diffusion coating.

The wafers are then etched gently in a bath of dilute hydrofluoric acid to remove any oxides formed during the plasma etch step, rinsed with DI water, and finally spin dried.

Anti-reflective Coating

AR coating is deposited on the front surface of the wafers. The purpose of the AR coating is to reduce the amount of sunlight reflected from the finished cell surface. The AR coating is deposited via chemical vapor deposition (CVD) or by spraying the chemicals on the wafers and then baking. Both methods achieve similar outcome of enhancing the solar cells' output and giving them the distinctive dark blue color (for poly solar cells).

NOTE: Different manufacturers deposit and fire the AR coating using different process parameters and sequence order. This is an important process, nevertheless, so its proper design and execution will determine the final, most important aesthetic and performance aspects of the cells.

Printing (Metallization)

Several screen printing steps are used to apply the metallization on the front and back of the wafers. First, silver paste is printed on the top surface which then becomes the front metal pattern (top contacts, or fingers). The paste is dried and the wafers are flipped for printing the back surface with aluminum paste. After drying, silver paste is printed in special slots in the dried aluminum. Then the wafers are put in the firing furnace.

Metal Firing

Thus metalized on both sides, wafers are run slowly through an IR-heated furnace where the metal pastes on the top and bottom sides of the wafers diffuse into the substrate, to make an electric contact with the p-n junction and the back surface. This step has critical process parameters.

Note: The firing of the front contacts is a very delicate process, where time and temperature are controlled to achieve the desired depth of penetration of the metal into the silicon surface. The depth of penetration determines the electro-mechanical properties of the finished cell. Specially designed automated printing-firing equipment is available for more precise and consistent process control.

Inspection, test, and Quality Control

Solar wafers and cells are inspected and tested at several stages of the process sequence. This is done by eye inspection, using magnification and other instrumentation. Electrical tests are also performed at some steps of the process. The final inspection is the most important step and must be performed by well trained and experienced operators.

Cell Flash Testing and Sorting

A certain percent of the completed wafers are placed on a test stand in the solar simulator and are illuminated for a period of time. I/V curve is generated for each cell and the output data are used to sort the cells into groups according to the I/V curve characteristics, prior to soldering and lamination into modules.

Solar cell Storage

The cells are finally loaded into cassettes, or coin-stacked (with protective material in between) and packed for ease of handling, transportation, or storage prior to laminating into solar modules or shipping to another location.

Silicon PV Modules Manufacturing

Once the materials—PV cells, laminates, glass, back cover, wiring etc.—have been received and gathered at the module production site, the module is assembled in the following sequence. See Figure 7-12.

Cell Sorting, Arranging and Soldering

Finished solar cells are flash tested and sorted by their I-V characteristics and power output. Cells that pass the test are placed in bins according to their performance and stored or taken to the module assembly area.

Wiring and Assembly

Cells are connected in a series circuit manually, or by a semi-automated soldering machine using solder coated metal ribbon (usually two in parallel) soldered to the top of one cell and to the bottom of the next cell. This process forms a string of cells which could be as long as desired but usually it is shaped to fit in the respective PV modules tray.

Electrical continuity and resistivity tests are performed on some modules to make sure that the bonds are good. "Pull" tests are done sometimes, to check the mechanical strength of the bonds. Thus, connected cells make a complete circuit (string), which is ready for lamination into a completed module.

Lamination

PV module laminators consist of a large-area heated metal platen mounted in a cabinet-like vacuum chamber. The top of the cabinet opens for loading and unloading modules. A flexible diaphragm is attached to the top of the chamber, and a set of valves allows the space above the diaphragm to be evacuated during the initial pump step and backfilled with room air during the press step. A pin lift mechanism is sometimes used to lift modules above the heated platen during the initial pump step, but most standard modules don't require it.

Laminators are available with two types of cover

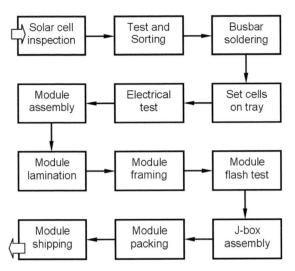

Figure 7-12. PV modules manufacturing process

opening systems: clamshell and vertical post. In the clamshell design, the cover is mounted on a hinge at the back of the laminator, which opens like the hood of a car. This leaves the laminator wide open on three sides, making it easy for an operator to load and unload modules manually. Automated belt-fed laminators, on the other hand, use the vertical post method, which lifts the cover horizontally above the process chamber.

Because the cover does not need to travel much for belt loading, the chamber opening and closing times, and resulting process steps (heating and vacuum pump down) are reduced. As a result, most high throughput module lines use belt-fed laminators with vertical cover lifts.

Temperature uniformities of ±5°C at the lamination point are sufficient for obtaining good laminations with acceptable gel content and adhesion across the module. While more uniform temperatures are available from some laminator suppliers, there is no real benefit to the module manufacturer.

A lay-up for lamination is prepared with clean top glass, and EVA film, onto which strings of wired cells are carefully placed. Sometimes the backing materials (Tedlar and back cover) are placed on top too, forming a complete module. Several lay-ups, each consisting of the above components are lined up in a large cabinet called a laminator. Using silicone vacuum blankets, the batch of lay-ups is heated and vacuum laminated at one time. After cooling, the modules are ready for use. This "batch" method is much cheaper than laminating one or two units at a time.

The excess lamination protruding from the edges of the modules is trimmed and terminal wiring is at-

tached to the end wires. In most cases, an aluminum extruded frame is assembled around the material and the unit is ready for shipping.

Note: Fully automated cell assembly and lamination lines exist today, but most low- to mid-volume assembly operations, especially those in Asia, still prefer manual lay-up and stringing operations, combined with low-throughput clamshell type laminators. This is due mostly to the availability of cheap labor. Although automating labor-intensive processes is no guaranty of a high quality product, it allows for much larger throughput and more efficient and robust quality control.

Modules Flash Testing and Sorting

After completing the assembly process by adding edge sealers, side frame, terminal box, etc., the modules are placed on a test stand; usually in a solar simulator (flasher). Here they are illuminated with a special type of light that resembles the solar spectrum at STC, for a period of time.

The temperature of the modules is kept at 25°C during the test by active or passive cooling. I-V curve is then generated and recorded for each module. The test data are used to identify, sort and label the modules according to their type, size, output, and other parameters. The modules that pass the test are packed up and shipped to the customers.

Certification Tests

All PV modules must be properly tested and certified before they are allowed on the market. For this purpose they are sent to test laboratories for official certification, which can be done by several authorized test labs. Only then can the modules be used in the destination country.

The testing process consists of a series of visual and electro-mechanical tests under the so-called *Standard Test Conditions (STC)*. These are the conditions (light characteristics, operating temperature and time) all PV modules are exposed to, in order to establish and certify their nominal performance parameters. PV modules are tested using test conditions accepted as "standard," or "standard test conditions" by the solar industry.

The STC test specs are as follow:
- Vertical irradiance E of 1000 W/m^2;
- Temperature T of 25°C with a tolerance of ± 2°C;
- Defined spectral distribution of the solar irradiance at air mass AM = 1.5.

There is also a number of temperature-cycling and mechanical (impact) tests performed on the tested modules, to ensure their mechanical and electrical integrity.

In the end, all that matters is the surface durability and output of the PV modules. Quality and reliability are also important, but these are built into the process and are not immediately obvious, nor can they be tested non-destructively.

So, accepting the final certification of the PV modules usually means that we trust the manufacturer. We trust that he has used the proper materials and procedures to ensure the quality and long-term reliability of the final product. Always remember that all we can measure at the time of purchase is the efficiency of the modules at that time. The answer to what will happen tomorrow, next year, or next decade is locked inside the module, to be revealed someday in the future.

Table 7-3 shows that mc-Si has the highest efficiency, except for some of the more exotic multi-junction solar products made out of III-V materials, the efficiency of which approaches 45% today. c-Si also has been around for the longest amount of time, thus it has the longest reliability track record.

Many of the PV technologies—thin film type especially—are too new on the market, so we must wait 10-20 years to see how they perform…especially under the extreme conditions in humid and very hot desert areas.

THIN FILM PV TECHNOLOGIES

Thin film PV (TFPV) technologies are a relatively new branch of the solar industry, which has grown much faster in popularity and size during the last several years than the other PV technologies. Since the active layers in TFPV cells and modules are deposited in the form of thin films, via thin film deposition methods, we refer to them as "thin film" PV products.

Thin films of special photovoltaic materials can produce solar cells or modules with relatively high conversion efficiencies, while at the same time using much less semiconductor material than c-Si cells. In addition, thin film equipment and manufacturing methods allow efficient, cheap, fully automated mass production which means lower price, and which is the main reason for their success lately.

TFPV technologies, however, have reduced efficiency (average 9-12%), which is expected to increase some in the future (in mass production mode), but it can never reach that of their cousins—the silicon PV cells and modules, and especially the III-V based PV cells and modules.

Because of this, and other issues, the TFPV sector is losing ground to the competition, a disparity that is expected to grow. Nevertheless, due to their versatility, TFPV products have become very popular for use in a number of applications. Recently they have gained a share in large-scale installations as well.

Some types of TFPV modules also show better efficiency under reduced solar radiation than the c-Si competition. This is very useful in many regions with cloudy climates, which could account for their quick rise in European and other world energy markets, although this trend is not consistent.

The major types of TFPV technologies considered for commercial and large-scale installations are:
- Cadmium telluride thin films
- CIGS thin films
- Amorphous silicon thin films
- Silicon ribbon
- Epitaxial silicon thin films
- Light absorbing dyes thin films

Table 7-3. Efficiency of different c-Si and thin film solar cells and modules.

Solar cell material	Cell efficiency η_z (laboratory) (%)	Cell efficiency η_z (production) (%)	Module efficiency η_M (series production) (%)
Monocrystalline silicon	24.7	21.5	16.9
Polycrystalline silicon	20.3	16.5	14.2
Ribbon silicon	19.7	14	13.1
Crystalline thin-film silicon	19.2	9.5	7.9
Amorphous silicon[a]	13.0	10.5	7.5
Micromorphous silicon[a]	12.0	10.7	9.1
CIS	19.5	14.0	11.0
Cadmium telluride	16.5	10.0	9.0
III-V semiconductor	39.0[b]	27.4	27.0
Dye-sensitized call	12.0	7.0	5.0[c]
Hybrid HIT solar cell	21	18.5	16.8

- Organic/polymer thin films
- Ink thin films
- Nano-crystalline cells
- Indium phosphide
- Single-junction III-V cells
- Multi-junction cells
  — Gallium arsenide based cells
  — Germanium based cells
  — CPV solar cells

The major TFPV technologies today are the cadmium telluride (CdTe), CIGS, and amorphous silicon (a-Si) thin films.

In more detail:

Cadmium Telluride (CdTe)

Cadmium telluride (CdTe) is a type of solar cell and module based on thin films of the heavy metal cadmium and its compounds, cadmium telluride (CdTe) and cadmium sulfide (CdS). CdTe is an efficient light-absorbing material, quite adaptable for the manufacture of thin-film solar cells and modules.

Compared to other thin-film materials, CdTe is easier to deposit in mass production environments and more suitable for large-scale production.

CdTe bandgap is 1.48 eV, which makes it almost perfect for PV conversion purposes. At 16.5% demonstrated efficiency in the lab, it is a candidate for a major role in the energy future. Mass production modules are sold with 8-9% efficiency. No significant increase is expected with the present production materials and methods, although manufacturing costs are down—at or below \$1.0/Wp.

With a direct optical energy bandgap of 1.48 eV and high optical absorption coefficient for photons with energies greater than 1.5 eV, only a few microns of CdTe are needed to absorb most of the incident light. Because only very thin layers are needed, material costs are minimized, and because a short minority diffusion length (a few microns) is adequate, expensive materials processing time and costs can be avoided.

The CdTe PV module structure, as shown in Figure 7-13, consists of a front contact, usually a transparent conductive oxide (TCO), deposited onto a glass substrate. The TCO layer has a high optical transparency in the visible and near-infrared regions and high n-type conductivity. This is followed by the deposition of a CdS window layer, the CdTe absorber layer, and finally the back contact.

For high-volume devices, the CdS layer is usually deposited using either closed-space sublimation (CSS)

Figure 7-13. CdTe thin film PV module

or chemical bath deposition, although other methods have been used to investigate the fundamental properties of devices in the research laboratory. In all cases, mass production and automation is possible, which is the greatest advantage of this technology.

The CdTe p-type absorber layer, 3-10 μm thick, can be deposited using a variety of techniques including physical vapor deposition (PVD), CSS, electrodeposition, and spray pyrolysis. To produce the most efficient devices, an activation process is required in the presence of CdCl2 regardless of the deposition technique. This treatment is known to recrystallize the CdTe layer, passivating grain boundaries in the process, and promoting inter-diffusion of the CdS and CdTe at the interface.

Forming an ohmic contact to CdTe, however, is difficult because the work function of CdTe is higher than all metals. This can be overcome by creating a thin p+ layer by etching the surface in bromine methanol or HNO3/H3PO4 acid solution and depositing Cu-Au alloy or ZnTe:Cu. This creates a thin, highly doped region that carriers can tunnel through. However, Cu is a strong diffuser in CdTe and causes performance to degrade with time. Another approach is to use a very low bandgap material, e.g. Sb2Te3, followed by Mo or W. This technique does not require a surface etch and the device performance does not degrade with time.

CdTe PV modules manufacturing is a sophisticated process—much more so than that of the conventional c-Si modules process, which uses simple 1970s manufacturing equipment, materials and processes. CdTe TFPV modules are manufactured with the help of modern, complex and expensive semiconductor type equipment and processes. Because of that, the precision and accuracy of the resulting process steps, and ergo the quality of the final product, are limited only by the quality of the materials and supplies, and the capabilities of the engineers, technicians and operators on the production lines.

CdTe thin-film solar modules are now being mass produced very cheaply, and it is expected with economies of scale that they will achieve the cost reduction needed to compete directly with other forms of energy production in the near future. Since CdTe thin film PV devices still have far to go to achieve maximum efficiencies, it will be interesting to see which materials and methods are most successful.

The most efficient CdTe/CdS solar cells (efficiencies of up to 16.5%) have been produced using a Cd2SnO4 TCO layer which is more transmissive and conductive than the classical SnO2-based TCOs, and including a Zn2SnO4 buffer layer which improves the quality of the device interface.

Future R&D efforts might bring additional efficiency increases, but transferring these in the mass production line is not that fast or easy. The big problem is that just because the efficiency is increased does not mean that the product is safe and reliable enough for long-term use under harsh conditions. So, in addition to the aggressive lab optimization efforts, long-term field tests are needed for every new and improved type of TFPV product.

Note: Remembering that some of the materials in the TFPV modules are poisonous, toxic, and carcinogenic (Cd, Te, In, As, etc.), we must not jump up and down excited about higher efficiencies. Let's do the proper testing before we fill the deserts with untested, questionable, and even dangerous products. Better safe than sorry.

CIGS

Early solar cells of this type were based on the use of CuInSe2 (CIS). However, it was rapidly realized that incorporating Ga to produce Cu(In,Ga)Se2 (CIGS) structure, results in widening the energy bandgap to 1.3 eV and an improvement in material quality, producing solar cells with enhanced efficiencies. CIGS have a direct energy bandgap and high optical absorption coefficient for photons with energies greater than the bandgap, such that only a few microns of material are needed to absorb most of the incident light, with consequent reductions in material and production costs.

The best performing CIGS solar cells are deposited on soda lime glass in the sequence—back contact, absorber layer, window layer, buffer layer, TCO, and then the top contact grid. The back contact is a thin film of Mo deposited by magnetron sputtering, typically 500-1000 nm thick.

The CIGS absorber layer is formed mainly by the co-evaporation of the elements either uniformly de-

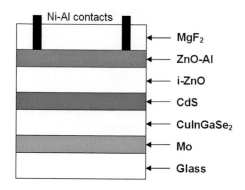

Figure 7-14. CIGS PV module

posited, or using the so-called three-stage process, or the deposition of the metallic precursor layers followed by selenization and/or sulfidization. Co-evaporation yields devices with the highest performance while the latter deposition process is preferred for large-scale production. Both techniques require a processing temperature >500°C to enhance grain growth and recrystallization.

Another requirement is the presence of Na, either directly from the glass substrate or introduced chemically by evaporation of a Na compound. The primary effects of Na introduction are grain growth, passivation of grain boundaries, and a decrease in absorber layer resistivity.

The junction is usually formed by the chemical bath deposition of a thin (50-80 nm) window layer. CdS has been found to be the best material, but alternatives such as ZnS, ZnSe, In2S3, (Zn,In)Se, Zn(O,S), and MgZnO can also be used.

The buffer layer can be deposited by chemical bath deposition, sputtering, chemical vapor deposition, or evaporation, but the highest efficiencies have been achieved using a wet process as a result of the presence of Cd2+ ions. A 50 nm intrinsic ZnO buffer layer is then deposited and prevents any shunts. The TCO layer is usually ZnO:Al 0.5-1.5 μm. The cell is completed by depositing a metal grid contact Ni/Al for current collection, then encapsulated.

CIGS solar cells have been produced under lab conditions with efficiencies of 19.5%, and lately modules with efficiencies of 15.7% were verified as well. Commercial, mass produced, CIGS PV module efficiency, however, is still lower than CdTe PV modules—and this will have a major impact on their future unless ways to increase their efficiency and reduce their costs are found soon.

CIGS TFPV modules have similar problems as those plaguing CdTe TFPV technologies. They have low efficiency, require larger mounting infrastructure, exhibit power loss under excess heat and have a significant annual degradation rate. Scarcity of materials and related toxicity issues are, as in the CdTe PV case, very important, but shoved on the back burner for now.

These issues must be evaluated from the point of view of large-scale installations, where thousands and millions of these modules will be installed. In these cases, reliability issues will cause mass failures and will delay further the development of these technologies.

Also, the small amounts of toxic materials in each module are multiplied mega times in large-scale installation and might become a substantial threat to the environment, the local wildlife and humans working and living in the area.

Special measures must be taken for proper disposal or recycling of these modules too. That fact should not be ignored, because it cannot be avoided and could become a great burden towards the end of life of large-scale TFPV power plants.

Amorphous Silicon

By definition, amorphous silicon (a-Si) is a classic thin film product produced via thin film processes based on depositing thin layers of silicon films on different substrates. Silicon thin-film cells are mainly deposited by chemical vapor deposition (CVD), typically plasma-enhanced (PE-CVD), using silane and hydrogen reactive and carrier gasses for the actual deposition.

Depending on the deposition parameters, the reactive species and the stoichiometry of the process, the reaction can yield different types of thin film structures. Some of these are: amorphous silicon (a-Si, or a-Si:H), or protocrystalline silicon (or nanocrystalline silicon, nc-Si or nc-Si:H), also called microcrystalline silicon.

These different types of silicon feature dangling and twisted bonds, which result in deep defects (energy levels in the bandgap) as well as deformation of the valence and conduction bands (band tails), which lead to reduced efficiency. Proto-crystalline silicon mixed with nano-crystalline silicon is optimal for high, open-circuit voltage PV cells production.

In all cases, solar cells and modules made from these materials tend to have lower energy conversion efficiency than those made from bulk silicon. They, however, have some important operating advantages, such as lower temperature degradation, which opens niche markets. They are also less expensive to produce than most other PV products, although the capital equipment expense is greater, due to equipment complexity.

a-Si has a somewhat higher bandgap (1.7 eV) than crystalline (c-Si) silicon (1.1 eV), which means that it absorbs the visible part of the solar spectrum more efficiently than the infrared portion. nc-Si has about the same bandgap as c-Si. That means that nc-Si and a-Si can advantageously be combined in thin layers, creating a layered cell called a "tandem cell," where the top a-Si cell absorbs the visible light and leaves the infrared part of the spectrum for the bottom cell in nc-Si.

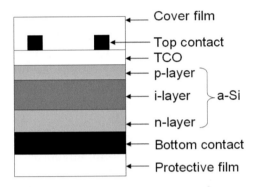

Figure 7-15. a-Si thin film structure

The biggest problem with the a-Si TFPV technology and a barrier to its success, however, is its low efficiency. Today's best cell efficiencies are about 12% in the lab, which is almost 50% lower than other PV technologies. Mass produced a-Si cells and modules are in the 6-9% efficiency range today.

Another problem with a-Si is its higher cost due to high initial capital investment—much higher compared with the competing PV technologies. Two proposed solutions to this problem are, a) higher manufacturing rates, and b) batch (simultaneous) processing of multiple modules. Good progress has been made in rates that are 3-10 times higher than those being used in present production, but all this is still on a lab scale and is yet to be proven on a large production scale.

On the positive side, while some of the more efficient cells and modules lose 20-30% of their output in the field with time, due to excess heat exposure, a-Si loses only 5-10%, due to its lower temperature coefficient.

Also, the active thin film structure is composed mainly of silicon films which have inert and homogeneous natures and show better chemical and mechanical stability than some of the competing thin films in case of an encapsulation failure. a-Si modules are also more resistive to the negative effects of shading in the field.

Of equal importance is the fact that a-Si PV modules do not contain any hazardous materials, which is paramount where large-scale PV installations are concerned.

These qualities put a-Si on the top of the list of PV technologies suitable for some specific applications. Even with low efficiency (well under 10%), a-Si thin film technology is being successfully developed for building-integrated photovoltaics (BIPV) in the form of semi-transparent solar cells which can be applied as window glazing. These films function as window tinting while generating electricity. It remains to be seen if the amount of generated electricity covers initial and operating expenses.

A triple-junction a-Si TFPV power plant has been operating near Bakersfield, CA, for several years now, and is providing proof of the excellent performance of this technology. The 500 kW grid-connects system has been performing well, meeting or exceeding its design goals.

Performance data from this larger-scale installation confirm data obtained from smaller a-Si systems and prove that this thin film PV technology can be successfully used in large-scale power plants, if the low efficiency can be justified.

Note: We review a number of other TFPV technologies in the "Future PV Technologies" section below.

FUTURE PV TECHNOLOGIES

A number of solar technologies promise to lead the way into the fossil-less future, but we cannot predict which will be the winner, how, when and where. Knowing the basic properties of the different technologies and the related energy market shifts, we can make some educated guesses.

With the available information today we clearly see silicon based solar cells and modules to dominate the solar energy field for the foreseeable future. The TFPV technologies are most likely to be used in different niche markets, where each occupies its own special place.

Below is a short list of most promising solar technologies.

Note: Much more detailed analyses of the most promising solar technologies are reviewed in our book, *Solar Technologies for the 21st Century*, published by The Fairmont Press, Inc., in 2013.

III-V Solar Devices

These are PV technologies based on the deposition of thin films of the III-V materials on different substrates. These are the most efficient solar power generators today.

The III-V PV devices are divided into single- and multi-junction, as follow:

Single-junction III-V Solar Cells

A number of compounds such as gallium arsenide (GaAs), indium phosphide (InP), and gallium antimonide (GaSb) have adequate energy band gaps, high optical absorption coefficients, and good values of minority carrier lifetimes and mobility, making them excellent materials for making high efficiency solar cells. These materials are usually produced by the Czockralski or Bridgmann methods, which provide high quality materials with increased efficiency and reliability, but at a higher price.

After silicon, GaAs and InP (III-V compounds) are the most widely used materials for single-junction (SJ) solar cells manufacturing. These materials have optimum band gap values (1.4 and 1.3 respectively) for SJ conversion of sunlight. The construction of solar cells made of these materials is similar to the regular single-junction c-Si solar cells we discussed elsewhere.

The major disadvantage of using III-V compounds for PV devices is the high cost of producing the materials they are made of and the related manufacturing processes. Also, crystal imperfections, including bulk impurities, severely reduce their efficiencies, so that only very high quality materials could be considered. Too, they are heavier than silicon, which requires the use of thinner cells, but they are weaker mechanically, so their design requires a delicate balance of thickness vs. weight.

The combination of high efficiency, high price, crystal imperfections intolerance, and mechanical weakness makes these devices useful for limited applications, where efficiency and overall behavior is more important than price. Thus they are not widely used in the general PV market, but still can be found in some important niche markets.

Gallium Arsenide Based Multi-Junction Cells

High-efficiency multi-junction GaAs cells were originally developed for special applications such as satellites and space exploration, but at present, their use in terrestrial concentrators might be the lowest cost alternative in terms of $/kWh and $/W. These multi-junction cells consist of multiple thin films produced via metalorganic vapor phase epitaxy. A triple-junction cell, for example, may consist of the semiconductors GaAs, Ge, and GaInP2.

Each type of semiconductor will have a characteristic band gap energy which, loosely speaking, causes it to absorb light most efficiently at a certain color, or more precisely, to absorb electromagnetic radiation over a portion of the spectrum. Semiconductors are carefully chosen to absorb nearly all of the solar spectrum, thus generating electricity from as much of the available solar energy as possible.

GaAs-based multi-junction devices are some of the most efficient solar cells to date, reaching a record high of 40.7% efficiency under "500-sun" solar concentration and laboratory conditions.

This technology is currently being utilized mostly in powering spacecrafts. Demand for tandem solar cells based on monolithic, series-connected, gallium indium phosphide (GaInP), gallium arsenide GaAs, and germanium Ge p-n junctions is rapidly rising. Prices are falling dramatically as well.

Twin-junction cells with indium gallium phosphide and gallium arsenide can be made on gallium arsenide wafers. Alloys of In.5Ga.5P through In.53Ga.47P may be used as the high band gap alloy. This alloy range allows band gaps in the range of 1.92eV to 1.87eV. The lower GaAs junction has a band gap of 1.42eV.

In spacecraft applications, cells have a poor current match due to a greater flux of photons above 1.87eV vs. those between 1.87eV and 1.42eV. This results in too little current in the GaAs junction, and hampers the overall efficiency since the InGaP junction operates below MPP current and the GaAs junction operates above MPP current. To improve current match, the InGaP layer is intentionally thinned to allow additional photons to penetrate to the lower GaAs layer.

In terrestrial concentrating applications, the scatter of blue light by the atmosphere reduces photon flux above 1.87eV, better balancing junction currents. GaAs was the material of the highest-efficiency solar cell, until recently, when Germanium-based MJ cells capped the world record at 41.4% efficiency.

Germanium-based Single- and Multi-junction Cells

Germanium (0.86eV band gap) is a semiconductor material, with properties far superior to other substrate materials used for PV cells and modules. It is ~40-50% more efficient than silicon and has a much lower temperature coefficient. It is several times more expensive than silicon, too, but with new superior slicing techniques, it can be cut into very thin wafers, saving a lot of material. This, combined with its higher efficiency and less degradation than silicon, could put it on the competitors' list within the next few years.

Germanium-based solar cells have been used mostly for space applications, but a number of manufacturers have geared up for mass producing them for high concentration HCPV and other high efficiency applications (the record is close to 45% efficiency).

Indium Phosphide Solar Cells

Indium phosphide is used as a substrate to fabricate cells with band gaps between 1.35eV and 0.74eV. Indium phosphide has a band gap of 1.35eV. Indium gallium arsenide (In0.53Ga0.47As) is lattice matched to indium phosphide with a band gap of 0.74eV. A quaternary alloy of indium gallium arsenide phosphide can be lattice matched for any band gap in between the two.

Indium phosphide-based cells are being researched as a possible companion to gallium arsenide cells. The two differing cells may be either optically connected in series (with the InP cell below the GaAs cell), or through the use of spectra splitting using a dichroic filter to produce different devices with wider market possibilities.

The presence of varying quantities of toxic materials in these devices must be considered when planning their use in large quantities.

Silicon Ribbon

Called EFG ("edge defined film fed growth"), this method, is not exactly a "thin film" process, as we know it, but the resulting material is thin enough, so it belongs in this category. Here, a graphite dye is immersed into molten silicon, making it rise into the dye by capillary action. It is then pulled as a self-supporting very thin sheet of silicon which hardens in the air above the dye. It can then be cut in different shapes and sizes for processing into solar cells and modules.

This method is more efficient than conventional c-Si ingot and wafer processes in terms of producing c-Si substrates of exact thickness and avoiding slicing it into wafers. Conventional processes waste 20-40% of the silicon material, use a lot of energy, and produce tons of hazardous waste materials.

Another similar process we need to mention here is called "dendritic web growth process." It consists of two dendrites, which are placed into molten silicon and withdrawn quickly, causing the silicon to exit and solidify as a thin sheet. A modification of this method now in use is called the "string ribbon method," where two graphite strings are used (instead of the dendrites) to draw the silicon sheet, which makes process control much easier. Again, the silicon sheet can be cut into different shapes and sizes.

Always, the silicon produced by these methods is multi-crystalline with a quality approaching that of the directionally solidified material. Although lab tests show efficiencies in the 17-18% range, solar modules made using silicon ribbons and produced by these methods, generally have efficiencies in the 10-12% range.

After the initial hoopla that silicon ribbon technologies would dominate the market, their share is quite small—less than 1% of total sales today—and does not seem to be growing. This is mostly due to the fact that the process is not easy to control, and the wafers' surface is not uniform enough, thus resulting in breakage, processing defects, and performance inefficiencies. The silicon sheets-forming process is also complex and uses a lot of energy, which makes it comparably more expensive than some of the other mass produced TFPV technologies.

Epitaxial Thin Film Silicon

The high cost of silicon material accounts for about half of the production cost of current conventional, industrial-type silicon solar cells. To reduce the amount of consumed silicon, the photovoltaics (PV) industry is counting on a number of options presently being developed. The most obvious is to move to thinner silicon substrates by producing thinner Si wafers, or shaving the thicker wafers, but this is proving hard to do for a number of reasons. A more feasible approach is the so-called epitaxial deposition of a thin film of silicon on a cheap substrate, thus creating efficient but cheap solar cells.

There are several approaches that can be used to create such a thin film cell:

Epitaxial Single Crystal sc-Si

To create an epitaxial thin-film solar cell on a cheap substrate we start with highly doped sc-Si wafers (e.g., from low-grade silicon or scrap Si material), and deposit an epi layer of Si by chemical vapor deposition (CVD). The resulting mix of a high quality epi layer and a cheap substrate is a compromise between high cost and efficiency, and yet offers a solution to gradual transition from a wafer-based (heavy material dependence) to a thin-film technology (less material and more sophisticated processing). This process is easier to implement than most other thin-film technologies today, but it remains to be seen if its efficiency and cost will be able to compete in the energy market.

Epitaxial Polysilicon Thin Film

To produce thin-film polysilicon solar cells, a thin layer (only a few microns) of polysilicon Si is deposited on a cheap foreign substrate, such as ceramic or high-temperature glass. These seed layers are then epitaxially thickened into absorber layers several microns

thick using high-temperature CVD with a deposition rate exceeding 1 m/min. Polycrystalline silicon films with grain sizes between 1-100 m appear to be particularly good candidates.

Good polycrystalline silicon solar cells can be obtained using aluminum-induced crystallization of amorphous silicon. This process leads to very thin layers with an average grain size around 5 m. This technology is still in R&D stages, but shows high cost-reduction potential and might become very important, especially in case of silicon shortage, or very high prices in the future.

Light-absorbing Dyes (DSSC)

These are special types of dye-sensitized solar cells, where a ruthenium metalorganic dye (Ru-centered) is used as a monolayer of light-absorbing material. The dye-sensitized solar cell depends on a mesoporous layer of nanoparticulate titanium dioxide to greatly amplify the surface area (200-300 m2/g TiO2, as compared to approximately 10 m2/g of flat single crystal).

Photogenerated electrons from the light-absorbing dye are passed onto the n-type TiO2, and the holes are passed to an electrolyte on the other side of the dye. The circuit is completed by a redox couple in the electrolyte, which can be liquid or solid. This type of cell allows a more flexible use of materials, and is typically manufactured by screen printing and/or use of ultrasonic nozzles, with the potential for lower processing costs than those used for bulk solar cells.

However, the dyes in these cells also suffer from degradation under heat and UV light, and the cell casing is difficult to seal due to the solvents used in assembly. In spite of these problems, this is a popular emerging technology with special applications and significant commercial impact forecast within this decade.

The first commercial shipment of DSSC solar modules was recorded in July 2009 from G24i Innovations.

Organic/Polymer Solar Cells

Organic and polymer solar cells are built from thin films (typically 100 nm) of organic semiconductors such as small-molecule compounds like poly-phenylene vinylene, copper phthalo-cyanine (a blue or green organic pigment), and carbon fullerenes and fullerene derivatives, such as PCBM.

Energy conversion efficiencies achieved to date using conductive polymers are low compared to inorganic materials. However, they were improved in the last few years and the highest NREL certified efficiency has reached 6.77%. In addition, these cells could be beneficial for some applications where mechanical flexibility and disposability are important.

These devices differ from inorganic semiconductor solar cells in that they do not rely on the large built-in electric field of a p-n junction to separate the electrons and holes created when photons are absorbed. Instead, the active region of an organic device consists of two materials, one which acts as an electron donor and the other as an acceptor. When a photon is converted into an electron hole pair, typically in the donor material, the charges tend to remain bound in the form of an exciton, and are separated when the exciton diffuses to the donor-acceptor interface.

The short exciton diffusion lengths of most polymer systems tend to limit the efficiency of such devices. Nanostructured interfaces, sometimes in the form of bulk hetero-junctions, can improve performance. Instability of the films, especially under harsh environmental effects is a major problem, which needs to be resolved, before full-scale implementation. Even with its advantages, this technology still has far to go to full market acceptance and serious deployment.

Ink PV Cells

A fairly new development, this light-activated power generating product is based on a unique and patented solvent-based silicon nanomaterial platform that can be applied like ink on any substrate. Developers claim that this approach has cost savings over traditional silicon products by using less silicon and having a more efficient manufacturing process as well as unique optical advantages.

This new technology consists of processing the quantum dots in the silicon "ink" in a way that makes it possible to use the old "roll-to-roll" printing technology used for printing on paper or film. Applying ink directly on any substrate (including a flexible one) allows applications such as tagless printing for clothing labels and portable chargers for consumer and military customers.

By controlling the sizes of the dots from 2 to 10 nm, the absorption or emission spectra of the resulting film can be controlled. This allows capture of everything from infrared to ultraviolet and the visible spectrum in between which is not possible with conventional technology.

The technology is also used as an efficient light source. By controlling particle size, you can produce light of any color or a combination of particle sizes that will give off white light. This application might provide additional, and possibly larger, markets for this technology in the near future—at least in some specialized areas.

Nano-crystalline PV Cells

These structures make use of some of the usual thin-film light absorbing materials, but are deposited as a very thin absorber on a substrate (supporting matrix) of conductive polymer or mesoporous metal oxide having a very high surface area to increase internal reflections. Hence, the probability of light absorption increases.

Using nanocrystals allows one to design architectures on the length scale of nanometers, the typical exciton diffusion length. In particular, single-nanocrystal ("channel") devices, an array of single p-n junctions between the electrodes and separated by a period of about a diffusion length, represent a new architecture for solar cells and potentially high efficiency.

We envision the development of this type of photo-conversion to be in the R&D labs for a while yet, but it opens new possibilities in areas where other technologies simply cannot compete, thus opening promising niche markets for these cells.

HIGH EFFICIENCY PV TECHNOLOGIES

There are types of solar cells and processes, which use the standard materials and process sequences with some modifications, which give them different (hopefully superior) qualities. One thing to note here is that when evaluating such technologies and devices, one must keep in mind that changing (even improving) one parameter alone is not enough; i.e., if the efficiency is increased at the expense of extreme sunlight degradation, then the "improvement" is worthless, and must be either abandoned or changed in a way that does not affect negatively the rest of the operating parameters of the final device.

The type of cells we are discussing here fall in the category of "high-efficiency" solar cells, simply because they are more efficient than the standard lot of similar materials and processes.

Following are some of the most efficient technologies today and most promising for large-scale power generation in the future.

CPV Solar Cells

Concentrating photovoltaics (CPV) is a branch of the PV industry, using special cells, optics and tracking mechanisms, developed in the 1970s by several companies under contracts and financing from U.S. DoE. Early CPV systems used silicon-based CPV cells, which had a problem with elevated temperatures.

These were later replaced by GaAs-based multi-junction cells, which have much higher efficiency, but still suffer from the effects of high temperatures. At first GaAs CPV cells were made by using straight gallium arsenide in the middle junction. Later cells have utilized In0.015Ga0.985As, due to the better lattice match to Ge, resulting in a lower defect density.

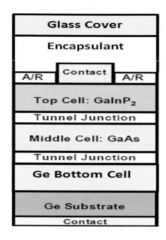

Figure. 7-16. CPV type solar cell

CPV cells are complex structures, consisting of many layers (some deposited, some diffused) in, or piled on top of, Germanium semiconductor material, which has more superior process and performance characteristics than silicon.

Current efficiencies for InGaP/GaAs cells are in the 40% range and constantly increasing. Research into methods to produce band gaps in the range between the Ge and GaAs is ongoing. Lab cells using additional junctions between the GaAs and Ge junction have demonstrated efficiencies above 41%. InGaP/GaAs CPV cells on GaAs substrate have demonstrated 42.3% efficiency.

CPV cells are mounted under lenses, which concentrate sunlight falling on the cells 100 to 1000 times.

This allows high efficiency and reliability, better land utilization, and other benefits. Cell-lens assemblies are mounted on trackers, which track the sun precisely through the day, providing the most power possible. Efficiencies of over 45% measured at the cell, and 30-35% measured in the grid are obtainable with these devices. We foresee CPV technologies as the primary choice for installation and use in large-scale power plants, especially those in desert regions.

HIT Solar Cells

The Hetero-junction with Intrinsic Thin layer (HIT) solar cell is composed of a single thin crystalline silicon

wafer, usually 4" or 5" square, sandwiched between two thin films of amorphous silicon (a-Si).

The structure of the HIT cell enables an improvement in the overall output by reducing recombination loss, which is the loss of electrical current that occurs when an electron and a hole (carriers) generated by impinging photons within the solar cell combine and disappear.

This effect is achieved by surrounding the energy generation layer of single thin crystalline silicon with high quality ultra-thin amorphous silicon layers.

Several manufacturers have developed special technologies for cleaning the energy generation middle layer and protecting it from damage during construction of the surrounding layer, with the result being an increase in the open circuit voltage from 0.718V to 0.722V. This also improves the efficiency and reliability of the solar cell by avoiding the usual problems related to doped and fired silicon solar cells.

This type of solar cell would be more expensive to produce, but might provide the much needed solution to the problems c-Si and TFPV modules encounter when exposed to extreme desert temperatures. Such cells theoretically would exhibit less heat sensitivity and power degradation, but, we have not seen proof of that happening as yet.

HIT cells have improved efficiency, over 22% in lab tests and around 19% in mass production, mostly due to the good a-Si:H/c-Si/a-Si:H hetero-interface of the HIT structure, which enables a high Voc (well over 0.7 V). This also results in much better temperature coefficient and better heat handling properties (theoretical values).

To reduce the manufacturing cost, numerous technologies are in R&D mode presently, with the goal to further improve the efficiency, and especially to increase the Voc of HIT solar cells, with the aim to achieve up to 25% efficiency in the near future.

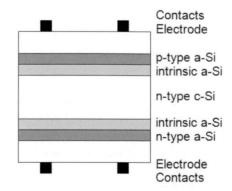

Figure 7-17. HIT solar cell

The commercial modules have the following key electrical and performance properties:

Cell Efficiency (n%)	19.7%
Module Efficiency (n%)	17.0%
Max. Power Voltage (Vpm)	55.8V
Max. Power Current (Ipm)	3.59A
Open Circuit Voltage (Voc)	68.7V
Short Circuit Current (Isc)	3.83A
Temperature Coefficient (Pmax)	-0.29 %/C
Temperature Coefficient (Voc)	-0.172 V/C
Temperature Coefficient (Isc)	0.88 mA/C

50% Silicon Use

The high costs of silicon used in photovoltaics dictate the price of conventional solar panels. Several new approaches are under consideration today, that promise great results with reduced amounts of silicon, in order to save on the expensive silicon material. The idea is to somehow change the shape of the active area of the solar wafers (and the resulting solar cells), or make them much thinner than the conventional technologies.

One such approach is to make solar cells that use half of the silicon in each solar wafer, by slicing the wafer into strips and using only half of these, while still generating 80-90% of the conventional solar cell's power. This is achieved by slicing the silicon wafer into thin strips and arranging these into a solar cell configuration by spacing the slices apart so that they only account for about half of the cell's total area. A clear plastic cover collects light from the entire cell surface and funnels it to the strips of silicon.

This approach would save money because the total cost of the most expensive material—silicon—is cut almost in half—and the device is much cheaper even after adding all other materials and manufacturing steps. The cost can be reduced further by using manufacturing equipment already developed for the semiconductor industry, which helps avoid expenses for customized equipment development and purchasing.

Also, in conventional solar cells, wires for collecting current are placed on top of the cell, where they block some of the incoming sunlight. Space and material can be saved if the wires are placed between the strips of silicon, where they block no light. The wires also don't need to be thin to avoid blocking light, thus they can be sized accordingly which might improve the collection of electrons from the adjacent silicon strips in the solar cell.

Back Mirror

Another new concept is under development, where the resulting solar cells are much thinner, have improved front surface texturing and added back surface mirror, all of which adds to efficiency increase. Higher efficiency is achieved, in addition to better texturing, mostly by means of trapping sunlight and keeping the photons inside the active material of the solar cell until their energy can be fully and efficiently used to free electrons and generate an electrical current.

Trapping light is greatly assisted by proper texturing of the front surface of the silicon wafer which is to become a solar cell. Effectively texturing the surface results in facets (pyramids, cones and such) that redirect incoming light, and even refracting. Instead of just passing through the silicon. Instead, the photons travel at least partially along the length of the silicon layer, and this way stay in the material longer. This way they have a much better chance of being absorbed by atoms in the material and free the electrons that generate current and conduct it into the outside circuit.

This phenomena can be assisted and amplified by adding a reflective layer at the back of the silicon layer. Thus created "photon mirror" will reflect the passing photons and will keep them moving around the solar cell still longer. This increases the chance of increasing the number of freed electrons, and from that the electric current in the outside circuit.

This design requires thinner silicon material, where the solar cell can be half its ordinary thickness, while absorbing the same amount of light. This results in the additional benefit of using much less of the expensive silicon material, which reduced the final costs of the devices. As an added benefit, impurities within the material can easily trap electrons before they reach the surface and escape to generate a current. Conventional silicon solar cells are approximately 200-250 μm thick, so the electrons have to deal with the impurities across the entire width of the material, which is a major reason for lower efficiency.

The new thinner cells are 100-125 μm thick, so the generated electrons have a shorter distance to travel, so they're less likely to encounter an impurity before they escape and do the work they were expected to do—generate electric current.

Cheaper and less pure silicon material could be used with this method as well, and lower-grade silicon is much cheaper and easier to make than the highly refined silicon ordinarily used in solar cells, which will reduce the final cost even further.

The battle for lower cost silicon material—which is still the preferred material for manufacturing solar cells—continues, so any effort in this area must be well worth it and encouraged. Silicon will be around for a long time yet, so the efforts to optimize the solar cells' materials and processes must continue, to achieve the best and cheapest possible combinations.

Buried Contact Solar Cells

The buried contact solar cell is another high efficiency commercial solar cell, this one based on a metal contact formed inside a laser-formed groove in the front surface. The buried contact technology has many advantages and avoids the problems of conventional solar cells related to the old screen-printed contacts method. This allows buried contact solar cells to have performance up to 25% better than commercial screen-printed solar cells.

A key high efficiency feature of the buried contact solar cell is that the metal is buried in a laser-formed groove inside the silicon solar cell, which allows for a large metal height-to-width aspect ratio of the contacts, which improves the device's operational characteristics, since a large metal contact aspect ratio allows a large volume of metal to be used in the contact finger, without having a wide strip of metal on the top surface. This allows a large number of closely spaced metal fingers, while still retaining a high transparency of the front surface.

For example, on a large area device, a screen-printed solar cell may have shading losses as high as 10-15%, while in a buried contact structure, the shading losses will be only 2-3%. These lower shading losses allow a higher level of photon collection (more photons hit the active surface), and low reflection (less photons hit the metalized front surface areas), which contributes to higher short-circuit currents.

In addition to good reflection properties, the buried contact technology also allows low parasitic resistance losses due to its high metal aspect ratio, its fine finger spacing, and its plated metal for the contacts.

The emitter resistance is reduced in a buried contact solar cell since narrower finger spacing dramatically reduces the emitter resistance losses. The metal grid resistance is also low since the finger resistance is reduced by the large volume of metal in the grooves and by the use of copper, which has a lower resistivity than the metal paste used in screen printing.

The contact resistance of a buried contact solar cell is lower than that in screen printed solar cells due to the formation of a nickel silicide at the semiconductor-metal interface and the large metal-silicon contact area. Overall, these reduced resistive losses allow large area solar cells with high FFs.

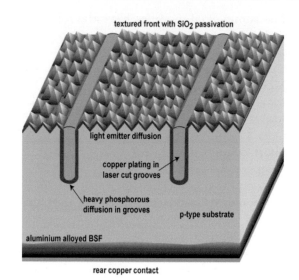

Figure 7-18. Buried contact solar cell

When compared to a screen-printed cell, the metalization scheme of a buried contact solar cell also improves the cell's emitter. To minimize resistive losses, the emitter region of a screen-printed solar cell is heavily doped and results in a "dead" layer at the surface of the solar cell. Since emitter losses are low in a buried contact structure, the emitter doping can be optimized for high open-circuit voltages and short-circuit currents. Furthermore, a buried contact structure includes a self-aligned, selective emitter, which thereby reduces the contact recombination and also contributes to high open-circuit voltages.

The efficiency advantages of buried contact technology provide significant cost and performance benefits. In terms of $/W, the cost of a buried contact solar cell is the same as a screen-printed solar cell. However, due to the inclusion of certain area-related costs as well as fixed costs in a PV system, a higher efficiency solar cell technology results in lower cost electricity. An additional advantage of buried contact technology is that it can be used for concentrator systems of up to 50x concentration.

PERL Solar Cells

There are several types of solar cells that can be built on silicon substrates, but which require much more expensive materials, sophisticated equipment, and elaborate processing. The passivated emitter with rear locally diffused cell (PERL) is one of these devices.

It is a fairly recent development of a solar cell design that uses micro-electronic techniques to produce solar cells of high efficiency, up to 25% in the PERL case.

Thinner silicon wafers (solar cells) are used in this

design, to optimize the efficiency, with 50 μm thick cells having the highest efficiency. Using such thin wafers in production is impossible due to breakage, so a compromise in the 150-200 μm wafer thickness area must be used.

One characteristic of this type of solar cell is the improved texturing of the active front surface area, which is done in the shape of inverted pyramids. If the base size of the inverted pyramid texturing is increased, then the effective surface area increases, which in turn increases absorption in the silicon structure. So a lot of effort is put in the optimized size and shape of the pyramids.

Bigger texture structures (> 10 μm) increase the optical absorption and the resulting cell efficiency. Using conventional chemical texturing, the maximum achieved base size of pyramid is around 9 μm. So, optimizing the pattern size is best done via photolithography methods, similar to those used in the semiconductor industry. This additional lithographical step adds an additional processing step. It has several advantages, which are worth the additional effort and expense.

Placing the metal contacts on top of the pyramidal structure is also critical and requires careful consideration and precise execution. The contacts placement will determine the path the carriers will travel to reach the contact, so the position of the contacts is critical for the overall electricity generation process. Properly executed front metal contact lines could be as thin as 2 microns, which would reduce the front surface shading to about 3% of the surface area. The reduced contact area reduces the total recombination at its proximity, which contributes to higher Voc of the device.

The passivated emitter is the high quality oxide deposited at the front surface of the cell, which significantly lowers the number of carriers recombining at the surface. The rear of the cell is locally diffused only at the metal contacts' areas, to minimize recombination at the rear while maintaining good electrical contact.

A major advantage of the PERL solar cell is the improved surface passivation of the front and rear cell areas via thermally grown oxide, and the respective dopants, which significantly reduces the emitter saturation current, thus leading to improved open-circuit voltage (Voc above 0.7V). This also reduces the recombination levels at the contact regions, due to suppressed minority concentration in these areas. PERL cells can reach 25% efficiency, but commercial types operate in the 19-20% range.

The PERL solar cells have not been proven reliable for large-scale operation under different environmental

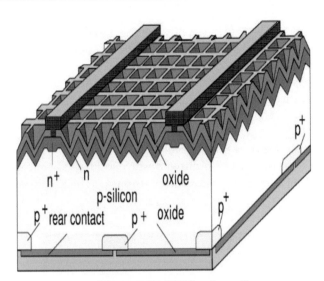

Figure 7-19. PERL solar cell

conditions. This, combined with their complex manufacturing process and higher cost will have to be tested by time and locations before this technology is fully established as a major competitor in the world's energy markets.

The PERL solar cells seem to be suitable for some special applications, such as solar powered vehicles and space applications, so we foresee their number increasing.

Rear Contacts Solar Cell

Rear contact solar cells eliminate shading losses altogether by putting both contacts on the rear of the cell. By using a thin solar cell made from high-quality material, electron-hole pairs generated by light that is absorbed at the front surface can still be collected at the rear of the cell. Such cells are especially useful in concentrator applications where the effect of cell series resistance is greater. An additional benefit is that cells with both contacts on the rear are easier to interconnect and can be placed closer together in the module since there is no need for a space between the cells.

Efficiencies of over 23% have been achieved with this cell design. These types of solar cells usually cost considerably more to produce than standard silicon cells, and are typically used in specialized applications, such as solar cars and for space exploration.

Passivated Rear Point Contacts Solar Cell

This type solar cell is even more specialized, due to its complexity, and although we don't see its use as such in the near-term energy markets, we do believe that some of its features (materials and processes) will be used in the future for developing new technologies. This alone is enough to put it in the category of most promising technologies for 21st century use.

Efficiencies over 22% have been reached with these cells, but due to their complexity and higher cost, their use is limited.

Selective Emitter Solar Cells

The "selective" emitter design solar cells concept uses laterally different emitter doping, which consists basically of:

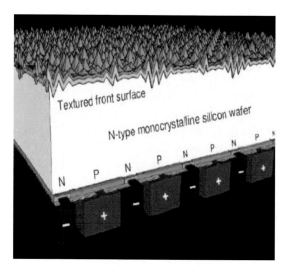

Figure 7-20. Rear contact solar cell

1. High doping under the front side metallization for low contact resistance between contact metal and semiconductor interface,

2. Lower doping between the contact fingers for a better short wavelength response due to less auger recombination and improved emitter passivation.

Unfortunately, most of these designs are very complex as far as the process is concerned, due to a number of masking steps required for achieving proper selective diffusion or emitter etch back.

In particular, such existing concepts are hard to implement into industrial production of solar cells. Nevertheless, several different approaches designed for industrial production are currently under development, and some are already being transferred to the industry.

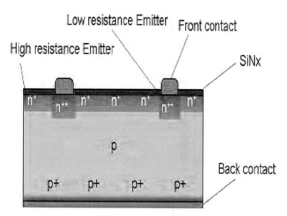

Figure 7-22. Selective emitter solar cells

A number of organizations are developing solar cells with this type of selective emitter, which allows the use of thinner emitters as needed to improve the short-wavelength spectral response.

We do foresee the use of this technology also limited to niche markets, but its components are valuable instruments in the battle for higher efficiency, reliability and cost effectiveness, which can be successfully used in the development of other solar technologies in the near future.

In summary, the above described high-efficiency solar technologies are good ideas that sound plausible, but most of them, due to their complexity and higher cost, need much more work to be made efficient, reliable and cost effective for use in the 21st century power fields. Nevertheless, some are already in mass production mode, but more importantly we do hope that the clever designs and approaches of these will be used in developing other solar technologies in the near future.

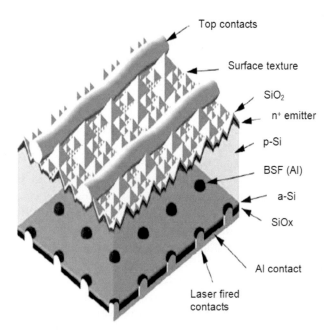

Figure 7-21. Passivated rear point contacts solar cell

HYBRID PV TECHNOLOGIES

These systems generate both photovoltaic and thermal solar power. The advantage is that the thermal solar part carries heat away and cools the photovoltaic cells. Keeping temperature down lowers the resistance and improves cell efficiency. Modified CPV systems have been tested, where the CPV cells are cooled by active flow of liquid, thus generating both heat and electricity.

The present-day hybrid solar technologies involving photovoltaics and heat generation can be divided into:

1. Photovoltaic-thermal hybrid, or PV/T, and

2. Concentrating PV-thermal hybrid, or CPV/T, which can be subdivided into:
   2.1 Concentrating photovoltaic/thermal hybrid, or CPV/T, and
   2.2 High-concentration photovoltaic/thermal hybrid, or HCPV/T.

A word of explanation and clarification:

PV/T is a combination of photovoltaic devices which incorporate both photovoltaics and thermal heating. These can operate at 1x (1 sun) to generate electricity, while running water through the system for use somewhere else.

CPV/T, therefore, is a combination (hybrid) of PV concentrators which concentrate the sunlight 10-100 times onto a receiver tube covered with solar cells mounted on a receiver tube through which water or cooling liquid flows. The trough and receiver track the elevation of the sun all day (one-axis tracking), and thus the device generates more DC power per active area, and the water/coolant are at much higher temperatures than their PV/T cousin.

HCPV/T is a combination of PV concentrators (mirrors or Fresnel lenses), which operate on higher magnification (100-1000 x) and track the sun in x-y direction (two-axis tracking) all through the day. These devices produce more power per area than their CPV/T cousin, while the water/liquid temperature is controlled as needed to optimize the efficiency of the device.

Below, we take a very close look at PV/T device operation, and their thermo-dynamics and static behavior, both of which could be used with some translation and extrapolation for use with CPV/T and HCPV/T devices. These calculations are absolutely necessary for the proper design of the devices and the power fields in which these are supposed to operate.

The hybrid photovoltaic-thermal (PV/T) system basically consists of an array of PV cells positioned directly on top of the absorber plate of a conventional forced circulation type solar water heater—similar to those we see on the roofs of houses.

PV/T Solar Collector

A practical, functional PV/T system consists of:
- Transparent cover allowing sunlight to pass towards the absorber and to create the effect of a greenhouse. It is composed of one or more glass or plastic panes.
- PV cells for the production of electricity.
- Absorber plate for transferring heat to the water or in the tubes built into it.
- Frame, for protection of the whole of these elements.
- Insulator, which allows limiting the losses by conduction through the walls' back and side.

The schematics of the PV/T collector are shown in Figure 7-23. The top cover is represented by a glass sandwich that includes PV cells. The cell area can cover the entire active surface or can be distributed in a grid where the spacing between adjacent columns and rows allows a direct gain of solar radiation to the absorber plate.

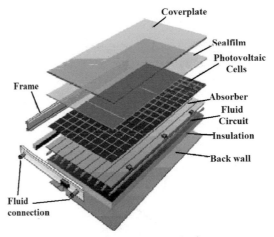

Figure 7-23. Hybrid PV/T

Different configurations of PV/T collector can be created by changing the cell area density, to balance electricity and thermal energy output of the system.

Concentrating PV-Thermal (CPV/T) Hybrid Technology

PV cells can generate more electricity if more sunlight is concentrated on their surface. The common silicon, or thin film solar cells can withstand 2-3 times the normal solar intensity, but their temperature increases with concentration, so they lose efficiency. Special materials and processes are used to manufacture special CPV and HCPV solar cells.

These, combined with thermal power generation form the foundation of CPV-T and HCPV/T technology, as follows:

CPV/T

The concentrating photovoltaic hybrid (CPV/T) hybrid technology is a fairly new development, and although a number of companies are working on this concept, full commercialization and large-scale deployment are still distant.

The CPV/T equipment usually consists of a parabolic trough lined with a mirror surface focused on a bank of solar cells mounted on a heater tube through which flows water or some liquid. When the sun hits the trough, its mirror surface reflects the sunlight directing it to the cells, which generate DP power at 15-20% efficiency. When the cells heat up under the sun's rays, they transmit the heat to the tube they are mounted on and heat the liquid running thru the tube.

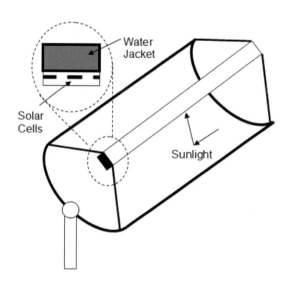

Figure 7-24. CPV/T one-axis tracker

The above dynamic and static calculations for the generic PV/T device could be used here too, with because the basic glass-absorber-cells-liquid configuration exists here in almost identical fashion. Their properties and behavior are similar, except for different results, due to much higher solar illumination, DC voltage and temperatures.

HCPV/T Hybrid

Other configurations of photovoltaic and thermal power generation are possible. One of these is the HCPV/T hybrid DC power and hot water generator.

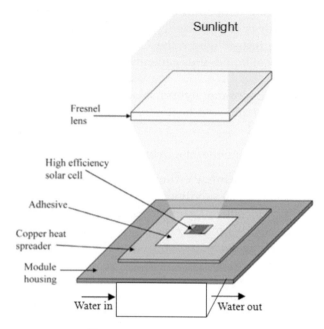

Figure 7-25. HCPV assembly

HCPV/T hybrid consists of a Fresnel lens that captures the sun rays and redirects them onto a high efficiency solar cell. The cell is mounted on a heat sink through which water or cooling liquid flow. This way, while the cell captures the sun's rays and generated DC power, it also heats water or coolant liquid.

The water flowing through the heat sink then has two important functions,

a)   it heats to high temperature and can be used for a number of industrial or domestic purposes, and

b)   the cooling water cools the solar cell so that its efficiency and reliability are increased.

This is the most efficient solar system in existence today, with great potential in the future. Its overall efficiency can exceed 65% under full sunlight. And because the solar cell assemblies are mounted on a tracker, electric power and hot water are produced non-stop all day.

Other PV Hybrids

PV cells and modules can be also used in combination with other technologies. There are, for example, PV and natural gas hybrid power plants, where solar power is used to compliment the natural gas power generation, especially during peak noon hours. PV can be also combined with wind power generators to produce a more consistent output.

One of the most practical hybrid applications—especially in developing countries—is the PV-diesel hybrid model.

Diesel-PV Hybrid

While the above described renewable hybrid schemes are possible from a technological point of view, they are not expected to take a significant part in the energy markets anytime soon. There are, however, other hybrids involving renewable energy generation that are more promising in the short term. One of these is PV assisted diesel power generation.

Diesel generators are a vital source of electricity in many parts of the developing world. They are widely used to provide back-up in areas where grids are intermittent (India is a primary example) and are used as the main source of power in remote locations.

There is an estimate of over 500 GW of diesel power generated around the world, and 30-40 GW of new diesel capacity are added annually. Diesel is an expensive fuel, and at a cost of over $4 per gallon in most African countries, this translates into a daily wage for many people. This is also equivalent to $0.40-0.50 per kWh, which price varies with the price of crude oil.

Solar power alone won't solve all the problems in these areas, but combined with diesel generators it could provide a welcome relief for many of these people. Sure enough, there is a growing interest in PV-diesel hybrid systems since the combination of PV with existing diesel generating facilities offers the chance of retaining the reliability of diesel while significantly reducing the amount they spend on fuel.

Of course there are problems related to technological and financial issues. There are still very few examples of PV-diesel hybrids.

On the technology side, the initial challenge is for the PV-diesel hybrid market to find the proper solution for each case, since the installation and optimal operation of each system requires significant technical consideration. The main challenge is stability, where in a diesel power generation, the diesel is providing the exact amount of energy needed. In these cases, power load and generation are in balance and no special equipment or effort is required to maintain this balance.

With PV added to the diesel grid, the existing load demand is not changing, but the power generation is now fluctuating due to the solar variability. Now it becomes much more difficult to balance the power load and its generation.

The controller "talks" to the diesel generators, analyzing their status and controlling the input of PV at an optimum level depending on certain load conditions. In all cases, the diesel operation must remain stable and robust, while controlled by the additional, complex, and expensive control equipment.

There are controllers designed specifically to facilitate the feed-in of PV to a diesel-based system, but these have limited field track records.

On the financial side is introducing a capital-intensive technology (new solar installation) into an established market that doesn't favor, or can't afford, such a solution. The financing could come either directly from the customer, who benefits from saving on diesel and electricity costs; or, lease-back financing is also a possibility, but somebody has to take on the risk of user. The initial expense and ongoing O&M costs remain a serious consideration in many parts of the world, and will take time to resolve.

What will make the PV-diesel hybrid concept ultimately sink or swim on the long run will be the extent to which solar and other companies are able to come up with practical commercial models that work for capex-sensitive businesses.

It seems that ultimately it will be the business model not the technology that will determine the destiny of the PV-diesel hybrids.

Nevertheless, experts predict some big strides forward for PV-diesel systems in the near future, with the Kenyan government soon expected to invite tenders for the retrofitting of PV to all of the country's existing diesel power stations. A number of other African countries are also set to follow its lead.

Other areas which could see an uptick in the use of PV-diesel hybrids are the telecoms industry, farms and safari eco-lodges.

It's a small step for solar, a big step for the African and Asian continents.

PV-hydrogen Hybrid

PV-hydrogen is a futuristic hybrid, representing an energy cycle where solar power is used to convert water to hydrogen and oxygen. Hydrogen is then stored for use by a fuel cell to produce electricity during cloudy days and/or at night.

The PV-hydrogen hybrid energy cycle can use any type of solar modules, and has numerous benefits. It is pollution free, as the only effluent from this cycle is pure water. The hydrogen powered energy economy is more stable than conventional energy economies, so this cycle can be especially beneficial in developing countries.

The major benefit of the PV-hydrogen energy cycle is that it could generate power 24/7, which is currently not possible with solar or wind energy alone. This technology promises to open new doors for the solar energy power generation of the future. Eventually, hydrogen generated by sunlight could replace fossil fuels for transportation and electricity generation.

Today, however, powering a car engine or a fuel cell by hydrogen is far cheaper with natural gas than using solar energy to split water. In the near future, solar power would be able compete with natural gas as a way to make hydrogen, if and when

a)   solar power efficiency gets above 25 percent,

b)   natural gas prices exceed a certain threshold, and

c)   the PV-hydrogen technology is fully developed, tested, and optimized.

PV-wind Hybrid

The futures of solar and wind energy are intertwined. Since they very conveniently generate power at different times of the day and night, their output can be combined to produce a constant level of output. This will reduce the uncertainties in variability of solar and wind power plants.

Note: We take a closer look at the combined PV-wind hybrid power generation below.

PV-wind hybrid systems offer several advantages over either single system. In different parts of the U.S., wind speeds are low in the summer, while at the same time the sun shines brightest and longest. On the other hand, wind is strong in winter when less sunlight is available. Because the peak operating times for wind and solar systems occur at different times of the day and year, hybrid systems are more likely to produce a steady level of power.

Some PV-wind hybrid systems are stand-alone systems and operate "off-grid," while others—usually much larger—are hooked into the grid. Adding a diesel generator, or other energy source, makes the system more compete and ensures uninterrupted 24/7 operation.

Figure 7-26 represents such a hybrid power generating system, consisting of PV, wind, diesel, and hydrogen power generators, combined with fuel cells and battery storage. This is a complete system for use under all conditions that could provide uninterrupted power on a 24/7 basis.

During times of normal solar and wind activities, the generated power could be used to feed the load, while some of the excess is stored in the batteries. An-

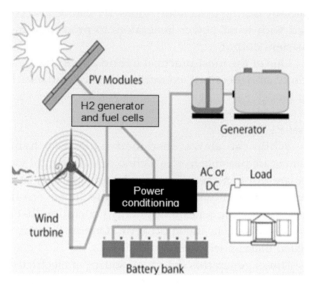

Figure 7-26. PV-wind-diesel generator hybrid system

other portion of the excess power could be converted into hydrogen and stored for future use. Depending on location and use, the solar and wind power generation combined could satisfy 60-80 percent of the total load.

During times of low solar and wind activities, the batteries would kick in and provide the needed power. When the batteries run low, the hydrogen could be fed into the fuel cells to generate power. The diesel generator could kick in to assist these activities, and or to provide full power if and when needed.

Such a complete system would be quite expensive today, but the technology is available, and as fossil and electricity prices go up, we will see many combinations and permutations of these hybrids in the future—especially in the developing countries.

SOLAR ENERGY USE

Solar energy is all around us (during the day), but it has to be captured and used in a certain way. Once we have solar cells and modules, we need to arrange them properly, to generate the maximum amount of electric power in each location and situation.

For that purpose, we arrange the solar modules in different types of arrays and systems.

PV Arrays and Systems

Typically, PV modules are used in combination with a number of similar modules to create an "array." A PV array is simply a group of PV modules installed and wired together as a group, to produce a certain amount of power.

PV arrays come in many forms, shapes and sizes, and each array can be fixed, sun-tracking with one axis of rotation, or sun-tracking with two axes of rotation.

Fixed PV array is the cheapest and most common on the PV energy market. It is not as efficient as the tracking array, but it is the simplest, the cheapest, and the one requiring least maintenance.

A fixed array consists of a number of modules that are mounted permanently on a solid frame (usually steel or aluminum angles) which is cemented, or otherwise solidly fixed into the ground or on a structure.

Tracking arrays are also mounted on a frame, but the frame is heavier and is mounted on a pivoting pedestal, that allows it to rotate on top of it, thus following the sun all through the day.

The tracking frame moves with the help of one motor-gear assembly for a one-axis system, or via two motor-gear assemblies for a two-axis system. A controller, which knows exactly where the sun is, sends a signal to the motors, which then activate the gears to move the frame into position.

The function, advantages and disadvantages of fixed vs. tracking arrays are quite important and must be considered in any PV project.

Tracking arrays generally have higher comparative efficiency, because they are positioned most advantageously towards the sun. One-axis tracking arrays keep closer to a 90-degree angle than fixed arrays during the day and usually generate 20-30% more power. Two-axis tracking arrays are always at 90 degrees towards the sun, thus receiving the maximum amount of sunlight at all times. They also generate the largest possible amount of electric power from sunrise to sundown.

Figure 7-28 shows the great difference between tracking and non-tracking modules and arrays. Two axis trackers deliver the maximum amount of output possible as they follow the sun on its journey across the sky. Some tracking systems also benefit from the use of very efficient (almost 45%) CPV solar cells, which are also very durable, but unjustifiably expensive for use in non-tracking systems.

CPV tracking arrays' output is several times higher than that of conventional Si or thin films non-tracking arrays.

The Participants

There are a number of actors and stakeholders participating in the solar activities, and their contribution to achieve cost reductions vary substantially. Cost reductions will require further leveraging of key stakeholders, whose functions are continuously evolving and may overlap with other categories. Also, achieving the objective requires the engagement of other groups.

The key participants and stakeholders in the solar game today are given below.

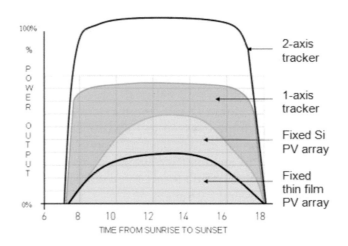

Figure 7-28. Performance of tracking vs. non-tracking PV arrays

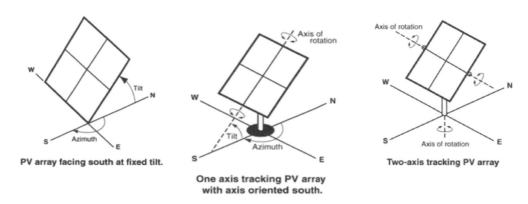

Figure 7-27. Types of PV arrays

Manufacturers

These are hundreds of companies worldwide that manufacture solar materials, cells, and modules. Some of these are small mom-and-pop shops, while others are multi-million and billion dollar enterprises. The manufacturing space changes periodically with the changes in the solar industry and the global financial situation.

Hundreds of solar products manufacturers and service companies popped up and then went away bankrupt or were swallowed by larger companies during the solar boom and bust of 2007-2012. The field is becoming more level now, dominated by large companies in all aspects of the solar energy business.

Developers/Installers

The greatest engagement in customer acquisition—especially in distributed residential and commercial installations—is required from developers and installers. Installers often create a suite of standard system designs to apply to sites with common parameters, with the goal of lowering costs, especially when designing systems for new leads. Most installers consider this a sound business practice that will endure, get optimized, and become more prevalent in the future.

Developers at times reach out to and partner with big-box retailers and national corporations, such as Lowes and Home Depot, mainstream consumers become increasingly exposed to PV. Also, corporations and national organizations, such as Honda and NREL, are beginning to include PV options as benefits (Solar-Benefits Colorado; Honda-SolarCity partnership, etc.) to their employees and customers.

Service Providers

The next-highest required engagement is from service providers. For example, over 40 information technology and software development companies are developing tools to reduce customer acquisition costs. The DOE SunShot Incubator Program, and other such tools, were developed to support multiple projects to reduce the costs. Some examples are the online quoting system (EnergySage.com), advanced siting tools (Simuwatt), and solar sales software (Genability).

Many maintenance companies are also offering their services in long-term solar fields maintenance contracts.

Utilities

The utilities play a major role in facilitating the development of residential and commercial solar projects. They hold the key to determining (albeit arbitrarily at times) who is qualified to participate in bidding and when and how to allow interconnect assistance. In essence, the utilities have the first and last word in the solar projects' development. We take a close look at the developments in this area later in this chapter.

Government

Among the actors/stakeholders with lower (yet extremely important) engagement are federal, state and local governments, and the regulators. These usually determine the business models via rules and regulations, such as subsidies, net metering, community solar, and third-party ownership, which affect the industry and its customer base.

The federal government plays a key role providing financing, subsidies, and research and development (R&D) to support innovative, game-changing technologies that help reduce soft costs, as seen with information technology/software product development through the DOE SunShot Incubator Program.

The state and local governments, acting through their respective branches with the help of the regulators, determine the who, when, and where fate of solar projects in their jurisdiction.

Solar Costs

The costs related to solar power projects' initiation and development are many and determine the opportunities for increasing solar installations in the U.S. These costs are usually classified as hardware costs and soft costs.

The hardware costs are mainly the cost of PV modules, support structures and related hardware. These costs are set by the national and international markets, and vary according to the development in these areas.

Hardware prices have dropped precipitously during the last several years and are now at, or close to, rock bottom at approximately $0.50-0.60 per installed watt in some areas.

The soft costs are the largest part of the cost structure, and consist of the following elements:

Permitting, Inspection, and Interconnection (PII) is done by state and local governments, so they have the biggest role in determining and improving the PII processes and reducing associated costs. The PII procedures are faced with the challenge of streamlining bureaucratic hurdles and reducing fees to cost recovery levels, while ensuring safety and reliability. This is a battle that is raging on several fronts.

Because the utilities determine and oversee interconnection requirements and processes, their willing participation is needed to overcome interconnection challenges and reduce costs. While the federal government cannot establish regulations directly, it can play a key role by supporting regulatory jurisdictions and the utilities via funding, research, and suggested guidelines.

Installation Labor. Labor costs vary from location to location. Equipment providers and EPC contractors play a critical role in reducing installation labor costs by offering innovative labor-saving products. Providers of different components and services must collaborate as needed to develop streamlined and integrated systems that can provide competitive advantages, resulting in labor and overall cost savings.

Financing. Of course, the financial community and the governments have a large role to play in determining and reducing solar projects' financing costs.

Solar Financing Close-up

The financing of solar projects is the key to their expansion. It depends on the following factors:

- The federal government's role is to enable opportunities through legislation [e.g., master limited partnership (MLP)] or regulatory (e.g., IRS letter rulings for REITs, etc.) actions. Another opportunity for the federal government exists around broader tax benefit monetization rule clarity, providing a stable accounting foundation underneath third-party financing. Continuous improvement on tax benefit monetization clarity will likely be critical to many solar business activities beyond today's third-party financiers. The federal government can also play a role in enabling residential solar loans through standard real estate lending vehicles in light of the conservatorship role the federal government holds over underwriters Fannie Mae and Freddie Mac.

- State and local governments also play a significant role, directly in the case of state "green" bonding programs or municipal bonding on MUSH buildings, as well as through collaborative actions between municipalities and their municipal utilities, such as community solar. State legislatures can also enact some policies that would otherwise be handled by PUC discretion.

- Banks must provide securitized debt offerings, institutional investors must participate in direct financing or securities purchases, and the real estate lending financial ecosystem—particularly underwriters (e.g., Fannie Mae and Freddie Mac)—must allow and help facilitate PV value to be included in

appraised value and be part of common real estate lending products. Such underwriting is perceived to be key to the greater penetration of mortgages and home equity loans for solar PV.

- Large public corporations must move into direct, full residential project third-party financing, likely through mergers and acquisitions of existing successful PV development companies.

- Financial community funding efforts also include advancement of crowd sourced funding and increased financing of community solar. A reconceptualization of offtaker (e.g., PPA or lease counterparty) credit and/or new financial structures may be needed to enable the undefined commercial solution and realize roadmap targets.

- Real estate and construction companies have a strong voice within the real estate finance community. Demand for appraisal value to include PV could go a long way toward realizing standard mortgages and home equity loans inclusive of PV value. In addition, providing solar-ready homes can increase the opportunity set for mortgages to cover PV capital cost.

- Developers, such as SolarCity, Verengo, and Sungevity, along with lease/PPA-provision companies, such as Clean Power Finance and SunRun, have led the transition to third-party financing.

  Further action is required in the areas of standardization of contracts, greater availability of payment performance/default and technical performance data, and standardization of project credit reviews.

- Third-party financiers play a key role in opening the lease and PPA market to subprime credit customers who are usually unable to receive financing.

- Utilities enable the business models that are allowed in a region, notably third-party finance, but they also enable new utility business models that can improve utilities' desires to participate in distributed PV. Most importantly, they influence the rate for solar power sold on the grid, such as innovative rates that move beyond simple net energy metering and capture the value of the solar power for the utility and the broader grid.

These rates play a profound effect on the economic proposition and therefore also the desire for the customer to attain solar financing or pursue solar at all.

Case Study: Soft Costs Reduction Roadmap

Solar installations consist of a number hardware of hardware elements: PV modules, support structure, inverters, wires, and lots of bolts and nuts. The costs of these materials are dictated by the respective markets, but in general are more the stable and predictable of the bunch.

The non-hardware components are those that are not related to the hardware and are referred to as "soft," or "business process" costs. These costs include permitting, inspection, interconnection (PII), overhead costs, installation labor, customer acquisition, and financing.

The soft costs in the U.S. were in the $3.00-4.00 per installed watt range until recently, but have dropped to just below the $2.50/W level today. Yet, because the industry lacks standardization, the soft and all other costs vary wildly from state to state, vendor to vendor, and case to case.

NREL Soft Costs Reduction Program

In the fall of 2013 The Energy Department's (DOE) National Renewable Energy Laboratory (NREL) issued a new report, "Non-Hardware ('Soft') Cost-Reduction Roadmap for Residential and Small Commercial Solar Photovoltaics, 2013-2020." (4) It was funded by DOE's SunShot Initiative and written by NREL and Rocky Mountain Institute (RMI).

The report builds off NREL's ongoing soft-cost benchmarking analysis and charts a path to achieve SunShot soft-cost targets of $0.65/W for residential systems and $0.44/W for commercial systems by 2020.

As it can be seen in Tables 7-4.a and b, the soft costs of residential PV installations dropped in parallel with the total costs of the total PV system installation cost—from $3.32 in 2010 to $2.25 in 2014—but still remain just over 50% of the system's total cost.

At the same time, mid-size the commercial PV installations dropped from $2.64/W in 2010 to $1.76 in 2014. This drop, however, still leaves the soft costs at about 50% from the total system install cost. The decrease is also smaller than that of the drop in soft cost of residential installations.

Note: There is a gap in this roadmap, where installations in the 5-100 kW range are left out in the cold, so to speak. We must assume that the soft costs of these installations are somewhere between the costs of the installations up to 5 kW and those over 100 kW, but more detail is needed here.

The soft cost reduction trend might continue, but at a slower pace. Although the NREL report and the roadmap make sense, we just don't see how a soft cost

of $0.65 and $0.44/W for residential and commercial installations respectively could be achieved at all—ever! Labor costs alone, in our estimate, will keep the total soft cost of small PV installations well above $1.00 for a long time to come.

And the other soft costs—see the Other Soft Costs items in Tables 7.4.a and b—are presently even greater than the installation labor, but are expected somehow to go to near zero—an absolute impossibility under the present conditions!

On the installation labor side, the installers have to buy tools and vehicles, pay salaries, insurance, taxes, many daily operational expenses—including rework and other extra expenses—and still have some profit left for unforeseen and future expenses.

The installers are not getting rich even under the present $2.50 soft cost prices, so we just don't see what they can cut to bring these down to $0.65/W. It seems like a practical impossibility.

As an example, under NREL's guidelines, an average 5kW residential PV installation would have a total soft cost of about $3,250. The soft cost for a similar commercial PV system would be even less—about $2,200. This is $1,000, or about 1/3, less than the soft cost of residential installations.

So, the contractor, who is assuming the major part of the responsibilities and all risks associated with the project, will bring his crew of 3-4 installation technicians to the site and start work. They will spend two-three days working on this project, starting with making the necessary inspection and measurements, preparing the roof, installing the supports and the PV modules, wiring them and the inverter, and testing and optimizing the system. This is lot of work, most of which is dangerous (working on roofs with electrical equipment), which requires training, expertise and experience.

At an average hourly pay of $15-20/hr., the salaries and benefits of the crew, in addition to the other compa-

Figure 7-29. Rooftop installation

Table 7-4.a. Residential PV Soft-Cost Reduction Roadmap (up to 5 kWp size)

	2010	2011	2012	2013	2014	2015	2016	2017	2018	2019	2020
Customer Acquisition ($/W)	$0.67	—	—	$0.53	$0.49	$0.45	$0.41	$0.36	$0.28	$0.19	$0.12
PII ($/W)	$0.20	—	—	$0.18	$0.16	$0.15	$0.13	$0.11	$0.10	$0.06	$0.04
Installation Labor ($/W)	$0.59	—	—	$0.51	$0.46	$0.42	$0.36	$0.30	$0.24	$0.19	$0.12
Other Soft Costs ($/W)	$1.86	—	—	$1.30	$1.14	$0.97	$0.82	$0.68	$0.56	$0.48	$0.37
Financing (WACC %-real)	—	—	9.9%	9.4%	8.8%	8.2%	7.7%	7.7%	4.8%	3.4%	3.0%
Total Soft Costs ($/W)	$3.32	—	—	$2.52	$2.25	$1.99	$1.72	$1.45	$1.18	$0.92	$0.65
Total System Costs ($/W)	$6.60	—	—	$4.99	$4.49	$3.99	$3.49	$3.00	$2.50	$2.00	$1.50

Table 7-4.b. Commercial PV Soft-Cost Reduction Roadmap (over 100 kWp size)

	2010	2011	2012	2013	2014	2015	2016	2017	2018	2019	2020
Customer Acquisition ($/W)	$0.19	—	—	$0.15	$0.13	$0.10	$0.08	$0.08	$0.08	$0.05	$0.03
Installation Labor ($/W)	$0.42	—	—	$0.33	$0.30	$0.25	$0.20	$0.16	$0.12	$0.09	$0.07
Other Soft Costs + PII ($/W)	$2.03	—	—	$1.53	$1.36	$1.22	$1.08	$0.90	$0.72	$0.53	$0.34
Financing (WACC %-real)	—	—	8.6%	9.5%	9.2%	8.2%	7.9%	5.1%	4.4%	3.9%	3.4%
Total Soft Costs ($/W)	$2.64	—	—	$1.98	$1.76	$1.54	$1.32	$1.10	$0.88	$0.66	$0.44
Total System Costs ($/W)	$5.96	—	—	$4.03	$3.64	$3.24	$2.84	$2.44	$2.05	$1.65	$1.25

ny expenses (offices, support personnel, transport, etc.) would get very close to the total NREL soft cost estimate of $3,250 for residential installations. These same expenses would easily exceed the $2,200 soft costs allocated to commercial installations by NREL. This leaves no room for the other soft costs expenses, so the project and the contractors would certainly go under financially.

The two-prong question is, Would the contractors be able to cover all their expenses on residential installations, and who would be willing to work on commercial installations at such low prices?

Nevertheless, the commercial PV roadmap offers a more certain path to SunShot soft-cost targets, although additional reductions of $0.11/W and 1.1% WACC beyond the current-trajectory reductions are required.

Customer acquisition also has a relatively certain path, although reaching the 2020 target hinges on the highly uncertain market penetration of improved site assessment and design cost-reduction opportunities (CRO), in addition to advanced customer acquisition tools that couple well with market-expanding ("new markets").

In the area of installation labor, commercial PV is more amenable than residential PV to streamlined installation practices, thus achieving the SunShot target by 2020 is more certain; provided that standardization, universal adoption of integrated racking and other innovations are implemented.

Commercial financing exhibits a similar level of challenge to reach the roadmap's weighted average cost of capital (WACC) target as residential financing. However, the commercial financing path requires the highly uncertain implementation of an undefined host-finance CRO (e.g., special rooftop property rights/easements or energy service agreements) as well as highly uncertain expansions of green bond programs and commercial property assessed clean energy (PACE) financing.

The soft costs in the NREL report are focused on permitting and interconnection, which are significant when measured in terms of dollars-per-watt, and are very important for they pose significant market barriers which slow PV deployment. The achievement of this target would depend on the government and state officials, the regulators, and the utilities coming up with improved, speedy and cheap permitting and interconnection procedures, which is not easily done.

Regardless of the specific path taken to achieve the SunShot targets, the combined and coordinated efforts of the stakeholders are required. The new NREL report shows that the required participation of each entity varies substantially by soft-cost-reduction category. It also emphasizes that this is not a done deal, and that the different roles and responsibilities will be complementary and will evolve over time.

Amazingly enough, the soft costs are a major part of the total costs in all residential and small commercial solar installations, but not so much in large commercial and utilities type PV projects. They have remained stubbornly high in recent years despite impressive hardware-costs reductions. To reduce the total cost of systems and installations, aggressive soft-cost-reduction pathways must be developed, which is also needed to achieve the SunShot Initiative's PV price targets.

Soft costs account for more than 50 percent of total installed residential solar costs and about 40 percent of commercial solar costs. The report points to strategies for dealing with market barriers and decreasing costs across four key areas: customer acquisition, permitting, inspection, and interconnection, installation labor, and financing.

We now know that installation labor, as well as permitting, inspection, and interconnection are facing the most uncertain near-term paths toward achieving the NREL roadmap targets. By leveraging proven methodologies adapted from the semiconductor and silicon PV industries, and considering comprehensive findings from market analysis and interviews with solar industry soft-cost experts—including financiers, analysts, utility representatives, residential and commercial PV installers, software engineers, and industry organizations— the roadmap attempts to identify specific cost reduction opportunities, but more details are needed to put the entire picture together and achieve the goals.

NREL suggests ways to decrease residential customer acquisition expenditures by using software tools, which reduce total time spent on site, and designing templates to reduce system design costs, as well as leveraging consumer-targeting strategies to increase the number of leads generated.

The NREL report is a good step forward—one of many needed to bring PV installations under a common denominator and bring their costs down. It is the first quantitative national roadmap that targets soft-cost-reduction opportunities for solar technologies, but it stops short of pointing a way to standardization of the procedures on all levels.

Note: The example of the global semiconductor industry could be used here to point out the difference between the two industries (semiconductor and solar) and to determine what is needed in the solar field. The semiconductor industry went through a similarly winding path, until solid standards and regulations were established and implemented. That changed the game worldwide. The new standards and regulations provided stability to the process equipment design and the related process steps, as well as to the final product quality, while allowing intellectual property retention.

Clearly, the solar industry is far from this stage and that much more work and refinement is needed to outline a sure path forward standardizing the production process and the final product—including the soft costs.

As any first-time trial-and-error attempts, the NREL report only scratches the surface and leaves many questions unanswered. Let's hope that a number of more serious steps in all areas of interest would be taken soon to optimize and standardize the procedures, while bringing the U.S. solar costs down.

Home PV Pricing

Solar (PV) penetration is increasing rapidly in the U.S., but proper valuing of homes with PV systems is lagging and remains a serious barrier to PV deployment. Houses with roof PV systems should be evaluated according to the size, quality, and the added benefits to the homeowner.

Unfortunately, due to ignorance and negligence, some appraisers do not consider the PV additions as value added items and ignore them partially or completely. Instead, alternative methods of valuing of PV homes are being developed across the U.S. These include the use of a total income approach, based on the value of PV energy produced over the lifetime of the system. Another alternative valuation method uses the replacement cost of an equivalent PV system.

These approaches are finding recognition, but the drivers underlying PV home premiums are still not well understood, let alone standardized. These limitations deter most appraisers from assigning value to PV systems, who just don't understand the technology and are afraid of making mistakes.

Recent analyses have confirmed that a premium for residential PV systems can exist in the marketplace and that PV systems have a real-life value, therefore their contribution to home values must be considered in all cases by all appraisers.

PV premiums of homes in California, for example, are correlated with PV systems' size and, where the estimates for larger PV systems are greater, while those for older PV systems (of the same size) are smaller.

On average, each 1.0 kW increase in size is equal to $5,000-5,500 increase in home value. At the same time, each year of systems age reduces the price by approximately 9% annually. Because of that, the total value of older PV systems (over 5 years of age) is substantially lower than the new systems of the same size.

Using different calculation methods produces different valuation results—some larger and some smaller, than the more simplistic "size-age" practical valuation method. In any case, some across-the-board and across-country standardization is needed in this area too. The methods used in the U.S. might not be applicable in Europe, or Asia.

Other, less tangible, factors that play a major role in the valuation of homes with installed roof PV systems could be:

- The buyers' intentions, where "green" motivated people are willing to pay much more for PV system additions in support of the "green" cause.

- In some cases, transaction and other costs are avoided by purchasing a home with already installed and operational PV systems, and those are not incorporated in the cost estimates.

- The residential electricity retail rates from area to area vary, and often determine the price, where home buyers in areas with higher rates are willing to pay a higher price for an existing PV system.

- Finally, the quality of the PV system—actual performance, esthetics, and reliability—are also significant factors that usually affect the valuation and the final price.

There are many factors to consider, but there is lack of understanding and/or lack of interest to get deep into it on the part of the evaluators. So, the only solution is standardizing the valuation methods, but that is a long-

term project. Until then, the homebuyers and seller must get educated on the different issues related to PV system operation and ownership, so they can make the right decisions.

Property Taxes

On the positive side, a property tax exemption for solar power systems in California (set to expire in 2016) was extended to 2025 by a bill as part of the annual state budget. The wording of SB871 extends the period during which property taxes will not be applied to "active solar energy systems," which includes PV and solar water heaters.

This same issue—whether or not property taxes could be applied on solar power systems depending on the perceived value added to the property—is also on the agenda in Arizona. In May, 2014 a reinterpretation of Arizona's property laws was introduced to mean that that solar systems leased from a third-party owner (such as Solar City or other large installers) would be taxed with due property taxes.

If this happens, the entire U.S. solar industry could be vulnerable to such alterations or reinterpretations to property tax law, since this is easier to do than making new laws that single out solar power systems for levying property taxes. Sneaky, eh?

The tax law changes proposed in Arizona are another symptom of the fight between the solar industry and the U.S. utilities, which see their monopoly over the electricity market crumbling.

"The extension of the exclusion does not take funds away from any jurisdictions where taxes are currently being collected, nor does it have an impact on the general fund. But the exclusion will reduce wholesale solar electricity costs for utility customers, and it reduces barriers to accessing solar for customer-sited projects." Simply put, many homeowners would not choose to install solar if faced with a property tax reassessment, according to SEIA.

Another recent piece of legislation, Assembly Bill 2188 (AB2188), would cut the "soft costs" for solar by streamlining the permitting process for residential solar power systems in the state. In brief, AB2188 could help end the bureaucratic nightmare that the solar installations approval process has become.

California is no doubt taken the lead in these areas, so we are expecting more good news.

SunShot Program

The SunShot program is a U.S. DOE brainchild, with a specific goal in mind: to assist in making the cost of solar-energy-generated electricity to be competitive, without any subsidies, with the cost of fossil fuel-generated electricity by 2020. The 2020 target is five cents per kilowatt-hour generated by PV power systems, without any subsidies. This represents approximately $1.00 per watt to install large, utility-scale, PV power plants, or $1.25/W for commercial rooftop PV, and $1.50/W for residential rooftop PV. This is about a 75 percent cost reduction from the cost at the program's 2011 launch.

For solar thermal plants, the SunShot program targets unsubsidized cost of six cents per kilowatt-hour. That translates into $3.60 per watt for the installation of solar thermal (CSP) systems, provided they offer up to 14 hours of thermal energy storage (TES). To achieve this ambitious goal, the SunShot program supports research in photovoltaic (PV) solar and concentrating solar power (CSP), along a number of disciplines in basic research and applied sciences.

There is only one problem...money. The U.S. Senate has been slow in passing a budget during the previous several years, so the program is operating under "continuing resolutions." This means that the funding is flat and even then there are uncertainties related to future budget limitations and cuts.

Another problem is that the low hanging fruit—or about half of the total—has been already picked, so the final phases (the remaining 50%) would be much more difficult to achieve. At the same time, the U.S. DOE is also more than halfway toward the final target of its cost cutting program for solar energy. So the most important components of the country's 2020 plan are already accomplished and the future path is clear for full implementation.

The SunShot Initiative is intended to make solar installations cost competitive with the fossils by 2020, which is to be achieved through a combination of technology advances, soft costs decreases, and (very importantly) by optimizing the energy use by businesses and households.

The SunShot program was launched in 2010 with a $12 million tranche of funding to help cut red tape in the industry. These are the so-called "soft costs" resulting from finance, permitting, marketing and other peripheral services that account for more than half of overall costs of each U.S. solar installation.

These "soft costs" are one of the reasons the solar installations fail to keep pace with the rapidly falling cost of the PV technology. And yet, there are still realistic expectations for the crucial final savings to come mostly from both hardware improvements and soft costs reductions.

Recently, the costs of solar installations have come down dramatically, mostly due to sharply reduced costs of PV modules. Utility-scale solar plants are now at less than $2/Watt installed cost. This versus $4.0-5.0/Watt in 2010—a 50% reduction in 3-4 short years, to be sure. The goal now is $1.0/Watt installed, but this is a dream that might be spoiled by increasing PV modules and labor costs.

There is no magic here. It all boils down to old fashioned American ingenuity and hard work, with the majority of gains coming from innovation across the value chain. Many more innovations in hardware—not just in PV cells and modules—are possible in the path to improving the efficiency and reducing the final prices of solar installations. There are many other opportunities as well, such as new customer education, acquisition and alternative finance methods.

All these activities combined will bring solar in the U.S. to a more competitive level by achieving SunShot's goal of reducing the cost of solar installations. So, promoting more Sunshot-like programs is one of the best things DOE can do to support solar development in the country, and although the future of the program is uncertain from year to year, it charts the way to success.

The Government Assistance Programs

Since the first decade of the 2000s, the U.S. government and many states have introduced a number of programs designed to assist and support the growth of renewable energy in the country. Some of these programs have been very successful, while others have either failed or produced minimal results.

Some of the most successful programs follow:

RPS

The renewable portfolio standard (RPS) policies and mandates were put in place in about 30 U.S. states, to ensure the growth of solar, wind, and other renewable technologies. It is basically a regulation that requires the utilities to increase the production of energy from renewable energy sources such as wind, solar, biomass, and geothermal. For the most part, the RPS have played a critical role in driving renewable energy deployment over the past decade.

Recently, however, fierce debates have arisen regarding the cost of RPS policies. A number of proposals have been introduced to repeal, reduce, or freeze existing requirements.

The key facts and issues related to RPS compliance and applications, according to Berkley and NREL analyses, are as follow:

- Among the states with RPS, the estimated RPS compliance costs over the 2010-2012 period were equivalent to roughly 1% of retail electricity rates. Substantial variation exists across different states and from year to year.

- Expressed in terms of the incremental (or "above-market") cost per unit of renewable generation, average RPS compliance costs during 2010-2012 ranged from -$4/MWh (a net savings) to $44/MWh across states.

- Methodologies for estimating RPS compliance costs vary considerably among different utilities and states. A number of states are in the process of refining and standardizing their RPS compliance costs estimation methods.

- Utilities in eight states assess surcharges on customer bills to recoup RPS compliance costs, which in 2012, ranged from about $0.50 to $4.00/month for average residential customers.

- Cost containment mechanisms incorporated into current RPS policies will limit future (increase in) compliance costs. In the worst case, the increase would be no more than 5% of average retail rates in many states and 10% or less in most others.

- Although typically not considered within utilities' estimates of net compliance costs, a number of states have independently estimated the value of RPS benefits associated with avoided emissions as follow:
  $4 to 23/MWh from renewable generation,
  $22 to 30/MWh from economic development, and/or
  $2 to 50/MWh from wholesale electricity price suppression.

RPS are now seen as controversial and the utilities are rethinking their stand on these. In some cases they are ready for a battle with the renewables.

RECs

Renewable Energy Certificates (RECs), also known as green tags, renewable energy credits, renewable electricity certificates, or tradable renewable certificates (TRCs), are tradable, non-tangible energy commodities in the U.S. They are basically an official proof that 1 MWh of electricity has been generated via eligible renewable energy resource.

The RECs certificates can be sold, traded, or bartered, while the owner of the REC can claim to have purchased renewable energy. They represent the environmental attributes of the power produced from renewable energy projects, but are sold separately from commodity electricity.

Note: Solar renewable energy certificates (SRECs) are RECs that are specifically generated by solar energy.

Traditional carbon emissions trading programs use penalties and incentives to achieve established emissions targets; RECs simply incentivize carbon-neutral renewable energy by providing a subsidy to electricity generated from renewable sources.

The key point here is that the energy associated with an REC is sold separately and can be used by another party. The consumer of an REC receives only a certificate.

In states that have an REC program, a green energy provider, such as a solar or wind power plant, is credited with one REC for every 1 MWh of electricity it produces. This incidentally is the amount of energy an average residential customer in Arizona consumes in a summer month.

A certifying agency gives each REC a unique identification number to make sure it is accounted for, and that it doesn't get double-counted. The green energy is then fed into the electrical grid (not to be used locally), and the accompanying REC is then sold on the open market.

Once fed into the grid, thus generated renewable energy mixes with the conventionally generated energy already circulating in the grid. This means that purchasing an REC is the same as purchasing an official claim that the REC owner consumed energy from the renewable portion of the energy in the grid. Because of that mix-up, REC purchase does not affect how, or how much, renewable energy was actually generated. RECs only affect, or rather represent, how the renewable energy was distributed.

A few states have established solar renewable energy certificates (SRECs). The solar set-aside establishes a separate market for SRECs that encourages the inclusion of solar technology in the renewable energy mix.

This differs from the REC multiplier approach used by some states in which an REC from solar might count 2-3 times as much as any other REC.

Multipliers have had limited impact in promoting solar technology since most REC buyers will find it easier to source 2-3 times their REC needs from the economics and scale that come with wind farms.

With a separate market for SRECs, states are able to ensure that a portion of their renewable energy comes from solar.

After the SRECs introduction, states with less solar insolation, like New Jersey and New York, have had amazingly high success in promoting solar energy through the RPS—in many cases much higher than states such as Texas, with a generic REC market, or such with REC multipliers.

ITC

Another significant federal program is the Business Energy Investment Tax Credit (ITC). It is a federal corporate tax credit applicable to commercial, industrial, utility, and agricultural sectors. It basically encourages the use of technologies like solar water heating, solar space heating, solar thermal electric power, solar thermal process heat, solar hybrid lighting, photovoltaic power, wind power, biomass, geothermal electric power, fuel cells, geothermal heat and direct use, CHP/cogeneration, and micro-turbines.

The ITC program is coordinated by U.S. IRS and DOE and was expanded by the American Recovery and Reinvestment Act of 2009 until 31 December 2016. It is largely responsible for the quick development of a number of renewable projects around the country. It was especially beneficial for the development of solar and wind power technologies and projects.

With generous up to 30% tax credits, it allowed many solar, wind, and other renewable companies to develop, upgrade, and optimize their technologies. Many of these are now growing, while others went bankrupt during the 2011-2013 downturn.

In any case, many companies would not survive, and many projects will not be built when the ITC is terminated in December 2016. This date is not very far off, and its approaching has already affected the industry. One of the most affected right now is the timeline for solar energy development at the utility-scale: early site identification and selection, surveying, permitting (often multiple rounds of back-and-forth) and construction.

Developers with long ramp-up periods are having a hard time planning and developing their projects under the threat of ITC cancellation. The effect was actually obvious in such solar, wind, biomass and geothermal projects even 2 years before the expiration. Some didn't get start of construction guidance until it was too late to ramp up. Now, even though Congress extends the PTCs, those projects can't get access.

So the renewable energy industry is hesitating and unless ITC is extended many projects would be cancelled. On the other hand, according to the industry experts, extension of the ITC would spur an additional 4-5 GW of solar capacity and that much wind during 2017 and 2018. This would also create tens of thousands of additional new domestic jobs.

All these programs—RPS, REC, SREC, and ITC—however, are temporary crutches, assisting the renewables to stand on their own feet. The success of these programs will, to a large extent, determine the future of the renewables in the U.S. The fact that the utilities are waging a war on some of these programs is not a good sign and jeopardizes the renewables' future.

The New Trend

In the US, utility-scale PV installations increased from just 5% of total annual PV installations in 2008 to 54% in 2012, which was the first time when utility-scale PV made up the largest segment of the U.S. PV market.

The U.S. large-scale solar market has been booming, making it the third largest end market in the world. In 2014 the U.S.' PV installations pipeline surpassed 43 GW, due to double-digit annual demand growth, which is exceeded only by China and Japan for now.

Importantly, a large percentage of projects could come under threat of being cancelled outright if they look likely to miss deadlines. An unspecified number could also be delayed while new financing structures are obtained that restore IRR levels, post-ITC regression.

Mega-scale PV projects—those above 100 MW—are now dominating the utility-scale market in the US. The 10 largest PV projects in 2014, with over 5 GWp of new capacity coming online by 2016 are in that category. These are, however, the exception.

The new trend, taking over the U.S. energy market now, is shifting away from mega-scale, to smaller, easier to handle, below 30 MW in size solar projects.

During 2013-2014 the number of such projects increased by 33%, which accounts for over 2,100 projects. More precisely, large-scale PV projects in the 10-20 MWp range now dominate the solar installations pipeline, which in most cases is stimulated by state-based renewable portfolio mandates.

Solar installations are also becoming increasingly viable, due to the rapid decline of PV modules and overall systems pricing, and other recent improvements.

Large companies such as SunEdison have been focusing for some time on smaller projects below 50 MW, mostly because these have shorter planning phases. These smaller projects are also easier to finance and can often be constructed within several months, which usually reduces highs and lows of quarterly revenue streams.

The trend to smaller projects is also motivated by an upcoming deadline that requires projects to be qualified for the full U.S. Investment Tax Credit (ITC) of 30% by the end of 2016. The ITC will be significantly reduced or even dropped in 2017, which will cause another shift in the solar industry. Until then, the trend is clear—10-30 MWp solar installations are the new game in town.

2016 is upon us, however, yet only less than 9% of the entire 43 GW U.S. PV solar installations are under construction, and nearly 5% were classified as delayed, or with uncertain completion dates.

Over 63% of these projects are in the "planned" stage, while 24% are in the "pre-planning" stages.

For these projects to qualify for the 2016 ITC deadline, however, they need to significantly increase the speed of their development. New solar projects need to be added too, in order to achieve the state mandate goals.

2017 will bring significant changes to the U.S. solar industry, which in most cases won't be in a positive direction. The 30% ITC has been a major driver in the large-scale solar projects developments, and many developers will have a hard time undertaking large-scale solar projects...unless something replaces the ITC, or new and more efficient materials, technologies, and processes are introduced.

SOLAR ISSUES

Solar power generation has a number of problems, some of which are quite serious. As any maturing technology, solar is going through ever changing "growing pains." Here we present some of the major issues of today that are hindering solar's progress.

Residential solar power systems are small and simple to operate. Although they suffer from similar problems as their larger cousins—the large-scale power plants—they are easier to take care of.

Large solar power plants are quite different from their residential cousins, and have a number of technical problems, which must be considered in all engineering calculations, plant design and operation activities.

The first and most important thing to keep in mind is that most large-scale solar power plants are located in desert areas, where they are exposed to the sun's blistering heat and UV radiation, freezing nights, flash floods, and severe sand and hail storms.

While most rooftop solar installations are somewhat shielded from the extremes, desert installations are fully exposed to the whims of nature. Because large desert installations are a fairly new phenomena, desert solar installations of any type must be considered endangered and treated that way until proven reliable and safe enough for long-term use.

Considering these and other facts from our experience, we discuss below some of the most significant issues experienced by solar (PV) power plants of all types today.

Temperature Coefficient

In hot desert-like regions, where sunlight is most abundant, the solar installations are most efficient—more sunlight, more power. The problem is that as the ambient temperature rises, so does the temperature of the PV cells and modules. The resistivity of the materials increases proportionately with the temperature, and according to Ohms law, the solar (PV) cells' efficiency drops significantly with rise of temperature too.

There is 0.4-0.6% drop in output per degree C increase of ambient temperature, above the standard 25°C.

The output of solar panels is measured by the manufacturer under lab conditions at 25°C, and that number is used to determine the nameplate of the PV modules. This is usually the maximum power a solar module can generate, and this number is usually used to calculate the maximum output (nameplate) of a solar plant as well.

But field conditions are much different from those in labs. During the high noon hours, the temperatures in the desert areas (measured in the solar panels) rises to nearly 200°F (93.3°C). This is about 70°C above the standard temperature, which means that the affected solar cells and panels could lose about 30-35% of their

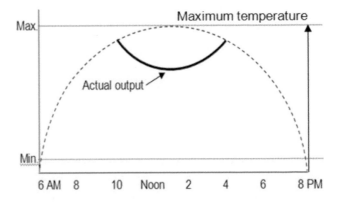

Figure 7-30. Reduced output during hot summer day in the desert

efficiency during the most solar-generation productive hours of the day.

70 degrees x 0.5% power loss = 35% efficiency loss

As can be seen in Figure 7-30, the output of the solar cells and modules starts to drop as the temperature approaches the maximum for the day. It then starts to recover as the temperature goes down, but never reaches the absolute maximum measured at the 25°C temperature in the lab.

This is a major problem with all PV technologies installed in the world's deserts. The actual power output during the hottest part of the day, especially in summer, drops significantly due to increased internal resistance of the materials with increase of temperature.

This also means that the solar panels exposed to such elevated temperatures never reach their maximum (nameplate) output.

So, taking as an example our 1.0 GW solar power plant, we see that during summer days (at full sunshine and with no clouds) we will get close to, yet below the 1.0 GW nameplate output of the power plant before and after the noon hour.

As the air temperature rises during the noon hour, so does the internal temperature of the solar panels, and we can see their output dropping 10, 20, 30% for awhile and then increasing again as the temperature decreases.

The nameplate power output is never reached during the hot summer months—the most productive months for solar energy generation—even for a short time. Because the generated power is much less before and after the noon hour, at the end of the day we will probably get an average of 450-550 MW, or 45-55% of the nameplate (1.0 GWp) power output.

This also means that when adding the rest of the hours of the 24-hour-day cycle, the overall efficiency of the system will be reduced to 15-20%, or the average daily output of our 1.0 GW PV power plant would be a mere 150-200 MW.

But this is not the end of the solar problems—it is only the beginning. Keep reading...

Annual Power Loss

Another serious issue of solar (PV) installations is that they lose about 1.0 % of their efficiency (and with that 1% of the projected power output) with every year. This is due to deterioration of the base materials—EVA sealants, solar cell materials, and the related inter-phases—abnormalities that contribute to increased resistance etc. malfunction mechanisms.

Figure 7-31 follows the operation of a 10 MWp solar power plant during its 30-year lifetime operation. What we see is that, at an estimated 1% annual power output loss, the plant will drop about 30%, or 3 MW in power output by the end of its useful life.

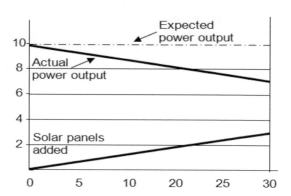

Figure 7-31. A 30-year operation of a 10 MWp solar power plant

If the expected 10 MW nameplate output is to be maintained for the duration, then about 100 kWp of new solar panels must be added annually to the tune of about $100,000 (parts and labor) each year.

An uninterrupted 10 MWp power output can be achieved only by adding more solar panels to compensate for weather variations and other inefficiencies. This might require a large number of additional solar panels, BOS equipment, and labor to keep the entire system operating at a steady 10 MWp power output at all times.

This also means that the uninterrupted operation of a 10 GW solar power complex, such as China is planning to install and operate soon, will require about 100 MWp of solar panels to be added every year, or 3 GWp for the 30-year operation. At about $1.00-2.00/Watt installed for parts and labor, plus a considerable expense for adding more land for the additions, this additional expense alone might amount to $4-6 billion. This is no small change, and would eat a considerable part of the profits.

Power Variability

A 1.0 GW coal-fired power plant can produce 1.0 GW electric power anytime it is required to do so on 24/7/365 basis (save for short scheduled maintenance periods). A similar size solar power plant has variable output that depends heavily on the local weather, the different seasons, time of day, temperature, and many other factors.

In reality, a solar power plant rated at 1.0 GWp output can produce 1.0 GW of power only for a very short time during the noon hours…and only if the sun shines brightly for the duration. At any other time—including early mornings, late afternoons, cloudy days and night hours—the solar plant is operating at reduced power output and even idling. In most cases it even uses energy to stay idle.

Figure 7-32 shows that the maximum power (nameplate capacity) of a non-tracking solar module is generated only for a short time during the high noon hours—and only at full sunshine. The daily average, therefore, is much lower (over 50% lower) than the advertised nameplate. When adding the night hours of inactivity, the output drops to 15-25% of the nameplate capacity.

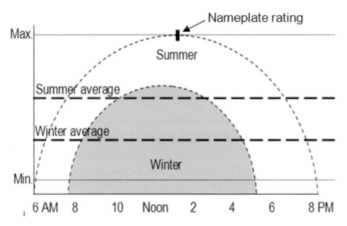

Figure 7-32. Daily solar power generation, summer vs. winter

Breakdowns and Failures

Once the power plant is installed, we start its operation, Here we start experiencing problems with equipment breaking and failing. This is a serious component of owning and operating a solar (or any other) power plant. Like any other type of equipment, solar panels and BOS equipment will malfunction and break down over their 30-year lifetime. No one can expect that all this hardware left to the elements will operate flawlessly and without failure all the time.

Sure enough, experience shows that 10, 20, even 30% of the hardware will experience some sort of problem—be it reduced output or complete failure—at sometime during the 30-year operating cycle. Repairs and replacements must be done at cost for materials and labor, which could be quite expensive.

There is an additional loss to be expected when the plant is shut down for maintenance. It might take several hours and even days for scheduled and unscheduled maintenance. In some cases the equipment has to undergo visual checks and test procedures of the suspected components to assess their status and complete the repairs. During all of these procedures, the power generation will be shut down, incurring further financial loss.

O&M Extras

During 30 years of non-stop operation the power plants undergo scheduled periodic maintenance (PM), which is included in the overall plant budget. There are a number of occasions, however, when different components malfunction or fail between PM procedures. In such cases, the plant is shut down and the faulty components are repaired or replaced at another additional cost for materials and labor.

This is true for PV, CPV, and CSP power plants, but the problems each technology presents are different. Nevertheless, the problem is here to stay and we cannot ignore it. What this means is that, everything considered, the new PV power plants are:

- The new installation is producing the maximum power possible now—albeit much less than the nameplate—in the 15-20% capacity range.

- At 15-20% total capacity factor, our 100 GW nameplate PV installation in the desert will actually generate the grand total of about 15-20 GW daily.

  Translated into gigawatt hours (GWh), at 100% capacity, our 100 GW power plant will generate 100 GW x 24 hrs = 2,400 GWh, or 2.4 TWh. At 15-20% capacity, however, the daily output of this same power plant would be 15-20% of the 100% nameplate total, or about 350-450 GWh.

- This also means that we have not replaced 100 average coal-fired power plants with solar power, but only one or two. This is a big difference, which needs to be well understood.

- As time goes on, the total output of the desert PV installations will be dropping about 1% annually, which, together with the other O&M issues, means that by 2030 our 100 GW desert power generation would have dropped to 75 GW nameplate output or less.

- At 15-20% capacity factor, this is about 240-300 GWh daily power generation.

- At that time, the performance warranty of the PV modules would have expired and whatever happens is at the owners' expense.

- Under these circumstances, it is unlikely that aging PV power plants at a declining state of readiness and performance would be able to stay in business very long, so after 25-30 years of non-stop operation, they have to be either rebuilt—basically starting from scratch—or decommissioned.

Starting from scratch is very expensive, and new calculations have to be done—including the old plant and decommissioning, which has to be done in all cases—to figure out the feasibility of a new installation.

Solar Power Fields Decommissioning.

At the end of its 25-30 years non-stop operation, every power plant is tired and ready to retire. This, however, is not an easy or inexpensive undertaking. Permits have to be obtained, and the old hardware has to be disassembled and removed from the site. After that, it is sorted and loaded on trucks for transport for disposal as waste, or to a recycling facility for processing.

Some of the components in solar power plants contain hazardous materials, which must be removed by special teams following special and very expensive procedures. Then it is recycled properly or disposed of at hazardous waste sites. This is always an expensive undertaking and should be included in the power plant's budget. If it is not, somebody will end up on the wrong end of the financial calculations or even worse—criminal court.

During the 30 long years, plants change ownership, manufacturers and insurance companies go out of business, and even in the best of cases, plant owners end up eating part of the cost of decommissioning.

When all equipment is removed, the land is bulldozed and otherwise treated to return it to its original condition as much as possible. This effort might require decontamination of the land, hazmat checks and tests, land surface rearrangement, adding grassy areas and trees, etc.—another set of lengthy and expensive procedures.

Silver Metal Dependency

The price of solar panels determines their marketability and the overall growth of the solar industry. The price depends on a number of factors, some of which were discussed in detail above. Access to affordable raw materials is one of the major solar power price-structure factors.

Silver metal is a key component of most solar panels used today. It is used in a form of paste in the manufacturing process of silicon and other solar cells.

Silicon solar panels predominate in solar installations worldwide with about 85% of market share.

An average photovoltaic (PV) solar panel contains about 20 grams of silver. About 40 tons of silver metal, equivalent to 40 million grams, or 1.4 million ounces, are needed to build a 1.0GW PV power plant.

The amount of silver used by the solar industry has been increasing steadily during the last several decades. In 2000, for example, 1 million ounces of silver were used for PV panels manufacturing.

By 2008 this amount had increased to almost 20 million ounces, which is a 20-fold increase in only eight years. During the next 5 years the global use of silver metal for PV panels tripled to nearly 65 million ounces in 2013.

Estimates now are that the solar market will use 100 million ounces in 2015, which represents about 10% of the total global silver market. The trend will continue, and most likely increase significantly, as long as the overall cost of solar electricity gets more competitive with fossil fuel-based energy sources.

The price and availability of silver metal suitable for PV panels manufacturing is something that needs to be understood and considered carefully when analyzing the supernormal growth estimates of solar installations.

Here is what we see happening:

Silver Prices

Silver metal prices could become a real stumbling block in the continued explosive growth of PV installations around the world. Silver is a crucial component of solar panel production, so a price of $20 per ounce (oz) is much more economical than $50 per oz. Since each PV panel contains about 1.25 oz of silver, the silver price difference is more than double in the latter case.

This means that the PV panels made with $50/oz silver will be $30-40 more expensive, a difference passed on to the consumer IF they accept it. Since price is the key determining factor in solar marketing, the price difference will decide who can afford the higher prices. Such price increase would surely slow down the global solar markets development.

Silver Has Multiple Uses

Its primary and most important use in the global economy is that of a monetary metal standard. It is also used in huge quantities as an industrial metal. This dual use can be good or bad, depending on which side of the equation you sit.

People concerned with global inflation usually look to hedge against inflation and would be more amenable to higher silver prices since precious metals historically retain their value and serve as a means to retain purchasing power. Investors, therefore, will have the final word in setting silver metal prices.

The solar industry does not control the silver market, so silver prices could increase dramatically for sustained periods of time. In such case solar will suffer a severe blow. Silver price of $100/oz. would mean doubling and tripling today's PV panel prices.

For example, a 100W PV panel with 20 grams of silver may cost $50 if the silver price is $20/oz. The same PV panel would cost $80-90 if the silver price is $50/oz. This is almost double the price, and would certainly have a long lasting devastating effect on the solar industry.

An increase of silver metal prices to $100/oz would double the price again, so our 100W PV panel will now cost nearly $200. Such dramatic increase would shut down the global solar industry for good.

Presently there is no substitute to silver metal in PV panels that can provide equal performance. There is a lot of R&D in progress, and a number of breakthroughs are showing promise to lower the overall cost of PV panels. Such developments might offer some breathing room should silver metal prices rise dramatically, but, a) it takes a long time for significant breakthroughs in the R&D labs to get to market and there are no such possibilities at present, and b) the global silver metal prices might explode and continue climbing to unsustainable heights for a number of reasons and there is simply no way for any breakthroughs to compensate sharply rising silver metal prices.

Silver Metal is a Finite Commodity

As such, it is in limited quantities, spread all over the world. The principal sources of silver today are ores of copper, copper-nickel, lead, and lead-zinc. The silver metal is produced as a byproduct of electrolytic copper refining, gold, nickel, and zinc refining, and from lead ores containing small amounts of silver. Commercial-grade fine silver is at least 99.9% pure, and purities greater than 99.999% are available, some of which are used in the solar industry.

The major silver ores mining countries are Peru, Bolivia, Mexico, China, Australia, Chile, Poland and Serbia. The top silver-producing mines are Cannington in Australia, Fresnillo in Mexico, San Cristobal in Bolivia, Antamina in Peru, Rudna in Poland, and Penasquito in Mexico. Tajikistan is known to have some of the largest silver deposits in the world.

According to the U.S. Geological Survey, in 2013, the United States produced approximately 1,090 tons of silver with an estimated value of $840 million. Silver was produced at 3 silver mines and as a byproduct or co-product from 39 domestic base- and precious-metal mines.

Alaska is the country's leading silver-producing state, followed by Nevada. There were 14 U.S. refiners that produced commercial-grade silver, with an estimated total output of 2,500 tons from domestic and foreign ores and concentrates, and from old and new scrap.

The total U.S. silver metal natural reserves are estimated at 25,000 tons, or about a year's worth of global silver production.

In 2013, the global silver metal production was estimated at 26,000 tons, with Mexico producing about 20% of the total, followed closely by China.

The BIG problem: The total global silver metal reserves are estimated at 520,000 tons, so at the present rate, the world has exactly 20 years of silver deposits left.

Let's assume that new deposits are found—just like with the unprecedented increase of the U.S. fossils reserves lately—increasing the global silver metal supplies to 40-50 years. Even then, silver is limited in quantity and prices will go up as its quantity decreases.

We must consider these restrictions every time we make big plans for the future.

Logistics

There are a number of additional problems that must be seriously considered as far as silver production and use are concerned:

a. Silver has to be mined in some difficult locations, and is actually getting harder and more expensive to extract in some places. As a result, the silver market was in a supply deficit of 113 million ounces in 2013, which is the widest deficit since 2008. Similar, and even larger, deficits are expected in the future too.

b. Electric power and fuel prices are inevitably going up, which increases the costs of silver mining and refining operations. As the energy costs go up, the price of silver goes up too.

c. China exported over 100 million ounces of silver annually in the 1990s, but recently it quickly reversed course. Now China is net importer of silver, where approximately 100 million ounces per year are imported and used mostly in PV panels. This puts significant pressure on the global silver market and if the trend continues, it might bring large silver metal price increases and quicker depletion of the silver metal reserves.

d.   Nevertheless, the price of silver has remained fairly consistent even in the wake of consistent supply deficits, which means that the PV industry and the silver market are in balance and working together to offer compelling opportunities along their respective value chains at both current and higher prices...for now.

As the future of solar energy become more ubiquitous and indispensible, however, silver stands to benefit disproportionately against the backdrop of a slowly improving global economy looking for a quick fix for its energy problems. As a result, silver prices will go up and solar energy might be the big looser in the not-so-distant future.

Silver Recycling

Fortunately, there is a light at the end of the tunnel...although we are not sure if it is the end of the tunnel, or a train approaching. The good news is:

a.   Silver in PV panels and other products can be partially recycled, thus reducing the "silver peak" threat. Silver, however, is a reactive material, so it can be only partially recycled. Still, enough silver can be recovered from PV panels to account for at least 20-25% of total use. This, however, is an expensive process, so the economic conditions have to be just right for undertaking a massive silver recycling effort.

b.   Many of the large-scale PV installations of today will be ready for recycling 25-30 years from now, which is just about the time when silver might become scarce and expensive enough. The recycled silver metal might extend the time of its depletion.

c.   The amount of silver used in PV panels can be significantly reduced by new technologies, and/or replaced by other materials. This is a long-term effort, but we count on it as the most promising option.

Drastically reducing (or better yet, eliminating) the use of scarce and expensive silver is the only way to avoid a silver metal supply crisis, thus ensuring unrestricted global expansion of solar power generation.

Standardization

The present-day solar industry is in its infancy, no doubt, consisting of a hodge-podge of different raw materials, equipment, processes, and products. There are some standards, such as testing the performance of the newly made solar panels, but this is far from being sufficient. It is even farther from determining the efficiency and reliability of the panels during their 30 years non-stop operation under different environmental conditions.

Standardization of all materials and steps of the manufacturing process—similar to that in place in the semiconductor industry—is absolutely necessary, if the solar industry is to rise to the highest level of quality and performance. Implementation of materials and operating standards will eliminate the shady operations from flooding the markets with cheap products of questionable quality. It will also give customers the peace of mind that their investment won't end up in the garbage bin after 5-6 years of on-sun operation.

SEMI's Example

For over 40 years, semiconductor device manufacturers, major fabs, original equipment manufacturers (OEMs), and related organizations around the world have relied upon the Semiconductor Equipment and Materials International (SEMI) international standards to optimize the production process, reduce costs, and spur innovation.

Standards have had enormous impact on the semiconductor industry and are now also making valuable contributions to other areas, such as flat panel display, micro-electromechanical systems, photovoltaic, and high-brightness LED (light-emitting diode) industries.

For example, SEMI has 305 pages compilation of official terms, definitions, abbreviations, and acronyms. (6) This work alone, as compared to the non-existence of universal terminology in the solar industry, is a great achievement. It is also a good, and absolutely needed start—step one—in achieving global standardization of the solar industry.

SEMI has relied on thousands of industry volunteers worldwide to help develop a wide range of standards. Over 4,600 volunteers representing over 1800 companies work in 21 global technical committees and over 200 task forces to find solutions to common technology challenges.

This is a great accomplishment, which the solar industry is very far from. Instead, there are continuous wars in the solar arena, and they do not seem to be going away anytime soon.

On the international front there are solar products import-export wars between different countries, and lawsuits among solar companies on intellectual property and other issues. In the U.S., there are wars between the utilities and the solar installers, in which the customer is the victim.

There is also confusion and hesitation at the government level as to the role of the solar industry, which determines the support the industry gets at the end of the day. The originally enthusiastic support is now gradually diminishing, so it is clear that government priorities are changing and solar is once again on the back burner. Been there, seen that...

It is, therefore, quite clear that unless the major confusion and the resulting discrepancies are resolved, the solar industry will not be able to follow the example of the semiconductor industry. The solar industry needs a closer cooperation with SEMI, or even better, establishing its own SEMI-like standardization body to develop, implement, and supervise industry standards. Short of that, the global solar industry will continue to go in circles, suffering from a number of ills.

Case Study; Solar Glass Standards

A solar glass certification developed in 2002, was designed to guarantee the quality of the glass and the glazing that is used as a transparent cover for *solar thermal* collectors. **Note:** *Thermal collectors*; not *PV modules*.

Over the last several years, more than 250 glass types made by different manufacturers have been measured and approved for use in solar **thermal** applications. This particular certification was explicitly developed for solar **thermal** applications and yet it became widely used in the manufacture of **PV modules** too. Why? Nobody really knows. Scientific conclusion, no?

The glass was there, had a label that certified it for *solar thermal* use, so let's use it for PV modules too. It is close enough. Its thermal properties might be acceptable, although there is no proof of that.

But how about its other physical and optical properties? Do they match what is required for efficient, safe, and reliable use of PV modules? The configurations and applications of thermal vs. PV solar are different, so we must agree that the results from the standardization and certification tests of solar thermal glass are not directly transferable to PV modules.

Nevertheless, the global solar industry has been using this erroneous glass standardization and certification for its PV modules. Millions of these modules were manufactured, sold, and are in use on countless roofs and large-scale installations in the U.S. and worldwide

with the wrong glass certification. How is this for a large-scale discrepancy?

Realizing the gap in standards between thermal and PV glass, which leads to erroneous conclusions in some cases, the PV industry adapted a special glass certification in 2012. The new certification fits the needs of the PV industry much better, so a dedicated PV solar glass certification is now available.

The PV module *cover glass* certification process consists of three performance characterizations:

- Optical transmissivity,
- Incident angle modifier (IAM), and
- UV degradation level.

The results obtained with different representative glass types, such as float glass, anti-reflective-coated glass, and rolled glass with different structures are quite different. The intricacies of the different types must be well understood, evaluated, and used in the most appropriate sites and locations, where they would perform most efficiently and reliably.

As yet, the tests and measurements used during the certification process are still debated, so it will take some time before we have a unified standard for PV glass certification for use worldwide.

Nevertheless, we now have a cover glass standard, which—as imperfect as it might be—is a good step ahead, and which cannot be said for other materials and process steps of the overall solar products manufacturing and use process.

Note: The new glass standard has limited (or no) requirements for mechanical and thermal stability, both of which are absolutely necessary for efficient, safe, and reliable field operation. The standard must be revised to include these parameters, if it is to be of consequence.

Most importantly, the new cover glass standard for PV panels is not widely accepted or implemented as yet. Instead, different companies use different materials that they think might do the job. Thus, the confusion in the solar industry continues.

As a result, some manufacturers use thermal glass, while others think that construction quality glass is best. These glass types, however, are manufactured, installed, and used according to the needs of the construction industry, which has different requirements. The construction glass, for example, has certain physical properties that make it less breakable, but the optical properties are usually not optimized for use in PV modules. Such glass would certainly fail the certification tests under the new PV glass standard.

The random, non-standardized choice and use of materials and processes during PV products manufacturing is presently the norm.

Meanwhile, billions of PV modules have been installed with different combinations and permutations of cover glass types and other materials. How good or bad this practice is will be determined after 10-20 years of non-stop field operations. Only then will we be able to count the total number of PV modules manufactured with inappropriate materials.

At that future time we will be able to analyze and count the number of PV modules with reduced efficiency, those with broken or hazed cover glass, and/or other failing components—depending on type of materials used by the different manufacturers.

Solar Power Plants Permitting

All residential solar installations follow an established protocol, which consists of obtaining permits for their construction and use. This is not a very hard undertaking and is usually done by the installers.

Large-scale solar power plants, however, are a totally different matter. Here large land areas are usually involved that need special environmental assessment and special permits. This is not an easy, quick, or cheap process. It may take years and millions of dollars to go through all necessary steps, and even in the best of cases, the chances of success are still low.

Since large-scale solar projects are located far away from populated centers, and from the power grid, special power lines and substations must be permitted and built to connect to the grid. This is another long and expensive process, which may or may not succeed in the end.

Special efforts today promise facilitating and speeding of the permitting processes, but it might take some time before this issue is fully resolved.

The U.S. Utilities

The growing problem in the U.S. solar sector is one that was least expected—the U.S. utilities. After embracing solar power in the first decade of this century, they are now making a U-turn and blaming it for many of their problems.

Barclays bank analysis of the U.S. energy sector in the summer of 2014 brought up some disturbing news for the utilities. They saw solar as a rising competition for the utilities, with the biggest danger coming from residential solar installations.

The analysts saw this as such a big problem that

it forced them to re-evaluate the high-grade corporate bond market for the US electric sector to "underweight." Solar is killing the utilities (profit margin), they thought...

The reason, "...a strong indication that solar PV-generated power coupled with energy storage (solar-plus-storage) presents a long-term disruptive risk to utilities."

Barclays' analysts think that the 19th century type establishment of the U.S. utility industry is not ready for the inevitable long-term challenges presented by increased use of solar power. The bank analysts see a real problem, because the utilities just don't see, and refuse to accept, let alone adapt to, the upcoming changes.

Presently, the greatest problem for the utilities is residential solar installations, which is trending towards adding energy storage, a.k.a. "solar-plus-storage." The new solar-plus-storage systems are basically a conventional solar PV array mounted on a house roof that feeds excess power into a bank of batteries, or other forms of energy storage devices.

But wait, if the excess generated power can be stored and used later on, then what do we need the utilities for?

This is a really big problem for the old, all-powerful industry which has never been faced with competition before. And now the utilities have to learn how to play the new game, which they thought would be played by their rules.

They really did think so! In the mid-2000s, the utilities made all kind of rules and regulations as of who can apply for solar installation permit and interconnect, how the applications were to be filled, how much the applicant must pay and on, and on. And they had fun doing it. There were meetings, conferences, media releases, and official regulations on the subject. All that worked for awhile, but now the game changed all of a sudden.

The U.S. utilities had the very unique and unchallenged positions as:

- The one and only undisputable provider of electric power in the country, and

- One of the most reliable long-term investments in the world.

That is how it was! Now, the invincible castle that the utilities built during the last century is crumbling. With $10,000 any Joe customer (in some parts of the country) can install a solar-plus-storage system and simply disconnect from the grid, thus saying goodbye to the monthly electric bills forever.

Unbelievable but true, this amazing development is just now starting and promises to grow exponentially in the years to come. Enough off-the-grid Joes can put the local utility out of business. It is that simple, and that ominous. This is something that the utilities surely did not foresee when they agreed to play the solar game and set the solar game rules so carefully in their favor.

So now, the utilities are threatened by the declining cost of solar (photovoltaic) power generation from house roofs, which when coupled with on-site energy storage makes the electric grid and the utility obsolete.

Since the utilities did not foresee this development at the initial stages of the game, they missed the opportunity to do something when they had a chance. The problem is that they still don't look at the problem from the proper perspective. It seems now that they saw solar as a temporary development; another short revival by a few enthusiasts which will go away as it did so many times before. Well, that was the wrong assumption.

Because of that, they underestimated the solar potential and totally failed to foresee the effects of the new trends. They ignored the possibilities of the new disruptive technologies and processes, the falling costs of solar products and installations, and the rapidly increasing deployment of residential energy storage.

The new, low cost residential PV systems with energy storage are the most significant—permanent and increasing—challenge to the utilities century-long reign over the U.S. energy sector.

The utilities still control the large-scale solar power generation by their 20th century rules and regulations, and many other tricks they have up their sleeve. That domain is unchallenged until a new disruptive technology or method comes along. Here again, they should be very careful, because the American entrepreneur's spirit is alive and ready for the challenge.

The utilities are helpless when confronted by the possibility of millions of Americans with housetops equipped with solar panels, the power from which could be stored for later use. The new solar-plus-energy-storage systems are capable of providing uninterrupted power on 24/7 basis without grid connection.

This is a true game changer. The day is coming soon when homeowners won't need the utilities at all. They will be able to generate enough power during the day, via solar and wind, and store the excess for use at night, or at periods of high demand.

This is a very real competition for the utilities. It is a truly cost-competitive substitute for grid power never available before. And all this is happening under their noses, without them being able to do anything about it. It is the beginning of the 21st century residential solar revolution.

Here is where the famous *American ingenuity* kicks in, and where average people are thinking of ways to beat the utilities at their own game. The game is just now starting, but the long-term prospects are convincingly in favor of the people.

The Rocky Mountain Institute in Colorado issued a report in the winter of 2014, calling the new development "possibility of defecting from the grid." Using the combination of solar power and electrical energy storage is now possible in five large regions in the continental U.S. and Hawaii, where solar-plus-storage has already reached grid parity. Even in New York state with its scarce sunlight, solar power with energy storage will reach grid parity by 2025.

Note: As a clarification of the new situation: Joe Customer-turned-energy-revolutionary has several options. He can:

- Install a small solar array on the roof and use the utilities maintained grid for free to play with his new toy while saving some money, or

- Install a larger solar array and eliminate most of his monthly utility bills, still using the grid for free, or

- Install a solar array large enough to eliminate all monthly electric bills, still using the grid for free, and/or

- Install even larger array, which generates extra power and then add energy storage (battery bank) to store the excess. Now Joe Customer can pull the plug on the grid and be totally free from monthly electric bills.

Wow! Freedom of choice is what America is all about, and people will not hesitate to make the appropriate choice at the right time. And this is the problem Barklays bank analysts see.

Barclays analysts even recommend that investors should move away from holding bonds in utilities in those 5 regions where grid parity is near reality.

Instead, they recommend investments in energy markets where this competitiveness is still in the distant future...for now.

The bank sees the energy market and the utilities as being blind, or refusing to see the reality, thus ignoring the disruptive risks the new solar technologies and developments bring. We see how it would be hard for a 20th century industry to think in terms of 21st century technological advances and large-scale social developments.

Although this revolution hasn't yet begun, the sparks are flying and millions of Americans fed up with the utilities games will not hesitate to pull the plug and be free. Maybe not tomorrow, but once the revolution starts it may explode overnight into a non-stoppable movement.

To remain competitive in the long run, though, the utilities need to find and implement new business models. They need to prepare for the day when cheap solar panels and storage batteries would dominate the residential energy market. They must be prepared to accept 80% self-sufficient and off the grid (ex)customers, thanks to the $10,000 solar-plus-battery systems on their roofs. That day is coming soon!

So, Mr. Utility, it is time to think in terms of the new solar energy reality—that of energy being generated by the people for the people. This will be the end of your kingdom. We, however, will not decapitate you, as people did in the past with their monarchs. Instead, we will only demote you from a rule-maker to a servant of the people. It is only fair, and we will all live happily thereafter.

The Solar Wars

The revolution has not started yet, but there are isolated conflicts through the energy environment. Arizona is a place where the utilities were successful in slapping their solar customers with additional charges for using their grid. This might be fair, but it was only the initial salvo, and the battle is heating. The utilities will fight for their survival the only way they can—exerting their monopoly on the masses.

The battle is spreading over the country too. Another sunny state in the U.S., Hawaii, where the solar electric industry was the hottest construction trade until recently, is on the warpath with the utilities. So far the score is utilities 1, solar 0.

During January-June, 2014, the permits for new solar (photovoltaic) system installations dropped by nearly 45 percent compared to 2013. At the same time, the total project value declined by more than 50 percent. Coming from the white-hot annual growth from 2007 to 2012, some of the top solar installation firms saw their

sales drop off by more than 90 percent over the same period.

The new energy revolution is happening before our eyes with solar losing the first battles by a huge margin. The significant decline is being blamed primarily on interconnection challenges. In other words, the utilities are not allowing any additional solar installations to have access to the grid.

In the absence of viable energy storage technology, home owners have no choice, so the utilities are winning for now. They are creating a behavioral impact, uncertainty and doubt, which are seeding hesitation in the minds of investors and potential solar systems owners.

The American Electric's PV division is more direct in stating that the decline is due almost totally to restrictions the utility has placed on the approval of PV systems for installation on the busiest, so-called "saturated" areas of the electric grid. Because the utilities control the net energy metering (and interconnection) approval process, very few solar installations are going up in Hawaii.

The reduced grid access and high volume of PV installations are in conflict, impacting the outdated electric grid which the utility has not updated to accommodate the renewables influx.

In summary, solar in Hawaii is on hold until the 19th century electric grid is brought up to 21st century standards, which may be someday soon, or never. So solar will resume its growth at that time, or...never. This is not good incentive for solar investors.

This signals the approach of a new era, that of post-net energy metering. Since it would cost billions to update the electric grid, the average Hawaiian customer hasn't many choices. For now they would put up with the utilities' monopoly and the related restrictions. Someday soon, however, they could break lose by installing their own PV system with energy storage. Then, they will say goodbye to the utilities...forever.

Grid Defection

Hawaii is not the only state that is going through solar growing pains. Distributed electricity generation (small solar PV installations on housetops) is rapidly spreading and getting much cheaper. The main barrier to its full expansion is the resistance of the utilities.

Energy storage is the solution to bypassing the utilities all together.

Energy storage technology is also going through growing pains, but prices are going down due largely to mass production of batteries for electric vehicles. As a result, energy storage for solar power is shaping up as a feasible—and very attractive—alternative for many home owners.

As a matter of fact, it is already becoming reality in some areas where the new technology is reducing the electric bills and eroding utilities' sales and revenues. The ultimate goal of any customer is to disconnect from the grid and manage its own energy needs.

This is what we call "grid defection."

This dream is becoming technologically feasible. Working together, solar and energy storage open new possibilities. They can make the electric grid optional for many customers, regardless of the utilities' actions or lack thereof. Equipped with a solar-plus-storage system, individual customers become a mini-utility. They can take or leave traditional utility service with their "utility in a box" approach.

This represents a fundamental shift from the "as usual" 20th century way of thinking of the utilities, and represents a huge challenge for them. Solar-plus-storage puts the power in the customer's hands…literally. It enables customers to cut the cord to their utility partially, or entirely, as they wish.

This development promises to see a considerable number of utility customers considering grid defection by 2020-2025. This will lead to significant revenue losses, especially during the pre-defection period, when the utilities enter a panic mode. The long-term customer defection will mark the end of the traditional utility-regulatory business models of today.

Some utilities will be able to adapt, and even see the new solar-plus-storage systems as an opportunity to add value to the grid and enhance their business models. The question is how would they do it? They have no flexibility under the existing regulatory frameworks, so a brand new regulatory landscape must be developed—one that is flexible enough to adapt and tap into the new energy sources and add value to build the electricity system of the 21st century by providing the best service at lowest cost possible.

Can they do it? When and how? These are the questions before all U.S. utilities. The clock is ticking…

The Utilities' Solution

In the summer of 2014, the largest utility company in Arizona, APS, serving over a million customers in the state, announced its decision to install 20 MW of solar on its customers' rooftops by the end of 2015.

APS will do this on its own—a trial experiment, if you will. If the plan is approved, 3,000 customers will receive a $30 credit each month for the 20-year life of the program. This adds up to about $7,200 for the duration. And get this: there is no upfront cost to the lucky customers.

This move is under the umbrella of APS' "AZ Sun" program, which allows the utility the right to install 200 MW of community solar in Arizona. The program has added about 170 MW total capacity since 2010. The new 20 MW capacity would result in a 10% increase in residential rooftop PV's contribution to APS' grid.

Recently, APS also filed a plan to install a 20 MW ground mount PV plant, so the Commission must decide which plan is best.

APS has been the center of growing controversy in the state, with solar advocates accusing APS of trying to kill solar. In 2013, more than a thousand people attended a rally as the corporation commission deliberated a proposal by APS to charge fees for net metering. And in 2014, a reinterpretation of property law was put forward by the Arizona Department of Revenue that would tax homeowners for their PV systems. That move was credited to Arizona lobbying group with APS members on the board.

Now, the fairness of the new plan is questioned, since it seems highly uncompetitive. APS is basically installing new solar systems on residents' roof for free. The customers get $30 a month as a roof-rental, APS pockets the rest. Clean and easy way to make money with solar energy. How can anyone compete with this?

The question now is whether APS is trying to help its customers, improve its bottom line, or kill the solar industry by unfair competition. The new plan surely stacks the deck in favor of APS which has unlimited resources. In this case, it introduces new rates-based solar with a guaranteed rate of return. Hard to duplicate.

The APS' motives, in light of ongoing disputes, are suspicious. Why would APS, which had clearly opposed solar until now, suddenly be looking to install its own solar systems? Here, the proverbial, "If you cannot beat them, join them" comes to mind. That way, the APS will be doing its own business, while the local solar companies would be trying to find a way to compete.

APS officially considers rooftop solar installations as damaging to its bottom line, so this latest tactic is seen as a "Trojan Horse" by some. If the new 20 MW APS solar project succeeds, it might lead the way for other utilities to follow APS' example. The other local utility, SRP, is not regulated, so it can install as much solar via "APS model" as it wants and needs. That will surely put the local solar industry out off business. And the solar game in the U.S. would be changed in favor of the utilities, if enough of them follow APS' example.

On the other hand, "fair play" is not in capitalism's play book, so APS has the right to run its business as it pleases, as long as it is within the law. Solar is just another game, which anyone is allowed to play, so why can't APS become a solar company, if it wishes? APS surely has an advantage, since it holds the interconnect rights, but who can stop them from using it?

Who knows, this twist of events might turn out for the best in the long run. APS might actually help the local solar companies by hiring them to install the new rooftop solar systems. In the process, some of the broken relations could get mended, while at the same time adding much needed solar power to Arizona's rooftops.

Solar Power and Energy Security

Access to cheap energy is essential to the proper functioning of our economy. The problems with fossil fuel production, transport, and use have led to numerous and significant vulnerabilities. Threats to our energy security come from all fronts—political instability of energy producing countries, attacks on supply infrastructure and transport routes, manipulation of energy supplies, increasing competition over energy sources, as well as accidents and natural disasters.

The recent Fukushima nuclear accidents in Japan have shed light to the extent our energy systems are vulnerable to disasters of any sort. Climate change is adding to the number of potential risks with pending weather and climate extremes. All these are serious threats to our energy security, which demand equally serious responses.

The only way to protect us from the negative effects of external energy risks is the development of a safe and efficient internal (domestic) energy system.

One way to do this is to develop a renewable energy sector that can help us meet the goals of energy supplies and reduce greenhouse gas emissions. This way, we not only secure our energy sources, but also limit future extreme weather and climate impacts.

Investing in renewable energy now is the only sure way to a stable long-term energy security.

The problem is that solar is still an immature technology with a number of issues, as can be seen in this text. There are still major stumbling blocks before solar can fulfill its destiny as a pillar in our energy supply and energy security.

Some of the most serious problems encountered today in the solar energy sector, which do now, and would eventually impact our energy security, are:

- Solar costs have been, and still are, so low because of extreme competition driven by huge government subsidies in China, the U.S. and Europe. The "solar wars" among the producing nations will eventually end, and the prices will inevitably go up. How high is anyone's guess...

- The efficiency and reliability of solar installations is improving, but they are still far from being as efficient and reliable as the fossils. Solar depends on nature's whims, and on the quality of the products, none of which is completely under control at present.

- Energy storage is the Achilles Heel of solar energy. It is badly needed, but is in its infancy. It will take many years for it to mature and become even closely practical. It will take even longer for it to become cheap and reliable enough for everyday use in large-scale solar installations.

- But even if and when all technologies mature, solar power will always remain a variable power generator, because you cannot store energy that you don't have. When there is no sunshine there is no power. Period. And we cannot control sunshine—not even in the sunniest places on Earth.

- Solar installations require a lot of land. While some countries, like the U.S., have a lot of empty and cheap land, most countries in Europe and Asia are land locked and offer limited land for solar installations. Because of that, we foresee an increase in land prices in many places, which in turn will hurt solar growth. Eventually, even the most expensive land will be used, at which point solar progress will cease in the affected areas.

- Government support is what catapulted solar into the present unprecedented, mind boggling, levels of development. Without the billions of dollars spent (and still being spent) on solar R&D and projects, the U.S. solar industry would've been still in the closet, where it spent the last 50-60 years of its life. Recently, the U.S. and EU governments started getting more fossil-oriented, which threatens solar to be pushed on the back burner...again!

- And very importantly, as someone said, "The difficulty lies not with the new ideas, but in escaping the old ones." We are entering a new era, where new ideas are badly needed, but the good ones are few and often rejected.

The biggest problem with the new energy technologies presently is that they are not allowed to develop due to their complexity and higher cost. The few who know exactly what can be done to fix the problems are not powerful enough to have their voices heard. So, the old energy technologies and their owners and proponents rule the energy field.

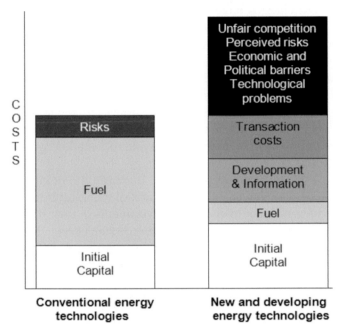

Figure 7-33. The new energy technologies vs. the old

Figure 7-33 shows the basic advantages and disadvantages of the conventional vs. the new and developing energy technologies. It is clear that the conventional energy generators (coal and natural gas) have a great advantage of being easy and cheap to install. The fuel is the greatest expense, but since we now have a lot of it, it is no longer of great concern to the rulers of the energy sector.

It is also clear from the same figure that the new and developing energy technologies are more complex and expensive. Generally speaking, they have more problems—all of which makes them not so convenient to use at this time.

Unless the new technologies (solar and wind especially) find ways to avoid the barriers and stand on their own feet—soon—they might be shoved back in the closet—something that has happened several times since the 1970s.

One must be blind not to see that solar and wind power are the key to our future energy security. Yet, they are vulnerable to the powerful (and at times destructive) forces of our capitalist system, which requires making a quick buck today, and forbids worrying (let alone doing much) about tomorrow.

In summary, we have the solutions to our energy security, and even our energy independence, but not the will to fully implement them.

INTERNATIONAL DEVELOPMENTS

The changes in solar development worldwide since 2008 have been extraordinary. We actually have an entire book on the subject, *Solar Technologies for the 21st Century*, published by Fairmont Press in 2013. In it we describe in great detail the changes—good and bad things happening since 2008—and what is to be expected in the near future.

We are now living in that near future and see the changes continuing, albeit at a much slower pace. There are no more Solyndra's, or mass bankruptcies, but the turmoil is still upon us and is more obvious in some countries than others, as we will see below.

Germany

Germany led the world in the number and size of solar installations until recently. This is a truly amazing accomplishment, considering the fact that Germany is NOT a solar country. As a matter of fact, the solar insulation in the majority of the German countryside is like that of Maine—partially cloudy and foggy most of the time at best.

Germany went through a period of hesitation during 2012-2013, when government regulations threatened to curtail the solar growth in the country. Still, they set a new record for solar PV generation in June of 2014.

With a solar PV capacity of 36 GW in 2013, and an increase well over 36.5 GW by the summer of 2014, Germany is again a world leader in solar energy installations. And now the country continues to work hard towards the goal of having 35% of its total energy capacity generated from renewable sources by 2020.

Get this: Germany plans to have 100% of its economy powered by renewables by 2050. Is this even possible?

Germany is a mostly cloudy and foggy place—especially during the fall and winter months, so is there even enough solar insolation available to produce so much power?

Nevertheless, as if to prove the point, Germany set a new record for solar production and consumption on June 9, 2014 with a total solar generation of 23.1 GW. This was about 50% of all electricity demand during that day. Amazing!

But wait; there is a catch. The record was obtained during a summer holiday period, during which less energy was used, as most businesses were closed and the sun was at its annual peak. The much lower energy demand and bright sunshine during the weekend were responsible for the good outcome.

Still, this is a plausible achievement, but the specialists advise that the next step for Germany—if it is to achieve its goals—is to focus on building its energy storage capabilities. Solar and wind alone just won't do the job, due to their inconsistent power generation. Energy storage will help to take further advantage of such record solar power generation, since it can be used during days of inclement weather and at night.

Energy storage systems are very expensive today, thus it is not economically feasible to store large quantities of electric power. The future forecast, however, predicts that the cost for storage systems will drop enough in the near future, to allow Germany to remain at the top of solar generation and use for a long time.

The broader picture reveals that the number of workers employed by the German PV industry fell by almost 50% in 2013, according to the German energy and economics ministry. The PV industry's workforce numbers in Germany fell from 100,300 people in 2012 to 56,000 in 2013.

Around 371,400 people were employed in all aspects of renewable energy generation in the country. Of this, 70.4%, or 261,500 jobs, can be directly attributed to the influence of the country's renewable energy act (EEG). The statistics also reflect a previous (albeit temporary) boom period for unsustainable PV growth.

A report put together collaboratively by the ministry of energy and other organizations—including the Centre for Solar Energy and Hydrogen Research Baden-Württemberg, and the German Institute for Economic Research—claims that the decline in job numbers was a consequence of lowered targets for PV installation. As a result of a managed decline and gradually falling PV system costs, a "boom" scenario like 2010-2012 is unlikely to happen again

In the future, Germany is looking for managed "expansion corridors" for renewable energy capacity which would add a degree of security to the industry and the job markets.

Changes to the EEG are likely to have a further impact on the industry over the next several years. Developers and investors are looking for ways to change their approach to large-scale projects as subsidies are further reduced, while manufacturers are increasingly looking to export their products to overseas markets to reduce their dependence on the fluctuating German renewable energy markets.

The EEG changes include the idea of applying surcharges for self-consumed PV energy, to cover grid and service costs, which are presently free to all solar users. These charges have proved controversial also because the heavy industrial energy users are exempted from the charges for economic reasons.

New business models are also emerging in the domestic market, such as optimizing self-consumption from residential rooftop PV installations. Also, the first subsidy scheme for lithium-ion battery storage systems for residential PV plant owners was launched in 2013, resulting in 4,000 new storage systems installed by May 2014.

Note: German people are increasingly looking to become independent of Russian fossil fuels and from the grid, which could also bring investments in their homes in line with Germany's wider energy transition.

Germany's national development bank KfW reported that nearly $90.82 million was given out in low interest loans by the development bank during 2013. Around $13.76 million was awarded in grants.

PV system owners could double the value of their system by storing some of the energy produced by the PV panels, and more people are considering the benefits of adding energy (battery) storage.

It is also possible to gain subsidies for retrofits, provided the PV system was installed after December 2012. More than double the amount of battery storage could be paid for each year in grants by redistributing the money saved by utilities in peak load shaving, for example.

The figure of installed systems is in line with predictions and research made by analysts. The subsidy program has been a success in terms of raising public awareness of storage systems, even though the actual uptake under the program is still limited.

Today two thirds of all German PV installers offer battery energy storage options in combination with solar system installations.

New Solar Charges

In 2013 the German cabinet approved a new charge on self-consumed solar power. It taxes solar installations owners for using their own solar generated electricity. Now they are required to pay a $0.06 for each kWh they use (that is not sent into the grid). This tax applies only to new rooftop installations above 10 kWp in operation since August, 2013.

The new surcharge is calculated as part (about 70%) of the total of the charges for 6.2 kWh to customers drawing power from the network—which self-generating-and-using consumers have been exempt from thus far. Fair, no?

According to solar industry insiders, however, this is like taxing people for using food they grow in their own backyards, with the proceeds being used to subsidize farmers. Since many of the solar owners chose to go solar for environmental reasons, the new taxation is viewed as an aberration and counterproductive to the proliferation of solar power generation.

The argument is that solar installation owners who consume their own solar power already relieve the energy transition budget because they have waived the feed-in tariff payments. They reduce the load on the distribution grid, since less power has to be transported from place to place. And since the major argument of the utilities is that the private solar installation use the grid for free, this new situation is somewhat of a misnomer.

The German government is also discussing plans for a similar charge on existing solar installations, but it is unclear if and when these plans are to be implemented. Probably soon after the protests against the present taxation die down.

The German solar industry trade association BSW-Solar is taking legal action over the reforms to the German Renewable Energy Act (EEG); the levy on self-consumption of PV. It described the changes as "impediment to the further expansion of solar energy in the country."

In BSW Solar's view, the cost of moving Germany's energy industry away from coal and nuclear should be more widely distributed and that a more significant financial contribution should come from fossil-fuel based industries than the reforms allowed for.

The changes in fact penalize environmentally friendly producers of solar power. It would be analogous to charging private gardeners a tax for the vegetables they harvest themselves.

The European Photovoltaic Industry Association (EPIA) also weighed into the argument, "The final outcome of such legislative process still remains unclear, but what is clear is that this ongoing uncertainty and this succession of turnarounds creates and nourishes a negative climate for investments, which impacts the development of PV in Germany."

The battle in Germany on the solar front is well underway. It will get more fierce with time, and we can't even venture a guess who is going to win.

Masdar PV

Masdar PV owned by Mubadala Development Co of the UAE, was a turnkey client of Applied Materials a-silicon technology. Production started in 2009 with great fanfare and hopes for world domination. Since then, however, the plant has supplied various sizes of a-Si modules to both ground-mount and BIPV projects.

It is one of the last remaining a-Si thin-film manufacturing plants in production globally. It is also the latest in Germany to close with the loss of 160 jobs at Masdar PV in Ichterhausen.

Masdar PV was expecting to become a major supplier for PV power plant projects in the MENA region as well as continue to develop its BIPV business; both sectors have suffered from long cycle times and limited project activity to date.

Alas, it was not to be! Masdar PV would handle all existing contracts until finally closing the facility in early 2015.

So, the solar sector in Germany is full of good and bad news, with the bad news predominating recently. Germany needs a lot of power, but has no natural resources and is tired of depending on Russian imports, so it is forced to look into the renewables—solar and wind especially. As a consequence, we foresee another wave of vigorous development in the German renewable energy sector.

UK Solar

The UK is another country in this world that is not blessed with much sunlight, so much so that one always thinks of the UK as a foggy, drizzly, rainy, cloudy and cold place. The solar irradiation over most of the country is only a fraction of the sunlight in the U.S., and many times less than the desert areas of the Southwest.

So how is it that it is the first country in Europe to officially issue a dedicated solar PV strategy? The UK's Department of Energy and Climate Change (DECC) published its version of the Solar Strategy in 2014, designed to implement a drastic change in deployment of solar energy in the country.

It encourages the deployment of rooftop solar, especially the underperforming commercial-scale sector, targeting one million solar homes by 2015. The UK is set to overtake Germany in 2014-2016 as the largest European market for rooftop PV deployment.

According to the energy ministry: "We have managed to put ourselves among the world leaders on solar and this strategy will help us stay there. There is massive potential to turn our large buildings into power stations and we must seize the opportunity this offers to boost our economy. Solar not only benefits the environment, it has the potential to create thousands of jobs across the region and deliver the clean and reliable energy supplies that the country needs at the lowest possible cost to consumers."

But the UK ground-mount solar sector is facing many challenges. DECC admits that the ground-mount pipeline had been much stronger than anticipated in government modeling and that the level of take-up has the potential to affect the financial incentives budget under the levy control framework. DECC is basically concerned that, left uncorrected, the large-scale solar could erode the approval rating of the sector overall.

The new strategy also fails to identify any specific targets for solar deployment, using the same figures published in the electricity market reform document of 10-12 GW of PV to be deployed by 2020. The strategy establishes 20 GW by 2020 as a possibility, albeit with the caveat, "the ability to achieve such a high level of deployment will be predicated by a number of factors. These include available budget within the levy control framework." It's the old story sung with a new melody.

Most importantly, DECC is working on allowing permitted development rights for roof-mounted solar up to 1 MW as well as simplifying the ROO-FiT application process. It is also considering changes to the financial support available to the sector to encourage further deployment.

There is still a lot of work to do in developing solutions to some of the barriers, but the excitement around the new found energy source is contagious and all issues will be sorted and addressed in the near future…sunlight permitting!

UK Update 2014

In the summer of 2014, the UK DECC confirmed that it is planning to remove support for solar under the Renewable Obligation (RO) scheme from 1 April, 2015. It outlined plans to entirely remove solar support under the RO for developments over 5MW in 2015/16. One step ahead, two back…

The government considers it necessary to take action to control the costs of large-scale solar PV to ensure it is affordable in the context of the RO and the Electricity Market Reform (EMR). So, solar projects over 5 MW will be able to apply for support under the Contracts for Difference (CfD) auctions as of October 2014.

The solar industry is concerned with the solar's classification as a "mature technology" and investors have responded negatively to the shorter, 15-year period, CfD scheme. This development would certainly slow down the falling cost of solar installations. It is a surprise that the solar industry has been singled out for such harsh treatment. It seems like the UK government has redefined the energy mix in favor of fossils, by hiding behind the lame excuses of "budget management."

DECC pushed the industry to work with communities to popularize solar, and now the new changes will knock the industry's extraordinary progress back. They will actually reduce healthy competition in the renewables sector. To reduce the pain, DECC is considering grace periods for projects which qualify.

In addition to removing support for ground-mount solar over 5MW, the government is also considering to extend the digression period for roof-mounted solar under the feed-in tariff, which is a market that DECC has targeted as critical in the UK's Solar Strategy.

According to DECC, "In order to support rooftop deployment, we are consulting today on splitting the current 'digression band' for projects over 50kW under FiTs into two: one for standalone, one for non-standalone. In other words, tariffs for building-mounted solar panels would reduce at a slower rate than for ground-mounted solar panels, so giving rooftop-mounted schemes access to more of the financial support available through FiTs."

These events actually show the immaturity of the solar industry in the country, and its dependence on government whims. Even if things turn out for the best this time, there is always a next time, when the government can change its mind and hurt the industry.

The worst part is that this is not an isolated case. We see similar teeter-totter in other EU countries and the U.S. as well. All this points to the fact that it will take a lot of time and effort before the solar industry can stand on its own feet.

Italy

The solar game is also changing in Italy. Today some PV plants owners in Italy are walking away from their existing feed-in tariffs (FiTs) in favor of power purchase agreements (PPA) as the market enters a new phase—that of a new market for subsidy-free solar that is emerging in the country.

Some solar power plants that have been receiving very good FiT from the third and fourth Conto Energia are switching to a PPA instead. This is a good sign for the solar market since the PPA business model works very well…if it is well designed and supported.

Analysts are expecting around 1 GW to be installed in 2014 with some of this already online. But the most exciting news is that 300-400 MW of *subsidy free* solar would be installed in the country under the current support scheme.

The market is heading towards rooftop self-consumption, as demonstrated by the latest IKEA solar project. The second, even larger and more important, solar market expected to start in Italy in 2015-2016, is the installation of subsidy-free ground-mounted solar installations with PPAs.

GSA, the entity which purchases electricity from the PV plants, has developed a simplified purchase and resale arrangement for projects between 1 and 10 MW with a tariff that varies with location. These zonal tariffs are set monthly by a market regulator, which is a good indicator of a maturing solar industry.

Although the Italian solar market is not expected to reach the 2011 heights, it is progressing and is now able to function in a sustainable way, which will improve in the future. Good news, no doubt.

Crime Wave

Sicily receives more sun and wind than any other part of Italy, so it became one of Europe's fastest growing renewable energies areas over the past decade. As soon as the Italian government began offering billions in subsidies to the sector, Sicily's crime families got busy.

In 2013, Italian prosecutors discovered an unfolding plot similar to "The Sopranos" as the authorities swept across Sicily in the latest wave of sting operations. Some of these revealed years of deep infiltration into the renewable-energy sector by Italy's rapidly modernizing crime families.

Most surprisingly, Suntech, the world's largest solar panel maker, had to restate more than two years of financial results because of allegedly fake capital put up to finance new solar plants in Italy. The discoveries follow the pattern of "eco-corruption" cases in Spain, where a number of companies stand accused of illegally tapping state subsidy money too.

One third of solar and wind farms were seized by authorities in 2013. Over $2 billion in assets have been seized and dozens of alleged crime bosses have been arrested, together with many corrupt local councilors and mafia-linked entrepreneurs.

The MO of the crime families was to target a piece of land suitable for wind or solar plants, then pressure the landowners to sell or offer long-term leases at low prices. Corrupt local officials were also enlisted to speed the permitting process, which usually takes 3-5 years.

With land ownership and permits in hand they would approach foreign investors attracted by the Italian government's green subsidies program. Some foreign investors did not know what they were getting into, while others just didn't want to know.

How the solar and wind projects proceeded after that depended on the mafia bosses' preferences, but in all cases they kept control over them. They also attempted to control the rest of the wind and solar business in the area.

Some project owners refused organized crime's attempts to control their businesses by using mob-connected developers or making customary payments of 2-3% of profits. Many of these businesses were targeted and came under attack.

Starting in 2007 arsonists set fire to a wind farm, causing $4 million in damage. In 2009, the Terrasi crime family attempted to block a new wind farm by claiming ownership rights to the land, which eventually resulted in 14 arrests and forcing the project owner and his family into two years of police protection.

The difficulty of doing business in Sicily eventually forced many companies to focus overseas, moving to invest in renewables in the United States, South America and North Africa.

It's not only the criminal infiltration; it is also the corrupt bureaucracy that makes it extremely difficult and expensive to do business in Italy. In October, 2013 the entire city council in Reggio Calabria, in southern Italy, was suspended because of alleged links to the Ndrangheta crime family. Italian prosecutors are still investigating suspected mafia involvement in multiple renewable-energy projects from Sardinia to Apulia.

As a result, foreign investment in the Italian renewables sector was hurt in the heart of the debt crisis. Italy was also ranked as the euro zone's most corrupt economy, together with Greece, by Transparency International. As a result, the foreign investment in Italy dropped to $87 billion during 2007-2012, compared to investments of $183 billion in the Netherlands, $289 billion in France, and $502 billion in Britain during the same time period.

Under the new anti-mafia Sicilian government that came to power in November, 2013, construction of most new renewable projects has been stopped. The new government is looking for ways to ensure that the mafia is purged out of the renewable industry before allowing fresh projects to go forward.

Criminal organizations caused major damage to

the Italian renewables energy industry. Italy lost a big opportunity for development, and the region lost the chance to profit from it. The Cosa Nostra got fully involved in the renewable energy game early on, because it was getting more profitable due to billions in government subsidies.

This, of course, cast a shadow over the entire Italian renewables energy sector, and is foreshadowing a massive challenge ahead for the entire European continent. A crime wave from Spain to Sicily is like a cancer growing in the middle of the newest and most promising industries.

Europe has many energy problems, and renewables are firmly imbedded in its future. But the single biggest barrier to achieving the goals might be overcoming the investment-inhibiting, market-distorting power of corruption lurking in the renewable energy sector.

China

The Chinese government has a different set of problems. After generously pumping billions of dollars into domestic solar products manufacturers during the last 5-6 years, it is now excluding over 80% of the remaining manufacturers, including the giants Shunfeng and LDK Solar, from any domestic support.

The government's plan was to throw as much money as possible at the solar manufacturers, in order to assist them in their ambitions to take over the world's markets. Now, after the open resistance from the U.S. and EU countries, China manufacturers are shrinking their ambitions and the government is shrinking its support, while trying to limit its losses.

The modified plan now is to curb oversupply and increase the quality of the Chinese solar manufacturing sector, thus making it more competitive with the world-class producers. So basically, after throwing millions of junk-quality products at the world's markets, China manufacturers are just now going to look into automating labor intensive processes, and focusing on quality instead of quantity.

Five hundred applications were received for entry in the list of "photovoltaic manufacturing industry norms conditions," which is basically a request to continue receiving state-run tenders and other support mechanisms. A final list of 109 approved companies (1/5th of the total applicants) was released in December, 2013 as a result of an "expert review" investigations and analysis.

The expert review criteria analyze a number of factors, including minimum conversion factors for solar cells, minimum manufacturing capacity for polysilicon,

Figure 7-34. China's labor intensive assembly operations.

ingot, wafer and ingot producers and other factors. The existing and future environmental conditions were taken into account as well.

The remaining 400 companies that did not make the list will not benefit from any domestic policy support, and are on their own. They cannot take part in domestic tenders, and/or benefit from export tax rebates. They basically must find a way to continue operation after being thrown overboard, after enjoying life on the mother ship for the last 5-6 years.

Senior officials at China's solar industry claim that the new measures amount to a "cull." The new situation will result in many producers being eliminated rather than acquired, as expected and planned initially. This will bring a lot of pain for thousands of laid-off workers, but may be a good thing for the industry in the long run, because many of these companies jumped on the bandwagon only because of the subsidies and cheap labor, and remain technologically and economically uncompetitive.

The newly instituted application process will be repeated every six months, which will give some companies time to upgrade their facilities, equipment, and processes, in order to meet the government criteria.

Update 2014. In June 2014, another 52 Chinese firms were added to the list of government-approved solar manufacturers. So now we have 161 in the "cull list" to receive favorable treatment and generous state support. The aim now is to introduce standardization within China's PV industry, especially in the key areas of R&D, quality control, and environmental practices.

Although initially it seemed as if the Chinese government is cutting the cord on failing manufacturers, now it appears unlikely that they would be pushed out of business. This adds uncertainty to the significance of the list. Companies on the list, however, will certainly be able to more easily access state support and tax benefits, which task might prove more difficult for the others.

One of the companies on the new list stated, "The purpose of PV manufacturing industry norms conditions" is to improve and standardize the photovoltaic manufacturing industry, eliminate backward production capacity, accelerate the PV industry to upgrade and develop the industry entry threshold. It has introduced strict quantitative requirements on corporate R&D, technology, quality control, marketing, production capacity, energy consumption and environmental protection."

So China is entering a new phase of PV industry development—that of officially supervised development of new standards and quality control procedures. This could be good for China and the world.

Legal action

But the Chinese government's and solar manufacturers' problems are not over. They are still haunted by SolarWorld's petition (several now), the last one of which claims that there are intentional and organized anti-competitive activities by Chinese and Taiwanese PV manufacturers, to be investigated in the U.S.' solar market.

This latest petition is designed to close a loophole in trade duties on Chinese solar imports introduced by the US government in 2012. It addresses a glaring loophole used by many Chinese PV modules manufacturers, who evade U.S. import duties (around 31%) by buying solar cells and other components from their Taiwanese and other third countries manufacturers. These imports (of questionable quality we must add) are then assembled in PV modules and sold as "Made in China" product.

SolarWorld had filed anti-dumping and anti-subsidy cases with the U.S. International Trade Commission and the U.S. Department of Commerce, explaining how the loophole is allowing many Chinese manufacturers to continue selling their PV modules at much lower cost in order to grab more market share.

SolarWorld is supported in this action by the Coalition for American Solar Manufacturing, with its 241 members, but some industry organizations, such as the Solar Energy Industries Association (SEIA), the U.S.' largest solar trade organization, oppose SolarWorld's latest move against the Chinese.

SEIA reasons that litigation is the wrong approach, since it is a blunt instrument and sometimes a knife with two sharp edges that can cut in both directions. It insists that litigation alone is incapable of resolving most of the complex competitiveness issues existing between the U.S. and Chinese solar industries. It is the opinion of SEIA that this conflict is damaging to the U.S. solar industry, and that is time to end this conflict and enter into friendly negotiations instead.

SEIA proposes a mutually satisfactory resolution, where both parties recognize their common and personal interests and agree on the solutions. There is no indication that proposal will be put forth in any meaningful way anytime soon.

So, the battle between the giants will be ongoing for quite awhile. Whether this is this good or bad for the U.S. and global solar industries is hard to predict. In any case, such actions and counteractions are expected to remain and even expand in the future.

Hy-Ref Technology

In the summer of 2014, China's Beijing Municipal Government entered officially into collaboration with IBM, focused on deploying its Hybrid Renewable Energy Forecasting "HyRef" technology. HyRef is a program developed to tackle air pollution, energy optimization for local industries, and to optimize renewable energy use and forecasting.

IBM launched its HyRef technology platform for renewable energy forecasting in 2013, employing its "Deep Thunder" computational super computers for high-resolution, micro-weather forecasts for regions of interest. When fully developed, the program will allow utilities and industries to implement and manage increasing levels of renewable energy into grids, thus reducing dependence on fossils.

All of IBM's 12 global research labs involved in the "Green Horizon" project will partner with Chinese government, academia, and industries in this 10-year effort to combat China's growing pollution problem due to significant economic growth over the last two decades.

The Hy-Ref program will tackle the problem with a range of policies that include the major adoption of renewable energy. The latest focus on distributed (solar) energy use in highly populated urban regions in the eastern part of the country is one of the primary goals of the program.

The Chinese government is targeting 13% of consumed energy to be provided by non-fossil fuels by 2017, which endeavor requires new renewable energy grids and energy efficiency. At the same time, the city of Beijing alone is planning to invest over $160 billion in air quality improvements with the aim of reducing fine particulate matter (PM 2.5) by 25% by 2017.

As part of the joint program, IBM and the Beijing Municipal Government are working on a system that is expected to pinpoint the type, source and level of emissions in the city, and predict air quality issues 72 hours in advance with high accuracy and street-scale resolution.

IBM's HyRef technology has already been tested on 30 wind, solar and hydro power sources around the country, with the largest being the Zhangbei Demonstration wind turbines project at the State Grid Jibei Electricity Power Company Limited (SG-JBEPC) in the northern province of Hebei.

As a result, SG-JBEPC has been able to integrate about 10% more alternative energy into the national grid, due to prediction accuracies of 90% proven on Zhangbei's wind turbines. IBM expects to deliver similar results on a much larger scale, which would help China in the long run.

Spain

Spain is the best illustration of the changing solar situation in Europe. Solar installations were popping up faster than mushrooms after rain in Spain in the mid-2010s. The frenzy lasted several years and suddenly the global financial crisis took Spain by surprise and put an end to the solar progress. An era of digress followed, and is still underway.

IKEA

As a sign of the times, in 2013 IKEA decided to get in the solar business and install a 10 MW solar plant on the roofs of its Spanish stores. But then—while part of the project was still under construction—IKEA changed its mind, decided to cut its loses and get out of the solar business altogether. In a split second, IKEA's—and maybe all of Spain's solar future—were put on hold as a result of new government regulations.

The losses imposed by these new regulations are too great for IKEA's 10 MW plant in Cuenca, Spain. The €65 million project, funded largely by the Santander banking group, was cancelled in mid-stream and the keys were handed back to the bank, in response to the Spanish government's cutting support to new solar projects.

IKEA has been praised for its efforts to install solar on its stores' rooftops and plans to sell panels in its UK stores, but IKEA of Spain decided to cut its losses after the retroactive changes took hold.

Without much fanfare, the government announced that the old and proven feed-in tariff (FIT) system is no longer feasible, and instead introduced payments that

cap the profit of any solar park at 7.5% of the initial investment. The IKEA case is the first casualty of this change, and could be a sign of things to come.

The new changes in solar policy are estimated to cut income from small to medium solar installations by an average of 25%. Larger solar installations will lose as much as 50% of the budgeted income. With already thin profit margins, 25-50% loss of income will make most solar installations money losers.

This is having a serious effect on the Spanish solar industry, and the investors are pulling out of the solar market in droves. No project can be profitable with 25-50% reduction in income. Some people will be able to renegotiate their loans with the banks, and some might be successful. For larger projects, however, this will not be possible simply because the income cuts are too great to justify, so many solar projects will be shut down and go bankrupt as a result.

There are already other examples of solar projects failing, and more are expected to fail.

The banks don't have many choices, so some might decide to recover their money by continuing to operate some of the solar installations, but there is no guarantee that they can succeed. This, in turn, is increasing the overall cost of finance for solar projects, where the usual 4-6% interest is now almost double at 9-10%.

The ripple effect of the new changes is affecting not only project developers, owners, and investors but is also forcing the country's PV manufacturers to reduce output. This will bring layoffs and increase solar products prices. In the end, the situation will make it impossible to manufacture or sell any solar products in Spain.

With the domestic market gone, the remaining manufacturers will have to compete on the global solar market...which is not an easy thing to do. Many solar installations will remain half-completed while many of the existing ones will be sitting idle.

Solar Police

There is another sign of the times in Spain, where widespread cheating has been the norm for awhile now. A number of large solar installations operators have been under-reporting their power use and making other "adjustments" to their records to avoid taxes. Many large solar projects have been penalized by large fines and some were shut down.

Now, the small residential solar owners are in trouble too. There is a potential for €60 million in fines for "illegally" generated solar power by some owners of residential solar installations. A new decree in the Spanish Energy Law allows inspectors to "raid" and inspect solar installations that are suspected of under-reporting or not reporting solar power use.

The plan boils down to raising cash to help pay off the "solar tariff deficit," or the difference between the cost of operating the national power grid and the money it generates. The new law aims to raise money for tackling a €26 billion debt to power producers which the state help build up over the years in regulating energy costs and prices.

All that the inspectors need to enter the property and conduct the investigation—anytime of day or night—is an administrative authorization. So now Spanish solar installations owners have an additional problem—they have to be ready for a knock on the door in the middle of the night from the "solar police."

If an owner denies entry, a court order is issued that will allow the inspectors access to the property with the help of the local police. Police officers are allowed to seize any documents they see fit to prove energy consumption irregularities and even seal off entry to the property.

This sets a precedent for obliging citizens to let inspectors enter a private residence on a whim. It also raises a serious doubt about whether this move is constitutional, and yet it is the law of the land for now.

The result of all this is that generating your own solar energy is a risky business in Spain. It exposes all solar owners, who already have a number of problems, to additional and quite serious risks.

Only 5-6 years ago Spain flung billions of pesos at solar companies and subsidized in a big way a large number of solar projects. In a reversal of roles, it is now in dire straights financially and is slapping a large fee on people who generate power from the sun for personal consumption.

In other worlds, Spain is punishing its middle class, which represents most of the people with solar panels on their roofs, and which benefits the large energy companies. And in yet other words, the Spanish government is taxing the sun and the people who dare to use it for their own power generation.

This makes independent power generation by households one of the biggest financial and personal risks Spaniards can currently take at a time when cutting costs is the key to the survival of the solar energy sector.

The broader implications of the introduction of "solar police" rules in Spain is that Europe and the world are carefully watching the developments, and the repercussions could be quite serious for the entire global solar industry.

Legal Battles

Finally, the solar battles in Spain were escalated in the summer of 2014 to the international arena when Spanish solar companies and investors filed a complaint with the European Commission. The law firm Holtrop SLP representing 1,500 renewable energy investors, delivered the complaint to the Directorate General for Energy at the European Commission.

The complaint claims that the changes, passed into law will limit the profits from solar projects to 7.5% before tax and around 5.0% after tax. Further revisions, to be done periodically according to the new law, are very likely to reduce profits even more. Added to investment and profits cap, punitive charges on self-consumption and other reforms to the ways payments are calculated were also incorporated into the new law.

The new law basically restructures the entire solar industry in the country by introducing a new and very complex system. It seems less about energy policy and more about recovering some of the sector's losses, creating a budget deficit of over $34 billion.

The standards and parameters of the new law are contained in 1,760 pages and described in great detail. This, according to the lawyers, is one huge smokescreen intended to make it difficult to initiate and proceed with court actions, and making sure that the law lasts at least two-three years during the anticipated court action. Certainly, the government goal of postponing the deficit problem as long as possible would be achieved.

The law, however, does not help the fledgling Spanish solar industry, because, in addition to short-changing the solar industry, it simply avoids tackling the real problems of the energy market, like restructuring the wholesale market.

A number of local governments of Murcia and Masdar have separately challenged the law as well. The problem is very likely to spread even beyond Spain. The impact of changes in Spain and the outcome of the court cases would inevitably reach further into the EU. As a matter of fact, the new Lithuanian energy tax law uses Spain as an example to follow. It explicitly referred to Spain's new energy laws and copied them as much as possible. Other countries are set to follow this example too.

The Spanish government—which supported and built the solar industry in the country in the recent past—now stands accused of killing it. It is intentionally and severely weakening the sector with new laws intended to curb future expenses and litigation at the expense of the solar industry.

This is one amazing development, the outcome of which cannot be predicted. Whatever happens, however, will have a resounding and lasting effect on the EU's and world's solar sector.

Japan

Japan's energy future is unclear too. Nuclear energy is on trial, and so is solar. After the Fukushima Daiichi nuclear accident, Japan is looking for new energy supplies, and lo and behold, solar and wind fit the bill perfectly...on paper. It is much more difficult in reality, but that will not stop the Japanese. Especially not now, when they have serious energy problems.

Even before the Fukushima accident, in 2011, the Japanese government introduced an incentive program to buy back electricity generated from smaller PV systems—those under 500 kW capacity. The program was actually geared mostly for PV systems much smaller than that, focusing on small backyard type systems and residential rooftop installations.

As a result of the incentive program, the installed PV capacity in Japan almost doubled in annual capacity, from 5% to a 9%.

In 2012 the government introduced a new, much more serious, FiT program that covered all types and sizes of solar installations. The new program caused the annual solar installations to increase to an average of 32% annually to date.

Around 35GW of large-scale PV projects in Japan were approved for FiT accreditation in the first two years of the program. Most projects, however, are not even close to breaking ground, which brought a number of unanswered questions and rising suspicions.

This anomaly grew to the point where the Ministry of Economy, Trade and Industry (METI) started an investigation into the status of the solar industry. As a result of the study, in October 2013, METI set two deadlines to be met in 2014:

Phase 1. Projects without land rights or equipment accreditation documents in place could lose their FiT approval by March 2014 after a series of hearings, and

Phase 2. Projects with land rights or equipment accreditation in place were given until the end of August, 2014 to obtain the additional permits and certification.

This action applied to nearly 750 projects in different stages of readiness. Cancellations were on the Japanese agenda before, but for a first time ever a wave of official project cancellations was announced in June, 2014.

As a result, 144 solar projects that missed the March deadline were notified that they have lost their FiTs, after failing to show the necessary equipment accreditation or land rights. These 144 projects account to nearly 300 MW of solar capacity.

Another batch of about 300 projects were also notified to get their documentation in order or face a similar fate in August, 2014.

The reasons for this discrepancy and the cancellations are many. They range from stakeholder disagreements, to attempts to use agricultural land (very difficult thing to do in Japan,) and many cases where the construction costs are so high that the projects look unrealistic.

The case of mass cancellations and many to come soon, clearly indicates the immaturity of the solar industry in Japan. It shows that people are intentionally or unintentionally misunderstanding, misusing, or abusing the system.

This is confirmed by the fact that there were a number of cases in 2012 when several solar projects were found to be simply bad. In the extreme, there were even fraudulent cases, where applications were submitted for plots of land picked at random and some from Google Maps.

Softbank, Japan, however, is coming to the rescue of the FiT debacle, offering assistance to solar companies struggling to meet the August deadline. Softbank is led by a millionaire, clean energy advocate, Masayoshi Son, who has already developed over 60 MW solar projects in Japan and is planning more.

Mr. Masayoshi Son's actually plan is to install close to 300 MW of PV power plants by end of 2015, and while at it he is offering his and Softbank services in sharing knowledge and offering support to Japanese solar developers in anyway possible to meet the deadline and make their projects reality.

A rich man doing a noble deed, no doubt, but at this pace of development Japan needs several thousand such men with knowledge and banks in their pockets to achieve the goals set by the government.

Japan has been in a tight place many times before, and has always found a way out. This time won't be different, but the path to getting there would be quite long and difficult.

Case Study: Koriyama PV Plant

One of the developments, paving the way to the new renewables energy future in Japan, is a new 50 MW solar power plant in Koriyama, Fukushima Prefecture. A Japanese company, Hybrid Service, supplier of toners for printing and LED lighting, has acquired land rights and feed-in tariff (FiT) approval for the new 50 MW solar power plant. Hybrid Energy, a subsidiary of Hybrid Service, is the project developer.

This development follows a conference on community power which took place in Fukushima in 2012, which announced the goal of powering the area 100% from renewable energy sources by 2040.

Hybrid Energy is a newcomer to the solar industry, which started installing residential solar systems in 2009, and then got into commercial solar systems in 2012. Nevertheless, Hybrid Energy received accreditation from Japan's Ministry of Economy, Trade and Industry (METI) and reached an agreement for interconnection to the regional grid, operated by Tohoku Electric Power.

Annual output of the PV plant is expected to be about 50,000 MWh. Note that a similar solar installation in the Arizona desert would generate at least 2 times this amount of power due to the greater sun intensity and daily/seasonal duration.

The Hybrid Energy's 50 MW project is one of the largest announced for the region, along with a 26.2 MW plant on a former golf course site by a subsidiary of steelmaker and heavy industrial corporation JFE Holdings.

But wait; a quick calculation shows that several hundred such solar installations will be needed to replace the capacity of the six nuclear generators of the damaged Fukushima nuclear power plant.

The new solar installation sounds good, but is there enough land, money and political will for such an endeavor?

Quick Analysis

The Fukushima nuclear power plant was one of the largest nuclear power stations in the world. It had a generation capacity of 7,500 MWe with about 880,000,000 MWh net annual electricity production.

So the 50 MWp solar power installation in the Fukushima prefecture, scheduled to generate about 50,000 MWh of electric power, is only 1/7,200 of the power generation capacity of the old Fukushima nuclear power plant.

That means that Japan's Ministry of Economy, Trade and Industry will need to build 7,200 such solar power plants (50 MWp each) to replace the Fukushima nuclear plant. If only ¼ of its power generation were used for the region, then Japan needs to build 4,400 such solar power plants (50 MWp each). Nope, there is not enough land in the area for such an undertaking.

Under the cloudy and foggy local skies, the new solar plant will be idling, or operating at reduced capac-

ity about 70-90% of the time. And of course, it will be shut down during the night, and rainy or foggy, days.

The annual power generation from the 50 MW solar power plant is estimated at an average of 2.74 hours daily operation (50MWp x 2.74hrs. x 365 days = 50,000 MWh annually).

At the price of $10 million for the land, plus about $200 million for the installation, the entire project comes to a total of $210 million. At an average income of about 10 cents per kWh of generated power, the plant would gross about $5 million annually. Operating expenses, financial obligations and such would suck half of it, so the net profit would be about $2.5 million annually.

At this rate, the new 50 MWp solar installation would need almost 50 years to repay the initial $210 million investment.

Anyone interested in investing in it? Nope, not many wise investors would jump at this chance, if they really know what they are doing.

However, with some ingenious subsidies, loan guarantees, and financing schemes propped by the government, the plant will become a reality very soon and more solar plants might be built, and yet we just don't see the region 100% solar powered anytime.

The new solar installation close to the destroyed nuclear plant is only a poor reflection of what was…and a quick peek into the uncertainties of Japan's energy future. But let's not forget that these are the first baby steps of solar power in Japan.

Just like in the U.S., solar power was shoved in the closet a long time ago and taken out occasionally for obligatory dusting. Now, however, the Japanese are feeling a real need for power and since they are poor on energy, solar and wind are out of the closet and put on the energy pedestal, where they belong.

A long and difficult road awaits solar and wind in the land of the Rising Sun, but if anyone can do it, the Japanese can.

Update 2014

Following the Fukushima accident the number of new solar (PV) installations increased dramatically. Japan showed promise to become the most aggressive solar country ever. Unfortunately, the euphoria did not last long…

Recently, sharp rises in consumer electricity prices, caused in part by a backlog of planned PV projects, could put Japan's solar industry in danger of losing public support.

Consumer electricity prices have risen 20% in some parts of Japan since the introduction of the feed-in tariff (FiT) in mid-2012. This makes the possibility of a public backlash to the price hikes a real possibility. The price increases and other problems had spilled over into the political and social life of the country, making it extremely dangerous for solar, which until now has experienced very high approval levels from politicians and customers alike.

So far the Japanese public has been strongly anti-nuclear and pro-renewable energy, but the electricity price hikes have proved extremely unpopular with Japanese consumers. The price rises are partially due to high cost of PV installations on Japan's aging grids.

The PV projects are operated independently of one another by 10 different utility companies, which create chaos and backlog. Yet. the government has continued approving solar projects at a quick pace.

This happened in Germany and other EU countries recently, so a "rebound" effect is to be expected as the public opinion is making a 180-degree reversal, which threatens to harm the solar industry there.

In the fall of 2013, the Japanese government was investigating the status of nearly 750 large-scale, FiT-approved PV projects that had applied for accreditation and land use permits but had yet to be built. The potential projects' developers were given deadlines of March or August, 2014 (depending on the status of their projects) to obtain the necessary paperwork needed to proceed with construction.

In the summer of 2014, the Japanese Ministry of Economy, Trade and Industry (METI) cancelled over 140 of these projects, since they have missed the March deadline. Another 300 PV projects are expected to be cancelled in August, which leaves about half of the original projects active.

Over 65 GW PV capacity had been given equipment accreditation thus far, further contributing to the backlog. This is a large number of projects, so the concern is that if all of these projects get approved and constructed, the energy surcharge levied on consumers would raise their monthly electricity bill sharply. This would inevitably trigger a backlash on the part of the consumers.

Complications come from the fact that equipment accreditation requires the use of materials and machinery from a list of approved suppliers, as part of the process of obtaining FiT approval in Japan. That is further complicated by obtaining proper land use and grid interconnection permits.

Japan has no "upper limit" in the amount of PV installations, so the government cannot refuse legitimate equipment, land and interconnect accreditation requests, even if the situation is critical and getting more so.

Presently, the only option for the Japanese government to remedy the situation is to revoke FiT approvals for a large number of projects, which could prove difficult. They can only cancel the applications where something is obviously wrong. Otherwise, they would lose in court, since Japan's law makes it very difficult to take away something that has been legally approved.

The next several years would see a large amount of PV deployed in Japan, no doubt, but as the planned end of the FiT approached, things are expected to change. The FiT might be cut to around 10% at the end of March 2015, which combined with the public's backlash against solar, might significantly delay the development of their solar industry.

Summary

The U.S. and global PV industries are going through major, unprecedented, unexpected, and hard to predict changes. Nevertheless, there are factors and developments that we need to point out in an attempt to present a comprehensive summary of the status quo and the developments in the sector. The predictions are based solely on our previous experience in the field, and analyzing the available facts and data.

The following bullet list is a condensed version of our understanding of the different drivers and developments in the PV industry and the global energy markets.

- Photovoltaic (PV) power generation is the most versatile and promising of all renewable technologies. It also has the potential of unlimited growth in efficiency and size, and in cost and price reduction as well.

- PV system components (PV modules and BOS) prices are now at record low levels, and expected to fall further before recovering and rising again during 2016-2020.

- PV installations' labor costs are slowly going down, and are expected to continue the trend until 2020.

- The supply side is recovering from negative operating margins during 2011-2012, with the suppliers playing a major role in the global PV deployment by controlling the quantity of PV modules that they can (or are willing to) ship, and where:
  — The manufacturers need some stability to re-

cover from the pricing fall in 2011-2013,
- Competition will increase as the dust settles, and prices will start rising again,
- Development of new technologies and processes will continue, but slowly,
- No major PV technology developments are expected in the marketplace until 2016.

- PV installations' growth during 2014-2016 depends on:
  - The ability of the users to implement large ground-mount projects,
  - The utilities' resistance to distributed PV power generation,
  - The 2014-2016 winners will be the cleverest and most risk averse players.

- The number of approved projects will determine the growth of the PV sector:
  - Net-metering, interconnection, finance conditions, and storage requirements will play a major role in which projects are approved and completed during 2014-2016,
  - The PV industry is no longer governed completely by the policy makers, and yet any changes in regulations affect the growth of the PV sector,
  - The residential market will see increased resistance from the utilities as they are beginning to blame it for some of their problems.

- The global PV market in 2014 is estimated to reach nearly 50 GWp, where:
  - Most of the new projects in the key markets are large-scale power plants,
  - The residential segment in Europe no longer dominates the sector's growth. A similar decline in the residential sector growth is anticipated in the U.S.

- Technology disruption is on the back burner now but will be back during 2016-2020:
  - Suntech and Yingli unsuccessfully tried to push their Pluto and Panda technologies to the market a few years ago. Their failure is a lesson for the rest.
  - Cast-mono ingot process was the disruptive technology of 2008-2010, but is dead now.

- Thin-film PV was considered a key disruptive technology until recently, but this is not the case today:
  - CdTe is taking the role of a specialized PV technology for use in certain locations,
  - CIGS technologies have the greatest disruption potential starting in 2016,
  - Hanergy's CIGS efforts are the center of attention, following the failure of its a-Si plans a few years ago,
  - If and when successful, Hanergy's CIGS plans will have a disruptive effect on the PV markets and the PV equipment supply chain.

- Promising developments 2016-2020:
  - Directional solidification furnaces for multi c-Si ingot production,
  - Slurry recycling during wafer slicing,
  - Fine-line and double-printing of c-Si cells,
  - High-efficiency multi c-Si developments,
  - n-type material use,
  - In-house power-supplies for polysilicon plants,
  - Continuous mono silicon pulling,
  - Optimized diamond wire sawing,
  - Use of thinner wafers,
  - Optimized FBR process,
  - Micro-inverters,
  - SIGS process developments,
  - Energy storage technologies,
  - Smart grid applications.

- Major obstacles to the potential 150 GW utility global PV market remain, as follow:
  - There is no comprehensive PV technology roadmap. Those who aligned with the European PV technology roadmaps of the past, for example, are today counting their losses and will stay away from any roadmaps for a long time. You can fool me once…
  - Standardization of the solar industry products and processes is badly needed
  - There is a lack of constructive consensus from the major players,
  - Utilities' resistance to residential PV is increasing and might lead to drastic changes,
  - A finite limit to how many large projects can be interconnected due to logistics,
  - Grid upgrades and capacity limits are a key challenge in most areas,
  - Low cost finance is not readily available under the present conditions,
  - Capex is still a weak point, and will be until market leaders return to profitability,

— End-market demand will change during 2014-2016, forcing changes in the industry,

— The methods of forecasting PV demand are inadequate and create confusion,

— LCOE is an oversimplified measure of PV technology's performance and overall potential.

— A Beijing mandate to facilitate financing, supply and completion of 14 GW of solar PV capacity in 2014 outweighs the LCOE calculations used as the basis of PV end-market sizing.

— Similarly so, the California's mandate of 30% renewables by 2020 is a key to pushing PV installations higher than anyone could've predicted...and regardless of LCOE calculations, as used to compare PV performance to that of the fossil fuels.

• The U.S. PV market is one of the top three global PV markets and is expected to grow further during 2014-2016. Various barriers at the state level, such as net-metering, interconnection, utilities' cooperation, and storage requirements will determine its future.

• Europe has gone from 75% to 25% of global PV demand
   — The European PV markets are stagnating
   — Government policies and regulations remain barriers in the EU PV market

• Europe, the U.S. and India continue investigation into wrongful foreign PV components imports.
   — The U.S. and European investigations will have the largest impact on the global supply/demand balance as they represent access to 15% and 25% of global demand, respectively.
   — The new, 2014 US investigation will linger on for the best part of 2014, with the outcome virtually impossible to predict.
   — The Europeans will have to revise the floor pricing and quota levels, but no great changes are expected in the European PV industry.
   — India has not yet progressed as originally planned and hoped. Because of that, the effects of its anti-imports actions will have no impact on the global supply/demand balance.

• The global PV market will not change much in 2014, with the U.S. leading the world, and Europe likely to rebound in 2015-2016 with much higher

PV demand than the emerging regions.
   — Latin America, Caribbean, and Eastern Europe's PV markets will continue to gain traction, but very slowly, as regulatory structures are still being adapted.
   — Government targets in some Asian countries are creating a rapid market pull.
   — Asia Pacific and Central Asia's major players are China, Japan, Australia, India and Thailand. All other countries in the region fall into the "emerging" category.
   — Most Japanese manufacturers have allocated most of their production to the domestic market. The extra demand will be supplied mostly by China and Taiwan.

• The global PV manufacturing segment is now about:
   — Incremental steady volumes,
   — Production efficiency,
   — Product and services quality improvements,
   — Supply chain cost reduction,
   — Increased pressure on outsourcing partners
   — Economy of scale,
   — Reduced costs of final product.

• Steady global drive to stability in the sector precludes any full-blown efforts to bring disruptive technologies or methodologies on-line...at least in the near future.

• No major changes in the PV technology and market status quo are expected at least until 2016.

• During 2016-2020 the remaining major players will intensify the efforts to distinguish themselves from the rest by introducing new and more competitive PV technologies and market approaches.

• U.S. energy security will increasingly depend on the renewable technologies, with a major jump in their importance during 2020-2030. Yet, there is no official road-map or halfway decent plan for getting there on time...for now.

Update 2014

The solar gold rush of 2007-2012 added 30 GW of PV capacity worldwide in the 2011-2012 time period alone. There was over 100 GW of solar (PV) installations at the end of 2012. Another 39 GW of PV power were added in 2013, 12 GW of which was in China. Additional

46 GW are being installed around the globe in 2014 and 56 GW are expected to be added in 2015. This brings the total global PV capacity to roughly 250 GW by 2015.

These numbers are also expected to grow through 2020 and 2030, mostly due to the fact that solar PV is already competitive without subsidies in 20 global markets and more are likely to follow.

Experts claim that rooftop solar in Germany is already cost-competitive with some power generating technologies, such as combined cycle gas turbines. It is also expected that by 2030 the PV utility-scale power plants in southern Germany will operate below the average LCOE (levelized cost of electricity) as compared with the fossil power generation.

In other words, solar has a chance to become cheaper than coal in Germany and many other places. Small rooftop PV systems will be able to compete with some (but not all) power generators too. Onshore wind and brown coal, hard coal, and CCG power plants are estimated to be more expensive to operate than solar in the near future.

Deutsche Bank (DB) stresses a number of reasons for the continuing solar success, as follow:

- The U.S. based distributed generation business models are set to become more pervasive in international markets and act as a significant growth catalyst in European markets that have significantly scaled back subsidies.

- Financing costs and availability for the solar sector are set to improve from 2014 on, and noted that sufficient access to low cost financing has been a significant constraint inhibiting the growth of global solar sector so far.

- DB expects downstream solar companies to participate in the "gold rush" to acquire solar customers at an accelerated pace. "Just like upstream/midstream solar companies participated in the gold rush to add manufacturing capacity during the 2005-07 timeframe, we expect another gold rush to add recurring MW over the next 2-3 years "until the (US investment tax credit) expires around 2016).

- While the past 5 years were above module cost reduction, the next 3 years (2015 and beyond) would be about reductions in the balance of systems costs. This includes the cost of inverters, hardware, customer acquisition and financing costs.

Goldman Sachs joins in the success predictions by finding that the renewable energy market is "incredibly compelling," and thus has made a commitment to invest $40 billion in renewable energy in the coming years.

- Investors clearly see the end of coal as the king of power generation, but warn that solar will become a major power ONLY IF and WHEN:

  — The overall solar costs (product, installation, and operation) decline to an acceptable and sustainable level,

  — The efficiency and reliability of solar installations improves to world class standards, and

  — Energy storage becomes readily available, efficient, cheap, and reliable

Only when all these requirements are met, will solar (and especially large-scale installations) be able to reach grid parity without subsidies. And "without subsidies" is a key factor in this equation that is ignored by many.

No doubt, solar power has been very successful lately. We have witnessed its incredible growth—almost 50% increase per year on average between 2000 and 2012. Yet, even after all this unprecedented growth, solar still accounts for just 0.6-0.7% of total global electricity generation.

This is impressive for solar power, yet still a very modest number as far as total global power generation is concerned. This number—regardless of the impressive gains—cannot be counted on to save the world from jumping off the post-fossils energy cliff.

If we must depend on solar for a major part of our energy supply, then we must see a 500-1000 times increase in solar installations worldwide by 2050.

WIND POWER

Wind power is a powerful energy source that has been used through the centuries to provide mostly mechanical power to a number of agricultural and commercial operations. These include powering sailing ships and mill wheels. The sophistication of wind power equipment has increased so much that today's amazing wind giants—the wind turbines—are powering entire cities. We see these overwhelming structures in many places, waving their giant arms as if to greet us and bid us well.

There is a lot of wind available in many areas of the U.S. and the world, where wind energy is waiting to be harnessed efficiently, economically, and safely. Wind power is also one of the cleanest technologies around. We'll see how clean it is below, but now we'll take a look at wind power in its present state of development.

Wind power is basically the conversion of wind energy into a useful form—mostly electric power—which is used locally, or sent into the national grid.

Figure 7-35. Wind power plant.

Large wind farms are of great interest to the power industry. This is mostly due to the fact that these power plants produce a large amount of electricity, which translates into an equally large share of revenue. The large-scale power plants consist of hundreds of individual wind turbines which are connected to the electric power transmission network.

Offshore wind farms can harness more frequent and powerful winds than are available to land-based installations and have less visual impact on the landscape, but construction costs are considerably higher. Offshore installations, however, are not readily accessible for maintenance, and the violence of the sea is a determining factor in the daily O&M.

Waiting for calm weather can result in delays in installation and maintenance schedules. In these cases, offshore turbines could be idling for long periods, reducing energy production and profits.

Large land-based wind farms don't have these problems, but face other obstacles, such as permitting, land availability, right-of-way (for accessing the power grid), and inconsistent wind velocity. Small land-based wind farms can provide electricity to remote locations and the local utility companies could buy surplus electricity produced by these as well.

Wind power is an alternative to fossil fuels, and unlike fossils it is renewable, widely distributed, and clean. It produces no greenhouse gas emissions during operation and uses little land. Although there are some environmental effects, these are much less problematic

than those from other power sources.

Wind power made significant progress in the U.S. lately, but its growth has been slowed for a number of reasons. Worldwide, Denmark is generating 25% of its electricity from wind, and another 83 countries are using wind power on a commercial basis. In 2010 wind energy production was over 2.5% of total worldwide electricity usage, and growing rapidly at more than 25% per annum.

Wind power levels are fairly consistent on a year-to-year basis, but have significant variations on a hour-to-hour daily basis. This creates problems when wind power is a major contributor to a local electrical system. As the proportion of wind contribution increases, new approaches for using wind power need to be implemented.

Power management techniques such as having excess capacity energy storage, geographically distributed turbines, dispatchable backing sources, exporting and importing power to neighboring areas, or reducing demand when wind production is low, can greatly mitigate these problems. Mixing wind and solar generation is another feasible approach, for the peaks of wind and solar compliment each other. This could be used to level the energy production and provide power in peak periods. Weather forecasting and management also permit the electricity network to be readied for the predictable variations in production that might occur.

One of the biggest problems wind power faces, despite its general acceptance by the public, is that the construction of new wind farms is not welcomed in most neighborhoods, due to aesthetics. The NIMBY (not in my back yard) sentiment is hindering the spread of the gentle giants around the U.S. and most of the developed world. This issue can be resolved only by locating large wind farms away from populated centers, which usually increases the final cost of wind power.

Wind Power Equipment

We saw some large equipment in use at coal and nuclear plants, so at a glance a windmill won't impress us much. Yet, a close look reveals a sturdy giant with very impressive electro-mechanical structure. Hundreds of feet tall, wind turbines are noticeable from a great distance. When these giants are working side-by-side, they look like something out of this world. Standing close by, looking at their long arms swinging high in the air—200 feet or more—whistling and whooshing nonstop, one gets an out-of-this-world sensation.

An even closer look reveals some technological miracles that make this giant operate smoothly and generate huge amounts of electric power. The large wind turbines, using about 1,000 square feet of land each, generate about 2 MW of power each. This is equivalent to the power generated by 15,000-20,000 solar panels installed on 10-15 acres of land. And today there are even larger off-shore wind turbines generating as much as 5 MW of power. True monsters...

Take a closer look at this promising technology.

The Wind Turbine

A wind turbine is an electro-mechanical device that is designed to convert the kinetic energy of the wind first into mechanical energy, which can be used to drive machinery. Most often, however, it is used to generate electricity.

Today's wind turbines are manufactured in a wide range of vertical and horizontal axis types. Small wind turbines are used for water pumping, battery charging, auxiliary power on boats, and such. The large wind turbines are usually used for grid-connected electric power generation, and their use is becoming increasingly important for commercial electricity generation.

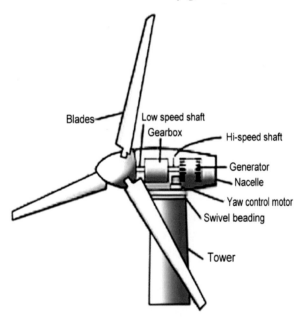

Figure 7-36. Wind turbine detail

Modern wind turbines have a number of key components:

- The structural support of the wind turbine includes the tower and rotor yaw mechanism.

- The rotor supports and controls the turbine blades for efficient and safe operation, to catch the maximum amount of wind and rotate the main axle.

- A gearbox or continuously variable transmission are attached to the main axle and convert its low-speed (but powerful) rotation to high-speed (low-resistance) rotation as needed for generating electricity.

- An electric generator and control electronics convert the rotational energy of the axle into electricity, which is sent into the grid.

The turbine blades are designed to rotate around an either horizontal or vertical axis, with the horizontal type being much older and more common.

Operation

Wind turbines are designed to capture the wind and convert its horizontal motion into rotational energy. Wind rotates the blades, which in turn rotate an electrical generator to produce electric current.

A computer connected to a number of sensors reads and analyzes the wind speed and other weather conditions around the wind turbine. It then makes decisions on the function of all active components and sub-systems based on the gathered weather data and the wind speed and variability.

The computer constantly reads the blades' speed and selects which generator to use and when and how, based on its analysis of the weather and wind data.

For example, at low wind speeds of less than 5 meters per second (m/s), the blades of an average wind turbine rotate at 15-16 rpm. This led to a 200 kW electric generator.

At high wind speeds over 5 m/s, the blades of the same wind turbine rotate at over 20 rpm, which motion is converted to 1800 rpm by the gearbox and the motion is transferred to a 900 kW electric power generator.

The computer decides which generator will operate at what blade speed and when it will make contact with the power grid to send the generated electricity.

The electricity generated by the chosen generator is conditioned by special on-site equipment to follow the grid sine-wave and is synchronized with the grid power via power controllers prior to sending it into the grid.

The computer shuts down the generator when the blade speed doesn't match its requirements or when it is disconnected from the grid.

Wind turbines usually operate at an absolute maximum 1000 rpm blade speed, which is limited and controlled by air brakes. The brakes slow the blade rotation to prevent damage, and are designed to stop the rotation totally at high wind speeds to prevent an out-of-control condition and potential breakdown and other electro-mechanical failures.

Types of Wind Turbines

There are number of types of commercial wind turbines in use today. Some of those used in today's wind power plants follow.

Horizontal-axis

Horizontal-axis wind turbines (HAWT) as in Figure 7-36 are the most common types, and the one we review in more detail herein. These turbines are designed with the main rotor shaft and electrical generator at the top of a tower, and operate by pointing into the wind.

Smaller turbines are pointed by a simple wind vane, while large turbines generally use a wind sensor coupled with a servo motor. Larger turbines have a gearbox, which turns the slow rotation of the blades into a quicker rotation that is more suitable to drive an electric generator.

Since a tower and the casing produce turbulence, the turbine blades are usually positioned upwind. Turbine blades are stiff, which prevents mechanical stress and fatigue of the material and also keep the blades from bending and smashing into the tower during high winds.

Downwind versions are used for small turbines, since they don't need an additional mechanism to keep them in line with the wind, and because in high winds the blades can be allowed to bend which reduces their sweep area and thus their wind resistance.

Large wind turbines used in wind farms for commercial production of electric power are usually three-bladed and pointed into the wind by computer-controlled motors. The high tip speeds of large turbines can reach over 320 km/h (200 mph), and are of high efficiency and low torque ripple, which contribute to reliable and safe operation.

The blades are usually colored white for increased visibility by aircraft and birds.

A gear box is used to step up the speed of the generator, but some designs use direct drive of an annular generator. Some models operate at constant speed, but more energy can be collected by variable-speed turbines which use a solid-state power converter to interface to the transmission system.

All turbines are equipped with protective features to avoid mechanical or electrical damage at high wind speeds. This is achieved by feathering the blades into the wind which ceases their rotation, and can be supplemented by brakes for even safer operation.

Vertical-axis

Vertical-axis wind turbines (or VAWTs) are different from the horizontal turbines in that the main rotor shaft is positioned vertically. The advantage here is that the turbine is fix-mounted and does not need to be reoriented into the wind to be effective. This is particularly advantageous on sites where wind direction is highly variable.

With vertical axis blades as in Figure 7-37, the generator and gearbox can be mounted on the ground, using a direct drive from the rotor assembly to the ground-based gearbox. This increases the mechanical stability of the system, and also reduces some expensive structural elements (like 300 feet of stairs in the horizontal axis type). It also makes maintenance and repairs much easier, faster, cheaper, and safer.

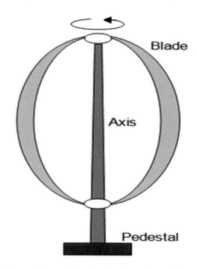

Figure 7-37. Vertical axis win turbine

Key disadvantages include the low rotational speed with the consequential higher torque and hence higher cost of the drive train, and the inherently lower power coefficient.

The 360-degree rotation of the aerofoil within the wind flow during each cycle and hence the highly dynamic loading on the blade, the pulsating torque generated by some rotor designs on the drive train, and the difficulty of modeling the wind flow accurately present challenges to analyzing and designing the rotor prior to fabricating a prototype.

Vertical wind turbines require different operating conditions than their horizontal counterparts. For example, when a vertical turbine is mounted on a rooftop, the building generally redirects wind over the roof, and this can double the wind speed at the turbine.

When the height of the rooftop-mounted turbine

tower is approximately 50% of the building height, it is near the optimum for maximum wind energy and minimum wind turbulence.

But wind speeds close to structures are generally much lower than at exposed rural sites, so a compromise must be found, because unpredictable turbulence and loud noise may occur in such cases.

Wind Farms

A wind farm is a group of wind turbines in the same location used for production of electricity. A large wind farm may consist of several hundred individual wind turbines, and cover an area of hundreds of square miles, but the land between the turbines may be used for agricultural or other purposes. A wind farm may also be located offshore.

Almost all large wind turbines have the same design—a horizontal axis wind turbine having an upwind rotor with three blades, attached to a nacelle on top of a tall tubular tower.

On a wind farm, individual turbines are interconnected with a medium-voltage (often 34.5 kV), power collection system and communications network. At a substation, this medium-voltage electric current is increased by a transformer to match the high voltage of the power grid.

Approximately 75% of the total cost of a wind farm is related to upfront costs such as the cost of the turbine, foundation, electrical equipment, and grid-connection.

A wind turbine is capital-intensive compared to conventional fossil fuel fired technologies such as a natural gas power plant, where as much as 40-70% of costs are related to fuel and operations and maintenance procedures.

In turn, a wind farm does not need any fuel to operate. It is also one of the cleanest and greenest power generating technologies available.

World's Wind Farms

Many of the largest operational onshore wind farms are located in the US:

- Alta Wind Energy Center in Kern County, California, is the largest onshore wind farm in the world at 1,020 MW, followed closely by the
- Shepherds Flat Wind Farm in Eastern Oregon at 845 MW, and the
- Roscoe Wind Farm in Texas at 781.5 MW.
- The Sheringham Shoal Offshore Wind Farm and the Thanet Wind Farm in the UK are the largest offshore wind farms in the world at 317 MW and 300 MW, followed by

- Horns Rev II in Denmark at 209 MW.

There are many large wind farms in operation or under construction around the world as well, including:
- The London Array offshore wind farm at 1,000 MW,
- BARD Offshore 1 at 400 MW,
- Sheringham Shoal Offshore Wind Farm at 317 MW,
- Lincs Wind Farm, Clyde Wind Farm at 548 MW,
- Greater Gabbard wind farm at 500 MW,
- Macarthur Wind Farm at 420 MW,
- Lower Snake River Wind Project at 343 MW, and
- Walney Wind Farm at 367 MW.

Wind Power Plant Design and Construction

Like any power plant, new wind power plants go through a number of complex, lengthy, and expensive procedures. Some of the elements are very similar to those discussed in the other chapters, so we will limit this discussion to the wind-particular issues.

This process is very similar to that used for solar power plants as well.

Site Assessment

Many factors related to the physical aspects of the proposed wind power plant—type of property, zoning regulations, transmission lines, environmental aspects, and the neighbors' perceptions about and acceptance of wind power—play a major role in designing a successful and profitable wind power plant.

All data related to environmental and weather conditions of the locale must be gathered, analyzed, and well understood to make an efficient, safe, and profitable wind farm design. These data include:

Land Considerations

Harnessing wind power has been done for many centuries, and can be a good investment today too. To get the most out of the investment, however, it is critical that turbines are located at the right place—a place with enough wind of the proper type is most suitable for commercial power generation.

As reviewed above, the turbines' placement depends on where and when the wind blows, the physical aspects of the property, zoning regulations, environmental concerns, and the neighbors' concerns about the physical aspects and noise of the turbines.

Proper site planning and adequate power plant design are absolutely critical for harnessing enough wind power, and efficient and profitable operation.

Zoning and Permitting

Wind turbines and wind turbine installation and operation are subject to local, state, and federal laws. Local zoning seems to be the most restricting, simply because of the presence of tall towers. Neighbors constantly complain about the unsightly view of these giants, which has led to numerous lengthy and expensive court battles.

Special permits must always be obtained from the local planning commission, so knowledge of the local zoning laws early in the development of the wind project can help decrease or avoid unnecessary delays and expenses.

Most states also regulate the size of the electric generating facilities. For example, any power plant larger than 25 MW in some states must receive a permit from the Public Utilities Commissions. This requires an elaborate environmental impact statement (EIS), which takes a lot of time and money.

State permits usually supersede local zoning and other laws. Local, state, and federal permitting requirements for wind projects change often, but the federal laws are always a priority. There are also several governing bodies that may need to issue an approval prior to construction, depending on location and size.

For example, the Federal Aviation Authority (FAA) has authority as far as structures close to airports are concerned. In particular, the FAA must permit every structure over 200 feet tall near an airport or within flight paths. Since many commercial wind turbines are 200 feet or taller, the FAA has the say in all cases.

Connecting to the Grid

Wind power plants usually connect the turbines output to the existing power grid, so the design engineers must make sure that the turbines are near a grid-connection point. Otherwise, they need to design an expensive transmission line and sub-station to connect to the grid. The proximity to existing transmission lines is critical for minimizing infrastructure requirements and keeping costs down. High voltage lines cost about $1 million per mile, so sites with good wind and adequate access to the grid can be very valuable.

Utilities usually restrict how close a turbine can be to power lines, and there are restrictions about proximity to airports and other facilities which must be also considered in the preliminary design.

Grid Operation

The power generated by wind turbines must be conditioned before plugging into the grid. Since induc-

tion generators are most often used for wind power, and they require reactive power for excitation, special substations are incorporated in wind-power generating systems, which include substantial capacitor banks for current conditioning and power factor correction.

Wind power is variable in nature, going up and down with the strength of the wind. The different types of wind turbine generators behave differently during transmission grid disturbances, so design engineers must pay special attention to the particulars of the site prior to installation. They use special grid codes that specify the requirements for interconnection to the transmission grid, which include the power factor, constancy of frequency, and dynamic behavior of the wind farm turbines during a system fault.

Grid operators also do extensive modeling of the dynamic and electromechanical characteristics of the wind power plant to ensure predictable and stable power during system operation.

All in all, it takes a village to operate a large wind farm.

The future of wind energy depends on finding ways to level the power output. This can be done by combining wind with solar or gas power generation or by adding energy storage to the wind farm setup. Either of these is an expensive undertaking, so a new technology breakthrough is needed in this area.

The Issues

Like any other power generating technologies, wind power has its own limitations and issues. Some of these are:

Environmental Concerns

Wind energy is considered a non-polluting energy generator, but the construction and operation of any power plant brings the risk of disrupting ecosystems on the site and beyond. Wind and solar power plants are no different.

In most cases, they are not allowed in protected areas, such as wetlands or other sensitive areas. They also should not be located in migratory bird flyways because, like any large structure, they can kill birds. Wind turbines in scenic areas have caused numerous lawsuits, so these must be avoided.

The Land

Wind turbines are tall, heavy structures that require a solid foundation. The land mass should be able to support the weight of the turbine, as well as the weight of construction equipment during the field assembly and installation.

Wind turbines also require a lot of open space to harness the power of the wind, in addition to access roads and support facilities, so large land areas are required for large-scale power generation. The wind turbine platform itself, even the largest available, occupies only a few square yards, so in most cases the surrounding land can be used for farming and ranching.

Site planning must consider possible changes of neighboring properties or buildings, and the related changes required by local zoning laws.

Safety

Working with, or on, the monstrous wind giants is a dangerous business. Skilled and well trained technicians and experienced engineers are in charge of the different operations during the entire cradle-to-grave process of the wind turbine. Safety is a priority in most wind power plant operations, but the dangers are constantly there.

Because it is human to err, and equipment does malfunction from time to time, a number of accidents are reported during all stages of the operation. Falling from the high towers is the most frequent cause of human injury, followed by a large number of incidents in which birds are killed by the rotating blades.

There are also documented cases of damage done by rotors breaking and flying off, as well as entire towers collapsing.

The Neighbors

The neighbors are one of the biggest enemies of newly proposed wind turbine power plants. Not-in-my-back-yard (NYMBY) is a movement that is taking the world over. Wind turbines are very tall and make a lot of noise, which is especially bothersome at night. The planning process must include the neighbors and listening to their concerns.

Turbines are best when installed where they will not be seen and heard by neighbors. Active general public involvement at the early stages of the wind project process is a must and is one way to increase the likelihood of a timely permit decision and reduce the possibility of protracted litigation.

The general public includes residents and members of communities near the wind development and community officials and representatives of various interests, including economic development, conservation and environmental groups. All these individuals must be involved in the decision making process from the very beginning.

Of critical importance to the wind farm's long-term success is getting enough information on the weather and wind conditions of the area.

Output Variability

Wind power is a variable power source—its intensity varies with the wind availability and speed. Wind turbines sit still when there is no wind to move them, and generate very little power at low wind speeds.

A new development is making energy specialists pause and wonder. The problems with wind projects lately, and falling solar panel prices, have prompted a number of wind companies to look into the competing technologies, and solar in particular.

The wind industry added more than 13 GW of capacity in 2012, but as a result of the fiscal cliff delays and hesitation, only 1/3 of that capacity was installed in 2013. At the same time, solar is going on unhindered—surging. U.S. solar installations are surpassing new wind additions for a first time.

The Solution

Solar is also a variable power source, but its peaks are different from those of wind power turbines. Too, it so happens that these two technologies compliment each other by providing a more stable power output. Because of that, solar is now considered wind's best friend and vice versa.

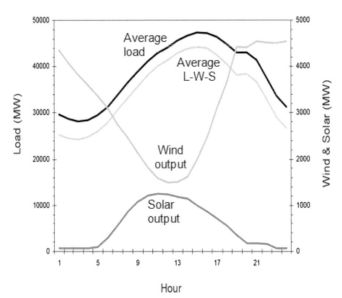

Figure 7-38. Simultaneous wind and solar generation

A marriage made in heaven. Wind and solar working hand-in-hand can provide a major portion of the peak loads of the national grid. Because wind still blows when the sun isn't shining, and vice versa, the combined power from wind-solar hybrids can provide more consistent power to the grid.

Companies like EDF Renewable Energy in San Diego, are developing solar-wind hybrid technologies and installing them in California's Mojave Desert. A 140 MW wind farm was constructed near a 143 MWp solar power plant, as an example of the best way to level the energy output of combined wind-solar power generation.

Solar and wind work together very well, so why not? Wind companies already have many in-house and field skills needed to develop energy projects. Many of the disciplines used in wind power installation are similar to those used in solar. GIS mapping, real estate contracts, transmission issues, and even the customers, are similar for both types of energy.

Because wind and solar go hand-in-hand so well, the future for wind-solar hybrids is just emerging. The addition of large-scale energy storage will provide a path to unlimited future growth.

Wind and Energy Security

The U.S. has immense offshore wind energy potential, due to strong, consistent winds along the lengthy U.S. coastline. Off-shore wind power potential of the U.S. coastal and Great Lakes waters is estimated at over 4,000 GW. Considering all restrictions, offshore wind could realistically supply nearly 2,500 GW of electricity, which is more than double our total combined generating capacity.

U.S. DOE expects wind energy to supply 20% of all U.S. electricity by 2030. Offshore wind could provide 54 GW of generating capacity, or 4% of the nation's total electricity use...if all present-day issues are resolved by then.

A robust U.S. offshore wind industry could generate tens of thousands of jobs and billions of dollars of economic activity. Much of this activity would be concentrated in economically depressed ports and shipyards, which could be repurposed to manufacture and install offshore wind turbines and equipment.

Lots of "would" and "could" still remain, mostly due to the major barriers to deployment of offshore wind power, such as the high costs of offshore wind power plants; long and uncertain permitting process; and technical challenges surrounding installation, grid inter-connection, and maintenance.

The majority of current offshore wind projects are in Europe, where more than 3 GW of capacity has been installed recently through more than 53 projects in several countries.

Reliability

Regardless of the achievements, the big question remains, "Can wind be relied on?" There's lots of wind, yes, but what if it dies down under a high pressure weather system, as often happens during some winter periods. What output can we expect from wind power at those periods?

For any other time of year in most places worldwide, wind power is a fairly healthy power generation resource. Combined with other power generators, like solar or natural gas, it could be quite reliable too.

There are several reasons to consider wind power a serious contributor to our national energy security:

- Wind turbines are assets, as opposed to nuclear reactors which depreciate in return value from day one.

- Wind turbines are much quicker and cheaper to deploy than most other power generators.

- Wind power fits the grid needs because it fits the market conditions as dictated by government subsidies and other measures that favor low carbon generation.

- Wind's only competitors are solar and natural gas-fired power plants. But these are actually wind's best friends, because together they make a good team and can be used to provide a steady power supply to the national grid.

- Wind power is spread over wider geographical areas, wider than the average anti-cyclonic high pressure system, so higher capacity values can be guaranteed. The more wind power we install over wider geographic areas, the higher power output we can expect.

Global Hubs

We already see a number of wind power "hubs" emerging around the globe. These are places where local conditions, financing and planning are more favorable for wind power growth. Some of these, such as those in the North Sea area, where plans for growing the integrated wind power market over a larger number of territories are developed to using wind power where it's most needed first. And if, at the end, "spare" wind capacity is available for the starving European power grids, then wind power is the way to go in parts of the continent.

The same is true for parts of North America, where there are documented favorable "wind corridors" in some areas. This is where wind power must be fully developed first, for it is the easier and cheapest way to implement this promising technology.

Wind Power Technology

It is becoming quite obvious that larger and higher wind turbines are the way to go—regardless of the problems this might bring. Such turbines are the best and most efficient (and profitable) way to capture more wind for power generation. Wind flow is much more regular, the higher the wind turbine blades are in relation to the land or sea surface. This stronger dependency of wind power determines the technology types to use, and is a major key to success of wind power in the future.

This, however, also increases the difficulties in installation and O&M operations. There are also increased risks of failures and damages resulting from high wind gusts. A number of giant turbines have collapsed and blades thrown far away, causing damage to properties and hurting people.

The Helpers

Wind power alone—no matter how good it gets—is not the solution to our energy security. As a matter of fact, it is just one of the many possibilities the future offers. When combined with other energy sources, such as solar, geo, tidal power and others, wind contributes significantly in generating a steady level of useful power. This way it becomes a major part of a solid energy security picture, where in the near future it (together with the other renewables) will be able to compete and even replace the fossils in base load power generation.

A number of wind-to-hydrogen projects are also testing the concept of using electrolysis of water by using spare wind (and solar) power to produce hydrogen gas that can be stored and burned later for power generation.

The more renewable energy technologies we develop, the more they can support each other while minimizing their respective weaknesses.

Improving and strengthening the transmission networks will also improve wind's reliability by getting "stranded" wind power (from distant and remote locations) to market. Constrained or curtailed wind power is surplus wind energy that could be captured for use as electrical power.

These are just a few ways that wind can be used to fill the gaps among the technologies. Some of these, or combinations of, will allow us to increase our renewable energy capacity as a whole. By doing this, we ensure our energy security at the same time.

The Counterview

The UK's Ministry of Defense (MoD) objects the to expansion of wind power in the country because, it claims that tall wind turbines with their long wavering hands interfere with radar's function and impede normal air flight operations.

The MoD has lodged objections to at least four onshore wind farms in the line of sight of its stations on the east coast of UK, because they make it hard to impossible to detect aircrafts flying in the area.

And not only do the turbines create a radar hole directly over a wind farm, but there is also a shadow beyond them that prevented low-flying aircraft being detected. The MoD is alarmed by the findings, since their impact is much greater than previously thought. Thus the more robust approach to wind turbine assessments prompted by the MoD officials.

The same objections apply to wind turbines in the North Sea, part of the massive renewable energy project undertaken recently. Wind turbines in that area would be directly in line with the three principal radar defense stations, Brizlee Wood, Saxton Wold and Trimingham on the Northumberland, Yorkshire and Norfolk coasts.

MoD experts claim that the turbines create holes in the radar coverage, which are so bad that most aircraft flying over these areas are not detectable. Amazingly enough, this occurs regardless of the heights of aircraft, the radar, or the turbines. DoE is alarmed by these findings and a fierce battle is ongoing. MoD has toughened its stand on the subject and is now objecting almost all wind farms in the line of sight of its radar stations. This puts millions of pounds of investment in wind farms in those areas at risk.

MoD is standing firm behind its findings in this case, and the Chief of the Defense Staff has given a firm direction that radar surveillance capability must not be degraded by any means.

The British Wind Energy Association is working with MoD and the government in solving the issue, but we know who is going to win. At the same time, the Department for Business Enterprise and Regulatory Reform has created an "Aviation Working Group," consisting of wind industry, MoD, and Civil Aviation Authority officials who are charged to solve the conflict.

This basically means that UK's MoD has the last word in deciding where a wind power plant can be located. It also sets a precedent to be followed by all countries. Who is next?

ENERGY STORAGE

We've seen that variability of solar and wind power plants' output is unavoidable and cannot be improved no matter what we do. This is because the variability depends on nature's whim, which we have no control over.

Variable and unpredictable output is the main obstacle in the progress of solar and wind energy generation today.

This will not change in the future, so wind and solar power generation might be combined with each other and/or other energy sources (i.e. geo-, bio-, ocean power, etc.) to smooth the variations and even match the peak load. But even achieving that will not provide a complete solution to our energy security.

During cloudy or rainy days the output of a solar power plant will be limited. Equally so, the output of a wind power plant is limited to times of wind activity. During days with no solar and no wind (and at night) we have no power generation, or very little at best.

The only way to rectify that anomaly in such cases is to provide *energy storage* for supplemental power generation. This stored energy can be used during periods of low energy generation, which is the only way to reduce the variability and inconsistency in the output of solar and wind power plants.

There are a number of potential energy storage solutions for use with solar and/or wind power plants. The available energy storage technologies under consideration and development for use now and in the future, are:

- Thermal storage
  - Steam accumulator
  - Molten salt
  - Cryogenic liquid air or nitrogen
  - Seasonal thermal store
  - Solar pond
  - Hot bricks
  - Fireless locomotive
  - Eutectic system
  - Ice Storage
- Electrochemical storage
  - Batteries
  - Flow batteries
  - Fuel cells
  - Electrical
  - Capacitor
  - Supercapacitor

  - Superconducting magnetic energy storage (SMES)
- Mechanical storage
  - Compressed air energy storage (CAES)
  - Flywheel energy storage
  - Hydraulic accumulator
  - Hydroelectric energy storage
  - Spring
  - Gravitational potential energy (device)
- Chemical storage
  - Hydrogen
  - Biofuels
  - Liquid nitrogen
  - Oxyhydrogen and Hydrogen peroxide
- Biological storage
  - Starch
  - Glycogen, and
- Electric grid storage

The most suitable and practical for solar- and wind power applications energy storage options are:

Thermal Energy Storage

Presently, this is the most widely used method of energy storage in thermal solar (CSP) plants. Heat is transferred to a thermal storage medium in an insulated reservoir during the day, and withdrawn for power generation at night. Thermal storage media include pressurized steam, concrete, a variety of phase change materials, and molten salts such as sodium and potassium nitrate.

The most widely used thermal energy storage technologies today are:

Pressurized Steam

Some thermal solar power plants store heat generated during the day in high-pressure tanks as pressurized steam at 50 bar and 285°C. The steam condenses and flashes back to steam, when pressure is lowered. Storage time is short—maximum one hour. Longer storage time is theoretically possible, but has not yet been proven.

Molten Salt

A variety of fluids can be used as energy storage vehicles, including water, air, oil, and sodium, but molten salt is considered the best, mostly because it is liquid at atmospheric pressure, it provides an efficient, low-cost medium for thermal energy storage, and its operating temperatures are compatible with today's high-pressure and high-temperature steam turbines. It is also non-flammable and nontoxic, and since it is widely used in other industries, its behavior is well understood and the price is cheap.

Molten salt is a mixture of 60% sodium nitrate and 40% potassium nitrate. The mixture melts at 220°C, and is kept liquid at 290°C (550°F) in insulated storage tanks for several hours. It is used in periods of cloudy weather or at night using the stored thermal energy in the molten salt tank to generate steam and turn a turbine, which in turn generates electricity. These turbines are well established technology and are relatively cheap to install and operate.

Pumped Heat

Pumped heat storage systems are used in CSP power plants and consist of two tanks (hot and cold) connected by transfer pipes with a heat pump in between performing the cold-to-heat conversion and transfer cycles. Electrical energy generated by the PV power plant is used to drive the heat pump with the working gas flowing from the cold to hot tanks. The gas is heated and pumped into the hot tank (+50°C) for storage and use at a later time. The hot tank is filled with solids (heat absorbing materials), where the contained heat energy can be kept at high temperature for long periods of time.

The heat stored in the hot tank can be converted back to electricity by pumping it through the heat pump and storing it back in the cold tank. The heat pump recovers the stored energy by reversing the process.

Some power (20-30%) is wasted for driving the heat pump and during the transfer and conversion cycles, but the technology can be optimized for use in large-scale PV plants.

In all cases, large heat energy losses accompany the energy storage processes. At least one third of the energy is lost during the conversion of stored heat energy into electricity. More is lost during the storage and following cooling cycles.

Batteries Energy Storage

No doubt, this is the most direct and efficient way to store a large amount of electricity generated by PV power plants. The generated DC electric energy is stored as DC power in batteries for later use.

There are several types of batteries, the most commonly used as follow:

Lead Acid Batteries

These are the most common type of rechargeable batteries in use today. Each battery consists of several electrolytic cells, where each cell contains electrodes of elemental lead (Pb) and lead oxide (PbO2) in an electrolyte of approximately 33.5% sulfuric acid (H2SO4). In the discharged state both electrodes turn into lead sulfate (PbSO4), while the electrolyte loses its dissolved sulfuric acid and becomes primarily water. During the charging cycle, this process is reversed.

These batteries last a long time, and can go through many charge-discharge cycles, if properly used and maintained. They are affected, however, by high temperatures, when the electrolyte can boil off and destroy the battery. Since there is water in the cells, the electrolyte can freeze during winter weather, which could destroy the battery as well.

Lithium Batteries

These are a mature technology, having been used widely for a long time in consumer electronics. They are actually a family of different batteries, containing many types of cathodes and electrolytes. The most common type of lithium cell used in consumer applications uses metallic lithium as anode and manganese dioxide as cathode, with a salt of lithium dissolved in an organic solvent. A large model of these can be used to store large amounts of electric power generated by a PV power plant, and due to their highest known power density they could be quite efficient—70-85%.

They are suitable for smaller PV installations, too, because scaling up to large PV plants would be a very expensive proposition.

Sodium Sulfur Batteries

These are high temperature, molten metal, batteries constructed from sodium (Na) and sulfur (S). They have a high energy density, high efficiency of charge/discharge (89-92%) and long cycle life. They are also usually made of inexpensive materials, and due to the high operating temperatures of 300-350°C they are quite suitable for large-scale, grid energy storage.

During the discharge phase, molten elemental sodium at the core serves as the anode, and donates electrons to the external circuit. The sulfur is absorbed in a carbon sponge around the sodium core and Na+ ions migrate to the sulfur container. These electrons drive an electric current through the molten sodium to the contact, through the electric load and back to the sulfur container.

During the charging phase, the reverse process takes place. Once running, the heat produced by charging and discharging cycles is sufficient to maintain operating temperatures and usually no external source is required.

There are, however, a number of safety and corrosion problems, due to the sodium reactivity, which need to be resolved before full implementation of this technology takes place.

Vanadium Redox Batteries

These are liquid energy sources, where different chemicals are stored in two tanks and pumped through electrochemical cells. Depending on the voltage supplied, the energy carriers are electrochemically charged or discharged. Charge controllers and inverters are used to control the process and to interface with the electrical source of energy.

Unlike conventional batteries, the redox-flow cell stores energy in the solutions, so that the capacity of the system is determined by the size of the electrolyte tanks, while the system power is determined by the size of the cell stacks. The redox-flow cell is therefore more like a rechargeable fuel cell than a battery. This makes it suitable as an efficient energy storage for PV installations.

A number of additional types of batteries are under development, and some show great potential for use in larger PV installations in the near future. Most batteries have problems with moisture, high temperature, memory effect, and use of scarce and toxic exotic materials, all of which cause longevity problems and abnormally high prices. If and when all these problems are resolved, the energy storage problems of PV power plants will be resolved as well.

Compressed Air Energy Storage

Compressing air into large high-pressure tanks is one of the most discussed and most promising energy storage methods for use with PV power plants today. It is a quite simple way of energy storage, using a compressor powered by the electricity produced by the PV plant compressing air into the storage tank. A lot of energy is lost by activating the compressor and heat is wasted during the compression process, so there are several compressing methods that treat generated heat so as to optimize conversion efficiency.

Some of these are as follow:

Adiabatic Storage

Adiabatic storage retains the heat produced by compression via special heat exchangers, and returns it to the compressed air when the air is expanded to generate power. Its overall efficiency is in the 70-80% range, with the heat stored in a fluid such as hot oil (300°C) or molten salt solutions (600°C).

Diabatic Storage

Here extra heat is dissipated into the atmosphere as waste, losing a significant portion of the generated energy. On removal from storage, the air must be re-heated prior to expansion in the turbines, which requires extra energy as well. The lost and added heat cycles lower the efficiency, but simplify the approach, so it is the only one implemented commercially these days. The overall efficiency of this method is in the 50-60% range.

Isothermal compression and expansion

This method attempts to maintain constant operating temperature by constant heat exchange to the environment. This is only practical for small power plants, which don't require very effective heat exchangers, and although this method is theoretically 100% efficient, this is impossible to achieve in practice, because losses are unavoidable.

There are a large number of other methods using compressed air, such as pumping air into large bags in the depths of lakes and oceans, where the water pressure is used instead of large pressure vessels. Pumping air into large underground caverns is another approach that is receiving a lot of attention lately.

Pumped Hydro Energy Storage

Pumped hydro energy is a variation of the old hydroelectric power generation method used worldwide, and is used quite successfully by some power plants. Energy is stored in the form of water, pumped from a lower elevation reservoir to one at a higher elevation. This way, low-cost off-peak electric power from the PV power plant can be used to run the pumps for elevating the water.

Stored water is released through turbines and the generated electric power is sold during periods of high electrical demand. This way the energy losses during the pumping process are recovered by selling more electricity during peak hours at a higher price.

This method provides the largest capacity of grid energy storage—limited only by the available land and size of the storage ponds.

Flywheel Energy Storage

Flywheel energy storage works by using the electricity produced by the PV power plant to power an electric motor, which in turn rotates a flywheel to a high speed, thus converting the electric energy to, and maintaining the energy balance of the system as, rotational energy. Over time, energy is extracted from the system and the flywheel's rotational speed is reduced.

In reverse, adding energy to the system results in a corresponding increase in the speed of the flywheel. Most FES systems use electricity to accelerate and decelerate the flywheel, but devices that use mechanical energy directly are being developed as well.

Advanced FES systems have rotors made of high-strength carbon filaments, suspended by magnetic bearings, and spinning at speeds from 20,000 to over 50,000 rpm in a vacuum enclosure. Such flywheels can come up to speed in a matter of minutes—much quicker than some other forms of energy storage.

Flywheels are not affected by temperature changes, nor do they suffer from memory effect. By a simple measurement of the rotation speed it is possible to know the exact amount of energy stored. One of the problems with flywheels is the tensile strength of the material used for the rotor. When the tensile strength of a flywheel is exceeded, the flywheel will shatter, which is a big safety problem.

Energy storage time is another issue, since flywheels using mechanical bearings can lose 20-50% of their energy in 2 hours. Those with magnetic bearings and high vacuum, however, can maintain 97% mechanical efficiency, but their price is correspondingly higher.

Hydrogen Generation (Storage)

Hydrogen, produced with excess power from solar or wind power generators, can be used as an efficient fuel during periods of high power demand. It is carbon-free and is converted back into water during the combustion process, thus we start with water and end with water.

We basically use the sunlight to generate hydrogen during the day, store it and then burn the hydrogen at night to generate heat and electricity, which process produces water again.

Excess hydrogen can be stored in tanks on site, or in underground caves for later use. Some of the excess hydrogen can be sold too.

This technology has a number of problems—related to the production, transport, and use of the excess hydrogen—but these will be solved eventually. When that is achieved, hydrogen generation via renewable energy sources has the potential to become industry standard.

Grid Energy Storage

Grid energy storage is large-scale storage of electrical energy, using the resources of the national electric grid, which allows energy producers to send excess electricity over the electricity transmission grid to temporary electricity storage sites that become energy producers when electricity demand is greater. Grid energy storage is a very efficient, albeit limited, storage method, which could play an important role in leveling and matching electric power supply and demand over a 24-hour period.

There are several variations of this method, all of which would require some modifications and upgrades to the national electric grid. This is billions of dollars expense that could be implemented in parallel with planned and pending smart grid upgrades.

One of the proposed grid energy storage methods, called "vehicle-to-grid energy storage system" uses modern (or future) electric vehicles that are plugged into the power grid and can release the stored electrical energy in their batteries back into the grid when needed. Millions of vehicles charged during low-power-use night hours, could be discharged into the grid at noon, or when needed, thus making a huge difference in the grid's peak hour response.

Far fetched, yes, but the future will demand many such ingenious approaches, if we are to strengthen our energy security and become energy independent.

In conclusion, there are a number of other energy storage methods such as fuel cells, new types of batteries, superconducting devices, super-capacitors, hydrogen production, and many other such innovative technologies under development, so the future looks bright in this area.

In practice, however, there are a few significant energy storage installations around the world, totaling over 2,100 MW, with the major technologies being:

Thermal energy storage—over 1,140 MW
Batteries energy storage—over 450 MW
Compressed air energy storage—over 440 MW
Flywheels energy storage—over 80 MW

Energy storage has many advantages, and could transform the electric utility industry by improving wind and solar power variability, availability and utilization, thus contributing to achieving energy security and environmental clean-up by avoiding the building of new power plants and transmission and distribution networks. Experts consider energy storage to be the solution to the electric power industry's issues of variability and availability, and the only way of opening wide the doors for wind and solar power use.

Complexity, safety, price and other restrains, however, will have to be worked out well before any of the large-scale energy storage methods become accepted reality.

Update 2104

Realizing the importance of energy storage for the future development of solar and wind energy in the U.S., DOE has included energy storage projects in a $4 billion loan guarantee program announced in July, 2014. The focus is on advanced grid integration and storage.

The new loan guarantee program is designed to support new renewable energy and energy efficiency projects in the U.S. in an attempt to avoid and/or mitigate GHG emissions, as part of Obama's Climate Action Plan.

The goal of the new $4 billion program is to support the development and deployment of near-market ready clean technologies, and finance projects that improve renewable energy variability and dispatchability, in an attempt to reduce grid congestion, and improve its control.

Some of the key areas of interest, identified by DOE at this time are: drop-in biofuels, waste-to-energy, enhancement to hydro power, and energy efficiency improvements. Different forms of energy storage are also priority of this round of loan guarantees.

Demand response and local storage are also some of the key features of the DOE's plan to allow access and grid compatibility for the different energy sources. This is the only way for solar and wind to compete and take a respectable place in the US power generation market.

DOE considers this a critical time to "address the effects of climate change and protect our children's future." This investment would also provide economic opportunities and jumpstart some low carbon and clean technology industries—another step in the right direction, no doubt! Technologically, however, we see many more rounds and many more billions of dollars spent before we see a true energy storage implemented at the large-scale solar or wind power plants.

100% Solar Power Dreams

To achieve 100% capacity (24/7/365 power generation capability), our 100 MW solar plant in a U.S. desert must generate:

100 MW x 24 hrs. x 365 days = 876 GWh/annum

This is the total annual power generation at 100% capacity.

Instead, even the most efficient solar power plants in the sunniest places in the world, would generate about 20% of this total power. This means that 80%, or over 700 GWh of energy storage would be needed for 24/7/365 operation—for just one 100 MW solar plant.

This represents 7,000 GWh energy storage for a 1 GW solar power plant, which is approximately what can be expected from an average wind farm too.

This is more energy storage than what is available in the entire world today.

Solar power plants in less sunny areas would require even more energy storage—double and triple this amount—in some cloudy areas of the world.

But installing and operating even the minimum required 700 GWh of energy storage—suitable to support 24/7 field operation of our 100 MW solar power plant in the desert, without using any other fuels—is unthinkable today.

This deficiency and the resulting delay in development of energy storage technologies is due to a multiplicity of unresolved technological difficulties and huge financial barriers still in the way of all types of large-scale energy storage.

So, the DOE loan guarantee program will no doubt bring us close to the day when we could rely on 7,000 GWh energy storage at each 100 MW solar or wind power plant. What a glorious day that would be—marking the end of our energy problems.

We know this will happen someday; we don't know when, but it surely won't be tomorrow or the day after!

OCEAN POWER

The world's oceans contain immeasurable amounts of never ending, renewable, and clean power, or rather potential energy. The key here is potential, because although the energy is there for the taking, taking it is not easy. Nevertheless, there are many efforts underway—large and small—to capture and use this energy.

There are basically four different ways to look at ocean energy. These are its tidal, wave, thermal, and salinity properties as energy sources.

In more detail, the different technologies that can be used to convert the energy contained in the oceans into useful electric power and other forms of energy can be described as follow:

Tidal Energy

One form of ocean energy is tidal energy. When tides come into the shore, they can be trapped in reservoirs behind dams. Then when the tide drops, the water behind the dam can be let out just like in a regular hydroelectric power plant.

For this to work well, however, a large dam wall needs to be constructed, and a great increase of the tides' height is necessary to generate enough power. An increase of 16 feet between low and high tide is the minimum needed for significant power generation.

There are only a few places in the world where such large tide changes can be found, and some power plants are already operating there using this idea. One such plant in France makes enough energy from tides to power 240,000 homes.

A tidal power generation method more practical for small applications is depicted in Figure 7-39. Here the tidal waves are directed into a channel (or tunnel), where they speed one way when the tide comes, and back the other way when the tide goes out. A turbine is located in the channel and its blades turn with the tides coming in and out. The rotational energy is converted into electricity by an electrical generator, and is used in the power grid or for directly powering electrical devices.

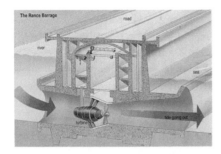

Figure 7-39. Ocean tidal engine

Wave Energy

Kinetic energy (movement) exists in the moving waves of the ocean and can be used to power a turbine, which in turn can generate electric power.

In Figure 7-40 the wave rises into a chamber, where the rising water forces the air out of the chamber. The moving air spins a turbine which can turn a generator. When the wave goes down, air flows through the turbine and back into the chamber through doors that are normally closed.

This is only one type of wave-energy system. Others actually use the up and down motion of the wave to power a piston that moves up and down inside a cylinder. That piston can also turn a generator.

Most wave-energy systems today are very small for the purpose of investigating their work function and future possibilities. Many can be used to power a warning buoy or a small lighthouse. Larger systems in the future could supply power to harbors, and coastal towns.

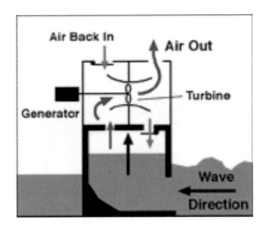

Figure 7-40. Ocean wave engine

Ocean Thermal Energy

The ocean energy idea uses temperature differences in the ocean. If you ever went swimming in the ocean and dove deep below the surface, you would have noticed that the water gets colder, the deeper you go. It's warmer on the surface because sunlight warms the water. But below the surface, the ocean gets very cold. That's why scuba divers wear wet suits when they dive down deep. Their wet suits trap their body heat to keep them warm.

Power plants can be built that use this difference in temperature to make energy, but a temperature difference of at least 38 degrees Fahrenheit is needed between the warmer surface water and the colder deep ocean water.

Cold seawater is an integral part of each of the three types of ocean thermal systems: closed-cycle, open-cycle, and hybrid-cycle. To operate, the cold seawater must be brought to the surface. The primary approaches are active pumping and desalination. Desalinating seawater near the sea floor lowers its density, which causes it to rise to the surface.

The alternative to costly pipes to bring condensing cold water to the surface is to pump vaporized low boiling point fluid into the depths to be condensed, thus reducing pumping volumes and reducing technical and environmental problems and lowering costs.

There are several types of ocean thermal energy systems.

Closed Cycle System

Closed-cycle systems use fluid with a low boiling point, such as ammonia, to power a turbine to generate electricity. Warm surface seawater is pumped through a heat exchanger to vaporize the fluid. The expanding vapor turns the turbo-generator. Cold water, pumped through a second heat exchanger, condenses the vapor into a liquid, which is then recycled through the system.

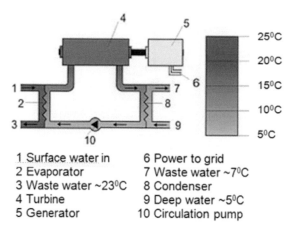

1 Surface water in   6 Power to grid
2 Evaporator   7 Waste water ~7°C
3 Waste water ~23°C   8 Condenser
4 Turbine   9 Deep water ~5°C
5 Generator   10 Circulation pump

Figure 7-41. Closed cycle thermal system

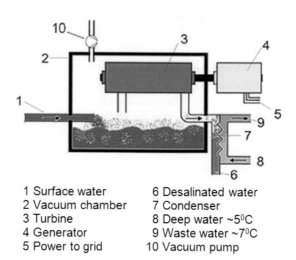

1 Surface water   6 Desalinated water
2 Vacuum chamber   7 Condenser
3 Turbine   8 Deep water ~5°C
4 Generator   9 Waste water ~7°C
5 Power to grid   10 Vacuum pump

Figure 7-42. Open cycle OTEC system

The first mini ocean thermal experiment in 1979 achieved the first successful at-sea production of net electrical power from a closed-cycle thermal system. The mini ocean thermal energy generating vessel was moored 1.5 miles (2.4 km) off the Hawaiian coast and produced enough net electricity to illuminate the ship's light bulbs and run its computers and television.

Open Cycle System

Open-cycle ocean thermal energy generator uses warm surface water directly to make electricity. Placing warm seawater in a low-pressure container causes it to boil. In some schemes, the expanding steam drives a low-pressure turbine attached to an electrical generator. The steam, which has left its salt and other contaminants in the low-pressure container, is pure fresh water. It is condensed into a liquid by exposure to cold temperatures from deep-ocean water. This method produces desalinized fresh water, suitable for drinking water or irrigation.

In other schemes, the rising steam is used in a gas lift technique of lifting water to significant heights. Depending on the embodiment, such steam lift pump techniques generate power from a hydroelectric turbine either before or after the pump is used.

The first vertical-spout evaporator to convert warm seawater into low-pressure steam for open-cycle plants was developed in 1984. Conversion efficiencies were as high as 97% for seawater-to-steam conversion (overall efficiency using a vertical-spout evaporator would still only be a few percent). In May 1993, an open-cycle ocean thermal plant at Keahole Point, Hawaii, produced 50,000 watts of electricity during a net power-producing experiment. This broke the record of 40 kW set by a Japanese system in 1982.

Hybrid Thermal Cycle Systems

A hybrid cycle combines the features of the closed- and open-cycle systems. In a hybrid system, warm seawater enters a vacuum chamber and is flash-evaporated, similar to the open-cycle evaporation process. The steam vaporizes the ammonia working fluid of a closed-cycle loop on the other side of an ammonia vaporizer. The vaporized fluid then drives a turbine to produce electricity. The steam condenses within the heat exchanger and could be used for desalination of sea water at the same time as well.

Ocean thermal energy conversion systems are being used in Japan and in Hawaii presently in some demonstration and small-scale projects.

Ocean Salinity Power

Salinity differential (or osmotic) power generation has been studied since the 1950s, but only recently was put into practice. The best place for this type of power generation is at estuaries, which is where river water mixes with seawater; usually along coastlines throughout the world. The salinity gradient between the two streams of water (fresh river water and salty ocean water) contains significant quantities of osmotic power. It is represented by the chemical potential in the differences in salt concentration between the two streams as they mix upon contact.

Enormous amounts of energy are released at that point, which can be captured and used to generate electric power. This process is defined as "power generation via transporting water through a semi-permeable membrane."

Westus (The Centre for Sustainable Water Technology) in the Netherlands is working on an osmosis based power plant using the Reverse Electrodialysis (RED) method of electricity generation. In it, fresh water from the Rhine river and saltwater from the North Sea are used in a type of battery containing two membranes permeable to ions (salt molecules), but not to water.

One of the membranes allows the passage of positively charged sodium ions into a stream of fresh water, while the other membrane allows the passage of negatively charged chloride ions into another channel of freshwater.

Thus separated and charged particles are captured by electrodes placed in both streams, and the DC electric current from these is conducted into a battery or (after conversion into AC power) into the grid.

Another team, Statkraft in Norway, is working on Pressure Retarded Osmosis (PRO), which also extracts electricity from salinity gradients. The PRO method utilizes a membrane, permeable to water, to draw fresh water into the concentrated salt water. This increases the pressure in the salt water chamber, which then drives a turbine to produce electricity.

In 2009, Statkraft built and started operation of the world's first osmotic power plant in Tofte Norway. This plant has a limited production capacity, around 4 kW, and is mainly used for R&D of the process, and data validation. This effort is expected to result in the construction of a larger commercial power plant of about 25 MW by 2015.

Although at a glance this technology has almost unlimited potential, it is still in its infancy and has a long way to go to full commercial deployment.

Environmental Effects of Ocean Power Generation

Power generation via tidal and wave power generating devices usually proceeds mostly without generating carbon dioxide, or any other harmful gasses or liquids. But as the number of wave and tidal projects in the world's oceans increases, attention is focusing on the negative effect power generating devices might have on marine life.

Ocean Environment

Any man-made equipment placed into the oceans creates some sort of interference with the marine environment. At the very least, the metals would have some chemical reactions, while the moving parts would affect it in another way.

For example, the presence of any electro-mechanical devices in the world's oceans would introduce some:

- Static and dynamic effects
- Chemical effects
- Acoustic effects
- Electromagnetic effects
- Energy changes and removal
- Many cumulative effects

This would result in some interference with:
- Near and far-field physical environment
- Habitat
- Invertebrates
- Migratory fish
- Resident fish
- Marine mammals
- Seabirds
- General ecosystem interactions

Figure 7-43. Ocean power generation

Noise, electromagnetic fields, mechanical damage etc. are possible effects suspected to affect marine life. Noise, for example, is known to confuse marine mammals, causing them stress and loss of orientation. Other marine species—sea turtles, crabs, sharks, skates, salmon and other fish—rely on Earth's magnetic fields for migrating and searching for food. The wave, tidal and hydrokinetic power devices, and the cables that bring the electricity they generate to shore, produce similar electromagnetic fields, which might confuse them.

Another thing that is known for sure is that fish swimming close to these devices might get hurt or killed by the moving parts—propellers, pistons, turbines, etc. Precautions have been taken to avoid collisions by installing special devices that emit noise to alert the potential victims, but it is not clear what other effects such noise makers would have.

We really don't know if the animals will be affected by all this or not. There's surprisingly little comprehensive research to tell us one way or the other.

Besides that (all of which still needs to be verified and taken care of) the ocean power-generating technologies are as clean as a whistle. They produce no pollution, nor do they cause any damage to the environment during operation.

On the front end, however, there are problems.

Large-scale power generating equipment from tidal and wave motion consist of very large pieces of metal, ceramics and plastics. The initial production of the different components and parts leave their environmental imprint. The equipment itself is made of different—usually expensive materials—and is of significant size. These components and parts are made of stainless steel, ceramics and other exotic and expensive stuff.

Here again, the metals were mined and refined in some Third World country, then transported, machined, or otherwise processed per spec into the needed materials. Then the finished parts were shipped to the tidal or wave power generating plant for final assembly.

How much pollution was generated during these steps? Mining, refining, transport, processing, more processing, and more transport require many sets of large equipment, facilities, and vehicles, all of which emit harmful gasses and liquids every step of the way.

And then, there is also considerable amount of disturbance to the ocean floor caused during the exploration and installation phases. Large pieces of equipment are brought in to dig deep holes in the ocean floor for the support structures. Large amounts of cement and other materials are then dumped in these holes to fix the structures in place.

These activities might go for days, weeks, and months during which time all marine life in the area would be killed or chased away. It cannot be even estimated how much damage is done during that time, but it is clear that the larger the project, the more damage would be inflicted to the ocean environment and local life in it.

And the damage to the local ocean flora and fauna continues and even magnifies with time, for the duration of the power plant life cycle.

GEOTHERMAL POWER

The term *geothermal* is from the Greek words *geo* (earth) and *therme* (heat), which put together mean Earth's heat. And this is exactly what geothermal power

generation is—using the earth's interior, which is naturally and eternally very hot.

This heat, geothermal energy, originates from the Earth's creation over 6 billion years ago. At the center of the Earth—Earth's *core*—which is over 4,000 miles deep, ferocious thermo-nuclear reactions are at play, and the temperatures reach over 10,000°F.

Like any heated body, the heat in the Earth's core is continuously emitted and flows towards cooler bodies in the surroundings. In some cases, the heated lava finds cracks and other openings and flows outward towards the surface, heating the rocks and earth as it cools on its way up.

The nearby layer of rock, the *mantle*, is heated red hot, and when temperatures and pressures become high enough, some mantle rock melts, becoming *magma*. During this process the magma becomes less dense and lighter than the surrounding rock, so it rises upward. This motion, or convection, makes it move slowly up toward the earth's crust, still hot but cooling down with every inch.

When hot magma gets near the surface, via volcanic explosions or through cracks in the ground, it flows down the hills like molten metal, which we know as *lava*. Often, however, the magma remains contained in pools at some distance below earth's crust. The pools of magma are heating nearby rocks and water in the water table or rainwater seepage. The rocks can get very hot—nearly 1000°F—and can keep the nearby water very hot too.

Some of the heated water flows up to the surface through cracks, where it forms *hot springs* or *geysers*. Most often, however, it stays in pools deep underground, where trapped in cracks and porous rock it forms natural formations called *geothermal reservoirs*.

Geothermal Electricity Generation

So, here we have a totally different type—truly clean and renewable—energy, ready to be extracted and used. It is truly renewable from our perspective, because when geo-thermal activities stop, all life on Earth will stop too. Period.

People have used the hot water on, or near, the Earth's surface for just relaxing and therapeutic activities since the beginning of time. But now, this free energy is used in other, much more creative and practical ways—mostly for heating buildings, and generating electric power.

The process of generating electricity from geothermal reservoirs is quite simple. We simply drill deep wells into the hot geothermal reservoirs, pump the water to the

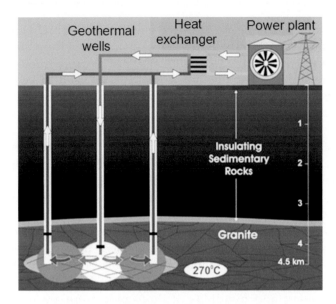

Figure 7—44. Geothermal power generation

surface and use it (and the *geothermal power in it*) to make steam. Thus produced steam is sent into a turbine, where it provides the force to spin the *turbine blades* and the *generator*, which produces electricity. The used, and cooled, geothermal water, exiting the turbine is then returned down the *injection well* into the reservoir to be reheated, thus the cycle can be repeated ad infinitum.

There are three kinds of *geothermal power plants,* depending on the temperatures and pressures of the underground geothermal reservoir, which can be divided as follows.

Dry Steam Reservoir

A "dry'" steam reservoir produces a lot of steam but very little water. The steam is piped directly into a *"dry" steam power plant* to provide the force to spin the turbine generator. The largest dry steam field in the world is The Geysers, about 90 miles north of San Francisco. Production of electricity started at The Geysers in 1960, at what has become the most successful alternative energy project in history.

Hot Water Reservoir

A geothermal reservoir that produces mostly hot water is called a "hot water reservoir" and is used in a *"flash" power plant*. Water ranging in temperature from 300-700°F is brought up to the surface through the production well where, upon being released from the pressure of the deep reservoir, some of the water flashes into steam in a "separator." The steam then powers the turbines.

Binary Systems

A reservoir with temperatures between 250 and 360°F is not hot enough to flash enough steam but can still be used to produce electricity in a *"binary" power plant*. In a binary system the geothermal water is passed through a *heat exchanger*, where its heat is transferred into a second (binary) liquid, such as isopentane, that boils at a lower temperature than water. When heated, the binary liquid flashes to vapor, which, like steam, expands across and spins the turbine blades. The vapor is then re-condensed to a liquid and is reused repeatedly. In this closed loop cycle, there are no emissions to the air.

Worldwide there are about 10 GW of geothermal power generated in over 20 countries, of which about 1/3 are generated in the U.S., which is equivalent to *not* burning over 60 million barrels of oil each year. The geothermal electricity generation is most prevalent where Earth's large oceanic and crustal plates collide and one slides beneath another. These are called subduction zones, with the best example being the Ring of Fire bordering the Pacific Ocean, the South American Andes, Central America, Mexico, the Cascade Range of the U.S. and Canada, the Aleutian Range of Alaska, the Kamchatka Peninsula of Russia, Japan, the Philippines, Indonesia and New Zealand.

Active geothermal energy sources are also found where these plates are sliding apart, such as in Iceland, the rift valleys of Africa, the mid-Atlantic Ridge and the Basin and Range Province in the U.S. There are also places called "hot spots," which are fixed points in the mantle that continually produce magma to the surface. In these cases the plate is continually moving across the hot spot, forming strings of volcanoes, such as the chain of Hawaiian Islands.

The countries currently producing the most electricity from geothermal reservoirs are the United States, New Zealand, Italy, Iceland, Mexico, the Philippines, Indonesia and Japan, but geothermal energy is also being used in many other countries.

Other Uses of Geothermal Energy

Geothermal power can be used even when the water is not hot enough, for applications other than generating steam and electricity.

Building heating uses the Earth's heat (the difference between the earth's temperature and the colder temperature of the air) which is transferred through buried pipes into the circulating liquid and then transferred into the building.

Building cooling is used during hot weather, where continually circulating fluid in the pipes "picks up" heat from the building—thus helping to cool it—and transfers it into the earth.

The main non-electric ways for using low temperature geothermal energy are *direct use* and *geothermal heat pumps*.

Direct Use

Geothermal waters ranging from 50°F to over 300°F, are used directly from the earth in domestic and industrial applications, such as soothing aching muscles in hot springs and health spas (balneology); growing flowers and vegetables in greenhouses (agriculture); growing fish, shrimp, abalone and alligators to maturity (aquaculture); pasteurizing milk, drying onions and lumber and washing wool (industrial uses).

Space Heating

Space heating of individual buildings and entire population centers is the most common and oldest direct use of nature's hot water. Geothermal *district heating* systems pump geothermal water through a *heat exchanger*, where it *transfers* its heat to clean city water that is piped to buildings in the district.

There, a second heat exchanger transfers the heat to the building's heating system. The geothermal water is injected down a well back into the reservoir to be heated and used again.

The first modern district heating system was developed in Boise, Idaho. In the western U.S. there are 271 communities with geothermal resources available for this use. Modern district heating systems also serve homes in Russia, China, France, Sweden, Hungary, Romania, and Japan. The world's largest district heating system is in Reykjavik, Iceland. Since it started using geothermal energy as its main source of heat, Reykjavik (once very polluted) has become one of the cleanest cities in the world.

Geothermal heat is being used in some creative ways. For example in Klamath Falls, Oregon, geothermal water is piped under roads and sidewalks to keep them from icing over in freezing weather. In New Mexico and other places, rows of pipes carrying geothermal water have been installed under soil, where flowers or vegetables are growing. This ensures that the ground does not freeze, providing a longer growing season and overall faster growth of agricultural products that are not protected by the shelter and warmth of a greenhouse.

Geothermal Heat Pumps

We all know that deeper into the ground the temperature is relatively stable compared to the surface air temperature. Using this free energy, geothermal heat pumps (GHPs) are used to take advantage of the stable earth temperature (about 45-58°F) just a few feet below the surface. Thus obtained heat can be used to keep household and commercial indoor temperatures comfortable.

The GHPs simply pump and circulate water or other liquids through pipes buried in a continuous loop (either horizontally or vertically) underground, next to a building to be heated.

Depending on the weather, the system is used for heating or cooling.

Ground heat pumps (GHPs) use very little electricity and are very easy on the environment. In the U.S., the inside temperature in more than 300,000 homes, schools and offices is kept comfortable by these energy saving systems, and hundreds of thousands more are used worldwide.

The U.S. Environmental Protection Agency has rated GHPs as among the most efficient of heating and cooling technologies.

Shallow Ground Heat

This is one of the most cost-effective methods from an initial capital point of view. The earth's temperature a few feet below the ground surface is relatively constant everywhere in the world (about 45-58°F), while the air temperature can change from summer to winter extremes. Shallow ground temperatures are not dependent upon tectonic plate activity or other unique geologic processes, thus geothermal heat pumps can be used to help heat and cool homes anywhere and very cheaply.

Summary

Thousands of megawatts of power could be developed from already-identified hydrothermal resources. With *improvements in technology*, much more power will become available. Usable geothermal resources will not be limited to the "shallow" hydrothermal reservoirs at the crustal plate boundaries. Much of the world is underlain (3-6 miles down), by *hot dry rock* which in most cases contains no water, but packs a lot of heat.

The U.S., Japan, England, France, Germany and Belgium are experimenting with piping water into this deep hot rock to create more hydrothermal resources for use in geothermal power plants. As drilling technology improves, allowing us to drill much deeper, geothermal energy from hot dry rock could be available anywhere. At such times, we will be able to tap the true potential of the enormous heat resources of the earth's crust

Advantages of Using Geothermal Energy

There are a lot of advantages to using geothermal heat for making electricity, and cooling/heating homes and businesses. Geothermal power plants, like wind and solar power plants, do not burn fuels to make steam to turn the turbines. Generating electricity with geothermal energy helps to conserve nonrenewable fossil fuels, and by decreasing the use of these fuels, we reduce emissions that harm our atmosphere. There is no smoky air around geothermal power plants, and in fact some are built in the middle of farm crops and forests, and share land with cattle and local wildlife.

For over ten years, Lake County, California, home to five geothermal electric power plants, has been the first and only county in the U.S. to meet the most stringent governmental air quality standards in the U.S.

The land area required for geothermal power plants is smaller (per megawatt) than for almost every other type of power plant. Geothermal installations don't require damming of rivers or harvesting of forests—and there are no mine shafts, tunnels, open pits, waste heaps or oil spills.

Geothermal power plants are designed to run 24 hours a day, all year, providing consistent and more reliable power than any other technology. A geothermal power plant sits right on top of its fuel source, so there is no need for mining, processing, transport, etc. operations. It is resistant to interruptions of power generation due to weather, natural disasters or political rifts that can interrupt mining, processing and transportation of fuels. Geothermal "fuel"—unlike the sun and the wind—is always on, and the economic benefits remain in the region, and no fuel price shocks are expected.

Geothermal power plants can have modular designs, with additional units installed in increments when needed to fit growing demand for electricity.

And finally, geothermal projects, with all their benefits, could help developing countries grow their energy system efficiently and without pollution. Installations in remote locations can raise the standard of living and quality of life by bringing electricity to people far from "electrified" population centers.

So if this is such a wonderful energy source, then why is it not used more frequently around the world? There are a number of reasons, some of which are unavailability of equipment, experienced staff, and infrastructure issues. All these hinder the proper and efficient location, installation and operation of geothermal plants across the globe. Not enough skilled manpower and availability of suitable-to-build locations (in addition to funding barriers) have been identified as the most serious problems in adopting geothermal energy globally.

Getting geothermal energy requires installation of power plants and other expensive equipment as needed to pump up hot water from deep within the earth and converting it into useful electric power. All this requires a huge one-time investment, a certified installer and the relocation of experienced staff to the plant location. Then, thus generated electric power must be delivered to the customers, who are usually far away, so another great expense is needed for adding electricity infrastructure that must be set up to move the power. This alone often makes the entire project prohibitively expensive and limits its financing opportunities.

Geothermal energy is only practical in regions with hot rocks below the earth and can produce steam over a long period of time. To identify such places, great research is required which is done by the companies before setting up the plant. This initial investment (time and money) is significant and only a few can afford it, since it increases the total expense in setting up a geothermal power plant.

There is also a potential of higher level of corrosion in some sites, due to dissolved minerals in the water used in the pipes and the power plant. This might increase the maintenance cost significantly and reduce the profits accordingly.

Because geothermal sites penetrate miles below the Earth's surface, the drill holes and hot rock reservoirs often contain toxic gases which can escape to the surface, and there is a fear of toxic substances being released into the local atmosphere. The toxic gasses can also travel deep within the earth and contaminate the water table, all of which is a liability to the investors and owners.

There is also a danger that geothermal sites can run out of steam over a period of time due to drop in temperature, shifts in the Earth layers, and/or if too much water is injected to cool the rocks. Since these are unpredictable events in most cases, the risk of financial loss for investors is high.

Geothermal Power Use

The United States leads the world in geothermal electricity production with about 3.0 GW of installed capacity generated at 77 geothermal power plants. The largest concentration of geothermal power plants in the world is located at The Geysers geothermal field in California. The Philippines is the second highest producer of geothermal power in the world, with close to 2.0 GW of capacity, which is nearly 20% of the country's electricity generation.

There are at least 1.5 million geothermal wells is Texas alone, most of them small and on private lands. The state looks like a Swiss cheese from that point of view. How many more holes can be drilled, and what is their use is uncertain, but geothermal power is increasing in importance, so expect to hear more good news about geothermal power in Texas in the coming years.

There are nearly 11 GW of geothermal power generated in 24 countries today, which generate over 65 GWh of electricity annually. This is about a 20% increase in geothermal power online capacity since 2005, and is expected to grow to nearly 20 GW by 2020. There are a number of projects presently under consideration, often in areas previously assumed to have little geothermal value, so these numbers seem plausible and reachable.

Al Gore said that Indonesia is on a path to becoming a super power in geothermal electricity production. It is not exactly clear what superpower means in this case, but Al has never been wrong before, so we will watch for the rise of Indonesia's super power.

Several countries (El Salvador, Kenya, the Philippines, Iceland, and Costa Rica) presently generate more than 15% of their electricity from geothermal sources.

Enhanced geothermal systems, several kilometers in depth are operational in France and Germany and some are being developed or evaluated in several other countries. The advance of these types of systems might bring geothermal power to places that were not considered possible until recently.

Note: Iceland has one of the best and most productive geothermal installations in the world. The total potential for electricity production from the high-temperature geothermal fields in Iceland is estimated at about 1500 TWh, which translates to energy generation of 15 TWh per year over a 100-year period. The electricity production capacity from geothermal fields presently is only about 1/10 of that, or 1.3 TWh per year.

Does this mean that Iceland has a 1,000-year supply of geothermal energy at the present level of use? Maybe, or maybe not... It all depends on many factors, some of which are not under the control of the Icelanders, and some of which might change the picture entirely and very quickly if and when nature decides to intervene.

Cost of Geothermal Power Generation

The cost of geothermal power installations arise from:
- Planning and design
- Exploration
- Permitting

- Well drilling
- Power plant construction
- Power plant operation and maintenance
- Site decommissioning

The cost of building a geothermal power plant **heavily weighs toward the initial expenses. This is different from other power plants, where fuel supplies, transport and disposal, are a major factor.** Here, well drilling and pipeline construction start the entire process, followed by resource analysis of the drilling information. Next is design of the actual plant, if the drilling information shows favorable conditions for geo-power generation: large quantities of high-temperature rocks at a reasonable depth.

Power plant construction is usually completed concurrent with final field development. The initial cost for the field and power plant is around **$2500 per installed kW** for large installations (over 1 MW installed capacity) in the U.S. It is somewhat higher—$3000 to $5000/kW—for a smaller power plant and in other countries that need to import expertise and technology.

Operating and maintenance costs range from $0.01 to $0.03 per kWh. Most geothermal power plants can run at greater than 90% availability, which means that they are fully operational (and at 100% capacity) 90% of the time. This is remarkable, and should be entered as best of all power generators.

Running at higher availability of 97-98%, however, can increase maintenance costs, so only higher-priced electricity justifies running the plant 98% of the time as needed to recover the higher maintenance costs.

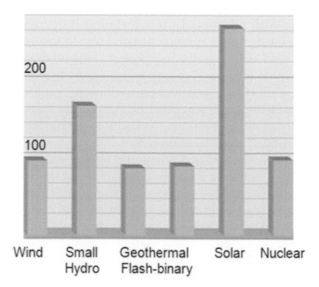

Figure 7-45. LCOE cost in $/MWh of generated electricity

Recent reports suggest that geothermal power may actually be cheaper than all other fuels, including coal. The U.S. stimulus package that includes $28 billion in direct subsidies for renewable energy, and additional $13 billion for research and development, gave renewable energy sources—geothermal power included—a shot in the arm, so we expect some great things to happen soon in this area of the energy industry.

Estimates from Credit Suisse estimate geothermal power costs at 3.6 cents per kilowatt-hour, versus 5.5 cents per kilowatt-hour for coal. This, however, does not include the risks and a number of assumptions that must be made when considering a new geothermal power plant.

Nevertheless, the basics are there; starting with the tax incentives, which save about 1.9 cents per kilowatt-hour, but they won't last, although the stimulus bill extended them through 2013.

Table 7-5. Installation cost of conventional geothermal power plant (in $/kW installed)

Stage	$/kW	%
Exploration	14	0.4
Permitting	50	1.4
Steam Gathering	250	6.8
Exploratory Drilling	169	4.6
Production Drilling	1,367	37.5
Plant & Construction	1,700	46.6
Transmission	100	2.7
TOTAL	**3,650**	

The development of geothermal energy requires the consideration and evaluation of factors such as site (geography), geology, reservoir size, geothermal temperature, and plant type. The majority of the overall cost is typically attributed to construction of the power plant, due to the high cost of raw materials including steel. The second highest cost are the intensive processes, exploratory and production drilling, which together cost as much as plant construction.

Geothermal power production is relatively capital-intensive with high first-cost and risk, but it has fairly low operating and maintenance (O&M) costs and a high capacity factor. This makes it one of the most economical baseload power generation options available. A number of factors contribute to the cost of developing a geothermal power plant, with the power conversion technology having some, but not great effect on the final cost.

Low-temperature reservoirs typically use binary power plants, while moderate- to high-temperature reservoirs employ dry steam or flash steam plants, based on whether the production wells produce primarily steam or water, respectively. The different technologies show similar construction costs and O&M expenses.

Remember, the significant up-front capital requirements, pervasive resource and development uncertainty, and long project lead times bring greater risk-related mark-up, than with renewables and traditional alternatives. Added to current economic conditions, this mean private firms seeking to develop geothermal projects may face greater difficulties obtaining the requisite capital for exploration and development. Industry analysts suggest that financing is still available, but the terms will be less attractive to investors and developers.

Most cost estimates rely on the *levelized cost of energy* (LCOE), which represents the total cost to produce a given unit of energy, usually in $/MWh. This figure assumes that the money to build a new geothermal plant is available at reasonable interest rates.

This, however, is not the case today. Although there is interest in geothermal energy generation, the general consensus is that it's difficult to get to and that it's expensive to produce. Therefore, only very low-risk and outstanding in their profitability projects can get even credit card interest rates, according to insiders. This means that the up-front costs are too high for a profitable project. So, utilities and other companies prefer to spend their money on lower front-end costs, like natural gas powered plants, which are cheap to build but relatively expensive to operate in the long run, due to the cost of fuel to run them.

Natural gas LCOE is averaged at $0.052 per kWh (vs. $0.036/kWh for geothermal power) but is finding greater use because it can be deployed anywhere. This is a great advantage over most other location-specific technologies, including geothermal power.

Only 13 U.S. states have identified significant geothermal resources, which severely limits its usefulness, since it is not possible to transport it like gas or oil. Nevertheless, there are estimates that the U.S. has over 30 GW of geothermal energy supplies that could be exploited using today's technology. These supplies are also estimated to last indefinitely, but, of course, all this is to be tested, verified, and proven.

Environmental Effects of
Geothermal Power Generation

The environmental problems caused by building a geothermal energy plant start while looking for a suitable site. During exploration, researchers will do a land survey, which may take several years to complete. To extract from a profitable heat generating site, we must find certain hot spots within the Earth's crust, which are most often around volcanoes and fault lines. But who wants to build their geothermal energy plant next to a volcano? So we go in search of similar conditions elsewhere.

Some areas of land may have the sufficient hot rocks to supply hot water to a power station, but many are in harsh areas of the world, near the poles or high in the mountains. So, the search continues for a long time, with people and machinery crawling all over the place, digging and drilling and causing some environmental damage in the process.

And of course some damage to the local environment and life would be done during the construction of the production facility. Large pieces of equipment will be brought in to level the place and drill deep holes. Lots of pipes will be delivered and laid in the exploration and production holes.

During the well drilling and operation stages, toxic gasses can escape from the site because it is hard to control what's happening deep under the surface. Things also change down there, so although there were no problems today, tomorrow things change and toxic gas leakage is detected.

These gasses can escape into the local air or penetrate deep into the ground where they could contaminate the water table. Always, the environment and all life in it could be damaged.

Another danger of geothermal energy is the likelihood of recurring earthquakes. People who live in areas of production report that there is an increasing number of earthquakes in the area. Although they are low-level quakes no stronger than magnitude 4, they are damaging homes and foundations.

Increased volcanic activities could bring the eruption of an old or a new volcano during hot rock drilling. Such an eruption could devastate the local flora and fauna in an instant. This, of course, is not as dangerous as a nuclear power plant explosion, but nevertheless, those unfortunate to be nearby during the eruption will pay with their health or lives.

Finally, after years of generating power, in many cases the site "cools down" and production stops. At that time more environmental damage will be done by the shutting down and decommissioning of the power plant. More large equipment arrives at the site and starts demolishing buildings, digging holes and levelling the ground. All of this is at the expense of the local environment.

The good thing is that in most cases, the geothermal site—unlike coal mines and gas and oil fields—can be restored to near-perfect (original) condition.

Author's note: Most importantly, looking at the larger picture, we can easily see that injecting large amounts of cold water into any hot body (including the Earth) day after day has limits—quantity and time limits. Where these limits lie is uncertain, but we should not be as naïve as to think that we can continue doing this indefinitely, and even increase the volume of cold water injected in the Earth's hot rocks, without any negative effects sometime in the future.

Like any other technology, when geothermal is applied in small quantities—as it is today—it cannot do much harm for a very long time. When deployed on a very large scale, however, things will surely change, and negative effects will follow…as they always do in such cases!!!

It is a great mistake to take examples of old, small, good and reliable production geothermal wells and extrapolate the results to large and very large-scale power generation. This is like comparing apples and oranges… or worse!

The energy in Earth's core is huge, no doubt!, but it is not unlimited. Cooling it down by pumping lots of cold water in/on it, without considering the possible consequences, might be the dumbest and most dangerous thing we've ever done. It is like cutting the branch we are sitting on, just to warm our hands temporarily. And this is why we do not consider geo-thermal power a renewable power. It is renewable while it lasts, but then it is gone, which happens sooner or later at each particular location. So how renewable is that?

BIO-ENERGY

Bio-energy technologies are most controversial, because, although bio materials are a plentiful and renewable energy source, they are affected by increased draughts. There is also global land and resource overuse and abuse, where the demand outstrips the supply in many cases, thus creating huge problems. On top of that, millions of tons of greenhouse gas (GHG) emissions are produced by using (burning) large quantities of bio-materials (wood and other vegetation).

Bio-energy, in the general meaning of the term, is renewable energy that is contained and can be extracted from materials created by natural biological sources. In more specific terms, biomass (from which we get bio-energy) is any organic material which has stored sunlight in the form of chemical energy. Other bio-energy sources are bio-fuels and their derivatives, which are produced from biomass.

As a fuel, biomass includes wood, wood waste, straw, manure, sugarcane, and many other byproducts from a variety of agricultural processes. Biomass (wood and its byproducts) are extensively used in Third World countries for cooking, heating, and industrial processes. Biomass can also be burned to generate electricity.

In 2010, there were about 35 GW of globally installed bio-energy electricity generation plants, of which 7 GW was in the U.S.

Note: There is confusion about *bioenergy, vs. bio-fuel,* which at times are used interchangeably. In technical terms, biomass is natural organic matter, which is used *as is*. It is usually burned directly, without any processing, or any modifications. Biofuels, and their sub-components, on the other hand, are usually derived from processing biomass and other organic or inorganic materials.

More specifically, biomass is usually mostly naturally grown, or with little human assistance, while bio-fuels need 100% human intervention, a lot of equipment and energy, and complex processes.

Biomass

Biomass is the common name for all organic materials classed as renewable energy sources such as wood, crops, and all types of animal, vegetation and other biological wastes. Biomass is a type of renewable energy source, and while it is replenished by nature in most cases, when used it is by no means green, or clean, as far as the environment and the fight against climate change are concerned.

The amount of energy released when a given unit of fuel is combusted is referred to as the *energy content* of that fuel. For example, the energy content of wood is generally in the range of 6 to 18 megajoules per kg (MJ/kg) of wood, depending on the moisture content of the wood.

Freshly cut wood could have as much as 60% moisture and would have a relative low energy content (e.g., 6 MJ/kg), whereas oven-dried wood with close to zero moisture content could have up to 18 MJ/kg. The average commonly used energy content value for commercial wood is 14-16 MJ/kg.

The energy content in some types of biomass, as well as their other properties can be seen in Table 7-6.

Table 7-6. Energy content in different fuels

Fuel	Energy
Crude oil	42 MJ/kg
Coal	27 MJ/kg
Natural gas	18 MJ/kg
Paper	17 MJ/kg
Dung (dry)	16 MJ/kg
Straw (dry)	15 MJ/kg
Wood	15 MJ/kg
Domestic waste	9 MJ/kg
Grass (fresh cut)	4 MJ/kg

Note: 1kW electricity = 3.6 MJ energy

The biomass production rate in terms of the quantity of biomass that can be grown on a parcel of land per unit time, which is generally given as kilograms of biomass per hectare per year (kg/ha/yr) varies greatly depending on the crop, soil type, availability of water, and moisture content of the crop.

Some representative yields of biomass products in kilogram/hectare/year are:

Sugar cane	35,000 kg/ha/yr.
Wood	20,000 kg/ha/yr.
Wheat, rice, and sorghum	15,000 kg/ha/yr.

*Note: 1 hectare is 10,000 m^2 or about 2.5 acres.

For liquid biofuels, the energy content is usually given in terms of megajoules per liter (MJ/L). Representative values for the energy content of crops grown for biofuels are:

Gasoline	35 MJ/L
Sunflower oil	33 MJ/L
Castor oil	33 MJ/L

The biomass production rate for plants to be grown for biofuels, given as liters per hectare or L/ha, however, varies greatly. Some representative values for plants grown as biofuels are:

Castor oil	1,413 L/ha
Sunflower	952 L/ha
Soybean	446 L/ha

Example

Assuming one crop per year, a hectare of castor beans would produce about 1,400 L of castor oil (with 40 MJ/kg. energy content), which can generate 56,600 MJ of bioenergy.

At the same time, given an average yield of 15,000 kg/ha/yr, and an energy content of 15 MJ/kg, the amount of energy that can be expected from a hectare of cultivated, mature, wood farm would be about 225,000 MJ/ha.

So, we get 56,600 MJ/ha from castor beans and 225,000 MJ/ha from natural wood—or about 4 times more. The difference here is that a castor beans farm can be harvested every year to produce this same amount of energy, while a mature wood farm needs 20-30 years to be harvested once.

This basically means that a hectare of castor beans would produce 5-8 times more energy that natural wood within the 20- to 30-year time period.

Note: 225,000 MJ is about the same amount of energy contained in 7.5 tons of coal, which is about one mechanized shovel worth of coal. All this means that natural wood has some serious competition today.

Today, the biomass power generating industry in the United States consists of approximately 11 GW of summer-only operating capacity supplying power to the grid, providing about 1.4 percent of the U.S. electricity supply.

The 140 MW New Hope Power Partnership is the largest biomass power plant in North America. It uses mainly sugar cane fiber (bagasse), and recycled urban wood as fuel to generate enough power for local industrial operations and to supply electricity for nearly 60,000 homes.

The biomass power generation reduces the import of 1 million barrels of oil annually, and by recycling sugar cane and wood waste, preserves landfill space in the local urban communities.

Let's take a closer look at the different types of biomass and bio-fuels.

Solid Biomass

Biomass is the common name for organic materials used as renewable energy sources such as wood, crops, and waste. Solid biomass materials include a number of natural and man-made products, such as wood, sawdust, grass trimmings, domestic refuse, charcoal, agricultural waste, nonfood energy crops, and dried manure.

Biomass is not to be confused with biofuels, which are usually liquid and are a product of the organic material in the biomass. Biomass, then, refers only to the organic matter contained in plant materials which can be used as a renewable energy source in a number of different ways. Raw biomass in a suitable-to-use form, such as firewood, can burn directly in a stove or furnace,

as needed to provide heat or generate steam. When raw biomass is in a different form, such as sawdust, wood chips, grass, urban waste wood, straw and other agricultural residues, the typical intention is to process it in order to increase the density of the "as is" biomass.

This process sometimes includes grinding the bulk to an appropriate particulate size to produce "hogfuel." Depending on the densification type, the final product can be 1 to 3 cm in size. It can then be further concentrated into a fuel product for ease of transport and use. The current processes produce wood pellets, cubes, or pucks of condensed biomass materials. The pellet process is common in Europe, and is typically a pure wood product.

Commercial densification operations are large in size and are compatible with a broad range of input feedstocks. The resulting densified fuel is easier to transport and feed into thermal generation systems such as boilers.

Table 7-7. Comparative value of biofuels (in MJ/kg)

Bio Fuels	Energy Content
Coal	28
Commercial wastes	16
Domestic refuse	9
Dung, dried	16
Biogas	55
Newspaper	17
Oil waste	42
Straw, bailed	15
Sugar cane residue	17
Wood, green	6
Wood, air-dried	15
Wood, oven-dried	18

One of the advantages of solid biomass fuel is that it is often a byproduct, residue or waste product of other processes, such as farming, animal husbandry, and forestry. This means that this type of fuel does not compete for resources with food production, although this is not always the case as we saw in 2008 when the ethanol production in the U.S. exploded and then imploded.

Different biomass materials contain different energy levels. Some examples are shown in the table opposite.

There are a number of materials that qualify as solid biomass. Below, we will take a look at several of these and their use as energy sources.

Note: Although biomass is classified as clean and renewable energy source, it is marginally so. The debate of its environment impact continues, and its renewable status depends on many factors.

1 m³ wood chips (.7 ton)	= 3.6 GJ	= 1 MWh		
1 ton air-dry wood fuel	= 11 GJ	= 3 MWh		
1 ton fire-dried wood	= 18 GJ	= 5 MWh		
1 barrel of crude oil	= 6.1 GJ	= 1.7 MWh	= 1.7 m³ wood chips, or 0.5 m³ wood chips	
1,000 m³ natural gas	= 36 GJ	= 10 MWh	= 6 barrels of oil, or 3 tons air-dry wood	

- In order to produce energy from biomass, for example, the organic matter must be burned in some way. This, in addition the energy used and emissions released during its growth, releases large amounts of carbon dioxide into the air; unlike the use of solar, wind, and other cleaner energy sources.

- The effect of increasing long-term draughts on biomass availability is also putting its renewable status in question.

Nevertheless, biomass is officially classified as a carbon neutral fuel, due to the total cradle-to-grave carbon cycle of the biomass. The carbon cycle means that while the crop grows it will absorb carbon dioxide, releasing it back into the atmosphere when burned.

In practice lots of carbon dioxide is emitted when burning biomass, which cannot be compensated by any means at the time of the emissions. Mass firewood burning during some periods of the year causes extreme pollution conditions in some locations, so we must always keep watch on the negative effects of biomass use on the environment.

Case Study: Straw to Energy

Straw from agricultural crops is the new hope for energy security in Germany. Until now straw, including corn stover and other agricultural byproducts, have been underutilized as a biomass residue and treated as waste material.

A total of 30-40 million tons of crops straw are produced annually in Germany, and most of it could be used sustainably for energy or fuel production. This could provide 2 to 3 million average households with electricity. At the same time it could provide 3 to 5 million households with heat.

There is a rise in the cultivation of winter wheat, rye and winter barley in Germany which produce approximately 30 million tons of straw per year. Granted, not all parts of the straw can be used, straw is also heavily used as bedding and feed in livestock farming, and some straw must be left in the fields for soil humus and nutrients balancing. Even then, about half of the 30 megatons—or about 15-20 million tons—of straw are available for ener-

gy production every year in Germany alone.

The straw can be used for generating electricity and/or heat by burning it in specially designed furnaces. Such effort is underway currently in Denmark, which is considered to be the world leader in straw-based energy production.

This work started in the late 1980s, when a plan was introduced in Denmark, ensuring the generation of about 5-6 GWh equivalent of energy per year. This is equivalent to the power of several large solar power plants...generated solely by straw burning and processing.

This number could be multiplied dozen of times if U.S. straw-to-energy production were considered, and several hundred times if the entire globe participated in such programs.

Straw could be a major participant in the future energy mix, by providing a large amount of additional energy. It would also contribute about 80% to GHG emissions reduction, as compared to fossil fuels in the generation of heat, combined heat and power generation, and/or as a second-generation biofuel production.

Different countries have different climate and economic conditions, which would determine the amount and method for straw production and use for energy. The German conditions, for example, are most favorable for using straw in combined heat and power generation.

Straw is best and most efficiently used in large district heating stations and/or combined heat and power stations. The technology still needs to be perfected for most efficient, cheap, and clean operation.

The Germans jumped very high (and fell very hard) when they raised solar energy on the pedestal of an energy savior a few years back. We hope and pray that they don't jump too high on the straw-to-energy solution too, because the fall might be even harder.

Overuse of straw and other agricultural byproducts, could eventually lead to thousands of acres of dead agricultural land, and/or other such disastrous results.

Firewood

Firewood is one of the oldest energy sources. It has been used for a number of purposes—cooking, heating, etc.—since the beginning of time. Until the middle of the 19th century the major energy sources were coal and firewood. Even when coal became a major energy source, firewood was still the next in line, mostly due to its unlimited supply from forests. At the time wood accounted for over 90% of U.S. energy generation, mostly for home heating and cooking.

An average of 18 cords (stack of 4x4x8 feet) of wood per year were needed to heat an American house in the mid 1800s. Consumption of fuel wood in this country reached a peak of almost 150 million cords in 1870, since at that time wood was the primary fuel.

This is always a great waste of energy, since it is 2-3 times the heat energy equivalent of the amount of heat energy used to warm a typical home today. This is mostly due to the inefficiency of the old fireplaces, which had less than 10% efficiency rating. They simply swallowed three quarters of all wood used at the time (approximately 75 million cords of wood) to produce little heat.

Woodstoves are about four times more efficient, but were not popular in the Americas at that time for convenience reasons. Wood stoves became fashionable in North America in the mid 20th century, and while wood burning efficiency increased, wood consumption increased as well.

The net effect of this move towards better energy efficiency was more energy use and more pollution emitted.

Wood also powered the growing industries of the 19th century. This included steamships and trains running on wood fire, which burned nearly 8 million cords of wood a year at the time. The iron and steel industry were the major industrial users of wood.

Firewood stoves are fashionable today, and are selling at a rate of 200,000 per year in the U.S. Over 100 million cords of firewood are used annually for home heating alone, which accounts for only a few percent of the total primary energy input in the country.

Wood and wood residue provide almost 2 quads of total energy, most of it in the wood products and paper industries. In these industries biofuels provided about half of the total energy needs.

If the use of wood and biomass increases to the very maximum, the total annual potential of biomass energy generation could provide only about 1/4 of the U.S. primary energy input. There is no estimate of the negative effects of such a move, but we are sure that they would be many and quite significant.

Firewood Fuel

Firewood is one of the oldest energy sources. As a matter of fact, the discovery of fire by burning wood is regarded as one of humanity's most important advances. The use of wood as a fuel source for heating and cooking is much older than civilization, and existed as early as the Neanderthal age.

Even today, firewood is one of the world's greatest energy sources (in terms of number of people affected and Btus used daily). Yes, brushes, branches, twigs and stumps of trees are burned by millions of people in the developed and developing countries for many reasons.

In terms of Btus generated, fire wood exceeds the power generation from the other sources during certain time periods.

Figure 7-46. Daily supply of firewood as seen in Africa.

Wood fuel may be available as firewood, charcoal, chips, sheets, pellets, and sawdust. The particular form used depends on the source, quantity, quality and application. In many areas, wood is the most easily available form of fuel, requiring no tools in the case of picking up dead wood. A few tools are needed to gather larger quantities of wood.

On an industrial scale, specialized equipment, such as skidders and hydraulic wood splitters, are used to mechanize production. Sawmill waste and construction industry byproducts also include various forms of lumber tailings.

Wood fuel today is used mostly for cooking and heating, and in some cases for fueling steam engines and steam turbines to generate electricity. Wood is also used indoors in furnaces, stoves, or fireplaces, or outdoors in larger furnaces, campfires, or bonfires.

While burning wood in the developing countries is mostly by choice, over 600 million people in Sub-Saharan Africa, 800 million in India, 500 million in China and 100 million in South America are using firewood for their daily needs. Firewood use in these cases is by necessity, for these people have no other choice.

Deadwood is collected around the roadways and forests, and many live trees are cut for this purpose as well. As time goes on, the deadwood gets used, and more and more trees are cut from the forest, thus helping in the propagation of the deforestation phenomena in some parts of the world.

Consumption of firewood is estimated at approximately 800 kg. per person per year). This comes to only 2-3 lbs. per person per day, but it is a large amount of wood—especially in areas that have been cleared from trees and vegetation for years. This creates great problems, including deforestation and tensions, especially in the more densely populated areas in a number of countries.

Let's assume for the purposes of this rough calculation that 500 kg. of firewood are used per person annually by approximately 2 billion poor people worldwide. Firewood has energy content equivalent to 0.35 tons of oil per ton of firewood, and assuming that firewood stoves and fire pits burn at maximum 15% efficiency, we can conclude that to replace the firewood used worldwide, we would need over 150 million tons of oil a year. This is nearly a quarter of U.S. oil consumption.

The U.S.' 350 million inhabitants use 4 times the energy used by the 2 billion poor around the world put together. The average American uses over 20 times more energy than the average person in the developing world.

Firewood is also used in large quantities in the rural areas of developing countries. The 1973 oil crisis brought out an entire firewood burning industry in the US, complete with most efficient wood burning stoves and processes. There are also many fireplaces in US cities that use natural wood to heat homes during the cold months.

Firewood use in the developed countries is not of necessity. It is a type of luxury, or convenience. Cuddling with a book by a roaring wood fire is part of the American dream, and people very seldom pass the opportunity. Firewood use has increased lately, so now there are many winter days with local wood-burning ban, due to increased particulate or CO_2 pollution.

Of course, with every wood fire there is smoke—lots of it. Toxic gasses belch in the air, in sometimes enclosed quarters, poisoning the inhabitants, and causing a number of health problems.

Believe it or not, wood burning is one of *the largest energy generators and largest polluters* in the world. It is the least known and discussed subject in the energy field, and it is also one that most energy principles prefer to ignore. It is of no consequence to them, or their business, and there is not much they can do about it, so why bother? Because of that, the entire issue has fallen between the cracks.

The majority of the wood-burning people happen to be the poorest segment of the world's population, and since there are few business opportunities in this area due to lack of money, the wood burning is left alone and will continue this way for a long time.

The extensive wood use is creating a problem big enough to classify wood burning as a "non-renewable energy," simply because the rate of use is much higher than the replacement.

We hesitate to call it renewable, because excessive use of wood in some parts of the world leads to uncontrollable and extensive destruction of forests and vegetation in entire regions. This is not something we should encourage. Instead, we should find ways to replace it with more sustainable and less damaging energy systems.

One simple 100 Watt solar panel, installed on a hut's roof could solve the energy problems of an entire Sub-Saharan family.

A hundred PV modules in the village might prevent 100 families from wasting time gathering wood 2-3 hours every morning. This would save a lot of trees and eliminate a lot of smoke emission.

Multiplied by the millions of wood burning families, this might save millions of tons of wood, and that many toxic gasses from being released in the atmosphere. Doing this would be the greatest environmental success ever! But this is not going to happen anytime soon, for the reasons we point out above—the main of which is indifference.

Large areas of bare land, where forests and woodlands were before can be seen for miles at a time in certain areas in Asia, South America, and Africa. These bare areas, in addition to reducing the world's forests, create additional problems, such as soil erosion and are becoming increasingly troublesome in a number of countries.

Wood burning cannot be eliminated completely, for it has some useful purposes, but providing alternative energy sources for people in places such as India and Sub-Saharan Africa, who are forced to burn wood daily, would help their development and contribute significantly to cleaning the environment from harmful gasses and the related effects.

Figure 7-47. Deforestation in progress

Some additional ways wood and wood products are used for energy today are covered below.

Sawdust

Sawdust, or wood dust, is a byproduct of cutting, grinding, drilling, sanding, or otherwise pulverizing wood, and is therefore composed of fine particles of wood. It is also the byproduct of certain animals, birds and insects which live in wood, such as the woodpecker and carpenter ants. It can present a hazard in manufacturing industries, especially in terms of its flammability. Sawdust is the main component of particleboard.

A major use of sawdust is for particleboard; coarse sawdust may be used for wood pulp. Sawdust has a variety of other practical uses, including serving as a mulch, as an alternative to clay cat litter, or as a fuel. Until the advent of refrigeration, it was often used in icehouses to keep ice frozen during the summer. It has been used in artistic displays, and as scatter. It is also sometimes used to soak up liquid spills, allowing the spill to be easily collected or swept aside. As such, it was formerly common on barroom floors, and used to make Cutler's resin. Mixed with water and frozen, it forms pykrete, a slow-melting, much stronger form of ice.

Sawdust is also used in the manufacture of charcoal briquettes, the invention of which is credited to Henry Ford who made some from the wood scraps and sawdust produced by his automobile factories.

Burning sawdust to generate electricity is not very common, but is still used in some special situations, mostly in developing countries.

In addition to the pollution problems described above, at sawmills, sawdust burners generate a lot of toxic gasses. Sawdust also may be stored in large out- door piles causing harmful leachates into local water systems, thus creating an environmental hazard. This problem has become a serious problem for small sawyers and environmental agencies, putting them in a deadlock.

Questions about the science behind the determination of sawdust being an environmental hazard remain for sawmill operators (though this is mainly with finer particles), who compare wood residuals to dead trees in a forest. Technical advisors have reviewed some of the environmental studies, but say most lack standardized methodology or evidence of a direct impact on wildlife. They don't take into account large drainage areas, so the amount of material that is getting into the water from the site in relation to the total drainage area is minuscule.

Other scientists have a different view, saying the "dilution is the solution to pollution" argument is no longer accepted in environmental science. The decomposition of a tree in a forest is similar to the impact of sawdust, but the difference is of scale. Sawmills may be storing thousands of cubic meters of wood residues in one place, so the issue becomes one of concentration.

But of larger concern are substances such as lignins and fatty acids that protect trees from predators while they are alive, but can leach into water and poison wildlife. Those types of things remain in the tree and, as the tree decays, they slowly are broken down. But when sawyers are processing a large volume of wood and large concentrations of these materials permeate into the runoff, the toxicity they cause is harmful to a broad range of organisms.

Charcoal

Charcoal is a type of man-made fuel, made from wood or vegetation and animal matter, via special process, where wood is burnt in a controlled presence of oxygen. The main reason for producing this type of fuel was, and still is, to convert unusable materials (even waste materials) into useful fuels that are easy to transport and use.

Charcoal is soft, brittle, light, with a dark grey to black appearance, very similar to coal. It consists mainly of carbon and ash, which remain after the water and other volatile compounds were removed from the raw materials. Its combustion properties are also close to those of coal.

In the past, charcoal was produced by piling wood logs, or other combustible materials, in a conical pile, which was covered to restrict contact with the air. A small opening at the base was used to control the intake of air, which was directed through a central air shaft serving as a flue. The fire was started at the bottom of the flue, and let to gradually spread upwards, which process took days at a time to complete.

Slow combustion is the key to producing quality charcoal. Small-scale production yields 50% by volume, or 25% by weight, of charcoal, which produces heat equivalent to the heat produced by the starting raw materials. Today charcoal is produced by a special carbonization process, where small pieces of wood, sawdust, or other waste materials are burnt under controlled conditions in cast iron retorts, which produce charcoal of different quality and for different purposes.

When charcoal is made at 300°C, it is brown, soft, friable, and readily inflames at 380°C. When it is made at higher temperatures it is hard and brittle, and does not fire until heated to about 700°C. Charcoal production is extensively practiced for the recovery of byproducts that have some useful heat content, such as tree branches, wood shavings, sawdust, etc.

There was massive production of charcoal during the last several centuries, which supported a large industry employing hundreds of thousands of workers in Central Europe and the UK. Most of the energy for iron smelting in the 1800s came from charcoal, which was made out of 1.5 million cords of wood per year. The charcoal used in iron smelting totaled some 700,000 to 750,000 tons annually. This is close to an estimated 750,000 tons of charcoal used every year by "outdoor and tailgate chefs" in the barbeques in this country alone.

Thousands of acres of forests were cut, which created a major deforestation in those areas. Large wooded areas were cut and regrown cyclically, to provide a steady supply of charcoal at all times.

Over-exploitation and lack of new supplies, and the increased demand for charcoal created a supply and demand disequilibrium, which facilitated the switch to coal and later on to fossil fuels for domestic and industrial use.

Charcoal is still an important energy source worldwide. Brazil uses charcoal in 45% of its iron smelting; 3.6 million tons of charcoal are used annually in Brazil alone, and consumption is increasing. Charcoal is also a major fuel in less developed countries such as Ghana and Kenya, where 250-300,000 tons are used every year.

Biochar

Biochar is charcoal which is used in special occasions, such as a soil additive and amendment. It is also a possible source of carbon sequestration, so it has the potential to help mitigate climate change, via carbon sequestration.

When mixed with soils, biochar can increase soil fertility, increase agricultural productivity and provide protection against some foliar and soil-borne diseases. Biochar is a stable solid, rich in carbon material that can endure in the soil for thousands of years.

Biochar made from agricultural waste can substitute wood charcoal. As wood stock becomes scarce, this alternative is gaining ground. In some parts of Africa, for example, biomass briquettes are being marketed as an alternative to charcoal to prevent deforestation associated with charcoal production.

Bagasse

Bagasse is the dry, fibrous matter that remains after crushing sugarcane, blue agave, or sorghum stalks, and extracting their juice. It is mostly used in the production of biofuel, paper products, and some special building materials.

Sugarcane and other plants are taken to a processing plant, where the juice contained in them is extracted for use in foods.

Sucrose accounts for little more than 30% of the chemical energy stored in the mature sugarcane plant, while about 35% is in the leaves and stem tips, which are left in the fields during harvest. Another 35% of the energy in the plant is in the fibrous material (bagasse) left over from crushing and pressing the sugarcane stalks, which can then be used for making fuel gas.

Bagasse is often used as a primary fuel source for sugar mills, where it is burned in large quantities to produce enough heat to run an entire sugar mill. The energy in biogas can be also used to provide both heat energy used in the mill, and electricity, which is typically sent into the electricity grid. This allows the plants to be energetically self-sufficient and even sell surplus electricity to utilities.

An average sugar- or ethanol-producing plant could produce 500 MW electricity for self-use, and 100 MW for sale. The sale of power is expected to boom as new regulations force the utilities to pay "fair price." This type of power is also especially valuable to utilities because it is produced mainly in the dry season when hydroelectric dams, and the electricity produced by them are running low.

Estimates of the potential power generation from bagasse in Brazil range from 1,000 to 9,000 MW depending on technology. Higher estimates assume gasification of biomass, replacement of current low-pressure steam boilers and turbines by high-pressure ones, and use of harvest trash currently left behind in the fields.

Presently, it is economically viable to extract about 288 MJ of electricity from the residues of one ton of sugarcane, of which about 180 MJ are used in the plant itself. Thus a medium-size distillery processing 1 million tons of sugarcane per year could sell about 5 MW of surplus electricity. At current prices, it would earn $18 million from sugar and ethanol sales, and about $1 million from surplus electricity sales. Not bad.

With advanced boiler and turbine technology, the electricity yield could be increased to 648 MJ per ton of sugarcane, but current electricity prices do not justify the necessary investment. Presently the World Bank would only finance investments in bagasse power generation if the price were at least $0.068/kWh.

In many other countries (such as Australia), sugar factories significantly contribute "green" power to the electricity supply. In the U.S., for example, Florida Crystals Corporation, one of America's largest sugar companies, owns and operates the largest biomass power plant in North America. The 140 MW facility uses bagasse and urban wood waste as fuel to generate enough energy to power its large milling and refining operations as well as supply enough renewable electricity for nearly 60,000 homes.

Researchers are also working with cellulosic ethanol, to optimize the extraction of ethanol from sugarcane bagasse and other plants viable on an industrial scale. The cellulose-rich bagasse is being widely investigated for its potential for producing commercial quantities of cellulosic ethanol. For example, Verenium Corporation is building a cellulosic ethanol plant based on cellulosic byproducts like bagasse in Jennings, Louisiana.

Bagasse is being sold for use as a fuel (replacing heavy fuel oil) in various other industries too, including citrus juice concentrate, vegetable oil, ceramics, and tire recycling. The state of São Paulo, Brazil, uses 2 million tons, saving about $35 million in fuel oil imports.

Bagasse burning is environmentally friendly compared to other fuels like oil and coal. Its ash content is only 2.5%, vs. 30-50% of coal-fired power plants, and it also contains very little sulfur. Since it burns at relatively low temperatures, it produces little nitrous oxides, which has a dual effect of reducing some of the worst pollutants (acid rain especially), and allowing to introduce ways to reduce nitrous oxides generation (which is not possible at high sulfur levels).

The resulting CO_2 emissions are equal to the amount of CO_2 that the sugarcane burnt in the power plant absorbed from the atmosphere during its growing phase, which makes the process of cogeneration greenhouse gas-neutral.

All in all, sugarcane and bagasse are shaping as an integral part of our energy future. The only problem with sugar cane is the need for a lot of water during the growing process. With increasing threat of global draughts, this might become a stumbling block for this promising energy source.

Wood Gas

Wood gas is a synthetic gas fuel which can be used to fuel furnaces and stoves. It can also be used as vehicle fuel instead of gasoline, diesel or other fuels. Wood and other biomass materials are gasified within the oxygen-limited environment of a wood gas generator to produce hydrogen and carbon monoxide.

These gases can then be burnt as a fuel within an oxygen rich environment to produce carbon dioxide, water and heat. In some cases, this process is preceded by pyrolysis, where the wood or biomass is first converted to char, releasing methane and tar rich in polycyclic aromatic hydrocarbons.

The quality of wood gas varies a great deal, depending on the raw materials and the process equipment. Staged gasifiers, for example, where pyrolysis and gasification occur separately (instead of in the same reaction zone), can be made to produce essentially tar-free gas (less than 1 mg/m³). Single-reactor fluid-bed gasifiers, on the other hand, may exceed 50,000 mg/m³ tar.

The heating value of wood is typically 15-18 MJ/kg, but these values vary from sample to sample. The heat of combustion of *producer gas* (wood gas used for car engine fuel) is rather low compared to other fuels. Producer gas has a heating value of 5.7 MJ/kg versus ten times that—55.9 MJ/kg for natural gas, and 44.1 MJ/kg for gasoline.

The chemical composition of producer gas, shown in Figure 7-48, can also vary with type of raw materials and process. We see here that its low heating value is due to low content of hydrogen and methane, which are the only combustible gasses in the mix.

Gas	Content
N_2	50.9%
CO	27.0%
H_2	14.0%
CO_2	4.5%
CH_4	3.0%
O_2	0.6%

Figure 7-48. Producer gas content (by volume)

Nevertheless, producer gas has been used in the past and a day may come soon, when it might be the only vehicular fuel left in the world.

The questions waiting for an answer today are:

- Can wood be considered a renewable energy source?

  — What should be done to ensure that it remains so, if it still is?

- Can wood be considered for use as replacement fuel for coal, oil, and natural gas?
  — What steps should be taken to make sure that it fulfills its role when the time comes?

In other words, humanity is fully determined to extract and burn the fossil fuels as quickly as possible. Since the fossils cannot be replenished, we need to think of their replacement by some other fuels. Wood seems to be an ideal replacement…if and when we find efficient and sustainable methods of doing so.

Can wood fill this need? Is it possible that people in the 22nd century and beyond will go back to using wood as their primary energy source? "History repeats itself," they say. So maybe we will make a full circle back to the 17th and 18th centuries, when wood was the only fuel.

But how would we control the deforestation? How would we make sure that the forests are not wasted in a short time? What about the great GHG pollution coming from wood burning?

We don't have the answers to these questions, so all we can do is take a close look at them and hope that we will eventually find answers.

Domestic and Agricultural Refuse

Also called garbage, domestic refuse is a waste type consisting of everyday items that are discarded by the public and end up in municipal dumps, where they rot and emit all kinds of gasses.

Wood waste is a large part of this mass. Grass, leaves, brush and branches from backyard maintenance comprise a significant amount of domestic waste, which also ends up in municipal dumps to add to the emissions.

Agricultural waste, consisting of dead vegetation and small animals, is added in large amounts to the pollution emitters on a daily basis. Manure, of course is an inevitable sign of large animal herding, and can be found in large amounts in the fields and on cow and pig farms.

Large parts of this refuse can be used for biogas generation, but some must be disposed of at municipal dump sites, landfills, and land spreading (in special circumstances).

In many cases, municipal solid waste can be used to generate energy. Several technologies have been developed that make the processing of solid waste for energy generation cleaner and more economical than ever. This includes landfill gas capture, combustion, pyrolysis, gasification, and plasma arc gasification methods.

Older waste incineration plants emit high levels of pollutants. Recently, new technologies have significantly reduced this type of pollution. For example, EPA regulations in 1995 and 2000 under the Clean Air Act reduced emissions of dioxins from waste-to-energy facilities by more than 99% from 1990 levels. At the same time mercury emissions have been reduced by over 90% as well. These improvements allowed waste-to-energy source to have less environmental impact than almost any other source of electricity.

Agricultural refuse, mostly in the form of manure and feed scraps, is organic matter that can be used as organic fertilizer in agriculture, where it contributes to the fertility of the soil by adding organic matter and nutrients.

Table 7-8. Husbandry biofuels potential

Animal	Manure		Biogas	
	Kg/day	MJ/day	m^3/day	MJ/day
Cows	40	62	1.2	26
Hens	0.19	0.9	0.18	0.6
Pigs	2.3	6.2	0.18	3.8

Table 7-8 shows that each cow produces 40 kg. manure each day, which converted into energy is 26 MJ. Pigs and hens produce much less, but their numbers are much greater, so the total amount of their daily manure production is most likely close to that of cows. This large quantity of manure can be processed in digesters to produce methane gas, in a process similar to that of producing biogas.

This is another significant energy source that is worth taking a closer look at, which in addition to providing energy, would also eliminate some environmental problems that large animal farms are creating.

Biomass Power Generation

In all cases, the organic matter in biomass products must be burnt, or processed, one way or another to produce heat or electric energy. Although the burning and processing of biomass emits carbon dioxide, it is still classified as a carbon neutral fuel—which is even better than wind and solar...in theory.

This is because of the carbon cycle, which means that while the crop grows it will absorb carbon dioxide, releasing it back into the atmosphere when burnt; or, CO_2 in + CO_2 out equals zero, or near zero.

There is only one problem: in most cases the biomass is grown in one place and burned in another. This creates zones of conflicting interests resulting in environmental damage at the burning, or processing sites.

This and other factors lead to the conclusion that biomass is actually a serious contributor to climate change. Other, somewhat unrelated, factors play a significant role, and complicate the situation. For example, biomass overuse can contribute to global warming as a result of "carbon leakage." Deforestation is one case of carbon leakage by reducing the world's total carbon absorption capacity, thus disturbing the natural equilibrium of carbon dioxide between the atmosphere, biosphere, geosphere, and hydrosphere.

There are also different types of energy use and the related pollution taking place during the planting, maintaining, harvesting, transporting, and processing of crops and biomass materials. With renewable energy sources such as solar, wind, and geothermal, the only carbon based energy used will be to manufacture, transport, and construct the system. And the amounts are fairly low, which is not the case with biomass.

So, keeping in mind that biomass is still much cleaner than the use of fossil fuels, it is not perfect in that respect, so improvements and even new alternatives must be sought. The major culprit is the use of biomass, which in all cases involves burning, which emits considerable amounts of pollutants, such as particulates and polycyclic aromatic hydrocarbons.

Modern biomass pellet-fired boilers generate much more GHG pollutants than coal, crude oil or natural gas boilers.

Pellets made from agricultural residues are usually much worse than wood pellets, producing much larger emissions of dioxins and chlorophenols.

Nevertheless, estimates show that biomass fuels (when considering the entire growing-to-burning process) have significantly less impact on the environment than fossil based fuels, due to the CO_2-in and CO_2-out concept.

Taking this into consideration, the global warming potential (GWP), which is a combination of CO_2,

methane (CH_4), and nitrous oxide (N_2O) emissions, and the energy balance of the system, need to be examined using a life cycle assessment. This takes into account the upstream processes which remain constant after CO_2 sequestration, as well as the steps required for additional power generation.

For example, *black carbon*—a pollutant created by incomplete combustion of fossil fuels and biomass—is possibly the second largest contributor to global warming. A recent study of the giant brown haze that periodically covers large areas in South Asia determined that it had been principally produced by biomass burning and to a lesser extent by fossil-fuel burning. Researchers measured a significant concentration of carbon, which is associated with recent plant life rather than with fossil fuels.

Forest-based biomass use has recently come under fire from a number of environmental organizations, including Greenpeace and the Natural Resources Defense Council, for the harmful impacts it can have on forests and the climate. Greenpeace recently released a report entitled "Fuelling a BioMess," which outlines their concerns about forest-based biomass.

Because any part of the tree can be burned, the harvesting of trees for energy production encourages whole-tree harvesting, which removes more nutrients and soil cover than regular harvesting, and can be harmful to the long-term health of the forest. In some jurisdictions, forest biomass is increasingly consisting of elements essential to functioning forest ecosystems, including standing trees, naturally disturbed forests and remains of traditional logging operations that were previously left in the forest.

Recent scientific research indicates that it can take many decades for the carbon released by burning biomass to be recaptured by re-growing trees, and even longer in low productivity areas. Also, logging operations may disturb forest soils and cause them to release stored carbon.

This means that we are in trouble now and that we have a pressing need to reduce greenhouse gas emissions now—in the short term—to mitigate the effects of climate change. Because of that, a number of environmental groups are opposing the large-scale use of forest biomass in energy production.

Finally, there is another serious issue that needs to be resolved. It is the difficulty of assigning a value to the contribution and the damage of biomass growing, harvesting, transport and processing in the different areas of the globe.

Money is paid sometimes (carbon tax) for excessive pollution and the related damages to properties and people in areas hurt by it. Is it possible, or fair, to assign monetary value to environmental damage and human suffering? These cases are no accidents, so can they be considered as planned and premeditated violation of environmental and human safety laws?

All this is to be considered in the equation of our energy security. How much can we count on natural wood, agricultural products and wastes as reliable energy sources? As we have seen, this is a complex issue with many ramifications, each of which we must understand and consider thoroughly.

Wood Power in EU

The EU countries plan to get 20% of their energy from renewable sources by 2020. Thus far, results show that the target will be missed by a mile if they rely on solar and wind alone. To meet the 2020 target, the EU is creating a new sort of energy business, that of wood power generation. Until recently, electricity from wood was on a small scale; consisting mostly of scattered waste-wood recycling operations.

Scandinavian pulp and paper mills, for example, use power stations which burn branches and sawdust residues to produce steam and electricity. Later, co-fired processes were used with marginal results.

Then in 2011, RWE, a large German utility, converted its Tilbury B power station in eastern England to run entirely on wood pellets—a common form of wood used in industrial processes. The plant, however, caught fire and the experiment died with it.

Another British company, Drax, one of Europe's largest coal-fired power stations, converted three of its six boilers to burn wood. It is scheduled to start operation in 2016, when it will generate 12.5 TWh of electric power annually. This energy is expected to get a subsidy (renewable obligation certificates), $68 for each MWh generated. This amount will be paid on top of the market price for electricity.

As a result, when operational Drax's wood-fired power plant will be getting $750 million a year in subsidies for using wood and biomass. This alone is more than its entire 2012 pretax profit of $250 million. Not bad...and if it is done right, it will be a precedent for others to follow.

The problem is that with incentives like these, many European firms are scouring the Earth for wood. Presently Europe consumed 13 million tons of wood pellets already, and with this new development, the European demand will rise to 25-30 million tons annually by 2020.

But Europe does not produce nearly that much timber, so a hefty chunk of it will come from imports of wood pellets, which have already risen by over 50%. Respectively, the global wood chips trade driven by Chinese and EU rising demand, could rise 5-6 fold from 10-12 million tons a year now to 50-60 million tons by 2020. Most of the new demand will be filled by new wood-exporting businesses that are now booming in western Canada and the American south. Here go the American forests in exports to EU and China wood-fired power plants.

This is a new industry "invented from nothing," the insiders call it. Business is brisk, because prices are going through the roof. Wood is not a commodity, officially speaking, so there is no single global price. As a result of the new developments, wood-pellet prices index published by Argus Biomass Markets rose from $152/ton in 2010 to $175/ton at the end of 2012. Prices for hardwood from western Canada rose about 60% since the end of 2011.

This new situation is putting pressure on wood producers and users. As a result of the increased demand and rising prices, 20 large sawmills making particle board for the construction industry were forced to shut down and went bankrupt in Europe during the past five years. Higher wood prices are also hurting pulp and paper companies, so the production of paper and board in Europe is now about 10% below its 2007 peak.

In Britain, a number of furniture-makers, and other wood users, complain that competition from power producers might lead to their collapse too. Their only hope is for the subsidies benefiting the power companies to be reduced or removed, which would reduce the use of wood for power generation.

Subsidizing biomass energy is seen as "an efficient way to cut carbon emissions," so these collateral damages are acceptable...if the policy were beneficial overall. We just don't know if it is or not.

The wood-to-energy processes use a lot of energy during their life cycles, which produce significant quantities of carbon. First in the supply chain is the process of making pellets out of wood, which involves grinding up the wood into a pulp to be pressurized and formed into pellets. Then it pollutes again in the power station, where it emits much more GHGs than any other fuels.

Overall calculations show about 200 kg of CO_2 are emitted by the wood (and the related processes) as needed to provide 1 MWh of electricity. So a 1 GW wood-fired power plant, would emit about 2 billion kilograms of GHGs annually—not a small amount, and something to be considered in making future plans.

Update 2014

Recent scientific studies reveal that the idea that *carbon in managed forests offsets carbon released by power stations* is an oversimplification. Carbon neutrality, actually, depends on the type of forests and raw materials used. What matters most are the local climate and environmental conditions, how fast the trees grow, and whether woodchips or whole trees are used in the power generation processes.

In 2011, the European Environment Agency stated, "…the assumption that biomass combustion would be inherently carbon neutral…is not correct…as it ignores the fact that using land to produce plants for energy typically means that this land is not producing plants for other purposes, including carbon otherwise sequestered."

Other scientists have estimated that if whole trees are used to produce energy, as they sometimes are, to simplify the process, the carbon emissions increase a lot as compared with coal, which is the dirtiest fuel by far.

The increase is quite significant and deserves a second look—about 80% over 20 years and 50% over 40 years. There is basically no carbon reduction for at least 100 years, when the replacement trees have grown up and are ready to repeat the cycle. But we're trying to reduce carbon levels now and cannot wait another 100 years.

In any case, the estimates and calculations on this matter need to be proven and verified, else we might get stuck with a process that does not reduce carbon emissions, does not encourage new energy technologies, and is basically incapable of doing the job we expect it to do.

Any miscalculations might affect our energy security by leading it in the wrong direction and at the end causing more harm than good.

BIOFUELS

Biofuels are types of fuels derived from biomass and other organic raw materials.

Biofuels can be divided into solid, liquid, and gas fuels.

- *Solid biofuels* are usually found in nature, and include biomass products (wood, sawdust, grass trimmings, domestic refuse, charcoal, agricultural waste, nonfood energy crops, and dried manure). We looked closely at some of those above, so it suffices to say that the current commercial solid biomass processes are used worldwide to make a number of products from biomass that are conve-

nient for use. These include wood pellets, cubes, or pucks, which are most common in Europe, and are typically a pure wood product processed in some form for commercial distribution and use.

Other types of densification produce products that are larger in size than a pellet, and that are compatible with a broad range of input feedstocks. The resulting densified biofuels (in the form of logs and other shapes) are easier to transport and feed into thermal generation systems, such as boilers and cooking stoves.

- *Liquid biofuels* include bioethanol, biodiesel, etc.—organic compounds in liquid form that have a high heat value. These fuels are produced from different raw biomass materials using special processes.

- *Gaseous biofuels* are often byproducts from the processing of solid or liquid biofuels. These include a number of bio-gasses that can be burnt onsite or delivered to other locations for burning.

Since we already took a look at the solid (biomass) fuels, we will now review the liquid and gas based biofuels, which can be divided into:

- *First generation biofuels* are made from the sugars and vegetable oils found in energy crops (sugarcane, soy, corn, etc.), which can be easily extracted using conventional technology.

  These are also called *conventional* biofuels, and include well-established processes that are already producing biofuels on a commercial scale. These biofuels, commonly referred to as first-generation, include sugar- and starch-based ethanol, oil-crop based biodiesel and straight vegetable oil, as well as biogas derived through anaerobic digestion. Typical feedstocks used in these processes include sugarcane and sugar beet, starch-bearing grains like corn and wheat, oil crops like rape (canola), soybean and oil palm, and in some cases animal fats and used cooking oils.

- *Second- and third-generation biofuels* are made from ligno-cellulosic biomass or woody crops, agricultural residues or waste, which makes it harder and more expensive to extract the required fuel.

  These are also called *advanced biofuels*, and are basically conversion technologies, which are still in R&D, pilot or demonstration phase.

Some of the second- or third-generation biofuels are ready for reclassification to the first generation category. These include hydro-treated vegetable oil (HVO), which is based on animal fat and plant oil, as well as biofuels based on ligno-cellulosic biomass, such as cellulosic-ethanol, biomass-to-liquids (BtL)-diesel and bio-synthetic gas (bio-SG).

This category also includes novel technologies that are mainly in the R&D and pilot stage, such as algae-based biofuels and the conversion of sugar into diesel-type biofuels using biological or chemical catalysts. The lines between the second- and third-generation technologies are fading, as the state of the art changes and provides more options.

Recently the boundaries among the biofuels of the first, second and third generation have become hazier, with many gaps and overlaps contributing to the confusion. In some cases the same fuel might be classified differently depending on technology specifics and its level of maturity, as well as its heating value, GHG emission balance, and the feedstock used in making the distinction.

In a more practical manner the biofuels could be divided into the following categories.

- Commercially available biofuels today are:
  — Ethanol from sugarcane and corn
  — Biodiesel, via trans-esterification
  — Biogas, via anaerobic digestion (biogas)

- Biofuels in early commercial stages today are:
  — Hydro-treated vegetable oils
  — Bio-methanol
  — Bio-gas reforming (H_2 gas)

- Biofuels in demonstration stages today are:
  — Cellulosic ethanol
  — BtL diesel from gasification
  — Biobutanol and DME
  — Pyrolysis-based fuels
  — Bio-SG (biogas)

- Biofuels in some type of R&D stage today are:
  — Biodiesel from microalgae,
  — Furanics (novel biofuel),
  — Hydrogen, novel bio-generation
  — Sugar based hydrocarbons
  — Gasification with reforming (H_2 gas)

Biofuels made from processing biomass are considered much cleaner than petrol/diesel alternatives. Theoretically they could be considered carbon neutral, since the biomass they were made from absorbs roughly the same amount of carbon dioxide during growth, as when burnt. While this is true, the fact that vegetation is grown in one place, biofuels are processed in another and burnt at a third, contributes to the regional inequality of global pollution. The air at the first place (the agricultural fields) is usually clean, while that at the processing plants and the burning site is badly contaminated.

On top of that, biofuels are responsible for other environmental inequalities. In many cases, large areas of forest are cut down to make space for the plantation of biofuel suitable crops. This deforestation not only harms the carbon cycle, but also harms surrounding civilizations/tribes who live off the forest.

Many environmentalists argue that biofuel is a disaster in the making, and doesn't offer a significant positive long-term environmental impact. No doubt, biofuels have some drawbacks, which—together with the benefits—must be thoroughly understood, carefully evaluated, and efficiently and safely implemented in the global energy future.

Biofuels can contribute to global warming as a result of "carbon leakage," an example of which is large-scale deforestation taking place in some areas of the world. It is a cause of carbon leakage, as we are reducing the world's carbon absorption capacity, disturbing the natural equilibrium of carbon dioxide between the atmosphere, biosphere, geosphere, and hydrosphere.

We must also take into account the energy involved during the planting, maintaining, harvesting, transporting, and manufacturing of the crops. And let's not forget the mountains of fertilizer, pesticides, and other chemicals that are used during the crop planting and growing cycles.

Water usage is another troublesome issue, especially with large-scale biofuel crops growing in desert-like areas. These crops require significant amounts of water several months at a time, year after year. Water is becoming a precious commodity in many areas, so it remains to be seen how much of it could be dedicated to growing biofuel crops.

And then, the processes involved in biofuels production are energy guzzling operations, which also emit a large amount of GHGs.

Some of these process are:

- *Hydrothermal processing*—a chemical process where biomass can be processed in a liquid media (typically water) under pressure and at temperatures between 300-400°C. The reaction yields oils and residual solids that have a low water content, and a lower oxygen content than oils from fast pyrolysis. Upgrading of the so-called "bio-crude" is similar to that of pyrolysis oil.

- *Pyrolysis oil* can be produced by fast pyrolysis, a process involving rapidly heating the biomass to temperatures between 400-600°C, followed by rapid cooling. Through this process, thermally unstable biomass compounds are converted to a liquid product. The obtained pyrolysis-oil is more suitable for long-distance transport than, for instance, straw or wood-chips.

  As a by-product, bio-char is produced that can be used as solid fuel, or applied on land as a measure of carbon sequestration and soil fertilization. The oil can be processed in ways similar to crude oil, and several research efforts are currently undertaken to upgrade pyrolysis oil to advanced biofuels.

- Dimethylether (DME) is another biofuel that can be produced from methanol through the process of catalytic dehydration or it can be produced from syngas through gasifying ligno-cellulosic and other biomass feedstocks. Production of DME from gasification of biomass is in the demonstration stage, and the first plant started production in September 2010 in Sweden (Chemrec, 2010). DME is the simplest ether and can be used as a substitute for propane in liquefied petroleum gas (LPG) used as fuel, and it is a promising fuel in diesel engines, due to its high cetane number.

- Biobutanol is used as a fuel in a number of applications, including unmodified internal combustion engines. It has a greater energy density (29.2 MJ/l) and is more similar to gasoline than ethanol, and could thus be distributed through existing gasoline infrastructure.

  Biobutanol can be produced by fermentation of sugar via the acetone-butanol-ethanol (ABE) process using bacteria such as Clastridium acetobutylicum. Demonstration plants are operating in Ger-

many and the US and others are currently under construction. Biobutanol can be produced from the same starch and sugar feedstocks that are used for conventional ethanol. In addition sugars can also be derived from lignocellulosic biomass, using the same biochemical conversion steps required for advanced ethanol production. This underlines the need for enhanced research into the biochemical conversion of biomass to sugars.

- *Solar bio-fuels* are produced by processing biomass into syngas using heat generated by a concentrating solar plant, thus potentially improving the conversion efficiency and providing higher GHG emission savings. More demonstration plants and further research are needed to make the process more efficient and allow for commercial-scale operation.

We will now take a close look at the key biofuels, remembering that some of them are well established and widely used, while others are still in R&D or small-scale production phases.

For the purposes of this text, we will review the biofuels that have the most practical value.

To start, we'll take a look at some of the biofuels currently in use:

- Bioethanol
- Biobutanol
- Biodiesel
- Bioethers
- Biogas
- Syngas, and
- Algae fuels

Bioethanol

Bioethanol, ethanol, or ethyl alcohol (C_2H_5OH) is a clear colorless liquid. It is biodegradable, low in tox-

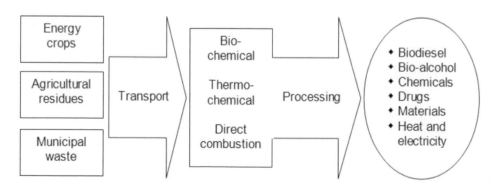

Figure 7-49. The bio-mass cradle-to-grave process

icity and causes little environmental pollution if spilt. When burning, ethanol produces carbon dioxide and water. Bioethanol is actually ethanol, but produced from bio materials; instead of from petrochemicals. It is the fuel substitute mixed with gasoline, used in the U.S. and other countries for passenger and commercial vehicles.

Ethanol is a high octane fuel and has replaced lead as an octane enhancer in petrol. Ethanol oxygenates the gasoline fuel mixture so it burns more completely and reduces polluting emissions. Ethanol fuel blends are widely sold in some states, mostly in the summer months, with the most common mix being 10% ethanol and 90% petrol, also called E10. Vehicle engines run well on this mixture, without the need of any modifications. Some hybrid vehicles can run on up to 85% ethanol and 15% gasoline blends, called E85.

In Brazil 25% ethanol is added to all gasoline mixes, resulting in E25 fuel mixture. Millions of flex-vehicles in Brazil can also run on pure ethanol, E100. As a matter of fact, Brazil has been running on ethanol since the 1970s, and why this practice is not widely implemented in other countries—including the U.S.—is a mystery, which we have been trying to solve for decades. It seems to be the trees that prevent us from seeing the forest...

Ethanol can be produced by fermenting biomass (types of vegetation and agricultural crops), or from petrochemicals by a reacting ethylene with steam. The so-called *energy crops*, are the main sources of sugar required to produce bioethanol commercially. These crops are grown specifically for bioethanol production, and include sugar cane, corn, maize and wheat crops, waste straw, willow and poplar trees, sawdust, reed canary grass, cord grasses, Jerusalem artichoke, myscanthus and sorghum plants. Solid municipal wastes are another possible source of ethanol fuel stock, but these are still in the R&D stages.

Bioethanol is produced from biomass by the hydrolysis and sugar fermentation processes of biomass wastes which contain a complex mixture of carbohydrate polymers from the plant cell walls known as cellulose, hemi cellulose and lignin. It is also produced from the energy crops, containing starch or other carbohydrates.

To break these into sugars, the raw materials are treated with acids or enzymes. This initially reduces the size of the feedstock and opens the plant structure, making it more susceptible to the fermentation process.

where
A   is kilograms of material needed to produce one liter ethanol, and

B   is square meters of land needed to produce one liter ethanol

Table 7-9. Ethanol production

Material	Ethanol Yield	
	A	B
Cassava roots	0.18	0.05 - 0.4
Maize grain	0.36	0.03 - 0.2
Sugar cane stalks	0.07	0.04 - 1.2
Sweet potato roots	0.12	0.1 - 0.5
Wood products	0.16	0.02 - 0.4

Note: Some biomass types contain mostly cellulose $(C_6H_{10}O_5)_n$, while the most used energy crops contain starch $(C_6H_{10}O_5)_n$. Note that the chemical formulas of cellulose and starch are basically the same, but that their behavior is totally different. Starch lends easily to chemical treatment when converted into sugars, while cellulose requires much more aggressive and expensive processing. This is because corn's molecule is actually one long chain of glucose molecules held together loosely, so adding enzymes is all that is needed to break the chains and separate the individual glucose molecules (sugar).

Also note in Table 7-9 that it takes much less sugar cane (0.07 kg) to produce 1 liter of ethanol, versus 0.16 kg wood that is needed for the production of the same amount of ethanol. This is mostly due to the fact that cellulosic materials (wood is composed mostly of cellulose) contain less glucose.

Cellulose is made out of similar long chains of glucose molecules, but the bonds between these are what make the big difference. The bonds (links) holding together the glucose molecules in the chain are very different in orientation and behavior, so fewer enzymes are capable of breaking down the long chains in the cellulose molecule.

Enzymes work in a *lock and key* system, where each enzyme is effective exclusively with a particular molecule, so the right enzyme is needed to build or degrade an organic molecule biologically.

Sugarcane, sugar beets and several other energy crops, on the other hand, are quite unique in that they contain *pure sugars* in their molecule, which are very easily extracted my mechanical or chemical treatment.

As it can be seen in Figure 7-50, the processing of sugar cane is much simpler and cheaper than that of cellulosic materials, such as switchgrass, which is becoming a favorite. In all cases, however, the goal is to convert the raw biomaterials into sugar—whatever it takes.

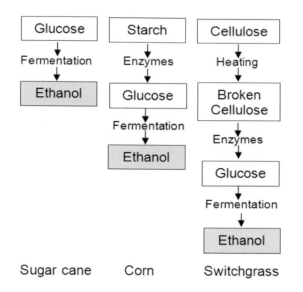

Figure 7-50. Bio-ethanol processing from different biomass materials

The cellulose and starch components are hydrolyzed (broken down) by enzymes or dilute acids into sucrose (type of sugar) in a solution. With time, heat, and yeasts the mass is then fermented into ethanol. The lignin and other waste products in the biomass are normally separated, and some are burnt as a fuel for the ethanol production plant's boilers.

There are three main commercial methods of extracting sugars from biomass and fuel crops: a) concentrated acid hydrolysis, b) dilute acid hydrolysis, and c) enzymatic hydrolysis.

Concentrated Acid Hydrolysis Process

The Arkanol process works by adding 70-77% sulfuric acid to the biomass that has been dried to a 10% moisture content. The acid is added in the ratio of 1.25 acid to 1 biomass and the temperature is controlled to 50°C. Water is then added to dilute the acid to 20-30% and the mixture is again heated to 100°C for 1 hour. The gel produced from this mixture is then pressed to release an acid sugar mixture and a chromatographic column is used to separate the acid and sugar mixture. The sugar mixture is then fermented with the help of enzymes or acids and the resulting dilute alcohol is distilled to pure bioethanol.

Dilute Acid Hydrolysis

The dilute acid hydrolysis process is one of the oldest, simplest and most efficient methods of producing ethanol from biomass. Dilute acid is used to hydrolyze the biomass to sucrose. The first stage uses 0.7% sulfuric acid at 190°C to hydrolyze the hemi cellulose present in the biomass. The second stage is optimized to yield the more resistant cellulose fraction. This is achieved by using 0.4% sulfuric acid at 215°C. The liquid hydrolates are then neutralized and recovered from the process.

Enzymatic Hydrolysis

In this process, instead of using acid to hydrolyse the biomass into sucrose, enzymes are employed to break down the biomass in a similar way. However this process is very expensive, and is still in R&D and early stage development.

Bioethanol in the U.S. is produced mostly from corn, via the following processes:

Wet Milling Process

In this process, the corn kernels are first steeped in warm water (thus the name) to break down the skin, and soften the kernel and the proteins, to release the locked-in starch. The corn is then milled in a mechanical mill, where germ, fiber and starch are produced. The germ is extracted and used to produce corn oil, while the starch fraction is centrifuged and then left for saccharifcation, which produces a wet cake of gluten material.

The ethanol is extracted from the gluten by a fractional distillation process. The wet milling process is used in large-scale bioethanol production plants, which produce hundreds of millions of gallons of ethanol every year.

Dry Milling Process

The dry milling process involves cleaning and breaking down the corn kernels into fine particles using a hammer mill process. This creates a powder with a course flour-type consistency. The powder contains the corn germ, starch and fiber. To produce a sugar solution, the mixture is then hydrolyzed or broken down into sucrose sugars using enzymes or a dilute acid.

The mixture is then cooled and yeast is added to ferment the mixture into ethanol. The dry milling process is used in lower volume factories, which on average produce less than 50 million gallons of ethanol a year.

Bioethanol from sugarcane is the easiest to process, since the sugarcane contains free sugars, which only need to be squeezed out from the stalks and fermented into alcohol.

Sugar Fermentation Process

The hydrolysis process breaks down the cellulosic part of the biomass or corn into sugar solutions that can then be fermented into ethanol. Yeast is added to the solution, which is then heated. The yeast contains an enzyme called invertase, which acts as a catalyst and helps to convert the sucrose sugars into glucose and fructose (both $C_6H_{12}O_6$).

The chemical reaction is:

$$C_{12}H_{22}O_{11} + H_2O = C_6H_{12}O_6 + C_6H_{12}O_6$$

Sucrose   Water   Fructose   Glucose

The fructose and glucose are produced in the presence of invertase (a catalyst). These sugars then react with another enzyme called zymase (also contained in the yeast) to produce ethanol and carbon dioxide.

The chemical reaction of this process is:

$$C_6H_{12}O_6 = 2C_2H_5OH + 2CO_2$$

Sugars   Ethanol

The fermentation process takes around three days to complete and is carried out at a temperature of between 250°C and 300°C. All we have to do after that is separate the alcohol from the mixture.

Fractional Distillation Process

The ethanol, which is produced from the fermentation process described above, still contains a significant quantity of water, which must be removed. This is achieved by using the fractional distillation process. The distillation process works by boiling the water and ethanol mixture. Since ethanol has a lower boiling point (78.3°C) compared to that of water (100°C), the ethanol turns into the vapor state before the water and can be condensed and separated.

Since bioethanol is a primary source of biofuels in North America, many organizations are conducting research in that area. The National Corn-to-Ethanol Research Center (NCERC) is a research division of Southern Illinois University Edwardsville dedicated solely to ethanol-based biofuel research projects. On the federal level, the USDA conducts a large amount of research regarding ethanol production in the United States. Much of this research is targeted toward the effect of ethanol production on domestic food markets. A division of the U.S. Department of Energy, the National Renewable Energy Laboratory (NREL), has also conducted various ethanol research projects, mainly in the area of cellulosic ethanol.

In More Detail

Glucose (a simple sugar) is created in the vegetation by photosynthesis.

$$6CO_2 + 6H_2O + light \rightarrow C_6H_{12}O_6 + 6O_2$$

During ethanol fermentation, glucose is decomposed into ethanol and carbon dioxide.

$$C_6H_{12}O_6 \rightarrow 2C_2H_5OH + 2CO_2 + heat$$

During combustion ethanol reacts with oxygen to produce carbon dioxide, water, and heat.

$$C_2H_5OH + 3O_2 \rightarrow 2CO_2 + 3H_2O + heat$$

After doubling the combustion reaction because two molecules of ethanol are produced for each glucose molecule, and adding all three reactions together, there are equal numbers of each type of atom on each side of the equation, and the net reaction for the overall production and consumption of ethanol is just: light → heat

The heat of the combustion of ethanol is used to drive the piston in the engine by expanding heated gases. It can be said that sunlight is used to run the engine (as is the case with any combustion-based energy source, as sunlight is the only way energy is added to the planet, and geothermal energy, which comes from the heat already present inside the earth).

Glucose itself is not the only substance in the plant that is fermented. The simple sugar fructose also undergoes fermentation. Three other compounds in the plant can be fermented after breaking them up by hydrolysis into the glucose or fructose molecules that compose them. Starch and cellulose are molecules that are strings of glucose molecules, and sucrose (ordinary table sugar) is a molecule of glucose bonded to a molecule of fructose. The energy to create fructose in the plant ultimately comes from the metabolism of glucose created by photosynthesis, and so sunlight also provides the energy generated by the fermentation of these other molecules.

Ethanol may also be produced industrially from ethylene. The addition of water to the double bond converts ethene to ethanol:

$$C_2H_4 + H_2O \rightarrow C_2H_5OH$$

This is done in the presence of an acid which catalyzes the reaction, but is not consumed. The ethene is produced from petroleum by steam cracking.

When burning in a pure oxygen environment, ethanol produces large amounts of CO_2 and H_2O.

$$C_2H_5OH + 3O_2 \rightarrow 2CO_2 + 3H_2O$$

However, when ethanol is burned in the atmosphere rather than in pure oxygen, other, much different and dangerous chemical reactions take place, since there are different gasses in the atmospheric air, such as nitrogen (N_2).

$$C_2H_5OH + 5O_2 + N_2 \rightarrow 2CO_2 + 2NO_2 + 3H_2O$$

During burning in this air mixture, ethanol produces lots of nitrous oxides—also a major air pollutant—300 times more dangerous than CO_2. This makes burning large quantities of alcohol a very dangerous affair, as far as the environment and global warming in particular are concerned. It appears also that equal amounts of CO_2 and NO_2 gasses are generated during the bioethanol burning process, which gasses are known to produce unwanted environmental effects.

Example: Starting with, say 100 lbs. of starch or cellulosic materials, we get about 120 lbs. of glucose after adding the enzymes to the mix. After fermentation, we measure about 60 lbs. of bioethanol and 60 lbs. of CO_2.

Note that there is almost as much CO_2 produced during this process as there is ethanol. Still, the process is considered CO_2 neutral, because the CO_2 released during the conversion process was actually absorbed by the plants during their growth in the fields.

But since the nitrogen in the air is also oxidized during the ethanol burning process, we now have 30 lbs. of CO_2 and 30 lbs. of NO_2. Considering the fact that NO_2 is 300 times more damaging than CO_2, we arrive at the conclusion that thus burned alcohol does at least 150 times more damage than burning equal amounts of other fuels that produce only CO_2. Not a good thing…

Another problem, related to environmental issues, that we need to consider is the actual *location* of the different operations. The CO_2 in question was absorbed in the fields where the biomass was grown, which usually are several hundred miles away from the processing plants. During their growth, the plants absorbed a lot of CO_2, while at the same time releasing a lot of oxygen, thus making the air clean and fresh in that location. Of course, there is no trace of NO_2 in these fields.

Fast forwarding to the place where bioethanol is burned, most likely by cars in large populated centers, where the entire amount of CO_2 stored in the biomass (plus equal amounts of NO_2) are released, both of which pollute the air at that location. If you happen to be in Los Angeles, or Beijing, on a late summer afternoon, you'll see and smell a lot of these polluting gasses in the air. At times you cannot see more than a few feet ahead, or take a good breath, from the dense smog in these places,

some of which is caused by ethanol burning.

The effects of this local gas pollution disequilibrium are open to debate, as is the entire effect of CO_2 on the global environment, but it is a proven fact that some areas have totally different air quality than others.

It is possible to minimize the CO_2 and NO_2 emissions, and significantly improve the greenhouse gas profile of ethanol, but we need to be aware of, and deal squarely with the issues at hand. Basically, we have a good thing in bioethanol fuels that comes with some bad consequences, which we just need to take a close look at and deal with properly and efficiently.

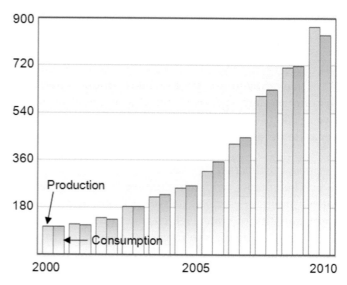

Figure 7-51. U.S. Bio-ethanol production and use (in thousand barrels/day)

Case study 1. Sacramento Bio-ethanol Production and Power Cogeneration Plant.

Sacramento ethanol and power generation plant was one of the first attempts for large-scale bio-ethanol production from cellulosic materials in the U.S. It was initiated around 1992, but it took 2 years to obtain a construction permit. At the time, it was not only the first commercial cellulosic biomass-to-ethanol plant, but also was a key part of the solution to the rice straw disposal problem facing California's rice growing industry in the face of regulations banning most field burning of such agricultural residues. This process would eliminate the post-harvest open field burning of rice straw on some 40,000 acres, approximately 15 percent of the total Sacramento Valley rice under cultivation.

Win-win situation, no doubt…at least on paper. In practice things are much more difficult—from political, regulatory, and technical points of view. The plant was to be built on 90 acres land in order to convert over 400 tons per day of rice straw and other cellulosic agricultural residue into approximately 35,000 gallons per day of fuel grade ethanol, using patented concentrated acid hydrolysis technology. This is 12 million gallons bio-ethanol obtained by processing 132,000 tons of rice-straw per year. Thus produced bio-ethanol would be used in tandem with natural gas to generate about 150 MW electricity as a base-load power plant.

The plant was essentially two separate yet linked projects with different owners united by a contractual arrangement, sharing a site and various operational synergies, including process heat and power supplied to the ethanol plant by the cogeneration plant, shared water supply and waste disposal provisions, etc. It's complicated…

Partnership issues flared up, delayed the project and finally caused the permit to expire. Five years later the owner applied for a new permit, which was granted in 2000. The struggles continued, but the plant was never built.

The plant construction did not go forward mostly due to complexities with the joint venture and multiple parties, who could not divide the pie to everyone's satisfaction. The technological approach was also unproven and somewhat too advanced for the time, which complicated things further.

Nevertheless, there were a number of lessons (albeit expensive) learned from this experience. It was an early test case of the California regulatory process for permitting a bio-refinery facility, where bio-alcohol production would be combined with electricity generation. As such, the project could be considered a partial success.

Some of the successfully executed tasks and lessons learned were:

- Extensive environmental analysis was conducted and complex mitigation measures considered on a full range of issues, including air quality, water supply and water quality, hydrology, and biological resources. Issuance of an air quality permit for the entire project was based on emission offsets to be obtained via the discontinuation of rice straw burning resulting from use of rice straw as the ethanol plant feedstock.

- Flood plain concerns resulted in modifications to the facility site plan. Original plans to use groundwater wells were changed to use of surface water; water supply arrangements included mitigation measures at the Sacramento River water intake to protect salmon. Various other mitigation measures were adopted involving several different endangered species found on the site.

- The unique features of the project, combining rice straw to ethanol production and electricity cogeneration, posed a number of considerations not previously encountered in the California regulatory proceedings.

- Reliability, or lack thereof, of the unproven cellulosic ethanol production process affected both the cogeneration performance and emission offset viability of the power plant. Various issues associated with the feedstock supply plan based on the yet-to-be-demonstrated use of rice straw were addressed as well.

Case study 2. Gridley Ethanol Plant

The Gridley Ethanol Project was another attempt, designed as a potential solution to the rice straw disposal problem in the Sacramento Valley region of California, which became acute with legislative mandates to significantly reduce the amount of rice straw burning after the fall rice harvest.

The Rice Straw Burning Reduction Act of 1991 (AB 1378) mandated a reduction in rice straw burning by the year 2000 to no more than 25% of the planted acreage. The California rice straw burning phase down has proceeded as required by the statute, with growers burning less than the statutory limitations. Other open-field burning laws and regulations further limit the actual rice straw acreage burned annually.

Despite the ongoing reduction of rice straw burning, no alternative market or disposal option sufficient to handle the quantities of rice straw being produced was available at the time. As a result, very large volumes of this material continue to accumulate without a viable market alternative to dispose of the rice straw. This could eventually render useless thousands of acres of rice lands, since in these hard clay-pan soils, no other crops have been successful. Production of bio-ethanol from rice straw was seen as a potential solution.

The original concept of the Gridley plant involved application of an unproven enzymatic hydrolysis process to produce ethanol. Lignin remaining from the hydrolysis process was to be utilized as combustion fuel for firing the facility's boiler for the production of steam and electricity to be used on site, with excess steam potentially used by adjacent facilities. Excess electricity would be supplied to the municipal utility and/or sold to the grid.

The actual Gridley plant planning began also in 1994 and was authorized in 1996 by NREL. Phase 1 started with initial screening of the technical and economic feasibility of a commercial rice straw-to-ethanol facility in the Gridley area. Phase II was to acquire financial and site commitments, perform pilot plant studies of the technology at NREL, prepare a preliminary engineering package, evaluate the economics and risks, and finally to prepare an implementation plan to commercialize the process. Phase II was to lead to a "go/no go" decision regarding the construction of the plants.

Here again, there were many partners, and sure enough, in 1997 the original conversion technology developer withdrew from the project. Since Phase I tasks had been completed and a rice straw-to-ethanol facility appeared feasible, a new partner was chosen to provide the conversion technology, which was basically also acid hydrolysis and fermentation, with lignin as a co-product.

During the progress of Phases I and II, it was determined that project economics with the then-current state of conversion technology would be enhanced by converting the bio-ethanol production plant into a cogeneration facility. It was then to be sited next to an existing biomass power plant in the region, which uses orchard prunings and forest wastes as feedstock. This co-location would reduce the costs and improve the efficiency of both the power plant and the proposed ethanol facility.

Construction of the new plant was projected to commence in early 2002 with operations to begin in late 2003. The collection and processing of rice straw became a paramount consideration, since infrastructure to harvest rice straw for use in the plant was virtually nonexistent. Processing of the rice straw for use as feedstock (i.e., grinding) presented technical challenges due to the high silica content of rice straw.

Rice straw supply studies indicated that the rice straw would cost over $30.00/bone dry ton (BDT) to be delivered to the facility. This did not include the grinding and processing of the rice straw at the facility. To produce the 23 million gallons of ethanol would require 300,000 dry tons of rice straw (some of which could be provided by orchard and forest wood wastes).

On top of that, there were environmental permitting and impact assessment studies that indicated some potentially higher costs than originally anticipated. Wastewater from the plant would have to be discharged to the local municipal wastewater treatment plant, which alone would cost several million dollars. Also, in order to discharge to the wastewater plant, an expensive

wastewater pretreatment adding several million dollars to the operating cost, would be necessary. ...*and* additional air emission control equipment would be needed to complete the environmental safety, which was not previously anticipated.

This, combined with the technical uncertainties of the two-stage dilute sulfuric acid conversion technology, led to a conclusion that the acid hydrolysis technology was not financially viable for use. A decision was made to investigate the use of a gasification technology to create syngas that could be converted to ethanol or other fuels.

This evaluation, done in 2002, indicated that switching from the dilute sulfuric acid process to a gasification process could have a number of advantages, such as a) increased yields of ethanol, with associated reductions in feedstock and other operating costs per gallon of ethanol produced, b) lower capital investment cost, c) fewer air emissions and wastewater effluents, and d) reduced feedstock requirements, which better fit the initial needs of the area for disposing of a critical mass of rice straw.

The plant was also to be moved again to its original location, due to a new industrial site availability, shorter transportation hauling distances from the rice fields, significantly reduced wastewater disposal costs and available infrastructure to better support the proposed facility.

NREL continued to finance and support the plant and its new gasification technology. A pilot facility was used for a proof-of-concept. The results were encouraging, and the Gridley plant was able to get funding from the U.S. Department of Energy. In December 2006 the plant management issued a request for proposals to construct and operate a thermo-chemical conversion system using rice straw to produce bio-ethanol and electricity.

The plant was awarded also a CEC grant in April 2007 to demonstrate an integrated biofuels and energy production system. The project was geared to support the construction, demonstration and validation of a cost effective and energy efficient biomass conversion system. But the conflicts among the partners, shifting locations, and changing processing technologies took their toll and, when the farmers refused to participate due to high transportation costs, the plant was put on hold...indefinitely...after millions of taxpayer dollars were spent in the process.

In summary, there were a number of failed bio-ethanol conversion plants in the U.S. during the last 2-3 decades too, but that didn't stop their progress. In the summer of 2013 there were 211 bio-ethanol operational plants in the U.S., producing the grand total of 13.5 billion gallons of bio-ethanol every year, mostly from corn.

What the future holds for bio-ethanol is anyone's guess, but we venture say that it won't go much farther for awhile—and while natural gas prices are so low. It will, however, be the fuel of choice for electricity generation and transportation in the distant future, when oil and natural gas prices are high enough and the fossils are depleted. Make no mistake about it!

Biobutanol

Biobutanol is a type of a 4-carbon alcohol that is produced from a number of biomass feedstocks via fermentation. A large variety of biomass types can be used in this process: corn grain, corn stovers, and many other feedstocks. As in other processes, these are processed into sugars, and the special microbes of the *Clostridium acetobutylicum* species, are introduced to the sugars, which are then broken down into various alcohols, including butanol.

Biobutanol can be produced by fermentation of sugar via the acetone-butanol-ethanol (ABE) process using bacteria such as *Clastridium acetobutylicum*. It can also be made using Ralstonia eutropha H16, a process requiring the use of an electro-bioreactor, and the additional input of carbon dioxide and electricity.

The difference between biobutanol and ethanol production is primarily in the fermentation of the feedstock and minor changes in the distillation setup and process parameters. The feedstocks are the same as for ethanol: energy crops such as sugar beets, sugar cane, corn grain, wheat and cassava, prospective non-food energy crops such as switchgrass and even guayule in North America, as well as agricultural byproducts such as straw and cornstalks.

In addition, sugars can be derived from ligno-cellulosic biomass, using the same biochemical conversion steps required for advanced ethanol production. According to industry experts, existing bioethanol plants can cost-effectively be retrofitted to biobutanol production.

Unfortunately, high alcohol concentration makes the butanol mixture toxic to these microorganisms. This condition made the fermentation process expensive and impractical when compared to petroleum costs.

New technological advances have improved the efficiency and reduced the cost of the fermentation process. Today, genetically engineered processes are making it possible for the most efficient microbes to withstand even the highest alcohol concentrations. This allows large quantities of biobutanol to be produced commercially.

As with the case of bioethanol, biobutanol can be prepared easier and directly from a ready source of sugar, such as sugarcane, but production from crop wastes and energy grass is possible too. It can also be made entirely with solar energy, from algae, called Solalgal Fuel, or special diatom materials.

Butanol has a total of four different isomers, but only three are used commercially: n-butanol, isobutanol, tertbutanol. They all have multiple uses in industrial and consumer products. The market for n-butanol and isobutanol is over 7 billion lbs. annually, and the producers of biobutanol have a captive market already.

N-butanol finds applications as production intermediate for a number of chemicals, such as butyl acrylate, butyl acetate, dibutyl phthalate, and also as an extractant for antibiotics, hormones, vitamins. It is also an ingredient in perfumes, degreasers, repellents, and cleaning solutions

Iso-butanol finds applications as a paint solvent, ink ingredient, gasoline additive, derivative ester precursor, viscosity reducer in paint, automotive polish, and paint cleaner additive.

Tert-butanol finds application as a perfume ingredient, gasoline octane booster, paint remover ingredient, solvent, as well as a synthesis intermediate of methyl tert-butyl ether (MTBE), ethyl tert-butyl ether (ETBE), and tert-butyl hydroperoxide (TBHP).

Biobutanol has a greater energy density (29.2 MJ/l) and is more similar to gasoline than ethanol, so it is used in internal combustion engines, primarily as a gasoline additive, or a fuel blend with gasoline. The energy content of biobutanol is only 10% less than that of regular gasoline, while the energy density of ethanol is over 40% lower. Biobutanol is also more chemically similar to gasoline than ethanol, so it can be integrated into regular internal combustion engines much easier than ethanol.

Biobutanol has the potential to reduce carbon emissions by 85% as compared to gasoline, which makes it a superior alternative to gasoline and to the gasoline-ethanol blended fuels. It also can be produced from feedstocks which do not compete with food. For example, algae biomass and waste wood particles can be converted to biobutanol, with the advantage that some of these require only a 10th of the land resource needed to grow corn.

Nevertheless, most of these achievements are in small-scale or R&D lab settings for now. Because of that, the commercial future of biobutanol is…still in the future.

Fuel	Energy density	Air-fuel ratio	Specific energy
Gasoline and biogasoline	32 MJ/L	14.6	2.9 MJ/kg air
Butanol fuel	29.2 MJ/L	11.1	3.2 MJ/kg air
Ethanol fuel	19.6 MJ/L	9.0	3.0 MJ/kg air
Methanol fuel	16 MJ/L	6.4	3.1 MJ/kg air

Table 7-10. Energy content of key fuels

Biodiesel

Biodiesel is a substance of pure or somewhat modified vegetable oil, or animal-based fat, that is used to power diesel engines. It has a technical definition of *mono-alkyl* (methyl, propyl, or ethyl) *ester* of long-chain fatty acids, which are produced from vegetable oils or animal fats. Soybean methyl ester is one of the pure biodiesel varieties. It is made from soybeans, has an average molecular weight of 292.2 and is comprised of: Palmitic acid $C_{15}H_{31}CO_2CH_3$, Stearic acid $C_{17}H_{35}CO_2CH_3$, Oleic acid $C_{17}H_{33}CO_2CH_3$, Linoleic acid $CH_3(CH_2)_4CH=CH-CH_2CH=CH(CH_2)_7 CO_2CH_3$, and Linoleic acid $CH_3(CH_2CH=CH)_3(CH_2)_7 CO_2CH_2$.

Since biodiesel is used to power expensive diesel engines in passenger and commercial cars and trucks, it must meet the strictest quality specifications of ASTM D 6751, and be compatible in any blend with petroleum diesel fuels.

Biodiesel is typically made by chemically reacting (trans-esterification) of vegetable oil or animal fat feed-

stock, lipids, vegetable oil, and animal fat (tallow) with alcohol producing fatty acid esters.

Recycled oil is processed to remove impurities such as dirt, charred food, and water from cooking, storage, and handling. Virgin oils are refined to commercial grade purity (not to a food-grade level.) A degumming process step is used when needed to remove phospholipids and other plant matter via different refinement processes.

Excess water is removed to prevent triglycerides from hydrolysis during base-catalyzed trans-esterification process, thus preventing the formation of salts of the fatty acids (soaps) instead of producing biodiesel. The acids present in the oil are either esterified into biodiesel, esterified into glycerides, or removed through neutralization with bases.

Trans-esterified biodiesel is a mix of mono-alkyl esters of long-chain fatty acids, with the most common form being methanol converted to sodium methoxide to produce methyl esters. These are commonly referred to as fatty acid methyl ester, or FAME.

Methanol is used to produce fatty acid ethyl ester, or FAEE biodiesel, but other alcohols such as isopropanol and butanol can also be used. Using alcohols of higher molecular weights improves the cold flow properties of the resulting ester, at the cost of a less efficient trans-esterification reaction.

A lipid trans-esterification production process is used to convert the base oil to the desired esters. Any free fatty acids (FFAs) in the base oil are either; converted to soap and removed from the process, or they are esterified, which yields more biodiesel, via an acidic catalyst. After this processing, unlike straight vegetable oil, biodiesel has combustion properties very similar to those of petroleum diesel, and can replace it in most current uses.

Glycerol is a by-product of the trans-esterification process, where 1 ton of biodiesel also produces 100 kg of glycerol. Crude glycerol contains 20% water and catalyst residues and has no practical use today. Research is underway to find use for it as a chemical building block in some products, such as epoxy resins.

Biodiesel is considered environmentally friendly, because its ozone (smog) forming potential, as well as its CO, particulate, and total hydrocarbons content are about 50% less than that of conventional diesel fuels. Sulfur emissions are essentially eliminated with pure biodiesel, and so are the unburned hydrocarbons, carbon monoxide, and particulate matter, all of which are typical for standard diesel fuels.

The bad news is that NO_x emissions from B100 bio-

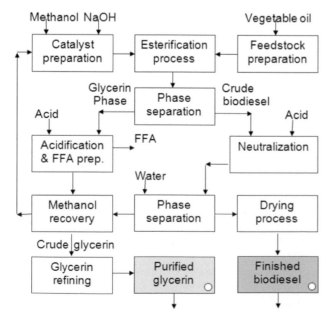

Figure 7-52. Biodiesel production process

diesel are 10% higher than those from standard diesel, which—as we saw above—is 300 times worse for the environment than the same amount of CO_2. This serious issue needs to be resolved, if biodiesel is to become accepted as an environmentally safe alternative to petrodiesel.

The good news is that biodiesel's lack of sulfur content allows the use of NO_x control technologies that cannot be used with conventional diesel engines. There are also some additives developed lately that can reduce NO_x emissions in biodiesel blends.

Basically, biodiesels reduce the health risks associated with standard diesel fuels. Biodiesel emissions show decreased levels of polycyclic aromatic hydrocarbons (PAH) and nitrated polycyclic aromatic hydrocarbons (nPAH), which have been identified as potential cancer causing compounds. In health effects testing, PAH compounds were reduced by 75-85%, while benzo(a)anthracene was reduced by roughly 50%. Targeted nPAH compounds were also reduced dramatically with biodiesel, with 2-nitrofluorene and 1-nitropyrene reduced by 90%, while the rest of the nPAH compounds are reduced to minute (trace) levels.

Biodiesel is extensively used in European and other countries around the world. Pure biodiesel, used in standard diesel engines, is different from waste oils (used for cooking) that are sometimes used to fuel *converted* diesel engines. Note the term *converted*, because using waste vegetable oil in a standard diesel engine would cause damage that might require expensive repairs.

Biodiesel can be used pure, or blended with petrodiesel. It can be also used as a low carbon alternative to heating oil. Blends of biodiesel and conventional hydrocarbon-based diesel are most commonly used in the retail diesel market.

The "B" factor system is usually used to reflect the amount of biodiesel in the respective fuel mix, as follows: 100% biodiesel is labeled B100, 20% biodiesel in 80% petrodiesel is labeled B20, 5% biodiesel in 95% petrodiesel is labeled B5, and 2% biodiesel in 98% petrodiesel is labeled B2.

Blends of 20% biodiesel and lower are used in standard diesel equipment with only minor modifications, but in some cases it can violate the manufacturer's warranty. Biodiesel used in its pure form (B100) usually requires major engine modifications.

Advanced Biodiesel

Several processes are under development that aim to produce fuels with properties very similar to diesel and kerosene. These fuels will be blendable with fossil

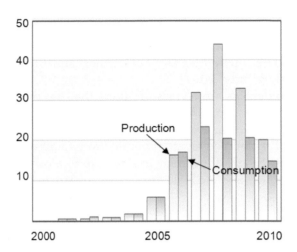

Figure 7-53. U.S. Biodiesel production and use (in thousand barrels/day

fuels in any proportion, can use the same infrastructure and should be fully compatible with engines in heavy duty vehicles. Advanced biodiesel and bio-kerosene will become increasingly important to reach this roadmap's targets since demand for low-carbon fuels with high energy density is expected to increase significantly in the long term.

Advanced biodiesel includes:

- Hydrotreated vegetable oil (HVO) is produced by hydrogenating vegetable oils or animal fats. The first large-scale plants have been opened in Finland and Singapore, but the process has not yet been fully commercialized.

- Biomass-to-liquids (BtL) diesel, also referred to as Fischer-Tropsch diesel, is produced by a two-step process in which biomass is converted to a syngas rich in hydrogen and carbon monoxide. After cleaning, the syngas is catalytically converted through Fischer-Tropsch (FT) synthesis into a broad range of hydrocarbon liquids, including synthetic diesel and bio-kerosene.

Advanced biodiesel is not widely available at present, but could become fully commercialized in the near future, since a number of producers have pilot and demonstration projects underway.

Biodiesel has great potential as a plentiful, clean and cheap fuel. With more improvements it might be one of the energy sources that will carry us through the energy transition gap of the 21st and 22nd centuries.

Bioethers

Bio-ether is a derivative from bioethanol, which

is obtained from the distillation of energy crops and sugar beet. The best known and widely used fuel ethers are MTBE (methyl-tertiary-butyl-ether) and ETBE (ethyl-tertiary-butyl-ether).

Bioethers are a class of ethers that are produced from biomass materials, energy crops, and such. They are classified as organic compounds that are characterized by an oxygen atom bonded to two alkyl or aryl groups. Ethers are similar in structure to alcohols, and both ethers and alcohols are similar in structure to water.

Figure 7-54. Chemical formulas of basic fuels

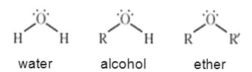

The difference here is that in the alcohol molecule, one hydrogen atom of a water molecule is replaced by an alkyl group, while in the ether molecule, both hydrogen atoms are replaced by either an alkyl or aryl group.

In the presence of acid, two molecules of an alcohol may lose water to form an ether. In practice, however, this bimolecular dehydration to form an ether competes with uni-molecular dehydration to give an alkene. Bimolecular dehydration produces useful yields of ethers only with simple, primary alkyl groups such as those in dimethyl ether and diethyl ether. Dehydration is used commercially to produce diethyl ether.

The most practical method for making ethers is the Williamson ether synthesis, which uses an alkoxide ion to attack an alkyl halide, substituting the alkoxy (−O−R) group for the halide. The alkyl halide must be unhindered (usually primary), or elimination will compete with the desired substitution.

Fuel ethers can be produced from a mixture of both petrochemical and agricultural feedstocks. In all cases, the building blocks for fuel ethers are isobutylene or isoamylenes compounds, reacted with methanol or ethanol.

A complete bio-chemical process involves using *bio-ethanol*, which is derived by a fermenting process from wheat, sugar beet and other agricultural products, and is the major feedstock for the production of ETBE or TAEE (tert-amyl-ethyl-ether).

Bio-methanol, which is also derived from biomass, is the second feedstock used in the production of MTBE or TAME (tert-amyl-methyl-ether).

Isobutylene is yet another feedstock used in both

MTBE and ETBE production, but it is derived from fossils, natural gas, or as a byproduct of petroleum refining.

Similarly, isoamylenes used in the production of TAME or TAEE are byproducts of petroleum refining.

Bioether production facilities are typically located close to a refinery with a fluid catalytic cracker unit, or in a chemical plant with a steam cracker. Large-scale "stand-alone" units are also in operation lately, but these are based mostly on fossils, using either butane isomerisation/dehydrogenation technology (where both the butane and the methanol are derived from natural gas), or by dehydration of tertiary butyl alcohols.

Figure 7-55. Bio-ether production scheme

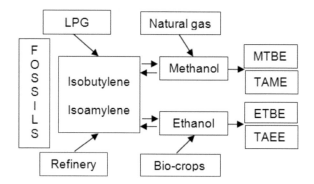

Ethers are pleasant-smelling, colorless liquids. They are less dense than alcohol and are less soluble in water. Ethers are relatively unreactive and have lower boiling points than their cousins, the alcohols. They are used widely as solvents for fats, oils, waxes, perfumes, resins, dyes, gums, and many hydrocarbons. Vapors of certain ethers are used as insecticides, miticides, and fumigants for soil.

Ethyl ether is used as a volatile starting fluid for diesel engines and gasoline engines in cold weather. Dimethyl ether is used as a spray propellant and refrigerant. Methyl *t*-butyl ether (MTBE) is a gasoline additive that boosts the octane number and reduces the amount of nitrogen-oxide pollutants in the exhaust. Its chemical formula is CH_3CH_2—O—CH_3CH_3.

Bioethers are one of the biofuels of choice today in Europe, with over three quarters of all bioethanol being used as bio-ether (ETBE). European energy policy is promoting the use of bio-fuels for transportation. Bioethers and bioalcohols are used as blending agents for enhancing the octane number. By changing the properties of common fuels, they make gasoline work harder, help the engine last longer, and reduce air pollution. Development of renewable fuels needs both knowledge of new thermodynamic data and improvement of clean energy technologies. In this context, the use of ethanol of vegetable origin in its manufacture process increases the interest of ETBE or bio-ETBE as an oxygenated additive.

Used Cooking Oil

There is a lot of oil—of vegetable and animal origin—used for cooking in restaurants and fast-food chains worldwide. Options for disposal of used cooking oil and grease are limited in most places. Disposal is difficult, because used cooking oil is usually in a liquid, or semi-solid, form, and most waste disposal regulations restrict the disposal of liquids in landfills.

Other disposal methods can also be problematic, for example: open burning of used cooking oil causes black smoke, which is prohibited. Using it as fuel in most standard heating systems will cause black smoke and soot too, and may even damage the system. Pouring used cooking oil down the drain can clog pipes and damage wastewater or septic systems, so it is a major no-no.

In addition to using it in pet food preparations, one of the best ways of disposing of and actually using cooking oil is to burn it in some approved incinerators. Another way is to use it as a fuel in modified diesel engines. Since the modifications and the supply might be questionable, the best way is to refine it into biodiesel for use in standard diesel engines.

For example, McDonald's UK is recycling 100% of its used cooking oil for biodiesel to be used as fuel in fuel delivery trucks. This is an emission savings of more than 3,500 tons of CO_2, or the equivalent of 1,500 cars being removed from the road each year.

In the USA, McDonald's is implementing a bulk cooking oil delivery and retrieval program, which includes collection of waste cooking oil in a separate tank, which is periodically taken back to a distribution facility where it is sold for re-use to a variety of vendors, including biodiesel companies. Nearly 9,000 U.S. McDonald's restaurants are enrolled in this program, and additional franchises are being converted for this service.

This means that the average participating U.S. restaurant recycles nearly 1,500 gallons of used cooking oil per year, which together with eliminating large amounts of plastic and corrugated paper packaging, this eliminates the pollution created during the manufacture of the packaging, and keeps additional packaging waste from going into the landfill.

Today there are small, portable, biodiesel generators which can process used cooking oil into biodiesel. These can be used at home, or commercially, and can produce 20 to 200 gallons of biodiesel per day.

Additionally, there are recipes for home-made biodiesel, in which vegetable oil (including used cooking oil) is mixed with sodium hydroxide and methanol in a blender. After blending, the mixture is left to separate, and the top layer can be used as biodiesel. It's easy to do, but the quality of the biodiesel is questionable, and this process cannot be used for large-scale fuel production.

One problem is that biodiesel has a limited shelf life. Some oils contain the antioxidant tocopherol, or vitamin E (e.g., rapeseed oil), and they remain usable longer than biodiesel or other types of vegetable oils. Biodiesel's stability can diminish after a week or two, and is usually unusable after 1-2 months. Higher temperature also affects negatively the fuel stability by denaturing it.

Biogas

Biogas is a type of gas-fuel produced by the breakdown of organic matter (biomass and other organic materials) in the absence of oxygen. It is considered a renewable energy source, which can be produced from regionally available raw materials and recycled vegetation and animal waste, all of which makes it environmentally friendly and CO_2 neutral.

Anaerobic digestion or fermentation of biodegradable materials used in the bio-gas production such as manure, sewage, municipal waste, green waste, plant material, and many crops, also has the advantage of reducing the overall amount of final waste from these materials. Biogas generated from digestion is comprised primarily of ~60% methane (CH_4), 30% carbon dioxide (CO_2), with small amounts of nitrogen (N_2), hydrogen (H_2), oxygen (O_2), hydrogen sulfide (H_2S), moisture, and siloxanes.

Biogas is most often produced as landfill gas (LFG) or digested gas in *biogas plants*. These are simple anaerobic (oxygen-free) digesters, designed to treat farm wastes or energy crops, such as maize silage. Some can process biodegradable wastes including sewage sludge and food waste into biogas as well.

The process requires an air-tight tank, where the biomass waste is *digested*, or transferred, into methane gas, thus producing renewable energy that can be used for heating, electricity, and many other operations, including fuel for car and truck engines.

Large-scale *landfill gas* is produced by wet organic waste decomposing under anaerobic conditions, where it is covered and mechanically compressed by the weight of the material that is deposited above. This material prevents oxygen exposure thus allowing anaerobic microbes to thrive, and the gas builds up with time. It is slowly released into the atmosphere if the landfill site has not been engineered to capture the gas, or in containers for later use if the design allows.

Table 7-11. Approximate content of household wastes

Material	Typical weight %
Ash	25
Carbon	25
Water	20
Oxygen	18
Hydrogen	3
Chlorine	0.7
Nitrogen	0.6
Sulfur	0.003

Landfill gas is hazardous because:

1. It is explosive when mixed with oxygen. Even a small amount, about 5% methane, can create an explosion.

2. Methane from landfill gas is 20 times more potent as a greenhouse gas than carbon dioxide, so if landfill gas is allowed to escape into the atmosphere, it contributes to global warming.

3. Landfill gas contains volatile organic compounds (VOCs) which contribute to the formation of photochemical smog—another bad environmental polluter.

In most cases, thus produced gas mixture is not good enough for use as fuel gas for car engines and other machinery; i.e., H_2S in it is corrosive enough to destroy the internal components of a car engine or power plant. Because of that, thus generated raw biogas must be purified, to remove the contaminants.

When the contaminants are scrubbed out, more methane per unit volume of gas is available for burning, and less harmful gasses are released.

There are several methods of refining and upgrading biogas:

a. water washing,
b. pressure swing absorption,
c. selexol absorption, and
d. amine gas treating.

The most practical of these is water washing, where high pressure gas flows into a process column in which the carbon dioxide and other impurities are removed by cascading water running counter-flow to the gas. This method produces over 98% pure methane, but some 2% of the total methane content is lost in the system. It also takes about 5% of the total energy content in the gas to run the scrubbers.

Thus produced upgraded biogas can be compressed, like natural gas, and used to power cars and trucks, and to fuel a number of important applications, and almost anywhere natural gas is used. Biogas, for example, has the potential to replace around 15-20% of vehicle fuel in some European countries, and since it also qualifies for renewable energy subsidies in some of these countries, interest in it is increasing.

Another major use for biogas is the so-called, gas-grid injection, where biogas is mixed into the natural gas distribution grid network. This way the gas is delivered to customers, where it can be used in domestic or commercial power generation (heat or electricity). One thing to note here is that typical energy losses in natural gas transmission systems are 1-2%, while the energy losses from a large electrical system range from 5-8%.

The difference is that methane gas, which is usually lost in the natural gas distribution system, is many times more environmentally damaging than any of the losses in the electrical system. This leads us to the conclusion that we need to separate the energy conversion efficiency losses from the actual physical loss of physical materials in the energy generation networks.

Bio-synthetic gas

Bio-synthetic gas, Bio-SG, or biomethane is actually methane (as in natural gas) derived from biomass, similar to landfill gas and even using the same source. Here, however, thermal equipment and processes are used to speed up the process, instead of letting the natural processes do the job.

The first such demonstration plant producing biomethane thermochemically out of solid biomass started operation in late 2008 in Güssing, Austria, and another plant is operating in Gothenburg, Sweden.

The increased use of natural gas vehicles (NGV) during the last decade is reaching over 25% and even more of the total vehicle fleet in some countries like Armenia and Pakistan. These vehicles can also be run on biomethane, derived from anaerobic digestion, or gasification of biomass.

Several routes to fuels and additives at different commercialization are available, including hydrothermal processing, pyrolysis oil, dimethylether (DME), biobutanol, and "solar" fuels (these produced with the help of solar as energy input).

Syngas

Syngas is a mixture of carbon monoxide, hydrogen and other hydrocarbons, which is produced by partial combustion of biomass. It is emitted during processes, such as the making of charcoal, where the biomass combustion is done with a controlled amount of oxygen. The gasification process usually proceeds at temperatures greater than 700°C, but due to insufficient oxygen, it does not convert the biomass completely to carbon dioxide and water. Instead, it removes enough of these to convert the raw biomass into a lighter combustible fuel, accompanied by release of syngas.

The resulting gas mixture, syngas, is more efficient than direct combustion of the original biomass, since more of the energy contained in it is extracted and contained in the resulting biogas.

The wood gas generator is a wood-fueled gasification reactor, that can feed directly an internal combustion engine. It can also be used to produce methanol, DME and hydrogen, or converted via the Fischer-Tropsch process to produce a diesel substitute. A mixture of alcohols, used for blending into gasoline can be produced too.

Lower-temperature gasification, as that used normally for co-producing biochar, results in syngas polluted with tar. After refining, syngas may be burned directly in internal combustion engines, turbines or high-temperature fuel cells.

There are a number of biogases available commercially today, depending on the raw materials and processes used in their production. They all, however, have similar composition and energy value, and all have to be refined and otherwise processed for convenient storage, transport and use.

A number of other biogases are in R&D mode, or are produced on a small commercial scale. These are the second- and third-generation biogases, which we will review below.

Algae Biofuel

Algal biofuel is a fuel derived from processing bio-materials as an alternative to fossil fuel. This process uses different types of algae as its raw materials. There is an ongoing effort by several companies and government agencies to develop the process in order to reduce capital cost and operating expense as needed to make algae fuel production commercially viable.

Algae materials release significant amounts of CO_2 when burned, but that amount is compensated by the CO_2 taken out of the atmosphere during the algae growing process. The world food crisis is increasing the interest in algae-culture in the production of vegetable oils, and biofuels on land unsuitable for agricultural crops.

Algal fuels can be grown with minimal impact on fresh water resources, and some can even be produced by using ocean and wastewater. Most algae products are biodegradable and harmless to the environment in any form and shape.

Algae cost more per unit mass, about ~$5000/ton, mostly due to high capital and operating costs. This is compensated by the fact that they can yield up to 100 times more fuel per unit area than most other crops. There are estimates that for algae fuel to replace all petroleum fuel used in the United States, it would require 15,000 square miles of land.

But, as with many other potential energy sources, these estimates remain on paper and far from any sizeable commercial application. And like the other renewables, biofuels production relies heavily on government support in the form of grants, tax and production tax credits. This shows the immaturity of this technology, and as promising as it sounds, we still have to wait awhile to see it compete shoulder-to-shoulder with the big guys—coal, oil and gas—and even with the renewables (solar, wind, and geothermal).

Algae grown in ponds of wastewater treatment plants can be harvested and processed into biofuels and ethanol. After all the hoopla, the enthusiasm died off, and the potential of harvesting oil for biofuels from waste pond algae is still far from any practical, let alone large-scale commercial, application.

Of great advantage, in addition to its projected high yield, algae-culture, unlike crop-based biofuels, does not interfere with food production. This is because no agricultural products can be grown in municipal waste plants, waste ponds, etc., so algae does not require farmland or fresh water for its growth.

A number of companies are looking into algae bioreactors for various purposes, including scaling up biofuels production to commercial levels; but again, the work is still confined to small test sites and university labs.

In recent years, several novel biofuel conversion routes have been announced, such as the conversion of sugars into synthetic diesel fuels. These include:

- Use of micro-organisms (yeast, heterotrophic algae or cyanobacteria that turn sugar into alkanes),

- Transformation of a variety of water-soluble sugars into hydrogen and chemical intermediates using aqueous phase reforming, and then into alkanes via a catalytic process,

- Use of modified yeasts to convert sugars into hydrocarbons that can be hydrogenated to synthetic diesel.

Unfortunately, these processes, as promising as they sound, have not been able to produce commercial products yet. The research continues.

Second-generation (Advanced) Biofuels

Second generation biofuels are made from ligno-cellulosic biomass or woody crops, agricultural residues or waste, which makes it harder and more expensive to extract the required fuel. These are produced from biomass materials, which are available in large quantity for mass biofuel production, and whose impact on GHG emissions, on biodiversity and land use is well known and acceptable.

Most second-generation biofuels are under development in universities, private and government labs and facilities. The development of cellulosic ethanol, algae fuel, biohydrogen, biomethanol, DMF, BioDME, Fischer-Tropsch diesel, biohydrogen diesel, mixed alcohols, and wood diesel are prime examples of the fuels developed, or under development, in these labs.

Cellulosic ethanol production is one of the great hopes of future biofuels. It is produced from non-food crops, wood waste, or inedible waste products and does not divert food away from the animal or human food chain.

Ligno-cellulose is the "woody" structural material of plants, which is the main energy source for the second generation biofuels. This feedstock is abundant and diverse, and in some cases (like citrus peels or sawdust) it is in itself a significant pollutant causing serious disposal problems.

Producing ethanol from cellulose, however, is a difficult technical problem to solve, which requires a lot of energy too. In nature, ruminant livestock (such as cattle) eat grass and then use slow enzymatic digestive processes to break it into glucose (sugar). In cellulosic ethanol laboratories, various experimental processes are being developed to do the same thing, and then the sugars released can be fermented to make ethanol fuel. In 2009, scientists reported developing (using "synthetic biology") "15 new highly stable fungal enzyme catalysts that efficiently break down cellulose into sugars at high temperatures," adding to the 10 previously known.

The use of high temperatures has been identified as an important factor in improving the overall economic feasibility of the biofuel industry and the identification of enzymes that are stable and can operate efficiently at extreme temperatures is an area of active research.

Research conducted at Delft University of Technology, Holland, has shown that elephant yeast, when slightly modified, can also produce ethanol from inedible ground sources, such as straw and other mostly cellulosic products.

The recent discovery of the fungus *Gliocladium roseum* points toward the production of so-called myco-diesel from cellulose. This organism, recently discovered in rainforests of northern Patagonia, Argentina, has the unique capability of converting cellulose into medium-length hydrocarbons typically found in diesel fuel.

Scientists are also working on experimental recombinant DNA genetic engineering organisms that could increase biofuel potential. Others have developed a technology to use industrial waste gases, such as carbon monoxide from steel mills, as a feedstock for a microbial fermentation process to produce ethanol.

Virgin Atlantic joined Lanzatech, New Zealand, to commission a demonstration plant in Shanghai for the production of aviation fuel from waste gases from steel production. Scientists in Minnesota have developed co-cultures of *Shewanella* and *Synechococcus* that produce long-chain hydrocarbons directly from water, carbon dioxide, and sunlight. The technology received ARPA-E funding.

Global companies such as Iogen, POET, and Abengoa are building refineries that can process biomass and turn it into ethanol, while companies such as DuPont, Diversa, Novozymes, and Dyadic are producing enzymes which could enable a cellulosic ethanol future.

Note: The conversion of cellulosic matter into alcohol has been a major effort of global proportions since the 1970s. The author had a chance to work at a South American government cellulose and paper pilot plant, experimenting with converting Brazilian jungle biomass into paper and biofuels. The results were promising, but the processes were extremely energy hungry and inefficient, which made the final product not practical, or profitable enough, for commercial applications.

Fast forward 30 years, and we still have problems with producing bioethanol from cellulosic materials. As a matter of fact, this was confirmed recently by the Appeals Court, which ruled in 2013 that the EPA's blending targets for the advanced biofuels, cellulosic ethanol, were simply infeasible.

The EPA had demanded that between 2010 and 2012, 20 million gallons of cellulosic ethanol be produced. However, very little cellulosic ethanol has been produced, and even less blended into commercial fuel, to date. The cellulosic ethanol pilot plants and startups have been simply unable to provide significant stock to blenders.

The EPA claims that it has the authority to enforce blends based on the 2007 Energy Act passed under the Bush Administration, which promoted biofuels (and corn ethanol) growth. But the Appeals court rejected that argument calling the decision to enforce targets on refiners, who are customers of the fuel producers rather than the producers themselves, "a bizarre and unprecedented government effort." Wow!!! Who'd ever think that biofuels can trigger a political war…?

It is apparent that although significant technological advances of late have introduced some improvements, a steady, fully efficient, and profitable large-scale commercial cellulosic bioethanol conversion process is still in the distant future.

Actually, it is not just the process, but also the process costs that are creating the problems. In our expert opinion, cellulosic biofuels will become a commercial reality on a large scale only when more energy efficient processes are developed and these are combined with using cheap energy sources (i.e. solar or wind power). At that point only the logistics (harvesting, transport, waste disposal, etc.) would be left to resolve.

Cellulosic Matter Conversion—Exercise

Remember that the cellulosic matter conversion objective is to:

a) develop the process of converting hardwood (trees mostly) and some otherwise useless agricultural wastes into biofuels via the cellulosic-to-biofuels conversion process,

b) expand the production into all available U.S. paper manufacturing plants, and

c) sell thus produced biofuels on the U.S. and global markets.

Let's start with the process.

• *Availability of raw materials.*

Keep in mind that the goal of this project is to use hardwood tree trunks, branches, and other inedible agricultural waste material. The idea here is that anything, even huge tree trunks, can be used, since they are grown in huge quantities and are easier to procure. This makes the supply chain (at least in theory) more controllable and ensures a steady supply of raw materials.

But the big question here is how much wood would Chuck need to chuck in order to chuck enough biofuels?

About 80-100 gallons of biofuels can be produced from a ton of hardwood.

Regulated, or man-made, forests can produce about 3-5 tons of hardwood per acre, which need 30-40 years to get to harvest. So, we can count on 250-500 gallons of biofuels from an acre…but only every 30-40 years.

Planned forestation and deforestation might provide enough materials, which combined with seasonal agricultural wastes would provide some quantity of biofuels. But how much and at what cost?

The Energy Independence and Security Act of 2007 mandates greater use of liquid biofuels for domestic transport. It requires a gradual increase in use of liquid biofuels by American fuel producers and set a standard of 9 billion gallons of renewable fuel in 2008, rising to 36 billion gallons by 2022.

Of this total, 21 billion gallons (or about 2/3 of the total) must be advanced biofuels, which includes cellulosic biofuels, or about 16 billion gallons annual production capacity, according to the Act.

16 billion gallons of cellulosic fuels are to be produced by 2022 in the U.S.

A quick calculation reveals that this large amount of biofuels would require about 32 to 64 million acres of mature forests…every year. Since the forests can be harvested only 30-40 years, this means that we need to have about 1,000-2,000 million acres forest reserves to support the cellulosic biofuels production.

But the U.S. doesn't even have so much forest lands (estimated total is about 750 million acres of forests on the continental U.S. territory), so we must assume that the rest would be procured from the agricultural fields… if we have enough, and provided that the process works and that it is efficient and profitable enough.

The Act also creates a new set of subsidy incentives, adding a $1.04 per gallon subsidy for cellulosic biofuels and decreasing the corn ethanol subsidy from $0.51 to $0.45. So is this what the big business billionaires are counting on? If so, then we agree; how can they go wrong with the U.S. government on their side?

Unfortunately, the U.S. government tends to change its policies and direction from time to time to fit the given situation, as we have seen many times in the past. The recent boom in oil and natural gas production promises a drastic change in direction this time too.

And the Energy Act has a number of gaping holes in it, full of ambiguities and limitations. Although the overall concept is to promote cellulosic biofuels, Congress appeared conflicted, and limited the sources from which the feedstock could be obtained. For example, Section 201 of the act includes a very restrictive definition of "renewable biomass" for the purposes of meeting the goals of the Renewable Fuel Standard.

Very importantly—and something that could throw a wrench in the overall cellulosic biofuels strategy—is the fact that no biomass from forests on federal lands is allowable. Also, only "planted" trees in private forests are allowed. Thus, naturally regenerated private forests, even areas of active management, cannot be included. What? What does that leaves us with?

Slash and materials from pre-commercial thinnings are allowed, but only from nonfederal lands. Because only wood from nonfederal, private planted forests is allowed, much of the U.S. forest estate is not available for biofuel feedstock.

Logging residue (25% of total tree harvest) could be used for biofuels production, but the cost would increase significantly due to transport and high handling costs.

So, suddenly, we see that even if the cellulosic process such as KiOR, fueled by billionaires' money is perfectly successful, and as a consequence all paper mills are converted to cellulosic biofuels production, the enormous amount of hardwood and cellulosic wastes needed to produce the needed biofuels is not available.

What about the 2007 Security Act mandate of 16 billion gallons of cellulosic biofuels production by 2022? Well, in 2014-2015 we were well below 1 billion gallons, so we can only guess how and when we'd catch up.

Costs

The production of cellulosic biofuels requires a lot of energy, expensive chemicals, and labor. The production of biofuels from wood biomass requires about 60% more energy that what it can generate when used. Basically, a gallon of biofuel (gasoline) needs about 1.6 gallons of crude oil or equivalent fossil energy. This also means that the production of a gallon of biofuel emits over 1.6 times (or more) the GHG pollution emitted by the fossils.

Recent studies estimate the cost of cellulosic ethanol from the existing commercial plants to be in the $2.00-$2.50 per gallon range, including everything except incentives. This compares to the current cost of $1.20-$1.50 per gallon for ethanol from corn.

Solar and wind might eventually provide cheap and clean energy, but that power is not always available, so cellulosic biofuels cannot compete with the conventional energy sources at the current state of the art. Some form of subsidy and/or policy assistance is also needed for now.

Figure 7-56 shows that cellulosic gasoline costs more than the conventional and even some of the other alternative fuels. At $150 per barrel, compared to $50 for corn ethanol, it would take a long time until it becomes a main stream product that everyone rushes to buy it.

Fuel	$/bbl	%GHG
Cellulosic gasoline	150	-50
Soybeans diesel	100	-50
Corn ethanol	50	-20
Coal gasification	40	80
Shale oil	60	75
Tar sands	25	25

Figure 7-56. Cost and pollution estimates

However, if the technology is fully developed, and when the fossils become more scarce and their prices high enough, then cellulosic gasoline and diesel might find a broad market. Until then, however, we doubt that anyone would prefer them before the cheaper and more readily available conventional fuels.

On the other hand, Figure 7-56 also shows that cellulosic gasoline generates much fewer GHG pollutants than the conventional, and even some of the other renewable, fuels. So if the price of pollution is calculated in the long-term survival equation, then the cellulosic biofuels make sense and take a priority in the world energy markets.

Unfortunately, we are not at this stage yet, and truly, it is not clear when (if ever) we will embrace in practice (not just talk about) this new way of life.

Instead, we see a drastic increase in natural gas, tar sands and shale oil shale use in the U.S., and significant increase in coal-fired power plants around the world. This trend will continue for a number of decades, which makes the advance of cellulosic biofuels a thing of the future...the distant future.

Case Study: KiOR Inc.

Dutch biofuels startup Bioecon and Khosla Ventures launched a joint venture called KiOR, with the intent to commercialize Bioecon's process for converting agricultural waste directly into biofuels, or "biocrude" as they call it. This is basically starting with a mixture of small hydrocarbon molecules in the agricultural wastes that can be processed into gasoline or diesel in existing oil refineries.

The KiOR process boasts numerous advantages over other methods of producing biofuels. For example, it could (eventually) prove relatively cheap; it relies on a nontoxic catalyst; it taps into the present fuel-refining and transportation infrastructure; and it produces clean-burning fuels that can be used in existing car and truck engines. Note the "eventually" condition. Anyone could also "eventually" move the Earth, given a long enough lever and a fulcrum.

KiOR Inc. recently announced $100 million in equity-related financing in two separate private placement transactions, as needed for expansion of production capacity in Columbus, Mississippi. The Columbus II Project, with the planned technology enhancements for both Columbus facilities is expected to achieve overall positive cash flow from operations for KiOR.

So, with the help of venture capital and government support, the project (presently $50-60 million in the red) will be able to continue operation for sometime to come. But as Solyndra's example shows, money alone cannot solve all the problems presented by the technically and logistically complex technologies.

Solyndra had a lot of money—$750 million to be exact—but took the wrong direction with the design, manufacturing, and marketing of their products and the rest is history...after leaving a half-billion-dollar hole in taxpayer pockets.

Could KiOR prove to be a repetition of Solyndra's case in the biofuels field, where a new and relatively undeveloped technology is taken to market prematurely? The world's richest investors are putting their money and reputations on the line.

They made money in the Silicon Valley, but can they do the same in the Energy Valley? Can their expertise in hi-tech electronic and computer related ventures be translated into the energy sector? One time an expert, always an expert...maybe, but it wouldn't hurt to keep in mind that energy generation is a different animal, and that energy ventures are much different than any other.

Biofuels Production and Use

Biofuels are increasing in volume and importance as the financial and energy crisis grow, and as the environmental issues take a front stage. Biofuels are considered by some as the solution to all of these problems (to one extent or another), so their use has increased from almost zero 20 years ago, to over 3.0% of the total road transport fuel use in some European countries. For example, in the UK 400 million gallons of biofuels are produced and sold annually.

At the same time, the U.S. produces over 5 billion gallons of ethanol, and about 2 billion gallons of biodiesel. Brazil also produces over 5 billion gallons of ethanol, and China follows with 1 billion. India is next with 500 million, followed by Germany, France, and Russia with about 200 million each, and Canada and South Africa with about 100 million each annually. Many other countries also produce biofuels, but in much smaller volume.

It is expected that by 2020 over 10% of the energy used in road and rail transport in European countries will come from renewable sources. This is equivalent to replacing 4.3 million tons of fossil oil each year. Conventional biofuels are likely to produce between 3.7 and 6.6% of the energy needed in road and rail transport, while advanced biofuels could meet up to 4.3% of the UK's renewable transport fuel target by 2020.

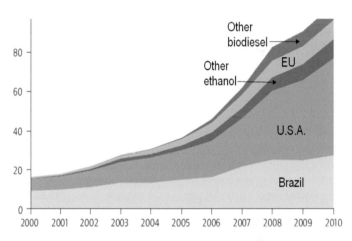

Figure 7-57. Global biofuels production in billion liters

The world leaders in biofuel development and use are Brazil, the United States, France, Sweden and Germany. At the same time Russia, which has over 20% of world's forests is one of the largest biomass (solid biofuels) suppliers, and is also looking into the possibility of increasing its share in biofuels production.

The transport sector presently is the largest consumer of fuels—fossil and biofuels—and will most likely remain in the lead for the foreseeable future. It is expected that the biofuels in 2050 will be used mostly for road transport too (over 50%), with 25% used for jet fuel, and the rest for shipping and other commercial purposes.

International organizations, such as IEA Bioenergy, were established by the OECD International Energy Agency (IEA), to improve the cooperation and information exchange between countries with bioenergy programs, and assist in the research, development and deployment of biofuels.

Table 7-12. Total vehicle fuel use in 2050 (estimate)

Fuel type	2050
Biofuels	27%
Diesel	23
Gasoline	13
Jet fuel	13
Electricity	13
Hydrogen	7
Heavy fuel oil	2
CNG, LPG	2

The UN International Biofuels Forum, formed by Brazil, China, India, Pakistan, South Africa, the United States and the European Commission has similar agenda and goals.

Table 7-13. Global bio-ethanol production (in million liters)

Area	2008	2010	2012
N. America	36,000	51,600	54,600
S. America	24,500	26,000	21,300
Europe	2,900	4,250	5,000
Asia/Pacific	2,750	3,100	4,000
Africa	65	130	235
World total	**66,000**	**85,000**	**85,000**

The global bioethanol industry continues to be a bright spot in the global energy sector and the overall world economy, and continues to grow. It is presently supporting nearly 1.5 million jobs and contributes over $300 billion to the global economy.

The future looks bright for biofuels, and bio-ethanol in particular.

Cost of Bio-energy Sources

The diversity of raw materials and processes for making biofuels is staggering. We can divide the sector into biomass and biofuels, although there is some intermingling possible.

Biomass Costs

Biomass can be subdivided into wood products from forests and other (agricultural crops waste materials, etc.). Forest wood for use as an energy generator and other purposes can be subdivided into:

a. Direct use of wood (as forest tree trunks and branches), and

b. Byproducts from wood and wood processing.

As can be seen in Table 7-14, there are a number of products and uses of wood and wood byproducts. Putting a value on each product and use is beyond the scope of this text, but in all cases the price of products and uses is determined by the market supply and demand dynamics. The availability and location of the wood products also has a determining role in the final price structure.

The harvesting and marketing of forest biomass in Kenya, for example, would be different from that in Sweden. In Kenya there are no laws, nor is there any significant enforcement of wood products harvesting and sale. Because of that, wood harvesting and use in forests and fields is basically free for the locals, but is sold at a certain minimal price in the large cities.

In Sweden, on the other hand, the wood harvesting process is heavily regulated and enforced, which will determine the market price, which varies with location and the seasons. And no doubt, wood products' market prices in Sweden are much higher than those in Kenya.

The price of biomass products is also determined by their specific heat value in relation to their mass. For example, wood pellets are the densest wood product with the highest heating value per ton of material. On the other side of the spectrum, wood shavings are the least dense material and although the heat value per ton is almost as good as that of the wood pellets, it will take almost 4 times the volume to generate equal

Table 7-14. Biomass origins and uses

Forest biomass				
Wood			Residues from wood based industry	
Forest residues	Stem wood	Whole trees	Dry residues	Wet residues
Pretreatment				
Solid biofuels				
Wood chips			Pellets	Residues
Bio-heating and bio-electricity systems				
District heat & power	Micro grids		Central heat & power	Stove

Table 7-15.
Bulk weight and heating values of biomass products

Wood products	Bulk weight of dry matter in ton / m^3	Heating value in MW/h per ton dry matter.
Pellets	0.70	4.80
Wood chips	0.45	4.70
Forest residues	0.40	4.40
Stem wood	0.45	4.40
Sawdust	0.40	3.00
Shavings	0.25	4.40

heat value.

As it can be seen in Figure 7-58, the cost of the biomass materials depends on a number of factors, based on materials and labor conditions. In addition to the major cost of the biomass material on the field, storage and transportation costs will also affect the cost. The farther from the source the processing facility and the power plant are, the higher the cost of transport would be. This, of course, will increase the cost of the final product as well.

Figure 7-58. Biomass production steps (in \$/1,000 tons/year)

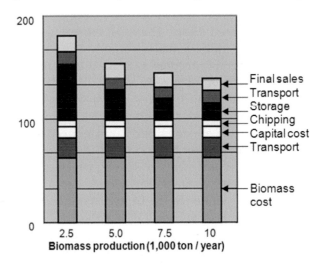

Another factor, determining price is the biomass quality; especially that destined for the generation of bioenergy. The European Committee of Standardization (CEN) is working on standards for fuels and has established three main categories for biomass quality:

a) *Pure biomass*, which comprises pure wood and other biomass;

b) *Contaminated biomass*, which is fuel that cannot be burnt without meeting the demands of waste combustion set by the EU and other governments, and

c) *Hazardous biomass*, e.g. CCA impregnated wood, which can only be burnt using a special combustion and cleansing technology.

Obviously, the price will be different for pure vs. contaminated biomass, while hazardous biomass will have a totally different price schedule, which might be on the negative side.

Table 7-16.
Average yield from different energy crops perunit area

Crop	Liters per hectare
Sugarcane	4,900
Sugar beet	4,000
Palm oil	3,600
Switch grass	2,850
Corn	2,600
Rapeseed	1,700
Soy beans	700

The price of the different energy crops also depends on the type and location. For example the price of a ton of sugarcane would be different from that of a ton of corn, and very different from a ton of bagasse. One of the reasons for this is the fact that sugarcane yields almost twice the amount of fuel from a hectare than corn, and 5 times that of bagasse. The cost of the land and amount of effort is reduced by the fact that sugarcane requires less land to produce equivalent energy.

Another important factor to consider in the price formation scenario of energy crops used for biofuels is the ease of processing. Sugarcane is the easiest and cheapest to process into biofuel, because it is pure sugar, which requires fewer process steps. This would raise its price significantly, as compared to the rest.

Due to the scarcity of wood and wood products in many places around the globe, import-export activities have increased during the last decade. Typical CIF prices for sea transport, including loading costs, ocean freight and insurance, are in the range of €100-120 per ton for wood pellets with heating value around 17 GJ/ton.

At this price, wood pellets are imported in the EU from Canada, United States, Baltic States and Russia. Large-scale freight volumes from 10,000 to 50,000 tons make wood products import profitable, since the cost for ocean fright is estimated at \$7.5/MWh.

The average international prices for the common

biofuel sources in 2007 were as follow:

Table 7-17. Average prices (in $/ton)

Bio-crude	167
Corn oil	802
Cottonseed oil	782
Crude palm oil	543
Maize	179
Peanut oil	891
Poultry fat	256
Rapeseed oil	824
Soybean oil	771
Sugar	223
Tea seed oil	514
Waste oil	224
Wheat	215
Yellow grease	412

The production of biofuels also varies significantly with location and raw materials types and availability. A municipal waste biogas generating plant would have a different price tag and production cost than an algae biofuel power plant.

As an example, a production plant of 200,000 tons of biodiesel per year has been estimated at approximately $500 million, or approximately $2,500 per produced ton. $150 million of the total is for the construction of an oxygen generating facility. So if the plant is located near an oxygen generator, the construction costs would be significantly reduced.

The production cost of biodiesel has been estimated at $500-600 per ton of diesel, or about $0.60-0.70 per liter of petro-diesel equivalent.

Renewable Fuels Program

Increasing crude oil prices and concerns about unreliable crude oil imports have forced us to consider self-sufficiency as the best, if not only, sure way of achieving energy security. The combination of all possible energy solutions, such as solar and wind, together with the other renewable technologies (geo-, ocean, bio-fuels, etc.) offer a real possibility to offset the increasing demand for energy.

Especially important to our energy security is the potential of providing renewable transportation fuels (bio-fuels and renewable hydrogen), as needed to replace imported crude oil.

The Renewable Fuel Program, Title XV of the Energy Policy Act of 2005 (EPAct—P.L. 109-58) was created to substitute increasing volumes of renewable fuel for gasoline, with EPA as the statutory authority for administering the program.

The act set a target production volume of 7.5 billion gallons of renewable fuels for 2012. The 2007 Energy Independence and Security Act (EISA) expanded the program to cover transportation fuels in general, extended the program to calendar year 2022, and increased the target volume to 36 billion gallons renewable fuel annually (857 million barrels annually or 2.3 million bbl/d).

Under current EPA rules, ethanol is blended up to 10% by volume in retail gasoline (E10) and up to 85% in E85 fuel for use in flex-fuel vehicles. In 2010, the EPA partially granted Growth Energy's waiver request application submitted under section 211(f) of the Clean Air Act. The partial waiver allows the sale of gasoline that contains ethanol up to 15% by volume (E15) for use in 2007 and newer model year vehicles.

EPA denied the waiver to use E15 in vehicles older than model year 2000, and is deferring a decision on using E15 in model years 2001 through 2006. The E15 fuel must be sold from a separate pump, as is E85. The new ruling appears to be at odds with the EPA's 2005 provision that limits the proliferation of boutique fuels. EPA also specifies the annual volumes for cellulosic biofuel, biomass-based diesel, advanced biofuel, and total renewable fuel requirements.

Some federal subsidies and tax breaks favoring ethanol production were reduced by Title XV of the Food, Conservation and Energy Act of 2008 (P.L. 110-246). The ethanol blender tax credit of $0.51 per gallon, which applies to all ethanol blends, was reduced to $0.45 per gallon in 2009 under Section 15331 of the act. The $0.54 per gallon import tariff on ethanol, which effectively offsets the blender tax credit when imported ethanol is blended into gasoline in the United States, expired at the end of 2010 under Section 15333 of the act.

During periods of strong motor fuel demand conditions, such as summer months and others, ethanol could be one of the means of extending the volume of refined transportation fuels. Ethanol could also displace refined fuels successfully (as is the case in Brazil), albeit with the help of government subsidies in the beginning.

Thirteen billion gallons of renewable fuels were produced in 2010, which represents more than 9% of the total gasoline consumption in the country. If the renewable fuel volume mandate is met by 2022, and with $0.45 per gallon subsidy, the goal of 857 million barrels could represent a $16.2 billion annual subsidy, displacing 564 million barrels of refined gasoline (equivalent).

Renewable fuels are our greatest hope for securing

our energy future—especially in the critical motor fuel supply sector—which is negatively affected by increasing crude oil prices, import irregularities, and shortages.

Biofuels Production Issues

Presently there are a number of technical, social, economic, and environmental issues, related to the production and use of biofuels. Most of these have been discussed in the media and the scientific journals, and include the effect of moderating oil prices, the *food vs. biofuels* debate, poverty reduction potential, carbon emissions levels, sustainable biofuel production, deforestation and soil erosion, loss of biodiversity, impact on water resources, as well as energy balance and efficiency, to mention a few.

Scientists warn that not all biofuels perform equally in terms of their impact on climate, energy security, and ecosystems, and suggest that environmental and social impacts need to be assessed throughout the entire life-cycle, prior to making a decision on which biofuel to use, when and where.

The fact that some biofuels are made by using huge amounts of fossil power distorts the "renewable" energy concept and reduces the "renewable" benefits.

Using fossil energy in the production of bio-fuels and in any other renewables is a regular practice. The renewables already use some energy during different stages of their cradle-to-grave life cycle, so the additional use of fossil energy must be avoided. At the very least, it must be accounted for and entered in the overall energy-environment-energy security equation.

The trend today points to the development of biofuel crops that require less land (or waste land) and use fewer resources than current biofuel crops do. Biofuel from algae is one of the solutions, since it uses unprofitable land and wastewater from different industries.

Different algae are able to grow in wastewater, which does not affect the land or freshwater needed to produce current food and fuel crops. Also, algae are not part of the human food chain, so do not take away food resources from humans.

Nevertheless, the effects of the biofuel industry on food production and the environment in general are still being debated. A recent study shows that biofuel production accounted for up to 30% increase in food prices at the height of the biofuels craze in 2008. Market-driven expansion of ethanol in the U.S. increased corn prices by more than 20% in 2009, as compared with the prices of ethanol production during previous years. This

changed the direction of the basic research, forcing the development of biofuel crops and technologies that will reduce the impact of a growing biofuel industry on food production and cost.

This prompted some drastic reactions, such as the 2012 decision of the United States House Committee on Armed Services to put language into the 2013 National Defense Authorization Act that would prevent the Pentagon from purchasing biofuels that offered improved performance for combat aircraft.

Developing biofuel crops that are optimized for the local climate and market is the best solution to this problem. Using specific local biofuel crops voids the need to transport fossil fuels from faraway places for processing and use. The problem is that many areas of the globe are unsuitable for producing the most efficient energy crops, since they require large amounts of water and nutrient-rich soil.

So biofuel crops, such as corn and sugarcane, are impractical in these places and must be grown in different regions. This complicates the picture, and creates a several-fold disequilibrium, where rich nations produce the crops to be used by poor nations. It also creates an environmental disequilibrium, where CO_2 absorbed in one part of the world is released in another part.

And speaking of CO_2...

The Biofuels Effects

It could be argued that biofuel is a product of solar energy, as the sun is needed to grow the biomass crops, which can then be manufactured into usable fuel. It is also true that naturally grown biofuels are near-carbon neutral, simply because the CO_2 emitted during their burning was absorbed during their growth cycle. It is also undeniable that the use of smartly made biofuels can help to reduce the costs associated with the purchasing of mainstream fuels such as gasoline and diesel.

Although *biofuel* sounds like a modern-day invention, we have used this type of fuel since the discovery of fire. If firewood can be classed as a biofuel, as wood is a biomass product, then biofuels are the oldest type of energy used for domestic and industrial applications. It is the material that was burnt to released energy in the form of heat since man walked on this Earth.

The types of biofuels mentioned above (biodiesel, biogas etc.) relate to the modern-day uses of biomass as a fuel energy source.

We, however, should not get carried away and exaggerate the importance of biomass and biofuels as future energy sources. To provide biofuel for every car, truck, bus, plane, boat, and factory across the globe would require a colossal amount of land to be used for the plantation of renewable biomass crops. This would also result in the deforestation of the world, as a direct result of biomass plantation becoming out of hand.

Electric cars powered from the electricity generated by renewable energy sources would be the best option, however, implementing this on a global scale will be a very challenging task too. Instead, we need to be looking into all possible methods for powering vehicles and machinery to reduce the future impacts that biofuel, and/or any one single technology usage may pose on the environment.

After all the praises of the benefits of biomass and biofuels, we must also take a close look at the negative effects these have on the different aspects of our daily lives.

Energy Use

Biomass is undoubtedly a *renewable* source of energy, although its renewable nature is heavily dependent on weather, economic, and socio-political conditions. Its seed-to-energy cycle inevitably involves the consumption of some fossil fuels at one or another point and time of the cycle. How much is used depends on the type of biomass and the production methods, but it usually includes fuels consumed by farm machinery in land preparation, planting, tending, irrigation, harvesting, storage, and transport.

There are also great quantities of fossil feedstocks for chemical products such as herbicides, pesticides, and fertilizers used during the growth cycle. And, of course, a lot of energy is required for plowing, seeding, growing, harvesting, and processing the bioenergy crop into a usable biofuel.

Energy requirements are higher for some crops because they require greater use of machinery and a higher level of water and chemical inputs (fertilizers and such). For many energy crops, the energy ratio, or the amount of useful bioenergy the particular crop can produce, compared to the fossil fuel consumed for feedstock production, could be very high.

Table 7-18 shows that hydropower is the most energy efficient power generator, producing 250 units of electric power for each unit of fossil fuel input. Biomass produces 30 units, while biofuel crops produce only 5 units of energy for each unit of fossil fuels used during their life-cycle.

Some crops, such as poplar, sorghum, and switchgrass grown in temperate climates have energy ratios of up to 30. This simply means that each Btu used in the bio-crop cycle generates up to 30 Btus of energy from the

Table 7-18. Energy ratios for different technologies

Technology	Ratio
Hydropower	250
Wind power	35
Biomass waste	30
Nuclear	15
Bio-fuel crops	5
Coal	5
Solar PV	5
Natural gas	5
Fuel cells	3
Crude oil	3

respective biomass crops. In tropical climates with good rainfall these ratios could even be considerably higher, because of higher yields and less labor and energy-intensive agricultural practices.

The energy ratios are much lower for crops that require much labor and mechanization, or yield a relatively small proportion of usable bioenergy feedstock per unit of plant matter produced. Some oil crops (like soybean), however, could have an energy ratio close to 1, which means that the amount of energy this crop produces is equal to the energy it required during its entire seed-to-fuel life cycle. For some crops, and in some cases, the energy ratio is so low that we need to think twice about using them as energy sources.

At the processing facility, more energy is used during the different steps of the conversion process and more polluting gasses, liquids and solids are generated. In all cases, massive amounts of fossil fuels are burnt, which further reduces the energy in/energy out ratio. And, of course, significant pollution is created in this process as well.

Emissions

Another factor to consider in the energy ratio estimates is the pollution created during production, transport, and processing of the crops. In most cases heavy machinery is used to cultivate the bio-crops, which uses a lot of fossil fuels and emits a lot of GHGs.

After harvesting, the crops are loaded on large trucks or trains to be delivered to the processing facility. Even if the processing facility is a mile away, which is rarely the case, the crops must still be loaded on vehicles and transported to the intake elevator.

Bio-energy is hailed as carbon neutral, and although at first glance this is so, this generalization needs a closer look. Carbon emissions from biofuels are emitted at all steps of their life cycle.

These can be reduced by:

1. Using renewable energy instead of fossil fuel energy, and
2. Increasing the amount of sequestered carbon.

The net carbon benefit is calculated by comparing the reality with what would have happened if fossils were used all through the life cycle instead. The amount and type of fossil fuel that would otherwise have been consumed, as well as the land that would otherwise have been used, must be entered into the calculations.

The relative carbon intensity must be assessed on the basis of the emissions associated with the biofuel crop production and the efficiency of the energy technology in which the biofuel is used.

As can be seen in Table 7-19, all biomass materials have fairly large carbon content (carbon-nitrogen ratio). In most cases, the carbon is released in one or another form, but most likely as CO and CO_2 during the digestion, harvesting, transport, burning, and/or otherwise processing of these materials.

Table 7-20 shows the energy conversion efficiency of different fuels during power generation. Natural gas has the highest efficiency, since about 45% of its energy input is converted into electric power. Biomass averages 20% efficiency, or only 20% of the available energy contained in it (and everything else considered) can be

Table 7-19. Carbon content in biomass materials

Type of Material	Carbon-Nitrogen Ratio
Alfalfa hay	18 : 1
Bagasse	150 : 1
Cichen manure	25 : 1
Clover	2.7 : 1
Cow dung	25 : 1
Cow urine	0.8 : 1
Grass clippings	12 : 1
Kitchen refuse	10 : 1
Lucerne	2 : 1
Pig droppings	20 : 1
Pig urine	6 : 1
Potato tops	25 : 1
Sawdust	200 : 1
Seaweed	80 : 1
Straw	200 : 1
Sewage sludge	13 : 1
Silage liquor	11 : 1

converted into useful electric power. The rest is emitted as waste energy and GHG gasses.

Table 7-20.
Efficiency and CO_2 emissions of different fuels (per kWh)

Technology	% Efficiency	Gr. CO_2 emitted
Diesel generator	20%	1,320
Coal steam cycle	33%	1,000
LNG combined cycle	45%	410
Biogas digester and diesel generator	18%	220
Biomass steam cycle	22%	100
Biomass gasifier and gas turbine	35%	60

Diesel power generators seem to be the most polluting, while biomass is the cleanest—assuming that the bioenergy crop is not counting the pollution emitted during the previous steps of its life cycle. It must be grown and harvested in a carbon-neutral manner, and there is no carbon net change in/on the crop fields and in the soil over the course of the complete crop growth and harvesting cycle.

In practice, the carbon in and on the land (in the soil) changes significantly, depending on how the biomass is produced and what would have happened in other cases. Taking as an example clearing the jungle forests, which leaves a bare land for growing energy crops, we see that the bare site has lost its carbon-reducing value and cannot be readily regenerated.

The carbon emissions from site preparation alone (trucks, bulldozers, etc.) could exceed the savings from biofuel use, and could be greater than the carbon emissions from a fossil-fuel cycle generating the same amount of power.

The justification for the land clearing activities, from an environmental perspective, is that long-term use of biofuels produced at the cleared land will compensate for the increase in CO_2 at the initial stages. This is a frequently used model for the production of energy and non-energy biomass, under which millions of acres have been deforested worldwide. The overall global effect of this massive land clearing is yet to be determined, let alone extrapolated into implications for the future.

Clearing large areas of natural forests, some of which are used for energy crops planting, means that the CO_2 sequestered in the natural forest will be released eventually during burning of the biomass materials at an amount that depends on the type of the forest. A rough figure of 300 metric tons of carbon per hectare (tC/ha) has been suggested, which is a significant, but reasonable, number.

As biomass feedstock is grown and harvested in cycles, carbon will be held on the land, partly compensating for the carbon released when the natural forest was cut down. Averaged over a single growth cycle, a typical amount of carbon sequestered on the land might be 30 tC/ha, so this leaves a 270 tC/ha balance unaccounted for. If the purpose of a bioenergy crop is to displace fossil fuels to reduce carbon emissions, it will achieve this by compensating for this 270 tC/ha difference derived from the above estimate over a period of roughly 40-45 years.

When environmental and social considerations (preserving habitat and protecting watersheds), are taken into account, the total of all considerations might increase the number of years and even outweigh any carbon-saving benefits.

If bioenergy crops are developed on unproductive land (degraded or abandoned land), then that land and the local environment could benefit tremendously from eventual crops re-vegetation. The degraded land is most likely carbon-starved, so the crop field will store some CO_2 in the soil and other below-ground biomass. This would compensate for some of the initial carbon emissions.

So, the overall effects of using wasteland for energy crops show definite benefits in,

a)   Providing immediate measurable carbon and other local ecosystem benefits, and

b)   Displacing fossil fuels for power generation in the long run.

Soil

Biomass crops are very similar to other crops, as far as managing soil, water, agrochemicals, and biodiversity are concerned. Following good crop management practices, while taking into consideration the specific technical and environmental challenges, is the key to success. None of this is new, or different from conventional agricultural methods of crops management.

One difference here is the fact that biomass plants are often harvested completely, unlike their cousins, large parts of which are left in the field. Bio-crops are usually chopped down to the roots, and sometimes are even harvested with the roots. This leaves very little organic matter or plant nutrients in the soil of the harvested field. This is a potential problem, since it would create soil problems especially in the developing world

where the soil depends on recycling crop wastes and manure, rather than on the use of expensive fertilizers. In such cases, biomass production could contribute to dramatic deterioration of soil quality.

To maintain the soil quality, clever farmers would choose to keep sufficient plant matter on the land for the soil's sake, at the expense of reduced crop yields. As another alternative, farmers might allow some nutrient-rich parts of the plant, like small branches, twigs, and leaves to decompose in the field. Some feedstock nutrient content can be recovered from the conversion facility in the form of ash or sludge, and trucked back to the fields to be used for soil improvement, instead of being put in a landfill. This, however, is an expensive undertaking and is rarely used.

Hydrology

Bioenergy crops usually consume more water than many food crops, and some energy crops like sugarcane consume immense amounts of water. These thirsty crops cannot be grown in most regions, simply because there is not enough water to support them. They also create huge problems with the local irrigation systems and water wells wherever they are grown.

An amazing fact: Three quarters of our fresh water—which is only 5 percent of the entire Earth's water supply—is used for crops irrigation and livestock farming.

Sugarcane and other water-hungry crops are known to lower the water table, or reduce stream flow, making local irrigation systems less reliable. For this reason alone, many agricultural communities have resisted the introduction of some energy crops and tree plantations.

A number of practices, such as growing tree crops without undergrowth, or planting species that do not generate adequate litter, reduce the ability of rainfall to replenish groundwater supplies, thus further increasing the local water overconsumption.

Water scarcity and the desertification of agricultural lands in the U.S. are proceeding at a frightening pace. Irrigation, for example is quickly depleting our Ogallala Aquifer on the Great Plains. The annual water consumption is about 1.3 trillion gallons faster than it can be replenished by rainfall. How long will that water be available for irrigation?

Biodiversity

Different farm crops have different, but significant, effects on the local ecosystem by enhancing or suppressing the biodiversity of the region. Energy crops have similar effects, but provide even more biodiversity, which is often closer to a natural habitat than other crops.

By enhancing biodiversity some gaps between fragments of natural habitat can be filled. For this purpose, in Brazil for example, environmental regulations require 25% of the plantation area to be left in natural vegetation. This helps to preserve biodiversity and provides other ecosystem benefits, where bioenergy crops can also serve as corridors between adjacent natural habitat areas for the benefit of all wildlife.

On the other hand, energy crops, like many other industrial crops, have been slowly escaping from the cultivated area and growing uncontrollably in the natural areas at the expense of the local species. There are many such examples, where crops in various regions have reproduced widely beyond man-made plantations to harm the local species.

Another problem is that growing energy crops at the same field for extended periods of time (monoculture) could have the effect of an incubator for pests or disease, which can then spread into natural habitats.

The complexities of bio-crops management, and the lessons learned from the past, suggest that growing energy crops is a delicate science that needs to be further understood, developed and implemented in an organized manner. Shortcuts, and hasty actions—like the 2007-2008 biofuels boom and bust cycle—might prove too expensive in the long run.

Meat Production

A lot of water, fuels, and other resources are used to grow meat for human consumption. Watering and cleaning the animals and the facilities consumes huge amounts of water, which usually end up in a cesspool nearby. Worse, it can drain into the water table or nearby water bodies.

Raising one pound of meat requires 2,000 gallons of water and produces 58 times more GHGs than growing 1 pound of potatoes.

It takes 7,000 pounds of grain to feed an animal that produces 1,000 pounds of meat.

It takes 16 times more energy (mostly fossil generated) to produce six ounces of meat than to produce a cup of broccoli, a cup of vegetables, or a cup of rice. **These numbers deserve our attention.**

Each year, ten billion bushels of corn are grown in the U.S. alone, 60 percent of which is used to feed cows and other livestock. The meat of these animals is the major ingredient in processed and fast foods, which in turn are the major cause of obesity and diabetes.

Beef meat also produces a lot of greenhouse gases, because of the methane and nitric oxide produced by cow flatulence. Grains and grass ferment in cows' stomachs producing large amounts of methane gas, which is then expelled by flatulence.

Thus released gases from millions of animals in the U.S. alone have to go somewhere…yes, you guessed it, so it represents one of the largest contributors to global warming.

Just think, if we all skipped just one processed meat meal each week, this would be the equivalent to taking 3/4 million cars off the road.

But the meat industry lobby is very powerful and would not allow that. When the USDA suggested "Meatless Mondays," the National Cattleman's Beef Association lobbied the government to retract their recommendation. And, yes, you guessed right again; there is NO Meatless Monday, Tuesday, or Wednesday in the U.S., or anywhere else…yet.

We usually don't need the government or the lobbyists to tell us what to do, do we? Meat eating and overeating is also a huge health problem in the U.S., so skipping a meat meal once or twice a week (or even more) might have a tremendously positive effect on both our environment and personal health.

Summary

Solid biomass use around the world is projected to continue at the present levels at least for the foreseeable future. The negative effects of that use are well known, but there is very little that can be done to reduce them. This is because the major use of solid biomass in the form of firewood is for providing basic, and often the only, human comforts, which cannot be denied to the needy people of the developing world.

Liquid and gaseous biofuels production is also expected to rise significantly. Developing countries are serious about improving their life style, so they demand a lot of energy for power generation and vehicle fuels.

Biofuels derived from organic matter will play an ever increasing important role in providing fuel and reducing CO_2 emissions in the U.S. transport sector, thus enhancing our energy security and helping to clean the environment. They could provide 20% of total transport fuel by 2025 and 30% by 2050, thus particularly significant role in the replacement of gasoline, diesel, kerosene and jet fuel.

Sustainably produced and efficiently used biofuels could avoid the generation of around 2.1 gigatonnes (Gt) of CO_2 emissions per year. For this to happen, however, most conventional biofuel technologies need to improve conversion efficiency, cost, and especially their overall sustainability.

Excess deforestation and using fossils as energy sources to produce biofuels are unsustainable long-run approaches.

Sustainable land-use management and proper certification schemes, together with support measures that promote "low-risk" feedstocks and efficient processing technologies, are the key factors in ensuring fast and efficient development of large-scale biofuels production.

Meeting the biofuel demand by 2050 would require around 65 exajoules of biofuel feedstock, occupying around 100 million hectares. This is in direct competition with the need for land and feedstocks to meet the rapidly growing demand for food and fiber around the world. An additional 80 exajoules of biomass will be needed by 2050 for generating heat and power.

The required 145 exajoules of total biomass for biofuels, heat and electricity from residues and wastes, along with sustainably grown energy crops could be achieved only by the implementation of sound policy framework and by the application of efficient equipment and processes.

Trade in biomass and biofuels will become increasingly important as well, especially in the effort to supply biomass to areas with very high levels of production and/or consumption. Scale expansion and efficiency improvements will reduce biofuel production costs over time. In a low-cost scenario, most biofuels could be competitive with fossil fuels by 2030. In a scenario in which production costs are strongly coupled to oil prices, they would remain slightly more expensive than fossil fuels.

The total biofuels production costs from 2010 to 2050 are estimated in the range between $11 trillion to $13 trillion, while the marginal savings or additional costs compared to the use of gasoline/diesel are in the range of only +/-1% of total costs for all transport fuels.

The United States produces 48 billion liters of biofuels annually. Of this amount only 20 million are produced from cellulosic materials (trees, grass, corn stalks, etc.) There is an estimate of 150 billion liters of bio-ethanol to be produced by 2025, 60 billion liters of which would be from cellulosic-ethanol.

The increased biofuels production, transport and use will also generate millions of jobs around the world, which will contribute to the economic development of a number of developing countries.

Advanced biofuels are the most promising, but least developed of the bunch. They require deployment on a large scale, which requires additional and quite substantial investment in their future development, demonstration, and commercial implementation. Special attention to, and support for, commercial-scale advanced biofuel plants is a must to their timely success.

Support policies should be developed and focused on incentivizing the most promising and efficient biofuels in terms of GHG avoidance. The policy framework must ensure that food security and biodiversity are not compromised, and that only social impacts are the result of the expansion efforts.

Note: The author was part of an R&D team in a South American government R&D lab which specialized in the production of cellulosic bio-products (paper and biofuels). The results of this effort, after several years of intense work, led to the conclusion that it is technologically feasible to produce cheap bio-ethanol and other bio-fuels from different types of tree and brush materials, as demonstrated recently by some companies.

The problem, however, is the high cost of the conversion processes and the huge amount of water used in these types of processes. Water is not readily available in many parts of the world, so this issue must be addressed and resolved first.

The energy input-output balance for most advanced fuels is also questionable, since they require a lot of energy, which is usually generated by fossils.

And then, there is the huge amount of waste products generated during the processing of the raw biomass. Since the useful content of the cellulosic materials is extremely low, the remaining waste comprises 90-95 percent of the raw materials input. This means that very large quantities of raw materials are hauled in, processed at a great expense of energy and chemicals, and then dried and stored in large storage bins, or hauled away again as waste or for other uses.

There are also rivers of chemicals that are prepared, used, and disposed of at the mass production level. This is a very large amount of solid and liquid waste products, so if the production increases to the estimated levels, we will eventually have a waste conundrum.

Some of the wastes could be burned for heat generation at the biofuel processing plants, but the huge amount of smoke, soot, and GHGs would make this process one of the dirtiest and therefore objectionable from an environmental point of view.

Generally speaking, there are a number of hard-to-deal-with problems—especially at the large scale of production anticipated in the next several decades. Bio-fuels, however, are a very promising and important part of our energy future, so solutions must be found.

So, bio-fuels are needed, no doubt, and if properly done can produce significant social and environmental benefits. Proper crop management and processing can help us ensure our energy security, and could also provide excellent carbon pollution reduction.

The key to success here is the implementation of proper and balanced soil and crop management techniques, which must be adapted to the local conditions. This way, greater crop efficiency and environmental benefits could be achieved.

Land selection is critical too, where the best case scenario is to use land that is abandoned or in some way is in degradation. Energy crops compete for land with food production, livestock grazing, and firewood gathering practices, so engaging and educating the local communities in the proper development and use of energy crops is another key to success.

All in all, lots of uncertainties lay ahead, and lots of battles will be fought on different fronts, but one thing is for sure: energy crops have an important role to play in our energy and environmental futures. But as we saw during the uncontrolled boom-bust of 2007-2008, they can also create large-scale problems on a global level.

Using the lessons of the past, we must make sure that future energy crop growing and processing methods are properly designed and implemented in a manner that would benefit the environment and those who are affected.

HYDROGEN ECONOMY

Hydrogen is expected to be one of the drivers of our future energy economy in more ways than one. It might help increase the use of solar and wind power by providing energy storage, which these variable energy sources need badly. And as importantly, hydrogen also might replace crude oil in powering our transportation sector, thus eliminating our dependency on fossil imports.

As it flows free in the mixture of other gasses in the air around us, however, hydrogen is useless for all practical purposes. To become a real fuel and be able to release its energy, it must be extracted from the air (or somehow created from other materials), collected, concentrated, processed, stored, delivered, and used as high energy gas, liquid, or solid.

In its pure, concentrated form hydrogen can be used as an energy source suitable for power generation and/or fueling transportation vehicles.

For the sole purposes of electric power generation it could be classified and successfully used as *energy storage technology* more than anything else. Its value is even greater in the transportation sector, where it could be used in gaseous or liquid form to power all kinds of vehicles.

Solar and wind power plants of the future will be equipped with hydrogen generators and hydrogen storage facilities. Hydrogen is one way of making wind and solar power generators viable, by reducing their variability (fluctuation with solar and wind availability) by storing the excess power these generate. Thus generated and stored hydrogen can be burned to produce electricity in times of low power generation (when no sun or wind are available) and at night.

The Most Possible Future of Solar Hydrogen

Mass hydrogen production is one of the possible ways to reduce the use of natural resources and stop the rampant air, water and soil pollution. Hydrogen is a nontoxic, clean-burning fuel which can be produced from water and sunlight. Thus produced "solar hydrogen" can then be used in every application where conventional fuels are used—including all residential, industrial and transportation applications.

Solar hydrogen is basically hydrogen gas produced by using solar energy (as a primary energy source) or one of its derivatives—renewable forms (secondary energy source) such as wind, ocean waves, falling water, and biomass. Hydrogen can then be used on site as needed, or it can be transported for use anywhere in the country and the world.

As a bonus, by replacing fossil fuels with solar hydrogen, we can move from an unsustainable, dirty, and wealth-depleting economy, to a clean, environmentally friendly, sustainable, wealth-enhancing economy.

Another great benefit that accompanies the production of solar hydrogen is the generation of mass quantities of oxygen as a byproduct of the water molecule splitting process. There are other benefits too, as we will see below.

Solar energy, albeit abundant in many areas of the world is inconveniently remote and inaccessible in large quantities. It is also variable, so we cannot count on it alone for providing reliable power. Generating hydrogen by splitting water via solar energy generated at large-scale solar fields in the high deserts, and storing it for use as primary energy source when solar is unavailable, might be one of the most promising solutions to the issues at hand.

Hydrogen is an efficient fuel. It is carbon-free and is converted back into pure water during the combustion process. This means that we start with water, use the sunlight to split it and generate hydrogen during the day, store it, and then burn the hydrogen at night to generate heat and electricity, which process produces water...again.

This is an almost perfect closed-loop system. A type of a practical perpetuum mobile machine, using only the sun's free energy to assist in the never-ending and very efficient water-to-hydrogen conversion process, where very little energy and water are lost and where the action never stops...as long as the sun shines and the wind blows.

Hydrogen Production

There are several different ways to produce free hydrogen gas suitable for practical use as a fuel. Hydrogen fuel can be produced by:

1. Electrolysis of water,
2. High temperature pyrolysis of natural gas, or
3. Steam reforming or hydrocarbons from petroleum or coal.

Water electrolysis via solar and wind power is the most promising from an environmental point of view, because when fossil fuels are burned to free hydrogen, they contribute to greenhouse gas emissions. Not to mention that we cannot count on them to be around forever.

So the true hydrogen economy of our "clean-energy" future depends on the use of renewables for hydrogen generation.

Using special catalysts created by chemists today, sunlight can turn water into hydrogen. When scaled up, this process could make solar hydrogen feasible.

There are, however, still many *IFs* and issues to be resolved, mostly related to the type, quantity, and cost of solar and wind power to be used, the conversion of hydrogen gas into liquid, as well as its storage and transport.

In addition to the technical issues, the hydrogen conversion process requires expensive equipment and processes, as well as a lot of energy. So it remains to be seen how far this technology can go at the present state of the art and what will follow.

The overall water-to-hydrogen-to-heat-to-electricity process goes something like this:

1. During the day:

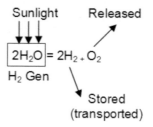

2. During the night:

$$2H_{2} + O_{2} = 2H_{2}O + heat \rightarrow electricity$$

The generated water, together with the sunlight used for splitting it are the only materials and fuel, respectively, needed to generate and store hydrogen, which could be used later to generate power. This is clean and abundant energy in a carbon-free, natural and renewable cycle.

Hydrogen can be liquefied and transported to the point of use just like the other fuels. It can be also stored and transported over long distances. Hydrogen can be also stored in large quantity in underground reservoirs or in containers to be used as needed by industrial enterprises, households, power stations, motor cars, and aviation.

The Challenges

Despite the advantages, water electrolysis and hydrogen/oxygen fuel cell technology still face enormous challenges. For instance, the electrodes used usually in water electrolysis equipment are currently coated with platinum, which is not a sustainable resource and is quite expensive.

Researchers are investigating the employment of nanomaterials with a large reduction in the amount of precious metal needed. Indeed, the main advantage of these materials used as catalysts that splits the water molecules using cobalt phosphate is that it is far cheaper and more abundant compared to expensive metals such as platinum.

Another problem is the intermittent nature of PV electricity (or solar heat), due to interruptions, like clouds and during nighttime, as discussed in more detail elsewhere in this text.

These anomalies create several problems for the H2 generation via solar power. For example, during cloud cover the hydrogen production would decrease or stop, which would shut down the hydrogen generation process. This is unacceptable because that will cause Ni dissolution at the cathode, which could damage it since it is driven to more positive potentials by short-circuit with the anode.

This and other shortcomings of the technology can be alleviated by using stored heat or electricity as energy sources, which would provide constant energy supply. The Ni cathodes eventually could be re-activated by coating them with a thin layer of different, more active and more stable materials.

Another great problem to be considered is the fact that lots and lots of electrical or thermal energy are required to break the H_2-O bond in water molecules, in order to free the hydrogen in it.

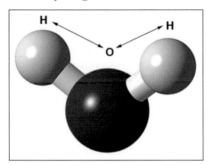

Figure 7-59. Water molecule

The simple reaction,

$$2H_2O \text{ (liquid)} = 2H_2 \text{ (gas)} + O_2 \text{ (gas)},$$

requires almost 3 times more energy (electric power or heat) than it can generate by burning the thus obtained hydrogen. This is not a good account of energy efficiency, but the usefulness of hydrogen electrolysis comes when no other ways of storing excess energy are available.

Photovoltaics, thermal solar and wind power generators will benefit tremendously when the hydrogen production technology is fully developed and implemented in large-scale power generation fields. And vice versa: hydrogen will benefit from the solar and wind power, which could reduce its cost and increase its market applications.

Cost

The problem with using hydrogen today for everyday purposes is that it costs about $6-8 to produce one kilogram of liquid hydrogen, which contains about the same energy content as one gallon of gasoline. Since gasoline is 2-3 times cheaper (after subsidies), producing hydrogen makes no economic sense at this time.

Once the infrastructure is developed, however, and equal subsidies applied, then hydrogen might be able to compete...someday. Also some day, when produced in large quantity, sometime in the distant future, its production cost would go down significantly. At that point, combining hydrogen with free oxygen (air) in fuel cells would make an extraordinarily convenient green form of power generation since there is a lot of energy produced by this reaction and the waste byproduct is pure water.

Platinum Depletion

For the hydrogen economy plan to work on a large scale, it will require millions of fuel cells, which use platinum metals. At that time, larger reserves of platinum-group metals must be mined as needed to build the requisite number of fuel cells.

The United States Geological Survey today estimates the total global reserves of platinum that can be mined at roughly 100 million kilograms, which is just enough to retrofit the 4 billion automobiles worldwide with fuel cells at the current 22.6 grams platinum needed per hydrogen-powered vehicle.

Soon enough after that, there would be no more platinum...

If and when platinum is depleted, we will have to find another material, or move mining operations off-planet, where there is evidence of much higher densities of platinum and platinum group metals (cobalt, nickel, ruthenium, rhodium, palladium, osmium, iridium) than are available on Earth.

Another pipe dream, yes...for now! It might, however, become a full reality when future generations run out of fossils, decide to shut down all their nuclear plants for fear of accidents, when water is too scarce and no longer available for hydro-power generation, and when the atmosphere will not accept anymore GHGs from fossil power plants.

Then, many things we consider impossible pipe dreams today might become life-and-death reality... Hydrogen-powered economy is one of those possibilities.

Hydrogen Storage

Once generated, hydrogen can be stored in its gaseous form, but it is more convenient to convert it into a liquid form. In this state, hydrogen can be transported, stored and burned either as free hydrogen or linked to other molecules like ammonia (NH_3), methanol (CH_3OH), methylcyclohexane (C_7H_{14}), or in a solid, such as a metal hydride.

Although liquid hydrogen (LH_2) has the highest energy density, it is disadvantaged by the energy required (approximately 30% of its combustion energy) for conversion to liquid and keeping it at a very low temperature of 20 K. These are energy wasting processes that cannot be avoided, if we have to use liquid H_2 fuel.

Producing large quantities of LH_2 is a relatively new process, the long experience from the space sector, and transporting liquid natural gas (LNG) are directly applicable for the new technology, since the storage techniques of hydrogen as a gas are similar to those of natural gas. LH_2 can be stored in storage tanks, underground in confined aquifers, or in abandoned natural gas reservoirs.

Most rock formations used for underground storage are sealed by water in their capillary pores, which reduces overall leakage and yet some losses are inevitable. The losses from underground reservoirs are on the order of 2-30% per year.

Another storage method is to convert (combine) LH_2 into Methylcyclohexane, which is liquid at ambient pressure and temperature. In this new liquid state (consistency and specific weight similar to diesel fuel) the LH_2 can be stored and transported in the usual oil product carriers. At the point of use it can be converted back into its original state. This process, however, requires specialized equipment, processes, and chemicals. Still, it could become important for a smooth energy-system infrastructure transition from crude oil to hydrogen.

LH_2 also can be stored as a solid in the form of solid, rechargeable metal hydrides. This method offers favorable volumetric hydrogen packing although the weight of the non-saturated compound needed for storage of hydrogen is rather high. However, it also requires special equipment, processes, and materials.

Hydrogen Distribution

In the future transition period, gaseous hydrogen can be transferred via existing natural gas pipelines, and delivered to natural gas fueled appliances in home areas and in industry. Some modifications to appliances would be necessary for added safety, but they are not seen as a show stopper.

Existing gas pipe networks also need retrofitting to be optimized for use with hydrogen. Hydrogen has a volumetric heating value of approximately one-third that of natural gas, so it requires at least 3 times higher flow rate and compressor power at given pipe diameter and energy flow as needed to keep the system under a minimum of 60-70 bar positive pressure.

Transportation of LH_2 can be done most conveniently and cheaply by pipelines, but wherever these are not available, then pressurized or cryogenic containers are used.

A gaseous hydrogen pipeline system of some hundred kilometers and a vacuum-jacketed LH_2 system of some hundred meters are in operation today, the latter in the Kennedy Space Center.

In the transition period, hydrogen can also be mixed with natural gas. As a matter of fact, the "town gas" in the recent past, containing up to 60% hydrogen, was transferred via regular gas lines.

Here again, we are talking about things that might happen in the near future, but most likely it would take many decades before these technologies become an everyday reality. Still, dreaming is good and is one way to prepare us for the future.

Fuel cells seem to be the most promising—in the short term—for quick implementation in the present energy sector, and during the transition to hydrogen economy.

Fuel Cells

The key element of an efficient, future electricity generation is the fuel cell, for which hydrogen is the best fuel. Since these are devices which transform chemical energy directly into electric energy, hydrogen generation is not subjected to Carnot cycle limitations. It is also done silently without moving parts and with high efficiency.

Efficiencies of future fuel cells like molten carbonate cells operating at 600-800°C with total energy efficiencies of 60-80% with waste heat at 600°C make the fuel cell concept a potential candidate for use in power stations, air- and road-transport, and space vehicles.

Advances in catalyst and electrode technology and cost reductions will have to be made, however, before fuel cells can seriously be considered for general use.

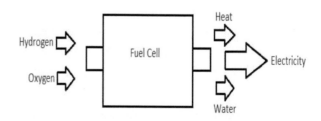

Figure 7-60. Fuel cell process

Notwithstanding, several power plants around the world have already been using different fuels (which is one of the great advantages of fuel cells.)

Types of Fuel Cells

There are many types of fuel cells, with the main difference among them all being the type and composition of the electrolyte layer. Fuel cells are, therefore, usually classified by the type of electrolyte they use, and each type comes in a variety of sizes. Other distinguishing elements are the composition, shape and size of the cell components, which would also determine its type and application.

Herein we will review several of the major and most promising fuel cell designs: a) phosphoric acid fuel cells, b) proton exchange membrane fuel cells, c) solid oxide fuel cells, and d) molten carbonate fuels cells.

Phosphoric Acid Fuel Cells

Phosphoric acid fuel cells (PAFC) are a type of fuel cell that uses liquid phosphoric acid as an electrolyte, and due to their simple and practical construction, they were the first fuel cells to be commercialized in the 1970s. Their quality and efficiency have improved significantly and have made them a good candidate for stationary power generating applications.

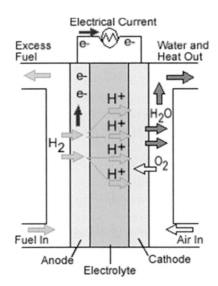

Figure 7-61. PAFC fuel cell

The chemical reactions for a PAFC system can be expressed as follow:

Anode reaction: $2H_2 \rightarrow 4H+ + 4e^-$
Cathode reaction: $O_2(g) + 4H+ + 4e^- \rightarrow 2H_2O$
Overall cell reaction: $2H_2 + O_2 \rightarrow 2H_2O$

The electrolyte here is highly concentrated liquid phosphoric acid (H_3PO_4), which is saturated in a silicon carbide material (SiC). The electrode is made out of carbon paper coated with a finely dispersed platinum catalyst. The normal operating temperature range is about 150 to 210°C, at which point the expelled water is converted into steam and can be used for air and/or water heating.

Note: Generating both hot air and water is known as combined heat and power process, which increases its practical efficiency up to 70%.

At lower temperatures phosphoric acid is a poor ionic conductor, and serious CO poisoning of the platinum electro-catalyst in the anode can occur. Nevertheless, these cells are much less sensitive to CO than PEFCs and AFCs.

PAFCs are also CO_2- and CO-tolerant (of concentrations of about 1.5 percent). This broadens the choice of fuels they can use, which includes gasoline if the sulfur is removed prior to injecting into the fuel cell.

A major disadvantage is the low power density of the cell, and the aggressive (corrosive acid) electrolyte in its structure.

PAFC have been used mostly for stationary power generators with output in the 100-400 kW range and for powering large vehicles such as buses.

The best known manufacturers of PAFC fuel cells are UTC Power, and Fuji Electric. India's DRDO is developing PAFC for air independent propulsion in their Scorpène class submarines, and the Indian Navy has requested a fully operational PAFC system for implementation by 2014.

Proton Exchange Membrane Fuel Cells

The proton exchange membrane fuel cell (PEMFC) design is based on a proton-conducting polymer membrane, as the electrolyte between the anode and cathode. Initially, this design was called *solid polymer electrolyte fuel cell* (SPEFC), before the key proton exchange mechanism was well understood.

The chemical reaction in this cell starts on the anode side, where hydrogen diffuses to the anode catalyst where it later dissociates into protons and electrons. The protons often react with oxidants which become multi-facilitated proton membranes, which allow the protons to pass through the membrane to the cathode. The membrane (which at this point is electrically insulating) does not allow electrons to pass through, so they are directed into an external circuit through wires, which can be connected to the grid, or used directly to power appliances and other electrical devices.

If a wire is connected from the anode to the cathode, the cathode catalyst (oxygen) reacts with the incoming electrons which have traveled through the external circuit, and the available protons, thus forming pure water.

Other hydrocarbon fuels can be used in this type of fuel cells, which include diesel, methanol, chemical hydrides, and others. In all cases, the waste products from all these types of fuel cells are CO_2 and H_2O.

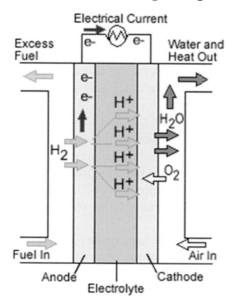

Figure 7-62. PEM fuel cell

A high temperature PEMFC consists of a bipolar plate electrode, with included in-milled gas channel structure. It is fabricated from conductive composites, enhanced with carbon containing materials (graphite, carbon black, carbon fiber, and/or carbon nanotubes) for more conductivity. The membrane is made of porous carbon paper.

The key components of a PEMFC are a) bipolar plate, b) electrode, c) catalyst, d) membrane, and e) wires and necessary hardware. These can be made of different material. For example, the bipolar plates can be made of materials such as, metal, coated metal, graphite , flexible graphite, C-C composite, carbon-polymer composites, etc.

The membrane electrode assembly (MEA) is the heart of the PEMFC, and is usually made of a proton exchange membrane sandwiched between two catalyst coated carbon papers.

Platinum and/or similar types of noble metals are usually used for catalysts in PEMFC devices, while the electrolyte could be a polymer membrane.

Water and air management in PEMFCs is critical.

The membrane must be hydrated, which requires water to be evaporated at precisely the same rate that it is produced—not an easy task. The problem is that if water is evaporated too quickly, the membrane dries and the resistance across it increases. This will eventually crack it, thus creating a gas "short circuit," where hydrogen and oxygen combine directly. This unwanted process generates heat, capable of damaging the fuel cell.

On the other hand, if the water is evaporated too slowly, the electrode floods, preventing the reactants from reaching the catalyst, which usually stops the reaction. Electro-osmotic pumps can be developed and used to control the flow, since a steady ratio between the reactant and oxygen is necessary to keep the fuel cell operating efficiently.

Temperature management is critical in this type of fuel cell, since a constant temperature must be maintained throughout the cell to prevent destruction of the cell through thermal loading. This is not easy, because the particularly challenging $2H_2 + O_2 \rightarrow 2H_2O$ reaction is exothermic and generates a large quantity of heat within the fuel cell.

Durability, or service life , and special requirements for some types of cells requires more than 40,000 hours of reliable operation at a temperature of −31 to +104°F. At the same time, automotive fuel cells require at least a 5,000-hour lifespan, which is equivalent to 150,000 miles operation under extreme temperatures. Unfortunately, the service life of present-day PEM fuel cells is about 7,000 hours under normal cycling conditions. Another problem to be resolve is the requirement that car engines must be able to start reliably at −22°F, and have a high power-to-volume ratio of typically 2.5 kW per liter, which is a problem with the present PEMFC state of the art.

Another problem here is that some non-PEDOT cathodes have very limited carbon monoxide tolerance, which narrows the area of their applications.

The cost of this type of fuel cells in volume production (projected to 500,000 units per year) was estimated at $49 per kilowatt, while the goal is $35 per kilowatt. This almost 25% cost reduction is needed for PEM fuel cells to compete with current market technologies, including gasoline engines. Companies working on this effort are Ballard Power Systems and Monash University, Melbourne.

Solid Oxide Fuel Cell

Solid oxide fuel cells (SOFC) use a solid material, most commonly a ceramic material, such as yttria-stabilized zirconia (YSZ), instead of an electrolyte. Because

SOFCs do not have any liquids, they have no limitations as to their size or shape, and are often manufactured as rolled tubes. However, they require much higher operating temperatures in the 800-1000°C range, but can be run on a variety of fuels, including natural gas.

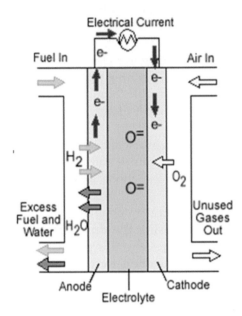

Figure 7-63. SOFC fuel cell

The chemical reactions for an SOFC system can be expressed as follow:

Anode Reaction: $2H_2 + 2O^{2-} \rightarrow 2H_2O + 4e^-$
Cathode Reaction: $O_2 + 4e^- \rightarrow 2O^{2-}$
Overall Cell Reaction: $2H_2 + O_2 \rightarrow 2H_2O$

During the SOFCs operating process, the negatively charged oxygen ions travel from the cathode to the anode. This is opposite to the case with other fuel cells, where positively charged hydrogen ions usually travel from the anode to the cathode. Here, oxygen gas fed through the cathode reacts with electrons to create oxygen ions, which then travel through the electrolyte to react with hydrogen gas at the anode. This reaction at the anode produces electricity and water as by-products.

Carbon dioxide is usually a by-product, depending on the fuel, but in all cases, the carbon emissions from an SOFC system are less than those from a fossil fuel combustion plant.

SOFC systems can run on fuels other than pure hydrogen gas, but since hydrogen is necessary for the reactions listed above, all fuels must contain hydrogen atoms. The fuel also must be converted into pure hydro-gen gas, for proper and efficient cell operation.

SOFCs are capable of internally reforming light hydrocarbons such as methane (natural gas), propane and butane. Heavier hydrocarbons including gasoline, diesel fuel, jet fuel and biofuels can serve as fuels in a SOFC system, but in these cases, an external reformer is required.

Another advantage of these systems is that waste heat from SOFC operations can be captured and reused, increasing the theoretical overall efficiency of the devices to as high as 80-85%.

The major disadvantage of using SOFC systems is their very high operating temperatures. Also, carbon dust builds up on the anode with time, which slows down the internal reforming process. The "carbon coking" issue can also be resolved by using copper-based cermet (heat-resistant materials made of ceramic and metal), which reduces coking and the loss of performance.

Note: The high operating temperature is largely due to the physical properties of the YSZ electrolyte, since as temperature decreases, so does the ionic conductivity of YSZ. So, to obtain optimum performance of the fuel cell, a high operating temperature is required.

Some SOFC manufacturers have developed methods of reducing the operating temperature of their SOFC system to 500-600°C. This can be done by replacing the YSZ electrolyte with a CGO (cerium gadolinium oxide) electrolyte. The lower operating temperature allows using stainless steel instead of ceramic, as the cell substrate, which reduces cost and start-up time of the system.

Another serious disadvantage of SOFC systems is their slow start-up time, needed to reach operating temperature. This makes them less useful for mobile, temporary, and low temperature applications. Nevertheless, the high operating temperature eliminates the need for expensive precious metal catalysts, like platinum, and that reduces the initial cost.

Molten Carbonate Fuel Cell

Molten carbonate fuel cells (MCFCs) also require very high operating temperatures, over 650°C, similar to SOFCs. Here lithium potassium carbonate salt used as an electrolyte liquefies at the high temperatures, thus allowing free movement of charges (negative carbonate ions) within the cell.

MCFCs are also capable of converting some fossil fuels to a hydrogen-rich gas in the anode, eliminating the need for an external hydrogen generator, while the reforming process creates CO_2 emissions. MCFC-compatible fuels include natural gas, biogas and gas produced from coal. The hydrogen in the gas reacts with carbonate ions from the electrolyte to produce water, carbon dioxide, electrons and small amounts of other chemicals.

The electrons travel through an external circuit creating electricity and return to the cathode. There, oxygen from the air and carbon dioxide recycled from the anode react with the electrons to form carbonate ions that replenish the electrolyte, completing the circuit.

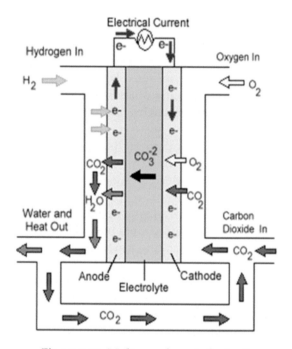

Figure 7-64. Molten carbonate fuel cell

The chemical reactions for an MCFC system can be expressed as follow:

Anode Reaction: $CO_3^{2-} + H_2 \rightarrow H_2O + CO_2 + 2e^-$
Cathode Reaction: $CO_2 + \frac{1}{2}O_2 + 2e^- \rightarrow CO_3^{2-}$
Overall Cell Reaction: $H_2 + \frac{1}{2}O_2 \rightarrow H_2O$

MCFCs hold several advantages over other fuel cell technologies, including their resistance to impurities. The main one is that they are not prone to "carbon coking," which is carbon build-up on the anode that results in reduced performance by slowing the internal fuel reforming process. Therefore, carbon-rich fuels like gases made from coal are compatible with the system. The Department of Energy claims that coal, itself, might even be a fuel option in the future, assuming the system can be made resistant to impurities such as sulfur and particulates that result from converting coal

into hydrogen.

MCFCs also have relatively high efficiencies. They can reach a fuel-to-electricity efficiency of 50%, considerably higher than the 37-42% efficiency of a phosphoric acid fuel cell plant. Efficiencies can be as high as 65% when the fuel cell is paired with a turbine, and 85% if heat is captured and used in a combined heat and power (CHP) system.

MCFC disadvantages include slow start-up times because of their high operating temperature. This makes MCFC systems not suitable for mobile applications, and this technology will most likely be used for stationary fuel cell purposes.

Another serious problem here is the cells' short life span. The high temperature and carbonate electrolyte lead to corrosion of the anode and cathode materials with time.

These factors accelerate the degradation of MCFC components, decreasing the durability and cell life. Researchers are addressing this problem by exploring corrosion-resistant materials for components as well as fuel cell designs that may increase cell life without decreasing performance.

Generally speaking, the fuel cell market is also growing at a healthy pace and according to insiders, the stationary fuel cell market is predicted to reach 40-50 GW by 2020.

Hydrogen Uses

Hydrogen is a carbon-free fuel with excellent combustion properties. This makes it a clean, universal fuel for use in industry, households, power stations, road vehicles, boats, trains, and aviation. The hard part is producing it, but once that is resolved (by using solar, wind etc. renewable energy in the future), using hydrogen in fuel cells would provide a number of benefits.

The specific properties of hydrogen induce new end-use equipment, processes and techniques that can be used for a number of practical applications. Fuel cells are just one, but a very important part on the top of the list of possible applications, which we will review below.

Upon combustion, hydrogen produces heat and clean water. Two pounds of hydrogen release about 120,000 Btu heat, or about as much as a gallon of gasoline. Or, 10 million tons of hydrogen per year can fuel 25 to 30 million hydrogen-fueled cars, or enough to power 6 to 8 million homes. Not bad, keeping in mind that 10 million tons of hydrogen could be easily generated in the Arizona or Nevada deserts, where sunlight is as abundant as the sand.

Table 7-21. Fuel cell types and applications

Fuel Cell Type	Temperature Electrolyte / Charge Carrier		Applications
Phosphoric Acid (PAFC) and Polymer / Phosphoric Acid	150–200° C		Distributed power Transportation
	H_3PO_4, Polymer/H_3PO_4 / H^+		
Polymer Electrolyte Membrane (PEMFC)	50–100° C		Distributed power Portable power Transportation
	Perfluorosulfonic acid / H^+		
Direct Methanol (DMFC)	50–100° C		Portable Power
	Perfluorosulfonic acid / H^+		
Alkaline (AFC)	25–75° C, 100–250° C		Portable Power Backup Power
	Alkaline polymer, KOH / OH^-		
Molten Carbonate (MCFC)	600–700° C		Distributed power
	$(Li,K,Na)_2CO_3$ / CO_3^{2-}		
Solid Oxide (SOFC)	500–1000° C		Electric utility Distributed power APUs
	Yttria–Stabilized Zirconia $(Zr_{.92}Y_{.08}O_2)$ / O^{2-}		

The practical implementation and application of new hydrogen based economy, however, is not simple, or cheap.

The production of 10 million tons of hydrogen annually, using the present state of the art as needed to run a hydrogen based U.S. transportation sector, would require:

- 35 coal/biomass gasification plants, similar to today's large coal fired plants,

- 25 large nuclear plants making only hydrogen, and

- 15 medium-size power plants, using oil and natural gas in multi-fuel gasifiers and reformers.

A quick glimpse in the future reveals that to provide the entire U.S. transportation system with (practical) hydrogen fuel we will need:

- 250,000 small neighborhood-based hydrogen electrolysis generators to fuel cars and trucks, or several hundred large-scale hydrogen generators near solar or wind power plants, and

- 20,000 interstate hydrogen vehicle refueling stations around the country for passenger and commercial vehicles refueling.

This basically means replacing and/or retrofitting most of the existing gas stations with new high pressure storage tanks, compressors, pumps, etc.—equipment needed to serve the hydrogen powered vehicles. This would come at an enormous expense, in addition to a number of other serious problems during the transition period.

Yet, the most attractive part of this pipe dream is the possibility of using free solar and wind energy. The huge unused desert areas in Arizona, New Mexico, Utah, Texas, and Nevada offer unlimited solar energy, which can be easily captured and converted into gaseous and liquid hydrogen to be used in the U.S. and abroad. There is also an unlimited supply of wind energy at many different areas around the country that are suitable for hydrogen production.

Fantasy? Yes…for now. There are still technological issues to resolve and equipment costs to reduce, but this is still what future generations will have to consider doing after we burn all the fossils. They will be forced into it, so it will be done one way or another…else the economy will stop dead in its tracks.

Some uses of hydrogen are:

Catalytic Burners

Hydrogen, having lower ignition energy than some conventional fuels, including natural gas, burns smokeless when in contact with a suitable catalytic surface. Its burning temperature is quite low too—between 100 and 400°C. This then makes it preferred technology for generating low temperature heat in residential space heaters, cooking devices and industrial dryers and heaters.

Catalytic combustion has the advantages of higher safety level and very high efficiency of up to 99% with negligible emissions of nitrogen oxides. Its use is limited by the complexity of the delivery networks and the catalytic conversion devices.

Liquid Hydrogen Technologies

Super/hypersonic aviation and direct use in jet engines at super/hypersonic speeds is considered, and could be advantageously—if not necessarily—powered with liquid hydrogen because of its:

1. Higher (three-fold) gravimetric heating value compared with kerosene, and

2. Cooling capacity, which can be used to cool critical parts (turbine inlet rim, wing leading edges, passenger cabins, etc.), since LH_2 cryogenic storage temperature is approximately –252°C.

As an added benefit, cooling of the wing skin (by stored-in-it liquid hydrogen) induces laminar flow and therewith considerably reduces drag by up to, theoretically, 30% (aka "laminar flow control").

The safety of a liquid hydrogen fueled airplane is debatable, but we'd be quite concerned with the additional safety mechanisms and procedures needed. Fueling, storing, and maintaining the pressure and the super-low temperature in the fuel tanks could prove to be tricky, especially in the learning stages of implementing this new technology.

Hydrogen Powered Vehicles

Fuel cells are important for use in vehicles, where presently their *tank-to-wheel* (TTW) efficiency is greater than 45% at low loads, and about 36% during a normal driving cycle load.

Note: The comparable value for a diesel vehicle is 22%, and Honda has a demonstration fuel cell electric vehicle, the Honda FCX Clarity, with a 60% tank-to-wheel efficiency.

The problem with this approach is that there are serious energy losses due to fuel production, transportation, and storage. Here we talk about power plant to wheel (PPW) energy conversion and efficiency; i.e., fuel cell vehicles running on compressed hydrogen have a power-plant-to-wheel efficiency of 22%, if the hydrogen is stored as high-pressure gas, and 17% if it is stored as liquid hydrogen.

Fuel cells cannot store energy like a battery, except as hydrogen, but in some applications, such as stand-alone power plants based on discontinuous sources such as solar or wind power, they are combined with electrolyzers and storage systems to form an energy storage system.

Most hydrogen, however, is produced by steam methane reforming, so most hydrogen production emits carbon dioxide. The overall efficiency (electricity to hydrogen and back to electricity) of such plants (known as *round-trip efficiency*), using pure hydrogen and pure oxygen can be "from 35-50 percent," depending on gas density and other conditions. While a much cheaper lead-acid battery might return about 90%, the electrolyzer/fuel cell system can store indefinite quantities of hydrogen, so it's better suited for long-term storage.

Solid-oxide fuel cells produce exothermic heat from the recombination of the oxygen and hydrogen. The ceramic can run as hot as 800 degrees Celsius. This heat can be captured and used to heat water in a micro combined heat and power (m-CHP) application. When the heat is captured, total efficiency can reach 80-90% at the unit, but does not consider production and distribution losses. CHP units are being developed today for the European home market.

Generally, while fuel cells are efficient relative to combustion engines, they are not as efficient as batteries, due primarily to the inefficiency of the oxygen reduction reaction, or the oxygen evolution reaction, should the hydrogen be formed by electrolysis of water. They make the most sense for operation disconnected from the grid, or when fuel can be provided continuously.

For applications that require frequent and relatively rapid start-ups, and where zero emissions are a requirement as in enclosed spaces such as warehouses, and in operations where hydrogen is considered an acceptable reactant, and if exchanging batteries is inconvenient, PEM fuel cells are an attractive choice.

Hydrogen Storage

The main problem of hydrogen utilization in vehicles is storage due to the low volumetric energy content of hydrogen, which is about one-third that of gasoline. Even liquid hydrogen is far less dense than hydrocarbon fuels. This translates to less miles per gas tank, which is unacceptable to most people.

As we saw above, there are several different storage and distribution technologies available today. Some are more mature than others, and all are considered for future mass hydrogen storage.

Hydrogen storage and distribution technologies today (and tomorrow) are:

- Gaseous hydrogen in gas lines, underground storage, or in pressure vessels,

- Liquid, cryogenically stored hydrogen, or in the form of hydrides (methylcyclohexane), and

- Solid matter; hydrogen chemically bonded in metal.

The low ignition energy of hydrogen and its wide ignition range of hydrogen/air mixtures make gas turbines as well as piston engines well adaptable for hydrogen combustion. The ignition within a wide range of non-stoichiometric air/fuel mixture permit combustion with high amounts of excess air at full-load operation with low nitrogen oxides emission.

A number of liquid hydrogen cars have been built and operated by BMW and DFVLR (Deutsche Forschungs-und Versuchsanstalt fur Luft-und Raumfahrt), and many universities and research groups. There are also vehicles powered by a version of the methylcyclohexane technology built and operated by the Paul Scherrer Institute in Switzerland.

Experimental vehicles with hydride storage have been built and operated by Daimler Benz and others. These vehicles show promising performance characteristics, but the necessary infrastructure for their refueling and service is not there, so their future is uncertain.

It should be emphasized here that the problems of the necessary infrastructure development are of major concern, due to serious technological, logistical, and economic factors. They are at least as difficult and important as the development of the different components of the new technology, so all this must be considered when discussing the options.

Practical Examples

In the last ten years, much hype has been associated with the hydrogen topic. Questions have been raised about the claim that hydrogen is an economically viable fuel for transportation because of the cost and greenhouse gases generated (from fossils) during its production. Also considered are the low energy content per volume, the weight of the container, the cost of the fuel cells, and the cost of the related H_2 fuels infrastructure.

The price of solar electricity has fallen significantly, and the pace of innovation in both fields has been so intense, so that a number of unexpected practical hydrogen based applications have emerged. Some examples follow:

Enel, Italy, started operation of a 12 MW H_2-powered electricity plant in Venice recently. The plant is fueled by hydrogen byproducts from local petrochemical industries. The turbines were specially designed to resist embitterment from hydrogen, but in any case the only emission of hydrogen combustion is water.

In 2009 the Austrian companies Fronius, Bitter and Frauscher successfully presented Riviera 600, the first electric boat powered by solar hydrogen fuel cells. The concept is that of a self-contained energy supply provided by hydrogen simply obtained by photovoltaic electrolysis of water.

The team "Future Project Hydrogen" has created budget calculations for the generation of hydrogen on site by use of photovoltaics under the premises of 10 boats for commercial use, or for boat rental. With a range of 50 miles (with a full hydrogen tank) and having been awarded a safety certificate by Germany's TÜV, the boat is 18 feet long, 6 feet wide and weighs about 3,000 lbs. Its 4 kW continuous-power electric motor has twice the range of conventional battery-powered boats.

The 47% efficiency of the noise-free fuel cell engine should be compared to the 18%-20% efficiency of a conventional (steel body) internal combustion engine.

The main economic advantage compared with conventional electric boats is the fact that no time has to be spent charging the batteries. For conventional electric boats, 6-8 hrs. of charging gives just 4-6 hrs. of use. The hydrogen-powered electric boat requires only the time that it takes to change the cartridge—5 minutes maximum.

The hydrogen boat's fueling system consists of a 20 kg cartridge that can be charged with up to 0.7 kg of hydrogen kept at 350 bar. Refueling is done using a standard filler coupling plus a simple exchange of an empty cartridge for a full one.

All three companies involved in the "Future Project Hydrogen" are based in Austria, very close to each other. Scientific and technical advice was provided by the Technical University of Graz, whereas the project was realized with support of the European Union regional programs and further funding from one of Austria's regions. The first 600 Riviera boat is commercialized at 150,000 V, with the first exemplars to be delivered to customers in early 2010.

The filling ("Clean Power") station uses PV modules integrated in a 250 m^2 flat roof, connected to an electrolytic cell. Even at Austria's cold latitudes, the station is capable of affording an annual yield of 823 kg hydrogen, equivalent to 1100 cartridges with a 27,200 kW hrs. energy content. This is enough hydrogen to run a boat non-stop for 80,000 km.

The installation is simple, thanks to the "container construction" design that can be carried out simply and quickly. The filling station consists of electric power charger, hydrogen, and payment units.

For comparison, storing power in batteries over long periods of time is linked to huge losses due to self-discharge (5-10% per month), while the energy density is a fraction of that for hydrogen, which means that by storing energy in the summer in a battery of the same capacity, one would have no more energy available in winter.

Another example that shows that photovoltaic renewable hydrogen is far from being solely a research topic is given by the world's first underground pipeline supplying H$_2$ to customers in the Italian city of Arezzo. At present, the pipeline serves 4 companies and the HydroLAb with a main channel of around 2000 feet where the whole network is around 3,000 feet. Four goldsmith companies use it for industrial and energy needs via four 5 kW fuel cells and two 1 kW fuel cells at the HydroLAb, the laboratory for hydrogen and renewable energies

The aim was to set up a completely off-grid testing lab for technologies in the renewable energy sector, collecting data to test solar energy technologies linked with hydrogen production and use. Hence, solar panels provide electricity, solar thermal vacuum tube panels provide heat for room heating and feed a 5 kW solar cooling machine in order to get zero emission air conditioning in summer.

Waste water is recycled through a special remediation dry technique and rain is collected and stored as well. The technologies implemented are continuously monitored, with the aim of further optimization in view of widespread commercial application in the building industry.

These are very encouraging developments, no doubt, but it could be decades before we see an American fishing boat with a twin diesel engine power plant going deep-sea fishing, powered by hydrogen. But it will happen...someday.

This is the humble beginning of a promising technology. Or, it might just be the end of a good idea that was not properly funded and developed. Only the future will tell...

Hydrogen Economy

Hydrogen has a number of advantages:

1. Hydrogen promises smooth transition from the existing dirty, inefficient, and unsustainable energy systems of power generation, storage and transport, to clean, efficient, renewable energy.

2. Hydrogen energy generation, storage and transport systems are fully developed today, do not need special materials or processes and can be implemented immediately.

3. The use of hydrogen as an energy source is ecologically safe, because there is virtually no pollution generated in the process, unlike most of the conventional energy generators and their supply chains.

4. Solar hydrogen generated power could be quite cheap, around 8-10 cents/kWh.

5. Hydrogen is a convenient renewable energy source which can be distributed all over the world, including Third World countries, where such implementation is badly needed.

Hydrogen can be cost competitive with gasoline. This estimate is based on the use of solar thermal energy to generate hydrogen, mass production of the necessary equipment, fully developed infrastructure, and subsidies equivalent to those provided to the oil, coal, and utility industries.

Large-scale production could deliver hydrogen at under $1 per GGE. Likewise, given the low cost of producing electrical power from existing hydroelectric dams, the use of this electricity to produce hydrogen by electrolysis would cost less than gasoline.

Hydrogen holds the world's record for burning faster (more explosive) than any other fuel and at leaner (lower) fuel-to-air ratios than the ratios of the hydrocarbon fuels. These characteristics allow engines that burn hydrogen to operate more efficiently than those that depend on other fuels.

Virtually every existing engine application from lawn mowers to automobiles and locomotives can be fueled with hydrogen, and benefits include more power and longer engine life.

But perhaps the most astonishing benefit of burning hydrogen in a conventional engine is what we might call minus emissions—that is, the exhaust pipe releases cleaner air than that which enters the engine. Atmospheric levels of carbon monoxide, tire particles, hydrocarbons, pollen, and diesel soot are reduced as air is cleaned by the hydrogen flame. The pollutants are substantially converted into harmless gases.

Kits for high-efficiency combustion of hydrogen can be retrofitted on most engines in the global fleet of motor vehicles. Ordinary engines that have been converted to operate on hydrogen show no sign of metal embrittlement or other degradation after decades of pollution-free service.

Engine oil stays clean, spark plugs last much longer, and degradation as measured by corrosion and wear on piston rings and bearings is greatly reduced. These vehicles can therefore last longer, run better, and clean the air. And this transition will facilitate the distribution of hydrogen for the advent of fuel cells on a commercial scale [see "The Fuel Cell Future," THE WORLD & I, April 1994, p. 192].

Solar hydrogen can make any country energy-independent and pollution-free as far into the future as the sun will shine, Illustratively, just a small portion of North America can produce enough solar hydrogen to supply all the energy needs of the whole continent. Currently, Canada, Germany, Japan, Saudi Arabia, and Russia lead the world in developing plans to employ solar hydrogen. And several major auto manufacturers have experimental fleets of hydrogen-powered vehicles.

In this manner, if solar hydrogen is used to provide energy-intensive goods and services to the world's population, it will facilitate wealth addition as opposed to wealth depletion that results from burning fossil resources. The harder we work in the solar-hydrogen economy, the more goods and security everyone can have, resulting in a lower rate of inflation and less reason for conflicts and strife. These circumstances can create a global incentive to work for higher sustainable living standards, for both present and future generations.

Yes, agreed! The above is the truth and hydrogen has a place in our energy sector as a replacement for the fossils. It could eventually replace crude oil for use in the transport sector, thus ensuring our energy security.

100% replacement of crude oil by hydrogen would bring us energy independence.

Yet, we are not even near that point. So let's get more realistic:

Practical Hydrogen Economy

A quote in the media from an unknown source: "With most hydrogen today being produced from hydrocarbons, the cost per unit of energy delivered through hydrogen is higher than the cost of the same unit of energy from the hydrocarbon itself." Duh!

Energy economy powered by solar hydrogen generation is the best, and for now the only, way to bring us to energy independence. However, this requires a lot of effort and money, as well as complete control of the production, storage, distribution, and utilization of hydrogen.

This could happen someday by:

1. Establishing renewable power fields, where generators convert solar energy and its derivatives, such as wind, falling water, wave motion, and biomass resources into electricity and hydrogen to be used as storage of energy or for transport to be used at remote locations.

2. Surplus hydrogen, and that for night power generation, could be stored in depleted natural-gas and oil wells and similar geological formations.

3. Existing infrastructure of electricity grids, natural gas pipelines, highways, and rails to distribute hydrogen and electricity from renewable resources could be used with little modification and expense.

4. Installation of kits that enable the current internal combustion engines of motor vehicles to operate with directly injected hydrogen, landfill gas, or gasoline. A vehicle converted to operate on hydrogen cleans the air through which it travels.

5. Introduction of automobiles that run on hydrogen-based fuel cells, to replace the wasteful and dirty internal combustion engines.

6. Installation of reversible electrolyzers and fuel cells in homes and businesses, where these could produce electricity during off-peak hours, to be converted to storable hydrogen. Thus stored hydrogen can be used to make electricity at peak hours.

Obviously, hydrogen production and use has problems, and although it is considered one of the cleanest energy technologies, hydrogen also has a negative environmental side as well.

Environmental Effects

Although much cleaner than the fossils, the environmental effects of hydrogen production which uses a lot of external power (fossils mostly), transport (via pipes), storage (in underground caverns, or pressurized tanks), and power generation (by direct burning, via fuel cells, or other means) is measurable and when the entire life cycle is considered, significant.

All energy sources used in the hydrogen gas production, mostly fossils, have their own environmental imprint, as discussed in previous chapters, so just by using them we are contributing to the complexity of the already complex environmental picture. Also a lot of energy is needed to free hydrogen from water molecules too. A lot of equipment has to be manufactured, installed and used in the process.

So we need to add here the impact of the solar and wind power equipment (which we discussed in detail in the previous chapter). This is in addition to the pollution created by the hydrogen gas generating and burning equipment (electrolyzers, fuel cells, and such).

The electrolytic cells equipment for mass production of hydrogen is made of different—usually expensive materials—and is of significant size too. These systems are contained in quartz, or stainless steel vessels, with a ceramic micro-porous separator (diaphragm) in the middle, while the electrodes are usually made of nickel, with Pt coated cathode, and MgO_2 coated anode.

And where did all this equipment come from? The metals were mined and refined in some Third World country, and then transported, machined, or otherwise processed per spec into the needed materials. Then the finished parts were shipped to the hydrogen producing plant for final assembly.

Exotic, expensive stuff that requires a lot of energy (mostly fossil) to produce, manufacture, and transport. All these processes are accompanied by serious pollution, including carbon dioxide, particles, and gas, liquid, and soil pollutants.

Once the hydrogen gas is generated, it has to be transported to a storage or liquefying facility, using steel pipes laid on, or in, the ground. Sometimes the hydrogen is sent into underground storage for later use, but more often it is sent to a compressor setup to be pressurized and liquefied in special cylinders. Thus filled cylinders are loaded on trucks or trains to be transported to the user location, which may be hundreds or thousands of miles away.

How much pollution was generated during all these steps? Mining, refining, transport, processing, more processing, and more transport, burning, and other steps require many sets of large equipment, facilities, and vehicles, all of which emit harmful gasses and liquids every step of the way.

Note: Fuel cells are thought of as more environmentally friendly than most other electricity generating technologies, and at the back end they are. The produced power is free of any pollutants, resulting in only pure, clean water as a byproduct.

Because of that, fuel cell power plants qualify under several environmental certifications established by the government, such as the Leadership in Energy and Environmental Design (LEED) program and Renewable Energy Standards (RES). DFC power plants also have been designated as "Ultra-Clean" by the California Air Resources Board (CARB), and exceed all 2007 CARB standards. FuelCell Energy's power plants eliminate emissions generated by fossil-fuel-based backup generators.

Another benefit of fuel cell technology is the fuel flexibility and diversity. In addition to hydrogen, renewable biogas, produced by industrial, agricultural plants and wastewater treatment facilities, can be used. Fuel cell power plants can harness the methane in this byproduct, and use the gas to power the system in lieu of hydrogen, or natural gas, making it a renewable energy source. In many places where digester gas production volume is variable, fuel cell plants can operate with automatic blending with natural gas or other fuels.

But looking at the front end of the process shows that the environmental effects of the initial production and transport of the fuel cells components and parts—just as in any other power generating technologies—are related to large gas emissions and waste that have significant environmental impact.

Raw Hydrogen Gas Emissions

The good news is that a 100% efficient system of producing (via solar energy), storing (underground), and transporting (via existing pipelines) hydrogen, theoretically speaking leads to no unwanted emissions of any type. But such a 100% efficient system does not exist, nor is it practically possible, for it would be exceedingly expensive.

So in reality—and this is very, very bad news—it is estimated that around 15-25% of the total hydrogen mass would escape into the atmosphere during the different stages of its cradle-to-grave life cycle.

This leads us to the wicked side of hydrogen production and use. Scientists estimate that if hydrogen fuel cells replace all of today's combustion technologies, the escaping hydrogen would double or triple the total hydrogen in the atmosphere.

This would have a great impact on the global environmental dynamics, for this excess of hydrogen would oxidize in the stratosphere, causing unpredictable and uncontrollable cooling of the stratosphere region, which in turn would create more clouds. What these clouds would do is uncertain, but the uncertainty of some unimaginable harm befalling us is there. And the numbers are too large to ignore.

Estimates show that the extra hydrogen will lead to up to an 8% rise in ozone depletion at the North Pole and up to a 7% at the South Pole. The excess cloud cover would delay the breakup of the polar vortex at the north and south poles, thus increasing the holes in the ozone layer; making them larger and longer lasting. These are significant amounts of damage that is capable of negatively affecting our already fragile climate system.

Another unknown effect of hydrogen production and use, and yet a very important one, is the possibility of excess hydrogen absorption into the surface waters and topsoil. It is conceivable that this process could equal all anthropogenic (man-made) emissions.

Although these possibilities depend on a number of additional, and mostly unknown, factors and quantities, they must be researched and well understood, before a large quantity of free hydrogen gas is, willingly or unwillingly, released into the atmosphere.

So, the transition to a fossil-free future, where hydrogen is expected to play a significant role, is not trouble-free. As such, we must see hydrogen as a friend and foe. It can help us with our daily energy needs, but large quantities of free hydrogen floating in the atmosphere might bring us more troubles than we can handle.

Be careful, people! Not all things that seem harmless are totally harmless.

METHANE HYDRATES

Here we discuss energy resources that have been around for billions of years, but are yet to be discovered and explored. These are the mysterious and little known *methane hydrates*, also called *hydrocarbon clathrates*—natural gas-like, intermolecular compounds that occur naturally in submarine continental margins and regions and predominantly in the arctic permafrost. They are also expected to be found within medium- to large-sized icy moons of the outer solar system and in the polar regions of Mars.

It has become increasingly evident that there are huge amounts of naturally occurring gas hydrates all over the globe. These are important components of the shallow geo-sphere and are of societal concern in at least three major ways: resource, hazard, and climate (in that order).

There are two main reasons that make gas hydrates attractive as a potential resource.

- The enormous amount of methane that is apparently sequestered within clathrate structures at shallow sediment depths of 2000-6000 feet and even deeper below the Earth's surface and the ocean floor, and.

- The wide geographical distribution of the gas hydrates, which are found all over the globe.

The energy potential of methane hydrates is considerably greater than that of the other sources of gas fuels such as coal beds, tight sands, black shales, deep aquifers, and natural gas.

The resource potential of marine gas hydrate is yet to be fully assessed and ascertained, but considering the possibility of enormous gas reservoirs, gas hydrates will continue to attract attention until their development potential is measured.

Methane hydrates are considered a major potential source of hydrocarbon energy and could be important in meeting natural gas demand in the future. The hydrates and any free gas trapped below the hydrate stability field may provide a significant hydrocarbon resource in the future.

Background

Natural gas hydrates are a curious kind of chemical compound called a *clathrate*. Clathrates consist of two dissimilar molecules mechanically intermingled but not truly chemically bonded. Instead, one molecule forms a framework that traps the other molecule.

Natural gas hydrates are modified ice structures enclosing methane and other hydrocarbons that melt at temperatures well above normal ice.

At 30 atmospheres pressure, methane hydrate begins to be stable at temperatures above 0°C and at 100 atmospheres it is stable at 15°C. This behavior has two important practical implications:

First, they form in natural gas pipelines and are a nuisance to the gas companies. The gas in the pipelines must be dehydrated thoroughly to prevent methane hydrates from forming in high pressure gas lines.

Second, methane hydrates are stable on the sea floor at depths below a few hundred meters and are solid within sea floor sediments. Masses of methane hydrate "yellow ice" have been photographed on the sea floor. Chunks occasionally break loose and float to the surface, where they are unstable and effervesce as they decompose. The emitted gas can be burnt.

Figure 7-65. Chunks of burning hydrates

The stability of methane hydrates on the sea floor has a whole raft of implications.

• They constitute a huge energy resource, which if carelessly disturbed might suddenly destabilize, triggering submarine landslides and huge releases of methane, and

• Methane is a ferociously effective greenhouse gas, since it oxidizes fairly quickly in the atmosphere, where it could contribute to global warming. Large methane releases may explain sudden episodes of climatic warming in the geologic past.

Methane Hydrate Properties

Oceanic methane hydrate and associated gas deposits have been formed due to gas flow in the sediments, and the subsequent development of a gas-fluid interface. This is the cause for the methane to be concentrate in one place—probably the coldest spots in the gas flow. The remainder of the gas flow has been most likely dispersed near its source of production without any significant concentrations of either natural gas or hydrate.

Methane hydrates have unique properties that are quite different from any other energy sources. They are stable under low temperatures and high pressures. The hydrate stability zone in marine environments is a function of the water depth, the seafloor temperature, and the geothermal gradient.

Any changes to the temperature and/or pressure, both at the surface and in the area adjacent to the hydrate, will affect the thickness of the stability zone.

Although temperature and pressure are the main controls in the formation of gas hydrates and the thickness of the hydrate stability zone, other factors such as gas chemistry and gas availability will also alter the thickness and location of the hydrate stability zone.

Methane hydrate has a very high concentration of methane gas. One cubic meter block of methane hydrate in its frozen state can release about 160 cubic meters of gaseous methane upon melting.

The growth and stability of the free-gas zone beneath gas hydrate related bottom-simulating seismic reflectors is investigated using analytical and numerical analyzes to understand the factors controlling the formation and depletion of free gas. Gas forms across a thick zone because the upward fluid flux is relatively low and because the gas-water solubility decreases to a minimum several hundred meters below the seabed.

The decomposition of gas hydrates is due to change in the pressure and temperature regime in the hydrate stability zone. If the heat transport and the pressure change processes are fast compared with pore pressure dissipation processes, the excess pore pressure and reduction in effective stress can result in violent action and counteraction.

During exploitation, hydrate dissolution and dissociation may significantly alter the structure and mechanical properties of the marine sediments and the subsequent softening. Decrease of the shear strength is considered as the main driving factor of sediment deformations and slope instabilities.

We know little about the hydrates, and less about possible ways of (safe) exploitation of this huge natural energy resource. Nevertheless, we are acutely aware of the associated dangers. Work site collapses, accompanied by loss of life and equipment, and enormous methane gas releases, are some of the early warning signs. The warning is clear, "Proceed with great caution."

Methane Hydrates Status

The Department of the Interior's U.S. Geological Survey (USGS) and Minerals Management Service

(MMS) have investigated and reported the available gas hydrate resources in the country. The USGS estimate of undiscovered technically recoverable gas hydrates in northern Alaska represents the most robust effort to identify gas hydrates that may be commercially viable sources of energy.

Despite a lack of a production history, USGS reports a growing body of evidence indicating that some gas hydrate resources, such as those in northern Alaska, might be produced in the near future with existing technology and despite only limited field testing.

Methane hydrates are widespread in sea sediments hundreds of meters below the sea floor along the outer continental margins and are also found in Arctic permafrost. Some deposits are close to the ocean floor and at water depths as shallow as 150 m, although at low latitudes they are generally only found below 500 m. The deposits can be 300-600 m thick and cover large horizontal areas.

A nearby deposit nearly 500 km in length is found along the Blake Ridge off the coast of NC at depths of 2000-4000 m. Gas hydrates trapped in the marine sediments require multifaceted, extensive efforts to bring them into world energy balance. Government organizations, laboratories and the oil industry have recently taken up projects to explore gas hydrates in India's offshore areas. However, participation from the academic institutions is negligible due to lack of collaboration among them.

Most methane hydrate deposits in the US are located in the Alaskan Outer Continental Shelf. Additional deposits are onshore in northern Alaska, in the Gulf of Mexico, and on the western and eastern outer continental shelves.

The USGS estimates 200,000 trillion cubic feet (TCF) of methane hydrate in the U.S. alone, compared to the estimated 1,400 trillion cubic feet of recoverable natural gas reserves.

Or, the U.S. has almost 150 times more methane hydrate deposits than natural gas.

If these deposits are properly and safely exploited, then the U.S. could count on enough energy supplies for the next several centuries.

But there is a big *IF*...that must be resolved first if we don't want to risk an environmental disaster on an unprecedented scale.

Mappings by the USGS of offshore North and South Carolina reveal the possible existence of a 1300 TCF methane hydrate deposit. If this is true, and if it could be extracted safely, it would represent a 700% increase in the current natural gas deposits in the U.S. At current consumption rates, this would be a 70-year reserve of natural gas.

Sandstones are considered superior reservoirs because they have much higher permeability than shales, which can be nearly impermeable. The marine shale gas hydrate reservoir may host hundreds of thousands of trillion cubic feet (TCF), but most or all of that resource may never be economically recoverable. It is likely that continued research and development efforts in the United States and other countries will focus on producing gas hydrates from arctic and marine sandstone reservoirs.

USGS issued a resource assessment of Alaska's gas hydrates, estimating that 85.4 TCF of natural gas can be extracted—an amount that could heat more than 100 million average homes for more than a decade. This does not mean that commercial production of these resources is imminent, but it does suggest a vast natural gas resource trapped in Alaska's gas hydrates.

At the same time, the USGS assessment indicates that the North Slope of Alaska may host about 85 TCF of undiscovered technically recoverable gas hydrate resources. According to the report, technically recoverable gas hydrate resources could range from a low of 25 TCF to as much as 158 TCF on the North Slope.

Of the mean estimate of 85 TCF of technically recoverable gas hydrates on the North Slope, 56% is located on federally managed lands, 39% on lands and offshore waters managed by the State of Alaska, and the remainder on Native lands.

- *Recent MMS assessment of undiscovered in-place gas hydrate resources in the Gulf of Mexico estimated their volume at over 21,000 TCF.*

- *The U.S.' average annual consumption of natural gas is about 25 TCF.*

Does this mean that we might have natural gas for the next 1,000 years? Just from the Gulf of Mexico and that much more from other locations?

Globally, the amount of gas hydrate to be found offshore along continental margins probably exceeds the amount found onshore in permafrost regions by two orders of magnitude, according to one estimate. With the exception of the assessments discussed above, none of the global gas hydrate estimates is well defined, and all are speculative to some extent.

Efforts to explore the development of gas hydrate

resources are already underway in energy resource poor countries such as Japan and India. Worldwide estimates of methane hydrate deposits reach the overwhelming number of 400 million trillion cubic feet—far outdistancing the 5500 trillion cubic feet of proven worldwide gas reserves.

Environmental Impact of Methane Hydrates

Today's world is facing two different but related and increasing environmental problems: global warming and air pollution. Both of these are linked to the heavy use of the fossil fuels. Global consumption of fossil fuels—coal, natural gas and crude oil, is staggering and is increasing at an alarming rate. The quadrupling of oil prices in the 1970s and rapid increase in recent years, the growing awareness of energy-related pollution, and the possibility of climate change have all contributed to a re-evaluation of energy use. This has also increased the environmental awareness of the masses.

The USGS estimates that there is twice the amount of carbon to be found in methane hydrate deposits as there is in all other fossil fuels combined. However, this estimate is made with scant information, and is just a guestimate.

Nevertheless, methane gas, contained in methane hydrate deposits is a very powerful greenhouse gas. Warmer Arctic temperatures could result in gradual melting of gas hydrates below permafrost, and warming oceans could cause gradual melting of gas hydrates near the sediment-water interface.

Although many news reports have presented this as a potential catastrophe, USGS research has determined that gas hydrates are currently contributing to total atmospheric methane and that a catastrophic melting of unstable hydrate deposits is unlikely to send large amounts of methane into the atmosphere anytime soon. *Anytime soon* are the key words here. Yet, the danger is there for us to assess and contain, instead of disregarding it. It is too big to be ignored! Anytime soon could come sooner than expected and then it might be too late to do anything about it.

Basically, methane hydrates sediments are sensitive to temperature and pressure changes. They can rapidly dissociate with an increase in temperature or a decrease in pressure—none of which we can control. This dissociation produces free methane and water, the huge quantities of which must be considered as a great risk of large quantities of GHG gasses in the atmosphere.

The conversion of solid sediment into liquids and gases could create a loss of support and shear strength. These can cause submarine slumping, landslides or subsidence that can damage production equipment and pipelines. With that, the methane release will escalate too without a chance to stop it.

The process can escalate to unimaginable size with untold damage to the environment and bringing devastation to coastal areas. This is a nightmare scenario, which, as unlikely as it might seem, must be taken into consideration when doing anything close to disturbing the sleeping giants.

Possible Effects on Climate Change

Hydrates may affect climate because when warmed or depressurized, they decompose and dissociate into water and methane gas, one of the greenhouse gases that warms the planet. Methane is a greenhouse gas. Discharge of large amounts of methane into the atmosphere would cause global warming. It has been well documented that methane levels in the atmosphere were lower during glacial than during interglacial periods. In fact, methane is many times more effective as a greenhouse gas than is CO_2.

Therefore, if the flux of methane to the atmosphere from dissociating hydrates is sufficient in quantity, this methane can cause excess global warming. Specifically, as the Earth warms, increasing bottom water temperatures could cause gas hydrate disassociation in many marine shelf locations. This gas hydrate disassociation would cause further warming due to the greenhouse effects of the gas which is released, such as methane releases from hydrates to atmosphere if sea level drops, and sea-level raises relatively warm ocean water to cover cold Arctic strata.

The resulting breakdown of stable gas hydrates within the sediment releases gas into the atmosphere. Methane releases from hydrates to atmosphere if sea-level rises. The dissociation of gas hydrates during deglaciation has therefore been linked to the ending of ice ages during the last the last few millions of years.

Recent studies show significant amounts of methane gas already rising from the seafloor of the Arctic Ocean. This is most likely due to methane gas released from melting methane hydrate deposits on the sea floor. This is one of the few locations on Earth with permanently elevated methane content.

In December, 2013, methane levels over 2400 ppb (parts per billion) lasting several months were measured in the Arctic region. This is 5-6 times higher than the maximum methane levels measured on Earth, which when spread over such large areas represent a very dangerous development in progress.

Methane levels well over 2660 ppb have been

measured in that area as well and must be taken as an initial warning. If the warming/melting of the methane hydrates on the ocean floor continues, and the emissions increase, they might become the second most serious GHG polluting source on Earth. That might bring us much closer and faster to a real global environmental disaster.

Conclusion

Methane gas hydrates are a major energy source of enormous potential. The advantages and disadvantages can be summarized as follow:

1) Gas from methane hydrate may be a significant clean energy source. It is now recognized that there are huge amounts of natural gas, mainly methane, tied up in gas hydrate globally, which could provide energy for centuries to come.

2) There are presently serious production problems. Gas hydrates are located deep underground and under the ocean floor. The major challenge is the fairly unstable formations, the exploitation of which resembles working in an underground mine dug into a sand dune—collapse of the walls is to be expected at any moment.

3) Methane hydrate may play a role in climate change. Methane is a strong greenhouse gas, so its escape to the atmosphere from natural gas hydrate could result in global warming.

Sub-seabed methane within the continental margin sediments is produced primarily by microbial or thermogenic processes. In the microbial process, organic debris are decomposed by a complex sequence (methanogenesis) into methane, by bacteria in an anoxic environment. Organic matter is composed of carbon, hydrogen and phosphorus in the ratio of 106:16:1, and decomposition results in production of methane.

Methane hydrates are also common in sediments deposited in high latitude continental shelves and at the slope and rise of continental margins with high bioproductivity. High biological production provides the organic matter buried in the sediment, which during early diagenesis and after exhausting oxygen, sulfate and other electron acceptors, eventually generates methane through fermentative decomposition and/or microbial carbonate reduction.

Methane hydrates located in the shallow submarine geosphere are part of a finely balanced system in equilibrium with all its components such as sediment, pore water, fluid flows, pressure, temperature, overlying water, hydrate.

Removal of any one component of this equilibrium may destabilize the whole system leading to irreparable damage. The destabilizing factors may be either natural perturbations or perturbations associated with exploitation. Studies have indicated that methane hydrates have the potential to affect global climate and the geological environment on a catastrophic scale.

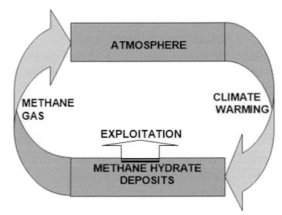

Figure 7-66. The vicious cycle

In conclusion, we now know about the existence of large amounts of methane hydrates—their properties, possibilities, and issues—which lead to the conclusion that they might be a significant part of our new energy future, and that of future generations. This is a comforting thought, which beats the alternative of leaving no energy sources at all for those following us, leaving them in complete darkness and an energy-deprived existence.

The methane hydrates use as an energy source, however, promises to be a huge, complex, expensive, and very dangerous undertaking. Presently we have no technology or process efficient and safe enough to warrant large-scale development of this unusual energy resource.

The safety issues surrounding large-scale exploration deep under the ocean floor are significant. Here, immense forces, most of which we are not that familiar with, nor equipped to handle, could damage equipment, hurt people, and even unleash a global disaster.

What this tells us is that there is hope for an energy bonanza at which we must take a close look, but approach very carefully.

Any delay might bring us to a fossil-poor existence, while any hasty move might result in costly and irreversible human disaster and lasting environmental damages. While we must be excited over the possibility of another energy source with great potential for supplying energy for centuries to come, we must be also aware of the real dangers involved.

Nevertheless, the availability of such a great amount of methane hydrates all over the world gives us hope that future generations will not be totally fossil-less. It will be up to them, however, to find the best and safest ways to exploit this unconventional fuel.

Notes and References

1. DOE, http://www.energy.gov/science-innovation/energy-sources/renewable-energy/solar
2. EIA, http://www.eia.gov/renewable/data.cfm#solar
3. National Geographic, http://environment.nationalgeographic.com/environment/global-warming/solar-power-profile/
4. NREL, Non-Hardware ("Soft") Cost-Reduction Roadmap for Residential and Small Commercial Solar Photovoltaics, 2013-2020. Kristen Ardani, Dan Seif, Robert Margolis, Jesse Morris, Carolyn Davidson, Sarah Truitt, and Roy Torbert. http://www.nrel.gov/docs/fy13osti/59155.pdf
5. METHODOLOGY FOR PRIORITIZATION OF INVESTMENTS TO SUPPORT THE ARMY ENERGY STRATEGY FOR INSTALLATIONS, by George Alsfelder, Timothy Hartong, Michael Rodriguez, and John V. Farr Center for Nation Reconstruction and Capacity Development United States Military Academy, West Point, NY 10996 http://www.usma.edu/cnrcd/CNRCD_Library/Energy%20Security%20Paper.pdf
6. SEMI, Compilation of terms http://www.semi.org/en/sites/semi.org/files/docs/CompilationTerms0614.pdf
7. Accommodating high levels of variable generation. NERC, April 2009
8. *Photovoltaics for Commercial and Utilities Power Generation*, Anco S. Blazev. The Fairmont Press, 2011
9. *Solar Technologies for the 21st Century*, Anco S. Blazev. The Fairmont Press, 2013
10. *Power Generation and the Environment*, Anco S. Blazev. The Fairmont Press, 2014

Chapter 8
Future Energy and Energy Security
(Our Survival Now and Tomorrow)

True "energy security" means: a) ensuring present-day energy supplies, while at the same time, b) preserving the environment, and c) planning the post-fossils energy future—in that order and with equal importance!

—Anco Blazev

BACKGROUND

The above statement reflects the one and only true meaning of energy security: providing energy today, without harming the environment, while making sure that future generations benefit from the gifts of the Earth as much as we do—simple, logical, and fair.

This is so, because it would make no sense to have a lot of energy today by destroying the environment and suffocating us all to death in its pollution. We cannot separate the environment from energy for obvious reasons, most of which are discussed in this text. The environment is very seriously affected by energy production and use, and so it is an integral part of anything related to energy, thus directly and indirectly affecting our long-term energy security.

We also cannot turn our backs on future generations for the sake of our "now" energy security. It is not fair to use all the fossils now and leave nothing to those who come after us. That would be criminal.

Note: We see a strong connection between our present energy security, global environment, and future energy supplies. We view these factors from different points of view as a repeating argument in this text. This cannot be avoided, if we are to get a complete picture of the situation. There is simply no way to look at these three variables as separate entities, thinking that by taking care of one of them our job is done, then turning our backs on the rest and feeling good about ourselves.

In this text we present a detailed view of the present energy and energy security developments, tasks, and issues, always keeping in mind that today's energy security is ONLY ONE of the three factors discussed above that define our way of life today and tomorrow. In more detail:

Short-term Energy Security

Today we are preoccupied with, and focused intensely on, our present energy security. The goal is to account for and eliminate the short-term risks at all cost, to produce and use as much energy as needed for normal functioning of our economy and society.

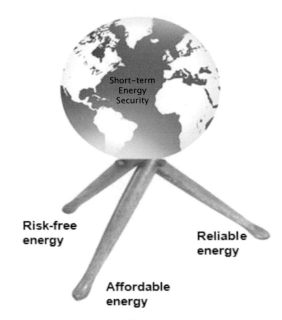

Figure 8-1. Our short-term energy security

Short-term energy security risks affect the energy supplies, which could cause interruptions in the power generation and transportation sectors.

The goal is to ensure reliable and affordable energy at reduced or no risks. Here we see the global energy security sitting on three legs—those of ensuring risk-free, reliable, and affordable energy. It is that simple. Or is it?

- *Leg One.* Risk-free energy is a pipe dream simply because there are too many players involved in the world's largest, most important, and most profitable (for some) energy markets. In capitalist terms, this is a prime terrain for fierce competition, shady deals, blatant cheating, and unrestrained crime. All of this brings risks—a great number of very serious internal and external risks, some of which we discuss in this text.

- *Leg Two.* Reliable energy supplies can be achieved only if a) all risks have been dealt with, sorted and eliminated, and, b) when the reliability of the production and delivery of sufficient amounts of energy supplies has been guaranteed. This is another pipe dream, because the external risks abound, the internal risks are growing, and the production and delivery amounts cannot be guaranteed forever for many reasons.

- *Leg Three.* The hope for affordable (and decreasing cost of) energy supplies is a thing of the past. At the present-day supply and demand ratio, global energy costs will continue to rise in unpredictable fashion. They are expected to go much higher as energy supplies dwindle and global competition increases.

Long-term Energy Security

Considering the above facts, we end up with another three-legged stool supporting the world. It represents the world's present and future energy security and environmental health. Enough energy for today, a healthy environment, and enough energy for tomorrow must be the goal. It sounds so simple and straightforward that one must wonder why is it even mentioned.

A closer look at Figure 8-2 reveals that all three legs of the world's energy security, which are equally important for the long-term survival of the human species, are wobbly. They all need repair—each with its own serious problems and urgent requirements, as follow:

- *Energy today.* Today this is the priority, so all countries are working hard on fixing the problems of the present-day energy supplies. Some are more successful than others, but all are racing towards achieving the same goal—providing enough energy for their people now. Today. That task, however, is becoming more difficult and more expensive to accomplish.

- *The global environment.* This is the worst and most dangerous part of the energy equation, since the

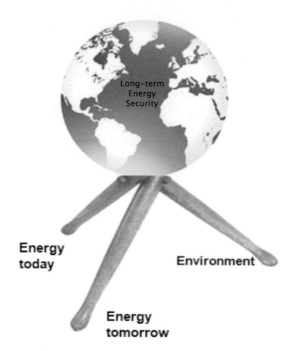

Figure 8-2. Our long-term energy security.

global environment is badly broken. Some of the key issues need to be addressed immediately and urgently, but the results thus far are not encouraging, and things seem to be getting worse.

- *Energy tomorrow.* This is also a very serious and even tragic situation, because we have not even admitted that there is a problem. Not yet...not really, and for sure not officially! But there is a problem...a big one. The way we use energy today will seriously affect future energy supplies ...unless we do something to prevent it.

"Drill, Baby, drill," and "Burn, Baby, burn" is the global *modus operandi* today, and the trend extends well into the future. The negative effect of our actions is evident now, but the full impact will be felt by those who come after us. It is hard to imagine what the world will look like 100-200 years from today, because it is hard to imagine life without fossil fuels. Certainly, the post-fossils reality will present serious challenges, if we don't change the way we produce and use our energy supplies.

Energy security means ensuring energy resources today and tomorrow while reducing the excess environmental pollution.

All three legs of the global energy security stool are equally important, and in need of fixing. We know exactly what the problems are, but there are no easy, fast, or cheap solutions, so we are ignoring some of them, only delaying the inevitable.

We all see the problems, and sporadic efforts are underway to solve at least some of them. We are not going to give up, and are sure that we (the world in general) will eventually solve the biggest and most urgent problems. We just don't see it happening now, or anytime soon...and the clock is ticking.

The longer we delay, the more wobbly the energy security legs become. If we delay long enough, one of the legs might just break off, plunging the whole world into energy and/or environmental oblivion.

We can clearly see this happening at the present, with the entire world slowly headed down the cosmic emptiness. Common sense dictates taking urgent and serious action, but common sense is not that common today, and is often overridden by personal, political, and corporate interests.

ENERGY SECURITY TODAY

To get a better picture of our near- and long-term future, we need to analyze our present energy situation in even more detail.

Figure 8-3 clearly shows that our energy security dictates our energy independence, which in turn directly affects our national security—which are all directly

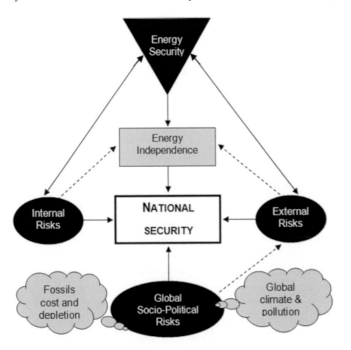

Figure 8-3. Energy-related dependencies

dependent on a number of internal and external factors and risks. What happens around the world affects us here at home one way or another.

Sometimes we see these effects as increases in the price at the gas pumps, and sometimes as increases in the military budget. Sometimes we feel them in a more direct and painful way, like the 911 attacks and their results.

What we see lately, is a split in the general population and the scientific community's opinions and actions, where some people clearly see and acknowledge the new energy security risks around us—including increased terrorist activities, pending energy sources depletion, increasing energy demand, unreasonable energy waste, and disastrous environmental damage. Some people don't see any of this (or pretend not to), while others just don't care.

This is normal for a population as diverse and opulent as that in the U.S. It is also normal to see the glass as half empty or half full. It all depends on which side of the issue you sit.

So, oil company executives see the new situation one way, while the unemployed single parent, sees it totally differently. The executives' primary concern is to increase corporate profits—at all cost—while the single parent struggles to pay ever-increasing energy bills.

As the battle on the energy front continues daily, how does it relate to our energy security and independence?

Let's take a close look at the three key energy related variables: energy security, energy independence, and energy survival:

ENERGY SECURITY, ENERGY INDEPENDENCE, AND ENERGY SURVIVAL

There is a lot of talk today in Washington, DC, and other high places about our energy present and future. Some of the issues are obscured by political, some by financial, and some by technical misunderstanding. This then is a good time and place to clarify the meaning of energy security, energy independence, and energy survival (this one is quite important, but usually ignored). In brief:

- *Energy security* is ensuring an uninterrupted energy supply today and tomorrow, by employing efficient supply methods, complete with accident and crime prevention measures.

- *Energy independence* is producing our own energy, so we don't have to import fuels and energy products.

- *Energy survival* is ensuring that we have enough energy now AND in the near future, so that neither we nor future generations end up in an energy blackout. This is an obscure subject, somewhere on the back burner.

Energy security, energy independence, and energy survival, while meaning different things, are interwoven in an integral way. They must be addressed as equal in importance and weight. One cannot exist without the other, especially in the long run. If one is ignored, our present and/or future energy balance and energy security would be at risk.

We know very well that we are using far too much energy and that at the rate we are using fossil fuels, they will not last long. Yet we continue the fossils extermination at an accelerated pace, silently ignoring the grim reality of pending fossils doom.

At the same time, the implementation of the renewable energy sources around the globe is moving at a snail's pace due to political hesitation, corporate grid, and many other socio-economic reasons. So what are we to do?

Let's take an even closer look at these three vitally important concepts:

Energy Security

There are many reasons why energy security is so important for providing normal (or rather opulent) life in our society. Without energy, or in a case of temporary interruption of the energy supply system, we would be vulnerable to a number of internal and external threats.

Energy security is one of the pillars of the national security and efficient economic development of modern societies.

Without energy the industrial machine stands still and does not produce anything. Products that are produced, cannot be moved to the markets. Armies cannot get to the battlefield. It is imperative for each nation to have enough energy to function properly. Interruption of the energy sector brings a number of undesirable consequences, external problems, and internal turmoil.

We'd divide the present-day risks to our energy sector, and our energy security to internal and external:

Internal Risks

The internal risks of our energy security are based and conducted. on our national territory. They can also be divided into natural (caused by natural disasters and such) and man-made (caused intentionally or unintentionally by human activities).

Unforeseen natural events and man-made accidents in mines, transport routes, refineries, and power plants threaten to disrupt the energy supply chain. Mine accidents still happen in the U.S. and around the world. They take lives and shut down mining operations temporarily or permanently. Railroad cars and transport ships' accidents and spillage can close a transport route indefinitely and do significant damage to the environment at the same time. Power plants are vulnerable to internal sabotage and terrorism, and such risks in the U.S. are increasing. Some of the accidents of the past have resulted in loss of human life as well. Hurricanes damage and shut down refineries and power plants every year.

As the state of the art of the energy technologies matures and we learn how to handle the natural and man-made accidents, some of the internal risks become more manageable. With that, the number of mine, railroad and transport ships' accidents is significantly reduced. Refineries and power plants, because of their complex infrastructure, on the other hand, are vulnerable to the force of hurricanes and other natural events, and not much can be done to improve that situation.

The most dangerous risks in the energy sector are those posed by incidents, accidents, and failures with nuclear power plants. Just mentioning Chernobyl and Fukushima brings gruesome memories and forces us to imagine what would happen if a similar accident happens nearby. Nuclear accidents are an extremely rare occurrence but when they happen, the local devastation is total and permanent, and their effects are felt over broad areas—even across continents.

This is a critical component in considering our energy security and safety. One single nuclear plant accident cannot only disrupt the energy supply, but even destroy the entire area and kill and sicken thousands. This issue needs to be seriously consideration as we review our energy options for the 21st century.

Internal terrorism is one way to gain control and damage a refinery or a power plant. There are a number of reported incidents, and even a greater number unreported. Today, computers run everything—including our energy infrastructure. Every step of our energy supply chain, from mining operations to transporting, processing, refining, and using energy sources is monitored and/or controlled by computers. That creates a serious problem—that of computer hacking and terrorism.

In 2012, a backdoor in a piece of industrial software used to control power plants allowed hackers to illegally access a New Jersey power company's internal heating and air-conditioning system. One of the viruses was accidentally discovered after an employee called in an IT technician to troubleshoot the USB drive. A simple virus check discovered sophisticated malware, capable of doing a lot of damage to the plant equipment.

Since the computer safety technology is fairly new, the defense mechanisms have not been fully developed and/or deployed, as was in this and many other cases. The workstations lacked backup systems, so they were lucky to discover the malware before it was activated.

Another intentional malware attack, spread by a USB drive, affected 10 computers in a steam turbine control system of the power plant. The incident resulted in downtime for repair of the impacted systems, which delayed the plant restart by three weeks.

The Stuxnet worm and the Flame malware were developed by the US and Israel to spy on, and even control, critical systems in power plants. In the summer of 2012, these programs successfully disabled an entire enrichment facility in Iran. These programs relied on USB drives to store the commands, propagate attack codes, and carry intercepted communications over computer networks. Microsoft has patched these vulnerabilities on Windows computers, but there are additional steps to be taken by power plant and other energy sector operators.

Of course, we cannot leave out the humbling experience of the massive computer security breach in December 2013 by foreign hackers who stole the identity and other private information of Target and several hundred thousand other U.S. companies' customers.

If hackers can get into one of America's largest and most prestigious chain-stores' computer systems, getting into the utilities' control systems is only a step away.

As a matter of fact, during 2012-2013, critical control systems in two US power plants were found infected with computer malware, spread intentionally by a computer virus brought in the plants via USB drives. The infected computers controlled critical systems controlling power generation equipment.

Intentionally planted malware poses a real threat by allowing the attackers to disable key equipment, thus disrupting its normal operation, or destroying it altogether. What a disaster that would be if allowed to shut down the cooling system of a nuclear plant!

But it is not just computers that are the internal terrorists' targets. In 2013, for example, a well organized and executed sniper attack on power grid components shut down the power in a large part of the Silicon Valley area and threw residences and businesses into darkness for several hours and even days in some areas.

The sniper shot at several transformers in a local high voltage substation which malfunctioned as a result and caused the power grid in a large part of the San Jose-Fremont area to shut down. Repairs were begun almost immediately, but it took some time to figure out what had happened and to repair the damages.

These examples are cautionary tales and lessons in safety procedures, as well as a call to action by owners and operators of critical energy infrastructure. New, 21st century security policies must be developed and implemented, in order to maintain up-to-date antivirus mechanisms, and manage system patching and the use of removable media.

Other, site-specific protection measures must be implemented for ensuring the energy infrastructure's physical safety. There are weak points throughout the power system—from power generation to delivery—which need to be evaluated and protected from physical and cyber attacks. This is not a simple or cheap task, but it must be done, if we don't want to risk waking up someday in darkness… or in the midst of a nuclear disaster.

Overall, internal risks are unavoidable, but we are well positioned to anticipate and respond in most cases. The internal security situation is basically under control for the most part, and we just have to add the necessary cyber and physical protection to key components, be more careful, and not let our guard down.

The external risks are more serious and complicated.

External Risks

The U.S. has many enemies. Traveling around the world, one can see the signs of animosity towards the U.S. in almost every country. Some hate us because we have done bad things in the past. Others hate us because they are jealous, and yet others hate us for the sake of hating. Either way, each of these groups has their own reasons and ways to inflict damage and pain to the mighty USA.

Terrorist attacks targeting oil facilities, pipelines, tankers, refineries, and oil fields are referred to as "industry risks," because they are part of daily life in the energy sector. The energy infrastructure is extremely vulnerable to external sabotage, with oil transportation, and its exposure at the five ocean chokepoints, at the top of the risks list. The Iranian controlled Strait of Hormuz is a prime example of a chokehold, where one attack on a Saudi oil field, or on tankers in the Strait of Hormuz,

Figure 8-4. Terrorist attack on French oil tanker *Limburg* **in 2002**

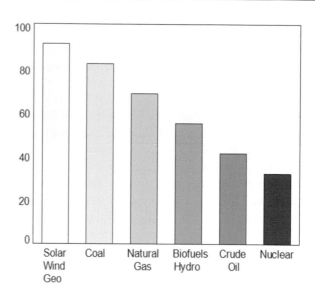

Figure 8-5. Energy security risks and dependencies (100 is risk-free)

could disturb the oil supply. A prolonged conflict in the area, would surely throw the entire world energy market into chaos.

> *External risks have always been, and will continue to be, a problem. Foreign terrorists spend a lot of time planning attacks on U.S. properties and personnel, around the world and at home. This is the price we pay for being the world's greatest power and a shining example of a functioning democracy.*

International terrorism affecting the world's energy reserves is of great concern lately as well, as evidenced by NATO leaders meeting in Bucharest in 2008, where international terrorism against energy resources was one of the key subjects. The group discussed the possibility of using military force to ensure the energy security of the region. One of the possibilities discussed included strategic placement of NATO troops in the Caucasus energy fields to police and protect oil and gas pipelines from terrorist attacks.

The U.S. energy supply and energy infrastructure are in the terrorists' eyesight too, and attacks are, no doubt, planned daily in terrorist dens worldwide. But how can they do that? We are too far away, and too powerful, to invade by sea or air. So, they've physically attacked people and structures, but recently they have been choosing different weapons and changing the battlefield tactics.

The Risk Levels

So let's take a look at the risk levels of the different energy sources.

Figure 8-5 is a rough summary of the level of risk the different energy sources present to our energy secu-

rity. From left to right, we see a decreasing level of risk-free operation of the different energy sources.

- Solar, wind, and geo-power are the most risk-free energy sources, simply because the fuels (sunshine and wind) are readily available and they and their technologies do not depend on external factors. There are, of course, some inherent problems, such as the variability of sunshine and wind currents, and internal risks, such as materials availability and cost, but these are few and are easily manageable.

This makes these energy sources virtually risk-free and highly reliable *ad infinitum*. We foresee them becoming major power generators as their technologies mature and the fossils become more expensive. They are the best (if not the only) bet for future generations.

- Coal is also relatively risk-free as far as its production, distribution, and use are concerned. Apart from its excess pollution, and mine accident problems, coal is, and will be, readily available and cheap enough for reliable and risk-free operation for a long time.

Its depletion in the more distant future, however, is imminent, so it is only a temporary solution to present-day energy issues. Let's not forget that someday soon—most likely during the next century—there will be no more coal on this Earth.

- Natural gas presents an increased level of risks,

due to its nature and the related production and distribution issues. The transport of natural gas, via pipelines and different land- and ocean-based vehicles increases the risks of natural accidents and criminal activities.

More importantly, the U.S. natural gas distribution infrastructure—the thousands of miles of gas pipelines running under our streets and homes—is old. In some places pipelines are over 100 years old, and are developing leaks and creating all sorts of problems. The incidents of gas line failure are increasing and becoming more violent. The last and most revealing example is the gas explosion that destroyed two apartment complexes is New York City and killed 8 people.

Natural gas is also a temporary solution to our energy problems. No matter how exiting the latest natural gas bonanza in the U.S. might be, we should not forget that its quantity is limited and that sooner or later—and surely sometime this century—it will be completely depleted.

- Biofuels and hydro power share the same unfortunate dependency on natural events—extended draughts and uncontrolled climate changes and weather patterns threaten the reliability of these energy sources. Biofuels also compete with the world's food supply, and as the global population increases, the competition between crops for biofuels and food will grow too.

Cellulosic biomass has great potential as a future energy source, but it is too expensive for now. However, it will become more reliable and cheaper in the future, as the technology improves.

- Crude oil, and its imports in particular, present a major risk to our energy security. Half of the oil used in the U.S. is imported. It comes from some of the most volatile and unfriendly areas of the world, and is transported through some of the most dangerous chokepoints. There is a constant threat to production levels in some countries, in addition to the risks with the distribution channels; thus, its import is a risky proposition.

Most importantly, let's not forget that crude oil production is entering a period of decline. Prices will keep rising, and quantities will diminish steadily. One thing is certain: There will be no more crude oil at sometime this century.

- Nuclear power presents serious dangers and risks to our energy security and personal safety. Computer warfare is the name of the game today, so a terrorist sitting in a cave in Afghanistan, or in a high-rise in Beijing, could gain access to the computer controls at a power plant or refinery and simply put them out of commission, or worse.

Nuclear reactors are potential terrorist targets, since they are not designed to withstand attacks by large aircraft, rockets and other air-born weapons. A well-coordinated attack, using powerful weapons, could damage the reactors in a nuclear plant, which would in turn have severe consequences for human health and the environment.

A recent study concluded that such an attack on the Indian Point Reactor in Westchester County, New York, could result in high radiation within 50 miles of the reactor, and cause 44,000 deaths from acute radiation sickness. An additional 500,000 long-term deaths from cancer and other radiation-caused illnesses would be expected as well.

Terrorists could also target a spent fuel storage facility by using high explosives delivered by ground or air vehicles. This would also result in radiation contamination of the immediate area. A terrorist group may infiltrate the personnel of a nuclear plant and sabotage it from inside. They could, for example, disable the cooling system of the reactor core, or drain water from the cooling storage pond. An internal attack is perhaps the most likely, and most dangerous, terrorist attack on a nuclear-power reactor.

There is also an inextricable link between nuclear energy and nuclear weapons, which pose the greatest danger related to nuclear power. The problem is that the same process used to manufacture low-enriched uranium for nuclear fuel, also can be used for the production of highly enriched uranium for nuclear weapons.

Expansion of nuclear power generation could, therefore, lead to an increase in the number of rogue states with nuclear weapons, produced at their "civil" nuclear programs. This is exactly what we are seeing play out in Iran and possibly North Korea. The use of nuclear power would, at the same time, increase the risk of commercial nuclear technology being used to construct clandestine weapons facilities, as was done by Pakistan in the recent past.

Today, the majority in Japan is against nuclear power. As a result, Japan's nuclear plants are shut down and the GHG emissions increased by 15% as more fossils are used for power generation in Japan since Fukushima.

One more, even much smaller, nuclear accident on the island of Japan will surely put an end to nuclear in Japan forever. To prevent that, Japan's nuclear plants spent billions of dollars for upgrades with the latest hardware and security measures, to protect against natural disasters and eliminate man-made accidents.

But even without any additional accidents, nuclear power in Japan will be on the back burner for a long time. Its destiny around the globe is also uncertain, and will be determined mostly by its safety record.

Environmental Impact

The environment is part of the external risks related to energy security simply because it is the media we live in. It is the air we breath, the water we drink, and the land we walk on. When we talk about energy security we must always *first* consider the environment and the impact on it that our actions could bring.

This is simply because it won't do us any good to have plentiful energy resources if we cannot breath the air, drink the water, or use the land. We see a glimpse of these abnormalities in the dense Beijing smog, the water in some U.S. states that is contaminated with fracking liquids, and the disappearing land mass in the Maldives and other places.

Coal-fired power plants belch millions of tons of GHG gasses, hydro racking activities pollute the fresh water supplies in large areas, and rising ocean levels threaten to submerge portions of inhabited land mass.

Many of these activities—many of which are done for energy security's sake—result in property damage, make people sick, and even kill many of us.

Energy security obtained at the expense of humans' safety is a partial security. It benefits some people, while hurting others. This situation is unsustainable and must be corrected before we reach a "point of no return."

Easier said than done…there are many interests to protect and a lot of money to be made by maintaining the status quo. Because of that, the talks continue, and many volumes are written on the subject, but the actions do not follow in most cases. For now…

Fossils Depletion

Another external factor of great importance, and even greater consequences, is the pending depletion of global fossil reserves. For over a century now we have been relentlessly digging and pumping huge amounts of coal, crude oil, and natural gas, day and night, nonstop. It is quite obvious that all measures discussed and

planned will not be enough to reduce the onslaught of fossils.

The rate of exploitation of the fossils is actually increasing in most areas of the world. This is especially true in the developing countries, many of which are determined to achieve a Western-like lifestyle in the near future. This effort requires a lot of energy, and they are determined to get it one way or another.

The present-day approach to achieving energy security ("drill, Baby, drill") promises energy poverty and even total lack of fossil fuel reserves in the 22nd century and beyond.

One doesn't have to be a genius or a specialist on the matter to see that the fossils are of a limited supply. This means that by the end of this century, and even before that, there will be no more crude oil. The sucking sound of empty natural gas reservoirs will follow soon thereafter, and coal reserves depletion will be next.

The fossil reserves are limited. They will not last forever. Most of them will be depleted within the next several decades. And then what…?

Energy Independence

Depending for energy (oil in particular) on the whims of the dictators of politically unstable countries is the major energy security risk at the moment. Our economy is still exposed to manipulation of energy supplies, such as the OPEC-orchestrated oil crisis of 1973, and the extraordinary jump in oil prices in 2008.

Energy independence is impossible without 100% domestic energy production. We will not see energy independence while importing any quantity of crude oil.

In 1973 the American government realized that we have a big problem and started a feverish effort to develop new energy sources to obtain energy independence, or rather to ensure an acceptable level of energy security.

Soon after the crisis was averted, the energy security drive gave way to making a quick buck by importing cheap Arab oil again. The lesson of 1973 was not learned!

The efforts to implement new, renewable energy sources fizzled as soon as OPEC lifted the embargo and oil prices dropped. The energy security effort was shelved until the next energy crises…and the next…and the next.

It has been proven time after time that oil imports are vulnerable to intentional or unintentional disruptions. These could be due to in-state conflicts, exporters' interests, and/or non-state players (terrorists and others) targeting the supply and transportation of oil resources.

The 1973 oil embargo is a good example of how oil supplies can be used against the U.S. The Arab nations decided to punish the U.S. for its support of Israel during the Yom Kippur War and turned the spigot off.

This was also the case during the economic negotiations during the Russia-Belarus energy dispute in 2007-2008. As a result, Putin shut down the Druzhba pipeline, which supplies oil to Germany, Poland, Ukraine, Slovakia, the Czech Republic, and Hungary, causing several days' extreme anguish and led to power failures and misery in these countries.

Wars, political conflicts and many other factors, such as strikes, can also prevent the proper flow of energy supplies. Venezuela is a prime example of everything that can go wrong with the oil supply chain. First it was the nationalization of the oil industry, followed by strikes and protests. Several years after the fact, Venezuela's oil production is yet to recover fully.

Since the nationalization of oil, Hugo Chávez threatened to cut off supplies to the United States time after time. He was holding American oil exports hostage and used them extensively to further his ideas and political agendas. His death in the spring of 2013 marks a new era in the relations between our two countries, but the long-term results are predicted to remain fluid and fluctuating. Exporters like Venezuela are driven by economic and political incentives, which at times force them to limit their exports. In other cases, export revenues are used to finance terrorist groups like Hamas and Hezbollah.

Amazingly, we (the U.S.) are paying the expenses on both sides of the conflicts. Saudi Arabia gets $150 billion annually from U.S. oil exports, about $4 billion of which goes to the Wahhabis who train their people to fight us. Then we pay to defend ourselves. Reasonable...?

There is also increased competition for energy resources worldwide, due to the population and standard-of-living increases in India, China, and other developing countries. These create energy price rises, which can get to high extremes like the unprecedented and unforgettable jump of oil prices to $180/bbl in 2008.

Thus increased competition over energy resources may eventually lead to the formation of security pacts between the major powers, to enable an equitable distribution of oil and gas. If and when this happens, however, it will affect positively the developed economies, while the developing countries will continue the daily struggle to supply fuels to power their homes and fledgling economies.

Concerns with the pending oil peak are lurking on the horizon too. This means reduced amounts of oil and higher prices, which will create another wave of competition among the major importers.

One big problem with the energy independence movement is that it has become a mantra of environmentalists and others interested in pushing their agendas, renewable energy, climate change, etc. It is also used by interest groups as a weapon for defending technological and economic reasons for using expensive and unreliable energy sources that are unable to compete on the market with oil and gas, in order to justify subsidizing them.

A close look at the energy independence situation reveals that all interests groups think that theirs is the only way of achieving it. The fossils are indispensible for providing reliable and cheap power, so they must be supported. Wind and solar are the solution for the future, so they must be supported unconditionally too.

The problem is that the support structure is limited and crumbling, so the different energy sources must find a way to support themselves while taking the country closer towards energy independence. This, however, requires a thorough understanding of the issues on both sides of the debate and coordinated efforts; both of which are missing today.

As things are going today, we can only talk about energy independence as a measure of our dependence on foreign fossils imports.

The Energy Independence and Security Act of 2007

For the purpose of ensuring our energy security and independence, the U.S. Congress passed The Energy Independence and Security Act of 2007. It is an Act of Congress addressing concerns with the energy policy of the United States. It was originally named the Clean Energy Act of 2007.

In brief, the Act consists of:

• Title I-Energy Security Through Improved Vehicle Fuel Economy, which requires increase in fuel economy standards for passenger cars, and established the first efficiency standard for medium-duty and heavy-duty commercial vehicles.

It is estimated to save Americans a total of $22 billion, by the year 2020, and to show a significant reduction in emissions equivalent to removing 28 million cars from the road. Title I is responsible for

60% of the estimated energy savings of the bill.

- Title II: Energy Security Through Increased Production of Biofuels, which contains the first legislation that specifically requires the addition of renewable biofuels to diesel fuel.

  Biomass-based Diesel, fuel must be able to reduce emissions by 50 percent when compared to petroleum diesel. Biodiesel is the only commercial fuel that meets this requirement thus far.

- Title III: Energy Savings Though Improved Standards for Appliance and Lighting, which contains standards for ten appliances and equipment.

  When fully implemented, the Title III modifications will void the burning of millions of barrels of oil, which will save millions of dollars, and contribute to cleaner environment.

- Title IV: Energy Savings in Buildings and Industry, which establishes new initiatives for promoting energy conservation in buildings and the industry.

  When fully implemented, Title IV modifications will reduce the energy used in federal buildings by 30 percent by the year 2015.

- The Act also provides: awards for developing a hydrogen economy; funding of R&D of renewable technologies; expanding research of carbon sequestration technologies; new training programs for "Energy efficiency and renewable energy workers"; new initiatives for highway, sea and railroad infrastructure; small businesses loans toward energy efficiency improvements; modernization of the electricity grid to improve reliability and efficiency via Smart Grid technologies; new federal standards for drain covers and pool barriers; and exclusions for people who have UV sensitivity that can be triggered by the higher UV radiation of the CFs.

A solid step or two forward, no doubt. There are already some good results to be reported from the above measures. And yet, energy independence is far from being a reality in the U.S. and most other countries.

Energy Survival

This is the hardest component of the energy dilemma. It is the skeleton in the closet, which no one wants to talk, let alone do something, about. Energy survival refers to the continuance of the energy supplies in the future, so that the coming generations do not experience a sudden and irreversible lack of fossil fuels. Such a plunge in the dark is absolutely unacceptable, and so it is our responsibility to do everything possible to avoid it.

In addition, excessive use and abuse of fossil fuels today is bringing increased emissions, which are causing adverse environmental effects. Continuing at this rate of use of fossils will put the world's population in increasing danger of land loss, air pollution, global warming, and many other negative effects. The combination of these effects puts human wellbeing and health in danger, so we cannot continue doing nothing to solve the problems.

It is obvious, from many facts cited in this text, that not much is done today to ensure the energy survival of the next generations, but these are serious issues that deserve our full attention.

Here is why:

1. We know and agree that *energy security and safety* are a must, for they affect us directly. Because of that all parties are working together and doing everything possible towards those goals.

2. We also know that *energy independence* is important for maintaining our way of life, so we are doing everything possible to ensure it too.

3. *Energy survival*, however, is not on the list of our priorities. It does not affect us directly today, because it is a long-term problem. It does not generate profits, and since we don't have to deal with its consequence directly, it is kept hidden in the closet of our guilty collective subconscious. This makes the subject quite uncomfortable for discussion, because our behavior today is similar to the way irresponsible parents spend their children's inheritance on frivolous and expensive life style, totally disregarding the future generations' needs.

As a matter of fact, most of our actions in ensuring energy independence, such as the increased exploitation and use of fossils (increased use of coal in China, and fracked oil and gas in the U.S.), are severely damaging to the integrity of the long-term energy survival of the affected countries and the entire world. These increased activities and fossils use deplete the limited resources, and pollute the environment excessively, leaving a deadly inheritance to the next generations.

We are only postponing the inevitable transition to a non-fossil economy, because we are too comfortable with the way things are now, and we are not willing to sacrifice for the sake of future generations. So as things are going today, we will be out of oil, gas, and coal by the mid-to late 21st century. What are the people living then going to do in a fossil-less society? What would they think of us?

There is a chance that by then all energy sources will be replaced by new and renewable such. Using solar, wind, biomass, and nuclear energy might be enough to substitute for all that the fossils are providing us today. These replacements, combined with more efficient use of residential and commercial energy, might be just enough to provide everything the society needs. If that happens, then our sin of indiscriminately abusing the precious energy sources will be washed away and forgotten…maybe…

It is more likely, however, that there would be problems during the delayed transition, resulting in energy- and fossil-based consumer goods shortages. It is difficult to predict when, and what exactly can be expected at that time, but looking around us today and seeing fossils in everything around us, we doubt that the life style of those who follow in our steps would be similar to ours.

One could argue that it might take longer than a century, but whatever it takes, the fossil supplies are limited, so they will be gone eventually. If not this century, for sure during the next century they will be no more. Sooner or later, one of the next generations is going to wake up in a fossil-less world. Not a drop of oil, a whiff of natural gas, or a piece of coal will be left on the face of the Earth. Just imagine what a very cold, dark and sad day that would be.

There will surely be many days before that, when the to-be-affected people will gather together to figure out what's happening with their energy sources and reserves. And they'll need a solution in hand by the time the last piece of coal and the last drop of oil go up in smoke. If they have not found substitutes by then, their lives will be quite miserable, and even in great jeopardy of extinction.

Present generations are obviously not concerned very much with the approaching end times of fossil energy. The U.S. energy economy, by 2020 at least and beyond, is based totally on fossils. The world as an entity is mining more coal than ever, pumping more oil than ever, and increasing gas and oil fracking to the highest levels imaginable. The use of all these precious fuels is increasing dramatically, especially in the developing countries, and the trend is growing.

We have already dug and pumped out at least half of the total fossils nature prepared for us during the billions of years in the past, and we will dig and pump out the rest within this century. That will leave the next generations with next to nothing in ready-to-use fossil reserves.

This makes our energy survival a short lived enterprise of ensuring enough energy now, but with no provisions for the future. Is there any way to change that? If there is, we don't see it. Capitalism dictates living well today, for tomorrow will take care of itself…somehow. We just don't feel responsible for whatever might happen in the future; good or bad.

SOCIO-POLITICAL CONCERNS

Energy and the related energy issues are no longer the sole concern of oil companies and governments. The general public is affected more and more by the developments of late in the field, so it is taking a major part in the energy dialog.

The Energy Debate

We must bring up the energy scenarios and the ensuing debate again, because it is quite important for the future of our energy security. It has many facets, as the role (and visibility) of the renewable energy sources is overwhelmingly becoming more important.

To clarify the situation, we show in Figure 8-6 what the experts expect to see by the mid-21st century in the global energy field and especially that of power generation. Obviously, energy use around the globe is increasing due to growing demand. The rate of increase is expected to keep up with the demand; with coal as the dominant power generator and major polluter. Natural gas' role globally will increase too, approaching, and in some cases surpassing, that of coal.

Hydropower is also expected to grow, as are nuclear and renewables power generation, but at a moderate (and unpredictable) pace and in a limited number of countries.

Here we must emphasize the fact that the future of the fossils (coal and natural gas) is quite clear, and we don't expect anything to get in their way in the short run. Their long-term future, however, is also very clear, but not that bright.

At this pace of exploitation, the fossils are doomed to complete extinction by the end of this century…or sooner!

The situation with hydro, nuclear and the renewables is more volatile, and many good or bad surprises are expected:

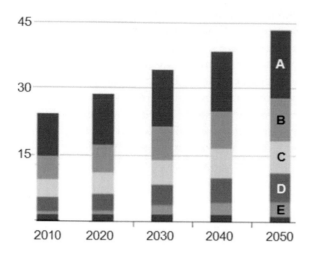

Figure 8-6. Global power generation by source (in trillion kWh/annum)
Legend:
A Coal
B Natural gas
C Hydropower
D Nuclear
E Renewables

- Hydropower depends heavily on water availability, which is getting scarcer. Periodic draughts of extended duration and severity are expected to aggravate the water scarcity around the globe, so hydropower might become an unwilling victim of the global warming.

- Nuclear's future, as predicted in Figure 8.6 is possible ONLY IF no new Fukushima accidents occur anytime soon. One more such accident could put the entire global nuclear power fleet on standby and the entire nuclear industry on its knees for an undetermined amount of time, or in a worst-case scenario, forever.

- The progress of the renewables (solar and wind especially) depends heavily on the above developments and many other factors. While coal and natural gas are abundant and cheap, the renewables will be treated as an expensive novelty and will not be given a chance for mass increase. Only when the fossils become scarce and expensive (most likely by the end of the century) will solar and wind be able to dominate the energy sector.

Government support and subsidies are also critical for the renewables, especially at this early stage of their development. Without government assistance, the renewables will stagnate until conditions change.

- The other renewables (biofuels, ocean wave energy, etc.) will continue growing modestly too, but only to pacify the environmentalists and meet the minimum government mandates. Their future also depends heavily on the availability and pricing of the fossil fuels. Government assistance is of utmost importance as well.

The Debate

As discussed above, the fossils rule for now, but their end is approaching, so the energy debate is mostly about fossils vs. renewables. Why, when, who, how much... all questions and issues that prevail in the debate. Our energy security, the environmental issues, economic development, and the variables, are also included in the debate, so it is quite a complex mixture of science, technology, politics, economics, and legal issues.

The number of interested parties is growing too, and when it comes to power generation they are now cleanly divided into three major groups. The major groups in the present-day energy debate are:

1. Those who are strongly against renewables,

2. Those who are strongly pro renewables, and

3. Those who don't care, or don't want to get involved.

In more detail:

- *Against renewables.* The most striking case against renewables we've ever seen, was quite eloquently expressed in an article titled, "Big Green, Not Big Oil, is the Enemy," published in the summer of 2013 in "the commentator state of the art." In it, the authors provide provocative ideas, information, and misinformation showing the renewables as a cancer growing within our society—a cancer ready to devour us, our values, and society as we know it today. (13)

"...because of angst-ridden theoretical speculation—Note: not empirical science—the modern green agenda has effected an intellectual disconnect. It is a disconnect that has seen eco-theories eclipse energy realities such that national leaders, industry executives and even reasonable people are not engaging in rational debate let alone action," the authors claim.

"There is no contest from an economic point of view. Solar, wind and other alternatives favored by the greens are not and will never be viable. From a thermodynamic point of view they will never amount to much more than one percent of world

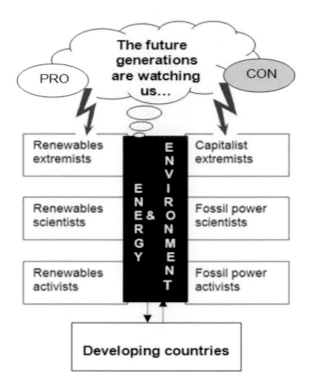

Figure 8-7. The energy and environment battlefield

energy demand without massive and unsustainable government subsidies," is the authors' opinion, so forget about renewables and focus on oil and gas.

And they continue, "The reality is, however, that it's the well-funded lobby of Big Green that has the greatest influence on government and government policy—and presents the greatest threat to the economy."

Sure enough, they make it quite clear, "Let's get it straight, Big Green is Big Business. Next time you are moved to pitch your 'hard-earned' cash into a tin shaken before you while being regaled with a nonsense message—e.g. that carbon dioxide is a 'pollutant'—you might like to bear that in mind."

As it stands, today's prevailing cultural 'green agenda' is nothing less than a case of anti-intellectualism, costing the earth, in pursuit of the illusory."

OK, we must agree that the renewables are not much now, and will never amount to much, if the author's point of view is accepted as the status quo. We must also agree that the Big Green lobby has done a lot of damage (remember Solyndra and its lobbyists), and that carbon dioxide is not a polluter for those who don't

see or feel its effects directly.

But the authors' conclusion, "The future of oil and gas is not solar and wind; the future of oil and gas is oil and gas," is a blatant denial of the fact that gas and oil—regardless of their potential, quantity, and present worth—are only a temporary solution to our energy problems.

It took Nature billions of years to create and store the fossils in the Earth. It took us—people—only a century to use half of them. How long will it take us to use the rest? Not very long, for sure!

If we do not have enough renewables by the time the inevitable "fossil-less" reality arrives, we and future generations will be faced with unprecedented challenges. Horrific fights for survival could erupt around the world, the outcome of which is unpredictable.

Maintaining that oil and gas is the only way to go is a good example of capitalist extremism (making a buck today, to heck with tomorrow), and a blatant refusal of the fact that oil and gas cannot last forever. It is self-centered egoism thinly veiled in economic justification.

No matter how well oil and gas serve us, or how much of them we have, they are a finite energy source, and they are killing us. Too, they will be depleted soon—20, 30, 50, or 100 years from today at the present level of exploitation. Then what?

In contrast, sun and wind will always be here. So it follows from a practical point of view (forget politics and science) that we must be careful when we imagine using fossils forever. It simply cannot happen, regardless of how much we like the idea, so the authors are barking up the wrong tree.

- *Pro-renewable* advocates can also be divided in several groups:
  — The extreme tree-huggers,
  — The caution-driven realists, and
  — The "let's wait and see" pragmatists

This is a large group in the U.S., where 80% of the people polled in 2014 have approved of renewables, while at the same time, 80% are against use of fossils.

Each of these affiliations has its problems stemming from misunderstandings of the technical details. The overwhelming conclusion is that we must be well educated on all issues, and very careful when discussing the different aspects of energy and energy security.

This is especially important for those who are given the privilege of making decisions on behalf of all of us—namely politicians and big company executives.

- *The agnostics*; those who don't care and/or don't want to get involved in any of these issues, can also be divided into several groups:
  — Those who are badly misinformed,
  — Those who don't have access to information, and
  — Those who refuse to even consider the issues.

This might be the largest group in the energy debate, since it includes hundreds of millions of people in Third World countries who have not even heard of the issues, or are too busy with daily survival to worry about the pending energy issues.

Each of the sub-groups has its own problems that cannot be resolved overnight. In all cases, reliable information, delivered in a believable way, is critical for educating these people. Armed with it, they can then make their conclusions and decisions.

Changes in the energy markets today are reflected by a) energy efficiency, b) energy intensity, and c) the energy-GDP ratio in the different countries, as follow:

Energy Efficiency

Energy efficiency is the new game in town. It is expressed in reducing the amount of energy required to provide products and services. Insulating a home, for example, reduces the energy use by requiring less heating and cooling to maintain a comfortable temperature inside.

Compact fluorescent lights (CFL) or natural skylights reduce the amount of energy required to light a room by 75%, as compared with using traditional incandescent light bulbs. CFL also last 5-10 times longer than their old incandescent cousins.

Great energy efficiency increase is attributed to new and more efficient technologies and production processes, and/or the application of proven methods of eliminating energy losses.

The key motivations to improve energy efficiency are reducing energy costs to the owner and cost saving to customers. Energy efficiency is also one of the solutions to reducing carbon dioxide emissions.

According to the experts, improved energy efficiency in buildings, industrial processes, and transportation could reduce the world's energy needs by 1/3 in 2050, and help control global emissions of greenhouse gases by an equal amount.

Energy efficiency is also seen as a pillar in national energy security, because it reduces the energy imports from foreign countries. Equally important, it also slows down the rate at which domestic energy resources are used and depleted.

Figure 8-8 needs some clarification. The U.S. is perched high on top of the world as far as its high productivity is concerned, but at the same time it is near the bottom of global energy efficiency. Could it be that Bangladesh is more energy efficient than we are? Yes; about three times more efficient. Why? It is our way life in the U.S. We are very smart and productive, but do not know how (or don't care) to conserve energy at home, or at work. Bangladesh, on the other hand, has no energy reserves, and the people are not accustomed to wasting energy, nor can they afford to.

And speaking of energy efficiency:

The U.S. Energy Efficiency Syndrome

Although the energy debate in the U.S. is escalating, and there are some good results to show, we don't expect any major developments on that front in the near future.

This is due to two main factors:

- *Infrastructure*. The way the U.S. roads and transport infrastructure are designed and built is very different from those in most other countries. The vast territory of the 48 continental states is scarcely populated, so public transport is not the best option. Although there are trains, planes, and buses running all over the place, they account for a very small percentage of human transport. Instead, most people prefer to drive from place to place, and even from state to state.

Mass transport systems exist in a few major cities, but in most U.S. cities and rural areas it is non-existent. Because of that, people drive everywhere around town in their own cars.

The interstate highway system is so well developed that one can make 100-200 miles drive in a car faster than taking a train, bus, or even a plane in many cases (not to mention that this is a much more convenient and pleasant way to travel). So millions of people do just that—drive and drive!

Nevertheless, it would not be that difficult or expensive to modify the national mass transportation system, so that more people use it. Trains and busses crisscross the cities and the country half-empty, so filling them would boost our energy efficiency. This, however, is unlikely to happen anytime soon, because of the other, even more important factor—our life style.

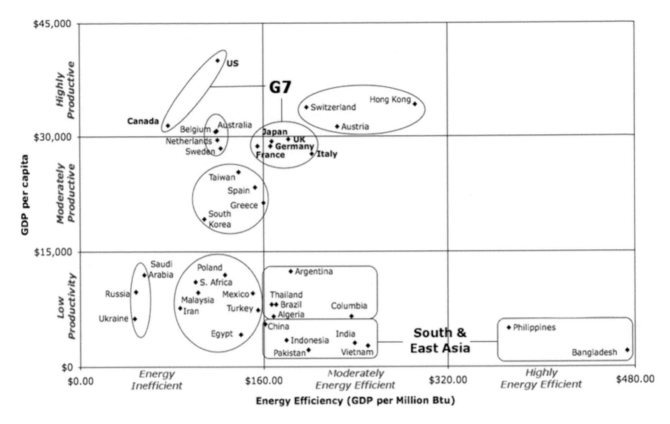

Figure 8-8. Energy efficiency vs. GDP

- *Our Way of Life* is very different from that of any other humans on Earth. We are used to having our cake and eating it too. Like spoiled children, we want everything to be done our way, and done now or sooner. Amazingly, we usually get it that way too.

These high expectations and the resulting high standard of living come at the cost of long hours of stressful work. However, the culture has become so entrenched in this system that most people feel there is no alternative. After all, it is a small price to pay for our almost unlimited freedoms and comforts.

The Major Change?

For generations, Americans have enjoyed living in large houses with large appliances, and large brightly lit rooms maintained at constant temperature throughout the year. Since the 1950s, Americans have been driving large cars, trucks, and SUVs without regard to price or anything else. The automobile is a symbol of freedom and independence which are two of the highest values for Americans, so any thought of altering our transportation system is viewed as a threat to our most cherished values.

Americans insist on doing whatever, whenever at a whim. It is unreasonable to expect Americans to live any other way unless something drastic happens to force a major change.

But what major changes could we expect? Even during the Great Depression many people drove cars, so we see cars as a key indicator of our way of life. It is how it is. This means that we will hang onto the car driving habit as a necessity until the last drop of oil is gone. Therefore, only when oil is totally exhausted could we talk about a major change in the energy sector.

Also during the Depression, FDR authorized electric power subsidies, which indicates the importance of this energy source. Today we can't even imagine life without electricity, so it is to be expected that we will burn coal and natural gas until they too are gone. Only then will we take steps towards implementing major change.

Europeans and Asians are accustomed to living in small, crowded, dimly lit rooms, with shoebox-size refrigerators, air-dry clothes lines all over the place, and many ride bicycles to work. Most have accepted the idea of living with ever-increasing energy sacrifices, while others are even contemplating life in strictly regulated eco-villages.

Well, that's perfectly OK, if this is what they want to do. We will let them. This, however, is not going to happen on any large scale in the U.S. Not now, not ever... at least not until the major changes are forced upon us.

The conservative thinkers of today must understand that "our way of life" cannot be changed, and work around it.

It matters not what our opinions of the status quo are, or how much better the alternatives might be; it is what it is and making any significant changes is not an option. It is the American way of life and people do not like change.

While a small number of free thinkers in Portland, Oregon, might consider living in an eco-village, going to Tyler, Texas, with that idea might get you in big trouble with the locals. Just think of a Texas farmer living on a 500-acre farm in a 5,000-square-foot house, with walk-in refrigerator/freezer and two air conditioning units on the roof. Then take a look at his equipment barn full of huge tractors and other machinery—F-150 truck, Escalade luxury SUV, a Softail Harley Davidson in the three-car garage—and his twin-engine fishing boat tied to the lake pier.

Now imagine this same guy riding a bicycle to the local bar at night, or to church on Sunday. How about moving out of the 3-bedroom home into a studio apartment with a shoebox-size refrigerator, 25 W lights, no air conditioner, and air-dry clothes line in the bathroom. Not a chance! This is not going to happen. Not in America. Not now, not later, not ever... unless and until the big major change comes.

So perhaps we should focus on greater energy efficiency instead? Yes, it is good to talk about it, so don't stop. You, however, cannot take my four-bedroom 3,000-square-foot home with two full-size refrigerator/freezers and 5-ton air conditioner on the roof away from me. And I will also continue driving my 8-cylinder SUV, my 1100 cc. motorcycle, and 30-foot diesel motor home anytime and everywhere I want. Our way of life encourages this type of behavior and only a major change could make a difference.

But let's continue the energy efficiency talk; it is a good thing, if it is not pushed down our throats. Instead, it must be somehow woven in the American fabric without sudden changes of our daily routines. The politicians know this very well, since they live in even bigger houses and drive even bigger cars and campaign busses. They are very careful not to step on any toes too, because one wrong step could cost them their political

future, together with their large homes and cars.

So, the energy efficiency, environmental protection and the need for major change debate in this country continues, but the changes that are planned or expected will have to be carefully crafted and even more carefully implemented. They must come slowly and without much sacrifice on the part of the people and their way of life. How far we can go that way is uncertain, but there are few alternatives. And that's the honest (albeit politically incorrect) truth!

Energy Intensity

Energy intensity measures the energy efficiency as compared to the economy of a nation, calculated as units of energy per unit of GDP. High energy intensities indicate a high cost of converting energy into GDP, while low energy intensity means lower cost of converting energy into GDP.

For example, 1 million Btus consumed with an energy intensity of 8,553 produced $116.92 of GDP for the U.S. between 2000 and 2010. At the same time, 1 million Btus of energy consumed in Bangladesh with an Energy Intensity of 2,113 produced $473 of GDP. This is more than four times the effective US rate. This, however, is an extreme case that does not set a precedent.

Energy intensity can be used as a comparative measure between countries, whereas the change in energy consumption required to raise GDP in a specific country over time is described as its energy elasticity.

Energy intensity is different in different world regions. For example, it is three times higher in CIS than in European countries. Energy intensity in OECD Asia and Latin America are about 15% above those in Europe, while North America stands 40% higher but remains below the world average. India is on a par with the world average, with energy intensity levels 60% higher than in Europe.

The high energy intensity in the CIS, the Middle East, China and other Asian developing countries is due to the predominant use of energy-intensive industries and relatively low energy prices.

Figure 8-9 shows that the U.S. energy intensity is in decline (a good thing). From 1950 to 2010, energy intensity in the U.S. decreased by almost 60% per GDP dollar. There was a period of sharp increase around the 1970s, although energy prices fluctuated only about 3% from year to year.

After the 1973 Arab oil embargo, energy prices rose significantly above the previous years, which lead to changes in the national energy policy. New vehicle efficiency standards were established, and consumer at-

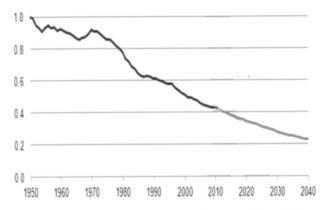

Figure 8-9. U.S. energy consumption per dollar GDP

titudes began to change as well. Since 1973, U.S. energy intensity has declined at a rate closer to 2% per year, and is expected to continue through 2040.

Surge of the knowledge-based economy, expressed in the rise of computer hardware, software and digital technology from 1991 to 2000, global economic productivity increased without parallel increases in energy use. New energy production technologies and use improved energy efficiency in almost every aspect of the economy. At the same time, the energy-intensive industries in the U.S. declined as well, with continuing structural changes in the economy.

Global energy intensity declined at an average annual rate of almost 1% percent in the 1980s and 1.40 percent in the 1990s. From 2001 to 2010, the rate dropped to 0.03 percent. This was a period of decline, when the developed countries restructured their economies with energy-intensive heavy industries accounting for a shrinking share of production. Global energy intensity increased 1.35 percent in 2010, reversing a broader trend of decline over the last 30 years.

In contrast, China's energy intensity declined 4.52% annually between 1981 and 2002, and a staggering 15.37% between 2005 and 2010. But that fell short of the government's incredible goal of 20%. The main reason for this shortfall was that over half of China's $630 billion stimulus plan was invested in infrastructure development, which drove up energy consumption.

The latest economic recovery has led to an increase in total energy consumption per unit of GDP, for the first time in more than 20 years (+0.5%) in the developed countries.

The European Union accounts for the highest increase in energy intensity, about 2.5% as compared to a 1.7% average annual decrease before the crisis. The poor EU performance was due to the industrial sector, where

energy consumption did not decrease at the same pace as the value-added segment due to lower efficiency.

The key drivers of the present and expected decline patterns in the U.S. include:

- Residential energy intensity, measured as delivered energy used per household, is expected to decline about 27% by 2040.

- Commercial energy intensity, measured as delivered energy used per square foot of commercial floor space, is expected to decline about 17% by 2040.

- Industrial sector's energy intensity, measured as delivered energy per dollar of industrial sector shipments, rises above its 2005 level initially owing to the 2007-09 recession but is expected to decrease 25% below its 2005 level by 2040.

- Transportation sector's energy intensity is more difficult to measure because of the multiple modes of transportation. Light-duty vehicles are by far the largest energy consuming part of the sector. Light-duty vehicle energy intensity, which is measured as their consumption divided by the number of vehicle-miles traveled, is projected to decline by more than 47% by 2040 from their 2005 value.

Things are looking good in the U.S., from that perspective at least.

Energy Use Intensity

Energy use intensity (EUI) is a measure of energy use per unit area of a building or business. As one of the key metrics in the efficiency mix, EUI basically expresses a building's energy use as a function of its size, function, and other characteristics.

Note: EUI is not to be confused with the (national) energy intensity we discussed above. EUI is a measure of a single unit—building, business, or home.

For most property types, the EUI is expressed as energy per square foot per year. It's calculated by dividing the total energy consumed by the building in one year (measured in kBtu or GJ) by the total gross floor area of the building. This can also be calculated from the energy use information in the monthly utility bills.

Generally speaking, a low EUI signifies good energy performance. A church campus, for example, uses much less energy compared to a hospital. The intensity of energy use, and the time periods are what makes the difference. Certain property types, however, will always use more energy than others.

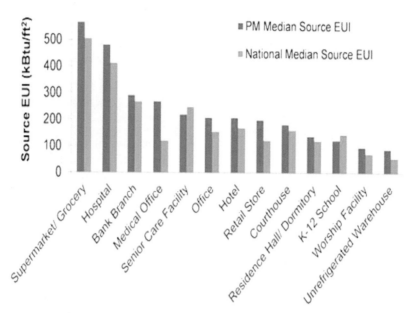

Figure 8-10. EUI of different building/business types

It is obvious from Figure 8-10 that supermarkets and hospitals (usually 24 hrs./day operation) are using much more energy than churches with their 2-hour weekly use, or warehouses with dim and partial lighting. Because of that, the efforts to increase energy efficiency must be focused on the large users.

Note: Why a bank branch would use that much energy is a mystery to us, but the figures are right, so we must assume that banks have some operations that require a lot of energy.

Energy and GDP

The amount of energy that a nation uses is related to its productivity, as expressed in its GDP and its population, which of course drives GDP. Therefore, to compare nations with different outputs and populations, it seems wise to divide both consumed energy and GDP by population.

Figure 8-11 shows unequivocally the wide range in power use (in kW per capita) between the lowest consumer (India) and the highest (USA). The GDP per capita is directly proportionate to the energy use per capita. In other words, the more money people make, the more energy they use.

Half of the rural areas in India don't have access to electric power, so the locals do with whatever they can, burning wood and coal, and many businesses generate electricity via diesel generators. In stark contrast, each U.S. citizen needs an average more than 10 kW at any mo-

ment to keep our way of life going. This includes all types of energy use in our daily lives: transportation, work, food, housing, leisure, etc. This begs the question: How much of this energy is wasted?

There is a rough overall correlation between GDP/capita and energy/capita. This is because higher productivity per individual will cause a higher energy need for each. But comparing a developing country with a highly developed one is like comparing apples with oranges. A more fair way to do this is to compare the U.S. and Japan, where the productivity per capita is comparable and are the best in the world from a productivity point of view. Yet Japan uses only half of the power used in the U.S.

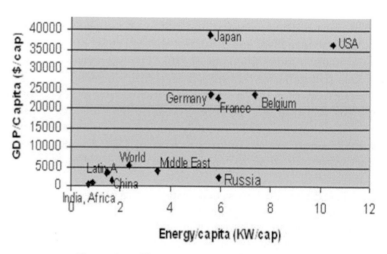

Figure 8-11. Energy use vs. GDP per capita

Because of that, Japan emits 9.5 tons of CO_2/capita, vs. 19.7 tons per capita in the U.S., which roughly corresponds to the 2:1 energy consumption ratio of the two countries. Remember that the global per-capita CO_2 emission is only 4 tons per year—2.5 times less than the Japanese and almost 5 times less than the U.S. For some developing countries this ratio goes up to 20 times less than the U.S.

If we use energy as frugally as the Japanese, we would not need any crude oil imports. This is one (the only) way to obtain energy security and energy independence, while significantly lowering the total GHG emissions at the same time.

Note: After the 2011 Fukushima disaster, however, Japan shut down its nuclear plants and replaced the lost power by coal and natural gas-fired such. This, of course, increased the air pollution significantly, so the emission ratio has changed, but is still lower than the U.S.

The U.S. energy use per capita was constant from 1990 to 2007 and began to fall after 2007. Energy use per capita continues to decline as a result of improvements in energy efficiency, reflected in the use of new more-efficient appliances and CAFE standards. We have been also slowly introducing changes in the way we use energy in the form of energy conservation. As a result, under the present trend, the U.S. population is expected to increase 21% by 2040, but the energy use will increase only 12%. The energy use per capita is expected to decline 8% during the same time.

The U.S. CO_2 emissions (in 2005 dollars of GDP) have tracked closely with energy use. With lower-carbon fuels accounting for a growing share of total energy use, CO_2 emissions per dollar of GDP are expected to decline more rapidly than energy use per dollar of GDP, or about 2.3% per year for a total of 56% below their 2005 level by 2040.

This is good and probably the most that can be expected from U.S. consumers, which is actually much better than the rate of energy use and CO_2 emissions expected from other, especially the developing, countries. This, however, is an unfair comparison, due to the much different (even opulent) lifestyle in the U.S., which makes a huge difference in every area of daily life.

Note: There is no official measure of the level of opulence (or extreme life styles) for different countries, but we don't have to dig too deep to see the striking difference. Just one example would illustrate the great life style chasm that exists between life in the States and other parts of the world.

During the 2014 Ebola epidemic events in Africa, thousands of people, including doctors, were infected and many died. In the best cases, the infected people—the lucky ones—got a bed and some food, but not much more than that. There were simply no adequate facilities and services to help them. They were basically left to fight the vicious Ebola on their own with little outside help. Almost half of the infected died, and the rest were left struggling in their misery for weeks and months.

At the same time, three American doctors were infected by the virus. With that news, the world stopped and held back its collective breath. Literally. The media were brimming with concerns about the American doctors' status and following their progress step by step. An intense and very expensive effort started, to save the lives of the Americans at all cost. No amount of effort or expense was going to delay, let alone stop, the operation.

A private jet was equipped with special isolation tents and specialized equipment, attended by dozens of specialists onboard and on the ground. The doctors were carefully isolated in special suites, placed in the isolation tents in the jet and flown to specialized hospitals in the U.S., which are equipped for, and specialized in, handling acute infectious cases. Several doses of a new experimental vaccine were administered with the hope of stopping the virus and speeding the recovery period. The evacuation and treatment operations took several weeks at a cost of millions of dollars and ended successfully with saving the Americans' lives, who were up and well within a week or two.

Looking at the two different scenarios we clearly see that Americans have enormous advantages over the locals in any African village. Money, expertise and preferential treatment are always available when our wellbeing is at stake. This includes the advantages we have in the energy sector, where many African villages don't even have access to electricity. During the 2014 Ebola disaster in Africa, many (mostly make-shift) hospitals didn't have the basic necessities to keep the patients alive. Some didn't even have electricity and/or running water. Thousands of people died from this lack of the most basic patient care needs.

At the same time, most Americans have access to unlimited electric power, water, health care and all the benefits that come with life in America, including unlimited resources in efforts to save American lives overseas. No wonder the developing nations are trying to catch up with us at all cost. Who can blame them?

Personal Safety and Comfort

We know now that the people in the developing world have as a goal to improve their lifestyle to match that of the developed countries. But what does the average person in Los Angeles, Tokyo, or London think about energy, energy security, energy prices, the environment, etc.? How are these things affecting our lives?

Figure 8-12 was compiled from different sources published in different developed countries. The concerns of the people in undeveloped countries go along the same line, but only to a point. Nevertheless, personal safety is shaping up as the most important concern of all people worldwide.

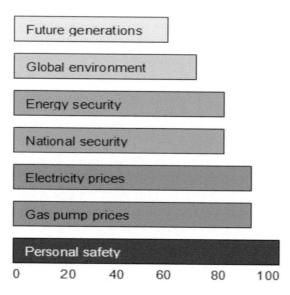

Figure 8-12. Global concerns (100 is most concerned)

Part of the energy survival goal is making sure that all people are safe and living in relative comfort—without dangers and threats of any sort—including having enough energy for daily activities.

Note: The concept of "comfort" varies widely from country to country, so we will just have to imagine the difference between feeling comfortable in a 4-bedroom house with central air conditioning and two cars in the garage in the U.S., vs. a hut in a fishing village in Thailand, or a camel herder's tent in the Sahara.

It is even harder to imagine the enormous disparity future generations would feel when comparing our wasteful life style with that of a their potentially fossil-less society. The safety of all people then would be affected too, due to lack of readily available and cost-effective fuels, fossil-based commodities, medications, and chemicals.

Today's safety and comfort could also become a thing of the past. The people of year 2222 would only imagine what an 8-cylinder gas guzzling car looks like, or what it feels like to roam the world's oceans on cruise ships.

The personal-safety-come-first fact was confirmed by the recent Fukushima nuclear plant accident. The devastation was so great that now the majority of Japan's citizens are against using nuclear power. Even if that would cost them in terms of finances and personal comfort.

So, the most environmentally minded nation in the world—the Tokyo Protocol instigator and host—trashed the Protocol in an instant and fired all its coal-burning power plants after the nuclear accident. Environment be still...personal safety comes first. Period!

As a result, the GHG emissions in the country increased by nearly 15%, and the protests against nuclear power continue. This means that Japan is never going to be able to restart its entire nuclear fleet and meet the Kyoto Protocol agenda.

But the environment and everything else is very low on the Japanese people's priority list. This is how important personal safety is—it is on the top of list. It is powerful feeling that must be experienced to be fully understood.

In all cases, personal safety and comfort should not be sacrificed at the expense of achieving energy security.

And there are many negative sides of the energy sector that are threatening the personal safety and comfort of people even today. It is not just the fear of nuclear accidents. Beijing citizens are afraid that the air they breathe is killing them and insist that air-cleaning measures be taken soon. People on the Maldives fear that their villages will be flooded someday soon, and that the land that feeds them will end up in the ocean.

Why should our efforts to achieve energy security cause such extensive human suffering?

As part of our safety and comfort, we are also concerned with the prices at the gas pumps and the monthly electric bills. These days, we must drive everywhere we go. Without vehicle fuel, or if its price becomes extremely high, we will be stuck, or greatly inconvenienced.

The same is true of the electricity we use. Everything today is run by electricity, so any shortages or high prices will have a serious negative effect on our lives.

These are major priorities of the average person here, until we hit a Fukushima-like threat. Then all these concerns become irrelevant or at least secondary. Personal safety becomes priority one.

How fair or unfair this is, is another subject, which we will let the statisticians answer, but the answer will not change the way things are now. Our primary concern is always our safety, and that of our family and friends. The rest varies in importance and changes from place to place and time to time.

NIMBY

We know that personal safety and comfort have a price and a limit. NIMBY (not in my backyard) is a powerful movement in a number of countries—including the U.S. that is trying to establish and enforce those limits. In the U.S. it started with opposition to nuclear power, a part of this movement since day one, starting with the first reactor, Fermi 1, to be built close to Detroit in 1957, where the United Autoworkers Union led a fierce anti-nuclear opposition, resulting in cancellation of the plans.

Then came the decision of PG&E to build the first large, commercially viable nuclear power plant in the U.S. at Bodega Bay, close to San Francisco. This was a very controversial project and serious conflict with local citizens began in 1958. After several years of non-stop protests and strikes, the NIMBY movement won in 1964. Plans for a power plant at Bodega Bay were abandoned. Another set of attempts to build a nuclear power plant in Malibu, California, failed under similar circumstances.

The awareness about the danger of nuclear power which started the NIBMY movement was fueled by a number of little known nuclear accidents during the 1960s. Those began with a small test reactor exploding at the Stationary Low-Power Reactor Number One in Idaho Falls in January 1961, that was followed by a partial meltdown at the Enrico Fermi Nuclear Generating Station in Michigan in 1966.

Environmentalists see the advantages of nuclear power in reducing air pollution, but are in general critical of nuclear technology on the grounds of its safety record. The major concerns with nuclear power were and still are about the possibility of nuclear accidents, nuclear proliferation, high cost of nuclear power plants, the possibility of nuclear terrorism, and the problems with radioactive waste disposal.

Because of that, nuclear power has been unable to face the opposition successfully thus far, and as things are progressing, it has a long way to go before achieving mass acceptance in the U.S. and the world. The success (or not) of the ongoing work at the destroyed Fukushima nuclear power plant in Japan will have a lot to do with determining the future of global nuclear power, so we will be watching carefully.

The NIMBY movement is now spreading and growing as a mass opposition to the expansion of hydro racking activities. This strong opposition has led companies to adopt a variety of public relations measures to educate, explain, and reduce fears about hydraulic fracturing. The admitted use of military tactics to counter drilling opponents and other such abnormal behavior have been in the center of the media reports. There is a report of a senior executive of a large oil company saying, "We need to download the U.S. Army Counterinsurgency Manual, because we are dealing with an insurgency."

Another executive confirmed that his company uses psychological warfare operations veterans in dealing with the NIMBY crowd. According to him, "...the experience learned in the Middle East has been valuable, when dealing with emotionally charged township meetings and advising townships on zoning and local ordinances dealing with hydraulic fracturing."

Police departments are increasingly forced to deal with intentionally disruptive and violent opposition to oil and gas development in different parts of the country. In March 2013, ten people were arrested during an "anti-fracking protest" near New Matamoras, Ohio. They illegally entered a well development zone and locked themselves to drilling equipment. A drive-by shooting at a well site in Pennsylvania was recorded, in which an individual shot a rifle at the drilling rig, shouting profanities at the site and fleeing the scene. There were cases of gas pipeline workers finding pipe bombs at pipeline construction sites, which would cause a disaster if not discovered and detonated in time.

Government Politics and Policies

Figure 8-13 shows that the primary concern of any government is national security. The other priorities vary and the rhetoric on the subjects continues at a variable volume according to current events. Politicians make appealing speeches, company officials promise to put personal safety and the environment on top of their list of priorities, but actions speak louder and compromises are usually the result.

There is actually some progress in these areas, but corporate priorities on making money can overshadow those of providing personal and environmental safety (which are interrelated). As a result, there are a number of accidents and incidents which damage properties and hurt humans. Some of these are well documented, but corporations refuse to admit fault or take responsibility.

Even worse, many companies pollute our soil, our water, and our air without blinking an eye. They justify these actions as, "done for the common good," and usu-

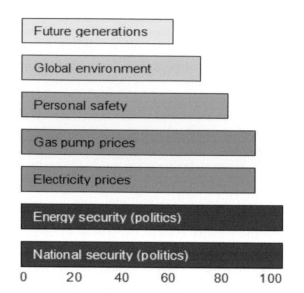

Figure 8-13. The government's concerns (100 is most concerned)

ally with the support of the governments. But as they say, "all politics are local," so the common good stops at home.

When a sinkhole full of green bubbling liquids shows up in my backyard, my water is contaminated, and the air is full of poisons, then all politics—including concerns with national security, energy, environment, and everything else—go out the window.

But the politicians and corporate executives sit on the other side of the issues, since they see the problems as small, isolated incidents. They look instead at mass production of energy as the only solution to our energy problems. The rest matters little or none. The race to dig out and pump up as much fossils as possible is on, whatever the consequences...until the last drop of oil and chunk of coal are burned.

Related environmental and health issues are discussed on all levels, and some action is taken here and there, but the coal burning in China, India, and even environment-conscious Japan, Germany, and other countries is on the increase. How long can the Earth's atmosphere take the billions of tons of GHG gasses sent down its throat (and that of its population) is uncertain, but nothing good is expected to come from this race. And what of the future generations? Well, they can wait...we have too many problems to worry about now.

The unintended consequences of the Fukushima disaster point the way to solving the energy problems in anticipation of the bad things to come (nuclear plants accidents and fossils depletion in particular.) The Japanese government, many companies, and the public are look-

ing to replace nuclear power and the fossils with alternative energies. The island of Japan is poor on fossils, its nuclear energy is misbehaving, and there is not enough land for bio-fuels and other alternative energy sources.

But Japan has a huge coastline, which can be used for wind power generation. It also has many unutilized residential and commercial roofs, most of which can be used for mounting solar panels.

Although the sun and wind are not as reliable as nuclear energy, they are much less dangerous. They are also less polluting than coal-fired power. Because of that, the country is headed in the direction of the alternative energies, but it will be a long and difficult road. For Japan it is the only long-term solution to their energy, environmental, and health problems.

The world should pay close attention to the developments in Japan, because what happens there might point the way to the future of energy and energy security.

National Energy Policy

The energy policy of the United States has been, and still is, a work in progress that changes with current events. Looking back, it seems to take a sharp turn after each major development during the last 40 years.

It started just about 40 years ago with the 1973 Arab oil embargo, which was a slap in the face that woke up the American public and policymakers to the fact that our energy security is totally insecure. For the first time since WWII we saw our nation attacked directly and mercilessly by foreign powers. It became quite obvious that we can be hit where it hurts most—by depriving us of life giving energy supplies—and that there is nothing we could do.

Lesson learned? Partially so! Soon after the Arabs decided to let us of the hook, we went back to our old ways of life and doing business. Except now we would sleep with one eye open and watch for another Arab oil embargo, adjusting our national energy policy according to the developments.

Some examples of what happened and didn't happen through the years:

- Nixon vowed to make sure that gasoline and diesel will never exceed $1.00/gallon. Instead, we paid $6.00 per gallon for diesel in Bakersfield, California, in 2008.

- Carter proclaimed that the U.S. will never again import as much oil as it did in 1977, but no comprehensive long-term energy policy has been proposed, albeit the discussions continue.

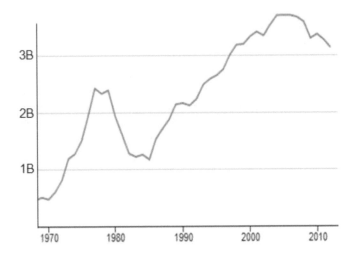

Figure 8-14. U.S. oil imports (in billion bbls/year)

- The U.S. now imports several times the amount of crude oil it imported in the 1970s, and the use of crude oil in the U.S. has increased well above that level as well.

- The federal energy policies have been dominated by crisis-management thinking, and knee-jerk reactions that favor expensive quick fixes and single-shot solutions—Solyndra style—which for the most part focus on masking the problem with the "band-aid" approach, while ignoring market and technology realities.

The government role is to provide stable rules in support of basic research while leaving the American enterprise machine alone, thus allowing good USA-style entrepreneurship and innovation to lead the markets.

Instead, Congress and administrations have time after time introduced energy policies that seem politically expedient, but with a short life on the energy markets. The trend was started by President Nixon who called for urgent "energy independence." This was the wake-up call for the U.S. policymakers, and a number of measures were taken to save energy and put energy production and use under some resemblance of control.

Some of these measures were successful, some not so much:

- In 1974 the national maximum speed limit was reduced to 55 mph through the Emergency Highway Energy Conservation Act. Soon after the "energy emergency" was gone, the speed limit was restored to 65 and 75 mph, and even higher in some states. Everything went back to "normal"; 8-cylinder cars and trucks were still the predominant vehicles on U.S. streets.

- Year-round daylight savings time began in Jan. 1974, in an attempt to save some energy, but due to wide public protests, pre-existing daylight savings rules were restored in 1976.

- The U.S. established the Strategic Petroleum Reserve (SPR) in 1975 and President Ford signed a bill creating the first fuel efficiency standards, which required auto companies to double fleet-wide fuel use averages by 1985.

- The Energy Policy and Conservation Act (EPCA) of 1975 were enacted and expected to achieve a number of goals, such as: increasing oil production through price incentives, establishing a Strategic Petroleum Reserve (SPR), and increasing automobile fuel efficiency.

- EPCA sought to roll back the price of domestic crude oil. Old oil was to be priced at the May 15, 1973, price plus $1.35 per barrel. New oil and stripper oil prices were set at the September 30, 1975, new oil price less $1.32 per barrel. In addition, the released oil program was dropped. It also authorized the creation of the Strategic Petroleum Reserve for the storage of up to 1 billion barrels of oil for emergency use only.

- The Corporate Average Fuel Economy (CAFE) standards were established at that time. They mandated increases in average automobile fuel-economy. These regulations set a corporate sales-fleet average of 18 mpg beginning with the 1978 model year, and established a schedule for attaining a fleet goal of 27.5 miles per gallon by 1985.

The CAFE standards apply separately to domestic and imported sales fleets. That program has been relatively successful, but its effect is minimal due to the extensive use of low mpg vehicles (large trucks and SUVs) in the country.

- In 1977, President Carter called the energy crisis the "moral equivalent of war" and established the U.S. Department of Energy (DOE), which consolidated several energy-related entities, and functions of the federal government into a single, Cabinet-level organization.

Until the 1970s, energy and fuel-specific programs were handled by several federal departments: The Department of the Interior managed most federal programs affecting the coal and oil industries, the Federal Power Commission regulated natural gas prices, and the Atomic Energy Commission (AEC) oversaw the development of nuclear power.

Post-1970 events accelerated the reorganization of the energy-related programs of the federal government. AEC was replaced with two new agencies; the Federal Energy Administration (FEA) was created in 1974 to administer programs that included crude oil price and allocation, and the SPR replacement of natural gas and oil with coal, and energy conservation.

One of the key functions of the DOE originally was to formulate comprehensive energy policy, but energy management in general became the primary role of the agency in the early 1980s. Energy shocks and public sensitivity during the last 2-3 decades have molded DOE into the country's energy manager. Energy conservation, alternative fuels, oil consumption reduction, national security, and energy prices are now DOE's major concerns.

- The second oil crisis hit the U.S. and the world in 1979 when the Shah of Iran was overthrown and the Iranian revolution began. The "tractorcade" (encircling of the U.S. Capitol by 3,000 tractors) called the attention of policymakers to the need for ethanol production and use. Now ethanol is part of the energy mix and its role is becoming increasingly important.

- The Three Mile Island nuclear plant accident around that time changed the plans for building more nuclear plants, and the U.S. nuclear industry has been in decline ever since.

- President Carter showed his support for renewables by installing a bank of solar panels (hot water heaters) on the roof of the White House in 1977. That experiment did not last very long and the panels were removed during the next presidency.

Another set of solar panels was installed on the roof of the White House (PV modules) in 2013 by President Obama. These events demonstrate the on-again/off-again fate of solar development in the U.S.

- Three Energy Policy Acts were passed later on in 1992, 2005, and 2007, which include many provisions for conservation, such as the Energy Star program.

- A set of energy development programs, complete with grants and tax incentives for both renewable energy and non-renewable energy were also enact-

ed. Most of these are now terminated, or changed to a lesser function.

- The latest developments, starting with the financial crisis in the U.S., and the subsequent global economic slow-down during 2008-2012 brought about the same jerk-knee reaction from the federal and state administrations. Billions of dollars were thrown at energy companies and projects, many of which were proven to be a waste of time, effort, and money. What government officials and their advisers at DOE were thinking is beyond our comprehension, but we saw similar awkward decisions and money waste during the renewables revival of the late 1970s, so the latest frenzy was not totally surprising.

Now, however, the plans to build mega-power solar and wind installations in the U.S, although still in the news, are on the government's back burner. Fossil energy—crude oil and natural gas—are abundant in the U.S., so we will use every last chunk and drop of these while we can, instead of leaving some to the next generations.

The Present-day Energy Status

Officially, today's U.S. energy policy is determined by federal, state, and local entities. It is geared to address issues of energy production, distribution, and consumption, such as building codes, gas mileage standards, pollution, etc. It also includes legislation, international treaties, subsidies and incentives to investment, guidelines for energy conservation, taxation and other public policy approaches and techniques. State-specific energy-efficiency incentive programs also play a significant role in the overall U.S. energy policy.

The U.S. had refused to endorse the Kyoto Protocol, preferring to let the market drive CO_2 reductions to mitigate global warming by means of CO_2 emission taxation. The major (unofficial) reason is that any emissions restrictions might hurt the related industries and cripple the national economy.

President Obama has proposed an aggressive energy policy reform, including the need for a reduction of CO_2 emissions, with a cap and trade program, which could help encourage more clean renewable, sustainable energy development. The new energy policy, however, is only partially viable now due to more urgent issues, and congressional responses.

President Obama also set a goal of 20% renewable energies to be implemented in the U.S. government installations (including DOD) by 2020. The US government is the largest energy consumer in the country, so the energy use reduction targets (which only apply to electricity consumption) will be phased in gradually.

Agencies must draw no less than 10% of their electricity from renewables by 2015, 15% by 2017, 17.5% by 2019 and no less than 20% by 2020. The different departments are required to do so by funding and installing their own generation through on- or off-site renewables. They can also purchase from third-party-owned clean energy plants built at their request, or renewable power from the grid, or renewable energy certificates. All power also must be generated by renewable sources less than 10 years old.

The target was mentioned in previous climate action speeches by the president but is now set in motion following his Presidential Memo. The US government is also looking to facilitate the work of developers to install renewable energy generation projects on federal land. A government auction for available federal land in Colorado in 2013 did not result in any bids, however.

Update 2014

The U.S. installed 1,330 MW of solar PV in Q1 2014, up 79% over Q1 2013, making it the second-largest quarter for solar installations in the history of the market. Cumulative operating PV capacity stood at 13,395 MW, with 482,000 individual systems on-line as of the end of Q1 2014. Growth was driven primarily by the utility solar market, which installed 873 MWdc in Q1 2014, up from 322 MWdc in Q1 2013.

Q1 2014 was the first time residential PV installations exceeded non-residential (commercial) installations nationally since 2002. For the first time ever, more than 1/3 of residential PV installations came on-line without any state incentive in Q1 2014. During the same time we saw school, government, and nonprofit PV installations add more than 100 MW for the second straight quarter.

74% of new electric generating capacity in the U.S. in Q1 2014 came from solar. PV installations are up about 40% over 2013 and nearly double the market size in 2012.

Q1 2014 was the largest quarter ever for concentrating solar power due to the completion of the 392 MWac Ivanpah project and Genesis Solar project's second 125 MWac phase. With a total of 857 MW (AC) expected to be completed by year's end, 2014 will likely be the largest year for CSP in history.

Great achievements, no doubt, yet solar and wind power generation are still a very small part of the energy mix in the country. Combined, they provide about 2-3% of the total power used on a daily basis.

The Renewables' Share

As Table 8-1 shows, the renewables, which category includes all types of renewable energy technologies (hydro, solar, wind, bio-, and geo-power), are about 10% of the total energy mix, but solar and wind are only about 2-3% of the total.

We discussed some of the issues that are stopping the full development of solar and wind power, such as variability, lack of storage, high price, increasing competition and resistance, which means that the renewables have a long way to go before they can compete with the conventional (fossil) fuels.

One of the major obstacles in the renewables' progress is the fact that the fossil energy sources are abundant and cheaper. This is causing a catch 22 scenario in the U.S., where we are using a lot of fossil fuels, because they are abundant and cheap, which is stifling the progress of the renewables. At the same time, we are emitting a lot of GHGs which are choking the environment and causing global warming.

To top it off, we are using such great quantities of fossil fuels (and even exporting some) that they will not last very long. So it seems like we are living in a temporary bonanza, indulging in energy gluttony in total disregard of tomorrow.

Regardless of the total amount of available fossil reserves, we must remember that they are finite and will be depleted soon—well before we have found a replacement.

Table 8-1. U.S. power generation by technology (2013)

Source	# Plants	Capacity (GW)	Annual energy (billion kWh)	% of annual production
Coal	557	336.3	1,514.04	36.97
Natural Gas	1,758	488.2	1,237.79	30.23
Nuclear	66	107.9	769.33	18.79
Hydro	1,426	78.2	276.24	6.75
Renewables	90	218.33	5.33	5.33
Petroleum	1,129	53.8	23.19	0.57
Misc	64	2	13.79	0.34
Storage	41	20.9	-4.95	-0.12
Imp-Exp	-	-	47.26	1.15
Total	6,997	1,168	4,095	100

Table 8-2 points the way to a sustainable future. The only problem is that the decision makers do not follow that path. They are, instead, gearing the economy for using and exporting more fossils in the near future.

Table 8-2. Sustainable National Energy Plan

Year	Coal	NG	Oil	Ren	GHG
2010	100	100	100	2	100
2015	95	95	95	5	95
2020	90	90	90	10	90
2025	85	85	85	15	85
2030	80	80	80	20	80
2035	75	75	75	25	75
2040	70	70	70	30	70
2045	65	65	65	35	65
2050	60	60	60	40	60
2055	55	55	55	45	55
2060	50	50	50	50	50
2065	45	45	45	55	45
2070	40	40	40	60	40
2075	35	35	35	65	35
2080	30	30	30	70	30
2085	25	25	25	75	25
2090	20	20	20	80	20
2095	15	15	15	85	15
2100	10	10	10	90	10

But what is really needed? Very simple. We need to ensure a sustainable energy future for the U.S., which can be easily achieved by:

• 50% reduction in fossil fuels use NOW,

• To be replaced by a 50% increase of renewables, which will result in

• 50% reduction in GHG emissions, and

• If all goes per plan, then by 2100 we will be on the path to a sustainable and clean future.

But first, we need to start this process, and achieve the goals set for 2050-2060. If this plan does not go as scheduled in Table 8-2 by then, and instead we continue the present rate of fossils exploitation and GHG emissions, it might be too late to go back…

Energy Policy and Strategy

The link between policy and energy is a critical as-pect of developing new methods for the electrical grid and our energy security in general. Through various directives, government policy has the power to regulate the types of energy sources utilized in the market as well as the different characteristics produced by those sources. Policies can also influence consumer preferences based on different directives and initiatives.

Government policy and other political factors significantly affect energy regulation and energy industries. Policies can be negatively enforced or positively enforced to help prevent or promote different results. For example, taxes and regulations could be used to reduce pollution and encourage the development of new technologies by shifting behaviors. The different policies and directives can be inspired by various influences.

The political landscape certainly influences the types of energy used within a country as well as any advancements to the grid through economic incentives. These advancements due to economic incentivizing are called induced innovation, which has a substantial role in reducing costs for energy and increasing research into different sources of production.

In more general terms, government policy has consequences on pricing and production, since changes in the capacity and generation of energy are driven by price. These changes are ultimately shaped by government policy and initiatives that greatly influence not only electricity producers but also consumers.

The impact of government policy and directives such as the NetZero program is significant and drives social behaviors, business practices, and energy initiatives for the national energy grid; yet, it is possible for the relationship to be symbiotic. The energy sector can also help to influence government policy, at times creating an interactive rather than a reactive relationship.

Considering the looming energy crisis, depleting natural resources, and pending threats of natural disasters or attacks, it is clear that there should be a focus on altering the energy grid to mitigate these concerns. There are several alternatives to developing answers to these pressing questions with the objective of building a stronger, smarter, and cleaner, more energy-efficient electric grid.

The basis of pricing as a means to influence consumer preferences and energy source investment can also factor into government policy. By influencing behavior in the social and business sectors it will help reinforce government policies focused on advancing the energy grid. The economic inefficiencies in the market may put pressure on politicians to alter their ideology that significantly influences policy proposals with regard to energy.

Further, pricing policy and strategy can provide incentives for producers to invest in research and development of new technology which in turn alters buying practices for consumers with more choices to meet their preferences. The relationship between government policies, initiatives, directives, and the energy grid is a critical factor in developing a methodology to optimize efficiency and mitigate many of the concerns facing the U.S.

With a looming energy crisis and other threats to the power grid, energy is and will continue to be a pressing policy issue. Understanding the relationships between policy, energy sources, and outputs is imperative to building economic solutions on national and global levels. Proper and timely implementation of these solutions should be a priority, and not hindered by third-party (personal and/or corporate) interests, which might twist the decision makers' thinking and influence their actions.

Executive Order 13514

Executive Order (E.O.) 13514, Federal Leadership in Environmental, Energy, and Economic Performance, was signed on October 5, 2009, by President Obama. It expanded upon the energy reduction and environmental performance requirements of the previous E.O. 13423. Surely, the Administration sees a strong relationship between energy production and use, and the environment. This is a good step ahead toward an energy efficient and environmentally friendly future.

E.O. 13514 states that the U.S. government is the largest consumer of energy in the country, with over 500,000 large, old, and mostly energy-inefficient buildings. E.O. 13514 sets numerous federal energy requirements, starting with a mandate for 15 percent of existing federally owned or leased buildings to meet Energy Efficiency Guiding Principles by 2015.

This means that at least 75,000 buildings are to be converted to energy efficient structures by 2015. A tall order. The annual progress is to be made toward 100 percent conformance of all federal buildings, where 100% of all new federal buildings are to achieve zero-net-energy status by 2030.

A "zero-net-energy building" according to E.O. 13514 is "a building that is designed, constructed, and operated to require a greatly reduced quantity of energy to operate, meet the balance of energy needs from sources of energy that do not produce greenhouse gases, and therefore result in no net emissions of greenhouse gases and be economically viable."

The executive order also states the general direc-tion the Administration is taking the country, "…the Federal Government must lead by example … increase energy efficiency; measure, report, and reduce their greenhouse gas emissions from direct and indirect activities … design, construct, maintain, and operate high performance sustainable buildings in sustainable locations; strengthen the vitality and livability of the communities in which Federal facilities are located; and inform Federal employees about and involve them in the achievement of these goals."

In October 2013, President Obama also signed a Memorandum under his Climate Action Plan, directing the federal government to consume 20 percent of its electricity from renewable sources by 2020. This is more than double the current level. Another tall order.

As part of this effort, federal agencies will also identify formerly contaminated lands, landfills, and mine sites that might be suitable for renewable energy projects, in order to return those lands to productive use with minimal fossil energy use.

To improve the federal agencies' ability to manage energy consumption and reduce costs, the Memorandum also directs them to use Green Button (a tool developed by the industry) which provides utility customers with easy and secure access to their energy usage information in a consumer-friendly format. This will provide transparency and will help further in managing the effort.

This is a great step ahead for the U.S. energy policymakers. Unfortunately, it is relevant only to the government complex. Similar patchworks of energy and environmental guidelines exist here and there across the national economy, but they are much more sporadic, and not as clearly defined. The final decision in executing the civilian energy policies in most cases, is left to the individual states and locals, which would make it hard to predict how the entire program will develop. A more unified, countrywide federal program is needed, but we will not hold our breath.

THE U.S. MILITARY

"Going forward, there should be no doubt: the United States of America will continue to underwrite global security." The National Security Strategy, 2010

The US Department of Defense (DoD) is the largest oil-consuming government body in the world. In 2004, during the Iraq war, US military fuel consumption increased to 144 million barrels annually, or about 40 million barrels more than the average peacetime military usage.

This is enough fuel for 3,300,000 cars to drive around the world not once, but twice. It is close to 400,000 barrels of crude oil per day, almost as much as the daily energy consumption of Greece and more than many other countries.

DOD purchases more light refined petroleum products than any other single organization and many countries.

The US DoD spent $8.2 billion on energy (fossil fuels mostly) alone in fiscal year 2004. In 2005, it bought even more, about 130 million barrels of fuel at a cost of $8.5 billion, or about $65 per barrel. Shortly after that the price of oil doubled, affecting the budget significantly.

This great expense makes the American GI the most energy-consuming soldier ever on the global battlefields. An estimated 40 million gallons of fuel are used during a 20-day period of combat operations in Iraq. This is equivalent to the total gasoline consumed by all Allied armies combined during the four years of World War I.

Note: During WWII, the 3rd Army led by General Patton had about 400,000 men and used about 400,000 gallons of gasoline a day to transport the troops and for other purposes. By today's standards, this is a humble one gallon per solder per day. The modern U.S. Army had about 1/3 that number of troops in Iraq, and yet used more than four times as much fuel during the same time period than General Patton's men.

About 12 gallons of fuel per solder were needed every day to support the U.S. operations in Iraq.

According to Defense Logistics Agency spokespeople, in the mid-2010s the U.S. military was using 10-11 million barrels of fuel each month to sustain operations in Afghanistan, Iraq and worldwide. This is 330,000-360,000 barrels per day, or more than double the amount of oil used in the Gulf War!

According to other sources, nearly 2.0 billion gallons of fuel were consumed within the U.S. Central Command's area of responsibility during Operations Desert Shield and Desert Storm, between August 10, 1990 and May 31, 1991. This makes a total of 44.8 million barrels, or 150,000 barrels of fuel a day during those 295 days of war operations.

During the same time, Saudi Arabia and the UAE supplied about 1.5 billion gallons of fuel without charge, but other countries (Bahrain, Egypt, Oman and Qatar) charged for the fuel they supplied.

Just think what this world, and the military conflicts, would be today if we had no crude oil or electricity.

Fast forward to 2014…it now takes 22 gallons of fuel per soldier per day to support combat operations. This is 22 times more fuel used per solder than during WWI and a 175% increase over the Vietnam War. So we went from 12 times to 22 times increase of fuel use per soldier, or almost doubling the fossil fuels use within the last 10 years.

The Air Force alone consumes 2.5 billion gallons of aviation fuel a year, which is about 30% of the total fuels used by DoD, which makes it the largest single-body petroleum consumer in the world. In contrast, the entire country of Greece, and many other nations, consume much less oil.

Note: Each U.S. soldier on the battlefield carries 60-70 rechargeable batteries on his/her back at all times. The batteries power different gadgets that the soldier also carries, such as GPS, lasers, communications, night vision, etc., etc.

The batteries must be charged daily or weekly, which requires energy right then and there. If there is no energy at the right moment, the gadgets won't work and the soldiers' mission and lives would be in jeopardy.

The Crude Oil Dilemma

Simply put, it takes a lot of energy to move and operate the military machine, to ensure our national security. Energy, energy, energy. It is all about energy… actually, mostly about oil. Figure 8-15 is an oversimplified schematic representation of the U.S. military machine. The most remarkable thing here is that crude oil is in its very foundation. Without sufficient oil supplies, the entire military structure will collapse like a house of cards. Or it would rather sit still and helpless, like the proverbial duck on ice.

Really! The military is driven by oil and its derivative—no oil, no movement, no action. How is a military, equipped with the latest equipment and gadgets—most of them crude oil guzzling beasts—going to function without crude oil? How would the equipment and soldiers move without gasoline, diesel, jet fuel, and lubricants?

Yes, this is a simplistic view of a huge and complex system. We are far from being military experts, but one doesn't need to be an expert to see that if the daily crude oil supplies and the refined fuels stop, the Army trucks will be parked, the F-18 fighter jets and Apache helicop-

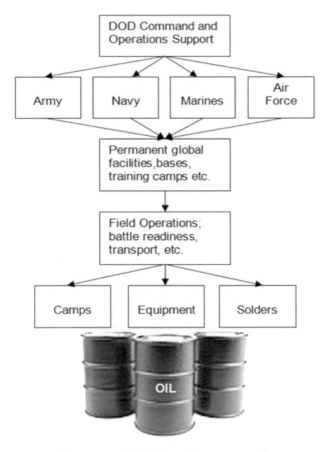

Figure 8-15. The U.S. military pyramid.

ters grounded, the battleships anchored, and the M1A3 Abrams tanks will sit still in their hangars.

Only God knows what else could happen, or not happen, in such a case since only crude oil can make them move.

What one needs to imagine are the thousands of people, equipment, weaponry, and vehicles standing behind each of the squares in Figure 8-15. Some of the equipment, weaponry, and vehicles are as large as a sky-scrapers, some are as fast as lightening, and some can blow an entire town out of existence in a split second. No other army, or natural force on this Earth, can do as much damage at a whim as the U.S. military machine.

Think "huge" and "awesome," since these two words summarize what we are talking about. It is not easy to imagine the whole thing, no matter how hard we try. It is also hard to imagine the immense amount of fuel that is needed to roll, fly, and float the millions of pounds of equipment, weaponry, vehicles, ammunition, and people around the world every day, all day and night.

For that purpose, there are thousands of special-ized trucks, trains, airplanes, and ships that haul non-stop millions of gallons of gasoline, diesel, jet fuels, nuclear materials, lubricants and other energy products around the globe every day. These and all other pieces of equipment and vehicles count on the prompt delivery of fuels every day. It has to be done on schedule, or they won't budge.

It is truly amazing, but a fact, that without the daily (and hourly) fuel refills, the military machine will come to a grinding halt.

In such a case, all war operations will be severe-ly constricted, and most will freeze. All military plans and programs will sit idle until a compromise solution is found. Nothing in the military machine can move, or operate properly, without timely (fossil) energy input. Crude oil is the main energy source (fuel and lubricants) used by all DoD branches, so it is what allows the daily operations to proceed as planned.

Since we import over half of the crude oil we use, our ener-gy security, and in extension our national security, are less than 50% ensured.

What does that mean to the future of our energy (and national) security? The DoD generals must answer this question, for it is too much for us to even grasp.

This is simply impossible, you say, so why fix it if it ain't broken? Maybe so...for now. But crude oil is getting scarce, prices are going up, and a day is coming soon when oil will be extremely expensive, and then it will be no more. Crude is a finite commodity that cannot and will not last forever.

The DoD generals see these problems clearly, since they pay the increasing energy bills and are re-sponsible for the battle readiness and the soldiers' lives. They take the real and perceived problems seriously and have been taking decisive action for awhile now, as summarized in their energy plans. Is this enough, is another question that we cannot answer, but which they are working on.

A closer look reveals...

The DOD Energy Policy

The U.S. Department of Defense (DoD), in terms of money, technology, and energy use, is larger, much more advanced, and many times more powerful than that of any other country in this world. As such, it is also held responsible for its actions. In response to the government's energy efficiency and environmental impact management efforts, the DOD has established its own energy plans with respective goals and accountability measures.

The U.S. DoD energy security goals are based on:
— Reducing energy consumption
— Increasing energy efficiency across platforms and facilities
— Increasing use of renewable and alternative energy sources
— Ensuring access to sufficient energy supplies
— Reducing adverse impacts on the environment
— Changing the overall energy culture

The energy management approaches in these plans are excellent and deserve to be reviewed and adopted by corporations and governments alike. Some of these should be urgently implemented in the overall U.S. economy, but there is not much chance of that happening anytime soon. The U.S.' economy is simply too large and diverse, but more importantly, it lacks the discipline needed to execute some of the basic energy use restric-

tive measures in a coordinated way.

Since the DoD is smaller and much more organized, its energy plans are easier to design and implement. It also enjoys a huge advantage of operating under strict discipline at all ranks. This way, once the plans are agreed upon by the top commanders, they are quickly and immaculately executed through the ranks—no questions asked, no debates, no protests, and no delays!

Some of the recent results from the DoD's energy efficiency improvement tactics are reflected in Table 8-3. It is obvious here that DOD takes the energy management effort seriously. There are still unreached goals and targets to meet, but the effort is well underway and DOD is determined to succeed in completing its goals.

In more detail:

• At the end of FY 2009, DoD was managing 1.93 billion square feet of facility space (EO 13423 defined goal-subject facilities) and had spent $3.6 billion on facility energy alone. In addition, DoD spent $9.6 billion on fuel for vehicles (non-fleet and fleet) and other equipment. This included jet fuel, avia-

Table 8-3. U.S. Department of Defense Energy Plan execution summary (2011)

GOALS	UNITS	ENTITY	FY2011 Achieved	FY2011 Target
Reduce facilities energy use intensity per FY 2003 baseline Per EI&SA*, 2007, #431	BTUs of energy consumed per gross square foot of facility space in use.	DOD	-13.30%	-18%
		Army	-11.80%	
		Navy	-16.90%	
		Marines	-9.40%	
		Air Force	-16.30%	
Consume more electric energy from renewable sources. Per EPA**, 2005, #203	Total renewable electricity consumed as a percentage of the total facility electricity use.	DOD	3.10%	5%
		Army	0.50%	
		Navy	1.00%	
		Marines	1.20%	
		Air Force	6.00%	
Reduce potable water water use intensity relative to FY2007 baseline. Per Exec. order 13423	Gallons of water used per square foot of facility space in use.	DOD	-10.70%	8%
		Army	-10.30%	
		Navy	-4.00%	
		Marines	-22.70%	
		Air Force	-13.10%	
Produce or procure more energy from renewable sources. Per U.S. Code Title 10, #291(e)	Total renewable energy (electric or non-electric) produced or consumed as a percentage of total facility electricity use.	DOD	8.50%	25% by 2025
		Army	4.30%	
		Navy	20.60%	
		Marines	0.60%	
		Air Force	7.10%	
Reduce petroleum use in non-tactical vehicles relative to FY2005 use. Per EI&SA, 2007, #142, Exec. Order 13514 #2(a)	Gallons of gasoline equivalent of petroleum fuel consumed.	DOD	-11.80%	12%
		Army	-10.30%	
		Navy	-26.00%	
		Marines	-16.60%	
		Air Force	-8.30%	

Notes: * Energy Independence & Security Act
** Energy Policy Act

tion gasoline, Navy-special fuel, automobile gasoline, diesel-distillate and liquefied petroleum gas (LPG)/propane.

- DoD delivered more than 209,000 billion British thermal units (BBtus) of energy to its EO13423 goal-subject facilities during FY 2009. This was a 1.3 percent increase over the FY 2008 amount. About 94 percent of the energy went to the military departments.

- About 79 percent of the delivered energy to DoD facilities in FY 2009 was from electricity and natural gas. In addition to using fuel oil, coal and purchased steam, DoD facilities also used a small percentage of LPG (liquefied petroleum gas)/propane and renewable energy sources. In FY 2009 DoD used renewable energy for 3.6 percent of its delivered electricity.

- In FY 2009 all DoD facilities had an overall energy intensity level of 104,527 Btus per gross square foot (GSF), which was a 1.1 percent increase over the 103,692 Btus/GSF level for FY 2008.

- When renewable energy is measured per 10 USC section 2911(e), DoD procured or produced 6.8 percent of its electric consumption from electric renewable sources in FY 2009.

- The Army had 67 active renewable energy projects operating in FY 2009. Of the total, 42 were generating electricity qualifying for credit toward the renewable energy goal and nearly all the energy produced was used on-site in federal Army facilities. The Army consumed 2.1 percent of its electricity from renewable sources in FY 2009.

Figure 8-16. Mid-flight refueling of U.S. fighter jets.
How many gallons of jet fuel does this operation take?
How many times a day is it preformed across the globe?

- The Air Force purchases Renewable Energy Certificates (RECs) to help achieve its renewable energy goals and continues to pursue the development and installation of renewable energy, with 5.8 percent of total electric consumption from renewable sources in FY 2009.

- The Navy consumed 0.6 percent of its electricity from renewable sources in FY 2009. These sources include wind and solar electric generation, but do not include the Navy's Naval Air Weapons Station (NAWS) China Lake geothermal site, whose 270 megawatts (MW) production capacity is not directly consumed by the Navy.

- The Navy uses thermal energy from the waste heat of six cogeneration systems to further meet energy intensity reduction goals. Cogeneration credits account for six percent of the energy intensity reduction, the largest single technology contribution. The most recent addition, a 39 MW cogeneration plant in Yokosuka, Japan, came on line in November 2008 to contribute 2.5 percent of the reduction to the Navy's overall energy intensity.

- DoD continued efforts in FY 2009 to acquire alternative fuel capable vehicles and provide the necessary supporting infrastructure. DoD acquired 105 neighborhood electric vehicles; received 863 low-speed or mini-utility vehicles, 150 hybrid electric vehicles, and 1,485 E-85 alternative fuel capable vehicles; and ordered 800 low-speed electric vehicles (LSEVs). DoD also completed the infrastructure for 16 E-85 and/or B-20 alternative fueling stations.

- DoD continues to make progress installing cost-effective renewable energy technologies and purchasing electricity generated from renewable sources (solar, wind, geothermal, and biomass). In FY 2009, 3.6 percent of DoD's electrical consumption came from renewable electricity, exceeding the EPAct 2005 goal of 3 percent and improving on the 2.9 percent achieved in FY 2008.

- DoD produced or procured 6.8 percent of all electricity from renewable sources. This is 3 percent less than the 9.8 percent reported -for FY 2008 because of a change in the calculation method. In FY 2008 and earlier, the calculation compared all renewable energy (electric and non-electric) produced or procured to total electricity consumed. Had FY 2008 been reported, limiting renewable energy to only renewable electricity sources, the FY 2008 achievement would have also been reported as 6.8 percent.

- The Army purchased 148,000 MWh of electricity qualifying toward the renewable energy goal. A large portion of the electricity was a direct purchase from a two megawatt photovoltaic array at Fort Carson, Colorado. Other sources included Renewable Energy Certificates (RECs) purchased by Fort Lewis, Washington; Fort Carson, Colorado; and the Pennsylvania Army National Guard. The Army also purchased a substantial amount of energy from renewable municipal solid waste plants at Redstone Arsenal, Alabama, and Aberdeen Proving Ground, Maryland.

- The Air Force acquires the most economical renewable energy available on the market, either by acquiring bundled renewable electricity, or through RECs. Bundled renewable electricity purchases represent 39 percent of the renewable purchases; RECs represent the remaining 61 percent.

- In 2010, U.S. armed forces consumed more than five billion gallons of fuel in military operations. The number one factor driving that fuel consumption is the nature of today's defense mission. Twenty-first century challenges to U.S. national security are increasingly global and complex, requiring a broad range of military operations and capabilities, which means a large and steady supply of energy.

Wow. Impressive! And this is only part of the long list of energy related projects and operations. There are surely many other applications of energy around the globe of which we are not aware and from which information cannot be released to the general public.

Operational Energy Strategy

In focusing on energy for the war fighter, the goal of "Operational Energy Strategy" is to ensure that the armed forces will have the energy resources they require to meet 21st century challenges.

"Our dependence on foreign oil reduces our international leverage, places our troops in dangerous global regions, funds nations and individuals who wish us harm, and weakens our economy." From: Risks to National Security, U.S. Military Advisory Board report.

The DoD energy strategy outlines three principal ways to a stronger and more efficient armed forces:

- More fight, less fuel: Reduce the demand for energy in military operations. Today's military missions require large and growing amounts of energy with supply lines that can be costly, vulnerable to disruption, and a burden on war fighters. The Department needs to reduce the overall demand for operational energy, improve the efficiency of military energy use to enhance combat effectiveness, and reduce military mission risks and costs.

- More options, less risk: Expand and secure the supply of energy to military operations.

  Most military operations depend on a single energy source, petroleum, which has economic, strategic, and environmental drawbacks. In addition, the security of the energy supply infrastructure is not always robust.

  This includes the civilian electrical grid in the United States, which powers some fixed installations that directly support military operations. The Department needs to diversify its energy sources and protect access to energy supplies, to have a more reliable and assured supply of energy for military missions.

- More capability, less cost: Build energy security into the future force. Current operations entail more fuel, risks, and costs than are necessary, with tactical, operational, and strategic consequences. Yet the Department's institutions and processes for building future military forces and missions do not systematically consider such risks and costs.

Fuel convoys like that in Figure 8-17 haul thousands of gallons of gasoline, diesel, jet fuel, and lubricants to U.S. military bases and field operations around the world. Each of the fuel convoys in war time operations (like during the Iraq war) supplies fuel to move the military hardware around the battle arena for several days or weeks. The same amount of energy could

Figure 8-17. U.S. Army fuel supply convoy.

supply power to half a dozen Sub-Saharan countries for a month or more. This number multiplied several times is what the entire military machine needs to operate efficiently.

The Department of Defense still needs to integrate operational energy considerations into the full range of planning and force development activities. Energy will be, in itself, an important capability for meeting the missions envisioned in the QDR and the National Military Strategy.

Military installations should also have the ability to "island" themselves from the power grid in order to support their strategic mission. Ideally, a military base can switch off of the energy grid, and still manage to power essential operations of the installation. Note that the social implications of islanding will not be addressed but they are profound.

This is important in the event of a physical or cyber attack on either the energy grid or our energy resources. Should this happen, the Army, must be able to keep key functions alive until power is re-established. Currently the military is mainly buying its energy commercially, having outsourced most of its power production facilities.

However, there is no current plan or funding to harden these outside facilities and the supplying grid. To island itself, an installation would almost certainly have to move some kind of energy source onto the installation itself, whether that is solar, wind, biomass, a nuclear generator, or any other kind of renewable energy source.

Critical national security and homeland defense missions are at risk of extended outage from failure of the grid. Currently the Army and the Department of Defense have implemented goals and policies to facilitate energy security and other NetZero energy initiatives.

Note: A Net Zero Energy Installation (NZEI) is an installation that produces as much energy on site as it uses, over the course of a year. To achieve this goal, installations must first implement aggressive conservation and efficiency efforts while benchmarking energy consumption to identify further opportunities. The balance of energy needs then are reduced and can be met by renewable energy projects.

From a solely ROI perspective these alternative energies are not cost efficient when compared to fossil fuels. The argument must be made here that the value of alternative energy is directly related to our energy security and good environmental stewardship—regardless of price and any other difficulties.

Renewables might be expensive and a hassle, but are the only viable energy sources in the long run.

The problem with the military energy use is that the huge machinery used in field operations requires petroleum products to move efficiently. It is not possible (at least for now) to run Abrams tanks or fighter jets with solar power, or any other fuel. Using biofuels is the best (if not only) option for replacing the fossil fuels in the future, but that in itself is problematic and requires much more effort.

So, DoD is faced with a real dilemma: while reducing energy use, or using renewables, on military bases is possible, fueling the U.S. war machine is not that easy. It is a thirsty beast that is built to use a lot of fossil fuels—crude oil derivatives mostly. Replacing these on mobile units would not be easy or fast...

For now DoD is doing all that is in its power to save energy, whenever possible.

DoD Net Zero Energy Initiative

The DoD Net Zero Energy Installation Initiative is an effort to increase the energy independence of installations by offsetting total annual energy use through on-site energy production. The Army goals for this initiative are for five Army installations to be net zero (outside energy independent) by 2020, 25 installations by 2030, and all Army installations by 2050.

Some of Army's renewable energy efforts in support of this initiative are a large concentrated solar system at Fort Irwin, California; geothermal steam resources at Hawthorne Army Depot, Nevada, replacing 800 petroleum-fueled non-tactical on-post vehicles with neighborhood electric vehicles; and development of consolidated waste-to-fuel projects at several locations. McGuire Air Force Base, New Jersey, is implementing four measures (including day lighting) to render a 30,774-square-foot facility the first energy-neutral facility in the Air Mobility Command.

Water Conservation

In FY 2009 DoD facilities had an overall water intensity (consumption) level of 57.1 gallons per gross square foot (GSF), which was a 1.1 percent decrease over the FY 2008 level of 58.1 GSF. Although the Army had an increase in water intensity compared to FY 2008 (58.2 compared to 54.0), the other military departments and DoD components (particularly the Defense Commissary Agency) had lower water intensity levels, which resulted in DoD meeting its water intensity reduction goal of 4.0 percent for FY 2009.

All of the military departments continue to install water-conserving toilets and urinals, and low-flow faucets and showerheads. Some installations have instituted aggressive leak detection surveys and followed up with repair programs for leaky valves and damaged pipelines, which have significantly reduced water consumption (as much as 20 percent at one location). All facility projects executed by the Army Corps of Engineers follow the International Plumbing Code, which prescribes water conserving fixtures.

The Air Force is implementing stronger conservation measures to reduce landscape irrigation, encourage low-water plantings, and repair leaking water and steam lines. The Defense Commissary Agency requires low-flow toilets and urinals with electronic flush sensors for new and renovated commissaries. Proposed landscaping for new DeCA facilities is closely reviewed during all phases of the design for low maintenance and watering requirements and includes requirement for xeriscaping and drip-versus-sprinkler irrigation systems.

Metering of Electricity Use

DoD has identified 37,493 buildings requiring either standard or advanced metering. The Navy accounts for 70 percent of this total (26,311), with meters installed in 64 percent (16,929) of its buildings in FY 2009. This effort, combined with that of the Air Force (93 percent complete), Army (44 percent complete) and the other ten DoD components (44 percent complete), brought the total number of metered buildings in DoD to 23,674, or 63 percent of all identified buildings in FY 2009.

Federal Building Energy Efficiency Standards

In FY 2009, 99 percent of DoD's new building designs included provisions to make them 30 percent more energy efficient than the American Society of Heating, Refrigerating and Air Conditioning Engineers (ASHRAE) 90.1-2004 standard.

The Army Corps of Engineers continues work with the DOE and the Office of the Army Assistant Chief of Staff for Installation Management to develop design guides for implementing building efficiency standards mandated by the EPAct 2005. The Corps has completed prescriptive design guides for battalion headquarters buildings, permanent party barracks, training barracks, and tactical equipment maintenance facilities.

Four of the most prevalent types of buildings were constructed in conjunction with Army troop stationing actions. Use of these design guides will help new building designs be 30 percent more energy efficient than ASHRAE standards without having to model each indi-

vidual project.

The Air Force initiated 98 new building designs in FY 2009; 100 percent will be life-cycle cost effective and meet the goal to exceed the ASHRAE efficiency standards by at least 30 percent.

The Navy is expecting 99 percent of the FY 2010/2011 military construction facility designs to meet or exceed federal standards and achieve life cycle cost effective sustainable designs.

The Defense Logistics Agency is redesigning a Child Development Center project in Columbus, Ohio, to ensure that it will be at least 30 percent more efficient than ASHRAE standards, and that it will be eligible for LEED Silver status.

The Missile Defense Agency is presently planning three military construction building projects (Redstone Arsenal, Alabama and Fort Belvoir and Dahlgren, Virginia). These buildings are designed in accordance with EPAct 2005, EO 13423 and the Whole Building Design Guide, and each building will be LEED certified.

The TRICARE Management Agency's Bureau of Medicine and Surgery (BUMED) is working with Naval Facilities Engineering Command (NAVFAC) to ensure that all new design work will exceed the ASHRAE efficiency standard by at least 30 percent where achievable. Ninety-three percent of new BUMED building designs started since the beginning of FY 2007 are expected to exceed the ASHRAE standard by 30 percent.

Note: Although the above figures are somewhat outdated, they clearly show the trend towards energy efficiency improvements in the U.S. military. These measures are saving a lot of energy and are diverting tons of fossil fuels from use on military bases.

Energy Innovation

A letter issued in 2012 and signed by more than 350 U.S. veterans, including retired generals and admirals, urges Congress to support the Pentagon's initiative to diversify its energy sources, limit the demand, and lower overall energy costs. This effort is needed, if the military is to be able to deploy clean energy technology to reduce its dependence on fossil fuels, while at the same time contributing to our energy security and strengthening national security.

The concern here is that some Congressmen are working to restrict the department's ability to introduce energy innovation by using advanced biofuels programs. In the long run, this would hurt DoD's capacity to reduce its dependence on oil prices, which might negatively affect its operational effectiveness.

Some congressional amendments, if adopted,

would limit and even bar the DoD from purchasing alternative fuels, which could also affect the fuels used to power all (including unmanned) vehicles for training and military operations.

The U.S. military is the world's largest consumer of liquid fuels, and according to the experts, needs over 22 gallons of fuel per soldier per day to support combat operations.

For every $10 increase in a barrel of oil, the DoD pays an additional $1.4 billion annually which comes at the cost of daily operations and battle readiness.

The Pentagon uses nearly 400,000 barrels of crude oil daily, spending over $16 billion in 2013. During 2011 and 2012, the department had $3 billion in unanticipated fuel costs, which additional budget burdens adversely affect military readiness and put troops overseas at imminent risk.

Most people agree that both the DoD and the nation as a whole must reduce their expenses and dependence on foreign oil. In response, DoD is trying to become more energy efficient by testing different advanced biofuels in its ships, planes, and land vehicles.

Alternative fuels are seen as the energy products to be used when they become readily available and cost-competitive with conventional fuels.

The so-called "second generation" biofuels can be produced from domestic biomass sources, for "as is" use. It will require no changes to current engine design, and would provide the same or better performance than conventional fuels.

The fight for the future of the U.S. military battle readiness (energy availability) continues…

Since 2007 a provision of the Energy Security and Independence Act, or Section 526, has allowed the Pentagon to develop advanced biofuels and other clean and renewable transportation alternatives. These are domestically produced fuels from non-food agricultural feed stocks, such as camelina and saw grass, that could be used in existing engines in fighter jets, land vehicles, and ships. They are, theoretically, supposed to provide similar power and performance as the traditional fuels used today in these areas.

The Navy and Air Force have successfully tested and certified some of these fuels, along with similar efforts by some commercial airlines, but the new defense authorization bills would repeal or weaken the provision for such fuels. They would also send a negative signal to the biofuels industry, which could result in adverse impacts to U.S. job creation, competitiveness, and overall rural development efforts.

The military's energy policy and investments have encouraged a domestic industry that is creating jobs and businesses in America. For these and other reasons, groups such as Airlines for America, the Farm Bureau, the Advanced Biofuels Association, the American Security Project, and others support the Defense Department's renewable energy programs and policies.

Table 8-4 shows the plan to gradually reduce energy consumption, which is still on the agenda.

Table 8-4. Energy consumption per square foot of federal buildings (as compared to 2003 energy consumption)

Fiscal Year	Percentage Reduction
2006	2
2007	4
2008	9
2009	12
2010	15
2011	18
2012	21
2013	24
2014	27
2015	30

The Problem

Section 526 makes the achievement of the energy plans more difficult by prohibiting the DoD and other federal agencies from purchasing some alternative fuels. This is because some of the alternative fuels have a higher (life cycle) carbon footprint thus emitting even greater emissions than conventional fuels. They also cause other problems such as land misuse and food price increases.

The existing provision has never prevented the defense department from acquiring fuel that it needs to operate, nor has it required the purchase of more expensive alternative or experimental fuels. Since these are needed for future applications, the DoD supports the goals and intent behind the current law and resists the proposed changes.

Most critically, the existing law has helped DoD to meet the mission needs, while the proposed one threatens to reduce and even eliminate the gains in the alternative energy use areas of its activities.

The reduction in use of biofuels and other alternative fuels is a step back and would mean that the DoD will continue its total dependence on fossil fuels and their unpredictable nature—from supply and cost points of view. This, in turn, would create additional uncertainties in the DoD's daily operations, and affect negatively its battle readiness.

How about that? Now, this is something that must change, but it probably won't happen anytime soon. It is just another barrier in the path to achieving DoD's energy goals.

The Environment

One significant achievement of DoD's policies is the full recognition of the direct connection between energy today and tomorrow, and the environment. Secretary of DoD, Mr. Panetta drew a clear line between environmental, energy use and our national security, since their relationship is important, as recently established in the Pentagon strategy.

"In the 21st century, reality is that there are environmental threats that constitute threats to our national security," Mr. Panetta stated in 2013, stopping short of naming the individual threats, which are many. Instead, Mr. Panetta laid out a strategic framework for how the military thinks about and acts on long-term environmental and energy issues. All this at a time of austerity, and when the military's alternative energy investment and achievements are under assault by members of Congress.

Never mind the hard times. Mr. Panetta confirmed the military services' commitment to add 3 GW of renewable energy in the coming years, emphasizing the military's history of anticipating changing trends in the global markets. Tomorrow is knocking on the energy door and we must answer.

Security implications of climate change have been a special interest of Mr. Panetta's policies too. Rising sea levels, extended droughts, and more frequent and severe natural disasters around the world are seen by him as a signal to increase the energy conservation effort by the U.S. military and a new and increasing demand for humanitarian assistance from the U.S. military.

Note: Melting polar ice caps are prompting competing claims by different nations in the mineral-rich Arctic, so Panetta calls for the U.S. to ratify the U.N. Convention on the Law of the Sea, which is the international treaty governing activity in the world's waters. It has been endorsed by U.S. military leaders, business groups and environmentalists, but has been stalled in the Senate for years by opponents who claim that it is a threat to U.S. sovereignty.

The U.S. is the only industrialized nation that has not approved that treaty as yet. There are many reasons for that, but the reality is changing daily and needs a clear and decisive action, and soon.

DoD Roadmap to Energy Security

The U.S. DoD has a seemingly firm grasp on the energy reality in its ranks, and its operations at home and abroad, as follow:

- Energy security and climate change goals should be clearly integrated into national security and military planning processes.

- DoD should design and deploy systems to reduce the burden that inefficient energy use places on our troops as they engage overseas.

- DoD should understand its use of energy at all levels of operations. It should also know its carbon footprint (or boot print) and strive to reduce it.

- DoD should reduce its use of energy at all installations through aggressive pursuit of energy efficiency, smart grid technologies, and use of electrification and renewables in its vehicle fleet.

All this sounds good, and the U.S. DoD is on the right path thus far, but the implementation of all these plans and ideas requires a lot of money. So, the question is how will all this happen in light of the decreasing DoD budget and the ongoing political constipation in Washington, DC?

Of course, as far as DoD is concerned, national security is its number one long-term goal, with the national energy security in the center of its short-term concerns.

Our national security depends heavily on energy, which is why our energy security is a high priority to DoD and U.S. politicians.

Lack of energy is a direct and serious threat to our national security and our way of life—simple as that. Hence the extra effort in ensuring energy supplies at all cost. It takes a lot of energy to get energy. For that purpose we police the world at all times, making sure parts of the world don't blow up, and that the energy supply

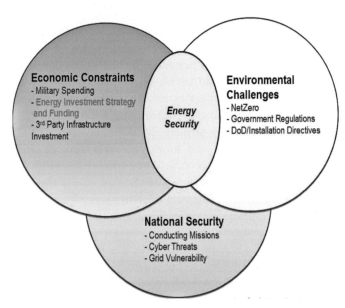

Figure 8-18. Energy security via DoD

import-export channels are clear of obstructions.

The economic constraints and the environmental challenges are another set of significant factors that affect our energy security. They are aggressively addressed because they also affect negatively our national security. So, as DoD is moving forward to a new era of energy dependency (or rather less dependency), we must remember that tanks, jet planes, patrol boats, and super aircraft carriers do not run on words and promises. They require millions of gallons of fossil fuels…every day…day after day. These fuels transport our solders and power all military hardware.

Fossil fuels ensure the proper function of the military machine, which ensures our energy security, which ensures our national security. No fossils, no security—energy, national, or otherwise!

So, the DoD is planning to save some energy, which it is doing quite successfully thus far. Nevertheless, its worldwide operations will still swallow a significant portion of our national (and global) fossil fuels in the process. There is no way around it for now.

In summary, we clearly see that the U.S. DoD understands and appreciates the role energy plays in its operations. The U.S. generals feel the pain of high fuel prices, and are worried about fuel supply disruptions. Both of these are potential barriers in the drive to more powerful, mobile, and versatile global fighter units.

To anticipate and control the threats to its energy budget and supplies, the DoD is actively pursuing a

well designed short-term plan for ensuring its energy supplies and the national energy security. The long-term plan of achieving full energy independence is also in the planning stages, but it is too far-fetched to become a reality anytime soon. Certainly not in our lifetime, since there are some serious problems in its way, which even the mighty U.S. military cannot overcome.

Note: The proverbial "giant with feet of clay" comes to mind here, except that the feet of the U.S. military are made out of oil. It (the giant) runs fast and tirelessly on his feet of oil, but as soon as the oil runs out, the giant will not be able to move an inch. No more running. Instead, the giant will be lying in the dust and the era of U.S. military might and total superiority will come to an abrupt end, all because of lack of oil.

For now, the U.S. military machine's energy supply is secure, and all the DoD generals have to worry about is the energy supplies of today, while shyly looking at tomorrow. This is where things might get very complicated and even dangerous, if DoD does not anticipate and control the upcoming energy debacle; that of global demand increase and pending depletion of fossil energy sources.

And speaking of fossils…

THE FOSSILS

Obviously, fossil fuels (coal, natural gas, and crude oil) are still a huge part of our military machine, but they are also an integral part of our economic complex. As a matter of fact, we dare say that they are (presently) its foundation. No fossils—no economic development. It's as simple as that. From the looks of things today, they will remain so for the foreseeable future, although their roles are changing, which complicates the picture.

So let's take a closer look at the new energy situation from the fossils perspective and discuss its effects on our economy, our energy security, and our lives in general.

Figure 8-19 shows that world energy use—especially in the large developing countries—is steadily increasing. There is an attempt (albeit not very successful) to reduce fossils use across the globe. The fossils are still the preferred energy source of the developing world, and they have few other choices. As the population in those countries increases, so does the demand for more energy, which only cheap fossil fuels can satisfy presently. So, that trend will continue for the foreseeable future.

Note: the black bars denote historical use while the lighter colors depict present use of different energy

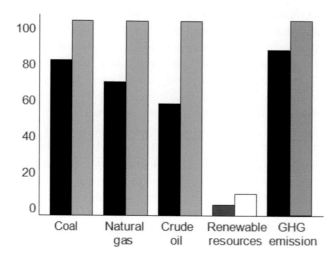

Figure 8-19. Relative changes in global use of energy resources (in %)

sources and GHG emissions. The trend is obvious and undeniable; all types of energy use are on the increase. With that, emissions are increasing proportionately.

The overall effect of this phenomena, in addition to faster depletion of global fossil reserves, is an increase in global GHG emissions. So, while the U.S. and other developed countries are working on reducing their contribution to global emissions by reducing fossils' use, the increased pollution of the environment from developing nations will continue unhindered. That, in turn, increases global pollution.

Here we must note that the U.S. is at present directly contributing to the global emissions increase by exporting large amounts of coal, natural gas, and refined petrochemicals to the developing world. The plans are for the U.S. exports of coal, LNG, and refined petrochemicals to increase incrementally in the near future. This strategy is amazingly short-sighted!

The Ignorance Factor

Before we dive into this complex matter, we need to clarify the situation from a technical point of view. Most of the discussion on energy subjects requires a minimal understanding of the science and engineering of the different materials, equipment, products, and processes involved.

This is a problem in many cases, where politicians, regulators, and even energy company officials lack in-depth knowledge (and at times even basic understanding) of the scientific, technical, and engineering disciplines involved. This technical ignorance (for lack of better words) often leads to misunderstandings and introduces errors in the decision making process.

Most politicians, regulators, lawyers, investors, and corporate heads—holding social, political, legal, or finance degrees—do not have technical background (physics, chemistry, and such). These people are mostly confused by the complexity of energy subjects, due to their limited scientific and engineering knowledge.

Even worse, the principals often rely on outside technical consultants and lobbyists, who may or may not have the necessary technical knowledge themselves. On top of that, often the consultants (and surely the lobbyists) have their own agendas, further complicating the picture and leading to confusion and more wrong decisions.

There are a number of cases where the lack of technical understanding has led to waste of time and money. We will cite only one, which reflects the situation quite well.

Case Study: Solyndra

One example of shallow technical understanding, influencing a major political decision is the $500 million taxpayer dollars wasted on the failed Solyndra company in 2011. Solyndra was a solar (PV) modules manufacturing company with brilliant technical staff and even more "brilliant lobbyists." What these two groups concocted was a gem of a team effort, geared primarily to get as much money as possible from gullible investors, and ignorant and/or negligent government bureaucrats.

Solyndra's engineers designed and installed fancy, unconventional, and extremely expensive custom-made production equipment, which produced some equally fancy, unconventional, expensive, and useless product prototypes. Then they printed some sleek product manuals, explaining in layman's terms the advantages of their new, shiny, "innovative" technology.

The lobbyists took the impressive brochures to Washington, DC, and being technically ignorant (and morally challenged) themselves, pushed them and their agenda onto the U.S. Department of Energy (DOE) and the White House, having almost unlimited access to both.

So, the self-serving agenda of Solyndra's lobbyists, combined with the negligence of the DOE managers and the ignorance of the U.S. government bureaucrats, resulted in $500 million poured into Solyndra's pipedream venture.

Solyndra's managers did not waste any time to frivolously spend all that money on large salaries and bonuses, brand new and totally unnecessary buildings, special but equally useless equipment, etc—all this at a time when they could not even show that their product worked.

Unfortunately for all involved, the Solyndra product was doomed from the very beginning. Their technology consisting of thin films deposited on the inside of long glass tubes, was sophisticated and innovative all right, but produced an absolutely impractical and prohibitively expensive product. The final product—cylindrical solar panels—in addition to the complex and expensive manufacturing process, required special mounting in specially selected areas painted white. Even then it still could not produce half of what the competition averaged at the time.

The reliability of Solyndra's evacuated glass tubes-made-into-solar panels was an even bigger problem. Glass-to-metal weldments in the long tubes would leak under mechanical and thermal stress, breaking the vacuum and allowing the elements to destroy the fragile thin films inside the glass tubes. So, in addition to getting less power per unit area, these installations would quickly deteriorate and lose power as well.

Solyndra produced worthless technology that had no chance of market success. All this was very, very obvious to those who knew the principles of solar power generation, thin film processes, and the overall energy market conditions. These are not new technologies, and DOE has hundreds of engineers who are specialists in these areas.

Why DOE's managers and the government's bureaucrats did NOT see the problems with Solyndra's technology in time, and instead recommended it for half-billion dollar funding, is beyond us. Technical ignorance, negligence, corporate greed, and political maneuvering are most likely the reasons why this case became a text book example of how not to do things. Or could it be that all participants collectively closed their eyes, justified by their ignorance and blinded by the complexity of the shiny new technology, so they could rush to finance one of the most embarrassing projects in the solar industry history?

Unfortunately, Solyndra is not the only such example. There were many similar, albeit on smaller-scale, cases where technical ignorance, combined with political maneuvering and corporate greed funded failing companies and ideas, wasting many millions of taxpayer dollars. We take a very close look at all these failures in our other books on the subject, see references 15, 16, and 17.

But let's get back to subject at hand. We've discussed the fossils (coal, natural gas, and crude oil) in detail in the previous chapters, emphasizing their utmost importance to our past and present-day wellbeing and comfortable life style. We also discussed the damages they cause and the possibility of fossil-less existence for the future generations in our previous publications, see reference #17.

Now we will take another close look at the fossils from our energy security's point of view.

Crude Oil

Crude oil is the Achilles heel of our energy security, since we import more than half of what we use. The imports amount to millions of barrels daily, coming mostly from countries with shady reputations and even shadier futures. How this, and everything related to crude oil, affects us and our energy security is the question of the 21st century. We have some of the answers, but many important questions are still left unanswered. The answers to the most important questions depend on who is answering them:

1. How much crude oil do we have left in the ground?
   a. In the U.S.?
   b. In the world?
2. How much of what remains can we pump up and use now and later?
3. How does oil quality change with well depletion?

Tackling the first question, we encounter the concept of crude oil's "proved reserves," which is an estimate of the total quantity of crude oil that "may reasonably be recovered from known reservoirs" now or in the near future with the existing technologies, and under the related conditions of the present-day life.

The proved reserves estimates change from source to source and from year to year, according to available information, which is constantly updated. This information includes discoveries of new fields, and/or reassessments of the potential of the existing ones. The combined sum of these is considered in the new estimate calculations.

The second question takes us to the concept of recoverable oil, where market prices and technology improvements play important roles in determining what's reasonably recoverable and what is not. These concepts vary with time and as the different technologies develop. Basically, there might be a lot of oil discovered in one spot, but some of it might be either too deep, or for some other reason too difficult to extract. That portion of the new discovery would be considered un-recoverable and subtracted from the total "available" estimate amount.

This is tricky, because there is a gray area between recoverable and un-recoverable oil deposits. As oil drilling and production technologies improve, many of the un-recoverable deposits today are likely to become recoverable tomorrow. Nevertheless, in most cases drilling and production work requires much more effort and expense today than it did in the past. This is an indication of an overall trend towards hard times coming upon the oil industry, which will very likely continue.

For example, 100 years ago wherever one poked a stick in the Saudi Arabian desert sands, high quality crude oil gushed 200 feet in the air. That was the time of "easily recoverable" crude oil.

During the last several decades things changed slowly, but steadily. The same oil wells that gushed up with the poke of a stick are almost dry now. In most cases oil has subsided many miles deep underground, thus making it very hard to pump up.

The simplified answer to the third question is that oil quality from each well changes over time, usually for the worse. Since most major wells are very old, we now see much poorer quality coming from the same wells that previously produced high quality crude oil.

The crude oil remaining deep under the surface is gummy and dirty; contaminated with dirt and unwanted, harmful chemicals. So additional and expensive processing steps must be considered at the refineries to bring the poor quality oil to world class levels.

These issues are carefully kept in the background by the oil companies and producing countries. It is their secret. They don't want us to know that the end of the unlimited high quality oil era is upon us, because that might cause worldwide panic. Panic is not good for business, so the status quo must be preserved.

Thus far, the secret is well kept, and with the new oil discoveries around the world, the urgency is gone (moved forward to the near future), and will remain out of the public eye for some time to come. Nevertheless, this does not change the situation much. It is normal today to dig 3, 5, and even 10 miles below the surface to find oil. And once you find the oil, it is no longer just a matter of pumping it up and refining it. Today's dirty oil requires special procedures to make it suitable for the refinery process. In many cases new oil has to be hydrofracked out of the rocks, which requires huge amounts of water, chemicals and energy, and which creates a big mess all over the place. All these unorthodox procedures make production costs rise significantly, since more time, effort, chemicals, and energy are required to extract and refine the oil.

As a matter of fact, it is only thanks to new and improved technologies that we are even able to get enough oil today and refine it at a reasonable cost. The expanding application of horizontal drilling and hydraulic fracturing in shale and other "tight" (very low permeabil-

ity) formations has opened new opportunities for new crude oil discoveries, drilling, and extraction. The new technologies now play a key role in their ability to reach deep in the ground and extract oil from places that were unthinkable just a few years back.

Rising oil demand and prices play a big part too, since they provide incentives for producers to explore and develop additional resources in very hard to work places, which is becoming increasingly important as the conventional oil supplies dry off slowly.

This is a natural self-correction imposed by the markets, where the new and more difficult to extract resources are developed for consumers in response to varying price signals, with the help of the new technologies.

There are conflicting opinions on these issues simply because people tend to look at them from different points of view, influenced by personal and corporate interest. So, if you ask an oil executive about the present and future oil status, they will paint the rosiest picture imaginable. A truck driver, paying $800 to fill up his rig's fuel tanks with diesel, on the other hand, would be mostly negative. The remaining majority would be somewhere in the middle, but most would express concerns about future oil price fluctuations, availability, the environmental and other oil- and fuel-related issues.

Oil Availability

As Figure 8-20 shows, the proved fossils, and crude oil and natural gas reserves especially, keep on increasing. This is not because new oil is created (there is no such thing as "new crude oil" anywhere on Earth). It is instead the ancient fossils that we are just now discovering and learning how to extract.

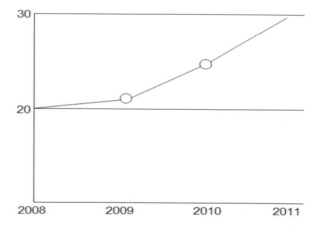

Figure 8-20. U.S. crude oil "proved" reserves (in billion barrels)

The only reason we have "new" oil reserve discoveries is because we are getting better—via new equipment and processes—in their detection and exploitation.

In 2010 alone, combined new technologies and high market prices boosted proved reserves of crude oil in the U.S. by more than 12 percent, and natural gas by nearly as much over the previous year estimates. So at that time, the U.S. had about 25.2 billion barrels of proved crude oil reserves, or 12 percent higher than the estimates a year ago and 50% higher than the increase estimated in 2008-2009.

In 2011, estimates jumped again to 29 billion barrels, or about 15 percent higher than the previous year.

Isn't this great? And surely it is...on the surface. But looking at the grand scale of things we see a big problem. Provided that we can pump all this oil, and do it at the present rate of use, we see that, even in best of best of cases, we can only count on our total reserves for a short time.

Get this: during the same time period, 2010-2011, the U.S. imported an all-time record of about 11 million barrels of crude oil per day.

Just as we were discovering, drilling, and pumping record levels of oil, we imported about 20% more oil than we usually import. But that was a record, a glitch, so let's leave it alone. Instead, let's look at today's average 8 million barrels of crude oil imports in the U.S.

If we multiply the 8 million barrels of oil imports by 365 days in the year, we see that the U.S. imports about 2.92 billion barrels annually. Assuming that we can use all U.S. crude reserves at will, we then see that by dividing the 29 billion barrels of total U.S. oil reserves (per 2011 estimates) by the 2.9 billion annual oil imports (which is about the amount of oil we use from domestic production), we have exactly 10 years of oil supply left on our territory.

Since we use only half of what we have, we then have 20 years of guaranteed oil reserves. Some may argue about the simplicity of this calculation. Agreed, but in the best of cases, the U.S. has no more than 25 years of guaranteed oil supplies. That is the very maximum according to the data available today.

Adding a greater margin of error and allowing for additional discoveries, takes us to 30 years of oil supplies.

In all cases, at the present rate of use, the U.S. will run out of crude oil by the end of the century.

As a matter of fact, a more precise estimate of the depletion of U.S. crude oil is around 2060-2070 time period. Then what? Well, at that time we would become 100% dependent on oil from Arab and other unfriendly states. What will happen to our SUVs, motor homes, eighteen wheelers, agricultural equipment, passenger planes, and cruise ships?

More urgently, how would the military trucks, ships, and jets operate around the world to defend us? On solar power? Corn biofuels? Maybe, but highly unlikely. Would they all be grounded, thus forcing the U.S. military to switch from global policeman to a domestic security guard? What would that do to our energy and national security? The answers to these questions are being tackled by politicians and high ranking DoD officials as we speak. Or are they just talking about it?

We will just have to wait to learn about the solutions they propose, if and when they agree on them and decide to let us know.

Maybe natural gas will solve the immediate mid-term problems, since we have a lot of it...

Natural Gas

The proved natural gas reserves in the U.S. have been rising even faster than oil. As a result, we now have over 320 trillion cubic feet (about 9 billion cubic meters) of proved natural gas reserves. There are unofficial estimates that the total (unproven) natural gas deposits are as much as three times this amount.

This is a lot of energy, that if used properly would serve us many decades, but that is not to be. The use of natural gas has increased significantly lately, so that in 2012 the U.S. consumed about 29 trillion cubic feet (TCF) of natural gas, which is projected to grow to about 32 TCF annually during this decade. An additional 10-15% of the total will be exported then too.

Again our simplistic estimates (based on other estimates) might not be precise, but look at the official numbers provided by the U.S. EIA: total proved reserves = 320 TCF vs. 32 TCF annual use. A 3rd grader could easily figure out that we have proved natural gas supplies for exactly 10 years.

Even when allowing a significant margin of error and other modifications, we still see the end of the natural gas as we know it today within the next 20-30 years max. Amazingly, most people don't know or don't care about this, so the trend will continue unhindered.

Everything considered, and under the best conditions possible, we will run out of natural gas by 2050-2070.

If you want to get technical, then note that the EIA estimates are in "wet" natural gas, which is the natural state of the commodity. When refined for commercial use, where water vapor and contaminants are removed, the final product is "dry" gas. In the process it loses some volume, so in reality we have much less natural gas available for commercial use than the official EIA estimates of raw gas.

In the best of cases—doubling and tripling new discoveries—natural gas deposits will be depleted well before the end of the century. End of the line! No gas, no oil...unless some drastic measures are taken...soon...now!

We don't see many, if any, such signs, so we must assume that the present trend will continue at least until oil and gas prices get sky high. Then, and only then, we may see some drastic measures taking place. Let's hope, for our energy security's sake, that it is not too late to do the right thing.

Until then, however, it will be the usual, "drill, Baby, drill," until it is all gone. Energy security now before anything else! Tomorrow?...we'll see. Most likely when "tomorrow" comes, we will switch from exporting to importing natural gas. That will keep us going for awhile until the global supplies' prices increase prohibitively. At that point we will enter into another energy crisis, the solution to which is to be provided by future generations, so they need to be ready.

The Bonanza

The present-day energy bonanza in the U.S. and some other countries is due mostly to new discoveries and the use of new technologies in the natural gas and oil exploitation fields. The new natural gas influx comes mostly via hydro racking of natural gas trapped in rock formations deep underground.

These deposits were not considered a viable source of gas until recently and were ignored. The technology was not sufficiently developed as well, which, together with low energy prices kept the shale gas safe underground.

Note: Shale formations are fine-grained sedimentary rocks that contain large amounts of crude oil and natural gas. The production of natural gas from shale formations has rejuvenated the natural gas industry in the United States and many other countries, to where we are now into a new phase of energy bliss.

Table 8-5 shows a staggering amount of natural gas available around the world. These, of course, are rough estimates, provided by different organizations at different times. Although we trust that they are close to the re-

ality, we must remember that estimates are just that—a bunch of numbers that depend on a number of things, all of which could change overnight.

Table 8-5. Global natural gas proved reserves (in billion m³).

Country	Total	Country	Total
Russia	48,600	Malaysia	2,350
Iran	33,600	Norway	2,313
Qatar	25,100	UAE	2,250
Turkmenistan	17,500	Kazakhstan	1,900
United States	9,100	Uzbekistan	1,841
Saudi Arabia	8,200	Kuwait	1,798
Venezuela	5,524	Canada	1,754
Nigeria	5,246	Egypt	1,656
Algeria	4,502	Libya	1,539
Australia	3,825	Netherlands	1,416
Iraq	3,600	Ukraine	1,104
China	3,100	India	1,075
Indonesia	3,001	Germany	550
ROW	remainder	Total world	220,000

These numbers can easily change in each direction as time goes on, so we must be cautious when relying on any estimates. As inaccurate as the estimates might be, however, we still must agree that there is a huge amount of natural gas available underground around the globe.

Assuming that the U.S. natural gas resources are depleted before the end of the century, we will be able to find reliable import sources and keep the price low enough. This, of course, would add another burden to our energy security, but even in the worst of cases, natural gas will be available at least for a while longer. At that point our politicians will be fully awake, and under increasing public pressure, will finally take serious measures to build a bridge to the fossil-less future.

LPG

Liquefied petroleum gases (LPGs) are mixtures of propane, ethane, butane, and other gases that are produced at natural gas processing plants and refineries. LPG is a convenient way to transport natural gas around the country and the world. It is pumped into special pressurized containers, which are loaded on trucks, trains, and ships. Its use is increasing as the demand from the energy-hungry world increases.

Propane is the main component in natural gas production, as one of the by-products of the new shale rock deposits exploitation. It is an energy-rich gas, with the chemical formula C3H8 and as such is one of the major liquefied petroleum gases (LP, or LPG).

Propane and other liquefied gases, including ethane and butane, are separated from natural gas at natural gas processing plants. They can be used/burned "as is," or can be separated from each other at the refineries for different uses. Additional large quantities are also separated from crude oil at oil refineries and all can be mixed for commercial use.

The amount of propane recovered from natural gas and crude oil is roughly equal, and constitutes a large amount of all gas produced around the world. Under normal pressure and temperature, propane is a volatile gas. At higher pressure and/or at lower temperatures, however, it becomes a liquid (LP)—at which point it occupies 270 times less volume than as a gas. The pressurization process, however, requires large and expensive equipment, which uses large amounts of energy.

Nevertheless, pressurization is required in order to "bottle" and transport LP to areas where no pipelines are available. In this liquid state, it can be stored and transported in special containers all over the world. It can be then used as gas when released from the pressurized containers at the destination point.

Propane accounts for less than 2% of all energy used in the United States, and yet it has some very important uses. For example, propane is the most common source of energy in rural areas that do not have natural gas service. In many homes, propane is used for heating the premises, hot water, cooking and refrigerating food, drying clothes, fueling gas fireplaces and barbecue grills.

On farms, propane is used to dry corn, to power farm equipment, and to power irrigation pumps. Businesses and industry use propane to run fork lifts and other vehicles and power equipment.

Only a small fraction of propane is used for transportation, and yet it is the second largest alternative transportation fuel in use today.

Propane is also used as a replacement for gasoline in fueling fleets of vehicles used by school districts and cities, government agencies, and taxicab companies. Hot-air-balloon enthusiasts use propane to heat the air in the balloons to makes them rise and travel long distances.

The LP role in providing a balance in the world energy scenario is increasing, as many European and Asian countries rely more on LP imports. With that, the role of the U.S. as a net energy exporter is increasing as well.

LP is a clean burning fuel, which is why it is preferred fuel for indoor equipment such as fork lifts. Its clean burning properties and its portability also make it popular as an alternative transportation fuel. However, it emits significant amounts of CO_2 and other GHG gas-

ses, thus polluting the atmosphere.

And as the quantity of natural gas and LPGs burned in the U.S. and around the world increases, we foresee a significant increase of GHGs coming from them too.

The U.S. is planning to significantly increase the use of natural gas at home as replacement for coal and vehicle fuels. U.S. exports of LPGs are also on the rise, all of which means increased energy security (albeit temporary).

Propane prices occasionally spike, usually during the winter months, sometimes increasing far beyond the normal supply/demand price fluctuations. This is mostly due to the logistics, driven by the difficulty of providing and maintaining the regular supply during the peak heating seasons. It is also because there is no ready source of incremental production when supplies run low, since propane is produced at a relatively steady rate year-round by refineries and gas processing plants.

Because of that, propane suppliers are forced to pay higher prices as propane markets gets higher in the winter months due to dwindling supply and logistic constraints. In these cases, the higher propane prices are passed on to consumers. Imports are too expensive to offer much cushion for unexpected demand increases or supply shortages.

Note: The black bars denote historical use, while the lighter colors depict present use of different energy sources and the total GHG emissions in the country. Compare this with the data in Figure 8-19, which shows a significant increase of energy use and pollution emissions in the global energy sectors today.

Figure 8-21 shows that while the use of coal and oil in the U.S. is decreasing significantly, there is significant

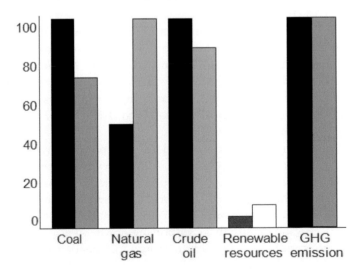

Figure 8-21. Relative changes in use of energy resources in the U.S. (in %)

increase of natural gas use. The renewables are also on the rise, but their contribution is still quite small, and projected to remain so for the foreseeable future.

The overall effect of this change—transition to natural gas—is that we are gradually replacing coal and crude oil with natural gas, which means that natural gas will be depleted much faster. Lower GHG emissions from the switch to natural gas are expected too, but it is hard to determine the long-term effect of this trend on the global environment. There are too many players and factors to consider, as we have discussed in the previous chapters.

Coal

Presently, the U.S. has all the fuel needed for power generation and heating (coal, natural gas, and crude oil). This way, we can truthfully proclaim energy independence in the domestic power generation sector. We have enough coal in existing and newly discovered deposits to last us over a century, even if we switched to using 100% coal as fuel for power generation. In fact, we have so much coal and natural gas that we are increasing the exports to a number of countries around the world.

Looking to the future; we have enough domestic gas and oil to last us at least 40-50 years, and coal for over a century.

Basically, we can celebrate energy independence in the domestic power generation and heating area of our economy. But we still import over half of the crude oil we use on a daily basis, so our energy independence is far off, and as things are developing, it seems to be getting even further off.

This is simply because crude oil drives the entire economy. Without crude oil to power the mine production equipment and unit trains, coal could not be dug out and transported to the power plants. Digging wells, pumping, transporting, and refining natural gas also cannot be done without crude oil to drive the process equipment and vehicles.

So crude oil is essential for the production and transport of all energy sources. Although we import only 50% of the oil we use, we cannot run the U.S. economy without the imports. Crude oil is, therefore, critical to our energy security and puts our energy independence in a checkmate situation. Importing crude oil is a major issue standing in our path to energy independence.

Our energy security is badly compromised, and energy independence is impossible, while importing crude oil.

Oil imports represent a large gap in our energy supply and a threat to our energy security, so replacing the oil imports with domestic liquid fuel (other than crude oil) should be on the top of our agenda. If we succeed in such a conversion, we would enjoy 100% energy independence.

One possible way to reduce and even stop oil imports, is by converting large quantities of coal (while they last) into liquid fuels.

While this is not the ultimate long-term solution (coal will be gone by the end of the century too), it would at least allow us to break away from the imports. Then we could use the money we save for developing new renewable and other energy technologies.

We have enough coal for that purpose, and the conversion processes are well established, so it is only economics (or a serious national or global disaster) that separates us from the day when coal (or rather its liquid derivatives) will become a major transportation fuel.

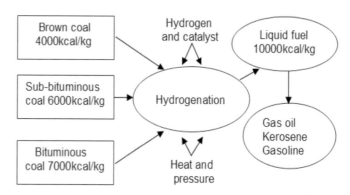

Figure 8-22. Coal liquefaction—a sure way to reducing oil imports.

Coal consists mostly of carbon atoms with very little hydrogen atoms in its molecule. The process of liquefying solid coal into liquid fuels, therefore, consists of simply increasing the number of hydrogen atoms in the new liquid fuel's molecule.

After removing the harmful hetero-atoms sulfur and nitrogen (S and N), most of the ash must be removed from the liquids. The final product is a liquid hydrocarbon fuel with proper H/C ration that can be refined into other fuels and products, or used as is for burning in power plants, and/or for powering vehicle engines.

As a bonus of this process, the heating value of thus produced fuels is 2.5 times higher than the input materials (coal). Basically, 1 lb. of processed coal would produce the energy of 2.5 lbs unprocessed.

Theoretically, coal liquefaction is a rather simple process of converting coal into liquid by adding a few hydrogen atoms. In practice, however, it involves a lot of factors, and uses complex and expensive equipment.

Although coal's physical structure is pretty simple, its liquefaction chemistry is very complex, where free radicals are produced by thermolysis of the covalent bonds of the coal molecule. Such reactions are prevalent in coal liquefaction, where thus formed radicals undergo various separate reactions. These could be decomposition to lighter components, stabilization by hydrogen transfer from hydrogen sources, and/or retrogressive reactions to heavier fractions.

Several chemical and physical factors affect the performance of the coal liquefaction process. The particular coal properties and structure, type of solvents and catalysts used, pressure, temperature, and the concentration of the coal-solvent process mix, are just some of the most important factors.

The Coal Liquefaction Process

Coal liquefaction actually refers to a number of sophisticated processes for producing liquid fuels from solid coal. There are two major categories: indirect and direct coal liquefaction processes.

— The indirect liquefaction (ICL) process involves gasification (special type of burning) of coal to a mixture of carbon monoxide and hydrogen (syngas). The gas then undergoes special (Fischer-Tropsch) process steps, designed to convert the syngas mixture into liquid hydrocarbons.

— The direct liquefaction (DCL) process converts coal into liquids directly, without the intermediate step of gasification. This is done by breaking down its molecule with the help of solvents and/or catalysts in a high pressure and temperature environment.

Liquid hydrocarbons generally have a higher hydrogen-carbon molecular ratio than coal, due to either hydrogenation (adding hydrogen atoms) or carbon-rejection (reducing the number of carbon atoms) methods used during the different types of coal liquefaction processes.

Coal liquefaction is a complex, high-temperature/high-pressure, process, which requires large capital investment and huge amounts of energy. This makes coal liquefaction practical and cost-effective only if and when oil and LNG prices are high. Since this is not the case presently, coal liquefaction remains a pipe dream, splashed over the pages of this text for awhile longer.

Its time will come though, and as global political and economic situations deepen into full-blown disasters. Coal liquefaction products might become a primary fuel, which together with the renewables could lead the way to our long-term energy independence.

The Liquefaction Process Factors

There have been many attempts to achieve the ultimate goal of correlating the physical and chemical properties of coal with liquefaction reactivity. But it is difficult to determine what property of coal is most important in the liquefaction reaction, due to coal's inherent inhomogeneous and complex chemical structure.

Historically, the reactivity has been estimated by the yield of some solvent soluble component in the type of coal at a set time and at fixed reaction conditions, although this is insufficient to compare the reactivity of the different types of coals.

Recently, more sophisticated reactivity measures, based on a combination of equilibrium and rate of liquefaction regimes have been used. Since no single property of coal's structure adequately predicts liquefaction conversion, correlations with combinations of several coal properties can be used.

The most important coal properties, as far as the efficiency of the liquefaction process is concerned are: type of coal, sulfur content, reactive macerals, volatile matter, vitrinite reflectant, and the atomic ratio of hydrogen/carbon (H/C) in the coal coming from different geological regions.

Coal properties and the related analyses that determine the coal liquefaction process efficiency, can be summarized as follow:

• Coal rank is considered as a major factor in the liquefaction reactivity and efficiency. Coal rank is defined by several coal properties such as volatile matter, fixed carbon, calorific value, vitrinite reflectance, carbon and hydrogen content in the coal, etc.

The initial coal chemical structure is a very important factor in determining coal's rank and in determining the liquefaction process. To predict the chemical reaction that may occur in the liquefaction process, we have to know the chemical structure of all materials involved in the chemical structure of the coal in question.

The structure of the coal molecule is dependent on the origin of the coal precursor and its rank. Higher ranks of coal contain a more condensed aromatic ring in the coal molecule. Anthracite, for example, has more a condensed level of aromatic rings in its molecule structure than bituminous, which level decreases with lower coal ranks.

Higher rank coals have more phi-phi bonding (condensed bonding) inter molecule. This type of bonding is very strong and only breaks during the liquefaction process in the reactor. Whereas, the weaker bonding such as hydrogen bonding and charge transfer bonding, break before that—in the preheating step.

High coal ranks have higher bonding strength, which makes it more difficult to break the coal molecule, and which results in lower liquefaction yields.

A few coal properties do not provide enough information in predicting liquefaction yield, but at least they can be used as precursor indicators in estimating the final liquefaction reactivity and efficiency.

Coal rank can be determined by several methods such as vitrinite reflectance, ASTM method (using fixed carbon and calorific value in certain basis). Generally, higher ranks of coal produce lower yields of liquefaction, since the higher rank means the increase of vitrinite reflectance and lower H/C ratio as well as lower volatile matter. The distillate yield of liquefaction also generally increases with increasing H/C ratio and decreased vitrinite reflectance.

Another property of higher rank coal that can reduce liquefaction yield is a more condensed aromatic ring in the coal molecule.

- Coal Maceral is an important coal property, since it indicates the origin and type of coal derivatives. It is based on the fact that coal contains three maceral groups; i.e., vitrinite, exinite (liptinite),and inertinite.

    Vitrinite and exinite are very reactive macerals, whereas inertinite is an un-reactive maceral. Therefore, coal that contain higher vitrinite and exinite is preferable in coal liquefaction. Higher reactive maceral contained in coal produces higher oil yield during coal liquefaction.

    Other elements in coal include carbon, hydrogen, nitrogen, and sulfur which are determined via laboratory analysis.

- Carbon and hydrogen are the most important parameters in coal property, since they indicate the yield of liquefaction by determining the ultimate H/C atomic ratio.

- Nitrogen content in coal is not a significant factor in the coal liquefaction process, since it is removed from the mix via hydro-treatment. The nitrogen content in product oil is one of the important parameters since it reduces the stability of thus produced oil by producing harmful gum deposits.

- Higher sulfur content in coal is desired in coal liquefaction process, since it enhances the reactivity of many catalysts.

- Oxygen content is obtained by subtracting the sum percentage of carbon, hydrogen, nitrogen, and sulfur content from the total.

Other important coal properties of importance to the liquefaction process include: moisture, ash content, volatile matter, and fixed carbon.

- Moisture content of coal does not affect the liquefaction process, since the slurry dewatering step is very effective in removing moisture from coal.

- Ash content affects the liquefaction process, since ash and catalyst accumulate on the bottom of the reactor and can slow, or even stop, the reaction. This can decrease the reaction volume in the reactor, resulting in a pressure drop and decrease of residence time of the slurry in the reactor. These solid materials can also deposit in the connecting pipes and reduce the flow of the slurry in down stream units, which also interferes with the overall process efficiency.

    The maximum limit of ash content of in liquefaction process is 10%, but lower ash content is preferable for proper and efficient liquefaction process.

- Volatile matter and fixed carbon are also the parameter of coal property that can be used to predict the level of the liquefaction yield. It is derived from aliphatic hydrocarbon. The higher the volatile matter content, the more aliphatic hydrocarbon is present in the coal molecule, resulting in a more reactive and efficient coal liquefaction process.

- The fuel ratio of coal is calculated as fixed carbon divided by volatile matter (FC/VM). The higher the fuel ratio, the lower the coal reactivity is, which results in lower liquefaction yield.

All in all, most of the coal types in the U.S. can be successfully used to produce liquid fuels, although the processes and yields might be different. The technology is available, and only the higher cost of processing is delaying the inevitable. Coal liquefaction is an option that will be used as soon as crude oil prices go high enough, and/or when the coal liquefaction process costs are reduced.

Considering the fact that crude oil imports are the main reason for our "energy dependence" on other countries, coal liquefaction process might be the determining factor to freeing ourselves from the imports burden. By doing this, we can be energy independent for a long time, since we have a lot of coal deposits.

Until then, however, we will be struggling with overseas oil imports and the unfriendly and unreliable exporters.

Now we'll take a look at one less known fossil fuel of potentially great importance for our energy future. These are the so-called methane hydrates.

Methane Hydrates

Methane hydrates, also called "natural gas hydrates" (NGH), or simply "gas hydrates" are, according to some, the solution to our long-term energy and environmental problems. Others, however, think that they are the devil in disguise that will bring our civilization to a disastrous end. The subject is somewhat obscured today by lack of information, and/or misinformation from interested parties, but will grow in importance as time goes on.

No matter what the future holds, this is an extremely important, albeit not well known phenomena, so we will take some more time dwelling on it in this text:

Methane hydrate is an ice-like crystalline; its molecular structure is similar to ice except that there is a methane molecule trapped within the hexagon/pentagon cage of ice. Methane hydrate exists in metastable equilibrium with its marine environment, but is affected by changes in pressure and temperature. When it dissociates (due to either lowering pressure or rising water temperature), the ice cage melts and the methane gas escapes.

Methane hydrates are usually found in the ocean sediments, deep under the sea floor of the continental shelf/slope. A fraction of methane hydrate resource is also found in the polar permafrost.

The presence of huge amounts of methane hydrates around the globe is the best kept secret of the energy industry. It is well known by many scientists, but some of them see it as a competitor, while others are too scared, to take it out of the closet. And there are some good reasons for that. The science behind this awesome energy source is too complex and overwhelming, while the logistics are even more so. And since the fossils still reign (and pay the bills) most of the professionals prefer to stay on the sidelines as far as methane hydrates and their future is concerned.

Nevertheless, some of those involved in the energy field are convinced that the methane hydrates are the energy source of the future. Others say it is an expensive exercise in futility, and yet others warn us that it is the most dangerous effort ever to be undertaken.

…and all this fuss over a single, otherwise humble, compound composed of methane (CH_4)—the simplest of all hydrocarbon molecules—buried deep under the ocean floor and in the permafrost.

NGH Global Supply

In the near future, the enormous importance of this fuel as disruptive technology will become evident, due to its excellent qualities and amazing quantities. Even if the potential of damages is proven, it will not override the potential of having unlimited supplies of energy for centuries to come. It will only make people more careful in their approach to this new and not well known energy source.

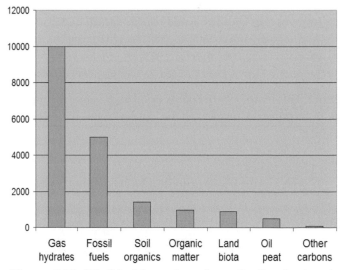

Figure 8-23. Worldwide carbon deposits (in gigatons)

Note the enormous supply of natural gas hydrates. More than twice the amount of ALL other carbon deposits on Earth put together. So maybe future generations will not miss the readily available coal, oil and natural gas that we will burn completely by the end of the 21st century.

If they are smarter and more ingenious than we, they will be able to find a way to extract, process, and use the huge NGH deposits and live happily thereafter for several centuries. Who knows…by then other energy sources might be available too, and the world will continue on without energy shortages and environmental problems. Let's hope…for their sake. Until then, we have a lot of issues to resolve. Just look below.

Environmental Impact of NGH Extraction and Use

Here is where things get sticky. Very sticky! The immense volumes of gas and the richness of the deposits make NGH deposits a strong candidate for development as an abundant energy resource. But they consist mostly of methane, a strong "greenhouse" gas, which in its raw form is many times more effective than carbon dioxide in causing climate warming.

Recent investigations indicate that some NGH deposits are unstable mechanically, and may cause landslides on the continental slope, if carelessly disturbed. A large NGH "avalanche" could result in an immense amount of methane to be released very quickly. What would be the consequences of such "explosion" of GHGs in the atmosphere? What would that do to our unstable environment, the GHG concentration in the atmosphere, and global warming?

> *Mining the NGH deposits at almost impossible depths and freezing temperatures would require special and expensive production, transportation, and storage equipment and facilities.*

In all cases, the transition to NGH won't be easy, quick, or cheap. Many issues need to be resolved before any large-scale work can be undertaken reliably and safely.

The Other Issues

Just a short list of potential NGH-related problems:

- There is an immediate problem with the huge amount of NGHs around the world. As the oceans warm due to the overall global warming, the trapped methane hydrates deposits will warm and start releasing methane in uncontrolled ways. Any activities in those deposits might accelerate the melting and methane release process. There are already some data showing that this unprecedented release has started.

- Raw methane gas is one of the worst environmental polluters, for it is over 40 times more efficient than

CO_2 as a greenhouse gas, so a significant release would increase global temperatures which would, in turn, melt more methane hydrate…*ad infinitum*. The vicious cycle could expand as it feeds on itself and accelerate the release of even more methane… until Earth's inhabitants find themselves in the worst global environmental disaster ever.

- Although we see a great opportunity of potential energy in the methane hydrate deposits lying idle on the bottom of the world's oceans, they are hard to get to and even harder to extract. Short of miraculous technical breakthroughs sometime soon, any of the present technical solutions would require a lot of time and money to develop fully on a mass scale. The methane hydrate deposits are usually deep in the ocean floor, where tremendous pressures and extremely low temperature would present the technological challenge of the century.

- Finally, it will take time for the oil companies and governments to agree on property rights and methods of exploitation. This, in many cases, will require global consensus on how to proceed, lots of money for R&D, and equipment and production sites development—all of which requires sustained efforts on all levels.

The way politicians and governments cannot agree on the simplest things today, it would be unreasonable to expect any serious action in the short term. So the great methane hydrates deposits will have to wait for future generations to get to them, and we are sure they will, after we burn all the fossils during this century.

Methane Hydrates' Future

Methane hydrates production has a huge potential, but very humble beginning with not very impressive results thus far:

- The world's first commercial production of useful methane gas from methane hydrate sites has a history of almost 20 years. The site is located at Messoyakh, Russia, where methane hydrates are mined directly from the permafrost in an effort to develop suitable and efficient technologies for methane extraction from the permafrost.

- An international onshore methane production test was carried out at the Mallik site in the Mackenzie Delta in the Northwest Territories of Canada in 2002. This area—damp in summer and frozen in winter—is a piece of land in the middle of nowhere. In the surrounding areas, the ground is permanently frozen to a depth of approximately 600 meters. The methane hydrates in the area are stable in their frozen state deep under at a low-temperature and high-pressure.

The test was a collaborative effort among Japan, Canada, the USA, India, and Germany, including seven research institutes. The "hot water circulation method" was selected for producing methane gas from methane hydrates. Here, hot water was heated up to 80°C and was fed into test wells to heat methane hydrate layers in their resting place at approximately 1,100 m below ground. The temperature and speed of the circulating hot water was controlled to be around 50°C when it came in contact with the methane hydrate layers.

This test produced approximately 470 m³ of methane gas over a five-day test period. However, it showed that the energy used to extract methane was many times higher than what the hydrates can generate. So, another heating method must be used. For that reason, the tests were continued in research laboratories.

- U.S.-Japan venture announced recently a successful extraction of methane from methane hydrates. An estimate of 100 billion cubic meters of methane can be extracted from the Gulf of Mexico and 1.1 trillion cubic meters from the Nankai Trough in Japan. This deposit alone could supply 100% of Japan's energy needs for over 10 years…if and when an efficient extraction methods are fully developed and implemented.

- China is in the midst of a 10-year research program, geared towards developing economical methods of extraction of methane from methane hydrates. Guangzhou Marine Geological Survey Bureau collected samples of high purity gas hydrates over nearly four months of surveys done by drilling 23 wells in the ocean floor off south China's Guangdong province. Two gas hydrate layers with a thickness of 15-30 meters were found at a depth of 600 to 1,000 meters below the ocean floor.

This methane hydrate reserve spans about 35 square miles, and holds the equivalent of about 150 billion cubic meters of methane. That's the size of the major conventional natural gas field in China's top gas province, Sichuan. This is nothing to sneeze at, and the Chinese are anxious to get their hands on it. We would be surprised if they don't come up with an efficient extraction technology in the next few years.

During the last two decades, ocean exploration has revealed an abundant fuel source of methane hydrates stored in deep ocean sediments in many places around the world. The majority of methane in ocean sediments has a biogenic origin, as a by-product of marine biogeochemical processes that include photosynthesis and oxidation. The rest of the known methane deposits have a thermogenic origin.

Methane is an ideal fuel that burns and produces some CO_2 but no other air pollutants. The amount of methane that is trapped in hydrate is perhaps 3000 times the amount contained in the atmosphere. The chemical energy stored in all known methane hydrate deposits is estimated to be two or three times that of all other fossil fuels combined. In principle, it is most desirable to develop the technology to exploit this enormous energy source in the near future.

But methane hydrates could also be at least partially responsible for the present-day global warming. Methane hydrates are sitting in their icy graves in a solid state. You'd think that they can be mined just like coal. Wrong! It is impractical and inefficient to mine methane hydrates this way because they are so deep in fragile and unpredictable geologic layers under a deep ocean.

The extraction of methane from this frozen solid state requires dissociating methane hydrate *in-situ*, right in their nests, and pumping the emitted methane gas through wells and production systems.

This is not easy, simple or cheap. The methane hydrates are stable at low temperature and under high pressure, so in order to break them out of their comfort zone, we must increase the temperature and/or decrease the pressure. Once methane gas is generated from the dissociated methane hydrates, then the usual methods, equipment, and facilities employed for the development of natural gas can be used.

But how do you do this at 5,000 feet below the ocean floor?

How do you increase the temperature, or decrease the pressure of solid frozen mass deep under the ocean floor, without spending a fortune, blowing yourself to pieces, getting buried alive, and/or poisoning the entire Earth's atmosphere in the process?

These questions separate us from untold energy riches, but they remain unanswered.

There are truly huge amounts of these fossils deposited deep under the ocean floor all over the globe. We see their presence today too; even without moving a finger. As the global climate temperature increases, the dissociation of oceanic methane hydrate increases. The process is accelerated by fluctuations in ocean currents and shifts in major warm current systems, such as the Gulf Stream migrating towards the continental slope of the northeastern United States/Canada.

The dissociative forces cause outgassing of the surface layers of the methane hydrate deposits, which is accompanied by the release of huge volumes of microscopic methane bubbles into the ocean waters. Methane oxidation bacteria and other microbes will convert the bulk of CH_4 into CO_2 and generate heat.

Both the CO_2 and any residual CH_4 will escape into the atmosphere while the heat warms the interior ocean (predominantly at intermediate depths). The warmed oceans will then evaporate more water vapor into the atmosphere, thus augmenting the enhanced greenhouse effect. This will cause further temperature increase, which will then accelerate the dissociative processes further, thus creating a non-stop chain reaction the final result of which is environmental disaster.

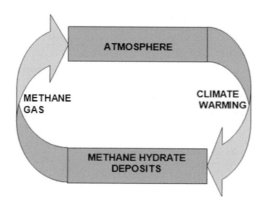

Figure 8-24. The vicious cycle

Noticeable melting of the ice cap and glaciers caused by gradual climate warming might be the most significant event of the 21st century. There are billions of tons of methane hydrates locked in the permafrost and deep in the world's oceans. These fossils have been kept locked in their watery-icy grave through the millennia by the low temperatures prevailing deep below the ocean floor.

As warmer ocean currents hit the deposits, the methane hydrates will thaw slowly at first, with a small release of methane gas in the water and the atmosphere. As a matter of fact, we might be at that first stage already, for there are signs of increased methane outgassing in some places around the world.

As the climate continues to warm, the speed of thawing of the permafrost will increase, and this is when things will get very dangerous. At some point in climate warming cycle, the chain reaction will enter a non-reversible and quickly accelerating speed, where the permafrost will melt at an ever increasing pace.

At such a time, the methane hydrate mountains will start collapsing and releasing immense amounts of methane gas. The resulting reaction cannot be stopped and will keep accelerating until all methane hydrates are completely dissociated and evaporated into the atmosphere. This will surely mark the end of the world… unless we find a way to harness their power.

If we spend enough time, effort, and money on developing a practical and economic process for methane extraction from methane hydrates, we might just bypass the darkness of the post-fossils era coming quickly upon us.

The extraction process development effort, however, in addition to billions of dollars, also requires the full participation of a) the U.S. government, and b) the oil companies. The government has the resources, and the oil companies have the know-how. Only this type of focused cooperation could bring successful and economically feasible extraction processes in time. All partial efforts, without the full and unreserved involvement of the above mentioned parties, would be a waste of time and money.

The other major fossil, which we have not discussed yet—radioactive ores—power our nuclear power plants. Yes, we don't think of nuclear as being powered by fossils, but where do you think uranium came from? It is a metal-like ore, created millions of years ago and stored in limited quantity in the Earth. We dig it out of the ground and process it just like the other fossils, so if it looks and behaves like a fossil, it must be a fossil.

NUCLEAR ENERGY

The U.S. nuclear industry today, aside from the fossils issues, reminds us of a giant coming from a long and bloody battle. For a long time he was the champion of the people. The knight in shining armor, whom everybody praised as their savior from the clutches of energy depravity and misery.

There were several minor battles in the 1950s and 1960s when the nuclear giant got a bloody nose, but they came and went without much ado. Only his pride and confidence were slightly hurt. But then came the big battles that he could not possibly win.

The first defeat was at Three Mile Island, then Chernobyl took him by surprise, and that took a long time to recover from (partially). Then the battle at Fukushima marked a major defeat, to where now his once brilliant future looks full of doubts and uncertainties.

Today the global nuclear giant is lying flat on the ground, tired, bloodied, and licking his wounds, while watching a bunch of kids run in circles around him. "Renewables will beat you. Solar and wind will take your place. You are out, Old Man, Nuclear," they shout incessantly.

What is a giant to do?

U.S. Nuclear Power

Nuclear power is still a major provider of electricity in the U.S. Of the 4300 billion kWh of electricity generated in the U.S. in 2012, 1640 TWh (38%) were from coal-fired plants, 1277 TWh (30%) from gas, 800 TWh (19%) from nuclear, 298 TWh from hydro and 141 TWh (13%) from wind and solar.

There were about 100 commercial reactors in 2013, of which 65 are pressurized water reactors (PWR), and 35 are boiling water reactors (BWR). These are installed and licensed to operate at 65 nuclear power plants, the combined output of which is over 100 GW. These power plants produced about 800 TWh of electricity in 2013, which was nearly 21% of the nation's total electric energy generation that year. This also represents 30% of the world's nuclear power generation—the largest supply of commercial nuclear power in one single country.

Globally, there are about 440 operational nuclear power reactors in 31 countries, but not all of these produce electricity. In addition, there are approximately 140 naval vessels using nuclear propulsion in operation, powered by some 180, albeit much smaller reactors.

Not considering the contribution from military nuclear fission reactors, presently nuclear power provides about 6% of the total world's energy needs, and generates about 375 GW of electric power, which is 13% of the world's electricity generation.

The U.S. nuclear power industry dates back to the 1960s, when most of the nuclear power plants in the country were built. Following the Three Mile Island accident in 1979, and due to changing energy industry and economic conditions, most planned nuclear projects were canceled. Nuclear power development in the U.S. came to a screeching halt almost 40 years ago.

All 100 nuclear reactors in operation in the U.S. were completed prior to 1977. No new nuclear power plants have been built in the United States since.

Over 100 plans for new nuclear power reactors, many of which were already under construction, were abruptly canceled in the late 1970s for many reasons. This led to confusion and chaos in the U.S. nuclear industry, and bankrupted several companies. The U.S. nuclear industry went into limbo followed by a period of stagnation and hesitation.

There was talk of a "nuclear renaissance" in the mid-2010s, which was supported by the Nuclear Power 2010 Program. A number of new applications for nuclear reactors were filed, but the new economic challenges, and the 2011 Fukushima nuclear accident put an end to those.

In 2013, four aging reactors were decommissioned due to high maintenance costs and pressure from natural gas price competition. These were San Onofre 2 and 3 in California, Crystal River 3 in Florida, and Kewaunee in Wisconsin. At the same time there are efforts to also close Vermont's Yankee and New York's Indian Point in Buchanan. Other nuclear power plants are also on the termination list of environmental, local activists, and anti-nuclear groups.

The Future of Nuclear Power

Nuclear is down, but not out. The fight is far from over, and as a matter of fact it might be just now starting.

Here is a brief summary of developments in the nuclear industry in the U.S. and around the world.

United States

Recently, the NRC approved the construction of four new reactors at existing nuclear plant sites. Virgil C. Summer Nuclear Generating Station, units 2 and 3 in South Carolina and Vogtle Electric Generating Plant, units 3 and 4 in Georgia started construction in March 2012. A new reactor at the Watts Bar Nuclear Generating Station in Tennessee is also under construction after it was halted in 1988. This is a total of five new nuclear reactors to be completed before 2020.

By the end of 2013, the planned additions to US nuclear power amounted to 7.7 GW of summer generating capacity, while planned retirements of nuclear facilities equaled 1.2 GW. Loss of nuclear generating capacity is expected to be offset by the five new nuclear reactors currently under construction, with a proposed combined capacity of more than 5 GW.

All five reactors are to be built at existing nuclear power plants, which means that no new nuclear power plants are planned in the U.S. at least until 2020.

Europe

While nuclear in the U.S. is mostly stagnating, the International Energy Agency (IEA) expects global nuclear capacity to rise more than 60% by 2035. With many new projects underway or being planned in the UK, France, Finland and across Eastern Europe, the European nuclear industry might be number one in nuclear number power then.

There are plans to build a number of new nuclear power plants in Europe:

- One in France by 2016
- Two in Belarus by 2020
- Two in Romania by 2020
- One in Bulgaria by 2022
- Two in Finland by 2024
- Two in Slovakia by 2025
- Three in the UK by 2025
- Two more in the UK after 2025.

China

China's nuclear plans and dreams are even bigger than the European's. China is actually aspiring to become the world's largest user of nuclear power by 2050.

Mainland China has 20 nuclear power reactors in operation, 28 under construction, and more about to start construction in the near future. China has become largely self-sufficient in reactor design and construction, as well as other aspects of the fuel cycle. For now it is using western technology while adapting and improving its own.

China's nuclear energy policy calls for implementing closed fuel nuclear power cycle. When the technology is perfected, it will be offered globally by exporting the Chinese nuclear technology, including entire rectors and components in the supply chain.

A number of reactors are planned to be built, including some of the world's most advanced, to give more than a three-fold increase in nuclear capacity to at least 58 GWe by 2020, then some 150 GWe by 2030, and much more by 2050.

The push for increasing nuclear power generation in China is mostly due to the increasing population and the public's concern with air pollution from coal-fired plants. So nuclear is one solution.

Japan

Since the Fukushima nuclear accident in 2011 Japan has become a test case for the global nuclear industry. Nuclear power generation in Japan went from 30% to zero the day after Fukushima. There were no people hurt or killed by the radiation, although 1600 died during the following months from heart and other stress-related problems. Three years after the accident, nuclear power in Japan is still zero.

Most of the world followed Japan's example in 2011. Nuclear plants in Germany and other countries also went dark at that time. Unlike Japan, most of the shut down plants were restated shortly thereafter. Japan, however, is holding back on restarting its nuclear power fleet. Public opposition, political pressure, and technological issues are keeping all 48 nuclear reactors in the country idling.

Unfortunately, Japan has no natural energy resources, but needs a lot of power, so it was forced to import huge amounts of coal, natural gas, and crude oil to keep the power generation going. They imported $266 billion in fossil fuel in 2013 alone.

The GHG emissions in the country increased during the same period by about 15%. The original Kyoto Protocol host country is failing its own mandate now and for some time to come. Renewables are only 1% of the total power generation mix now, and it would take years and many more billions of dollars to add a significant amount of renewables. Plus, Japan doesn't have unlimited land mass or sunshine for wind and solar power generation, so that option is off the table.

Now Japan is faced with a dilemma: to continue importing and burning fossils at a great cost to the budget and the environment, or to restart its nuclear plants. What would it be? Whatever happens in Japan will be followed anxiously and studied for many years by the nuclear industry. The developments in Japan might just determine its future...

This is an unfair situation, to be sure. Nuclear is doing everything right 99.9999% of the time, but one single error of statistically unimportant magnitude changes everything. Then nuclear power is back on the defendants' bench; it is again guilty until proven innocent.

While nuclear power is no doubt a significant factor in our energy security arsenal, it is also a great risk to it and our lives in general. Is the risk worth the benefits? Are we willing to sacrifice our health and lives for ensuring our energy security? The answers depend on the developments in Japan and/or any new developments in the global nuclear power sector.

India

India is also counting heavily on nuclear power for meeting its future energy needs. As a matter of fact, there is a talk about an upcoming nuclear revolution in India. The government has a vision of becoming a world leader in nuclear technology due to its expertise in fast reactors and the thorium fuel cycle.

India has a lot of thorium deposits, so it is counting on developing this new (yet to be proven) technology, to utilize the enormous natural resources. The new thorium technology works by circulating a solution containing thorium in a jacket around a core containing a small amount of fissile material (uranium or plutonium). As the decaying core material bombards the surrounding thorium it breeds more uranium 233, creating heat. This heated fluid is passed through a heat exchanger to convert water to steam and, via a turbine, electricity. The bred uranium 233 is then passed into the core, more thorium is added and the cycle continues.

The thorium nuclear cycle is heralded as the safest nuclear power generation, and since India has a lot of it, if the process works, India might meet its nuclear vision goals. For now, however, the thorium process is still in development in the research labs.

The latest achievement in India is the 2013 startup of a Russian-built nuclear power plant (does Chernobyl ring loudly here?) at Kudankulam. It took 12 years to build the plant and start the first reactor at 40% budget overrun. The plant's total capacity is 2 GW, which makes it the largest in India, and the operators claim that it is one of the safest nuclear power facilities in the world. Its safety is expressed in mechanical strength, where it could resist even the strongest of tornadoes or even direct impact by an aircraft.

The problem here is that this is a little consolation for the locals, since Chernobyl was also quite safe from outside attacks, but it blew from the inside. As a result, the new nuclear power plant experienced almost 2 years delay, due to NIMBY protests and demonstrations at the site. People feared another nuclear disaster in an area which was impossible to evacuate. Law suites reached India's Supreme Court, which finally ruled in favor of the plant construction. Safety was not a consideration for the court; the larger public interest was, so the locals (the affected minority) lost.

Slowly but surely, India is (at least in official statements) on the way to a huge nuclear power expansion. The new Russian-built power plant in Kudankulam is just a start. According to nuclear power insiders, India has official plans to build nearly 500 GW of new nuclear power capacity by 2050. This is more than the entire nuclear power capacity in the world today, which is under 400 GW.

Is this Indian magic? Let's see: If each new nuclear power plant in the 2050 plan is 1 GW in size, then they have to build at least 1 new nuclear power plant every month for the duration. Since they have not even started yet, this whole thing seems quite unrealistic. Or is it?

The World Nuclear Association records show that in the decade of the 1980s, 218 new nuclear power reactors were started up around the world, or an average of one every 17 days. During this decade one new nuclear plant came online every 77 days in the U.S. alone. So what India plans is not impossible, generally speaking. It is, however, highly improbable, under the economic conditions and logistics problems in the country.

Over 500 million people in India do not have access to electricity—twice the entire population of the U.S. or the EU—so we see clearly the need for electricity as a driving force. But even if India installs that much nuclear power, it does not have the infrastructure (power grid, substations, etc.) needed to deliver the power to most areas of the country.

The combination of these possible, but improbable, scenarios suggests that India must think of nuclear power as only one of the solutions. Looking for other means of adding more electric power in the country might be a better way to utilize already scarce resources.

Conclusions

Increasing world population and rising demand for energy is reflected in over 60 new nuclear power plants under construction worldwide, most of them in Asia. While the discussions on Fukushima, or no Fukushima continue, 160 new nuclear power reactors with a total net capacity of nearly 200 GW will be operating in the near future globally. The problem we see is that some of these new power plants use outdated Russian hardware, installed in countries with shaky political and economic systems.

Yes, there are many problems. The memories of Chernobyl and Fukushima raise big questions about nuclear power's viability, safety, price, etc. that cannot be ignored. The images are engraved in people's minds and bring questions, the answers to which will determine the future of the nuclear dreams.

Before nuclear goes to "business as usual," however, people need to know:

- Is the risk of nuclear accidents still alive and is it growing?
- Is Fukushima going to be the last nuclear accident ever?
- Is the nuclear waste storage problem resolved?
- Is the nuclear supply chain (equipment and services) capable of providing adequate support?
- Is the cost of nuclear power going to remain compatible?

The world is carefully watching the developments in Japan and other countries, and the questions remain:

— Will Japan restart its nuclear power plants,

when and how?
— Are Fukushima's problems with continuing radiation release going to be resolved anytime soon, when and how?
— What are the major European countries going to do with their existing nuclear and future nuclear plants?
— Will the biggest, and currently only hope of permanent nuclear waste disposal site—the Onkalo storage site in Finland—be successful?
— Why did the construction cost of the new Olkiluoto nuclear plant in Finland escalate from the original estimate of €3.3 billion to well over €8.5 billion?
— Are the above problems isolated cases, or are they new trends of hesitation, uncertainty, and ever increasing costs in the nuclear power industry?

These are very serious questions, which went away for awhile after Chernobyl, but popped up again after the Fukushima nuclear disaster in 2011. Many of these have been on the front pages of the international news, and there is no sign of them going away unanswered this time.

Complete and timely answers to these questions in terms of physical achievements and in gaining the support of the customers (and investors) is the key to the future of the global nuclear industry. Due to the complexity of the situation, however, we don't see how anyone could even begin to answer these questions. Any serious attempt would be a very risky undertaking, and an official response would be professional or political suicide in most cases.

Any new nuclear accidents, operational glitches, price increases, or unresolved issues might renew and increase the concerns, adding to the list of questions, and eventually changing the nuclear industry's course drastically.

The NIMBY movement is also alive and well in Europe, Japan, China, the U.S. and almost everywhere nuclear plants are operating or planned. The movement is actually growing in most countries, so the nuclear industry has to tiptoe around it, watching very carefully each step it takes. It is trying to avoid any serious battles with the NIMBY crowd, because it is bad publicity and because it is very hard to kick against the goads of public opinion.

Nuclear vs. Solar

Comparing the solar and nuclear industries is like comparing apples and oranges. The nuclear industry's Goliath is sitting silent for now, but a blow aimed at solar is expected any moment. We can't know how this trend will play out.

Figure 8-25. Nuclear vs. solar

Money and effort spent on developing the nuclear energy industry vs. money and effort spent on solar energy further reinforce the image in Figure 8-25. The U.S. nuclear industry was once the champion of the energy sector and the pride and joy of the U.S. government.

Just consider how the 2005 U.S. Energy Bill continued the amazing, uncompromised, and all encompassing government support in subsidies and technical and financial assistance for the nuclear power industry that started in the 1950s. Since then billions upon billions of taxpayer dollars have been spent on basic and specialized nuclear research, which was justified during the Cold War with the development of all kinds of nuclear weapons. The effort was then switched to nuclear use for power generation, which is basically a fancy way to boil water.

The U.S. government started developing nuclear energy as a weapon out of necessity, but ended up adopting nuclear power generation as the great promise of the energy future. In the process, they showered it with gifts and benefits. We should mention here the billions of dollars spent on the failed Yuka mountain nuclear waste storage facility. Also, more billions were spent on the failed NIF laser fusion facility. There are also the countless billions spent (and still being spent) on cleaning old nuclear research and waste materials storage facilities. The government-sponsored, multi-billion dollar development and servicing of the nuclear industry effort continues.

Subsidies for nuclear power continue, as follow:
— $5.7 billion were set aside for production tax credits

— $4.8 billion represent various subsidies, including R&D, tax breaks, loan guarantees, and risk insurance.

— $1.5 billion, "plus some additional unspecified sums" for an Ohio nuclear plant to produce hydrogen fuel by 2021.

— $500 million for the U.S. DOE to build new nuclear plants and its Generation IV program.

— And "out of budget" expenses, such as the equipment, effort, and expense needed to build and maintain the extensive and complex nuclear materials and waste transport routes and storage facilities, complete with anti-terrorism protection. How many billions of dollars are spent here? Who knows? There are so many agencies and activities involved that it is hard to keep track.

— There are also unlimited loan guarantees for all types of risks that are estimated at above 50% of the total. For example, 98% of the $600 billion government accidental insurance for worst-case accident recovery effort, are to be paid by taxpayers. This is easy to estimate; we, the taxpayers, will end up with a Fukushima-like bill of about 590 billion dollars. Not to mention the property damage and casualties, if it ever comes to it.

— But even more importantly, the nuclear industry could have never, ever, in a million years, been able to develop the processes, build the power plants, and step up on its own feet without governmental help. As a matter of fact, the modern nuclear industry still cannot function properly without government's help. So, in addition to direct subsidies, R&D, radioactive materials transport, waste storage, etc., operations are done under the supervision and/or with the help of the government.

— There are not enough private investors willing to put their money in such a risky business. As a consequence, not many private insurance companies would insure nuclear power plants. Here again, the government steps in and provides whatever is needed to keep the industry going.

Now compare this with the help provided to the development of renewable energy sources. "Sporadic, inefficient, and insufficient" would summarize the expense and the effort, which fades in comparison with the past, present, and future nuclear power benefits.

But could these two rivals find a common ground and go hand-in-hand into the energy future? Possibly, but not likely. The possibility comes from the fact that solar power plants could reduce the power load during peak periods, which helps grid management.

It is not clear that this helps the nuclear plants, since they operate at full speed most of the time. So the general attitude of nuclear plant operators would be that of indifference, or frustration.

Case Study: Combined Nuclear-Solar Power Generation

Still, this might be a good time to work together. Since nuclear, nor solar can provide the entire power load of the country, combining forces might be the best, if not only, approach for the future.

No matter how much we like solar, it is only a part of the solution.

Remember that the total nameplate capacity of the 100 U.S. nuclear plants, operating at 90% capacity, is over 100 GW. Combined, they generate about 800 TWh of electricity every year. So, what would it take to provide 800 TWh of electric power in the U.S. using solar panels?

First, solar (PV) power plants operate at 20-25% of their nameplate capacity (vs. 80-90% of their nuclear cousin). This is because the sun shines only part of the day, and because they do no provide maximum power all through the day even in mostly sunny days.

Second, we need over 10,000 acres to build a 1.0 GW solar power plant in the Arizona desert. Considering this estimate, we would need over 1,000,000 acres to install the 100 GW worth of solar panels. But at 20-25% power generating capacity, we would need 4-5 times that much land and panels. Or 4-5 million acres of desert land are needed to generate 800 TWh solar power to match the U.S. nuclear power output.

Third, since the sun does not shine at night, we will need an additional 30% of electricity generated by solar power plants during the day to be stored for night use. This is brings us to a total of 7-8 million acres.

We need a total of about 11-12,000 square miles, or an area of about 100 miles by 120 miles in size to generate all electric power the U.S. needs.

So, imagine most of the desert between Las Vegas and Phoenix packed with solar panels. Then consider the fact that all U.S. nuclear reactors could fit into one square mile. Finally, imagine the logistics (sweat, dollars, and tears) of maintaining the solar panels baking under the broiling Arizona desert sun in the 12,000-square-mile area, and you will get a good idea of the great difference between nuclear and solar power. Surely, the nuclear power operators sitting in their comfortable chairs in air conditioned control rooms appreciate it.

Looking closely, we see that nuclear is too dangerous and solar requires too much land and effort. So what are we to do? Doing it all with nuclear would be exceedingly dangerous, doing it with solar would be foolish. How about combining these two power sources?

For example, Palos Verdes nuclear plant located 50 miles from Phoenix, Arizona, is one of the largest such plants in the country. It hums happily non-stop during the entire year, until the broiling summer days invade the valley. In most of the exceedingly hot afternoons, May through October, electric demand in the area exceeds the nuclear power plant's capacity.

A number of peaker power plants, powered by diesel or natural gas, kick in during the hot summer afternoons to help the nuclear power plants keep up with demand. The peakers are usually idle until called into action by hot weather or in emergencies. Even with their help, the power grid is still close to the breaking point during most summer afternoons, and the prices jump up 2-3 times the regular rates.

So obviously, nuclear power cannot do it alone, conveniently, solar is most productive during the sunniest (and hottest) summer days, so if properly designed and implemented, it could take large part of the extra load during hot summer afternoons. This additional power would prevent the peakers from starting up and burning extra quantities of fossils. The power grid would take a break, and the electric power prices won't go up so high.

Figure 8-26 is a hypothetical scenario, showing a 1.0 GW nuclear power plant operating at maximum capacity on a summer day in the Arizona desert. Close by a smaller, 500 MW solar plant is cranking some power too. The nuclear power plant alone is capable of keeping the power demand met most of the time, but just before noon the residential and commercial air conditioners kick in at full power and start consuming megawatts of power in anticipation of another blistering hot summer day in the Valley of the Sun.

With temperatures approaching 110°F in the shade, the total power demand increases steadily, and eventually surpasses the maximum power level provided by the nuclear plant. At 1 p.m. the local utility starts charging peak rates, which are 2-3 times higher than off-peak rates, but that would not reduce the demand by much. Life without air conditioning during Arizona's summer afternoon is a death wish, so every single home

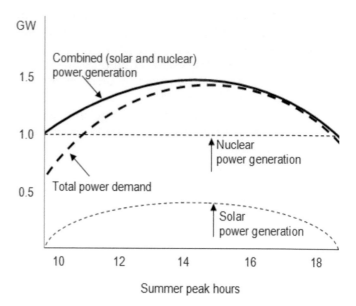

Figure 8-26. Combined solar and nuclear power peak power generation

and business building has a large air conditioning system. Some of the commercial air conditioning systems consist of multiple units, some bigger than a truck and consuming more power in a day than entire villages in Africa use in a year.

All this cooling and fanning requires a lot of energy, so the power grid reaches full capacity around noon. It would shut down at this point, if only power from the nuclear plant were available. Instead, it starts sucking electric power from other states. In most cases, even that is not enough, so around 1 p.m. the local peakers (diesel or natural gas power plants on standby for emergency use only) fire up and start generating the additional 500 MW electric power to keep the grid operating and thousands of air conditioning units running.

The local solar power plant, which has been ramping up slowly, reaches its maximum output at this time (maximum solar insolation) and provides the additional 500 MW needed to keep the locals cool and happy. Now the peakers can slow or even shut down, saving money and fuel, and sending less GHGs into the atmosphere.

Figure 8-26 shows combined solar-nuclear power generation as an almost perfect fit, but it's far-fetched as a standard approach. Too much controversial politics and conflicting private interests are involved in this process, so although we think this is the best way to build a bridge to the fossil-less future, it is not happening—not on a large enough scale to make a difference.

Still, common sense dictates that the intense summer heat, caused by too much sunlight could be successfully counteracted by generating equal amounts of electric power from the same sunlight. Common sense and future generations are asking, "How can you ignore such a winning combination?" But common sense and the future generations are absent for all practical purposes, so the decisions are often based on political and corporate interests. Because of that, both nuclear and solar power are paying the consequences.

HYDROPOWER

Hydropower is one of the oldest forms of power generation. It has been used since the beginning of time to propel grain grinding and other devices. Its use for large-scale electric power generation dates back to the beginning of the 20th century, when a number of hydro dams were constructed in the U.S. Most of these still provide a significant portion of the total energy used in the country.

Hydroelectric power provides almost one-fifth of the world's electricity, with China, Canada, Brazil, the United States, and Russia as the largest producers of hydropower today. One of the world's largest hydro plants is the Three Gorges on China's Yangtze River. The water reservoir for this facility is 75 miles long, and required flooding of thousands of acres of agricultural land and villages, together with their cultural and ethnic effects.

The Three Gorges water storage reservoir filling started in 2003, but the power plant did not start full operation until 2009-2010. The flooding of the reservoir area was a horrifying experience for thousands of locals, who were forced to resettle. The world watched while thousands of acres were submerged forever under the rising waters, bringing total destruction of the region.

The Three Gorges dam structure is 1.4 miles wide, 607 feet high and contains a number of new and sophisticated features, including 22.5 GW non-stop hydropower generators.

There are only two other large hydrodams generating over 10 GW in operation worldwide, Itaipu Dam at 14 GW, located at the Paraná River on the border between Brazil and Paraguay, and Guri Dam at 10.2 GW, located on the Caroni river in Venezuela.

Large-scale hydroelectric power stations are more commonly seen as the largest power producing facilities in the world. Some of the water storage reservoirs are bigger than most cities, and the hydropower plants on these are capable of powering entire cities as well. Their power generating capacity is also many times the installed capacities of the largest nuclear power stations.

Their number, however, is limited due to limited land area and other resources needed for such huge undertakings. Instead, smaller dams and hydropower generators are more feasible.

Figure 8-27 represents the potential hydropower around the world vs. operating hydropower plants, and those under construction, or planned. It is easy to see from this image that there are not many plans for increasing hydropower generation in North America, and few operating such, although the hydro potential is undeniable.

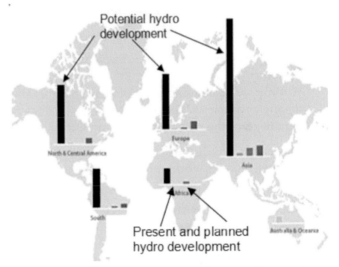

Figure 8-27. Hydropower potential around the world.

When most people in the U.S. hear the world "hydropower," they immediately think of the huge Hoover dam, and the Three Gorges dam images of human tragedy pop up too. And yes, some hydropower plants can be very large, but there are some very small too. These take advantage of water flows in rivers, streams, municipal water facilities, or even irrigation ditches.

There are also "dam-less" hydropower generators, where channels, pipes, or run-of-river facilities channel part of a stream or a river or stream through a powerhouse before dumping the water back where it came from.

Whatever the method, hydroelectric power is much easier to obtain and more widely used than most people realize. In the U.S., for example, every state uses some sort of hydropower for electricity generation; some more than others.

Hydropower is clean and safe, and usually costs much less than most of the conventional energy sources. The states of Idaho, Washington, and Oregon get the majority of their electricity from hydropower, and as a result their energy bills are much more stable and lower than the rest of the country.

Another advantage of hydropower is that most hydropower facilities can ramp the power level up and down from zero power to maximum output and back. This property is very important in power grid operations, and so hydropower is used to provide essential back-up power during many special events, including major electricity outages or power disruptions.

The biggest hydropower plant in the U.S. is the Grand Coulee Dam on the Columbia River in northern Washington. It, and several smaller dams, provide more than 70 percent of the state's electricity.

Still, hydropower in the U.S. has a number of serious problems:
1. Hydropower, amazingly, is not even considered officially as a renewable energy source, so it does not benefit from the respective government subsidies.

2. The existing hydropower infrastructure, dams and transmission networks, are getting old and in need of serious and expensive repairs.

3. Building large hydro dams requires huge land mass for their water reservoirs. Such land is not readily available, and flooding it causes enormous environmental damage and human suffering.

4. Constant draughts of long duration are drying the dams' water storage, especially in the southwestern part of the U.S. and some countries around the world. Since the water levels are low, the generated electric power levels are low. This makes the entire hydro power system (at least in those parts of the country and the world) unreliable in the long run.

As Figure 8-28 reveals, hydropower generation in the U.S. has fluctuated through the years, depending on a number of factors, some of which we discuss in this text. The renewables (solar, wind, bio-fuels, etc.), on the other hand have been steadily increasing, and if the trend continues will overtake hydropower.

And as a matter of fact, in April 2014 the U.S. Energy Information Administration (EIA) announced that the annual non-hydro renewable generation in this country surpassed annual hydropower generation for the eighth month in a row. Under the present water conditions, this might be a steady trend.

Until recently, hydropower was the largest source of renewable energy in the U.S., since over 7 percent of America's electricity comes from water-driven turbines. Only a decade ago, hydropower accounted for three

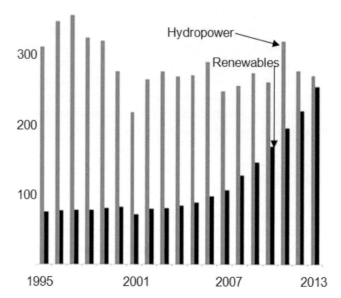

Figure 8-28. U.S. hydropower and renewable power generation (in million MWh)

times as much power generation in the U.S. as non-hydro renewable sources (wind, solar, biomass, geothermal, landfill gas and municipal solid waste). What happened in such a short time to change all that can be explained by the stagnation of hydropower vs. the rapid grow of the renewables, led by solar and wind power.

The renewables growth reflects ambitious and successful government policies such as state renewable portfolio standards and federal tax credits, as well as declining costs of the technologies. At the same time, hydropower received little attention, and even less support, with the overall trend pointing toward non-hydro renewable generation.

Adding insult to the injury, EIA projects that non-hydro renewables will generate more than twice as much power as hydropower by 2040. Assuming the continuation of tax credits and other policies that support non-hydro renewables, they could increase even more and faster.

Hydropower capacity has increased about 1% over the past decade, while at the same time wind capacity has increased almost 10 fold.

Nevertheless, hydropower is a viable alternative in most of the North American continent. If the full potential of hydropower is captured, we could count on it to bring at least 500-600 GW of new electric power generation to the continent.

This still requires taking the above factors under consideration and solving the issues.

Hydropower Development

Hydropower fell out of favor during the late 20th century, due mostly to the disruptive ecological and social effects of large dams' water reservoirs. The Three Gorges Dam in China is a text book example of large-scale property destruction and human misery caused by large water reservoir creation. It is a good example of how not to do things in the future.

Is this what's hindering new hydropower in the U.S.—the negative public opinion? According to the National Hydropower Association, 78% of U.S. voters agree that hydropower is cleaner than the conventional energy sources, while 77% believe that it is environmentally friendly, and 74% insist it is also reliable.

Most U.S. voters also favor increasing the effort to maintain existing hydropower plants and add new hydropower generation. In support of the efforts towards ensuring energy security, 77% of voters favor investing in federal hydropower facilities and hydropower R&D. At the same time, 74% favor providing government credits and incentives to hydropower, similar to those provided to other renewables.

Overseas, hydropower is going through a present-day revival as some international institutions are in support of its full development by using less invasive approaches. The World Bank, for example, is trying to find solutions to economic development of smaller hydro dams in developing countries, which avoid adding substantial amounts of carbon to the atmosphere. Implementing small-scale hydropower projects that cost less, and have much less negative impact than those of huge hydro dams, is one of the proposed solutions.

'Low head', "small-scale," or "distributed," hydropower generation is one of the keys to unlocking hydropower's potential around the globe.

Such small projects could be installed on rivers and streams all over the world, where they could provide much needed power to local villages or the power grid.

There is huge undeveloped and underdeveloped hydropower potential in some parts of the U.S. as well, where hydropower can be generated by water dropping 10-20 feet.

For example, there are presently over 80,000 dams in the U.S. erected and used for different purposes, but which do not generate any power. The U.S. Department of Energy suggested recently that we could add up to 12 GW of capacity by retrofitting the largest of these dams, which would boost the country's total hydropower capacity by an additional 15 percent. That's like adding a dozen large coal-fired plants without the pollution and the hassles coal brings.

Additional 5-10 GW hydropower can be installed in manmade waterways and at existing dams that do not generate electricity.

Another estimate claims that there is a grand total of over 70 GW of potential small hydropower plants that can be built in the wet areas and the rivers of the U.S. This is nearly the amount of the currently installed hydropower capacity. Such undertaking, however, requires the full support and direction of the U.S. government.

There are a number of reliable, cost-effective low head hydropower technologies available today, and new are being developed constantly. With some minimal technical and financial assistance from the U.S. government, thousands of low drop hydropower plants can be developed around the country to produce clean power with minimal environmental effects.

As with nuclear power, hydropower also has an Achilles Heel. It is the unpredictable nature of rain fall and snow cover. Without enough and consistent water coming from the sky, hydropower cannot be relied on. The future looks bleak from this point of view, at least in areas where serious draughts of long duration are converting the water reservoirs into mud puddles.

Tidal Power

Tidal power plants, using the power of the in- and out-going tides involve a dam, or a barrage, built across a water inlet. Sluice gates, which are devices used to control water levels and flow rates, installed on the barrage allow the rising tides to pass through a turbine system installed on the barrage structure.

This way the unlimited, truly renewable, power of the ocean tides can be converted into steady electrical flow into the power grid.

There are currently several tidal power barrages operating in the world. The largest is the Sihwa Lake Tidal Power Station in South Korea, which has rated capacity of 254 MW. The second and oldest is in La Rance, France, with 240 MW capacity rating. The Annapolis Royal, Nova Scotia, Canada, is a distant third largest with only 20 MW. Then come the Jiangxia Tidal Power Station in China at 3.7 MW, Kislaya Guba, Russia at 1.7 MW, and the 1 MW Uldolmok Tidal Power Station in South Korea.

A great disadvantage to tidal power is the mass extermination effect a tidal power plant could have on plants and animals in the local estuaries. Tidal barrages can change the tidal level in the basin and increase the turbidity, or the amount of solid matter in suspension in the water. They can also negatively affect navigation and recreation, including damaging boats and hurting people caught in the tidal currents.

There are no tidal plants in the U.S., and only a few sites where tidal energy could be produced economically. Other countries, like France, England, Canada, and Russia have much more potential to use this type of energy in the future.

There are also other ways to extract energy from the world's oceans. Capturing the energy stored in ocean waves, currents, and/or the temperature gradient of the ocean body are some of the most promising technologies to be developed in the future. There are a number of successful tidal- and wave-power plants around the world, but the high cost of construction, maintenance expenses, and logistics are preventing these technologies from expanding quickly—for now!

The ocean contains enormous amounts of energy, most of which is simply wasted. With some effort, a significant part of this energy can be captured and become a significant addition to global hydropower generation, increasing the renewables fraction of the total energy mix.

Pumped Storage

Pumped storage is another not-so-new technology that has a new meaning and added importance. As solar and wind power plants grow, so does their variability problem. Imagine a huge solar power plant idling during the rainy season? Or a wind power plant sitting in the midst of mid-day, low-wind conditions.

What good are zillions of watts of installed capacity sitting idle? Here is where pumped storage comes to the rescue. It is just another type of hydroelectric power plant that can store, and then generate electric power when, and as much as, needed.

When excess power is available at the solar or wind power plants, it is sent into the electric generators of the hydropower plant. The power in the generators spins the turbines backward, causing them to pump water from a river or lower reservoir to an upper reservoir. The gravitational power of the water is stored as potential energy, ready for use when needed.

When time for more power comes, the stored water is released from the upper reservoir and flows back down through the turbines, activating the generators and producing electricity. This is an elegant solution to a huge problem, and might provide a solution to the variability issues with large-scale solar and wind power generators.

Political Currents

Hydropower in the U.S. already has quite a few problems, a major one of which is the dependency on Congress to pass a number of laws in its support. It is urgently needed, and has been already submitted to the U.S. Congress, to officially include hydropower in the renewable energy category. This way, hydropower could count on more government support in all areas of its development and implementation. But this doesn't seem to be happening…

Another law that would help the development of new hydropower projects is to require (some) utilities to get a significant chunk of their future electricity from hydropower. That's a much more contentious undertaking, which also might take many years to become a law under the present political situation in Washington. If and when implemented, however, it might become a major contributor to adding significant electric power to the U.S. energy mix.

Another problem is the ongoing droughts in the U.S. Because we cannot change the drought patterns even if new laws are passed, we need to work around this by focusing on areas that are hindering hydropower development presently. In that respect, the U.S. Congress must be given credit for passing two hydropower bills in the fall of 2013. These had a strong bipartisan support and mark a step in the right direction for both the U.S. Congress and hydropower.

The first set of bills, H.R. 267 and S. 545, would speed licensing time for small hydropower projects, which is presently a huge hurdle. The other set of bills, H.R. 678 and S. 306 would make it easier for the Bureau of Reclamation to add hydropower to many of the canals, pipelines, aqueducts, and other man-made waterways that it already owns, operates, and maintains. This would also facilitate the development of hydropower infrastructure in man-made waterways.

A serious hydropower renaissance push would require new and more efficient technologies as well, which is what some Congressmen have been promoting with the "Marine and Hydrokinetic Renewable Energy Act." This Act would provide research money and direction into ways to tap energy from rivers and streams via "dam-less" hydropower. This approach doesn't involve damming up the flow for the creation of large reservoirs, which would be a major boost for hydropower in general.

These are encouraging steps towards reviving hydropower as a major energy source in the U.S., but the battles are just now starting. On the downside, some environmental groups and locals like the Native American tribes are pushing hard to demolish a number of dams in the Pacific Northwest, some with power generating plants, in order to rescue depleted fisheries and restore natural habitats. Who is right and who is wrong remains to be seen, but the battles pro- and con-hydropower in the U.S. continue. Stuck between merciless draughts, politics, and local resistance, the future of hydropower in the United States is not clear. With that, its role in solving the problems with our energy security are also yet to be resolved.

SOLAR ENERGY

Solar energy generation (PV especially), due to its flexibility, adaptability, ease of use, and lower prices, is likely to remain as the most widely used energy source of the coming centuries. Solar thermal technologies, using the sun's heat to heat water that turns a turbine to generate electric power, have seen a steady decline recently. Their major disadvantage is the need for huge amounts of water, which is just not available in the deserts, where this technology is most efficient. Because of that, we see their use diminishing even further, which is why we will focus our attention to the photovoltaic (PV) end of the solar industry.

Solar photovoltaic technologies are very flexible and can be used in small and large quantity in many different areas of our personal and business lives. They don't need water or much maintenance, so they can fill many niches and gaps that other technologies cannot.

Sunlight—albeit with different intensity—is free energy that is available in every corner of the world and can be captured and used at will. For this reason, we consider solar the energy of the future, and predict its unprecedented expansion in the near future.

Background

In recent years, solar and wind power have been among the fastest-growing sources of power generation in the U.S. and the world. Their promise as the energy technologies of the future is indisputable, yet a number of serious questions loom over their future. Some of these affect their short-term survival, while others will determine their development in the more distant future.

For now, the key questions are related to the uncertainty of government support and international disputes. Government support is vital during the infancy stage of these technologies. Will federal incentives that are important to the growth of solar and wind power continue? Most importantly for the long-term future of these technologies, what happens if and when those incentives are discontinued—as happened in most European countries recently?

International developments in the renewable energy sector are also problematic as the major powers of the world disagree and flex muscles about unfair trading practices. The ongoing solar energy trade war between the U.S. and EU, on one hand, and China, on the other is threatening to bankrupt many solar companies in all countries involved. If the present-day trade war drags on for a long time, it will eventually cause the entire industry value chain to collapse. How this scenario ends will have profound effects on the world's solar power growth.

Still, the renewables (solar and wind especially) are out of the closet forever. They have been finally recognized as the most practical and probable future energy providers.

Solar will never be shoved back in the closet again.

Solar still faces a number of problems. One is the expiration of the very important investment tax credit (ITC) program on December 31, 2016. The ITC was created as part of the stimulus package, and the industry is lobbying for a one-year extension, but that is unlikely to happen.

Whatever happens in 2016, the ITC will expire someday, which will have a significant compressing effect on the amount of renewable solar energy that gets financed. While extending the ITC would allow the U.S. solar business to grow 30-50 percent faster, it will stagnate, or worse, without the ITC.

Some other technologies, like fuel cells and small wind turbines, have access to similar tax credits through the same time period, and are also threatened with a slowdown in its absence. Their contribution to the total power load, however, is quite small, so the impact won't be as strong.

Another major federal program, the provision of loan guarantees to aid large renewable energy projects, ended in 2011. That program became controversial after Solyndra, the first solar recipient, filed for bankruptcy, leaving taxpayers potentially liable for more than $530 million.

While the loan guarantee program helped finance emerging technologies and "higher-risk projects," the grant program aided "extremely low-risk projects" using off-the-shelf technology. Chinese imports also benefited tremendously from these perks during the 2007-2012 wind and solar boom-bust in the U.S. and Europe.

Why that was allowed is a mystery, but at the very least we must learn from the past and not repeat our mistakes.

The continuation of federal incentives is important and like all forms of energy—including fossil fuels—solar also relies on a complex web of state and federal credits and aid. More established technologies, including oil, gas, coal and nuclear power, are still taking advantage of incentives that were established in the beginning of the 20th century. One important federal incentive for oil and gas drilling, for example, has been in place since the 1910s, at a cost of over $6 billion annually, though its days are numbered, if the Administration has its way.

It is clear that it is time to make a large investment in renewables, like we have invested heavily in conventional power industries (coal, oil, natural gas, hydro and nuclear) for a long time now. For the new technologies (solar, wind, bio-, and geo-power), trying to establish themselves and reduce their costs, government incentives usually are make-or-break situations, as it was for the conventional energy industries in the beginning of the 20th century.

Unlike the established oil and gas industries, where those companies could still make money even if they didn't have some of the assistance they currently have, the renewable energy industries won't develop properly, and some may even die as happened a number of times in the past. The perpetual threat that the incentives and all other government support will expire soon makes financing of projects and long-term planning in general difficult to impossible. Lack of certainty is a serious issue, so companies hope to be able to rely on the federal grant programs, but are forced to plan for life without subsidies and grants.

The costs of solar have dropped substantially in recent years, but not enough to compete with conventional energy sources, even when including the present-day tax credits, subsidies and grants.

In addition to the political uncertainties, low natural gas prices and the renewed interest in gas, have made it tough for solar to compete. It will become even harder if the investment tax credit ends.

The solar industry is globalized to the point where China, for example, can manufacture high-end PV panels at competitive prices, but it needs inexpensive, high quality American or EU silicon. At the same time the U.S. and EU also benefit from cheaper Chinese and wind and solar energy products.

We are used to the ups and downs of the renewables sector, created by government on-again, off-again policies and international developments, but it seems that solar will survive this time. A decrease or demise of any of the alternative energy sectors, however, would be very bad news for our energy security and for the well-being of the future generations.

Solar Technologies

A number of solar technologies show great promise for deployment in the future and have a great potential in contributing to our energy security.

Thermal solar technologies are making some progress, but we don't see their future as particularly bright due to, a) the huge amount of land needed for their deployment, and b) the large quantity of water that is needed for their operation. Because land and water are in limited supply, we will not spend much time on these technologies here. The reader can find detailed information on solar thermal technologies in our other books; see reference 15 and 16.

Photovoltaic technologies are the most promising of all renewables. They are numerous and very flexible in their designs, manufacturing and applications. Because of that, we see them as major tools in our energy security toolbox, which is why we focus on them in this text.

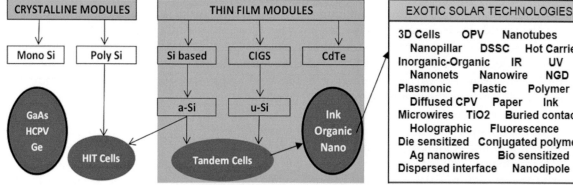

Figure 8-29. Major PV technologies

The key PV technologies presently are:

Crystalline silicon (c-Si) PV technologies:
- Single crystal silicon
- Multi-crystalline silicon solar cells
- Poly-crystalline silicon solar cells
- Quasi-crystalline silicon solar cells
- Super mono-crystalline silicon solar cells
- Microcrystalline silicon solar cells
- Silicon ribbon solar cells
- Silicon foil solar cells
- Silver silicon solar cells
- Emitter wrap-through silicon solar cells
- 3-D silicon solar cells

and

Thin film solar technologies:
- CdTe thin film modules,
- SIGS thin film modules
- Alpha silicon (a-Si) modules
- Ribbon silicon (thin film) modules

c-Si, CdTe, SIGS and a-Si technologies are the most established and promising, at present and most used in residential, commercial and large-scale installations.

The Solar Developments

PV manufacturing in emerging regions became bleak in 2013 after the silicon PV manufacturing capacities in China increased disproportionately during the 2007-2012 solar boon. That resulted in over supply, followed by equally disproportionate decreases of PV module prices. This unexpected event resulted in severe profit margin loss for most global manufacturers, and also contributed to feeding illusions for unreasonably cheap solar installations.

The momentum didn't last long. Few companies could withstand the pressure, and the abnormal situation culminated with numerous company closures and insolvency filings. Most expansion and upgrade plans were put on hold for the duration.

The overwhelming misconception that the flood of cheap Chinese PV modules will continue forever kept the frenzy alive. U.S. and EU installers would not consider anything but Chinese-made PV modules during the latest solar boom. Today they are looking for similar deals too, but these are becoming few and far between.

During 2007-2012 U.S. installers imported thousands of container loads full of cheap (and cheaply made) Chinese PV modules. The race was on, with price

as the only consideration. Ultimately, this abnormality escalated into global trade disputes, followed by confusion, price increases, questions about the quality of Chinese-made products, and overall decline in interest in solar.

Reduction in investment, and lack of capital spending in the PV industry supply chain in 2013 was a result of that confusion. It had a significant impact on the development of new products, which was most evident in the failure of the major equipment manufacturers to launch new products or programs lately. There was, for example, a 35% decline in new solar products advertised online in 2013 as compared with 2012.

The biggest problem the U.S. solar industry is facing today is the relentless competition from fossil fuels. Based on general observations and experts' opinions, we now see a long-term trend developing that will affect the solar industry profoundly.

— Considering the dismal investment in the industry of late, investors are losing interest in solar too fast to sustain development.

— More importantly, the incremental steps taken lately—as evidenced by the new products revealed in 2013—are too small to ensure quick development of solar energy resources around the world.

— Energy storage is the key to the successful future development of solar technologies. It is critical for the full integration of solar energy in the global power network. Solar energy cannot develop properly as a "variable" power source, because as such it will continue to be treated as a step child by utility companies and investors. Energy storage is showing very little progress at the present.

— New, more significant and disruptive materials, processes, products, and services are badly needed if solar energy is to outgrow its infancy. It must abandon the focus on shiny toys, and spectacular but unsubstantiated performance promises. Instead, solar must take large steps towards increasing power output, reliability, and decreasing product and service costs.

— The utilities are the largest users of solar energy products, since there are already many GWs of solar power generated worldwide, with many more planned. The utility type solar power generation is expected to grow even faster, so the utilities are in fact shaping the future of solar energy. They, however, are now seeing solar energy as a competitor and are starting to kick back. It will do us very little good to improve the solar technologies if the utilities refuse to use them, or make their use more difficult.

And speaking of the utilities:

The Utilities Games, 2014

Most investor owned utilities (IOU) in the U.S. are not happy with the way things have been going lately in the solar energy areas. The IOUs, regulators, installers, and customers have different ideas on the issues, and confrontations are becoming the norm.

California has the largest amount of installed solar power in the U.S. Since California usually leads the way in many areas—including solar energy—the developments there are very important, because many states are likely to follow its example.

Edison Electric Institute (EEI), the association of U.S. shareholder-owned electric power utilities, recently published a startling report calling our distributed energy resources (DER) "disruptive technologies" that threaten the very existence of today's U.S. electric utility industry.

Note: DER refers to small- to medium-size solar installations on residential and commercial rooftops. Their number in California has grown dramatically, thus the utilities' concern.

The authors of the EEI report, and the commentaries that followed, view solar primarily as a threat to their business model (and their bottom line)—a serious threat, no less. The "disruptive distributed"…sounds like a rap song, but in reality it is a serious business dilemma, that is just now hitting the media, and promises to be in the headlines for a long time.

And as a matter of fact, in 2014 the departing commissioner Mark Ferron from the California Public Utilities Commission (CPUC), which regulates the utilities in the state, issued a warning and voiced fears in his final report, that the utilities usually would consider solar energy only if forced to do so.

Mr. Ferron said that while "there is no better place to be than California when it comes to energy and climate policy, the CPUC needs to watch the utilities' legalistic, confrontational approach to energy regulation very, very carefully." Not very, but very, very carefully, is his suggestion, which means that things are quite serious.

Commissioner Ferron praised the California utilities for being "orders of magnitude more enlightened than their brethren in the coal-loving states, although," he continues, "I suspect that they would still dearly like to strangle rooftop solar if they could."

Another escalation: from just watching the utilities' anti-solar actions very, very carefully, he is warning the CPUC to take the necessary measures to prevent the utilities' attempt to strangle the solar industry. Murder in the making…?

Mr. Ferron thinks that the relationship between utilities and CPUC is more "cat and mouse" than a partnership, and advises CPUC to be "one smart and aggressive cat." This fear and advice coming from an energy insider dealing with the utilities on a daily basis explains why most solar industry professionals blame the utilities for being the biggest obstacle to the growth of solar in the country.

Quite obviously, utilities representatives say all the right things in public and on TV, while in real life and in regulatory proceedings they are doing everything possible to stop, or at least curtail, solar's advances. The excuse they give most often is that they're not anti-solar, but that they're just looking out for the rights of their customers.

Lately the utilities have been saying openly that solar, and especially the distributed energy resources (DER) represented by thousands of roof-top solar installations, are interrupting their usual MO. This is forcing them to spend a lot of time and money on the problems brought upon the grid and their daily operations by the new influx of solar DER of roof-top residential and small commercial solar installations.

The problems utilities quote often are that DER installations: a) use the power grid infrastructure for free, and b) interfere with normal operations due to solar's variability and need for additional maintenance.

All this, of course, is true and is no doubt costing the utilities additional effort and expenses in controlling and maintaining the grid, and adjusting power generation to compensate for solar's variability. Eventually, this additional work and expense increases the cost of electric power and decreases the utilities' profits.

In fairness, we must agree that the California utilities have seen a huge increase in rooftop solar, with growth rates at over 40% a year during the last several years. There were over 400 new solar installations in San Diego in the month of July, 2012, and over 700 installations in February, 2013. Somebody has to pay the additional expenses these are causing for grid operation and maintenance.

Since the utilities are stuck between the CPUC, the solar installers, and the customers, they are trying to figure out a way to get out of the tangle by stopping, or at least reducing, solar power installations by introducing new rules. In most cases, however, this maneuvering doesn't help either the solar or the non-solar customers.

Sand Diego Gas and Electric (SDG&E), for example, has a top tier rate of around $0.30-0.32 per kWh (as in hot summer peak-power afternoons when the A/Cs come on). At the same exact time, SDG&E is using much lower $0.08-0.10 per kWh cost of electric power generated by DERs and other renewable sources. The end result is that the utility charges their non-solar customers the difference of about $0.20-0.26. This, of course, is an unfair and unsustainable situation that needs to be resolved.

The CPUC is playing an arbitration role in the solar price wars, and in 2012 rejected a rate raise proposal from SDG&E for a network use charge, which would've made PV customers pay additional charges for using the electric distribution grid. But the utility will try again... you can bet on it.

Over 95% of the electricity customers do not have solar on their roofs, and many can barely pay the existing bills, and yet get charged the difference for solar which they simply cannot afford. At the same time, the customers with solar systems on their roofs get paid several times less the going rate. That difference, however, is obviously not enough, so the utilities want to change the game.

The California utilities will continue trying to change the rules of the game to ensure their bottom line. A new cost-benefit analysis from the CPUC on net energy metering is setting the stage for the next battle.

The utilities are stuck between a rock and a hammer—the regulators and the customers. They play cat and mouse with the regulators, and are so good at it that at times it is not clear who is the cat and who the mouse.

The customers, playing the role of a sitting duck, are split into two distinct categories: a) the minority DER solar owners, who get paid much less than the utility rates, and b) the majority poor, non-solar, customers, who shoulder the bulk of the additional charges.

Judging from the lessons of the past, it seems likely the cat and mouse game between the utilities and CPUC will continue for a long time. The sitting duck customers, watching carefully but not being able to do much, will eventually lose the battle. And as usually happens, the biggest losers will be the poorest among us.

More Utilities Games

Another U.S. utility has joined the cat and mouse game with its customers, in a trend started by Arizona's APS in 2013. The largest municipally owned utility company in the U.S., CPS Energy in Texas, is planning to add connection fees and monthly charges to its rooftop solar customers as well.

The plan is to expand the city's rooftop solar program by retaining net metering, developing a solar leasing scheme, building a community solar project, and adding another $21 million to its solar rebate program.

But this will come at a price for the residential rooftop solar customers. CPS's services will cost the solar customers a connection fee of $450 plus monthly charges of $1 per kW generated. The monthly charge is estimated to be set at a maximum of $17.50 per customer over the life of the program.

Commercial customers in Texas will be charged on a case-by-case basis, dependent on system size and other factors.

Now the average household system would have to pay around $5 per month based on the size of a typical residential system. This is similar to the charges implemented by Arizona's APS public utility. The utilities seem united in their decision to shape the game as they feel needed, and there is not much the customers can do.

As a result of the CPS's decision, the combined one-time charge plus the average monthly fee over a year adds to about $510 as needed to permit, connect, and use a rooftop PV system for the first year of operation.

CPS justifies the move as necessary to recover its cost of service. It costs a lot of money to build, maintain, and upgrade the power grid, and since the solar customers use the grid electricity—both for sending power into it and for using its power during hours when their systems are not generating—they must pay their share of the expenses.

During the solar boom-bust cycle of 2007-2012, most utilities saw solar as a passing fancy that would soon fade, as it had before. But this time, surprisingly, solar was able to step up on its own feet, and there is no going back. The advances of late and the support solar enjoys on all political and social levels are undeniable, so it is here to stay...unless the utilities have anything to say about it.

Many utilities are taking action to defend their old and proven business model and their profit, only the methods they use are different. Some utilities prefer a slow and frictionless approach to manage the situation, while others confront the solar invasion head-on.

Solar 2025

On the bright side, and looking from a technical point of view, the new solar products, methods for harvesting and storing solar energy are becoming very advanced, efficient, and cost effective. Because of that, solar is still growing and is projected to become the primary source of energy in the near future.

In the summer of 2014, Thomson Reuters analysts tried to identify and predict the rapid progress and emergent trends in scientific research and technological innovation in the solar industry. They concluded that as the world's population grows and global energy needs increase, renewable energy (solar and wind especially) will remain high on the priority lists.

The good news in 2013-2014 included many companies working on new, or improving existing, solar products. Here is a list of some of the new products revealed recently:

- Q.PRO-G3 series PV module by Hanwha Q CELLS is a lighter and slimmer design that offers superior linear performance and extended 12-year product warranty.

- Series 3 CdTe thin-film PV module by First Solar offers harsh environment capability and conversion efficiencies of 13%. A new record of 21% was announced in the summer of 2014.

Note: Here we must mention that regardless of the performance, CdTe PV modules contain a substantial quantity of toxic, carcinogenic, heavy metal cadmium. Since this is fairly new technology, we must beware of the negative effect of so much poisonous materials spread over large land areas.

- 2nd gen AC PV module by Canadian Solar is a follow-up of the original AC module. It has captured the attention in the US market, but has less success elsewhere.

- Fully integrated EVA cross-linking metrology solution by Meyer Burger and LayTec is designed to eliminate the need for destructive off-line testing.

- SmartWire Connection Technology by Meyer Burger produces fine line Cu interconnects with up to 2,000 contact points per cell, adding 5% power gain and reducing production costs.

- Cluster Controller by SMA Solar, designed for monitoring and control of decentralized PV power plants of up to 1 MW capacity using maximum of 75 string inverters.

- 2nd gen micro-inverter by Enecsys is a new iteration designed to handle higher output modules. It claims over 33% higher power output and is scalable to 300W AC.

- 60-cell PDG5 frameless module by Trina Solar is designed for harsh environmental conditions. It is equipped with PID-free attributes for utility-scale applications and is IEC 61730-2 certified.

- Gallium arsenide nano-material by Sol Voltaics's claims to offer low-cost ink process for solar cells mass production, although a product is as yet unavailable.

- InPassion's ALD system by SoLayTec offers superior Al_2O_3 deposition in mass production of solar cells.

- Perovskite materials and the related processes have shown significant advances in material efficiency gains in the short period of their development. Their use in silicon cells is most promising, where they could increase the cell reliability and efficiency 3-5% or more.

A close look at these developments and improvements, reveals that most of them are quite sophisticated and interesting from a technological point of view. Unfortunately, they represent very small steps in the race with the competition. They also need a lot of time for entering the mass production regime. More importantly, they need a lot of time for efficiency and reliability testing for their long-term use in different environments.

But some new developments in the PV technology sector are expected to allow its quick growth in the main and niche energy markets. The progress would come from some hi-tech areas like chemical bonding research, photo-catalysts, three-dimensional nanoscale heterojunctions, etc. Fabrication of novel heterostructures, such as CO_3O_4-Modified TiO_2 nanorod arrays with enhanced photo-electrochemical properties, as well as modifying donors behavior in bulk-heterojunction solar cells promise a 10-fold increase in efficiency in the future.

At the same time, new production methods, such as mesoscopic oxide films sensitized by dyes, and quantum dots, will increase significantly the present solar conversion efficiency rate of 8-9%. Optimized chemical bonds in PV materials and new photosynthetic processes will make solar energy available when needed—even under low solar insolation and at night (don't hold your breath). New materials such as cobalt-oxide and titanium-oxide nanostructures, photo-catalysts and 3D nanoscale heterojunctions will dominate the different niche markets.

All these developments promise to bring solar out of its perception as just an environmentally conscious undertaking and to spread its use by the masses around the globe. Solar no longer would be a privilege for the rich people in the rich countries, but will be available for the last villager in Asia and Africa.

Solar equipment will become much more efficient and reliable. Thus generated energy will be stored and used during cloudy days and at night. Solar thermal and PV technologies from new materials will heat buildings, and provide hot water and electricity in homes and commercial and manufacturing facilities alike.

These futuristic technologies look good in the lab (and even better on paper), but they have a long way to go to the marketplace. Many of them will find specialized niche markets where they can thrive while being further developed and optimized. Where and what these markets will be is another big subject that needs an entire book to be fully addressed.

Government Subsidies

2017 will be the year that will determine the future of solar power in the U.S. Especially that of large-scale, utility type installations. It is when the investment tax credit (ITC) will drop from 30% to 10% or less. Since the ITC is a buffer—a guaranteed profit margin for any utility solar power plant—its absence in most cases would mean no profit. Because of that alone, the experts predict that there will be almost no new utility-scale solar in the U.S. in 2017 and beyond…unless something drastic happens.

The ITC reduction is going to impact the utility-scale sector in a major way, so the participants are getting ready. We will see a huge boom in solar installations in 2015 to be completed in 2016, in order to benefit from the present 30% ITC. Then, a complete collapse of the utility-scale solar developments will follow as of January 1, 2017.

There are some solar projects with signed PPAs as of 2018 and 2019, which are likely to complete construction in 2016, in order to get the ITC, but they face a gap period until the PPA kicks in. For that, plan to bridge the few difficult years. There are two ways to do that:

— A "merchant nose" is a solar project, which counts on selling wholesale power to potential buyers for a few years and until the PPA kicks in. This is risky proposition that requires a lot of effort, in the hope of getting good returns until an appealing long-term PPA is offered by the utility. Merchant nose power plants are already in operation in the US. An example is First Solar's 18 MW merchant plant in Texas in 2014. First Solar, however, is a rich company that can afford this and much more in terms of financing. Not many companies could follow its example.

— A "bridge PPA" is another way, which involves a short-term PPA that might last a few years with a different utility and at a lower rate before the full, longer-term PPA kicks in. This is another risky proposition, since signing a PPA with any utility is a major undertaking. Signing a short-term PPA is not a normal way of conducting large-scale solar business, so we would not advise using this business model, unless you have deep pockets to cover the losses while waiting for PPA. Many projects have failed and many companies have gone bankrupt while waiting for a favorable PPA.

So, 2017 is the start of a new era—that of the U.S. solar power industry standing on its own feet and running the marathon. The only other option is collapsing in the dirt, tired of the long battle against the other, more established, energy sources.

Just like mother eagle in Figure 8-30, who is planning to kick her fledgling off the cliff to teach it how to fly, the U.S. government is ready to kick solar power off the ITC cliff. Let's hope the U.S. solar power industry finds enough strength and ingenuity to fly.

Figure 8-30. The U.S. Government vs. the solar industry movement.

Solar Politics

Since the mid-2010s solar has been very skillfully used as a weapon by political party candidates, as the "solar theme" predominates in political campaigns, especially in the western states. The battle is especially noticeable in election years, but the mud slinging does not help the solar cause.

There were a number of "solar teams" running for seats in different local and state offices in these states, which appealed to the solar crowd, so this approach worked at first. Soon, however, it became apparent that the "solar thing" was nothing but a means to achieve the goal. Once in office, the "solar" candidates realized the complexity of the situation and in most cases abandoned their dedication to furthering the "solar goals" they promised during the campaigns.

On the federal level, there are a variety of obstacles to the progress of solar power. Federal, state and local rules, regulations, incentives, and taxes on new solar projects vary from location to location. Whatever incentives are available at one time or location, for example, are not available at others. In most cases they are sporadically designed, and unpredictably phased in and phased out.

The absence of suitable power infrastructure, and universal feed-in tariffs makes it impossible to integrate excess solar power generated by home solar systems into the grid. This is simply because the U.S. utilities are a one-way-street; they are experts at generating and sending power to customers, but are not set up, and still don't know how, to take it back from the customers. Neither do they want to, unless forced. Here is where the federal government must step in and introduce standards that are acceptable by all parties.

But the biggest and most dangerous setback of solar progress in the U.S. is coming as soon as January 1, 2017. That day will mark a cataclysmic shift in the U.S. solar energy market. It is when the government subsidies are planned to end for good, or are to be significantly reduced. It is when the investors will see no way to make money from their investment in solar energy projects. It is when the U.S. solar industry will either sink or swim.

Such a catastrophe could be prevented, but the government is not interested in solving the solar dilemma. What is needed is a comprehensive nationwide renewable energy program. It must include universal technical, logistical, and financial standards, serious funding for basic solar R&D, and most importantly, clearly outlined, and totally predictable incentives for adoption of solar power and other renewable energy technologies across the nation.

Instead, our energy politics point toward a meaningless "all-of-the-above" energy strategy—the government is washing its hands of maintaining a balance in the energy sector.

Under the new "all of the above" policy, the oil companies are pushing increases in domestic natural gas and oil exploitation, while using the renewables for publicity stunts. Many congressmen do not believe that government has a role to play in promoting renewable energy development, so the oil companies have *carte blanche* for the foreseeable future in most states.

Note: California is an example of how rapidly renewable energy can be adopted when local, state, and federal policies provide the needed push. But this is not likely to happen in most other states, and things in the renewable energy sector there may deteriorate soon.

So, the U.S. solar industry is facing several major hurdles: a) severe competition from the cheaper fossil fuels and Chinese imports, b) the utilities, with their inadequacy to handle the increase of solar power generated by their customers, and c) dysfunctional government politics that threaten the development of renewable energy. Republicans' victory in 2014 will make things worse for solar.

WIND ENERGY

Recently there were two world records documented in the wind power generation sector. Spain's wind farms generated more energy than any other energy source during 2013, which is a remarkable event since it has never happened there before. Wind now has a 21.1 percent share, at 55 GW total power generation, in Spain.

Other remarkable news coming from Spain's wind energy sector is that the price of electricity dropped from $150 per MWh down to $7 per MWh in the span of several short years—a jump of more than 20-fold within 5-6 years. Although not everything is peachy in Spain's wind power sector, this is a good example of the progress that wind could make under appropriate circumstances.

During several months in 2013, Denmark generated more than half of its energy from wind, for a total of 54.8 percent of the total power generation—another world record. In December, 2013, for example, wind power met the country's entire energy demand during several days (of low power demand). Over the course of the year, wind produced one third of the total power used in the country. Remarkable.

There is good news at home too. In 2012, the U.S.'s wind energy capacity surpassed 60 GW, which is enough to power 15 million homes. Wind power added more power to the national grid than any other source, including natural gas. Another record that went virtually unnoticed...

In Texas, during the cold spell in January 2014, wind energy helped the grid to meet the demand increase. Several power plants were shut down at that time, and wind energy from Western Texas' wind farms helped avoid dangerous blackouts in other parts of the state. This is a result of Texas having more wind power plants than most other states.

Most importantly, the average prices of wind power are at a record low of 4 cents per kW/hour—more than 50 percent less than 3-4 years ago. As a confirmation of the upward state of wind power, Warren Buffett invested $1 billion for the installation of 1.0 GW wind turbines in Iowa. If this is the total investment, and if the project went as planned, it means that it cost about $1 per watt installed—another record for power plant construction, which beats all other records.

Figure 8-31. Off-shore wind power plant

It looks good on paper, but for wind power in the U.S., the situation is still precarious. The federal production tax credit, which has provided incentives for wind farm operators to produce power since 1992, expired at the end of 2012. Congress has extended it in the past, most recently in 2009 as part of the federal stimulus package, but no more... So wind is now on its own, and the 2015-2020 will be a trial period for wind power. Estimates call for minimal wind power plant installations in the United States during that period, according to industry specialists.

For renewable energy developers, including wind, the threatened expiration of incentives may even have a few benefits. For example, the wind deadline at the end of 2012 caused customers to accelerate decisions.

So, 2012 was a banner year for wind, and maybe the largest thus far in the U.S. in terms of number and size of installations. But that is history, and the treat of a dramatic fall-off looms over the entire alternative energy industry now.

Provided that Congress renews the production tax credit for another four years, wind power will probably reach 6 percent of the country's electricity supply by 2020, according to industry insiders. This is up from 3 percent thus far.

Background

Wind power generation is the cleanest technology with a bright future, while at the same time the conventional power generators (coal, natural gas, crude oil, and nuclear power) are going through critical cyclical changes. The most obvious of these is the increased awareness of damages done by these energy sources. It is estimated, for example, that $500 billion in premature deaths, asthma, emphysema, heart disease, cancer and other health problems are caused by coal use. The hidden costs from damages caused by hydro racking, oil use, and nuclear accidents are several times that much.

These costs are invisible, because they are usually paid by different methods—in most cases by taxpayers. This cost is in addition to over $50 billion in annual subsidies (and many other expenses) going to fossil and nuclear companies, which are also paid by the U.S. taxpayers.

Wind energy gets no such breaks. As a matter of fact, Congress is still to renew the Production Tax Credit (PTC) for wind, which is one of the few tax incentives that support the clean energy industries. Without PTC, wind would have a harder time competing with the conventional power generators.

In addition to providing cheap and clean energy, wind is responsible for the creation of 80,000 jobs in the U.S. during installation and operation. At the same time, ¾ of the wind turbine parts are manufactured domestically, which also creates jobs in many local industries.

The global wind energy industry has been going through the ups and downs of the energy sector during the last several decades, but saw a sharp rise since 2004. In 2007, its best year yet, the wind industry installed close to 20 GW worldwide—an increase of 31 percent over the previous year.

The U.S., China and Spain led the progress at the time, bringing the total global installed wind power capacity to about 95 GW. The top five countries in terms of installed capacity were Germany (22.3 GW), the US (16.8 GW), Spain (15.1 GW), India (7.8 GW) and China

(5.9 GW).

The ratio is changing today, with China taking the lead with over 91 GW—an almost 20-fold increase in less than a decade—while the other countries are growing at a much slower pace. The fast trend of wind power growth in China is expected to continue through 2020.

Table 8-6a. Global wind power capacity in 2013 (in MW)

Country	Wind capacity
China	91,424
United States	61,091
Germany	34,250
Spain	22,959
India	20,150
United Kingdom	10,531
Italy	8,552
France	8,254
Canada	7,803
Denmark	4,772
Rest of world	48,351
Total	**318,137**

Note: In 2011, the 781 MW Roscoe Wind Farm in the U.S. was the world's largest onshore wind farm. The Thanet Wind Farm in United Kingdom was the largest offshore wind farm at 300 MW, followed by the 209 MW Horns Rev II in Denmark. The United Kingdom remains the world's leading generator of offshore wind power, followed by Denmark.

In terms of economic value, the global wind market in 2007 was worth about $37 billion in new power generating equipment, and attracted $50.2 billion in total investment. Europe was the leading market for wind energy, and new installations represented 43 percent of the global total.

The U.S. was the second largest wind power market in terms of cumulative installed capacity base and annual capacity addition in 2012, but the economic slowdown and the nation's lack of long-term policy certainty have caused a 43 percent drop in installations.

By the end of 2012 the nation recorded annual wind power installations of 13.1 GW, and reached a cumulative installed capacity of 60 GW. The experts predict that growth in U.S. wind power installations will be slow up to 2020, given the expiration of the Production Tax Cred-

it (PTC) after 2013.

The extension of the PTC is a political decision that will impact the wind energy market significantly. The situation is getting even worse, due to the low cost of natural gas for power generation found in the U.S. presently.

In any case, the U.S. wind market will not fall to zero, as there are other support mechanisms for wind in the U.S., including the state-based renewable portfolio standards which require an increasing proportion of electricity consumed to be renewable. This would force some utilities and municipalities to add wind energy to their energy mix. Wind will benefit from the state standards, no doubt, but it is unclear where, when, and how much.

GW vs. GWh

Here is a concept that needs to be understood, if we are to make the right decisions. Installed capacity (in GW) is one thing; it shows how much maximum power can be produced at any time. The actual wind-power generated electricity (in GWh) tells us how much electric power was generated during each hour. It depends on a number of factors and variables. Some of these are wind availability, equipment availability, grid operation and use efficiency.

Wind power generation basically depends on the availability of wind at the wind power plant location. Because the availability and speed of the wind vary with the seasons and the weather, wind is considered a variable power generator, a condition causing problems for the utilities.

It is obvious from Table 8-6b that the U.S. leads the world in the amount of generated electricity, even though China has the largest installed wind capacity. This is actually a significant lead—about 35%, everything considered—that shows that the U.S. is serious about wind power and knows how to generate and use it.

This discrepancy is mostly due to superior conditions in the U.S., as far as the efficiency of design, installation, operation and use of wind power plants and the related infrastructure. The power grid infrastructure makes a huge difference, and is where the U.S. has a significant advantage.

China is the most ambitious of all wind game players; the government plans to order its electrical companies to source up to 15 percent of all of their power from wind turbines by 2020. But China has a big problem—a large number of wind power plants were installed in areas with no access to the grid, so the generated power is partially used, or not used at all.

Table 8-6b. Wind-power generated electricity in 2012 (in TWh)

Country	TWh	%
United States	120.5	26.2
China	88.6	19.3
Germany	48.9	10.6
Spain	42.4	9.2
India	24.9	5.4
Canada	19.7	4.3
UK	15.5	3.4
France	12.2	2.7
Italy	9.9	2.1
Denmark	9.8	2.1
Rest of world	67.7	14.7
World total	**459.9**	**100%**

Note: There are already a number of ghost cities in China, and following that pattern, China is now adding ghost wind farms to the picture.

The countries in the European Union had 117,289 MW of installed wind power in 2013, or almost as much as the U.S. Denmark generates more than a quarter of its electricity from wind, while 83 countries around the world are using wind power on a commercial basis.

Presently there are about 200,000 wind turbines generating electricity around the world on 13,600 large and small wind farms. About 950 off-shore wind power plants generate over 280 GW of electric power.

In April 2013, California set a record in wind power production with 4.2 GW peak power produced at a time when the entire California power system used 24 GW of electricity. This means that wind provided about 17.5 percent of the state's entire power needs at the time... albeit for a short while.

Another 135 GW of potential wind production await development and connection to the grid, according to industry data, while the U.S. DOE reports suggest that by 2030, about 20 percent of America's energy could be generated by wind.

Unfortunately, there hasn't been a lot of investment in the U.S. power grid for the last two decades, and the overall expansion of wind power is hindered by lack of grid lines in the areas of the country with great wind resources. The grid was built a long time ago to service coal, nuclear and hydropower plants without wind or solar power energy in mind. Since wind power plants are located in remote areas, and operate intermittently, it is difficult and expensive to get the electricity from those remote areas to the national grid.

There is no money for grid upgrades, but at the same time the U.S. government subsidizes the fossil industries to the tune of up to $50 billion annually.

Similar investment in clean energy would put the renewable energy markets on their feet, ensure our energy independence, and speed up the fight against climate change. But this is not possible without sufficient political will, so the abundant natural renewable resource will remain under-utilized awhile longer.

Regardless of all obstacles, the U.S. and global wind power industries are growing, and the future looks bright. According to new studies wind can provide half of the world's energy in the future.

About 4 million turbines installed at strategic places around the world, could generate about 8.0 terawatts of energy annually, which is more than half of the power fed into the global power grids.

Of course, this is not easily done. Some of the major obstacles would be:

— The need for over 10 million acres of suitable land,
— Thousands of miles of new transmission lines, and
— Billions of dollars in new investment.

Where would the wildlife and people who now live on these 10 million acres go? And/or how would they get along with the huge monsters swinging their giant hands day and night and making dreadful noise. Where would the raw materials for 4 million windmills come from (there is already scarcity of some special materials)? How long would it take to install this large amount of windmills and related distribution lines? Where the money for the land and the lines come from? These are questions for future generations to answer.

The idea of half of the world energy being provided by wind power is a bit far fetched, but doable at its extreme. The reality is somewhere in between, and the choice is ours. We must agree that wind is not the most reliable energy source...for now at least. Sometimes it blows, sometimes it doesn't.

So yes, there are reasonable concerns about the use of wind power as a reliable energy source, but it is not the only one we can rely on. Wind can be very successfully used in combination with other renewable energy sources, such as solar, hydro, or ocean power.

The successful development of energy storage solutions would be the ultimate solution to wind's problems, making it less variable. But that development is also in the future.

The Wind Power Future

In 2012, Greenpeace International and the Global Wind Energy Council released a bi-annual report on the future of the wind industry. It shows that wind power could supply up to 12% of global electricity by 2020, creating 1.4 million new jobs and reducing CO_2 emissions by more than 1.5 billion tons per year, more than 5 times today's level.

By 2020, the IEA's New Policies Scenario suggests, total capacity could reach 587 GW, supplying about 6% of the total global electricity. The GWEO Moderate scenario suggests that it could even reach 759 GW, supplying 7.7-8.3% of global electricity supply. The Advanced scenario suggests that with the right policy support wind power could reach more than 1,100 GW by 2020, supplying between 11.7-12.6% of global electricity, and saving nearly 1.7 billion tons of CO_2 emissions. And by 2030, wind power could provide more than 20% of global electricity supply. The potential is undeniable, as are the barriers. Yet, with over 300 GW installed capacity in 2013, wind power is now a major source of clean power and a major contributor to emission reductions.

Wind power also saves lots of fresh water which the other energy sources can use to generate electricity.

Note: A large nuclear plant uses a million gallons fresh water a minute. Got that? One million gallons every minute is pumped up into the reactors for cooling the core. This is 1 x 60 x 24 x 365 = 525.6 billion (with a B) gallons annual water use. Some of this water is returned (hot) from the water body it came from, increasing its temperature on the long run. But a significant portion of the cooling water evaporates in the heat exchangers and goes up in the clouds to change that environment as well.

Wind uses no water, and causes no negative effects on the atmosphere, so this alone is a unique, attribute, which is increasing in importance as the world's draughts continue. Only wind and solar (PV) power possess this advantage, making them even more attractive options in an increasingly water-constrained world.

Wind power provides a number of environmental and social benefits, such as zero CO_2 emissions, zero water use, no air pollution, and no soil or water contamination—all at no additional cost.

Wind power is an indigenous energy source, which could be particularly useful to countries poor in natural resources, and/or burdened with large fossil fuel import bills. Wind power is now competitive in an increasing number of markets, even when competing against heavily subsidized "conventional" energy sources.

No doubt, wind energy is safe, useful and practical, so it is going to play a major role in our energy future, but for wind to reach its full potential it needs help. The level of uncertainty surrounding the wind energy sector is still great, so the specialists foresee several different scenarios for the wind industry's future. Each of these scenarios depends on factors related to government decisions, investment scenarios, energy needs and prices.

Wind Power Developments

Technologically, wind is an intermittent and unreliable power source, so the utilities are rethinking their acceptance of wind power in their power generation mix. New technologies, such as large-scale energy storage, are needed to stabilize wind power generation. These technologies, however, will not be available or economically feasible anytime soon, so the only way to remediate the wind's intermittency is to combine it with other power sources, such as solar, hydro, and others.

Financially, wind power is subject to governmental policies and regulations, and since both the global political and financial sectors are still in turmoil, wind's future is full of uncertainties and risks. The long-term success of the wind industry requires stable, long-term policy, sending a clear signal to the investment community that it has the governments' support. In any case, wind power is a valuable constituent of our future energy security.

U.S. DOE Wind Program

The latest U.S. Department of Energy (DOE) Wind Program, developed in cooperation with the U.S. wind industry, is revisiting the 2008 DOE "20% Wind Energy by 2030" report. In it, DOE is developing a renewed vision for the U.S. wind power research, development, and deployment.

The new initiative includes:
— Analysis of the U.S. wind industry progress and the effect of recent trends,
— Analysis of wind power costs and benefits affecting the nation's energy sector,
— A roadmap to optimizing the levels of wind power in the country.

The objectives of the initiative, in addition to analyzing the current status of the wind industry, are to:

— Provide leadership in developing a practical long-term vision for the U.S. wind power industry,

— Analyze a range of attainable industry growth scenarios,

— Provide best available information to address stakeholder concerns, and

— Provide objective and relevant information for use by policy and decision makers.

The U.S. DOE is developing a "Vision for the U.S. Wind Industry," with the realization that it is only as powerful as the key players behind it. The DOE Wind Program has over 150 participants from different wind energy companies and organizations. These include equipment manufacturers; environmental organizations; federal, state, and local government agencies; research institutions and laboratories; associations; and foundations.

The different Wind Vision Task Forces within the group have been organized around key topics to ensure breadth and depth in the approach to the future of wind power in the U.S. Each task force is led by one of the Wind Program's nine funded labs and is working to address key industry issues or considerations.

Task Force topic areas include wind plant technology, manufacturing and logistics, operations and maintenance, project performance and reliability, transmission and integration, wind power project development (siting and permitting), scenario modeling, market data and analysis, offshore wind, and roadmaps.

This is a sign that wind power in the U.S. has (some) government support that it could count on for growth.

European Wind Power

During the latest solar boom-bust cycle of 2007-2012, most European leaders took pride in thinking that they lead the way towards the era of renewable power generation and low-carbon emissions. Now the game has changed to where they seem more worried about the cost of the fuels than anything else.

Between 2005 and 2012, gasoline prices fell by 66% in America, even after the sharp spike during 2008-2009. At the same time gasoline prices rose 35% in Europe. This means that while the U.S. is forecast to become a net exporter of energy in the next few decades, Europe's dependence on foreign fuels and electric energy will only grow.

This threatens Europe's competitiveness and in-creases worries that energy-hungry European factories are packing up and fleeing to the U.S. and other countries with cheaper and more reliable energy. So, confusion is setting up among European leaders who don't know how to handle the developing (and ever changing) energy market.

The most difficult aspect of the problem is the growing web of national incentives, where power companies chose to trade even more energy in wholesale markets. German taxpayers, however, do not want to subsidize French or Polish companies, so the controversy deepens.

And there are the technological issues. For example, on Sunday, June 16, 2013 Germany's solar panels and wind turbines generated a record 60% of Germany's electricity…albeit it was on a slow weekend afternoon. At the same time, France and Belgium were generating lots of nuclear power that could not easily be cranked down, so for several hours during that remarkable afternoon, power generating companies had to pay customers to take their surplus power.

Negative wholesale prices have become more common as European countries turn to renewables too. Germany became a particularly interesting example after the forced shutdown of its nuclear power. Such events, and the addition of a lot of renewable power, have changed the way power companies do business. At times Germany generates too much power, as during the afternoon of June 16th, while at others it sucks power from nuclear plants across the border in France, Poland, and other neighbors. German customers still are faced with the risk of blackouts when the sun does not shine and the wind does not blow.

Still, Germany is shaping up as the leader of the pack today—energy and otherwise. As Europe's biggest energy consumer and largest producer of renewable energy, its power grid would be at the heart of any upgraded pan-European network. Inevitably, Germany will have the last say in gas and electricity distribution plans.

Many of the German neighbors, such as Poland and the Czech Republic, however, are starting to complain that power surges from Germany are playing havoc with their grids too. This is a strange, and we must add unexpected and unplanned for, consequence of subsidized renewables.

Another unexpected and unwanted effect is that now governments have to pay the power companies to produce electricity from fossil fuels to ensure that back-up power is available. All this because the sun doesn't shine all of the time, the wind doesn't blow all of the time, and we haven't figured out how to combine these as a best solution for uninterrupted power generation.

Now Europe is switching to burning more heavily polluting coal instead of the cleaner and more flexible gas, because coal is cheap, the gas market is far from liquid, and the carbon-emissions system is broken.

Another oddity of late is that Germany pays some of the lowest wholesale prices for electricity in Europe, yet at the same time offers some of the highest retail prices. Consumers are forced to pay a number of unusual charges such as network fees, taxes, and ever-growing charges for subsidizing renewable energy projects.

The German energy system also encourages heavy users, making them exempt from some charges. And while politicians are in disagreement on how to proceed, the energy system will be going as is for a while longer. Abandoning the renewables is not one of the solutions, as the Spain case shows. Spain has huge renewable power from sun and wind, so much so that now government subsidies are putting a large hole in the country's ailing economy. Spain, however, cannot sell much of this bounty because there is no adequate distribution network to its neighbors.

If all the European grids were linked up properly, like in the U.S., they would form a large integrated energy market. The peaks and valleys of power demand and supply could be easily handled, as in the U.S. This would have required forward thinking. For example, instead of installing solar panels in cloudy Germany, they would be much more efficient in sunny Greece. It's the same with wind power generation, which could be concentrated in the windiest parts of the continent, instead of here and there according to the availability (or not) of grid connection.

Cross-border cooperation has been discussed since the creation of the European Union, but full integration is far off. Eastern Europe still has "energy islands," especially the ex-Soviet Baltic states, vulnerable to blackmail from Russia. Done properly, an integrated EU energy market could favor the transition to renewable power, enhance security and promote cheaper energy.

What does all this mean for European wind power? It simply puts it in a very disadvantageous position. Wind power plants are small in size and built on remote locations, usually far from the power grid. This causes fragmentation of the wind industry, making it inefficient and the generated power expensive.

As we enter the era of cheap gas, and until Europe integrates its power grid, wind power on the Continent will grow, but much slower than in the past.

China Wind

China is entering a new phase of serious plans for reducing its carbon footprint and increasing electricity production in rural areas. According to the specialists, China has doubled its cumulative wind capacity each year between 2006 and 2011, at a compound annual growth rate of 76 percent.

China, number two market after the U.S., and together with the other top markets Germany, the UK, Italy, Spain and India, accounted for 75% of global installed wind capacity in 2012.

China's plan has been focused on two goals: a) to support the development of a Chinese manufacturing base for wind turbines and other equipment which can then be exported, and b) to connect more renewables to the grid to reduce carbon emissions from electricity generation.

The progress of the Chinese wind power-related equipment manufacturing and increased exports are a testament that the latter goal has been achieved, where several Chinese manufacturers are now global players and Chinese wind products exports are unrivaled. The U.S. helped China to further its wind manufacturing by purchasing billions of dollars of Chinese-made windmills during the wind boon period of 2007-2012.

That growth began to slow in 2012 as some aspects of the nation's wind power strategy were proving challenging and the global markets started to slow. The biggest problem with wind (and all renewables) in China is connecting the new power plants to the grid. Because of that, it has been unable to fully accommodate the already installed wind capacity. This is especially serious in remote areas where wind yields are at their highest.

So, wind power plants output is often disconnected from the transmission system and simply wasted, when supply exceeds demand, or in cases of system congestion. This is a major technological problem and a major economic dysfunction and disincentive for any wind project. Because of that, only the most economically favorable projects are financed today, usually with a lot of government help.

China's wind industry will continue growing, but in smaller steps at least until the power grid infrastructure is upgraded to integrate existing and new wind generation. This effort, however, might take several years, or even decades, which will certainly limit the size of the Chinese wind power market as compared to the recent past glory days.

This unexpected (due to poor planning) obstacle in the internal wind power development is forcing the Chinese wind equipment manufacturers to focus on increasing their exports to keep the production plants' doors open. This in turn might result in another cheap Chinese products dumping, similar to the solar panels dumping in the near past. So, we should be looking for some significant price competition and even global price wars over the next few years and until the global wind power markets stabilize.

The experts' estimates are that the Asia-Pacific wind power market, led by China, India, Korea, Thailand, and the Philippines, will continue to grow slowly by 2020. Budding wind markets in Argentina, South Africa, the Philippines, Ukraine, Brazil, Korea and Mexico are expected to expand rapidly during the same time.

Combined Wind and Solar

The future of the renewable energies is not clear, mostly because of capital investment uncertainties, and the competition from the conventional fuels. But even more importantly, they are seriously disadvantaged and cannot compete well because of their variability. As discussed above, the utilities are concerned with the variability problem (solar and wind in particular), so a lot of work and money will go into finding solutions.

One of the immediate solutions might be seen in combining wind and solar operation. One example is the wind and solar plant by Element Power in New Mexico. According to insiders, it happened almost by accident.

There were plans for two different projects—wind and PV—and they were, as a matter of fact developed separately. The demand for renewables in that part of the country is immense, and it just happened to be in a place where both solar and wind can be operated at maximum capacity. The synergy is not just in sharing land and resources, but in conveniently spreading out peak time power generation, which provides much more power than wind or solar can produce by themselves.

The wind power plant is composed of 28 Vestas V100 turbines, 1.8 MW each, which were installed with labor, services and materials from the local community. This cooperation always brings prosperity to the locals, which helps build trust between the parties.

The solar plant is a 50-megawatt PV installation, using conventional solar modules. The solar plant also has a PPA with the local utilities, so the power produced by both plants will be pumped into the local power grid, via a nearby substation, which was built for this purpose.

The fact that there is excellent solar radiation, and good winds at the right times, enables both wind and solar, to be cost-competitive sources of energy for the local utilities. Combining these two power generating sources

is also a big plus which brings more profit and contributes further to the success of the project.

Transmission system operators are always concerned with managing wind and solar variability, but since these two plants are near each other the task is somewhat simpler. Since wind and solar generate at different times, there is alternating of the generated power; i.e., wind feeds the transmission lines 30-40% percent of the time, but is usually weak during the mid-day hours. Solar kicks in exactly during those hours, thus generating power when it is most needed—during the noon peak hours.

In general, the combination of wind and solar power generation reduces the variability, which these power sources experience when operating separately. This is a good solution to a problem that cannot be solved by any other means (for now). It would be welcomed by grid operators too, if and when implemented in large-scale installations.

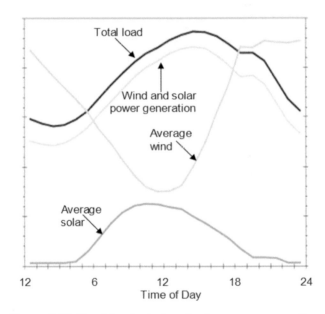

Figure 8-32. Combined wind and solar power generation

Solar energy generation could also be combined with other energy sources when constant output is the goal. Addition of solar power to conventional power plants is one approach. Using energy storage devices (batteries or water storage) is another.

A more natural and most efficient way, however, is combining PV power with wind at locations especially chosen for this purpose. At such locations PV power is complementary to the output of wind generation, since it is usually produced during the peak load hours when wind energy production may not be available. Variability around the average demand values for the individual characteristic wind and solar resources can fluctuate significantly on a daily basis.

However, as illustrated by Figure 8-32, solar and wind plant power generating profiles—when considered in aggregate—can be a good match to the load profile and hence improve the resulting composite capacity value for variable generation.

In this example, the average load (upper line) is closely followed during the day by the average output from the combined wind and solar generators (the second from top line) during the same time. This average is created regardless, and because, of the fluctuations of the individual wind and solar power generators. This is a marriage made in heaven, and this combined power generating combination will work well if wind and solar power outputs can be matched as closely as this one.

Although there are areas in the US and abroad that match this wind and solar profile, the combined effort is usually hard to execute, because the best places for wind and solar are at different locations, often miles apart. There is also lack of infrastructure at some of the most suitable locations, which requires expensive new grid lines, so combining wind and solar is a costly undertaking.

Because of that, it will require great effort to implement large-scale "wind-solar load matching" schemes anytime soon with the existing technologies. Having as a goal the matching of wind and solar power outputs will force us to find the most suitable locations and appropriate technologies for this match. This won't happen overnight, but if we approach this solution seriously, we will have a large-scale power output—nationally—that matches the grid power loads.

Solar and wind combinations are not uncommon in residential and small commercial applications. The combo solar panels and small wind turbine, for example, are used to maximize local resources, provide power for remote locations, pump water, run appliances. etc.

On a larger scale, China's biggest power network operator, State Grid Corp, installed and is operating a 140-megawatt combined wind-solar hybrid project, where 100 megawatts of wind and 40 megawatts of PV solar are installed and operating in tandem since January, 2012. It also has a 20-megawatt battery storage capability, which provides even more stability to the hybrid power generation.

This is the largest on-line utility-scale solar-wind hybrid project in the world, but we are sure that it is not the last.

A smaller example is the Western Wind's fully integrated 10.5-megawatt hybrid system in Arizona, which consists of five 2.0 MW Gamesa turbines and a 500 kWp c-Si PV array. There are a number of plans to develop similar projects—either new installations or retrofitting existing solar with wind power generation capabilities, to reduce the harmful variability issue.

Presently enXco's Pacific Wind/Catalina Solar project is the largest wind-solar combo in the US. It has seventy 2 MW REPower turbines, combined with the largest Solar Frontier CIGS solar field in the world. It will be used to test the effectiveness of feeding the grid simultaneously with both energy resources.

We do foresee the wind-solar hybrid generation as one of the most reliable and widespread alternative energy generators in the 21st century. The combined action of these two power sources, with added energy storage, will eliminate (or significantly reduce) the generated power variability, which is one of the major barriers to the development of large-scale power fields today. The wind-solar hybrid is also the surest way to successfully compete with the conventional energy sources in the near future.

The New Problem

While solar and wind could work together very well in the fields, thus reducing the overall variability of the two energy sources, some of the major solar and wind power plant owners are on the war path. EDF and E.ON are two of the largest offshore wind project developers in Europe. In the summer of 2014, both backed the UK Department of Energy and Climate Change's (DECC) plans to cut support for large-scale solar under the country's Renewable Obligation program (RO).

Currently, large-scale solar shares subsidies with offshore wind and a number of other renewable technologies. The new RO states that beginning in 2015, solar projects over 5MW will only be able to apply for support through the new Contracts for Difference scheme (CfD). This puts solar in direct competition with onshore wind, thus the friction. Wow! The most promising solution to the variability problems of wind and solar out the window! Just like that.

It appears that the unprecedented success of solar installations on the Continent has forced the government into considering solar as a threat to the balance of the Levy Control Framework (LCF). So, removal of large-scale solar would leave more money in the LCF for offshore wind and other projects.

EDF and E.ON, of course, support measures to control budget spending on rapidly deployable technologies such as large-scale solar PV projects, in order to ensure cost control for the government and affordability for consumers. All this for fairness sake, according to EDF, "Effectively managed LCF is important to ensure that some technologies are not unfairly disadvantaged through other technologies deploying rapidly." Got it; wind is afraid of solar. This is a new twist, which we did not see coming (not a good start for a promising long-term relation). Let's hope it is the exception.

So the inside battle has started, and all parties in the UK renewable field are in agreement; solar has no place in the government subsidy programs. The best possible solution for reducing the variability of solar and wind power generators just went down the drain... at least in the UK.

ENERGY STORAGE

So close and yet so far, energy storage is the key to serious development of solar and wind power in the U.S. and the world. It is badly needed to eliminate the variability and reliability of these technologies.

Energy storage technologies seem so simple at first glance, but their full development and implementation have proven that to be wrong. After many years of talks, they are still on the design boards and many companies are still working towards their commercialization.

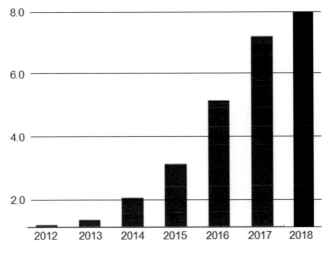

Figure 8-33. PV energy storage (in GW)

Energy storage can be done using different technologies, but none has proven cost-effective and/or reliable to date. There are efforts to develop and prove some of the existing technologies, as well as developing new such, but these are far behind the estimates.

While there are some successes with demo equipment and small energy storage sites, the success is far from proven. This is especially true for large-scale ener-

gy storage. There is a lot of equipment needed to store 10 MW of electricity, for example, and even more is needed to store 100 MW. This is also a very expensive proposition—from the initial investment, to maintenance, and end-of-life disposal.

The technology to cover all aspects of energy storage is still immature, unreliable, and extremely expensive, so in our estimate, large-scale, cost effective and reliable energy storage technology will not be available for the next several decades. Because of that, solar and wind will have to proceed and survive without it. This won't be easy, because the utilities simply cannot easily handle the variability and intermittency of solar and wind power without energy storage and back-up options.

Energy Storage Options

The reliability and flexibility of our future electric distribution system depends largely on the effective application of renewable energy and energy storage devices. Renewable energy storage lets electric energy producers send excess electricity over the electricity transmission grid to temporary electricity storage sites that become energy producers when electricity demand is greater, optimizing the production by storing off-peak power for use during peak times.

Many renewable energy technologies such as solar and wind energy cannot be used for base-load power generation as their output is much more volatile and depends on the sun, water currents or winds. Batteries and other energy storage technologies therefore become key enablers for any shift to these technologies.

The power storage sector generally includes traditional batteries, but also covers hydrogen fuel cells and mechanical technologies like flywheels, water storage, etc., that are straight potential replacements for batteries. More and more research is also conducted in the field of nanotechnology as ultra-capacitors (high energy, high power density electrochemical devices that are easy to charge and discharge) and nano-materials could significantly increase the capacity and lifetime of batteries.

The replacement of fossil fuels as the primary sources of energy for electricity generation and transportation will take place over the next few decades. At that time, growing penetration of renewable energy sources and a shift, hopefully, to plug-in hybrid electric vehicles (PHEVs) and all electric vehicles (EVs) will require a much more dynamic electric infrastructure.

Beginning with the U.S. DoE "Grid 2030 Vision" report, energy storage emerged as a top concern for the future. In 2007 the DOE convened an Electricity Advisory Committee (EAC) to make recommendations for an energy road map for the U.S. including energy storage. The EAC produced a report to the U.S. Congress, which provided a road map development of storage technologies and goals for storage deployment in the U.S. grid over a 10-year period.

Globally, other nations like Japan and Germany have been working to make larger amounts of energy storage a vital part of their energy plan. Japan has a near-term target of 15% storage in the grid with Germany planning 10% compared to just over 2% in the U.S.

Electric Energy Storage Devices

There are several categories to consider when classifying energy storage units: size and weight, capital costs, and life efficiency. Size and weight of storage devices are important factors for certain applications. The influence of size and weight with respect to output energy density as well as the values of various storage devices can be seen in Figure 8-34a.

However, the electrically rechargeable types, such as zinc-air batteries, have a relatively small cycle life and are still in the development stage. The energy density ranges reflect the differences among manufacturers, product models and the impact of packaging.

While capital cost is an important economic parameter, it should be realized that the total ownership cost is a much more meaningful index for a complete economic analysis. For example, while the capital cost of lead-acid batteries is relatively low (as shown in Figure 8-34a), they may not necessarily be the least expensive option for energy management due to their relatively short life for this type of application.

Note: The costs in Figure 8-34b have been adjusted to exclude the cost of power conversion electronics. The cost per unit energy has also been divided by the storage efficiency to obtain the cost per output (useful) energy. Installation cost also varies with the type and size of the storage. The information in the charts is to be considered a guide, not detailed data.

Efficiency and cycle life are two important parameters to consider along with other parameters before selecting a storage technology. Both of these parameters affect the overall storage cost. Low efficiency increases the effective energy cost as only a fraction of the stored energy could be utilized. Low cycle life also increases the total cost as the storage device needs to be replaced more often. The present values of these expenses need to be considered along with the capital cost and operating expenses to obtain a better picture of the total ownership cost for a storage technology.

— Advanced battery technologies have been the main focus in distributed storage systems

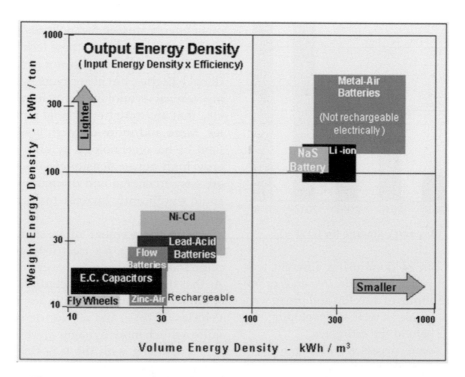

Figure 8-34a. Storage devices' weight per unit energy density (U.S. DOE).

followed by flywheels, super-capacitors, and compressed-air energy storage. Studies conducted in the past to determine the effectiveness of fast-response batteries and flywheels in frequency regulation applications showed these systems could affect frequency control with approximately 40% less energy as compared to fossil fuel plants because of the very fast response time (cycles versus minutes to respond).

Based on these findings small size (up to 20-30 MW) energy storage plants are being built in the U.S. and other countries. In the wind turbine industry super-capacitors have been widely adopted for powering the pitch control of turbine blades and offer back-up energy to safely shutdown a wind turbine if loss of power occurs. Super-caps are being applied as well with small solar arrays to insure smooth power flow as clouds pass over.

Lithium-ion batteries, which have achieved significant penetration into the portable/consumer electronics markets and are making the transition into hybrid and electric vehicle applications, have opportunities in grid storage as well. If the industry's growth in the vehicles and consumer electronics markets can yield improve-

ments and manufacturing economies of scale, they will likely find their way into grid storage applications too.

Developers are seeking to lower maintenance and operating costs, deliver high efficiency, and ensure that large banks of batteries can be controlled. As an example, in 2009, AES Energy Storage and A123 Systems announced the commercial operation of a 12 MW frequency regulation and spinning reserve project at a substation in the Atacama Desert, Chile. Continued cost reduction, lifetime and state-of-charge improvements, will be critical for this battery chemistry to expand into grid applications.

— *High-speed flywheel energy storage systems* consist of a massive rotating cylinder (a flywheel attached to a shaft) that is supported on a stator by magnetically levitated bearings. To maintain efficiency, the flywheel system is operated in a vacuum to reduce drag. The flywheel is connected to a motor/generator that interacts with the utility grid through advanced power electronics. This is an expensive, but functional system with unlimited applications.

Some of the key advantages of flywheel energy storage are low maintenance, long life (20-30 years or

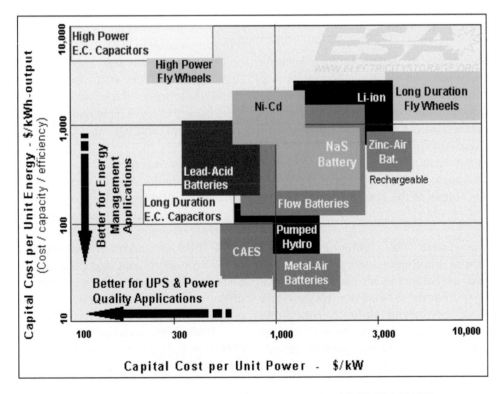

Figure 8-34b. Storage systems $/output vs. unit's $/kW (U.S. DOE)

tens of thousands of deep charge-discharge cycles), and negligible environmental impact. Flywheels can bridge the gap between short-term ride-through power and long-term energy storage with excellent cyclic and load following characteristics.

Currently, high-power flywheels are used in many aerospace and UPS applications. Today 2 kW/6 kWh systems are being used in telecommunications applications. For utility-scale storage a "flywheel farm" approach can be used to store megawatts of electricity for applications needing minutes of discharge duration. Currently several "flywheel farm" facilities are in the planning or construction stages to sell regulation services into open ISO markets.

— *Compressed-air energy storage (CAES)* uses off peak electricity to compress air into either an underground structure (e.g., a cavern, aquifer, or abandoned mine) or an above-ground system of tanks or pipes. The compressed air is then mixed with natural gas, burned, and expanded in a modified gas turbine. In a conventional gas turbine, roughly two thirds of the power produced is consumed in pressurizing the air before combustion. CAES systems produce the same amount of electric power as a conventional gas turbine power plant using less than 40% of the fuel.

Recent advancements in the technology include above-ground storage in empty natural gas tanks and "mini-CAES," a transportable technology that can be installed at or near individual loads (e.g., on urban rooftops). The first commercial CAES was a 290-MW unit built in Hundorf, Germany, in 1978.

The second commercial CAES was a 110-MW unit built in McIntosh, Alabama, in 1991. Several more CAES plants are in various stages of the planning and permitting process.

— *Grid energy storage.* The challenge in developing a more intelligent electricity network (smart grid) is balancing all of the variables associated with dynamic load control powered from an ever increasing variable (renewable energy) sources. This "balancing act" can be made simpler with small amounts of energy stored throughout the grid.

A specific example of storage in a smart grid is the concept of placing small amounts of energy storage (1-2 hours) on the feeders of residential areas. American Electric Power applied this idea by developing the community energy storage (CES) concept. CES units are placed at the very edge of the grid allowing for ultimate voltage control and service reliability.

As more and more sophisticated electronic loads, computers, appliances, etc. are added by customers who demand greater service reliability, new even larger loads will be added randomly in the grid. On top of these changing load patterns more and more solar arrays on rooftops will introduce a growing amount of energy flowing back into the grid when solar generation exceeds the power demand of the specific customers.

Today, a neighborhood with a significant number of solar roofs generates a fair amount of energy that dissipates back into the utility network during solar peak periods. Since the solar peak precedes the customer load peak by two to three hours each work day, it is desirable to store that energy or use when the load grows later in the day.

Current renewable energy storage systems have varying degrees of variability and uncertainty, and the output characteristics of the associated technologies vary substantially. However, research in this field is advancing rapidly. It is likely that a solution for small-scale renewable energy storage will become a reality within the next few years.

Reliable, compact, and cost effective storage systems are needed in the future to store and release energy on demand at solar and wind power plants. The future smart grid systems will also demand reliable energy storage solutions, which can be provided in large quantities on a grid level. This makes grid energy storage the most practical and urgent.

Grid Energy Storage

Grid energy storage seems to be the most practical solution to all energy storage problems. It is not simple or cheap to do, but we see an emerging market in the U.S. for grid energy storage devices, processes, and services. Their full and timely development might be a shot in the arm for the old and tired U.S. power grid and its owners—the utilities. It might provide the frustrated utility operators with a new opportunity for increased profits and much-needed recognition.

Note: This is a good time and place to express our gratitude to the thousands of utilities' and grid operators' managers and employees who are on the short end of the stick often, being blamed for things for which they are not personally responsible. These tireless servants of the old and outdated national grid, serving millions non-stop, are doing their best to keep power flowing to our homes and businesses day and night. This is not easy, and has been made even harder lately with the ever changing demand, supply, and new regulations, so we must give them due credit for their hard work now and in the future.

The need for new grid energy storage technologies has created a number of separate and specific market niches, which are expected to grow and develop during the next few years. These new opportunities will help the industry development and will quickly make way for more substantive market developments that may exceed $100 billion in the future.

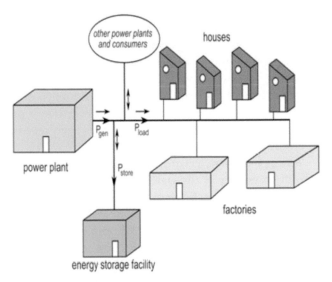

Figure 8-35. Energy storage scheme (the dream)

Sounds easy and elegant, right? Excess electricity is sent into some sort of storage facility and is kept there until needed. The problem is that while water, air, or dirt can be stored in pools, tanks, or piles, electricity does not lend its self to storage. At least not easily or cheaply.

The grid power storage schemes of the future can be built to have a single function or to provide a combination of services. They can consist of electrical storage hardware such as batteries, or pressurized air which can be released at will. For now, however, due to the high cost of such facilities, grid storage must use more indirect ways and schemes.

These could be commercial and industrial energy management systems, that can provide commercial and industrial customers with a method to reduce electricity costs incurred by time-of-use pricing, or by demand changes. At the same time, this system can also be set up to provide distribution benefits by providing distribution grid support and peak load management.

Industry insiders, forecasting the adoption rates, feasibility and overall volume of installations for facilities that provide either single or combined applications, predict the near-term grid storage market to grow by 50% by 2020. This upsurge in installation will be primarily driven by storage facilities and schemes able to perform multiple applications.

More than 90% of all operating grid storage facilities at that time will be able to serve more than one application. This is so because ultimately it all boils down to the financial considerations. Facilities providing combined solutions will produce the best economic results in the short- to mid-term by increasing opportunities for revenue and value, thereby maximizing their inherent financial viability. This will be critical for early adoption, as economic feasibility is currently the most significant barrier to the development of grid storage projects.

While the installation of facilities providing multiple applications will experience the most growth, we see a number of limited long-term opportunities in multi-application installations. These are mostly niche markets that are expected to reach saturation by 2020.

The gradual decline in installation potential for combined application facilities in association with further decreases in technological and economic barriers will provide ideal conditions for an increase in the installation of facilities optimized for individual applications.

Pumped-storage hydroelectricity is presently the largest-capacity form of grid energy storage available. It accounts for 99% of bulk energy storage capacity worldwide, with around 130 GW. The energy efficiency of this method is 70-75%, so 20-25% of the originally generated electric power is lost from pumping the water uphill.

Solar and wind power plants can benefit from grid energy storage, since it would improve their variability. It is one method that the operator of an electrical power grid can use to adapt energy production to energy consumption, which usually vary over time. This increases the overall efficiency and lowers the cost of energy production, while at the same time facilitating the use of intermittent energy sources, like solar and wind power generators.

Another way to achieve grid energy storage is to use a sophisticated smart grid communication infrastructure to enable efficient demand response. The final effect of these technologies is to shift energy usage and production on the grid from one time (and/or place) to another.

Collectively, the market size for individual and combined application grid energy storage facilities is expected to grow to $70-$90 billion by 2020. Participants will need a solid understanding of the technologies and the market forces at play to optimally position themselves to take full advantage of the vast potential in this nascent sector.

None of the solutions will be as simple as constructing a new building and stuffing it with electrical equipment. Much more sophistication and imagination are needed to bring energy storage (energy conservation) to fruition. There are already such attempts underway in California and other U.S. states, which are watched carefully by other states and most countries.

According to a 2012 U.S. DOE report: "Modernizing the electric system will help the nation meet the challenge of handling projected energy needs—including addressing climate change by integrating more energy from renewable sources and enhancing efficiency from non-renewable energy processes. Advances to the electric grid must maintain a robust and resilient electricity delivery system, and energy storage can play a significant role in meeting these challenges by improving the operating capabilities of the grid, lowering cost and ensuring high reliability, as well as deferring and reducing infrastructure investments. Finally, energy storage can be instrumental for emergency preparedness because of its ability to provide backup power as well as grid stabilization services."

Groups engaged in the development of grid energy storage in the U.S. include: the Office of Electricity Delivery and Energy Reliability, ARPA-E, Office of Science, Office of Energy Efficiency and Renewable Energy, Sandia National Laboratories, and Pacific Northwest National Laboratory.

The Future of Energy Storage

In addition to grid energy storage, a boom in the use of electric vehicles (EV) will help push the overall energy storage market to over $50 billion by 2020, according to the experts. Demand for storage technologies in different transportation applications is foreseen to grow faster than that for consumer electronics, becoming a $21 billion market by 2020.

Note: EVs, smart phones, and other devices are included in the overall grid energy storage scheme, because they can be charged at night, or during hours of low power demand, store the energy in their batteries, and use it during the day. With millions of these around the globe, this is not a small thing to consider.

Electronics will remain the largest single market for storage in 2020, with the automotive market well on its way to displacing consumer electronics as the biggest

user of energy storage. This will lead to further scale and a new round of cost reductions, which will impact larger stationary applications as well.

In transportation, the electric and hybrid vehicles represented the biggest opportunity. A modest sales volume of about 500,000 such vehicles represents over $6.5 billion of energy storage use. The US is expected to be the largest customer for electric vehicles storage technologies, due to subsidies. It won't take much longer for China to take over, soon after the U.S. subsidies dwindle.

In consumer electronics, smart phones will remain the strongest market, growing 15% by 2020 to reach $10 billion. The market for stationary applications, including residential solar, is expected to grow to over $3 billion by 2020. Solar integration appears to be the biggest opportunity for the stationary segment, with forecasts suggesting a growth from $0.1 billion today to over $1.5 billion by 2020. This would be as a consequence of robust downstream industry, and strong government policies encouraging solar energy storage, as seen recently in Germany and the U.S.

Energy storage, combined with different ways to co-produce energy is the best way forward for the renewable technologies. Solar and wind power plants can be combined with coal and gas power generation to supplement their daily output in peak hours or as needed.

The winning power co-production scheme is combining large-scale solar and wind power generators, where these two power sources take turns and compliment each other in providing a more stable power output. A properly designed, built, and operated energy storage facility, would serve as a buffer for times of low solar insolation and/or lack of wind.

This entire scheme plugged into a Smart Grid system would be the most efficient form of energy generation, storage, and use known to man. It is also something that would be able to take us successfully through the transition to the post-fossils period.

THE POWER GRID

Electric power is essential to life in the 21st century, and is the foundation of our energy security. Economic prosperity, national security, and public health and safety cannot be achieved and maintained without it. Communities and countries that lack electric power, even for short periods, have trouble meeting basic needs for food, shelter, water, law and order.

A prerequisite for an efficient and practical power

plant, in addition to its ability to generate power, is to be able to transfer it from the source into the national grid, to be used where needed. So the grid is the life's blood of our nation, and we need to know all there is to know about it.

In 1940, 10% of energy consumption in America was used to produce electricity. In 1970, that fraction was 25%. Today it is 40%, showing electricity's growing importance as a source of energy supply. Electricity has the unique ability to convey both energy and information, thus yielding an increasing array of products, services, and applications in factories, offices, homes, campuses, complexes, and communities.

The national power grid accomplishes that by its thousands of miles of power lines, substations and other equipment, which convert the raw power coming from thousands of different power plants around the country to voltages and frequencies that are adequate for transmission and use by all.

Looking at the map in Figure 8-36 we see that the national electric grid, just like the human body's circulatory system, carries life-sustaining energy to all the different parts of the country. Without it, most critical activities would simply stop. The economy would come to a hard and very expensive stop. Life as we know it would cease to exist. The consequences are impossible to even estimate, but nothing good can be expected to result from a drastic interruption in the national energy supply. Even a temporary disruption would be catastrophic, as demonstrated during blackouts in New York, California, and other states in the recent past.

Figure 8-36. The US Power Grid

The U.S. Power Grid

The economic significance of generating, distributing and using electricity in the U.S. today is staggering. It is one of the largest and most capital-intensive sectors of the economy. Total asset value is estimated to exceed $800 billion, with approximately 60% invested in power plants, 30% in distribution facilities, and 10% in transmission facilities.

Annual electric revenues—the nation's electric bill—are about $300 billion, paid by America's 140 million electricity customers, which includes nearly every business and household. The average price paid is about 10-15 cents per kilowatt-hour, although prices vary from state to state depending on local regulations, generation costs, and customer mix.

There are more than 3,100 small, medium, and large electric power utilities in the US.

Of these:

- 213 stockholder-owned (large) utilities provide power to about 73% of the customers

- 2,000 public utilities run by state and local government agencies provide power to about 15% of the customers

- 930 electric cooperatives provide power to about 12% of the customers

Additionally, there are nearly 2,100 non-utility power producers, including both independent power companies and customer-owned distributed energy facilities. Add to this thousands of newly created small power producers with several kW of solar power on their roofs, and you see a huge picture of producers, users, and middle men.

The bulk power system—the large power plants—consists of three independent networks:
- Eastern Interconnection,
- Western Interconnection, and
- The Texas Interconnection.

These networks incorporate international connections with Canada and Mexico as well. Overall reliability planning and coordination of this dynamic system are provided by the North American Electric Reliability Council (NERC). The Council, a nonprofit organization, was formed in 1968 in response to the Northeast blackout of 1965.

NERC's mission is to ensure the reliability of the bulk power system in North America. This is done by creating and enforcing reliability standards, annual assessment of the system reliability, and monitoring the bulk power system through system awareness. NERC also educates, trains, and certifies industry personnel.

NERC's area of responsibility spans the continental United States, Canada, and the northern portion of Baja California, Mexico, including users, owners, and operators of the bulk power system, which serves more than 340 million people. NERC is subject to oversight by the Federal Energy Regulatory Commission and governmental authorities in Canada.

It all starts with electric power being generated in some sort of power plant somewhere in the country.

Power Generation

America operates a fleet of power plants, mostly thermal (coal, natural gas, and diesel fuels) with average efficiency of around 45%. Amazingly, the power generation efficiency has not changed much since 1960s, mostly because of slow turnover of the capital stock and the inherent inefficiency of central power generation that cannot recycle heat efficiently, wasting it instead. Nuclear and hydro plants are more efficient, since the losses are limited, but they are much more expensive to build and maintain.

Power plants are generally long-term, long-lived investments, with the majority of existing US capacity 30 years old, or older. The aging power plant fleet and grid infrastructure are another huge burden to our crumbling national infrastructure—perhaps the largest and most urgent of all, that needs to be addressed immediately.

The power plants can be divided by their daily function into:

- Baseload power plants, which are run all the time to meet the minimum power needs, and which actually provide the bulk of the bulk.

- Peaking power plants, which are run only to meet power needs at maximum load (known as peak load, or peaker plants), and

- Intermediate power plants, which fall between the two and are used to meet baseload, intermediate and emergency power load needs, whatever they may be.

Coal accounted for over 39% of U.S. electricity production in the country, supplying 16.5 quadrillion Btus of energy to electric power plants in 2013. This is nearly 92% of coal's contribution to energy supply, where the utilities buy more than 90 percent of all coal mined in the U.S.

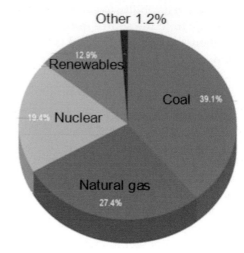

Figure 8-37. U.S. power generation in percent of total (2013)

Recently we saw a significant drop (to 39.1% in 2013) from 53% total contribution of coal in 1997. This is due to the fact that natural gas power plants are slowly but steadily replacing their coal-fired cousins. There are many reasons for this switch, some of which are justifiable and some not, but the trend is expected to continue *ad infinitum*.

There are over 600 coal-fired power plants amidst the mix of over 20,000 individual power generators in the U.S., each with a nameplate (generation capacity) of at least one megawatt. These energy facilities typically combine heat and power generation and achieve efficiencies of 40% to 55%.

A shift in ownership is occurring from regulated utilities to independent, competitive suppliers. The share of installed capacity provided by these competitive suppliers has increased from about 10% in 1997 to nearly 45% today. Recent data suggest that this trend is continuing up, and even increasing fast, as so many new solar and wind power plants are added daily.

Cleaner and more fuel-efficient power generation technologies are becoming available too. These include combined cycle combustion turbines, wind energy systems, advanced nuclear power plant designs, clean coal power systems, and distributed energy technologies such as solar, wind, and combined heat and power systems.

Because of the expected near-term retirement of many aging plants in the existing U.S. fleet, growth of the information economy, economic growth, and the forecasted growth in electricity demand, we face a significant need for new electric power generation. In this transition, local market conditions will dictate fuel and technology choices for investment decisions. Capital markets will provide the financing, while federal and state policies will affect siting and permitting.

This is an enormous challenge that will require a large commitment of technological, financial and human resources in the years ahead. The discussion on these issues has started, but the actual work is still in the planning stages. It will be interesting to see how the effort develops; how extensive it might be, where the money would come from, etc.

Solar, wind and other renewable power generating sources are constantly added to the already complex power generation and distribution system. Although they contribute to the overall national power generation, their location, size and volatile performance present a set of unprecedented problems for the utilities and the national grid. In the best of cases, it will take a long time for the new energy sources to be fully integrated into the energy sector. Meanwhile, they seem to make the bad situation with the national grid even worse.

Presently the utilities are doing their best to accommodate the additional unconventional load, but there are a number of problems related to power interconnection, distribution, and grid control resulting from overload and variability. These problems are caused by the new solar and wind power installations and as a consequence, the utilities are beginning to resent the newly created problems. Since the utilities are responsible for the proper grid function and maintenance, they are fighting against the new forms of renewable power added to their grid and using it for free.

These, and many other, issues must be resolved before the new renewable energy sources could take their place in, and are accepted as part of, the national power generation and distribution systems.

Electricity Pricing

Based on the ever evolving growth of new technologies such as the smart grid, micro grids, and renewable energy production, new pricing strategies are being developed to capture the dynamics of supply and demand in this market. This electricity pricing strategy development is one of the most important and complex aspects of the present-day energy debate.

The dynamic pricing strategies that are used to offer customers shifting prices depend on several internal and external factors. Dynamic pricing makes value and cost of energy use transparent to consumers which enables them to determine when cost exceeds value, which in turn enables the adjustment of energy consumption and production needs.

As smart grid and other technologies emerge, the necessity of new and efficient means of energy pricing become urgent. There has been an array of attempted pricing and forecasting models by utility companies and regulators—usually designed to increase end use efficiency, energy conservation, capacity utilization, savings, in the short term, and to ensure energy security, in the long run.

A price based energy network architecture is based on marginal costs, consumer supply and demand, and the effects of deregulation and competition within a market.

There are many different factors that impact electricity pricing, and which change with advances in technological capacity. These factors affect pricing schemes no matter which pricing model is implemented. Prices fluctuate by location—city, state, country, region—and by input source (petroleum, oil, coal, solar, etc.).

Key factors that affect the price of electricity and energy include:

- Peak and demand rates are vehicles that help the power companies pay their bills and make capital investments to meet the present and anticipated peak demands (by charging the customers as needed).

- Alternative energy technologies can help reduce the rates (especially the peak rates), but despite advances in recent years their use is limited due to high implementation costs, uncertainty in their long-term operation, legal constraints, and recently due to conflicts with utilities operations (it costs too much to operate the grid with many renewables hooked into it).

- The state and local taxes imposed on utilities change from state to state and from time to time, affecting energy rates.

- The power generation or input source determines the price of energy delivered to the customers. A mix of energy sources is ideal for trimming peak rates, where, for example, base-load (gas-fired power plant) is assisted by a solar power plant, which delivers maximum power during the peak hours.

- Environmental considerations complicate the situation and limit the choice of the utilities, forcing them to use technologies with a smaller carbon footprint (wind and solar), which are more expensive to use.

- The capital expense and maintenance of the transmission network is of great importance, as is the distance from the energy source. For example, greater losses are experienced over long-distance power transmission, also making rates higher.

Many factors affect energy rates. The overall picture is complex, and the expenses for producing and maintaining energy supplies are great.

Utilities are caught in the middle between the customers (who demand lower prices) and the regulators (who order lower prices). So, the utilities are forced to kick back by increasing prices in whichever area of their activities they can. Recently, the utilities have been trying the increase the prices charged to residential solar power owners to compensate for those owners' free use of the power grid. The utilities claim that the increased use of the power grid by thousands of new residential solar installations is forcing new capital investment and additional maintenance. True, but the customers do not like the utilities' dictatorship and are rebelling against it.

Power Grid Security

The power grid, with its exposed, unprotected, and vulnerable infrastructure is an attractive target for vandals and terrorists. Attacks on the national power grid can be divided into two distinct categories:

- Cyber attacks, which are usually conducted from remote computer terminals that are geared to interfere with the proper and safe power generation and distribution, and

- Physical attacks on the power grid infrastructure.

- Cyber attacks on power plants have been successfully attempted on several occasions with varying results. Always, they have shown that a remote attack, resulting in great damage is possible. While still a real threat, cyber attacks seem to be properly addressed and well within our ability to control them...at least for now.

Physical attacks on the grid, however, are not so common and are not so easy to control. The grid is spread over thousands of miles across the entire country. There are over 2,000 power plants and 4,000 substations. How do you protect all of them day and night?

There is also reluctance on the part of some of the principals to even discuss the issues. Thus, there is clear and present danger of physical attacks on the power lines, power plants, and sub-stations around the U.S.

The potential of great damage as a result of physical attacks on the grid became reality in the spring of 2013 after the attack on a major substation in California. The April, 2013, attack on PG&E's Metcalf transmission substation started when somebody entered an underground vault and cut telephone cables leading to the substation. Several attackers then fired more than 100 shots at the substation wires and equipment, causing millions in damage.

Quick response from PG&E personnel averted a massive blackout, but it took almost a month to repair the damages.

Note: The Metcalf substation is one of the several extremely critical, strategically positioned, power distribution centers in the country. Located in the middle of the Silicon Valley in California, it serves several million people and many businesses, some of which are major names in the hi-tech industry. The selection of this particular target and the precision of the attack shows that the attackers are well trained and determined to do the most harm. This was not a midnight target practice by a handful of teenagers.

According to insiders, this was the most significant incident of domestic terrorism resulting in a successful attack on the U.S. power grid infrastructure ever. It was done by a well trained and well coordinated group of people. Energy experts fear that the 2013 attack was a dress rehearsal, and that it is only a glimpse of what could happen.

Amazingly, the attack was (intentionally or unintentionally) hidden from the public until it was finally publicized by *The Wall Street Journal* almost a year later. Several senators sent a letter to the Federal Energy Regulatory Commission and the North American Electric Reliability Corporation, stating, their concern that voluntary measures may not be sufficient to provide a reasonable response to the risk of physical attack on the electricity system. The senators also warned that this attack is a wake-up call to the risk of physical attacks on the grid.

The letter was followed by a meeting of several other senators, and industry and government officials to further discuss the adequacy of voluntary measures in protecting the national power grid.

The lawmakers may know something we don't, but apart from isolated vandalism, this is the first major attack on a key substation that got close to throwing part of the Silicon Valley into darkness. It is also the only one, to our knowledge, that has led to such serious damages to the infrastructure, which required major and lengthy repairs.

The utilities, and the national grid in extension are already heavily regulated, so it is not clear how additional regulations will add to national power grid safety. The House Energy and Commerce Committee insist that protecting the grid remains a top priority and that the work on mitigating all emerging threats is ongoing.

The attack had the potential to cause a lot of damage, and the perpetrators have not been identified yet, so the threat cannot be ignored. These people may strike at anytime anywhere in the country, so something must be done. But what…?

Power Quality and Reliability

Power quality is an important parameter of electric power, with the importance increasing as more electronic devices are becoming more sensitive to power fluctuations. High quality power means no voltage fluctuations or harmful signals intermixed with the base power. This is a big concern for today's power grid and the sophisticated loads it serves.

It would make no sense to have a lot of power available for use, if it is of poor quality. Computer equipment, which runs everything today, is very sensitive to power variations. The ubiquity of computers in today's manufacturing and service environments means high power quality is very important to most commercial and industrial firms as well as the average homeowner.

Alternating current, which is the predominant way of transporting and delivering electric power, can be illustrated as a sinusoidal wave, as shown in Figure 8-38. Over time, the voltage oscillates between a positive and negative value around the average voltage levels.

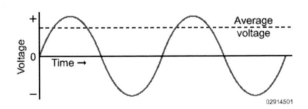

Figure 8-38. AC power wave form.

Although alternating current is established as the world's standard, it still has unresolved problems and issues. These are important, as far as power quality is concerned, in light of the contribution of the new power generating plants.

Below is a list of conditions and characteristic for AC power transmission and use:

- *Voltage sags and swells.* The amplitude of the wave gets momentarily smaller or larger because of large electrical loads such as motors switching on and off.

- *Impulse events.* Also called glitches, spikes, surges, or transients, these are events in which the voltage deviates from the curve for a millisecond or two (much shorter than the time for the wave to complete a cycle). Impulse events can be isolated or can occur repeatedly and may or may not have a pattern. The largest voltage glitch, or surge, is caused by a lightning strike.

- *Decaying oscillatory voltages.* The voltage deviation gradually dampens, like a ringing bell. This is caused by banks of capacitors being switched on by the utility.

- *Commutation notches.* These appear as notches taken out of the voltage wave. They are caused by momentary short circuits in the circuitry that generates the wave.

- *Harmonic voltage waveform distortions.* These occur when voltage waves of a different frequency (some multiple of the standard frequency of 60 cycles per second) are present to such an extent that they distort the shape of the voltage waveform.

- *Harmonic voltages.* These can also be present at very high frequencies to the extent that they cause equipment to overheat and interfere with the performance of sensitive electronic equipment.

Other power quality problems may also be considered reliability problems because they occur when the transmission system is not capable of meeting the load on the system. They can be described as follow:

- *Brownouts.* These are persistent lowering of system voltage caused by too many electrical loads on the transmission line.

- *Blackouts.* These are, of course, a complete loss of power. Unanticipated blackouts are caused by equipment failures, such as a downed power line, a blown transformer, or a failed relay circuit.

- *Rolling blackouts.* These are intentionally imposed upon a transmission grid when the loads exceed the generation capabilities. By blacking out a small sector of the grid for a short time, some of the load on the grid is removed, allowing the grid to continue serving the rest of the customers. To spread the burden among customers, the sector that is blacked out is changed every 15 minutes or so—hence, the blackouts "roll" through the grid's service area.

All of these issues have a negative effect on the normal daily operations. The energy sector is changing by the addition of PV power generation from millions of rooftops and larger solar and wind installations, which affect the energy sector and the grid in particular. The task the utilities are tackling today (on their time and money) is to figure out the contribution of all these factors and what can be done to reduce their negative effects.

This is a complex and pertinent question that can be answered by the energy experts, and which they are surely working on as we speak. One of the most discussed solutions is the implementation of the new and upcoming "smart grid."

THE SMART GRID

The "smart" grid concept, vs. the old, tired, and not-so-smart national grid is supposed to fix all problems and make the power grid operate like a Swiss watch. Since we have to fix the old and tired power grid anyway, it makes sense to replace it with something better, instead of just patching up the old one.

How would we do that, how long is going to take, and how much would it cost? Those questions must be addressed first, and many companies are working on them.

The U.S. Department of Energy (DoE) defines a smart grid as follows:

A smart grid uses digital technology to improve the reliability, security, and efficiency of the electricity system, from large generation through the delivery systems to electricity consumers.

Smart grid deployment covers a broad array of electricity system capabilities and services enabled through pervasive communication and information technology, with the objective of improving reliability, operating efficiency, resiliency to threats, and our impact on the environment." *(Department of Energy, 2012)*

The Smart Grid tailors to the demands of the 21st century electricity consumers by optimizing energy efficiency and combating new security challenges. To accomplish this, a Smart Grid has to rely on new digital technology, specifically focused on communication between the consumer and the producer.

There are six major characteristics of a smart grid, according to DOE. The smart grid:
• Enables informed participation by customers

• Accommodates all generation and storage options
• Enables new products, services and markets
• Provides power quality for the range of needs
• Optimizes asset utilization and operating efficiency
• Operates resiliently to disturbances, attacks, and natural disasters

The concept of the smart grid arose from the need to diversify energy resources to promote energy security and meet the demands of consumers. As more developments emerged, the scope of the smart grid expanded to drive a new electrical system. These drivers include the environment, system reliability and operational excellence.

These drivers push the smart grid into being adaptable, to better meet the new demands of the consumers and environmental challenges. The success of a smart grid system relies on meeting changing and evolving demands and solving a number of implementation issues.

The smart grid works as a multi-way communication system integrating the consumer (and its electric and electronic needs) into the distribution grid, ultimately communicating with the supplier. For the success of this system, the consumer needs an interface in which they can communicate with the network. The suppliers also need to be able to respond to these changing demands quickly as well as forecast future demands to meet power requirements.

The power supply derives its power from not just one main power source but rather many distributed sources to better meet individual demands and increase reliability in the system. This would also decrease transmission losses—one of the largest losses in the electric grid.

In the past, power systems were based off huge power generation systems such as a coal power plant, and distributed through a traditional hub and spoke method. This centric system caused power to be transported over long distances to reach consumers within the region. The transportation of energy created large losses of electricity through transmission, and increased the overall price for electricity.

When this distribution system failed, for example a storm disrupting power lines, there was no way to reroute the power, so customers had to wait until the power crews physically restored the power lines. This has significant impacts on all customers, from residential users to power-dependent industries and businesses such as grocery stores, hospitals, and factories. Some of the larger entities have their own generators, which are used in emergencies, which is a great expense in initial cost, maintenance, and fuel.

In the smart grid, power generation will retain the traditional system while at the same time adopting smaller power generators distributed throughout the system. These integrate within a new and more flexible distribution system, bringing more reliable power and reducing transmission costs. Power generation sources would be more evenly distributed throughout the system as the prevalence of wind and solar power generation increase.

Energy could be supplied through different points within the system, acting as a failsafe in case a power failure; i.e., a nuclear power plant emergency shutdown. Consumers would also be able to better manage the type and amount of power they would be receiving, meeting their interests and creating a market for certain types of power.

For example, one consumer might prefer carbon neutral (solar) power and be willing to pay more for it. Power suppliers would then meet this market demand and produce more carbon neutral power to take advantage of the economic surplus.

The main driver of functionality of a smart grid system is an efficient distribution management system.

Such efficient (advanced) distribution management system for the smart grid starts with data acquisition and supervisory control. The current dispatch and system operation systems operate with insufficient or minimal data and on manual or analog systems. Operators must work with multiple interfaces and systems, making individual decisions that restrict the efficiency of the system.

Actual hands-on experience of grid operators is critical for proper operation and safety of the system. By creating a new (semi-automated) supervisory control and data acquisition system, most of the inefficiencies could be reduced or eliminated. This will allow the operating costs to be reduced while efficiency, reliability, and asset utilization can be maximized. With an efficient distribution management system on one hand, and a consumer interface on the other, the power grid can meet the changing demands of consumers while maximizing cost and operational efficiency.

Demand response is the key to an efficient smart grids and remains one of the most important elements in the new smart grid systems design.

Note: Demand response is the actual use of electricity by consumers in response to changes in pricing controls or other direct and indirect methods that change the consumer consumption patterns. Effective demand response means effective and efficient utilization of power sources and assets.

As consumers get integrated into a distributed management system, the power generation and the distribution network have to be able to successfully communicate with the consumer and meet their power demands. This dynamic change illustrates the challenges that load management has faced since the early 1980s. Load management takes power strategies such as peak shifting and direct load control and implements them to manage power supplies so producers are able to meet the high demands of consumers.

With demand response, system operators are able to use control methods that consumers can respond to, changing their level of demand. This changes the dynamic of the load whereby system operators will be able to monitor and control the consumer demanded load through direct or indirect methods such as pricing controls to meet electricity needs and maximize the utilization of assets.

The Smart Grid Players

Presently, most medium- and low-voltage networks cannot be remotely observed and controlled, which is a big problem that causes inefficient power use and money losses. When fully developed and implemented, smart grid components will eventually solve that problem. Or at least this is the goal of the industry specialists and the companies involved.

Various companies are already developing different technologies aiming at creating and supporting different smart grid network segments. However, some of these developments are either futuristic, or are based on supposed technological possibilities, rather than on a sound problem analysis and a coordinated, truly structured and realistic, smart grid approach.

The major players in the new smart grid game are: ABB, Accenture, BPL Global, Echelon, Freescale, GE, Holley Metering, Moxa, RuggedCom, Siemens, State Grid Corporation of China, Wasion, XD Electric, and XJ Group.

Many other, not so major companies worldwide are working feverishly on developing new components and gadgets for the smart grid technologies, which is leading to new inventions and concepts that will prove to be helpful in the future.

In the recent past, a great variety of sensors, protocols, communication equipment and the like have been designed to support the move toward smart grids.

However, because of differences of opinion and lack of standardization, most of them have not found wide application.

Standardization of smart grid concepts is step number one on the way to solving power distribution and use problems.

Lack of standardization can be blamed for a number of false steps and inefficient solutions in the past. A new "smart grid" gadget is useless unless it provides sizeable solutions to at least part of a big problem.

In other words, there is too much technology push—too many shiny toys—and too little market pull. There are cases where manufacturers of unsuccessful technologies blame network operators as conservative (for not accepting their solutions) instead of improving the price and performance ratio of their products. This reflects the existing anomaly and further hampers a real take-off of smart grid concepts in the energy markets.

In the longer term, smart grid technologies will play an important role in maintaining reliability of supply and improving sustainability. The complexity of power distribution increases, as new solar and wind power plants are connected into the grid. This also applies to small generators, such as rooftop PV and small windmill installations, where smart grid solutions are needed to support these developments by continuously monitoring and controlling their interaction with the grid and the main power generators.

"Smart grid" is a fairly new concept, with growing importance. It is a common denominator for a wide range of developments that make power generation and the related medium- and low-voltage power distribution grids, as well as the use of electric power in general, much more intelligent, efficient, safe, affordable, and flexible than they are presently.

The smart grid vision is becoming clearer, and ever greater efforts will be spent on developing appropriate smart grid technologies in cooperation with commercial energy companies, other grid operators, and suppliers. These efforts will also help to increasingly focus discussions between regulators and the government on the future energy supply and the role of smart grids in it. The energy future looks bright, and smart grids will play a significant part in it.

In some respects the U.S. military might be a step further in developing a "smart grid" system, albeit on a smaller and more manageable scale. Nevertheless, its example must be studied and well understood, to avoid some mistakes in the design and implementation of a national "smart grid" system.

Military Micro-grids

Traditionally, if the power goes out at a military base, each building within the base will switch to a back-up energy source, most often provided by a diesel generator. The military isn't crazy about this setup, in part because this is an expensive solution, and also because generators can fail to start—especially bad news for base hospitals and other critical operations. And if a building's backup power system doesn't start, there is no way to use power from another building's generator.

In addition, most generators are oversized and use a lot of dirty and increasingly expensive fuel. So the military is working on ways to change that by connecting clean energy sources like solar and wind in micro-grids that can function when commercial power service is interrupted.

A new $30 million initiative could transform the way U.S. military bases deal with power failures. Termed SPIDERS—short for Smart Power Infrastructure Demonstration for Energy Reliability and Security—the three-phase project will focus on building smarter, more secure micro-grids on military bases and other facilities, that make use of renewable energy sources. Sandia Labs has been chosen as the lead designer and technical support for the SPIDERS project.

Diesel generators are used all the time to provide emergency power to buildings, but they are usually not interconnected with alternative energy sources like solar, hydrogen fuel cells, etc., as needed for a stable and significant energy source. It's a real integration challenge, according to Sandia engineers working on the SPIDERS program. So, Sandia is working to set up a smart, cyber-secure micro-grid that will allow renewable energy sources to stay connected and run in coordination with diesel generators, which can all be brought online as needed. The new system is expected to not only make the military's power more reliable, but to lessen the need for diesel fuel and reduce its carbon footprint.

The project is being funded and managed through the Defense Department's Joint Capability Technology Demonstration with the support of the U.S. Department of Energy.

Someday soon, the departments hope to use the SPIDERS plan for civilian facilities like local hospitals and other applications as well.

Micro/Smart Grids

An important concept to understand here is that of the micro grids, which are small electrical distribution networks that can be operated in islanded mode or interconnected with the mains. The key point being that a micro grid can be self sufficient. (5)

Micro grids can pull power from the main grid during peak times if needed, but can be mostly self-sustaining. The energy market in developed nations mostly relies on macro grids. Treating the energy market as a system, there are many concerns currently associated with the structure.

When dealing with such large areas, the macro grid makes it difficult to forecast consumer demand and produce the required energy. Another disadvantage of the macro grid is that it monopolizes power. This leads to higher prices for the consumers.

Arguments can be made that a move to micro grids from traditional large power distribution networks (macro grids), would be a return to the past. The first power grids created were often very isolated from one another. Each was self-sustaining with no reliance on other grids.

One of the first examples of a micro grid was Thomas Edison's Manhattan Pearl Street Station in 1882. However, the era of isolated systems was short lived in large part due to advances in alternating current (ac) technology. The first system to utilize ac technology was implemented in 1896. This system carried electricity from a hydro station in Niagara Falls to Buffalo, NY.

Current examples of micro grids can be seen in the island of Lemnos and in the English Borough of Woking. Lemnos uses wind turbines, solar panels, and diesel generators to create 14.84 MW of power. Woking uses different forms of energy production including combined heat and power plants and fuel cells to generate 2 MW of power.

Utilization of micro grids offers many advantages such as cost reduction for the consumer, renewable resource utilization, improvement in reliability, and a reduction in negative effects of the environment due to existing power generation.

As stated earlier, the micro grid in Lemnos uses wind turbines and solar panels to create energy. This is effective in a micro grid due to the size of the system. Many of these renewable resources are not currently able to generate large wattages based on the current technologies. This makes it difficult for a wind turbine plant to supply enough energy for a large geographic region. However, utilization of wind power in a small town is much more feasible considering this micro grid would be able to meet consumer energy demands.

Micro grids can also increase reliability in the system. They have the potential to not only turn on when local supply needs more energy, but micro grids can also be used to help reduce an excessive load in the macro grid. The utilization of micro grids during these peak times would decrease the potential for blackouts.

Not only can a micro grid provide all of the electricity and heating needs for local customers, but it can also reduce overall emissions during the energy production process. Micro grids make up a part of a larger, more effective power supply and distribution network known as Smart Grids.

The Smart Grid Markets

The smart grid, although a fairly new development, is finding enthusiastic support in many developed and developing countries. It is a technically complex and financially expensive undertaking, and yet there are takers, as follow:

The U.S.

In December 2007, Congress passed, and the President approved, Title XIII of the Energy Independence and Security Act of 2007 (EISA). EISA provided the legislative support for DOE's smart grid activities and reinforced its role in leading and coordinating national grid modernization efforts. Key provisions of Title XIII include:

— Section 1303 establishes at DOE the Smart Grid Advisory Committee and Federal Smart Grid Task Force.

Section 1304 authorizes DOE to develop a "Smart Grid Regional Demonstration Initiative."

— Section 1305 directs the National Institute of Standards and Technology (NIST), with DOE and others, to develop a Smart Grid Interoperability Framework.

— Section 1306 authorizes DOE to develop a "Federal Matching Fund for Smart Grid Investment Costs."

Office of Electricity (OE) is the national leader, partnered with key stakeholders from industry, academia, and state governments to modernize the nation's electricity delivery system. OE and its partners identify research and development (R&D) priorities that address challenges and accelerate transformation to a smarter grid, supporting demonstration of not only smart grid technologies but also new business models, policies, and societal benefits.

OE has demonstrated leadership in advancing this transformation through cooperative efforts with the National Science and Technology Council (NSTC) Subcommittee on Smart Grid and the Federal Smart Grid Task Force.

"The National Science and Technology Council Subcommittee on Smart Grid: Chaired by the Assistant Secretary for OE and the National Director for Smart Grid at NIST, the Subcommittee is promulgating a vision for a smarter grid including the core priorities and opportunities it presents; facilitating a strong, coordinated effort across federal agencies to develop smart grid policy; and developing A Policy Framework for the 21st Century Grid which describes four goals the Obama Administration will pursue in order to ensure that all Americans benefit from investments in the Nation's electric infrastructure: better alignment of economic incentives to boost development and deployment of smart-grid technologies; a greater focus on standards and interoperability to enable greater innovation; empowerment of consumers with enhanced information to save energy, ensure privacy, and shrink bills; and improved grid security and resilience."

The above is a long and winding statement, which, in our opinion, reflects the long and winding road in front of the US smart grid. And please note that the statement is calling for a "smarter" grid, while what we need is a totally "smart grid." This might mean that the government bureaucrats are thinking of implementing a "smarter" grid first, before jumping to the "smart" grid version—which only makes the road longer and more winding.

We saw similarly long, winding, confused and even misdirected statements in the 1970s, when solar energy was hailed as the "savior" of the American Dream, and where the US government was taking charge of its development. Solar was hailed as the only way to a sustainable energy future, so lots of money was thrown at solar companies then (some of it at the wind, as happened today)…until the government changed, and things went downhill from there.

Solar energy was brought out and shoved back into the closet several times since then, and it is where it was until recently, when it was taken out, dusted off and hailed again as the only way to energy independence… until the new government takes power and solar is shoved back in the closet…again. That may not be so easy or fast this time around, but the signs unequivocally point in that direction.

Is the same fate awaiting the US smart grid? Nobody knows, but due to the great amount of effort and money involved, and judging from the lessons of the past, we should not count on the government bureau-

crats—self-proclaimed "smarter" grid saviors—to lead the smart grid effort to a successful end.

Personal interests, changing political winds and everything else that is awkward with politics and politicians today, makes us believe that the government should stay out. Instead, we should let the US capitalist system take over and dictate the rules and the effort. Keeping the government bureaucrats in a sub-serving role, where they belong, and where they can be controlled, is the best, if not the only, way to manage this, and any other, serious effort.

One example of ill-conceived (albeit well intentioned) intervention of the government is the Department of Energy's Smart Grid Investment Grant Program. Grants ranging from $500,000 to $200 million are to be issued for deployment of smart grid technologies, most of the recipients of which seem to be working on smart metering solutions (since it is the easiest, and most accepted by the public, smart grid related technology.)

In addition, IRS issued a guidance in 2010 providing a safe harbor, under which the $3.4 billion of federal Smart Grid Investment Grants (SGIGs) issued under the American Recovery and Reinvestment Act of 2009 (ARRA) will not be taxable to corporate recipients.

Great news, right? But here is the catch; energy conservation, derived from the full implementation of smart meters (as encouraged by the government grant and tax programs) means lost revenue to utilities at a time when the utilities are in the midst of expensive changes and are not ready for additional revenue losses. The lost revenue, forced by the government programs is simply untimely and is not aligned with the objectives of most utilities and their shareholders.

In addition, long-established regulations have favored supply-side resources over energy conservation, so utilities have been encouraged to add new generation because they earn a rate of return on investments on their assets, and mainly the power generation, transmission and distribution infrastructure. Now, without proper warning and preparation, the utilities must abandon the profitable business model and gear for energy conservation, which simply translates to loss of revenue.

This misalignment might cause delays and even failure of some energy-efficient technologies and services in the U.S. and reflects the disconnect and fragmentation in the sector. Some of the disconnect is caused by the inadequacy of the government bureaucrats, who talk the talk, but seem more interested in their agendas than anything else. This anomaly is fueled by the ignorance of the regulators as well, which completes the circle of incompetency, and which will make smart grid implementation that much harder.

The bureaucrats do not take into account the fact that the technologies that promote energy conservation have evolved much faster than the regulations that govern utilities and other suppliers. This creates an imbalance, which will continue until new regulations, promoting the implementation of demand response and other energy conservation programs are implemented. To cross the divide seamlessly, appropriate and timely regulations and other policies should be properly structured to justify and encourage full utility participation and investments in demand response.

Table 8-7. Smart grid investment (in millions)

Country	Total
China	$7,500
USA	7,000
Japan	850
S. Korea	850
Spain	800
Germany	400
Australia	360
UK	300
France	260
Brazil	200

The work has started, no doubt, and some money has been spent, but the effort is still full of gaps, fragmentation and misdirection. To coordinate and encourage the activities, new legislation must be approved, and regulation must be implemented to create incentives for utilities to reconsider investing in generation, transmission and distribution assets and promoting energy conservation and demand response.

Standardization of the "smart grid" equipment and procedures is lacking and is a must for proper advance of the technologies involved. There must be also a balance between the approaches, and the utilities must be encouraged to use demand response as part of the generation portfolio and modus operandi. Some type of shared-savings program that allows utilities to participate in the savings customers receive from reducing their energy usage is needed as well. This will persuade them to promote energy conservation because they will benefit, as well. In addition, there should be a penalty if a utility does not encourage customers to reduce their energy usage.

Another way to encourage utility participation would be that a utility gets compensated for a portion of its avoided supply costs obtained through demand response and other related programs.

So, the future of the smart grid in the US is not clear. Looking to the history lessons of the past: solar was the hope for energy independence in the early 1970s. After 40 years, we know what solar can do, but we still don't know what to expect from it.

Smart grid technologies are headed in the same direction. Right now they are changing at an exponential pace—ten years ago we didn't even have smart phones, and the term "social media" was still in the making. Today, these are everywhere and we cannot imagine life without them.

So how could we—at the beginning of an era—even imagine where, or what, the "smart anything" might be in 2015...2020...2050 and beyond. It is not possible, but we can say with certainty that there will be progress, with heavy emphasis on customer (partially virtual) services.

The utilities are well aware of the present and pending socio-demographic changes. They see the new generation of informed and active young customers, as well as the large retiring workforce, where by 2020 half of the utility workforce will be retired as well. And these people know way too much...so nothing can be kept hidden.

Customers are accepting new technologies quickly as evidenced by the fact that, according to some statistics, the volume of text messages now exceeds that of phone calls. But just think—10 years ago text messaging did not even exist...

In the 21st century, information is the foundation, customer acceptance is the route, and integration is the vehicle.

The existing global database is enormous, and the data markets are growing exponentially. Because of that, all participating systems must be integrated in parallel to receive and process the data for maximum efficiency and lowest possible cost. Smart grid is the best, if not only, solution to all these problems.

No matter how you look at it, at least the direction is clear: the smart grid and its smarter virtual applications are here to stay. Some key things haven't even been invented yet, but we know that whatever is coming in the area of smart grids will have a large virtual component with a high level of self sufficiency and reliability. The utilities must be given a chance to get ready for the new order of things. It won't be easy for them, especially while carrying the baggage of century-long traditions being managed by personnel more than 50 years old.

Our fully implemented and integrated, smart grid based "green" energy future seems attractive, but we cannot even guess how and when it will arrive. We just have to go step by careful baby step toward its proper and complete development and implementation in our lives.

California Edison's Smart Grid Roadmap

Southern California Edison (SCE), which serves about 13 million people in the Golden State, is expected to make more changes to the grid and its overall power infrastructure in the next ten years than in the previous century. Some of these changes are related to old hardware upgrades and new smart grid introductions. It is a good example of a big investor-owned utility that's taken a leading role in the smart grid era.

There is already an impressive list of smart grid accomplishments, which mark the beginning of a transformation that is going to bring many changes. That comes with a long list of projects needed to meet the myriad challenges. And with the projects come the problems.

The Developments

One of the top concerns of the California utilities today is how to manage the growing share of intermittent renewables (solar and wind) powering its grid. The recent rise of distributed energy resources (DER), represented by a growing number of solar roof systems, has complicated the situation even further. Its solution is becoming extremely important and even urgent.

State mandates call for utilities to get 20 percent of their power from green energy today, and that figure is set to grow to 33 percent by 2020. Some of that power comes from stable sources like geothermal or small hydropower, but an even greater share is coming from new (small and large) solar and wind power installations, and much more renewable power is to be added during the next decades.

The variable power sources are creating problems for the utilities, that have traditionally delivered power from central generation plants to end customers.

The present-day issues and solutions before the utilities, brought about by the new technologies and approaches, can be summarized as follow:

- Solar and wind power are different from the baseload sources (fossil power plants) because they cannot be dispatched when needed. Sun doesn't shine all the time and wind doesn't blow 24/7. Instead, the utilities are forced to react to the whims of Nature which drive the solar and wind power generation. This is a big problem with limited solutions at present.

- Many of the generation resources are connected directly to the distribution networks (power lines to homes and businesses), thus bypassing the (long-distance) transmission lines altogether. The distribution network, however, is part of the entire grid system and was not designed to operate independently and/or handle two-way power flow. This prevents the central grid control structure from managing this extra power, so a new system control must be developed and implemented.

- Plug-in electric vehicles are another headache for the utilities. They use enough power when plugged in (at unusual hours), to amount to an entire household's power use, to which the grid must react and adapt.

- Millions of new smart meters deployed recently represent another disrupting factor, because they force the utilities to collect and process more data than usual. The new meters also enable customers to sign up for time-of-use rates, instead of the old way of paying flat fees for power used. This created more work for the utilities and is forcing them to base the energy use on real-time feedback from thousands of home energy management platforms.

- The old traditional customers are now sophisticated "transactive energy" players, who in a sense are competitors since they generate electricity and, a) either send it into the energy markets, or b) use the energy to take advantage of price fluctuations. What is a mother utility to do when the kids are grown, make their own power, and use it as they wish?

Note: This reminds us of the old days when AT&T would not allow people to have their own telephones. The mother company had to bring the telephone unit and install it before activating the number. Monthly charges included phone unit rental. Fax lines had to be registered and paid separately too. The author was even threatened with a lawsuit in 1976 for installing and using a fax machine without AT&T's blessing. Now, only 40 years later, most people don't even have home phones anymore. Wireless is the new game in town and AT&T had to adapt to it, which, albeit under duress, they are doing quite well.

- Customers now have a chance to make energy decisions based on economic choices. This brings a new level of unpredictability to the utilities' (bottom line) equation. As the renewables increase in number, so will the utilities' headaches, while their profit margins shrink.

- Battery-based, grid-scale energy storage is presently on the drawing boards as one way of balancing the grid. Two Department of Energy smart grid stimulus grant-funded projects are fueling the drive. The energy storage effort is focused simultaneously on the large-scale and small-scale energy storage systems. This will increase the independence of the individual power generators, giving them a choice of when and how much energy to use or send into the grid.

- SCE is working with A123 Systems on an 8-megawatt, 32-megawatt-hour lithium-ion battery storage option in the Tehachapi mountains. The goal is to stabilize and integrate large-scale wind power generation into the grid.
  — SCE has undertaken a number of other projects, including a California Energy Commission (CEC)-funded partnership with GRIDiant and New Power Associates, which analyzes transmission and distribution systems for a grid serving about 275,000 customers.
  — SCE is also a leading player in a DOE stimulus grant-funded project to deploy synchrophasors, which are devices that monitor transmission lines by collecting data in sub-second intervals, across the entire western United States.

- SCE is integrating four different configurations of batteries (ranging from a 2-megawatt substation battery to smaller residential energy storage units [RESUs]), into its smart grid demonstration project in Irvine, California. This is an $80 million, Department of Energy grant-funded project designed to test the interplay of energy-smart homes, solar panels, grid batteries, plug-in car chargers, grid voltage management, self-healing circuits, and communications and controls networks in a single neighborhood. The list of partners, and potential designers and manufacturers of the related equipment includes: Boeing, General Electric, SunPower and Space-Time Insight, to name a few.
  — SCE is also testing a variety of "smart inverters" and other devices which could help integrate roof solar power into the grid by allowing each rooftop solar system to better manage its interaction with the grid. This has been successfully tried in Germany, where solar inverter regulations and requirements were put in place recently, to manage the massive share

of distributed solar power and the grid interactions. Ultimately, using digital devices on the grid can help record and even manage the variable flow of distributed energy resources and customers' energy use. The challenges, however, are great—from hooking up smart meters to distribution management systems, to integrating transmission grid measurement units into the utility's operations.

The future role of energy storage is undeniable, but the type and size of the systems will depend on the respective economics, reliability, and efficiency factors. CPUC recently decided that it would require at least 50 megawatts of energy storage for the Los Angeles region power mix by 2020. This may or may not happen, but if it is not done right, it might distort the market without solving the energy storage problems.

At the same time, the utilities are spending a lot of effort and money on keeping the old grid humming nonstop 24/7/365. Replacing old power poles and wires, transformers and other components of the crumbling electric infrastructure costs SCE over $3-billion-per-year in capital investment. The number is tripled when the efforts of the other California utilities are added. Utilities in other states are in the same situation, and like it or not must dedicate a lot of effort and money to fixing the old grid and its components. This alone makes any investment in smart grid initiatives involving the old grid imprudent.

With all this considered, we must conclude that:

- The U.S. power grid is not ready (in its present state) for smart grid additions and improvements. It needs a major upgrade and overhaul to become "smart grid" compatible;

- The most advanced and most promising smart grid technologies are not ready for mass-deployment, since most of them are still in R&D labs; and

- Most of the smart grid technologies that are commercially ready and available are not approved by the California Public Utilities Commission, since they require a lot of testing and verification of the cost-benefit and efficiency equations involved with their energy market deployment.

So, the smart grid is a great idea and would work wonders for the grid, the utilities, and their customers, but its full implementation is still far in the distant future. Partial additions and improvements are underway, so we wait to how these work out and how they affect the existing grid and the overall energy situation in the country.

One of the biggest problems that the smart grid is bringing to the national power grid, to our energy and national security, is increased vulnerability to outside attacks. While now the terrorists must penetrate the security around, and physically blow up a substation to disrupt power to a large area, in the near future they can do this with a computer from thousands of miles away. This way, not one but several substations can be disabled, and many power plants can be shut down by skillful computer terrorists. This is not very likely, but a possible and very scary, scenario.

Smart Grid Cyber Security

Smart grid activities are shaping as most promising, as far as ensuring the future grid functionality. From a technical standpoint this is a doable, exciting and profitable effort, but there is a big problem—the new smart grid opens the power grid to cyber attacks.

Since the smart grid will grow big no matter what, ensuring its security is going to be a very big business. Unfortunately, this big business will not make money for the utilities that buy it; the majority of cyber-security investments will go to meet regulations and make the investors happy. The major effort and money will go into justifying the extra costs for preventing grid infrastructure attacks, and the related failures, accidents and incidents.

There are estimates that the $120 million spent on cyber-security products and services in 2011 will double by 2016 and triple by 2018. This number will grow over $6 billion, spent mostly by the utilities and the U.S. government, by 2020 to secure the grid and avert the consequences of a successful cyber-intrusion. Worldwide, this number could grow to $80 billion in the same time frame.

Cyber security costs exceed the total spending on the smart grid itself. This is truly big business that is taken very seriously by all involved.

It is a very risky business too, since the grid must be protected from any eventuality, such as minor data theft to taking over grid assets, causing blackouts, or damaging substations and power plant equipment. A lot of research and trial-and-error will go into making sure this is done right.

A number of cyber-attacks and hacks at major U.S. organizations and installations have elevated the issue to the White House level. As a consequence, the Obama administration has issued an executive order demanding that all industries involved in the smart grid and power generation come up with a plan to deal with cyber and other threats to the grid's integrity.

How far and how fast this effort will develop remains to be seen, but it will not be a transparent process, since one of the key tenets of cyber-security is that it is most secure when nobody knows how secure it is. We surely will not learn much about the specifics of discovering, isolating, and eliminating any new intrusions and attacks. And since the methods of the intruders change daily, there will be a lot of changes and adjustments done to the cyber-security infrastructure as well.

The utilities are interested in and required to secure their infrastructure by all means available. The effort will spread on all levels from installing fences and locking gates and doors to ensuring the safety of the smart meters and computer networks, and to implementing the latest cyber-intrusion detection software and schemes.

This effort will require new products and support services, similar in size and form to what the banking and telecommunications industries have embraced during the last 20-30 years. It was a long and expensive route for them too, and even after all this time and expense, the banks are still not as secure as they wish to be.

The cyber-security spending on this continent is driven by the North American Electric Reliability Corporation (NERC)'s Critical Infrastructure Protection (CIP) requirements. In the U.S. and Canada these rules are implemented and enforced by stiff fines. There are cases of up to $1 million-per-day fines for utilities and other entities that do not meet the security guidelines.

The Department of Energy is dangling a $4.5 billion carrot in front of the cyber-security and wanna-be-such companies. The DOE stimulus grants for a number of projects also require cyber-security compliance, as outlined and coordinated by the U.S. National Institute of Standards and Technology.

Securing the new IT infrastructure of the power grid against cyber-attack, for example, will be big business, although it makes no money for the utilities. It is mostly about meeting regulations, satisfying shareholders, and trying to justify the costs based on what it's trying to prevent.

Putting a dollar figure on averting the consequences of a successful cyber-intrusion in most cases is simply impossible. Utilities have to worry about everything from data theft to a full-blown attempt to take over grid assets, to cause blackouts, or to overload generators.

North America accounts for over 40 percent of the

global cyber security market, with Europe at 30 percent and Asia-Pacific at 17 percent. China is expected to overtake the U.S. as the largest market by 2020, leaving the Asia-Pacific region with 35 percent of the market share. China, Japan, South Korea, and Russia will see growth rates of 40 percent and up in the next decade.

International Smart Grid

Around the world, governments are working on their own rules and regulations for ensuring the safety of their respective power grids. The European Union's cyber-security agency, ENISA, for example, is building a framework that is focused on a risk-based approach with a certain degree of freedom for the utilities and all companies involved, in order to do their job more promptly and efficiently.

Europe

Europe's Smart Energy Collective is a sector-transcending cooperation involving a wide range of companies working for smart energy and smart grid implementation. Members include: ABB, Alliander, APX Endex, BAM,DELTA, DNV KEMAEnergy & Sustainability, Efficient Home Energy, Eaton, Eneco, Enexis, Essent, GEN, Gemalto, Heijmans, IBM, ICT Automatisering, Imtech, KPN, Nedap, NXP Semiconductors, Philips, Priva, Siemens, Smart Dutch, Stedin, and TenneT.

In May 2012, members of the Smart Energy Collective in Europe have approved the second phase of the smart grid initiative, which involves the development of five large-scale smart grid demonstration projects in the Netherlands.

Schiphol Airport, ABB, Siemens offices, and several residential districts were the chosen sites for smart grid implementation and tests. To make this possible, an intelligent energy system is designed that uses a combination of innovative technologies and services. It will enable the owners to control energy use and keep energy costs low with a comparably high level of reliability.

This effort would be an important step towards standardization too, in addition to just testing the smart grid operation in actual use. A survey during the first phase showed that there are more than 6,000 relevant standards that play a role in the new technologies, which according to the group, will be introduced to the market over the coming years in one or another form.

Separate work groups have been established for standardization, market mechanisms, services and business cases, privacy and security, and infrastructure, in order to establish a solid foundation for the design of the five demonstration projects, as well as for future proj-

ects. These will serve as the foundation for developing a broad, international, cyber security network on the continent.

Asia

Asia is quickly becoming a major player, and the center of global smart grid activity. The combined smart grid market in China, Japan and South Korea is estimated at over $10.0 billion, with an estimated increase to over $30.0 billion by 2020, according to industry specialists.

As many Asian countries are irreversibly becoming the predominant smart grid markets, the competition and the positioning of different vendors is increasing as well. Lack of clear understanding of the energy scenarios (present and future) in the major Asian countries (China, Japan, and South Korea) is a major obstacle that hinders progress and needs to be resolved first. The energy, political, and socio-economic conditions in these countries are quite different, so no unified approach is expected soon. This will also delay the process and make it more expensive.

It is widely expected, however, that the smart grid markets in Asia will move forward at a breakneck pace during the next decade or two. The developed countries on the continent are already positioned for the race with over $45 billion in funding, earmarked by the respective governments and utilities across China, Japan and South Korea. The majority of funding and related opportunities, of course, is located in China and its booming energy market.

A level of uncertainty and secrecy is still distorting the smart grid vision, so determining the trends and establishing a clear path of energy policies and currents will allow much faster implementation of smart grid technologies in these countries. If and when some of these problems and uncertainties are lifted, development of the Asian smart grid markets will proceed very fast too.

China

The growth of the smart grid market in Asia is characterized by the special needs of each country and its energy demands, as well as the local utilities and existing grids' specifics. For example, the smart grid investment in China is focused on transmission and distribution automation as needed to support the plans for a new power grid (planned to be developed) and robust renewable energy infrastructure (planned to be built.)

China is aiming to become a world leader in smart grid technologies in the next decade. The "Strong and

Smart Grid," an 11-year plan revealed in 2009 outlines the ambitious steps to get there. It involves all aspects of the power grid, including increase of power generation capacity, implementation of smart meter programs, emphasis on large-scale renewable energy, and a large transmission lines and substation build-out.

China, however, has basic needs for a better and more extensive power grid. Nevertheless, today, the smart grid plan is still alive and well. It might be like putting the cart before the horse, so the State Grid Corporation of China, which is actually one of the largest utilities in the world and the executor of China's smart grid plans, is already in phase two of the three-phase program. It is the construction phase, which is scheduled for completion in 2015.

New transmission lines are a major focus (and a big problem) for the state grid in the construction phase, which is struggling to meet the growing energy demands of the rising middle class in the east and south of the country. So, in reality, China is building its power grid first and before jumping into the smart grid aspects of it. You must have a grid first, before you can make it smart.

The problem is that most coal, hydro, wind, and solar energy resources in China are thousands of miles away from the populous east and south urban centers. High voltage, HV (under 300 kilovolts), extra-high voltage, EHV (300 kilovolts to 765 kilovolts), and ultra-high voltage, UHV (765 kilovolts and up) lines are being installed currently. There are plans for at least one 1,000-kilovolt UHV AC and/or DC lines installed annually until completion. Overall transmission line investments estimates are approximately $280 billion, equivalent to the combined market cap of ABB, GE, and Schneider Electric put together.

China is adding so much new transmission capacity and so many power lines that it could build three quarters the length of a new American transmission grid in just five years. When the dust settles, there will be over 125,000 miles of new 330-kilovolts-and-up transmission lines built, for a total of 560,000 miles of transmission lines, compared to 160,000 miles of transmission lines presently in the U.S. What...?

At a cost of $1.05 million per mile for UHV transmission line and equipment, each UHV line requires billions of dollars to build, and State Grid put in a staggering $80 billion investment into the 25,000 miles of UHV lines to be built by 2015.

A new Chinese business model (in terms of quality) is apparent here. A 1,250 miles, 800-kilovolt UHV DC line recently constructed that has an incredibly low 5 percent line loss rate per 1,000 miles and a high 6.4 GW transmission capacity. It is also 30 percent cheaper than a 500-kilovolt EHV DC or 800-kilovolt UHV AC line of the same length. The plan is for the UHV lines to have 300 GW of transmission capacity by 2020, roughly split 60 percent AC and 40 percent DC.

This innovative, brisk and competitive business environment as seen in the transmission grid build-out is indicative of the upcoming smart grid market in China. High-quality goods, competitive costs, and a well-built relationship with State Grid are needed to win a contract for most companies. Fierce vendor competition, due in part to State Grid's competitive construction procurement process, is also well underway. All projects costing over $300,000 are required to go through an open bidding process that aims to enforce fairness and transparency. The State Grid still holds the reigns tightly on choosing project developers. It has the final say-so and does a rough 45/45/10 split when evaluating the major factors of quality, cost, and bankability of each company.

The new developments, together with the promise of fewer power shortages and a stable energy supply base are ushering in the next era of smart grid opportunities in China. Smart meters and renewable integration are already big businesses, and a new, more efficient, substation infrastructure is expanding the substation automation equipment market. The need for better monitoring equipment has risen too and is promising decrease in the interruption duration index and improved power quality to its customers.

State Grid has earmarked over $40 billion toward these smart grid technologies to be spent by 2016, with smart meters alone being a $2.5-$3 billion annual market. Substation automation technologies are of special interest and there are plans for installing 75 new digital substations for 63 to 500 kilovolt systems during this decade.

While this number is small compared to the existing 40,000+ substation base in China, State Grid intends to include digital technology in all new substations. Foreign companies such as BPL Global have been expanding their substation operations (equipment and services) in China, which has been met with stiff domestic competition. The substation market offers promising growth over the next ten years, so the competition is growing proportionately.

The transmission grid build-out also has an impact on technologies at the distribution level and downward. China plans to build 36 million new urban homes by 2015-2016, and all will be equipped with modern building automation and smart meter technologies.

The coming years promise to create a new and vibrant building automation market, but for the time being, the market continues to focus on meeting demand shortfalls, blackouts, and other key infrastructure problems.

We expect to see an exciting shift toward technologies at the distribution level and downward in the next five to ten years, as China upgrades and expands its transmission grid and improves its generation sources. China is set to become a leading smart grid market in this decade. We will be watching, as the distribution grid build-out and digitization will be the major indicator of China's smart grid progress. The lessons learned by companies working in China can then be transferred to the smart grid development projects in the U.S. and Europe.

The Rest of the World…

The shutdown of many nuclear plants in Japan and around the world after the 2011 Fukushima nuclear accident created a need for demand response, energy management and smart meter deployments in the country and the world. Smart grid and its safety is now on the agenda much more than before. Money is available, the need is there, the customers are willing, so only time separates Japan and many other countries from the full implementation of efficient smart grid concepts.

We would not be surprised if Japan leads the world in this area in the near future, and even becomes the first country to claim complete smart grid deployment. It is much smaller, more developed, and more intensely technological than China, so it seems that it has more of a chance to cross the finish line first.

The South Korean market is quite different. It is the country with the most reliable grid in the world, and is looking seriously into developing the next-gen smart grid technologies and components (hardware and software) across all segments. At this point the emphasis is global export of these technologies, but the trend for domestic use is increasing too. If the focus is shifted to more domestic use, then South Korea might beat both Japan and China to the finish line.

In summary, the global smart grid future is bright. It has the potential to develop into one of the most important parts of the world's energy markets, thus bringing us closer to the overall goals of achieving energy security and cleaning the environment.

The 21st century will bring a lot of developments in this area, and the Asian countries will be the first to claim smart grid implementation and use.

TRANSPORTATION SECTOR

"Over the past four decades, eight presidents tried to reduce America's oil dependence, but despite some tactical progress, were unsuccessful, and each passed to his successor a country that was more, rather than less, dependent on oil. One of the reasons for this failure is that they all focused on reducing the level of oil imports when the real problem is oil's status as a strategic commodity which stems from its virtual monopoly over transportation fuel."
— *The United States Energy Security Council, 2014*

We can add a lot to this statement, but it says it all—we use a lot of crude oil and have allowed it to run essentially any and all aspects of our personal and business lives. No exception. Crude oil powers everything from our electric shavers, laptops and cars, to huge factories, cruise ships and fighter jets.

Transportation is especially, directly, and heavily dependent on it. Just think that 99% of all vehicles and moving equipment on Earth run on gasoline, diesel, jet fuels, and different fuel blends, all of which include crude oil products.

The energy markets encompass, and in one or another way affect almost every area of our daily lives. One must live in a cave in a far, faraway place to not feel the presence of, and the need for, energy in his/her life. Amazingly, crude oil is the driving force behind it all. It is what brings the other energy sources and makes the entire energy market work. Everything around us—directly or indirectly—needs crude oil or its derivatives to function.

Without crude oil, the economy, and life as we know it, stops!

Don't think so? Take a little walk with us…

Walk in the Future…

It is not very hard to imagine a world without crude oil, since it drives us and everything around us. Just imagine for a moment a catastrophic event that cuts all global crude oil supplies. No oil comes from the oil wells and no oil products come from the refineries. The world will not end, but it will feel like it, at least during the initial stages, and until some solutions are found.

This oil-free picture has no cars and trucks on the highways, no trains on the railroads, no boats in the water, no airplanes in the sky. Without gasoline and diesel, all forms of powered transportation stop—even electric cars, because they depend on the power generated by power plants, which are driven by fuel delivered by diesel-powered vehicles. No diesel, no fuel, no electricity, since coal from the mines cannot get to the power plants without diesel to drive the unit trains.

Those who get to work won't be able to do much work, because there would be no lights, no air conditioning, no power for the computers and the electric motors driving the machinery. Agricultural production would stop dead in its tracks, because it is totally dependent on diesel-powered equipment, in the fields, during transport, and in the processing plants.

Renewable energy will keep the lights on...not! The operating solar and wind power fields need replacement parts and human intervention, none of which would be available, so they would start failing one by one. The entire power grid would shut down, because the intermittent (solar and wind) power generation cannot keep it going. Yes, nuclear and natural gas power plants could continue operating (at least for awhile) but they cannot keep the entire power grid operational. The grid overload would cause massive blackouts and myriad of disasters. Sooner or later the nuclear power plants would also shut down due to lack of uranium (which is transported overseas) and the situation would only get worse.

But the most dangerous part of such a scenario is that the entire U.S. military machine would stall. All trucks, tanks, fighter boats and planes, drones and everything else that moves would be grounded. Iraq, Iran, Ukraine, and all other places would become unreachable and the U.S. would stop being the global policeman and peacemaker. The consequences of such development on a global scale cannot be predicted, but no good can come of it.

Internally, the police force would be grounded too, since their cars and SWAT vehicles need a lot of diesel and gasoline. With reduced counter-terrorism capabilities, and lack of quick emergency response, the country would become much more vulnerable to internal terrorist attacks. From here, things would go from bad to worse *ad infinitum*.

Agreed: this entire scene is fantasy—a waste of words and a highly unlikely scenario. This is because we are awash in oil and gas today. So not to worry, all is well, the experts say...today! The present fossil fuels bonanza will keep us going for several decades, so pay no attention to such absurd scenarios.

But common sense dictates that crude oil is here today, gone tomorrow. There is no way around it. A day like the one described above *will* arrive, unless we take

steps *now* to avoid it. Back to the 18th century we'll go. Because of that, the national and global energy markets, and those of crude oil especially, deserve our full and undivided attention.

So let's start from the beginning of the energy markets as we know them today. No, they did not start with the Big Bang, but developed with time and were dormant for millennia. Finally the energy balance was disturbed and they were brought to life.

In the Beginning...

Although energy products and services have been sold, bought, and traded since the beginning of time, the beginning of the energy markets, as we know them today, started in the 1970s with a big bang.

The U.S. and most of the other developed countries were asleep at the wheel and misjudged the global energy situation. One beautiful fall day we saw ourselves cornered by an oil embargo that we did not expect and which cost us dearly—a lesson we will never forget and one that serves us quite well. We are better prepared now, and this disaster should never happen again.

It all started in the fall of 1973, when the U.S. woke up with an Arab oil embargo, which slowly spread across the world. This meant that we no longer controlled the type and amount of energy we use. Factories shut down, gas pump lines went around the block, and the country's economy was faced with a serious risk of collapse.

Almost overnight, in the fall of 1973 the energy markets were clearly outlined, and energy security became the main preoccupation and goal of all developed countries.

The new reality became painfully obvious. While we were enjoying the cheap fossil supplies, most of them were controlled by foreign states. The wicked OPEC cartel had control over the global energy markets and our energy security was in the hands of the oil cartel, which included some enemy states.

This was not a new thing; it was something that the politicians simply ignored, thinking that since they control the purchasing end of the deal, they can also control the supply end. Wrong! Wrong! Wrong! Lesson learned? Partially so, but good enough...for now!

The "Seven Sisters " term was used since the 1950s to describe the seven oil companies of the "Consortium for Iran" cartel which dominated the global petroleum industry at the time. The Anglo-Persian Oil Company (now BP); Gulf Oil, Standard Oil of California (SoCal), Texaco (now Chevron); Royal Dutch Shell; Standard Oil of New Jersey (Esso) and Standard Oil Company of New York (Socony) (now ExxonMobil) were the seven pillars of the oil cartel at the time.

The Seven Sisters controlled around 85 percent of the world's petroleum reserves, but since the 1960s the dominance of these companies had declined due to the increasing influence of the OPEC cartel and state-owned oil companies in emerging-market economies, like Venezuela and Nigeria.

Everything changed in 1970s as the influence of OPEC grew substantially and the repercussions of the 1973 oil crisis shook up the global energy markets. The U.S. and EU politicians woke up from their decades-long slumber and started talking about energy issues, energy security, and developing energy market plans. They started creating national energy policies that encourage the development of an energy industry in a safe and competitive manner. This included the development of renewable technologies like solar and wind power generation.

The global energy markets were born, albeit under duress. They then developed as the major vehicle ensuring the achievement of the energy security goals. Unfortunately, the renewables were put back in the closet, when the oil crisis subsided. They were brought out several times since the 1970s, only to be shoved back in the closet following the original pattern.

Today most countries, including the U.S., have given up on the idea of achieving complete energy independence. This is mainly due to the fact that they lack one or another energy source needed for their ensuring 100% of their energy supplies. Instead, they are focusing on ensuring some resemblance of energy security.

That, however, is not an easy task...even for the U.S. Just take a close look at the daily problems in ensuring a safe passage of the oil tankers headed to the U.S., while going through the world's choke points. And the worse is still ahead...with fossil fuel price hikes first on the global energy horizon.

The Present Situation

In 2014, U.S. EIA demonstrated the profound nature of the rise in oil and gas production over the last decade: The proved reserves of U.S. oil and natural gas rose by the highest amounts ever recorded since 1977. This is stunning news, considering the general trend of decreasing production between 1980 and 2000, which now has been reversed.

Proved reserves of crude oil are now at their highest levels since 1991

"Proved reserves" are estimates of oil and natural gas deposits that may reasonably be recovered from known reservoirs under existing economic and operating conditions. The estimates change from time to time, of course, and depend on factors such as discoveries of new fields, reassessments of existing reserves, and the availability of new detection and recovery technologies. The overall extraction prices, which depend on the new technologies, play the most important roles in determining what's reasonably recoverable as well.

According to EIA: "The expanding application of horizontal drilling and hydraulic fracturing in shale and other 'tight' (very low permeability) formations, the same technologies that spurred substantial gains in natural gas proved reserves in recent years, played a key role." This pretty much explains what accounts for the record growth in reserve estimates of late.

Rising oil prices are to be credited in part, too, since they provided incentives for producers to explore and develop additional resources. This is a natural self-correction, since when the energy markets are allowed to work unobstructed, the new resources are usually developed in response to price signals. Today, both technology and the market's price signals combine to boost proved reserves of crude oil by more than 12 percent, and natural gas by nearly as much.

There were already some impressive reserves estimates issued in 2009; a record 22.3 billion barrels in crude oil proved reserves, which went to 25.2 billion barrels in 2010 and kept increasing with time.

These are truly extraordinary developments, worth celebrating especially when considering the economic benefits. Thousands of jobs, affordable and abundant energy supplies and new revenues for government through taxes and royalties. Those are achievements, which we all must applaud, no doubt.

All this helps in furthering our energy security goals too, but we must keep the distant future in mind too. We must never, ever, forget that no matter how much oil and gas we have, these are commodities with finite reserves. Sooner or later they will be depleted, so we must be mindful of the way we exploit and use them.

U.S. Transportation Sector

According to the U.S. EPA, the U.S. transportation sector includes air, ground (non-automobile), water transportation and transportation cleaning equipment. The sector is now classified under the 2007 North American Industry Classification System (NAICS) as:

Air Transportation (EPA Section 481)

Industries in the Air Transportation subsector provide air transportation of passengers and/or cargo using aircraft, such as airplanes and helicopters. This subsector distinguishes scheduled from nonscheduled air transportation. Scheduled air carriers fly regular routes on regular schedules and operate even if flights are only partially loaded. Nonscheduled carriers often operate during nonpeak time slots at busy airports. These establishments have more flexibility with respect to choice of airport, hours of operation, load factors, and similar operational characteristics. Nonscheduled carriers provide chartered air transportation of passengers, cargo, or specialty flying services. Specialty flying services establishments use general-purpose aircraft to provide a variety of specialized flying services.

Scenic and sightseeing air transportation and air courier services are not included in this subsector but are included in a special Subsector 487, Scenic and Sightseeing Transportation and in Subsector 492, Couriers and Messengers.

Although these activities may use aircraft, they are different from the activities included in air transportation. Air sightseeing does not usually involve place-to-place transportation; the passengers flight (e.g., balloon ride, aerial sightseeing) typically starts and ends at the same location. Courier services (individual package or cargo delivery) include more than air transportation; road transportation is usually required to deliver the cargo to the intended recipient.

Rail Transportation (EPA Section 482)

Industries in the rail transportation subsector provide rail transportation of passengers and/or cargo using railroad rolling stock. The railroads in this subsector primarily either operate on networks, with physical facilities, labor force, and equipment spread over an extensive geographic area, or operate over a short distance on a local rail line.

Scenic and sightseeing rail transportation and street railroads, commuter rail, and rapid transit are not included in this subsector but are included in Subsector 487, Scenic and Sightseeing Transportation, and Subsector 485, Transit and Ground Passenger Transportation, respectively.

Although these activities use railroad rolling stock, they are different from the activities included in rail transportation. Sightseeing and scenic railroads do not usually involve place-to-place transportation; the passengers trip typically starts and ends at the same location. Commuter railroads operate in a manner more consistent with local and urban transit and are often part of integrated transit systems.

Water Transportation (EPA Section 483)

Industries in the Water Transportation subsector provide water transportation of passengers and cargo using watercraft, such as ships, barges, and boats. The subsector is composed of two industry groups: (1) one for deep sea, coastal, and Great Lakes; and (2) one for inland water transportation. This split typically reflects the difference in equipment used.

Scenic and sightseeing water transportation services are not included in this subsector but are included in Subsector 487, Scenic and Sightseeing Transportation. Although these activities use watercraft, they are different from the activities included in water transportation. Water sightseeing does not usually involve place-to-place transportation; the passengers trip starts and ends at the same location.

Truck Transportation (EPA Section 484)

Industries in the Truck Transportation subsector provide over-the-road transportation of cargo using motor vehicles, such as trucks and tractor trailers. The subsector is subdivided into general freight trucking and specialized freight trucking. This distinction reflects differences in equipment used, type of load carried, scheduling, terminal, and other networking services. General freight transportation establishments handle a wide variety of general commodities, generally palletized, and transported in a container or van trailer. Specialized freight transportation is the transportation of cargo that, because of size, weight, shape, or other inherent characteristics requires specialized equipment for transportation.

Each of these industry groups is further subdivided based on distance traveled. Local trucking establishments primarily carry goods within a single metropolitan area and its adjacent nonurban areas. Long-distance trucking establishments carry goods between metropolitan areas.

The Specialized Freight Trucking industry group includes a separate industry for Used Household and Office Goods Moving. The household and office goods movers are separated because of the substantial network of establishments that has developed to deal with local and long-distance moving and the associated storage. In this area, the same establishment provides both local and long-distance services, while other specialized freight establishments generally limit their services to either local or long-distance hauling.

Transit and Ground Passenger Transportation
(EPA Section 485)

Industries in the Transit and Ground Passenger Transportation subsector include a variety of passenger transportation activities, such as urban transit systems; chartered bus, school bus, and interurban bus transportation; and taxis. These activities are distinguished based primarily on such production process factors as vehicle types, routes, and schedules.

In this subsector, the principal splits identify scheduled transportation as separate from nonscheduled transportation. The scheduled transportation industry groups are Urban Transit Systems, Interurban and Rural Bus Transportation, and School and Employee Bus Transportation. The nonscheduled industry groups are the Charter Bus Industry and Taxi and Limousine Service. The Other Transit and Ground Passenger Transportation industry group includes both scheduled and nonscheduled transportation.

Scenic and sightseeing ground transportation services are not included in this subsector but are included in Subsector 487, Scenic and Sightseeing Transportation. Sightseeing does not usually involve place-to-place transportation; the passengers trip starts and ends at the same location.

How many vehicles are we talking about? Many millions, to be sure! And this does not even include the private mopeds, motorcycles, cars, SUVs, trucks, vans, off-road, racing, and many other types of ground, water, and air vehicles Americans use daily. Nor does it include millions of private business vehicles; rental cars, executive cars, ambulances, industrial self-propelled equipment, etc.

The U.S. has the largest passenger vehicle market of any country in the world. There are over 275 million passenger vehicles registered in the country.

This number has increased steadily every year since 1960, indicating a growing number of vehicles per capita. All vehicles on the U.S. roads put together come to more than one vehicle registered to every American.

The number of commercial and business vehicles, combined with those used for special purposes (military, police, emergency services, etc,) is another large number, so we are looking at a huge number of vehicles that move many miles every day on their different tasks and destinations. They all have one thing in common; their thirsty engines need a lot of crude oil-based fuels every day. Without it, the army of vehicles will just stop.

The transportation sector consists of billions of tons of metal, plastics, and glass that move and provide us with all we need in our daily lives. Most of this movement requires crude oil and its derivatives: gasoline, diesel, jet fuels, different mixes, and lubricants. No oil, no movement. The consequences of immobility would be awesome and in some cases, life threatening.

Where do all these fuels, that we depend so much on, come from? Yes, Saudi Arabia, Iraq, Venezuela, Nigeria, and several other similarly unstable and/or unfriendly nations.

So, since we import half of the crude oil we use, at least half of our transportation sector depends totally on imported fuels coming from shaky political regimes. In case of a problem overseas, at least half of our transport network would stop moving. Which one would it be? There would be enough for the military, police, and emergency services vehicles, but how about the rest? How would we proceed with daily life?

One good thing to consider in such a case would be that if we don't have enough crude oil to run the billions of vehicles worldwide, the global environment would take a deep breath of clean air for a first time since the 1950s. Good for the environment, but devastating to our energy and national security, with the world economy taking a great hit.

OUR ENVIRONMENTAL FUTURE

What does the environment have to do with our energy and energy security? Nothing, if you are an oil company executive with six-digit salary, sitting in a plush office on the 20th floor of the corporate headquarters in Houston, Texas. But it is everything, if you live in Shanghai and breath the dirty air all day, every day. Or if you are a subsistence farmer in the Maldives, with your house 6 inches above the ocean level, which is rising daily, threatening to flood it and your fields.

What is energy security worth, if you lose your health and your livelihood? Without these, the locals must move away and look for new places to live and new fields for crops. A lot of misery is awaiting millions of people directly affected by the changing climate and rise in ocean level—some of which is due to rising air pollution levels.

Of course, thousands of people living next to the many new hydrofracking wells also have a lot to say about the health and other problems caused by contaminated ground water from the drilling sites. As the hydrofracking industry expands, so do the problems surrounding it. Thousands of acres of productive land are being poisoned and sacrificed for the sake of energy production. Fair? You be the judge!

There is another group of people who complete the picture—the wasteful users of energy, most of whom (this author included) live in the abundance and excesses of the American-type lifestyle. This group is only a fraction of the world's population, but uses the largest part of its energy and natural resources, and creates the greatest amount of global pollution.

These are the people who crank their heaters to turn the house into a sauna in winter, and who lower the A/C thermostat to make the interior feel like a freezer in the summer. They also drive huge, 3-ton SUVs, which get 5 miles per gallon going downhill, and emit more GHGs than a fleet of taxicabs in Hong Kong.

We also have 30-story cruise ships leaving dozens of ports on the Pacific and Atlantic oceans with thousands of happy tourists cruising in circles to nowhere and back while wasting thousands of tons of diesel and spreading its particulate pollution over the world's oceans.

We cannot ignore the enormous U.S. military machine with its megatons of iron—trucks, tanks, jet fighters, submarines, battleships, aircraft carriers, etc.—using huge amounts of fuel on a daily basis. The U.S. military hardware is spread all over the globe, so fuel is shipped and used all over the place too, creating large GHG pollution over the entire Earth. The amount of daily energy used by the U.S. military is greater than that used by many countries in a month.

This is only a small list of excess energy use and misuse that we see around us daily; yet, most of us close our eyes, refusing to see the reality, and hoping for the best.

"The era of procrastination, of half measures, of soothing and baffling expedience of delays, is coming to a close. In its place we are entering a period of consequences. We cannot avoid this period; we are in it now," said Winston Churchill in 1936.

Although this was said almost 80 years ago, Churchill's words sound even more relevant today. Was Mr. Churchill clairvoyant? Was he talking about something else? Or was this problem too obvious even to those who lived over three-quarters of a century ago?

Whatever the case, it is more than clear now that we are in such a period of half measures and procrastination. The fossil energy resources are being depleted very quickly and our environment is getting dirtier by

the day. Although there are talks on all levels, the situation is only getting worse…

But this was not supposed to be so! According to the Kyoto Protocol, an official commitment of 7% decrease of GHGs was agreed upon, but something went wrong, and things are just not going as planned. The Kyoto Protocol is an official international agreement on the environment linked to the United Nations Framework Convention on Climate Change. It was agreed to by a number of countries, and was adopted in Kyoto, Japan, on 11 December 1997, and entered into force on 16 February 2005. The detailed rules for the implementation of the Protocol were adopted at COP 7 in Marrakesh, Morocco, in 2001, and are referred to as the "Marrakesh Accords." Its first commitment period started in 2008 and ended in 2012.

All parties committed to set internationally binding emission reduction targets, recognizing that the developed countries are principally responsible for the current high levels of GHG emissions in the atmosphere. Since the developed countries have benefited from over 100 years of unrestricted industrial activity, the Kyoto Protocol places a heavier burden on the developed nations under the principle of "common but differentiated responsibilities."

In December 2012, the "Doha Amendment to the Kyoto Protocol" was adopted in Doha, Qatar. It includes:

- New commitments for Annex I Parties to the Kyoto Protocol who agreed to take on commitments in a second commitment period from 1 January 2013 to 31 December 2020;

- A revised list of greenhouse gases (GHG) to be reported on by parties in the second commitment period; and

- Amendments to several articles of the Kyoto Protocol, which specifically referenced issues pertaining to the first commitment period and which needed to be updated for the second commitment period.

During the first commitment period, 37 industrialized countries and the European Community committed to reduce GHG emissions to an average of five percent against 1990 levels. During the second commitment period, Parties committed to reduce GHG emissions by at least 18 percent below 1990 levels in the 8-year period from 2013 to 2020; however, the composition of parties in the second commitment period is different from the first.

The United States of America refused to be part of this agreement for many reasons, but we should take credit for the fact that today our GHG emissions—Kyo-

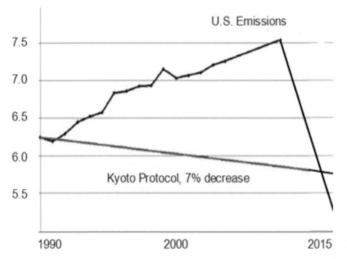

Figure 8-39. GHG emissions U.S. vs. Kyoto in giga-tons of CO₂ equivalent

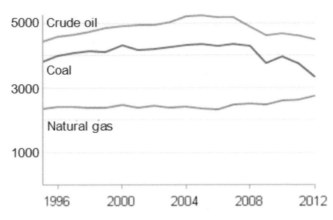

Figure 8-40. U.S. Fossil GHG emissions (in millions of tons per year)

to, or no Kyoto—have been reduced significantly all across the country. Regardless of the reasons—some of which have nothing to do with intentional reduction of GHGs—the U.S. is one of the few countries that can report reduced GHG emissions. We are now well within, and even better than, the Kyoto GHG limits.

At the same time, the country of the Kyoto Protocol—Japan—is now in dire straights and is cranking the coal- and oil-fired power plants, thus increasing the GHG emissions by over 15%. This is the case in many European and Asian countries too, which only confirms the fact that any agreements like the Kyoto are just a piece of paper. In most cases they are just a political move, designed to pacify the populace. Long-term activities and the results thereof speak louder and nullify paper agreements.

In 2012, the U.S. industrial complex emitted about 5.5 billion tons of CO_2, while the Kyoto target puts the cap on U.S. emissions at 5.8 billion tons annually. This unexpected (and we dare say unplanned) environmental compliance makes the U.S. the first major industrialized nation in the world to meet the United Nation's original Kyoto Protocol 2012 target for CO_2 reductions. And we didn't even sign the darn thing...

The reduction in GHG emissions in the U.S. is due mostly to the decrease of coal use for power generation, as it was slowly replaced by natural gas.

The steady increase of natural gas use, however, leads to increased emissions, rivaling those of coal. Yet natural gas is hailed as "cleaner" technology. How clean is clean? Does total amount count?

Industry insiders predict a 60% reduction of GHG emissions by 2016, due to the coal-to-gas conversion of older power plants. But how can we count on reduced emissions when the number of power plants increases. Less emissions per unit, multiplied by increasing number of units is a larger number than what we started with, according to simple math estimates. And natural gas has other problems too, which are as serious as those of coal. Not to mention that we cannot count on natural gas to last forever and use it as if its supply is endless, which it is not.

There was also some reduction in use of crude oil for transportation, due to a number of (mostly economic) reasons. Oil prices went up during the 2008-2012 economic slow down, and people simply stopped driving. Some also bought smaller and more economic cars, thus the decrease in oil use. Further decrease in emissions from the transportation sector is expected in the near future for these and other reasons.

But how much good does all this do for the world as a whole? The U.S. still emits about 5.5 billion tons of poison in the air every year, which is 5.5 billion tons too much to live with in the long run. In 10 years this is 55 billion, and in 100 years, 550 billion tons. This huge amount of poison is choking our atmosphere and creating all kinds of problems worldwide. But this is only part of the total picture.

The worse part is that while the U.S. and other countries are trying (somewhat successfully) to reduce energy use and GHG pollution, China, India and many other developing countries continue to increase their CO_2 emissions in their attempt to develop their industries and provide better lives for their citizens. As a matter of fact, China now leads the world with about 10.0 billion tons of annual GHG emissions, an amount quickly increasing with the addition of new coal-fired plants, millions of vehicles on the roads, etc. The U.S. is second, with slightly over 5.5 billion tons, or half as much, and dropping.

So what can we say? Don't the developing countries deserve the lifestyle we enjoy here in the West after a century of unrestricted fossils exploitation and use? Do we have a right to put limits on their energy use, CO_2 emissions and industrial development?

Later, we will take an imaginary walk into the not-so-distant energy-environmental future. But before we do that, let's take a close look at an experiment designed to predict and optimize that future. It was a "man vs. nature" trial done 20+ years ago in the Arizona desert, called Biosphere II.

It was a sensation at the time, aimed to prove the feasibility of living in a closed, self-sustained environment and learning how to live more efficiently with less energy and outside input. All this was supposed to help us live better and with less waste on Earth, as well as prepare us for life in space.

Biosphere II Experiment

Biosphere II was an expensive experiment conducted during the early 1990s in the Arizona desert. It consisted of several structures, spread over 40 acres of land, with the main building several stories tall and over 3 acres in size. It was an all-glass structure, designed and built as a totally closed and self-supporting ecological system.

Its construction and operation cost about $200 million, including the land, and peripheral support research greenhouses, test modules, and living accommodations for support staff.

The basic idea was to see if such systems could be built on other planets and to research and some of the environmental problems on Earth.

All work was to be done by several people living totally isolated from the outside—including air, food, and everything else—thus simulating, as closely as possible, the natural cycles on Earth.

Food was to be grown inside the structure, liquids were to be recycled, and air was supplied by specially designed and enclosed air-lungs; huge structures that moved up and down as needed. No outside input of materials was allowed. Heating and cooling water circulated through independent piping systems, and the building also used passive solar input through the glass panels covering most of the facility.

Electrical power, however, was supplied into Biosphere 2 from an onsite natural gas energy center. This

Figure 8-41. Biosphere II campus

raised many questions about the validity of the experiment from the very beginning, since there is no outside electric power on Mars, nor is it readily available in case of a global environmental disaster.

Nevertheless, there was enough money raised, so the experiment proceeded as designed. There were several areas in the structure based on biomes, an agricultural area, and human living and working space to study the interactions between humans, farming and technology with the rest of nature.

The representative biomes were:
- 1,900 square meter rainforest,
- 850 square meter ocean with a coral reef,
- 450 square meter mangrove wetlands,
- 1,300 square meter savannah grassland,
- 1,400 square meter fog desert,
- 2,500 square meter agricultural system,
- Human habitat, and
- Below-ground support infrastructure.

The agricultural area of Biosphere II was planted by the crew a year before sealing the structure, for growing and processing food, so that there would be a supply of food grown inside when the experiment (and full closure) began. This was another red flag, since the Martians might not allow us to go there to prepare for more permanent residence. Nor would Nature give us enough warning when deciding to shut down the Earth experiment.

But the experiment continued as designed... Several week-long periods of simulated full closure data were gathered on agricultural operations and productivity, and allowed the crew to get used to their workload. The structure was completely sealed from the outside with several researchers of different specialties living inside for the duration. The first stage of the project was conducted by a crew of eight people from 1991 to 1993.

The sealed nature of the structure allowed scientists to obtain some information on the continually changing chemistry of the air, water and soil contained within the closed space. At the same time, the health of the researchers inside was closely monitored by a medical team.

The effort produced results that are somewhere between controversial and contradictory. That stage was followed by a six-month transition—or hesitation—period during which researchers were not sure what to do next, so they conducted unplanned research and executed some building and system engineering improvements.

Then a second stage with a crew of seven people was conducted for a much shorter time between March 1994 and September 1994. Now, in addition to the inside problems to be resolved, an external dispute over management of the financial aspects of the project ended the effort prematurely.

Columbia University took management of the facility for research and as a campus in 1996 and changed the virtually airtight, materially closed structure designed for closed system research, into a "flow-through" system. The closed-system research was abandoned.

More realistic tests were performed by manipulating carbon dioxide levels for global warming research, by injecting desired amounts of carbon dioxide and induced venting as needed to simulate natural events. This lasted until 2003, when several years went by with the Biosphere's fate in the balance. In 2006, the area was slated for redevelopment into a planned community and the surrounding land was sold to a home developer to build homes and a resort hotel. The University of Arizona took over the research at the Biosphere II in 2007, supported by private gifts and grants.

Presently the Biosphere II is a tourist attraction with several tours daily. Simultaneously, scientists are conducting research projects, including the terrestrial water cycle and how it relates to ecology, atmospheric science, soil geochemistry, and climate change. During a visit in 2013, the author saw deteriorating flora and fauna and bleak signs of life, which did not appear to be of any real use to science or anything else.

In Summary

Although Biosphere II did not answer the $200 million question it was supposed to answer, it revealed and confirmed several important things related to the natural environment, and humans, mainly:

- All things environmental are interconnected and very complicated, and

- Humans are the biggest problem in the natural environment.

There were a number of design flaws that showed how little we know about materials and their interactions. Oxygen in the structure ran out only a short time after it was sealed. Excess fertilizer in the agricultural area absorbed most of the available oxygen and released CO_2, which created a living hell for the inhabitants.

Outside air was pumped in, which basically annulled the validity of the entire project and caused a public outcry. Most of the animals and insects died as a result of the CO_2 fluctuations...the multi-million dollar experiment quickly started losing scientific value.

Other problems appeared when overstocked fish clogged the filtration system; abnormal condensation made the "fog desert" too wet; morning glories overgrew the "rainforest" and blocked out access to light for other plants. Local ants were unintentionally sealed in the structure too, and their population grew so fast that an explosion of ants and cockroaches took over.

Most importantly, humans had the greatest problems, and were driven to the edge of their physical and emotional limits. This brought some abnormal decisions and actions, which caused disagreements and even fights among the crew members.

Problems continued later on too. Members of the first crew, for example, vandalized the structure and invalidated the second stage of the project by opening airlock doors and emergency exits, leaving them open for fifteen minutes. Glass panes were also broken, causing 10% of the biosphere's air to be exchanged with the outside during this time.

Soon after that, one of the members just left the Biosphere, to be replaced by another scientist, and later another was replaced as well. Then, completing the list of human issues plaguing the project, the ownership and management company of the Biosphere was officially dissolved in 1994. The second stage—and the entire project was ended prematurely on September 6, 1994.

The Biosphere II mini-mission was too short, poorly designed and even more poorly executed to provide any value of meaningful closed-area life, including agriculture or animal husbandry. No data were gathered that might have been useful in estimating whether the Biosphere itself was capable of sustaining eight people for two years.

The Biosphere II experiment did not provide any useful data on the main objectives, living in space, and helping the environment. There were, however, many questions arising from this experience and a number of lessons of how *not* to do things. The question, however, still persists. Can a closed system exist on Earth, or in space? Unfortunately, nobody else has tried to duplicate the Biosphere II experiment in an attempt to answer this question.

In conclusion, we can draw a parallel between what happened in Biosphere II and what is happening with the global environment today. The number of issues, and their seriousness, is growing. People continue misunderstanding, or intentionally ignoring, the issues. But the worst part is that we all disagree on how to solve the issues at hand.

Mishandling the issues negatively affects our energy supply, continues to pollute the environment, and makes the global climate change and its consequences even more serious. The big difference between the failed Biosphere II and our failing efforts is the fact that the scientists in the Biosphere locked up in the structure had several exit doors, while we here on Earth don't have any. We have no place to escape to, if and when things go deadly wrong, and they *are* headed in that direction… If we are smart enough, we will start thinking of finding or building exit doors too.

Walk In The Future…

By now we know all there is to know about energy (or at least its most important components), and how it affects our environment. The overwhelming fact is that we use fossils more than any other fuels, while at the same time we blame them for most of our energy and environmental problems. It might do us some good to take a quick glimpse at how life might be as the fossil fuels begin to vanish.

Let's take an imaginary walk into the period of transition to a fossil-less energy existence—life during the hard years of the energy transition.

We reached the oil peak. Natural resources and fossils in particular became too difficult and too expensive to extract and needed to be replaced by something cheaper, better and more plentiful. That was time for hard, sometimes life-changing, decisions, following decades of indecision and half measures (which we are clearly living with today).

Amazingly enough, after all the years of talks on the subject, our sophisticated and advanced society woke up one day in the 22nd century with no substitutes for coal, oil, and natural gas. Even more amazing, we were not even paying much attention to the approaching end of the fossil

fuels when suddenly faced with jumping off the proverbial fossil-less energy cliff.

So, imagine that in the beginning of the 21st century, oil peaked at just over 74 million barrels per day (mbpd) and even increased slowly. There was some additional oil that brought the total to about 84 mbpd (equivalent), but it was not the conventional crude we are accustomed to; it was rather a number of unconventional hydrocarbons, such as natural gas and coal derivatives, "extra heavy" oil, synthetic oil from Canadian tar sands, biofuels, and some renewables like solar and wind.

Crude oil production dropped soon after that, mostly because the oil fields were getting depleted quickly, and new deposits had become progressively smaller, of lesser quality, and much harder (and more expensive) to exploit. The decline rate accelerated over 5-10% per year as of 2025. By 2050-2075 crude oil was on the endangered species list.

Natural gas was hailed as the energy problem solver (albeit temporary), but it peaked around 2050 too, declining sharply soon thereafter—a short-lived glory, that brought with it a number of environmental disasters, human suffering and death.

Coal was more plentiful but its use decreased, while the production of cheaper natural gas increased. When the prices of natural gas and crude oil went sky high, coal was the only fossil on the menu of the global energy industry. Since its use increased several fold, coal was also depleted shortly after 2100.

Figure 8-42 depicts our vision of the global energy future, where the depletion of oil, gas, and coal will be complete by the end of the 21st century.

Cruise ships were grounded. All unnecessary air travel ceased.

The American Dream downsized to the max. By the end of the century, electricity prices sky-rocketed and even midget electric cars became luxuries.

Then the transition to non-fossil fuels started, slowly and painfully when the extraction of methane hydrates

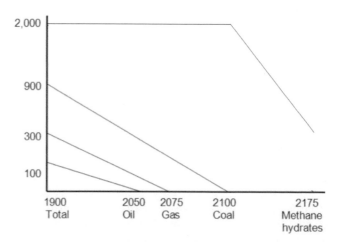

Figure 8-42. Depletion of fossil reserves (in billions of tons)

from the permafrost was fully developed, with other sources (advanced peat, shale, etc. new fossils production) so the world had another type of fossils for another century.

Methane hydrates were abundant globally, keeping the world power plants and transport running, but that bonanza, as the others, lasted only several decades.

People had enough electricity, and were able to drive electric cars, recharged by electricity from the new methane hydrate power plants. Cruise ships were modified to use the new methane gas from the new deposits, but even that energy source was exhausted by the end of the 22nd century. There was just no way around it; this is a hungry and needy world that needs to be fed, clothed, and transported every day. And because to most people only today matters, we are not used to worrying about tomorrow. Not really. We talk about it, and find ways to make money on the subject, but we do not like to sacrifice for sake of tomorrow.

By the 2250, all fossils on Earth were gone. A real transition from fossils to non-fossil types of energy became an absolute and unavoidable reality. Little was done in terms of preventing major energy transition disasters until then, so the work finally began in earnest. Amazingly, as many times before, during extreme duress great things happen.

The world's economies were forced to live within shrinking, instead of expanding, energy budgets. Having gone through the cycle of commodity price spikes several times, all indications pointed to economic destruction, leading to supply destruction, leading back to price spikes, and continuing the cycle. The entire world economy was in a free fall and down-spiraling for a long while.

The first consequence was the end of globalization as we know it. The days of shipping raw materials to Asia, then shipping back ready-to-use (albeit of low quality) gadgets were gone. People had to learn how to make things with their own hands—just as our predecessors did long ago—and how to live within the local boundaries. Every nation and locale was on its own.

Americans are less than 5% of the world population but are accustomed to consuming 25% of its energy and goods, so they were hit the hardest. The American dream was reduced to a daily fight for survival of the fittest. Production of all goods was localized; food and other goods traveling an average 1,000 miles before reaching the markets was no longer feasible. Foods would now come from nearby fields.

High prices and lack of fertilizers forced the use of organic crops and alternative growing methods. The good old days of using 10-15 calories of fossil fuel input (diesel fuel and fertilizers) for every calorie of food were gone.

Bedroom communities, 50 miles from the industrial centers were converted into self-sustaining independent entities with specialized local economies. The long daily drive, and sitting in idling cars was seen only in old movies. Instead, mass transport vehicles and moving walkways were the new transportation methods.

Renewable energy was the only solution and a lot of ef-

fort went into its final development and implementation. But the big question remained; "Why that was not done ahead of time?" The answer was obvious: Human nature dictates to take care of me first, and worry about today only. And so, year after year the inevitable was swept under the rug, until it became a reality. Then things changed overnight and something had to be done quickly.

The Environmental Connection

We don't need to dig too deeply to imagine the future. The signs of increasing global devastation are quite obvious today around the world, and we can see them here in the U.S. as well.

In the near future we will see much more of that and some new and more ominously negative events taking place, such as ocean levels rising—where Venice, the Maldeves, the Florida coastline, parts of Louisiana and the Netherlands are flooded and remain under water for good. With their homes and businesses inundated by the rising waters, the population of these areas will be forced to migrate.

Many parts of the world will be hit by unprecedented draughts and others by killer floods. Worse, excess methane hydrate exploitation will accelerate global warming when the permafrost starts melting, and ocean drilling of those enormous fossil reserves expands. The unpredictable origin of these resources could bring accidents, the destructive nature of which could dwarf that of the Chernobyl and Fukushima disasters.

Transitions are usually hard, and very often tragic, events. The fossil-less energy transition would be the most serious global crisis ever. The consequences will be dramatic and even tragic, but can eventually have a very positive outcome too.

It all depends on which path the governments chose. Solar, wind, geothermal and ocean resources could provide the total energy the world consumes every day, if and when the people of the 22nd century and later learn how to harvest them, and if their 25th century descendants follow through.

They could build energy efficient, carbon-free economies, put an end to climate change and avert resource wars. They would have healthier food and a safer, more resilient and equitable, world where all nations and peoples live in peace and harmony. This is a dream after all, and in dreams everything is possible, so dream with us…

But Back to Our Environmental Present

We see a lot of changes in the energy markets and the global environment. What are these changes, and what do they do? Energy production and use have been blamed for environmental disasters such as rising ocean levels due to global warming. At the same time, environmental disasters have been blamed for erratic energy use such as the Fukushima accident forcing the entire country of Japan to shut down the clean energy generating nuclear power plants and crank up the coal-fired power plants which are now belching 15-20% more pollutants into the atmosphere.

So, is this a catch 22 situation that feeds on itself and propagates beyond our understanding, let alone our ability to control it.

The Future of Climate Change

We all know that fossils create excess pollution and that they must be replaced by clean burning renewable technologies. The greatest argument against doing this is that there is not enough money to expand the effort as needed. Recent estimates show that we need over $550 billion every year from now until 2030 (or a total of over $10 trillion) to implement the transition to 100% sustainable energy. This is needed not only to preserve the precious fossils from extinction, but to also keep the global temperatures below a 2°C rise.

To just initiate such a program, doubling the present level of investment is needed at the very minimum. This would scale-up renewable energy to about 25-30% of the total energy mix. At the same time, it would reduce carbon emissions to an acceptable level, thus helping mitigate the catastrophic impact of climate change.

According to the experts, most of the initial investment would have to go towards solar, wind and hydropower, which are the "low hanging fruits" of the energy market. Such development would also create nearly 1 million jobs by 2030. It would also save $100-200 billion a year in health related costs, as a result of shutting down fossil power plants.

The greatest challenge to the concept of increasing the share of renewable energy today is financing the end users of these technologies. With other words, the renewables market is primed, and the customers are willing and waiting, but have no money to buy its products.

The June, 2014 Renewable Energy Map issued by IRENA (a consortium of 130 states and the European Union, working on cooperation in renewable energy) shows that global solar deployment needs to increase by a factor of 12, which would provide $740 billion of annual savings on environmental costs from burning fossil fuels. This alone would pay for the investment costs required to reach the 30-35% renewables fraction of the energy mix.

Unfortunately, the U.S. and the rest of the world,

are not ready for this undertaking. Unless the majority of world's governments step up their efforts, this is just not going to happen. We need to act now, because "later" might be too late.

During 2014, the European Investment Bank provided $1 billion in support of renewable energy development. At the same time, $219 million was invested by the Danish Climate Investment Fund from various pension funds, and $8.4 billion was invested in renewables by Norway's oil-generated sovereign wealth fund, Government Pension Fund Global. Not a bad start, but far from the $550 billion needed to start the jump to renewables process.

While IRENA and some of the EU states show a united front in the area of renewables, there are disagreements around the globe that are causing confusion, delays, and even conflicts—all hindering the overall effort.

The Global Division

There is a serious problem of trust among nations that is reflected in ineffective international environmental agreements. It is the mistrust between the old industrialized nations of the West on one side, and the newly developing countries, like China, India and Brazil on the other.

The developing economies want the West to take care of the climate problem, since it created it. In addition, they want the West to compensate them for their existing and future problems related to environmental changes.

This is a hidden debate that makes difficult things even more difficult. At a recent conference in Warsaw, an agreement had been reached on a climate deal, when a Fiji delegate issued an objection, protesting a single formulation in the treaty document. "There is absolutely nothing to write home about at the moment," he said. That annulled the entire agreement and derailed the negotiations.

The most important demand of the developing nations is to be given financial aid when struck by natural disasters resulting from climate change, since climate change is resulting from Western countries' created pollution.

The developing nations demand a new institution to be created that would manage the financial aid awards and distribution, while the West wants it to be handled by the existing framework. The fear is that they would be held personally liable if a new category and managing institution are created.

A new Warsaw Mechanism for Loss and Damage

was agreed on to be renegotiated at the 22nd climate conference in 2016. The Warsaw treaty also states that the losses and damage caused by weather catastrophes go beyond the scope of adaptation. Ultimately, the agreement delays any real decision.

A new universal climate agreement is intended to be agreed on in 2015 in Paris, which will include concrete goals for curbing CO_2 emissions. It will also seek to limit the rise of global temperatures to 2 degrees Celsius maximum. So this is work in progress, but as Warsaw experience shows, Paris 2015 will have to deal with the same issues that separate the two blocks—the developing and developed countries.

For the time being, six UN funds support some poor countries in tackling climate change. The Western countries promised in Warsaw to provide larger contributions to the funds, with Germany serving as one of the leading donors. A preliminary agreement calls for the industrialized countries to provide $100 billion per year to developing nations, starting in 2020. The intent of this assistance is to help mitigate the effects of climate change. A working group was established at the Warsaw conference that is charged with developing the new finance plan. One of the intended uses of the funding is reforestation projects worldwide. More exploration and research are needed to determine the extent to which planting trees benefits the climate.

The world is split on the most important issues, which prevents agreement. Right now the issue is the amount and timing of payments to developing nations. They want at least $70 billion per year starting in 2016, but the industrial nations had paid only $10 billion annually during the 2010-2012 time period.

There are no commitments for any payments between 2013 and 2019, except Germany's contribution of about $2.5 billion and Europe's total contribution $600 million to the "Fund for the Poorest Countries." The total amount might rise to $4 billion by 2020, but these are still preliminary plans.

The developing countries are frustrated by how slow things are moving, and especially by the lack of solid commitment on the part of the EU and US. Whereas earlier international drafts mention "commitments" by the developed countries, now the language is changed to read "contributions" to the cause of global environmental health. Not a small change, thus the frictions.

So the world is divided in two polarized groups—the developed and developing countries. Each has its own problems and suggested solutions to a common goal, but there is no common agreement on how to proceed. It is the age-old battle between the haves and the have-nots. The gap is getting wider and there is no bridge, nor even an agreement to build one. Because of that, we see lots of conferences coming and going without any concrete results.

So, we have problems. But the world has always had problems. What makes today different from yesterday? Different from 100, or even 1,000 years ago? The biggest difference we see is that in addition to the natural problems, which existed then and exist today, man has added a series of new, man-made problems.

The natural and man-made problems have different origins and behaviors, which is complex and hard to understand, let alone control. When the natural and man-made problems are mixed, however, the situation becomes many times more complex and impossible to control.

Let's take a look at these problems:

MAN-MADE PROBLEMS

Not all problems related to environmental pollution are man-made. As a matter of fact, Nature is a significant contributor to the developing energy-environmental picture, and we will take a look at that below.

Man-made problems, however, are not to be ignored for they are a major contributor to everything related to energy and the environment, and affect our energy security in a big way.

Air Pollution

We discussed the issues of air pollution in detail in the previous chapters, so we will only summarize them here:

- The most important pollutant—CO_2, SO_2, and particulate matter emitted from power plants—is responsible for a major part of the global air pollution, and which is blamed for the climate change.

- The GHG gases emitted from natural gas-fired plants are about 2 times less than those from coal-fired plants, but the increased use of natural gas around globally means that no significant reduction in total GHG emissions is expected anytime soon.

- Exhaust pollution from transportation vehicles is another major issue, that is held responsible for air quality deterioration in major cities.

- Acid rain, which results from excess air pollution is also hurting people and damaging buildings and infrastructure.

- Excess particulate matter from motor vehicles (diesel mostly), volcanic activities, dust storms, etc. causes a number of environmental problems, including creating a dust cover on glacier ice, which makes it melt faster than predicted.

- Methane gas is emitted by natural gas production, distribution, storage, and processing sites. It is over 20 times more potent as GHG than CO_2, so even though it is emitted in smaller quantities, its contribution to air pollution is quite significant.

- Methane gas is also emitted in huge quantities by animals on feeding farms.

Case Study: Cattle GHGs

In a 2012 report on U.S. greenhouse gas emissions, the U.S. Environmental Protection Agency (EPA) reported that the U.S. cattle fleet emits over 140 million metric tons of methane annually.

Nope, this is not a typo; cattle fleets emit as much GHGs (mostly methane gas released from both ends of the animals) as more than a dozen coal-fired power plants put together.

Methane happens to be 23 times more potent than carbon dioxide as a greenhouse gas, and the agricultural sector (not just the cattle fleet) is responsible for over 8% of the total U.S. GHG emissions, so this is serious.

Enteric fermentation from ruminant animals is the largest anthropogenic source of methane in the U.S. The natural gas systems are the second largest source with 127 million metric tons annual GHG emissions.

In 2006, the United Nations Food and Agriculture Organization released a report on environmental issues related to livestock which recommended cutting in half the "environmental costs" of livestock, which includes greenhouse gas emissions, and land and water degradation caused by these activities.

According to the UN environmental specialists:

"When emissions from land use change are included, the livestock sector accounts for 9% of CO_2 deriving from human-related activities, but produces a much larger share of even more harmful greenhouse gases. It also generates 65% of human-related nitrous oxide, which has 296 times the Global Warming Potential (GWP) of CO_2. Most of this comes from manure."

As the economies of developing nations improve, their populations eat more meat. The UN projects meat production to grow from 240 million metric tons in the early 2000s to nearly 500 million metric tons by 2050.

The milk output is expected to almost double from 580 million tons to over 1,000 million tons during the same time period.

The sharp increase in demand for meat and milk will require more GHG-emitting cattle worldwide, so we can expect an equal increase of GHG gasses. So, the present 140 million tons of methane emitted annually by the U.S. cattle fleet will double and triple during the next several decades.

This, added to the total world cattle fleet means that we are to expect over 1 billion tons of methane to hit the Earth's atmosphere every year from 2050 on.

Presently 21% of global fossil-related GHG emissions comes from natural gas, and 44% from coal—both burned for power generation. 35% of the global GHG emissions come from crude oil, most of which is used in transportation and some for power generation.

An average U.S. coal-fired power plant emits about 2,000 lbs. of CO_2 per each MWh, so an average 1.0 GW power plant at 80% operational capacity will emit about 14 billion lbs. of CO_2 annually.

This is about 7 million tons annually from our average coal-fired power plant—and—the global cattle fleet has the potential of generating the GHG emitted by more than 140 such coal-fired power plants.

Remember that methane is 23 times (and nitrous oxide is 296 times) more potent than CO_2 as far as its GHG potential is concerned. So if we multiply the above numbers by 23 and 296 respectively we come up with an astronomically high GHG amount, equal to that emitted by 3,200 coal-fired average size power plants, which is close to the global GHGs emitted by power generation emissions around the world.

The world will not reduce—let alone stop—meat and milk consumption, so we must brace for a significant growth in the global cattle fleet. With it, we will see a huge influx of GHGs which will do significant damage to the quality of our environment and will contribute to serious deterioration of the air we breathe.

Soil and Water Pollution

Oil spills, chemical releases and spills, and a number of other soil- and water-polluting activities abound today in the U.S. The oil and gas bonanza-turned-energy-frenzy is causing damage on a large scale. Most of it remains unreported, but we see bits and pieces of the negative effects brought upon the locals by the oil and gas drilling rigs and production wells.

Oil Contamination

Oil companies reported over 1,000 oil spills in North Dakota in 2011 alone. This is only a small portion of the many accidental spills going unreported there, and nationwide. But this is not a secret. Everybody, including government officials, knows this, but not much is done for many reasons.

The law is not perfectly clear on these issues, so the boundaries are often blurred and willingly crossed by some of the participants.

Some of the reported oil spills include an Exxon-Mobil pipeline that ruptured and spilled over 45,000 gallons of oil into the Yellowstone River, near Billings, Montana. The pipeline has been transporting tar sands oil from Alberta, Canada; a low-grade, dense, more toxic and corrosive type of oil than the normal crude oil types. Amazingly, the regulators were unaware that the pipeline had ever carried tar sands oil, but the evidence after the spill was reported pointed in that direction. This might have been the cause of the accident to begin with. Either way, now the river and the locals must deal with 45,000 gallons of oil floating in the river.

Another great example of contamination due to faulty design and construction of an oil well is the BP Gulf blowout disaster in 2010. Enough is known about this unprecedented case, so we won't spend more time on it. It is, however, a good example (on a large scale) how even the best plans fail. It also reminds us that oil leaks are quite possible anywhere, anytime, and that many of the small ones will remain willingly or unwillingly undiscovered and/or unreported.

Chemicals Contamination

Storage of waste water from hydrofracking sites is currently under the regulatory jurisdiction of the different states, many of whom have weak to nonexistent policies protecting the environment. So, the amount of chemical contamination from drilling waste liquids increased more than 5,100 percent over the past decade, partially due to the fact that the regulations have not kept pace. We now have official estimates that more than 550,000 tons of such chemicals were used, and some were dumped in local rivers, streams, or directly into the soil in 2013 alone.

Drilling and oil production operations, combined with process chemicals and waste liquids hauling, go on 24/7/365 non-stop at thousands of sites in the Dakotas and other states.

Chemical additives are used in the drilling mud, slurries and fluids required for the fracking process. Each well produces millions of gallons of toxic fluid containing not only the added chemicals, but other naturally occurring radioactive material, liquid hydrocarbons, brine water and heavy metals.

Fissures underground created by the fracking process can also create underground pathways for gases, chemicals and radioactive material. These are difficult to trace, and impossible to control and maintain. The industry denies that such things could occur, pointing to the sophistication and safety features of the new technology.

Because of this opposition, the claims of the locals that land and water have been contaminated are filed as unreasonable and unsubstantiated. One such community is Pavillion, Wyoming. Here the local claims were also ignored until the U.S. EPA and USGS recently confirmed what the residents had been claiming for awhile. There is a significant evidence that hydrofracking fluids and gasses had contaminated their soil and groundwater.

In another incident, EPA, initially under an emergency administrative order, forced three major oil companies operating in the area to reimburse the city of Poplar, MT, for damages of their water infrastructure, which resulted in expenditures needed to remediate the damages from drilling contamination.

Of course, the oil companies appealed the EPA order, but were eventually forced by a federal judge to comply and rectify the violations. What is fair is fair, but that still doesn't rectify all the problems in the area, because the contamination could pop up anytime in the same or nearby areas. Then, the locals are back to square one—fighting in court with the oil companies.

Earthquakes

Reports of earthquakes in areas that have never had such events are increasing, mostly in areas with heavy hydrofracking activities. This is another problem related to deep-well oil and gas drilling operations. It is no wonder that things underground would change for the worse when large tunnels are dug out, flooded, and aggressively exploited.

Scientists call the earthquakes caused by the injection of fracking wastewater underground during well exploitation "induced seismic events." In other words, we are now dealing with unprecedented numbers of small earthquakes all over the country. Most of these events are small ground shakes as far as magnitude is concerned (although some measure well over 5 on the Richter scale.

This adds more uncertainty to the overall hydrofracking and fossils exploitation equation. Here we have millions of gallons of toxic chemicals underground and even more waste liquids stored above ground, combined with unprecedented earthquake activities. This is what nightmares are made of, and it understandably adds to the anguish of the locals. It also makes the long list of fossil fuels' externalities even longer, adding another layer of risks and dangers.

NATURAL EVENTS

There are places around the world, and some states in the southwestern U.S.—Arizona, Nevada, New Mexico—that have a few natural problems. There are no earthquakes, hurricanes, floods, tsunamis, land slides, or other natural disasters to speak of in this part of the world. Hot summers, dust storms, and brush fires are the few seasonal events people worry about.

On the other hand, there are areas, that are threatened periodically by a number of natural disasters. The situation is made worse by the ongoing environmental and climate changes, which bring a new dimension to the threats at hand—not unlike the multiple whammy at Fukushima's nuclear power plant.

Although the threats in the most affected places are different, they have one thing in common: climate change is making them more serious and difficult to manage.

Some examples of the world's most dangerous places:

Tokyo, Japan, with its 38 million inhabitants is under constant threat of earthquakes, monsoons, river floods and tsunamis. The Tokyo-Yokohama region is considered the riskiest place to live in the entire world. The locals are fully aware of the danger and are expecting a very large earthquake to hit the place any day, and yet they prefer to live there and take their chances.

Tsunamis are another big danger Japan must live with. It is an island, fully exposed to the ocean whims. Tsunami risks abound along its long coast line, with its many populated areas. Many urban centers are located along or near the Ring of Fire, the active faults of the western Pacific, with their never-ending grumbling and shaking activities. The Great Kanto Earthquake of 1923 devastated Tokyo and Yokohama, killing over 142,000 people. Another one is predicted to hit the area some time this century.

On top of that, today Japan must worry about nuclear radiation spreading from the Fukushima accident. Tokyo is not very far from the damaged nuclear power plant, and the continuing radiation leaks from it are making people in the capital—as well as in most of Japan—quite nervous.

Manila, Philippines, is located just off the Philippines' earthquake trench, and is one of the most dangerous places one could live. Severe earthquake risks and high wind speeds are a constant reminder of the dangers. Recently, one of the most severe typhoons, Haiyan, swept the country and destroyed several central islands, ruined the coastal city of Tacloban, and killed thousands in its path.

Los Angeles, CA, is located on top of the San Andreas Fault. This makes it one of the most earthquake-prone cities in the world. Fifteen million people living in the area daily face the possibility of an earthquake. Several earthquakes have hit Los Angeles through the years, some causing major damage and death.

The Big One, as they call the future great earthquake, is predicted to happen within this century, at which point not much of Los Angeles and the surrounding area would be left standing. Many other populated centers in California will be affected to different degrees. Amazingly, most people living there just shrug their shoulders when reminded of the danger. They are not well prepared, and yet not worried, which would suggest that they rely on luck to survive the Big One.

As the environmental and climate changes progress, we witness increased numbers of natural disasters, which also deliver more punch and devastation. How far these changes will push the natural disasters is unknown, but the people who live in the endangered zones must be prepared for the worst. Some of the natural events we must deal with today are mentioned below.

Droughts

Droughts around the globe are affecting the climate (or is it the other way around—hard to tell), so that agricultural crops, animals, and many people are affected. It is so bad, in fact, that even the Global Drought Monitor website crashed in 2013, displaying a morbid sign, "IMPORTANT NOTICE: 19th November 2013." The server running the UCL Global Drought Monitor has malfunctioned beyond repair.

"We are currently exploring possibilities and will provide an update on the future of this service as soon as possible. Please accept our apologies for any inconvenience." A sign of the times…?

This sounds like the end of the world, but is actually just a reminder of how futile human efforts are to control Nature. Even efforts to monitor Nature are doomed to fail…

Seriously, periodic droughts are not uncommon, and we often hear the "seven-year drought" theories. While there is something to that, Nature is much more complicated than putting labels on her actions.

The seven-year drought in Arizona, for example, is going on over ten years now, and is actually getting worse, with no signs of change.

At the same time, California seems to be entering its seven-year drought period, that seems to be worse than any seen before. And yes, there were many such droughts in its past, so we should not be surprised, nor attribute it to any one factor.

Some droughts today are caused by global warming (according to the scientists). Elevated temperatures increase the evaporation of water (evapo-transpiration) from land and water surfaces and from plants due to evaporation and transpiration.

This leads to increased drought in many areas, which felt worse in the drier regions of the world. There, evapo-transpiration produces long periods of drought, which are basically lower water levels of rivers and lakes. Eventually, the groundwater level drops too, which brings reduced soil moisture in the agricultural areas.

As a result, there has been a significant decline in precipitation in the tropics and subtropics since the early 1970s. At the same time, southern Africa, the Sahel region of Africa, southern Asia, the Mediterranean, and the southwestern U.S. are getting noticeably drier. Even the relatively wet areas are experiencing longer dry conditions.

Worse, there is a noticeable expansion of dry areas worldwide, not only in the Saharan desert sands. There are estimates that the amount of land affected by drought will grow, and water resources in these areas will decline over 30 percent by 2050.

These changes are thought to occur, at least partly, due to an expanding atmospheric circulation pattern known as the Hadley Cell. Here, warm air in the tropics rises and loses moisture to tropical thunderstorms, which usually dissipate over the ocean or over wet areas which don't need that much water—sometimes causing huge floods.

The dry air from the tropics descends in the subtropics as more dry air, creating more drought conditions. As jet streams continue to shift to higher latitudes, and storm patterns shift along with them, semi-arid and desert areas are expanding steadily.

Once flourishing gardens are being converted into parched lands with little hope of recovery. The big problem is that droughts in key states usually mean reduction of agricultural production. This results in higher food prices and deficit of agricultural products. From an energy security point of view, extended droughts mean reduction of the amount of hydropower and biofuels we can produce. Of course, energy prices will increase accordingly too.

One of the best examples of the negative effects of extended droughts on energy supplies can be seen in the sugar cane fields of Brazil. Brazil runs on ethanol, which comes from sugar cane. Sugar cane is the most efficient biofuel raw material, but it needs a lot (with a capital L) of irrigation water.

Brazil, fortunately, has a lot of water (mostly from rain), which allows it to grow a lot of sugar cane. This in turn has allowed it to process it into ethanol, which fuels 90% of its transportation sector since the 1970s…until now.

The persisting global climate change, accompanied by extended drought periods are threatening to put Brazil's transport system back on crude oil dependence. One of the worst droughts in three decades slashed Brazil's northeastern sugar cane crop in 2013-2014 by as much as 30 percent in some areas and even more in others.

Brazil's north and northeast regions account for only about 10 percent of national cane output, but the crop is an important source of sugar and ethanol at home and abroad when the main center-south crop is idle between harvests. The Alagoas region, which begins harvest in September, typically crushes cane to produce sugar and ethanol through March, or about six months, just as the center-south starts up harvest of the new crop.

This time, cane processing operations ended sooner simply because there was not enough cane to crush. The northeast produced only 52 million tons in 2013, down 15% from the record 62 million tons in 2012. Sugar output from the region also fell by 15% from the over 4 million tons produced in 2012.

Some states saw a 20-25% drop in cane output as a direct result of the fact that the cane fields received only 25-30% of the normal rainfall in 2013. This was the worst crop since 1983. When cane and sugar production is down, ethanol production suffers, and gas pump prices go up accordingly. If the drought continues, Brazil may have to import ethanol (and more crude oil) in order to supply its gas stations.

California's Drought

California is in the midst of an unprecedented drought. The state is not a stranger to dry conditions, but in terms of severity and length, the present 7-year drought is unprecedented. New estimates show that the drought is so severe that it's causing the ground to rise. What?

Scientists estimate that 63 trillion gallons of water have been lost since 2013 alone. So what happens when 63 trillion gallons of water disappear? A lot. This is about 240 billion tons of water that normally fills lakes and rivers. As the water evaporates, the ground shifts… in an upward direction. There are measurements in California's mountains, where huge ex-lake bottom areas have lifted as much as half an inch.

California's water supply is in three large reservoirs, all of which are at roughly 30 percent capacity. Other, smaller, water reservoirs are doing better. The statewide lowest average of 41 percent was recorded in 1997, when another devastating drought struck the state. The present drought is nearing the 1997 record, and as things are going, it will set new records.

Figure 8-43. Lake Oroville…err, stream Oroville, more precisely.

Lake Oroville is (was) one of California's largest water reservoirs. Now it is a skinny steam flowing where the lake bottom was…a near-record low. So what does that do to the state's energy production and the overall economy? What is a state running out of water to do?

In August, 2014, The California Public Utilities Commission (CPUC) ordered all state water companies under its jurisdiction to provide direct notice to their customers of mandatory water use restrictions. These are accompanied by potential fines in case of non-compliance with the State Water Resources Control Board's Emergency Regulation for Statewide Water Conservation.

Starting in February, 2014, CPUC adopted drought procedures for water conservation, rationing, and service connection moratoria for regulated water utilities. In April, 2014, Governor Brown issued an Executive Order to strengthen the state's ability to manage water in drought conditions.

Then in July, 2014, the Water Board adopted an Emergency Regulation that prohibits the use of drinking water for outdoor landscapes in a manner that causes runoff; the use of a hose without a shut-off nozzle to dispense drinking water to wash a motor vehicle; the application of drinking water to driveways and sidewalks; and the use of drinking water in a fountain or other decorative water feature, except where the water is part of a recirculating system.

Violation of the prohibited actions brings a fine of up to $500 for each day in which the violation occurs. Additionally, all CPUC jurisdiction water utilities are ordered to comply with the Water Board's requirements in implementing either a) mandatory outdoor irrigation restrictions or b) mandatory water conservation measures.

Utilities must include notice of the implementation of either the mandatory outdoor irrigation restrictions or mandatory water conservation measures as required and as part of the required customer notification the CPUC required in its last order.

The CPUC-regulated water utilities are to implement water conservation measures consistent with the Water Board's mandate. CPUC will closely monitor the water utilities' progress in encouraging water conservation and consider further action if warranted. CPUC-regulated water utilities will also take steps to monitor and promptly inform customers about leaks and to work with customers to stop leaks and water waste, consistent with best practices being implemented by other California water utilities.

The Future of California

In anticipation of increasing drought conditions, the city of Carlsbad, California, received final approval to build a $1.0 billion desalination plant. It took 12 years of planning and over 6 years of permitting battles with the California Coastal Commission, State Lands Commission and the Regional Water Quality Control Board, but now the city has a 30-year Water Purchase Agreement with the San Diego County Water Authority for the entire output of the plant and is ready to go.

Construction on the plant and pipeline is already underway and the plant will be delivering water to the businesses and residents in San Diego County by 2016. When operational, the plant will use huge amounts of electricity for the desalination process, to pump, process and deliver 50 million gallons of drinking water every day, which will be sent to local homes and businesses.

This amount, however, is only 5-6% of the total water demand of the service area, so it is a little bit more than the proverbial drop in a bucket. The new desalination project will also increase the water bills to each household by $5-6 every month. Is it worth it? In the words of a city official, "When you turn the water tap on and no water comes out, price becomes insignificant."

So, is it worth it? This is the question California must answer soon, because as the draught continues, additional desalination plants are scheduled to be built.

Obviously, California cannot exist on desalinated water, so the government is taking all feasible steps to reduce water use, including curbing outdoor water use, all through the drought crisis. At the same time, Governor Brown called on all Californians to (voluntarily) reduce their water use by 20-30 percent and to eliminate all kinds of water leaks and waste.

All this is fine and will help with managing the drought, but what if it continues for 7 more years? What if it never goes away? What will happen to the Golden State? What will the Silicon Valley crowd do? Where would the millions of people in San Francisco and Los Angeles go? There are no clear answers.

How does all this relate to our energy security? It does so directly and profoundly with long-term negative consequences. As the drought intensifies, it dries lakes and rivers, which in turn reduces hydropower generation. This forces other base-load power plants (usually fossil-fired) to increase power generation with predictable results.

More importantly, the drought also brings the water table in the state lower and lower. Since all thermal power plants (coal, natural gas, nuclear, and concentrated solar) use millions of gallons of water daily for cooling, they are directly and negatively affected. It costs more to pump water from deeper under ground, and some power plants might have to be shut down for fear of complete depletion of precious local water resources.

What then? Reducing the power generation and shutting down power plants would add to the misery in the most heavily populated, and richest state in the nation. The consequences of energy reduction would be felt heavily by the entire state economy. The end results are unpredictable, but the longer the drought goes on, the more disastrous the results will be.

Sooner or later, the present draught will end, so the state can recover and get back to normalcy. But climate experts predict that other draughts, even more severe and of longer duration, are expected in the future. It is not certain how much the state can take and how long it could fight, but the choices are few.

Floods

Globally, floods are becoming more frequent and dangerous. In the U.S., floods have always been a way of life in some flood-prone areas. Some floods in the past were exceptionally damaging, so in response to heavy flood damages in the 1960s, the National Flood Insurance Program (NFIP) was established. It is a federal program created by the U.S. Congress through the National Flood Insurance Act of 1968.

The program enables property owners in participating communities to purchase insurance protection from the government against losses from flooding.

The intent was to reduce future flood damage through community floodplain management ordinances and provide protection for property owners against potential losses through an insurance mechanism that requires a premium to be paid for the protection.

The NFIP is meant to be self-supporting, though in 2003 the GAO found that repetitive-loss properties cost the taxpayer about $200 million annually.

As the global climate changes, and flood risks increase, the national budget simply would not be able to support the expected flood of floods predicted to hit our rivers and shore communities.

Earthquakes

San Francisco, 1906—an earthquake of 7.8 magnitude destroyed the city. Chile, 1960—the largest 9.5 magnitude earthquake ever recorded in the 20th century. Scientists say that another big one is overdue, so it is just a matter of time, before we awake someday to see images of destruction in major cities.

The quake expected to hit the Pacific Northwest of the United States is estimated to pack a power punch of a 9.0 or more on the Richter scale. It would also cause a massive tsunami like the one that hit Indonesia and India in 2004. Tsunami activity has been recorded and evidenced on the coast of Japan showing that about every 300 years these large quakes occur. The last one hit in 2011, causing a lot of destruction and irreparable damage to the Fukushima nuclear power plant, which is still a major disaster zone.

Volcanoes

There are a number of these sleeping giants around the world. Yellowstone National Park is a beautiful place on the surface, storing underneath a massive reserve of magma, hundreds of miles in diameter, waiting to blow.

If just one of the underground calderas exploded, it would set off a chain reaction creating a big hole where Yellowstone sits now. It would also bury most of the continental U.S. and parts of Canada and Mexico in a blanket of ash.

Such an explosion would cause massive destruction of farm production in the U.S. Some even predict an unprecedented type of winter weather, where the sun would be blotted out from the landscape. This event could be apocalyptic. But we don't know when it will happen, so we carry on as usual.

Tsunamis

The U.S. east coast could someday become the biggest surf ever seen. There is a volcano on the Canary Island called Cumbre Vieja that is expected to explode in the near future, causing a massive landslide as a major part of the island collapses and slides into the Atlantic Ocean. The displacement of the huge land mass would cause a massive wave surge towards the American eastern coastline. Waves of over 75 feet could be expected to wash inland for miles, taking everything in their path.

Tornadoes

In 1925, Missouri, Illinois, and Indiana were hit by a massive tornado that ripped quickly across the three states. This tornado was much more fierce than any tornado known to man. It cleared a 200-mile path, destroying buildings and farms and causing massive damage to everything in its path. There were more than 700 people killed and 2,000 injured. This was the strongest and costliest tornado ever, which even raised the classification in tornado ratings to a class five.

Regular tornadoes are harmful enough, but a tornado like that one is especially dangerous, for if it struck a major population center, it could wipe out the entire area and the people in it. And there is no stopping such a monster...

Storms

We have seen a lot of storms and hurricanes in the north Atlantic, some of which cause significant damage such as Katrina and Sandy. With the changing environmental conditions, there is a chance of three of more such storm fronts to combine and grow into a massive category five or greater event.

Falling in full force upon the mainland U.S. or the Caribbean Islands, the damage would be catastrophic. Storm surges, flooding, and tornado activity would devastate any area the storm contacts. The economic impact could cause damage many times worse than hurricane Katrina.

Though unlikely to happen, changing weather patterns create unpredictable combinations, and just one of these storms could bring any nation to its knees.

Freezes

Many blizzards have hit the northeastern United States, accompanied by several feet of snow fall. These disasters are expected to grow in severity and length as the global climate changes. An extreme blizzard of extended duration could paralyze a major population center, taking a toll on the population and its agricultural sector.

Massive blackouts, transportation and communication interruptions, fuel shortages, and loss of major public services would result. Such a freeze is not likely, but planning is of paramount importance.

Heat Waves

Heat waves are a serious problem in the Southwest U.S., where the temperature in summer reaches 120°F. Without air conditioning, life in the desert areas would slow significantly and might even cease altogether. Interruption of electrical services during the summer months, caused by storms, or other natural events could be catastrophic for these areas.

The Southwest is also suffering from prolonged and very serious droughts. The lakes and rivers are drying, and the entire area—including its economy and agriculture—is suffering. California's unprecedented 7-year drought is devastating the state's agricultural and fishing industries. Hydropower generation is dropping too, contributing further to the problems in the Golden State.

Gas Anoxia

As discussed in this text, there are massive reserves of methane hydrates (natural gas trapped in icy form) under the ocean floor. These deposits have been laying dormant in their icy graves for millions of years, without causing any harm. When disturbed by volcanic activities or landslides, however, they emit huge quantities of natural gas.

Because the gas is toxic, large emissions could cause serious damage. As a matter of fact, they have been blamed for the mass extinction of wildlife and humans. For example, a methane gas pocket erupted in Lake Nyos in Africa in 1986, in an event called "limnic eruption." The gas spread on land, killing 1,700 people and 3,500 livestock in nearby towns and villages.

There are similar methane hydrate deposits in the ocean floor along the California coastline, where a major earthquake could trigger a massive release of the gasses. This would add to the devastation of the earthquake, so people must be aware of such possibility. Otherwise, the limnic eruption would have horrifying end results.

While all these events are possible, albeit not very likely and limited in their amount of damage, there are others that are much more far-reaching. One of these is the possibility of a strong solar storm hitting the Earth.

Solar Storms

Solar storms are powerful electromagnetic discharges created constantly on the sun's surface. Most are small in energy content and speed and do not travel too far be of any danger to us. Too, most of the dangerous ones are shot away from Earth.

From time to time, however, a series of powerful solar flairs are shot in our direction, creating enormous danger. The waves of solar radiation pack an huge punch.

So, if a powerful solar storm hits the Earth head-on, anything in its way might be evaporated with extensive damage to hundreds of miles around the epicenter. The super-powerful electromagnetic discharge could wipe out many satellites circling the Earth and cause a global communication blackout within seconds.

A powerful solar storm could knock out power plants and their grid, essentially freezing all commerce, travel, finance, and human services. Food would spoil, banking would cease, mass panic would ensue.

Strong solar storms have hit the Earth in the past, but the results cannot begin to compare to what could happen under the same circumstances today. *Now* we rely on power grids and telecommunication satellites in every area of our lives. These networks are most vulnerable to the damaging effects of a powerful solar storm.

Solar storms reaching Earth in the near past include:

The Carrington Event of 1859

The Carrington Event was a solar flare (or coronal mass ejection) that impacted the Earth's magnetosphere and induced the largest known solar storm to date.

Numerous sunspots were observed on the Sun in late August, 1859, and just before noon on September 1, the first observations of a huge solar flare were made. It caused a major coronal mass ejection (CME) to travel directly toward Earth, followed by a second that took only 17.6 hours to reach Earth, instead of the usual 3-4 days. The second CME moved so quickly because the first one had cleared the way of the ambient solar wind plasma.

On September 1-2, 1859, the largest geomagnetic storm ever recorded occurred. Aurorae were seen around the world, those in the northern hemisphere even as far south as the Caribbean. Those over the Rocky Mountains were so bright that their glow awoke people who thought it was morning. People in the northeastern US could read a newspaper by the aurora's light. The aurora was visible as far from the poles as Cuba and Hawaii.

Telegraph systems all over Europe and North America failed, telegraph pylons threw sparks, telegraph operators experienced electric shocks. The telegraph network was so charged that it continued to send and receive messages even though the power supplies were turned off.

The Quebec Solar Storm of 1989

A severe geomagnetic storm—the result of a coronal mass ejection on the Sun's surface several days earlier—caused the collapse of Hydro-Québec's electricity transmission system. The storm began on Earth with extremely intense auroras at the poles, which could be seen as far south as Texas and Florida. Some people thought this unusual light was a first-strike nuclear attack, while others considered it to be associated with the Space Shuttle mission STS-29, which had been launched earlier.

The burst caused short-wave radio interference, including the disruption of radio signals from Radio Free Europe into Russia. Around midnight, an ocean of charged particles and electrons in the ionosphere flowed from west to east, inducing powerful electrical currents in the ground that surged into many natural nooks and crannies.

Some satellites in polar orbits lost control for several hours. GOES weather satellite communications were interrupted, causing weather images to be lost. NASA's TDRS-1 communication satellite recorded over 250 anomalies caused by the increased particles flowing into its sensitive electronics.

Today…

In July, 2012, a massive, and potentially damaging, "Carrington-class" Solar Superstorm shot near Earth, missing it at the last moment. According to NASA, there is a 12% chance that a similar event may happen to Earth in the next ten years, and greater chance in the more distant future.

In June, 2013, a joint UK-US research team used data from the Carrington Event to estimate the current cost of a similar event to the U.S. at $0.6-2.6 trillion. It is hard to make these assumptions and estimates, because they are directly related to the intensity of the solar storm and the impact location. But for the sake of argument, let's imagine a huge solar flare strike over New York.

First, since everything is driven by electricity and computers, there would be instant darkness and silence over the entire area. Power grid, computers, and telecommunications would all be down—no lights, no fuel, no transportation, water and food shortages, and lots of chaos.

This is not a scenario we can plan for or prevent. Because of the concerns that utilities have failed to set protection standards and are unprepared for a severe solar storm such as a Carrington Event, the Federal Energy Regulatory Commission (FERC) has proposed a ruling that may require utilities to create a standard that would require power grids to be protected from severe solar storms. It is not clear how this would be done on a large scale, but with today's technological advances solutions are always available. Is money also available for such an undertaking?

One area of extreme vulnerability is our large-scale solar and wind power plants. Sprawled over thousands of acres in open desert land, they would act like huge antennae, attracting the charged particles of a strong solar storm and getting obliterated in the process. In a split second, billions of dollars and years of hard work would go up in smoke. Cleanup of the resulting molten mass of metal and glass, mixed with silicon and other chemicals (some toxic) would be a very expensive undertaking. Yet, there has been no effort to address a potential disaster and take some preventative measures.

In contrast, the Nuclear Regulatory Commission has begun a phased rule-making, published in the Federal Register, to examine the sufficiency of cooling systems of stored spent fuel rods of nuclear power plants. Other systems, considered vulnerable to long-term power outages from events such as space weather, high-altitude nuclear burst electromagnetic pulse or cyber attacks, are also evaluated, to design and implement preventative measures.

Summary

All these events and risks affect our energy supplies, our energy security, our national security, and our way of life. Since we have no way of predicting, let alone controlling, most of them, we are exposed to a number of serious dangers.

The U.S. has a number of organizations and procedures for handling some of the more common natural disasters. Most of these deal with managing the after-effects of events. We know what to do when a storm or hurricane hits a populated area or when a neighborhood gets flooded. Yet, we can only react. We also know how they affect our energy supplies, our power generation, and the supply of motor fuels, so we respond accordingly. So far we have been successful in handling most natural events, but how about those with which we have no experience? How would we get electric power and fuel to run the emergency equipment and transport emergency response teams? There are partial solutions, like stand-by generators in hospitals and power plants, but there is no complete solution to date.

OUR ENERGY SECURITY

This subject requires in-depth understanding of the technical, political, financial, and logistic issues. To start with, we must look back on our history, and especially the global industrial development during the last 200-250 years.

The Industrial Revolutions

The world had a long way to go before getting to the present level of technological development and prosperity. It started with the first industrial revolution (IR-I), which took place between 1760 and 1840.

IR-I was most famous for the development of new, more efficient, manufacturing processes. This included going from all manual production to using machinery and machine tools in the chemical, iron, textile, and other production processes.

IR-I began in Britain and spread to Western Europe and the U.S. within a few decades. New and more efficient methods of using energy were the key factor in the success of the industrial development in this period. Among the most important developments was the improved equipment and methods of using water power as well as increased use of steam power (by burning coal) to drive machinery and processes. This change alone provided large quantities of energy for added efficiency of the industrial and agricultural sectors.

The most important changes during IR-I were possible because of the revolutionary transition from wood, and biomass, to coal use.

IR-I was a major turning point in history, affecting almost every aspect of daily life in one way or another. It contributed to unprecedented income growth—the first time in history—when the living standards of the masses began to increase significantly. This was the precursor of our modern capitalist economy.

While economists argue about the actual results, they all agree that IR-I is the most important event in the history of humanity since the invention of the wheel and the domestication of animals and plants.

The first industrial revolution evolved smoothly into the second industrial revolution (IR-II), which started around 1840-1860. Technological and economic progress continued with the increasing use of better and more efficient steam-powered boats, ships and railways. Progress was especially noticeable in the large-scale manufacturing of machine tools and the increasing use of machinery in steam-powered factories.

The real change started when technological and economic progress gained momentum with the development of the internal combustion engine and electrical power generation. These technologies opened new horizons for manufacturing and transport that brought unprecedented comforts in the home and amazing efficiency and quality improvements in the global industrial complex.

Crude oil and electric power were the drivers of IR-II. Natural gas joined the party later. The party is still on…we just don't know how long it will last.

Then began industrial development as we know it today. Gas and diesel powered vehicles increased in number like mushrooms after a pouring rain. New roads and highways were constructed to accommodate the new transportation system, which added another element of efficiency to the global industrial growth.

Electric power plants and distribution networks delivered electric power to homes and businesses. The increased use of electric power alone is credited with the majority of the technological developments and benefits we enjoy today.

Now, we are on the verge of transitioning to industrial revolution three (IR-III). It is marked by the transition from the traditional fossil-fuels of IR-II to more efficient and cleaner means of energy generation—the renewables. A transition away from the fossils is coming sooner or later, no doubt. We are actually already in it and the signs of change are all around us, but this is only the beginning.

Today we also see another, equally important development (IR-IV), we call the "green energy revolution" (GER). It is actually a gradual transition from IR-III, marked by the distinct difference in efficiency and environmental attributes of the different energy technologies and sources, as well as the increasing competition for dominance among them.

This time, however, we don't see a smooth transition from one to another type of energy source or technology, nor do we see a clear distinction between any of these. Instead, all technologies and energy sources with their mixed-up properties and corresponding ups and downs and cons and pros will compete. They are all also undergoing major changes during their development within the energy mix. Yet, they are all moving in parallel as if in a race for dominance of the global energy future.

Some of the new technologies are very efficient and cheap, but polluting, while others are not as cheap or efficient, but are much cleaner. Coal is cheap, but very dirty. Natural gas is cleaner, but more expensive and still dirty. Solar and wind energy are more expensive, but very clean. Bio-fuels are somewhere in between. Because of these differences, some people prefer one type, while others prefer another.

The most important aspect of the new revolution is the transformation we see in all sectors. Coal, for example, is cleaning up its act by conversion into cleaner fuels via gasification. Natural gas is contributing by increasing its efficiency via combined cycle power plants and running vehicle engines. At the same time, wind and solar are optimizing their performance reliability and efficiency, while getting the price down.

Finally, we also see far into the future another energy revolution (IR-V), which we call the "post-fossils" energy revolution, or PFR. This is time in the future when the fossils would be depleted and forgotten. What comes in their place is anyone's guess, but we must assume that the renewables will play a major role in the energy mix at that time.

There is a lot of activity in the energy field. Can we find a way to mix and match the needs and requirements with the properties of the different technologies and energy sources? This is what the energy markets must do, if we are to jump over the fossil-less future cliff.

And speaking of energy markets…

Energy Markets

The energy markets are related to anything and everything energy: energy supplies, energy equipment, energy products, energy professionals, energy and energy related services, and energy use. This includes the entire cradle-to-grave process of production, distribution, and use of energy sources.

Note: The subject of energy markets is so large that it requires a book of its own, so we will only touch on the key basics here.

Energy markets start with someone needing energy. This entity directly or indirectly commissions someone else to find the energy, which initiates the energy supply phase of the process. Thus discovered energy sources must be extracted, which is the second phase. The extracted energy products are then sent to a power plant, or fuel processing facility, which initiates the third phase of the energy cycle. The final stage of the energy cycle is the delivery of the actual energy products to the end user via power grid, pipelines, gas stations, etc.

In all stages of the long journey of each energy source, there are thousands of people using their hands and brains with the help of complex pieces of equipment. This is an army of engineers, technicians, operators, managers, consultants, specialized professionals, and laborers.

Each step of the process is also assisted by another army of investors, bank executives, government officials, regulators, insurance agents, lawyers, security experts, and many types of consultants.

All of these people and their activities drive the global energy markets. In this text we cover a number of these activities.

The entire cradle-to-grave energy production and use process (which is a major driver of the global energy markets) consists of:

Energy Source Production
- Request and need for energy
- Initial investment and legal work
  - Private banks and investors
  - Government finance
- Energy source locating and discovery
  - Equipment and services
  - Operations
- Energy source development
  - Mine construction
  - Equipment and services
  - Operations
  - Risks

Electric Energy
- Energy source procurement
- Energy generation
  - Power plants construction
  - Equipment and services

- Power plants operation
- Power plants decommissioning
- Energy distribution
  - The National Electric Grid
  - Equipment and services
  - Operations
  - The smart grid
- Risks

Vehicle Fuels
- Energy source procurement
- Energy source processing
  - Refineries construction
  - Equipment and services
  - Operations
- Fuels transport
  - Pipelines
  - Trains, trucks, barges
- Retail distribution and sales
- Risks

Petrochemicals (commercial use)
- Energy source procurement
- Petrochemicals processing
  - Production facilities construction
  - Operations
  - Final products transport
    - Pipelines
    - Trains, trucks, barges
- Retail distribution and sales
- Risks

Conventional energy sources
- Firewood
- Coal products

Environmental Aspects
- Environmental markets
- CO_2 trading
- Government and legal action
- Risks

Financial Aspects
- Private investments
- Government finance and subsidies
- Energy stocks, options, and futures trading
- Risks

Energy market transactions
- Domestic energy markets
  - Production issues

— Transport issues
- International energy markets
  — Trade
  — Global shipping
- Security
- Risks

Political and Regulatory Aspects
- Government direction
- Political decisions
- Regulatory compliance

Legal Ramifications
- Policy challenges
- Lobbying issues
- Lawsuits

Future Energy Concerns
- Fossil-less lifestyle options
- The role of the renewables
- New technologies and approaches

We will be discussing these subjects and sub-subjects in future texts, since we just don't have enough space to cover them all in detail here. For the purposes of this text, we review the energy markets in terms of global and domestic developments and the related risks, as follow.

Table 8-8. Major components of energy production and use.

The global energy markets include:
- Production of Energy Sources
- Transport of Energy Sources
- Imports and Export of Energy Sources
- Energy Security, and
- Energy Trading

The domestic energy markets include:
- Production of energy sources,
- Power plants construction and operation, and
- Vehicle fuels processing and distribution

International Energy Markets

Energy production, distribution, and use is a huge international business. All countries are involved in some sort of energy buying, selling, and use. The global energy markets have entangled the entire world in a delicate web.

The main aspects of the international energy market are:

- *Production of Energy Sources* is a huge market, which includes mining, drilling, and extraction (digging and pumping) and the related operations in many countries. Coal, uranium, natural gas, and crude oil are produced in varying quantities at different times during these operations. Production schemes are complex and depend on a number of geo-political and socio-economic factors. We address the basics of these in this text and in full detail in our other books on energy and environmental subjects. See references 15, 16, and 17.

- *Transport of Energy Sources* is another significant market. It consists of transport of raw materials from the production sites to processing facilities, power plants, and other consumers. The transport of huge amounts of crude oil, coal, and natural gas from exporting to importing countries across the world's oceans is another big market.

- *Imports of Energy Sources* is an international affair of great proportions. Most countries import crude oil, natural gas, coal, and electricity daily. This has triggered the formation of several international commodity markets where these products are bought, sold, and traded.

- *Energy Security* requirements of the production, transport, distribution, and use of energy sources have created another big international market. Here, products and services designed to ensure the safe and reliable production and delivery of energy sources are provided by specialized companies all over the world.

- *Energy Trading* is a big business, which consists of trading (physically and on the commodity market). The physical trading is the actual sales and purchase of electricity and fuels (crude oil, natural gas, and uranium). This trade is conducted among countries and different companies daily. Commodity trading is done via several crude oil and other energy markets, where individuals, companies, and countries buy and sell stock and commodity futures.

Domestic Energy Markets

Each country has its own internal energy demand and supply requirements and needs. Some of the developing countries produce and/or import enough energy to fulfill the bare necessities of their populations. Most of the people in these countries don't have access to the official energy sources (electricity, gasoline, etc.), so they have developed and use their own energy sources (wood, coal, etc.). These countries could survive a global energy crunch, since they are disconnected from the international energy markets.

Other countries, like the U.S. which consumes about ¼ of the world's energy every day, consider energy a driving force behind their economic success. The U.S. and most of the OCED countries would not be able to survive a global energy crunch. At the very least, they would have to lower and otherwise drastically change their way of life and business.

The main aspects of the domestic energy market are:

- *Production of Energy Sources* domestically includes mining, drilling, and extraction (digging and pumping) and the related operations. Coal, uranium, natural gas, and crude oil are produced in varying quantities at different times during these operations in different areas of the country.

  The production and distribution schemes are complex and depend on a number of geo-political and socio-economic factors. They also vary from country to country, and even in different areas within a country.

- *Power Plants Construction and Operation* is a booming business in many countries. Power plants provide power that makes or breaks economies, so the importance of their construction and operations cannot be understated. Activities in this area have created a large market of primary and secondary products and services.

- *Vehicle fuels* processing and distribution can trigger the development of huge infrastructures for processing and distribution companies and facilities across the entire country. A number of thriving industries can be developed around these activities as well.

Energy Markets

As we saw above, the energy markets are related to anything energy: supplies, products, services, and use. This includes the production of energy sources, the imports of such, and using them for power generation and transportation fuels.

The energy markets determine our energy security strategy and directly affect our national security.

In financial terms, energy markets are commodity markets that are especially designed for, and deal exclusively with, the trading of energy related products and services; i.e., the energy market is a place or service, where energy products and services are bought and sold.

In the broader sense, energy markets involve—directly or indirectly—the entire process of locating, constructing, producing, transporting, processing, supply chain management, and distribution and use of energy materials and products, and the related services.

In more detail, the global energy markets (anywhere energy is produced, used, or in anyway involved) are shown in Table 8-9.

All products and services that are produced for, or from, energy sources are integral parts of the global energy markets. Coal, natural gas, crude oil and its derivatives; gasoline, diesel, heating fuels, electricity, as well as the renewable energy products and services are the foundation of the energy markets. From these, a number of industries develop, or branch out to form the entire network of products and services that comprise the global energy market in its entirety.

Table 8-9 is a partial list of the key products and services that comprise the global energy markets. A complete discussion on the subject deserves a book of its own, to do it justice. We have plans for such a book, where we can take a closer look at the U.S. and global energy markets. For now, we will settle for a general discussion on the subject.

Energy drives our economies and provides our necessities and comforts. No energy, no economic development, no comforts. It is as simple and as serious as that.

Energy Security Markets

Energy security...national security...? Are these terms related, and if so, how? To begin with, we must emphasize that it is widely believed that there are a number of risks created by America's energy policies and practices. Some of these constitute a serious threat to our national security—from military, diplomatic, and economic points of view.

We cannot get too deep in this subject here, but it suffices to take a close look at the U.S. military's presence around the world and try to figure out what portion of this presence is geared towards ensuring our energy security. How much energy is needed to run the daily operations on land, sea, and air, to achieve a certain level of energy security?

A significant portion of this force is geared towards protecting the crude oil transport channels. Huge amounts of energy (electricity from coal and natural gas,

Table 8-9. The major components of the global energy market.

Mining (coal, uranium, silicon, special metals, etc.)
— Exploration and location services
— Engineering design services
— Mine equipment manufacturing
— Mine construction
— Mine operation and maintenance
— Transport services

Natural Gas and Crude Oil Production
— Exploration and location services
— Engineering design services
— Well equipment manufacturing
— Well construction
— Well operation and maintenance
— Transport services

Hydro- and Geo-power
— Exploration and location services
— Engineering design services
— Production equipment manufacturing
— Site construction
— Site operation and maintenance

Biofuels
— Exploration and location services
— Engineering design services
— Production equipment manufacturing
— Agricultural field operations
— Processing operation and maintenance
— Transport services

Solar and Wind Power
— Exploration and location services
— Engineering design services
— Production equipment manufacturing
— Production operation and maintenance
— Transport services

Power Plants (all types)
— Exploration and siting services
— Engineering design services
— Power plant equipment manufacturing
— Power plant construction
— Transport services
— Power plant operation and maintenance

Transportation Sector
— Vehicles and equipment manufacturing
— Fuel transport and storage
— Mass transport
— Personal transport

Government/Military
— Contracting services
— Technical services
— Equipment manufacturing
— Civilian transport services

Peripheral Products and Services
— Commodity trading
— Physical and cyber security
— Political and regulatory services
— Safety and environmental services
— Legal and financial services

crude oil, and nuclear power) are needed daily to ensure the efficient function of the military machine with protects our energy supplies and safety.

The energy security markets ensure energy security at home in terms of political, diplomatic, and military actions. For the U.S., this translates into, "utilizing everything in the U.S. government's power to ensure reliable and cost-effective energy supplies." We have enough coal and natural gas for now, so the internal energy security is covered. Crude oil is the only missing piece in the overall energy security puzzle, so it is one of the key priorities of our international policies.

Energy security starts at the energy production sites, protects the import-export transport routes, watches over the power plants, and ends up with the final, safe, efficient, and cost-effective distribution and use of the energy sources.

The world relies on energy supplies to fuel the basic requirements of our lifestyle, it is indispensible, and it has become the backbone of most countries' economic development, which in turn determines the growth of the world's economy.

The developed countries need a lot of energy to maintain their lifestyle, while the developing countries require increasing amounts to advance their national development. In all cases, energy is in the focal point of most governments. Energy has an extremely high priority for the U.S. government. Due to increasing risks and threats across the energy supply (crude oil transport) routes, some of our energy resources have become more vulnerable, so we have stepped up security measures in these areas.

To protect power plants and energy resources, many countries are focusing on securing them from internal and external physical attacks and/or cyber attacks. These attacks are getting more sophisticated and more violent in nature, and are carried out by skilled terrorist groups and cyber criminals, who use increasingly modern tools and methods.

The need to protect the energy sources production, transport routes, and power generation facilities has created a new type of global market, the so-called "energy security market."

A number of measures are available to protect energy resources and routes against threats of attack on supply and transport infrastructure, accidents, natural and unnatural disasters and rising terrorism and cyber attacks. Various intelligent security solutions which enable the providers to integrate, collect and analyze the security networks through the data generated by their control and data acquisition grids are under development.

The basic energy security segments, on which the security solutions are focused, are physical security solutions and network security solutions. These can be further divided by types of products and services, and by the different world regions.

Global energy security markets can be identified

and sorted by the revenues and the trends in each of the principal and the related sub-markets, as shown in Table 8-10.

It is obvious from this long (yet incomplete) list that different physical and network solutions, or combinations and permutations of these, would be employed in different situations and varying regions. This makes all solutions complex and expensive, but changing world conditions demand such responses. As a matter of fact, we are just now scratching the surface of the upcoming physical and cyber battles on the global energy markets.

Because of that, we see the energy security markets growing exponentially across the globe in the near future. Even now these markets are estimated in the billions of dollars, but these numbers will grow exponentially. As the competition for energy and terrorist threats around the world increase, we will hear much more about the energy security markets.

Energy Security Markets Today

As with all products and services in a capitalist society, energy security is defined by the marketability of its products and services. These basically involve providing a degree of security and safety of daily operations to ensure reliable and affordable energy supplies.

Ensuring an acceptable level of energy security is

Table 8-10. Energy security market segments

Physical security solutions and services:
- Microwave intrusion detection
- Perimeter fencing and IR fields
- Anti-terrorism methods
- Secured communications
- Video surveillance and CCTV
- Transportation security
- Detectors and access control
- Air and ground surveillance
- Costal surveillance
- Over and underwater surveillance
- Fire detection and alarm systems
- Personnel tracking and RFID
- Identity and access management
- Building management systems
- Scanning systems
- Biometrics and card readers

Network security solutions and services:
- Distributed denial of service (DDoS) protection
- Computer networks firewalls
- Antivirus/malware programs
- Intrusion detection systems (ITS)
- Intrusion prevention systems (IPS)

- Unified threat management (UTM)
- Security information and event management (SIEM)
- Disaster recovery services
- Incident management programs
- Supervisory control and data acquisition (SCADA)
- Chemical, biological, radiological, nuclear and high-yield explosives (CBRNE)
- Hazardous materials (HAZMAT)

Professional services:
- Risk management services
- Anti-terrorism action groups
- System design and consulting
- Compliance, policy and procedures consulting
- Audit and reporting

Region:
- North America (NA)
- Europe (EU)
- Middle East Africa (MEA)
- Asia-Pacific (APAC)
- Latin America

a complex, never-ending, and very expensive activity. It involves thousands of specialists in physical and cyber security, engineering, computer science, insurance, legalities, political and regulatory issues, and many other areas. These activities require the use of sophisticated equipment and instrumentation, special facilities and infrastructure, and a lot of expenses. They combine to comprise the energy security market.

The global energy security market is expected to reach nearly $100 billion by 2020. North America is expected to own over 30 percent of the global share of this technology-based market with myriad new and untested products and services.

The energy security market is driven by political pressure, government regulations, terrorist threats, and other factors related to ensuring the security of the energy sector. Increased activities of late in these areas are contributing to rapid growth of North America's energy security market.

Rising investments in energy resources and infrastructure around the globe are triggering a sharp demand for providing their long-term security. This is not an easy task, and often requires the most advanced technological solutions.

The effort of providing security to the entire energy system is complicated by the fact that there are many different components, most of which are spread over large and often remote areas. These are the elements of the national power grid (high power transmission lines) with thousands of miles of wires stretched around the country and thousands of substations installed far away from populated centers.

There are also thousands of miles of pipelines, collection and distribution centers, and gas and oil transport infrastructure crisscrossing the country—mostly in remote areas. To ensure the safety of these installations, and avoid terrorist attacks, surveillance along with remote monitoring technologies are increasingly deployed.

Video cameras, intrusion detection, supervisory control and data acquisition (SCADA), aerial and underwater surveillance are being used by the power companies to protect their infrastructure from attacks and natural disasters.

New and emerging technologies have been focused on in recent years, such as unmanned air and water vehicles, and underwater surveillance systems deployed to secure the perimeters in remote areas of installations.

Note: Power generating facilities and the power transmission infrastructure have been classified by government agencies as primary vulnerable targets of terrorists and cyber attackers. This has led to industry-wide push to deploy technology and methods in answering the call, which is fueling the growth of the market.

Governments in the developed countries, including the U.S., have developed laws and regulations that define security standards, which need to be deployed by the owners and operators of power plants and the related infrastructure. This includes facilities that develop, generate, and store nuclear energy and materials, and all oil and gas energy, thermal and hydropower and renewable energy entities.

Ensuring the physical security of these facilities requires a number of sophisticated of technologies and approaches. These include, but are not limited to, microwave intrusion detection, perimeter fencing, IR isolation fields, video surveillance, motion detectors and other access control devices, air and underwater surveillance, fire detection and alarm systems, building management systems, and remote hazardous and nuclear materials detection technologies.

The energy security market is projected to grow, but political instability and economic stagnation might reduce the estimates in the near future. Nevertheless, it is a needed measure to protect our energy sources and way of life.

The experts forecast the global energy security market to grow 6.5 percent annually over the period 2013-2018. One of the key factors contributing to this market growth is the increased occurrence of cyber threats. The global energy security market has also been witnessing an increase in the construction of smart grid projects. The unpredictability of cyber attacks poses a real threat to the growth of this market, so a lot of work is done in that area too.

Energy security is one of the new comers in the energy field, and as such it is immature and in many cases the results are uncertain. Nevertheless, the field is growing in importance, which will force the energy security market to mature quickly.

The key companies dominating this space are ABB Group, Aegis Defense Services Ltd., BAE Systems PLC, Cassidian SAS, Elbit Systems Ltd., Ericsson A.B., Flir Systems Inc., Honeywell International Inc., and Siemens AG.

Other companies directly and indirectly involved in the sector are Acorn Energy Inc., Agiliance Inc., Anixter, Inc., HCL Infosystems Ltd., Intergraph Corp., Lockheed Martin Corp., Mcafee Inc., Moxa Inc., Northrop Grumman Corp., Qinetiq Group, Raytheon Co., Safran

S.A., Symantec Corp., Thales Group, Tofino Security, and Tyco International Ltd.

The number of participants is growing, and the effort is becoming very important and has high visibility.

Criminal Energy Markets

Criminal activities in the global energy markets are becoming more prevalent. Some of the greatest violations of today are in the energy sector, as follow:

Nigeria's Bunkering Market

Illegal oil bunkering is big business in many, mostly poor and developing, countries. It is a criminal activity undertaken by individuals or organized crime groups, which consists of illegally tapping into crude oil lines and storage tanks. Thus bunkered oil is sold (also illegally) to other not-so-legal enterprises.

Illegal oil bunkering is a particularly huge, and well developed "business," in Nigeria. It is growing and taking an increasingly heavy toll on the national economy, the local environment and the people. Illegal oil bunkering in the Niger Delta region is growing in leaps and bounds, because there is no law enforcement and corrupt administrations benefit from it.

According to insiders, Nigeria loses about $20 billion annually to crude oil theft. An estimated 30,000 to 300,000 barrels of oil per day are stolen from the country's oil reserves. There was an estimate of $100 billion lost in illegal bunkering during 2003-2008.

In 2001, the government set up a Special Security Committee on Oil Producing Areas, aimed at identifying the organizers of the illegal oil bunkering and implementing solutions for resolving the problem. A report issued by the committee noted that, "...a major threat to Nigeria's oil industry arises from activities of a 'cartel or mafia,' which comprises highly placed and powerful individuals within the society, who run a network of agents to steal crude oil and finished produce from pipelines in the Niger Delta region."

According to the committee, those responsible for halting or diverting oil production and preventing free traffic on the waterways could be enjoying the patronage of some retired or serving military and security personnel. Vessels used in the bunkering deals were often seized by the army and navy, but their cargoes were unaccounted for and usually went unreported.

In 2008, the government appealed to the Group of Eight (G8) nations for assistance with solving the bunkering problem. The call for help was then repeated at the United Nations later on that year, and again in Washington late in 2009.

The international community made efforts to help Nigeria fight the ongoing, decades-long problem, but they were frustrated by top military personnel and politicians, who benefit from the illicit oil business.

Top military personnel have been openly accused of involvement in illegal oil bunkering in the Niger Delta, but there has been no action taken on any level. Further investigation revealed that military personnel were actually supporting and even managing the illicit oil bunkering. Some of the military officers linked to the criminal activities had been retired, while others are still in the service, so the trend continues. One way to stem the tide would be for the troops in the Niger Delta to be changed periodically. This would at least curtail the rate at which Nigeria's oil is stolen, since it takes time to organize new bunkering operations.

Youths in the region have been accused of complicity in the crime, but they can only do "bucket bunkering," which is small-scale theft, since they don't possess the technology required for larger operations.

Instead, according to WikiLeaks, politicians and military leaders, not the youths or militants, were responsible for the majority of oil thefts in Nigeria. An unnamed Nigerian official, a member of a government panel on troubles in the Niger Delta, implicated a late general and a former vice-president as being the biggest names behind the thefts. The military wants to remain in the Niger Delta because they profit enormously from money charged for escorting illegally bunkered crude and from money extorted in the name of providing security on the roads. The foot soldiers are not the only ones who profit; the commissioner of police, the director of the State Security Service and the military, all benefit from the illegal activities.

In 2010 vessels carrying 724 metric tonnes of stolen crude were impounded, and about 6,000 illegal refineries across Niger Delta were destroyed. About 200 persons suspected in illegal bunkering were arrested. Oil companies' workers were also accused of cooperating with thieves, but the oil workers' unions absolved their members from the crime, pointing out that they have been helping, instead, in the fight against this crime by raising alarms over the increasing rate of crude oil theft and pipeline vandalism.

The unions also point out that the volume of oil stolen in Nigeria is more than double the total production of some countries—a crime that could destabilize any government. As unstable the Nigerian government is, more destabilization could lead to a total collapse.

The only way to remediate the quickly worsening situation is for the federal government to get to the root of the issue by finding the real culprits and bringing them to justice. Sure enough, Nigeria legalized bunkering in 2014 in an effort to stop the illegal activity and increase revenues.

In response, the Nigerian National Petroleum Corporation (NNPC) has (finally) made plans to start a bunkering business of its own. This way, NNPC would bypass, or make much more difficult, the illegal activities, thus capturing a huge business that was lost until now. The only problem is that NPCC doesn't know how, and is not equipped, to do this work. So now they are looking for the expertise of reputable bunkering firms. The study is underway, and very soon the huge bunkering business will be taken advantage of...legally?

The effort has the support of the government's security agencies. Initially, it will consist of allowing vessels operating in Nigerian waters to refuel locally instead of going to neighboring countries or using illegal oil supplied by bunkering operations.

This is a good step ahead in crime fighting, but a little too late for the environment. The damage to the Niger Delta is done. The ground is soaked in oil, and the water bodies in parts of the country are flooded with it. This is permanent devastation on a grand scale. It is not clear how or when the region's environment would recover, if ever.

Solar Energy Crimes

Criminal activities in the global energy sector are not limited to conventional fuels. Increased solar power development worldwide has opened doors to new riches for individuals, corporations, and organized crime.

In Spain, for example, the regulators shut down several dozen solar power plants during 2012-2013, the owners of which were misreporting their profits and underpaying their taxes. This is a new trend of cheating that was discovered in Spain, but is also prevalent in many other countries.

The situation is worse in Italy, where the mafia had found ways to penetrate the solar energy sector to run its operations within. We described these events in detail in the previous chapters, so here we will only conclude that even clean solar cannot be kept clean from crime without intense security measures from within and without.

Power Plants Security

Power plants as the source of energy generation must be protected from physical as well as cyber attacks for smooth functioning of the national grid. Protecting power plants is the responsibility of a number of gov-

ernment authorities, and power plant owners themselves. For this purpose, they are working together on implementing both physical and cyber network security solutions.

These solutions include surveillance systems, different sensors, perimeter security, secured communications, intrusion detection, transportation security, and special detection and access control systems.

Also, under consideration and development are network security systems such as SCADA, firewalls, antivirus and IDS/IPS systems. These are designed to detect and prevent all kinds of intrusions to secure the normal operation of power plants.

The energy security market is broadly segmented according to the types of network security and/or the physical security systems and solutions. This includes services provided by government organizations and private companies to ensure our energy security. The focus today is on the important drivers which are driving the energy security market and the major factors in it.

The companies providing security services usually work closely with government agencies, regulators, and power plant operators to provide solutions to secure the premises from physical and cyber attacks.

The Major Risks

Crude oil is of special concern, since we import a large portion of it for producing motor fuels and heating oil. These imports have a number of risks, which endanger our energy and national security.

The major risks that affect our energy security addressed in this text, and which are on the forefront in the fight for energy independence, are:

- Dependence on oil weakens international leverage, and undermines foreign policy objectives

- Endless dependency entangles America with unstable and even hostile regimes, further complicating our foreign politics and policies.

- Importing millions of gallons of crude oil daily carries a high price tag—over $200 billion annually go in the coffers of others, some of whom use it to fight against us.

- Inefficient use and overreliance on oil burdens the government and military apparatus, and undermines combat effectiveness, resulting in increased financial burden and loss of life.

- Dependence on fossil fuels (and imports) undermines our economic stability, which is critical to achieving our national security objectives.

- A fragile domestic electricity grid makes our national and military installations, and their critical infrastructure, unnecessarily vulnerable to deliberate or accidental incidents.

Continuing business as usual (by burning large quantities of fossils) is very dangerous, because of the converging national security risks of both energy demand and climate change.

Some of the major factors in this equation are:

- The market for fossil fuels is shaped by declining supplies and increasing demand, so continuing our heavy reliance on these fuels is, in the long run, a security risk of great proportions.

- Regulatory frameworks driven by increasing climate change concerns will steadily increase the costs—economic and geopolitical—of fossil fuels.

- Destabilization driven by ongoing climate change and rising fossil fuels use adds significantly to the uncertainties of the future energy markets.

Confronting these converging risks is critical to ensuring America's secure energy future. Due to the destabilizing nature of increasingly scarce resources, the impacts of energy demand and climate change could increasingly drive military missions in this century.

Experts recommend that the first priority for the U.S. Administration is to fully integrate energy security and climate change goals into our national security and military planning processes. Consistency with emerging climate policies should shape America's energy and national security planning; the U.S. should not pursue energy options inconsistent with the national response to climate change. Diversifying energy sources and moving away from fossil fuels where possible is critical to future energy security.

While the current financial crisis provides enormous pressure to delay addressing these critical energy challenges, experts warn against delay. The economic risks of this energy posture are also security risks.

As economic cycles ebb and flow, the volatile cycle of fuel prices will become sharper and shorter, and without immediate action to address our nation's long-term energy profile, the national security risks associated with the nation's and the military's current energy posture will worsen.

The U.S. consumes 25 percent of the world's oil production, yet controls only 3-5 percent of an increasingly tight liquid fossils (crude oil mainly) supply.

Oil is traded on the world market, and the lack of excess global production makes that market volatile and vulnerable to manipulation by those who control the largest shares. Reliance on fossil fuels affects our national security, largely because nations with strong economies tend to have the upper hand in foreign policy and global leadership.

Recently, the U.S. has entered an energy renaissance where we produce a lot of oil, and some of our old bad habits in energy waste persist and even worsen.

We all know that reasonable and efficient use of energy and astute energy conservation are the keys to long-lasting energy prosperity, but U.S. automakers continue to produce gas-guzzling SUVs, vans, and trucks, and residences and office buildings are cooled and heated at luxurious levels.

The Fossils

Figure 8-44 tells us that the U.S. is one of the few countries in the world that has enough coal and natural gas—and even exports significant quantities of those. As a matter of fact, 10-15% of the annual coal and gas production is exported (or planned for export) mostly to Europe and Asia.

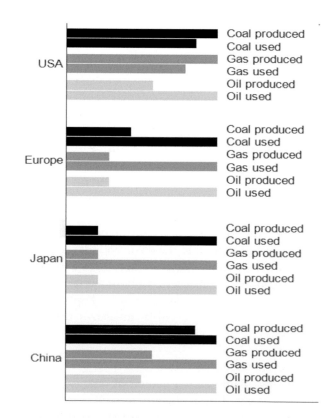

Figure 8-44. Global fossils production and consumption (in %)

That trend will continue with gradual increase of exports as some countries in Europe and Japan are shutting down their nuclear plants. The threat from reduced Russian energy exports is also looming over Europe after the events in Ukraine, so the Europeans will depend more and more on other imports—an increasing portion of which will come from the U.S.

Even if the European relations with Russia were patched up and nuclear plants in Japan were restarted, U.S. exports will grow, because the newly developing countries need more energy. As their economies improve and their populations grow, they have more money for purchasing imported energy.

One equally important fact to be considered in the overall picture of production and use of fossils is the amount of proven reserves of each type globally and in the different countries. To get a complete picture of the situation, we must consider the ratio of the amount of proven reserves vs. their present and future production rates. The results may shock you.

These factors are important, because the fossils in a country with very high production rates and limited proven reserves might not last too long. On the other hand, limited reserves in a country with low production rates might last much longer, but these countries usually like to export the amount they don't use, so the exploration rates are high in all scenarios.

Proven U.S. fossil reserves explored at the present rates might last another half century or so, but they will be severely depleted by the end of this century and completely disappear shortly thereafter.

*Crude oil is imported in high quantities and at great risks, which represent **immediate** threats to our energy security and our national security. Not even partial energy independence can be achieved under these conditions.*

*Crude oil is also limited in total quantity of available global reserves, which represents a serious threat to our **long-term** energy security.*

So what are we to do?

Personal Energy Use

There is no free lunch in the land of plenty, but we are quite spoiled anyway—most of us living in full comfort for which we are willing to pay, while reserving the right to complain. The average American family pays many bills every month, so we have plenty to complain about.

Fuel and repairs for 2-3 cars per household, rent or mortgage for a 2-3 bedroom house and all other household related bills, food, transportation and medical insurance, eating out, clothing, entertainment...

Energy bills—electricity and gasoline—are a significant part of the American family's budget, comprising about 1/4 of the monthly bills' total.

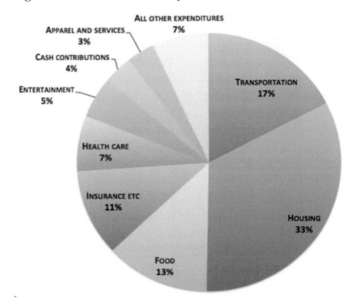

Figure 8-45. American family expenditures, in % of personal budget.

Note in Figure 8-45 that almost 1/5 of the budget of an average U.S. family is spent on vehicle fuel for personal use. Fuel used for powering motor vehicles is essential for the social and economic life of the country. It is indispensible for the function of our society, because we all commute long distances to work, school, the store, etc. There is no other way to move around, especially in rural areas.

Many people believe that their car is the largest energy user and the greatest source of air pollution for which they are personally responsible. In fact, the average *home* uses more energy and emits more than twice the carbon dioxide of the average car.

This is because most of the energy consumed in our homes is produced by burning fossil fuels like coal, oil, and natural gas. This pollution is actually a hidden cost for the energy we use, over and above the $250 billion Americans spend each year on their home energy bills.

Most American homes are very large and much more luxurious as compared to homes in other parts of the world—including those in the developed countries in Europe and Asia. There are large rooms with special lighting, several TVs and stereos, heaters, air conditioners, computers, dishwashers, clothes washers and driers, refrigerators, stoves and ovens, garbage disposals, and other appliances and gadgets plugged in the electrical outlets sucking power day and night.

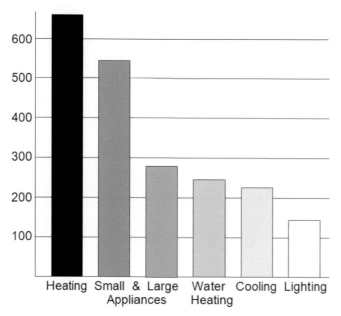

Figure 8-46. Residential energy bills (in $ per annum)

The average use of energy in American homes comes to over $2,100 annually! This amount increases in the hot desert areas and the cold northern states, due to excess cooling and heating bills.

Uninterrupted electric energy supply is provided by thousands of power plants run primarily by coal, natural gas, nuclear, hydro, solar and wind power. Natural gas in gaseous or liquid form is also supplied to many homes and businesses for heating and cooking.

Uninterrupted and *cheap* are the keywords describing the American power supply. Anything less than that is unacceptable, and the utilities, regulators and politicians are constantly reminded of their importance.

Much of the energy used in American homes, however, is wasted or used unnecessarily. Lights and heaters left on in empty rooms are mindless wastes of energy and are common events in the U.S. There are few excuses for this, because there are a variety of proven, widely available products on the market (heating, cooling, appliances, windows, lighting, etc.) designed to drastically reduce the home's energy bills, and the accompanying pollution.

In addition, many of these products actually improve the comfort and livability of our homes. Unfortunately, we don't have time in our busy schedules to

include those, so they remain on the back burner.

But as fuel and energy costs become more expensive, families' budgets must be adjusted. Fuel price fluctuations affect the lives of all Americans, and some balance must be maintained in order to keep things running smoothly.

On personal level, we must consider the type and amount of energy we use, and the related pollution emissions, in light of: a) our personal budget, and b) our national energy security.

Of course, the personal budget takes precedence in most cases, which is a problem when discussing energy security. While Americans are willing to accept some personal sacrifices, these cannot come at the expense of our comfort level, or the ever tightening personal budgets. Especially not during the hard economic times we have been enduring lately. So national energy security can wait, as far as the average American is concerned.

Technology and Energy

Above we took a look at the role of personal energy use and that of the government and military. Another significant group needs to be included here—the business community (small, medium, and large companies and corporations) and their contribution to the energy picture and our energy security. They are, after all, also great users of energy, and even more importantly, providers of technology and services in the energy areas.

Technology is a determining factor in the economic development of the world. The 20th century overemphasized the importance of technology by bringing us some amazing developments that literally changed all aspects of our lives.

It is becoming more important today, and will become even more so during the latter part of the 21st century, when new technological developments will improve existing ones on a daily basis.

Hydrofracking is one such technological development that has allowed production of huge quantities of oil and natural gas from places that were unthinkable just a decade ago. It also promises even further advances in the future. But that development has brought a lot of damage to property and has hurt wildlife and humans. Nevertheless, the expansion of this technology continues, and the damage increases with it.

The fracking industry does not acknowledge the damages, attributing them to unrelated accidents. This is not a new approach, since there have been a number of such debacles in U.S. history, where companies cause damage but refuse to take responsibility. Let's take a look at the past, in an attempt to see the similarities with today's developments:

Unleaded Gasoline Debacle

Car engines at the beginning of the 20th century were performing poorly and making lots of knocking noises. GM scientists found the answer to better performing and quieter car engines in the early 1920s by adding the lead compound, tetraethyl lead, to gasoline. The resulting fuel was marketed by GM and Standard Oil under the name Ethyl fuel.

In 1924, however, several workers died from a form of sudden lead poisoning at the Standard Oil facility in New Jersey, The symptoms of lead poisoning were unmistakable and undeniable, since they all became delirious and violent just before their deaths. Other workers had suffered similar deaths at a DuPont Company plant, but both companies tried to keep word of the fatalities from getting out.

A number of complaints had already been voiced by public health reformers. A public warning was issued by scientists claiming that innocent bystanders by the hundreds might simply fall down dead on the nation's sidewalks if motorists began using leaded gas. This was, of course a gross exaggeration, and yet it struck a responsive chord with the public.

The press and the public were in a frenzy, but Standard Oil had already put Ethyl on the market based only on the results of its own recent tests with the substance. No long-term tests were even planned, let alone executed.

There was no official federal body in the 1920s with powers to investigate the manufacture and distribution of a new industrial product. The public uproar, however, was so loud that it triggered a quick response from local health boards, and after a short fight any further sales of Ethyl was discontinued until unbiased tests could be conducted.

Scrutinizing important technological developments before they had reached the market, and were perhaps halted, was almost unheard of at the time. DuPont's "Better Living Through Chemistry" slogan had convinced the nation that new chemicals would bring prosperity, and leaded gasoline was one of those miracle compounds, so the affected companies stood behind their product regardless of the fatalities.

Finally, in May 1925 the surgeon general called for a conference of experts and interested parties to consider the pros and cons of this new fuel compound. The corporate interests dominated the hearings, for they had controlled all the research and testing of the new gasoline.

They tried to prove that the men who had died from handling Ethyl had been exposed to concentrations far greater than the motoring public would ever see in using leaded gasoline.

The opposition pointed out the known dangers associated with lead poisoning and cited the tremendous health risks of even tiny amounts of lead being discharged in automobile exhaust fumes. They were, however, overwhelmed by the corporate lawyers who called Ethyl "a gift of God" due to its potential to improve automobile performance. Finally, the panel agreed to lift the ban on the sale of leaded gasoline.

This marks the beginning of a reoccurring practice of allowing unsafe products to enter U.S. markets. Lead, nicotine, asbestos, to mention a few, have hurt and killed thousands.

This was a huge victory for the large corporate interests, which would be allowed to sell their poisonous product for the next 45 years. They kept did so unhindered until the 1960s, when scientific evidence made it clear that airborne lead was a serious health hazard. Once again efforts were underway to outlaw lead in gasoline, and finally federal restrictions governing the lead content of motor fuels came into effect in the 1970s, and lead in gasoline was no more.

Lead exposure causes a wide range of illnesses in adults and poses especially high risks for children, affecting their neurological development, growth and intelligence. That was proven even in the 1920s, but corporate interests drowned the voices of caution at the time. They knew that leaded gasoline, although it was good for car engines, was very bad for people; and yet, they only emphasized the good side of the issue and covered up the downside.

Leaded gasoline is still in use in some developing countries, where, especially in large cities, it is still posing grave health risks to people, especially small children.

Radium Poison

Also in the 1920s, the Waterbury Clock Company was making glow-in-the-dark watches that were fashionable at the time. Mostly girls were employed to use their keen eyes and nimble fingers to paint tiny numbers on the watch dials.

The process required the women to press their brushes between their lips before dipping them in the radium-laced paint to give their small brushes a nice, fine point. The gritty-textured paint made their mouths glow in the dark, but this didn't bother the girls. They went as far as painting their dress buttons and fingernails, and rings on their fingers with the radium paint.

But the party ended when some of the girls got sick, and some even died. Evidence was mounting that this radioactive element, radium, could be the cause for the illnesses and deaths. Amazingly, the best workers—those that produced the most watch dials—became ill and/or died first, most likely because they absorbed the most radium

The company was such a success, and was making so much money, that things were run as usual even when young women working on the radium assembly lines in other places began to develop horrific symptoms.

The company did not want to hear, let alone admit, that radium was the cause. Even after it was forced to pay retribution to the victims sometime later, a company spokesman announced that this human disaster was caused by the fact that they hired weak women and invalids, who could not handle the work pressure and became ill or died as a consequence.

The dial painters were finally compensated, but they had to be willing to keep quiet about their experiences and stay out of court. Such stonewalling of occupational accidents or diseases is not uncommon.

Coal miners with black lung disease endure similar silencing tactics, as did asbestos workers with silicosis. Things are much better today, but there are still many cases that we just don't hear about. This is because most of them are settled ahead of trial, since the companies don't want to risk a huge financial loss and negative social exposures, while the affected families try to avoid the emotional trauma of prolonged trials. The money also comes handy in covering expenses and other family needs.

There is a pattern here that goes like this: first, a group of workers get sick; then someone makes the work-illness connection; and finally advocacy groups attempt to get societal recognition, worker protection and compensation.

The recognition and remediation of occupational accidents or diseases in the past, usually went through these stages. In many cases it was accompanied by corporate cover-ups and social resistance at each step, which made the results unpredictable.

Today, awareness has increased, and reporting procedures have changed for the better, so we don't expect to see such blatant violations of workers' rights. But old habits don't go away easily, so we still see abnormali-

ties, even though the consequences of the mistakes of the past are still around.

We should add to the list of lessons learned (or not) the widespread use of lead pipes in the beginning of the 20th century, which led to numerous deaths. Although lead's toxicity was suspected, and even scientifically proven, the large corporations were able to put the concerns on hold until they made enough money.

There is also the equally broad use of asbestos and nicotine, the harmful effects of which were covered up for a long time. Only after many lawsuits were filed, and many politicians became involved, were the issues brought out in the open and resolved by banning or restricting the use of asbestos and nicotine.

Why would anything be different today? The corporate world's main goal is to make money—whatever it takes. So why would fuel companies stop fracking … just because several people lost their homes, health, or even lives? Why would large companies manufacturing solar panels containing cadmium and other toxic materials stop selling them …just because there is a clear and present danger of environmental contamination in the future?

That's just not going to happen soon, because capitalist enterprise and personal interests dominate the decision making process. Too, it takes a lot of work and long time to pass and enforce a law that hurts large corporate interests.

Today's Situation

How do the above examples relate to today's energy and energy security realities? Energy security is meant to ensure the supply of energy, which in turn is supposed to improve life. What we see today is a race for more and cheaper energy—at all costs. What we need to watch for in the future, therefore, is a repetition of the above-mentioned abnormalities, where political inefficiency (for lack of better words) and corporate greed result compromise the health and lives of some of the participants. So, while ensuring energy security for some, we cause harm to others…

A close look at today's energy industry reveals a number of incidents in coal mines and oil fields, where corporations tried to cover up accidents…remember the BP actions during and after the Deep Horizon blow-up? Or the Exxon Valdez spill? Or the variety of coal mine accidents and miners' lawsuits? No need to go in detail; the overwhelming conclusion is that we still have problems with transparency and fairness. Corporate and personal interests more often than not come before the common good.

As an example, take a look at the complaints of late from people living next to hydrofracking sites. They will tell you how aggressive the oil companies have been in getting what they want, how fraudulent their approaches have been, and how blatantly they have been disregarding the privacy and rights of the land owners. There are hundreds of such cases, which is only the beginning of the problems. Soon after the well is drilled, the local water is unusable, and even flammable. Sink holes full of bubbling toxic chemicals pop-up in people's back yards, and many other hard-to-believe stories. At the same time, the industry denies most of these claims and goes to court to fight them.

While the companies of the 1920s could be partially excused for reasons of lack of scientific facts and technical ignorance (science was not as developed at the time) there are no plausible excuses today. The state of the art in all areas of our industrial and medical complexes is developed to the point where we know all there is to know about, and can easily prevent, industrial accidents and occupational diseases...when we want to.

The question is DO the large companies really want to do this? And that brings us back to our example with hydrofracking. The oil and gas well owners know that there is no way to seal 100% the walls of every well. They know that from time to time a well wall will collapse, crack, or somehow become damaged enough to allow oil, gas, and/or the fracking fluids to travel through the ground in whichever direction. But even if the well wall holds, the fracking fluids are pumped into the lose soil and rock formation. There is no containment, no structure, no control of any sort deep underground. So what is to stop the fluids from popping up on the surface in whichever direction they chose. This could easily contaminate the local water table, and/or blow a hole in the unsuspecting farmer's property to flood it with poisons.

Similarly, some solar companies use toxic, carcinogenic heavy metals and chemicals in their solar panels, which are then installed by the millions in U.S. deserts. Most of these are new technologies that have not been thoroughly tested for long-time performance in severe desert conditions, so we don't know when these millions of panels will spill their poisons on the desert floor and contaminate the air, ground, and water in the area.

Tons of Cd, Te, As, In, Cu and other toxic chemicals are now baking under the desert sun while the U.S. regulators are sitting on the sidelines waiting for something to happen before they act. What might happen is too scary and tragic to even consider, but it might be a close repeat of the large-scale lead pipes and asbestos insulation debacles of the past.

These companies, similar to the companies involved in hydrofracking (owners and suppliers) would deny that their products are capable of damaging property or hurting people. They most likely will not admit fault in any case, even if and when the unthinkable happens. Like the hydrofracking enterprises, they will continue to operate as usual even in the face of undeniable clams of damaged properties and ill people.

The only difference is that the solar companies release the contents of their products, unlike the hydrofracking companies and suppliers who refuse to let the public know the chemical contents of the millions of gallons of proppants they pump into the ground daily.

How can we justify achieving energy security at the cost of people's property and lives? Isn't there a better way? Here again, the federal government is late in providing the necessary solution, except that today there is no excuse.

Unlike the 1920s, there is a huge government bureaucracy now that is in charge of controlling these things. DOE, EPA, NIH, BLM, etc., entities, employing thousands of well educated and knowledgeable people, are supposed to assist the technological developments, with people's safety as priority number one. In some cases, the priority is shifted to number 2, 3, or beyond, or totally ignored. Hydrofracking is a case in point.

These agencies are also well equipped with the best and latest scientific equipment, capable of analyzing and finding solutions to all pertinent problems. There are no excuses today.... But as in the past, things move slowly on the federal level, so we will probably see much more damage and hurt before decisive action

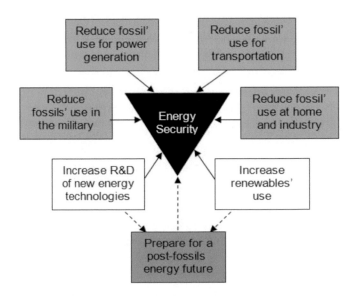

Figure 8-47. Present-day energy security goals

is finally taken in the oil and solar fields. We hope we don't see too much damage and hurt, in the name of energy security, before improvements are made.

No government action can stop the fossils from being depleted, whether it will be tomorrow, or next century, a day will come when the fossils are no more. Let's not be like the proverbial frog who stays in the slowly heating water until it boils to death. The poor thing doesn't know any better. We do know what is coming, so let's prioritize our goals in the drive towards a fossil-less future. At the very least, we should lay the foundation for the transition. We cannot continue living the careless and wasteful lifestyle of spoiled Americans.

It is unsustainable because end of the fossils that allow our life style is coming. It is unfair, because developing countries are watching us and want to emulate our life style. Do we have the right to deny them the luxuries we have enjoyed for so long, and continue to enjoy today? Why should they sacrifice, if we don't? And it is quite obvious that they have no intention, at least for now, to make any sacrifices. On the contrary; they plan to increase the use of fossil energy with time, which will have devastating effects on our environment and will bring the fossil-less reality much closer.

More than anything, it is not fair to future generations, who by no guilt of their own will be totally deprived of the benefits of the fossils. They not only will not have the luxury to burn them, but fossils won't be available for basic needs like lubrication, plastics, medicines, fertilizers, and so many daily necessities.

So what can be done? Looking at Figure 8-47 we see that we must:

1. Reduce the use of fossils in the power generation sector by increasing the use of other energy sources—preferably renewable.

2. Reduce the use of fossils in the transportation sector by increasing the vehicles' efficiency and switching to renewable energy sources.

3. Reduce the use of fossils in our homes and in the industrial sector by increasing the efficiency of our equipment and processes, implementing energy conservation measures, and switching to renewable energy sources.

4. Reduce the use of fossils in the military sector by increasing the efficiency of fossil use, by introducing energy saving procedures, and switching to renewable energy sources.

5. Clearly, renewables are the key to a smooth transition to a fossil-less energy future, so we need to work on their optimization and full deployment in all areas of our lives.

6. And finally, R&D of new energy technologies will play an ever increasing role in helping us enter into the post-fossils realty without major shocks and disasters.

THE FUTURE OF ENERGY SECURITY

We agreed that the electric energy supply in the US is guaranteed for now, due to large coal and natural gas deposits The United States, however, still imports about 50 percent of its petroleum supply, but it also exports some of it, so for now we depend on imports for less than 50% of our overall crude oil use. Yet, petroleum product imports cost the United States roughly $200 billion annually. This does not include the peripheral expenses and hidden costs, like the military patrols in and around the transport choke points and all over the world.

But there are alternatives. Filling our gas tanks with ethanol is one of the cost-effective ways of lowering U.S. dependence on petroleum imports and saving lots of money. The Energy Security and Independence Act of 2007 (EISA) seeks to do just that; it promotes the use of biomass to displace imported petroleum in meeting the U.S. demand for liquid transportation fuels.

By 2022, EISA would require the use of 36 billion gallons of biofuel in the United States, up from the current level of roughly 10.5 billion gallons. This will also reduce the amount of GHGs emitted by fossil fuels. Each gallon of gasoline burned releases over 20 pounds of carbon dioxide, and the transportation sector emits about 17 percent of the total U.S. emissions of GHGs.

Ethanol derived from biomass is expected to lower GHG emissions by 50 to 90 percent per gallon of gasoline displaced. This would be a major step in achieving our energy security objectives, but it is just one possibility.

Electricity use (and misuse) are other potential fossils- and money-saving undertakings. We all use electricity for computers, heating and cooling our homes, etc. Since most of the electricity used in the U.S. is generated by domestic energy sources, there is no energy security issue associated with it.

Unfortunately, about half of the electricity used in day-to-day activities comes from coal—about a billion tons of coal are burned annually—releasing about 2 billion tons of CO_2. This is about 30-40% of the total U.S. annual GHG emissions.

The alternative energy sources could replace a major part of the coal-fired power generation, displacing the fossils and reducing GHG emissions. Therefore, the use of the alternatives enhances our energy security.

The use of biomass to replace coal in generating electricity, for example, can achieve an 80 percent reduction in GHG emissions relative to burning coal, or a reduction roughly 2.5 times greater than its use in producing cellulosic ethanol.

This will contribute to cleaner air and less damage to property (caused by fossil emissions), which will ultimately result in improved human health.

The systemigram in Figure 8-48 was published by a West Point team (5) and illustrates the complexities associated with obtaining and maintaining total energy security from DoD's point of view. It shows the kind of dependencies that exist and need to be understood in order to develop a systemic solution to this problem. It will require coordination between elements of the US government, the military and private corporations. Beyond that, even the societal and economic effects of any proposed solution will need to be understood and quantified.

The assumption made here (which is possible but not very likely in today's socio-political climate) is that national energy security is a product of total and willing systematic partnership of key institutions, coordinated by government instruments, and supported by third-party (private) financing.

"Is this even possible?" one could ask after taking a close look at Figure 8-48.

Case Study: DoD Energy Security Model

What we have today in the power generation sector is unsustainable in the long run.

As Table 8-11 shows, almost half of our electric

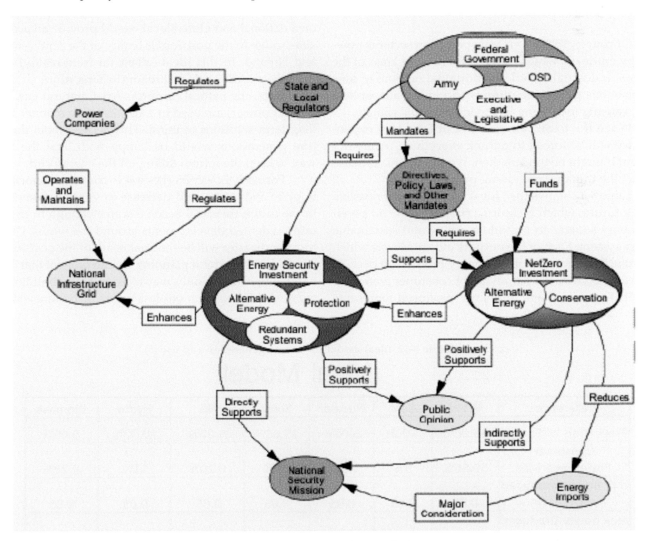

Figure 8-48. Energy security through the DoD

energy is generated by coal, which is simply killing the environment. We are the most powerful and influential economy in the world, and our example in environmental unaccountability is being followed by most of the other countries.

Table 8-11.Present-day model of power generation and industrial fuels use.

Source	Use %
Coal	45
Gas	20
Nuclear	15
Oil	10
Hydro	5
Wind	2
Solar	2
Biomass	1

Of course, 95% of our transportation sector is powered by crude oil, which is also used in other areas of the national industrial complex. Additional millions of tons of poisonous gasses are pumped up in the atmosphere daily, causing increasing havoc in the global climate.

When the fossils are gone, or at least very expensive, we will be forced to rethink everything we are doing, but it might be too late then. Wouldn't it make sense to start the thinking and doing processes now?

Table 8-12 shows the "Ideal Model" of our possible energy future, which includes a proper mix of all possible energy sources to provide a robust and sustainable energy system. Each of the outputs indicate levels which optimize the efficiency of the energy production cycle in order to maximize the parameter of consumer preferences met. The percentage of energy produced represents the optimal mix of energy sources and the amount of each energy source that should be produced in a certain time period.

Looking at the ideal model we see:

Table 8-13. The ideal model of power generation in the DoD study

Source	Use
Wind	35%
Solar	35
Hydro	10
Coal	5
Nuclear	5
Gas	5
Biomass	5

Wow! This works, and if and when implemented on a national and global level would provide an acceptable bridge to the post-fossils reality of the 22nd century and beyond. In this ideal (albeit far-from-reality) case we have a sustainable future for the long run.

Here, the natural resources (coal, natural gas, and nuclear power) are used in a conservative manner with long-term solutions in mind. This is one model the future generations would be happy with, and the only way to keep them from falling off the energy cliff.

For now, however, the fossils continue to increase in price and usage, and decrease in volume. Sometime in the future they will become scarce enough to trigger unprecedented developments around the world. Disaster after disaster will be the final result of the post-fossils era. Only with careful planning we can prevent that from happening, and the only, way to do this is by building an energy bridge to the fossil-less future using renewables.

Table 8-12. Ideal model for energy generation and use (5)

Ideal Model

Preferences	Wind	Coal	Nuclear	Solar	NG	Hydro	Biomass
Proportion of Energy	35.00%	5.00%	5.00%	35.00%	5.00%	10.00%	5.00%
Consumer Preference Met	59.85%	5%	0.25%	39%	0.20%	11%	4.26%
min power produced per unit >=	0.05	0.05	0.05	0.05	0.05	0.05	0.05
max power produced per unit<=	0.35	0.35	0.35	0.35	0.35	0.35	0.35
Total Preference	118.36%						

Energy Waste

Before we start getting serious about energy efficiency, which is one of the precursors of our energy security, we must take a close look at some of the obvious waste patterns of our capitalist society. Here are a few examples:

Packaging

What does packaging have to do with energy security? It is actually one of the most visible and noticeable symbols of energy and environmental impact of the 21st century. Plastic bags, containers, wrappers of all sorts and sizes are all around us. We see them around the house, the park, in the rivers and the oceans.

Packaging is credited with heavily contributing to our economic development and raising our standard of living. One look at the items in a U.S. supermarket, as compared with the street market in most African and Asian countries, gives you a good idea of the stark difference that packaging makes.

In the U.S., with our sophisticated packaging, storage, and distribution systems, only 2-3 percent of the food is wasted before it gets to the consumer. In developing countries, on the other hand, 40-50 percent of food shipments are spoiled because of inadequate packaging, storage, and distribution systems.

On the consumer side, however, this ratio is reversed; 30-50% of the food from U.S. consumers ends in the waste baskets, while in the developing countries only 2-5% of the food total is wasted by the consumers.

So how can people without efficient cooling and packaging of food items be so efficient in their use? How can they do without packaging, and why are we so wasteful having good packaging options?

Packaging plays a growing role in the global economy, especially in the developed countries where it is not only for protection of goods from contamination, but also used to convey information about contents, preparation, and use, but in most cases it also keeps would-be tamperers from poisoning or otherwise hurting the customers.

At the same time, however, packaging—its manufacture, transport, use, and disposal—uses enormous amounts of energy and is causing an environmental disaster. It is the largest and fastest growing contributor to one of the most troubling environmental problems—garbage. In the U.S., packaging accounted for more than 30 percent of the municipal solid waste stream in 1990.

Where is all this packaging going? In this country, most packaging and other waste is buried in landfills. But even with its abundance of open land, America is running out of room for its garbage. One quarter of the country's municipalities have already exhausted their landfill capacities, so that more than half the population lives in regions with lack of landfill capacity.

Even though the environmentally sound alternatives to burying garbage—recycling, reuse, and energy recovery—are well developed, our disposable and throwaway society is still generating more garbage from packaging than the entire rest of the world.

Packaging is not the only culprit in the solid waste crisis, but it is the most highly visible component in a garbage dump. It also directly involves the customers, who use it for a very short time. Its short lifetime exacerbates the problem, since the useful lives of most packages are very limited.

While some packaging materials may last as long as several years (i.e. paint and food cans, and reusable canisters), the useful lives of most others can be as fleeting as a few minutes, or hours. Think of the hamburger wrapper, or a new gadget in a large cardboard box received in the mail.

We deal with a huge volume of different types of packaging daily that goes into the solid waste stream. This is a big problem, which also presents an opportunity, since even very small improvements in packaging, due to its huge amount, can make a real difference in the magnitude of the garbage crisis. So packaging offers this unique opportunity for individuals and companies to assume leadership roles in environmental responsibility.

Industry's response to the environmental challenge of excess packaging has been focused on recycling and packaging materials reduction. But this is not enough, because these are complex issues that demand a systemic, integrated approach based on comprehensive analysis. Long-term vision and implementation of innovative solutions are also some of the tools we must use.

Life-cycle analysis is another tool, which needs to be applied to every product we consume. Think of millions of dirt-cheap gadgets coming from China enclosed in large packages made of heavy-duty plastics. How much material and energy was use to make these packages? How much energy was used to manufacture, use, and dispose of them? What is their overall benefit to our well-being?

Life-cycle analysis is an important step toward understanding the full energy and environmental implications of packaging choices.

Using this technique, we consider the energy use and environmental implications integral parts of each product. Considering the energy used during the entire product cycle—from raw material acquisition, design, manufacturing, transportation, to final use and disposal of the packaging—will allow us to see the other side of packaging. That's the side that affects our economy and contributes to great energy waste. This will allow us to make better packaging choices.

Our energy security will not be achieved by solving the packaging problems, but present-day packaging is one of the elements of our way of life that is indicative of our societal problems in need of fixing. Less waste means more and cheaper energy for all.

And speaking of waste…

Food Waste

Because we cannot live without food it is on our government's priority list, and a lot of effort (and energy) goes into ensuring the food supply on the American tables. This comes at a heavy price.

Getting food from the farm to our dinner table eats up 10 percent of the total U.S. energy budget, uses 50 percent of U.S. land, and swallows 80 percent of all freshwater consumed in the country.

Yet, 40 percent of food in the United States today is unused and/or wasted. This is over 20 pounds of food per person every month ending up in waste baskets and eventually in municipal garbage dumps. Amazing numbers…

Today, the average American consumer wastes 10 times as much food as people in Southeast Asia and Africa, and 50 percent more than we wasted in the 1970s.

This not only means that Americans are throwing out the equivalent of $165 billion of different food items each year, but that the uneaten foods must be collected, transported, and left rotting in landfills. Food waste is the single largest component of U.S. municipal solid waste, where it also accounts for a large portion of U.S. methane emissions.

Organic matter (mostly food leftovers) in municipal waste dumps accounts for 16% of U.S. methane emissions and contamination of 25% of all fresh water supplies in the country.

Not to mention the huge amounts of energy, chemicals, and land that are wasted in this process.

There is work underway in Europe in search of better understanding of the drivers of the problem of food waste, and identifying potential solutions for it. In 2012, the European Parliament adopted a resolution to reduce food waste by 50 percent by 2020. It also designated 2014 as the "European year against food waste."

An extensive U.K. public awareness campaign called "Love Food Hate Waste" has been conducted over the past five years too. As a result, over 50 of the leading food retailers and brands have adopted a resolution to reduce waste in their own operations, as well as upstream and downstream in the supply chain. As a result of these initiatives, during the last five years, household food waste in the United Kingdom has been reduced by 18 percent.

The situation is somewhat different in various countries, where America is in a class of its own. Here, food represents a small portion of most Americans' budgets, which makes the financial cost of wasting food low enough to be considered a regular, more convenient, mode of life.

Then there is the basic economic truth of advanced capitalist society, that if consumers waste more, industry sells more. This affects the entire supply chain—from the fields to the supermarkets. Here, any waste downstream translates to higher sales for the upstream companies.

Overcoming these and other related challenges is not easy, and will require the total cooperation of the government, consumers, and businesses. But this effort can start only after everyone understands and agrees that this is a problem. Only then, can the problem of reducing food waste be raised to a higher level of priority.

Reducing waste food losses in the U.S. by just 15% could feed 25-30 million people in Africa every year. It could also save millions of barrels of oil wasted during food production, storage, transportation, preparation, and disposal.

One simple solution (one of many that the government can undertake) is to standardize and clarify the meaning of expiration date labels on food packages. All foods have recommended use periods stamped on their packages. These stamps are hard to read and are easy to misunderstand. This is causing tons of foods to be thrown out due to misinterpretation.

Expiration date clarification on food packages could prevent about 15-20 percent of wasted food in U.S. households.

Businesses must understanding the extent of their own waste streams and adopt best practices. For example, a U.S. food chain saves over $100 million annually after an extensive analysis of the freshness, shrink, and customer satisfaction in the perishable foods department.

The consumers can reduce their waste by getting better educated on the basics of foods: shelf life, when and how the different types of foods go bad, as well as buying, storing and cooking food with waste reduction in mind.

Implementing efficient food waste reduction strategies, can bring tremendous social and economic benefits. It can reduce hunger, save energy, and reduce environmental pollution. Are we ready for it?

Sustainable Living

Sustainable living is a lifestyle that attempts to reduce the carbon footprint and minimize the use of Earth's natural resources. This is most often done by alternative methods of *transportation*, energy consumption, and special diets.

Sustainable living aims to structure human life in ways that are consistent with sustainability, or making sure that biological and all other life-supporting systems remain diverse and productive. This ensures their natural balance and preserves humanity's symbiotic relationship with the Earth's natural resources, ecology and cycles. Sustainable and ecological living is interrelated with the overall principles of sustainable development, which is a form of sustainable sustainability.

The concept of sustainable living is changing in the 21st century. Now it is focused—in addition to reducing carbon footprint, etc.—on shifting to a renewable energy-based, reuse/recycle economy and a diversified transport system.

The trend in sustainable living today is eco-villages. These are intentional communities focused on becoming socially, economically and ecologically sustainable and/or perhaps independent from the conventional societal life support systems.

The new era eco-villages usually vary in size from 20 to 200 people, and some are much smaller. Larger eco-villages of up to 2,000 individuals exist as networks of smaller sub-communities. Some eco-villages have grown by the addition of individuals, families, or other small groups who are not necessarily members settling on the periphery and participating in the eco-village community's activities.

Much larger eco-villages are also being developed in the midst of large metropolitan areas. These function as independent patches of alternative living in the midst of a conventional life style. Parts, or even entire large cities in Europe and Asia are using eco-village principals and trending towards sustainable living.

One example of a sustainable living patch (mini eco-village) in the midst of a large city is Christiania, or Freetown Christiania—a self-proclaimed autonomous neighborhood of about 850 residents—in the Christianshavn neighborhood in Copenhagen, Denmark. Christiania covers about 85 acres within the city limits and is officially regarded as a large commune with its own unique legal status. The commune is regulated by the Christiania Law of 1989, which transfers parts of the supervision of the area from the municipality of Copenhagen to the state.

Christiania has been a source of controversy since its creation in a squatted military area in 1971. Its cannabis trade was tolerated by authorities until 2004, which is when the conflicts escalated. Since then, measures for normalizing the legal status of the community have led to frequent police raids and ongoing negotiations.

During the height of the conflict, the neighborhood was closed by residents in the spring of 2011, while discussions continued with the Danish government as to its future. It is now open again and has resumed its normal daily operations. Conflict with the official powers is not a good, sustainable, or exemplary lifestyle, so we just have to wait and see what will happen at Christiania in the long run.

The practical, 21st century eco-village lifestyle requires a shift to renewable technologies, which approach is complex and expensive. This means that the concept can be successful only if the new lifestyle and the surrounding environment are attractive to (and can be afforded by) the local culture. Only then can the eco-village concept be maintained in the short run and adapted as a necessary and best life style by the generations to come.

Case Study: Ten Thousand Villages

Ten Thousand Villages is an all together different concept of collective sustainable living on a global scale. It is a nonprofit fair trade organization that produces and markets handcrafted products. It consists of disadvantaged artisans from more than 120 artisan groups in more than 35 countries.

As one of the world's largest and oldest fair trade organizations, Ten Thousand Villages cultivates long-term buying relationships in which artisans receive a fair price for their work and consumers have access to gifts, accessories and home décor from around the world. As part of the "global village" this concept opens new opportunities for people who are looking for a way out of the 20th century type of job market.

Large or small, the eco-village concept is growing and pointing towards a way of life, which actually reminds us of the communist system, which was designed and developed in a similar, self-sustaining, closed-loop environment. The basic concept was good—equality and opportunities for all—and the system could've been successful, but human greed, ignorance, and stupidity took over. After a short period of misery and agony for the masses, while the rulers twisted the rules and piled on wealth, communism fell with a crash. The effects of the 70 years of life in a closed society still reverberate around Russia and Eastern Europe, so we must be careful when considering new closed-loop systems. While the new "eco-village" concept might be a good one, we must remember the lessons of the past.

The eco-village concept, however, is contrary to the American way of life, where freedom from any organized bondage, rules, and regulations is priority number one for most free-thinking Americans. The American Dream is contrary to living in closely regulated communities. Because of that, we don't expect any great movement in that area in the U.S. anytime soon.

The Future Energy Dilemma

We have discussed a number of plausible scenarios for achieving acceptable energy security. While some of these approaches of energy generation and use, combined with energy efficiency improvements, are possible on theory, they also represents a great effort at a great expense. Presently, investment in all forms of alternative energy and energy efficiency is well below what is required to replace the conventional fuels, avert energy disasters, and remediate dangerous global warming.

The IEA, in its Energy Investment Outlook report published in the summer of 2014, puts a $53 trillion (with a T) price tag on the investment needed to significantly change our path to energy and environmental disasters. Meeting the rising global energy demand by 2035, nearly $48 trillion of investment would be needed. That means total spending on renewables and energy efficiency of $6 trillion and $8 trillion respectively are needed by 2035.

But even if this effort is successful, it will bring higher levels of warming that are assumed safe. The IEA's "450 Scenario," charts a path that could give the world a 50% chance of staying within the 2 degree warming limit. For that, the total investment needed has to be increased by another $4.8 trillion.

Spending on energy supplies would be $39.4 trillion, slightly less than the $40 trillion envisaged under the IEA's new policy scenario total. More of this investment must be spent on more costly low carbon energy sources, like solar, which would require $4.2 trillion of additional investment by 2035.

For this target to be achieved, investment in low carbon energy technologies must be increased to about $900 billion annually through 2035. Thus far we have seen that figure reaching only $260 billion—3.5 times below the mark.

Energy efficiency measures, implemented in buildings, transportation, and across industries, is essential to achieving the goals. The total price tag for this effort is estimated at $14 trillion. Significantly increased investment in energy technologies and efficiency measures is no doubt needed. Thus far national policies have been unable to achieve the necessary shift in investment flows. But this is not just a matter of getting the money. It is also about deciding how that capital is allocated and used.

The series of talks on climate change scheduled to take place in Paris in the summer of 2015 will be crucial to answering some of these questions. They are also expected to chart the path to global energy security and cleaner environment. In anticipation of these talks, the IEA's report stated, "Clarity of intent and, to the extent possible, stability of implementing measures are essential to build the confidence of investors to realize the energy transition."

What's in the Future?

We cannot predict the future of global energy, but we can easily see into the future of fossil fuels:

- *Short term*: quickly increasing production of crude oil, natural gas, and coal. Fossil exports from producing countries will increase as needed to feed increasing global energy needs and glut.

- *Mid-term*: gradual depletion of fossils. Crude oil will become increasingly harder to produce, and its quality will deteriorate simultaneously. Prices will increase gradually and availability will become sporadic.

- *Long term*: total depletion of all fossils—crude oil, natural gas, and coal—is inevitable sometime in the 22nd century. Coal might last longer, but will eventually be depleted too.

What are we to do? At this point, there are no clear solutions. We hear a lot of talk and different (mostly biased) suggestions. Capitalism still works! Making a buck today and living better than yesterday takes priority until some global events change the balance.

So, how about methane hydrates? Could they provide the bridge to the fossil-less future; giving us more chance to get ready for the inevitable?

The Fossil Energy Alternative

There are many suggestions of how the energy future will develop, but most are biased. Since we have no crystal ball, or any biases, we see the following as the most plausible way to enter the energy future of the 22nd century.

Figure 8-49 shows the most likely (or desirable) scenario of our energy future, where we dig and pump out the fossils as fast as we can, but even with the latest technologies and new reserves discoveries, they won't last more than a century…if that long.

After that, oil companies could focus on the huge methane hydrates reserves, since digging and pumping fossils from the ground is what they do best. Not to mention, this is the most profitable way of energy mass production.

The only problem with methane hydrates is that they are so explosive. The reserves are unstable and an earthquake at the right time and place could unleash a global disaster with untold consequences.

Using the renewables is the safest way to proceed. They will be around to fill the gap, but unfortunately (as things look today), their participation will be limited to completing the energy balance gaps. Nevertheless, the more reasonable, desirable, and most practical, scenario for our energy future is that depicted in Figure 8-50, where the fossils are used in a planned, gradual manner, while allowing for the renewables and other alternative technologies' development and implementation.

This way we could eliminate the effort and expense needed to develop technologies for extraction of unstable methane hydrates and the potential of a global environmental disaster. The renewables offer the only way to ensure steady, uninterrupted energy supply for future generations.

The Future Developments

The worldwide amounts of carbon bound in natural methane gas hydrates (NGH) is conservatively estimated to total twice the amount of carbon found in all known fossil fuels on Earth. NGH deposits contain immense amounts of methane gas, with major implications for energy resources and climate.

The natural behavior and managed control of NGH deposits as well as their impact on the environment, are interrelated, very complicated, and extremely poorly understood.

Realizing the incredible size of methane hydrate deposits in marine sediments, and their importance as a major future energy source, a number of scientific organizations are focused on the investigation of methods for their exploration. There are selected areas where hydrates are known to be common, and where their influence on energy resources, climate, and seafloor stability can be easily (or more easily) analyzed and documented.

Results of recent investigations indicate that methane hydrates possess unique acoustic properties, which could facilitate their exploration and make their exploitation easier. Efforts in this area, however, are sporadic—marking the baby steps of the industry—but the work continues.

Due to the recent fossil bliss in the U.S. and other countries, the future direction of research and exploitation of the NGH's natural energy bonanza is totally unpredictable. This should not make us forget that we have a great resource at our disposal, not if but when we run out of the conventional fossils.

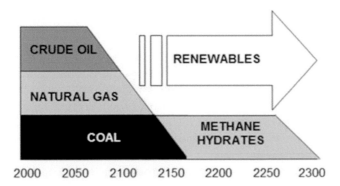

Figure 8-49. The world's energy future—most favorable scenario

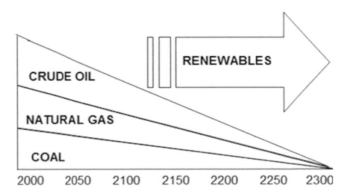

Figure 8-50. The world's energy future—the preferred scenario.

The Energy Bridge

In this text we discuss the global industrial revolutions—IR-I, IR-II, IR-III of the past, and IR-IV, or the Green Energy Revolution, of the present. But there is another one coming, IR-V, or the "post-fossils revolution (PFR). We don't know how exactly it will develop, but we know it is coming, because the fossils are a finite commodity.

How long would the fossils last? What will we do without fossils? Would nuclear and hydro be able to sustain their share in the energy mix now and later? What if they do not? What if instead we see more nuclear plant accidents? What if hydro power drops in production due to extended draughts? What would replace the 20-30% total power generation loss from these energy sources? Would we ramp up coal-fired power generation and build more coal plants? How long would that period last? Or would we rely more on the renewables?

There are a lot of unanswered questions in our energy and energy security future. For now, however, we have adopted the conventional "we will cross that bridge when we get to it" attitude. There is no bridge right now, nor are there any serious plans to build one. So how will we cross the gap into a non-fossils existence?

*The only feasible and sustainable long-term approach is to make sure that we build a **solid energy bridge** to our energy future by slowly and steadily replacing the fossils with renewable energy sources.*

If we build a bridge (using renewables) to bypass the effects of a fossils-less energy future, we might be prepared for the inevitable post-fossils reality. There are no indications of building such a bridge, so with every day we get closer to the crossing point when it will be too late.

Figure 8-51 shows the weak point of today's energy policies and approaches. We are focused on the present so much that we have lost all perspective of the post-fossils future. The energy gap is large and we cannot avoid it. We even make it larger by the day, by exporting some of our precious fossils to many countries, some of which are our enemies.

A bridge of renewable technologies must be built eventually to prepare us for the time when the fossils become too expensive and disappear.

International Energy Security

Energy security is important to all countries, but it has different connotations for different regions. While energy security in Saudi Arabia means protecting the

Figure 8-51. The energy bridge to the fossil-less energy future.

crude oil reservoirs and ensuring plentiful power generation for the population, in Sub-Saharan Africa it is reduced to gathering enough firewood for cooking the next meal.

The developed countries are especially affected by the continuous supply of great quantities of energy sources, and it seems that the more developed a country is, the more important energy security becomes. It is extremely important in North America. We cannot even imagine a day when flipping the light switch will not turn the lights on, or, God forbid, waiting in a gas line. So, energy security should be at the top of our government's agenda.

We discussed the energy security issues in this text and agreed that there are many, but the greatest one for the U.S. is the continuous and reliable supply of crude oil. We have enough coal and natural gas to run our power generation sector for a long time, but we still import half of the crude oil we need.

The transportation sector depends heavily on the supply of crude oil, so the risks to our energy security in that area are numerous and some are very serious. We spend a lot of money and effort on ensuring a reliable supply of crude oil, but it comes from unreliable sources and travels through some even more unreliable routes, so the risks are great.

Nevertheless, the U.S. is fortunate in that it is only crude oil that partially jeopardizes our energy security. The situation is much more serious in other developed countries.

Europe

The political, economic, and social relationship between Europe and Russia has deep roots that are unlikely to be severed anytime soon. The natural gas consumer/supplier relationship between Russia and Europe, however, is fundamentally negative and has been a subject of debate and frustrations on the part of all parties involved.

A critical natural gas "partnership" between the EU and Russia occurred in 2000 when the president of the EU Commission, declared an intent to double EU gas imports from Russia. As oil and gas prices steadily rose, however, so did the tensions between these "partners."

The EU began publicly calling for security of supply, while Russia complained about the need for security of demand. This tension culminated in 2006 when Gazprom cut off its supplies to Ukraine during natural gas pricing disputes in the middle of a severe winter weather.

This action compromised European supplies as well at that time, because about 80% of European natural gas imports from Russia flowed through Ukraine.

Tensions escalated again in 2009 when Russia announced that it would not ratify the Energy Charter Treaty with Europe. This document was created following the Cold War to support European energy security. It was signed by Russia in 1994, so its failure to ratify the treaty sent a strong and irreversible political message.

Following these incidents, Europe and Russia have been implementing (but mostly talk and political posturing) limited measures to diversify supply and markets respectively.

Things became very tense again in 2014, when Europe protested Ukrainian take-over and took measures to punish Russia for this unprovoked act of war. Putin's response actions are unpredictable, to say the least, so Russia's status of potentially unreliable energy supplier was quickly changed to that of absolutely unreliable in the long run.

The Russia-Ukraine conflict is not going away, so the concerns over a possible cut in Russian natural gas to Europe remains and even increases. The Europeans are not going to turn a blind eye to Russia's aggressions, so they are getting ready to change energy strategies to mitigate the risks of losing Russian supplies.

They are taking a closer look at Turkey, which has the potential of becoming an energy hub for Caspian and Middle Eastern oil and gas to Europe. Azerbaijan has significant fossil resources and is willing to supply Europe. Its location is also a natural conduit for oil and gas exports to Europe from other energy-rich countries, like Kazakhstan and Turkmenistan.

In the short term, Turkey and Azerbaijan are shaping up as important partners in supplying energy to Europe in its quest for energy security (or at least some semblance of it). These countries have been acting upon the promise of European exports by working on two strategic energy projects: the Baku-Tbilisi-Ceyhan (BTC) oil pipeline and the Baku-Tbilisi-Erzurum (BTE) natural gas pipeline.

When fully implemented, these developments can significantly weaken Russia's grip on Europe's energy markets—even providing somewhat more diversified and reliable energy supplies to the continent.

Those projects in Russia's periphery, however, can only be achieved with unwavering American and European backing. Sure enough, most U.S. administrations have considered European energy security as part of the U.S. national interest, and have been supporting the diversification of Europe's natural gas supplies through the new channels.

The BTC and BTE pipelines were part of the U.S. political support and played an important role in their realization. The only problem now is that regardless of how successful these projects might be, the amount of natural gas delivered by them is much less than the present-day Russian exports. The Crimean crises forced the U.S. to renew its commitment to increasing reliable energy supplies to Europe via these and other projects.

One of these is the Trans-Anatolian Gas Pipeline (TANAP) project, which as an important part of the Southern Gas Corridor, can deliver natural gas from Azerbaijan through Georgia and Turkey to Europe. It is expected to carry 23 bcm (billion cubic feet) after the second phase is completed in 2023, and 31 bcm after the third phase ending in 2026.

The Southern Gas Corridor is vital for Europe's energy security, but it is also important to Turkey's energy security and economic development. Turkey also is trying to reduce its reliance on Russian and Iranian gas imports by increasing imports from Azerbaijan. So, the additional 6 bcm per year provided by TANAP will reduce Turkey's dependence on Russian and Iranian energy, while at the same time significantly reducing the total annual natural gas costs.

The overall result from these efforts is increased energy security for the European continent through diversification of its energy supplies via adding new suppliers and routes. This will also help accelerate the economic development of all countries involved.

The problem here—as history tells us—is when relations between Europe and Russia warm again, plans for energy supplies diversification will be put aside... again. So, the uncertainty of European energy security will remain uncertain.

Germany

Germany set a new record for solar PV generation in early June of 2014. With a solar PV capacity of about 35.5 GW in 2013, and an increase to over 36.3 GW by the end of April 2014. Germany is a world leader in solar energy installs. The goal is having 35% of Germany's total energy capacity generated from renewable sources by 2020 and 100% by 2050. Wow!

Germany set another record for solar production/consumption on June 9, 2014 with a total solar generation of 23.1 GW, which actually meets 50% of its electricity demand. Albeit for one single day, Germany demonstrated that the goal is achievable.

The analysts forecast that the cost for storage systems (which are key to the progress of solar power) will drop in the near future, which means that Germany is sure to remain at the top for sometime.

The bad news is that in the summer of 2014 the German parliament approved controversial reforms to the country's energy policy. Starting in 2015 PV systems over 10kW in size will be subject to an initial 30% surcharge, rising to 35% in 2016, and 40% in 2017. At the same time major industrial energy users will pay a 15% rate, since a higher rate would put them at a competitive disadvantage compared with the European competitors.

Residential solar systems of under 10kW are exempt, but this is one-fifth of the country's present and planned PV expansion, which is likely to be dominated by mid-scale systems. This could leave small- and medium-sized businesses exposed to the EEG surcharge while the fossil industry is exempt.

The new "sun tax" is making the cost of solar power artificially high, and the suspicion is that the large utility companies are behind this development. Renewable energy groups have promised to challenge the reforms in court, so the battle for the German rooftops is just starting.

Japan

Japan's energy situation can be summarized in a few short sentences:

- *Japan has no significant fossil resources, so it imports most of its fossil energy.*

- *Nuclear energy was to be the solution, but Fukushima changed all that in 2011.*

- *Now Japan's energy future is uncertain...nuclear, or no nuclear, CO_2 or no CO_2?*

This is a sad story of one of the world's most advanced societies falling victim to its great technological advancements. Nuclear power and Japan have a long, happy history and were looking forward to an even brighter future.

In March of 2011, a powerful earthquake and a huge tsunami ravaged the eastern coast of Japan, heavily damaging the Fukushima Daiichi nuclear plant and sweeping away entire villages.

Some radiation is still being released three years after the accident, with no prospects of full containment anytime soon. The clean-up process, undertaken by 4,000 workers on daily basis is expected to take 30 years.

An area of about 30 square miles around the damaged plant will be totally contaminated for a long time. The monetary damages from clean-up and recovery efforts are in the billions and rising. In addition to radiation contamination of air and water is the increased carbon emissions resulting from increased use of coal-fired power plants.

As the nuclear plants have been shut down in Asia and Europe, their electric power generation was replaced by increasing the output of hundreds of coal-fired power plants. That, of course, increased the CO_2 emissions to where Japan—the Kyoto Protocol host country—became overnight one of the world's greatest CO_2 emitters.

The real damage to the world's nuclear industry is much more serious. The severity of that accident forced many people to re-evaluate their feelings about nuclear energy, and it is now viewed as a mass killer. Japan, Germany, and others shut down all their nuclear power plants for inspection. Many plans for new nuclear power plants are now on hold.

After the nuclear accident, the Japanese government started looking into the options, with the renewables being the obvious choice. A number of wind and solar power plants were installed and plans for many more were made. But something went wrong in the solar energy sector and as a consequence many of Japan's approved but not-yet-built solar PV projects from the 2012 fiscal year were cancelled.

Nearly 2.0 GW of cancellations were announced by the government in the summer of 2014. This is about 10% of all solar projects at the time, which brings a number of questions. On top of that, hearings are now expected to be held for another 3.0 GW of similar solar projects.

A special task force, investigating the solar projects approved in 2012, has found that nearly 50%, or about

9.0 GW of all 2012 solar projects approved in 2012 were already completed, or had produced the necessary documentation to go ahead with construction. But 5.0 GW, or about 25%, of these projects are questionable, so the investigation continues.

Some of the official reasons for solar projects cancellations point to grid connection issues. This is a legitimate concern that has become a stumbling block to much development in the remote northern island region of Hokkaido and the equally remote sub-tropical southern island region of Okinawa. It is unclear how many projects were cancelled for that reason, and more importantly, why were they approved to begin with?

Experts suspect that another reason for the cancellations is the fact that the lucrative FiT rates offered in 2012 through Japan's tariff program, made it very easy to obtain FiT approval. This allowed for substandard or impractical solar projects to be approved. Another problem under investigation is the fact that some developers applied for projects in 2012, but postponed development until the cost of PV panels went down, to maximize their profits. Some of these developers and projects are now under investigation.

So, with its nuclear fleet on stand-by and the solar industry taking baby steps, Japan is cranking up the fossil power generators, and with that the CO_2 emissions continue rising. With its foggy coastal areas and limited land area, Japan is not a "solar" country *per se*, so solar is not the immediate solution to its energy woes. Wind energy looks more promising, but it also is going through growing pains.

In best of cases, wind and solar cannot provide a significant amount of power to replace the need for nuclear power in the country. No matter what happens, Japan is faced with serious energy problems, which will be making headlines for a long time to come.

India

Coal and diesel drive India's economy. India, with its increasing energy deficit, is the fourth biggest energy consumer after China, the US and Russia. The energy split in the country is as follows: 55% of all energy comes from coal, 30% crude oil, 8% natural gas, 5% hydro electricity, 1% nuclear energy, and 1% biomass, wind, and solar power. Electricity generation in India has been hindered by domestic coal shortages, so India relies on coal imports for most of its electricity generation.

In 2013, India imported about 145 million tons of crude oil, 16 Mtoe (million tons oil equivalent) LNG, and 95 Mtoe coal. This is total of over 255 Mtoe of fossil energy, or nearly 45% of the total primary energy consumption. About 70% of India's electricity generation capacity is from fossil fuels, causing huge GHG emissions, which are also on the increase.

Even with all these imports, India still has one the worst electrical generation and distribution systems in the world. Blackouts are daily occurrences in most regions, and long-term power interruptions are to be expected at anytime. That has led to the development of the world's largest distributed power generation, where people use small diesel generators to power their homes and businesses during the periodic blackouts. Most of India's crude oil imports go to power this home-grown industry, and what a waste of crude oil and money this is—not to mention the pollution and health hazards that come with it.

Things are getting even worse. India's economy and populations are growing, as is the demand for more energy. As the trend continues, the dependence on energy imports is expected to exceed 55% of the country's total energy consumption by 2030.

India has ambitious plans to expand its renewable (wind and solar) and nuclear power industries. Wind power development is the one of the best developments in India's energy market, but it is still a very small part of the total energy picture. There are five nuclear reactors under construction (third highest in the world) and plans to construct 18 additional nuclear reactors by 2025.

If the future plans are to be judged by the developments in the solar sector, however, we see big problems in India's energy future. Tata Solar came up with a report in the summer of 2014, claiming that solar energy in India has the potential to become a viable alternative to fossil fuels. Tata estimates that up to 145 GW of solar, or about 13% of India's energy generation, can be deployed in the next ten years as solar becomes increasingly competitive…and IF the country follows the most appropriate strategies.

The report compared four scenarios for India's solar future; ultra-mega projects, utility-scale projects, large rooftops, and small residential rooftops. Comparing different scenarios for various sizes of solar installations across India, the report predicts solar prices will drop by 5% by 2024, while coal imports are expected to rise by 12%, and domestic coal prices are to increase by 7%.

Solar is expected to achieve parity with domestic coal soon, and storage technology will be cost competitive with imported coal by 2017…maybe. Good forecast, IF the country follows the most appropriate strategies in creating a market, which relies on reliable policies, like net metering, and creating consumer finance options. Again, IF…

The present reality, however, does not match the expectations. After several months of trade dispute over anti-dumping duties on PV modules manufactured in China, US, Malaysia and Taiwan, India gave up and decided to focus on growing its own solar industry. But the current state of India's PV manufacturing is an indication of a big problem in the making.

In 2014 there were 15 solar cells manufacturing companies in India, and 48 with PV module production capabilities. Only 12 firms had equipment capable of producing both solar cells and modules. But 6 of the 15 solar cells manufacturing companies, and 14 of the 48 PV modules manufacturers were in idle mode with no existing or planned production at all for the duration.

Amazingly, the larger PV manufacturers in India, such as Moser Baer, IndoSolar, Tata Solar, Vikram Solar, Warree Energy and XL Energy were among the number of idling facilities.

XL Energy, for example, has 80 MW solar cells production capacity, and 210 MW of PV modules capacity, but is producing neither. Another large player, Warree Energy, has a large, 250 MW of PV module production capacity that is also idling. Vikram Solar, with 150 MW PV modules production capacity produces only 75 MW, or 50% module production capacity. Another giant, IndoSolar, has 450 MW of modern, automated nameplate capacity, but only 80 MW was produced, or utilization rate being only less than 18%. The situation is even worse with the major player Moser Baer. With solar cell capacity of 200 MW the production lines are totally idled. From the 230 MW PV modules production capacity, only 80 MW were produced, or a maximum utilization rate of less than 35%.

The industry leader, Tata Solar, with 180 MW production capacity was producing 70 MW of solar cells, or a utilization rate of around 39%. At the same time, the company was producing 200 MW PV modules, or 100% utilization. Still, very far from the 145 GW projected in the above report.

The smaller module manufacturers seem to be relatively better off than the larger players. Gautum Solar, Modern Solar, Shan Solar, Sova Power, Topsun Energy and Photon Energy Systems are running at high utilization rates of above 80%, and several are at full capacity.

Still, the acute shortage of solar cells suggests that a significant part (over 70%) of all solar cells are purchased from abroad (mostly China and Taiwan). This means that imports of millions of solar cells and panels are in India's future, if the goal is to be even partially met.

The overall energy situation in the country is crying for a general reorganization and consolidations in the different sectors. Although the government now is promising help for the solar industry, it is not certain how and when the discrepancies would be resolved. Our experience with working in India shows that things there change very slow. And when they finally change, it is not always for the better.

Note: India is home of the mighty Thar desert, which takes more than 10% of the country's territory. It is one of the world's largest deserts, with enough daily solar insolation to power the entire world several times over. The Thar Desert is rich with incredible ethnic, cultural, and wildlife diversity, but high tech is not part of its wealth. This is pretty much the situation in the entire country too, where with the exception of a small minority, most people are basically uneducated and barely survive on basic skills.

India can use the Thor Desert and other natural resources it has available on its territory, but new approaches and government policies are needed. New educational initiatives, R&D incentives, lower cost of finance, and other such government-sponsored programs are needed to provide support for India's fledgling solar, and other, energy sectors.

Qatar

Qatar is a small oasis in the blistering desert, rich in oil and sand. Thus far it is relying on oil exports for its existence. So far so good. Lots of money is pouring in every day, and where there is money, there are usually good things happening. Now Qatar is planning to use its sand dunes to offer something very interesting that deserves a close look. It is the perfect 21st-century city in the desert—already in full construction mode.

Starting from scratch, on a sweltering strip of desert land, the future home of the 2022 World Cup is slowly rising from the ashes (uh, sand dunes). The new city, Lusail, is a vision of the government of Qatar and its developers. This is an old concept that nobody has ever tried because of the prohibitively high price tag. But Qatar's sheikhs have the money and would like to show off by building Lusail, an ambitious planned city on 28 square miles of waterfront desert along the Persian Gulf.

The compact city consists of a commercial district, a lagoon, four islands, two marinas, an upscale shopping mall, a hospital, a zoo, two golf courses and housing for over 250,000 people. There is also an 86,000-seat soccer stadium, surrounded by a moat, where the final game of the 2022 FIFA World Cup would be played.

The city's skyline is distinguished by four 75-story-high commercial towers, surrounded by five-star hotels, an entertainment district and the Marina Mall's

600,000-square-feet retail palace. The mall is a cluster of five interconnected pods, shaped like giant boulders, with cantilevered white roofs to repel the heat. Its architects want the space to evoke desert canyons. A canal runs through it, and waterfalls splash throughout.

Mass transport is provided via a light-rail network, a water-taxi system and a network of underground pedestrian tunnels. All energy and communications services, as well as the entire transportation system will be run from a computerized command center. Computers will drive the entire infrastructure, automatically adapting it to ever-changing traffic, weather, and emergency conditions.

The infrastructure itself is a new concept too. The city's gas and water pipes, as well as all electric wires are placed in an underground network of tunnels, thus allowing for their maintenance without disrupting aboveground activities. A network of surveillance cameras, monitored around the clock, will keep the streets and buildings safe too.

Solar power will be used to cool the buildings with chilled water, which will be pumped through a vast network of pipes. Another new concept, which, at least on paper, seems cheaper and more energy efficient than electrical air conditioning.

One of the most innovative concepts is that of public and residential toilets being fitted with special sewer lines. These are not your usual, gravity driven sewer pipes. Instead, an efficient network of pneumatic tubes, like those we see in bank's drive-up windows will transport human waste to its final destination for processing.

Presently, over 20,000 workers, mostly foreigners, are building Lusail, which is estimated for completion by 2019, three years before Qatar is scheduled to become the first Arab country to host the World Cup.

The idea is for Lusail to become the first truly "smart" city in the world. One thing the smart computers won't be able to adjust, however, is the ferocious desert heat that will plague the 2022 games, but that too is being scrutinized. You never know, money talks, so we would not be surprised to see some 21st century, air conditioned soccer stadiums rise from the sand too. For now, the plans are for stadiums to contain open-air cooling systems that will keep World Cup players and fans comfortable enough.

But there are other problems too. There are talks about the World Cup being moved to another country, and not just because of the heat. FIFA, soccer's international governing body, has been reconsidering Qatar because of allegations of bribery, use of slavery during construction, the country's condemnation of homosex-

uality, and, of course, the blistering summer heat, predicted to reach 120 degrees Fahrenheit during the 2022 games.

Figure 8-52. The futuristic City of Lusail, Qatar

This level of sophistication and futuristic solutions to urban problems can be achieved by Qatar, because, a) it's being built from the ground up, and b) money is not a problem. The estimated cost of $45 billion for the City of Lusail is a drop in the bucket for the oil-rich country which gets millions for their oil every day.

We also can't help it but wonder what Qatar is going to do after their oil is gone. They would have to reconsider life in general. Perhaps as a sign of things to come, the City of Lusail website crashed 9 out of 10 times we attempted to load its images. The future is not hard to predict; in many cases, just looking at the present gives you a good idea of things to come.

SUMMARY

Finally, we present below a list of facts that affect the energy sector, from an energy security point of view. These conclusions are based on our experience, resulting from extensive work and research in the energy sector.

Energy Independence

First, let's clarify and expand on the energy independence issue, which we have been avoiding in this text. The reason for that is that it is difficult to even think about energy independence under the present state of affairs. And the way all things energy are going today, it is even harder to think of it when looking toward the future.

It is simply unthinkable for any country to achieve energy independence until 100% of energy used domestically is produced domestically. No country has achieved this, and those that are close to it might achieve it but only temporarily.

Energy independence requires 100% domestic energy production and 100% energy security.

Looking at Figure 8-53 we see that the domestic electric power in the U.S. (generated by coal and natural gas) is more than secure. We have so much of these fossils today that we are even exporting a lot of them.

Crude oil, however, is a different case. We don't have enough, so we import over 50% of what we need. Every day there are convoys of oil tankers headed to the U.S.—5, 6, 8, 10, even 12 million tons per day have been unloaded at U.S. oil terminals in the past

Energy security is unachievable and unthinkable while consistently importing such large quantities of crude oil.

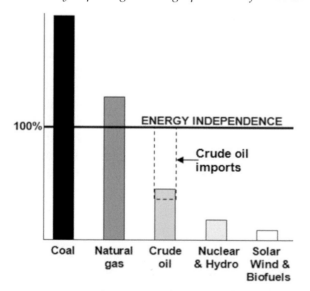

Figure 8-53. U.S. Energy independence...not!

Our power generation sector meets and exceeds the requirements for 100% energy security and is also well above the requirements for energy independence. The problem is that crude oil powers the majority of our transportation sector, and since we import about 50% of the total crude oil we use, our energy security and energy independence are affected.

Depending on such large oil imports from unstable countries is a serious problem. Any major world conflict

could reduce, or put an end of our imports. This would cripple the transportation sector, which will have lasting negative impact on the entire U.S. economy.

But this is only half the story. The other half, which is even worse, is that 40-50% of what we pay at the gas pump goes into the pockets of people who are either terrorists themselves, or are sponsoring terrorism. And most of the hostilities are focused on the U.S. and the developed countries.

So, by buying our energy security with oil imports, we are exposing ourselves to the whims of oil-producing monarchs and terrorists who control the oil flow. These are mostly people, with whom we have nothing in common. What they would do next is anyone's guess...

We need 50% more domestic production of crude oil to bring the energy import-export balance to zero. Only then can we talk about complete energy security and full (albeit temporary) energy independence.

What we have at this point—due to huge crude oil imports—is some sort of temporary "quiet before the storm," resulting in a false sense of safety reflected in an unsustainable, false sense of energy security. This "calm" could be interrupted at any point by another oil interruption like the 1973 Arab oil embargo. In the best of cases, how far would the dependence on fossils take us? As soon as the crude oil or natural gas reserves start diminishing, we would be back on the import bandwagon.

Of the other energy sources, the contribution of nuclear and hydro energy in the future is questionable too. One more Fukushima accident might shut down the entire nuclear fleet, or at least a portion of it. Yes, it can; just like it did in 2011 in Japan and Germany. Nuclear power is a good but unfaithful friend. It is not to be trusted, at least not until we are absolutely, positively sure that it is safe. This will not happen anytime soon, so beware of false assurances.

Hydropower, on the other hand, depends on the climate, and as things are going, we are witnessing a severe increase in draughts in parts of the U.S. and around the globe. These are already affecting the hydropower generation negatively. Hydropower is a good and safe energy resource, but it is unreliable in the long run.

Solar, wind, and biofuels are on the increase, but their contribution is still small enough to make a big difference in the short- to mid-term future. It will take a very long while, under the present global conditions, before they grow in size and significance in the energy sector.

Until some drastic changes occur, fossils are the major players in the energy markets. They (crude oil in particular), however, threaten our energy security and deprive us of energy independence.

Summary of Energy and Energy Security Facts and Factors

We are humbled by the depth and width of the energy and energy security sectors, and do not think that it is possible to provide a complete list of subjects, issues, explanations, and solutions, so the list below is a partial summary of these at best. It is, however, a good start of the discussion on these important subjects:

- Our energy security is determined primarily by the integrity of our domestic power generation and the transportation sectors, as follow:

  a) The power generation sector:
    — Presently our *power generation* is 100% secure, mostly due to use of abundant coal, natural gas, nuclear-, and hydro-power resources. We are even planning to increase the fossils exports, which is an ill-advised move that needs to be reconsidered for the sake of our (future) energy security.

  b) The transportation sector:
    — Our *transportation sector* is only 50% secure, due to unsecured imports of crude oil. Replacing crude oil imports with domestic production, such as biofuels, syngas, and fuels from coal liquefaction is the best way to 100% energy independence.

  Nevertheless, there are no official plans to reduce, let alone discontinue, the 50% crude oil imports anytime soon. This is a major breach of our energy and national security, which will only get more pronounced and serious with time.

- Fossils—natural gas and fossils—are presently the foundation of the power generation sector, and we have enough of them for now. The long-term energy security of the sector, however, is uncertain since the fossils' prices will keep rising and will be exhausted by the end of the 21st century.

- The present natural gas windfall is expected to generate a major portion of our electric energy, and partially fuel the transportation sector. This bonanza will last 40-50 years, and then natural gas will be exhausted. So measures must be taken for the long-term survival of the power generation and transportation sectors that depend on natural gas.

- Crude oil is the foundation of the transportation sector, 50% of which is imported, so the energy security of our transportation sector is questionable, since it depends on imports from unstable and unfriendly nations. The situation will become even more sensitive by the mid-21st century, when the global oil resources will be largely exhausted. At that time, their availability will be reduced, while prices will increase dramatically.

- Presently, the U.S. is NOT low on fossil fuels (coal and natural gas) *per se*, but only on transportation fuels (crude oil especially). Since crude oil is used primarily for transportation, and since efficient and cheap transportation is absolutely necessary for a healthy economy, another oil embargo would be devastating to our way of life.

- Oil shortages, and eventually running out of oil would have the most serious effects on the U.S. military. The military machine would come to a screeching halt. Planes would be grounded, warships will remain tied to the piers, trucks with provisions and ammunition won't reach their destination. The American soldier won't even be able to get to the battlefields, and the entire country and its interests would be vulnerable to enemy attacks.

- The fear of running out of fossils—and crude oil in particular—forces us to consider alternative fuels, like biofuels, synfuels, natural gas, shale oil, etc.
  — Biofuels are a promising fuel for the transportation sector's energy security, and their use will increase as crude oil supplies dwindle. New technologies—in addition to corn ethanol—must be developed for biofuels to compete. Cellulosic matter is the most promising alternative biofuel source.
  — Synfuels are another promising practical and important fuel. When fully developed, they should be cheap enough if, for example, can be made from cheap coal and natural gas. Synfuels, in theory, are one the few energy sources that have a chance to eventually compete with compressed natural gas.
  — Shale oil is also shaping up as a potentially promising new energy source for the transportation sector. Shale reserves in the U.S. are

enormous, and successful extraction technologies have been developed. Shale oil production is cheaper than synfuels, and will eventually become a major competitor.

- Another very important factor in the long-term energy security of the transportation sector is improved automobile mileage and more efficient use of cars and trucks. Car and truck manufacturers are working on more efficient engines, but this is not a priority, so they need a push from the politicians from time to time to come up with plausible solutions.

- Smarter driving habits by U.S. drivers is the best way to save fuel and protect the environment. Walking, biking, and mass transport are widespread in many countries, but getting Americans (this one included) out of their luxury cars won't be easy. No amount of regulation will make much difference until gasoline at the pump is $20-30 a gallon....or more!

- Hydrogen production is a promising energy alternative, but it is presently stuck somewhere between too complex and too expensive. It might become a viable energy competitor only when cheap energy sources for its production are implemented—a catch 22 situation. Some limited use in special areas is expected in the near future.

- Nuclear power use is limited by the danger of nuclear accidents and nuclear waste storage—both of which are real, but somewhat exaggerated. Nevertheless, because of these uncertainties and the lack of definite solutions, the short-term future of nuclear power is uncertain. In the long run, the uranium supplies will run out by the end of the 21st century, and if nuclear power survives by then (without any new major accidents) substitute fuel must be found.

- We also import a significant part of our nuclear material—some from unstable and unfriendly nations. This is not a big problem at the present, but will become such in the future as prices go up, the producing nations destabilize further, and global competition for these materials increases.

- Some of the more exotic energy sources, like geothermal, tidal energy, and ocean wave energy,

are limited for reasons of technological complexity and logistics. Their future will be determined largely by the level of their technological development and the price of the competing technologies and fuels.

- Solar energy is undergoing spectacular progress, but the jury is still out on its potential to compete with the major energy sources. Solar cells are the most flexible and most promising solar technology, since solar thermal is proving to be not so flexible, since it requires a lot of water. At the same time challenges—changing regulations, utilities resistance, lack of financing, and interconnect difficulties—are barriers in solar's long-term future.

- Wind power has seen even larger increase lately, and is proving to be a truly clean and cheap energy generating technology. The problems of intermittency and availability in predominantly remote locations are slowing its progress, as is the growing opposition from neighbors and environmentalists.

- Energy storage, which is the only way to reduce the intermittency of wind and solar power generation, is not going to be feasible for technological and financial reasons for some time to come. It is, however, essential for these technologies to compete with the major energy sources, and miracles happen....

- Hybrid cars have some promise in the future, but plug-in hybrids and all-electric automobiles not so much. Not in the U.S. Complexity, high cost of repairs and battery replacement, short range, and sluggish performance (per American standards) have been some of the reasons of their slow U.S. market penetration.

  As a confirmation, while Volts and Prias abounded at the International Auto show in Detroit in the past, there were few of these types of vehicles at the 2014 show, and even fewer interested visitors. Muscle and sports cars, large trucks, and SUVs took the spot light and are headed back to the American roads. Goodbye, Volts and Prias... until the next time oil rises over $150/bbl. Capitalism works...

- For the long-term energy security of the U.S., we must focus on improving the energy productivity, and optimizing the efficiency of the energy technologies. New energy use efficiency and energy conservation technologies and measures can yield the best returns at lowest cost and least effort.

- Energy efficiency and energy conservation are the greatest weapons we have to ensure both; our energy security and environmental integrity. These are, however, the hardest measures to implement in America, where everyone is chasing the American Dream, which requires a lot of energy. For such measures to work properly and efficiently, new regulation must be designed and implemented. Work in that area has started, but much more effort is needed before we can report great results.

- Very few of the publicly proposed solutions to increase energy efficiency and decrease CO_2 emissions have had a significant positive effect. They are usually discussed, but not implemented, or if implemented they have no realistic chance of making a big difference, because:
  — Internally, we are busy increasing our economic growth, and although there is decrease in CO_2 emissions lately, it is just not enough to offset the increasing global emissions, and
  — Internationally, we claim to be setting an example for the world to follow, but most nations can't, or don't want to, follow our example. Encouraging developing nations to consider coal-to-natural gas conversion might be the only productive effort in this area at this time.

- The nuclear disasters of Chernobyl, Fukushima, as well as the Exxon Valdez, and the BP Gulf oil spills were all quite devastating to the environment, wild life, and people alike. They are also good examples of the worst that can happen in the energy cycle. After a short-lived world-wide scare, and some temporary measures, there were some changes in the energy policy of some countries. But on the overall, things went back to "normal" as if nothing happened. This means that energy generation is a top priority for governments and the population alike. It also tells us that nothing can deviate us from the course we have chosen—accidents or no accidents.

- Global warming and the related climate change, caused largely by humans, are threatening large part of the world, and yet there are no significant efforts to resolve the underlying problems. This

is because the situation can be controlled only by means of inexpensive or profitable methods to reduce GHG emissions in the major polluting countries of this world.

Such methods are unavailable now and unlikely to popup any time soon. And since most countries are not willing to sacrifice, the status quo will be unchanged for a long while. That means higher CO_2 levels, more smog in the cities, increased global warming, and unpredictable climate change.

- As the U.S. is entering a new era of abundant (albeit temporary) energy availability, it fortunately is also reducing GHG emissions. This will reduce the total of pollutants emitted in the atmosphere, but it will not even make a dent in the total GHGs released globally. This is because many developing countries—China, India, Russia, and many others—are determined to catch up with the West. They plan to do this by adding thousands of coal-fired power plants and putting millions of new automobiles on their roads. These additions are increasing the global total GHG emissions significantly, thus making the efforts of the Western countries irrelevant.

- A number of less-known fuels have the potential of adding to our energy sources, thus improving our chances of ensuring stable energy supplies. Some of the most important are:
  — Coal liquefaction is a process capable of converting the single coal molecule into versatile hydrocarbons in liquid fuels. This process—when properly developed and implemented—could eliminate all crude oil imports. We have enough coal to last over a century, and are presently reducing its use in the U.S., so instead of exporting it (as planned presently) we could use the excess to make liquid fuels for our transportation sector and for other applications, as needed to replace crude oil imports. This is the fastest and most plausible approach to full energy security and long-lasting (albeit not eternal) energy independence.
  — Methane hydrates are natural gas in frozen state that are found in huge quantities on the bottom of the world's oceans. When the extraction processes are properly developed and implemented, methane hydrates have the potential of supplying energy for centuries to come. The extraction technology, however,

needs to be developed first and made absolutely fail-prove.

The list above reflects the major factors in the energy sector that affect our energy security. These are complex matters that we must understand and act upon, if life in this country (and on Earth) as we know it is to continue unchanged. Some of the developments often go against the public's beliefs and interests, and are also beyond a government's ability to control.

The most important of these is the fact that we are not preparing for the inevitable depletion of the fossil resources. Oil companies, encouraged by politicians, are drilling and digging out as much coal, natural gas, and oil as they can—as if there is no tomorrow...literally!

No doubt, looking at Figure 8-54a, the energy bonanza in some countries ensures economic boon, personal comfort, and safety to their people...while the fossils last.

Figure 8-54b, however, shows something we don't want to see. It is how our fossil-less (or pre fossil-less) future will look, if we continue the present rate of exploitation of these precious commodities. Coal, crude oil, and natural gas reserves will be largely depleted in the near future, and whatever remains will be difficult and expensive to produce.

Energy use as we know it today will be no more. Heating and cooling of homes will not be done with fossil fuels. Cars, trucks, planes, and boats will not run on gasoline and diesel. Industrial enterprises will not have huge—almost unlimited—quantities of electricity, coal, natural gas, and crude oil for their daily operations.

The question is: What will the fossil fuels be re-

placed with? Will the next generations have enough time to replace them with something, or must they revert to the stone age?

In the fossil-less society, energy security will no longer be a major concern, because unless each country has developed its own energy supplies by then, there will be no energy to compete for. Countries that do not develop internal energy resources, will have to live without reliable energy.

We have many other choices of energy resources, such as solar, wind, hydro-, geo-, ocean-, and bio-fuels, to mention a few. These must be developed and implemented much faster than the present plans allow, in order to fill the gap left by the depleted fossils of our future fossil-less existence.

Conclusions

"We live on a Smarter Planet, where data, mobile, social and cloud are transforming industries and allowing us to create value in new ways. The evidence is abundant and everywhere." From IBM Website

We cannot agree more with IBM's marketing gurus who have captured the essence of our modern lives in one sentence. This, no doubt, is a new world full of amazing inventions, exciting discoveries, shiny toys, and new developments that improve the quality of life around the globe. We are full of hope for the future.

On the practical side, however, we must ask what is driving all of this. The answer is over a century old: it is our old friends the fossils that are running our homes and businesses. Our national security, energy security, military operations, vehicle transport, power gen-

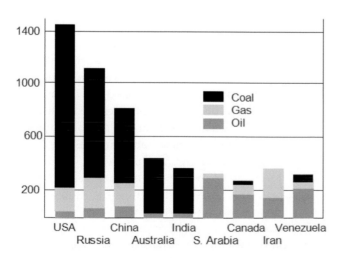

Figure 8-54a. Global fossil reserves (in billion bbl equivalent in 2013)

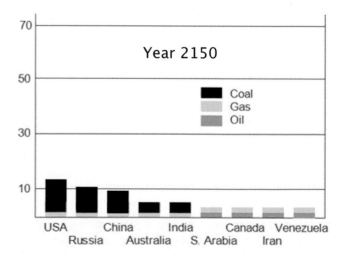

Figure 8-54b. The fossil-less future (in billion bbl equivalent by 2150)

eration, and entire economic development depend on fossils, with crude oil as the major driver of it all. Its availability (or lack thereof) and increasing price are the greatest threat to it all.

The Institute for 21st Century Energy is following America's Energy Security Risk Index. It shows that the U.S. reached 100% energy security risk in 1980. After a short remission during the mid-1990s, the risk index increased near 100% during the 2008-2012 economic crises. The risk index went over 100% for a first time in 2011.

The U.S. energy security risk index is now decreasing, and is expected to drop below 90% by 2020. Then, however, it will slowly increase to 100% by 2040 and will remain at that level, or higher. This tells us that we are entering a very dangerous, long-term period of energy insecurity, so America, tighten your belt and get ready for what's coming. It *is* coming.

It is not easy ensuring some resemblance of energy security with 50% of oil imports coming across the world (an effort requiring an army of fighter planes and battleships to ensure their safe arrival) every day. It is even naïve to talk about, let alone plan for, energy independence, or any semblance of such, under the existing, difficult, and unsustainable energy import-export conditions (crude oil in particular).

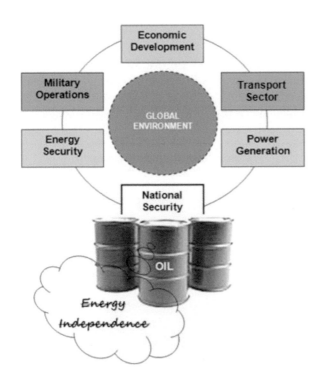

Figure 8-55. The energy-environment circle of the 21st century.

It is hard to imagine anything in our daily life that does not need some sort of energy input—cars, appliances, gadgets—all daily necessities and all running on fossil fuels. Without energy, our lives would become totally different; at the same time, all these things generate some sort of pollution at some point of their life cycle.

Excess energy use (founded on our old friends the fossils) puts the health of our environment in the center of the energy picture. How sustainable is this?

Right now the biggest problem for the U.S. energy future and energy security is ensuring steady crude oil supplies. This effort represents daily negotiations and plans for import quantities and prices, transport security, and refining and distribution logistics activities.

The other fossils—coal and natural gas—require less effort, but they will also become a problem—and maybe even more serious than crude oil—in the near future. Over a century old coal mines and coal-fired power plants provide most of the power we need to turn on our cars, appliances and gadgets. The electricity produced by this coal dug out from deep underground, accompanied by clouds of dust, grime, and choking air pollution, is what turns the lights on in the conference rooms where the new ideas for a bright future are born.

Thousands upon thousands of huge drill rigs burying pipes miles underground and pumping out precious natural gas to replace coal in the electric generation is the new phenomena in the energy picture. In all cases, electricity is what drives the sophisticated equipment needed to produce and run micro-processors and other parts that allow our computers to make our lives easier and bring the future closer.

Electricity also drives every piece of production equipment that makes the goods we use on a daily basis: cars, computers, food, clothing, shoes, chemicals… And of course, our cars, trucks, busses, trains, and planes would not move an inch without the crude oil derivatives in their fuel tanks. The scary part is that our military will not run, without adequate and energy input.

Without electricity and crude oil, the entire world's economy and transport would come to a grinding halt—especially those in the U.S. and the developed countries. The misery, suffering, and pain of the people would be immeasurable. Goodbye energy security, goodbye comfortable life style, goodbye unrealized dreams for a brilliant future.

It is that simple! No electricity and fossil fuels, no economic development, and no bright future! Within a very short time period we would be thrown back into the Dark Ages. At the rate we are using coal, natural gas, and crude oil, we will run out of these resources within the next several decades.

The concept of energy security would become semantics. The American Dream would become a daily nightmare, and those living in it would point a finger at us—the irresponsible people of the 21st century who in an amazingly selfish way deprived them of the chance to live a normal life.

There are ways to avoid such pessimistic outcome, but they require a lot of money and effort, none of which is in the official plans today.

If we don't start planning for the fossil-less tomorrow now, soon it will be too late to avert the upcoming energy doom.

We dig and we pump and we burn as much as we possibly can—and even more than we need. And although we clearly see the fuel gauge nearing the empty mark, we not only don't slow down, but we accelerate production and use of fossils.

The U.S. now has a lot of fossils (coal and natural gas). So much, in fact, that we cannot use them all. But instead of saving some for tomorrow, we are exporting a major part of our fossil reserves; coal, natural gas, and even refined crude oil products.

We are so focused on achieving energy security and economic prosperity today (now) that we are cutting the branch we are sitting on in the process.

After over 40 years of talking about it, we are no closer to energy security today than we were then. At the same time, we are getting farther from energy independence with every passing day. The dreams of yesterday are still dreams, and it seems that they will be dreams in the future as well.

There is a bar in Texas with a sign over the counter saying, "Free Beer Tomorrow." You go there tomorrow and the sign still says, "Free Beer Tomorrow." So you ask the bartender and he says, "Can't you read; the free beer is tomorrow."

Just like in the Texas bar where the future never comes, our energy independence has been coming for over 40 years, but it is yet to arrive...

Presently, we import about 50% of the crude oil we use to fuel our vehicles on daily basis. We also import 90% of the uranium ore we use in the nuclear plants. How that affects our energy security is not difficult to guess.

One does not need to be an energy expert to figure out that any unfriendly energy exporting government

can turn the export spigot off in a split second. A similarly unfriendly terrorist group could sabotage the transport routes and stop the crude oil imports.

What that does to our energy security depends on the situation, but under all circumstances the energy independence dream will remain forever a thing of tomorrow.

On a global level, we are getting even farther away from achieving any of the energy goals. This is mostly due to the increased demand for energy, growing population, and rising demand for fossils. None of these trends shows any signs of change, so we cannot be very optimistic about the world's energy future.

And the environment? We actually spent a lot of time on the subject in this and previous texts (see reference 17), so we would spare the reader another spiel about today's and tomorrow's environmental problems. We all know how bad things are getting...and this is just the beginning...literally!

It is hard to put all these factors and variables into one picture, but Figure 8-56 clearly shows the dependence between fossils availability, energy use, the environment, and our energy security. These are intimately, yet inversely, related, so as time passes, the energy use and the environmental problems increase, while the energy sources—the fossils in particular—decrease, approaching complete exhaustion. In time, our confidence in achieving an energy security balance also decreases proportionately.

It is obvious from Figure 8-56 that 15 years ago—around 2000—we were aware of energy problems and knew that our energy security was in jeopardy. After several wars, millions of dollars and thousands of sol-

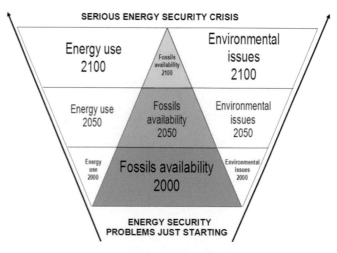

Figure 8-56. Fossils, environment, energy, and energy security.

diers dead or wounded, we now live in a world that is more dangerous than ever. Part of this is due to the speed with which the social media let us know what is happening even in the most distant and isolated places on earth. Be that as it may, it cannot be denied that energy has a lot to do with it. Was the war in Iraq started because of our intentions to control its vast oil reserves? Or was it just a good Samaritan gesture?

Different people have different opinions, but one thing cannot be denied: the oil fields of Iraq and Kuwait are soaked in American blood. There was more money spent and blood shed there than anywhere else since WWII. Why? There might be many reasons, but crude oil and our energy security always have something to do with it...

Try as we may, crude oil is not going to last forever, no matter how much effort we put into ensuring its supply. Someday the oil tanker convoys will run dry, because there will be no more oil to be pumped from the ground. So put yourself in the shoes of your great-great-great-great-grandchildren, looking back and analyzing our lifestyle and our actions. Looking at their bleak, fossil-less lives, they'll scratch their heads in disbelief.

None of us will live to see the energy poverty and the environmental disasters during the next centuries, but one thing is for sure: there will be serious problems around the world as a consequence of our actions or lack thereof, IF we don't change our ways soon.

"IF" is the key word here. There will be environmental, political, economic, and social disasters of great proportions IF we don't take serious measures to change the ways we:

- Exploit our natural energy resources, and

- Use all types of energy in our personal and business lives.

Could we change the way things are going? Yes, but it is very unlikely, at least as things are going now.

Here is a letter we need to send future generations so they will know what happened and what to expect:

January 1, 2020

From: The 21st Century Generation
To: All Future Generations
Re: Your Energy Security and Environmental Integrity

Dear Future Generations,

We are sorry for burning all the fossil fuels on Earth in the name of our energy security. We just couldn't agree on a better way to do it, and when we realized what was happening, it was too late.

We are also sorry for the environmental mess we left you, too. It was not intentional; we just had to survive the best way we could. Unfortunately, this is what it is!

Basically, you are on your own! We hope you are smart enough to handle the lack of fossils and the environmental mess we left you. Please don't hold it against us. Nothing personal, you know....

Wishing you best of luck. Really!

Sincerely,

The People of the 21st Century

Notes and References

1. Institute for 21st Century Energy. Solutions for Securing America's Future. http://www.energyxxi.org/energy-security-risk-index
2. ExxonMobil. U.S. oil and natural gas Industry sets reserves records. http://www.exxonmobilperspectives.com/2012/08/21/u-s-oil-and-natural-gas-industry-sets-reserves-record-3/?gclid=CJ-YlZSwwMACFZSFfgodJGUAuw&gclsrc=aw.ds
3. IEA. Energy Security. http://www.iea.org/topics/energysecurity/
4. Methodology for prioritization of investments to support the army energy strategy for installations, by George Alsfelder, Timothy Hartong, Michael Rodriguez, and John
5. Farr Center for Nation Reconstruction and Capacity Development United States Military Academy, West Point, NY 10996 http://www.usma.edu/cnrcd/CNRCD_Library/Energy%20Security%20Paper.pdf
6. DoD, Energy of the Warfighter. Operational Energy Strategy, 2010
7. DoD, Annual Energy Management Report, Office of the Deputy Undersecretary of Defense (Installations and Environment), 2010
8. Energy Pricing Report, 2013. Center for Nation Reconstruction and Capacity Development.
9. Derivation of power gain for three types of three dimensional photovoltaics cells, by Jack Flicker and Jud Ready. *Materials Science and Engineering*, Georgia Institute of Technology, Atlanta, GA.
10. *Natural Gas Hydrates: Overview*, Steven Dutch.
11. Energy game changers-IQ2 talks. http://www.energypost.eu/many-gamechangers-energy-sector/
12. *Journal of Energy Security*, http://www.ensec.org/
13. *Big Green, Not Big Oil, is the Enemy*, by Peter C. Glover and Michael J. Economides. http://www.thecommentator.com/article/3768/big_green_not_big_oil_is_the_enemy
14. U.S. Energy Security Council. http://www.usesc.org/energy_security/
15. *Photovoltaics for Commercial and Utilities Power Generation*, Anco S. Blazev. The Fairmont Press, 2011
16. *Solar Technologies for the 21st Century*, Anco S Blazev. The Fairmont Press, 2013
17. *Power Generation and the Environment*, Anco S. Blazev. The Fairmont Press, 2014

Appendix

ABBREVIATIONS

'	Minute
	Foot
''	Second
	Inch
α	angular acceleration
	Alpha rays
β	velocity in terms of the speed of light c, unitless
	Beta rays
γ	Lorentz factor
	Gamma ray
	Photon
	Shear strain
	Heat capacity ratio
Δ	A change in a variable
π	Energy efficiency
	Coefficient of viscosity
λ	Wavelength
ν	Frequency
Σ	Summation operator
σ	Electrical conductivity

ACRONYMS

a	Acceleration
AAO	Antarctic Oscillation
ABE	Acetone-butanol-ethanol (process used in the production of butanol)
ACC	Arizona Corporations Commission
	Air Combat Command
ACCS	Assured Combinable Crop Scheme
ACE	Area Control Error
AD	Anaerobic digestion
ADB	Asian Development Bank
AEE	Association of Energy Engineers
AEGIS	State-of-the-art radar and missile system
AEP	American Electric Power
AEO	Annual Energy Outlook
AETC	Air Education and Training Command
AEZ	agro-ecological zoning
AFB	Air Force Base
AFCEE	Air Force Center for Engineering and the Environment
AFCESA	Air Force Civil Engineer Support Agency
AFFEC	Air Force Facility Energy Center

AFIT	Air Force Institute of Technology
AFMC	Air Force Material Command
AFRETEP	African Renewable Energy Technology Platform
AFSO 21	Air Force Smart Operations for the 21st Century
AFSPC	Air Force Space Command
AFV	Alternative Fuel Vehicle
Ag	Silver
ALCC	Life Cycle Cost Annualized
ANL	Argonne National Laboratory
ANGB	Air National Guard Base
AO	Arctic Oscillation
AOGCM	Atmosphere-Ocean General Circulation Model
ARAR	Applicable or Relevant and Appropriate Requirements
AR4	Fourth Assessment Report (of the IPCC)
Ar	Argon
As	Arsenic
ASHRAE	American Society of Heating, Refrigerating and Air Conditioning Engineers
ASN(I&E)	Assistant Secretary for Installations and Environment
Au	Gold
AU$	Australian dollars

B	Boron
BA	Balancing Authority
BASF	Badische Anilin- und Soda-Fabrik, Ludwigshafen, Germany
BC	British Columbia
BMDS	Ballistic Missile Defense System
BBTU	Billion British thermal units
Be	Beryllium
BIG	Biomass integrated gasifier
BLM	Bureau of Land Management (USA)
Bio-SG	bio synthetic gas (also referred to as bio synthetic natural gas)
BOE	Barrels of oil equivalent
BOP	Balance of Power
BPA	Bonneville Power Administration
Br	Bromine
BRAC	Base Realignment and Closure
BTL	biomass-to-liquids
BTU	British thermal unit
BUMED	Bureau of Medicine and Surgery

c	Speed of light (300 million meters per second)
C	Carbon
	Celsius (degrees)
Ca	Calcium
CAFE	Corporate average fuel economy
CARB	California Air Resources Board
CCl4	Carbon tetrachloride
CCS	Carbon capture and storage
CC	Combined Cycle
CT	Combustion Turbine
CCSM	Community climate system model
Cd	Cadmium
CdTe	Cadmium telluride
CDC	Central Distribution Center
CDD	Consecutive dry days
CDM	Clean development mechanism
CDP	Carbon Disclosure Project
CEC	California Energy Commission
CEM	Certified Energy Manager
CEMT	DFAS Corporate Energy Management Team
CEQA	California Environmental Quality Act.
CER	Certified emission reduction
CERCLA	Comprehensive Environmental Response, Compensation and Liability Act of 1980
CES	Civil Engineering Squadron
CEU	Continuing Education Unit
CFC	Chlorofluorocarbons
CFE	National electric utility, Mexico
CFL	Compact fluorescent light
CH_2Cl_2	Methylene chloride
CH_3Br	Methyl bromide
CH_4	Methane
CMA	Court of Military Appeals
$CMIP_3$	Coupled Model Inter-comparison Project 3
CNG	Compressed natural gas
CNIC	Commander, Navy Installations Command
Co	Cobalt
CO_2	Carbon dioxide
CO_2e	Carbon dioxide equivalent
CONUS	Contiguous United States
COP	Conference of Parties
CPUC	California Public Utilities Commission
COR	Contracting Officer Representative
CPV	Concentrating photovoltaics (solar PV)
Cr	Chromium
CRA	Climate risk assessment
CRM	Climate risk management
Cs	Cesium
CSIRO	Commonwealth Scientific and Industrial Research Organization

CSP	Concentrating solar power (thermal solar)
Cu	Copper
d	Diameter
DASA(E&P)	Deputy Assistant Secretary of the Army for Energy and Partnerships
DASA(I&H)	Deputy Assistant Secretary of the Army for Installations and Housing
DASN(I&F)	Deputy Assistant Secretary Navy for Installations and Facilities
DBCP	Dibromochloropropane
DCB	Dichlorobenzene
DCMA	Defense Contract Management Agency
DDC	Direct Digital Controls
DDGS	dried distiller's grains with solubles
DDT	An environmentally-persistent insecticide banned for most uses by the U.S. EPA in 1972.
DeCA	Defense Commissary Agency
DeCAH	DeCA Design Criteria Handbook
DESERTEC	Foundation "Clean Power from Deserts"
DFAS	Defense Finance and Accounting Service
DFD	Defense Facilities Directorate (WHS)
DHHS	Department of Health and Human Services (USA)
DHS	Department of Homeland Security (USA)
DI	Deionized water
DIA	Defense Intelligence Agency
DIAC	Defense Intelligence Analysis Center
DJF	December, January, February
DLA	Defense Logistics Agency
DME	dimethylether
DoA	Department of the Army
DoAF	Department of the Air Force
DOC	Department of Commerce (USA)
DoD	Department of Defense
DOD	Depths of Discharge
DOE	Department of Energy (USA)
DOI	Department of Interior (USA)
DOJ	Department of Justice (USA)
DoN	Department of Navy
DOT	Department of Transportation (USA)
DRE	Destruction and removal efficiency
DR	Demand Response
DTSC	Department of Toxic Substances Control
DUSD(I&E)	The Deputy Under Secretary of Defense (Installations and Environment)
E	Energy
E/P	Energy/rated power
E85	85 percent ethanol fuel

EACC	Economics of adaptation to climate change	g	Standard gravity acceleration
ECD	Estimated Completion Date	Ga	Gallium (semiconductor)
ECIP	Energy Conservation Investment Program	GaAs	Gallium Arsenide (semiconductor)
		GAC	Granular activated carbon
ECMWF	European Centre for Medium-Range Weather Forecasts	GBEP	Global Bioenergy Partnership
		GC	Grand Coulee
EDF	Electricite De France	GCM	General Circulation Model
EIA	Environmental impacts assessment	GCOS	Global climate observing system
EISA	Energy Independence and Security Act	GDP	Gross domestic product
EJ	exajoule	Ge	Germanium
EMCS	Energy Management Control Systems	GEF	Global Environment Facility
EMSG	Energy Management Steering Group	GET-CCA	Global Expert Team on Climate Change Adaptation (of the World Bank)
ENSO	El Niño-Southern Oscillation		
EO	Executive Order	GFCS	Global Framework for Climate Services
EPA	Environmental Protection Agency (USA)	GHG	Green house gas (effect)
EPAct	Energy Policy Act	GNP	Gross national product
EPCRA	Emergency Planning and Community Right-to-Know Act	GPC	Government Purchase Card
		GPI	Genesis Potential index
EPEAT	Electronic Products Environmental Assessment Tool	GSA	General Services Administration
		GSF	Gross Square Feet
		GSHP	Ground Source Heat Pump
ERA-40	ECMWF 45 year analysis of the global atmosphere and surface conditions (1957-2002)	GT	Giga ton
		$GTCO_2eq$	Giga tons of carbon dioxide equivalent
		GTS	Global Telecommunication System
ESC	Engineering Service Center	GW	Giga-watt
ESCO	Energy Service Company	GWh	Giga-watt-hour
ESMAP	Energy Sector Management Assistance Program	GWP	Global warming potential
ESPC	Energy Savings Performance Contract	h	Height
ESPP	Energy Savings Performance Program	H	Enthalpy
ETP	Energy Technology Perspectives		as in pH, measure of strength of acidic or basic aqueous solution.
ETSD	Engineering and Technical Services Division		
		H_2	Hydrogen
EUMETNET	European National Meteorological services	H_2O	Water
		H_2SO_4	Sulfuric acid
EV	Electric vehicle	HCFC	Hydro chlorofluorocarbons
		HCL	Hydrochloric acid
f	Frequency, in Herts or s^{-1}	HDD	Heating degree days
F	Fluorine	HDI	Human development index
	Fahrenheit (degrees)	HEAT	Hands-on Energy Adaptation Toolkit
FAME	Fatty acid methyl ester	HF	Hydrofluoric acid
FAO	Food and Agricultural Organization	HFC	Hydrofluorocarbons
FDA	Food and Drugs Administration	Hg	Mercury
Fe	Iron	HHV	Gross calorific value
FEMP	Federal Energy Management Program	HP	Horse power
FES	Facility Energy Supervisor	HQ	Headquarters
FFV	Flex-fuel vehicle (multi-fuel use)	HQCC	Headquarters Command Complex (MDA)
FOB2	Federal Office Building #2 (Navy Annex)	HQDA	Headquarters Department of the Army
FT	Fischer-Tropsch	hr	Hour
FY	Fiscal Year	HVAC	Heating, Ventilating, and Air Conditioning
		HYBLA	Hybla Valley Office Building (WHS)

HVO	Hydrotreated vegetable oil	LED	Light Emitting Diode
		LEED	Leadership in Energy and Environmental Design
I	Current (in Amps)		
	Iodine	LFD	Lease Facilities Division (WHS)
IAEA	International Atomic Energy Agency	LFF-1	USMC, Facilities and Services Division Facilities Branch
IATA	International Air Transport Association		
IBEP	International Bio-energy Platform	LGE	Liter gasoline equivalent (energy content 33.5 MJ/litre)
ICAO	International Civil Aviation Organization		
ICBM	Intercontinental Ballistic Missile	LHPP	Large hydropower
ICPD	International Conference on Population and Development.	LHV	Lower heating value, or net calorific value
		Li-ion	Lithium-ion
IEA	International Energy Agency	LMP	Locational marginal price
IESP	Infrastructure Energy Strategic Plan	LRS	Load and Resource Subcommittee
IEO	International Energy Outlook	LPG	Liquefied natural gas (Propane)
IFAD	International Fund for Agricultural Development	LSS	Lean Six Sigma
		lt	Long ton
IFC	International Finance Corporation	LTC	Lithium Technology Corp
IFES	Integrated Food and Energy Systems	LUC	Land-use change
IFRC	International Federation of Red Cross	LULUCF	Land use, land use change and forestry
IGCC	Integrated combined cycle plant		
IGO	International government organizations	m	Mass
IIASA	International Institute for Applied Systems Analysis		Meter
		m^3/year	Cubic meters per year
ILUC	indirect land-use change	M&V	Measurement & Verification
IMCOM	(Army) Installation Management Command	MAJCOM	Major Command
In	Inch	MCL	Maximum contaminant level
IPCC	Intergovernmental Panel on Climate Change	MCLB	Marine Corps Logistics Base
IPPF	International Planned Parenthood Federation	MDA	Missile Defense Agency
IRENA	International Renewable Energy Association	MDA/DOH	MDA Office of Human Resources
ISCC	International Sustainability and Carbon Certification System	MDMS	Meter Data Management System
		MEA	Multilateral Environmental Agreement
ISSA	Inter-Service Support Agreement	MEDCOM	Medical Command (DoA)
		MEK	Methyl ethyl ketone
J	Joule	MENA	Middle East and North Africa
JJA	June, July, August	MEO	Most Efficient Organization
IRR	Internal rate of return	Mg	Magnesium
		mg/g	Microgram per gram
K	Potassium	mg/kg	Microgram per kilogram
kcal	Kilocalories	mg/l	Microgram per liter
km	Kilometer	mi^2	Miles square
km_2	Square kilometers	MHA	million hectares
kW	Kilowatt	MILCON	Military Construction
kWh	kilowatt-hour	min.	Minute
			Minimum
l	Length	MJ	Megajoule
LCA	Life-cycle assessment	mm/yr	Millimeter per year
LCC	Life cycle cost	MMA	March, April, May
LCCA	Life-Cycle Cost Analysis	MMBtu	Million British Thermal Units
LDCF	Least Developed Countries Fund	Mo	Molybdenum
LDE	Liter of diesel equivalent (energy content 36.1 MJ/litre)	LTC	Lithium Technology Corp
		MRI	Meteorological Research Institute, Japan

m/s	Meters per second	OMB	Office of Management and Budget
mt	Metric ton	OPEC	Organization of Petroleum Exporting Countries
MTOE	Million tons of oil equivalent		
mW	Milliwatt		
MW	Megawatt	P	Phosphorous
MWh	Megawatt-Hour, 1 million Watt-hours		Power (as in Watts)
			Pressure (as in lbs/in2)
n	number of observations or replicates in a statistical sample	P/E	Power to energy
		PACAF	Pacific Air Forces
N	Nitrogen	PAH	(or PNA) Polynuclear Aromatic Hydrocarbons
Na	Sodium	Pb	Lead
NAO	North Atlantic Oscillation	PBMO	Pentagon Building Management Office
NAM	Northern Annular Mode	PC1	Planning case
NAPA	National adaptation plan of action	PCS	Power conversion system
NaS	Sodium-sulfur	PCB	Polychlorinated biphenyls
NATO	North Atlantic Treaty Organization	PCE	Perchloroethylene
NAVFAC	Naval Facilities Engineering Command	PCP	Pentachlorophenol
NAVSTA	Naval Station	PENREN	Pentagon Renovation Office
NAWS	Naval Air Weapons Station – China Lake	PG&E	Pacific Gas & Electric
NCEP/NCAR	40-year reanalysis project, NOAA	pH	Level of acidity, alcalinity
NCS	National Climate Service	PH	Pumped hydroelectric
NDAA	National Defense Authorization Act	PHEV	Plug in hybrid electric vehicles
NGA	National Geospatial-Intelligence Agency	PH&RP	Pentagon Heating & Refrigeration Plant
NGO	Nongovernmental organization	PM	Program Management
NGV	natural gas vehicle		Periodic Maintenance
NH	Naval Hospital	PM10	Particulate matter less than 10 microns in diameter.
NIST	National Institute of Standards and Technology		
NMC	Naval Medical Center	PNA	Pacific North American Pattern
NMHS	National Meteorological and Hydrological Services	PNNL	Pacific Northwest National Laboratory
		PROMOD	Production cost modeling software by Ventyx
NNI	Net national income	RPS	Renewable portfolio standards
NNP	Net national product	PRECIS	Regional climate modeling system, UK Met Office
NOx	Nitrogen oxides		
NOOA	National Oceanic and Atmospheric Administration	POP	Persistent organic pollutants
		ppb	Parts per billion
NPDES	National Pollutant Discharge Elimination System	ppm	Parts per million
		PPP	Purchasing power parity
NPL	National Priorities List	Proalcool	Brazilian ethanol fuel program
NPP	Net Primary Production	PRP	Potentially Responsible Party
NREL	National Renewable Energy Laboratory	Pu	Plutonium
NSA	National Security Agency	PV	Photovoltaic
NSW	New South Wales	PVC	Polyvinyl chloride
nW	Nanowatt	pW	Picowatt
NWPP	Northwest Power Pool	PW	Petawatt
O2	Oxygen	Q	Heat (Btu, calories)
O3	Ozone	QSR	Quality Surveillance Representative
O&M	Operations and maintenance	Qual	Quality
OECD	Organization for Economic Co-operation and Development	r	Radius
		R	Resistance (in Ohms)

Note: In the table above, NO_x, O_2, O_3 are written with subscripts.

R&D	Research and development
RAF	Royal Air Force
RAP	Remedial Action Plan
RCM	Regional climate model
RCOF	Regional Climate Outlook Forum
RCRA	Resource Conservation and Recovery Act
RD&D	Research, development and demonstration
RDD&D	Research, development, demonstration and deployment
RDF	Remote Delivery Facility (WHS)
REAP	Reduced Energy Appreciation Program
REC	Renewable Energy Certificate
RED	Renewable Energy Directive
REM	Resource Efficiency Manager
RFF-PI	Resources for the Future—Policy Instruments
RFP	Request for Proposal
RI/FS	Remedial Investigation/Feasibility Study
RMCS	Refrigeration Monitoring and Control Systems
ROI	Return on investment
RPS	Renewable portfolio standards
RRSO	Roundtable on Responsible Soy Oil
RSA RSA	Insurance Group, Munich
RSB	Roundtable for Sustainable Biofuel
RSPO	Roundtable on Sustainable Palm Oil
RX5D	Yearly maximum precipitation in five consecutive days
RWQCB	Regional Water Quality Control Board
S	Sulphur
SAF/IE	Secretary of the Air Force for Installations, Environment and Logistics
SAL	State action level
SAM	Southern Annular Mode
SARA	Superfund Amendments and Reauthorization Act of 1986.
SARA	Title III—Emergency Planning and Community Right-to-Know Act of 1986
SBS	Sick building syndrome
SCCF	Special Climate Change Fund
SDD	Sustainable Design and Development
sec.	Second
SECNAV	Secretary of the Navy
SE ITP	Sustainability and Environment Integrated Product Team
SERP	Super Efficient Refrigerator Program
SFG	Senior Focus Group
SHPP	Small hydropower
Si	Silicon
SiO2	Silicon dioxide (sand)
SIOQ	Quality Assurance Division (NGA)

SIP	State Implementation Plan
SLBM	Submarine Launched Ballistic Missile
SNARL	Suggested No Adverse Response Level
SO_2	Sulfur dioxide
SON	September, October, November
SREC	Solar renewable energy credits
SRES	Special report on emissions scenarios, IPCC
st	Short ton
STLC	Soluble Threshold Limit Concentration
SWMU	Solid waste management unit
t	Time
	Ton
T	Temperature
TAR	Third Assessment Report (of the IPCC)
TCE	Tetrachloroethylene
	Tons of coal equivalent
TCLP	Toxicity Characteristic Leaching Procedure.
TCP	Tetrachlorophenol
TEPPC	Transmission Expansion Planning and Policy Committee
TFR	Total fertility rate
TLV	Threshold Limit Value
TMA	TRICARE Management Agency
TNT	Equivalent A measure of energy released during nuclear explosion
TOC	Total organic content
toe	Tons of oil equivalent
TSCA	Toxic Substances Control Act
TTLC	Total Threshold Limit Concentration
TW	Terawatt
U	Internal energy (molecular level)
	Uranium
UESC	Utility Energy Services Contract
UFRJ	Federal University of Rio de Janeiro
UK	United Kingdom
UKCIP	United Kingdom Climate Impacts Programme
UN	United Nations
UNCCO	UN Convention on Combating Desertification
UNDP	United Nations Development Programme
UNEP	United Nations Environment Programme
UNESCO	United Nations Educational, Scientific and Cultural Organization
UNFCCC	UN Framework Convention on Climate Change
UNIDO	United Nations Industrial Development Organization
UNWTO	World Tourism Organization

US	United States
USA	United States of America
US$	United States Dollar
USACE	US Army Corp of Engineers
USAF	United States Air Force
USAMRIID	United States Army Medical Research Institute for Infectious Diseases
USGBC	United States Green Building Council

v	Velocity in m/s, in/s, MPH
V	Potential (in Volts)
	Volume
VAV	Variable Air Volume
VOC	Volatile organic compounds

w	Width
W	Watt
	Work
WBG	World Bank Group
WCC-3	WORLD CLIMATE CONFERENCE-3
WCRP	World Climate Research Programme
WCSS	World Climate Services Systems
WECC	Western Electricity Coordinating Council
WFP	World Food Program
WHO	World Health Organization
WHS	Washington Headquarters Service
W/m²	Watts per square meter
WMO	World Meteorological Organization
WRAMC	Walter Reed Army Medical Center
WRMA	Weather Risk Management Association
WRMF	Weather Risk Management Facility
WU	Wage units

GLOSSARY OF TERMS

A

Acceptable intake (as related to sub-chronic and chronic exposure). Numbers which describe how toxic a chemical is. The numbers are derived from animal studies of the relationship between dose and non-cancer effects. There are two types of acceptable exposure values: one for acute (relatively short-term) and one for chronic (longer-term) exposure.

Accumulation: the build up of a particular matter, like pollutants, over time

Acetone. A widely used, highly volatile solvent. It is readily absorbed by breathing, ingestion or contact with the skin. Workers who have inhaled acetone have reported respiratory problems.

Acids. A class of compounds that can be corrosive when concentrated. Weak acids, such as vinegar and citric acid, are common in foods. Strong acids, such as muriatic (or hydrochloric), sulfuric and nitric acid have many industrial uses, and can be dangerous to those not familiar to handling them. Acids are chemical "opposites" to bases, in that they can neutralize each other.

Acid Rain: term applied to acid precipitation formed when emissions of sulfur dioxide (SO_2) and oxides of nitrogen (NO_x) react in the atmosphere with water and other compounds

Acid Rain Program: created under the Clean Air Act to reduce acid rain; employs a cap and trade framework to achieve SO_2 reductions. Specifically means the limited authorization to emit one ton of SO_2 during a given year.

Act. In the legislative sense, a bill or measure passed by both houses of Congress; a law.

Action level. A guideline established by environmental protection agencies to identify the concentration of a substance in a particular medium (water, soil, etc.) that may present a health risk when exceeded. If contaminants are found at concentrations above their action levels, measures must be taken to decrease the contamination.

Activated sludge. A term used to describe sludge that contains microorganisms that break down organic contaminants (e.g., benzene) in liquid waste streams to simpler substances such as water and carbon dioxide. It is also the product formed when raw sewage is mixed with bacteria-laden sludge, then stirred and aerated to destroy organic matter.

Activity (of a radioactive isotope). The number of particles or photons ejected from a radioactive substance per unit time.

Active solar. Using solar energy with the assistance of external power.

Acute hazards. Hazards associated with short-term exposure to relatively large amounts of toxic substances.

Acute Loading: a term that applies to the short-term build up of a pollutant and which suggests that, in the short-term, significant amounts of a pollutant can accumulate

Adjournment. The end of a legislative day or session.

Adverse health effects. Effects of chemicals or other materials that impair one's health. They can range from relatively mild temporary conditions such as minor eye or throat irritation, shortness of breath or headaches to permanent and serious conditions such as cancer, birth defects or damage to organs.

Advisory level. The level above which an environmental protection agency suggests it is potentially harmful to be exposed to a contaminant, although no action is mandated.

Aeration. Passing air through a solid or liquid, especially a process that promotes breakdown or movement of contaminants in soil or water by exposing them to air.

Aerosol. A suspension of small liquid or solid particles in gas.

Air pollution. Toxic or radioactive gases or particulate matter introduced into the atmosphere, usually as a result of human activity.

Air stripping tower. Air stripping removes volatile organic chemicals (such as solvents) from contaminated water by causing them to evaporate. Polluted water is sprayed downward through a tower filled with packing materials while air is blown upwards through the tower. The contaminants evaporate into the air, leaving significantly-reduced pollutant levels in the water. The air is treated before it is released into the atmosphere.

Alkaline (synonym basic, caustic). Having the properties of a base, a pH greater than 7. Usually used as an adjective, i.e. "alkaline soil."

Allowance: the term generally used to refer to the emission reduction unit traded in emissions trading programs; in the Allowance Loan: transaction wherein an owner of allowances, the lender, allows another party, the borrower, to use the allowances. The borrower customarily promises to return the allowances after a specified period of time with payment for their use, called interest. The allowances returned are not necessarily the exact ones loaned, but are allowances of similar vintage years

Allowance Loan Rate: payment for the lending of allowances over a specified period of time, calculated including the cost-of-carry charge

Allowance Tracking System (ATS): a computerized system administered by EPA and used to track the allowances and allowance transactions by all market participants

Allowance Transfer Form (ATF): official form used to report allowance transfers to the ATS. The ATF lists the serial numbers of the allowances to be transferred and includes the account information of both the transferor and the transferee

Alluvial deposit. An area of sand, clay or other similar material that has been gradually deposited by moving water, such as along a river bed or shore of a lake.

Alpha particle. A positively-charged particle emitted by radioactive atoms. Alpha particles travel less than one inch in the air and a thin sheet of paper will stop them. The main danger from alpha particles lies in ingesting the atoms which emit them. Body cells next to the atom can then be irradiated over an extended period of time, which may be prolonged if the atoms are taken up in bone, for instance.

Alternative energy. Energy that is not popularly used and is usually environmentally sound, such as solar or wind energy (as opposed to fossil fuels).

Alternative fibers. Fibers produced from non-wood sources for use in paper making.

Alternative fuels. Transportation fuels other than gasoline or diesel. Includes natural gas, methanol, and electricity.

Alternative transportation modes. Travel other than private cars, such as walking, bicycling, rollerblading, carpooling and transit.

Ambient air. Refers to the surrounding air. Generally, ambient air refers to air outside and surrounding an air pollution source location. Often used interchangeably with "outdoor air."

Amendment. A change or addition to an existing law or rule.

Ancient forest. A forest that is typically older than 200 years with large trees, dense canopies and an abundance of diverse wildlife.

Applicable or Relevant and Appropriate Requirements (ARARs). Federal or state laws, regulations, standards, criteria or requirements which would apply to the cleanup of hazardous substances at a particular site.

Apportionment. The process through which legislative seats are allocated to different regions.

Appropriation. The setting aside of funds for a designated purpose (e.g., there is an appropriation of $7 billion to build 5 new submarines).

Aquaculture. The controlled rearing of fish or shellfish by people or corporations who own the harvestable product, often involving the capture of the eggs or young of a species from wild sources, followed by rearing more intensively than possible in nature.

Aqueous. Water-based.

Aquifer. A water-bearing layer of rock or sediment that is capable of yielding useable amounts of water. Drinking water and irrigation wells draw water from the underlying aquifer.

Arms control. Coordinated action based on agreements to limit, regulate, or reduce weapon systems by the parties involved.

Arsenic. A gray, brittle and highly poisonous metal. It is used as an alloy for metals, especially lead and copper, and is used in insecticides and weed killers. In its inorganic form, it is listed as a cancer-causing chemical under Proposition 65.

Artesian well. A well that flows up like a fountain because of the internal pressure of the aquifer.

Asbestos. A general name given a family of naturally occurring fibrous silicate minerals. Asbestos fibers were used mainly for insulation and as a fire retardant material in ship and building construction and other industries, and in brake shoes and pads for automobiles. Inhaling asbestos fibers has been shown to result in lung disease (asbestosis) and in lung cancer (mesothelioma). The risk of developing mesothelioma is significantly enhanced in smokers.

Ash. Incombustible residue left over after incineration or other thermal processes.

Ask: the price a prospective seller is willing to accept (a.k.a. "offer")

Asthma. A condition marked by labored breathing, constriction of the chest, coughing and gasping usually brought on by allergies.

Atmosphere. The 500 km thick layer of air surrounding the earth which supports the existence of all flora and fauna.

Atomic energy. Energy released in nuclear reactions. When a neutron splits an atom's nucleus into smaller pieces it is called fission. When two nuclei are joined together under millions of degrees of heat it is called fusion.

Average Weighted Price: calculation used to determine price taking into account the quantity of allowances sold

B

Backfill. The word is used in two contexts; a) to refill an excavated area with uncontaminated soils; and b) the material used to refill an excavated area.

Background concentration. Represents the average amount of toxic chemicals in the air, water or soil to which people are routinely exposed. More than half of the background concentration of toxic air in metropolitan areas comes from automobiles, trucks and other vehicles. The rest comes from industry and business, agricultural, and from the use of paints, solvents and chemicals in the home.

Bases. A class of compounds that are "opposite" to acids, in that they neutralize acids. Weak bases are used in cooking (baking soda) and cleaners. Strong bases can be corrosive, or "caustic." Examples of strong bases that are common around the house are drain cleaners, oven cleaners and other heavy duty cleaning products. Strong bases can be very dangerous to tissue, especially the eyes and mouth.

Beach closure. The closing of a beach to swimming, usually because of pollution.

Bear Market: prolonged period of falling allowance prices.

Benzene. A petroleum derivative widely used in the chemical industry. A few uses are: synthesis of rubber, nylon, polystyrene, and pesticides; and production of gasoline. Benzene is a highly volatile chemical readily absorbed by breathing, ingestion or contact with the skin. Short-term exposures to high concentrations of benzene may result in death following depression of the central nervous system or fatal disturbances of heart rhythm. Long-term, low-level exposures to benzene can result in blood disorders such as aplastic anemia and leukemia. Benzene is listed as a cancer-causing chemical under Proposition 65.

Berm. A curb, ledge, wall or mound used to prevent the spread of contaminants. It can be made of various materials, even earth in certain circumstances.

Beta particles. Very high-energy particle identical to an electron, emitted by some radioactive elements. Depending on their energy, they penetrate a few centimeters of tissue.

Bid: price a prospective buyer is willing to pay

Bill. A proposed law, to be debated and voted on.

Billfish. Pelagic fish with long, spear-like protrusions at their snouts, such as swordfish and marlin.

Biodegradable. Waste material composed primarily of naturally-occurring constituent parts, able to be broken down and absorbed into the ecosystem. Wood, for example, is biodegradable, for example, while plastics are not.

Bioaccumulation. The process by which the concentrations of some toxic chemicals gradually increase in living tissue, such as in plants, fish, or people as they breathe contaminated air, drink contaminated water, or eat contaminated food.

Biodiversity. A large number and wide range of species of animals, plants, fungi, and microorganisms. Ecologically, wide biodiversity is conducive to the development of all species.

Biomass. Two meanings: a.) the amount of living matter in an area, including plants, large animals and insects; and b.) plant materials and animal waste used as fuel.

Bioremediation. A process that uses microorganisms to

change toxic compounds into non-toxic ones.

Biosolids. Residue generated by the treatment of sewage, petroleum refining waste and industrial chemical manufacturing wastewater with activated sludge.

Biosphere. Two meanings; a.) the part of the earth and its atmosphere in which living organisms exist or that is capable of supporting life; and, b.) the living organisms and their environment composing the biosphere.

Biosphere Reserve. A part of an international network of preserved areas designated by the United Nations Educational, Scientific and Cultural Organization (UNESCO). Biosphere Reserves are vital centers of biodiversity where research and monitoring activities are conducted, with the participation of local communities, to protect and preserve healthy natural systems threatened by development. The global system currently includes 324 reserves in 83 countries.

Biota. The animal and plant life of a particular region.

Biotic. Of or relating to life.

Birth control. Preventing birth or reducing frequency of birth, primarily by preventing conception.

Biotransformation. Transformation of one chemical to others by populations of microorganisms in the soil.

Birth defects. Unhealthy defects found in newborns, often caused by the mother's exposure to environmental hazards or the intake of drugs or alcohol during pregnancy.

Birth rate. The number of babies born annually per 1,000 women of reproductive age in any given set of people.

Bloc. A group of people with the same interest or goal (usually used to describe a voting bloc, a group of representatives intending to vote the same way).

Blood lead levels. The amount of lead in the blood. Human exposure to lead in blood can cause brain damage, especially in children.

Boring. Usually, a vertical hole drilled into the ground from which soil samples can be collected and analyzed to determine the presence of chemicals and the physical characteristics of the soil.

Borrow pit. An area where soil, sand or gravel has been dug up for use elsewhere.

Bottled water. Purchased water sold in bottles.

Broker: person who acts as an intermediary between a buyer and a seller, usually charging a commission

Brownfields. Abandoned, idled, or under-used industrial and commercial facilities where expansion or redevelopment is complicated by real or perceived environmental contamination.

Bubble: a regulatory term which applies to the situation when a company combines a number of its sources in order to control pollution in aggregate; under a bubble facility operators are allowed to choose which sources to control as long as the total amount of emissions from the combined sources is less than the amount each source would have emitted under the conventional requirement

Budget. A formal projection of spending and income for an upcoming period of time, traditionally submitted by the President or Executive for consideration and approval.

Budget reconciliation. Legislation making changes to existing law (such as entitlements under Social Security or Medicare) so that it conforms to numbers in the budget resolution.

Budget resolution. The first step in the annual budget process. This resolution must be agreed to by the House and Senate. It is not signed by the President and does not have the effect of law, but instead sets out the targets and assumptions that will guide Congress as it passes the annual appropriations and other budget bills.

Bull Market: prolonged rise in the price of allowances. Bull markets usually last at least a few months and are characterized by high trading volume

Bycatch. Fish and/or other marine life that are incidentally caught with the targeted species. Most of the time bycatch is discarded at sea.

Bycatch reduction device (bdr). A device used to cut bycatch while fishing. These gear modifications are most commonly used with shrimp trawls. They are also called "finfish excluder devices" (feds) or, when specifically designed to exclude sea turtles, they are called "turtle excluder devices" (teds).

C

Cadmium. Cadmium is a natural element in the earth's crust, usually found as a mineral combined with other elements such as oxygen. Because all soils and rocks have some cadmium in them, it is extracted during the production of other metals like zinc, lead and copper. Cadmium does not corrode easily and has many uses. In industry and consumer products, it is used for batteries, pigments, metal coatings, and plastics. It is used also to make solar cells and panels. Cadmium salts are toxic in higher concentrations.

Cairo Plan. Recommendations for stabilizing world population agreed upon at the U.N. International Conference on Population and Development, held in Cairo in September 1994. The plan calls for improved health care and family planning services for women, children and families throughout the world, and also emphasizes the importance of education for girls as a factor in the shift to smaller families.

Calendar. In the legislative sense, a group of bills or proposals to be discussed or considered in a legislative committee or on the floor of the House or Senate.

California Environmental Quality Act (CEQA). First enacted in 1970 to provide long-term environmental protection, the law requires that governmental decision-makers and public agencies study the significant environmental effects of proposed activities, and that significant avoidable damage be avoided or reduced where feasible. CEQA also requires that the public be told why the lead public agency approved the project as it did, and gives the public a way to challenge the decisions of the agency.

Call Option: a contract that grants the right to buy, at a specified price, a specific number of allowances by a certain date

Cancer. Unregulated growth of changed cells; a group of changed, growing cells (tumor).

Cancer risk. A number, generally expressed in exponential form (i.e., 1 x 10 -6, which means one in one million), which describes the increased possibility of an individual developing cancer from exposure to toxic materials. Calculations producing cancer risk numbers are complex and typically include a number of assumptions that tend to cause the final estimated risk number to be conservative.

Cap. A layer, such as clay or a synthetic material, used to prevent rainwater from penetrating the soil and spreading contamination.

Carbamates. A group of insecticides related to carbamic acid. They are primarily used on corn, alfalfa, tobacco, cotton, soybeans, fruits and ornamental plants.

Carbon adsorption. A treatment system in which organic contaminants are removed from groundwater and surface water by forcing it through tanks containing activated carbon, a specially-treated material that retains such compounds. Activated carbon is also used to purify contaminated air by adsorbing the contaminants as the air passes through it.

Carbon monoxide (CO). A very poisonous, colorless and odorless gas formed when carbon-containing matter burns incompletely, as in automobile engines or in charcoal grills used indoors without proper ventilation.

Carbon dioxide (CO2). A naturally occurring greenhouse gas in the atmosphere, concentrations of which have increased (from 280 parts per million in preindustrial times to over 350 parts per million today) as a result of humans' burning of coal, oil, natural gas and organic matter (e.g., wood and crop wastes).

Carbon tax. A charge on fossil fuels (coal, oil, natural gas) based on their carbon content. When burned, the carbon in these fuels becomes carbon dioxide in the atmosphere, the chief greenhouse gas.

Carbon tetrachloride (CCl4). A colorless, nonflammable toxic liquid that was widely used as a solvent in dry-cleaning and in fire extinguishers. It is listed as a cancer-causing chemical under Proposition 65.

Carcinogens. Substances that cause cancer, such as tar.

Carpooling. Sharing a car to a destination to reduce fuel use, pollution and travel costs.

Catalyst. A substance that accelerates chemical change yet is not permanently affected by the reaction (e.g., platinum in an automobile catalytic converter helps change carbon monoxide to carbon dioxide).

Catalytic Cracker Unit. In a petroleum refinery, the catalytic cracker unit breaks long petroleum molecules apart, or "cracks" them, during the petroleum refining process. These smaller pieces then come together to form more desirable molecules for gasoline or other products.

Caucus. A meeting of a political party, usually to appoint representatives to party positions.

Caustic. The common name for sodium hydroxide, a strong base. Also used as an adjective to describe highly corrosive bases.

Caustic scrubber. An air pollution control device in which acid gases are neutralized by contact with an alkaline solution.

Chlorinated herbicides. A group of plant-killing chemicals which contain chlorine, used mainly for weed control and defoliation.

Chlorination byproducts. Cancer-causing chemicals created when chlorine used for water disinfection combines with dirt and organic matter in water.

Chlorine. A highly reactive halogen element, used most often in the form of a pungent gas to disinfect drinking water.

Chlorofluorocarbons (CFCs). Stable, artificially created chemical compounds containing carbon,

Chlorine, fluorine and sometimes hydrogen. Chlorofluorocarbons, used primarily to facilitate

cooling in refrigerators and air conditioners, have been found to damage the stratospheric ozone layer which protects the earth and its inhabitants from excessive ultraviolet radiation.

Chlorobenzene. A volatile organic compound that is

often used as a solvent and in the production of other chemicals. It is a colorless liquid with an almond-like odor. It is toxic.

Chloroform. Chloroform was once commonly used as a general anesthetic and as a flavoring agent in toothpastes, mouth wastes and cough syrups. It is listed as a cancer-causing chemical under Proposition 65.

Chromated copper arsenate. An insecticide and herbicide containing three metals: copper, chromium and arsenic. This salt is used extensively as a wood preservative in pressure-treating operations. It is highly toxic and dissolves in water, making it a relatively mobile contaminant in the environment.

Chromium. A hard, brittle, grayish heavy metal used in tanning, in paint formulation, and in plating metal for corrosion protection. It is toxic at certain levels and, in its hexavalent (versus trivalent) form, chromium is listed as a cancer-causing agent under Proposition 65.

Chronic exposure. Repeated contact with a chemical over a period of time, often involving small amounts of toxic substance.

Class I landfill. A landfill permitted to accept hazardous wastes.

Clayoquot Sound. One of the last remaining unlogged watersheds on the west coast of Canada's Vancouver Island.

Clean Air Act Amendments of 1990: reauthorization of the Clean Air Act; passed by the U.S. Congress; strengthened ability of EPA to set and enforce pollution control programs aimed at protecting human health and the environment; included provisions for Acid Rain Program

Clean Air Act. A federal law passed in 1955 and extensively modified in 1970. It is enforced by the California Air Resources Board and the local air quality management or air pollution control districts, as well as by U.S. EPA nationally.

Clean fuel. Fuels which have lower emissions than conventional gasoline and diesel. Refers to alternative fuels as well as to reformulated gasoline and diesel.

Cleanup. Treatment, remediation, or destruction of contaminated material.

Clear cutting. A logging technique in which all trees are removed from an area, typically 20 acres or larger, with little regard for long-term forest health.

Clean Water Act. A federal law of 1977 enforced by U.S. EPA. A key provision is that "any person responsible for the discharge of a pollutant or pollutants into any waters of the United States from any point source must apply for and obtain a permit." This is reflected by the National Pollutant Discharge Elimination System (NPDES), through which the permits are issued by Regional Water Quality Control Boards. Permits are now being required for storm water runoff from cities and other locations.

Cleanup process. A comprehensive program for the clean-up (remediation) of a contaminated site. It involves investigation, analysis, development of a cleanup plan and implementation of that plan.

Clearing Price: price at which a buyer and seller agree to transact a trade

Climate change. A regional change in temperature and weather patterns. Current science indicates a discernible link between climate change over the last century and human activity, specifically the burning of fossil fuels.

Cloture. The formal end to a debate or filibuster in the Senate requiring a three-fifths vote.

Coastal pelagic. Fish that live in the open ocean at or near the water's surface but remain relatively close to the coast. Mackerel, anchovies, and sardines are examples of coastal pelagic fish. commercial extinction—the depletion of a population to the point where fisherman cannot catch enough to be economically worthwhile.

Collar/Zero-cost Collar: set of contracts used to hedge against the risk of prices moving in both directions; involves purchasing a call option and selling a put option. Option premiums in a collar that cancel each other out are "zero-cost" collars

Communities of color. Hispanic, black or Asian people or groups living together or connected in some way.

Community right-to-know. Public accessibility to information about toxic pollution.

Compact fluorescent light (CFL). Fluorescent light bulbs small enough to fit into standard light sockets, which are much more energy-efficient than standard incandescent bulbs.

Compost. Process whereby organic wastes, including food wastes, paper, and yard wastes, decompose naturally, resulting in a product rich in minerals and ideal for gardening and farming as a soil conditioners, mulch, resurfacing material, or landfill cover.

Combustion gases. Gases produced by burning. The composition will depend on, among other things, the fuel; the temperature of burning; and whether air, oxygen or another oxidizer is used. In simple cases the combustion gases are carbon dioxide and water. In some other cases, nitrogen and sulfur oxides may be produced as well. Incinerators must be controlled carefully to be sure that they do not emit more than the allowable amounts of more complex, hazardous compounds. This often requires use of emission-control devices.

Combustible vapor mixture. The composition range over which air containing vapor of an organic compound will burn or even explode when set off by a flame or spark. Outside that range the reaction does not occur, but the mixture may nevertheless be hazardous because it does not contain enough oxygen to support life, or because the vapor is toxic.

Comprehensive Environmental Response, Compensation and Liability Act of 1980 (CERCLA). Also known as Superfund, authorizes EPA to respond directly to releases of hazardous substances that may endanger public health or the environment.

Congressional Record. A document published by the government printing office recording all debates, votes and discussions taking place in the Congress; available for free inspection at all government document repositories, as well as in some major libraries.

Consent decree. A legal document, approved and issued by a judge, formalizing an agreement between DTSC and the parties potentially responsible for site contamination.

Confirmation Sheet: formal memorandum from a broker to a client giving details of an allowance transaction

Containment. Enclosing or containing hazardous substances in a structure to prevent the migration of contaminants into the environment.

Contamination. Pollution.

Contraceptive. Preventing conception and pregnancy.

Copper. Distinctively-colored metal used for electric wiring, plumbing, heating and roof and building construction, and in automobile brake linings. It is known to be toxic at certain levels.

Corrosivity. A characteristic of acidic and basic hazardous wastes.

The characteristic is defined by a waste's pH and its ability to corrode steel. A waste is corrosive if it has a pH less than or equal to 2.0 or greater than or equal to 12.5.

Cost-of-carry: out-of-pocket costs incurred while an allowance holder retains allowances for future transfer

Counterparty: the party opposite the buyer or seller in a transaction

Credit Risk: the financial risk that an obligation will not be paid and a loss will result

Creek. A watercourse smaller than, and often tributary to, a river.

Creosotes. Chemicals used in wood preserving operations that are produced by distilling coal-tar. They contain polycyclic aromatic hydrocarbons and polynuclear aromatic hydrocarbons (PAHs and PNAs) and so high-level, short-term exposures may cause skin ulcerations. Creosotes are listed as cancer-causing agents under Proposition 65.

Criteria pollutants. Air pollutants for which standards for safe levels of exposure have been set under the Clean Air Act. Current criteria pollutants are sulfur dioxide, particulate matter, carbon monoxide, nitrogen oxides, ozone and lead.

Critical mass. The minimum mass of fissionable material that will support a sustaining chain reaction.

Crop dusting. The application of pesticides to plants by a low-flying plane.

Cryptosporidium. A protozoan (single-celled organism) that can infect humans, usually as a result of exposure to contaminated drinking water.

Cumulative impact. The term cumulative impact is used in several ways: as the effect of exposure to more than one compound; as the effect of exposure to emissions from more than one facility; the combined effects of a facility and surrounding facilities or projects on the environment; or some combination of these.

Cyanide. A highly toxic chemical often used in metal finishing or in extraction of precious metal from ore.

D

Deferred Swap: a trade of one allowance for another in order to exchange the vintage years of the allowances; settlement occurs after more than 180 days

Degrease. To remove grease from machinery, tools, etc., usually using solvents. Aqueous (water-based) cleaners are becoming popular and are required in some parts of the state.

Deionized water. Water which has been specifically treated to remove minerals.

Demand Side Management (DSM). An attempt by utilities to reduce customers' demand for electricity or energy by encouraging efficiency. A term referring to the need (or demand) for power generation among a utility's customers.

Demersal. Fish that live on or near the ocean bottom. They are often called benthic fish, ground fish, or bottom fish.

De minimis risk. A level of risk that the scientific and regulatory community asserts is too insignificant to regulate.

Department of Toxic Substances Control (DTSC). A department within the California Environmental

Protection Agency charged with the regulation of hazardous waste from generation to final disposal, and for overseeing the investigation and clean-up of hazardous waste sites.

Designated Representative: for a unit account, the individual who represents the owners and operators of that unit and performs allowance transfer requests and all correspondence with EPA concerning compliance with the Acid Rain Destruction and removal efficiency (DRE). A percentage that represents the number of molecules of compound removed or destroyed in an incinerator relative to the number of molecules that entered the incinerator system. A DRE of 99.99 percent means that 9,999 molecules of a compound are destroyed for every 10,000 molecules that enter the system. For some compounds a DRE of 99.9999 is required.

Development. a) developed tract of land (with houses or structures); b) the act, process or result of developing.

Dewater. To remove water from wastes, soils or chemicals.

Diazinon. An organophosphate insecticide. It is used in agriculture, and for home, lawn and commercial uses.

Dibromochloropropane (DBCP). An amber-colored liquid used in a agriculture to kill pests in the soil. Inhalation of high concentrations of DBCP causes nausea and irritation of the respiratory tract. Chronic exposure results in sterility in males. Although not in use as a pesticide in this country since 1979 (until 1985 in Hawaii), it is found as a contaminant at many hazardous substances sites. DBCP is listed as a cancer-causing agent under Proposition 65.

Dichlorobenzene (DCB). A volatile organic compound often used as a deodorizer, and as a moth, mold and mildew killer. It is a white solid with a strong odor of mothballs. It is toxic and is listed as a cancer-causing agent under Proposition 65.

1,1-Dichloroethane. A colorless, oily liquid having an ether-like odor. It is used to make other chemicals and to dissolve other substances such as paint and varnish, and to remove grease. In the past, this chemical was used as a surgical anesthetic, but it is no longer used for this purpose. Because it evaporates easily into air, it is usually present in the environment as a vapor rather than a liquid.

Dieldrin. An insecticide that was used on crops like corn and cotton. U.S. EPA banned its use in 1987. It is listed as a cancer-causing chemical under Proposition 65.

Diesel. A petroleum-based fuel which is burned in engines ignited by compression rather than spark; commonly used for heavy duty engines including buses and trucks.

Diesel engine. An internal combustion engine that uses diesel as fuel, producing harmful fumes.

Dioxins. A group of generally toxic organic compounds that may be formed as a result of incomplete combustion (as may occur in incineration of compounds containing chlorine). RCRA regulations require a higher destruction and removal efficiency (DRE) for dioxins and related furans (99.9999 percent) burned in incinerators than the DRE required for most other organic compounds (99.99 percent). They are rapidly absorbed through the skin and gastrointestinal tract and are listed as cancer-causing chemicals under Proposition 65.

Dispatch: the ordered use of generation facilities by an electric power utility including which units will operate, when they will operate, and at what capacity

Double hulled tankers. Large transport ships with two hulls with space between them, protecting the cargo (in most cases, oil) from spilling in case of a collision.

Downgradient. The direction in which groundwater flows.

Dredge. A fishing method that utilizes a bag dragged behind a vessel that scrapes the ocean bottom, usually to catch shellfish. Dredges are often equipped with metal spikes in order to dig up the catch.

Driftnet. A huge net stretching across many miles that drifts in the water; used primarily for large-scale commercial fishing.

Dump sites. Waste disposal grounds.

E

Ecologist. A scientist concerned with the interrelationship of organisms and their environment.

Ecology. A branch of science concerned with the interrelationship of organisms and their environment.

Ecosystem. An interconnected and symbiotic grouping of animals, plants, fungi, and microorganisms.

Edge cities. Cities bounded by water, usually with eroding or polluted waterfront areas.

Effluent. Wastewater, treated or untreated, that flows out of a treatment plant, sewer or industrial outfall. Generally refers to wastes discharged into surface waters.

Electric vehicles (EV). Vehicles which use electricity (usually derived from batteries recharged from electrical outlets) as their power source.

Electrostatic precipitator. An air pollution control device that uses electrical charges to remove particulate matter from emission gases.

Emulsifiers. Substances that help in mixing liquids that don't normally mix; e.g., oil and water.

Emissions cap. A limit on the amount of greenhouse gases that a company or country can legally emit.

Endangered species. Species of wild life in danger of extinction throughout all or a significant part of its range.

Endocrine disruptors. Substances that stop the production or block the transmission of hormones in the body. energy conservation—using energy efficiently or prudently; saving energy.

Endosulfan. An insecticide used on vegetable crops, fruits and nuts.

Energy efficiency. Technologies and measures that reduce the amount of electricity and/or fuel required to do the same work, such as powering homes, offices and industries.

Enrolled bill. The final, certified bill sent to the President; House and Senate versions of a bill must match exactly in order to be enrolled.

Equity. In the environmental sense, the planned dispersement of toxic or waste facilities in regions throughout the socioeconomic strata.

Estuary. Areas where fresh water from rivers mixes with salt water from nearshore ocean. They include bays, mouths of rivers, salt marshes and lagoons. These brackish water ecosystems shelter and feed marine life, birds and wildlife.

Ethylene glycol. Used in the manufacture of a wide variety of industrial compounds and in certain cosmetics. It is used most commonly as an automobile antifreeze. It is toxic.

Eugenics. The study of hereditary improvement of the human race by controlled selective breeding.

Everglades. Large and biologically diverse wetland ecosystem in South Florida.

Executive session. A congressional meeting closed to the public (and the media).

Exercise Date (or Expiration Date): last day on which an option can be exercised

Exposure pathways. Existing or hypothetical routes by which chemicals in soil, water or other media can come in contact with humans, animals or plants.

Extraction wells. Wells that are used primarily to remove contaminated groundwater from the ground. Water level measurements and water samples can also be collected from extraction wells.

Exurbia. a.) the area of suburbs; b.) the region outside a city and its suburbs where wealthier families live.

F

Factory farming. Large-scale, industrialized agriculture.

Factory ships. Industrial-style ships used for the large-scale collection and processing of fish.

Fallout. The radioactive dust particles that settle to earth after the denotation of a nuclear device. It is also used to describe dust particles settling from smoke, etc.

Family planning. A system of limiting family size and the frequency of childbearing by the appropriate use of contraceptive techniques.

Fauna. The total animal population that inhabits an area.

Federal land. Land owned and administered by the federal government, including national parks and national forests.

Feasibility study. An evaluation of the alternatives for remediating any identified soil or groundwater contamination.

Feedlots. A plot of ground used to feed farm animals.

Fertility. The ability to reproduce; in humans, the ability to bear children.

Fertility rates. Average number of live births per woman during her reproductive years, among a given set of people.

Filibuster. A tactic used to delay or stop a vote on a bill by making long floor speeches and debates.

Fiscal year. A financial term referring to any twelve-month period, usually to set a budget. The federal government's fiscal year begins October 1.

Filter cake. A mixture of sediments that results from filtering and dewatering of treated wastewater.

Fisheries. An established area where fish species are cultivated and caught.

Fissile material. Material fissionable by slow neutrons. The fission process and the fissile isotopes are the source of energy in nuclear weapons and nuclear reactors.

Fission. The process whereby the nucleus of a particular heavy element splits into (generally) two nuclei of lighter elements, with the release of substantial amounts of energy.

Flammables. A class of compounds that ignite easily and burn rapidly. The Department of Transportation requires that Vehicles transporting flammables must have special markings (placards).

Flash point. The lowest temperature at which a liquid generates enough vapor to ignite in air. If a waste has a flash point of less than 140° F, then it is an ignitable hazardous waste.

Flora. The total vegetation assemblage that inhabits an area.

Florida Bay. A bay at southern tip of Florida which is bounded by the Florida Keys.

Fly ash. Non-combustible residue that results from burning fuels in an incinerator, boiler or furnace. It can include metal oxides, silicates and sulfur compounds, as well as many other chemical pollutants. It is fine ash carried along by flue gases that must be captured by some means before it reaches the mouth of the chimney.

Footprint. The outline of an area within which hazardous substances are suspected or known to exist.

Forest certification. A process of labeling wood that has been harvested from a well-managed forest.

Forests. Lands on which trees are the principal plant life, usually conducive to wide biodiversity.

Formaldehyde. A water-soluble gas used widely in the chemical industry and in the construction and building industries, largely in wood products and in foam insulation. It is also used in some deodorizing preparations, in fumigants and as a tissue preservative in laboratories. Formaldehyde is listed as a cancer-causing agent under Proposition 65.

Fossil fuel. A fuel, such as coal, oil, and natural gas, produced by the decomposition of ancient (fossilized) plants and animals; compare to alternative energy.

Fresh Kills. New York City's only operating landfill, located in Staten Island. Infamous as the largest landfill in the world.

French drain system. A pit or trench filled with crushed rock and used to collect and divert stormwater or wastewater. Most often, perforated piping at the bottom provides easy drainage.

Fugitive emissions. Releases of pollutants to the atmosphere that occur when vapors are vented from containers or tanks where materials are stored. They can also be caused by spillage during the unloading of vehicles, leaks from pipes and valves, and through equipment operation.

G

Gamma radiation. A high-energy photon (ray) emitted from the nucleus of certain radioactive atoms. Gamma rays are the most penetrating of the three common types of radiation (the other two are alpha particles and beta particles) and are best stopped by dense materials such as lead.,

Gas. Natural gas, used as fuel.

Gasoline. Petroleum fuel, used to power cars, trucks, lawn mowers, etc.

General Accounts: accounts in the SO_2 or NOx ATSs which were created after the initial allocation; general accounts can be opened by any individual and they are not automatically adjusted for compliance

Geophysical logging. A general term that encompasses all techniques for determining whether a subsurface geological formation may be sufficiently porous or permeable to serve as an aquifer. These techniques typically involve lowering a sensing device into a borehole to measure properties of the subsurface formation.

Geothermal. Literally, heat from the earth; energy obtained from the hot areas under the surface of the earth.

Gigawatt (GW). One thousand megawatts, or one billion watts.

Gigawatt hour (GW/h). One thousand MW generated or used in one hour

Gillnets. Walls of netting that are usually staked to the sea floor. Fish become entangled or caught by their gills.

Global Climate Change: change in the earth's climate; caused by increasing greenhouse gas (GHG) concentrations in the atmosphere; human activities considered to be major new source of

GHGs

Global warming. Increase in the average temperature of the earth's surface.

Golden Carrot. An incentive program that is designed to transform the market to produce much

Greater energy efficiency. The term is a trademark of the Consortium for Energy Efficiency.

Granular activated carbon (GAC). A form of crushed and hardened charcoal. GAC has a strong potential to attract and absorb volatile organic compounds from extracted groundwater and gases.

Grassroots. Local or person-to-person. A typical grassroots effort might include a door-to-door education and survey campaign.

Grazing. The use of grasses and other plants to feed wild or domestic herbivores such as deer, sheep and cows.

Green design. A design, usually architectural, conforming to environmentally sound principles of building, material and energy use. A green building, for example, might make use of solar panels, skylights, and recycled building materials.

Greenhouse. A building made with translucent (light transparent, usually glass or fiberglass) walls conducive to plant growth.

Greenhouse effect. The process that raises the temperature of air in the lower atmosphere due to heat trapped by greenhouse gases, such as carbon dioxide, methane, nitrous oxide, chlorofluorocarbons, and ozone.

Greenhouse gas. A gas involved in the greenhouse effect. A variety of gases including carbon dioxide, methane, and nitrous oxide; the buildup of these gases in the atmosphere prevents energy from the sun to escape back out into space, creating the "greenhouse effect"

Greenway. Undeveloped land usually in cities, set aside or used for recreation or conservation.

Groundfish. A general term referring to fish that live on or near the sea floor. Groundfish are also called bottom fish or demersal fish.

Ground-level Ozone: the occurrence in the troposphere (at ground level) of a gas that consists of 3 atoms of oxygen (03); formed through a chemical reaction involving oxides of nitrogen (NOx), volatile organic compounds (VOC), heat and light; At ground level, ozone is an air pollutant that damages human health, vegetation, and many common materials and is a key ingredient of urban smog.

Groundwater. Water beneath the earth's surface that flows through soil and rock openings, aquifers, and often serves as a primary source of drinking water.

Growth overfishing. The process of catching fish before they are fully grown resulting in a decrease in the average size of the fish population.

H

Habitat. the natural home of an animal or plant; or the sum of the environmental conditions that determine the existence of a community in a specific place.

Half-life. The amount of time that is required for a radioactive substance to lose one-half its activity. Each radioactive substance has a unique half-life. It is also used to describe the time for a pollutant to lose one half of its concentration, as through biological action; and the time for elimination of one half a total dose of a drug from a body.

Halogens. The family of elements that includes fluorine, chlorine, bromine and iodine. Halogens are very reactive and have many industrial uses. They are also commonly used in disinfectants and insecticides. Many hazardous organic chemicals—such as polychlorinated biphenyls (PCBs), some volatile compounds (VOCs) and dioxins contain halogens, especially chlorine.

Harpooning. A surface method of fishing that requires considerable effort in locating and chasing individual fish. Harpoons are hand-held or fired from a harpoon gun and aimed at high-value fish, such as giant tuna and swordfish.

Hazardous waste. Waste substances which can pose a substantial or potential hazard to human health or the environment when improperly managed. Hazardous waste possesses at least one of these four characteristics: ignitability, corrosivity, reactivity or toxicity; or appears on special U.S. EPA lists.

Haze. An atmospheric condition marked by a slight reduction in atmospheric visibility, resulting from the formation of photochemical smog, radiation of heat from the ground surface on hot days, or the development of a thin mist.

Health-based remediation targets and levels to which hazardous substances on the site will be cleaned up. These target levels are health-based, meaning that exposure to the hazardous substances at or below the target is not expected to present a significant health risk.

Health risk/endangerment assessment. A study prepared to assess health and environmental risks due to potential exposure to Hazardous substances.

Hearings. Testimony (sworn statements like those given in court) given before a Congressional committee.

Heavy metals. A group of elements (such as chromium, cadmium, lead, copper and zinc) that can be toxic at relatively low concentrations and tend to accumulate in the food chain.

Hedge: strategy used to offset investment risk. A perfect hedge is one eliminating the possibility of future gain or loss

Heptachlor. An organochloride insecticide once widely used on food crops, especially corn, but has not been in use since 1988. It is listed as a cancer-causing chemical under Proposition 65.

High seas. International ocean water under no single country's legal jurisdiction.

Highly migratory fish. Fish that travel over great areas.

Horizontal wells. Extraction and monitoring wells are typically drilled vertically. A horizontal well has the advantage of providing a large area of groundwater capture for a lower overall cost.

Hot spot criteria. Cleanup levels for small areas on the site that have particularly high concentrations of hazardous substances.

Household hazards. Dangerous substances or conditions in human dwellings.

Hydrochloric acid (HCl). Clear, colorless and acidic solution of hydrogen chloride in water often used in metal cleaning and electroplating. Many hazardous wastes contain chlorine compounds which create small amounts of hydrogen chloride when they are burned. This can contribute to the formation of acid rain. Regulations require that air pollution control equipment remove either 99% of the hydrochloric acid, or that the emissions contain less than four pounds per hour.

Hydroelectric. Relating to electric energy produced by moving water.

Hydrofluorocarbons (HFC). Chemicals used as solvents and cleaners in the semiconductor industry, among others; experts say that they possess global warming potentials that are thousands of times greater than CO2.

Hydrogeology. The geology of groundwater, with particular emphasis on the chemistry and movement of water.

Hydropower. Energy or power produced by moving water.

Hypoxia. The depletion of dissolved oxygen in water, a condition resulting from an overabundance of nutrients of human or natural origin that stimulates the growth of algae, which in turn die and require large amounts of oxygen as the algae decompose. It was the most frequently cited direct cause of fish kills in the U.S. from 1980 to 1989.

I

Ignitability. A characteristic of hazardous waste. If a liquid (containing less than 24% alcohol) has a flash point less than 140°F, it is a hazardous waste in the United States.

Immediate Settlement: conclusion of an allowance trade in which a party pays for allowances within days of the confirmation of the transaction

Immediate Vintage Year Swap: an trade of one allowance for another in order to exchange the vintage years of the allowances; settlement occurs within days (or at least less than 180 days)

Impoundment. A body of water or sludge confined by a dam, dike, floodgate or other barrier.

In-situ soil aeration. Applying a vacuum to vapor extraction wells to draw air through the soil so that chemicals in the soil are brought to the surface where they can be treated.

Incinerators. Disposal systems that burn solid waste or other materials and reduce volume of waste. Air pollution and toxic ash are problems associated with incineration.

Incompatible wastes. Wastes which create a hazard of some form when mixed together. This could be intense heat or toxic gases, for example.

Indicator chemicals. Chemicals selected from the group of chemicals found at the site and used for a public health evaluation. They are selected on the basis of

toxicity, mobility and persistence, and are thought to be the chemicals of the greatest potential risk.

Industrialized countries. Nations whose economies are based on industrial production and the conversion of raw materials into products and services, mainly with the use of machinery and artificial energy (fossil fuels and nuclear fission); generally located in the northern and western hemispheres (e.g., U.S., Japan, the countries of Europe).

Insecticides. Substances used to kill insects and prevent infestation.

Interim remedial actions (IRAs). Also known as Interim Remedial Measures Cleanup actions taken to protect public health and the environment while long-term solutions are being developed.

International Conference on Population and Development. A conference sponsored by the United Nations to discuss global dimensions of population growth and change in Cairo, Egypt in September 1994. The conference is generally considered to mark the achievement of a new consensus on effective ways to slow population growth and improve quality of life by addressing root causes of unwanted fertility.

International Planned Parenthood Federation (IPPF). An international organization made up of national level affiliates representing every region of the world. IPPF receives and distributes funds from international donor nations to its affiliates, who in turn provide services (prenatal care, contraceptive counseling and service provision, and other reproductive health services) within a country. Some national level organizations provide abortion services, others do not. IPPF sets and supports policies encouraging governmental provision of comprehensive reproductive health care.

Irritant. A chemical that can cause temporary irritation at the site of contact.

J

Joule; unit of energy measurement

K

Kilowatt (kW). One thousand watts of electric energy

Kilowatt hour (kW/h). 1000 Watts generated or used in one hour

Kyoto Protocol: an agreement under the UNFCC signed by 84 nations; establishes greenhouse gas targets ("budgets") and framework for implementation; the Protocol has been agreed to and signed by the U.S. and now awaits ratification by the U.S. Senate

L

Landfill. Disposal area where garbage is piled up and eventually covered with dirt and topsoil.

Landings. The amount of fish brought back to the docks and marketed. Landings can describe the kept catch of one vessel, of an entire fishery, or of several fisheries combined.

Land use. The way in which land is used, especially in farming and city planning.

Law. An act or bill which has become part of the legal code through passage by Congress and approval by the President (or via Congressional override).

Leachate. Typically, water that has come in contact with hazardous wastes. For example:

Water from rain or other sources that has percolated through a landfill and dissolved or carries various chemicals, and thus could spread contamination. Current landfills have systems to collect leachate before that can happen.

Lead. A heavy metal present in small amounts everywhere in the human environment. Lead can get into the body from drinking contaminated water, eating vegetables grown in contaminated soil, or breathing dust when children play or adults work in lead-contaminated areas or eating lead-based paint. It can cause damage to the nervous system or blood cells. Children are at highest risk because their bodies are still developing. Lead and its compounds are listed as a reproductive toxic substance for women and men, and a cancer-causing substance under Proposition 65.

Lead agency. A public agency which has the principal responsibility for ordering and overseeing site investigation and cleanup.

Lead poisoning. Damaging the body (specifically the brain) by absorbing lead through the skin or by swallowing.

Least-cost planning. A process for satisfying consumers' demands for energy services at the lowest societal cost.

Leukemia. A form of bone marrow cancer marked by an increase in white blood cells.

Life cycle assessment. Methodology developed to assess a product's full environmental costs, from raw material to final disposal.

Light pollution. Environmental pollution consisting of harmful or annoying light.

Lindane. Lindane (gamma hexachlorocyclohexane) is an insecticide, once used on fruit and vegetable crops. It is still used to treat head and body lice, and scabies. It is highly toxic to humans, freshwater fish and

aquatic life. It is listed as a cancer-causing chemical under Proposition 65.

Litter. Waste material which is discarded on the ground or otherwise disposed of improperly or thoughtlessly.

Logging. Cutting down trees for commodity use.

Long: a market position in which a party records (or anticipates recording) emissions less than its yearly emissions allocation, thus it has surplus allowances

Longlines. Fishing lines stretching for dozens of miles and baited with hundreds of hooks. longlines are indiscriminate and unintentionally catch and kill immature fish along with a wide variety of other animals in the Atlantic including tunas, sharks, marlins, sailfish, sea turtles and occasionally pilot whales and dolphins.

Long-term Forward purchase or sale of a specific quantity of allowances, with delivery and settlement scheduled for a specified future date, usually more than one year out

Low-emission vehicles. Vehicles which emit little air pollution compared to conventional internal combustion engines.

Low-impact camping. Camping that does not damage or change the land, where campers leave no sign that they were on the land.

Lumber. Wood or wood products used for construction.

Lung diseases. Any disease or damaging conditions in the lung or bronchia such as cancer or emphysema.

Lymphoma. A tumor marked by swelling in the lymph nodes.

M

Magnesium oxide. Also known as magnesia, magnesium oxide is used medicinally ("Milk of Magnesia"), industrially and in agricultural soil supplements. It is also used to enhance biological processes and to cleanup groundwater.

Magnesium. This light metal and its derivatives are used in aerospace alloys, in incendiary devices such as flares, and elsewhere. When scrap magnesium is thinly shaved or powdered, it is considered to be a hazardous waste, as it ignites easily and burns with an intense, white flame. It is also a nutritionally essential trace metal.

Majority leader. The leader of the majority party in either the House or the Senate.

Malthusian—based on the theories of British economist Thomas Robert Malthus (1766-1834), who argued that population tends to increase faster than food supply, with inevitably disastrous results, unless the increase in population is checked by moral restraints or by war, famine, and disease.

Malathion. Insecticide that, at high doses, affects the human nervous system.

Mammal. An animal that feeds its young with milk secreted from mammary glands and has hair on its skin.

Managed growth. Growth or expansion that is controlled so as not to be harmful.

Manatee. A plant-eating aquatic mammal found in the waters of Florida, the Caribbean, and off the coast of West Africa.

Marbled murrelet. A rare and imperiled bird that nests in ancient forests on the west coast of the U.S.

Marine mammal. A mammal that lives in the ocean, such as a whale.

Market Maker: an individual or company that maintains firm bid and offer prices in allowances by standing ready to buy or sell allowances at market prices

Mark-up. Action by a Congressional committee to amend and/or approve a bill; following mark-up the bill is "reported" out of committee and is ready for consideration by the entire House or Senate.

marsh—wetland, swamp, or bog.

Mass transit. Public transportation.

Maximum contaminant level (MCL). A contaminant level for drinking water, established by the California Department of Health Services, Division of Drinking Water and Environmental Management, or by the U.S. Environmental Protection Agency. These levels are legally-enforceable standards based on health risk (primary standards) or non-health concerns such as odor or taste (secondary standards).

Medfly. The Mediterranean fruit fly, a flying insect.

Megalopolis. A large city expanding so fast that city government cannot adjust to provide services (such as garbage disposal).

Megawatt (MW). One million watts.

Megawatt hour (MW/h). One million watts generated or used in one hour

Mercury. Also known as "quicksilver," this metal is used in the paper pulp and chemical industries, in the manufacture of thermometers, and thermostats, and in fungicides. Mercury exists in three biologically important forms, elemental, inorganic and organic. It is highly toxic and affects the nervous system, kidneys and other organs. It also accumulates in animals that are high in the food chain (predators). Organic mercury compounds are the most toxic, and transformations between the three forms of mercury do occur in nature.

Methane (CH_4). An odorless, colorless, flammable gas that is the major constituent of natural gas. It can be formed from rotting organic matter (i.e., trash in a landfill), and seep up through soils or migrate through underground piping to the surface. It also seeps up through the ground in areas that have shallow petroleum deposits or improperly abandoned oil wells, such as certain areas of the Los Angeles Basin. If it collects in a closed space and reaches certain concentrations, a spark can cause an explosion. It can also displace air and cause a suffocation hazard in low, enclosed spaces. This is one of the reasons landfill gas is collected and burned, sometimes for generation of electricity.

Methyl bromide (CH_3Br). The gaseous compound is used primarily as an insect fumigant; found to be harmful to the stratospheric ozone layer which protects life on earth from excessive ultraviolet radiation.

Methyl ethyl ketone (MEK). MEK is a flammable solvent that has many industrial uses, primarily in the plastic industry as a solvent. MEK is also used in the synthetic rubber industry, in the production of paraffin wax, and in household products such as lacquer and varnishes, paint remover, and glues.

Methylene chloride. A colorless liquid that evaporates easily. It has been used as a metal cleaner, paint thinner, in wood stains, spot removers, fabric protectors, shoe polish and aerosol propellants. Mild exposure can cause skin and eye irritation

Microgram per gram (mg/g). A measurable unit of concentration for a solid. A mercury level of 1.0 mg/g means that one microgram (one millionth of a gram) of mercury was detected in one gram of sample. It is equivalent to one part per million.

Migration. The movement of chemical contaminants through soils or groundwater.

Milligram per cubic meter (mg/m3). A unit of concentration for air contaminants. A mercury vapor level of 1.0 mg/m3 means that one milligram (one thousandth of a gram) of mercury vapor was detected in each cubic meter of sampled air.

Milligram per kilogram (mg/kg). A unit of concentration for a solid. A mercury level of 1.0 mg/kg in fish means that one milligram (one thousandth of a gram) of mercury was found in each kilogram of sampled fish. (A kilogram is 1,000 grams or approximately 2.2 pounds). Also equals one part per million.

Mining. The removal of minerals (like coal, gold, or silver) from the ground.

Minority leader. The leader of the minority party in either the House or the Senate.

Minuteman. An American-made ICBM; 500 Minuteman III ICBMs are deployed currently in the United States.

Mitigation. Actions taken to improve site conditions by limiting, reducing or controlling hazards and contamination sources.

Monitoring wells. Specially-constructed wells used exclusively for testing water quality.

Moratorium. Legislative action which prevents a federal agency from taking a specific action or implementing a specific law.

Mulch. Leaves, straw or compost used to cover growing plants to protect them from the wind or cold.

N

National Ambient Air Quality Standards (NAAQS): health-based standards for a variety of pollutants set by EPA that must be met by states across the country

National Pollutant Discharge Elimination System (NPDES). A system under the federal Clean Water Act that requires a permit for the discharge of pollutants to surface waters of the United States. In California, NPDES permits are obtained from the Regional Water Quality Control Board.

National Priorities List (NPL). U.S. EPA's list of the top priority hazardous waste sites in the country that are subject to the Superfund program.

Natural long: a party whose allowance allocation is greater than its actual emissions

Natural short: a party whose allowance allocation is less than its actual emissions

National Recreation Areas. Areas of federal land that have been set aside by Congress for recreational use by members of the public.

Negative Declaration. A California Environmental Quality Act document issued by the lead regulatory agency when the initial environmental study reveals no substantial evidence that the proposed project will have a significant adverse effect on the environment, or when any significant effects would be avoided or mitigated by revisions agreed to by the applicant.

Neutrals. Organic compounds that have a relatively neutral pH (are neither acid nor base), complex structure and, due to their carbon bases, are easily absorbed into the environment. Naphthalene, pyrene and trichlorobenzene are examples of neutrals.

Nickel. A metal used in alloys to provide corrosion and heat resistance for products in the iron, steel and aerospace industries. Nickel is used as a catalyst in the chemical industry. It is toxic and, in some forms, is listed as a cancer-causing agent under Proposition 65.

Nitrate. Formed when ammonia is degraded by microorganisms in soil or groundwater. This compound is usually associated with fertilizers.

Nitroaromatics. Common components of explosive materials, which will explode if activated by very high temperatures or shocks. 2,4,6-trinitrotoluene (TNT) is a nitroaromatic. Some are listed as cancer-causing chemicals under Proposition 65.

Nitrogen oxides (NOx). Harmful gases (which contribute to acid rain and global warming) emitted as a by-product of fossil fuel combustion. Gases produced during combustion of fossil fuels in motor vehicles, power plants and industrial furnaces and other sources; is a precursor to acid rain and ground-level ozone

Nitrogen oxides (NOx) Budget Program: a NOx cap and trade program adopted by 13 jurisdictions in the Northeast to address ozone transport in that region

Noise pollution. Environmental pollution made up of harmful or annoying noise.

Non-attainment pollutants. See "Criteria pollutants." If any of the criteria pollutants exceed established health-based levels in a given air basin, they are identified as "non-attainment pollutants."

Nuclear energy. Energy or power produced by nuclear reactions (fusion or fission). Alsop called nuclear power.

Nuclear reactor. An apparatus in which nuclear fission may be initiated, maintained, and controlled to produce energy, conduct research, or produce fissile material for nuclear explosives.

Nuclear tests. Government tests carried out to supply information required for the design and improvement of nuclear weapons, and to study the phenomena and effects associated with Nuclear explosions.

O

Offers: price at which someone who owns an allowance is willing to sell (a.k.a. "Ask")

Oil. A black, sticky substance used to produce fuel (petroleum) and materials (plastics).

Oil spills. The harmful release of oil into the environment, usually in the water, sometimes killing area flora and fauna. Oil spills are very difficult to clean up.

Old growth forests. See ancient forests.

Omnibus spending bill. A bill combining the appropriations for several federal agencies.

Operation Plan. A document submitted to DTSC that gives details about how a permitted hazardous waste facility is built, a detailed description of the hazardous waste operations, the plan to be used in case of emergency, and other plans. A DTSC facility permit requires that the reviewed and approved Operations Plan be followed. It is sometimes referred to as the "Part B" of the hazardous waste facility permit.

Option: a contractual right to buy or sell allowances at an agreed price; the option buyer pays a premium for this right. If the option is not exercised after a specified period it expires Option Premium: amount per share paid by an option buyer to an option seller for the option Out-of-the-money Call Option: term used to describe an call option whose strike price for an allowance is higher than the current market value

Organochlorides. A group of organic (carbon-containing) insecticides that also contain chlorine. These chemicals tend not to break down easily in the environment. DDT, Toxaphene and Endosulfan are all organochlorides.

Organophosphate. A group of organic (carbon-containing) insecticides that also contain phosphorus. Although they do not have a long life, some can be very toxic when first applied. Malathion and Parathion are organophosphates. Malathion is mildly toxic, and parathion is extremely toxic.

Out-of-the-money Put Option: term used to describe a put option whose strike price for an allowance is lower than the current market value

Over-development. Expansion or development of land to the point of damage.

Over-fishing. Fishing beyond the capacity of a population to replace itself through natural reproduction.

Over-grazing. Grazing livestock to the point of damage to the land.

Overpacking. Process used for isolating waste by jacketing or encapsulating waste-holding containers to prevent further spread or leakage of contaminating materials. Leaking drums may be contained within oversized ones as an interim measure prior to removal and final disposal.

Over-the-counter Market: Market in which allowance transactions are conducted through the direction interaction of counterparties rather than on the floor of an exchange

Oxidizers. A group of chemicals that are very reactive, often but not always supplying oxygen to a reaction. Some oxidation reactions can release large amounts of heat and gases, and, under the right conditions, cause an explosion. Others can cause rapid corro-

sion of metal, damage to tissue, burns and other serious effects. Examples of oxidizers include chlorine gas, nitric acid, sodium perchlorate, and ammonium nitrate.

Ozonation. Ozone reacts with volatile organic compounds (VOCs) to change them into chemicals which pose no potential threat to human health, by breaking them down to form carbon dioxide and water. This is done with an ozonation unit

Ozone. A naturally occurring, highly reactive gas comprising triatomic oxygen (O3) formed by recombination of oxygen in the presence of ultraviolet radiation. This naturally occurring gas builds up in the lower atmosphere as smog pollution, while in the upper atmosphere it forms a protective layer which shields the earth and its inhabitants from excessive exposure to damaging ultraviolet radiation.

Ozone depletion. The reduction of the protective layer of ozone in the upper atmosphere by chemical pollution.

Ozone hole. A hole or gap in the protective layer of ozone in the upper atmosphere.

Ozone Transport Assessment Group (OTAG): a multi-stakeholder workgroup convened to address problems associated with the long-range transport of ozone and its precursors; encompassed stakeholders in 37 jurisdictions

P

Paper. Thin sheet of material made of cellulose pulp, derived mainly from wood, but also from rags and certain grasses, and processed into flexible leaves or rolls. Used primarily for writing, printing, drawing, wrapping, and covering walls.

Paper mills. Mills (factories) that produce paper from wood pulp.

Paper products. Materials such as paper and cardboard, produced from trees.

Parathion and Methylparathion. Toxic insecticides.

Particulates. Small solid or liquid particles, especially those in the emission gases of incinerators, boilers, industrial furnaces or in exhaust from diesel and gasoline engines. Particles below 10 microns (10 one-millionths of a meter, 0.0004 inch) in diameter are considered potential health risks because, when inhaled, they are taken deep into the lungs. Regulations require that an incinerator emit no more than 180 milligrams of total particulates per dry standard cubic meter per minute.

Particulate pollution. Pollution made up of small liquid or solid particles suspended in the atmosphere or water supply.

Parts per million (ppm). A measuring unit for the concentration of one material in another. When looking at contamination of water and soil, the toxins are often measured in parts per million. One part per million is equal to one thousandth of a gram of substance in one thousand grams of material. One part per million would be equivalent to one drop of water in twenty gallons.

Parts per billion (ppb). A unit of measure used to describe levels or concentrations of contamination. A measure of concentration, equaling 0.0000001 percent. For example, One part per billion is the equivalent of one drop of impurity in 500 barrels of water. Most drinking water standards are ppb concentrations.

Passive solar. Using or capturing solar energy (usually to heat water) without any external power.

Pelagic species. Fish that live at or near the water's surface. Examples of large pelagic species include swordfish, tuna, and many species of sharks. Small pelagics include anchovies and sardines.

Pentachlorophenol (PCP). A petroleum-based chemical that is used as a wood preservative because it kills fungus and termites. It is toxic listed as a cancer-causing chemical under Proposition 65.

Perched groundwater. Water that accumulates beneath the earth's surface but above the main water bearing zone (aquifer). Typically, perched groundwater occurs when a limited zone (or lens) of harder, less permeable soil is "perched" in otherwise porous soils. Rainwater moving downward through the soil stops at the lens, flows along it, then seeps downward toward the aquifer.

Perchloroethylene (PCE). A volatile organic compound used primarily as a dry-cleaning agent. It is often referred to as "perc." It is toxic and listed as a cancer-causing chemical under Proposition 65.

Percolation. The downward flow or filtering of water or other liquids through subsurface rock or soil layers, usually continuing to groundwater.

Pesticide. A general term for insecticides, herbicides and fungicides. Insecticides kill or prevent the growth of insects. Herbicides control or destroy plants. Fungicides control or destroy fungi. Some pesticides can accumulate in the food chain and contaminate the environment.

Petrex Method. A method for collecting vapor samples from surface soil.

Petrochemicals. Chemical substances produced from petroleum in refinery operations. Many are hazardous.

Phenols. Organic compounds used in plastics manufacturing, tanning, and textile, dye and resin manufacturing. They are by-products of petroleum refining. In general, they are highly toxic.

Piezometers. Small-diameter wells used to measure groundwater levels.

Pilot study. A study of a possible cleanup alternative during the Feasibility Study for a specific site. It is used to gather data necessary for the final selection of the cleanup method.

Plastics. Durable and flexible synthetic-based products, some of which are difficult to recycle and Pose problems with toxic properties, especially PVC plastic.

Plume. A body of contaminated groundwater flowing from a specific source. The movement of the groundwater is influenced by such factors as local groundwater flow patterns, the character of the aquifer in which the groundwater is contained, and the density of contaminants. A plume may also be a cloud of smoke or vapor. It defines the area where exposure would be dangerous.

Plutonium. A heavy, radioactive, man-made, metallic element (atomic number 94) used in the production of nuclear energy and the explosion of nuclear weapons; its most important isotope is fissile plutonium-239, produced by neutron irradiation of uranium-238.

Poison. A chemical that adversely affects health by causing injury, illness, or death.

Poison runoff. See polluted runoff.

Polluted runoff. Precipitation that captures pollution from agricultural lands, urban streets, parking lots and suburban lawns, and transports it to rivers, lakes or oceans.

Pollution prevention. Techniques that eliminate waste prior to treatment, such as by changing ingredients in a chemical reaction.

Polychlorinated biphenyls (PCBs). A group of toxic chemicals used for a variety of purposes including electrical applications, carbonless copy paper, adhesives, hydraulic fluids, and caulking compounds. PCBs do not breakdown easily and are listed as cancer-causing agents under Proposition 65.

Polynuclear aromatic hydrocarbons (PAHs or PNAs). PNAs or Polynuclear Aromatic Hydrocarbons, are natural constituents of crude oil, and also may be formed when organic materials such as coal, oil, fuel, wood or even foods are not completely burned. PNAs are also found in lampblack, a by-product of the historic gas manufacturing process. PNAs are found in a wide variety of other materials, including diesel exhaust, roofing tars, asphalt, fireplace smoke and soot, cigarettes, petroleum products, some foods, and even some shampoos. PNAs tend to stick to soil and do not easily dissolve in water, and generally do not move in the environment. The test method used to analyze for PNAs detects seventeen different compounds. Of the seventeen, seven are suspected of causing cancer in humans.

Polyvinyl chloride (PVC). A plastic made from the gaseous chemical vinyl chloride. PVC is used to make pipes, records, raincoats and floor titles. It produces hydrochloric acid when burned. Health risks from high concentrations of vinyl chloride (not the polymer) include liver cancer and lung cancer, as well as cancer of the lymphatic and nervous systems. Vinyl chloride (not the polymer) is listed as a cancer-causing chemical under Proposition 65.

Population. Two meanings: a.) the whole number of inhabitants in a country, region or area; b.) a set of individuals having a quality or characteristic in common.

Post consumer waste. Waste collected after the consumer has used and disposed of it (e.g., the wrapper from an eaten candy bar).

Potentially Responsible Party (PRP). An individual, company or government body identified as potentially liable for a release of hazardous substances to the environment. By federal law, such parties may include generators, transporters, storers and disposers of hazardous waste, as well as present and past site owners and operators.

Power plants. Facilities (plants) that produce energy.

Power Pool: a situation where output from different power plants are "pooled" together, scheduled according to increasing marginal cost, technical and contractual characteristics (so-called must-runs), and dispatched according to this "merit order" to meet demand

Pretreatment unit. A wastewater treatment unit that is designed to treat wastewater that does not meet the sewage discharge standards so that it meets or exceeds those standards. Pretreatment units usually require a permit from a local agency.

Principal organic hazardous constituents (POHCs). Specific hazardous compounds monitored during an incinerator, boiler or industrial furnace trial burn. They are selected on the basis of their high concentration in the waste feed and the difficulty of burning them.

Proprietary information (trade secret). The Department will classify information as proprietary provided the owner demonstrates the following: the business has

asserted a business confidentiality claim; the business has shown it has taken reasonable measures to protect the confidentiality of the information both within the company and from outside entities; the information is not, and has not been reasonably obtainable without the business' consent; no statute specifically requires disclosure of the information ; and either the business has shown that disclosure of the information is likely to cause substantial harm to its competitive position, or the information is voluntarily submitted and its disclosure would likely impair the government's ability to obtain necessary information in the future.

Public estate. Area or plot of public land owned by community or government.

Public health. The health or physical well-being of a whole community.

Public land. Land owned in common by all, represented by the government (town, county, state, or federal).

Public participation plan. A document approved by DTSC that is designed to determine a community's informational needs and to provide a program for public involvement during facility permitting, site investigation and cleanup, or other similar activities.

Public transportation. Various forms of shared-ride services, including buses, vans, trolleys, and subways, which are intended for conveying the public.

Pulp. Raw material made from trees used in producing paper products.

Pump test. A field test by which a well is pumped for a period of time and data are collected for use in assessing characteristics of subsurface water-bearing zones, or aquifers.

Put Option: a contract that grants the right to sell, at a specified price, a specific number of allowances by a certain date

Q

Quench tower. A gas cooling and pollution control device in which heated gases are showered with water. Gases are cooled and particulates "drop out" of the gases. They can generate a waste called "quench tower drop-out."

R

Radiation. The process of emitting energy in the form of energetic particles (such as alpha particles or gamma radiation), light or heat. It also refers to that which is emitted.

Radioactive. Of or characterized by radioactivity.

Radioactive waste. The byproduct of nuclear reactions

that gives off (usually harmful) radiation.

Radioactivity. The spontaneous emission of matter or energy from the nucleus of an unstable atom (the emitted matter or energy is usually in the form of alpha or beta particles, gamma rays, or neutrons).

Radionuclides. Radioactive elements, which may be naturally-occurring or synthetic. They emit various types of energetic radiation— alpha and beta particles and gamma radiation. Their half-lives range from a minute fraction of a second to many thousand years. Certain radionuclides have valuable medical and industrial uses. One is used in home smoke detectors at an amount that can cause no harmful effects.

Radium. A radioactive element with a half-life of 1,600 years that emits alpha particles as it is transformed into radon. In the past, radium was mixed with special paints to make watch faces and instrument dials glow in the dark.

Radon. A gaseous, radioactive alpha particle-emitting element with a half-life of about four days. Radon exists naturally in many locations, and may present a serious health risk when it accumulates in basements or crawl spaces beneath homes. A cancer-causing radioactive gas found in many communities' ground water.

Rainforest. A large, dense forest in a hot, humid region (tropical or subtropical). Rainforests have an abundance of diverse plant and animal life, much of which is still uncatalogued by the scientific community.

Ranking member. The lead member of a Congressional committee from the minority party, usually chosen on the basis of seniority.

Reactive. A class of compounds which are normally unstable and readily undergo violent change, react violently with water, can produce toxic gases with water, or possess other similar properties. Reactivity is one characteristic that can make a waste hazardous.

Recess. Ending a legislative session with a set time to reconvene.

Recycling. System of collecting, sorting, and reprocessing old material into usable raw materials.

Reduce. Act of purchasing or consuming less to begin with, so as not to have to reuse or recycle later.

Refrigerants. Cooling substances, many of which contain CFCs and are harmful to the earth's ozone layer.

Regional Clean Air Incentives Market (RECLAIM): initiated in 1993; a set of market initiatives designed address air pollution in the Greater Los Angeles area of California; includes cap and trade programs for NOx and SOx.

Regional Water Quality Control Board (RWQCB). Agencies that maintain water quality standards for areas within their jurisdictions and enforce state water quality laws.

Remedial Action Plan (RAP). A plan that outlines a specific program leading to the remediation of a contaminated site. Once the Draft Remedial Action Plan is prepared, and approved by DTSC a public meeting is held and comments from the public are solicited for a period of not less then 30 days. After the public comment period has ended and the comments have been responded to in writing, DTSC may modify the Draft Plan on the basis of those comments before it approves the final remedy for the site (the Final RAP).

Remedial Investigation/Feasibility Study (RI/FS). A series of investigations and studies to identify the types and extent of chemicals of concern at the site and to determine cleanup criteria (Remedial Investigation), and to provide an evaluation of the alternatives for remediating any identified soil or groundwater problems (Feasibility Study).

Remediation. Cleanup of a site to levels determined to be health-protective for its intended use.

Resource Conservation and Recovery Act (RCRA)

Renewable energy. Energy resources such as wind power or solar energy that can keep producing indefinitely without being depleted.

Reservoir. An artificial lake created and used for the storage of water.

Responsible party. An individual or corporate entity considered legally liable for contamination found at a property and, therefore, responsible for cleanup of the site.

Resolution. A formal statement from Congress.

Retire (Allowances): to remove a portion of allowances from the market

Reuse. Cleaning and/or refurbishing an old product to be used again.

Rider. Usually unrelated provisions tacked onto an existing Congressional bill. Since bills must pass or fail in their entirety, riders containing otherwise unpopular language are often added to popular bills.

Riparian. Located alongside a watercourse, typically a river.

Risk assessment. A risk assessment looks at the chemicals detected at a site, the frequency and concentration of detected chemicals, the toxicity of the chemicals and how people can be exposed, and for how long.

Routes of exposure to people are generally through ingestion, such as eating, contact with the skin, or inhalation. The most significant potential routes of exposure are trough ingestion and contact with the skin. Based on the standard risk assessment guidelines established for use nationwide by U.S. EPA, exposures for an on-site resident are generally assumed to e daily contact over a 30-year period starting with children ages 0-6, and continuing from 6-30 years.

Rotary kiln. An incinerator with a rotating combustion chamber. The rotation helps mix the wastes and promotes more complete burning. They can accept gases, liquids, sludges, tars and solids, either separately or together, in bulk or in containers.

Run-off. Precipitation that the ground does not absorb and that ultimately reaches rivers, lakes or oceans.

S

Sagebrush Rebellion. A movement started by ranchers and miners during the late 1970s in response to efforts of the Bureau of Land Management (B.L.M.) to improve management of federal lands. While its announced goal was to give the lands "back" to the western states, its real goal—and the one it achieved—was to force the B.L.M. to abandon its new approach to public land management.

Salvage logging. The logging of dead or diseased trees in order to improve overall forest health; used by timber companies as a rationalization to log otherwise protected areas.

Sanitary landfill. A landfill which does not take hazardous waste, often called a "garbage dump." It must be covered with dirt each day to maintain sanitary conditions. The Integrated Waste Management Board regulates these facilities.

SARA Title III. Or the Emergency Planning and Community Right-to-Know Act of 1986. It requires each state to have an emergency response plan as described, and any company that produces, uses or stores more than certain amounts of listed chemicals must meet emergency planning requirements, including release reporting.

Scrubber: a pollution control technology utilized in power plants to remove pollutants from plant emissions

Secondary containment. A structure designed to capture spills or leaks, as from a container or tank. For containers and aboveground tanks, it is usually a bermed area of coated concrete. For underground tanks, it may be a second, outer, wall or a vault. Construction of such containment must meet certain re-

quirements, and periodic inspections are required.

Second-growth forests. Forests that have grown back after being logged.

Sediment. The soil, sand and minerals at the bottom of surface waters, such as streams, lakes and rivers. Sediments capture or adsorb contaminants. The term may also refer to solids that settle out of any liquid.

Seismic stability. The likelihood that soils will stay in place during an earthquake.

Selenium. This metal is a nutritionally essential trace element that is toxic at higher doses. High levels of selenium have been shown to cause reproductive failure and birth defects in birds.

Semi-volatile organic compounds. Compounds that evaporate slowly at normal temperatures.

Short: a market position in which a party records (or anticipates recording) emissions in excess of its yearly emissions allocation, thus it has a deficit of allowances

Short-term Forward: purchase or sale of a specific quantity of allowances at the current or spot price, with delivery and settlement at a specified future date, usually within one year

Sick building syndrome. A human health condition where infections linger, caused by exposure to contaminants within a building as a result of poor ventilation.

Silos. Fixed vertical underground structures made of steel and concrete that house an ICBM and its launch support equipment.

Silver. Silver is a metal used in the manufacture of photographic plates, cutlery and jewelry. Silver nitrate is used in an array of industrial chemical processes. It is toxic.

Silvex. A chlorinated herbicide.

Sinkhole. A depression formed when the surface collapses into a cavern.

Site mitigation process. The regulatory and technical process by which hazardous waste sites are identified and investigated, and cleanup alternatives are developed, analyzed, decided upon and applied.

Slurry wall. Barriers used to contain the flow of contaminated groundwater or subsurface liquids. Slurry walls are constructed by digging a trench around a contaminated area and filling the trench with a material that tends not to allow water to pass through it. The groundwater or contaminated liquids trapped within the area surrounded by the slurry wall can be extracted and treated.

Smog. A dense, discolored radiation fog containing large

quantities of soot, ash, and gaseous pollutants such as sulfur dioxide and carbon dioxide, responsible for human respiratory ailments. Originally meaning a combination of smoke and fog, smog now generally refers to air pollution; ground level ozone is a major constituent of smog. Most industrialized nations have implemented legislation to promote the use of smokeless fuel and reduce emission of toxic gases into the atmosphere.

SO_2 Allowance Auction: provided for in the Clean Air Act, the SO_2 auction is held annually by the US EPA; the auctions help to send the market an allowance price signal, as well as furnish utilities with an additional avenue for purchasing needed allowances

Soil borings. Soil samples taken by drilling a hole in the ground.

Soil gas survey. Soil gas or (soil vapor) is air existing in void spaces in the soil between the groundwater and the ground surface. These gases may include vapor of hazardous chemicals as well as air and water vapor. A soil-gas survey involves collecting and analyzing soil-gas samples to determine the presence of chemicals and to help map the spread of contaminants within soil.

Soil vapor extraction. A process in which chemical vapors are extracted from the soil by applying a vacuum to wells.

Solar energy. Energy derived from sunlight.

Solid waste. Non-liquid, non gaseous category of waste from non-toxic household and commercial sources.

Solid waste management units (SWMUs). Any unit at a hazardous waste facility from which hazardous chemicals might migrate, whether or not they were intended for waste management. They include such things as containers, tanks, landfills among others.

Solidification. Mixing additives, such as fly ash or cement, with soil containing hazardous chemicals, especially metals, to make it more stable. This process lessens the risk of exposure to the hazardous chemicals by making it less likely that those chemicals will move into and through surface or groundwater.

Soluble Threshold Limit Concentration (STLC). The limit concentration for toxic materials in a sample that has been subjected to the California Waste Extraction Test (WET), a state test for the toxicity characteristic that is designed to subject a waste sample to simulated conditions of a municipal waste landfill. If the concentration of a toxic substance in the special extract of the waste exceeds this value, the waste is classified as hazardous in California. This is distinct from the Total Threshold Limit Concentration (TTLC). The California Waste Extraction Test procedure is more stringent than the federal Toxicity Characteristic Leaching Procedure (TCLP).

Solvent. A liquid capable of dissolving another substance to form a solution. Water is sometimes called "the universal solvent" because it dissolves so many things, although often to only a very small extent. Organic solvents are used in paints, varnishes, lacquers, industrial cleaners and printing inks, for example. The use of such solvents in coatings and cleaners has declined over the last several years, because the most common ones are toxic, contribute to air pollution and may be fire hazards.

Soot. A fine, sticky powder, comprised mostly of carbon, formed by the burning of fossil fuels.

Speaker. The leader of the House of Representatives, who controls debate and the order of discussion; chosen by vote of the majority party.

Special Allowance Reserve: roughly 2.8 percent dangerous anthropogenic interference with the of the cap set aside each year to supply the climate system; established a framework for annual allowance auction agreeing to specific actions

Spotted owl. Reclusive bird, found in the American West, requiring old-growth forest habitat to survive.

South Coast Air Quality Management District (SCAQMD): the air pollution control agency for the four-county region including Los Angeles and Orange counties and parts of Riverside and San Bernardino counties

Sprawl. The area taken up by a large or expanding development or city.

Stabilization. Changing active organic matter in sludge into inert, harmless material. The term also refers to physical activities such as compacting and capping at sites that limits the further spread of contamination without actual reduction of toxicity.

State action level (SAL). The maximum concentration of a contaminant in drinking water that The California Department of Health Services considers to be safe to drink. Drinking Water Action Levels (ALs) are health-based advisory levels established by the Department of Health Services (DHS) for chemicals for which primary maximum contaminant levels (MCLs) have not been adopted. There are currently 36 ALs. ALs are usually expressed in parts per billion (ppb) or parts per million (ppm). Drinking water with concentrations of impurities greater than the state action level must be treated to reduce or remove the impurities.

State Implementation Plan (SIP). Mandate for achieving health-based air quality standards. The plan that each state must develop and have approved by the US EPA which indicates how the state will comply with the requirements in the Clean Air Vintage Year: represents the first year in which Act; each State's SIP is amended as they address the allowance can be used for compliance specific or new requirements such as the NOx reductions required in the NOx SIP.

State land. Land owned and administered by the state in which it is located.

State parks. Parks and recreation areas owned and administered by the state in which they are located.

Static stability. The likelihood that soils at rest will remain at rest.

Still bottoms. Residues left over from the process of recovering spent solvents in a distillation unit.

Stockpile. Nuclear weapons and components under custody of the U.S. Department of Defense.

Straddling stocks. Fish populations that straddle a boundary between domestic and international waters.

Strangle: sale or purchase of a put option and a call option on the same underlying instrument, with the same expiration, but at strike prices X equally out of the money.

Stratosphere. The upper portion of the atmosphere (approximately 11 km to 50 km above the surface of the Earth).

Strike Price (or Exercise Price): price at which Y the allowance underlying a call or put option can be purchased (call) or sold (put) over the specified period. Z

Strip mining. Mining technique in which the land and vegetation covering the mineral being sought are stripped away by huge machines, usually damaging the land severely and limiting subsequent uses.

Submarine Launched Ballistic Missile (SLBM). A ballistic missile carried by and launched from a submarine.

Subsidence. Sinking or settling of soils so that the surface is disrupted, creating a shallow hole.

Suggested No Adverse Response Level (SNARL). Drinking water standards established by the U.S. EPA, but not enforceable by law. SNARLs suggest the level of a containment in drinking water at which adverse health effects would not be anticipated (with a margin of safety).

Sulfur dioxide (SO_2). A heavy, smelly gas which can be condensed into a clear liquid; used to make sulfuric acid, bleaching agents, preservatives and refrigerants; a major source of air pollution in industrial areas. A gaseous pollutant which is primarily released into the atmosphere as a by-product of fossil fuel combustion; the largest sources of SO_2 tend to be power plants that burn coal and oil to make electricity.

Sump. A pit or tank that catches liquid runoff for drainage or disposal.

Super Efficient Refrigerator Program (SERP). An organization of 24 U.S. utilities that developed a $30 million competition to produce a refrigerator at least 25% lower in energy use and 85% lower in ozone depletion than projected 1994 models. The winning product, produced by Whirlpool, cut energy use by 40% in 1995.

Superfund Comprehensive Environmental Response, Compensation and Liability Act of 1980 (CERCLA).

Supply-side: a term referring to the generation (or supply) of power by a utility

Surge tanks. A tank used to absorb irregularities in flow of liquids, including liquid waste materials, so that the flow out of the tank is constant.

Sustainable communities. Communities capable of maintaining their present levels of growth without damaging effects.

Swap: an exchange of one allowance for another to exchange the vintage years of the allowances held in accounts Surface water. Water located above ground (e.g., rivers, lakes).

T

Table. In the legislative sense, an action taken to halt debate on a bill.

Tailings or Mine Tailings. Crushed waste rock deposited on the ground during mining and ore processing, including some of the rock in which the ore is found. Unless they are handled carefully, they frequently release contaminants. As they age under the effects of air, rainfall and bacteria, some oxidize to produce new toxic materials, such as sulfuric acid, that can leach out and poison streams, rivers and lakes.

Tap water. Drinking water monitored (and often filtered) for protection against contamination and available for public consumption from sources within the home.

Tax shift. Replacing one kind of taxes with another, without changing the total amount of money collected. For example, replacing a portion of income taxes with carbon tax or other pollution taxes.

Telecommuting. Working with others via telecommunications technologies (e.g., telephones, modems, faxes) without physically travelling to an office.

Tetrachloroethylene (TCE). Volatile organic compound that is commonly used as an industrial degreasing solvent. TCE affects the central nervous system and is listed as a cancer-causing chemical under Proposition 65.

Tetrachlorophenol (TCP). Tetrachlorophenol is a toxic fungicide.

Thermonuclear. The application of high heat, obtained via a fission explosion, to bring about fusion of light nuclei.

Threatened species. Species of flora or fauna likely to become endangered within the foreseeable future.

Three Gorges. A project along the Yangtze river in China to build the largest hydroelectric dam in the world.

Threshold Limit Value (TLV). Public health exposure level set by the National Institute for Occupational Safety and Health for worker safety. It is the level above which a worker should not be exposed for the course of an eight-hour day, due to possible adverse health effects.

Timber. Logged wood sold as a commodity.

Time Weighting: an investment strategy in which allowance purchases and sales are transacted over an extended period of time and in small increments, thereby eliminating risk associated with highs and lows in the market

TNT Equivalent. A measure of the energy released in the detonation of a nuclear weapon, expressed in terms of the quantity of TNT which would release the same amount of energy.

Toluene. A toxic volatile organic compound often used as an industrial solvent.

Tongass. A national forest in southeast Alaska comprising one of the United States' last remaining temperate rainforests.

Total Threshold Limit Concentration (TTLC). A test for the toxicity characteristic: If the total concentration of a toxic substance in a waste is greater than this value, the waste is classified as hazardous in California. This is distinct from the Soluble Threshold Limit Concentration, or STLC, which is concerned with only the soluble concentration.

Toxaphene. A chlorinated pesticide insecticide that was widely used to control pests on cotton and other crops until 1982, when it was banned for most uses. (In 1990, banned for all uses.) It was also used to kill unwanted fish in lakes. It is toxic to fresh-water and marine aquatic life and is listed as a cancer-causing chemical under Proposition 65.

Toxic. Poisonous.

Toxic emissions. Poisonous chemicals discharged to air, water, or land.

Toxic sites. Land contaminated with toxic pollution, usually unsuitable for human habitation.

Toxicity. Ability to harm human health or environment, such as injury, death or cancer. One of the criteria that is used to determine whether a waste is a hazardous waste (the "Toxicity Characteristic").

Toxicity Characteristic Leaching Procedure (TCLP). A federal test for the Toxicity Characteristic (TC). If the concentration of a toxic substance in a special extract of a waste exceeds the TC value, the waste is classified as hazardous in the United States (a "RCRA waste"). The extraction procedure is different from that of the California Waste Extraction Test (WET).

Toxic Substances Control Act (TSCA). A federal law of 1976 to regulate chemical substances or mixtures that may present an unreasonable risk of injury to health or the environment.

Toxic waste. Garbage or waste that can injure, poison, or harm living things, and is sometimes life-threatening.

Toxification. Poisoning.

Traffic calming. Designing streets to reduce automobile speed and to enhance walking and bicycling.

Trader: anyone who buys or sells allowances with the intention of making a profit

Transit. See public transportation.

Transportation. Any means of conveying goods and people.

Transportation planning. Systems to improve the efficiency of the transportation system in order to enhance human access to goods and services.

Trash. Waste material that cannot be recycled and reused (synonymous with garbage).

Trawls. Nets with a wide mouth tapering to a small, pointed end, usually called the "cod end."

Trawls are towed behind a vessel at any depth in the water column.

Trial burn. A test of incinerators or boilers and industrial furnaces in which emissions are monitored for the presence of specific substances, such as organic compounds, particulates, metals and hydrogen chloride (all specified by agency permits).

Trichloroethane. 1,1,1-TCA, or methylchloroform is used as a cleaning agent for metals and plastics. It is toxic.

Trichloroethylene (TCE). A volatile organic compound that is often used an industrial degreasing solvent. It is toxic and is listed as a cancer-causing chemical under Proposition 65.

Trip reduction. Reducing the total numbers of vehicle trips, by sharing rides or consolidating trips with diverse goals into fewer trips.

Tritium. A radioactive form of hydrogen with a half-life of 12 years. It emits beta particles. It is used to mark chemical compounds so that the structure or chemical activity can be determined. Also used in nuclear weapons research and construction. Small amounts of tritium occur naturally, and some exists as a by-product of previous nuclear testing and nuclear reactor operations.

Trolling. A method of fishing using several lines, each hooked and baited, which are slowly dragged behind the vessel.

Turtle excluder device (TED). A gear modification used on shrimp trawls that enables incidentally caught sea turtles to escape from the nets.

U

Unit Accounts: accounts in the SO$_2$ or NOx ATSs which hold allowances initially allocated to those sources required to participate in either the acid rain or OTC NOx programs; EPA adjusts these accounts for compliance each year

United Nations Framework Convention on Climate Change(UNFCC): a treaty signed in 1992 by 165 countries and ratified by 160 countries (including the U.S.); took effect in March 1994; set a target of stabilizing greenhouse gas concentrations in the atmosphere to a level that would prevent further increase of global climate warming.

Unsaturated zone. Underground soil and gravel that could contain groundwater, but lies above the aquifer. This is in contrast to a saturated zone, where the space between soil particles is filled with water.

Up-gradient. The direction from which water flows in an aquifer. In particular, areas that are higher than contaminated areas and, therefore, are not prone to contamination by the movement of polluted groundwater.

Urban planning. The science of managing and directing city growth.

Uranium (U). A heavy, radioactive metal (atomic number 92) used in the explosion of nuclear weapons (especially one isotope, U-235).

Urban parks. Parks in cities and areas of high population concentration.

Utilities. Companies (usually power distributors) permitted by a government agency to provide important public services (such as energy or water) to a region; as utilities are provided with a local monopoly, their prices are regulated by the permitting government agency.

V

Vanadium. A toxic metal that is both mined and is a by-product of petroleum refining. Compounds of vanadium are used in the steel industry, as a catalyst in the chemical industry, in photography and in insecticides.

Veto. A Presidential action rejecting a bill as passed by the U.S. Congress. The President can also effect a "pocket veto" by holding an unsigned bill past the signing period.

Video logging. A method for close-up inspection of the interior of a well or pipe by means of a color camera that can view the well casing and screen at 90 degrees to the well's axis.

Vinyl chloride. Vinyl chloride is widely used in the plastics industry in creating polyvinyl chloride (PVC). It is listed as a cancer-causing agent under Proposition 65.

Virgin forest. A forest never logged.

Viscosity. A measure of the ease with which a liquid can be poured or stirred. The higher the viscosity, the less easily a liquid pours.

Voice vote. A vote where members vote by saying either "yes" or "no" together; individual member's votes are not placed on record.

Void space. The space in a tank between the top of a tank and the liquid level. If the tank is used to store combustible liquids that easily evaporate, this space can fill with vapors which may reach explosive levels.

Volatile. Describes substances that readily evaporate at normal temperatures and pressures.

Volatile organic compounds (VOCs). Organic liquids, including many common solvents, which readily evaporate at temperatures normally found at ground surface and at shallow depths. They take part in atmospheric photochemical (sun-driven) reactions to produce smog.

Volatilization rate. The rate at which a chemical changes from a liquid to gas. It is also known as "air flux."

W

Warhead. The part of a missile which contains the nuclear explosive.

Waste. Garbage, trash.

Waste feed. The flow of wastes into an incinerator, boiler or industrial furnace. The waste feed can vary from continuous to intermittent (batch) flows.

Waste site. Waste dumping ground.

Waste stream. Overall waste disposal cycle for a given population.

Waterborne contaminants. Unhealthy chemicals, microorganisms (like bacteria) or radiation, found in tap water.

Water filters. Substances (such as charcoal) or fine membrane structures used to remove impurities from water.

Water quality. The level of purity of water; the safety or purity of drinking water.

Water quality testing. Monitoring water for various contaminants to make sure it is safe for fish protection, drinking, and swimming.

Watershed. A region or area over which water flows into a particular lake, reservoir, stream, or river.

Water table. In a shallow aquifer, a water table is the depth at which free water is first encountered in a monitoring well.

Well. A dug or drilled hole used to get water from the earth.

Wetland. An area that is regularly saturated by surface or groundwater and, under normal circumstances, capable of supporting vegetation typically adapted for life in saturated soil conditions; they are critical to sustaining many species of fish and wildlife, including native and migratory birds. They include swamps, marshes, and bogs, and may be either coastal or inland. Coastal wetlands are brackish (have a certain mixture of salt).

Wilderness area. A wild area that Congress has preserved by including it in the National Wilderness Preservation System.

Wildlife. Animals living in the wilderness without human intervention.

Wildlife refuges. Land set aside to protect certain species of fish or wildlife (administered at the federal level in the U.S. by the Fish and Wildlife Service).

Windpower. Power or energy derived from the wind (via windmills, sails, etc.).

Wise use movement. A loosely-affiliated network of people and organizations throughout the U.S. in favor of widespread privatization and opposed to environmental regulation, often funded by corporate dollars.

Woods Hole. A town on Cape Cod where several important ocean research institutes are located.

Work plan. The site work plan describes the technical activities to be conducted during the various phases of a remediation project.

X

x the horizontal axis in a plot or a chart

x 8-12 GHz frequency band

x-ray; a form of electromagnetic radiation with a wavelength in the range of 0.01 to 10 nanometers, and energies in the range 0.1 to 100 keV, which is shorter than the wavelength of UV radiation but longer than that of gamma rays.

Xylene. An aromatic hydrocarbon used in gasoline, paints, lacquers, pesticides, gums, resins and adhesives. It is toxic and flammable.

x-y tracking; frames with PV cells or modules tracking the sun in the x and y directions all day long

Y

y often used to denote year (i.e., 100 kWh/y means 100 kWh generated yearly)

y the vertical axis in a plot or a chart

y yard

Z

Zoning. The arrangement or partitioning of land areas for various types of usage in cities, boroughs or townships.

Zinc. A metal used for auto parts, for galvanizing, and in production of brasses and dry cell batteries. It is nutritionally essential but toxic at higher levels.

Index